Articular Cartilage Dynamics

David W. Smith · Bruce S. Gardiner
Lihai Zhang · Alan J. Grodzinsky

Articular Cartilage Dynamics

Springer

David W. Smith
Faculty of Engineering and Mathematical Sciences
University of Western Australia
Perth, WA, Australia

Bruce S. Gardiner
College of Science, Health, Engineering and Education
Murdoch University
Perth, WA, Australia

Lihai Zhang
Department of Infrastructure Engineering
University of Melbourne
Melbourne, VIC, Australia

Alan J. Grodzinsky
Departments of Biological Engineering, Electrical Engineering, Computer Science, and Mechanical Engineering
Massachusetts Institute of Technology
Cambridge, MA, USA

ISBN 978-981-13-1473-5 ISBN 978-981-13-1474-2 (eBook)
https://doi.org/10.1007/978-981-13-1474-2

Library of Congress Control Number: 2018953314

This Springer imprint is published by the registered company Springer Nature Singapore Pte Ltd.
The registered company address is: 152 Beach Road, #21-01/04 Gateway East, Singapore 189721, Singapore

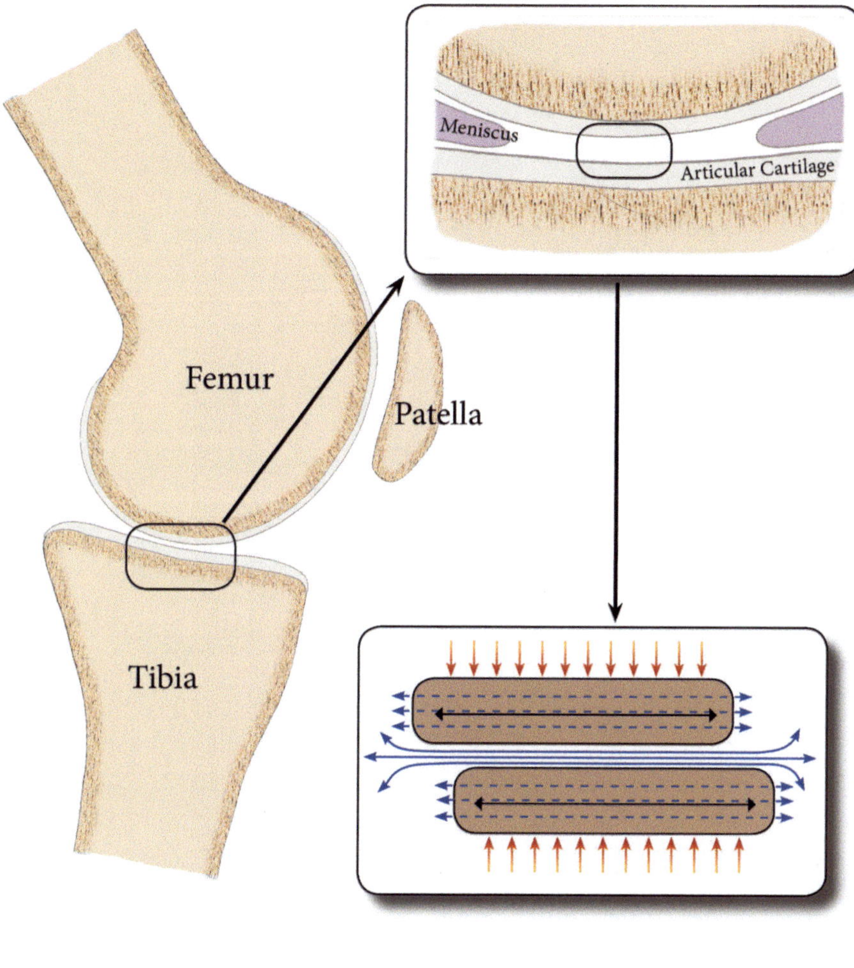
Femur
Patella
Tibia
Meniscus
Articular Cartilage

Foreword

Articular cartilage is found within diarthrodial joints covering the ends of long bones. The two primary mechanical functions of articular cartilage are to reduce the magnitude of contact stresses (the contact area of cartilage on cartilage is much larger than if bone directly contacts bone) and provide a low-friction bearing surface during motion (cartilage on cartilage has a much lower friction coefficient than if bone rubs directly on bone). The reduced contact stress and low friction at the contact interface together minimize internal damage to the cartilage extracellular matrix and resident chondrocytes, as well as minimizing wear damage at the cartilage surface. Fulfillment of these two primary mechanical functions provides the mechanical basis for the cartilage tissue to survive an extremely challenging mechanical environment.

Yet despite the best efforts by joint tissues to maintain themselves, many people develop disease states known collectively as osteoarthritis (OA). OA is a progressive and often disabling disease of the whole joint, involving multiple joint tissues. Over a period of years, the articular cartilage softens and degenerates—the hallmark of OA—leaving bare bone to rub painfully on bare bone. This results in substantial morbidity and much suffering in the general community, as well as generating substantial economic loss due to ill health and/or early retirement. To date, most drug treatments for OA are directed at symptomatic pain relief. It is unfortunate that current drug treatments have no demonstrable impact on altering the rate of structural change within osteoarthritic cartilage.

Even after an enormous research effort around the world over many decades, involving the expenditure of many billions of dollars, no single cause of osteoarthritis has been found. Both fundamental and clinical researchers have concluded that there is in fact no single cause and that OA disease states arise through a variety of pathways. Some osteoarthritic disease pathways to OA involve inflammatory initiators, while other pathways clearly involve mechanical initiators (e.g. joint trauma and overuse syndromes). A common view is that the onset of osteoarthrosis occurs when the recuperative processes within the cartilage and the surrounding tissues are pushed beyond their capacity to maintain tissue homeostasis, due to a patient-specific combination of genetic and environmental factors.

The evidence to date suggests that there are no easy answers and there will be no 'silver bullet' that eliminates OA. A more plausible scenario is that avoiding or managing established OA requires advice and treatment on a case-by-case basis—first identifying the specific combination of factors at work in any one individual and then instituting a patient-specific treatment regime. Undoubtedly, the most accurate assessment of disease potential and its management require an evaluation of patient-specific data in the context of a 'system-level' understanding of both normal cartilage physiology and OA cartilage physiology, gained from a deep and sustained integration of knowledge grounded in facts gathered over decades of basic biomedical science and clinical research.

It is our view that despite the intensive research efforts over decades, such a system-level understanding of cartilage is largely missing from the current cartilage literature. The focus of much basic research today remains on discovering the components in cartilage (e.g. signaling molecules, genes or protein expression profiles, or discovering the effect of 'gene silencing' or disabling protein activity). At a clinical level, there is a focus on high-level statistical correlations between system 'perturbations' and system outcomes (e.g. partial meniscectomy and incidence of OA, BMI and the incidence of OA, diabetes and the incidence of OA). Such 'discovery' approaches become problematic only because they completely dominate the current research landscape and, like all activities, are subject to the law of diminishing returns (i.e. successive studies yield diminishing increments to our knowledge). Practiced to excess, discovery, epidemiological and clinical research approaches deflect both attention and research funding from the development of integrated understandings of cartilage, which only serve to limit the potential benefits of the hard-won knowledge obtained from the previous discovery and clinical approaches. In short, our contention is there is currently a deficit of knowledge integration in cartilage research. An aim of this book is to help change this situation.

When integrative models of cartilage are employed, too often such models are far too simple. For example, articular cartilage is presented as a static tissue with a very limited capacity to repair—if it has any capacity to repair at all? From this simple conceptual model, it follows logically that cartilage inevitably 'wears out,' much like a pair of shoes wear out, and so patients are inevitably in need of joint replacement surgery (usually involving a metal implant of some kind). A large industry has grown up around joint replacement, partly because it is so effective.

But the same conceptual model leads to the conclusion that 'conservative treatment' of osteoarthritis is ineffective, and from this, it is a small step to conclude that conservative treatment is simply not beneficial. In practice, most people experience considerable inconvenience, suffering, and loss of income over one or two decades prior to joint replacement, and there is no effective treatment for OA during this long and debilitating period when inactivity, isolation and depression are all too frequent. So-named 'conservative treatments' may be able to reduce the high personal costs of this disease, but one barrier to the development of conservative treatments is conceptual. In their review of osteoarthritis, Vincent and Watt (2018) neatly sum up the core issue:

> Many clinicians regard OA as being simply a disease of 'wear and tear,' and by implication one in which disease modification is not possible. Such prejudices have led to significant academic apathy in this disease that is reflected not only in our poor understanding of disease pathogenesis, but also in the failure to classify the disease with greater precision and to develop sensitive tools for diagnosis and prognostic assessment.

Another barrier to the development of conservative treatments is the type of business model required to support conservative treatments. It is a truism to observe that much of the modern medicine is geared to acute care, rather than geared to disease prevention or chronic care models—but this should not diminish the importance of either prevention or chronic care. With healthcare systems geared to acute care, conservative treatments almost inevitably play a minor role among the therapeutic options in most advanced healthcare systems around the world today. With many vested interests within the healthcare systems, weak consumer knowledge and little bargaining power in their dealings with acute care specialists, potentially health insurers are the most likely vested interest group capable of leading a rebalancing of healthcare systems toward the prevention and conservative management of chronic diseases. Health insurers already reduced the premiums and so reward those who lead certain lifestyles, but if health insurers realized the potential advantages, quite possibly health insurers might in the future offer funding for 'integrative research' of the type presented here.

In this book, we argue for a new, deeper functional understanding of normal cartilage that is offered by the simple 'use and replace' model of cartilage described above. The theme throughout the book is that cartilage is not a static tissue, rather, it is a dynamic tissue that adjusts and remodels itself throughout the life to meet the changing environmental demands placed upon it. Unsurprisingly, articular cartilage is much like all tissues in the body [and yes, even brain tissue remodels, and in some parts of the brain, there is even neural cell proliferation (Boldrini et al. 2018; Sahay et al. 2011; Spalding et al. 2013; Pilz et al. 2018)]. However, articular cartilage tissue remodeling occurs over a longer timescale than many other tissues in the body (and often on a timescale much longer than busy modern lifestyles allow us!). For in the absence of a deeper understanding of cartilage tissue physiology, we are severely constrained in the formulation of strategies for maintaining healthy cartilage and for treating diseased cartilage. Without a deeper understanding, palliative care followed by joint replacement is all but inevitable. And in the absence of an integrated framework that explains how the cartilage 'works' as a tissue, discovery-based science is left to find new therapies through serendipity. But we know that ignorance of basic tissue and cell physiology is demonstrably an inefficient and very costly way to advance clinical science.

For these reasons, we aim to accelerate the process of developing a logical framework for understanding cartilage tissue by attempting to describe how we believe cartilage successfully operates as a tissue over a lifespan. To develop a more complete and hopefully more useful integrated framework for cartilage physiology we have to extrapolate beyond the limited experimental data currently available. The conceptual frameworks presented here can be interpreted as an extended hypothesis. We adopt a hypothetico-deductive approach, which enables us to formulate a more subtle and complete understanding of cartilage physiology that has been attempted heretofore. The risk in this approach is overreach. But the potential reward of this approach is that it offers new conceptual frameworks that are essential for the development of integrated models of cartilage behavior. By taking this riskier approach, we hope that new ways to rationally modify tissue responses to preserve cartilage tissue or reduce its rate of degradation will suggest themselves naturally to readers. And using this approach, basic biomedical science can be more rapidly and precisely be applied to clinically important questions.

To both interpolate between sparse data sets and extrapolate beyond existing data sets, we use our backgrounds in systems modeling of cartilage and related tissues, guided by our grounding in various fields of engineering, physics, chemistry, materials science, poroelasticity, and reactive transport. We freely apply engineering, physics, and chemical principles in the context of cartilage, gleaned from our experiments and a wide re-reading of the cartilage literature. Such knowledge helps to constrain the 'space' of acceptable solutions. We prefer a 'Bayesian approach' to inference (http://en.wikipedia.org/wiki/Bayesian_inference). In other words, our prior beliefs, developed in the engineering sciences, have strongly shaped our interpretation of the literature in order to identify what we believe is the most appropriate a priori models for cartilage (i.e. those interpretations consistent with established principles in the major fields of endeavor noted above). This book can then be interpreted as an attempt to develop rational a priori models (or 'constructs' as some prefer to call them) to progress our conceptual understanding of cartilage physiology.

For a 'Bayesian approach' to inference, such a priori models are crucial because they provide the basis for the interpretation of new data sets. For a Bayesian approach to science, knowledge is cumulative, with a posteriori models becoming a priori models in the next iteration. This fits nicely with an engineering approach to developing analytic and computational models of systems. This Bayesian approach may be contrasted with a 'frequentist approach' to inference, adopted by many scientists, where either a true or false inference is drawn on the basis of 'statistical tests' on data: see (http://en.wikipedia.org/wiki/Frequentist_inference). In this frequentist approach to science, a priori knowledge often plays no formal role in inference, as the underlying pattern is to be distinguished from 'random events' on the basis of statistical tests that have a pre-determined power and significance levels to control type I and type II errors. But the academic landscape is finally beginning to change, and Bayesian approaches are now becoming well-established, even in Statistics Departments, over the last two decades.

Of course, this does not mean that the extended hypothesis on cartilage physiology developed here is necessarily right—by definition, a Bayesian approach dynamically weights and updates hypotheses (a priori knowledge) given the new evidence and/or new beliefs. Explaining the scientific process, McIntyre (2017) comments on 'multilevel thinking':

> Similarly, the most powerful, versatile, and innovative scientific thinking uses varieties and hierarchies of models including precise mathematical models and computer codes in partnership with "conceptual" models that are fuzzier, more intuitive, and often not overtly mathematical—perhaps expressed more in words or pictures or even in intuitive feelings. Quite often, the intuitive models are wrong or partly wrong to begin with, until modified as a result of crosschecking against something more precise.

Precisely, the approach we believe is most beneficial and so efficient given our current state of knowledge. Our a priori 'multilevel' conceptual models of cartilage integrate many separate hypotheses (some of which are reasonably well established in the existing research literature), while developing others that are more speculative. This approach leads to some obvious conclusions about cartilage (often with the usual response by biologists being 'of course!'), along with some surprising conclusions and the seemingly inevitable responses from biologists and clinicians: 'do a (frequentist-designed) animal study or clinical trial to test its significance!'. But doing an animal study or clinical trial is not always possible or advisable. While some frequentist inference is clearly necessary (particularly, in the discovery-led phase of science), to rely on it as much as we do today with decades of research behind us leads to an inefficient and very costly way to progress science from our current state (e.g. many experiments duplicate previous work, or identify false leads that are pursued at considerable cost of limited time and resources) (Ioannidis 2005; Ioannidis 2016). Indeed, for complex systems, application of standard statistical tests can lead to unacceptably high rates of false positives and false negatives (e.g. see the NAS publication: *The Irreproducibility Crisis of Modern Science Report* https://www.nas.org/articles/nas_launches_new_report_the_irreproducibility_crisis). On the other hand, concepts and theories can organize and give meaning to large swathes of experimental data and help flag false interpretations, and by these means increase information gain and improve research efficiency.

Hopefully, government-sponsored funding bodies will come to realize and acknowledge these real costs to the research enterprise of 'business as usual' and begin to reduce their reliance on discovery-based biomedical research, rebalancing toward a more appropriate mix of theoretical biology and experimental research that reflects the maturity of biomedical and clinical sciences today. In the future, we expect funding bodies will offer greater support to research that organizes existing knowledge and research that has the goal of embodying the cumulative knowledge of that discipline area, providing a firmer platform for new knowledge creation.

While an essential part of organizing knowledge is about developing reductionist databases (e.g. UniProt), we need to aim higher in our 'knowledge organization' endeavor and improve the fidelity of both conceptual and quantitative models of biological systems. While the politics of negotiating change is tortuous, such a change in scientific methodology is all but inevitable as disciplines advance. Indeed, history shows science practiced in its mature phase is led by theory rather than by discovery (e.g. the Higgs boson was discovered theoretically—by Peter Higgs in the 1960s—as a particle in the Standard Model for particle physics. This theoretical prediction was confirmed experimentally in 2013 using the Large Hadron Collider at CERN). Interestingly, signs of a transition in the practice of biomedical and clinical sciences can be found at the US FDA, which is beginning collaborative agreements to develop new computational tools to 'virtually test' new drugs and medical devices (e.g. heart pacemakers) as part of their approval process (http://www.3ds.com/products-services/simulia/solutions/life-sciences/the-living-heart-project). Perhaps surprisingly, it is often regulations that most influence the pace of change, rather than rational debate about research funding models.

Despite such cultural differences in the practice of research, we believe the extended hypotheses championed here can provide a useful foundation for thinking about and formulating rational, coherent a priori functional models of cartilage physiology. Adopting this approach can help both wet-laboratory and computational biologists alike and eventually help speed translation into a range of rational therapies employed in the clinical sciences. Where possible, we explain the relevance of our extended hypotheses to clinical science, but in the near term, we expect that the extended hypotheses developed here will be most useful in interpreting new and existing experimental data, in formulating new testable hypotheses, and in helping to direct new experimental and theoretical studies toward the most relevant areas of missing information. We both hope and expect that the approach developed here can be built upon (and of course be modified over time) and that this process will yield additional value from the many billions of research dollars already invested in the cartilage research, as well as yield greater value from the expensive experiments and clinical studies to take place in the future.

This book is organized in the following way:

In Chap. 1, we introduce articular cartilage biology consistent with that found in the literature. It explains what cartilage is, how it works as a biocomposite material, how it deforms, and how cartilage depends on synovial tissue for its nutritional needs.

In Chap. 2, we explain how an aggrecan-collagen network forms in vivo and how they work together to create a functional tissue suited to its in vivo mechanical role. Under repeated cyclic loads, aggrecan, collagen and dozens of extracellular matrix components are continually damaged and so have to be repaired. A key focus is how aggrecan, hyaluronan, the collagen network, and the chondrocyte population together maintain homeostasis of the tissue in vivo. This helps establish a view of articular cartilage as a responsive tissue that can continually adjust to its changing circumstances. While much of this chapter is well supported by the research literature, we touch on some aspects that are more speculative (i.e. meaning there is paucity of direct experimental evidence to support them (e.g.) (i) collagen homeostasis, (ii) radial zonation of the extracellular matrix around chondrocytes, and (iii) ongoing chondrocyte proliferation and migration), but we attempt to make sense of them based on our knowledge of cartilage and in light of our understanding of physical and biological sciences.

In Chap. 3, we propose that articular cartilage is able to adapt to significant changes in its environment. This new view of cartilage is more speculative and really forms an 'extended hypothesis' that attempts to answer some of the key unknown questions about cartilage: (i) From where does the superficial zone in adult articular cartilage originate, (ii) what are the failure modes in articular cartilage, and (iii) is aggrecan and collagen really shed to the synovial fluid and what role does this play in maintaining articular cartilage health?

In Chap. 4, we explain the technical basics of our current understanding of cartilage lubrication. Clearly, lubrication of the joint is very important functionally, for without adequate lubrication, cartilage surfaces roughen, wear rates increase, and the cartilage tissue may progress toward osteoarthritis. We distill a complex set of ideas and principles into something that is coherent using the principle of effective stress and show that that weeping lubrication and a type of boosted lubrication together ensure articular cartilage minimizes wear.

Chapter 5 explains a number of 'systems approaches' for understanding biological tissues. This chapter contains a reasonably detailed account of the cell–cell signaling systems operating in cartilage and a briefer account of the intracellular signaling systems operating inside chondrocytes. It also contains higher-level conceptualisations drawn from other disciplines to aid our understanding of cartilage tissue at a system level.

Chapter 6 explains in detail the meaning of osmotic pressure, Donnan osmotic pressure, and effective stress. We explain the relationships between physical chemistry and articular cartilage mechanics. The chapter begins the process of building a rational, quantitative framework from first principles, which complements biological methods of cartilage investigation.

Chapter 7 presents a large deformation, continuum framework for quantitative modeling of cartilage, using poroelastic modeling as a starting point. This model can be extended as far as a research question might require (e.g. to poro-visco-elasto-plastic models with or without fiber reinforcement with reactive transport).

We hope you enjoy joining us on this demanding but exciting journey, and moreover, we hope that this book forms the basis of ongoing thought-provoking discussions about how cartilage functions as a dynamic tissue.

WA, Australia David W. Smith
WA, Australia Bruce S. Gardiner
VIC, Australia Lihai Zhang
MA, USA Alan J. Grodzinsky

References

Boldrini, M., Fulmore, C. A., Tartt, A. N., Simeon, L. R., Pavlova, I., Poposka, V., et al. (2018). Human hippocampal neurogenesis persists throughout aging. *Cell Stem Cell, 22,* 589–599.

Ioannidis, J. P. (2016). Why most clinical research is not useful. *PLoS Medicine, 13,* e1002049.

Ioannidis, J. P. A. (2005). Why most published research findings are false. *Plos Medicine, 2,* 696–701.

Pilz, G. A., Bottes, S., Betizeau, M., Jorg, D. J., Carta, S., Simons, B. D., et al. (2018). Live imaging of neurogenesis in the adult mouse hippocampus. *Science, 359,* 658–662.

Sahay, A., Scobie, K. N., Hill, A. S., O'carroll, C. M., Kheirbek, M. A., Burghardt, N. S., et al. (2011). Increasing adult hippocampal neurogenesis is sufficient to improve pattern separation. *Nature, 472,* 466–U539.

Spalding, K. L., Bergmann, O., Alkass, K., Bernard, S., Salehpour, M., Huttner, H. B., et al. (2013). Dynamics of hippocampal neurogenesis in adult humans. *Cell, 153,* 1219–1227.

Vincent, T. L. & Watt, F. E. (2018). Osteoarthritis. *Medicine, 46,* 7

Acknowledgements

The authors are especially grateful to Ms. Yvette Harrap, who has worked tirelessly to help bring this book to completion in numerous ways, including providing artwork for the front-piece, particularly in Chaps. 4 and 6. We are also grateful to Dr. Saeed Miramini who has done numerous numerical simulations and to Associate Professor Piaras Kelly, who kindly checked the equations for introductory remarks on large deformation poromechanics. The authors take responsibility for any remaining errors.

Contents

List of Figures

cartilage retrieved from an OA patient and stained (scale bar is 100 μm). Based on the calibration reported in the paper (see Fig. 2 in the paper), acridine orange stains approximately green for pH > 7.0, yellow for pH 6.0–7.0, light orange pH 5.0–6.0, and dark orange less than pH 5.0 (Konttinen et al. 2002). Note that the tissue pHs suggested by this staining are indicative only. Actual tissue pHs may be higher or lower, and the actual pHs for this specimen are in any case probably different to normal articular cartilage, as this specimen was taken from OA cartilage, even though it was graded as 'normal.' Nevertheless, it is clear there is a general decrease in pH with depth from the cartilage surface. It is also apparent that around individual chondrocytes (or around chondrocyte clusters), there is a local decrease in pH (as indicated by orange 'spotting' about the same size as chondrocytes within the ECM) 207

Fig. 2.61 Comparative expression of MMP genes in the fractured NOF (i.e. used as normal reference control cartilage) (filled boxes) compared with osteoarthritis (open boxes) in hip synovium and cartilage. The expression level of each gene was determined and normalized to the level of *18S* rRNA gene expression using ΔC_T [C_T(target gene) − C_T(18S)]. The box-and-whisker plot shows median values, and each section represents a quartile of the data. Samples on or below the dotted line have C_T(target gene) = 40 and are therefore not detectable (n.d.). (Davidson et al. 2006). Note that the top three MMPs (normalized median expression level above 10^{-4}) in normal cartilage are MMP1, MMP3, and MMP14, while in osteoarthritic cartilage the top five MMPs (normalized median expression level above 10^{-4}) are MMP2, MMP9, MMP13 and again MMP3 and MMP14. Note MMP mRNA expressions in the synovium generally remain fairly constant, though MMP2 and MMP13 are upregulated in osteoarthritis, suggesting the primary pathology is cartilage related. Perhaps synoviocytes are responding to the same signaling molecules and ECM damage fragments in the synovial fluid that the chondrocytes are responding to within the cartilage? 210

Fig. 2.62 Comparative expression of TIMP genes in the fractured NOF (i.e. used as normal reference control cartilage) (filled boxes) compared with osteoarthritis [cartilage] (open boxes) in hip synovium and cartilage. The expression level of each gene was determined and normalized to the level of 18S rRNA gene expression using ΔC_T [C_T(target gene) − C_T(18S)]. The box-and-whisker plot shows median values, and each section represents a quartile of the data (Davidson et al. 2006) 210

such as sugar molecules). In this example, the container is filled with solvent (e.g. water). The concentration of osmotically active particles on the left-hand side, c_1, is greater than concentration of osmotically active particles on the right-hand side, c_2 (which is our 'reference solution'). Our reference solution is taken to be at zero pressure (at the top of the container). The Pitot tube measures the mechanical fluid pressure in the left-hand side compartment. The pressure gauge has an impermeable sensor membrane, and so it measures the total stress [which is equal to (any) solid stress minus the fluid pressure, i.e. $\sigma_T = \sigma^s - p$]. For this example, the pressure gauge measures the fluid pressure (as solid stress is taken to be zero). The sign change arises because fluid compression by convention is taken as positive. The semipermeable membrane is so named because it is permeable to water, but it is not permeable to the colloidal molecules in solution. It is the semipermeable membrane that gives rise to an imbalance in momentum exchange between the molecules in the solutions, which results in a pressure jump at the entrance to the membrane pore. However, this pressure jump causes fluid to flow from right-hand side to the left-hand side of the membrane (down the chemical potential gradient of the solvent, water). This flow is transient and causes a mechanical pressure in the solution with more osmotically active particles. This pressure in the fluid is the osmotic pressure, i.e. $\pi_{os} = RT(c_2 - c_1)$. The transient fluid flow and resultant pressure difference across the membrane causes the fluid in the manometer (Pitot tube) to rise, the pressure gauge to read the fluid pressure, and the flexible semipermeable membrane to bend. In physiological experiments, the Pitot tube is often replaced by a more convenient 'wick needle' or 'glass micropipette.' All these devices measure the 'interstitial fluid pressure' (which is the net effect of osmotic and other mechanical pressures in the fluid between cells). Figure drawn by Yvette Harrap 431

Fig. 6.2 Donnan osmotic pressure arises when there is a difference in osmotically active particles [i.e. macromolecules (blue), Na ions (green), and Cl ions (red) in solution] across a semipermeable membrane, but in this case, water and ions in solution are all free to move through the membrane pores—only the (usually negatively charged) macromolecule is not. In Fig. 6.2, the macromolecule is free to move in the solution on the left-hand side. For Donnan osmotic pressure, ions redistribute in response to the negatively charged protein on the left-hand side. At equilibrium, electroneutrality is maintained

Chapter 1
Introduction to Articular Cartilage

1.1 Setting the Scene

Relative to other soft biological tissues, the magnitude of the mechanical loads experienced by articular cartilage is extreme. Peak cartilage-to-cartilage contact stresses reach up to 10 MPa or more (Adams 2006; Hodge et al. 1986; Anderson et al. 2008). Cadaver experiments demonstrate peak contact stresses around 5 MPa are likely during everyday activities such as walking (Marzo and Gurske-DePerio 2009; Hodge et al. 1986; Venalainen et al. 2016). In vivo, measurements suggest 5–6 MPa is experienced by large joints during gait (Morrell et al. 2005) and up to 18 MPa rising from a chair (Morrell et al. 2005). Numerical experiments suggest peak contact stresses while walking of between 4 and 7 MPa (Venalainen et al. 2016).

To put these numbers in perspective, 10 MPa is the stress exerted by 100 kg of weight-force (i.e. 1000 N) acting on an area of just one square centimeter (an area less than the size of your thumbnail). Placed in a similar mechanical environment, all other soft tissues in the body would be triturated (i.e. ground and pulverized) between the 'hammer and anvil' of articulating bones—yet articular cartilage survives and even thrives in this environment over many decades. To help appreciate just how demanding the joint environment is, an important point to grasp is that despite our modern technologies, no inanimate man-made substitute for cartilage has been found that can last in vivo for more than a few decades, despite enormous research effort to find such materials.

In such a hostile mechanical environment, damage to articular cartilage is unavoidable. It follows that articular cartilage *must* have evolved mechanisms that enable it to withstand repeated punishing cyclic loadings for years and decades on end. Given this, a priori we assert that articular cartilage must be equipped with powerful homeostatic processes to repair when damaged.

In this book, we attempt to define the processes that enable homeostasis of articular cartilage and to explain how tissue homeostasis operates. To do this, we want to first identify and then develop as detailed an understanding of these

D. W. Smith et al., *Articular Cartilage Dynamics*,
https://doi.org/10.1007/978-981-13-1474-2_1

maintenance processes as we possibly can. We want to explain how the coordination of maintenance processes can lead to a dynamic tissue that can adapt to its environment. We want to explain how the failure of one homeostatic process can lead to a 'cascading failure' that ends in tissue destruction.

Unfortunately to elucidate these homeostatic processes in cartilage is not an easy task. Some of these processes have not yet been identified, and many of the remaining maintenance processes are poorly characterized. For example, some dispute that chondrocytes in normal, non-osteoarthritic tissue can proliferate, while others maintain that the collagen network in adult tissue is essentially unchanging over time. Indeed, something as basic and fundamental as how the extracellular matrix 'self-organizes' itself and then maintains itself are only dimly perceived in the cartilage literature, with no coherent account to be found, which seems quite remarkable when one thinks about it.

Even when a homeostatic process is comparatively well understood, the processes involved are often not well-articulated and/or they are not quantified in the extensive research literature on cartilage. For example, though it is well known that aggrecan is mechanically and enzymatically degraded and replaced in cartilage, but when, where, and the details of how this occurs are not clearly known. For example, it is not known if there are different pools of aggrecan in articular cartilage with different turnover kinetics and/or different damage mechanisms, and it is not known how aggregate, a key molecule that helps retain aggrecan in cartilage and so gives the ECM its compressive stiffness, is maintained in vivo.

In a historical review of the 'understanding of cartilage,' Benedek observes (Benedek 2006):

> Three questions intrigued nineteenth century investigators: of what is articular cartilage composed, how is it nourished, and what keeps it from wearing away.

In pursuing our 'understanding of cartilage,' these and many other 'difficult questions' about cartilage homeostasis and adaption in vivo will be the central focus of this book.

To achieve our ends, we need to start at the beginning, first creating a suitable platform of knowledge from which we can develop our dynamic systems view of cartilage. Some readers may be unfamiliar with biology, and they may like to first read Appendix A. Appendix A reviews some of the fundamental chemical, physical, and biological concepts that apply to all cellular tissues. While basic, the review is uncommon because it collects together chemical, physical, and engineering principles and relates them to cell function.

For those who are familiar with biology but unfamiliar with cartilage tissue, this chapter describes some of the basic known facts about articular cartilage, drawn from the very extensive research literature on articular cartilage. We first describe where articular cartilage is found in the body, how it differs from other types of cartilage, its histological features, and its composition, and how it deforms under load. Throughout the book, we generally focus on human knee articular cartilage (as perhaps most published data pertains to this joint), but we also draw on data obtained from other joints (e.g. hip, ankle, and head of humerus), and we draw on data obtained from other species (e.g. mouse, rabbit, and bovine articular cartilage) as well as some data from bioengineered tissue scaffolds.

The human knee joint is just extraordinary to view while contemplating its function. How stable is it? What forces do knee joints transmit as joint angle changes? Why are the femoral condyles rounded while the tibial plateau is so flat? What functional role do the various ligaments serve? What are the crescent-shaped structures referred to as menisci, and what exactly do they do? In any case, how do the menisci survive given that on first impressions they are wedge-shaped in cross-section—do not wedges usually prevent relative motion—couldn't they jam joint rotation? What effect do the powerful muscles groups around the knee joint have on knee joint stability? What role do tendons (and the sesamoid bone) play?

Clearly, the knee joint is kinematically much less stable than the 'ball and socket' joint at the hip above or the talo-crural 'hinge' joint of the ankle below. Stability refers to the magnitude of the restoring forces when a joint is perturbed (or dislocated) from its normal operating position. Mathematically, stability is assessed from the eigenvalues of a 'stability matrix' (McGill and Cholewicki 2001), but actually assessing dynamic stability of a joint (i.e. during joint motion as muscle forces vary) is not straightforward. So, clinicians use some kind of simple clinical test. For example, they might assess knee joint stability based on response to a perturbation with the knee at rest. For example, with the patient in the supine position and the knee flexed, the clinician sits on the bed and while stabilizing the foot (usually by partially sitting on it), attempts to pull the tibia forward on the femur, and based on the movement so induced, assesses laxity of the cruciate ligaments.

What is clear is that in contrast to the hip joint, the tibial plateau in the knee joint offers only a shallow depression on top of the tibia for the menisci and extremities of the femoral condyles (in much the same way as the rounded head of humerus rests in a shallow depression called the glenoid fossa), so there is not much stability arising from the geometry of the knee joint (e.g. the resistance to sliding movement arising from kinematic relationships of tibia and femur is small). Perhaps unsurprisingly, both the knee and shoulder joints are classified as having low stability, based on the fact that they are comparatively more likely to 'sub-lux' (i.e. begin to dislocate) or even to dislocate under peak loadings experienced during daily activities, particularly if the joint has experienced previous trauma and the ligaments are lax. To help prevent the femoral condyles simply sliding off the tibial plateau, strong ligaments join the tibia and femur bones together (i.e. the medial and lateral ligaments and the anterior and posterior cruciate ligaments), while still allowing the normal flexion (bending) and extension (straightening) of the knee joint.

Strong muscles span across the knee joint to powerfully stabilize the knee while standing, walking, and running (joint stability due to muscle forces across the joint arises because a component of the muscle force acts to restore any (sliding) shear perturbation at the joint). Indeed, muscle forces across the knee joint are so large that while standing approximately the entire body weight acts across *each* knee joint (rather than (about) half the body weight as may be expected based on accounting for body weight alone). It is the muscle forces across the knee joint while standing that accounts for the other half of a body weight per knee joint, bringing the total to one body weight acting across each knee joint while standing (i.e. two body weights are measured across the knee joints while standing). And while walking on level ground or walking downhill, muscle activity increases, so the 'body weights' acting across each knee joint can be four times body weight for level walking or up to eight times body weight for downhill walking (Kuster et al. 1997; Winby et al. 2009).

Stress is defined as force per unit area. Let us put use some numbers to calculate stresses. Assume a large 100-kg person has say 1000 mm^2 of contact area in each knee joint (Kuster et al. 1997) and exerts a peak load of four body weights while walking. A simple calculation shows the peak average stress in each knee joint during level walking is about ($4 \times 100 \times 10/1000$ mm^2 equals) 4 MPa, while the average stress in each knee joint for downhill walking increases to about 8 MPa. As mentioned above, some have estimated contact stresses of 10–20 MPa for activities such as walking upstairs (Adams 2006). Similar stress magnitudes to those in the knee joint are estimated for articular cartilage in the ankle joint (e.g. 6.5 MPa during walking and higher during running) (van Dijk et al. 2010).

To develop an appreciation of the meaning of these stress numbers, we can confirm from the definition of stress that 10 MPa is the stress exerted by 100 kg of weight-force acting on an area of just 1 cm^2. Now, a stiletto heel is defined as having a heel diameter at the ground of less than 1 cm (i.e. 0.79 cm^2), so 100-kg weight-force acting on a single (large diameter) stiletto exerts a stress of about 12.7 MPa on the floor. To appreciate the implications of such a stress, we recall that stiletto heels often damage linoleum and wooden floors, leaving small lasting depressions in their surfaces.

But stress distributions on the cartilage surface are not uniform. Taking into account that contours of stress magnitude can be defined over the cartilage-on-cartilage contact area over the tibial plateau [see, for example, the experimental data presented by Marzo and Gurske-DePerio (2009)], then it is not difficult to estimate that peak contact stresses could be as high as 20 MPa during activities such as walking upstairs or downstairs, or up and down a hill. However in this context, we mention that experiments with normal cartilage rubbed back and forth on normal cartilage continuously for 1 h, under 17 MPa of compressive stress, have shown visible damage to the cartilage surface (McCann et al. 2009). Something extraordinary must be going on within the knee joint in vivo to prevent such damage rapidly accumulating, as people often walk downhill over long distances repeatedly loading their cartilage, and yet they do not develop diseased joints (or at least, diseased joints do develop within one hour). Comments (Mow et al. 1992):

> Thus the fundamental difference between wear of articular cartilage in vivo and wear of an engineering bearing is that in the biological system, there is a balance of mechanical attrition and biological synthesis and repair.

We will discuss repair and wear rates articular cartilage and the lubrication of synovial joints at length in Sects. 2.6 and 3.5 and Chap. 4.

We need to keep in mind that alterations in a person's activity level, weight, and gait pattern, or defects in the alignment, shape, or strength of bones, or changes in the ligaments or the strength of muscles, or changes in neural control, may all contribute to knee joint instability. And given the typically high stresses outlined above, it becomes clear that any joint instability may in turn create very challenging new stress conditions for articular cartilage, perhaps causing the joint to degenerate. For example, simply lengthening of the anterior cruciate ligament (e.g. as a result of trauma to the knee) may allow the femur to slide forward on the tibial plateau causing unusual shear stress on the surface of articular cartilage [which may lead to increased rates of chondrocyte apoptosis (Waller et al. 2013)]. Or damage to the meniscus may reduce the effective contact area and so concentrate contact stresses on the cartilage surface (Marzo and Gurske-DePerio 2009). A slight misalignment of the ankle can dramatically increase contact stresses and similarly for the knee. Though the geometry of adult joints does change gradually in adulthood (the area of tibial plateau increases with age at a rate of about 1.5% per year [see discussion in Sect. 3.8)], comparatively more rapid changes in a person's weight can lead to very significant increases in joint loadings, which in turn can lead to early onset osteoarthritis (Silverwood et al. 2015; Niu et al. 2017). Indeed, obesity is believed to be responsible for the increasing rates of joint replacement at earlier ages (Harms et al. 2007). There is a nonlinear relationship between BMI and risk of developing knee osteoarthritis (Mork et al. 2012; Zhou et al. 2014) (the relative risk of OA is reported to increase from 1.0 for a BMI 22.5 to 1.6 at BMI 25, 3.6 at BMI 30 to 7.5 at BMI 35). And with osteoarthritis and joint replacement at earlier ages, comes increasing rates of joint revision (Warth et al. 2014; Pabinger et al. 2013)—unfortunately, subsequent joint replacements are usually less satisfactory than the initial replacement. This largely explains the advice to the young and middle aged to delay joint replacement as long as possible, which clearly prolongs disability and suffering experienced by those with osteoarthritis.

Exactly how the structures in the knee become the coordinated functional unit we refer to as a 'normal knee joint' remains very poorly understood. For example, what is an appropriate answer to this apparently 'simple' question: How do menisci, joint capsule, ligaments, tendons, and muscles attain the right length and tension to facilitate normal knee movement, while minimizing instability and potentially damaging interactions between each of the tissues? Can we even begin to sensibly answer such a question?

In attempting to generally answer this question, clinical practice provides some important clues to this simple question. For when a normal joint is immobilized, the joint and surrounding tissues suffer 'disuse atrophy' (i.e. muscle bulk is lost, the amount of collagen in tendon and ligament reduces and may become disorganized, bone becomes osteoporotic, while articular cartilage 'thins') (Brandt 2003; Eckstein et al. 2006; Yasuda and Hayashi 1999; Udaka et al. 2008). Further, those joint tissues normally in tension (such as muscle, tendon, joint capsule, and ligaments) shorten with disuse (Harvey and Herbert 2002; Herbert and Crosbie 1997; Gardner et al. 2012; Young et al. 2016). Indeed, the tissues normally in tension shorten so much that within a matter of weeks the joint may be described clinically as 'frozen' (i.e. the joint offers noticeable resistance when subjected to manually applied bending moments). Avoiding so-called contractures by intermittent mobilization of the joint, either passively or actively, is a major focus of physical therapists, along with being an important focus of general nursing. For example, avoiding joint contracture strongly influences post-surgical care—e.g. it is now standard practice that patients become 'mobile' (i.e. get out of bed) as soon as possible following surgery (Buckwalter 1995).

At a high level, we can interpret these clinical observations in terms of two physiological processes operating continuously within normal diarthrodial joint tissues. The first process is a 'constitutive tendency' for the tissues normally in tension to shorten, effectively 'pulling' the joint together. The second process is the tendency for mechanical loading to damage and lengthen the tissues in tension, so loosening the joint and promoting its mobility.

We may postulate that normal joint homeostasis can be interpreted as a continual balance between the constitutive tendency for joint tissues to contract, 'pulling the joint together,' and the functional tendency for mechanical loading to lengthen joint tissues, 'tearing the joint apart.' This conception places mechanical loading in the role of being a fundamentally important part of normal joint homeostasis. The concept has been implemented in an adaptive tendon model (Young et al. 2016). Constitutive contraction of tissues is mediated by cells, while functional lengthening results from daily tendon usage. Unsurprisingly, overloading of tissue may lead to overt muscle, tendon and ligament tears, which generally required prolonged periods of reduced activity to enable the tissue to adequately repair. It is well known clinically that tissues comprised primarily of extracellular matrix (such a tendon, ligament and cartilage) are not well vascularized and comparatively slow to repair.

Skeletal tissues in compression, such as bone and cartilage, are usually more dimensionally stable than the tissues in tension, but disuse leads to tissue atrophy (e.g. both cartilage and bone 'thin') (Brandt 2003; Behrens et al. 1989, Eckstein et al. 2006). Normal levels of activity (and so normal skeletal loads) lead to healthy skeletal tissues by promoting an appropriate balance between the rate of damage and the rate of repair. Interestingly, only comparatively low levels of mechanical loading are required to maintain bone (e.g. only 10 load cycles is required at 2000 microstrain, though the number of cycles required increases as the strain magnitude decreases (e.g. about 10,000 load cycles is required at 600 microstrain) (Qin et al. 1998).

Excessive activity and joint loading may lead to the rate of damage exceeding the tissue's rate of repair, while overloading the joint may lead to cartilage trauma or to bone fractures. It becomes apparent that skeletal tissues in tension and compression have a 'normal operating range' for cyclic mechanical loading, over which the rates of tissue damage and repair can come to an appropriate and manageable balance. In fact, our daily mobility is so habitual and expected by our bodies that it is taken as a given, so now we actually require an appropriate 'dose' of mechanical loading each day (along with attendant tissue damage) to avoid contractures and so maintain skeletal tissue health (Eckstein et al. 2006). These observations and high-level conceptualization goes some way toward answering the question: How do menisci, joint capsule, ligaments, tendons, and muscles attain the right length and tension to facilitate normal knee movement, while minimizing instability and potentially damaging interactions between each of tissues?

But clearly there are many subtleties, for what may be an optimal mechanical loading dose for one joint may not be optimal loading for another joint, even in the same limbs (Munugoda et al. 2018). This disparity in optimal mechanical loading dose may be expected to increase over time, as different joints suffer different levels of joint trauma with increasing age and so accumulate different levels of tissue damage.

To satisfy ethical and many practical constraints when doing experimental research (e.g. obtaining ethics approval from an appropriately convened ethics committee, limited funding, personnel, equipment and resource constraints including finding suitable subjects or test materials, etc.) as far as possible, researchers like to 'keep it simple.' Consequently, biological researchers usually pose comparatively simple, well-defined research hypotheses and/or study an individual component of a synovial joint (e.g. cartilage). But as we have deduced above, the truth is cartilage performance is strongly dependent on the satisfactory performance of all the anatomical structures that go to make up a synovial joint. Interactions between joint tissues are crucially important, as well as interactions with blood plasma, lymphatics, and the immune system. So, when a meniscus or ligament within a joint, or a muscle or tendon across a joint, or a nerve to a joint, or bone supporting articular cartilage in a joint, or the immune system is either genetically abnormal or damaged, almost invariably this has important consequences for the long-term performance of cartilage within the joint.

For example, clinical experience has shown that significant damage to either the meniscus or cruciate ligaments usually results in subsequent damage to articular cartilage (Chinzei et al. 2018), which feeds back to further damage of the meniscus and cruciate ligaments. Or for another example, there is a little researched interaction between innervation and the immune response, which can impact joint health (Stangenberg et al. 2014). Or for another example, diabetes can lead to sympathetic nerve degeneration and 'Charcot joint' (Watkins and Edmonds 1983; Varma 2013). But in fact any cause leading to loss of joint innervation and so loss of pain sensation can lead to a Charcot joint, which usually results in rapid joint destruction due to abnormal joint mechanics and overloading of the articular cartilage. Clearly, pain is an integral part of a crucially important behavioral (and protective) feedback loop! Because studying these interactions between tissues is both more difficult and more time consuming than studying single tissues, not a lot is known about the interactions between joint tissues. What is known mainly arises from experience obtained in clinical practice, which necessarily takes a high-level, practical view of these matters. Perhaps needless to say, there is much to learn.

As we begin our relentless focus on articular cartilage, these macro-observations and high-level conceptualizations of joint behavior nicely set the scene for all joint tissues to be viewed as dynamic entities that maintain their own homeostasis within a dynamically varying mechanical environment—cartilage is but one tissue that has to continually adjust itself to its environmental conditions within the joint.

Indeed, these macro-observations and high-level conceptualizations of joint behavior are helpful in reminding us that we always need to keep in the back of our mind that articular cartilage functions as part of an extraordinarily complex mechanical system we know as a diarthrodial joint. For it is undeniably apparent that articular cartilage is dependent on many other tissues functioning normally in the joint, as they in turn are dependent on articular cartilage functioning normally in the joint (Falah et al. 2010; Brandt et al. 2006). In other words, articular cartilage is part of a 'joint system.' Indeed, systems within the joint system, set within the whole body system, will be a recurring theme of this book. This reminds us that just about everything is depends on everything else in the body, to a greater or lesser extent.

1.2 Where Is Articular Cartilage Found?

Diarthrodial joints enable joint mobility and so mobility of the organism. Diarthrodial (or synovial) joints are characterized by their joint capsule, which has a 'lining' of synovial membrane that produces synovial fluid—hence their anatomical description as 'synovial joints.' In synovial joints, synovial fluid directly bathes the surface of the articular cartilage covering the ends of long bones.

Articular cartilage in diarthrodial joints is made of a tissue known as 'hyaline cartilage.' Fresh normal hyaline cartilage has a glistening white appearance when viewed on the bone in vivo. Sliced away from the bone and held up to the light, hyaline cartilage appears to be translucent milky-blue, which is described as being 'glassy' in appearance. This simple observation is the basis for naming the tissue 'hyaline' cartilage, as hyaline is derived from the Greek and Latin words meaning 'transparent' and 'glassy.'

Hyaline cartilage may be contrasted with 'elastic cartilage,' which is found in the outer ear and epiglottis (see Fig. 1 in Krishnan and Grodzinsky (2018) for schematic illustration of the distribution of cartilage types in the human body). Elastic cartilage is similar to hyaline cartilage but takes on a yellow color due to the presence of many elastic fibers within the extracellular matrix. Elastic fibers are usually rare in hyaline cartilage. Hyaline cartilage may also be contrasted with 'fibrous cartilage.' Fibrous cartilage is found in non-articulating joints (e.g. the pubic symphysis joint), but fibrous cartilage is also found toward the center of menisci found in the knee joint. For dogs, cartilage accounts for less than 1% of the dry body weight, while articular cartilage accounts for less than 10% of the total uronic acid content, present mainly in cartilage throughout the body (Atencia et al. 1989) (disaccharides of glucuronic acid and (sulfated) galactosamine make up most of the glycosaminoglycans (GAGs), a major extracellular component of articular cartilage). So, articular cartilage is probably about 0.1% of the dry body weight.

Importantly, we note that when articular cartilage is given a chance to repair following significant tissue damage (e.g. when the hyaline cartilage in diarthrodial joints has been partially worn away or completely worn away to the underlying bone), fibrous cartilage repair tissue usually replaces normal hyaline cartilage.

Importantly, experience has shown that a fibrocartilage repair of articular cartilage is much less durable than normal hyaline cartilage (Falah et al. 2010).

Fibrous cartilage is distinguished by having a significant fraction of type I collagen forming the cartilage collagen network, while normal hyaline cartilage contains only type II collagen (i.e. there is no type I collagen normally in articular cartilage). We note that type I collagen is the primary tensile, structural extracellular matrix fiber found in all tissues of the body *other* than hyaline and elastic cartilage. For example, type I collagen is the primary structural collagen found in tendon, ligament, muscle, and bone. However, fibrocartilage with type I collagen is not well suited to the *high compressive loadings* experienced by articular cartilage (Falah et al. 2010). A specialized type II collagen has evolved for these particular mechanical conditions. Type II collagen will be discussed further in Sect. 1.4.

As mentioned previously, a particular focus of the book is the knee joint. Even a casual inspection of the anatomy of the knee joint makes it clear that the knee joint is surprisingly complicated (see, for example, Figs. 4 and 5 in Last (1948), showing anterior and posterior views of the flexed knee with anatomic details). It is immediately apparent that articular cartilage within this finely sculpted and tuned piece of 'machinery' will depend on the satisfactory operation of many separate components and tissues within and around the knee joint, including the joint's vascular supply, the ligaments, the crescent-shaped menisci, the joint capsule, the underlying bone, the tendon, and muscles around and across the joint, together with innervation of these structures.

To remind us of this important fact, some clinicians speak of the 'joint organ' (Loeser et al. 2012). An organ is a group of tissues that performs a specialized function. In this case, the function of the synovial joint is joint mobility, enabling locomotion.

1.3 How Is Hyaline Cartilage Described Histologically?

Chondrocytes are the only cell type found in normal cartilage (see Fig. 1.1). Under the light microscope, chondrocytes can be seen to be located in 'openings' in the collagen network of the cartilage extracellular matrix. The openings in the collagen network are called 'lacunae.' In vivo, the lacunae contain one or more chondrocytes, with more than one cell indicating chondrocyte proliferation (see Fig. 1.1).

We note that many lacunae containing more than one chondrocyte are often elongated [i.e. so they have an axis orientation (see Figs. 1 a–c in (Poole 1997)], which suggests the lacunae elongate to accommodate chondrocyte proliferation). The lacunae that contain more than one chondrocytes tend to have their long axis oriented vertically when in the deep zone and horizontally in the superficial zone (see Figs. 1.1, and 2.21 for a close-up of groups of vertically oriented cells in the deep zone and Fig. 2.41 for a close-up of groups of horizontally oriented cells in the superficial zone).

Chondrocytes are surrounded by the so-called pericellular matrix. Beyond the pericellular matrix is the 'territorial matrix' and more distant again that is the

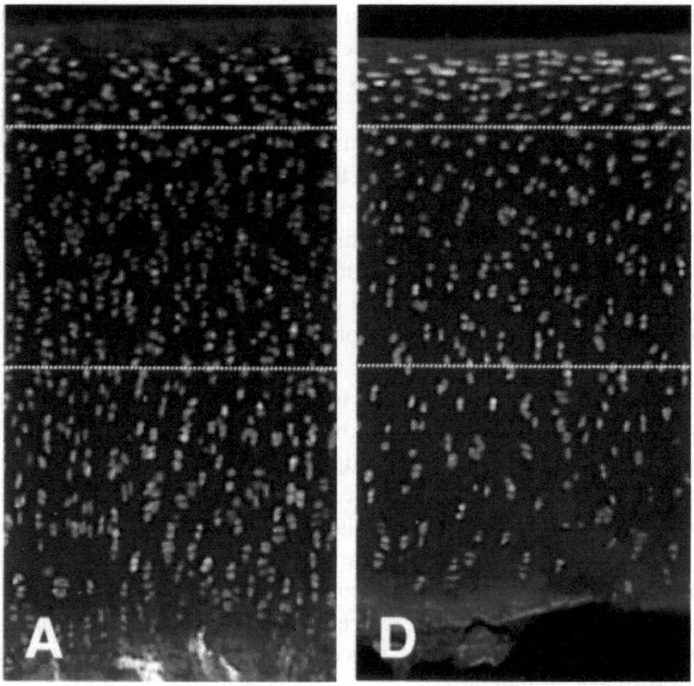

Fig. 1.1 Low magnification epifluoroscope images of bovine (1–2 years old) articular cartilage labeled for actin microfilaments (Langelier et al. 2000). Image **a** is for cartilage subject to load-bearing contact forces on its top surface (top of image), while Image **b** is for a more peripheral region of cartilage that experiences reduced load-bearing contact forces. Both images are segmented by horizontal lines into superficial, middle, and deep zones. Note that Image **a** has a higher cell density than Image **b**, indicating more highly loaded regions of tissue experiencing larger strains require (*and have*) greater proliferative capability and ECM repair capacity, so as to maintain the integrity of the tissue. Consistent with this interpretation, Image **a** *shows* greater ongoing proliferation than Image **b**, as a greater proportion of chondrocytes is arranged in 'proliferative *columns*' in both the deep and middle zones. In addition to differences in cell density and proliferative drive, centrally located (porcine) chondrocytes contain about twice as much aggrecan mRNA as lateral chondrocytes, and upon loading central chondrocytes increases aggrecan mRNA about twice as much as similarly loaded lateral chondrocytes (Bevill et al. 2009)

'interterritorial matrix' [see, for example, the labels shown in Fig. 1c of Poole (1997)]. Together, the pericellular, territorial, and interterritorial matrix are known as the extracellular matrix (ECM). Unusually for any tissue except cartilage, around 95% to 98% of adult articular cartilage tissue volume is ECM (Fig. 2.21 gives some sense of the volume of extracellular matrix served by a comparatively small number of chondrocytes).

Usually, four zones are identified in mature articular cartilage: (i) the superficial zone, (ii) the transitional zone, (iii) the deep zone, and (iv) the calcified cartilage zone [see zonation labels in Fig. 1.1 and the labels in Figs. 1a in Poole (1997)]. The transitional zone is sometimes referred to as the 'middle' or 'intermediate' zone, while the deep zone is sometimes called the 'radial' zone (due to the 'radial orientation' of the collagen fibers—where 'radial' is used in the sense of collagen fibers being oriented normal to the underlying subchondral bone). When greater specificity is required, the transitional and deep zones are sometimes divided into upper and lower transitional zones, and upper and lower deep zones.

As the name suggests, subchondral bone is that bone immediately beneath articular cartilage. The subchondral bone generally refers to the cortical (or dense) 'bone plate' on which the articular cartilage rests, but the cortical bone plate in turn rests on trabecular (or spongy) bone.

For humans, the proportion of each articular cartilage zone relative to the total cartilage thickness for normal human femur (hip and knee) cartilage is reported as being $7 \pm 1\%$ for the superficial zone, $19 \pm 7\%$ for the transitional (middle) zone, and $75 \pm 9\%$ for the deep (radial) zone (Changoor et al. 2011). We note that to classify the zones in the cartilage based on morphology and so arrive at this estimate, plane light microscope images were analyzed using the software Bioquant Osteo II, as explained further in Changoor et al (2011).

However, we need to keep in mind these are average estimates. We should also remember the 'zonal thicknesses' depend on the 'selection criteria' employed. The influence of selection criteria is perhaps most evident in (colorized) 'T2 MRI' [see, for example, Fig. 2 in Goodwin et al. (2004)], which depends on both the density and the orientation of collagen fibers. This image shows substantial variation in the thickness of the 'superficial' and 'deep' zones across the tibial plateau when 'collagen density and orientation' are used as selection criteria for zonation. We mention that the interpretation of Fig. 2 in Goodwin et al. (2004) is perhaps best done with the aid of Fig. 6 in Goodwin et al. (2004). Further information about T2 MRI and its sensitivity to both collagen density and collagen fiber orientation can be found in Gold et al. (2009), Xia (2013).

In the transitional zone of articular cartilage, the collagen network is 'interwoven,' but it also has observable 'arcade-like structures.' The collagen network arcades form as radially oriented collagen fibers, originating in the deep zone, 'turn over' in the transitional and superficial zones (see arcade-like surfaces/shapes shown in Fig. 1.2 and the sketch of these arcades in Fig. 2.16), thereby anchoring the superficial zone and transitional zones to the deep zone and ultimately anchoring the collagen fibers directly into the bone (Poole 1997; Clarke 1971, 1974; Lane and Weiss 1975; Hughes et al. 2005).

In the lower deep zone, thick collagen fibers (sometimes forming 'sheets' or 'ribbons' of collagen) are arranged predominantly normal to the subchondral bone surface (see the scanning electron microscope images shown in Fig. 1.2). However, between the thick collagen fibers and sheets of collagen, there remain regions of interwoven collagen network, which presumably help resist shear forces as well as help to retain the aggrecan in the deep zone collagen network (Poole 1997; Clarke 1974, 1971; Lane and Weiss, 1975).

The cartilage calcified zone (CCZ) lies between the 'tide-mark' (which is a thin region a few microns thick at the interface between the CCZ and the hyaline cartilage) and the 'cement line' (which is a thin region a few microns thick at the interface between the articular cartilage and the subchondral bone). The cement line

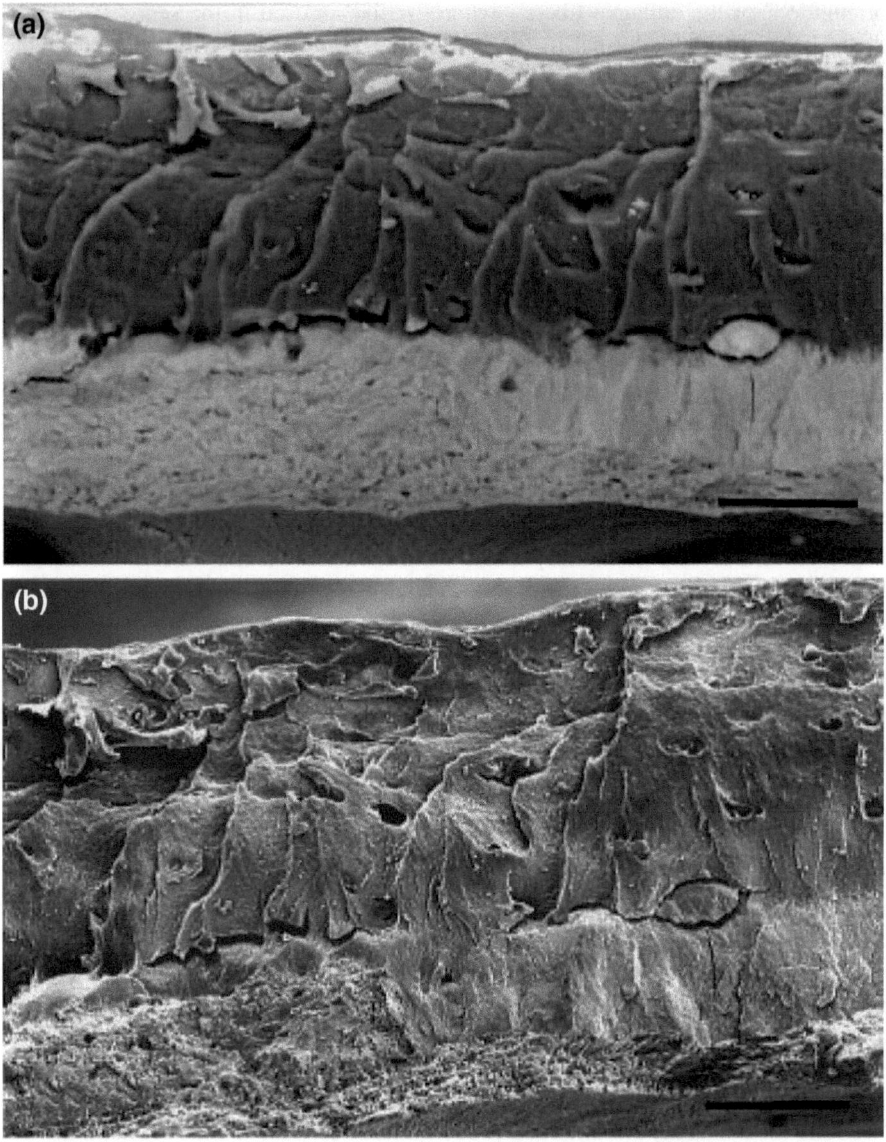

Fig. 1.2 Scanning electron micrograph of a freeze fracture showing normal human articular cartilage on osteoporotic subchondral bone in 76 year old (Li et al. 1999). Scale bar equals 100 μm. Image **a** shows sheet-like structures revealed by the freeze fracture preparation imaged in backscatter mode, while Image **b** shows the same surface in scanning electron mode (showing more clearly the irregularities in the 3D surface). Further high-resolution (7T) MRI of human knee tibial cartilage can be found in Goodwin et al. (2004)

is defined by a sharp demarcation between type II collagen in articular cartilage and type I collagen in subchondral bone.

The calcified cartilage zone is distinguished by containing inorganic mineral crystals of similar composition to bone. The crystals are mainly small, disordered crystals of calcium hydroxyapatite (i.e. $Ca_{10}(PO_4)_6(OH)_2$) (Duer et al. 2009; Zhang et al. 2012), together with other small, disordered calcium crystals [e.g. calcium carbonate ($CaCO_3$)].

The dissolution or precipitation of hydroxyapatite depends on the concentrations of calcium, phosphate, and hydroxyl ions in solution. The product of these ion concentrations, raised to appropriate powers, form the so-called ionic product ($I_p = [Ca]^{10}[PO4]^6[OH]^2)$). To decide if dissolution or precipitation is likely to occur, the ionic product is compared to the so-named 'solubility product' (K_{sp}), https://en.wikipedia.org/wiki/Solubility_equilibrium. The solubility product is a special kind of equilibrium constant, such that sparingly soluble salts in an aqueous solution are in equilibrium with the solid phase of a mineral (the activity of a solid mineral phase is by definition equal to one). For the mineral hydroxyapatite https://en.wikipedia.org/wiki/Hydroxylapatite, $K_{sp} = [Ca]^{10}[PO_4]^6[OH]^2 \sim 10^{-117}$ (Dawes 2003). Dissolution of a crystal occurs when the ionic product (I_p) of the calcium, phosphate, and hydroxyl ions in solution is less than the solubility product for hydroxyapatite K_{sp} (i.e. $I_p < K_{sp}$), while precipitation occurs when the reverse occurs (i.e. $I_p > K_{sp}$) (Dawes 2003).

We observe that for normal body chemistry, calcium crystals tend to dissolve in low pH environments (pOH reduces as pH increases, i.e. $pOH = pK_w - pH \approx 14 - pH$). The solubility of hydroxyapatite increases approximately tenfold for each unit decrease in pH (e.g. at pH 7, the solubility of hydroxyapatite is 30 mg/L, while at pH 4 it is 30 g/L) (Dawes 2003). So the formation of calcium crystals in tissues like cartilage is usually conditional on the tissue pH being only somewhat acidic (say pH 5.5) or higher. We note that if chondrocytes in the deep zone increase their metabolic rate and produce more lactic acid (via anaerobic metabolism), the increase in concentration of lactic acid has the effect of reducing the local pH (the pKa of lactic acid is 3.86), which will tend to dissolve calcium carbonate and apatite crystals locally.

We mention that in practice crystal formation often has some barriers to initial crystal formation (e.g. the surface energy of the crystal may be significant when the crystal is small). This means that $I_p > K_{sp}$ and no crystals form until I_p increases to a sufficiently high value (the solution is then said to be 'supersaturated'). This barrier to crystal formation may be reduced by so-named 'nucleators' (some nucleators are secreted proteins), which have the effect of reducing the surface energy and thereby promoting crystal formation.

The thickness of the calcified cartilage zone tends to decrease with age. In the head of humerus and femoral head articular cartilage, the CCZ is about 200 μm thick at age 20, but decreases linearly with age to about 100 μm at age 80 (Lane and Bullough 1980). However, the number of visible tidemarks in the CCZ increases with age, averaging about one at age 20, but increasing to an average of two at age 80 (some articular cartilages have several visible tidemarks). The presence of several visible tidemarks suggests that the CCZ thickness increases and decreases over time, probably in response to environmental stressors on cartilage tissue. It is noted that the CCZ may increase in early OA but decreases in late-stage OA (Hwang et al. 2008).

An examination of histological sections of articular cartilage [see, for example, Fig. 1a–c in Poole (1997)] reveals that normal adult hyaline cartilage contains no blood vessels within the tissue. Adult articular cartilage is thought to be avascular because the large compressive strains normally experienced by the tissue would inevitably mean that red blood cells in articular cartilage are triturated, along with any vascular tubes made of endothelial cells. This challenging mechanical environment requires that chondrocytes escape this fate themselves by secreting hyaline cartilage extracellular matrix, which is specially designed to survive repeated large compressive loadings experienced by diarthrodial joints.

Chondrocytes themselves are reported to have a small in situ compressive stiffness (e.g. $E_{\text{chondrocyte cytoplasm}} \sim 0.3$ kPa (Trickey et al. 2000) or $E_{\text{chondrocyte cytoplasm}} \sim 3$ to 4 kPa, $E_{\text{chondrocyte nucleus}} \sim 5$ to 6 kPa) (Ofek et al. 2009), with some reports up to 15 kPa) relative to that of the ECM. This suggests the compressive stiffness of chondrocytes is about two to three orders of magnitude less than that of the initial equilibrium compressive stiffness cartilage ECM (i.e. $E_{\text{whole cartilage tissue}} \sim 0.3$ to 5 MPa) (Ding et al. 1998a, b, Darling et al. 2010). Such an arrangement suggests that the chondrocytes are 'protected' from the mechanical stress borne by the extracellular matrix in the sense that most of the load goes around the chondrocytes through the stiff ECM (much as lines of maximum principal stress go around a tunnel opening). However, this measured stiffness may be an artifact of testing, and the range of chondrocyte stiffnesses is probably an open question. For example, consideration of force–stress balance across the cell membrane suggests higher chondrocyte stiffness. For GAG, aggrecan core protein and aggrecan are synthesized within chondrocytes (where they contribute to intracellular osmotic pressure), pass through the secretory pathway where they are assembled and modified (Prydz 2015), and then are secreted into the pericellular matrix that has a high concentration of aggrecan. If these newly synthesized molecules are in low concentration within the cell, then a great deal of work needs to be done at the cell membrane in order to secrete them. In addition, chondrocytes need to be capable of adapting rapidly to straining of the local ECM and to changes in local osmotic pressure within the ECM. In practice, this means that water needs to move easily between intracellular and extracellular compartments, so intracellular and extracellular osmotic pressures remain reasonably well balanced at all times.

Chondrocytes obtain energy and materials required to construct their cell bodies and ECM via the articular cartilage surface that is exposed to the synovial fluid, which may be some considerable distance away (i.e. two or more millimeters) from the chondrocytes located in the deep zone. The transport limitations of distant nutrient and oxygen supply are often evoked to explain why cartilage is very slow to heal. It is true that the better the blood supply to a tissue, the faster the rate of healing. For example, a small cut to the surface of the tongue—which has a very dense blood supply—usually heals within a day, while a small cut to the external skin—with a good blood supply—usually heals within a week. Without its own blood supply, there is no doubt that adult articular cartilage heals slowly (Wang et al. 2006) relative to the rates of healing observed in tissues with a good blood supply.

But we also note that at least some of the explanation for the poor rate of healing lies in the fact that the joint remains in daily use, so any repair to excessive mechanically damaged cartilage may be quickly undone by further excessive mechanical load. Indeed, it has been shown that joint distraction (i.e. using a metal frame around a joint to unload the whole joint for periods usually of 8–12 weeks) leads to a surprising amount of 'cartilage regeneration' (Intema et al. 2011; Mastbergen et al. 2013; van Valburg et al. 2000). And it has also been shown that the cartilage regeneration following distraction appears to be sustained for at least a period of 2 years (Wiegant et al. 2013). Such observations testify to the regenerative capacity of cartilage given suitable environmental conditions, despite the absence of a blood supply within articular cartilage.

Identifying the timescales for healing of cartilage relative to activity levels is clearly an important requirement for well-informed clinical advice on rehabilitation options following damage to articular cartilage. But beyond the general statement that articular cartilage is slow to heal relative to other tissues, there is frustratingly little objective clinical advice relating specific cartilage injuries and treatments (including physiotherapy and exercise levels) to healing times or damage progression. This 'knowledge gap' has spurred the investigation of MRI as a potential predictive biomarker (Wang et al. 2006; Hunter et al. 2011; Chu et al. 2014).

It is also well known that normal hyaline cartilage contains no nerves. The tissue is probably aneural for the same reason that it is avascular (i.e. it is very difficult for neural tissue to survive the mechanical strains experienced within the tissue). This means that for normal joints (with normal cartilage) knee pain cannot originate from the articular cartilage itself. In other words, joint pain usually arises from one or more of synovial, tendon, ligament, and bone tissues (which do contain pain-sensitive nerve endings). In contrast to hyaline cartilage, we note that fibrous cartilage is both vascular and does contain nerves (so the central region of the meniscus does have a blood supply and does contain nerves, and so it too, can also generate pain sensation).

Clearly, hyaline cartilage (containing types II and III collagen) is adapted to the high compression and low shear environment normally found in articular joints, while fibrous cartilage (containing types I and III) is suited to lower compression and higher shear environments. Fibrous cartilage appears to be particularly suited to conditions where there is large biaxial 'compression and tension' stress in the tissue, generating significant shear stress (as occurs in the menisci of the knee, the temporomandibular joint and when tendon wraps around bone).

1.4 What Is the Composition of Hyaline Cartilage?

The major components of hyaline articular cartilage in large joints such as the knee and hip are water (typically reported to be about 70% by weight per unit volume, but ranges between 65 and 80%), type II collagen (about two-thirds of the dry weight of articular cartilage is polymeric collagen (Eyre 2002) (about 20% by weight per unit volume), but this ranges between 55% and about 75% of the dry weight), aggrecan (about one-third of the dry weight (about 10% by weight per unit volume) but ranges between 15 and 40% of the dry weight), and chondrocytes (<2 to 5% by volume per unit volume). Collagen and aggrecan contents are typically higher for adults.

For normal adult human (hip) articular cartilage (age 40), the number of chondrocytes per milliliter is reported to be about 5 million (Bobacz et al. 2004) [n = 41 individuals, see Fig. 1 in Bobacz et al. (2004)]. For normal adult human (femoral) articular cartilage (age 20), the number of chondrocytes per gram of human articular cartilage averages about 8–9 million (or about 10 million per milliliter of cartilage tissue) (see Fig. 1 in (Barbero et al. 2004)). The chondrocyte number density remains constant between ages 20 and 39, but is reduced to about half that number between age 40 and 49 [similar to that reported by Bobacz et al. (2004)], and slowly declines thereafter with increasing age to about one-third of that at age 20 [n = 69 patients, see Fig. 1 in Barbero et al. (2004)].

The numbers of chondrocytes per milliliter (ml) of hyaline cartilage vary with depth in humans (Gilmore and Palfrey 1988). For adult (1 year old) bovine articular cartilage, the density of chondrocytes toward the surface of the cartilage is around 150 million cells per ml, but this decreases to around 50 million cells per ml at 1 mm depth below the cartilage surface (Jadin et al. 2005). Similar changes in chondrocyte density are observed in other species (Jadin et al. 2005). It is noted that a general decrease in chondrocyte DNA with increasing body mass of the animal accords with the generally higher chondrocyte densities observed in smaller animals that have thinner cartilage (Malda et al. 2013). Nevertheless, chondrocytes account for only a few percent of the total cartilage volume, the great majority of hyaline cartilage volume being extracellular matrix (ECM).

The genome is now fully known, resulting in the categorization of the 'extracellular matrisome' [see, for example, the review and tribute to Prof Dick Heinegard by Hsueh et al. (2014)]. The 'core extracellular matrisome' probably comprises about 300 proteins (Hynes and Naba 2012). Hynes and Naba report: '*In addition there are large numbers of ECM-modifying enzymes, ECM-binding growth factors, and other ECM-associated proteins.*'

After decellularization, treatment with surfactant, ultrafiltration, and using guanidine-HCl extracts, Hsueh et al. identified some 425 proteins in total (with about 51% of Gene Ontology Identifiers [GO IDs] indicating proteins associated with the extracellular matrix) from all human articular cartilage zones, hip and knee joints and for all physiological conditions examined (Hsueh et al. 2016). Typically, around 200 proteins were found in each articular cartilage zone—see Fig. 7 in Hsueh et al. (2016). No significant differences were discerned in joint type or physiological condition of the cartilage; however, aggrecan G1 and G2 domains, hyaluronan, proteoglycan link protein, matrilin, serine protease, chondroadherin, clusterin, SPARC, carbonic anhydrase, TIMP3, and emilin-1 were all significantly enriched in hip versus knee cartilage, while chitinase-3-like protein and MMP1 demonstrated significant enrichment in knee cartilage. A total of 198 proteins (47% of the total) were common to superficial, intermediate, and deep zones. Lists of ECM proteins found in human articular cartilage, along with further analysis, can be found in Hsueh et al. (2016).

Extracellular matrix biochemists are still trying to discover and define the roles of the many molecules making up the ECM. Hyaline cartilage contains a variety of signaling molecules (such as TGFβ, IHH-PTHrP, BMPs, and Wnts) and regulatory molecules controlling tissue degradation and repair such as TIMPS, fibronectin, fibulin, tenascin, collagens VI, IX, and XI, COMP, CILP, asporin, matrilins, WARP, osteoadherin. Additional regulatory molecules include heparin sulfate proteoglycans (HSPs—include membrane-bound HSPs such as syndecan and glypican, and secreted HSPs such as perlecan, agrin, and collagen XVIII). Further regulatory molecules include the family of small leucine-rich proteoglycans (known as SLRPs). In addition, there are many enzymes (such as ADAMTSs, MMPs, cathepsins, meprins, and lysyl oxidase), and many (structural) mers that form polymers within the ECM (such as tropocollagens, lecticans, linker protein, hyaluronic acid, chondroitin sulfate, keratin sulfate). See, for example, Fig. 1 in Heinegard (2009) for a schematic illustration of some of these molecules in the ECM.

Through all this biological complexity, there are two dominant structural molecules in mature articular cartilage ECM: type II collagen fibrils and aggrecan [see, for example, the schematic diagram as shown in Fig. 1 (Poole et al. 2002)]. Indeed, much of the complexity in the makeup of cartilage is functionally related to how to manage and sustain these two structural components, so that articular cartilage can perform its necessary mechanical functions under challenging environmental conditions.

We note here that type II collagen in humans is only found in articular cartilage and elastic cartilage (i.e. in the pinna of the ear and epiglottis), within the temporomandibular joint cartilage, within vertebral disks, within the tympanic membrane (i.e. the ear drum) and within membranes of the inner ear. Type II collagen is a highly specialized collagen for retaining highly charged proteoglycans such as aggrecan. Type I collagen simply cannot retain aggrecan as well as type II collagen. So when high compressive loads are encountered, and high concentrations of aggrecan are required to resist the high compressive loading, then a type II collagen network is employed, because type II collagen can more effectively retain the high concentrations of aggrecan required to resist the high compressive loading.

The meniscus of the knee contains type I collagen and a high concentration of aggrecan. In this example, type I collagen specializes in carrying large tensile (in this case 'hoop') loads generated within the crescent-shaped menisci upon joint loading. However for this particular example, we note that the 'fibrocartilage core' of the meniscus containing the type I collagen is surrounded by a 'shell' of hyaline cartilage. It is the shell of hyaline cartilage that retains the high concentration of aggrecan within the meniscus at a concentration appropriate for knee loadings, despite the presence of a fibrocartilage core to resist the meniscal hoop stresses during knee loadings. In other words, to serve the functional needs of the meniscus most effectively, the collagen type changes with region through the meniscus.

Given this structural arrangement to facilitate its function, it is worthwhile to reflect on the expected effect of removing all or part of the hyaline cartilage shell surrounding the fibrocartilage core of the meniscus (as sometimes occurs during partial meniscectomy), on the long-term functional performance of the meniscus. It is also interesting to note that when the meniscus is damaged, the tibial and femoral cartilages move closer together, leading to joint space narrowing (JSN is a characteristic radiographic sign supporting the diagnosis of osteoarthritis).

Interestingly, collagen type is also graded in the temporomandibular joint (TMJ) disk, which is functionally somewhat like a meniscus. The TMJ disk is reported to have four zones: (i) the articular zone (characterized by the expression of lubricin), (ii) the proliferative zone (characterized by the expression of SOX-9 and the absence of the expression of collagen type II), (iii) the chondroblastic zone (characterized by the expression of SOX-9, collagen types I and II, and Indian hedgehog), and (iv) the hypertrophic zone (characterized by the expression of Indian hedgehog, osteopontin, and collagen type X) (Wadhwa and Kapila 2008). The TMJ disk is classified as a 'fibrocartilage' containing collagen types I, II, and III (Wang and Detamore 2009; Wadhwa and Kapila 2008; Minarelli and Liberti 1997). Embryologically, TMJ cartilage is classified as a 'secondary cartilage' formed from periosteum or endosteum (while hyaline cartilage is a 'primary cartilage' derived from the anlagen, which forms as a 'template' for bone tissue formation). The thin proliferative zone separates the fibrocartilaginous fibrous zone at the surface (i.e. the articular zone with mainly type I and type III collagen) from the more hyaline-like mature and hypertrophic zones below (containing types I, II, and III collagens) (Wang and Detamore 2009; Minarelli and Liberti 1997).

We mention that in the most common TMJ disorder—disk displacement—the TMJ disk is dislocated anteriorly, and so 'retrodiscal tissue' moves anteriorly where it is then squeezed between the mandibular condyle and the mandibular fossa. As the retrodiscal tissue is highly vascularized and innervated, this squeezing causes considerable pain. However over time, this condition usually resolves as the retrodiscal tissues redifferentiate and adapt to become fibrocartilage, almost indistinguishable from the original disk (forming a 'pseudo-disk'). This resolution of a clinical condition clearly illustrates the capacity for adaption by fibrocartilage tissues.

We finally note that it has been observed there is a grading of collagen types in the walls of intervertebral disks. Remarks (Nerlich et al. 1998):

> As previously reported, we observed a distinct staining pattern for each collagen type analysed in the intervertebral discs of our series. Thus, collagen I is absent from the juvenile/adolescent nucleus pulposus and only faintly seen in the annulus fibrosus, mainly the outer annulus fibrosus. The fact that the annular collagen I staining is rather weak has also been reported by others, although biochemically collagen I represents the major bulk of annular collagen. The reason remains unclear, but it may well be that the closely parallel orientation of the annular collagen I fibers leads to some antigen "masking". Collagen II is the main collagen component of the nucleus pulposus, but it is also seen in the (inner) annulus fibrosus and a major constituent of the end-plate.

So, collagen type I is primarily found on the 'outside' of the annular fibrosis (of the intervertebral disk), which then grades into type II collagen on the 'inside' of the annular fibrosis and lines the end plate of each vertebra, and is found within the nucleus pulposus itself. Again, type II collagen serves the function of retaining the concentrated aggrecan found in the central region of the disk (i.e. within the so-called nucleus pulposus of the intervertebral disk).

Is there a gradient of collagen type I and type II through articular cartilage, as observed for these other tissues? No—normal adult human articular cartilage is entirely collagen type II; i.e. there is zero collagen type I in articular cartilage. This suggests that for articular cartilage, the functional benefit of increased strength (i.e. type I collagen) does not outweigh the functional benefit of retaining aggrecan (i.e. type II collagen), at any location throughout articular cartilage. This implies that the tensile strength collagen type II is sufficient in normal articular cartilage.

The 'type II collagen network' found in hyaline cartilage is itself not a simple molecule, but made up of a number of molecules. While adult hyaline cartilage contains up to 90% type II collagen, in addition to type II collagen there is about 1% type IX collagen covalently linked to type II collagen, 3% type XI, and up to 10% type III collagen (most type III collagen is covalently linked to type II collagen, but there is some type III collagen (e.g. 1%) that forms part of type IX collagen) (Eyre et al. 2006) (Fig. 1.3).

It is likely that in the transitional and superficial zones of articular cartilage, type III collagen is covalently cross-linked with type II collagen to strengthen any weakened type II collagen (Wu et al. 2010). Type III collagen is less prominent in the deep zone of the cartilage, grading to being particularly prominent in the superficial zone of the cartilage (Eyre et al. 2006). This gradation of type III collagen through the depth of the cartilage is consistent with the idea that collagen type III helps to repair damaged (type I and type II) collagen (Liu et al. 1995), and we later suggest that type II collagen damage increases toward the cartilage surface.

It becomes apparent that structure of type II collagen fibrils within the cartilage ECM is itself a complex system of molecules. Indeed, the structure of type II collagen revealed by a rotary shadowed electron microscope image is extraordinary—spiky projections with globular heads protrude along the length of the fibril (Vaughan et al. 1988) (see Fig. 1.4).

The structure of type II collagen fibrils is clarified by a very helpful schematic diagram by Kadler et al. (see Fig. 1.5) [and also see the schematic diagram by Poole et al. (1992)]. The type II collagen fibrils are composed of many aligned, staggered,

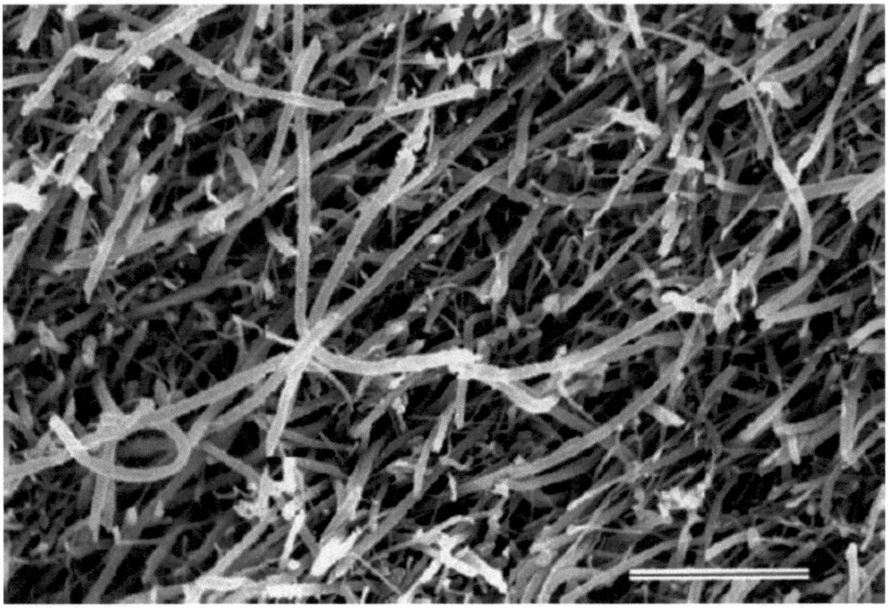

Fig. 1.3 Scanning electron microscope image of collagen network in [femoral] cartilage showing fibrillar entwinement in the extracellular matrix. Scale bar 2 μm (Nickien et al. 2013; Broom et al. 2001). See also Figs. 2, 4, 6, 7, and 8 in Nickien et al. (2013) and Figs. 3, 7, 8, and 9 (Broom et al. 2001)

type II collagen sub-units, polymerized around (a nucleating core of) collagen type XI (Kadler et al. 2008). The collagen type II fibril grows in diameter. When sufficiently large, collagen type IX then binds to the surface of the collagen type II fibril. Binding of collagen type IX probably prevents further growth of the collagen type II fibril diameter (this suggests that if collagen type II is to grow in diameter, the collagen type IX probably needs to be removed first). The Col3 'arm' (collagen sub-unit type III) of the type IX collagen projects out from the main type II collagen fibril, which presumably helps to increase the likelihood for mechanical entanglement of the collagen network with aggrecan and aggregate molecules in the pore spaces between the collagen fibrils. In other words, collagen type II is a multi-molecular fibrillar structure built up in layers—collagen type II binds to the nucleating core collagen type XI, while collagen type IX (as well as decorin and other SLRPs—see discussion below) binds to collagen type II, and finally aggrecan binds/interacts with collagen type IX.

The type IX collagen is a heterotrimer (made up of Col1, Col2, and Col3 segments). Type IX can bind integrins on the cell surface, and it has a proteoglycan binding site at the NC3 domain (i.e. at the site of the kink in the type IX collagen—see Fig. 1.5), though this binding site is reported to not be utilized in articular cartilage (Eyre et al. 2006). The type IX collagen molecule has a positive charges site (containing Zn^{2+}) on the NC4 terminal domain that binds negatively charged

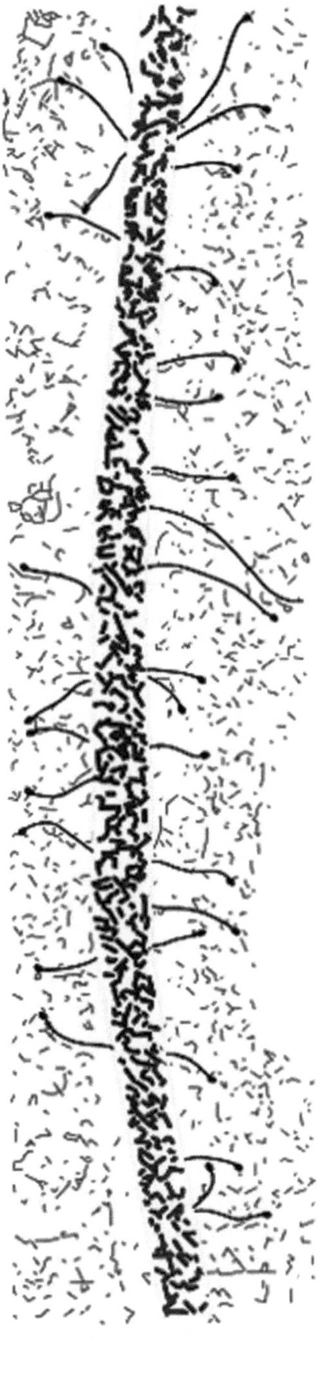

Fig. 1.4 Drawing of image shown in Fig. 3 of Vaughan et al. (1988) showing rotary shadowed cartilage fibril. The fibril in this shadow appears to be sheathed in collagen type IX. The globular amino-terminal domain (1) and a short non-collagenous domain (NC3), which gives rise to a prominent kink (2) in the otherwise relatively rigid collagenous portion, are notable features also visible in the molecules on the fibril surface (Vaughan et al. 1988). Note collagen fibril shown in Vaughan et al. (1988) is about 35 nm in diameter. Figure sketch drawn by Yvette Harrap. See also Fig. 3 (Broom et al. 2001), which shows an SEM image with the (tiny) globular amino-terminal domains collapsed into the collagen fibrils, and Fig. 1.5

molecules (Leppanen et al. 2007). In addition, the Col3 and Col2 sections of type IX collagen contain many arginine amino acids that are positively charged for the physiological pH in normal cartilage (a pK_a of 12.48 means arginine is positively charged at typical pHs in cartilage, i.e. pH 7 and lower). These positively charged sites can electrostatically bind to the negatively charged sites on GAG molecules (see Fig. 1.6; Kapyla et al. 2004). Explains (Horkay et al. 2012): '*These large aggrecan/hyaluronan complexes are enmeshed in a network of collagen, where the structural stability is enhanced by the ionic interactions between the negatively charged CS bristles and the positive charges on the collagen fibrils.*' It has been shown that the aggrecan keratin sulfate-rich region has moderate affinity for isolated collagen type II (Hedlund et al. 1999). The NC4 terminal domain sits at end of the Col3 arm of the type IX collagen (see Fig. 1.5), which is positively charged and binds GAGs, matrilins, osteoadherin, fibromodulin, heparin, COMP and other molecules ensuring the formation of a macromolecular network (Kalchishkova et al. 2011).

It has been shown using surface plasmon resonance that chondroitin sulfate E (i.e. chondroitin-4,6-sulfate) binds strongly to collagens II, III, and IX (as well as collagens I, V, and VII) (Munakata et al. 1999). We note in passing that the sulfation patterns of chondroitin sulfate (CS) change with age (Sauerland et al. 2003b) (e.g. '*additional tissue maturation-related changes can also be seen in the*

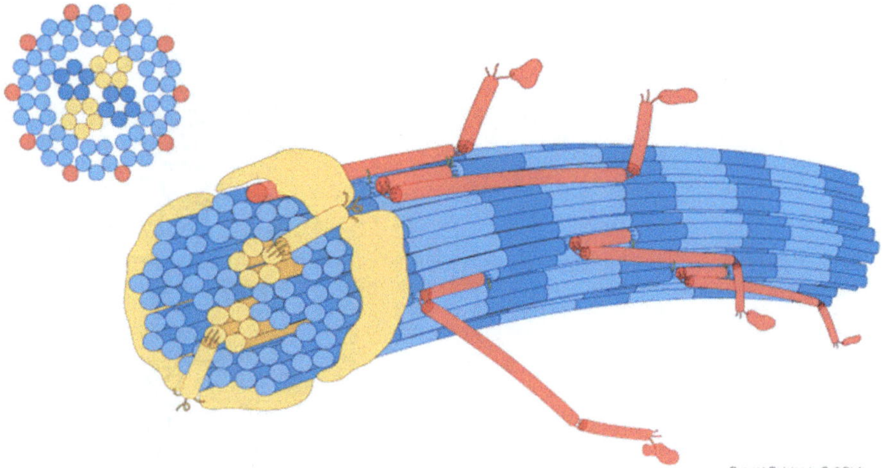

Fig. 1.5 Schematic of the 10 + 4 microfibril structure of a thin cartilage collagen fibril. A pair of collagen XI microfibrils comprise half of a four microfibril cores surrounded by ten microfibrils at the surface. The collagen XI/IX/II assembly is a cross-linked heteropolymer, as is V/I, and is an important component of the fibril assembly mechanism. Blue: collagen II molecules; yellow: collagen XI molecules; red: collagen IX molecules. The N-terminal thrombospondin-like domains of collagen XI (yellow) are shown extending from the core microfibrils onto the fibril surface (Kadler et al. 2008) (model kindly provided by Dr David Eyre, University of Washington, Seattle). Such a prototypical fibril is about 20 nm in diameter (Gottardi et al. 2016). Thicker fibrils are formed by winding two or more prototypical fibrils together (so a 100-nm fibril is formed from about 7 prototypical 20 nm fibrils)

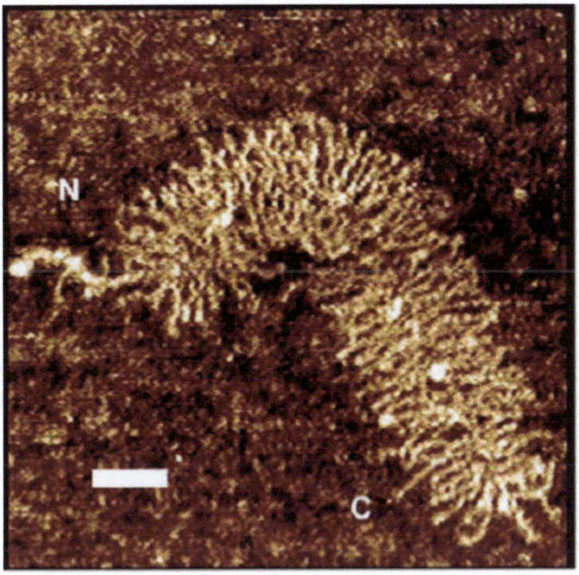

Fig. 1.6 Atomic force microscope image of the fetal aggrecan molecule (generated by AFM in 'tapping mode'), showing protein core with GAG molecules attached (via linker protein) clearly visible (N and C denote N-terminal and C-terminal regions of the core protein) (Ng et al. 2003). Scale bar approximately 50 nm

sulfation of the terminal GalNAc residues, which are almost exclusively 4-monosulfated in CS from embryonic and postnatal growth cartilage, but contain an additional sulfate (GalNAc4,6S) on ~50% of CS chains in adult human cartilage').

Importantly, the type II collagen fibril itself is reinforced and stiffened by aldol cross-linking reactions ['*bone collagen features both pyrrole and pyridinoline cross-links, whereas mature cartilage contains only pyridinolines*' (Robins 2007)]. These cross-linking reactions are catalyzed by lysyl oxidase (Eyre 2002; Eyre et al. 2010; Ahsan et al. 2005; Robins 2007). Remarks Eyre et al (2010): '*Different tissue types showed distinctive patterns of cross-linking chemistry, though all were essentially based on the reactions of peptide-bound aldehydes created from specific lysine and hydroxylysine side chains by the action of lysyl oxidase(s) during the assembly of collagen sub-units into fibrils*' (Eyre et al. 2008). We note that lysyl oxidase (LOX) is one of a number of enzymes in the LOX family (LOX, LOX1, LOXL2, LOXL3, and LOXL4) (Hulmes 2008, Barker et al. 2012). This cross-linking and consequent collagen fibril strengthening and collagen network stiffening develop over weeks, months, and years (Charlebois et al. 2004; Waugh et al. 2012). We note that it has been shown that hypoxic conditions can promote cross-linking via lysyl oxidase (and lysyl oxidase may be induced by HIF-1) (Makris et al. 2015). Lysyl oxidase activity may be modulated by fibromodulin (Kalamajski et al. 2016).

As chondrocytes are the only cell type within cartilage, these cells produce all the structural components required to form hyaline cartilage (e.g. type II collagen, GAGs, aggrecan, linker protein, and hyaluronic acid (HA)), as well as the regulatory molecules that control the polymer assembly process (e.g. some collagens,

SLRPs, COMP). In addition, chondrocytes produce many of the signaling molecules that control cell behavior within the tissue (e.g. PTHrP, IHH, BMPs, Wnts, and TGFs). Important collagen-related regulatory and structural molecules are the small leucine-rich proteoglycans (the so-called SLRPs) (Iozzo 1999). The SLRPs include decorin (Jungmann et al. 2012) and biglycan (Poole et al. 1996; Vynios et al. 2001), lumican, fibromodulin, as well as other larger SLRPs. The collagen fibril assembly regulatory roles played by the SLRPs are surprisingly varied. For example, SLRPs can bind to: (i) different types of collagens, (ii) toll-like receptors, (iii) cMET, EGF-R, and IGF-R, (iv) TGF-β, (v) LRP-1 and WISP-1 (so influencing Wnt signaling) and BMP4 (Merline et al. 2009). Lumican may be necessary for effective toll signaling (Amjadi et al. 2013), and in scleral tissue lumican forms disulfide linkages with aggrecan (which may provide a mechanism to help retain aggrecan in type I collagen tissues such as the sclera (Dunlevy and Rada 2004) and fibrocartilage). We also note in passing that asporin is a leucine-rich repeat protein that has a protein structure similar to decorin.

Other molecules involved in matrix formation and removal include collagen oligomeric matrix protein (COMP) (DiCesare et al. 1995), regulatory collagens [such as collagen XI (Kadler et al. 2008)], aggrecanases, cathepsins (Dejica et al. 2012), matrix metalloproteases (MMPs) (Visse and Nagase 2003) as well as tissue inhibitors of cathepsins (e.g. the various cystatins) and inhibitors of metalloproteases (e.g. the various TIMPS) (Baker et al. 2002). It is likely that all these molecules (and many others) play important roles in shaping the structure of the ECM around chondrocytes (see further discussion in Sect. 2.10).

Interestingly, it has been shown that there are different 'protein expression profiles' in cartilage from different joints (Hsueh et al. 2014). For example, matrilin 1 and epiphycan are only found in rib and tracheal cartilage, while asporin is particularly abundant in the meniscus (Onnerfjord et al. 2012). CILP, COMP, and asporin all play a role in cross-linking various components of the ECM together (Onnerfjord et al. 2012), with their expression profile depending on the type and amount of cross-linking required in that tissue [we note in passing that CILP is upregulated in early OA, and this may antagonize the actions of IGF-1 and TGFβ (Johnson et al. 2003)]. Hsueh et al. (2014) report that more proteins tend to be expressed in the superficial layers of cartilage. These differences in the protein composition of cartilage doubtless reflect specific functional roles for cartilage in different types of joints. This hints that 'cartilage is not just cartilage,' much as Castrol (and more recently dieticians!) tells us that 'oils ain't oils' (meaning that while articular cartilage appears to the uninformed to have much the same outward appearance, there are actually significant differences between articular cartilages, and so one articular cartilage is not necessarily interchangeable with another). We are really only beginning to appreciate the subtleties of different articular cartilages.

It is clear that chondrocytes are primarily responsible for maintaining the structural integrity of the cartilage tissue. In other words, we are always mindful that if the chondrocytes die for any reason, then cartilage inevitably degrades (and similarly for other tissues, e.g. if osteocytes die for any reason, then bone inevitably degrades). We next consider how the collagen and aggrecan behave together to form a biocomposite that is functionally suited to articular cartilage's challenging mechanical environment.

1.5 Cartilage Is a Biocomposite Material

We now focus on the two major structural components—collagen and aggrecan (or in normal in vivo cartilage, hyaluronan, and aggrecan combine to form 'aggregate')—and how they work together to create the biocomposite material that gives articular cartilage its distinctive mechanical properties.

Aggrecan is a proteoglycan with a molecular weight of about 2.0–3.0 MDa. Aggrecan comprises a protein core [made up of 18 exons (Watanabe et al. 1995)] with many (up to 100) glycosaminoglycan (GAG) side chains (Aspberg 2012; Lee et al. 2013; Sauerland et al. 2003b) (see Fig. 1.6 for schematic and beautiful AFM images of aggrecan shown in Lee et al. (2013). The overall dimension of a typical cylindrical solvated aggrecan is around 350 nm in length and 60 nm in diameter (Ng et al. 2003) (see also Fig. 1 (Nilsson et al. 2017), Fig. 3 (Han et al. 2011), and Figs. 1 and 3 (Roughley and Mort 2014) for schematic illustrations). The side chains are principally the polyanionic polysaccharides keratin sulfate and chondroitin sulfate [see, for example, Fig. 1 (Roughley and Mort 2014)]. Chondroitin sulfate comprises usually 30–40 (but up to 50) disaccharides of uronic acid (which has an ionizable carboxyl acidic group) and *N*-acetyl galactosamine (which has on average one ionizable sulfate group—and occasionally two ionizable sulfate groups) (Sauerland et al. 2003a). The saccharides are joined by alternating β1-3 and β1-4 glycosidic bonds. A single disaccharide about 1 nm in length typically contains two charged sites at physiological pH. We see that a 60-nm diameter for an aggrecan molecule is consistent with GAG chain lengths of about 30 disaccharides (i.e. two radii (of 30 disaccharides stretched out 'straight'—see Fig. 1.6) sum to the diameter of the aggrecan molecule).

The carboxyl and sulfate groups ionize under normal physiological pH conditions in the joint, giving rise to the 'fixed' negatively charged molecules in the cartilage ECM [see, for example, the illustrative sketches shown in Fig. 1 (Nilsson et al. 2017), Fig. 3 (Han et al. 2011), and Figs. 1 and 3 (Roughley and Mort 2014)]. This fixed charge density can largely explain the osmotic pressure of GAG and aggrecan solutions and in 'semidilute' states under physiological conditions (at least for cartilage bathed in 0.15 M NaCl salt solution and containing up to around 80 mg/ml of aggrecan) (Bathe et al. 2005) (see Chap. 6). We note here that this 'fixed charge' turns out to be very important in contributing to the compressive stiffness of the cartilage tissue.

Aggrecan itself is in turn non-covalently attached to hyaluronic acid (HA) to form a very large molecule known as 'aggregate' (Roughley and Mort 2014; Day and Prestwich 2002). There is a strong binding affinity of aggrecan for HA—of the order of 10^{-7} to 10^{-8} M. The aggrecan attachment to HA is non-covalently stabilized and strengthened by an additional molecule named 'link protein.' The aggregate molecules can measure up to about 1 μm or more in length, and so on a molecular scale they are huge, with a molecular weight up to about 300 MDa (Ng et al. 2003; Heinegard et al. 1978).

The aggrecan molecules may also be cross-linked together by various molecules in the extracellular matrix. For example, fibulin (Aspberg et al. 1999; Olin et al. 2001) and tenascin (Aspberg et al. 1997) may cross-link the G3 domains of several aggrecans (G3 domains are at the ends of the aggrecan core protein) [see Fig. 4 in Aspberg (2012)].

In addition to the direct (via electrostatic or non-specific molecular interactions) and indirect molecular adhesion (via other molecules such as tenascin) of aggrecan to the collagen fibrils, it is plausible to suppose that the enormous molecular weight and length of the aggregate molecules substantially increase the probability of mechanical retention of the aggregate molecules within the cartilage, via 'mechanical entanglement' of aggregate molecules in the type II collagen network (which we now know is 'spiky' due to the presence of type IX collagen on the surface of collagen fibrils and presumably aids mechanical aggregate retention, and we now know is also 'sticky' due to the presence of positive charges on the arginine amino acids along type IX collagen). Indeed, it is a basic principle of chromatography that the larger the molecule and the smaller the effective pore size opening, the more slowly the molecule migrates through the chromatographic column. We believe this offers a functional explanation as to why aggregate molecules have developed their unusually large size (up to 300 MDa) and type II collagen its distinctive structure (i.e. 'spiky' and 'sticky').

Further, we note that various cross-linking molecules such as tenascin and fibulin (and possibly including COMP, asporin, and CILP molecules) provide an additional level of 'chemical retention' by making the aggregate molecules even bigger in situ, with 'super-aggregates' aiding 'entanglement' of aggregate by completely enclosing numerous collagen fibrils making up the collagen network. Indeed, it is apparent in early OA that tenascin expression is increased (Pfander et al. 2004), presumably in an attempt to try and better retain the already depleted concentrations of aggrecan and aggregate within a damaged collagen network.

From the foregoing, it begins to become apparent that aggrecan is an extraordinarily 'slippery molecule' (Seror et al. 2011). When concentrated, aggrecan is extraordinarily difficult to confine and retain within a porous network. Indeed, the properties that give aggrecan its affinity for water and resistance to compression are the same properties that make it slippery and difficult to retain in a biological porous network. It then follows that anything that compromises the retention of aggrecan within the collagen network (e.g. any defect in aggrecan molecules themselves or defect in the link molecules or defect in the collagen network or depletion of type II collagen) is likely to contribute to the development of disease states associated with being unable to effectively retain adequate aggrecan within articular cartilage tissue. This immediately opens up the possibility for many genetic–environmental disease states.

Normally, strong double-layer repulsive forces between like-charged molecules keep them separated. The counterions keep the molecules electrically neutral overall; however, nanoscale separation of fixed charge and counterions in aqueous solution generate strong double-layer repulsive forces between the molecules (Chandran and Horkay 2012; Grodzinsky 2011) (and see Chap. 6 for further discussion). Most importantly, the large electrical and osmotic repulsive forces within the double-layer overlaps between aggrecans molecules provide a substantial fraction of cartilage tissue's resistance to compression (Basser et al. 1998; Bathe et al. 2005; Nap and Szleifer 2008).

In short, aggrecan expansion and aggrecan movement are impeded by the chemical attachment and mechanical entanglement of aggregate within a dense type II collagen network. But this same repulsive force generates a strong tendency for the aggrecan molecules to (slowly) disperse through a normal collagen network, away from regions of higher concentration toward regions of lower concentration. But because the migration of aggrecan is slow relative to the movement of water, the osmotically active counterions draw in water, which expands the biocomposite. Crucially, it is the differences in timescales required for migration of aggrecan/aggregate out of the collagen network relative to the much smaller timescale for migration of water molecules into the tissue (a difference in timescales of about three orders of magnitude) that leads to the sustained inflation the collagen network.

It is now very clear that one of the key roles of the type II collagen network is to retain aggrecan, enabling aggrecan concentrations to increase to high enough concentrations to resist large compressive loadings. At steady-state equilibrium, the rate of aggrecan production within the collagen network equals the rate of aggrecan loss from the collagen network. At steady-state equilibrium, aggrecan concentration is then related to the magnitude of the loading (which also influences the rate of aggrecan production and so the equilibrium) and the density of the collagen network (which also influences the rate of aggrecan loss and so the equilibrium).

In summary, aggrecan/aggregate in articular cartilage wants to imbibe water and osmotically expand, but this expansion is resisted by the collagen network, creating a local (osmotic) swelling pressure in the collagen network (see Fig. 1.7 and Chap. 6 for a discussion of osmotic pressure). A 'balloon-string model' of cartilage, shown in Fig. 15 of Nickien et al. (2013), also conveys the idea of collagen (represented by string) in tension and aggrecan (represented by balloons) in compression. If the collagen network was not present, aggrecan/aggregate would quickly disperse throughout the joint space (including the synovial fluid), and then aggrecan could offer no resistance to compression. However, resistance to osmotic swelling of the aggrecan is offered by the collagen network, which places the collagen network in tension, while the retained aggrecan is in compression. Without any load applied to the surface of articular cartilage, at steady state, the (nominal) compressive stress in the aggrecan and (nominal) tensile stress within the collagen network are equal and opposite. This means that cartilage tissue subject to no external loading is 'self-equilibrating' (Maroudas 1976; Basser et al. 1998).

To describe this situation in the language of material science, cartilage is a 'prestressed biocomposite' material. The amount of prestressing in the collagen network can be adjusted by adjusting the aggrecan concentration within the collagen network. We note that for structural engineering applications, prestressing is

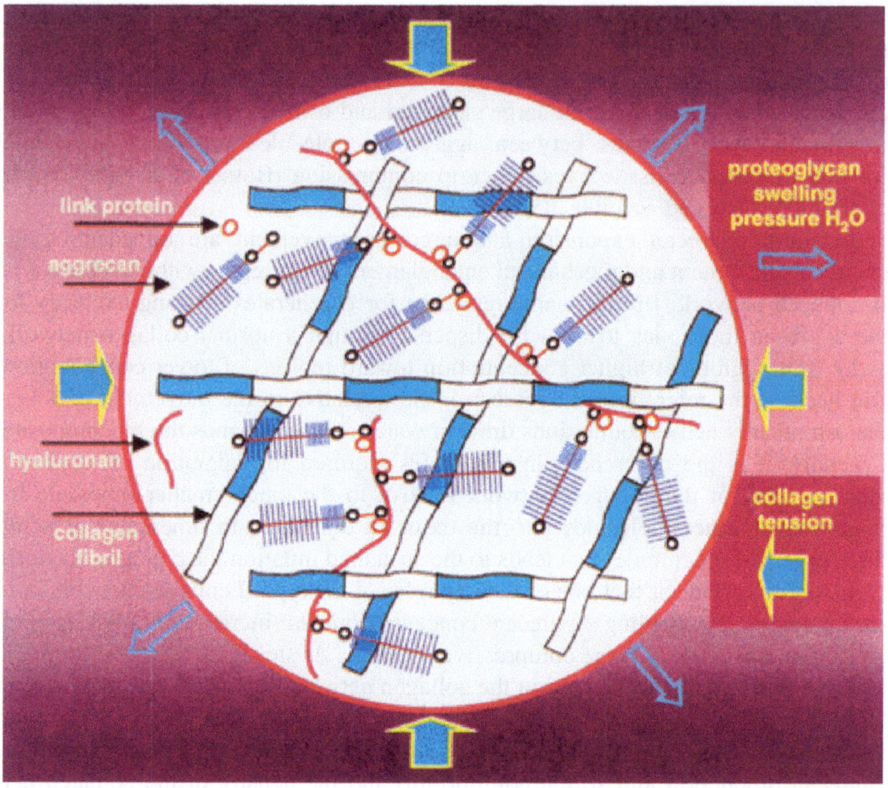

Fig. 1.7 Combined properties of collagens and aggrecan in cartilage (Kiani et al. 2002). See also Fig. 4 (Roughley and Mort 2014)

often employed in composite materials to minimize tensions in some material components, minimize deflections, and increase shear strength (e.g. prestressed concrete bridge). The steady-state equilibrium aggrecan concentration depends on the balance between the rate of production and its rate of loss from cartilage tissue (see Appendices E and F).

So, we now have new significant insights aiding our understanding of articular cartilage behavior. One of these insights relates to the mechanical properties of cartilage. As a result of the initial prestress in the collagen network, articular cartilage has an initial strain prior to any loading being applied to its surface. If this cartilage is now compressively loaded on its surface, the collagen network immediately beneath the applied load 'relaxes' and unloads, while the aggrecan immediately beneath the surface load is compressed [see Fig. 4 (Roughley and Mort 2014)]. The prestress has the curious effect of increasing the initial stiffness of the cartilage in both tension and compression (i.e. $E_{composite} > E_{col}$) (see Fig. 1.8 for explanation).

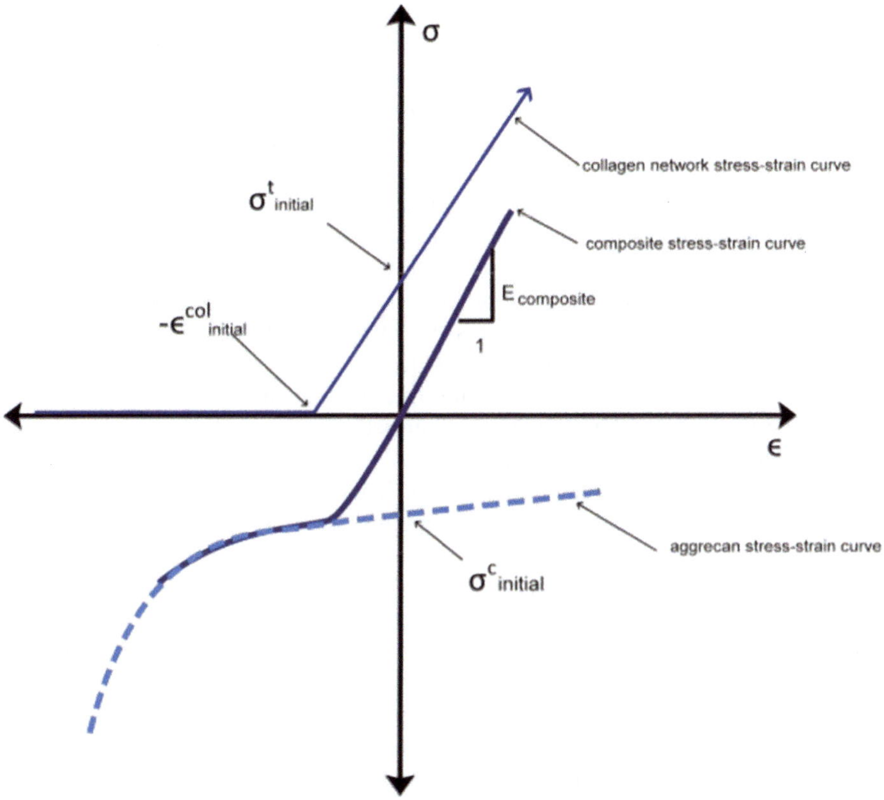

Fig. 1.8 Stress–strain behavior of the cartilage composite is the sum of the collagen network stress–strain curve (depicted here as bilinear for simplicity, but more generally is nonlinear) and a nonlinear stress–strain curve for aggrecan. Initially, there is a tensile prestress in the collagen network and a compressive prestress in the aggrecan network. With no load on the cartilage surface, these two stresses are self-equilibrating ($\sigma^{c}_{\text{initial}} = \sigma^{t}_{\text{initial}}$ at $\varepsilon = 0$). Interestingly, we see that upon loading of the cartilage surface, the initial stiffness of the cartilage composite is greater than either the stiffness of the collagen alone or the stiffness of the aggrecan alone (i.e. $E_{\text{composite}} > E_{\text{collagen}} > E_{\text{aggrecan}}$). At large tensile strains, the stiffness of the cartilage composite approaches the stiffness of the collagen network. For compressive strains greater than the initial collagen network strain ($\varepsilon^{\text{col}}_{\text{initial}}$), the cartilage composite assumes the compressive stiffness of the aggrecan. We note that a similar 'compression curve' is reported experimentally [see, for example, Fig. 8a (Chahine et al. 2004)]

Indeed in the limit of the collagen network being infinitely stiff, and in the limit of a uniform aggrecan concentration profile with depth, the surface of the cartilage would not deform until the initial prestress in the collagen network is exceeded. In this case, the surface of the cartilage would appear to be initially infinitely stiff (as load is applied with no surface deformation), and then at some load magnitude the surface suddenly becomes compressible (and deforms as more load is applied). To give a human scale analogy, the cartilage surface would be like a stiff 'crust' on the surface of a soft soil.

However, the real collagen network does have a finite stiffness, the aggrecan concentration profile with depth is not uniform, and so the initial stiffness of the cartilage due to surface loading is also finite (see Fig. 1.8). At some compressive strain greater than the initial strain in the collagen network, the collagen network becomes slack and carries no load. Note that if there is an increase in aggrecan concentration with depth (as is usually the case in real cartilage), this slack in the collagen network develops progressively below the cartilage surface as compressive loading increases. As this 'slack front' passes a certain depth below the surface of the articular cartilage, all the compressive load is then carried by the aggrecan between the surface and that slack-front depth (while collagen network remains embedded within aggrecan). The biocomposite modulus increases as the cartilage is further compressed, as the aggrecan concentration continues to increase as volumetric strain increases (see Appendix D).

Another significant insight into cartilage behavior is gained by observing the short-term effects of substantial changes in either the rates of aggrecan production or degradation. The effect of depleting the aggrecan that normally 'inflates' the collagen network in the cartilage biocomposite is perhaps most vividly illustrated by the injection of the enzyme papain into the veins of rabbits. The ears of some rabbit breeds have sufficient bending stiffness they normally stand erect. This 'bending stiffness' in the rabbit ear is due to the presence of sufficient aggrecan concentration giving the collagen network and aggrecan adequate prestress, enabling the rabbit ear to act as a (cantilever) beam. When the rabbit ear has an adequate/normal prestress, if the rabbit ear is bent (i.e. an external moment is applied to the ear) an internal moment is generated in the ear tissue. The internal moment is a 'resisting moment' (resisting the externally applied moment). The resisting moment is formed by the lever arm between the collagen network in tension on one side of the neutral axis within the ear and the unloading of the prestressed collagen network and aggrecan in compression on the opposite side of the neutral axis (the neutral axis is the line of no strain change). It is this resisting moment (or bending resistance) arising from the internal ECM stresses that enables the rabbit's ears to remain erect, despite an investigator bending the ear or gravitational and dynamic loading of the ear (e.g. due to head turning).

However following a papain injection, within 18 h the rabbit's ears lose their bending stiffness, as the aggrecan concentration in the ear cartilage is depleted by papain. Without adequate bending stiffness, the rabbit ears sit limply on each side of rabbit's head, much like the ears of spaniels. As time passes, the chondrocytes within the ears produce more aggrecan to replace that degraded by the papain, and the rabbit's ears gradually become erect again over the next 3–4 days (Thomas 1956). This clearly illustrates that the functional integrity of the biocomposite tissue to resist bending depends upon the mechanical properties of both aggrecan and collagen. Clearly, this example also highlights that both the rates of production and degradation of the collagen network and aggrecan/aggregate are crucial to the long-term structural integrity of biocomposite tissue known as cartilage.

For very large compressive strains, it is possible that the dense fibrous collagen network found in cartilage may become so densely packed together that it compresses onto itself (analogous to a 'pile of rope'), and this structure may carry some compressive load. However, it is difficult to believe that this 'pile of rope mechanism' is a normal load-carrying mechanism for a collagen type II network because repeated '(crossing) point or line contact' loadings between contacting collagen fibers would almost inevitably cause stress concentrations, which would locally damage the collagen type II fibrils (see Fig. 1.5) in the collagen network irretrievably. Contacts, knots, and kinks form stress concentrations in loaded ropes, and so these locations are usually the 'weakest link.' Indeed, standard rope care and safety demands ropes never cross under load, and one never places a load on a pile of rope for similar reasons. One can imagine that a specialized collagen molecule such as type IX collagen, protruding from the type II surface, is even more easily damaged by stress concentrations due to abnormal loadings.

We also note that aggrecan is electrostatically attached to and surrounds the collagen network, and so it is likely aggrecan nearly always inserts itself between the collagen fibers offering 'cushioning protection,' reducing the likelihood of stress concentrations in the collagen fibers and reducing the likelihood of abrasive wear if there is sliding (i.e. relative movement between the collagen fibers). It is more likely that the collagen fibers buckle, while maintaining regular fiber spacing between neighboring collagen fibers due to the presence of aggrecan as the 'spacer molecule,' helping to produce what is often described as a 'crimp' or 'wavy' collagen patterning [see Figs. 16 and 17 in Fick et al. (2010)]. So, the usual way for normal collagen network to undergo very large deformation is to form regular waves of collagen fibril deformations to form that are sometimes referred to as 'chevron patterning' [see, for example, Fig. 2.25 and Fig. 14 in Nickien et al. (2013)]. By reducing collagen fiber damage, such crimp-forming behavior helps preserve the structural integrity (i.e. load-carrying capacity) of the collagen network. For these reasons, it seems more likely that closing of the inter-GAG spacing between aggrecan (GAG) molecules accounts for most if not nearly all of the normal compressive strain deformation in cartilage tissue (rather than some other potential mechanisms, e.g. lateral deformation of the collagen fibers due to changing load, or osmotic redistribution of intrafibrillar water to the extrafibrillar compartment).

Some report that a compressive strain of about 20–25% is required before the collagen network through articular cartilage becomes completely slack (Basser et al. 1998). However, Inamdar et al. (2017) report collagen fibril tensile (pre)strain in normal bovine and (femoral) human articular cartilage are only about 1–2%. How can we reconcile these two findings? Assuming a uniform aggrecan concentration profile, a large compressive deformation of 25% before the collagen network becomes slack could mean the collagen network itself has compliance similar to that of aggrecan. Clearly, a biocomposite with these compliant properties does not result in a suitable material to serve the functional needs of articular cartilage, because such regular large strains in the tissue are more likely to result in tissue damage and aggrecan loss. Indeed, it is this 'softened state' that diseased cartilage often acquires, which contributes to its destruction through excessive straining (e.g.

excessive mechanical strain [see, for example, (Nguyen et al. 2010)], simply tears extracellular matrix molecules apart, abrades collagen fibrils and aggregate alike, and may lead to cell death).

To serve its functional purpose, more usually the collagen network in healthy cartilage is much stiffer in tension relative to the stiffness of aggrecan in compression. For example, the measured stiffness of collagen is up to 20–30 MPa (depending on collagen network geometry and preload) (Akizuki et al. 1986; Huang et al. 2005), while aggrecan is comparatively compliant (its compressive stiffness of aggrecan in normal adult cartilage is usually reported as being initially less than a few MPa (and usually less than one MPa). So the explanation for why a compressive strain of around 25% might be required to fully unload the collagen network is not to do with collagen network having a similar stiffness to aggrecan. Rather, the explanation, touched on above, is more subtle.

Normally, the aggrecan concentration profile is non-uniform with depth, so aggrecan stiffness is non-uniform, and consequently the deformation (or more accurately strain) from a surface loading through the depth of the cartilage is also non-uniform (see Fig. 1.10). So rather than the collagen network unloading all once (as would be the case for an initially uniform aggrecan profile, which implies all the resistance is aggrecan efflux is concentrated right at the cartilage surface), the collagen network in articular cartilage unloads gradually as the surface loading increases. In other words, it is as if exceeding a 1% or 2% strain threshold for collagen fibril unloading [as reported by Inamdar et al. (2017)] moves progressively to greater depth through the thickness of the cartilage as it is loaded. At a particular loading that has reached equilibrium, the collagen network is unloaded by the compressive strain up to a certain slack-front depth from the cartilage surface. In this way, it may take up to 25% average compressive strain to unload the collagen network over the full depth of the cartilage (Basser et al. 1998), i.e. from the cartilage surface up to the osteochondral junction.

We mention here that the compressive stiffness of aggrecan is strongly dependent on aggrecan concentration (and so on preload) (Smith et al. 2016; Huang et al. 2005). And importantly, we mention that the volume fraction of collagen (along with other proteins) influences the cartilage compressive stiffness very significantly (Basser et al. 1998; Maroudas et al. 1991, 1992). Space filling by the collagen network and other proteins in cartilage means that the actual aggrecan concentration in the extrafibrillar space (mg of aggrecan/ml of extrafibrillar volume) is significantly higher than the apparent aggrecan concentration in the cartilage tissue (mg of aggrecan/ml of cartilage tissue), which is the quantity usually reported in the literature (see Appendix D for further discussion). Technically, one refers to the 'apparent concentration' and the 'true concentration' of aggrecan—usually, it is the apparent concentration of aggrecan that is reported in the literature, and it is the true concentration of aggrecan than controls the compressive stiffness of articular cartilage.

The point to be made here is that the disparity in material properties between the collagen stiffness in tension and aggrecan stiffness in compression in normal articular cartilage is usually large, i.e. usually more than tenfold (Huang et al.

2005). This means that the initial strain in the collagen network is normally small (probably just a few percent), so at small compressive strains (e.g. a few percent compressive strain) the compressive surface load is transferred to the aggrecan and the collagen is completely unloaded near the surface. As the loading increases, so unloading of the collagen network due to increasing compressive strain progresses deeper from the cartilage surface. One effect of the load transfer to aggrecan as the collagen completely unloads is to then squeeze much more water from the cartilage composite (because the aggrecan is less stiff), causing more fluid to 'weep' from the cartilage tissue into the contact interface, thereby improving the functional performance of surface lubrication when cartilage rubs against cartilage in vivo (see further discussion Chap. 4). Let us now look more closely at the compressive loading of cartilage to better understand how intact articular cartilage deforms.

1.6 How Does Cartilage Deform When Compressed?

Cartilage is a porous material comprising solids bathed in an aqueous solution known as pore fluid. There are non-collagenous proteins, glycoproteins, proteoglycans, many small molecules, and salts in pore fluid, but they do not carry any mechanical load from one molecule to another through the pore fluid, at least under equilibrium (or steady-state) load conditions. The main structural solids within the solution are the collagen network and aggrecan (or aggregate when aggrecan is combined with aggrecan), which do carry mechanical loads from one molecule to another (through the fluid) in a steady-state (equilibrium) load state (see Sect. 6.4). But what happens in a transient (non-equilibrium) load state?

Importantly, it is a good approximation to assume that both the solids and the pore fluid (i.e. mainly water) in the cartilage are themselves incompressible. Incompressibility means that if the cartilage is abruptly and incrementally loaded in all-round compression, the cartilage volume must initially be unchanged. When cartilage is in this initial loaded, constant volume state, the cartilage is said to be 'undrained' because pore fluid has not had time to escape the tissue. Only when pore fluid begins to be expelled from the cartilage can it change its volume. As the water is being expelled by the increment of compressive loading and the cartilage volume decreases over time because of the compressive loading, the cartilage is said to be 'consolidating' and is 'partially drained'. When the pore fluid stops exuding, the cartilage is said to be 'drained.' Drained does not mean the cartilage contains no pore fluid, but it means that pore fluid drainage from the cartilage in response to an incremental loading is complete.

Importantly, it takes a finite time for consolidation of cartilage to occur, because it takes a finite time for the pore fluid in the cartilage to be expelled from the cartilage (i.e. in vivo, most pore fluid is normally expelled to the synovial fluid bathing the cartilage surface—a small fraction of pore fluid may be expelled through a normal osteochondral junction). This consolidation behaviour of loaded cartilage tissue is not unusual. In fact, all saturated biological tissues behave this way, as do many inanimate materials (e.g. indeed, the (original) theory for porous media was developed for saturated soils). How long it takes for cartilage to consolidate depends on the 'hydraulic conductivity' or 'permeability' of the cartilage, the length of the drainage path for the water to reach the cartilage surface where it can escape, the stiffness of the cartilage, and any resistance to fluid flow at the cartilage surface (e.g. due to the small gap size at a contacting cartilage surfaces).

Importantly, all else being held constant, decreasing stiffness, decreasing hydraulic permeability, increasing drainage path length, and the increasing resistance to flow at the cartilage surface, all increase the time it takes for the cartilage to consolidate to the same degree. This gives us significant insight into the mechanical behavior of cartilage, as we can now predict how the rate of cartilage consolidation will change as the cartilage itself changes its composition.

For normal knee cartilage (specifically patellafemoral groove cartilage) experiencing a significant 'step increment' in loading, it takes around one to three (and sometimes more) hours for all the pore fluid (water) to be expelled from the cartilage and for consolidation to be complete (see MRI measurements of cartilage deformation and fluid flow rates shown in Fig. 1.9). The time required for consolidation to a specified percentage of the initial volume or thickness may be chosen as the 'characteristic time for consolidation' of the cartilage. With our new found knowledge of what influences consolidation, it is clear that the characteristic time for consolidation depends on many cartilage and joint specific details, for example the joint geometry and contact area, cartilage thickness, concentration of aggrecan in the cartilage, presence and size of meniscus, collagen/protein volume fraction, surface roughness at the contact interface. But in simplified form, the characteristic time for consolidation is given by a simple expression, viz., L^2/c, (where L is the 'characteristic path length' (normally the maximum fluid path length) and c is the consolidation coefficient, which is the product of the equilibrium stiffness and the (hydraulic) permeability—see Sect. 7.2.1).

When a step increment in compressive loading is held constant for a long time, the compressive deformation of the cartilage increases over time as the water is expelled [see Fig. 1.9 and Figs. 2 and 3 (Herberhold et al. 1999)]. Eventually, no more water can be expelled, and the cartilage deformation settles to a new equilibrium (or steady state). This new equilibrium state is said to be the 'drained' deformation state, because all the water that is going to be expelled from the cartilage due to the applied loading has been expelled. Note that this does *not* mean that *all* the water has been expelled from the cartilage (and the cartilage is now 'dry'). In fact, the cartilage remains hydrated, and it remains saturated by the aqueous solution (so more water can be squeezed from it if the compressive load on the surface is again increased). The cartilage stiffness in this new drained state is the new equilibrium stiffness for the cartilage tissue.

When the compressive loading is incrementally stepped to higher and higher levels, the compressive deformation of the cartilage also increases again and again. For each step increment in load, there is zero initial deformation at constant volume (undrained state), then more water is expelled from the cartilage, and eventually the deformation settles as the cartilage again reaches a new, drained, equilibrium state.

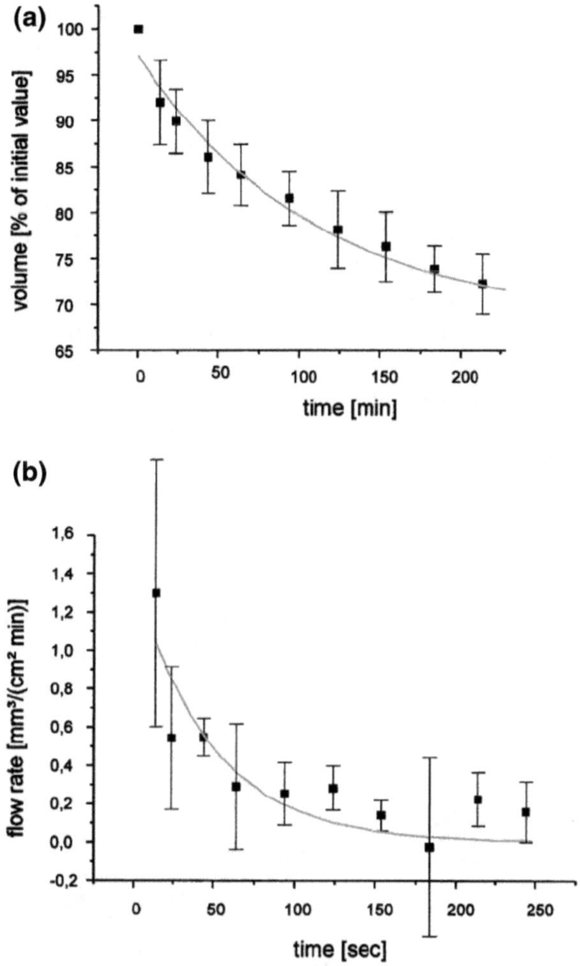

Fig. 1.9 In situ, volume changes and fluid flux during [static, normal] loading [of the patellofemoral joint at 60° flexion] with 150% body weight [measured maximum contact pressure 3.6 MPa]. The error bars show one standard deviation of the variability between individuals [there are six samples tested]. **a** Volume change of the patellar cartilage (3D analysis), **b** rate of fluid loss from the interstitial patellar cartilage matrix per square centimeter surface area (and fluid microns/s), based on the assumption that during compression the fluid flow will occur through the articular surface into the joint cavity. A 43% fluid loss was observed from the interstitial matrix after 3.5 h of static loading, the fluid flow rate per square centimeter articular surface being initially 1.3 ($\pm$ 0.5) mm^3/min cm^2 (0.217 $\pm$ 0.083 μm/s) for the first 14 min after loading, and 0.22 ($\pm$ 0.04) mm^3/min cm^2 (0.037 $\pm$ 0.007 μm/s) in the terminal phase of the experiment (>120 min) (Herberhold et al. 1999). *Note* The timescale on diagram (**b**) appears to be minutes rather than seconds, as indicated by the figure caption. In other words, even after 3.5 h of steady loading, the cartilage was still consolidating, albeit at a slow rate

The new (drained) equilibrium states of cartilage subjected to four incremental steps in surface loading are shown in Fig. 1.10 (Szarko and Xia 2012). Similar behavior is observed for human cartilage (Chen et al. 2001) and for bovine cartilage (Schinagl et al. 1996, 1997).

From a visual inspection of the images in Fig. (1.10), and also based on a quantitative analysis of this data (as shown in Fig. 1.11), it is apparent that the top of the cartilage experiences the most deformation (and strain) under a compressive loading, while cartilage in the deep zone experiences the least deformation (and strain). Such an outcome is usual following surface loading of articular cartilage. Consideration of strain change with depth (e.g. a very simple strain analysis shows the equilibrium change in height for the top half of the cartilage is much greater than the equilibrium change in height of the bottom half of the cartilage, even though the load is constant throughout the depth of the cartilage) shows that the compressive stiffness of the cartilage varies through its depth (see Fig. 1.11 for strain analysis with depth). We see that when the average tissue strain is 29% strain (see Fig. 1.11), the superficial zone strain is about double the average strain (at approximately 56%), while in the deep zone strain is only a small fraction of the average strain. When the average tissue strain is 29%, the superficial zone strain is about (55%/4.5%) twelve times greater than the deep zone strain (see Fig. 1.11).

These results are consistent with the earlier studies of Shinagl et al., who directly measured the compressive stiffness of thin slices of bovine cartilage as a function of

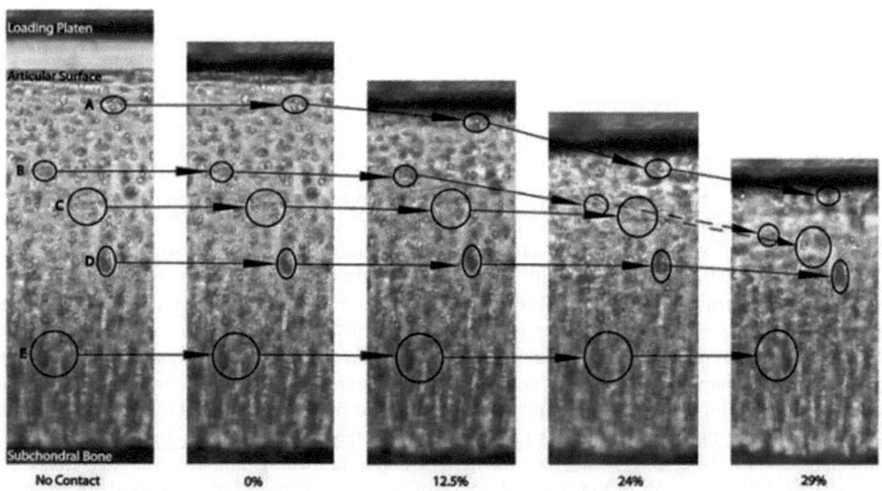

Fig. 1.10 Representative sample of (dog) cartilage with initial height 0.6 mm (shaved 0.1 mm thick) under various compression levels [zero strain, 12.5, 24, and 29% engineering strain (where the 'engineering strain' is defined as the change in specimen height divided by the initial specimen height)]. The circles correspond to cells that were tracked through the increasing compression in the superficial zone (0–60 μm) (A), transitional zone (61–180 μm) (B, C), and deep zone (181–600 μm) (D, E) 10× magnification (Szarko and Xia 2012)

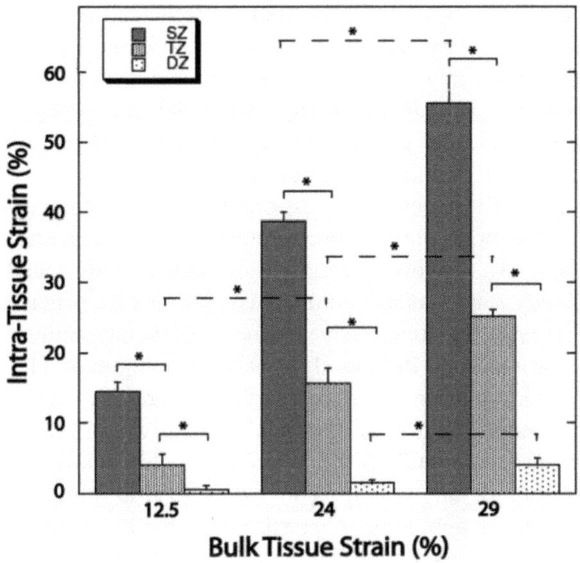

Fig. 1.11 Graph summarizing the depth-dependent properties of intratissue strain within the three bulk tissue strain levels (12.5, 24, and 29% [i.e. superficial zone, transitional zone, and deep zone (DZ)]). Significantly decreased intratissue strains were seen with increasing tissue depth. Intratissue strains also significantly increased with increasing bulk tissue strains (although the increase in the DZ was only significant upon 29% bulk tissue strain). The * symbol indicates significance at the $p < 0.05$ level (Szarko and Xia 2012). Based on reported stress–strain responses (Szarko and Xia 2012), it is interesting to note that the equilibrium cartilage stiffness in dog humeral head cartilage at 29% average Young's modulus (deep zone) is measured to be around 0.165 MPa (stress)/0.04 (strain) = 4.1 MPa

depth within the tissue and found that the stiffness of the deep zone cartilage was an order of magnitude higher than that of the surface cartilage (Schinagl et al. 1997).

This gradient in compressive stiffness is explained by a gradient in aggrecan concentration, which has a relative low aggrecan concentration at the surface of the cartilage (where the cartilage is less stiff) and a relative high aggrecan concentration at the osteochondral junction (where it is more stiff). We note again that the 'true concentration' of aggrecan (true concentration = mass of aggrecan/(n_{agg} V_T), where n_{agg} is the fraction of total volume accessible to aggrecan and V_T is the total volume) is larger than the commonly reported 'apparent concentration' of aggrecan (apparent concentration = mass of aggrecan/V_T). In the initial state, a gradient in aggrecan concentration nevertheless remains even when true aggrecan concentration is considered. However, this gradient in true aggrecan concentration decreases with increasing compressive load. For very large compressive loadings, the equilibrium true aggrecan concentration becomes nearly constant over the depth of the articular cartilage.

It is immediately apparent that chondrocytes located in the deep zone are somewhat 'protected' by the presence of more aggrecan, in the sense that the local strain they experience is less than the strain experienced by chondrocytes in the superficial zone. It is interesting to note that the difference in chondrocyte shape probably reflects this difference in strain environment. This difference in strain environment probably explains why the superficial chondrocytes tend to more 'pancake shaped,' while the reduced strain variation at depth enables the chondrocytes at depth to be 'rounder' in articular cartilage.

There are comparatively few compression studies that examine the deformation behavior of cartilage–bone plugs, while simultaneously measuring the behavior of bone and cartilage. An interesting example of such a study enabling quantitative insights into a steady-state dynamic compressive stiffness of cartilage of human tibial cartilage, together with the compressive stiffness of the supporting subchondral trabecular bone, is provided by Rohl et al. (1997) and Ding et al. (1998a, b).

A series of tests on normal and osteoarthritic cartilage were undertaken. For example, osteochondral plugs from the tibial plateau were obtained from nineteen human donors aged between 60 and 85 years (ten controls and nine with early stage osteoarthritis) (Ding et al. 1998a). The osteochondral plugs were 7.5 mm in diameter and about 11-mm long, consisting of about 2.5 mm of cartilage and 8.5 mm of bone (including about 1 mm of dense cortical bone and 7.5 mm of trabecular bone). A typical (unconfined) load–unload test loading (at a rate of 5 mm/min), with a load magnitude sufficient to cause 0.5% strain in the underlying bone, is shown in Fig. 1.12. We note in this example test result, 0.5% strain in the underlying bone results in a stress in the bone and cartilage tissues of around 6 MPa (so, the bone had a compressive stiffness of about 1200 MPa). We note in passing that the small strain stiffness of mineralized cortical (or dense) bone is usually around 10–20 GPa (i.e. about 10 times greater than measured for the osteochondral plug). However, the stiffness of trabecular (or spongy) bone is much less than that for cortical bone (in this case, it is around 1.2 GPa, i.e. about an order of magnitude less than the stiffness of cortical bone).

The average maximum steady-state dynamic (i.e. *non-equilibrium*) compressive stiffness for the control (i.e. normal) group, for cartilage from the medial compartment of the tibial plateau, was found to be 82 MPa for the cartilage (range 34–235 MPa) and 311 MPa for the underlying bone (range 53–887 MPa). For the same control group, the average maximum steady-state dynamic compressive stiffness of the lateral tibial plateau cartilage was found to be 67 MPa for the cartilage (range 16–170 MPa) and 292 MPa for the underlying bone (range 30–723 MPa). The early stage osteoarthritis group was found to have an average maximum dynamic steady-state compressive stiffness of 58 MPa for the cartilage (range 15–180 MPa) and 237 MPa for the underlying bone (range 53–887 MPa).

We also note that the stiffness of the cartilage is reported to be significantly correlated with the stiffness of the underlying bone for normal cartilage. Importantly, Rohl et al (1997) found that for normal cartilage and bone tested under the conditions specified are related by $E_{\mathrm{bone}} = 9.4(E_{\mathrm{cart}})^{0.99}$ ($r = 0.87$) (Rohl et al. 1997). That is for these test conditions, the stiffness of normal trabecular bone is about 10 times that of normal cartilage when 7.5-mm osteochondral plugs are tested

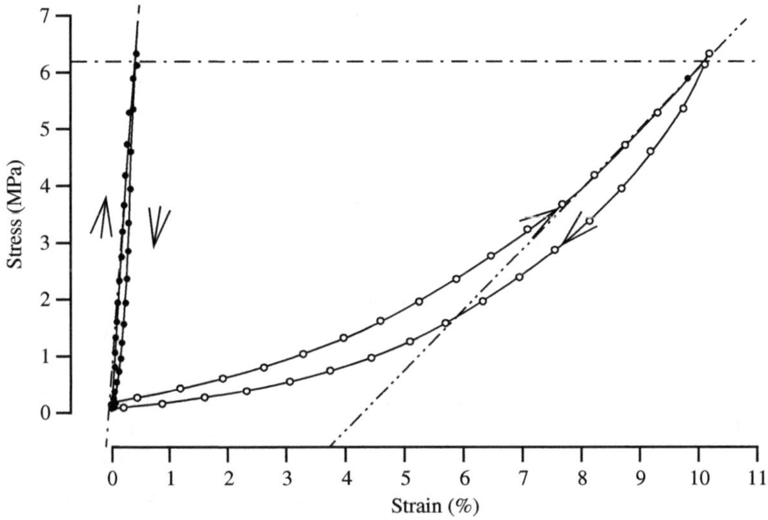

Fig. 1.12 A typical combined testing curve. [Osteochondral plugs consisting of] cartilage and bone were tested to a load resulting in 0.5% bone strain. The stiffnesses of cartilage and bone were determined as the tangent to the loading curves at 0.45% bone strain Rohl et al. (Ding et al. 1998a). Note that the osteochondral plugs tested are specimens 7.5 mm in diameter with a bone length 8.5 mm (which included 7.5-mm trabecular bone beneath the subchondral bone plate). The samples are tested unconfined, with a preload of 0.09 MPa. 20 cycles of preconditioning applied at 5 mm/s (i.e. around 0.02 and 0.03 cartilage strain per second, a complete load cycle took between 3 and 6 s, so preconditioning took between 1 and 2 min), prior to the test loading (a typical test loading stress–strain curve is shown above). Mean thickness of normal cartilage was 2.5 mm (range 1.7–3.6 mm). Mean thickness of early stage osteoarthritis cartilage is slightly thinner at 2.3 mm (range 1.8–3.0 mm)

at 5-mm displacement per minute (i.e. about 1.7–3.3%/s strain rate for cartilage tissue, with loading stopped at 0.45% bone strain).

However more importantly than the actual stiffness ratio (which is specific to these test conditions), a high correlation (a correlation being the geometric mean of two ordinary least squares regressions) between the YM of the two tissues means the other tissue's YM can be largely predicted if one tissue's Young modulus is known. In this case we have two different tissues, with working strains an order of magnitude different, yet the stiffnesses are highly correlated. This high correlation is consistent with a key principle of bone tissues, namely 'force (or stress) chain' adaption so as to maintain similar strains in bone at each level in the force chain. Bone adaption is a consequence of the Frost's well-accepted 'mechanostat hypothesis' for bone (Frost 1987a, b). However in this case, it is apparent that both bone and cartilage tissues in different individuals have also adjusted to maintain a similar 'stiffness ratio' ((trabecular) bone stiffness/cartilage stiffness), equal to about 10 (though this number only applies for test conditions). However, we expect that a wider principle to be drawn from this data is that because the two tissues are both driven by a common loading, each tissue in this 'force chain' normally adjusts its properties proportionally to the load magnitude it carries. Based on this principle, for example, we might expect that both cartilage and bone will similarly adapt when subjected to higher environmental loadings. We also note that the correlation between bone and cartilage was not significantly correlated for cartilage with early stage osteoarthritis (Ding et al. 1998a), but this would be expected if abnormal homeostatic processes operating in a disease state can no longer deliver normal correlations between properties across tissues subjected to the same force chain.

Importantly though, this is dynamic experiment that is clearly not at (thermodynamic) equilibrium (as evidenced by the significant hysteresis over the 20th load cycle, primarily due to water being imbibed and expelled over each load cycle). This means that the applied load is being shared between the pore water phase and the aggrecan solids in the cartilage, with the pore water giving an apparent stiffening effect to the cartilage. Our own poroelastic modeling of Fig. 3 of Rohl et al. (1997) shows that the cartilage is in a dynamic steady-state dynamic equilibrium (i.e. hysteresis is the same over each of the load cycles, so the system has reached a steady state by the 20th load cycle). Taking into account nonlinear stiffening due to aggrecan and nonlinear changes in hydraulic permeability, our model shows that on the mid-plane of the cartilage, about half the load is being carried by the pore water when at maximum load and about half the load is being carried by the aggrecan when at maximum load. However, about 96% of the 'total stiffness' (i.e. the slope of tangent approaching the end of load curve [see tangent shown in Fig. 1.12) averages about 75 MPa in Rohl et al. (1997)] may be attributed to horizontal straining of collagen (i.e. 72 MPa), driven by the sideways movement of the fluid phase due to its incompressibility (see discussion below), while (according to Appendix D) on average only about 4% of the stiffness is due to aggrecan stiffness (i.e. 3.0 MPa). We note here that if cartilage experiences large static loads for long periods of time (e.g. several hours), and the cartilage compression approaches up to 50% strain, then aggrecan stiffnesses can become very large to sustain the applied loading at 50% strain.

We may ask: What happens to the cartilage stiffness when it is loaded quickly? Consistent with the previous example, it turns out that as the frequency of compressive loading increases, the cartilage appears to become stiffer [this effect is sometimes referred to 'self-stiffening' (Nia et al. 2013; Sah et al. 1989)]. For short-term loadings in vivo, the main reason cartilage behaves the way it does is because the applied loading is largely carried by excess pore pressure in the pore fluid within the cartilage. Following application of a surface load, the load is shared between the fluid and solid phases within cartilage. At least in the short term (where short term means relative to the characteristic time for consolidation under the operating conditions in vivo), the load is mainly carried by the pore fluid pressurization. At high rates of loading, the pore fluid does not have time to leave the cartilage (because of its low permeability) and so most of the load is carried by the pore fluid (Sah et al. 1989), at least until sufficient time has passed for the majority of the pore water pressures to have dissipated (see Sect. 7.2.1).

During unconfined compression testing on cartilage explants, the stiffening effect is due to the combined effect of fluid not leaving and the so-called Poisson's ratio effect. If the pore fluid cannot leave and tissue volume is conserved, for small deformation theory this means the Poisson's ratio is close to 0.5 (note: the Poisson's ratio can be larger when deformations are large). Due to the so-called Poisson's ratio effect, upon axial compressive loading, radial and hoop stresses are generated in the collagen network, which resist the axial loading. This resistance to compressive axial deformation of a volume element occurs because it is necessarily accompanied by lateral expansion of the volume element, due to the constant volume constraint.

Consider an example—a cartilage sample is tested so that fluid can flow through the sides of a cartilage sample under conditions of 'unconfined compression.' When the rate of loading is sufficiently large, water does not have time to leave the test specimen and it remains in a nearly undrained state. Because water is nearly incompressible (this is a good approximation at physiological stresses), volume is conserved. Now, assume the cartilage sample is cylindrical and it is given a large compressive strain, of say 50%, in the axial direction, then the diameter of the cylinder must increase by about 40% (i.e. $\sqrt{2} - 1$) to maintain constant volume, and this lateral expansion of the cylinder places the cartilage network in tension; i.e. both tensile radial and tensile hoop stresses are generated by the compressive axial loading. For this example, the ratio of minus the radial strain to axial strain is then about 80%. How does this estimated ratio compare with experimental data? Impact testing on 2–3-year-old bovine cartilage (Flachsmann et al. 2001)—the rapid loading rate during impact means that there is little time for water to escape the tissue—has shown that the maximum radial strain in the superficial zone collagen is 60–80% of the strain in the axial direction (decreasing at higher stress magnitudes). So, the assumption of approximately constant volume appears to provide a reasonable explanation for the experimental observations reported in Flachsmann et al. (2001).

We also note in passing that based on these calculations, if the tensile stiffness of the collagen network in the superficial zone is 100 MPa in adult cartilage, and the ultimate tensile strain of superficial collagen in healthy adult human aged about 25–30 years is around 35% of the tensile stiffness (see discussion in Sect. 3.3), then impact loading leading to 50% strain in the vertical direction under undrained conditions is likely to exceed the tensile strength of the collagen in the superficial zone, causing fracture of the cartilage surface (Flachsmann et al. 2001). This expectation appears to be borne out in practice (Flachsmann et al. 2001). The result of overloading is cartilage splitting, with the split usually originating from the cartilage surface (see Fig. 2.27). The effect of cartilage splitting is graphically demonstrated in images shown in Figs. 4 and 5 of Workman et al. (2017) (we note in passing that the crack tip in the (Workman et al. 2017) images shows all the classic indicators of a plastic failure zone, e.g. multiple satellite cracks appearing near the crack tip, collagen fibers continuing to spanning the advancing crack. This is indicative of a material that has high fracture toughness).

For 'confined cartilage sample tests' (i.e. tests with zero lateral strain) and with no fluid flow through the confining surface, the initial apparent cartilage stiffness becomes theoretically 'infinite' (or at least exceedingly large relative to equilibrium cartilage stiffness). This is because water under physiological loading is well approximated as incompressible. Alternatively, if the same confined sample is loaded very slowly and the fluid can flow easily through the cartilage surface (i.e. a drained state is maintained at all time), the initial compressive stiffness is typically a fraction of 1.0 MPa. Importantly, we note that in vivo load conditions are usually somewhere 'in between' confined and unconfined test conditions. Along 'axes of symmetry' (e.g. often well approximated as being close to where contact stresses are largest), test conditions are more closely approximated by confined compression tests. Clearly, the magnitude of the stiffening effect can be difficult to quantitate without a relatively sophisticated computational model or experimental tests, as many factors can have a strong influence on the magnitude of the observed stiffening effect.

It is also interesting to note in passing that this tensile radial and hoop deformation under both undrained or partially conditions probably in large measure explains the functional significance of the collagen architecture toward the surface of the cartilage. The collagen fibers in the superficial zone are largely parallel with the cartilage surface. In particular, the vertically oriented collagen fibers in the deep zone of the tibial plateau (resisting aggrecan expansion and anchor articular cartilage to bone) curve over in the transitional zone to a variable degree. Curvature is abrupt for fibers along the axis of symmetry directly beneath the maximum (in vivo) contact stresses (fibers tend to crimp with drained vertical deformation), but fiber curvature increases radially, in sympathy with maximum principal surface strains (i.e. with radial surface displacements increasing with radial distance from the point of maximal load) (Figs. 5C and 6; Goodwin et al. 2004), in sympathy with meniscal displacement upon loading and the surface shear generated by the same loading and shear displacement. Meniscal translational displacement, subchondral bone orientation, load orientation, and regional strain differences largely explain further local differences in collagen fiber orientation. We note in passing so-named 'split lines' (i.e. the 'wheel-spoke' of surface 'gaping' following a pinprick of the articular cartilage surface, demonstrating residual tensile hoop stress in the cartilage surface) are oriented in the direction of maximum principal strain generated in the cartilage surface upon compressive loading (i.e. split lines are oriented in a 'radial direction' from the centre of the tibial plateau), are usually most prominent toward the edge of the tibial plateau, and are smaller and sometimes absent toward its center—as illustrated in Fig. 6 in Goodwin et al. (2004).

Figure 6 in Goodwin et al. (2004): *Drawing shows architecture of extracellular matrix of tibial plateau. Viewed from above (a), split lines in peripheral region of plateau are arranged in wheel-spoke pattern. After fracture sectioning in coronal plane, surface of anterior half is viewed en face (b) and then viewed at oblique orientation (c). In central portion of tibial plateau (1), cartilage has column-like structure that is perpendicular to underlying subchondral bone and curves at superficial level (2). At tibial eminence (3), cartilage maintains column-like structure, but instead of perpendicular alignment relative to subchondral bone, its alignment approaches weight-bearing axis of joint. At submeniscal region of joint*

(4), tissue curves both away from center of joint and in annular direction perpendicular to orientation of split lines in articular surface (5) (Goodwin et al. 2004).

To a lesser extent, 'stiffening effects' also occur because of (loading) 'rate effects' in the collagen network and between aggrecan molecules. It is reported that rate stiffening of the collagen network in adult bovine cartilage (16–24 months old) can result in an increase of measured Young's modulus of collagen from 35 MPa at 0.1% strain rate to 55 MPa at 80% strain rate, at 5% strain (Ahsanizadeh and Li 2015). Under normal operating conditions for cartilage in vivo, it is unlikely such stiffening effects have a significant effect on tensile stresses in the collagen network in the superficial zone cartilage; however, under impact loading conditions, such stiffening effects potentially offer some additional protection against cartilage splitting, providing there is sufficient ultimate tensile strength available in the collagen.

Clearly articular cartilage presents a very large range of possible compressive stiffnesses, ranging from a stiffness of less than one megapascal to potentially several hundreds of megapascals. All these stiffnesses are accessible through simple changes in specimen size, material properties and flow and loading conditions on the surface of the test specimen. These examples serve to highlight the extraordinary potential for a composite porous material such as cartilage to modify its structural behavior in response to changing environmental conditions and so to adapt. In other words, there is no simple way to characterize cartilage mechanical behavior in vivo; rather, a great deal of information about the cartilage is first required before any meaningful statements can be made.

Even precisely measuring the stiffening effect under in vivo conditions is currently challenging, and in some ways questions about stiffening and component responses have not been adequately addressed in the literature. With ethical constraints on the types of experiments possible and technology in its current state, this question is almost certainly most efficiently explored using a combination of experiments and computational modeling of cartilage. Experimentalists have investigated a number of comparatively simple experimental systems in the laboratory. A sinusoidal load of varying frequency applied to an osteochondral plug or cartilage explant under unconfined compression is a typical test that attempts to quantify the magnitude of the stiffening effect in cartilage.

One such test is shown in Fig. 1.13. From Fig. 1.13, we see that for cyclic loading frequency of 0.1 Hz, the initial (loading) stiffness increased from 15 MPa to 49 MPa at 16% strain, while at 1 Hz the measured initial stiffness increased from 16 MPa to 66 MPa at 13% strain, and while for 40 Hz the measured initial stiffness increased from about 29 MPa to 61 MPa at 9% strain. This data suggests that for these particular test conditions (which are not physiological), there is about a doubling of the measured loading stiffness as the load frequency increases from 0.1 to 40 Hz (Park et al. 2004).

It is also clear from Fig. 1.13 that there is load–unload hysteresis (i.e. the unloading curve always lies below the loading curve, so less mechanical work is recovered from the material than the work done on the material by the applied

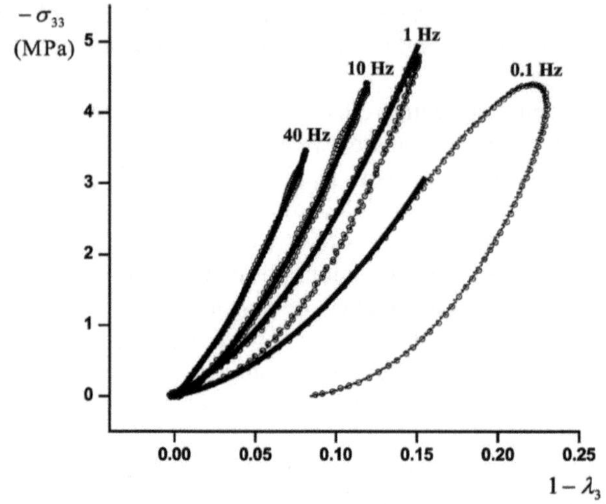

Fig. 1.13 Stress–strain response for a typical specimen [bovine cartilage (3 months to 3 years of age) explants plugs, 3 mm in diameter and 1.8 mm in height, which were tested under unconfined compression with impermeable platens and free draining sides], at various loading frequencies. Loads ranged from zero to a nominal 5.7 MPa, while the engineering strain in the load direction ($1 - \lambda_3$) ranged about 12% seating strain to total loading strain of 34% (a difference of about 22%, as shown above) (Park et al. 2004)

mechanical load). In other words, the hysteretic loop area represents an estimate of the mechanical energy loss (i.e. the mechanical energy dissipated as heat) within the material during a load–unload cycle.

We note that hysteresis is theoretically zero at the lowest frequency (i.e. theoretically, there is no energy loss for a 'quasi-static' process completed infinitely slowly so that the system is always in thermodynamic equilibrium). For practical tests on macroscale specimens (i.e. finite rate of loading on specimens at the millimeter scale), hysteretic loss becomes higher as frequency increases, but then the hysteretic loss decreases again with increasing load frequency to reach a minimum, usually around 1 to 10 Hz (i.e. at the frequencies containing the majority of the power in physiological loadings). We see that hysteretic loss is close to zero again at 10–40 Hz for Figs. 1.13 and 1.14. The precise value of minimum energy loss depends on the test specimen and test conditions.

For microscale test specimens (i.e. for specimens measured at the micron scale), hysteretic loss is also very low at very low frequencies, becoming higher at frequencies in the kilohertz range and finally decreasing again at the highest frequencies (see Fig. 1.14) (Nia et al. 2013). We note in passing that depletion of GAGs of cartilage causes a marked shift in mechanical energy loss as a function of kilohertz frequency for microscale tests (see Fig. 1.14).

It turns out that the hysteretic loss is greatest when the timescale of loading (i.e. period is the inverse of the frequency) is comparable to the characteristic timescale for drainage of pore water from the poroelastic material (Nia et al. 2011) (i.e. the

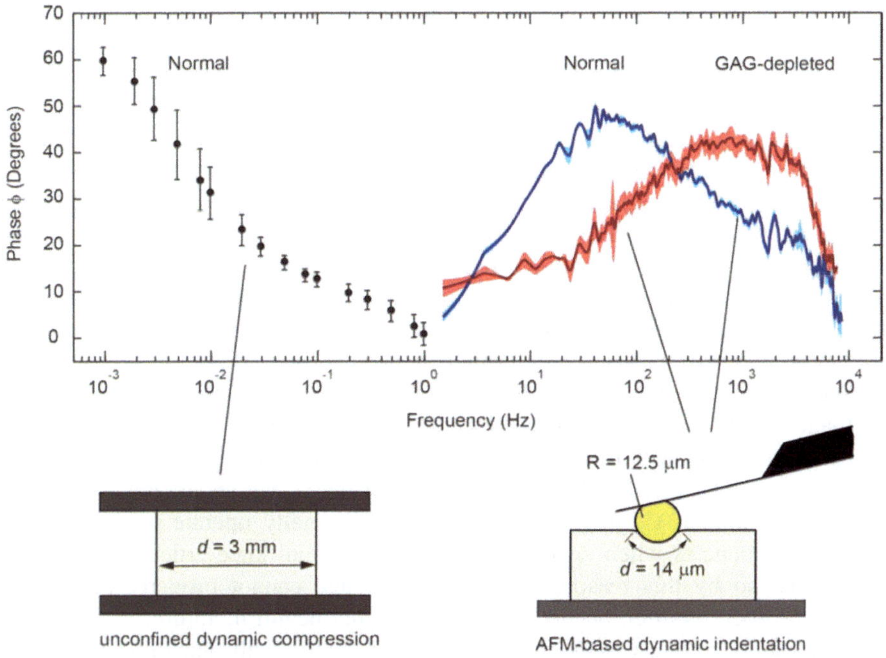

Fig. 1.14 Phase φ of the dynamic modulus [i.e. the inverse tangent of the ratio of the energy loss to energy stored in the cartilage] for both normal and GAG-depleted cartilage obtained by AFM-based dynamic indentation is compared with the phase angle from macroscopic unconfined compression tests using 3-mm diameter by 1-mm-thick cartilage disks. The length scales involved in the AFM-based dynamic indentation are: probe radius R = 12.5 μm, δ_0 = 2 μm, and probe tip-cartilage contact distance d = 14 μm. The relevant length scale for the unconfined compression is the diameter of the sample, d = 3 mm. As we observed a shift in φ in AFM-based indentation after GAG depletion, a similar shift in the phase φ is expected at physiological macroscales. However, due to experimental limitations in measurements at low frequency, this macroscale hypothesis has yet to be reported for sinusoidal testing (Nia et al. 2013). We note in passing that the shift in φ higher with aggrecan loss is due to an increase in the coefficient of consolidation (i.e. an increase in the product of permeability and equilibrium stiffness) with aggrecan loss. The increase in the coefficient of consolidation is due to the rate of increase in permeability eventually 'outpacing' the rate of decrease in aggrecan stiffness as aggrecan concentrations are lowered

characteristic timescale pore fluid drainage is L^2/c, where L is a 'characteristic length' and c is the 'coefficient for consolidation' (which involves the product of k, the permeability of the porous medium to pore fluid, and E, the equilibrium compression modulus—see Sect. 7.2.1 for further discussion on characteristic timescales for pore water drainage and cartilage consolidation).

It makes sense that hysteretic losses are reduced when there is a 'separation of timescales' (i.e. separation of load cycle period from the characteristic time for pore water drainage and consolidation)—for if the frequency is high (i.e. the load cycle period is significantly shorter than the characteristic time for drainage), the water does not have time to leave the cartilage, so viscous dissipation is reduced. On the other hand, if the frequency is low (i.e. the load cycle period is significantly longer than the characteristic time for drainage), there is a lot of time for the pore fluid to leave the cartilage, pore fluid velocities are consequently comparatively small, and so again viscous dissipation is minimized. This explains qualitatively the existence of maxima in hysteric losses in saturated porous media subject to cyclic loading.

We note that the frequency maxima shift lower (i.e. period increases) as the size of the specimen tested increases (along with L, and so the characteristic time for pore fluid drainage). In vivo, we have seen the characteristic time is of the order of two to three (or more) hours in normal human patellafemoral groove cartilage (see Fig. 1.9), so we expect that hysteretic losses would peak around a load frequency of 0.00014 Hz. Since normal load frequencies are of the order of 1 Hz, this suggests hysteretic losses in cartilage are normally very small.

We mention in passing this finding is at odds with common descriptions of cartilage functioning as a 'shock absorber,' since shock absorbers are normally designed to dissipate kinetic energy during dynamic loading, and to do this the period of loading and the characteristic time for the spring shock absorber system need to be similar. Cartilage clearly does not normally operate in the range of loading frequencies when energy dissipation in normal knee articular cartilage is prominent, so by this reasoning, articular cartilage cannot properly be called a shock absorber, as often occurs in the professional health literature.

Importantly, because standard poroelasticity theory (with tension–compression nonlinearity—see Fig. 1.8) can explain the data very well, this suggests that the primary mechanism for hysteretic loss within cartilage is the flow of pore fluid within the cartilage in response to excess pore water pressures that arise from surface loading, though no doubt some losses are associated with loading and unloading collagen fibrils and deformation of aggrecan.

We may also wonder about the geometry of cartilage contact in vivo. More specifically, we may ask: What are typical deformations and contact areas of knee cartilage during normal activities such as walking? Eight healthy subjects (six men and two women) aged 32–49 years (average BMI 23.5) were recruited. The knee was imaged while walking at 0.67 m/s using a fluoroscope at 30 frames a second (with this apparatus, cartilage thickness could be measured to within about 1.8% $\pm$ 1.6% error) (Liu et al. 2010).

One minute after starting walking, it is interesting to find that the peak load on the tibia moves forward 2.3 mm from heel strike to 20% stance, followed by a 2.0-mm posterior shift from 20 to 60% stance, and then forward again 4.67 mm from 60% to toe off (100%). Peak strain is estimated to increase from 8 $\pm$ 5% at heel strike to 23 $\pm$ 6% at 30% stance to around 16 $\pm$ 7% deformation at 30–60% stance, and at 80% stance increases again to 22%. Total contact area for a single joint (i.e. the sum of contact areas for medial and lateral compartments) was estimated simultaneously (error estimate about 14%) (see Fig. 1.15) (Liu et al. 2010).

In another set of experiments, eleven healthy subjects with average age 31 years (range 23–40 years), we asked to stand and then to make a single forward lunge on one knee and hold for 5 s at 30, 60, 75, 105, and 120 degrees of knee flexion, with their upper body held upright (Bingham et al. 2008). It is apparent from Fig. 1.16

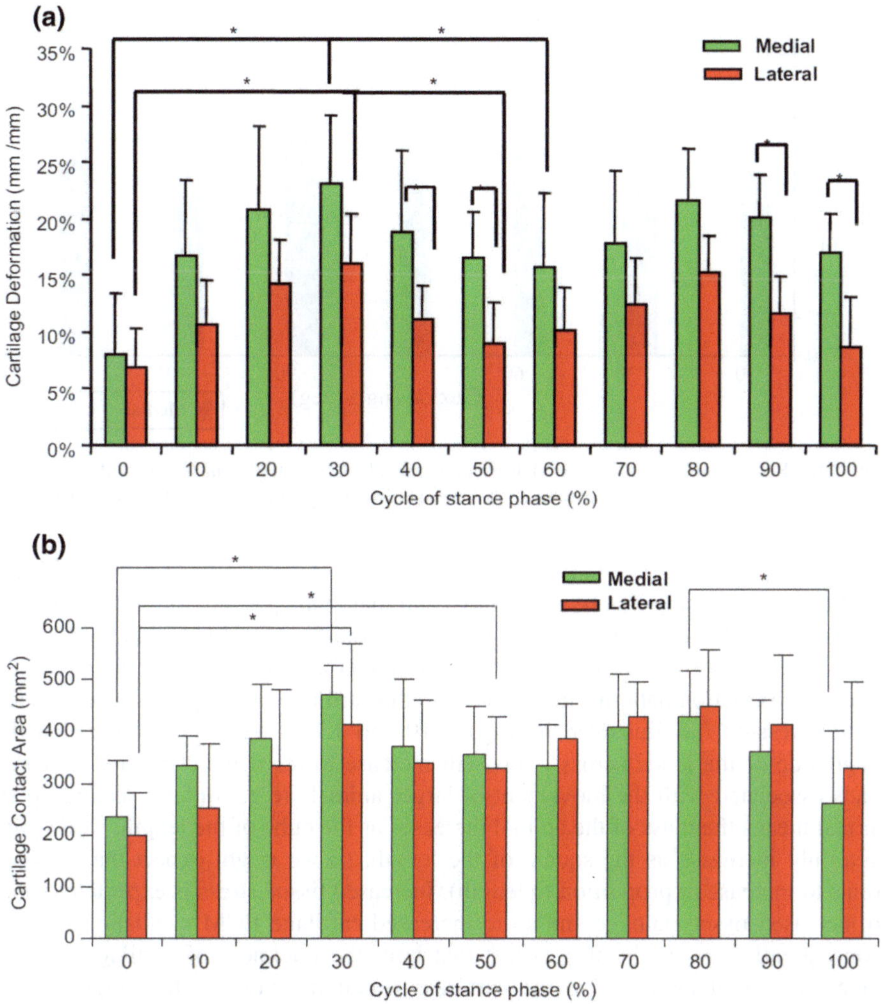

Fig. 1.15 Cartilage contact characteristics during the stance phase of gait: **a** peak cartilage contact ['deformation,' i.e. average strain across both cartilages in mm/mm)] and **b** cartilage contact area on the medial and lateral tibiofemoral compartments. Error bars indicate standard deviations. Asterisk denotes significant difference at $p = 0.05$. Because of the numerous comparisons, we only present the relevant comparisons, i.e. between the maximal and minimal values (Liu et al. 2010)

that on initial (i.e. undrained) loading, cartilage deformations are estimated to reach about 30% of the total cartilage thickness upon significant knee flexion [up from about 22% when standing: see Fig. 4 (Bingham et al. 2008)]. It is expected deformations would increase over time if activities were continued for up to a couple of hours (see Fig. 1.9).

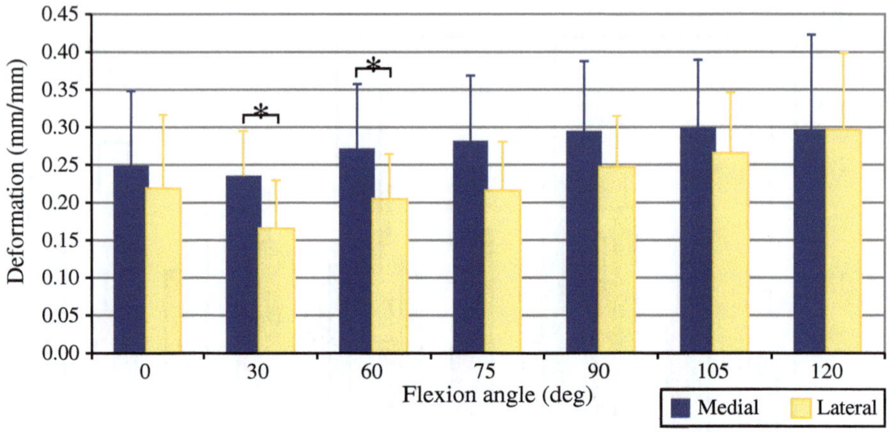

Fig. 1.16 Peak tibiofemoral contact deformation for all flexion angles in the medial (blue bars) and lateral (yellow bars) compartment (* denotes statistical significance for $P < 0.05$) (Bingham et al. 2008)

We note in passing that reptiles, birds, and mammals past and present, no matter their size, all have articular cartilage in their diarthrodial joints (and yes, even dinosaurs had cartilage in their joints) (Holliday et al. 2010). And while a broadly similar deformation patterns are observed in the cartilage of all species when loaded in compression (as illustrated by Figs. 1.10 and 1.11), we might expect that the magnitude of the loads across joints will increase to resist the static and dynamic loads associated with the movement of larger animals (e.g. crudely, scaling up an animal means the mass of the animal increases as the cube of the length, while joint area only increases as the square of the length, so we might expect the stress in joints to increase in proportion to length). Increased tissue stress is expected to lead to increased tissue damage and so to increased cartilage ECM synthetic capacity. Consequently, we might also expect that both the thickness of cartilage and the concentration of aggrecan in the cartilage would increase as the weight of the animal increases.

We do find that the thickness of femoral condylar cartilage increases with the size of the animal: In the rat the thickness is around 170 μm, in the rabbit it is around 350 μm, in the goat it is around 900 μm (Kamisan et al. 2013), while in humans it is around 2000 μm (Malda et al. 2012). Various studies have confirmed this general trend (Frisbie et al. 2006; Malda et al. 2013). In humans, a trend between the weight of the person and the thickness of the cartilage has also been reported (e.g. mean cartilage thickness of the knee is 1900 μm at 40-kg body weight but increases to 2800 μm for 80-kg body weight) (Shepherd and Seedhom 1999).

However, the (apparent) concentration of GAG appears to be not well correlated with the size of the animal. So in the rat knee, there are reported GAG concentrations of around 30–40 mg/ml (Lundberg et al. 1996), in rabbit knee around 28 mg/ml (Price et al. 1996), while in humans the GAG concentration in the knee is

50 mg/ml [based on the average of five reports, but may be 80 mg/ml, e.g. (Wedig et al. 2005)] while in horses it is around 50 mg/ml (Malda et al. 2012). A study of the femoral condyles of many different species revealed no well-defined trend between GAG concentration and the size of the animal (Malda et al. 2013). This is probably related to the fact that apparent GAG concentrations are reported rather than true GAG concentrations (see Appendix D for discussion of excluded volume effects of collagen and non-collagenous proteins on aggrecan concentration), as well as to differences in the contact stresses experienced by each joint, depending on the details of joint geometry (e.g. area and congruence of cartilage), on animal-specific joint muscle mechanics (hopping vs. walking on two or four feet involves quite different joint angles and loadings), and probably to other factors as well.

In summary, in the section we have seen that characterizing the deformation behavior of the cartilage biocomposite material clearly presents a challenge, but this behavioral complexity is also a valuable resource that enables articular cartilage to adapt its behavior to cope with a wide variety of joint microenvironments.

1.7 The Synovial Fluid and Synovial Membrane

The synovial fluid in diarthrodial joints is in intimate contact with the contacting surfaces of the articular cartilage and with the cells lining the synovial membrane (see Fig. 1.17). The contact of the synovial fluid with synovial membrane is extensive, with the synovial membrane extending into the suprapatellar pouch and to the 'infrahoffatic recess' (see Fig. 1.17). If these pouches (or recesses) become swollen with excess fluid and inflamed, the condition is known as bursitis.

The intima cells lining the synovial membrane are not an 'epithelium' [so, there is no basement membrane present (Simkin 2015)] (see Fig. 1.17). There are broad gaps between the synovial fibroblasts, and molecules readily move back and forth [see Fig. 1 in Smith (2011), Simkin (2012, 2015)]. Large molecules in the synovial fluid are retained by the extracellular matrix between the cells, and glycosaminoglycans are in relatively high concentration in the synovial ECM (Sabaratnam et al. 2007).

The intima cells lining the synovial membrane are specialized synovial fibroblasts (type B cells) with occasional macrophages (type A cells) interspersed between the type B cells. There appears to be a sub-population of type B cells that extend dendritic-like processes between the other lining cells (see Fig. 1.18; Kitamura et al. 1999). The specialized synovial fibroblasts secrete large amounts of hyaluronan into the joint space. The synovial fluid has a pH of about 7.4–7.7 (Jebens and Monk-Jones 1959). Additional immune cells (including CD4+ and CD8+ T cells) can be found in the sub-intimal tissue.

Small concentrations of GAG are found in the (rabbit) synovial tissue (0.55 mg/g). We note that this is much less (about 50 fold less) than the GAG concentrations found in (rabbit) cartilage (28 mg/g) (Price et al. 1996).

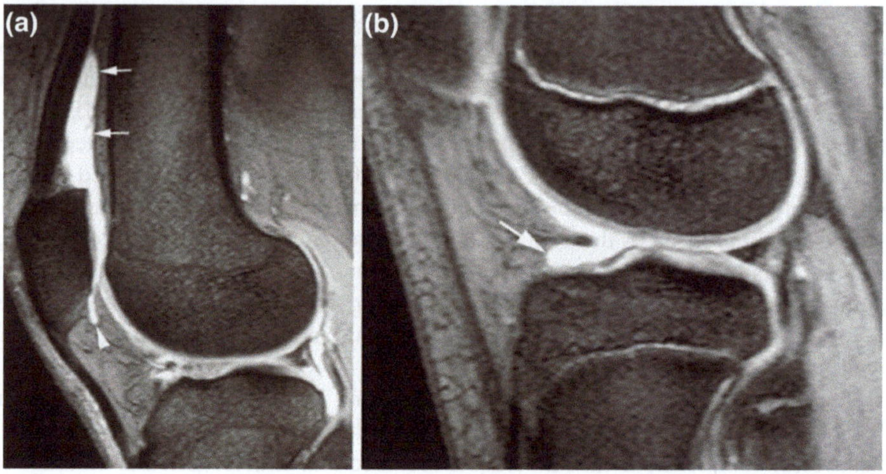

Fig. 1.17 MRI of anterior knee synovial recesses, showing synovial fluid (white). **a** Sagittal T2*W GE image showing synovial fluid in the suprapatellar pouch (arrows) and suprahoffatic recess (arrowhead). **b** Sagittal T2*W GE image showing synovial fluid in the infrahoffatic recess (arrow)

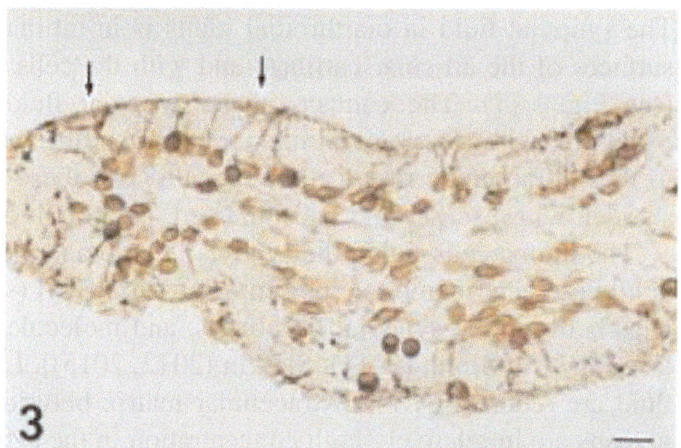

Fig. 1.18 Tip of synovial villus from a horse (scale bar 20 μm). Note region containing central vessel and PGP 9.5 stained 'dendritic cells' extending processes to the villus surface (Kitamura et al. 1999)

The 'synovial membrane' lines non-articular portions of the joint, including pouches and recesses, as well as intra-articular ligaments and tendons, but 'bursal membranes' are smooth rather than folded and do not have villi.

See Fig. 1 in Hirschmann et al. (2007)*: intima, divided in a lining layer and a lamina propria, and sub-intima [for bursal membrane].*

The sub-intimal tissue of the synovial membrane contains collagen fibers (mainly composed of collagen types III, IV, and VI, laminins, fibronectin, and proteoglycans (but not collagens type I and type II). The specialized collagen composition of the synovial membrane probably reflects the fact that it needs to be a very deformable tissue, which can accommodate large joint movements. It appears that collagen type III is preferred when deformations can be very large, being abundant in 'elastic tissues' (Gelse et al. 2003), and is particularly suited to forming long thin fibers (e.g. collagen III forms so-called 'reticulin fibers' supporting soft tissues) (Fessler and Fessler 1978).

'Synovial villi' protrude from the surface of the synovial membrane (see Fig. 1.18) (Kitamura et al. 1999). Importantly, the villi have a central blood vessel, but they contain no lymphatic vessels (Fisk 1952). The lymphatic vessels are to be found in the fibrous capsule or sub- (or deep) intima of the synovial membrane, usually some distance from the surface of the synovial membrane. The 'synovial fluid' originates from the plasma filtrate that emerges from the gap junctions between the endothelial cells of the blood vessels within the synovial villi, which then mixes with the hyaluronan secreted by the synovial fibroblasts to form synovial fluid. The transudate flows from the villi to the joint space, forms the synovial fluid, and then flows back through the synovial membrane, to the lymphatic vessels in the fibrous capsule or sub- (or deep) intima, creating a 'circulation of the synovial fluid' through the joint space and surrounding tissues. We begin to see the importance of the synovial fibroblasts not secreting a basement membrane. For it now becomes apparent that such a membrane would impede the circulation of the plasma filtrate from the synovial villi to the synovial fluid, and impede the removal of hyaluronan or tissue debris carried by the synovial fluid, as it flows back to the lymphatic vessels located in deeper synovial tissues.

Synovial fluid production is governed by the usual balance of mechanical and oncotic potentials as described by the revised Starling's law (see Levick and Michel 2010). So, for example, an increased production of hyaluronan by synovial fibroblasts will result in an expansion of the synovial fluid. Indeed, any process resulting in an elevation in the concentration of oncotically active particles in the synovial fluid will cause an expansion of the synovial fluid volume. Indeed, expansion of the synovial fluid due to increased concentration of oncotically active particles is usually what leads to the palpable swelling in a 'bursitis.' Synovial fibroblasts secrete various proteases, which degrade ECM components and create 'cell debris' in the joint cavity. If the concentration of proteases in the synovial fluid increases, the rate of production of osmotically active particles in the synovial fluid increases, so the volume of the synovial fluid expands. Clearly, intermittent fluxes of aggrecans and collagen fragments released to the synovial fluid from damaged articular tissue may transiently increase synovial fluid volume significantly—perhaps to such an extent, it becomes clinically apparent as a 'swelling of the joint capsule' (which clinicians call an 'effusion'). However normally, the synovial fluid volume of the adult human knee joint is small—generally less than 5 ml in volume (Heilmann et al. 1996; Kraus et al. 2007).

In the pig, synovial fluid is probably replaced every hour or less, as labeled albumin first appears in the thoracic lymph duct within 20 min of injection. The clearance half-life of the labeled albumin from the pig joint is 8.3 h (Jensen et al. 1993). For the rabbit, the half-life of albumin (MW = 67 kDa) is about 1.2 h, while for acridine blue in the rabbit (MW = 370 Da) it is just 0.2 h (Bajpayee et al. 2014). As might be expected, the half-life of hyaluronan (around 500 kDa) in the rabbit knee is considerably longer than albumin at around 31–38 h, meaning the hyaluronan is probably replaced every few days in vivo (Coleman et al. 1997). Indeed this rapid turnover makes intra-articular injection of therapeutics for OA pathologies is extremely challenging, as the injected drug is cleared in a matter of hours, leaving little time for the drug to diffuse into cartilage tissue (Evans et al. 2014).

Normally, synovial fluid is removed via lymphatic vessels mostly found in the sub-intima and fibrous capsule (Xu et al. 2003). It is the flux of new synovial fluid into the joint cavity from the synovial villi (and blood vessel contained therein) that brings with it the nutrients, oxygen, growth hormone, IGF-1 (produced by the liver), IGF-2 (produced by muscle), thyroxine, cortisol, sex hormones, and many other molecules that support chondrocyte vigor within the articular cartilage.

Indeed, we may poetically think of the synovial tissues as the 'handmaiden to articular cartilage.' In other words, the blood vessels in the synovial membrane are the source of all the nutrient and energy needs of the avascular articular cartilage. Transfer of nutrients and removal of water products between these two tissues occur via the synovial fluid. Mass transfer of molecules to and from articular cartilage occurs mainly by diffusion, but there is an interaction between pore fluid flow in and out of articular cartilage and mass transfer. Together with diffusive transport, advective transport plays some role in mass transfer (Gardiner et al. 2007; Zhang et al. 2007), and there is some evidence that sliding at contacting surfaces may enhance advective mass transport (Graham et al. 2018). This 'arrangement' between the two tissues clearly arises because the hostile mechanical environment of articular cartilage precludes blood vessels, compelling this tissue to be avascular, and so cartilage depends almost entirely on the synovial membrane for its requirements.

The total protein concentration of synovial fluid is usually about 0.25 that of blood plasma (i.e. 10–30 mg/ml, with coagulation proteins absent, so synovial fluid does not clot) (Faryna and Goldenberg 1990). The main protein in synovial fluid is albumin (55–70%), the balance of protein mainly being made up of alpha, beta, and gamma globulins (30–45%), including complement proteins. HA accounts for only 3–4 mg/ml, so proteins dominate by weight and have an important influence on the viscosity properties of synovial fluid (see Chap. 4). A detailed account of the proteins and signaling molecules found in synovial fluid can be found in Hui et al. (2012).

The synovial fluid has a glucose concentration of over 90% of the concentration found in the blood (i.e. normally glucose concentration is not more than 0.6 mmol less than the concentration in blood, which normally averages 5.5 mmol glucose, i.e. about 1 mg/ml). Chondrocytes also store glucose as glycogen, and they store triglycerides as fat globules in their cytoplasm. These stores presumably smooth any temporary shortfalls between energy requirements and energy supply by transport to chondrocytes, as glucose concentrations decrease with depth in cartilage (Spitters et al. 2013). Temporary shortfalls in oxygen supply are managed by anaerobic metabolism, which ensures an ongoing supply of ATP, albeit at reduced levels. It is noted that lactate concentrations in synovial fluid are normally that of blood plasma, but lactate concentrations are higher in the cartilage tissue (so, there is a flux of lactate out of articular cartilage toward the synovial fluid). Meanwhile, the flux of synovial fluid out of the joint cavity removes various cytokines and signaling molecules picked up from the joint space, along with debris from the breakdown of chondrocytes and cartilage ECM. These are transferred to lymphatic vessels located in the sub-intima of the synovial membrane.

Interestingly, it is known that an increased rate of synovial fluid turnover (induced by increasing the number of lymphatic vessels in the synovial membrane) can reduce the severity of inflammatory arthritis (Zhou et al. 2011), and conversely, reduced synovial fluid turnover (induced by a reduction in the number of lymphatic vessels in the synovial membrane) significantly worsens inflammatory arthritis (Guo et al. 2009). This suggests that synovial fluid turnover is an important regulator of concentrations of the various molecules in the synovial fluid. Assuming constant source strength and a variable sink strength, this experimental finding makes sense from the viewpoint of mass balance and the concentration of molecules sourced from the cartilage and surrounding tissues within the synovial fluid.

In vivo, the density of lymphatic vessels appears to be controlled by macrophages (Kataru et al. 2009), while the number of macrophages is controlled by the amount of debris to be removed from the joint space. This 'adaptive network' response serves to highlight that the synovial membrane is a dynamically adjustable tissue, changing its operational behavior in response to the needs of the articular cartilage (e.g. changing its structural form to enable more removal of debris when cartilage damage increases or when the articular cartilage requires more nutrients). The synovial membrane is not an inert tissue, but a tissue that adapts to the needs of the tissue it serves.

It becomes apparent that the synovial membrane and the articular cartilage normally work together, partnering each other to achieve articular cartilage tissue homeostasis. Co-cultures have been developed to explore these interactions (Beekhuizen et al. 2011). Undoubtedly, there are many additional partnerships between the synovial membrane and cartilage, via the synovial fluid, that ensures cartilage normally has access to adequate nutrients, energy, and oxygen sources. The sustaining ability of synovial fluid in vivo is evidenced by it being shown to be capable of supporting other tissues in addition to cartilage, e.g. tendon tissue (Lundborg and Rank 1978).

The fact that the synovial membrane is a dynamically adjustable tissue again reminds us that articular cartilage probably is too. The discussion so far has given a largely static view of articular cartilage tissue, as we have focused on what it is, how it deforms, and how it is nourished. Given this, in the next two chapters we want to develop the arguments supporting the view that cartilage is in fact a dynamically adaptable tissue, much like the synovial tissue that supports it. But before we do that, we want first to introduce osteoarthritis (OA).

1.7.1 What Are the Implications for Osteoarthritis?

What is osteoarthritis (OA)? Osteoarthritis is a common, chronic joint disease, causing joint stiffness, pain, and disability (normal and osteoarthritic cartilage is shown in Fig. 1.19). OA is often responsible for premature cessation of physical activities, early retirement from the workforce, and a great deal of morbidity in the general population, including co-morbidities such as anxiety and depression (Neogi 2013; Fernandes and Valdes 2015; Whittaker et al. 2015; Sharma et al. 2016).

The understanding and classification of osteoarthritis have evolved over time. In 1977, it was suggested by Mitchell and Cruess (1977) that the former classification of osteoarthritis as either: (a) 'idiopathic' (meaning of unknown origin) and (b) secondary (i.e. arising as a result of another disease) was obsolete! In 1977, they argued there was enough evidence to suggest osteoarthritis is primarily the result of: (i) abnormal concentrations of force on a joint or (ii) normal concentrations of force on abnormal cartilage matrix (Mitchell and Cruess 1977). Mitchell and Cruess (1977) developed a useful classification system for degenerative osteoarthritis based on the idea of an abnormal joint force or an abnormal joint cartilage [see Table 1 in Mitchell and Cruess (1977)]. This classification system is also helpful to us because it illustrates the extraordinary number of ways that a (late-stage) osteoarthritic phenotype may develop (Mitchell and Cruess 1977).

There are many pathways to osteoarthritis, which is perhaps not surprising given the extraordinary number of interdependences between tissues in synovial joints (which are invariably 'circular' interdependencies, i.e. meaning they involve a cascade of events that closes on itself), and because of the many homeostatic processes within and between joint tissues (which are themselves also invariably 'circular'). Of course, the joint is part of the whole animal, and then again there are many circular interdependencies between joint tissues and the rest of the body, and

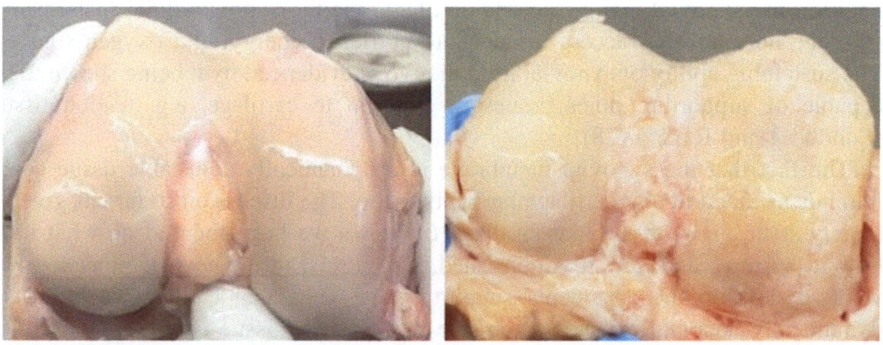

Fig. 1.19 LHS near-normal femoral articular cartilage (female 60 years, Collins Grade 1 (Muehleman et al. 1997); Grade 0 is normal while Grade 4 is end-stage OA. RHS osteoarthritic femoral articular cartilage (female 54 years, Collins Grade 3). We thank the Gift of Hope Organ and Tissue Donor Network (Itasca, IL) and donor families for human articular cartilage and Dr. Arkady Margulis for tissue procurement

again there are many homeostatic processes operating. Therefore, it is usually very difficult to identify a first cause (i.e. meaning that event most antecedent in time, which leads to the current state), because subsequent interactions (stretching over decades) obscure first cause(s). However, Vincent and Watt remark (Vincent and Watt 2018):

> Irrespective of how the disease is classified, the unifying aetiological factor in OA development is mechanical load – either abnormal load on a normal joint, or normal load on a joint that has lost its mechanoprotective mechanisms. This is perhaps most clearly illustrated in young individuals who have sustained destabilizing injuries to the joint (e.g. meniscal and cruciate ligament injuries). They exhibit an OA risk of approximately 50% within 10 years of injury. It is also the case that repetitive low-impact injuries, often occupational, are strongly associated with disease. Likewise, malaligned and misshapen joints are at increased risk of disease.

Indeed more than 40 years ago, Mitchell and Cruess (1977) present an original and still useful classification of degenerative osteoarthritis by causation (with examples). In a review of OA biological discoveries made in 2017, Appleton identifies three (early stage) 'osteoarthritic phenotypes': suggesting there may be: '*potentially critical differences between biological mechanisms of post-traumatic, age-associated, and metabolic phenotypes of OA,*' and urges researchers to add to this list of 'OA phenotypes,' suggesting possible phenotype candidates '*basic calcium phosphate, genetic and inflammatory*' as possible addition phenotype categories. However, we note that classification system shown in Fig. 1 of Appleton (2018) is broadly similar (though not as detailed) as that of Mitchell and Cruess (1977), and so that if one wants to add 'OA phenotypes' based on causation, then the more extensive list of Mitchell and Cruess's classification system above may prove to be useful starting point.

Note that there are many examples in Table 1 of Mitchell and Cruess's classification system that involve genetic causes (Mitchell and Cruess 1977) (this list has grown enormously in recent years as the costs of genetic testing reduced). Often they are consequences of other diseases [e.g. sickle cell anemia (itself a genetic disease), or diabetes (type II diabetes is one dimension of what is now referred to either as a 'systemic metabolic derangement' (Appleton 2018), or 'metabolic syndrome' (Courties et al. 2017), or some such similar name)], it arises from preexisting unfavorable conditions [such as obesity (now generally viewed as another dimension of 'metabolic syndrome')]. Or the cause is 'environmental' (e.g. excessive joint loading due to one's occupation, or due to an unfavorable topography for one's everyday living environment, or due to an obsessive commitment to one or more sporting activities, etc.).

What is now very clear is that both environmental and genetic factors can be important risks for developing osteoarthritis. Indeed, as for most multifactorial polygenic disease states, the conclusion from twin studies is that osteoarthritis seems to be around 40–60% due to environmental conditions and around 40–60% due to genetic predispositions. This range for attribution is relatively wide, probably because again attribution is blurred by both feedback interactions between genes and the environment (one example is epigenetics), and the fact that different environments realize different outcomes from the same genome, and so reveal different capabilities of an organism. That osteoarthritis is both environmental and genetic is a seemingly somewhat vague statement, but it is almost certainly very important when assessing risk factors for OA, and properly delineating environmental and genetic risks lie at the heart of accurate risk prediction.

But what about a modern definition of OA? Most modern definitions of osteoarthritis emphasize that there is a failure of repair processes to keep pace with damage processes, eventually resulting in 'tissue failure.' In 2011, the Osteoarthritis Research Society International, along with the US Federal Drug Administration (FDA), defined osteoarthritis as (Lane et al. 2011):

> Based on evidence to date, there was consensus that OA is usually a progressive disease of synovial joints that represents failed repair of joint damage that results from stresses that may be initiated by an abnormality in any of the synovial joint tissues, including articular cartilage, subchondral bone, ligaments, menisci (when present), periarticular muscles, peripheral nerves, or synovium. This ultimately results in the breakdown of cartilage and bone, leading to symptoms of pain, stiffness and functional disability. Abnormal intra-articular stress and failure of repair may arise as a result of biomechanical, biochemical and/or genetic factors. This process may be localized to a single joint, a few joints, or generalized, and the factors that initiate OA likely vary depending on the joint site. The complexity and variability of OA etiology suggests the need for patient-specific, etiology-based treatment.

OA is often described as a 'degenerative disease,' which implies a failure to repair 'wear and tear' tissue damage. The above definition of OA highlights 'a failure of repair of joint damage,' a recognition that OA may originate from 'stress' (not necessarily mechanical stress, e.g. it could be 'oxidative stress,' or 'metabolic stress,' or 'stress' from inflammatory cytokines) resulting from an abnormality in any joint tissue, and a recognition that there is an interaction between the local microenvironment (including both biomechanical and biochemical dimensions) and the genetic potential of an individual. In this context, we note Brandt et al.'s remark (Brandt et al. 2006):

> …the aetiology and progression of osteoarthritis should not be thought of as being invariably attributable to a single tissue, such as articular cartilage, but as possibly due to disease in any of the tissues of the affected organ, the diarthrodial joint, including the subchondral bone, synovium, capsule, periarticular muscles, sensory nerve endings and meniscus (if present). Supporting ligaments should be added to the list.
>
> Although reviews of osteoarthritis often contain a statement to the effect that it is not merely a disease of cartilage, the large amounts of time, money and brainpower that have been invested in attempts to develop "chondroprotective" drugs and efforts to find the best ways to image minuscule (and clinically meaningless) changes in articular cartilage and to identify biomarkers of cartilage damage in osteoarthritis are evidence that we really do not believe what we write.

In 2015, Osteoarthritis Research Society International (OARSI) proposed another definition of OA, viz.:

> Osteoarthritis is a disorder involving movable joints characterized by cell stress and extracellular matrix degradation initiated by micro- and macro-injury that activates

maladaptive repair responses including pro-inflammatory pathways of innate immunity. The disease manifests first as a molecular derangement (abnormal joint tissue metabolism) followed by anatomic, and/or physiologic derangements (characterized by cartilage degradation, bone remodeling, osteophyte formation, joint inflammation and loss of normal joint function), that can culminate in illness.

This definition highlights micro- and macro-injury, and again 'cell stress' which leads to net loss of ECM due to 'maladaptive repair responses.' This basically conforms to the idea of damage processes exceeding the capacity for tissue repair (though it is not explicitly stated as such), leading to net loss of ECM. However, it also suggests that the repair response may be 'maladaptive.' Exactly what 'maladaptive' means here is open to question, but it could mean repair is inadequate, it could mean that repair processes are abnormal (e.g. a tissue 'physiologic derangement' (e.g. a breakdown in the compartmentalization of signaling processes to each tissue), or a different genetic potential embodied as a 'molecular derangement,' or an abnormal chemical reaction embodied as a 'molecular derangement'), or it could mean that unusual repair pathways have been recruited (and they give an example as activation of pro-inflammatory pathways of 'innate immunity,' or it could be recruitment of macrophages to a tissue, e.g. synovium).

Importantly, we note that this 2015 OARSI definition of OA limits inflammatory pathways to those associated with innate (i.e. humoral or non-specific) immunity, rather than adaptive (i.e. cell-mediated or specific) immunity. This rules out septic arthritis and other types of arthritis such as rheumatoid or psoriatic arthritis that involve adaptive immune responses (e.g. rheumatoid arthritis involves auto-antibodies against rheumatoid factors, while psoriatic arthritis is a result of the autoimmune disease psoriasis). Innate immune responses (e.g. complement activation) are always 'on-the-ready' in all tissues as the critical first defense against microbial invasion (see discussion in Sect. 5.6.14), while neutrophils and macrophages are 'on-the-ready' in most tissues, but not cartilage (because it is avascular). The innate immune system can be triggered inadvertently by genetic mutations or by disease processes. However in the context of innate immunity, we mention that inflammatory mediators (e.g. cytokines such as IL-1, TNFα, chemokines) are part of the normal repair processes following microbial infection, and these same 'inflammatory molecules' may also be also part of (non-inflammatory) autocrine signaling involved in normal cartilage repair processes (Appleton 2018).

Metabolic syndrome is said to consist of '*four core features, variably defined, including hypertension, atherogenic dyslipidemias, visceral obesity, and insulin resistance*' (Appleton et al. 2017). These 'four core features' (three classified as disease states) have grown enormously as the quantity and composition of food in our diet have changed, raising questions about when does 'food' become an agent for disease. The key is to realize risk-free living is simply not possible—the problem is to quantify and so attribute that unavoidable risk. Again, because our genomes are unique, every person's risk profile is unique, so what may be deemed harmful for one person is not for another. Indeed, sensitively understanding and then deeming what may be harmful for a person is the 'art of being a physician.'

The 2015 OARSI definition of OA does not specifically mention 'metabolic syndrome' as a cause of OA, but presumably this is implied in word combinations like 'molecular derangement.' A link between metabolic syndrome and OA is presumably mediated via adipokines, including leptins (Abella et al. 2014; Calvet et al. 2016), or mediated by diabetes and/or dyslipidemias (Courties et al. 2017; Robinson et al. 2016). Metabolic syndrome-related OA is an active area of research given the increasingly prominent 'obesity epidemic' (Courties et al. 2017). Comments Appleton et al. (2017).

> Thirty years ago, Altman et al. told us that osteoarthritis is not a single disease. That 1986 description of OA as "a heterogeneous group of conditions that lead to joint symptoms and signs…" remains true today. But the simple recognition of OA as a group of related but distinct joint disorders among clinicians and researchers is hampered by the lack of a clearly accepted set of criteria to distinguish independent clinical OA phenotypes. Moreover, the description of these clinical OA phenotypes in molecular, anatomic, and physiologic domains remains a formidable, yet fundamental task before us in the field of OA research. Notwithstanding, the blanket term "OA" should no longer be used in isolation to describe the typical joint pathology and symptoms of the most common form of arthritis…
>
> The nature of the interaction between metabolic syndrome and OA remains unresolved. It is unclear whether the most important link is due to an influence of OA on metabolic syndrome (e.g., decreased mobility due to OA leads to obesity and therefore metabolic syndrome), vice versa (abnormal joint loading—with or without metabolic derangement—fuels OA pathophysiology), or if a common set of risk factors exist which drive both conditions in parallel.

It is extremely difficult to untangle events and identify first causes, as interactions are circular and complex. There are so many possible circular interactions that it is not difficult to show using (hypothetical) factorial experimental designs to test hypotheses and that it is not practical to delineate and quantitate the relative contributions of myriad factors in the causation of OA using clinical trials, because of the enormous number of clinical trials required to do so (and the even more enormous numbers of subjects with associated costs). In no time at all, the clinical trials involve more people than live on earth. We would suggest that the most cost-effective approach to improved classification of OA is combining basic and clinical experimental research with computational modeling of normal joints and disease processes. For then, a 'virtual (modeling) environment' can be employed to quantitate the likelihood of causes and outcomes for a potentially unlimited number of (patient-specific) scenarios.

Nevertheless as more becomes known, no doubt we can expect more intricate definitions of what constitutes osteoarthritis, as opposed to other disease entities. For example, Malemud (2015) has proposed that osteoarthritis is re-classified as a 'systemic musculoskeletal disease,' rather than a disease of synovial joints (Malemud 2015). Says Malemud (2015):

> Human osteoarthritis is a systemic musculoskeletal disorder involving activation of innate and adaptive immune systems accompanied by inflammation exemplified by the elevated production of proinflammatory cytokines, which play a significant role in the progression of the disease. The future of novel therapies for osteoarthritis should consider developing drug development strategies designed to inhibit proinflammatory cytokine-induced signal

> transduction. These strategies have been successful in the development of drugs for the treatment of rheumatoid arthritis....
>
> Therefore, the next period of basic and clinical OA research will be forced to take into account the mounting evidence which points to a targeted genomic approach to limit pro-inflammatory cytokine-induced changes in synovial joint immune-mediated inflammation....Finally, the future of OA therapeutic drug development will likely require the use of biologic approaches designed to alter the pathologic responses and activities of chondrocytes, synoviocytes and macrophages. Thus, drug development strategies such as those employed to successfully treat RA may also have to be aligned with those strategies that analyze the effects of these biologic drugs as inhibitors of OA chondrocyte, macrophage and synoviocyte MMP.

Now, both the innate and adaptive immune systems are included in this view of osteoarthritis as a systemic musculoskeletal disorder. We observe that first causes are not mentioned, and inflammation is said to play a significant role in osteoarthritis progression—this is hardly controversial in the sense that inflammation has been recognized for a very long time to play a role in OA progression.

We note that the usual way of defining disease, by signs and symptoms in patients, has fallen out of favor with the research community, as this approach does not accommodate new information about disease mechanisms (El-Tawil et al. 2016; Kraus et al. 2015). But understandably (and properly), clinicians often 'push back,' emphasizing clinically observable signs and reported symptoms (e.g. stiffness and pain), as these are the immediate issues faced by patients and their doctors. But this is really a case of different models and approaches for different questions.

How prevalent is osteoarthritis? Lawrence et al. report OA is the leading cause of disability and estimate the prevalence at 11.5% of the adult US population (i.e. 77% (>18 years) $\times$ 304 M = 234 M people are adults in 2008) (Lawrence et al. 2008):

> We estimated that among US adults, nearly 27 million have clinical osteoarthritis (up from the estimate of 21 million for 1995), 711,000 have polymyalgia rheumatica, 228,000 have giant cell arteritis, up to 3.0 million have had self-reported gout in the past year (up from the estimate of 2.1 million for 1995), 5.0 million have fibromyalgia, 4–10 million have carpal tunnel syndrome, 59 million have had low back pain in the past 3 months, and 30.1 million have had neck pain in the past 3 months.

The Arthritis Foundation reports that in 2016 approximately 31 M adults have osteoarthritis in the USA, with the prevalence increasing with age (<2% at 45 years and >80% at 75 years), https://www.arthritis.org/Documents/Sections/About-Arthritis/arthritis-facts-stats-figures.pdf. People (14 M) have symptomatic OA, with about half having sufficient progression for them to be eligible for knee replacement. The incidence (i.e. rate of newly diagnosed cases) of OA peaks in the age range from 54 years to 65 years. We note that the economic cost of OA in the USA is estimated to be $80B per year between 2008 and 2011.

Having defined osteoarthritis and considered its incidence and prevalence, we now discuss osteoarthritis in the narrow context of the information presented in this chapter. For example, as the discussion on equilibrium stiffness of cartilage in compression made clear, adequate aggrecan is essential for limiting the long-term (drained) deformations of cartilage to acceptable magnitudes. If the cartilage strain becomes too large, ECM molecules and polymers are increasingly damaged by large strains, and more chondrocytes undergo apoptosis (see discussions Chaps. 2 and 3). This is a fundamental (and probably eternal truth for articular cartilage tissue) that often seems to be forgotten in the excitement as new 'upstream' biologic complexities are discovered.

Detailed calculations on the compressive stiffness offered by aggrecan, taking into account the volume occupied by the collagen network and other proteins, and the measured concentrations of aggrecan in cartilage, demonstrate that aggrecan can largely explain the magnitude of equilibrium compressive stiffnesses of cartilage that is measured experimentally (see discussion in Appendix D). The importance of having adequate aggrecan to resist compressive loadings on articular cartilage is difficult to overemphasize. We note that maintaining adequate aggrecan concentrations in articular cartilage is not just a matter of synthesizing aggrecan, but also of the collagen network retaining aggrecan within the cartilage. Clearly, damage to the collagen network increases the loss of aggrecan, and this also increases strains within the tissue and so damage to the ECM and cells. No matter what else may be occurring in and around articular cartilage, it is certain that adequate fractions and proportions of both aggrecan and collagen are essential for normal articular cartilage function.

The 'stiffening effect' of the pore fluid on the material properties of cartilage is clearly of vital importance for its structural performance in vivo. The stiffening effect arises when cartilage is loaded over a timescale that does not leave adequate time for the pore water to escape cartilage tissue. It turns out how much stiffening is offered by water to cartilage under repeated compression loadings depends on many things including: (i) the drainage conditions at the cartilage surfaces, (ii) the length of drainage pathways, (iii) the permeability of the cartilage, (iv) the amount and distribution of aggrecan and collagen in the cartilage, (v) the load boundary conditions (confined, unconfined, or something else), (vi) the size of the joint, (vii) the load distribution acting across the cartilage contact interface, and (viii) its amplitude and frequency content.

The stiffening effect of cartilage can be large. The simple examples described in Sect. 1.6 show that depending on boundary conditions and loading rate, the stiffness of cartilage can vary from less than one megapascal to hundreds of megapascals, all obtained through simple time-dependent fluid flow. These examples serve to highlight the extraordinary potential for a composite porous material such as cartilage to change its structural response to changing loading conditions, or changing surface conditions, or changing composition of cartilage.

Indeed, these examples also serve to highlight the sensitivity of the material response to its load state. This sensitivity renders cartilage vulnerable to significant damage under certain conditions. It becomes clear the types of 'surface conditions' that help *preserve* cartilage tissue and minimize cartilage damage are similar to those of 'confined compression.' In vivo, preservation of cartilage tissue is promoted when congruence of the contacting cartilage surfaces is high, loading distributed over an extensive region of contact, leading to approximately confined compression conditions, and low or no flow through the cartilage surface (i.e. ensuring that an 'undrained state' lasts for a long time). Practically, this is achieved by having long drainage path lengths for the pore fluid, as well as having a small 'coefficient of consolidation.' A good example of a highly congruent joint with long drainage paths is the ankle joint. It is found the ankle joint has a relatively low incidence of osteoarthritis.

Conversely, surface conditions that increase the likelihood of cartilage damage and degradation of the ECM are conditions approximating 'unconfined compression.' This state is more closely approximated by concentrated loading, leading to less confined compression conditions beneath the loading. This state is more closely approximated by short drainage pathways, leading to rapid flows and rapidly approaching the 'drained state,' where there is then reliance on 'equilibrium aggrecan stiffness' to control load-induced compressive strain within the tissue. Rapid drainage is also promoted by a large 'coefficient of consolidation.' A good example of reduced drainage path lengths is meniscal damage in the knee joint, which is only accentuated by partial or complete meniscectomy. Given the information on loading deformation, strain, and cartilage stiffness in this chapter, it is entirely consistent to find post-meniscectomy joints have a much higher incidence of osteoarthritis.

In the context of joint loading leading to mechanical ECM damage initiating OA versus low-grade inflammation initiating OA (Robinson et al. 2016), it has been argued that while knee OA may be explained by excessive mechanical loading, because 'we do not walk on our hands,' the existence of OA in hands is suggested to weigh against mechanical factors initiating OA (Cicuttini and Wluka 2014). Says Cicuttini and Wulka:

> For decades, obesity has been recognized as a risk factor for hand OA. Given that we do not walk on our hands, this risk factor is circumstantial evidence against a mechanical pathogenesis for OA… Although mechanical factors are also involved in hand OA, the case for a systemic mechanism in the pathogenesis of this condition seems clearer.

We point out that argument like '*we do not walk on our hands*' is too coarse for an argument about the relative importance of mechanical loading in initiating OA in hand and knee joints, because contrary to what first impressions may be, relative to their size, hand joints are subject to large loads too. When considering joint loading, one way to take size into account is to consider contact stress, as contact stress normalizes the joint force by the appropriate articular cartilage contact area. Analysis shows that hand joints are in fact subject to large mechanical stresses, somewhat smaller but of certainly of similar magnitude to those experienced by knee joints during everyday activities. For example, Butz et al. say (Butz et al. 2012):

> The FEA [finite element analysis] illustrates the manner in which stresses may be expected to act at the surfaces of the MCP [metacarpophalangeal] joint primarily because the MCP is one of the most highly affected by arthritic degeneration. The highest predicted average stress of 2.5 MPa is similar to stresses experienced at other anatomical sites such as the hips and temporomandibular joints, where the stresses commonly range from 2 to 5 MPa….

> Though finger joints may not experience stress levels as consistently high as the hips throughout an average day, this analysis suggests that common activities may have a highly strenuous effect on the hand and should be taken into account when considering joint mechanics or the design of replacement prosthetics. In fact, given the large stresses that have been shown to occur at the MCP, it is not surprising that there is frequently a large incidence of fractures and breaks in MCP prostheses.

We observe that peak stresses (rather than average stresses) may be more relevant when considering the initiation of mechanically induced OA. Further, the timing and number of repetitions of some kinds of hand activities may be far more damaging to metacarpophalangeal joints (e.g. jar and tap opening and closing, key turning, weight carrying) than other hand activities (e.g. keyboard typing, pen grip). For calculated stresses experienced by hand joints during some of these activities, see Table 3 in Butz et al. (2012). This also needs to be taken into account, as fatigue damage of articular cartilage is a function not only of stress magnitude but also the number of repetitions (see discussion in Sect. 3.3).

One also needs to be taken into account that tendons and ligaments and other connective tissue membranes in the hand play important roles in distributing forces to joints and that any damage to these tissues may amplify stress experienced by articular cartilage. Further, any mechanical failure of subchondral bone, which may result from quite different loadings experienced by the hand (e.g. impact loading), also needs to be taken into account. All this implies that comparisons of stresses experienced by normal hand joints in young adults may not be representative of the stresses experienced by normal hand joints in older adults. Taken together, all this suggests that mechanics may well play a central role in the pathogenesis of hand OA, much as it does for knee OA.

Clearly though, low-grade inflammation may play a synergistic role with mechanical factors in the pathogenesis of hand OA and, in some people, may presumably play the dominant role. The problem is how to systematically delineate their relative importance. We would suggest that computational modeling of joints has an important role to play in such an endeavor.

In the context of osteoarthritis, particularly noteworthy is the complete reliance of articular cartilage on its surrounding tissues (Brandt et al. 2006). For example, articular cartilage is completely dependent on synovial tissues for its day-to-day metabolic needs (i.e. for the supply of nutrients and removal of waste products, including cell and ECM debris). If then, the synovial tissue is compromised in any way, particularly if the synovial tissue suffers from one of a number of disease states (inflammatory, infectious, degenerative, traumatic or hemorrhagic), then an osteoarthritic state is much more likely to develop in articular cartilage. Indeed, osteoarthritis is often a complication of joints suffering rheumatoid arthritis. Rheumatoid arthritis is an autoimmune disease of the synovial membrane that leads to articular cartilage being exposed to increased concentrations of inflammatory cytokines within the synovial fluid and to activated complement in the cartilage ECM (Nakagawa et al. 1999), which together with joint loading hastens the development of articular cartilage with an osteoarthritic phenotype. It is entirely possible that low-grade inflammation develops in synovial tissues for any number of reasons, and this may be involved in initiating OA (Schott et al. 2018).

Finally, we note the economist Paul Krugman once famously observed: '*Productivity isn't everything, but in the long run it's almost everything.*' This can be adapted to provide a profoundly important insight into cartilage and its longevity (and indeed osteoarthritis and other degenerative pathologies of aging): '*Damage and repair isn't everything, but in the long run it's almost everything.*' This highlights the essence of modern biological-based definitions of osteoarthritic phenotypes, whatever the primary cause of damage (mechanical or inflammatory) there is incomplete repair of tissue damage, and so damage accumulates over time.

Of course, the main practical difficulty is that much like productivity (which refers to how efficiently labor and capital 'come together' to produce outputs), damage and repair processes in cartilage are not easy to measure accurately and so to quantify. But this should not stop us trying to measure damage, cumulative damage, and repair, any more than economists should stop trying to measure productivity (they are both crucially important to our long-term health!). The fact is productivity changes in the workplace every single day, one step at a time, just like damage and repair processes change in tissues every single day, one step at a time.

But the difficulty of measuring and quantifying damage and repair processes serves to bring to the fore the important role to be played by theoretical biology in understanding the origins of osteoarthritis. We need theoretical biology to develop (abstract) constructs that help explain the processes of change in tissues that may be hidden from our initial view and so make sense of the world around us. In the following, we draw on the theoretical foundations of disciplines such as biochemistry, matrix biology, and materials science in our search for a fundamental understanding of tissue damage and tissue repair processes that are continuous in articular cartilage. By adopting a theoretical viewpoint and framework, by developing abstractions through reinterpreting existing experimental data, we arrive at some unusual and surprising models, statements, and conclusions. At least, they are surprising in the context of existing discipline dogmas. Let us now have a closer look at cartilage homeostasis.

Chapter 2
Cartilage Tissue Homeostasis

2.1 Introduction

We now want to begin to develop a holistic view of cartilage as a tissue that has an integrated set of homeostatic processes for maintaining itself. In the context of other tissues in the body, such a proposal is hardly controversial, for it is well known that there are myriad of homeostatic processes maintaining our bodies throughout all our tissues, which despite outward appearances of being unchanging are in a continual state of flux.

To give but one famous physiological example from cardiovascular homeostasis, consider blood volume. When the volume of blood in the cardiovascular system is too large, it causes increased stretching of the muscle in the heart wall. This causes the release of natriuretic peptides into the blood, which leads to the kidney tubules to increase excretion of sodium and water. This process (or sequence of events) decreases blood volume back toward its normal 'set point' volume (Suzuki et al. 2001). On the other hand, if the blood volume is too low, then renin and angiotensin increase, and aldosterone is released from the adrenal gland. This induces the kidney tubules to retain more sodium and water (Briet and Schiffrin 2010), and this process increases blood volume back toward its normal 'set point' volume. By these means, blood volume is regulated to a set range, and we say that blood volume is homeostatically controlled. This homeostatic control of blood volume then plays a role in controlling blood pressure homeostasis (Seeliger et al. 2005).

There are many more examples of homeostatic processes maintaining homeostatic conditions in the cardiovascular system. Some homeostatic processes involve remodeling of cardiovascular tissues themselves. For example, a change in blood pressure within an artery causes the artery wall to 'remodel.' The artery diameter becomes larger in response to an increased pressure, and the diameter smaller in response to a decreased pressure. Both processes act to change the wall stress

D. W. Smith et al., *Articular Cartilage Dynamics*,
https://doi.org/10.1007/978-981-13-1474-2_2

(which is closely related to the blood pressure in the vessel lumen) back toward the normal ‘set point’ wall stress (Liu and Fung 1996).

For all active homeostatic processes involving cells, a deviation from the normal ‘set point’ is detected by some cellular sensing system, which then sets in train events that return the system toward the normal set point (see further discussion Sect. 5.4). An important point to mention is that both sensors and the effectors may be multiple and distributed across one or more tissues. In other words, homeostasis is a distributed property of a tissue or system—there is usually no one thing or single mechanism about which we can say—here it is, the cause of homeostasis.

A further important and related point is this—even for the cardiovascular system, probably the most studied physiological system in the history of medicine and clinical science, which now has many well-established homeostatic processes identified, the homeostatic regulation of something as fundamental as blood pressure is still very much under debate (see for example Malpas 2009; Osborn et al. 2009; Evans and Bie 2016). This debate over control of blood pressure is driven by the complexity of homeostatic blood pressure control under different physiological conditions and in various disease states. In this context (i.e. after several centuries of intensive research studying the cardiovascular system and homeostatic control of blood pressure, this research has still not resulted in a ‘settled science’ for blood pressure control), it is then perhaps unsurprising that comparatively, not a lot is known about homeostasis of articular cartilage. And of course, what is known about homeostatic control of articular cartilage is also subject to debate.

Establishing the homeostatic processes operating in cartilage is not an easy task. One reason for this is that most of the reported experimental literature relate to the behavior of cartilage at ‘snapshots’ in time (tissues are usually fixed and examined under the microscope) (Bobacz et al. 2004), or over relatively short times with multiple ‘snapshots,’ compared to the timescales over which cartilage restores its normal ‘set point’ state (Li et al. 2013). This ‘separation of timescales’ leads to a hiatus in data because of the practical, ethical, and cost considerations involved in gathering the longitudinal data required to uncover more details about the homeostatic processes operating within cartilage. Nevertheless, many ingenious experiments have been carried out that give us important insights into the tissue processes going on within articular cartilage. Interpolating between the findings made in various experiments, and then abstracting them in the context of the behavior of other tissues, allows us to create a more complete picture of the homeostatic processes operating in cartilage.

As noted in the foreword, our methods for interpolating the data and inferring how cartilage works as a homeostatic tissue through abstraction are based on our background in systems modeling of cartilage and related tissues and are grounded in our background in physics and engineering principles. It has been developed from a wide re-reading of the literature on articular cartilage, with an ‘integrative systems framework’ in mind.

However, we again mention that we do not claim that our proposed view of cartilage homeostasis is the final word. Yet as for other tissues in the body (e.g. bone, tendon, muscle, kidney, liver, heart), we do think a ‘homeostatic view’ of cartilage physiology provides a powerful and rational basis for developing an abstracted and connected, functionally adaptive understanding of cartilage. The 'pay-off' for developing such an integrated dynamic view of cartilage is that this 'lens' can be applied to help interpret new experiments, formulate new hypotheses, guide the development of new studies directed at critical missing information, or correct what has been proposed here or by others.

2.2 Overview of Cartilage Homeostasis

Articular cartilage clearly maintains its functional integrity over many decades in a very challenging mechanical environment (Bader et al. 2011). But exactly how this is achieved is not well understood (Lee et al. 2006; Bader et al. 2011; Buschmann et al. 1995; Mott and Werb 2004). Here we describe some of the main dynamic mechanisms maintaining homeostasis of the cartilage extracellular matrix (ECM) (Videman et al. 1981; Karsdal et al. 2008; Ma et al. 2013; Zhu et al. 2008; Wolf et al. 2007; Loeser 2009). As part of this extensive package of ideas, we propose that there are also mechanisms maintaining homeostasis of the chondrocyte population density throughout the cartilage tissue (Jadin et al. 2006; Lee et al. 1993; Brisby et al. 2013).

We begin this chapter by describing the processes for extracellular matrix turnover. First we observe that the timescales for aggrecan and collagen turnover are significantly different, which suggests 'tissue homeostasis' involves a range of timescales. We also find in this chapter that the timescale for tissue turnover varies spatially with depth through the tissue. It turns out that the half-life for aggrecan turnover in the deep zone of the cartilage is surprisingly long, while the half-life for collagen in the superficial zone is surprisingly short. Homeostatic processes occurring over long timescales are clearly much more difficult to study due to practical constraints, so generally much less is known about them. For example, a lot is known about minute-to-minute and day-to-day regulation of blood pressure, but much less is known about month-to-month and decade-to-decade regulation of blood pressure. Next to nothing is known about month-to-month, year-to-year, and decade-to-decade regulation of articular cartilage homeostasis in an individual, as we only have a very limited number of snapshots and data points about the state of their articular cartilage. Much of what we know has to be inferred on the basis of animal data, but again there are many practical limitations on the experimental data available.

We propose here that the processes regulating chondrocytes numbers and their distribution play an integral part in maintaining homeostasis of the cartilage ECM over the long term (Jadin et al. 2006; Lee et al. 1993; Brisby et al. 2013; Ryan et al. 2009). For as cyclic strain damages the cartilage extracellular matrix, impact loading or repeated excessive strains inevitably lead to chondrocyte apoptosis (Kuster et al. 1997; Seol et al. 2012; Chen et al. 2001). ECM homeostasis and chondrocyte apoptosis are clearly related, for a reduction in chondrocyte number density reduces the capacity of the remaining chondrocytes to respond to changes in the ECM. A reduction in chondrocyte density may in turn lead to a depleted ECM and greater tissue strains (assuming the same loading), which further damages the extracellular matrix while simultaneously increasing the likelihood of chondrocyte apoptosis. Clearly, an absence of chondrocyte proliferation will create a positive feedback cycle leading to tissue destruction—we therefore need to look at the evidence for chondrocyte proliferation very closely.

Indeed exceeding the capacity of the tissue's homeostatic processes, possibly due to a dysregulation of a tissue's normal homeostatic processes (Loeser 2009), usually leads to a positive feedback cycle that ends in the tissue's destruction. Because there are many homeostatic processes operating in cartilage, each with important components, it follows there are many points for dysregulation of homeostatic processes in articular cartilage. This provides one plausible explanation for the clinical observation that there are many 'pathways to osteoarthritis.' It becomes apparent that even classifying the 'pathways to osteoarthritis' is difficult without knowledge of the key homeostatic processes operating in the tissue.

Almost certainly one of the foundational homeostatic processes in cartilage is maintaining chondrocyte density (Jadin et al. 2006; Lee et al. 1993; Brisby et al. 2013). When growth is complete, maintaining normal adult chondrocyte density clearly relies upon the balance between the rates of chondrocyte proliferation and chondrocyte apoptosis. Experiments show that the level of strain that mechanically causes chondrocyte apoptosis depends on a number of factors including load type and rate, load magnitude, load number and timing, chondrocyte location within the cartilage and the vigor of cellular repair processes (Loening et al. 2000; Hembree et al. 2007). Increasing the level of exercise (e.g. running) has been shown to increase chondrocyte proliferation (Brisby et al. 2013).

It has also been found experimentally that various systemic hormones, local cell–cell signaling molecules (i.e. signaling between chondrocytes), and ECM molecules (e.g. matrilins) can render chondrocytes more or less susceptible to mechanically induced apoptosis (Zhu et al. 2008; Lotz et al. 1999; Miwa et al. 2000; Goggs et al. 2003; Leighton et al. 2007). It has been shown that transcription factor concentrations also change within chondrocytes, which naturally alters the likelihood of chondrocyte apoptosis (Hashimoto et al. 2009a; Schipani et al. 2001; Ye et al. 2014).

In light of the effect of altering concentrations of signaling molecules and transcription factors has on the likelihood of chondrocyte apoptosis, we have to accept that critical homeostatic 'set points' for chondrocyte apoptosis are themselves dynamically adjustable. This is the case generally—homeostatic 'set points' vary to a greater or lesser extent over time as circumstances change—and sometimes they change quickly. Indeed, it is this adjustment of tissue homeostatic set points that enables tissue adaption to optimize the tissue to new environmental conditions (e.g. activity levels increase and are maintained). This means tissue 'set points' are only fixed in a relative sense (e.g. they adjust with the hormonal environment and with activity levels).

The cartilage tissue system clearly has a large number of mechanical and chemical inputs. The systems for integrating and assessing this environmental data are 'spatially distributed.' By spatial distributed we mean a cell's decision processes are distributed across the cell's nearby ECM, cell membrane, cell cytoplasm and cell nucleus. In other words, a cell's decision machinery is not located in a discrete structure, but rather it is spatial distributed.

Indeed, it is the spatial (and temporal) distributed nature of the complex networks of chemical reactions that underlie a cell's 'decision machinery' that makes the decision machinery difficult to understand, design experiments to test, numerically model, and ultimately difficult to predict. For example, compare the conflicting interpretations of experimental results obtained from transgenic mice reported by Yan et al. (2016) and Newton et al. (2018). Says Newton et al. (2018):

> Accordingly, the data should always be interpreted carefully and although the Cre-lox approach is one of the most powerful tools for examining the functions of gene products in target tissues, expression of Cre in non-target cells is a serious drawback and must always be taken into account...In conclusion, and in contrast to a recent report, elevated mTORC1 signaling in chondrocytes does not alter the proliferation or differentiation of growth plate chondrocytes. The described phenotype is likely caused by aberrant Cre-activity in other non-cartilaginous tissues early in development.

Interpreting data in distributed control/decision systems is not at all straightforward.

If we wish to reliably predict what a cell will decide to do in a given set of conditions, then clearly a rather elaborate 'systems network model' is required that reflects all the different processes that contribute to the cell's decision (see discussion in the following chapter). Unfortunately, such a numerical model is not yet available, for we do not know all the network components, and even if we did, we have no way of measuring all the relevant concentrations, binding constants, transport constants or the initial state. We are a long way from developing such a detailed predictive model that might replicate a cell's decision process.

But we do know some of the components in 'decision networks' that enable articular cartilage tissue homeostasis. For example, osmolality sensors, extracellular matrix component sensors, fluid flow sensors, strain sensors, and signaling molecule sensors (otherwise known as receptors) all signal directly through complex intracellular chemical networks that help a chondrocyte tune its 'homeostatic response' to the ECM and its environment.

Through all the complexity of a cell's 'decision machinery,' one basic decision that has two fundamental outcomes is straightforward to describe. Either a cell survives or it does not survive [i.e. a chondrocyte 'chooses' to undergo apoptosis or programmed necrosis, or it does not—we note that most cell death in cartilage injured by traumatic loading is due to cell apoptosis (Chai et al. 2007)]. For those cells that 'survive,' further basic cell 'decisions' relate to whether or not a cell proliferates, and to whether or not a cell migrates. Of course, many more additional functional and maintenance decisions may be added to these basic decisions (e.g. what molecules to secrete (based on a range of local environmental stimuli), when, where, and how much of them to secrete, what molecules and organelles to replace, recycle, degrade). We note that a cell's possible decisions are not unlimited—after all, normal (non-cancerous) cells in normal tissues are there to do a set number of tasks appropriate for that tissue—that is all.

Clearly regulating chondrocyte density is fundamental to cartilage homeostasis, for without cells the tissue ECM cannot be repaired. In the following, we propose that apoptotic chondrocytes in vivo are in fact replaced in a two-step homeostatic process: (i) proliferation of the remaining chondrocytes (to correct the diminished total tissue repair capacity that is the inevitable consequence of chondrocyte apoptosis) and (ii) migration of daughter chondrocytes to new spatial locations within the cartilage tissue (so as to appropriately match local repair capability with local repair needs).

Because a low aggrecan concentration is a key indicator of articular cartilage in need of repair, we further propose that proliferation becomes more probable in low aggrecan environments (i.e. in comparatively hypo-osmotic microenvironments), and further, other things being equal, that the usual direction of chondrocyte migration is 'down' an aggrecan gradient, in the direction of steepest descent. Because aggrecan is normally lost from cartilage surfaces, this means the direction of chondrocyte migration is generally toward a nearby cartilage surface.

Mechanical damage of extracellular matrix and chondrocyte apoptosis is usually more likely toward the surface of the articular cartilage, diminishing with distance from the cartilage surface (Saarakkala et al. 2010; Lin et al. 2004; Kurz et al. 2005). For example, following an extensive set of 'confined compression' experiments, the macroscopic effects of cyclic mechanical loading on articular cartilage are nicely summarized schematically by Lin et al. (2004) (see Fig. 2.1).

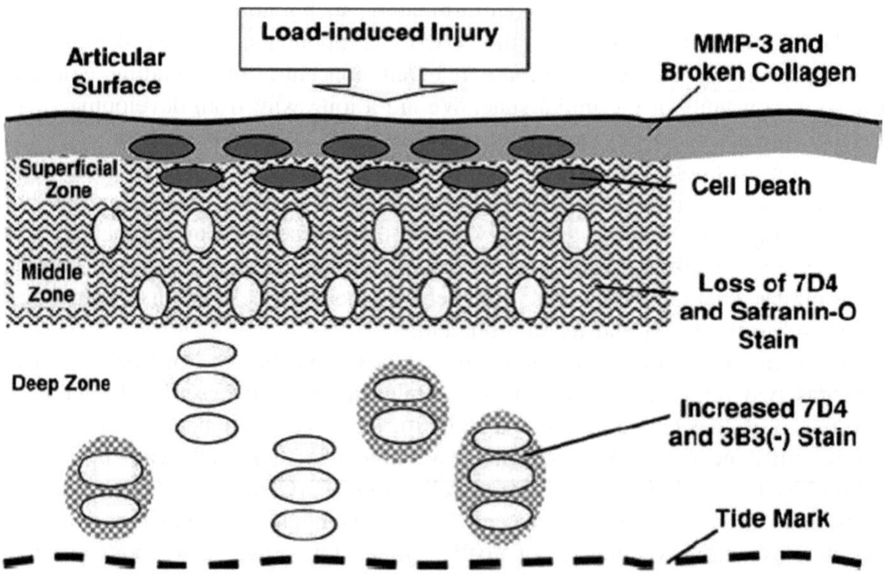

Fig. 2.1 Summary of in situ changes due to load-induced injury. The loss of 7D4 and safranin-O is located in the superficial and middle zones. Increased levels of MMP-3 are co-localized with cell death and broken collagen. The increased 3B3(−) and 7D4 staining in the deep zone indicates an alteration of GAG (Lin et al. 2004)

Lin et al. (2004) observed that following cyclic loading (loading directed normal to the cartilage surface, i.e. no shear load applied to the cartilage surface), there is graded 'damage response' from the top to the bottom of cartilage. Damage is clearly most extensive at the cartilage surface where mechanical strains are largest. At the cartilage surface, there is aggrecan loss, cell death, and collagen network damage. However, collagen network damage extends only a small distance from cartilage surface; cell death extends somewhat further into the cartilage tissue—while aggrecan loss extends even further into the cartilage tissue (see schematic Fig. 2.1). This suggests aggrecan loss is a sensitive indicator of cartilage damage. We note that aggrecan synthesis decreased in the zone of aggrecan loss, however in the deep zone, aggrecan synthesis increased (though some abnormalities in the synthesized aggrecan were detected). This suggests that deep chondrocytes support the secretion effort of more superficial chondrocytes, with the secreted aggrecan moving down the aggrecan gradient toward the superficial ECM and cartilage surface.

When homeostatic processes are operating normally in the cartilage, then an increased rate of mechanical damage (including cell death) should soon lead to an increased rate of repair [including cell proliferation (Xu et al. 2010; Akiyama et al. 2004; Brisby et al. 2013; Ryan et al. 2009)]—otherwise the cartilage will not survive long term. Given that mechanical damage is depth dependent, it follows that the rate of repair (or tissue turnover) should also be depth dependent. This leads us to expect the rates of ECM tissue turnover will decrease with distance from the cartilage surface. This line of reasoning leads us toward the conclusion that while the collagen network in the deep zone has a comparatively long half-life, collagen network in the superficial zone will have a much shorter half-life.

We also propose that chondrocyte proliferation and migration play roles during all the stages of articular cartilage's life history. For example, chondrocyte proliferation and migration play an important role during the extensive cartilage tissue remodeling and zonation that accompanies maturation of the animal (as the animal grows in size and sexually matures). Another example where chondrocyte proliferation and migration play central roles is (remodeling) adaption, which occurs as result of a prolonged change in environmental loadings (e.g. a previously sedate lifestyle is replaced by heavy manual exertion, or the reverse). And finally, there is chondrocyte proliferation and migration that occurs during normal environmental loadings, which is required for tissue homeostasis (see Babyn et al. 1996; van Turnhout et al. 2008, 2010; Iordan et al. 2010; Brisby et al. 2013; Jadin et al. 2006). In normal adult tissue, cell proliferation decreases with increasing distance from the cartilage surface.

Given that chondrocyte apoptosis increases as a result of significant mechanical loading, on this basis we might expect that the rate of proliferation would also be greater where mechanical loading is greater. This appears to be borne out in practice (see for example Fig. 1.1 and Ryan et al. 2009; Brisby et al. 2013).

'Modulation' of chondrocyte phenotype has been proposed previously (Aigner and Dudhia 1997). We propose here that chondrocyte proliferation and migration requires de-differentiation of the chondrocytes. In vivo de-differentiation is driven partly by falling osmolality of the extracellular matrix (Urban et al. 1993; Hopewell and Urban 2003; Palmer et al. 2001; Negoro et al. 2008), and partly by cell–cell signaling (e.g. Wnt signaling contributes to signaling between cells) (Chun et al. 2008; Urban et al. 1993; Akiyama et al. 2004). In other words as osmolar levels fall, chondrocytes de-differentiate from a fully differentiated chondrocyte with a 'rounded' morphology toward a less differentiated chondrocytes with a more 'spindle-shaped' morphology in cross section (see Figs. 1 and 2 in Horky and Tichy 2004).

While changing osmolality plays a role in this phenotype transformation, undoubtedly a number of different mechanisms operate simultaneously to effect a change in the gene expression profile of chondrocytes. For example, Plaas et al. (2011) argue that rising ADAMTS5 concentrations in the ECM lead to a switch in TGFβ signaling that induces a more fibroblast-like phenotype (see Fig. 5.4). A switch from 'canonical' toward 'non-canonical' Wnt signaling also leads to a more fibroblast-like phenotype (see discussion Sect. 5.6.3). Naturally, there are also less direct mechanisms, such as changes in the level of physical activity, which provide stimuli that activate multiple phenotype pathways that influence a range of cell behaviors such as proliferation (Brisby et al. 2013). For simplicity, here we focus on osmolality change as the primary driver for change, while being aware that a number of signaling mechanisms almost certainly contribute simultaneously to cause chondrocyte de-differentiation.

Upon lowering osmolality, de-differentiating 'rounded chondrocytes' first develop a polarized secretion profile (i.e. asymmetries develop in cytoplasm and the cell membrane enclosing the cell), while at lower osmolality cells begin to develop lamellipodia. Finally at low osmolality, chondrocytes take on fibroblast-like characteristics and become spindle-shaped. When spindle-shaped, a migrating (myofibroblast-like) population of chondrocytes extend processes and pull themselves through the damaged cartilage (Qiu et al. 2000) [e.g. toward a newly created cartilage surface (Lyman et al. 2012)], much as fibroblasts (Hinz et al. 2001) and tenocytes can do (Tsai et al. 2003).

Indeed in the reduced aggrecan environment found in the superficial zone of the cartilage, it is observed that local chondrocytes do become more spindle-shaped in cross section (see Figs. 1 and 2 in Horky and Tichy 2004). These chondrocytes proliferate more easily and normally extend filopodia and lamellipodia horizontally until they touch their neighboring cells, probably facilitating communication between them via gap junctions (Gonzalez et al. 2007). We note in passing that tenocytes, with a similar morphological phenotype, also have gap junctions. Indeed, cellular networks (functional syncytia) between fibroblast-like cells are not at all unusual (Langevin et al. 2004). Chondrocytes in the superficial zone and upper transitional zone appear to have greater probability of proliferating than more fully differentiated chondrocytes, which are more distant from the cartilage surface. We also note that the population of chondrocytes in the superficial and upper transitional zones have been identified as including a 'chondroprogenitor cell population' (Hayes et al. 2008; Dowthwaite et al. 2004; Alsalameh et al. 2004; Hunziker et al. 2007; Seol et al. 2012; Jadin et al. 2006).

Based on all these findings, we propose here that there is a continuum of 'differentiation states' for chondrocytes [from morphologically rounded to spindle-shaped chondrocytes in cross section (see Figs. 1 and 2 (Horky and Tichy 2004)), though we mention that a spindle-shape in cross-section may resemble a disk shape in 'plan view' (see Fig. 3 (Poole et al. 1984))], which map onto a continuum of 'proliferation states' (from little proliferation in the normal adult deep zone to comparatively high levels of proliferation in the upper transitional and superficial zone). In addition, these morphological and differentiation states map onto a continuum of 'migration states' (from nearly all cells in the deep zone being stationary, through to a relatively large proportion of cells capable of much faster migration in the transitional and superficial zones).

We propose that morphologically more rounded migrating chondrocytes can only migrate slowly, driven forward by a 'push migration mode' (usually in the vertical direction from the deep and lower transitional zones toward the upper transitional and superficial zones). At the other end of the continuum are spindle-shaped chondrocytes that can proliferate and migrate relatively quickly, in a 'pull migration mode.' The fast pull migration mode normally occurs in comparatively heavily damaged tissue (e.g. around splits or scratches on the cartilage surface) or in very diseased tissue (e.g. late-stage osteoarthritis) (Seol et al. 2012). Chondrocytes in the upper transitional and superficial zones may be in either the normal slower push migration mode, or in a faster pull migration mode, depending on local microenvironmental conditions (i.e. depending on the state of cartilage damage). There is again a continuum of states between the 'slow' and 'fast' migration modes and that chondrocytes in intermediate states can sometimes use elements of both slow and fast migration modes depending on the local environmental conditions.

In the following, we propose a detailed mechanism explaining how the slow, 'push-mode' migration operates. As part of the normal maintenance of cartilage ECM following normal rates of tissue damage, chondrocytes in the upper transitional zone and superficial zone usually slowly migrate down their aggrecan local gradients (Pabbruwe et al. 2009), removing the (usually damaged) collagen network in the path of the migrating cell. Importantly in this slow migration mode, we propose that the cell secretes generous quantities of newly synthesized procollagen (Eyre et al. 2006), which then has the telopeptides removed from procollagen and polymerizes behind the migrating cell to form a new collagen network as the cell passes by. We suggest here that so-called tapered chondrons (which are found in normal cartilage) develop around migrating chondrocytes as a consequence of cell migration-induced remodeling of the collagen network (Poole 1997).

Seemingly from 'nowhere,' we now have a lot of proposed mechanisms going on to sustain and maintain cartilage. Cartilage is often portrayed as an 'inert' and even a 'boring' tissue—not here, not now! But we are not yet finished—it gets even more intriguing. In this chapter, we further describe how we propose that extracellular matrix 'self-organizes.' Self-organization of macromolecules is driven by chemical reactions that minimize the partial molar Gibbs 'free energies' of each material component in the solution (see Appendix K). This remarkable self-organization process is remotely regulated by chondrocytes by altering their local chemical milieu in the pericellular ECM as well as expressing 'regulatory molecules,' thereby exerting control over the chemical reactions taking place more distantly in the ECM.

In other words, radial zonation of the ECM around discrete chondrocytes is fundamentally driven by secretion of ECM molecules by chondrocytes. The control the chondrocytes exert is limited to the timing and composition of the secretion profile (which may include regulatory molecules). Everything else happens based on (known) chemical and physical laws. For example, secreted molecules are then transported radially away from each chondrocyte, usually along their concentration gradient (or more accurately, along their chemical potential gradients). Differential transport rates for different species lead to different chemical milieu at different radial distances around the chondrocyte. What happens in vivo is somewhat analogous but more complex than separation of a mixture of chemicals along a chromatographic column. But it is analogous in the sense that differential transport rates in the column lead to spatial variations in chemical composition along a 'separation column' https://en.wikipedia.org/wiki/Van_Deemter_equation. Naturally, the different chemical milieu at different radial distances around the chondrocyte gives rise to some chemical reactions being favored over others.

The chemical reactions favored at different distances from a chondrocyte are so striking that they have been identified histologically as three distinct zones around a chondrocyte, namely (i) pericellular matrix, (ii) territorial matrix, and (iii) interterritorial matrix. In this chapter, we explain how a molecule's size and charge usually determine how far a molecule can be transported away the chondrocyte, which strongly influence the chemical reactions that are possible at different distances from the chondrocyte. We explain how the net charge on a molecule can be regulated by the interplay between the isoelectric point (pI) of the molecule and the pH variation with distance from the chondrocyte.

Then something interesting happens as the radial zonation around each chondrocyte begins to interact and overlap. ECM molecules begin to accumulate throughout the cartilage, and the boundary conditions at the surface of the cartilage tissue then become relatively more important. Because the osteochondral junction is normally a 'no flux' condition for most secreted molecules (meaning that molecules cannot escape from the cartilage through this surface to the underlying bone), while the 'free cartilage surface' is a 'concentration boundary condition' for any secreted molecule (meaning molecules can escape from the cartilage through this surface). As a consequence of the asymmetry of these two surface or 'boundary' conditions on articular cartilage in vivo, inevitably there is a net transport of ECM molecules (and cells) toward the cartilage surface.

As ECM molecules accumulate to higher and higher concentrations, this tendency to escape through the surface of the cartilage becomes larger and larger, and the transport (or 'flux') of molecules through the tissue and surface becomes larger and larger. The interaction between the initially radial transport of molecules and their movement toward the cartilage surface becomes more significant as ECM macromolecular concentrations increase. For aggrecan, this leads to the formation of preferred transport pathways (or 'channelization') of aggrecan movement through territorial zone matrix (around the interterritorial matrix). This channelization provides a mechanism for chondrocytes in the deep zone to quickly contribute to ECM molecules for maintenance of the transitional and superficial zones. This is discussed further in Chap. 3.

From the above introductory description, it becomes clear that our abstraction of cartilage homeostasis is really a package of ideas, from which we attempt to provide a coherent set of arguments and observations that support homeostasis of the cartilage ECM. Naturally, a coherent picture of tissue homeostasis includes homeostasis of the chondrocyte population in vivo. We hope the reader finds that the various strands of reasoning 'come together' to create a convincing picture of cartilage as a subtle tissue that maintains its equilibrium in the face of environmental challenges that generate ongoing damage. We believe this way of thinking about cartilage provides a rewarding perspective on existing experimental evidence —and in some cases, a fresh perspective on existing experimental evidence. Let us now look at cartilage homeostasis more closely.

2.3 Aggrecan: Homeostasis and Turnover of Extracellular Matrix

2.3.1 Oxygen and Aggrecan Production

The mechanical environment in diarthrodial joints is very different to the mechanical environment found in other soft tissues. Contact pressures in weight-bearing joints are routinely several MPa [and 10+ MPa at times, with reported stresses as high as 18 MPa for certain activities e.g. rising from a chair, climbing stairs (Morrell et al. 2005)]. Stresses of this magnitude preclude the usual soft tissue architectures seen elsewhere in the body (Elder and Athanasiou 2009; Lee et al. 2006; Milner et al. 2012). Consequently articular cartilage is avascular, alymphatic and aneural, with a very high fraction of volume occupied by a dense extracellular matrix (about 95–98%).

Without any blood supply (and so easy access to the abundant resources in blood that is offered to some tissues for rapid cell proliferation and regeneration e.g. mucosal surfaces, skin, bone marrow), hyaline cartilage is generally believed to have a more limited capacity to repair itself than many other tissues (Huey et al. 2012). However, limited capacity is quite different to no capacity. Chondrocytes are adapted to their special environment. Articular cartilage is supported by the resources that can be delivered via the synovial fluid, which proves to be sufficient to enable both extracellular matrix and chondrocyte turnover. In this context, we should keep in mind that the synovial membrane is a very dynamic tissue that can adjust its blood supply, cell population, and lymphatic drainage. By these means, synovial membrane can adjust the level of support and resources it offers to articular cartilage in a synovial joint. In effect, the function of blood vessels, which cannot be in the articular cartilage by dent of its challenging mechanical environment, has been 'handed over' to the synovial membrane, which is located to the 'side' of the mechanical load path through the joint, safely out of harm's way. In short, cartilage has 'outsourced' logistic support of itself to the synovial membrane, allowing its further specialization. It is the demanding mechanical environment experienced by articular cartilage that drives the particularly close 'partnership' between articular cartilage tissue and synovial tissue.

The flux of nutrients and oxygen from the synovial fluid into the cartilage may be 'transport limited' (e.g. Zhou et al. 2004). In other words, the rate limiting transport step for the supply of nutrients to the articular cartilage may not be within the synovial membrane, or the synovial fluid [which facilitates transport (Mobasheri et al. 2002)], but rather through the dense cartilage matrix. For example, the concentration of oxygen in the superficial articular cartilage is approximately 6 or 7% [46–53 mm Hg 'oxygen tension,' meaning partial pressure of oxygen (PO_2)], suggesting a low but perfectly adequate oxygen supply. But for deeper cartilage, it is only about 1% (8 mm Hg oxygen tension) and perhaps even less (Fermor et al. 2007; Henrotin et al. 2005), suggesting a permanently 'hypoxic' environment is normal (Fermor et al. 2007). Biochemists define a hypoxic environment as one in which mitochondria cannot generate ATP at their maximal rate due to limited oxygen availability, while physiologists define hypoxia by loss of cellular functions (Hockel and Vaupel 2001). Clearly, deep chondrocytes are still functional, so hypoxia in this case would refer to a relative increase in anaerobic metabolism required to maintain normal cell functions.

We note that atmospheric oxygen accounts for 21% of the total atmospheric gas pressure, and so has an oxygen tension of 160 mm Hg, while fully oxygenated arterial blood ejected from the heart has an oxygen tension around 95 mm Hg. Oxygen tension decreases with increasing distance from blood vessels, and there may even be a significant gradient of oxygen within a cell (mitochondria closer to the blood vessel and cell surface have higher oxygen tensions). Oxygen tension within mitochondria has to be greater than around 0.5–1 mm Hg for aerobic oxidative phosphorylation to operate at anywhere near maximal rate (Hockel and Vaupel 2001).

HIF-1α expression is essential to the survival of hypoxic cells in fully differentiated tissue (Schipani et al. 2001). Indeed, HIF-1α expression is necessary for regulating glycolysis under aerobic and anaerobic conditions. Interestingly and probably importantly, increased HIF-1α expression strongly increases both aggrecan and collagen type II production by epiphyseal chondrocytes, and '*HIF-1α null chondrocytes were unable to maintain ATP levels in hypoxic microenvironments*' (Pfander et al. 2003).

Chondrocytes respiring anaerobically produce only 2 ATP molecules and lactic acid from a single glucose molecule. This low ATP yield, relative to 30–36 ATP molecules typically produced per glucose molecule for aerobic metabolism, explains the comparatively low rates of metabolism by chondrocytes. The lactic acid produced by anaerobic metabolism lowers the pH of the cartilage ECM locally relative to that of the synovial fluid (Sengers et al. 2005; Milner et al. 2012). Generally, the pH falls with increasing depth in the cartilage (see Fig. 2.60) as the proportion of cellular ATP sourced from anaerobic metabolism increases. Importantly as chondrocyte metabolic activity increases, more ATP is required by the cell. If this increased rate of ATP production is sourced primarily from anaerobic metabolism, the pH of the cartilage ECM will be lowered further due to the increased rate of production of lactic acid. We note here that the increased metabolic activity of chondrocytes in osteoarthritis probably explains the fall in pH observed in osteoarthritic cartilage ECM (Konttinen et al. 2002).

2.3.2 Aggrecan Homeostasis

It has been shown that cyclic surface compressive mechanical loadings damage the extracellular matrix of articular cartilage tissue, causing accelerated loss of (damaged) ECM components (see for example Figs. 2.1, 2.2, and 2.3) and references in (Ko et al. 2013; Bader et al. 2011; Bellucci and Seedhom 2001; Melrose et al. 2008a; D'Lima et al. 2001; Jeffrey et al. 1997; Ewers et al. 2001; Thibault et al. 2002; Chen et al. 1999). The loss of ECM components continues for 5–6 days after the damaging loading (see Fig. 2.3).

For stress-controlled surface loadings, generally the faster the rate of loading, the larger the stress required to cause damage (Flachsmann et al. 2001; Quinn et al. 2001). For example, Flashsmann et al. found that splitting of the surface occurred at 15 MPa static load, but dynamic loading of 28 MPa did not cause splitting (note: measured dynamic strain was about half static strain). Quinn et al. found that for the

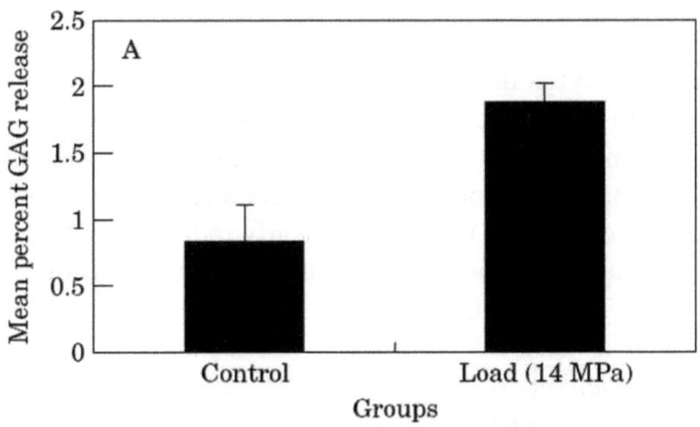

Fig. 2.2 GAG release in response to single impact mechanical loading of cartilage explants. Cartilage explants were loaded in unconfined compression to 14 MPa (resulting in about 30% impact strain) and cultured for 96 h. Cumulative sulfated GAG release was measured by 1,9-dimethylene blue assay. The results are expressed as a % of total explant GAG content per mg and represent mean values from eight explants in each group from three separate experiments (separate donors) (D'Lima et al. 2001)

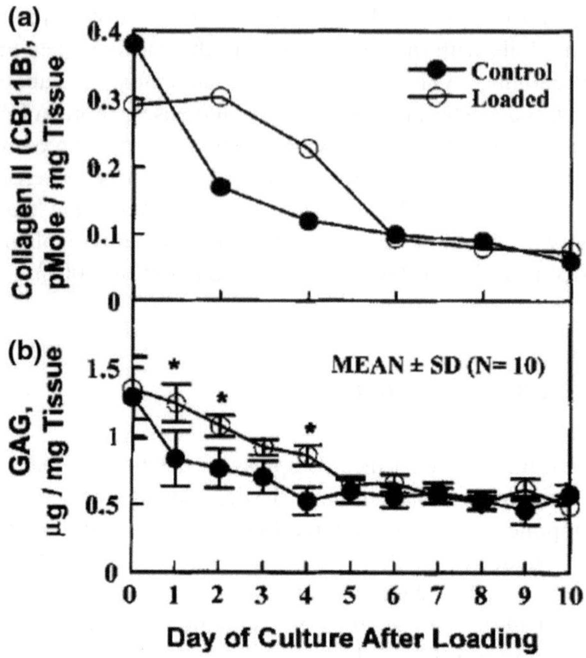

Fig. 2.3 Cartilage explants obtained from 1- to 2-year-old steers, subject to 40 cyclic loads over 20 min. Denatured type II collagen and **GAG** release to media during culture after loading: collagen II fragments (**a**) and **GAG** (**b**) released to culture medium after cyclic loading were assessed. Loading occurred on day 0 on these graphs and the day 0 point represents fragments found in the medium from the 48-h (**a**) or 24-h (**b**) period before loading. There is an increased release of denatured collagen type **II** and of **GAG** to the culture medium after loading compared to unloaded controls. *$p < 0.005$ comparing loaded to control (Thibault et al. 2002)

same strain rate of loading (3×10^{-5} strain per second), the amount of aggrecan produced by chondrocytes decreased with increasing load [3.5 MPa (unchanged relative to control) to 7 MPa (reduced to about 50% of control) to 14 MPa (reduced to about 20% of control)], but this effect was blunted by faster loading rate (0.7 strain per second), there being only slight reduction relative to controls for all loads (see Fig. 4 in Quinn et al. 2001). This data clearly indicates that the stiffening effect of pore fluid has generally beneficial functional implications for cartilage tissue.

Chen et al. (1999) found that damage to canine articular cartilage required repeated impacts of at least 2.5 MPa and a stress rate of 30 MPa/s at 0.3 Hz for 2 min or longer (Chen et al. 1999). Chen et al. (1999) also noted, following damage, that proteoglycan production was increased above control levels after short periods of loading (40–100% increase above controls after 2 min of 5 MPa impact loading at 0.3 Hz), but production decreased when the impact loading duration (and damage) became longer (20 and 120 min). Fibronectin production increased (20–50%) for all impact load durations (Chen et al. 1999)—fibronectin is a multi-functional glycoprotein that has multiple binding sites for a number of macromolecules and cell surface receptors (Makogonenko et al. 2002). Collagen network damage was visualized using a collagen denaturation antibody, which revealed damage at the cartilage surface after 20-min impact loading, and at the cartilage surface and close to the subchondral bone after 120-min impact loading [see Fig. 4 (Chen et al. 1999)].

This indicates a general finding: a small amount of dynamic loading can increase aggrecan production (see for example Fig. 2.7) but larger and more damaging loadings decrease aggrecan production (Chen et al. 1999; Li et al. 2013). Importantly, experimental data also shows that for physiologically relevant loading rates, slow rates of loading are much more damaging than faster loading rates (Quinn et al. 2001) (see Fig. 2.4). Mechanical damage to the ECM is generally not uniform, with mechanical damage being greatest where tissue strains are largest (i.e. this is usually directly beneath the loading and toward the cartilage surface) (Rosenzweig et al. 2012; Ewers et al. 2001; Chen et al. 1999; Lin et al. 2004; Quinn et al. 2001).

In addition to mechanical damage of the ECM, there is ongoing 'enzymatic damage' of the ECM. It is not known how much of the enzymatic damage is 'debridement' of aggrecan that has been mechanically damaged, but it is probable that debridement of mechanically damaged aggrecan (and collagen) is one of the main purposes of aggrecanase (and protease) activity in articular cartilage (Murphy and Nagase 2008). Indeed, much of the aggrecan lost by enzymatic digestion only removes 'end portions' of the entire aggrecan molecule (Struglics et al. 2009), which is consistent with the notion of 'trimming' of damaged portions from aggrecan molecules (see Fig. 2.13).

But it is the same compressive loading causing cartilage tissue damage that promotes the differentiation of chondrocytes and the formation and maintenance of the distinctive articular cartilage ECM (Juhasz et al. 2014). Aggrecan is secreted by chondrocytes (Karsdal et al. 2008). The secreted aggrecan replaces the aggrecan that is damaged by mechanical loading, it replaces the aggrecan that is removed by secreted aggrecanases and/or proteases, and it replaces the aggrecan that is continually transported through the cartilage tissue via its migration down its concentration gradient, eventually exiting through the cartilage surface (Belcher et al. 1997; Bayliss et al. 2000; Smith et al. 2016).

To maintain tissue homeostasis, usually the rate of aggrecan loss is equal to the rate of aggrecan synthesis, and so articular cartilage is in a dynamic state of equilibrium. However, if the rate of aggrecan degradation (and/or removal) consistently exceeds the rate of aggrecan production, the local aggrecan concentration decreases and the cartilage will 'soften' (a typical description of OA cartilage), and possibly become unsuitable for its mechanical purpose in vivo. So matching the rate of aggregate formation/repair to the rate of aggregate damage is vital for maintaining healthy articular cartilage.

Matching the rate of repair to the rate of damage is an example of cartilage tissue homeostasis. As noted in the introduction, cartilage homeostasis is the name given to all the processes that contribute to maintaining the cartilage in an equilibrium state after it is perturbed by environmental disturbances. Homeostasis usually

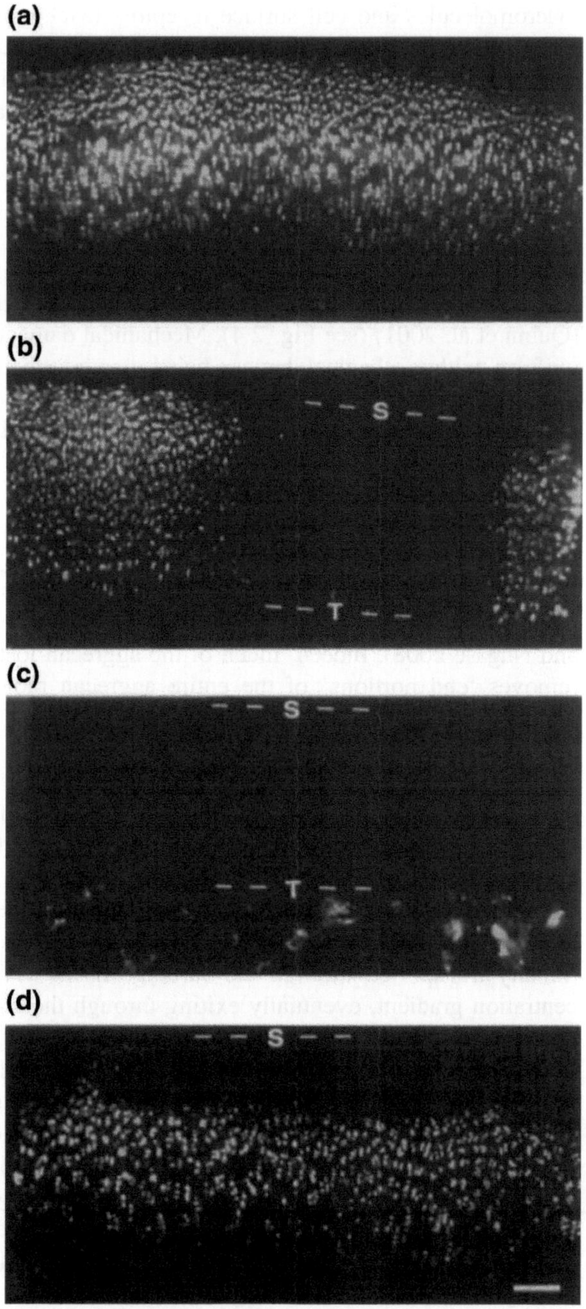

Fig. 2.4 Fluorescence microscopy of osteochondral tissue sections (−0.5 mm thickness) after incubation in fluorescein diacetate. Fluorescence signal indicates the locations of metabolically active cells 3 days after mechanical compression had been applied to osteochondral explants. Compression was characterized by: **a** 3.5 MPa peak stress at 3×10^{-5} strain rate per second; **b** 7.0 MPa peak stress at 3×10^{-5} strain rate per second; **c** 14 MPa peak stress at 3×10^{-5} strain rate; **d** 14 MPa peak stress at 0.7 strain rate per second. Bar = 200 μm (Quinn et al. 2001). This data illustrates that higher compressive stresses lead to more cell death, and for physiologically relevant loading rates, slow rates of loading are much more damaging than faster loading rates (compare **a**–**c** with **d**). This outcome can be explained by cartilage stiffening as the rate of loading increases, so reducing the amount of strain for a constant loading and so the amount of damage (see discussion on stiffening Sect. 1.6)

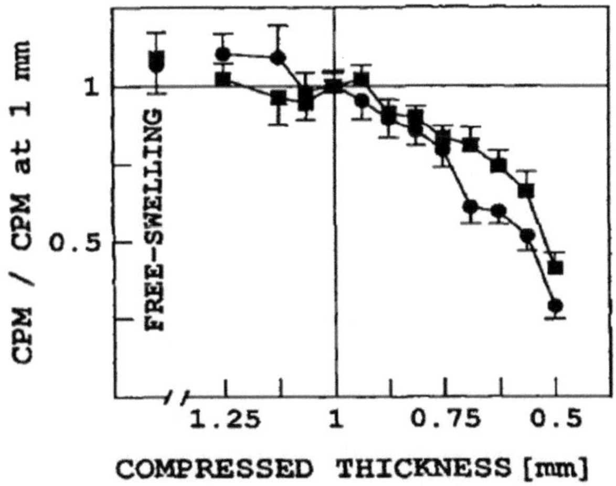

Fig. 2.5 Radiolabel incorporation during 12 h static compression. ^{3}H-proline (square) and ^{35}S-sulfate incorporation (dot) are expressed relative to incorporation into disks held at 1 mm (the original explant thickness); free-swelling controls were incubated in 24-well culture dishes. All points are mean ± SEM (n = 9–12) (Sah et al. 1989). Note the substantial decrease in aggrecan and collagen synthesis with sustained compression

involves negative feedback (which means a signal is generated based on a discrepancy between the equilibrium state (or 'set point') and the actual state). This discrepancy is detected by the chondrocytes. The negative feedback signal transduction within chondrocytes sets in train processes to restore tissue equilibrium.

For example, chondrocytes at each depth in the articular cartilage may have a preferred local aggrecan concentration (i.e. a set point aggrecan concentration). This set point concentration is determined by the resident chondrocyte's local environmental conditions. Known environmental conditions influencing the chondrocytes' preferred local aggrecan concentration include the oxygen and nutritional status of the tissue, and the concentration of various growth factors and cytokines, the recent local histories of fluid flow, hydrostatic pressure, and strain experienced by the tissue, and of course, the local ECM composition. The relative contributions of each influence on the aggrecan set point are not known as this stage, but it is almost certainly true that the relative contributions of each influence depend on the specific circumstance of the chondrocyte (i.e. contributions of each influence are nonlinear).

When cartilage is statically compressed, the local concentration of aggrecan around chondrocytes increases (see Appendix D). The concentration increases because the local mass of aggrecan remains constant (i.e. aggrecan is for practical purposes immobile over short timescales, e.g. seconds, minutes, or hours associated with normal daily joint loadings) while the static compressive volumetric strain reduces local cartilage tissue volume. By this means, tissue compression temporarily increases local aggrecan concentration in the pericellular ECM. It seems extraordinary, but chondrocytes can sense this temporary local increase in aggrecan concentration. They can do this by sensing local hyperosmotic conditions relative to their local osmotic set point (Schneiderman et al. 1986), but there may be other sensors and mechanisms in operation too (e.g. aggrecan-HA-CD44 signaling or integrin signaling, Rho-Rock signaling). The chondrocytes respond to static compressive loading by *reducing* aggrecan (and collagen) production (see Fig. 2.5) (Buschmann et al. 1996a; Schneiderman et al. 1986; Ryan et al. 2009). This effect is strikingly visualized by comparing Figs. 3.7 and 3.8.

However, chondrocytes respond quite differently to cyclic loading—cyclic loading *increases* aggrecan production (see Fig. 2.6) (Saadat et al. 2006). Given that repetitive mechanical loading damages the ECM including aggrecan, it is perhaps unsurprising that repetitive mechanical loading also stimulates chondrocytes to secrete more aggrecan (Grodzinsky et al. 2000). It has also been shown that chondrocytes can sense their environment and increase aggrecan synthesis in response to increased interstitial fluid velocity (higher fluid velocities are experienced at the 'radial edge' of the explant) (see Fig. 2.7) (Buschmann et al. 1999;

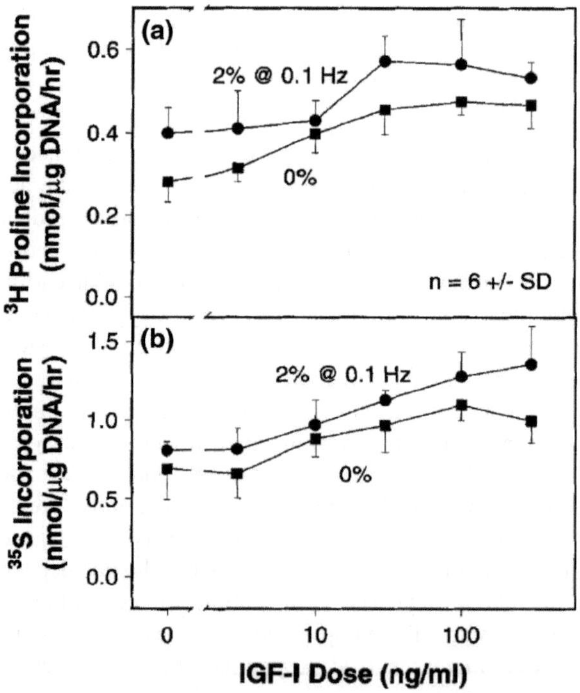

Fig. 2.6 Effect of dynamic compression on the dose response of IGF-I (for one- to two-week-old bovine cartilage obtained from femoropatellar groove). Shown here is incorporation of [^{3}H]-proline (**a**) and [^{35}S]-sulfate (**b**) into cartilage disks treated with 0, 3, 10, 30, 100, or 300 ng/ml IGF-I in the absence (square) or presence (dot) of a 2% sinusoidal strain imposed at 0.1 Hz for 48 h. Data is represented as mean ± S.D. for n = 4 samples (Bonassar et al. 2001). Notice that sustained cyclic loading (even at just 0.1 Hz), significantly stimulates aggrecan and collagen production

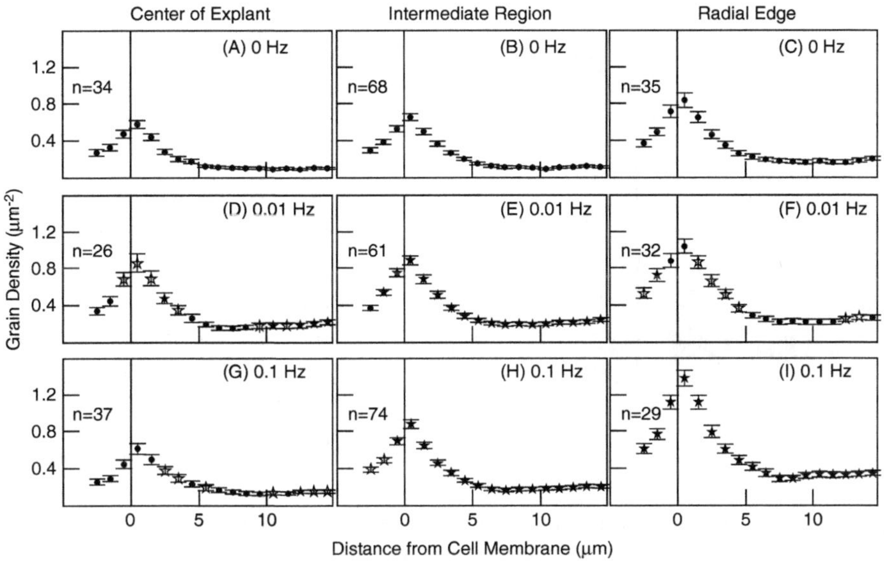

Fig. 2.7 Effects of dynamic compression on proteoglycan synthesis and deposition on a cell-length scale. Grain density is expressed as a function of distance from the cell membrane *d* averaged over all [radial directions (mean ± SEM, *n* = number of cells sampled)]. Frequencies of applied compression included 0 Hz (**a–c**), 0.01 Hz (**d–f**), and 0.1 Hz (**g–i**). Cells were sampled as a function of radial position *R* within explants, and include those derived from central (**a**, **d**, **g**), intermediate (**b**, **e**, **h**), and radial edge (**c**, **f**, **i**) regions. Values different from those for 0 Hz (by two-tailed Student's *t*-test; all observed differences were increases) are indicated by *q* ($P < 0.05$) or *w* ($P < 0.01$) (Quinn et al. 1998). 8 h ^{35}S pulse was applied at end of 24 h of cyclic loading. Notice that aggrecan production is significantly enhanced toward the outside of the cartilage disk, where fluid flow is greatest, particularly when the frequency increased to 0.1 Hz

Quinn et al. 1998) and in response to increased hydrostatic pressure (Toyoda et al. 2003; Parkkinen et al. 1993). We note here that the aggrecan gene contains a 'shear-stress response element' (SSRE) in its promoter. We also note in passing that tenocytes change their gene expression profile in response to fluid shear flow too (Maeda et al. 2011).

It is very likely that chondrocytes in different parts of the articular cartilage modulate their response to the various spatially dependent mechanical stimuli arising from loading and straining a saturated porous media (i.e. mechanical stimuli include volumetric strain and shear strain, fluid flow and hydrostatic pressure) (Nguyen et al. 2010). For example, see the remarkable data shown in Fig. 2.7. This data shows how newly synthesized aggrecan is radially distributed from the cell membrane, and how its relative concentration is influenced by its location across the articular cartilage tissue explant (note that more aggrecan is produced where pore fluid flow rates are higher at the 'radial edge'), and the loading frequency (note increasing aggrecan production from 0 to 0.1 Hz). This and other data have been

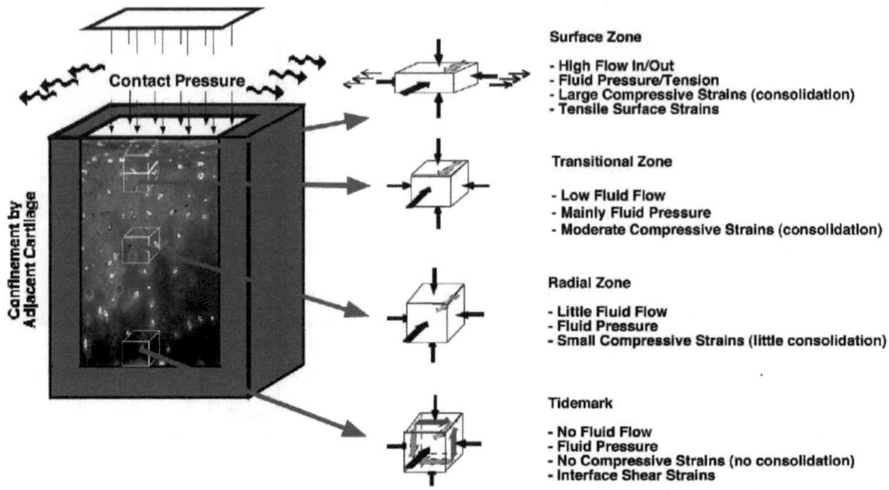

Fig. 2.8 Schematic representation of the in vivo mechanical environment of articular cartilage under intermittent joint loading and motion (Carter et al. 2004)

incorporated in a computational model of aggrecan in cartilage (Zhang et al. 2008a, 2009).

In another example, it has been shown that chondrocytes bearing the highest loads toward the center of the tibial plateau, contain about twice as much aggrecan mRNA relative to peripheral chondrocytes bearing lower loads (Bevill et al. 2009), and interestingly, chondrocytes from central regions responded to mechanical loading by producing mRNA about as twice as strongly relative to chondrocytes from peripheral regions.

In addition to lateral variations in cartilage load, there are depth-dependent variations in mechanical stimuli. Fluid flow, chondrocyte volumetric, and shear straining probably provide the main stimuli in the superficial and transitional zones of cartilage. Shear strain depends on regional shear stiffness over the cartilage depth [‘*normal adult femoral condyle cartilage tested with normal synovial fluid (SF), shear strain reached 5.6% near the articular surface and 2.8% overall, increasing markedly with degeneration to 28 and 4.1%, respectively*’ (Nguyen et al. 2010) (see also Figs. 4 and 5 (Nguyen et al. 2010))].

Hydrostatic pressure stimulus provides a general anabolic signal (Milner et al. 2012), which probably becomes more prominent further from the surface, e.g. perhaps being most important in the deep zone cartilage. This is because the tissue closest to the surface of the cartilage drains more quickly than tissue in the deep zone (see discussions Chaps. 4 and 7). The concept that mechanical stimuli vary spatially through the cartilage is nicely illustrated by Carter et al. (2004), Wong and Carter 2003)—see Fig. 2.8.

Importantly, we here note that aggrecan synthesis is strongly depth dependent, with more aggrecan being produced more distant from the cartilage surface (see Fig. 2.9). Though this may on first impression be surprising—for example, deep zone chondrocytes are further from nutrients and energy that is sourced from the synovial fluid through the cartilage surface. However, this arrangement makes sense on a couple of levels: (i) the greater distance that aggrecan is synthesized from a free surface, the more difficult it is for the aggrecan to escape from the cartilage tissue, and (ii) given chondrocytes that are relatively well protected from excessive strains in the deep zone, they are well placed to produce ECM components which then have more opportunities (and time) to construct (or repair) ECM along the path to the cartilage surface in any region closer to an exit surface (i.e. an aggrecan molecule produced in the deep zone may be used to form ECM in the transitional or superficial zones too). This is discussed further in Chap. 3.

In summary so far, it seems most likely that aggrecan synthesis in vivo is turned off by high local osmolality, as aggrecan is the main contributor to osmolality of the ECM (Urban et al. 1993; Hopewell and Urban 2003; Palmer et al. 2001; Negoro et al. 2008). So aggrecan production is regulated by its own 'feedback inhibition.' Aggrecan synthesis is also turned off by static loading, at least partly through changes in local osmolality. Interestingly, aggrecan synthesis may be turned back on by ERK1/2 inhibition (while chondrocyte proliferation marker PCNA, which increased by static loading, is also decreased by ERK1/2 inhibition) (Ryan et al. 2009).

On the other hand, dynamic mechanical loading of the cartilage causes cyclic strain, fluid flow, and hydrostatic pressure, which all promote aggrecan synthesis. On functional grounds, it is plausible that the mechanical drivers would promote aggrecan synthesis in order to regulate compressive straining of the tissue to acceptable levels, and in addition, to regulate tissue permeability to help regulate compressive straining to acceptable levels. So tissue volumetric compressive strain (and fluid flow) is regulated by feedback inhibition via aggrecan production. Because local compressive strain and tissue permeability are closely correlated to local osmolality, the osmolality set point works alongside the local strain set point and the local fluid shear-stress set point in chondrocytes to maintain local aggrecan homeostasis and tissue integrity. This group of homeostatic sensors is central to maintaining cartilage tissue integrity. In other words, if these sensor systems fail, then articular cartilage tissue will fail.

In addition to mechanical stimuli (cyclic strain and fluid shear stress), chemical stimuli are also sensed by chondrocytes. Chemical stimuli feed into 'decision processes' within chondrocytes about how much ECM to secrete. There is a large number of chemical stimuli, which may be arbitrarily divided into local and systemic (see for example Table 1 in Loeser 2003, Table 1 in Plaas et al. 2011 and Bader et al. 2011). However, probably the most important local chemical stimuli promoting matrix synthesis in vivo are osmotic stimuli, the IGFs, the TGFs (and related BMPs, e.g. BMP7) and possibly FGFs (e.g. FGF18). All these chemical stimuli elicit powerful anabolic responses. Importantly, we note that in mice withdrawal of TGFβ (e.g. by TGFβRII knock-down) leads to upregulation of MMP13 and ADAMTS5, which then progresses to an OA phenotype (Shen et al. 2013).

Using a reporter gene construct and isolated chondrocytes, increased osmolality (and so osmolality above the local set point for chondrocytes at that depth) cause chondrocytes to decrease promotor activity and decrease mRNA production (see Fig. 3 in Palmer et al. 2001). Truncation of part of exon1 of aggrecan core protein, known to be an important regulatory region, resulted in the loss of this downregulation. We also note in passing that osmolality is also an important determinant of chondrocyte volume (Xu et al. 2010), and integrins appear to play a role in signal transduction (Jablonski et al. 2014).

Maximum GAG synthesis rates have been reported to occur at about an osmolality of about 350–400 mOsm (Urban et al. 1993; Negoro et al. 2008) (for comparison, we note blood osmolality is about 290 mOsm). The existence of a maximum rate of synthesis implies there is initially a positive feedback process on aggrecan production up to an osmolality somewhat below the set point osmolality, and then a negative feedback process kicks in on aggrecan production at higher osmolalities (e.g. see Fig. 2 in Urban et al. 1993). Positive feedback implies increased concentration of aggrecan in the pericellular matrix induces more synthesis and secretion of aggrecan. Such positive feedback behavior is very interesting, as it helps to explain how chondrocytes produce so much ECM volume relative to the size of the cell volume, and why, as shown in Fig. 2.9, the amount of aggrecan secreted increases with depth (i.e. with increasing aggrecan concentration). Presumably the existence of this positive feedback on aggrecan production is one of a number of attractor mechanisms that help the tissue lock into the cartilage phenotype.

Further, it has been reported that the ‘set point’ for maximum synthesis rate may change with increasing osmolality. For example, Hopewell and Urban (2003) reports that aggrecan synthesis initially decreased when they exposed tissue to 550 mOsm, but after adaption over 24 h, synthesis rates increased and became largest at 550 mOsm (Hopewell and Urban 2003). Clearly much remains to be understood about chondrocyte ‘set points’—and what drives aggrecan production—it may be that aggrecan production is at least partially limited by mass transfer of resources to the cartilage surface, which limits aggrecan synthesis and keeps aggrecan concentrations below the target set-point? It may be that aggrecan concentration is the pericellular matrix is at least partially limited by aggrecan losses (e.g. mechanical damage and transport mechanisms lead to loss of aggrecan through the cartilage surface), which keeps aggrecan concentration in the pericellular matrix below the target set-point? We can get a flavor of the complexity of the regulation of aggrecan (and other molecules) by considering some additional influences and their interactions.

Key hormones carry information to all tissues of the body (i.e. to cartilage tissues), relaying information about the general state of the whole organism. Thyroxine sets the metabolic rate of cells throughout the body; growth hormone plays numerous roles in determining growth rates of the organism (Lewinson et al. 1994); IGF-1 provides information about the nutrient and energy state of the liver (and so its ability to support tissue maintenance, proliferation, and growth); the sex hormones [estrogen (Schicht et al. 2014), progesterone, and testosterone (Irie et al.

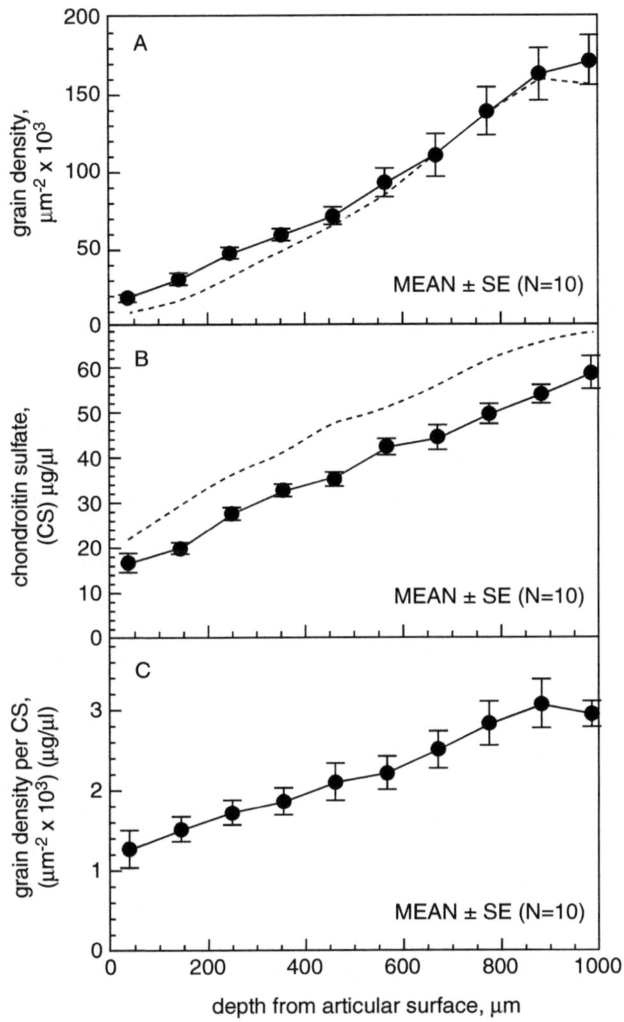

Fig. 2.9 Bovine animals 10–14 months old, cartilage from humeral head. **a** The RHT-derived chemographic grain density was obtained as a function of distance from the articular surface by analyzing digitized images similar to the photos in Fig. 2.1 using a method developed previously for quantitative autoradiography. **b** Analysis of papain-digested tissue sections from different tissue depth regions using the cationic DMB spectrophotometric agent is represented as equivalent chondroitin sulfate concentration determined by a chondroitin sulfate standard. Solid lines with data points and standard errors show the results of one experiment containing 10 specimens; broken line shows the results of an independent prior experiment also containing 10 specimens. Both grain density and equivalent chondroitin sulfate concentration increase with depth. **c** The ratio of grain density to equivalent chondroitin sulfate concentration as a function of depth found using the data from **a** and **b** of this figure. The ratio is not constant and is a function of distance from the articular surface, possibly because the binding chemistry in dense tissue sections of cationic species such as RHT and DMB may be different from that in dilute solution (Buschmann et al. 2000). We note that a similar depth-dependent aggrecan production profile is found in adult human articular cartilage (Maroudas 1975)

2005)] provide information about the maturity of the organism and stage of the fertility cycle in women; melatonin keeps internal time for a central circadian clock, which helps coordinate some functional activities in tissues; parathyroid hormone can reset peripheral clocks in bone and cartilage tissue (Okubo et al. 2015).

Some hormones may provide more specific information about the state of the organism. For example, information is conveyed about total body mass by leptins (Loeser 2003). Information about total muscle mass and levels of muscle activity is carried by IGF-II [IGF-II is correlated with muscle mass (Stewart and Rotwein 1996; Wang et al. 2005a)]. Indeed, IGFs are very important hormones stimulating a general anabolic response in cartilage tissue (see Fig. 2.6). IGF-II has essentially the same effect as IGF-I, binding the IGF-1R receptor (albeit binding less tightly than IGF-I).

In addition to all this hormonal information coming in to the cartilage tissue about the state of the organism, there are many additional regional signaling molecules [e.g. periostin (Chinzei et al. 2018)]. For example, some regional signals cause chondrocytes to reduce aggrecan synthesis, and indeed to degrade aggrecan in the ECM. Three of the most important (regional and local) catabolic cytokines are IL-1β, TNFα, and IL-6 (though there are several others that may play key roles under certain circumstances, e.g. activation of RAGE in aging or activation TOLL-like receptors may play a key role in joint infection). These catabolic cytokines are very powerful (particularly IL-1β and TNFα) and can be highly destructive of cartilage in a short period of time (see Sect. 5.6.6).

Of course chemical (hormonal, regional, and local) and mechanical stimuli (whole body, regional, and local) may interact to increase aggrecan production rates. For example, the anabolic response of IGF is enhanced when combined with cyclic loading (see for example Fig. 2.6) (Bonassar et al. 2000). But mechanical stimuli and chemical stimuli may also have opposing effects on aggrecan synthesis rates. Importantly, application of cyclic mechanical loading of sufficient magnitude has been shown to inhibit the cellular effects of IL-1β and TNFα, which promote tissue degradation (Torzilli et al. 2010; Li et al. 2013). For example, (Torzilli et al. 2010) demonstrated that light loading (e.g. 0.2 MPa at 0.5 Hz) had little effect on aggrecan and collagen loss from explants exposed to IL-1 at 10 ng/ml, while 0.5 MPa at 0.5 Hz reduced aggrecan and collagen loss by about half.

In a separate set of experiments, Li et al. (2013) exposed cartilage explants to cytokines in the support medium (specifically, the cytokine combination was rhTNFα at 25 ng/ml, rhIL-6 (50 ng/ml) and sIL-6R (250 ng/ml)) over the entire eight days of the experiment. Initial injurious loading of 50% compression was first applied at a rate of 100% strain per second. Then sustained mechanical loading over the next eight days was at 10%, 20%, or 30% cyclic strain, 0.5 Hz, for 1 h on 5 h off (40% duty cycle). The results show sGAG loss is reduced with increasing cyclic strain, relative to controls with initial injurious loading with continuous exposure to cytokines. Rates of aggrecan biosynthesis increased with increased cyclic loading strain up to 20% cyclic strain, but 30% cyclic strain leads to a slight reduction of synthesis compared to 20% cyclic strain.

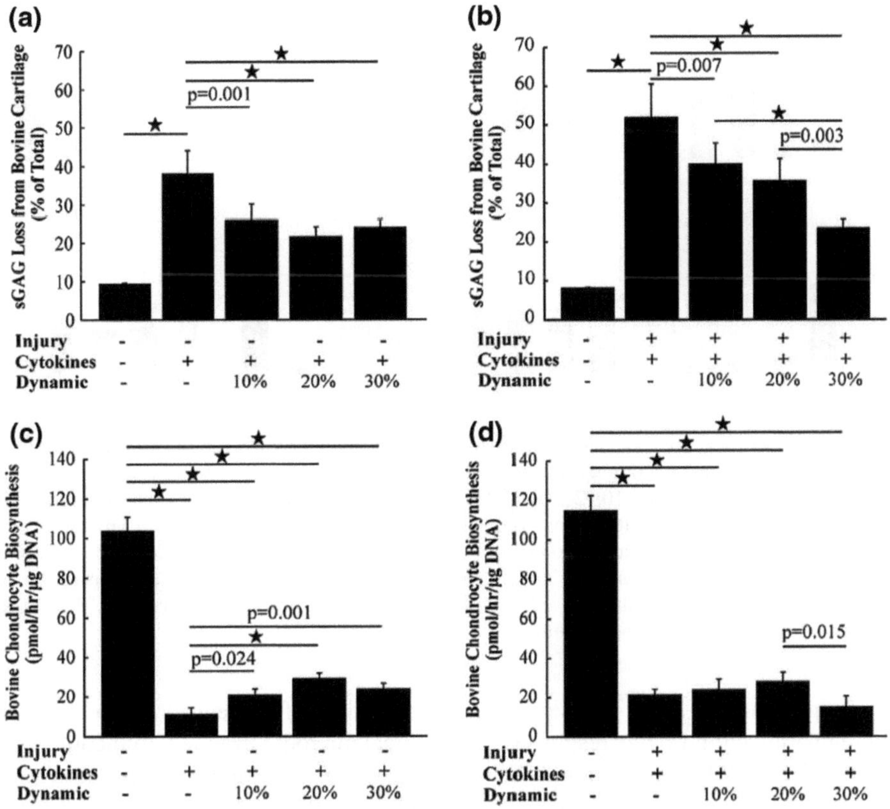

Fig. 2.10 **a, b** Cumulative sGAG loss from bovine cartilage to the medium in response to 8-day treatments without or with initial mechanical injury. **c, d** Bovine chondrocyte biosynthesis measured during day 6–8 as 35S-sulfate incorporation rate for the same cartilage disks as in **a** and **b**, respectively. Values are mean ± 95% confidence interval; **a, c** $n = 3$ animals (6 disks/animal); **b, d** $n = 4$ animals (4–6 disks/animal); $^{*}P < 0.001$ (Li et al. 2013)

Figure 2.11 is largely confirmatory of the aggrecan loss data shown in Fig. 2.10, showing that at the 48-h time point, aggrecan and collagen mRNA expression is progressively upregulated relative to control, at 10% cyclic strain, to 20% cyclic strain, but then mRNA expression is reduced at 30% cyclic strain (see aggrecan loss data Fig. 2.10a). Importantly, 10 and 20% cyclic strain is also observed to reduce NF-κβ. IL-6 and ADAMTS5 expression (but 30% cyclic strain greatly increased ADAMTS5 expression) (Li et al. 2013). Clearly too little and too much cyclic strain are both deleterious relative to the optimal cyclic strain.

Possible signaling pathway interactions explaining the protective effects of mechanical loading following cytokine exposure are depicted in Fig. 2.12. Clearly understanding and manipulating these interactions have important implications for potential strategies for minimizing tissue damage following injury (Bader et al. 2011).

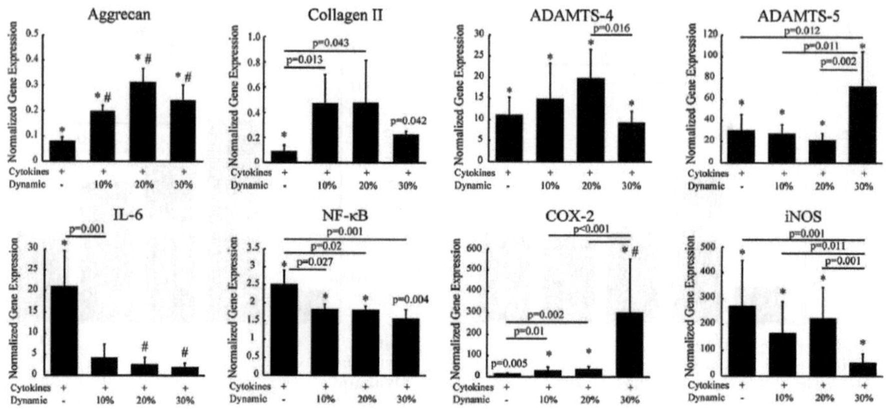

Fig. 2.11 The effects of dynamic compression on bovine chondrocyte gene expression after 48 h of treatment with exogenous cytokines (normalized cell expression on vertical axis, and no dynamic loading, strain of 10, 20, and 30% on the horizontal axis). For each condition, six cartilage disks from the same animal were pooled for mRNA extraction; $n = 4$ animals. Gene expression levels were normalized to that of the 18S gene and then normalized to the no-cytokine, no compression control condition which had an expression level = 1. Data is presented as mean ± 95% confidence interval, $^*P < 0.001$ compared with untreated control; $^{\#}P < 0.001$ compared with cytokines-alone treatment (Li et al. 2013)

It appears very likely that several ECM breakdown fragments are also involved in regulating chondrocyte responses to their local microenvironment (Vincourt et al. 2010). For ECM fragments and their 'damage-associated molecular patterns' (i.e. DAMPs as they are sometimes called) provide a local signal indicating the *specific* molecular-scale mechanical damage occurring to the ECM, or the particular enzymatic debridement of molecules following mechanical damage (Murphy and Nagase 2008). The molecular fragments carry important information to chondrocytes about the specific type of damage taking place in the tissue.

While many of these molecular fragment signals probably remain to be discovered, some are now known. For example, fibronectin fragments induce a strong decrease in ECM component synthesis (Homandberg et al. 1998; Ding et al. 2008), act as proteases themselves, and induce MMP production (Barilla and Carsons 2000). Short hyaluronan (HA) upregulates a number of genes involved in matrix remodeling and turnover, while long HA tends to suppress these genes (Knudson and Loeser 2002; Ohno et al. 2005). Mannosamine fragments from the degradation of aggrecan can inhibit IL-1β degradation of cartilage (Patwari et al. 2000), while a 32 amino acid fragment from the aggrecan core protein can stimulate the toll-like receptor 2, which accelerates cartilage enzymatic damage (Fosang 2015). Collagen fragments have been shown to influence chondrosarcoma behavior (Vincourt et al. 2010). Debridement of damage tissue may be an important first step before repair can commence, so a highly nuanced time sequence of chondrocyte secretions driving debridement, synthesis, and repair may evolve following damage. It is not

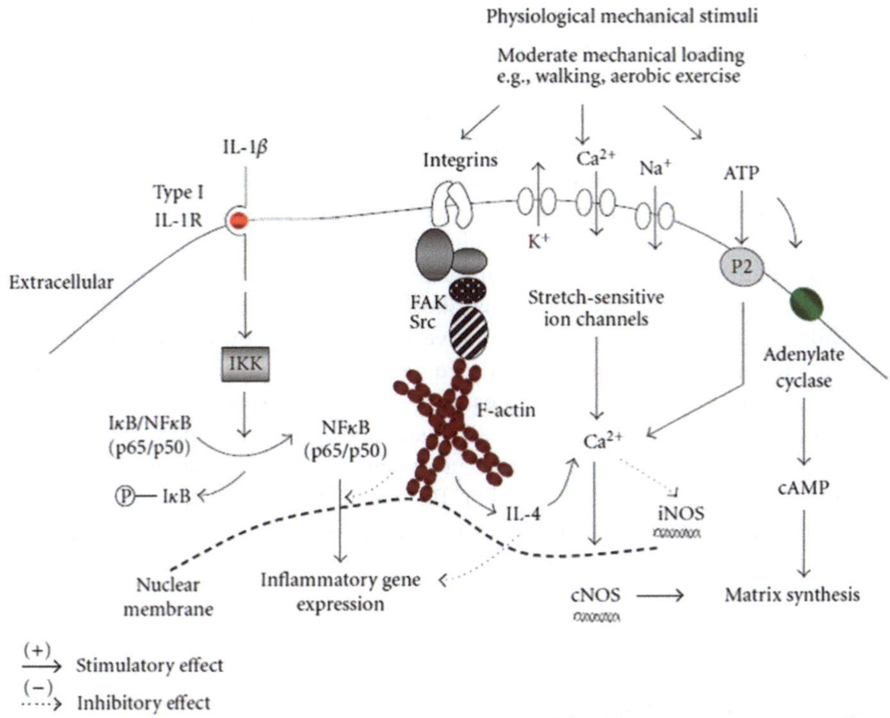

Fig. 2.12 Model depicting the potential protective effects of physiological mechanical stimuli in chondrocytes stimulated with interleukin-1β (IL-1β). Moderate mechanical loading induces a number of signaling cascades which leads to the production of extracellular matrix components. Mechanical loading will stimulate integrin-mediated release of interleukin-4 (IL-4) via actin cytoskeleton, mechanical perturbation of stretch-sensitive calcium or sodium channels, or stimulation of a purinergic pathway involving ATP release and subsequent purinoreceptor (P2) or cAMP activation. The loading-induced calcium may cause instability of inducible nitric oxide synthase (iNOS) mRNA or increase transport of interleukin-4 (IL-4), which blocks catabolic effects. In the presence of IL-1β, mechanical stimuli inhibit cytoplasmic dissociation of NFκB from inhibitory κB-α (IκB-α), which prevents nuclear translocation of the p65/p50 dimers and/or proteolytic degradation of IκB-α by IκB-specific kinases (IKK) or impair IκB-α degradation, thereby switching off transcription for the pro-inflammatory genes (Bader et al. 2011)

too fanciful to imagine a local healing process by local chondrocytes being directed by the types of ECM molecules being sensed by local chondrocytes.

We again mention that mechanical stimulation of cartilage tissue is necessary for cartilage tissue to maintain its resistance to compressive loading. For if a joint is immobilized, within a matter of weeks cartilage thickness is reduced by about half, the rate of aggrecan synthesis diminishes 40% and the aggrecan content may be reduced by half (Brandt 2003; Eckstein et al. 2006). This basal level of aggrecan production in the absence of mechanical stimulation is probably maintained by Indian hedgehog (Ihh) signaling via parathyroid hormone-related protein (PTHrP) (for further discussion see Appendix C). Following a period of immobilization, in a matter of weeks, ambulation restores cartilage thickness, normal rates of aggrecan synthesis and cartilage aggrecan content returns to normal.

Given all of the above, it is clear that mechanical and chemical stimuli (including growth factors, hormones, cytokines, metabolic signals, osmotic stimuli and ECM fragment stimuli), local, regional, and systemic, provide a rich and powerful set of inputs to chondrocyte sensors. We need to keep in mind that all these signals vary spatially and temporally in articular cartilage. The variation within cartilage tissue allows chondrocytes to respond locally with a highly nuanced regulation of aggrecan synthesis and degradation in response to their own environmental conditions, so maintaining aggrecan homeostasis throughout the whole tissue.

While we do not yet have a quantitative understanding of aggrecan in articular cartilage, the examples in this section give a sense of the potentially rapid changes in aggrecan production, the rapid rates of aggrecan loss upon immobilization of a joint and the rapid net aggrecan synthesis upon remobilization. This clearly demonstrates that articular cartilage is a dynamic tissue responding to its environmental stressors (Bader et al. 2011). But we would like to try and gain some quantitative understanding of the rates of aggrecan turnover during normal tissue maintenance. In the next section, we attempt to estimate the rate of aggrecan turnover in the human knee joint.

2.3.3 Aggrecan Turnover

We now turn to the difficult questions as to the rates of tissue turnover in articular cartilage, and to the states of aggrecan molecules when they finally exit the articular cartilage and enter the synovial fluid. The relative concentrations of aggrecan in different states may indicate the relative rates of mechanical damage versus enzymatic damage within normal cartilage tissue. We consider aggrecan turnover first.

Tissue turnover time is defined as the time to replace the ‘old material’ in a tissue with ‘new material’. So we may usefully speak of ‘cell turnover’, i.e. the time to replace all the cells in a tissue. For example, every 3–5 days all the epithelial cells lining mucosal crypts lining the gut are replaced. In this case, cells proliferate at the bottom of the crypt, mature as they migrate up the walls of the crypt, and exit through the ‘mouth’ of the crypt in a mature state (van der Wath et al. 2013). Because there is usually an orderly progressive movement of all epithelial cells from the bottom to the top of the crypt, when the last of the initially labeled cells exits the crypt mouth, all the epithelial cell in the crypt have been replaced, and the time required for this to occur is known as the crypt ‘turnover time.’

But if epithelial cells in the crypt migrate at different rates (some fast, some slow), then turnover time would be less clear cut. We need a different model to interpret tissue turnover. In this case, we could take a probabilistic interpretation and ask what is the half-life of the initial cells in the crypt (i.e. the time to replace half the cells in the crypt), much like we ask what is the half-life of a radioactive material. But it could be that epithelial cells have a diffusive transport component (which is random), and an advective transport component (which is like 'plug flow/ transport'). We see that 'tissue turnover time' is a theoretical estimate, which carries with it an implied model about the processes by which tissue replacement occurs.

Following from the general definition given above, 'aggrecan turnover' is defined as the time to replace all the aggrecan in the cartilage ECM. But we now realize that aggrecan turnover can be calculated in different ways depending on the processes that degrade/remove aggrecan molecules. Are aggrecan molecules degraded (which is similar to radioactive decay), or are they removed by diffusion transport and/or advection transport processes? To accurately calculate the tissue turnover rate, we really need to know exactly what happens to aggrecan molecules (which is not yet clear to us).

In the absence of this information, we could simply assume a mechanism for aggrecan replacement and calculate tissue turnover rate based on this assumption. So assuming the aggrecan produced in the deep zone of the cartilage were to simply to push all the old aggrecan ahead of it, to the shallower cartilage zones and finally out of the cartilage (i.e. resulting in advective 'plug flow'), then as for the mucosal crypt example above, when the last of the initial aggrecan molecules exited the cartilage, the time elapsed would be the aggrecan turnover time for the articular cartilage. So with this model in mind, one way to calculate aggrecan turnover is to divide the total amount of aggrecan in the articular cartilage ECM by the total rate of aggrecan synthesis by the articular cartilage ECM. This is not an uncommon approach to use.

However, it is likely while 'advective transport' of the aggrecan (i.e. movement of aggrecan by 'plug flow') does occur in adult tissue (see Sect. 3.4 and Figs. 3.6, 3.7, and 3.9), it is also observed that newly synthesized aggrecan molecules can diffuse among the existing aggrecan molecules (Buschmann et al. 1999, 2000). So aggrecan transport in the cartilage tissue is a mixture of both diffusive transport and advective transport. Examination of these figures suggests that in adult tissue, dominant transport modes may be different in territorial and interterritorial matrix. It is possible that advective transport may dominate diffusive transport through the contiguous territorial zones ('channelization') around neighboring chondrocytes in adult articular cartilage tissue, and vice versa for the interterritorial zones.

Aggrecan is synthesized throughout the depth of the cartilage, though the data suggests that considerably more aggrecan is synthesized in the deep zone rather than in the superficial zone—see Fig. 2.9. This makes sense from the viewpoint of aggrecan retention by the collagen network. Similar findings have been shown in human articular cartilage. For example, for a group of 16 normal and OA cartilage subjects with an average age of 64, Maroudas (1975) demonstrated a fourfold variation in the aggrecan production in the femoral condyle with depth. We note that the peak aggrecan production appears to have shifted in the elderly group from the osteochondral junction to the upper deep zone (Maroudas 1975).

However, the time required to replace half the aggrecan is sometimes calculated using a radioactive decay-type model, e.g. Maroudas (1975). This approach is most appropriate when aggrecan is randomly degraded (or produced) [e.g. by aggrecanases (or chondrocytes)], which is analogous to radioactive decay. The half-life of a molecule in a tissue is calculated by measuring the rate of removal of molecules over time (or the addition of sulfate 35), and fitting an exponential curve to the data (see for example Maroudas et al. 1998). This type of analysis is particularly suited to 'well-mixed' systems (e.g. the half-life of a reacting molecules is a fundamental performance characteristic in a 'mixed continuous batch reactor'). But clearly articular cartilage is not a well-mixed, continuous system. Rather articular cartilage is clearly variably mixed, continuous system that has well-defined heterogeneity (i.e. zones in the ECM) around chondrocytes.

Clearly both 'turnover time' and 'half-life' are good estimates only in so far as the in vivo processes of aggrecan replacement accords with the underlying assumptions of the model used to interpret the experimental data. Because aggrecan replacement in articular cartilage is actually a mixture of many different processes (e.g. fast and slow spatially distributed chemical reactions, including synthesis and degradation, in the pericellular and territorial, interterritorial zones, slow and rapid advective transport, slow and rapid diffusive transport), to interpret the rate of tissue turnover accurately requires a rather sophisticated reactive transport model. For this reason, neither the calculated times for 'aggrecan turnover' or 'aggrecan half-life,' which are both based on simple models of tissue turnover, actually represent the actual turnover of aggrecan in articular cartilage. The use of different methodologies for interpretation of experimental data at least partly explains some of the divergence in published estimates for aggrecan turnover (see below).

But is there a model of intermediate complexity, which can be usefully assist with at least a qualitative understanding for estimating tissue turnover more accurately than the simple methods described above? If the tissue is in homeostatic equilibrium, the rate of aggrecan replacement in the tissue may be estimated from either the rate of synthesis or the rate of degradation, because the two processes will be nearly the same at homeostatic equilibrium. Often the rate of sulfate incorporation is used to estimate the rate of production. But even at tissue equilibrium, one important confounding factor is that the newly synthesized molecule may go into one of two (or more) aggrecan 'pools.' For example, there may be an aggrecan pool with rapid turnover (e.g. in the pericellular matrix), and a pool with a moderate rate of turnover (e.g. in the territorial matrix) and another pool with much slower rate kinetics (e.g. in the interterritorial matrix) (see further discussion in Sect. 3.4).

Many tissues appear to have ECM components that have at least two pools—one pool with 'fast kinetics' and a second pool with 'slower kinetics.' For example, the rate of production of procollagen in Achilles tendon is about 1% per day, which is remarkable large, for if all of the procollagen were turned to tropocollagen and incorporated into tendon collagen fibrils, then an Achilles tendon could be completely replaced in just 3 months (Langberg et al. 2001). But in fact most of the procollagen is not incorporated into collagen fibrils—most of the procollagen is rapidly resorbed by the cells that produced them (fast turnover pool), with only a small fraction finding its way into the collagen fibrils (slow turnover pool) (Young et al. 2016), so the half-life for adult Achilles tendon collagen is measured in hundreds of years.

Much like tendon, experimental evidence supports the idea that aggrecan is quickly turned over in the pericellular region of chondrocytes, while the aggrecan in the territorial and interterritorial regions is turned over more slowly (Mok et al. 1994; Lohmander et al. 1973). It has been reported that the rate of production of aggrecan in newborn calf cartilage is about 2.9% (of total aggrecan content) per day (Sah et al. 1989), meaning for these conditions all the aggrecan could be replaced in as little as one month. But because of the existence of pools of aggrecan with faster and slower turnover, it has been estimated that the rate of synthesis of aggrecan by chondrocytes could be about 50% higher than the rate of turnover of aggrecan incorporated into the ECM (Maroudas et al. 1998).

For this reason, the half-life estimated using the 'total aggrecan produced' is often measured experimentally by the incorporation of the radionucleotide sulfate 35 (as this sulfate 35 is related to GAGs produced), while the half-life of the aggrecan actually incorporated in the territorial and interterritorial zones of the ECM may alternatively be measured using racemization. Racemization refers to the rate of transformation of *levo* (L) molecules into *dextro* (D) molecules of amino acids in the core protein of aggrecan, as racemization of amino acids requires longer timescales and so can measure processes that have longer timescales. In other words, short-term aggrecan turnover may have little interaction with and so effect on the kinetics of aggrecan pools with slower turnover.

It is known that the measured aggrecan turnover depends on many factors including the animal species, the joint type, the age of the animal and the depth beneath the cartilage surface (Maroudas 1975). In a careful experiment based on sulfate incorporation rates into GAGs articular cartilage, Maroudas et al. (1975) found the aggrecan turnover time depended on depth (see Table 4 in Maroudas 1975), but was about $(0.05/(3.3 \times 10^{-6} \times 24) = 630$ days (range 560–720 days) for cartilage from condyles of the human femur, and about 1600 days for the human head of femur. We note that the average age of the measured group was 64 years and included normal (6 people) and osteoarthritic cartilage (10 people). The time for half of aggrecan turnover was only 100 days in cartilage from a young rabbit head of femur, and 200 days in cartilage from an adult dog head of femur, which was similar to a human baby.

We show in Appendix E that estimated rates of aggrecan production by sulfate 35 incorporation (at 630 days; see previous paragraph) are very close to the estimated rate of aggrecan removal through the synovial fluid (at 640 days). Based on the rate of aggrecan removal via the synovial fluid, the estimated rate for aggrecan turnover in young adults is about 300 days (this estimate assumes an average aggrecan concentration in the knee cartilage for the young adult is about twice that for an old adult).

Later estimates of aggrecan half-life by Maroudas, based on racemization of aspartate in the proteoglycan of aggrecan, put the 'half-life' of aggrecan incorporated in the ECM of the head of femur at about 3.2 years (for the so-called large fraction) (Maroudas et al. 1998). This is similar (at least within 50%) to that estimated from the sulfate 35 incorporation experiment. But as explained above, racemization estimates give greater weight to the slower pool of aggrecan turnover in cartilage, than the fast pool. This difference in weighting given to the two aggrecan pools by the two different methods may explain the difference in estimated aggrecan turnover rates.

As mentioned above, it has been shown that the rate of synthesis of aggrecan, and so the turnover rate, varies strongly with age. For example, Verbruggen et al. showed that the rate of sulfate incorporation in a 20-year-old person is 2–4 times greater than for 45- to 60-year-olds (Verbruggen et al. 2000). Maroudas showed that the measured rate of sulfate 35 incorporation in a newborn femoral cartilage was five times the rate for a group of adults with an average age of about 65 years (Maroudas 1975). Bayliss et al. showed that cartilage from a 9-year-old person synthesized aggrecan much faster than cartilage from a 65 year old, and what is more that the aggrecan from old cartilage polymerizes into aggregate much more slowly than it does for young cartilage (Bayliss et al. 2000).

Following an examination of the literature, it becomes clear that the rate of aggrecan turnover decreases with age. The root causes explaining this are probably the endocrine system and decreased activity levels. The hormonal environment within the body changes markedly with age, which apart from directly influencing cell activity within the animal, also influences physical activity levels of the animal, and so the level of mechanical stimulation of the cartilage. Given that the rate of aggrecan turnover decreases with age, aggrecan has a longer time to be degraded in vivo by aggrecanases (which presumably would be particularly evident in the interterritorial zone, which probably has a much slower turnover time for aggrecan).

A decreased aggrecan turnover is reflected by a decrease in the ratio of the 'G3/G1 domains' found on cartilage aggrecan molecules in vivo as age increases. The G3 domains of the aggrecan molecule (see Fig. 2.13) are removed more quickly by aggrecanases than the G1 domains, and indeed the G1 domains may accumulate on 'aggrecan stubs' that remain attached to hyaluronan. Dudhia et al. report a 13-fold decrease in the G3 domains in the 65-year-old relative to that of a newborn (Dudhia et al. 1996). Taken together, this data suggests that age is a powerful influence on the rate of GAG and aggrecan synthesis. We also note that as age increases, not only the quantity, but the quality of the aggrecan decreases (e.g. the amount of glycosaminoglycan per aggrecan molecule decreases, and the sulfation becomes more variable and typically trends lower). Consequently beyond about 30 years of age, each aggrecan molecules secreted generate less osmotic pressure in the ECM, because it contains less glycosaminoglycan per aggrecan molecule (Ng et al. 2003).

It becomes apparent that there is no single number for aggrecan turnover in cartilage—it may be of the order of days to many months or years depending on: (i) the aggrecan's proximity to the cartilage surface (aggrecan at the cartilage surface is turned over in days, while the interterritorial matrix in the deep zone turnover may take many years), (ii) the joint type (knee cartilage aggrecan turnover is faster than hip cartilage aggrecan turnover), (iii) the age of the animal (in humans, there is probably around at least a fivefold difference between the very young and the very old cartilage turnover rates) and (iv) the species (small animals tend to have faster aggrecan turnover than larger animals, which may at least partly reflect differences in cartilage thickness).

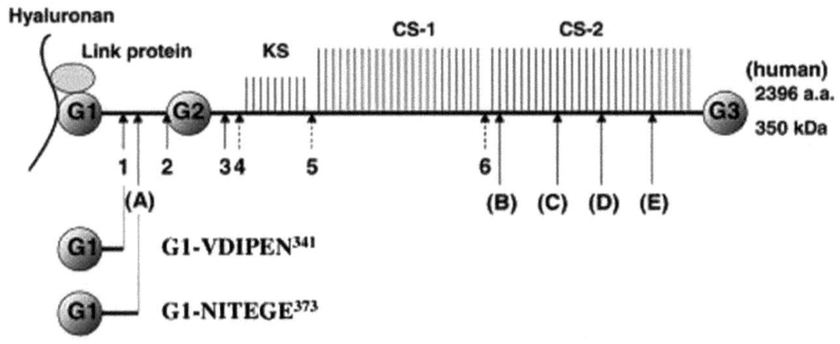

Sites cleaved by aggrecanases	
(A)	...RNITEGE^{373} ^{374}ARGSVIL...
(B)	(A) (G) (1480) (1481) (D) ...STASELE^{1545} ^{1546}GRGTIGI...
(C)	(1667) (1668) ...TTFKEEE^{1714} ^{1715}GLGSVEL...
(D)	(1771)(1772) ...QAPTAQE^{1819} ^{1820}AGEGPSG...
(E)	(V) (1871) (1872) ...EPTISQE^{1919} ^{1920}LGQRPPV...

Sites cleaved by MMPs	
1	(S) ...FVDIPEN^{341} ^{342}FFGVGGE...
2	...TAFTSED^{441} ^{442}LVVQVTA...
3	(V) (A) ...HAFCFRG^{666} ^{667}ISAVPSP...
4	not determined
5	not determined
6	not determined

Arthritis Research & Therapy

Fig. 2.13 Aggrecan cleaved by aggrecanases and matrix metalloproteinases (MMPs). Aggrecan core protein has three globular domains (G1, G2, and G3). The N-terminal G1 domain interacts with hyaluronan with the help of a link protein. G1-VDIPEN^{341} and G1-NITEGE^{373} are G1-bearing N-terminal products generated by MMPs and aggrecanases, respectively. Sites cleaved by aggrecanases are shown as (A)–(E), and sites cleaved by MMPs are shown as 1–6. The dotted arrows are sites predicted based on SDS-PAGE analysis of Little et al. (2002) and of Sandy and Verscharen (2001). *KS* keratan sulfate-rich region; *CS* chondroitin sulfate-rich region. Residues and numbering in parentheses indicate bovine sequences. From Nagase and Kashiwagi (2003)

We now turn to the difficult question as to the state of the aggrecan when it exits through the cartilage surface and finds its way into the synovial fluid. Does the aggrecan leave the cartilage intact, or is it first cut into pieces? After all, it is known that aggrecanases and MMPs may cut the aggrecan molecule at very specific places along the core protein in aggrecan (see Fig. 2.13).

The detection of intact and fragmented aggrecan in the synovial fluid is difficult because of the low concentrations present (aggrecan in synovial fluid is turned over quickly, while the flux of aggrecan into the synovial fluid is comparatively small), the large range of enzymatic reactions that can cleave aggrecan [e.g. at least theoretically, a fragment may be released by an aggrecanase and then be further 'processed' (i.e. degraded) by MMPs, or vice versa (Struglics et al. 2006)], and finally the potential for non-specificity of antibodies chosen to identify the characteristic epitopes of cleaved aggrecan. Interpreting the research data is further complicated by the use of different animal species (e.g. small rodents have different aggrecan dynamics relative to larger animals), some studies focus solely on disease states, while other studies focus on the development of antibodies and evolution of laboratory techniques (Struglics et al. 2009).

With this in mind, we consider the important paper by Larsson et al. (2009). For normal healthy cartilage and synovial fluid obtained from a group of 16 people with an average age of 27, they find that the median concentration of GAG in the synovial fluid is about 74 μg/ml (as estimated by alcian blue precipitation), while a median aggrecan concentration of about 120 μg/ml is estimated using an 1-F21 ELIZA and Western blot method (which detects the core protein of aggrecan in the keratin sulfate region close to the G2 domain) (see Table 3 in their paper and Fig. 2.14) (Larsson et al. 2009).

We can partially reconcile these two estimates using different measurement methods by first noting the aggrecan concentration in the synovial fluid as estimated from the GAG concentration in the synovial fluid. Assuming 75% of the aggrecan weight is GAG as determined by alcian blue, then 74 μg/ml is equivalent to about 100 μg of aggrecan per ml of synovial fluid, which is reasonably close to the estimated aggrecan concentration based on the 1-F21 ELIZA (Moller et al. 1994) and quantitative Western blot (120 μg/ml). If the aggrecan fragments were small and predominantly from the G3 end of aggrecan, then 1-F21 ELIZA would predict lower aggrecan concentrations than the alcian blue estimate, but this is not found to be the case. This strongly suggests that for normal cartilage, intact aggrecan is entering the synovial fluid intact.

ARGS fragments result from aggrecanase cleavage between 392 glutamine and 393 alanine in the interglobular domain of aggrecan (between G1 and G2 domains), releasing N-terminal 393 ARGS fragments (Sandy et al. 1992) (see site (A) Fig. 2.13). It has been found that this cleavage site is particularly deleterious to the mechanical properties of articular cartilage, much more so than cleavage along the core protein in the chondroitin sulfate 2 domain (Patwari et al. 2000). It has been shown that the preferred cleavage site for aggrecanases is mostly found toward the G3 end of aggrecan, in the chondroitin sulfate 2 domain (Little et al. 2007) (see cleavage sites for aggrecanases and MMPs shown in Fig. 2.13).

To avoid compromising the compressive stiffness of the cartilage, clearly a functional viewpoint would suggest that ARGS fragments should be minimized in healthy cartilage. Importantly, Larsson et al. (2009) found that the ARGS fragments are indeed in very low concentrations in the synovial fluid of normal cartilage (<1 picomolar). This means that ARGS aggrecanase fragments are not the normal way for aggrecan to exit healthy cartilage (i.e. again, the majority of the aggrecan is probably leaving cartilage and entering the synovial fluid intact).

Indeed, researchers looking at tendon have noticed that ARGS fragments from normal tendon are usually much higher in tendon than for normal cartilage (Rees et al. 2000). Higher rates of enzymatic-induced aggrecan turnover in tendon than cartilage explain the commonly observed side effects of general protease inhibitors such as Marimastat and Actinonin—both protease inhibitors lead to an accumulation of aggrecan in the tendon and painful tendinopathy (Waggett et al. 2004).

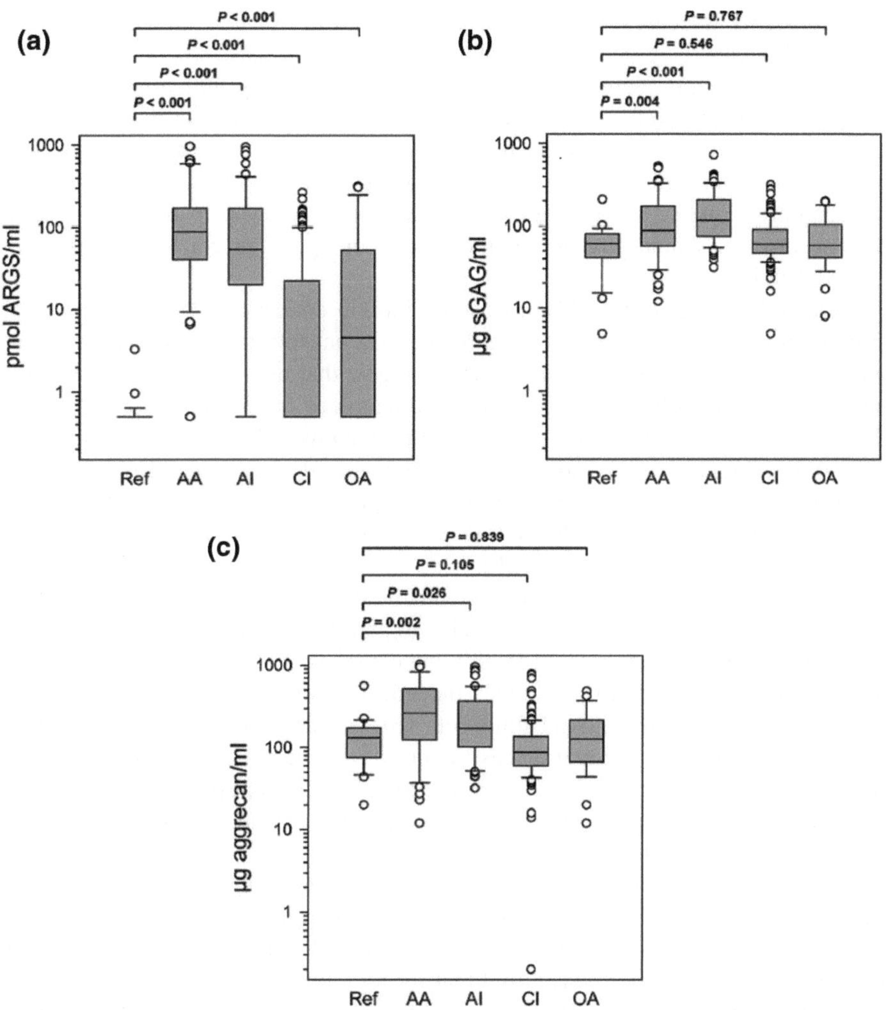

Fig. 2.14 Concentrations of **a** ARGS fragments, **b** sulfated glycosaminoglycan (sGAG), and **c** aggrecan in the study groups healthy knee reference, acute inflammatory arthritis (AA), acute knee injury (AI), chronic knee injury (CI), and knee osteoarthritis (OA). The boxes define the 25th and 75th percentiles with a line at the median, error bars defining the 10th and 90th percentiles and circles represents individual outliers. Note that in panel (**a**) the median level of the chronic injury group is the same as the lower limit of the box; 0.5 pmol ARGS/ml. After Bonferroni correction, *P* values below 0.013 are considered significant to retain the 0.05 overall significance level (Larsson et al. 2009)

However if the cartilage is acutely inflamed or injured, ARGS fragments in cartilage increase dramatically (×100), and the total rate of aggrecan release to the synovial fluid approximately doubles (×2) (Larsson et al. 2009). This is consistent with acute infection and injury raising cytokine expression (e.g. IL-1β and TNFα concentrations increase), resulting in the production of large amounts of aggrecanases that cleave aggrecan and release ARGS fragments.

ARGS fragments were elevated in chronic knee injury and osteoarthritis too (×10 or ×4 respectively), but are much less than for acute infection or injury. This is consistent with chronic knee injury and early stage osteoarthritis not being inflammatory in origin. Larsson et al. (2009) make the point that changes in ARGS concentrations in the synovial are much more sensitive way to measure cartilage disease processes than changes in total GAG or aggrecan concentrations in the synovial fluid.

Further Larsson et al. (2009) measured the G3 concentration in synovial fluid from both normal and diseased joints. If the G3 concentration in the synovial fluid were to rise in disease states, this would suggest that aggrecan being released by cleavage is sourced from newly synthesized aggrecan (as older aggrecan often has the G3 end already removed—see discussion above). However, Larsson et al. (2009) find the G3 concentration does not change, ruling out the ARGS fragments in disease states being generated by a pool of aggrecan with rapid kinetics. This strongly suggests that the ARGS fragments are being generated by cleavage of aggrecan bound to the existing ECM.

Given that aggrecan in normal synovial fluid of normal cartilage includes the keratin sulfate region of aggrecan (see above) and that ARGS fragments are in very low concentrations (see Fig. 2.14), this data implies that at least the majority of aggrecan is leaving the cartilage intact. Indeed, (Larsson et al. 2009) use the ratio of aggrecan fragments in synovial fluid detected as ARGS as 5% of the total (estimated using sGAG concentration) as the cut-off between normal and disease cartilage states (see Table 4 in Larsson et al. 2009). This is important because it means that in normal cartilage, aggrecanase cleavage is not the predominant way for aggrecan to be degraded.

This evidence is corroborated by a search for 'intact aggrecan' in the synovial fluid (Kobayashi et al. 2004). Defining 'intact aggrecan' to be aggrecan that is capable of binding HA, as well as having G1, G2 and keratin sulfate domains (Kobayashi et al. 2004), it was found that for a group of osteoarthritic patients, the mean intact aggrecan concentration was 30 μg/ml of synovial fluid. This demonstrates that even in osteoarthritic cartilage, a significant proportion (about one third) of the aggrecan exiting through the articular cartilage surface is actually intact aggrecan.

If enzymatic fragmentation is not the main way for aggrecan to exit normal healthy cartilage, that leaves either intact large aggregate molecules exiting the cartilage (which is probably less prominent given their enormous size) or mechanically driven fragmentation of aggregate into either smaller aggregate fragments and aggrecan (probably predominantly aggrecan). The stripping of aggrecan from HA by repeated cyclic loading is plausible, given that only non-covalent bonds attach aggrecan to HA, and this bonding is then strengthened by the non-covalently bound link protein.

Further, we note that pushing aggrecans together can result in entanglement of GAG chains, which combined with excessive deformation will presumably facilitate mechanical damage to GAG chains aggrecan molecules. Indeed, it has been shown experimentally that force-mediated detachment of aggrecan from the HA molecule is indeed possible, and that this bond represents the 'weakest link' in the 'molecular force chain' (Liu et al. 2006). Because detached aggrecan can reattach, it is possible that mechanical loading may detach aggrecan from HA, and as it makes its way toward the cartilage surface, it reattaches to another HA. How often such 'cycles of detachment and reattachment' may occur in vivo is unknown, but it is probable that this is a feature of normal aggrecan behavior in normal articular cartilage in vivo.

Based on the non-covalent bond energy of aggrecan to hyaluronan, we mention that Nap and Szleifer (2008) calculates that to avoid excessive repulsive forces between aggrecan molecules and disruption of the HA-aggrecan binding, the minimum spacing distance along hyaluronan between aggrecan molecules should be greater than 20.5 nM (at pH 5.6 and 50 disaccharides on the GAGs). However, the presence of link protein will reduce this estimated spacing further. Measured separation distances are usually between 15 and 40 nm, which is in agreement with this calculated estimate.

Taking this evidence into account, it appears as though aggrecan turnover in normal cartilage should probably not be estimated based on enzymatic cleavage of aggrecan (i.e. not using a simple 'radioactive decay'-type model), but rather aggrecan turnover should be estimated using an aggrecan production and transport model (which may include both diffusive and advective transport). It is likely that intact aggrecan molecules move down their chemical potential gradient and make their way to the cartilage surface. Newly synthesized or mechanically detached aggrecan molecules are most likely to be transported and may reattach to hyaluronan sites downgradient (how many times an aggrecan molecule may detach from hyaluronan and reattach to hyaluronan as it makes it way to the cartilage surface is unknown). For osteoarthritic cartilage, aggrecanase activity increases markedly, and then aggrecan turnover may be more closely approximated using a production and enzymatic cleavage model. Sophisticated computational models to more accurately estimate aggrecan turnover are clearly an open research direction.

2.4 Hyaluronic Acid: Homeostasis and Turnover of Extracellular Matrix

2.4.1 Introduction

Hyaluronic acid is a non-sulfated glycosaminoglycan molecule made of the repeating disaccharide unit—D-glucuronic acid and *N*-acetyl-glucosamine. This is contrasted with the repeating disaccharide in chondroitin sulfate, which is composed of D-glucuronic acid and *N*-acetyl-galactosamine. Hyaluronic acid (HA) is also known as 'hyaluronan,' as at physiological pH the counterion to ionized hyaluronic acid is most often a metal of a salt (e.g. the Na^+ ion) rather than the hydronium ion (H^+).

The sugars in the repeating disaccharide unit—D-glucuronic acid and *N*-acetyl-glucosamine—are covalently joined by alternating β1–3 and β1–4 glycosidic bonds. The number of repeating disaccharides can range up to 25,000 units or more, with a total length up to 10 μm or more (Necas et al. 2008; Matsumoto et al. 2009). The molecular weight of sugar molecule is about 400 Da (i.e. 800 Da per disaccharide unit), so the molecular weight of HA can range from 5000 Da up to an enormous 20 MDa. The HA in synovial fluid is typically about 3–4 MDa (i.e. about 5000 disaccharide units in length). Reviews on HA can be found in the following papers (Fraser et al. 1997; Necas et al. 2008; Bastow et al. 2008).

HA is synthesized at the cell membrane by the integral protein hyaluronan synthase (HAS), and directly secreted into the pericellular space (i.e. outside the chondrocyte). HA synthase has three isoforms that produce HA at different rates with different final lengths. The isoform expressed by chondrocytes may change with age, but the main HA synthase isoform expressed in normal cartilage is HAS2 (Moffatt et al. 2011). We note in passing that the production of HA via the cell membrane contrasts with the production of aggrecan and link protein, which are both synthesized in the endoplasmic reticulum and pass through the Golgi apparatus within chondrocytes. The Golgi and endoplasmic reticulum route allows for the synthesis of more complex molecules than direct production via the cell membrane into the ECM.

We know now that with high concentrations of aggrecan necessary to resist large compressive loads, it is extremely difficult to retain the aggrecan in cartilage tissue. Aggrecan maximizes its entropy in the aqueous solution by escaping from the articular cartilage tissue to the synovial fluid along its gradient of chemical potential. Various measures are employed by the cartilage tissue to help reduce the rate of aggrecan migration through the collagen network. As discussed above, these measures including the attraction of aggrecan to the type II collagen network (which is specially designed to retain aggrecan), the attachment of aggrecan to hyaluronic acid to form enormous molecules known as aggregate (e.g. 300 MDa), and the use of various 'cross-linking molecules' such as tenascin and fibulin (and quite possibly molecules such as asporin, CILP, and COMP) to form even larger 'super-aggregates' within cartilage.

2.4.2 Hyaluronic Acid Homeostasis

The homeostasis of HA is very poorly understood—few publications refer to this important topic directly. Indeed, it is reported that the regulation of HA appears to be 'multifaceted' (Recklies et al. 2001), whatever that may mean. The study of HA homeostasis is made more difficult by the fact that HA is not a protein, and so is not directly the product of a gene (so mRNA detection is not an option). Rather HA is a product of the integral membrane enzyme HA synthase. So while the biological effect may be related to the number of HAS molecules in the chondrocyte's cell membrane, the biological effect may also depend on direct regulation of the HAS enzyme itself.

Once HA is secreted directly into the pericellular space by HAS, it has a variety of options. HA may make its way into the territorial zone of the cartilage alone, or it may bind to secreted aggrecan and link protein in the pericellular space. HA (and aggrecan-link protein complex) may bind directly to the cell membrane. Indeed, there appears to be a competition between transport of molecules away from the cell membrane and competition for binding to CD44 receptors on the cell surface. The length and state of the HA appear to be important in this pathway competition (Knudson and Loeser 2002).

The main HA receptor on the cell membrane is CD44, but other 'hyadheren receptors' are ICAM-1 and RHAMM. Once bound to CD44, HA may be recycled through endocytosis, but there appears to be also a signaling pathway that helps to regulate the chondrocytes response to its local environment. In fact, it has been proposed that binding of HA to the cell surface helps regulate HA homeostasis (Knudson and Loeser 2002).

It has been shown that upon binding to CD44 on the cell surface, short HA and long HA bound to aggrecan have different signaling effects. Short HA upregulates a number of genes involved in matrix remodeling and turnover, while long HA tends to suppress these genes (Knudson and Loeser 2002; Ohno et al. 2005). A similar effect is found when CD44 is knocked down (using a microRNA). Perhaps unsurprisingly, a mechanism for HA-CD44 binding and then interacting with signal transduction pathways that influence the chondrocyte's general metabolism has also been proposed (Turley et al. 2002). It is noted that Plaas et al. (2011) present a convincing case that aggrecan-HA-CD44 signaling is central to maintaining normal TGFβ signaling, so that chondrocytes maintain a well-differentiated rounded chondrocytic phenotype. However, an increase in ADAMTS5 leads to a switch from normal TGFβ signaling (maintaining a chondrocytic expression profile) to abnormal TGFβ signaling (maintaining a fibroblast-like signaling profile with increased protease expression levels)—see Fig. 5.4. This theory places aggrecan-HA-CD44 signaling complexes in a central role for maintaining healthy cartilage.

What we are certain of is that a key mechanism to minimize aggrecan loss is the formation of the huge molecules known as aggregate (see schematic shown in Fig. 1 appearing in Roughley and Mort 2014). Consequently, we might expect that an absence of HA would lead to a reduction in aggrecan concentrations in cartilage tissue. While HA synthase knockouts are fatal, using conditional knockouts in Prx1-cre mice (which allows HA production to be turned-off some time after birth), it has been shown that failure to synthesize HA does in fact result in low aggrecan concentrations in the cartilage tissue (Matsumoto et al. 2009). This is reassuringly consistent with HA's role in retaining aggrecan in tissues, and particularly in cartilage where aggrecan concentrations are high (Bastow et al. 2008). In this context, it is interesting to again note that the mouse knockout for link protein (which strengthens the bonding of proteoglycans such a aggrecan, lumican, versican, brevican, and neurocan to HA) results in perinatal lethality (Czipri et al. 2003), while low concentrations of link protein result in 'genetic defects' of the heart (Wirrig et al. 2007). Though link protein may also be a signaling molecule, these results also suggest that the formation of proteoglycan aggregates is crucial for normal development of the organism.

Interestingly, it has been shown that despite its enormous size, HA can penetrate the superficial layer of cartilage [e.g. incredibly, HA penetrates about 300 μm into superficial zone cartilage over 48 h (Forsey et al. 2006)]. This suggests that one source of HA in the superficial zone of articular cartilage is in fact the synovial fluid itself (i.e. synovial membrane synoviocytes may supply the superficial cartilage zone with HA). In this case, it is possible that a flux of aggrecan and link protein from the deep zone into the superficial zone could meet a flux of hyaluronic acid in the superficial zone, forming aggregate in the superficial zone. How significant such *de novo* aggregate formation is for normal cartilage health is unknown.

HA may be degraded by hyaluronidases (Flannery et al. 1998). We also note in passing that it has been shown that 6 MDa hyaluronic acid injected into the synovial fluid is cleared from the rabbit knee joint with a half-life of about 13 h (Brown et al 1991). This clearance rate is probably somewhat slower than for the pig and human knee joints (i.e. $t_{1/2}$ is about 8 h).

2.5 Collagen: Homeostasis and Turnover of Extracellular Matrix

2.5.1 Introduction

We turn now to the collagen network in articular cartilage. We begin by describing the evolution of the collagen network structure in collagen, its functional role in retaining aggrecan and mechanisms for protection of the collagen network. We then move on to homeostatic mechanisms, and finally discuss collagen network turnover.

2.5.2 The Evolution of the Collagen Network

For adult articular cartilage, it is found that the collagen network is organized into characteristic zones, i.e. superficial zone, transition zone, deep zone, and calcified cartilage zone. The collagen network in the upper layers of the cartilage is often described as a 'Benninghoff structure' (Benninghoff 1925) (see original sketches by Benninghoff shown in Fig. 1 of Matcher 2015). This refers to collagen fibers being arranged in 'arcades' (see Fig. 2.15 and Fig. 6 in Matcher 2015), as first described by Benninghoff. The collagen fibers in the transitional zone 'bend over' to become

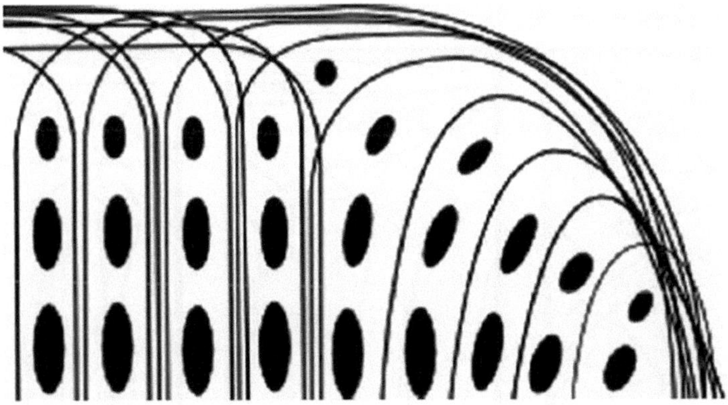

Fig. 2.15 Redrawn from Benninghoff who described this figure as: 'a schematic representation of the principal direction of the collagen fibrils in joint cartilage. The chondrons are represented as black elipses' (Nickien et al. 2013)

parallel with the cartilage surface in the superficial zone. Looking at Fig. 2.15, it is perhaps unsurprising that we later propose how the cartilage superficial zone develops through a mechanism involving collagen fibers 'bending over.'

The Benninghoff structure of adult cartilage is strikingly different to the more uniform collagen structure seen in very young, sexually immature animals (Rieppo et al. 2009; Hughes et al. 2005). In adult articular cartilage, not only do collagen fiber orientations change significantly with depth, creating well-defined zones in the cartilage (see Chap. 1), but the collagen fiber size increases significantly with depth (Changoor et al. 2011). In contrast, very young animals typically have relatively uniform collagen fibers, with the arrangement usually described as 'randomly' oriented (see Fig. 2.16). However, some of the collagen fibers appear to be oriented roughly parallel to the cartilage surface, probably reflecting the process of perichondrial cells laying down the initial cartilage template (van Turnhout et al. 2008, 2010). This collagen network structure found in the very young then 'evolves' as the animal grows in size and joint loads increase.

This 'evolution' of articular cartilage structure over time has been studied in detail in New Zealand white rabbits. Initially, articular cartilage is very thick, but this thickness decreases precipitously up to puberty (at between 3 and 4 months for the rabbit), mainly by cartilage anlagen 'resorption' associated with bone formation at the base of the cartilage (see Fig. 3.10) (Hunziker et al. 2007). However, the decreasing cartilage thickness is accompanied by cell proliferation and interstitial growth in articular cartilage near the cartilage surface. The cartilage thickness changes by several simultaneous processes (including resorption, bone formation, proliferation and surface growth, and tissue straining due to joint loading). These processes together 'shape' the articular cartilage to a coarse-scale congruence of opposing cartilage surfaces that determines its final thickness at around 8 months. Meanwhile, it is likely that 'compressive reorientation' of collagen fibers mainly explains how superficial zone cartilage develops post-six months in the rabbit, i.e. it

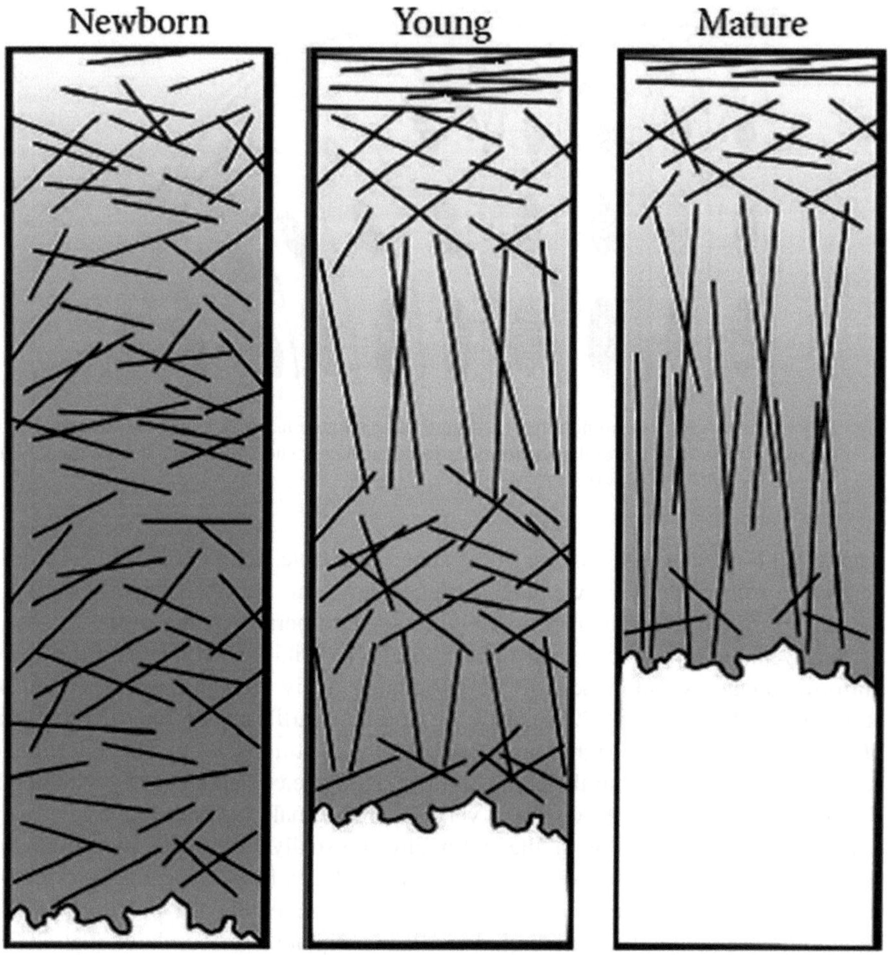

Fig. 2.16 Presentation of the collagen network organization in maturing articular cartilage based on articular cartilage from rabbit, pig, and sheep. When maturing, a non-organized collagen fibril network slowly forms into the traditional described zonations (superficial, transitional, and deep), and a Benninghoff-type arcade structure may develop in some animals (including humans). At the same time, cartilage thickness is decreased (Julkunen et al. 2013). See similar remodeling process reported for cartilage construct implanted in Gottingen minipigs (Paetzold et al. 2012)

is likely that the process of compressive reorientation shapes the finer-scale congruence of opposing cartilage surfaces (see later discussion).

For humans, the proportions of each cartilage zone relative to the total thickness for normal human femur hip and knee cartilage are reported as being 7 ± 1% for the superficial zone, 19 ± 7% for the transitional (middle) zone, and 75 ± 9% for the deep (radial) zone (Changoor et al. 2011). However, the fractions of cartilage thickness in each zone change dramatically for 'degraded' cartilage (note that the meaning of ‘degraded’ in this context is defined in Fig. 1 of Changoor et al. 2011). It is found that the relative proportions of each zone in cartilage classified as ‘degraded’ are: 10 ± 3% for the superficial zone, 47 ± 10% for the transitional zone, and 46 ± 10% for the deep (i.e. radial) zone. It is clearly apparent that there is a substantial expansion of the transitional zone at the expense of the deep zone in degraded cartilage relative to normal cartilage. These macroscale observations of structural changes in the collagen network of articular cartilage again attest to the dynamic shaping of the collagen network in articular cartilage throughout life.

For normal human cartilage, collagen fibril diameter increases with depth (Gottardi et al. 2016). ‘Prototypical’ collagen fibrils are 20 nm in diameter (see Fig. 1.5). Collagen fibrils increase in size as more prototypical fibers are wound together (with a pitch of at least 400 nm). So four prototypical fibrils wound together are about 50 nm in diameter, while seven prototypical fibrils wound together are about 90 nm in diameter (Gottardi et al. 2016). For knee cartilage, collagen network fibers in the superficial zone are about 50 nm in diameter, while

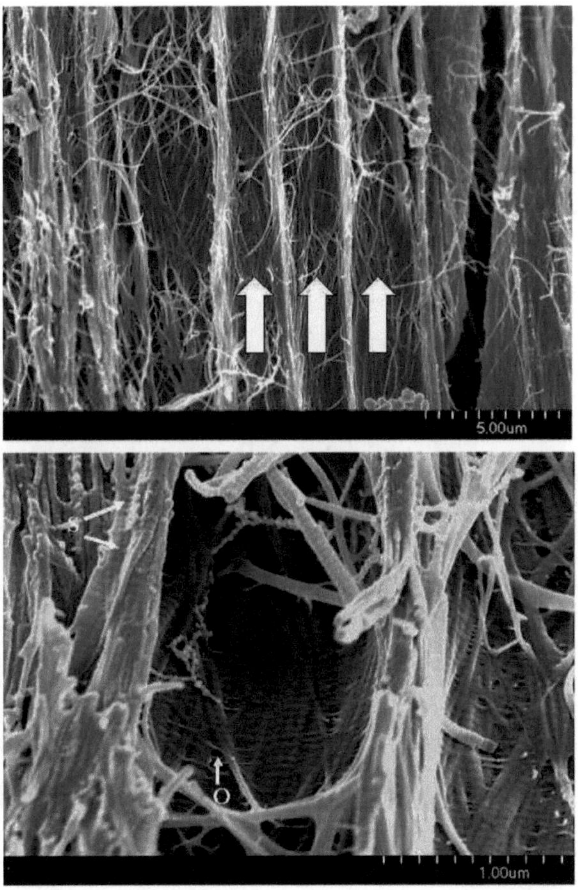

Fig. 2.17 The image shows the base of the radial zone, at the interface with the calcified zone in the rabbit (top image: white arrows highlight three parallel tubes). Bottom image: bases of tubular structures are seen, including orthogonally arranged fibrils apparently binding the 30-nm-diameter collagen fibrils lining the tubules’ walls (arrows). More than a few microns from the interface the collagen structure was entirely disrupted (Gwynn et al. 2002)

collagen fibrils in the deep zone are about 70 nm in diameter. For hip cartilage, collagen fibers in the superficial zone are about 30 nm in diameter, while collagen fibrils in the deep zone are about 90 nm in diameter (Gottardi et al. 2016).

It is noted that freeze fracture reveals significant species differences in the collagen arrangements within the deep zone (Kaab et al. 1998). Surprisingly, the collagen in the deep zone of some animals (e.g. mouse, rat, rabbit, sheep) is arranged in 'tube-like structures' of collagen, which contain columns of proliferating chondrocytes (ap Gwynn et al. 2002; Hughes et al. 2005; Kaab et al. 1998). In other animals (e.g. dogs, pigs, and humans), the deep collagen is arranged in sheet-like structures (Kaab et al. 1998). For cows, there is some disagreement as to the architecture (Jeffery et al. 1991; Kaab et al. 1998)—which may be partly explained by these authors examining different joints.

In the mouse, the chondrocytes are orientated into columns of cells, surrounded by territorial matrix that appears to wrap itself around the pericellular capsules (see Fig. 1 of Hughes et al. 2005). A similar structure is reported in the rabbit (see Fig. 2.17 showing the very well-developed collagen tubes). The collagen tubes presumably facilitate chondrocyte proliferation and fast migration along a straight, unimpeded, path from the deep zone directly to the transitional zone (ap Gwynn et al. 2002) (Fig. 2.18).

There are a number of images in the literature suggesting a 'sheet-like structure' for collagen in the deep zone of articular cartilage in humans (Kaab et al. 1998; Clarke 1971; Marzo and Gurske-DePerio 2009; Matcher 2015), but perhaps the most vivid images of sheets—or 'kelp-like ribbons'—of collagen in the deep zone rolling into arcades are to be found in Li et al. (1999) (see Fig. 2.19).

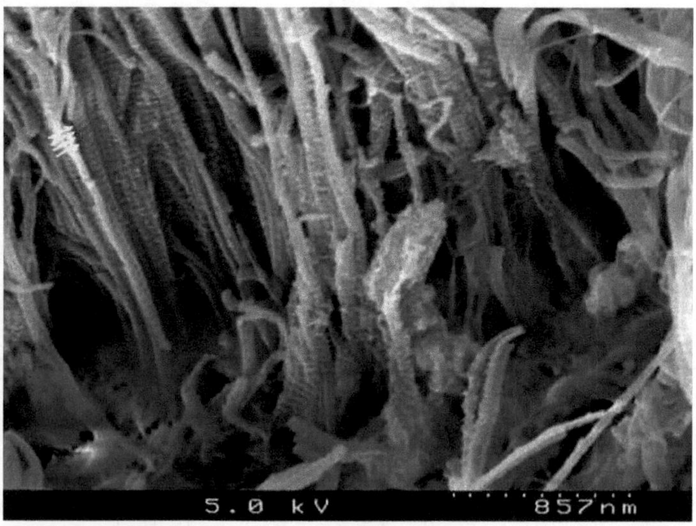

Fig. 2.18 In the rabbit, well-developed collagen tubes are apparent. Image shows close up of collagen tubes with inner circumferential collagen lining (Gwynn et al. 2002)

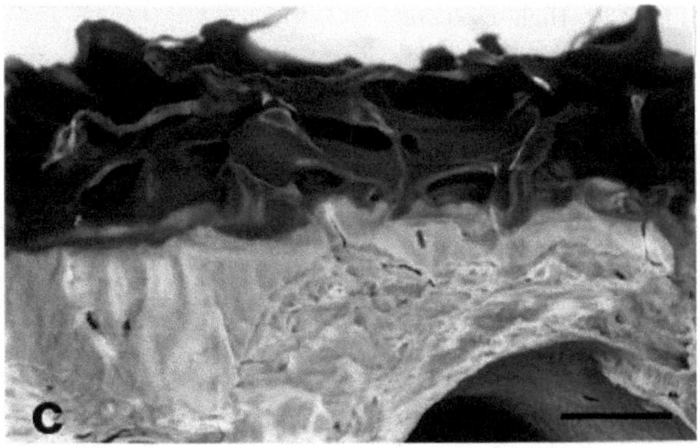

Fig. 2.19 Scanning electron micrograph of normal 76-year-old femoral head cartilage (imaged in backscatter mode), showing details of osteochondral junction (Li et al. 1999). Scale Bar 100 μm. Note the ribbon-like structures defining collagen sheets

While tube-like collagen structures have not been reported explicitly in human cartilage (as observed in other species) columns of chondrocytes are reported in the deep zone of articular cartilage (Hunziker et al. 2002). Based on images of newly synthesized aggrecan being transported en masse creating 'plumes' through the territorial matrix, it seems likely that these plumes represent preferential pathways (i.e. channelization) along which ECM components and chondrocytes may be transported (see for example Figs. 3.6 and 3.7). If aggrecan is transported en masse, then it seems very likely that other molecules would be transported along with the aggrecan (e.g. type II and type III collagen mers, hyaluronan, aggrecan-link protein, COMP, tenascin, fibulin, and others).

In the deep zone of human cartilage, it is plausible that with varying activity over a life span, these columns of proliferating cells ascend 'vertically' between parallel sheets or ribbons of collagen, migrating between these constraints in the direction of their local aggrecan gradient. These 'channels' probably represent preferred low resistance pathways to transport any new chondrocytes (and ECM materials) through the collagen network (Poole 1997; Changoor et al. 2011). This would suggest while humans do not have well-defined collagen tubes through the cartilage, human articular cartilage probably does have preferential cell migration and ECM transport pathways through the cartilage (e.g. perhaps these 'channels' represent contiguous territorial matrix that surrounds nearby chondrocytes). Presumably, these preferential transport routes serve a similar purpose to the well-defined cartilage tubes observed in other species (e.g. see Fig. 2.17 showing collagen arranged in tube structures in rabbit cartilage) (Fig. 2.20).

It is plausible that the significant differences in the deep zone collagen arrangement between species have a functional purpose. Indeed, it is likely that animals have different collagen structures depending on joint type and function. In this context, we propose that joints with well-developed tube-like collagen structures in the deep zone of their cartilage tend to have their cartilage subjected to less fatigue-induced damage during normal physiological use, relative to those joints with sheet-like collagen structures in the deep zone, which generally subject their cartilage to much more demanding physiological loadings. Those joints suffering a

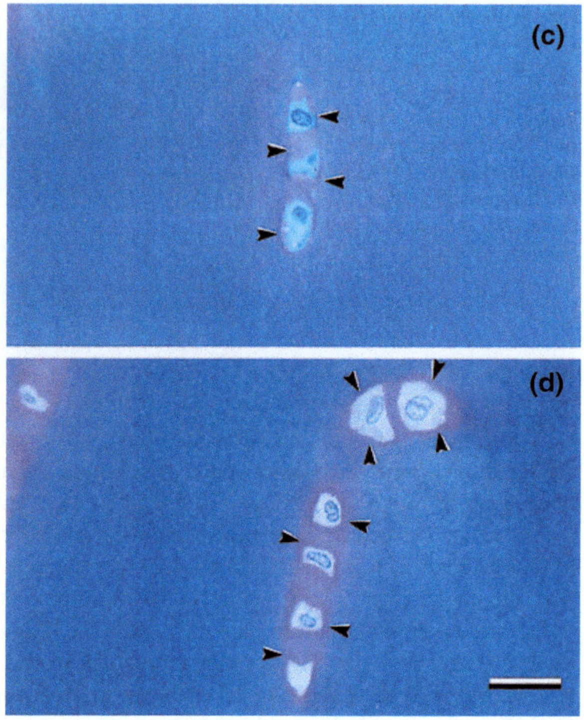

Fig. 2.20 High-resolution light micrographs of the deep (i.e. radial) zone in normal human articular cartilage. Images **c** and **d** illustrate typical chondrocyte clusters (chondrons) within the upper **c** and lower **d** halves of the deep zone. As must be apparent from their scant distribution, these cell clusters control huge matrix domains. The arrowheads in **c** and **d** mark the boundary between the pericellular/territorial matrix compartment and the vast interterritorial one. Semithin (1 μm thick) section of Epon-embedded tissue stained with Toluidine Blue 0. Scale bar = 20 μm (Hunziker et al. 2002)

large numbers of cyclic loads, with high normal and occasional high shear and/or twisting loads subject the collagen network in their cartilage to internal deformations that require sheets or ribbons of thick collagen fibrils to form, so as to adequately resist the imposed deformations. Indeed as we later discuss, normal joint forces are probably responsible for forming the sheets and ribbons to resist the imposed forces (see Sect. 3.3).

For example, knee joints in larger animals capable of endurance running (e.g. dogs and humans) subject their articular cartilage to a very demanding type of cyclic loading. Indeed, it is reported by Arokoski et al. (1996) that after long-distance running training of Beagle dogs (up to 40 km/day for 15 weeks), the superficial collagen network of articular cartilage was either deteriorated or the fibrils were reoriented due to an altered loading pattern. Kiraly et al. reports that many hours of cyclic loading (up to 20 h at 4.1 MPa and 0.5 Hz in unconfined compression) causes '*packing or condensation of the tissue*' (Kiraly et al. 1998). Kiraly et al. also report the thickness and birefringence of superficial, transitional, and deep zones changes (Kiraly et al. 1998). Indeed, both thickness and birefringence of the superficial zone increased (*e.g.* '*data suggest that the loaded superficial cartilage showed higher orientation of collagen fibrils or that the fibrils were packed more in a more parallel manner and closer to each other*'), while the converse holds for the deep zone) (Kiraly et al. 1998). In contrast, knee joints of small animals that usually run in short bursts (e.g. mouse and rabbit) have mechanical conditions that are much less demanding on the collagen network in the cartilage. However, small animals experience significant mechanical loads during rapid accelerations and so chondrocyte apoptosis is likely to be prominent, particularly in the superficial zone cartilage. Presumably this creates the need to efficiently replace apoptotic cells at the cartilage surface by proliferating cells more distant from the cartilage surface. Deeper proliferating cells experience smaller cell strains and so are protected by virtue of the distance from the cartilage surface. Small animals replace superficial zone chondrocytes by migration of daughter chondrocytes along collagen tubes running from the deep zone to the upper transitional zone.

Much like the 'tubed structures' in rodents and sheep, for dogs, pigs, and humans, sheet-like collagen structures also rise vertically from the subchondral bone, but turn over in the transitional zone to form arcade-like structures (Matcher 2015). This collagen structure firmly anchors the transitional zone collagen network directly to the subchondral bone (Goodwin et al. 2000). In dogs, pigs, and humans, the chondrocytes still proliferate and form preferred (tube-like) tracks through the interwoven collagen along contiguous territorial matrix, but their migration pathway is constrained by the collagen sheets in the interterritorial matrix of the deep zone (Goodwin et al. 2000). Unlike the collagen tubes seen in rodents and sheep, we speculate these preferred chondrocyte pathways in dogs, pigs, and humans are probably variable over time. It is possible that this variability in pathways from the deep zone to the transitional zone and from the transitional zone to the superficial zone enables chondrocytes to be directed to where collagen network damage is greatest (i.e. chondrocytes migrate through collagen damage).

The collagen network structures that develop in articular cartilage have a profound effect on the directional strength of articular cartilage. Here we note that if the cartilage is subjected to a severe non-physiological loading, the collagen network may well tear apart in very unusual ways. This is because the collagen fiber orientations have been shaped to most effectively resist physiological loadings, rather than applied non-physiologic loadings.

For example, if a so-called channel indenter is placed on the surface of the cartilage and then loaded (see Fig. 2.21a), it subjects the cartilage to a severe non-physiological loading. It is observed that upon load application, the cartilage bulges into the channel. In this case, the rapid loading of the channel indenter displaces the cartilage pore water centrally, creating a growing collection of pore water under extreme pressure immediately beneath the channel of the indenter. The pressure gradient between the highly pressurized water that collects in the cartilage and the low pressure at the cartilage surface causes the collagen network to bulge upward into the channel. As more load is applied, the fluid pressure gradient literally tears the collagen fiber network apart (revealing the 'horizontally' oriented interdigitating fibers in the superficial zone). Tearing at the interface between the superficial zone and the transitional zone is caused by the tensile strength of the collagen network at this location being exceeded by increasing excess pore fluid pressure gradients (see Fig. 2.21b).

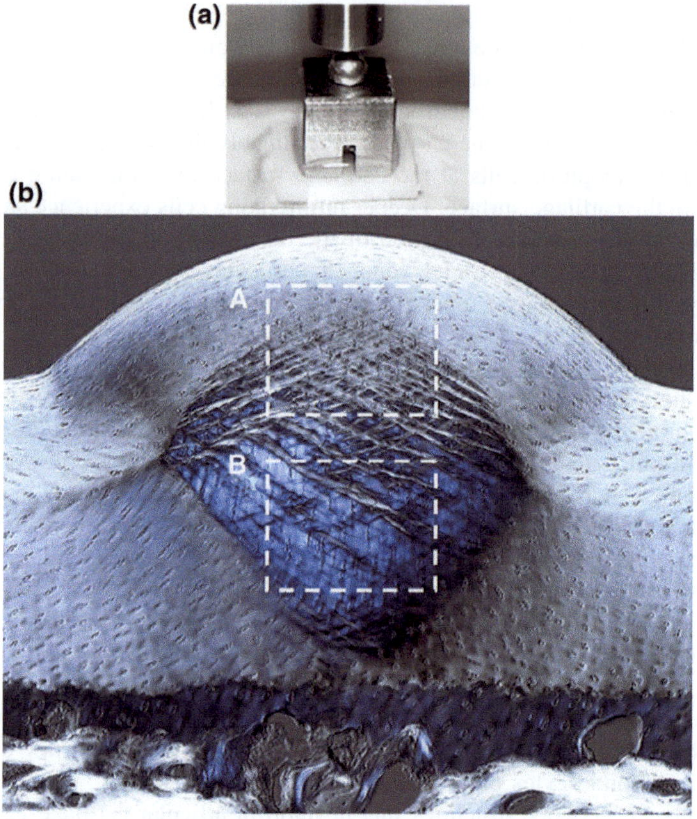

Fig. 2.21 **a** Channel indentor pressed into the surface of cartilage (Bevill et al. 2010). **b** High magnification differential interference contrast (DIC) optical image of typical channel relief zone cartilage microstructure for a surface-intact sample (note that cartilage bulges into channel as cartilage is loaded). Beneath the strain-limiting superficial layer, the periodicity and thickness of oblique shear bands within the cartilage bulge vary with depth. Near the STZ interface (region A), the bands are fine and relatively dense, but band thickness and spacing increase in the deep zone (region B) (*Note* This is a severe non-physiological loading)

2.5.3 Collagen Protection

On functional grounds, it is generally supposed that the evolution of collagen network structure in adult cartilage facilitates greater resistance to tissue deformation caused by physiological loadings in the adult animal (van Turnhout et al. 2011). Indeed, this is perfectly reasonable. However, the mechanisms by which such remodeling of the tissue occurs are not widely discussed in the literature (Changoor et al. 2011), yet plausible and helpful explanations for collagen network remodeling are now available.

For example, it has been shown experimentally that when collagen fibrils are not loaded (and not stretched), they are more likely to be degraded by proteases than when they are loaded (and stretched) (Nabeshima et al. 1996; Wyatt et al. 2009; Hadi et al. 2012; Flynn et al. 2010). Indeed, it has been shown that sufficient collagen stretch can even prevent collagen degradation (Flynn et al. 2013). Presumably this effect occurs because the stereochemistry of the surface molecules comprising the strained collagen fiber does not facilitate either recognition by MMPs, or facilitate their migration along the collagen fibril, or prevent 'loop formation' next to the MMP, or hinder unwinding of the collagen triple helix, or hinder interaction between the reactive sites of enzymes and the chain of amino acids in the collagen fibrils (Rosenblum et al. 2010b; Collier et al. 2011; Lauer-Fields et al. 2001; Flynn et al. 2010, 2013), or some combination of these factors. Probably for all these reasons, mechanical-induced strain protects collagen fibrils from enzymatic degradation.

In fact, it has been shown experimentally that as little as 0.1% strain can reduce collagenase activity on collagen fibrils, and as little as 1.2% strain can prevent individual collagen fibril degradation by collagenase (Flynn et al. 2013). Testing single collagen fibrils stretched between two glass microneedles (see Fig. 2.22) has demonstrated an 'exponentially decreasing relationship' between rates of degradation by collagenase with strain in the collagen fibril (Fig. 2.23).

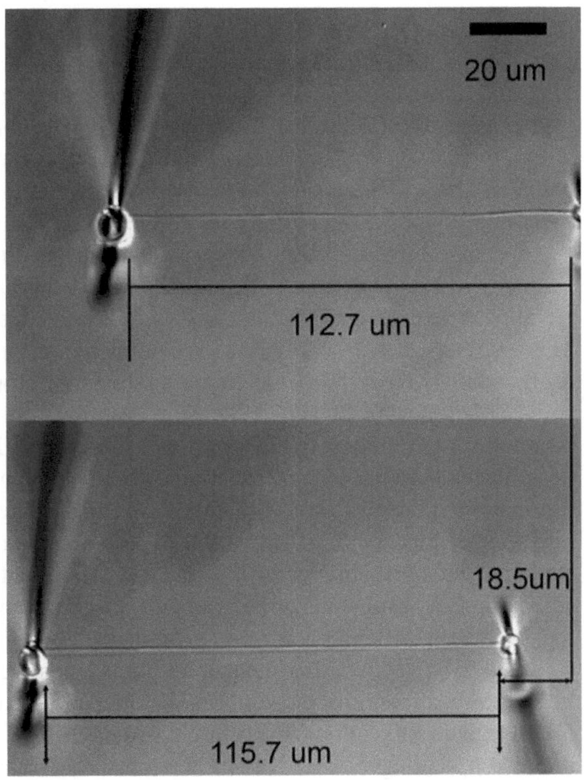

Fig. 2.22 DIC image sequence of mechanical testing. An isolated native type I collagen fibril from bovine sclera is epoxied to calibrated (57 ± 8 nN/μm) glass microneedles, submerged in buffer, and mechanically loaded. Top Gauge length (zero-load). Bottom High-load of 1054 ± 149 nN produces 2.6 ± 0.3% strain

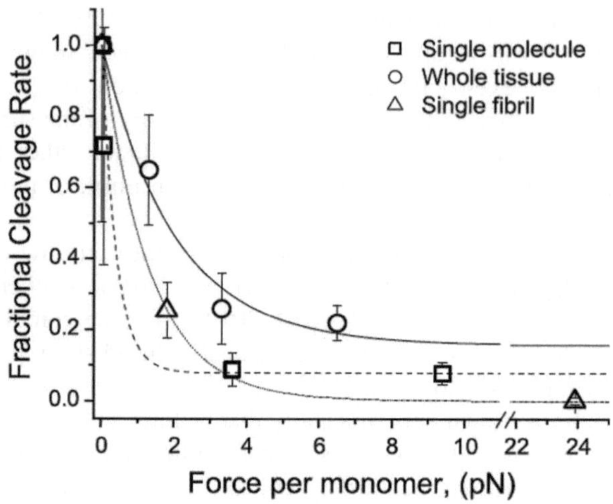

Fig. 2.23 Normalized fractional cutting rate versus applied force per monomer for three size scales. Data is normalized against maximum cutting rate, from the zero-load experimental series, and fit to a single exponential curve (Flynn et al. 2013). *Square* (Camp et al. 2011), *circle* (Zareian et al. 2010). Note that collagenase degradation of a single collagen fiber is close to zero from about 8 pN stress, while collagenase activity reduces to about 20% in whole tissue and 10% for procollagen

Comments (Flynn et al. 2013): '*[single fibril] strain levels are much lower than those found to reduce enzymatic susceptibility at the whole-tissue scale (4–8%). This difference is explained well by findings that molecular sliding and network realignment translates 5% tissue-level strain to only 0.5–1.5% molecular-level strain, mainly due to fibril network realignment*' (Flynn et al. 2013). In other words, collagen fibers in a collagen network may rotate as they stretch, thereby facilitating extension of the whole tissue with comparatively much less collagen fiber strain.

Of course how well tissue-level strain translates to fibril strain depends on the architecture of a particular collagen network in the tissue. For example, for human Achilles tendon, the fibrils are well approximated as bundles of single fibrils in primary fibers that probably run for centimeters in the same direction with little or no branching, so in this case tissue strain more closely approximates fibril level strain (Mehdizadeh et al. 2017). However for 'rat tail' collagen, the difference between rat tail tendon tissue strain and fibril strain is much greater. Similarly for the collagen network in cartilage, the relationship between cartilage tissue strain and fibril strain depends on the location of the collagen network sample and the direction of loading (e.g. we would expect a close approximation between tissue strain in the fibril directions in both the superficial zone or deep zone, but the association is probably more variable in the transition zone, where network realignment with loading is more prominent).

Practically, these experimental findings mean that collagen fibril retention is biased toward retaining those fibrils that provide greatest resistance to deformation for any given physiological loading regime. On the other hand, it means collagen removal by collagenases is biased toward those collagen fibers not carrying much load. Given sufficient time, this 'process bias' by collagenases exerts a noticeable 'selective pressure' on the total population of collagen fibrils. After some time, only those fibrils carrying significant load remain. Newly formed 'randomly' orientated collagen fibrils joining the existing fibril population are subject to the same selective pressures by collagenases. By this mechanism, the collagen network is gradually 'sculpted' into a preferred directional alignment of collagen fibers that optimally resists the imposed physiological loads (Hadi et al. 2012; Flynn et al. 2013). While not implementing a computational model with fibril degradation and formation (so mechanistically modeling this collagen fibril selection process), Wilson et al. did implement a simplified computational model in which collagen fibers were 'selected' to simply rotate toward the principal strain directions (Wilson et al. 2006). Interestingly and importantly, this algorithm resulted in a Benninghoff-type structure like those shown in Fig. 6 of Goodwin et al. (2004) and Fig. 2.15.

This sculpting effect (by proteases and load) is particular noticeable when the load direction and magnitude are constant. Importantly, we note that a constant magnitude and directed loading does indeed occur in all pre-stressed (biocomposite) materials, including cartilage. In cartilage, the 'radially oriented' collagen in the deep zone clearly resists aggrecan swelling in the radial direction, i.e. momentum exchange between aggrecan and collagen generates shear forces between the aggregate-aggrecan and collagen fibers resulting in large constant tensile forces in the deep zone collagen fibers (Smith et al. 2016), which is protective of the collagen network.

There are a variety of loadings that may occur on the surface of cartilage. A concentrated physiological contact load applied to the cartilage surface places the cartilage surface (as shown in Fig. 2.24a, b) in tension in the direction tangential to the surface (in the plane of the superficial zone collagen), but the cartilage surface is in compression in the direction normal to the cartilage surface. As evidenced by Fig. 2.24b–d, the compression strain causes buckling (i.e. crimping or 'chevron patterning') in the collagen fibers in the direction normal to the interface between the surficial collagen network and the underlying collagen network. The interface between these two regions is a zone of intense collagen fiber buckling and bending.

Based on the formerly described mechanism for collagen fiber reorientation, one would predict that given normal repetitive loading, collagen fiber orientations would over time become predominantly parallel to the surface in the superficial zone. In this orientation, the collagen fibers will resist in-plane tension induced by any shear stress on the surface of the cartilage due to normal physiological loads. Observations on the tangential orientation of collagen fibers in the superficial zone are consistent with tangential tensile stresses in this region (see Fig. 2.25).

While biased collagen network degradation by collagenases is probably one of the key mechanisms reshaping the collagen network in cartilage, it is almost certainly not the only mechanism reshaping collagen fibril architecture in cartilage. Indeed, there are likely to be several additional important mechanisms orienting the collagen to be parallel with the cartilage surface.

Because cartilage tissue is a saturated poroelastic biocomposite material, a change in surface loading induces excess pore pressure gradients through the tissue. These pressure gradients in turn induce fluid flow through cartilage tissue. For

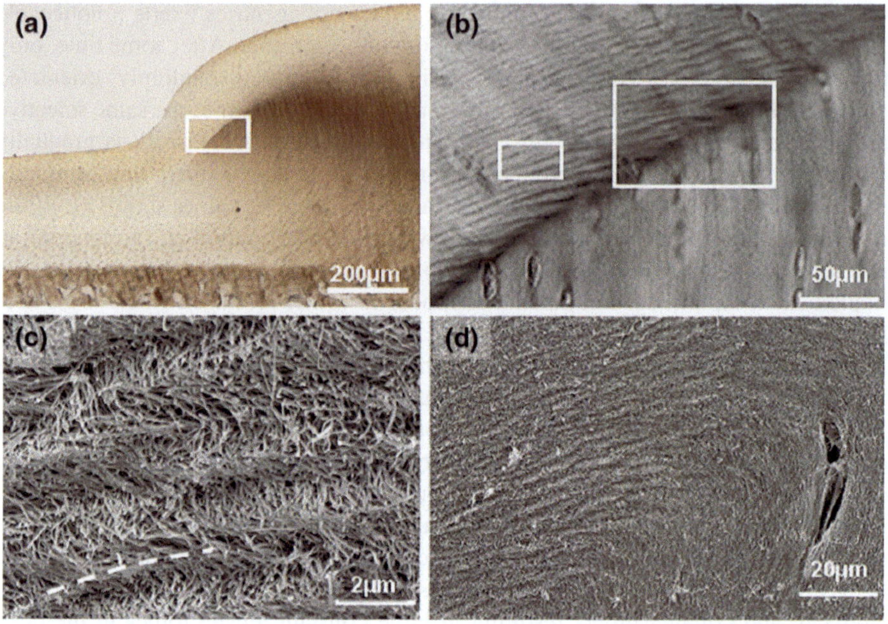

Fig. 2.24 Differential interference contrast (DIC) optical images of the deformed matrix near the edge of the indenter of intact samples tested in 0.15 M saline. The boxed region in **a** highlights the distinct shear boundary referred to in previous studies as the 'chevron discontinuity' 9, 10. **b** DIC image at higher magnification of the boxed region shown in **a**. Note the intense creasing in the upper deformed cartilage layer. **c** SEM image of the smaller boxed region in **b**. The dotted line highlights a force line of action aligned with the creases shown in **b** and reveals an in-phase fibrillar aggregation. **d** SEM image of the larger boxed region shown in **b**. Note the intense fibrillar aggregation, and its dissipation near the shear boundary (Bevill et al. 2010)

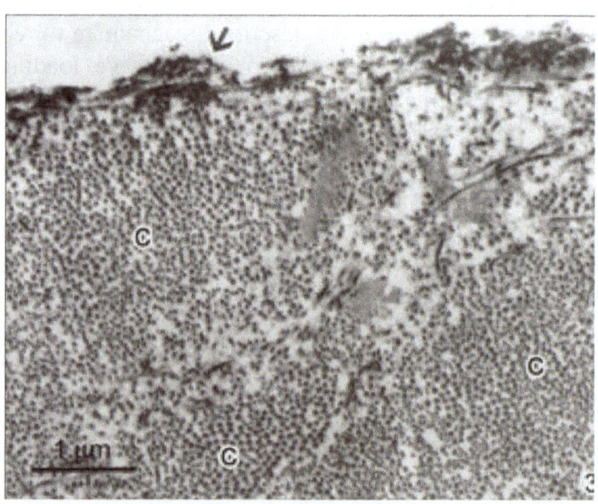

Fig. 2.25 Transmission electron microscope image of articular cartilage surface in a young dog. Chondrosynovial membrane (i.e. lamina splendens) (→) with fine aperiodic fibrils. Bundles of collagen fibers (c) matted together; image magnification ×14,000. Scale bar 1 µm (Horky and Tichy 2004)

transient normal and shear loadings tracking across the cartilage surface (as occurs in the diarthroidal joints as they flex while loaded), both pressure gradients and fluid flow within cartilage are dynamic. Normal physiological loading induces fluid velocity fields that often have fluid velocities that are mainly directed normal to the surface of the cartilage (because the pressure in the synovial fluid outside the contact zone is effectively zero, while being positive (or negative—upon unloading) within the cartilage. This situation creates a large alternating pressure gradient normal to the surface. It has been shown experimentally that fibroblasts producing ECM usually deposit collagen *normal* to the direction of fluid flow (Swartz and Fleury 2007; Ng and Swartz 2003). In other words, fluid flow back and forth across the cartilage surface may provide cues to resident chondrocytes that helps to orientate the deposition of collagen fibers in a direction tangential to the cartilage surface (as is observed in the superficial zone). Fibroblasts orientate themselves normal to this flow. If chondrocytes were to act the same way as fibroblasts, then in zones of substantial flow (i.e. the superficial layers) chondrocytes would also orientate themselves parallel to the cartilage surface. Indeed, this is what is observed.

The superficial zone is mainly dense collagen type II fibers lying parallel to the cartilage surface (Clarke 1971, 1974; Kobayashi et al. 1996). Importantly, the local fiber density in the superficial zone is high, which plays an important functional role of providing substantial resistance to aggrecan movement out of the cartilage close to the cartilage surface. This resistance to aggrecan movement supports an abrupt change in aggrecan concentration near the cartilage surface. Probably, an increased resistance to aggrecan movement close to the cartilage surface is necessary because when the tissue is compressed and pore fluid drainage occurs, aggrecan gradients to the cartilage surface become much larger than at equilibrium under no loading, providing a very powerful driving gradient (the gradient approaches a 'step-change' in aggrecan concentration at the cartilage surface). This large aggrecan concentration gradient promotes efflux of aggrecan from cartilage.

The resistance to aggrecan movement in the superficial zone, increasing toward the cartilage surface, is aided by the presence of a type III collagen network attached to the type II collagen network. The type III collagen network, with its fine fibrils attached to the exterior surface of type II collagen fibrils, is particularly well developed in the superficial zone of the cartilage (Wu et al. 2010), but is also found in the transitional zone of articular cartilage. The concentration distribution of *N-terminal propeptide of collagen type III* (PIIINP), indicative of type III collagen distribution in knee cartilage, is shown in Fig. 3 at http://www.biology-online.org/articles/articular_cartilage_collagen_irreplaceable/figures.html). The monoclonal antibody 4G9 specifically detects PIIINP. This antibody is seen to have high concentrations near the cartilage surface. Along with the dense collagen type II network toward the cartilage surface, the type III network further increases collagen density and reduces pore opening size, helping to retard the efflux of aggrecan through the cartilage surface.

Type III collagen mers also forms part of type IX collagen molecule (see https://chempolymerproject.wikispaces.com/Collagen-D-apml). The type IX collagen molecule has a positive charges site (containing Zn^{2+}) on the NC4 terminal domain that binds negatively charged molecules (Leppanen et al. 2007) like COMP (isoelectric point of COMP is pH 4.36 http://www.phosphosite.org/proteinAction.do?id=21867&showAllSites=true) and aggrecan (the isoelectric point of GAGs such as chondroitin sulfate is very low compared to the tissue pH, meaning they are negatively charged) (Leppanen et al. 2007). The NC4 terminal domain also binds matrilins, osteoadherin, fibromodulin, heparin, and other molecules ensuring the formation of a macromolecular network (Kalchishkova et al. 2011). Type IX collagen molecule also has an NC3 domain that may bind GAGs, as well as positively charged arginines along the Col3 domain of type IX collagen, which can interact electrostatically with aggrecan (see discussion Sect. 1.4). Clearly, binding between aggrecan, matrilins, osteoadherin, COMP, and other molecules to the collagen network increases the effective macromolecular network 'fibril' size and reduces the effective pore opening size of the macromolecular network. This decrease in effective pore opening size increases the resistance to transport of negatively charged molecules like aggrecan through the superficial zone.

The resistance to aggrecan efflux by the collagen network in the superficial zone is visually evidenced in histological sections of normal cartilage by the intense staining for aggrecan right up to the cartilage surface. To computationally model a localized 'higher resistance' to aggrecan transport close to the cartilage surface, one can define an increased resistance to aggrecan transport in the superficial zone. Or more simply, this increased resistance can be approximated using a so-called Robin boundary condition (Smith et al. 2016).

If it is true, the resistance to aggrecan efflux increases the closer to the cartilage surface (which appears to the case), and aggrecan production per unit volume increases with depth (as appears to be the case), this implies that at a long-term equilibrium, the true concentration of aggrecan is closer to constant throughout cartilage than it would have otherwise been (and would become nearly constant as the surface resistance becomes sufficiently large). Cyclic compression loading then decreases cartilage volume near the cartilage surface, which increases the true aggrecan concentration there, and locally steepens the aggrecan gradient near the surface, thereby promoting efflux of aggrecan at the cartilage surface.

Due to the difference in timescale between loading-induced aggrecan efflux and the time for the original concentration to be restored (aggregate transport through cartilage is slow), the amount of cyclic loading may have a significant influence on the local aggrecan gradient at the cartilage surface. This type of reasoning predicts that activity levels could be reasonably well correlated with the aggrecan gradient at the cartilage surface (say, when measured using MRI imaging). This is intriguing, as this means if a normal aggrecan gradient could be defined for normal cartilage, then a significantly lower aggrecan gradient at the cartilage surface implies an unhealthy activity level for the cartilage in that particular person. As articular cartilage is different in each person, this suggests that it may be possible to make person-specific recommendations about the activity levels (i.e. recommendations about number and magnitude of cyclic loadings) that their cartilage can manage without experiencing an abnormal risk of disease in the future.

Large cyclic tensile forces within the surface of the cartilage relative to its ultimate strength can cause fatigue damage of the collagen network, which could progress to splitting of the cartilage surface (Clarke 1971; Flachsmann et al. 2001;

Workman et al. 2017). This is more likely to occur when the menisci or their attachments are damaged, as the menisci distribute the loading over a larger area, reducing local stress, and so reduce the curvature (and so tension) in the superficial zone collagen fibers. Splitting of the cartilage surface is also more likely when there are significant shear forces acting on the cartilage surface, which is more likely with increased normal stress.

A combination of high surface curvature and high surface shear stress is also likely during impact loadings and can cause splitting of the cartilage surface in a single loading when the ultimate tensile strength of the superficial zone collagen is exceeded (see Fig. 2.26) (Huser and Davies 2006; Flachsmann et al. 2001; Ewers et al. 2001; Workman et al. 2017). We note in passing that in young adult humans,

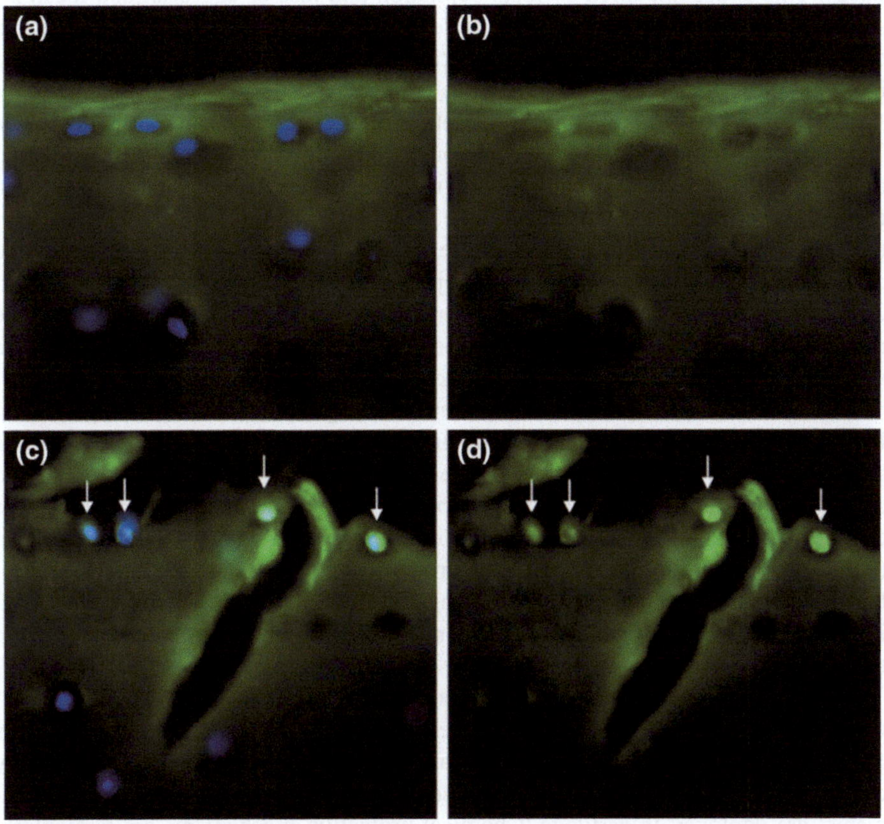

Fig. 2.26 Representative histological sections of cartilage explants taken from the interphalangeal joints of horses, incubated with the specific caspase-3 substrate PhiPhiLux G1D2 (green), and counterstained with DAPI (blue), after 12 h of culture. **a** and **b** are unimpacted controls, **c** and **d** suffered a single mechanical impact and show split in cartilage. **a** and **c** are merge images of DAPI and PhiPhiLux signals, **b** and **d** are PhiPhiLux G1D2 signal only. White arrows point to cells where caspase-3 is active. Original magnification 400 times (Huser et al. 2006). See also Ewers et al. (2001) for extent and depth of surface fracturing

the ultimate tensile strength of superficial zone cartilage is about 25–30 MPa, which decreases by about half at 60–80 years of age (Kempson 1982).

When a split penetrates the superficial zone of the cartilage and extends down into the transitional zone (Ewers et al. 2001)—or perhaps even extends into the deep zone of the cartilage (Huser et al. 2006; Ewers et al. 2001; Workman et al. 2017), this provides a path of reduced resistance to aggrecan and aggregate movement. In other words, a surface split creates a preferential transport pathway, enabling the aggrecan to exit the cartilage tissue more easily than it would have otherwise (Huser and Davies 2006). As aggrecan exits the tissue, local aggrecan concentrations adjoining the cartilage split fall, which leads to local softening of the cartilage tissue around a split. Local softening increases local strains, which may lead to chondrocyte apoptosis. It is noted that for early-to-middle-stage osteoarthritis, multiple splits often appear along the cartilage surface (Pearle et al. 2005). This is accompanied by local depilation and loss of aggrecan and localized softening of the tissue (Poole 1997; Pearle et al. 2005). Similar processes appear to occur around so-named partial thickness cartilage defects. An example can be found in (Hembry et al. 2001).

Collagen fibril size increases with depth in the cartilage (Changoor et al. 2011). This increase in fibril size can be functionally explained by the body forces acting on the collagen network generated by the aggrecan swelling and the aggrecan gradient pointing normal to the cartilage surface. The body forces means the collagen network has to carry increasing tensile load in the direction normal to the cartilage surface, which increases with depth from the cartilage surface (Smith et al. 2016). The densely packed small diameter collagen fibers in the superficial zone are regulated by the presence of type IX collagen, decorin and biglycan, and by a well-developed type III collagen network.

Collagen fibers subject to intermittent loading are vulnerable to proteases. Because surface loads are cyclic and transient, deformation of collagen fibers tangential to the cartilage surface in the superficial and transitional zone is also cyclic and transient. This means ‘strain protection’ against protease activity in the direction parallel to the cartilage surface in the superficial zone is only intermittent (particularly if the unloaded cartilage surface is concave—unloaded convex cartilage surfaces tends to maintain some collagen tension). How long ‘strain protection’ is offered following a transient strain event is not yet known, but clearly protection time must be finite.

How then to protect functionally important collagen fibers that are not continuously being strained? To reduce the rates of degradation of collagen by proteases to more manageable levels, when any part of the collagen network is temporarily unstrained and so vulnerable to proteases, various ‘chemical protection mechanisms’ are deployed in cartilage. Chemical protection mechanisms are particularly evident in the superficial zone, as the collagen network in this zone has a large proportion of collagen fibers oriented tangential to the cartilage surface, which may well be unstrained for extended periods of time (e.g. during sleep).

What molecules might offer chemical protection of the collagen network? Fortunately, it has been demonstrated that aggrecan itself is protective of collagen degradation (Pratta et al. 2003). But it seems unlikely the findings of Pratta et al. can be explained by the high concentrations of aggrecan preventing the migration of MMPs from the chondrocyte to the territorial and interterritorial collagen, for while it has been shown that high concentrations of aggrecan do reduce the diffusion rates in the pericellular matrix [the pericellular region normally has the highest concentration of aggrecan is in the pericellular region (see Fig. 2.7)], this high concentration of aggrecan does not prevent the migration of all kinds of molecules (structural, regulatory, and otherwise) from the pericellular region to the territorial and interterritorial zones. Indeed, if such transport were prevented, the cell could not manage to repair the ECM (see discussion on radial zonation of the ECM in Sect. 2.10). So this appears to rule out high concentrations of aggrecan preventing the migration of MMPs up to collagen fibrils.

As suggested by Pratta et al. (2003), a more likely explanation is that aggrecan binds to and protectively coats the surface of type II collagen (Pratta et al. 2003; Hedlund et al. 1999; Raspanti et al. 2008), or binds to the type III and type IX collagen on the surface of type II collagen, which protectively coats the collagen fibril. It has been shown that the aggrecan keratin sulfate-rich region has moderate affinity for isolated collagen type II (Hedlund et al. 1999). A model for aggrecan binding to collagen type II has been proposed [see Fig. 2.27 (Fosang and Beier 2011; Hedlund et al. 1999)]. Indeed, we have previously noted that the NC4 region of type IX collagen is positively charged, that the NC3 region of type IX collagen is capable of binding to GAGs, while the Col3 domain of type IX collagen contains many arginine amino acids that are positively charged at physiological pH in cartilage and so can interact with negatively charged sites on GAGs (see Sect. 1.4).

In other words, a multimolecular collagen fibrillar structure is built up in 'coats' or 'layers'—collagen type II mers polymerize around the nucleating core of collagen type XI, then collagen type IX mers polymerize (as well as collagen type III, decorin and other SLRPs—see discussion below), binding to collagen type II, and finally aggrecan (and other molecules such as COMP, matrilins, osteoadherin), bind to collagen type IX and type III on the fibril surface. Finally the whole collagen fibril and associated molecules are 'wrapped' in aggrecan/aggregate (and the whole 'parcel' tied together by molecules such as tenascin and fibulin). The two outer layers (i.e. (iv) aggregate-aggrecan (together with tenascin/fibulin), and (iii) type III and type IX collagen (and associated molecules like COMP, matrilins, osteoadherin), protect the collagen type II from degradation by MMPs. It is also possible that relative movement between the collagen fibril and aggregate-aggrecan-GAG molecules during deformation may 'sweep' any attached MMPs from collagen surface, but this remains to be discovered.

The collagen network superficial zone is more susceptible to degradation for a combination of reasons including: (i) relatively lower aggrecan concentrations, particularly at or near the cartilage surface, (ii) the superficial zone is exposed to greater amounts of proteases due to the continual flux of MMPs (produced deeper in the cartilage) toward the cartilage surface, and (iii) the fact that the collagen fibrils in the superficial zone are only subject to intermittent load-induced tangential straining (and may actually be in a compressed configuration when the unloaded

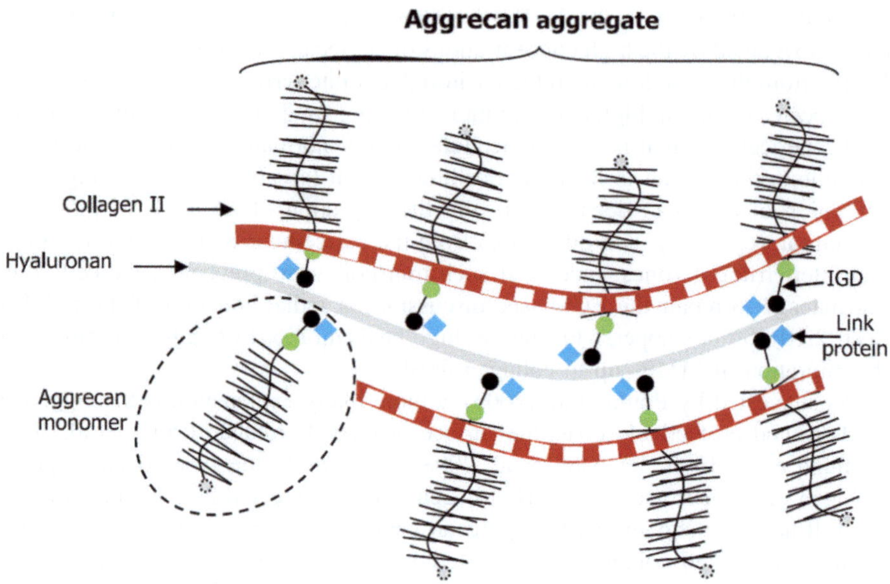

Fig. 2.27 Theoretical arrangement of aggrecan and collagen in cartilage. The aggrecan core protein contains three globular domains including a G3 C-terminal domain, a G2 domain (green), and an N-terminal G1 domain (black). Aggrecan is anchored in the cartilage matrix by binding to link protein (blue diamond) and hyaluronan to form large multimolecular aggregates. Type II collagen fibrils can bind aggrecan in the keratan sulfate-rich region. The figure shows a theoretical arrangement of type II collagen (red) and aggrecan aggregates in which the charged chondroitin sulfate chains of aggrecan protect collagen fibrils from cleavage by collagenases. Another theoretical possibility is that collagen II fibrils bound at the keratan sulfate-rich region obstruct access of aggrecanases to the cleavage site located in the aggrecan interglobular domain between G1 and G2 (Fosang and Beier 2011). Aggrecan keratin sulfate-rich region has moderate affinity for isolated collagen type II (Hedlund et al. 1999)

cartilage surface is concave; see for example Fig. 4.14). For all these reasons, additional chemical protective mechanisms are required for the superficial zone collagen network.

Superficial zone collagen is chemically protected by a number of molecules with comparatively high concentrations. For example, collagen is both regulated and protected by various proteoglycans including members of the ‘small leucine-rich proteoglycans’ (SLRPs) such as decorin and biglycan (Heathfield et al. 2004). Decorin and biglycan bind tightly to the collagen fibrils, inhibiting the further addition of collagen mers (Svensson et al. 1999), so regulating the collagen fibril diameter. Decorin and biglycan may also modify the operational efficiency of MMPs (Stuart et al. 2011). Importantly, decorin and biglycan concentration are found in the highest concentration in the superficial zone, their concentration decreasing with distance from the cartilage surface (Poole et al. 1996). This is consistent with their role in regulating and protecting the collagen network (Fig. 2.28).

Fig. 2.28 The micrograph clearly demonstrates the interfibrillar bridges (arrows) [note this image is of type I collagen, rather than type II collagen] formed by the decorin side chains. A few chains lie longitudinally along the fibril axis (Raspanti et al. 2008)

Interestingly, it has been shown that some molecules display a circadian variation in serum concentration (Markoulli et al. 2012). One such protein that appears to be undergoing circadian variations (at least based on measured serum concentrations) is cartilage oligomeric matrix protein (COMP) (Heinegard and Saxne 2011). COMP is a major glycoprotein in cartilage and synovial tissue (Geng et al. 2012), perhaps accounting for as much as 1% of the cartilage tissue wet weight. The COMP monomer is multidomain glycoprotein with MW of 97 kDa, while the functional COMP pentamer is about 550 kDa. It is known that COMP binds to the matrilins and collagen types I, II, and IX via calcium and zinc divalent ions (Mann et al. 2004; Rosenberg et al. 1998).

During the day, *serum* COMP concentrations are high (possibly because COMP is mechanically dislodged from cartilage and other tissues, e.g. tendon during daily activity), but concentrations fall during the night (Andersson et al. 2006a). Supporting this interpretation is the finding that exercise temporarily elevates serum COMP concentration. With the cessation of exercise, COMP concentrations return to their baseline so that there is no statistical difference after about 30 min (Andersson et al. 2006b). This data suggests that COMP becomes attached to ECM components when there is no cyclic loading, but when cyclic loading is applied to cartilage/tendons/ligaments (e.g. with exercise), the deformations in the tissue dislodge some of the COMP molecules from its attachment to ECM components. COMP then finds its way from articular cartilage tissue into the synovial fluid, and from their through the lymphatic drainage and into blood plasma (where the serum concentration is easily measured).

It seems plausible that tissue straining associated with exercise would also dislodge other protective molecules from the collagen network, including aggrecan and SLRPs, which then make their way toward the cartilage surface in increasing numbers. The effectiveness of exercise in dislodging protective molecules, which then migrate through the collagen network, will depend on the binding affinity of the molecule for the collagen type II surface (or binding affinity for 'coating molecules' such as type IX and type III collagens, or for HA in the case of aggrecan), the local collagen fibril strain, and how surface binding affinity alters with the local fibril strain (e.g. straightening of a curved collagen fibril, or closing of the 'branch angle' in the collagen network may cause high local strain and dislodge protective molecules). Of course the reduction in chemical protection of collagen fibrils is associated with increasing strain protection of collagen fibrils, so as one protective mechanism declines another protective mechanism increases, so overall collagen network protection may well be maintained relatively constant.

It is known that COMP knockout mice show no skeletal abnormalities during development (Svensson et al. 2002); however, defective COMP is associated with epiphyseal dysplasia and early onset osteoarthritis (as are defective matrilins and collagen IX). This suggests that COMP is in a global sense protective of the collagen network. While its absence is not particularly relevant during development (when protease activity in many parts of the skeletal system is minimal), its absence becomes relevant in the long run (and so becomes associated with cumulative damage to the tissue with aging).

Based on the description above and the limited experimental data currently available, it is plausible to suppose that one important role for COMP is protection of collagen type II from degradation by collagenases. COMP is found throughout articular cartilage (e.g. see the distributions of COMP in articular cartilage shown in mice [see Fig. 4c in Salminen et al. (2000) and in the horse: see Fig. 1 (Murray et al. 2001)], but this additional chemical protection is probably particularly important in the superficial zone where the protective mechanism of collagen fibril strain is intermittent.

2.5.4 Collagen Turnover

The turnover of collagen in articular cartilage is a complex story for several reasons. First, the shaping of the collagen network appears to take place in several stages over the life of the animal: (i) growth and remodeling of the collagen network as the animal grows to maturity, (ii) followed by a (depth-dependent) period of quiescence through adulthood (at least quiescent relative to the first stage), and (iii) an apparent 'reawakening' of remodeling (meaning more remodeling and to greater depth) with the onset of osteoarthritis, which is observed both in small animals and humans (Grimmer et al. 2006; Hotta et al. 2005; Ruiz-Heiland et al. 2012; Sandell and Aigner 2001; Eyre et al. 1980).

To explain the period of depth-dependent quiescence during adulthood, we first note the key fact that a high concentration of aggrecan suppresses chondrocyte proliferation (Xu et al. 2010). Further high concentrations of transcription factor

(TF) SOX9 in adult cartilage suppresses Wnt signaling (Akiyama et al. 2004), which in turn suppresses chondrocyte proliferation in adult cartilage. High concentrations of aggrecan and SOX9 (together with SOX5 and 6), and low concentrations of Ihh are characteristic of mature adult cartilage. Other effects like telomere shortening may play a role (Martin and Buckwalter 2001, 2002). It is plausible that together these factors largely explain the period of depth-dependent proliferative quiescence in healthy adult cartilage, at least for adult cartilage making up the lower transitional zone and deep zones. However, the upper transitional zone and superficial zone usually retain proliferative activity throughout adulthood, driven by PTHrP concentrations and mechanical damage as a result of cyclic loading. These more superficial regions have lower aggrecan and SOX9 concentrations, and are usually subject to the highest levels of ECM damage and cell death, and so these regions require ongoing repair, e.g. they require collagen network repair and they require ongoing chondrocyte proliferation (see further discussion Sect. 2.7).

Based on racemization of the amino acid aspartic acid, it is also of interest at this point to note that the half-life of collagen in adult articular cartilage has been estimated to be between 100 and 400 years (Eyre et al. 2006; Verzijl et al. 2000b). This may be compared to a half-life of 15 years for collagen in skin. (We note in passing that triglycerides stored as fat in adipocytes can be lost much more rapidly than collagen remodeling can remove excess skin (half-life of 15 years)—probably explaining why 'skin wrinkles' appear after weight loss). We note that similarly long collagen half-life has been estimated for collagen type I turnover in (horse) tendon (i.e. around 200 years (Thorpe et al. 2010)).

Based on carbon 14 measurements, again similarly long collagen type I half-lives appear to be indicated for human Achilles tendon (Heinemeier et al. 2013, 2018). We note that this very long estimated type I half-life for collagen fibrils sits in stark contrast to measured collagen type I synthesis rates in human tendon, being around 1% of total tendon collagen per day! It noted that this rate of type I procollagen production can increase fivefold upon training exercise, which clearly suggests a physiological role for the huge amount of procollagen synthesized, i.e. presumably this level of production is necessary to effect the repair of damaged collagen fibrils (Langberg et al. 2001).

How do we resolve this discrepancy between type I collagen turnover rates based on procollagen synthesis and fibril collagen replacement? One explanation for the difference between the observed very large rates of procollagen synthesis and very low rates of collagen degradation is that while collagen in cartilage, tendon, or ligament is continually damaged and repaired, usually only a small fraction of the collagen network is damaged [see Fig. 4 (Chen et al. 1999; Schechtman and Bader 1997; Young et al. 2016)]. And of that small damaged fraction, only a small portion of each collagen fibril is damaged (i.e. the damaged collagen fibrils are damaged only along part of their length) (see Fig. 2.29) (Provenzano et al. 2005). Consequently, only a very, very small fraction of the collagen network is in need of repair following daily damage events. Indeed based on such concepts, Young et al. (2016) developed a collagen remodeling model that predicted collagen fibril

Fig. 2.29 Collagen fibril and fiber damage [for a ligament] above the damage threshold. Hence, shortly after [5.8% stretch] injury some fibers are intact and appear normal, some fibers are completely ruptured and a 'gap' exists between ruptured fibers, and additionally some fibers are intact but contain ruptured collagen fibrils that contain a 'gap' between ruptured fibril ends. As such, if the tissue gap remains and is present when new tissue 'fills in' without substantial contracture, the repaired fiber or fibril would be longer than its pre-injury length and as such it will be more lax. Magnification 20,000× (Provenzano et al. 2005)

turnover rates consistent with those estimated experimentally (i.e. with an estimated half-life around 100–200 years) (Young et al. 2016).

Consider now the data on collagen damage resulting from cyclic loading of canine cartilage explants at 5 MPa for 120 min at 0.3 Hz (2160 cycles) (Chen et al. 1999). It is evident from the immunohistological straining shown in Fig. 4 in Chen et al. (1999) that damaged collagen is highly localized—in this case, primarily at the cartilage surface.

Figure 4 in Chen et al. (1999): *Immunohistochemical localization of mechanically induced collagen denaturation with use of COL2–3/4m antibody for* ***a****: non-loaded control (×100) (bar = 100 microns) and* ***b****: loaded sample (at 5 MPa for 120 min). Staining is increased in the superficial surface and deep zone in the core (direct loaded) region (indicated by arrows) (×100). The enlarged views (×200) of the* ***c****: non-loaded control and* ***d****: loaded sample show the surface injuries and damaged collagens piled up in several layers perpendicular to the surface of the loaded sample* (Chen et al. 1999).

These results are corroborated by Lin et al. (2004) (see Fig. 2.1).

Within the sub-fraction of damaged collagen network, only a portion of the collagen fibrils are damaged (see Fig. 2.29). Further, the damaged collagen fibrils are only damaged over part of their length (see Fig. 2.29). It seems likely that the regions of an individual collagen fibril most likely to be mechanically damaged in daily activities are those regions experiencing the highest stress concentrations (e.g.

at bends in fibril direction, i.e. regions of crimp) or branch points in the collagen network. Based on measured CTX-II fragments (CTX-II are cross-linked C-terminal telopeptides, or end fragments from tropocollagen, released through damage of the collagen network) in urine in adults between the ages of 18 and 40, the rate of collagen fibril damage declines about threefold over this age range (Struglics et al. 2015).

Note that the rate of collagen incorporation in the collagen fibril network may well be greater than the rate of removal by enzymatic degradation. This is because the collagen fibril network gradually expands as collagen fibrils mechanically fracture (see Fig. 2.29 and discussion Sect. 3.5). At steady state, the new collagen incorporated into the collagen fibril network as it expands is balanced by the removal of collagen fibril network through wear at the cartilage surface (see discussion Sects. 3.5 and 4.4). So the collagen fibril network probably has two pathways for collagen removal—enzymatic degradation (associated with debridement of damaged collagen) and abrasive wear at the cartilage surface.

We also observe that collagen repair of fractured collagen fibrils is effected through the 'self-organization' of tropocollagen mers (see discussion Sects. 2.6 and 2.10.5), requiring no direct involvement of cells. Crucially for the collagen network repair process to work, the key requirements are: (i) an adequate concentration of procollagen is maintained in the ECM (it then immediately becomes more clear why a higher rate of procollagen synthesis is maintained relative to that incorporated in the collagen fibril network) and (ii) that procollagen is converted into tropocollagen. There are a number of enzymes involved in the conversion of procollagen to tropocollagen. N-terminal telopeptides are removed from procollagen by ADAMTS2 and meprins, while C-terminal telopeptides are removed from procollagen by BMP1, tolloid-like protease, furins, and meprins (see discussion Sect. 2.10.5).

Understanding the need to maintain an adequate concentration of procollagen just in case collagen network repair is needed, shifts our focus from collagen network repair to procollagen synthesis by chondrocytes. It is clear that a sustained compressive load reduces procollagen synthesis, just as it reduces aggrecan synthesis (see Fig. 2.5). But it appears as though procollagen synthesis is fairly constant in articular cartilage (at least in a time-averaged sense) (Ackermann and Steinmeyer 2005), varying more with age [being higher in both the young and old (Hermansson et al. 2004) and increasing with tissue injury (Eyre et al. 1980)]. In other words, it seems most plausible that chondrocytes usually just continually produce more than enough procollagen to sustain whatever repair may be needed in the tissue, but if there is 'moderate' damage, local chondrocytes increase procollagen production (Hotta et al. 2005). In an ACL transection dog model, Eyre et al. (1980) report that the rate of radioactive proline incorporation in collagen is 10 times that of the control knee (Eyre et al. 1980). Indeed, under most circumstances the amount of procollagen used in repair probably accounts for only a fraction of the total amount of procollagen produced by chondrocytes. Like tendon, it is likely that the majority of procollagen synthesized throughout the articular cartilage is either uptaken by healthy chondrocytes, or 'lost' from the cartilage to the synovial fluid. It is unclear which of these pathways is more important.

As if these complications in understanding collagen turnover in articular cartilage are not enough, another complication is that while the collagen network is mainly made up of mainly type II collagen, there are actually three different collagens that go to make up 'type II collagen' (types II, IX, and XI). Further, the volume fraction of collagen in articular cartilage changes over time [some calculation can show the hydrated volume in calf cartilage aged two weeks is around 10% (Sah et al. 1989), while in adult human cartilage it is around 25–35% (see Appendix D)]. The composition of the collagen network changes between young animals and adults (Eyre 2002). In young animals, the amount of type II is <80%, while type IX is >10% and collagen XI is >10%. In adult animals, collagen type II is generally >90% while collagen IX is about 1% and collagen XI about 3%. This suggests that much more collagen type IX is required to cover the surface of small fibrils in young animals, while much less collagen type IX is required to cover the collagen type II fibrils in the mature collagen network found in adults, probably because the mature type II collagen fibrils are generally of larger diameter. Further in adults, type III collagen can be a significant fraction of the total collagen (e.g. up to 10%), particularly in the transitional and superficial zones (see Fig. 2.30) (Eyre et al. 2006; Aigner et al. 1993). The total amount of collagen may be fairly constant with age in adults (Maroudas et al. 1992).

Another significant complication is that collagen network repair probably occurs by several different mechanisms in articular cartilage (the efficiency of each repair method depending on procollagen being able to access the site of repair after being synthesized by a chondrocyte). For example, in Stages 1 (before sexual maturity) and 3 (aging animal) described in the introduction to this section, cartilage expresses type II procollagen as a (longer) type II isoform [the so-called A isoform, as part of what Sanchez et al. refer to as the 'chondroblastic phenotype' (Sanchez et al. 2017)], while in Stage 2 (adult animal) the cartilage expresses a (shorter) type II procollagen isoform (the B isoform). This differential isoform expression is regulated by signaling molecules in the extracellular matrix influencing the 'splicing machinery within chondrocytes (Matsushita et al. 2007).

It is likely that the longer type IIA mers have a lower isoelectric point (i.e. a lower pI), which presumably enables this isoform of procollagen to migrate long distances through the cartilage ECM even when the cartilage pH is toward the lower end of its range in normal cartilage. In this case, the pH of the ECM remains greater than pI of the procollagen molecule, and so the procollagen molecule carries a negative charge that facilitates its diffusive transport through negatively charged aggrecan. The ability of this procollagen isoform to migrate through the ECM to the repair site, where damaged collagen type II fibrils reside in the territorial and more distant interterritorial zones, is obviously crucial to fulfilling its function.

It is likely that in Stage 2, adult cartilage tissue is less metabolically 'stressed,' so the pH rises and the shorter type IIB mer, which has a higher isoelectric point, is then expressed. This works fine so long as the pH in the cartilage tissue remains high enough, for then distant damaged collagen network can still be accessed by procollagen as the B isoform. However, if the pH in the cartilage now falls, the type IIB mer movement becomes restricted, e.g. it may only access the pericellular

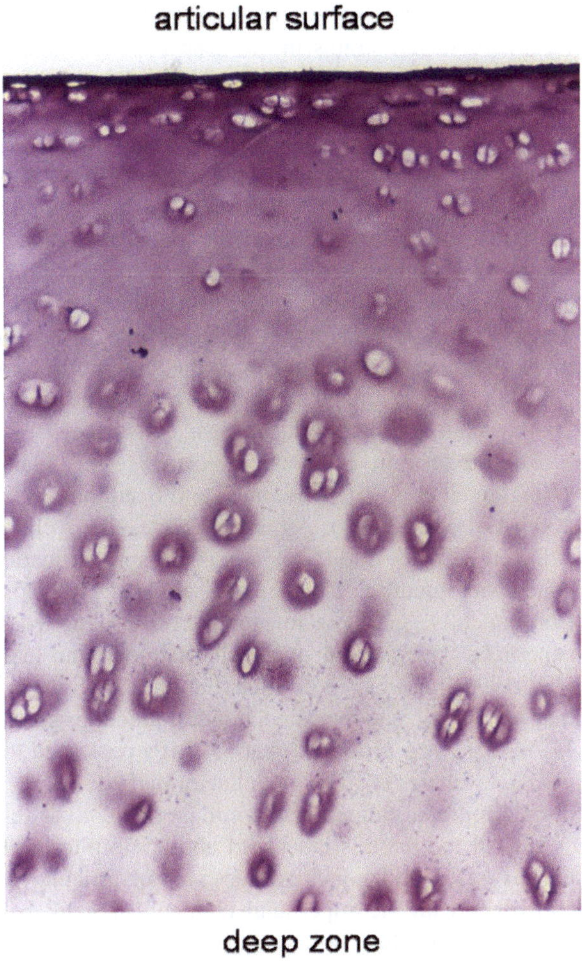

Fig. 2.30 Immunochemical localization of type II collagen in an intact articular cartilage region from the knee joint of a 59 year old female with familial OA. mAb 4G9 specifically sees the N-propeptide of collagen type III, which is concentrated in the surface zone and around chondrocytes at depth (Eyre et al. 2006)

matrix, because it is positive charged, which will cause the B isoform to bind locally rather than diffuse away to more distant ECM (see discussion in Sect. 2.10.5 for further details). Under these conditions, it seems likely that collagen type III may play an important role in the immediate repair of damaged type II collagen distant from chondrocytes. However, if the pH fall is substantial and sustained, the type IIA isoform can be again expressed (as it is in osteoarthritis), which probably is an adaption to restore collagen network repair capability.

It becomes apparent that the story of collagen turnover is likely to be as complicated as that for aggrecan.

To maintain collagen fiber stress at target levels (probably around about 1–2% strain, e.g. Inamdar et al. 2017), the collagen fiber diameter increases with depth in the adult, commensurate with the increasing expansion pressure generated by aggrecan/aggregate (Changoor et al. 2011). However, the collagen in the deep zone generally experiences less mechanical fatigue damage than collagen near the surface, as collagen strains in the deep zone are generally smaller than closer to the cartilage surface. In addition, the deep zone fibrils experience less protease degradation because the collagen fibrils at depth are more likely to remain constantly strained (rather than intermittently strained). The greater constancy of the strain environment protects the collagen fibers from protease degradation (Wyatt et al. 2009; Flynn et al. 2010). We note that beyond a certain threshold of collagen fibril mechanical strain (probably between 1 and 2% strain), collagen may be completely protected from the action of collagenases (Flynn et al. 2013). Because collagen in the deep zone is largely protected from both mechanical and enzymatic degradation, it is evident that collagen in the deep zone will usually turnover much more slowly than the collagen in the superficial zone. In other words, collagen at depth in adult cartilage may have an exceptionally long half-life (Verzijl et al. 2000a) relative to the collagen near the surface.

Repeated strain is known to cause fatigue damage to the collagen network (Weightman et al. 1978; Bellucci and Seedhom 2001; Chen et al. 1999; Kurz et al. 2005; Lin et al. 2004). As noted above, the cyclic shear and tensile strain in the collagen network due to physiological surface loading is usually least in the deep zone and greatest in the superficial zone close to the cartilage surface. Consequently, fatigue damage of collagen network is generally greatest in the superficial zone and upper transitional zone and least in the deep zone (Hollander et al. 1994) [though we do note that under extreme load conditions (shear) damage may also localize to the osteochondral junction (see Levy et al. 1996 and Chen et al. 1999) (e.g. with 2160 cycles at 5.0 MPa peak load (0.3 Hz) for 120 min)]. In other words, estimates of average collagen half-life for the whole tissue mask significant normal regional differences in collagen half-lives through the depth of the tissue.

For all the reasons discussed above, despite the long experimentally measured half-lives for collagen fiber turnover in articular tissue as a whole, this is still consistent with the superficial collagen network being dynamically adjustable in response to its changing mechanical environment (see discussion Sect. 3.5).

Indeed, collagenases (Heathfield et al. 2004) and mechanical loading on the superficial zone creates ongoing damage that requires collagen network repair—and collagen network formation (ongoing formation is principally in the transition zone) —to maintain the ongoing integrity of articular cartilage. And it is plausible that sometimes protease degradation in the superficial zone is significant, given that the flux of proteases toward the cartilage surface produced by chondrocytes lying in deeper tissues, and considering that the superficial zone cartilage is exposed to all the proteases that may enter via the synovial fluid. For example, proteases may enter the synovial fluid via the transudate from the blood, production by synoviocytes, and in the case of the knee joint, proteases may also be produced by the synovial, meniscal, and ligmentous tissues.

If collagen network repair is not effective, a 'vicious cycle' of tissue degradation can be established. Because damaged collagen network provides less resistance to aggrecan egress from articular cartilage, over time the local aggrecan concentration reduces and so the local compressive stiffness of the cartilage reduces (Kurz et al.

2005). This can lead to increased local strain, which increases rates of collagen fatigue damage, leading to further aggrecan loss and decrease in cartilage stiffness. This is a powerful positive feedback process that can lead to comparatively rapid deterioration of the collagen network putting articular cartilage tissue integrity at risk—unless the collagen network can be repaired!

Focusing on a fibril repair process, it is possible that the fatigue damage to the collagen network may lead to aggrecan loss as described above. Then a repair process follows. This presumably involves removal of collagen type IX from the surface of collagen type II at or adjacent to the fibril damage, thereby allowing MMPs to attach to the collagen fibril. This is followed by: (i) protease debridement of damaged collagen fibrils (perhaps reducing their diameter, trimming up ends or removing them altogether) (Flynn et al. 2013), prior to (ii) local collagen fibril repair (i.e. polymerization of collagen type II tropocollagen), and (iii) reattachment of type IX collagen to the surface. This is followed by the restoration of the normal local aggrecan concentration.

Collagen network degradation by collagenases and mechanical damage both point to collagen repair processes normally being most active in the superficial and transitional zones, and rather less active in the deep zone. Indeed, in normal cartilage this explains increasing amounts of degraded collagen type II toward the cartilage surface. This is evidenced experimentally by more collagenase and cathepsin K (collagen type II) neoepitopes (neoepitopes are new molecular shapes that are generated by collagenases and cathepsin K) emanating from chondrocytes and the ECM closer to the cartilage surface (see Fig. 2.31) (Dejica et al. 2012b), as well as by superficial and transitional zone cartilage being more disorganized than deep cartilage [see Fig. 2 (Matcher 2015)].

Greater amounts of type II collagen are usually produced in osteoarthritis (i.e. Stage 3) (see Fig. 3 in Hermansson et al. 2004). This increased rate of collagen type II production normally commences in the chondrocytes close to the cartilage surface and progresses to the deeper tissues as the disease progresses.

We observe that the findings of Hermansson et al. (2004) are consistent with the experimental findings of Eyre et al. (1980). Eyre et al. (1980) found that in a surgically induced arthrosis in dogs, collagen synthesis rates were increased tenfold above those found in controls (Eyre et al. 1980).

While having a better qualitative picture of collagen turnover, our quantitative understanding of collagen network turnover remains rather deficit. In Sect. 3.5, we use the available information and some informed guesstimates to very approximately estimate the total rate of collagen fiber loss from the surface of the whole human adult knee joint experiencing 1 million load cycles per year to be in the range 8 to 70 mg/year. For this analysis, the rate of surface collagen fiber wear loss is assumed to be equal to the rate of new collagen network formation, which occurs principally in the transitional zone cartilage, while the average rate of new collagen network formation in normal transitional zone cartilage is estimated to be equal to about 1% of the total collagen network content in the high load regions of knee per year (i.e. probably around 0.5% of the total collagen network content of articular cartilage in the normal adult knee).

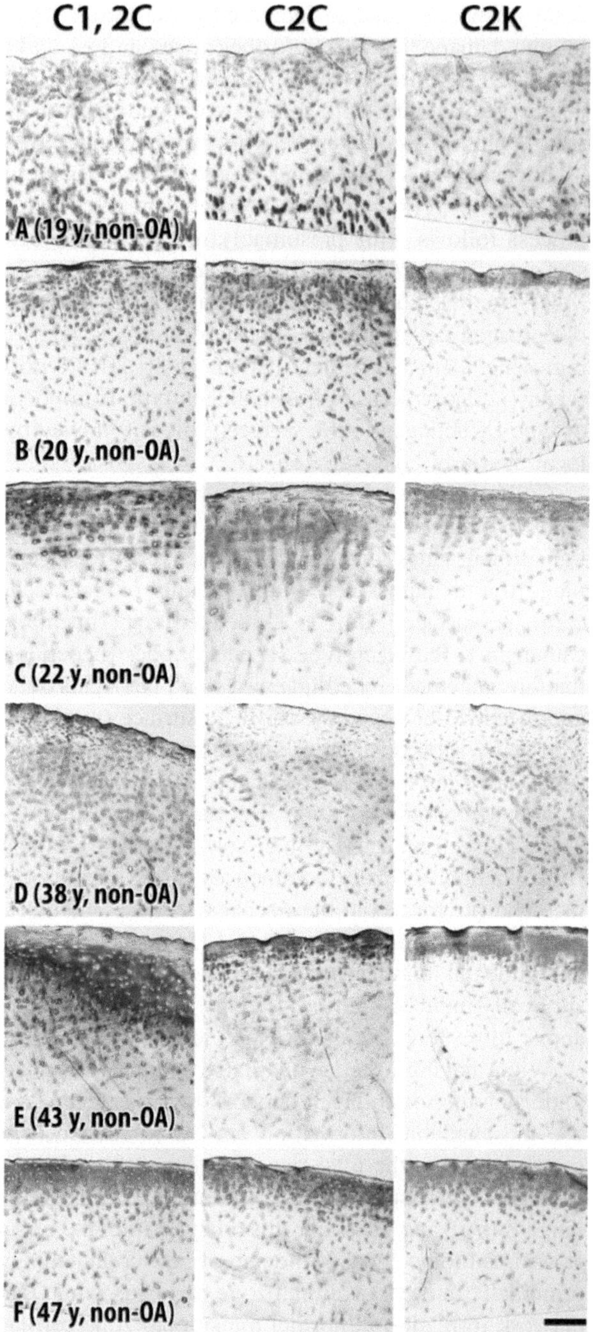

Fig. 2.31 Immunostaining of cartilage from young normal individuals. Sections of human femoral condylar cartilage from six non-OA individuals (ages 19–47 years), showing detection of neoepitopes generated by collagenases (C1, 2C, and C2C) and cathepsin K (C2K) in type II collagen. Specimens are positioned so that the articular surface is at the top of each figure. Bar = 200 μm (Dejica et al. 2012b)

2.6 Overview Collagen Network Repair Mechanisms

It is likely the collagen network in articular cartilage is created and repaired by at least four different mechanisms.

First, procollagen is 'activated' by protease action forming tropocollagen, which then self assembles to form *new* collagen fibrils, and continues the growth of collagen fibrils. This is how the collagen network is created. In addition, minor collagen fibril damage to existing collagen fibrils may be repaired by binding and cross-linking of types II and/or III collagen mers to the predominantly type II collagen network (Eyre et al. 2006; Videman et al. 1981; Wolf et al. 2007; Wu et al. 2010). Type III collagen is found predominantly on the outside of collagen type II fibrils (up to 10% of the collagen network may become type III collagen). Type III collagen repair process probably dominates in the superficial zone of the cartilage (Wu et al. 2010; Aigner et al. 1993). In the transitional zone, some fraction of repair is achieved by covalent linking of type III collagen, while in the deep zone the addition of type II collagen mers enables the fibers in the deep zone to grow in thickness over time and mature through cross-linking (cross-linking increases both strength and stiffness of the collagen fibers (Hansen et al. 2009; Tang et al. 2010; Silver et al. 2003)).

Second, while elastic straining of collagen fibrils is protective, larger strains combined with sufficiently frequent cyclic straining are themselves damaging [see Figs. 5 and 6 (Tang et al. 2010)], leading to fatigue damage. Assuming sufficient cross-linking (and so stiffness), and subject to sufficient stress-controlled loading ('stress-controlled' loading is normally the type of loading experienced by articular joints—in contrast to 'strain-controlled' loading), collagen fibrils may suddenly fracture, leading to a sudden drop in load carrying capacity, to near zero [e.g. see the sudden drop in load indicated by the dashed line in Fig. 5 (Tang et al. 2010)]. In the case of such random breaks in collagen fibrils, collagen network repair is mainly effected by debridement of the damaged collagen fibril, following by building new fibril from type II (and/or type III) tropocollagen mers, connecting together the broken ends of the damaged fibril. Note that repeated fracture damage and repair of this type drives gradual expansion of the collagen network, while maintaining structural integrity of the collagen network. This is probably the main repair process operating in physiologically normal cartilage.

Like the process for new fibril formation, this process of collagen network repair requires a steady concentration of procollagen in the ECM, which can be 'activated' by cleavage of telopeptides, and the tropocollagen used to build new collagen fibrils between the broken ends. A key requirement is an adequate concentration of procollagen needs to be maintained in the pore fluid of the ECM, so that sufficient procollagen can be converted into tropocollagen [which in addition to suitable proteases, may require the presence of procollagen C-proteinase enhancer (Vadon-Le Goff et al. 2011)]. Assuming there is sufficient procollagen throughout articular cartilage, this shifts the focus of collagen network repair from chondrocyte synthesis of procollagen (which is reported to be fairly constant in articular cartilage

(Ackermann and Steinmeyer 2005)) to the control of regulatory enzymes that convert procollagen to tropocollagen (via removal of N-terminal telopeptides from procollagen by ADAMTS2 and meprins and removal of C-terminal telopeptides from procollagen by BMP1, tolloid-like protease, furins, and meprins) (see discussion Sect. 2.10.3).

Understanding this process helps to explain the observation that procollagen production by chondrocytes (quite unlike aggrecan production) is reasonably constant as mechanical loading is decreased or increased, as this system of collagen repair simply relies upon having an adequate background concentration of procollagen in the ECM available for repair at all times—just in case should it be needed (i.e. the background concentration of procollagen needs to be above a threshold concentration that ensures adequate repair, while the particular concentration of procollagen in the ECM is less important, assuming it is above the threshold value).

Collagen fibrils subject to sufficient (strain-controlled) loading may show local plastic deformation of a collagen type II fibril [e.g. dash-dot line Fig. 5 (Tang et al. 2010)]. It is possible that plastic straining of the collagen surface may result in dislodgement of the collagen type IX (and other molecules) from the collagen surface. This is because plastic deformation changes the molecular structure of the collagen type II surface (see Fig. 6 of Tang et al. 2010 for schematic depiction). The change in stereochemistry of the surface presumably significantly reduces the affinity of the type IX and other molecules for the (plastically) damaged collagen surface, so type IX collagen no longer wants to bind to the surface.

See Fig. 5 in Tang et al. (2010): [Theoretical mesoscale elasto-plastic] *stress–strain curves of m × 5 collagen fibrils with cross-linking density beta = 20. The parameter m changes from 2 to 10. Different snapshots along the deformation histories of the 2 × 5 and 5 × 5 collagen fibrils.* See also snapshots of actual tropocollagen configurations at points (a–f), visualizing elastic extension and 'slip motions' in collagen fibrils as they plastically deform shown in Fig. 6 of Tang et al. (2010). Further details can be found in Tang et al. (2010).

Removal of type IX collagen from the type II collagen fibril surface should then allow MMPs [i.e. normally MMP1, but when cartilage is 'stressed' MMP1 is joined by MMP13 (and sometimes by MMP8)] to access the collagen type II surface, whereupon the MMPs debride the damaged regions of a collagen type II fibril. The debrided damaged surfaces allows for firm attachment of new tropocollagen mers to the existing collagen network. Then the repair by polymerization of new collagen fibril from type II tropocollagen can occur (see Sect. 2.5.3). As mentioned above, this repair process is maintained by having an adequate concentration of procollagen mers available in the ECM for repair. Collagen type II repair processes described above may be supported by type III collagen repair (perhaps by also forming a temporary 'scaffold' before more permanent type II collagen repair). Note that normal type IX collagen binds to the surface of collagen type II fibril, which then binds to aggrecan (see Sect. 1.4). Aggrecan's presence normally inhibits the degradation of a normal collagen type II fibril by MMPs (see Sect. 2.5.3).

When collagen network damage is (sufficiently) large, this leads to a local decrease in aggrecan concentration (as aggrecan either escapes from the collagen network toward the cartilage surface or forms a local with low constant aggrecan density, creating steep local aggrecan gradients around the damage). Such extensive collagen network damage leading to aggrecan loss is illustrated in Fig. 3.5c–f. A decrease in local aggrecan concentration and collagen damage stimulates nearby chondrocyte proliferation. After proliferating, a daughter cell may migrate toward the region of low aggrecan concentration. As chondrocytes migrate slowly through the damaged ECM, they degrade the collagen network in their path [see for example the description of fibroblast migration through a collagen network (Iordan et al. 2010)]. Then finally, a completely new collagen network polymerizes in the 'wake' of the migrating chondrocytes.

This more extensive collagen network repair mode related to chondrocyte migration (or 'collagen network remodeling mode') helps explain why the rate of collagen synthesis in cartilage is difficult to manipulate by mechanical loading (Wolf et al. 2007). In this case, repair intensity is related to the number of slowly migrating chondrocytes. Cell migration driven repair may normally be a significant repair process in the upper transitional zone of the cartilage (Aigner et al. 1993). But it is noted that this cartilage repair process takes considerable time to be effective. Consequently, C telopeptide collagen fragments may remain elevated for months or indeed for many years following cartilage injury (Lohmander et al. 1996) as chondrocyte migration is slow, and if damage is ongoing, then in some cases, C telopeptide collagen fragments may remain elevated for decades (Lohmander et al. 2003). The processes driving chondrocyte division and chondrocyte migration, and the rates of migration, are explained in more detail in Sect. 2.9.

We therefore envisage different collagen repair mechanisms dominating in different regions of articular cartilage at different times, depending on the type and severity of collagen network damage. It is noted that some types of cartilage damage are inherently difficult to repair in vivo using the three abovementioned repair mechanisms, because each of these three mechanisms are only effective when the damage is within the extent of the existing cartilage tissue itself. But cartilage sometimes splits in the direction normal to the cartilage surface, under impact loading or with increased surface shear tractions (Clarke 1971, 1974; Huser and Davies 2006; Workman et al. 2017). Once a split from the cartilage surface forms, it allows aggrecan to more easily gain access to the synovial fluid, and the cartilage ECM close to the split surface softens. In this case, a fourth mechanism for cartilage repair is required, one that can generate new 'cartilage tissue' external to the existing domain of collagen network.

This fourth (and final) collagen network repair mechanism, normally only occurring when there is very extensive articular cartilage damage, and involves either de-differentiated chondrocytes moving to the cartilage surface and exiting the cartilage tissue altogether, or mesenchymal pluripotent cells circulating in the blood finding their way into the synovial fluid and then differentiating into fibroblasts. Whatever their origin, these cells typically reside on the surface of the damaged tissue. For example, the cells attach to the sides of the cartilage split or the sides of (partial or full thickness) traumatic injuries, where they are partially protected from the sliding and rolling of the articular cartilage contact surfaces, which happens every time two articular cartilages (and menisci) transfer load from one to the other. Once the cells are in situ, they begin to produce new fibrocartilage tissue (not hyaline cartilage tissue).

The population of chondrocytes that migrates to the cartilage surface from cartilage ECM de-differentiates. These cells de-differentiate to such an extent that they now predominantly produce type I collagen, rather than type II collagen, and in this de-differentiated state these cells generate fibrocartilage. Fibrocartilage is known to have inferior functional properties relative to that of the hyaline cartilage that it replaces. This is primarily because collagen type I cannot retain aggrecan as effectively as collagen type II, and so the fibrocartilage cannot effectively resist large compressive loads as well as hyaline cartilage. Because aggrecan concentrations are too low, this leads to compressive loads being carried by the type I collagen fibrils. In this situation, compressive loading of collagen fibrils damages the collagen fibrils, and the substitute articular tissue usually fails in service.

We finally mention here that it has been shown that for large, full thickness cartilage defects that have bone rubbing on bone, joint distraction for a period of several months can lead to significant increases in cartilage volume (Yanai et al. 2005). However, this newly formed cartilage is again largely fibrocartilage, formed by the (fourth) mechanism described above, rather than normal hyaline cartilage (which is produced by the first three collagen network repair mechanisms mentioned above). Nevertheless, joint distraction is reported to result in significant functional improvements in knee joint operation (see discussion Sect. 1.3).

It is apparent that the sustained performance of cartilage over decades is only possible because of cartilage repair processes coordinated by chondrocytes embedded within the ECM (Bader et al. 2011). No other known tissue, either natural or a synthetic substitute, can withstand decades of repeated cyclic loadings (with tens of millions of load cycles per decade), as normal healthy cartilage tissue manages to achieve because of its repair capability. Indeed, most knee replacements loosen or wear out over a period of 15+ years, particularly in young active people. Taken together, this suggests that repair processes in normal cartilage need to rather more vigorous and effective than some researchers and clinicians may believe them to be.

2.7 Chondrocyte Proliferation and Apoptosis

The stress–strain state within in vivo cartilage is complex, and it is difficult to recreate similar conditions in the laboratory—so laboratory tests usually employ simplified stress–strain states.

For example, cartilage surface load is typically applied as an impact load, a sinusoidal load, or a steadily increasing ramp load. Usually, load is directed normal to the surface (i.e. it usually involves no surface shear loading). Testing is most often done on cylindrical explants under 'unconfined compression' (i.e. test specimen subject to zero lateral stress), and less often they are cylindrical specimens under 'confined compression' (i.e. test specimen subject to zero lateral strain). Actual load conditions in vivo are generally somewhere between these load conditions. It is useful to note in passing that 'lateral bulging' of a material in compression is reduced when the extent of the lateral load is large relative to its thickness (i.e. the approximation of confined compression becomes better), as is the case for most congruent joints (e.g. hip, knee (with meniscus intact), and ankle).

Excessive mechanical loading causes cell death in cartilage tissue (Kurz et al. 2005; Levin et al. 2005). Mechanical tests on newborn bovine cartilage explants have shown that apoptotic injury of cartilage begins at about 4.5 MPa (i.e. 30% strain) when subject to a single load application in 'unconfined compression' at a rate of 1 mm/s displacement (which held for 5 min (allowing pore fluid drainage) and 25 min off, and repeated six times on 1-mm-thick specimens) (Loening et al. 2000). The fraction of cells undergoing apoptosis increases to 50% of chondrocytes at 20 MPa (i.e. 50% strain) (Loening et al. 2000). The number of apoptotic cells peaked within one day. The experimental data obtained from newborn bovine cartilage appears to be consistent with data obtained from unconfined compression tests on adult human cadaver cartilage explants (aged between 18 and 45 years) (D'Lima et al. 2001). In these experiments, a single 14 MPa unconfined compression test (resulting in 30% strain held for 500 ms) leads to about 30% of chondrocytes undergoing apoptosis over 7 days (D'Lima et al. 2001).

Note that though the degree of apoptosis was 30% at 7 days, there was only 10% of apoptotic chondrocytes at 1 day (see Fig. 5 in D'Lima et al. 2001). This suggests that following traumatic injury to cartilage, there is a 'window of opportunity' in which the degree of apoptosis may be influenced. Indeed, it has been shown that following traumatic injury, appropriate mechanical loading can increase the biosynthesis of aggrecan and collagen while decreasing the amount of apoptosis, but above a threshold of mechanical loading, apoptosis is further increased (see Fig. 2.32) (Li et al. 2013).

Note that injurious loading in the experiments by Li et al. (2013) was 50% compression, loaded at a rate of 100 strains per second. Cytokines added to medium in some experiments were rhTNFα at 25 ng/ml, rhIL-6 (50 ng/ml), and sIL-6R (250 ng/ml). Cyclic straining was 10%, 20%, or 30% at 0.5 Hz (1 h on, 5 h off, i.e. 40% duty cycle) (Li et al. 2013). The results show that 20% cyclic strain approximately halves the amount of apoptosis, but 30% cyclic strain approximately doubles the amount of apoptosis (see Fig. 2.32) (Li et al. 2013).

Chondrocyte apoptosis may also be assessed using 'confined compression.' For example, 5-mm-diameter cartilage explants (explants had intact articular surface) obtained from young and mature bovine cartilage were tested at 1 and 5 MPa at a frequency of 0.5 Hz for 0.5, 1, 3, 6, and 16 h at 37 °C and 5% CO_2 (Levin et al. 2005). Mature and immature cartilage, subject to 0.5 Hz cyclic load for 6 h at

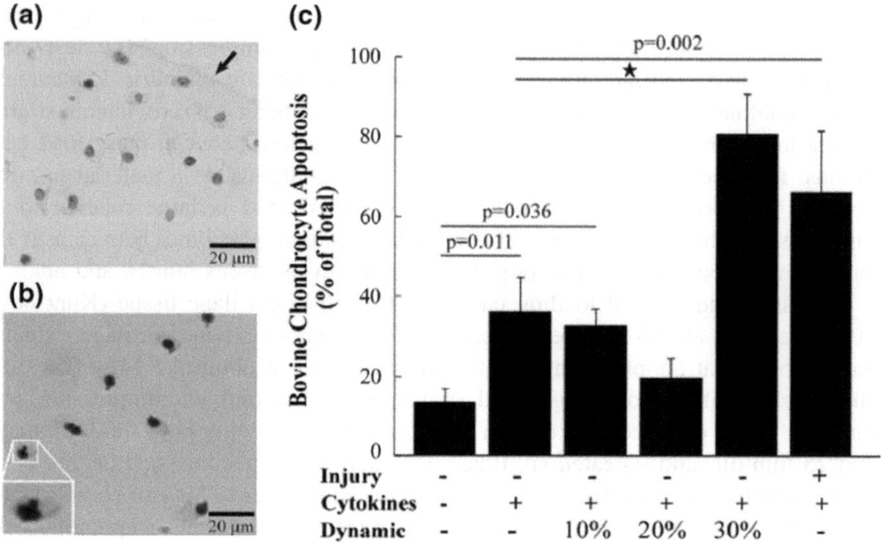

Fig. 2.32 Bovine [calf] chondrocyte apoptosis within the explants in response to 4-day treatments. **a** Representative image (40× objective) of histological sections from untreated control disks using hematoxylin staining for the nucleus. The superficial-most surface is visualized in the upper right corner (arrow). **b** Histological section from explant treated with the combination of cytokines (TNF-α + IL-6/sIL-6R) plus intermittent 30% dynamic strain amplitude. Image (40× objective) was taken from middle-zone cartilage: apoptotic cells displayed nuclear blebbing, a morphological marker of apoptosis. Insert: Higher magnification of nuclear blebbing. **c** Percentage of chondrocytes in histological sections that underwent apoptosis, quantified as the ratio of cells showing nuclear blebbing to total cell count; $n = 4$ disks per condition. Values are mean ± 95% confidence interval; $*P < 0.001$ (Li et al. 2013)

5 MPa peak stress, results in chondrocyte cell death within 300 μm (mature) and 600 μm (immature) of the cartilage (load) surface (see Fig. 2.33). It is noted that a similar depth of cell death occurred after just 0.5 h of loading at 5 MPa in mature cartilage (see Fig. 2.34). It is also clear from Fig. 2.34 that the depth of cell death increases with load magnitude, and usually with the duration of loading. Importantly, we note from Fig. 2.34 that half an hour of 1 MPa loading at 0.5 Hz in mature cartilage has little effect on the rate of apoptosis, however, for a load of 5 MPa for half an hour at 0.5 Hz, cell death increases substantially (up to 150 μm from the (load) surface) (Levin et al. 2005).

It is observed that there was little cell death if the load is reduced to 1 MPa for 1 h (and is about five times less than for 6 h compressed to 5 MPa load), but cell death increased substantially within the top 100 μm if the 1 MPa load continued for 16 h. There are some indications that the mature cartilage is more resistant to cell death than immature cartilage. The amount of cell death did not increase between 24 and 48 h after loading (Levin et al. 2005).

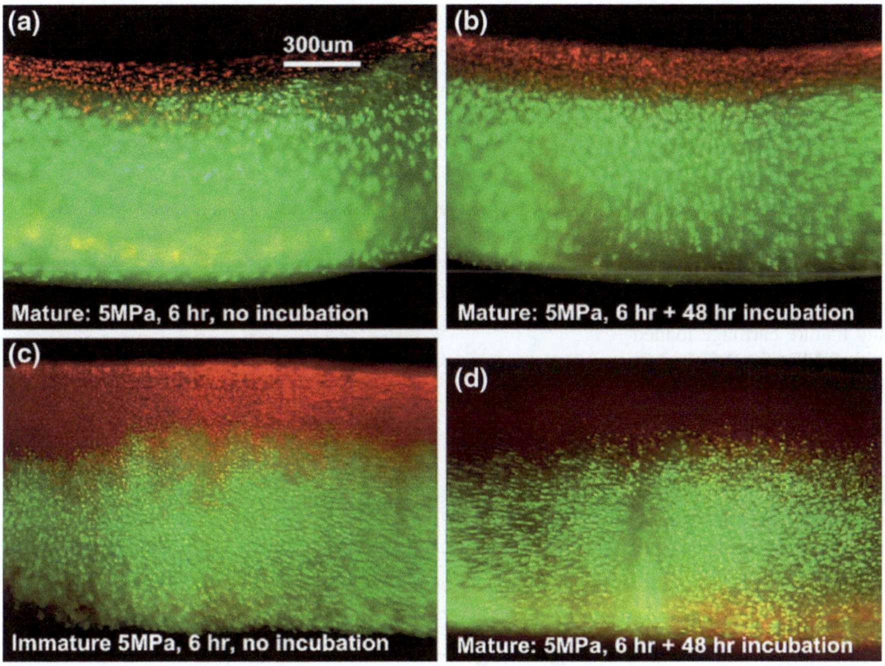

Fig. 2.33 Low-power fluorescent photomicrographs of cartilage loaded with 5 MPa for 6 h. Live cells were stained green by FDA, whereas dead cells were stained red by PI: mature cartilage **a** immediately after loading and **b** 48 h after unloading; immature cartilage **c** immediately after loading and **d** 48 h after unloading. Images **a** and **b** were taken from the same cartilage explant, so were images **c** and **d**. Bar = 300 mm (Levin et al. 2005). Note that for compressive surface loadings, the depth of cell death is usually less than the depth of the aggrecan depletion, but greater than the depth of collagen damage (see illustration Fig. 2.1)

Importantly, we note that for a given stress magnitude, cell damage and cell death are *reduced* when the strain rate is fast (see for example Fig. 2.4). This is because of the constant volume 'stiffening effect' of poroelastic materials like cartilage. Poroelastic materials exhibit near constant volume when the rate of loading is fast relative to the rate at which water can exit the cartilage (Levin et al. 2005) (see discussion Sect. 1.6).

Shear load applied to the surface of bovine cartilage explants (which induces shear straining through the depth of bovine cartilage explants) has been found to cause depth-dependent chondrocyte apoptosis (see Fig. 4.20) (Waller et al. 2013). It is reported that the amount of apoptosis depended on the type of lubrication (and so friction coefficient) at the cartilage surface, which is related to the shear load at the surface and the shear straining through the depth of the explant.

While the number of apoptotic cells depends on load type, magnitude, and frequency, and apoptotic set points in cells (which in turn are influenced by hormonal and local cytokine and growth factor signaling environments), experiments

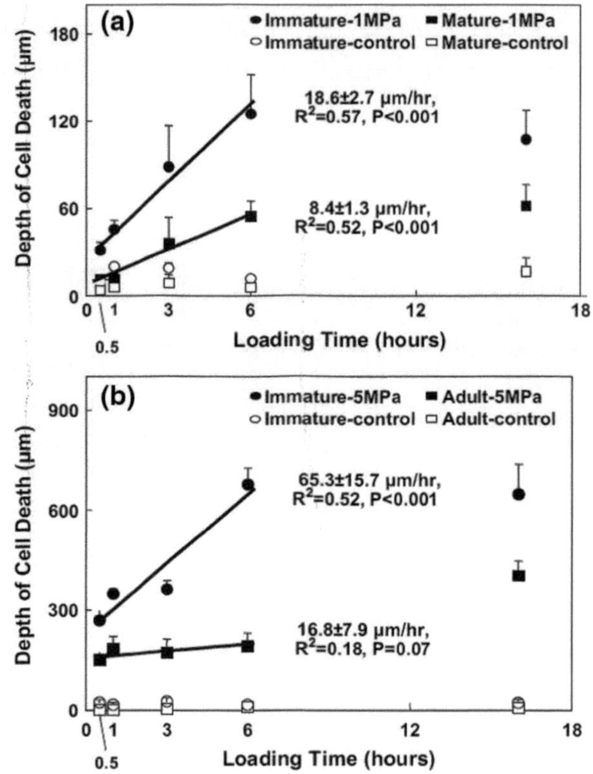

Fig. 2.34 Depth of cell death from the articular surface as a function of loading duration. **a** Immature and mature cartilage loaded with 1 MPa for 0.5, 1, 3, 6, and 16 h. Linear regression analysis (rate of increase in depth) found that the cell death in immature cartilage increased at a greater rate than that in mature cartilage. **b** Immature and mature cartilage loaded with 5 MPa for 0.5, 1, 3, 6, and 16 h. The rate of cell death in immature cartilage increased at a greater rate (slope given as microns increase in depth of cell death/ additional hour of loading) than the mature cartilage (Levin et al. 2005)

like these demonstrate that excessive mechanical loading is responsible for chondrocyte death in articular cartilage.

A fundamental hypothesis is that the primary activator of homeostatic systems within mechanical tissues (including articular cartilage) are changes in the mechanical properties of the tissue. So for example, if there is a substantial loss of aggrecan, remaining chondrocytes are in danger of apoptosis. If remaining chondrocytes do not have the potential to synthesize sufficient ECM to raise the local osmolality and so regulate the local mechanical microenvironment to the desired set point, then signaling systems will be activated so that chondrocytes are stimulated to proliferate (Xu et al. 2010; Akiyama et al. 2004; Brisby et al. 2013; Ryan et al. 2009; Barbero et al. 2004). An increase in chondrocyte density means that the total ECM production potential per unit volume of tissue is increased, which gives the tissue the capability of restoring its desired mechanical properties.

We observe an increase rate of chondrocyte in vivo proliferation is reported for (8–12 weeks old Sprague Dawley—body weight 200 g) rat knee articular cartilage subject to moderate running exercise over a 3-month period [see Figs. 2 and 3 (Brisby et al. 2013)], while an increase in proliferation marker PCNA is reported for static loading (40% strain) on 2–3 month-old bovine femoral condyle articular cartilage [see Figs. 2 and 5 (Ryan et al. 2009)]. Pairs, clusters, and ‘horizontal strings' of chondrocytes are common in the superficial zone of normal articular cartilage (Rolauffs et al. 2008), suggesting proliferation. And novel spatial patterns of chondrocytes are observed in osteoarthritis (e.g. double parallel strings of chondrocytes), with cell proliferation increasing superficial zone chondrocyte density in osteoarthritis (Rolauffs et al. 2010).

There can be no doubt about the importance of hormonal, regional, and local signaling environments for cartilage homeostasis (Zhen and Cao 2014), says Zhen and Cao:

> The indispensable role of TGFβ in maintenance of articular cartilage metabolic homeostasis and structural integrity has been well established. TGFβ stimulates early events in chondrogenesis, including chondrogenic condensation, chondroprogenitor cell proliferation and differentiation.

Chondrocyte proliferation is substantially downregulated when CREB signaling is blocked (Long et al. 2001). It is noted that CREB signaling is activated by ATP, PGE_2, and PTHrP, which are all stimulated by mechanically induced fluid shear stress (Ogawa et al. 2014). Chondrocyte proliferation also requires the activation of a number of different signaling pathways including: (canonical) Wnt signaling (via canonical Wnts 3 and 7, which activate 'frizzled receptors'), Smad signaling (via TGFβ 1 to 3 activating TGFβ receptors), IGF signaling (via IGF-I and II activating tyrosine kinase receptors), insulin signaling (via insulin receptors), and probably other signaling pathways that also activate AKT (e.g. integrin signaling) and ERK-mTOR. Barbero et al. also report that a mix of TGFβ, FGF-2, and PDGF-BB increased proliferation rates for isolated chondrocytes by about 3.7 relative to proliferation in control media (n = 69 patients) (Barbero et al. 2004), while Qu et al. (2017) report that chondrocytes exposed to PDGF-AB and protease increased relative proliferation rates several fold (Qu et al. 2017). We also note the importance of hormonal, regional, and local signaling environments regulating chondrocyte apoptosis. These include not only the molecules mentioned above—which tend to increase cell survival as well as increasing cell proliferation (and so decrease the likelihood of apoptosis)—but also other signaling molecules such as nitric oxide, FAS, and some kinds of Wnt signaling, which increase the likelihood of cell apoptosis (Lotz et al. 1999; Zhu et al. 2008; Lories et al. 2013).

Clearly, proliferation increases the total synthetic potential of the tissue, giving a chondrocyte population the capacity to produce the amount of ECM required to control the mechanical strains, fluid flows, and fluid pressures within the tissue under the applied loading. This is arguably the primary reason higher chondrocyte densities are usually found directly beneath higher mechanical stresses (see for example Fig. 1.1). To maintain chondrocyte homeostasis, rates of tissue damage including chondrocyte apoptosis must be matched by rates of tissue repair including chondrocyte proliferation.

Mechanical loading not only damages ECM molecules, it also damages molecules within chondrocytes. Exactly what damage is done inside cells is not well known, but it can include rupture of the cytoskeleton, rupture of internal cell membranes, and disruption of internal cell signaling. When chondrocytes experience deformations that are too large and/or too often, then chondrocytes undergo apoptosis (see for example Chen et al. 2001a, Kuhn et al. 2004; Zimmerman et al. 1988; Buckwalter et al. 2006; Waller et al. 2013). These conditions are most likely to occur in the superficial zone, despite chondrocytes being adapted to this mechanical environment. Apoptotic cell markers are observed in normal cartilage (Heraud et al. 2000; Sharif et al. 2004; Hashimoto et al. 1998a). For example, Heraud et al. report 2–5% of cells in a histological section in normal adult cartilage are 'TUNEL positive' for this marker of apoptosis (median age 77) (Heraud et al. 2000), while Sharif et al. report 0.6% of chondrocytes are 'TUNEL positive' for apoptosis in normal (e.g. samples obtained from patients with osteoporosis) human femoral and knee cartilage (median age 68), while the rate of apoptosis increases two- to fourfold in osteoarthritis (see Fig. 2 in Sharif et al. 2004).

Some reports have it that 'apoptosis rates' in a histological section of normal cartilage are as high as 5–11% (Sandell and Aigner 2001), but (Sandell and Aigner 2001) argue this rate is too high on the basis that the cartilage homeostasis could not be maintained if apoptosis rates were that high:

> Recently, it was reported that apoptotic cell death is a dominant event in the degeneration of osteoarthritic cartilage, although the results are not in good agreement: for example, cell death in cartilage samples ranged from 5 to 11% and in patients with OA, from 22 to 51% of all cells. We think it is very likely that these numbers are overestimates of the extent of apoptosis in cartilage, because if they are correct, other biosynthetic parameters of OA would be impossible; indeed even 'normal' cartilage would soon lose the capacity to undergo biosynthesis. In theory, a major degree of cell death would easily lead to a failure of turnover of the cartilage matrix, because chondrocytes are the only source of synthesis of matrix components in articular cartilage and there is no renewal of chondrocyte population. In our studies (T Aigner, unpublished findings), we have confirmed that apoptosis occurs in osteoarthritic cartilage, but at a very low rate with approximately 0.1% of the total cell population apoptotic at a given time point, indicating that the death of chondrocytes has only a limited impact on the pathology of osteoarthritis.

We agree that if apoptosis rates are high, then '*indeed even 'normal' cartilage would soon lose the capacity to undergo biosynthesis.*' Cartilage needs to maintain its biosynthetic potential to cope with the environmental demands made upon it. We also agree that the normal *rate* of chondrocyte apoptosis in normal cartilage is likely to be only a fraction of a percent of the total number of chondrocytes in a histological section, and somewhat higher in osteoarthritic cartilage (Sharif et al. 2004) due to softening of the ECM, larger strains, and more ECM and cell damage. It has been shown that chondrocyte apoptosis is associated with ECM damage (Thomas et al. 2007), and that chondrocyte apoptosis is associated with excessive or unusual loading (Waller et al. 2013).

But what does: '*we have confirmed a very low rate of approximately 0.1% of the total cell population apoptotic at a given point in time*' actually mean, when there is no time involved in a 'snapshot' of tissue (i.e. '*at a given point in time,*' no rate of apoptosis is actually specified by any fraction of chondrocytes expressing a marker of apoptosis)? This makes it difficult to argue that one marker with a high 'labeling fraction' is less valid than another marker with a low labeling fraction, only that each marker has different (time) dynamics. It does not mean that one marker is necessarily more accurate than another marker.

Nevertheless, we do believe that the rate of apoptosis must be only a fraction of a percent of the total number of chondrocytes, otherwise the chondrocyte turnover rate would be too high. For example, take the (Sandell and Aigner 2001) estimate of 0.1%, and assume that marker 'turns over' each day (i.e. a new set of cells display the marker each day), or each week or each month. If the marker turns over each day, then for a rate of apoptosis equal to 0.1%, and assuming there is no chondrocyte replacement, after 3 years there is only 33% ($=0.999^{(3 \times 365)}$) of chondrocytes remaining in the tissue. If the markers turned over each week, it would take 21 years to reach the same fraction (i.e. 33%), and if the marker turned over each one month, it would take some 84 years to reach the same fraction of chondrocytes remaining. This means that even for low rates of chondrocyte apoptosis (e.g. say an apoptosis rate equal to 0.01% per day without replacement, less than half the chondrocytes remain after 21 years), there needs to be chondrocyte proliferation taking place to maintain an approximately constant chondrocyte cell population in the human adult knee cartilage between 20 and 39 years of age (see Fig. 1 in Barbero et al. 2004).

Following chondrocyte apoptosis, some lacunas in the cartilage ECM are left empty (Gonzalez et al. 2007), more often in the superficial zone than further from the cartilage surface [see Fig. 2.35 (Aigner et al. 2001)]. The increased percentage of empty lacunae in the superficial zone is consistent with a higher rate of chondrocyte apoptosis in the superficial zone, as evidenced by a greater percentage of chondrocytes expressing the apoptotic marker caspase 3 in the superficial zone (see Fig. 2.36). But a small, relative constant percentage of empty lacunae in the superficial zone between 30 and 60 years of age [see Fig. 2.35 (Aigner et al. 2001)] suggests ongoing chondrocyte *proliferation* in the superficial zone throughout life.

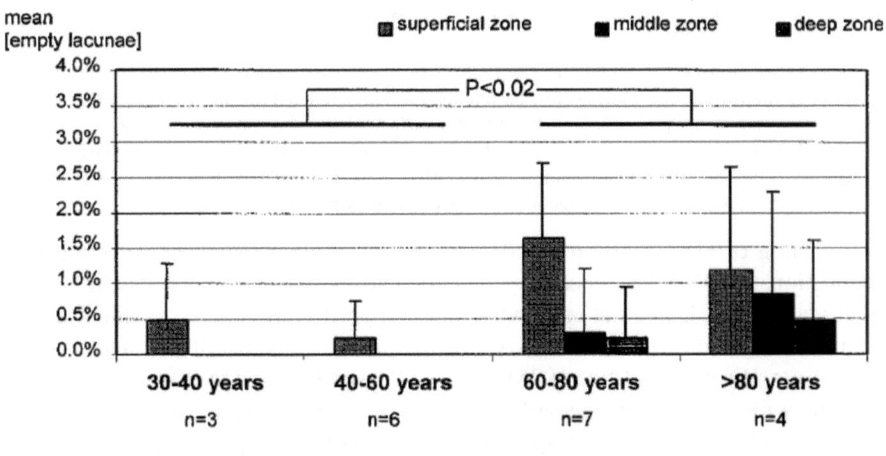

Fig. 2.35 Diagrams showing the mean ± SD percentage of empty lacunae found in the various age groups of normal cartilage samples (Aigner et al. 2001)

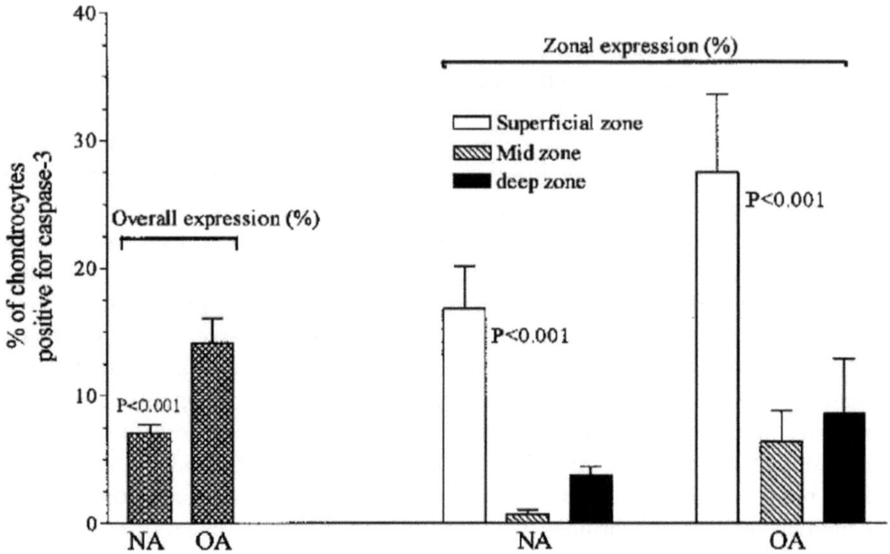

Fig. 2.36 Percentage of chondrocytes expressing caspase-3 in non-arthritic (NA) and osteoarthritic (OA) cartilage. Bars show the mean and SEM (Sharif et al. 2004). Note that caspase 3 is an enzyme that mediates apoptosis, and so is a molecular marker indicating an increased likelihood of apoptosis (see also Fig. 6 in Sharif et al. (2004) for data on pro-apoptotic marker p53)

Based on the concept of chondrocyte homeostasis, one might expect that chondrocyte proliferation rates are similar to chondrocyte loss rates (as a first approximation, the loss rate can be considered equal to the apoptosis rate). Observations of the molecular marker Ki67 in normal adult knee cartilage indicate that about 0.6% of chondrocytes in deep cartilage and 0.1% of chondrocytes in the superficial cartilage are proliferating (Muinos-Lopez et al. 2012) (see Fig. 2.37). We note remarks by Scholzen and Gerdes (2000):

> The expression of the human Ki-67 protein is strictly associated with cell proliferation. During interphase, the antigen can be exclusively detected within the nucleus, whereas in mitosis most of the protein is relocated to the surface of the chromosomes. The fact that the Ki-67 protein is present during all active phases of the cell cycle (G(1), S, G(2), and mitosis), but is absent from resting cells (G(0)), makes it an excellent marker for determining the so-called growth fraction of a given cell population.

Labeling of DNA using tritiated thymidine is another means of detecting DNA replication, as DNA is a necessary precursor to cytokinesis. Such labeling is normally taken to indicate that cell division or cell proliferation is proceeding (and will occur), though some point out DNA labeling does not necessarily mean cytokinesis will occur. Hirotani and Ito (1975) exposed 21 small cartilage 'blocks' taken from patients with 'secondary osteoarthritis,' and 20 small cartilage 'blocks' taken from patients with un-united intracapsular fracture (which are presumed to be normal cartilage), to tritiated thymidine for 4–20 h (immersed in Eagle MEM medium).

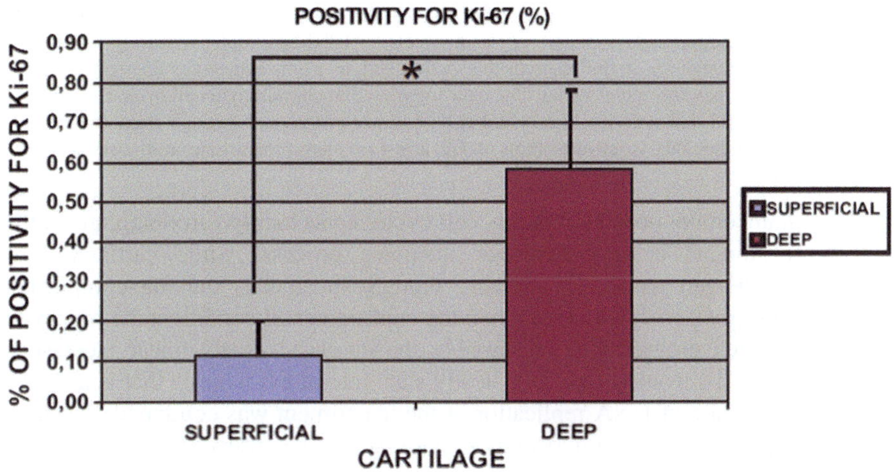

Fig. 2.37 Analysis of *Ki67* expression in histological sections of normal articular cartilage from human knee articular cartilage. Percentage of cells positive for *Ki67* in both cartilage biopsies. $^*p < 0.05$ (Muinos-Lopez et al. 2012) (inferred $n = 12$). Note in this experiment, cells were isolated by trypsin and collagen type IV digestion, so they are not normal chondrocytes within articular cartilage tissue

They found that only five of the cartilage osteoarthritic specimens had cells that were labeled with thymidine, suggesting that chondrocytes in 'normal' cartilage do not divide. However, the labeling period was very short, and in addition, the protocol described (4–20 h) suggests that the labeling time, in addition to being short, may have varied considerably. In other words, there may not have been enough time for some normal cartilage cells to undergo DNA replication and be labeled with such a short labeling 'pulse time.'

This appears to be borne out by the data presented in Jadin et al. (2006) (S phase DNA labeled for 6 days using BrdU) and (Lee et al. 1993) (S phase DNA labeled for up to 14 days using tritiated thymidine). Lee et al. (1993) examined 18-month-old steer cartilage. The samples tested were 5 mm × 5 mm full depth blocks from metacarpophalangeal joints, cultured in Dulbecco's minimal essential medium with 20% (v/v) fetal calf serum added. These experiments have tritiated thymidine labeling periods that ran for 1, 3, 5, 7, 9, and 14 days (rather than 4–20 h). After 1 day of labeling, there is hardly any detectable proliferation in normal cartilage, but after 14 days of labeling, about 4% of cells are labeled in normal cartilage (see Fig. 2 in Lee et al. 1993). This data suggests an average chondrocyte DNA replication rate for these test conditions of about 0.3% per day. As explained by Lee et al. (1993):

> [Tritiated thymidine] incorporation is an indication of DNA synthesis, but not necessarily of cell division. If mitosis does occur, it leads to the formation of two separately labelled nuclei adjacent to each other. This may be termed a labelled couplet. In control explants, labelled couplets first appears between 5 and 7 days. After this point, the number of

> labelled couplets rise linearly up to 14 days. Even at the end of the culture period, however, only 3% of labelled cells were present as couplets. A labelled couplet will only be visualised if both nuclei are in the plane of the section and, as a consequence, values presented may be lower than the actual value. The data suggests that chondrocyte cell cycle time is at least 5 days and the majority of labelled cells had not progressed through mitosis within 14 days. Whether this is an indication of G2 arrest or merely of a long and variable G2 phase is unclear.

A long, uncertain, possibly halting, cell-cycle seems hardly surprising, given that cell replication is a very resource intensive process, while cartilage is a resource-challenged tissue. Presumably though, most cells that have replicated DNA will progress to cytokinesis when the environmental conditions both demand and permit such progression. Presumably the observed nearly linear increase in labeling day 5–14 would come to a steady-state rate of cytokinesis that is a sizable fraction of the rate of DNA replication if the experiment was continued, and there were adequate tissue nutrition to support cytokinesis. We observe that a steady-state DNA replication rate of 0.3% per day and a steady-state cytokinesis (couplet formation) rate of ($0.3\% \times 3\% = 0.009\%$ per day) are not compatible with one another long-term, and the true in vivo steady-state rate of cytokinesis is probably somewhere between these two values (indeed Lee et al. (1993) explains in the extract above that the measured cytokinesis rate is probably a lower bound because of sampling bias).

We mention that Fig. 6 in Lee et al. (1993) demonstrates the importance of fetal calf serum in the explant support medium, as rates of thymidine labeling are significantly reduced without exposure to fetal calf serum. Indeed, fetal calf serum '*has become the supplement of choice for cell culture research, as it contains an array of proteins, growth factors and essential ions necessary for cellular viability and proliferation in vitro*' (Bryan et al. 2011). But it should be mentioned that it has been shown that human serum can substitute for calf serum, and it has a similar effect in supporting cell growth. Actually, human serum appeared to be somewhat better than calf serum when culturing human fibroblasts, as evidenced by shortening of the cell population doubling time (Witzeneder et al. 2013). Certainly, cartilage in vivo is exposed to the mix of nutrients present in their own serum (i.e. synovial fluid is an ultrafiltrate of the serum), so the proliferation data for cartilage exposed to fetal calf serum appears to be appropriate.

If we plausibly assume that the cytokinesis rate is about 3.3 times larger than the lower-bound steady-state rate of cytokinesis measured above (an assumption based on considering sampling bias introduced by not counting 'out of (sectioning) plane' cell divisions as couplets, the thymidine labeling rate being so much higher than the couplet formation rate, and taking into account the stimulatory effect of cut surfaces), this would mean the average rate of cytokinesis in adult bovine tissue is about ($0.3\% \times 10\% = 0.03\%$ per day). Even for this estimated average rate of cytokinesis, this implies that cell division could replace all the chondrocytes in the tissue within 9 years (i.e. about 11% of chondrocytes are replaced each year). Clearly though, this is an average value and there would be regional variations in the rate of cytokinesis depending on resources available for cell replication, growth factor stimulation, and the rate of damage to the tissues. We would therefore expect that some regions of cartilage tissue will have cytokinesis rates where every cell divides within a year (or alternatively, possibly fewer cells divide more often), while other tissue regions may have cells dividing every couple of decades.

Though another species, bovine cartilage is similar to human cartilage, and 18-month-old bovine aging is probably a reasonable approximation to teenager to young adult cartilage tissue in humans. Indeed, similar thymidine incorporation trends are reported by Benaroch (1990) for human femoral cartilage, ranging from very young to aged tissue—though the variability in thymidine incorporation per microgram of DNA for different people is notable in these experiments, with some older cartilage having greater rates of thymidine incorporation than cartilage from much younger people. We note that Jadin et al. (2006) found, while proliferation occurred through the full thickness of articular cartilage, most of the proliferation (about 3 or 4 times above the average) occurred in the top 25% for bovine articular cartilage between one to two years of age.

We also mention the estimated proliferation rate of 0.03% per day and the cytokinesis duration of (at least) 5–7 days measured by Lee et al. (1993) can be combined to estimate the total fraction of cells in some stage of the cell cycle other than the G0 phase as 0.03% per day $\times$ 5–7 days (the time required to complete the cell cycle) = 0.15 to 0.2%. This is the estimated fraction of cells would be labeled by Ki67 (Scholzen and Gerdes 2000). This estimated fraction appears to be reasonably consistent with the fraction of cells actually marked for cell division using ki67 reported by (Muinos-Lopez et al. 2012) for normal adult femoral cartilage (i.e. 0.1–0.6%—see Fig. 2.37).

Assuming chondrocyte density homeostasis, we observe that as a first approximation, the reported rates of chondrocyte proliferation markers also align approximately with the rate of apoptosis markers as reported by Sharif et al.—about 0.6% of chondrocytes have apoptotic markers in normal human femoral cartilage (Sharif et al. 2004)—providing we assume that the apoptotic TUNEL marker can label apoptotic cells in cartilage on average for 20 days. This is plausible, as macrophages cannot quickly engulf and remove apoptotic cells in cartilage, as they do in other tissues.

Despite this evidence, we note that some believe that apoptosis and proliferation is not a widespread phenomenon in normal articular cartilage (e.g. Aigner et al. 2001; Newman 1998) and of what proliferation there is, proliferation is most frequent close to the cartilage surface (Rolauffs et al. 2008). Furthermore, some data suggests chondrocyte numbers in normal cartilage decrease with age (see for example Fig. 2.38), together with rising numbers of empty lacunae. However, for adult humans, it is reported that total chondrocyte numbers are relatively constant to age forty (i.e. implying that for first twenty years of adulthood, apoptosis and replacement are balanced), and then their total chondrocyte number declines at about 2% per year (Bobacz et al. 2004), though (Vignon and Arlot 1981) report a 1% per year decline in chondrocyte numbers for human femoral cartilage from age 30.

In this context, it is relevant to mention the decrease in tritiated thymidine incorporation in isolated, cultured human chondrocytes with age (most of the tissue tested was obtained at arthroplasty). For these cells, thymidine incorporation per cell decreases some threefold between 15 and 75 years of age (see Fig. 3 in Martin and Buckwalter 2001 and Fig. 2A in Martin and Buckwalter 2002). Barbero et al. report a twofold decrease in proliferation capacity of isolated (femoral) chondrocytes in control media (i.e. with no additional growth factors) between 25 and 65 (see Fig. 1 in Barbero et al. 2004). Together with the estimate above for the cell proliferation rate in vivo, if about 11% of chondrocytes are replaced each year at age 15, then only 4% would be replaced in normal cartilage at age 75. So even though proliferation continues, it decreases with increasing age, while the decrease in apoptosis presumably lags somewhat, leading to a net loss of chondrocytes after age forty (i.e. suggesting apoptosis rates are somewhat higher than replacement rates after age forty) (Fig. 2.38 and Barbero et al. 2004).

We also observe that for osteoarthritic cartilage, chondrocyte numbers are reported to stay constant (see Fig. 2.38), even though the rate of cell apoptosis is expected to increase for the reasons described above. Steady-state chondrocyte number density and increased apoptosis are only compatible statements if there is increased chondrocyte proliferation. And indeed, chondrocyte proliferation is clearly evidenced in osteoarthritic cartilage (see for example Fig. 2 in Poole et al. 1991b; Rolauffs et al. 2010).

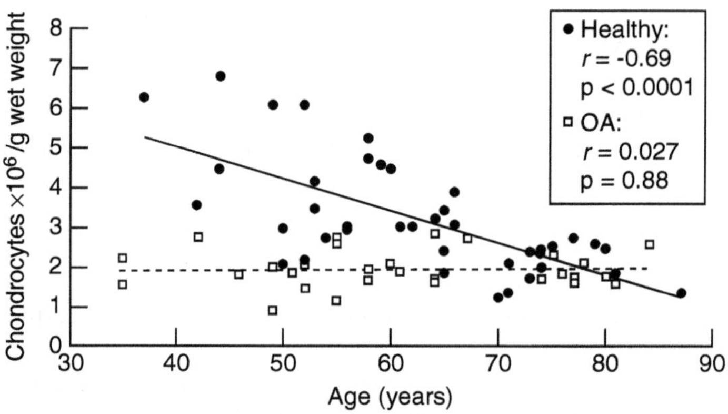

Fig. 2.38 Changes in the total chondrocyte numbers of healthy and osteoarthritic articular cartilage with increasing age. Chondrocytes were enzymatically released from their extracellular matrix in 0.2% collagenase B. Directly after digestion cell numbers were assessed using light microscopy. The values are given as chondrocytes 6×10^6/g wet weight. There was a decrease in chondrocyte numbers in healthy articular cartilage (black circles, black regression line) as a function of age ($n = 41$; $r = -0.69$, $p = 0.0001$). Cellularity in osteoarthritic cartilage (white squares, dashed regression line) showed a significant reduction in cell number compared with healthy cartilage ($p = 0.0001$), but no dependence on patients' age [$n = 30$; $r = 0.027$, $p = 0.88$ (i.e. no difference to the null hypothesis)] (Bobacz et al. 2004). Note this data implies that in normal knee cartilage there is a net loss in total chondrocyte number of around 2% per year from age 40, while Vignon and Arlot (1981) reports a chondrocyte loss of 1% per year for human femoral cartilage from about age 30

Taken together, all this data and reasoning help build confidence that chondrocyte proliferation continues throughout the life of articular cartilage. Clearly, the rate of proliferation varies throughout life, with very high rates of proliferation during postnatal development, but then slowing greatly after sexual maturity is reached, then decreasing with increasing age and increasing again in disease states like osteoarthritis. For normal adult human articular cartilage, we have made an admittedly very approximate estimate for the average rate of cell proliferation, i.e. 0.03% per day in early adulthood, and possibly 0.01% per day in late adulthood. This estimate will be revised when better data becomes available. We add that there is considerable variation from person to person (and from region to region within cartilage tissue) (Benaroch 1990).

We mention here that we have not considered loss of chondrocytes due to cartilage wear loss (see Sect. 3.5 for estimates of surface wear rates). It is probable wear also makes a reasonably significant contributor to chondrocyte loss. It is likely that cell apoptosis probably occurs before wear remove cells, so in this sense apoptosis probably dominates wear-related cell loss too.

Based on these estimates of cartilage proliferation rates, mitotic figures in normal cartilage are expected to be rare. For example, if the proliferation rate is 0.03% per day, and if mitosis takes about 1 h, then the frequency of mitosis is estimated to be about one cell per hundred thousand cells. While this is rare, it is not never, yet mitotic figures are almost never reported in normal adult cartilage in the literature. For example, Jadin et al. (2006) examined 6000 cells in adult bovine tissue and found no mitotic figures (i.e. there was only a 6% chance), though we note Muldrew et al. (2001) does report seeing mitotic figures in ovine cartilage.

How can this very infrequent observation of mitotic figures explained? It seems likely that in addition to having to examine a lot of cells to have a reasonable statistical chance of finding cells undergoing mitosis, additional factors may also play a role. It could be that the frequency of cell mitosis decreases further due to unfavorable cell-cycle conditions in the explant tissue after it is extracted from the living animal for testing, or it could be that normally cell division has a pronounced circadian variation (see Sect. 2.10.1). Indeed, it is known that mitotic figures are seldom seen even in skin tissue (which is among the most proliferative of the body's tissues) unless biopsies are taken at night time http://act.downstate.edu/courseware/histomanual/cell_division.html. To reason this in another way, it makes no sense for the 'delicate ballet' of mitosis to take place when an animal is under (possibly severe) environmental stress (e.g. when running, the repeated impact loading send shock waves through the skeleton, while simultaneously there is strong competition for serum resources from active muscle groups). Consequently, it makes sense for most cell divisions (including articular cartilage) to occur during sleep, the period for growth, renewal, and repair.

How can we explain the loss of 1–2% of chondrocytes per year after the age of 40 years in terms of the concept of chondrocyte density homeostasis? It is possible that the 'set point' for chondrocyte density (which is in any case a function of chondrocyte depth) is also a function of age (see Sect. 5.5 for a discussion of negative feedback and 'set points'). For example, an age-related decline in total chondrocyte density in normal cartilage may well be associated with a change in set points associated with declining systemic blood hormone concentrations.

For example, blood concentration of IGF-1 declines with age at about 1% per year—see Fig. 2.39). Further, blood concentrations of growth hormone decline with age. Growth hormone secretion rate declines exponentially from a peak of 150 μg/kg/day at puberty to about 25 μg/kg/day at age 55 (Hersch and Merriam 2008). Similarly for other hormones, e.g. thyroxine [T_4 declines about 0.2% per year (Lipson et al. 1979)]), and sex hormones (e.g. estrogen levels drop precipitously post-menopause in women and testosterone decreases at about 1% per year from age 40) (Nardozza Junior et al. 2011).

However, Fig. 2.38 also indicates that for an osteoarthritic population, chondrocyte numbers remain constant over decades (at about two million cells per gram of tissue wet weight) (Bobacz et al. 2004; Barbero et al. 2004). This suggests the cell number set point for chondrocyte homeostasis remains constant in osteoarthritis (i.e. so the rate of cell loss equals rate of cell proliferation). Perhaps this is driven by abnormal local production of IGF-1 by chondrocytes in osteoarthritic states among other things (i.e. systemic signaling factors controlling set points have been replaced by local signaling factors controlling set points). However, other evidence suggests chondrocyte cell numbers rise in early osteoarthritis before declining in late-stage osteoarthritis, so perhaps the constant cell number over decades masks variability in individuals at different stages of the osteoarthritic disease process. Indeed, it is possible greater set point control exerted by the tissue (and so decreased systemic set point control) leads to abnormal states (e.g. due to excessive local production of IGF in the tissue leads to chondrocyte 'cloning'—see Sect. 2.9.5).

Based on the evidence detailed above, it seems reasonable to suppose there is homeostasis of both the ECM and chondrocytes. However, the 'set points' for ECM components and chondrocyte cell density appear to be a function of both depth and age, and vary over time (i.e. the set points drift with changes in cell conditions). Both age- and depth-related changes of homeostatic set points may be driven by the decline in systemic hormone concentrations, as well as any local changes in concentrations of paracrine and autocrine signaling molecules.

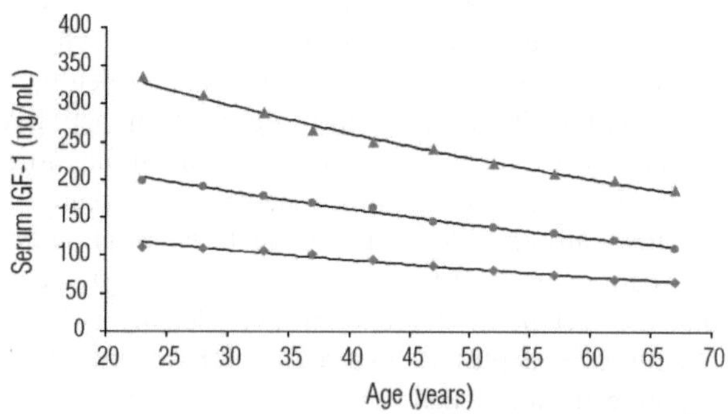

Fig. 2.39 Curves representing the 97.5th, 50th (median), and 2.5th percentiles of IGF-1 concentration as a function of age (Rosario 2010). Note median IGF-1 concentration declines at about 1% per year throughout adulthood

2.8 Chondrocyte Proliferation and Migration

We now focus on a process that facilitates chondrocyte homeostasis, namely chondrocyte migration. Chondrocyte migration helps facilitate matching of spatial mismatches in chondrocyte removal and proliferation. Let us assume that chondrocytes first proliferate then migrate (to assume the reverse would mean one region becomes depleted in chondrocytes while another region gains chondrocytes, which seems counterproductive). Then a logical deduction supporting chondrocyte migration in vivo follows—if cells proliferate and *do not migrate* then chondrocyte 'clones' form. Chondrocyte cloning means that a single chondrocyte proliferates and the daughter cells (clones) accumulate in 'multicellular clumps' within a large single lacuna. Chondrocyte cloning is very common observation in osteoarthritic cartilage (see Fig. 2.55). Chondrocyte cloning occurs because there is a defect in the mechanism for chondrocyte migration, at least there is a migration defect in advanced osteoarthritic cartilage when chondrocyte cloning is most frequently observed (see discussion Sect. 2.9.5 for further explanation).

See Fig. 8 in Poole (1997): *Type IX collagen distribution in spontaneously osteoarthritic canine cartilage showing (a) the reactive erosion front beneath the articulating surface [note chondrocyte clones in lacunae], (b) relatively normal chondrons near the interface with the reactive zone, and (c) a chondrocyte cluster near the eroding surface. Bars, 10* m (Poole 1997).

See Fig. 9C in Poole (1997): *Type VI collagen distribution in osteoarthritic tissue, (c) cell division and migration within the expanded microenvironment [chondrocyte cloning], and (d) clonal chondrocyte cluster formation. Bars 10 microns* (Poole 1997).

There is evidence for chondrocyte apoptosis (e.g. empty lacuna), and there is evidence for chondrocyte mitosis, e.g. multiple chondrocytes in 'columns' are common during development, while chondrocyte 'rows' or 'strings' become localized toward the surface of the cartilage in normal adult cartilage (i.e. parallel with the cartilage surface) (Noskina and Cole 2002; Rolauffs et al. 2008). There is no evidence of multicellular clumps of cells developing throughout the transitional and deep zones in normal healthy articular cartilage. In the superficial zone of normal healthy cartilage, 'strings' (which probably result from repeated oriented cell proliferation and migration), clusters (we note there are a few clusters that may resemble clumps with chondrocytes located around the perimeter, but this may occur by chance), pairs (i.e. these are probably recently divided 'couplets'), and single chondrocyte (non-dividing cell population) are reported by Rolauffs et al. (2008). However, we note cloning is observed in early osteoarthritis, and cloning is a hallmark of advanced osteoarthritis (see Fig. 2.55 and Poole 1997; Povysil et al. 2008). Cloning also occurs in response to some specific injuries (Havdrup et al. 1982; Muldrew et al. 2001). The logical explanation that is consistent with all the observations for normal cartilage is that following cell division, in healthy cartilage, chondrocytes migrate to new spatial locations in the ECM (see discussion on chondrocyte migration Sect. 2.9).

Based on this discussion, we now propose that apoptotic chondrocytes in vivo are in fact replaced by a two-step homeostatic process that ensures not only normal chondrocyte numbers, but normal chondrocyte density (relative to the local set point, which is probably controlled by local concentrations of chemical signals and the mechanical environment around chondrocytes): (i) proliferation of chondrocytes (which helps to correct the diminished total tissue repair capacity, which is diminished as a consequence of cell loss) and (ii) migration of daughter cells to new spatial locations within the cartilage tissue (so as to help spatially match repair capability to repair needs).

Because a low aggrecan concentration is a key indicator of cartilage tissue with damaged collagen network and/or low chondrocyte density and so in need of repair, we further propose that the usual direction of chondrocyte migration is down an aggrecan gradient. Because aggrecan is normally lost from cartilage surfaces, this means the direction of chondrocyte migration is generally toward a nearby cartilage surface. Chondrocytes in the superficial zone have the least protection to excessive mechanical straining, and so they are the most likely to undergo apoptosis due to mechanical causes (see Figs. 2.35 and 2.36). Even though superficial zone chondrocytes are better adapted to surviving large cyclic strains, the large cyclic strains take their toll (e.g. caspase 3 levels Fig. 2.36 are high and there are more empty lacuna Fig. 2.35), and so rates of apoptosis are relatively high. Apoptotic chondrocytes in the superficial zone can be replenished either by chondrocytes migrating from the transitional zone to the superficial zone, or by proliferation within the superficial zone. Both occur. When chondrocytes in the superficial zone do proliferate, the familiar 'columns' become oriented horizontally in 'rows,' parallel with the cartilage surface (see for example, the horizontal columns of proliferating cells evidenced in Fig. 2.40), or cluster or pairs (Rolauffs et al. 2008). We note that the proportion of cell arrangements differs between joints and across surfaces within joints (Rolauffs et al. 2008). It may be that high shear environments (e.g. femoral condyles, patella groove, distal humerus, radial groove) produce more 'strings' of chondrocytes, while low shear tends to produce clusters or pairs of chondrocytes (e.g. tibial surface, talar dome).

Now tenocytes in tendon also arrange themselves in 'chains,' just as superficial chondrocytes arrange themselves in 'strings' or 'rows' (Rolauffs et al. 2008), among other cell arrangements. Tenocytes have cell processes (or sheets) that extend and 'touch' processes from nearby cells (Waggett et al. 2006). The cell processes have connexin molecules between them, providing direct molecular communication pathways [tenocytes express both connexin 43 and 32 (Waggett et al. 2006)], just as chondrocytes do in the superficial zone of cartilage (chondrocytes express connexin 32, 43 and 45) (Knight et al. 2009). We note in passing that connexin 32 may be associated with anabolic processes, while connexin 43 and 45 may be associated with catabolic processes.

The tenocytes align themselves along the collagen fibers, which in turn aligns with the major principal stress in the tissue. Given these similarities, it is perhaps unsurprising that chondrocytes in tension in the superficial zone sheet take on tenocyte-like morphology and align themselves in rows (see Fig. 2.40). Indeed, it is

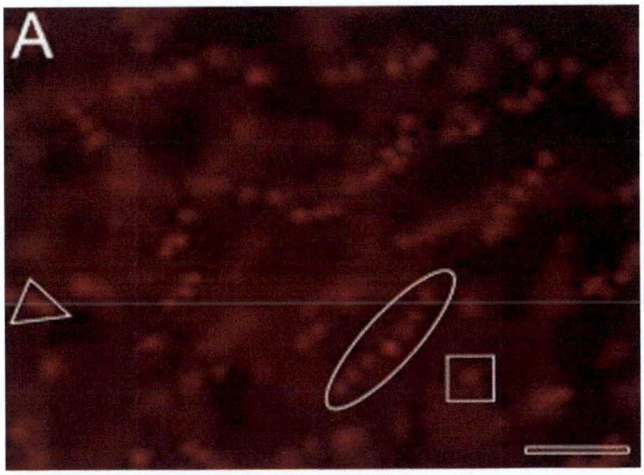

Fig. 2.40 Confocal fluorescence image looking down on the surface of human articular cartilage (Rolauffs et al. 2011). It is evident that there are many linear groupings of chondrocytes parallel with the cartilage surface. This suggests that chondrocytes in the superficial zone proliferate to form horizontal arrangements of daughter cells, unlike the deep zone where daughter cells form vertical columns of chondrocytes. Rolauffs et al. (2011) that changes in the angle of proliferating chondrocytes may indicate the onset of osteoarthritis (Rolauffs et al. 2011). See also Rolauffs et al. (2008, 2010)

reported that for about 2% of the normal population, chondrocytes at the surface of the superficial zone becomes so tenocyte-like they express some collagen type I (Roberts et al. 2009), though it is often said that no type I collagen is expressed in normal cartilage. On this basis, we might therefore expect that chondrocyte 'strings' or 'rows' would align themselves approximately along the first and second major principal tensile stress directions in the tangential plane within the collagen sheet in the superficial zone.

It is noted that increased variability of the angles between chains of chondrocytes (viewed from above) has been suggested as an early sign of osteoarthritis (Rolauffs et al. 2011). This makes sense given the propensity of cells to align themselves with the principle stresses, as increased tissue damage is likely to lead to greater tissue heterogeneity in load directions acting on the cartilage surface (and so a wider dispersion of angles around the first and second major principal stress directions). This will increase the variations in directions of collagen fibers, and so the directions chondrocytes choose to proliferate and migrate.

While chondrocyte migration has been often reported experimentally in vitro and in explant systems, chondrocyte migration in vivo has not been observed directly in normal cartilage tissue, so the idea that chondrocytes may migrate in vivo is controversial (Lyman et al. 2012). However, it well known that chondrocytes in cartilage tissue display all the morphological features of migratory cells. They can display filopodias, lamellipodia, centrioles, cilia, and have cytoskeletal fibers (e.g. smooth muscle alpha actin) associated with migratory cells.

However, we also propose that not only does chondrocyte migration enable a more homogenous distribution of chondrocytes and so aggrecan and other ECM molecule production throughout the cartilage tissue, but it also has important implications for maintenance of the *collagen network* within normal cartilage (see Sect. 2.9.4). Normally, mechanically induced damage of extracellular matrix is non-uniform in cartilage, being comparatively large toward the surface of the articular cartilage (superficial and upper transitional zones) and diminishing with distance from the surface (deep zone). Consequently, we propose that while the collagen network in the deep zone has a comparatively long half-life, the collagen network in the near surface zones has much shorter half-lives. Further when the collagen network is so damaged that it cannot retain aggrecan, chondrocytes migrate toward the areas of collagen network damage, and in the process of migrating, they replace the collagen network with new collagen network in their 'wake.'

When the collagen network is significantly damaged, it loosens the collagen network and local aggrecan concentrations are lowered, and over time cartilage tissue osmolality falls and approaches that of serum (i.e. isotonic saline). Under these conditions, chondrocytes develop lamellipodia and take on fibroblast-like characteristics. In this state, the migrating population of chondrocytes extend processes and begin to pull themselves through damaged cartilage (Qiu et al. 2000), possibly toward a newly created cartilage surface (Lyman et al. 2012), much as fibroblasts can do (Hinz et al. 2001) and much as tenocytes can do (Tsai et al. 2003). Indeed in the lower aggrecan environment found in the superficial zone of the cartilage, local chondrocytes become spindle shaped, and normally extend filopodia and lamellipodia horizontally so they 'touch,' which probably facilitates communication and exchange of materials between the cells (Gonzalez et al. 2007).

In Sect. 2.9, we explore the concept that there is a continuum of differentiation states for chondrocytes, and that these differentiation states map onto a continuum of migration states—from stationary chondrocytes through to fast migration. We propose that 'slow chondrocyte migration' is a 'push migration mode' (the usual migration mode in the vertical direction in deep and transitional zones). Migration may be in the horizontal direction in the superficial zone. At the other end of the migration continuum is a fast 'pull migration mode.' The pull migration mode is most often found close to cartilage surfaces, and usually only occurs in heavily damaged tissue, e.g. near splits in the cartilage surface, near partial-depth defects or in diseased tissue. We describe in detail how the push migration mode operates and review what experimental evidence there is supporting this migration mode.

If the concept of chondrocyte migration is unfamiliar, we note here that so-called tissue remodeling in growing animals provides further evidence supporting the proposal that chondrocytes migrate in vivo. For cartilage undergoes tissue remodeling as it is subjected to increasing mechanical loadings as the animal grows from newborn to an adult. In very young cartilage, the distribution of chondrocytes throughout the ECM and the ECM itself are rather uniform. But upon reaching sexual maturity, this uniform cell and ECM distributions have 'remodeled' to form distinct zones within the articular cartilage (described as radial, transition, and superficial zones) (Karlsson et al. 2009). In adult tissue, cell numbers and collagen organization and density then vary between zones. As chondrocytes are the only cells present in cartilage, these cells have to be the cells responsible for 'tissue remodeling' in articular cartilage.

A logical explanation consistent with these observations of tissue remodeling is that following chondrocyte division (which most agree is prevalent in young cartilage tissue), chondrocytes migrate to new spatial locations within the cartilage, and while migrating they remodel the collagen network to form the foundational collagen network, from which characteristic collagen structures within each zone are formed through the combined actions of proteases and cyclic mechanical loading. In other words, chondrocyte migration plays an important role in both assisting the remodeling of the ECM and in shaping the spatial distribution of chondrocyte density (Babyn et al. 1996; van Turnhout et al. 2008, 2010; Iordan et al. 2010; Hunziker et al. 2007).

2.9 Modes and Mechanisms of Chondrocyte Migration

Chondrocytes in damaged articular cartilage can migrate. It is well established that chondrocytes do migrate in a 'fast migration mode' when articular cartilage damage is severe. Here we also propose that chondrocytes can migrate in a 'slow migration mode,' when articular cartilage damage is much less severe. In fact, it seems likely that slow mode migration may be a normal part of collagen network repair mechanism in the upper transitional zone and superficial zones of articular cartilage.

We begin our discussion by first considering fast mode chondrocyte migration, and then review the evidence supporting a slow mode for chondrocyte migration. We then explain the relationship between the two migration modes and then consider in more depth how slow migration actually works and how it may play a role in tissue repair. Finally, we consider the consequences of imbalances in chondrocyte proliferation and chondrocyte migration, and how these outcomes are displayed in the characteristic appearance of degraded articular cartilage when examined histologically.

2.9.1 Fast Mode Chondrocyte Migration

ECM osmolality and cell–cell signaling molecule concentrations are continuously variable, and so gene expression profiles of chondrocytes are also continuously variable. This dynamically adjustable gene expression profile leads to a 'graded response' in the differentiation states of chondrocytes (Yasuhara et al. 2011; Nadzir et al. 2013). Graded states of chondrocyte differentiation are clearly evidenced morphologically (see for example Fig. 2.41) (Bush and Hall 2003).

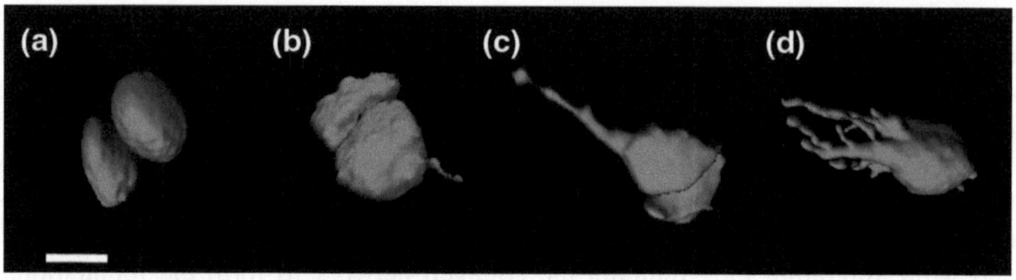

Fig. 2.41 Examples of morphologically abnormal human articular mid-zone chondrocytes. **a** Morphologically normal, spheroidal cell with relatively smooth surface, **b** cell with single small process (<8 μm), **c** single chondrocyte with single process more than 8 μm, and **d** single chondrocyte with multiple processes more than 8 μm (Bush and Hall 2003)

While changing osmolality plays a role in this transformation, undoubtedly a number of different mechanisms operate to effect a change in the gene expression profile. For example, Plaas et al. (2011) argue that rising ADAMTS5 concentrations in the ECM lead to a switch in TGFβ signaling that induces more fibroblast-like phenotypes (see Fig. 5.4). Others have suggested changes in ALK5/ALK1 ratios lead to changes in signaling that lead to more fibroblast-like phenotypes (van der Kraan et al. 2012). There may be changes in TGFβ concentrations, as well as changes in TGFβ signaling pathways. A switch from canonical toward non-canonical Wnt signaling may also lead to a more fibroblast-like phenotype (see discussion Sect. 5.6.3). Again for simplicity, here we focus primarily on osmolality changes as the primary driver for change, while remembering that a number of microenvironmental cues probably contribute to de-differentiation of chondrocytes.

In normal low aggrecan density ECM, less well-differentiated chondrocytes become more fibroblast-like in appearance (Yasuhara et al. 2011; Ma et al. 2013). Chondrocytes lose their rounded morphology and become more spindle-shaped. In very severely damaged collagen network that has very low concentrations of aggrecan, chondrocytes extend long processes and pull themselves through the ECM. This 'pull mode' for cell migration has been reported in a number of publications (Morales 2007; Chang et al. 2003; Lyman et al. 2012).

For example, using human osteochondral plugs (obtained at the time of surgery from osteoarthritic cartilage with 'normal' appearance—mean age of patients 62 years), and then introducing 2 mm deep partial thickness defects into the cartilage, it was shown that wound healing on the edge of the cartilage defect involved chondrocytes undergoing flattening, polarization of actin cytoskeleton, extension of pseudopods toward the defect and the extension of filopodia to interact with collagen fibrils (see Fig. 2.42 and description by Lyman et al. 2012).

Interestingly, Lyman et al. also investigate cell density adjacent to and distant (i.e. >300 μm) from the margin of the cartilage defect—these authors report that: '*Cell density decreased progressively in the 300-μm zone adjacent to the defect to an average of approximately 25% to 35% after 3 weeks*' (Lyman et al. 2012). Chondrocyte numbers close to the defect decline in the superficial and deep zones

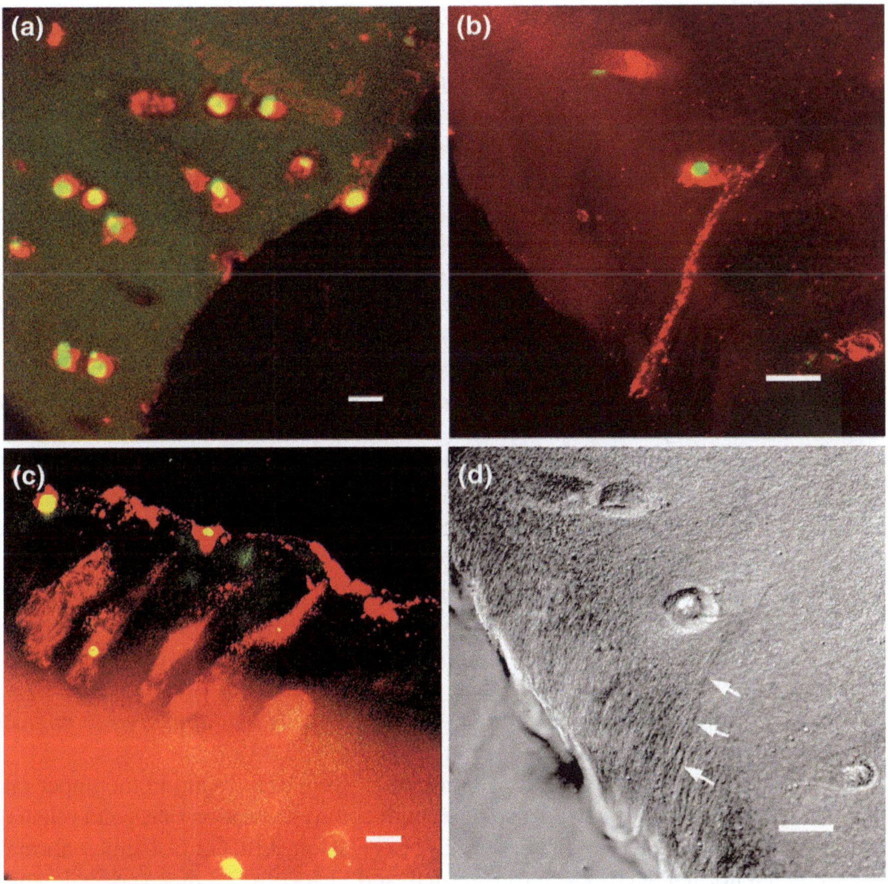

Fig. 2.42 Chondrocytes extend cell processes toward the edge of the defect. The red fluorescence is CD44 labeling of plasma membranes, and the green is Hoechst staining of nuclei (images are pseudocolor). **a** Confocal microscope image of a defect edge in the middle zone on the day of defect creation. Chondrocytes were rounded and found in abundance adjacent to the defect edge. Bar = 10 μm. **b** Confocal image of a defect edge in the deep zone at 3 days. A cell process extends from the chondrocyte to the fibrillated edge of the defect. Bar = 10 μm. **c** Confocal image at 4 weeks demonstrating multiple cells with extensions toward the defect and cells on the edge of the defect. Bar = 10 μm. **d** DIC image of defect edge in the deep zone at 3 days (corresponding to Fig. 2B). A small cleft extends from the chondrocyte to the fibrillated edge of the defect (arrows). Bar = 10 μm (Lyman et al. 2012)

to 10–25% of the original cell density. However, from two weeks onward, cell density begins to rebound quite strongly in the deep zone (clearly indicating proliferation of chondrocytes) (see Fig. 9 in Lyman et al. 2012). Cell density in the distant cartilage in the superficial, transitional, and deep zones cartilage remains relatively constant throughout the four-week experiment. Proliferation of chondrocytes adjacent to the partial thickness defect is consistent with the reports of Hembry et al. (see Fig. 6 in Hembry et al. 2001).

See Fig. 9 in Lyman et al. (2012): *Graphs comparing the change of cartilage cellularity adjacent and distal to the defect edges in 3 different zones of the explants. (**A**) The changes in cellularity of the superficial zone adjacent to the defect (diamond) and distal to the defect (square) are compared. Over 300 μm from the defect, the cellularity of the cartilage did not change significantly, while within 300 μm, the cellularity dramatically decreased during the first 2 weeks after the defect was created. (**B**) Small changes in the middle zone are seen. Again, greater than 300 μm from the defect, the cellularity was not affected dramatically by culture conditions. The cellularity adjacent to the defects was also unchanged. (**C**) Changes in the deep zone are demonstrated. These included a gradual decrease in the cellularity in the deep zone over the first 2 weeks. The cellularity of the deep zone distal to the defects, like that of the middle and superficial zones, remained relatively stable throughout the 4 weeks. In the area adjacent to the defect, the cellularity increased during the last 2 weeks ($n = 3$ for all points; mean ± SD is shown).*

In another example of fast mode chondrocyte migration, blunt impact trauma to mature bovine cartilage tissue stimulates a vigorous migratory and proliferative response by chondrocytes in the tissue (Seol et al. 2012). Both human and bovine cartilage subject to impact loading may lead to chondrocyte 'progenitor cells' migrating through the cartilage matrix, or exiting the cartilage (as demonstrated by Lyman et al. 2012). Chondrocytes that exit the cartilage may migrate (by extending long thin processes, i.e. fast migration mode) and proliferate on the cartilage surface (see Fig. 2.49). Seol et al. observe that: '*non-viable areas in cartilage injured by blunt impact were repopulated within 7–14 days by cells that appeared to migrate from the surrounding matrix*' (Seol et al. 2012) (Fig. 2.43).

Seol et al. (2012) report that 'progenitor cells' were only found in the upper one third of the cartilage. These cells displayed high colony formation rates, and putative stem cell markers were also increased, e.g. expression ABCG2 > 4 times normal, SOX9 by 3.8 normal, RUNX-2 by 11 normal, and FLK-1 by 2.8 times normal, while Notch-1 and CD44 were increased by 7.4 and 14 times normal, respectively, i.e. final gene expression profile are closer to MSCs than to normal chondrocytes. No progenitor cells were found in the lower two-thirds of the cartilage. This suggests that the proliferative cells in normal mature bovine cartilage are found in the upper one-third of the cartilage (which accords with the mobile α-actin-expressing cells being in the upper portion of cartilage). High-mobility group protein B1 (HMGB1—a DNA bending molecule) was identified as the primary chemo-attractant following blunt trauma (HMGB1 is released by cytolytic cells such as natural killer lymphocytes and macrophages). Blocking the RAGE receptor also significantly reduced migration. High levels of cytokine and chemokine expression were evident, which doubtless facilitated the progenitor cell population's rapid movement, but interestingly the progenitor cells also expressed high concentrations of PRG4 (which is a marker of chondrocytes in the upper transitional and superficial zones of articular cartilage). Similar results were found for human cartilage, suggesting that mature bovine cartilage is a reasonable model of adult cartilage.

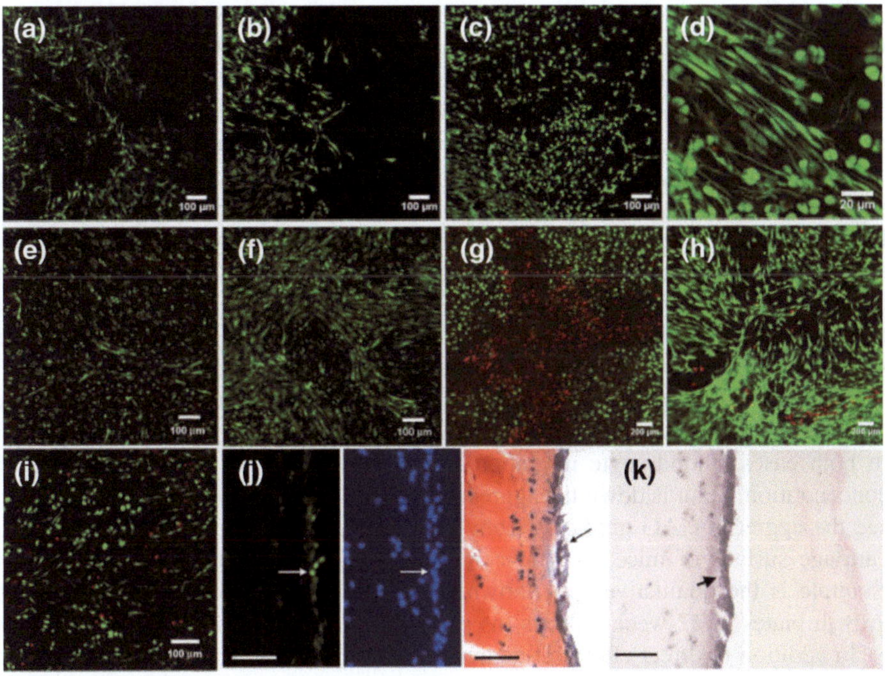

Fig. 2.43 Migrating cells on injured cartilage. **a–c** Confocal microscopy images showing live cells (green) in the same area within an impact site on the surface of an explant on day 7 (**a**), day 11 (**b**), and day 15 (**c**) after impact. **d** Higher magnification view showing the elongated morphology and dendritic appearance of the cells. **e**, **f** Live cells observed on a human talus specimen (obtained from a 36-year-old man) on day 6 (**e**) and day 10 (**f**) after impact. **g**, **h** Dead cells (red) and live cells in a bovine explant with a cross-shaped needle scratch. Images were obtained immediately after the injury (**g**) and 14 days later (**h**). **i** Appearance of migrating cells on the surface of cartilage dissected free from subchondral bone immediately after impact. **j** Left, Green immunofluorescence staining for proliferating cell nuclear antigen, showing positive cells (arrow) on the surface of a cartilage explant. Middle, Blue staining in the same section, showing all nuclei. Right, Surface migrating putative chondrogenic progenitor cells (arrow) in a consecutive section stained with Safranin-O–fast green. **k** Immunohistochemical staining for lubricin. Arrow (left) indicates strongly positive migrating cells in an impact site. Right, Negative control. Bars in **j** and **k** 100 μm (Seol et al. 2012)

2.9.2 Slow Mode Chondrocyte Migration

It is hypothesized that the primary driver for activation of homeostatic response systems in mechanical tissues such as articular cartilage is a change in the mechanical properties of the tissue as a result of mechanical loading (i.e. tissue damage). As part of the whole suite of homeostatic mechanisms, if there is a substantial collagen network damage and loss of aggrecan, then signaling systems are activated so that chondrocytes are stimulated to proliferate. Daughter chondrocytes then slowly migrate in the direction down local aggrecan gradients. In doing so they move to new locations, remodeling and repairing damaged collagen while migrating. In the following, we propose a mechanism for slow chondrocyte migration in vivo based on proteolytic degradation of the ECM along the 'leading edge' of the chondrocyte (in the direction of chondrocyte migration), and secretion of aggrecan and procollagen along the 'trailing edge' of the chondrocyte.

More specifically, we propose that falling concentrations of aggrecan stimulate remaining chondrocytes toward migration (Urban et al. 1993; Lories et al. 2013; Pabbruwe et al. 2009). Aggrecan concentration may decrease because of insufficient aggrecan production as a consequence of net chondrocyte loss (Lotz et al. 1999), or because of rapid enzyme degradation by secreted aggrecanases (Ilic et al. 2007), or because of collagen damage and excessive loss of aggrecan through the cartilage surface, or some combination of these factors.

We note that slow chondrocyte migration in the articular zone is in the opposite direction to chondrocyte migration in the growth plate in the same bone. However in both cases, chondrocyte migration is in the direction of decreasing aggrecan concentration, i.e. it is down the aggrecan gradient. For examples of such gradients, see the aggrecan/GAG gradients from bone and growth plate through to articular cartilage surface of mice, as shown in Fig. 4 of Mertz et al. (2012). Or another example is the qualitative data shown as the intensity of alcian blue staining of growth plates in 12-week-old rats (as shown in Fig. 3 of Bassett et al. 2006).

In addition to direct effects of aggrecan loss on chondrocytes through changing osmolarity [e.g. reduced osmolality promotes chondrocyte proliferation (Xu et al. 2010)] in vivo, reduced aggrecan concentrations also change the load-deformation response of the tissue, which will also contribute to the changing the gene expression profile of the chondrocytes. Clearly, chondrocytes are also modulated in their proliferation and migration through a variety of mechanisms, including cell–cell signaling molecules, e.g. concentrations of IGFs (Capito and Spector 2007), Wnts (Ma et al. 2013; Lories et al. 2013), TGFβs (Beier et al. 2001; Zhen and Cao 2014) and PTHrP and Ihh (Beier et al. 2001; Chen et al. 2008), all being particularly important (Zhong et al. 2015).

As the local aggrecan concentration in the tissue decreases the chondrocyte gene expression profile changes, and consequently so does the chondrocyte's phenotype (including cell morphology and secretion profile) (Urban et al. 1993; van der Windt et al. 2010; Aigner et al. 1993). This is not surprising given that osmolality influences SOX9 expression, probably the key transcription factor determining the chondrocytic cell type (Tew et al. 2009). In this context, we note that changes in cell morphology are strongly correlated with disease processes (see for example Fig. 1 of Sandell and Aigner 2001). Phenotype change is not unusual, rather, it is what cells do.

Because ECM osmolality and the cell–cell signaling molecule concentrations are continuously variable, so the gene expression profile of chondrocytes is continuously variable. It is proposed here that this continuously adjustable gene expression profile leads to a 'graded response' in the differentiation states of chondrocytes (Yasuhara et al. 2011; Nadzir et al. 2013). See also Fig. 2.41 (Bush and Hall 2003). Normally, chondrocytes in the deep zone are well-differentiated, rounded chondrocytes, while chondrocytes in the superficial zone are less well differentiated, more spindle-shaped, and fibroblast-like. Some of these fibroblast-like cells express smooth muscle α actin, a standard marker of TGFβ-induced conversion from fibroblast-like chondrocytes to myofibroblast-like chondrocytes (myo-cells have considerable contractile ability).

The superficial zone of cartilage has to sustain very large tensile strains as the surface tissue deforms under compressive loading. In this biaxial strain environment, chondrocytes take on a morphological appearance more like tenocytes (and very occasionally, a few chondrocytes may even express collagen type I in the most superficial layer of the superficial zone). This chondrocyte phenotype in the superficial zone is very different to the morphological appearance of chondrocytes in the deep zone (Yasuhara et al. 2011). This serves to remind us that chondrocytes in each zone are phenotypically different, as indeed they are phenotypically different in the lateral direction. In other words, the differentiation state of the chondrocytes alters as the ECM composition, chemical signaling, and the mechanical pressure and strain state change. Given the infinite variety of microenvironments, this serves to remind us that no two chondrocytes are exactly the same.

If chondrocytes do migrate in the direction from deep to superficial zones, they will progressively de-differentiate as the migration path encounters new microenvironments along the way. At high osmolality (i.e. high aggrecan concentration), chondrocytes are smooth and rounded (Nadzir et al. 2013; Hunziker et al. 2002). In this (pre-hypertrophic) differentiation state in normal adult cartilage, the majority if not all chondrocytes are stationary in the deep zone cartilage ECM. When aggrecan concentrations are sufficiently high, chondrocytes in the deep zone produce Indian hedgehog (Ihh). These pre-hypertrophic cells produce enzymes that degrade granulin epithelium precursor (GEP). Degradation of GEP blocks chondrocytes in the deep zone progressing to a hypertrophic state and then undergoing apoptosis (Feng et al. 2010). In normal adult cartilage, the deep zone chondrocytes 'stay put' while busily secreting large quantities of ECM components (see for example, the aggrecan synthesis profile over the depth of cartilage in Fig. 2.9). Clearly this ECM production in the deep zone provides the 'foundational ECM molecules' required for articular cartilage, thereby helping to maintain homeostasis of the cartilage tissue throughout its entire volume, as ECM components may polymerize anywhere along their migration pathway from point of production to the cartilage surface.

See Fig. 2a, b (Poole et al. 1984): *Transverse section (a) and diagrammatic representation (b) of the deeper layers of canine tibial cartilage showing concentrically, the chondrocyte (C), pericellular matrix (*), pericellular capsule (arrowheads), territorial matrix (TM) and interterritorial matrix (IM).* This image and schematic shows morphological appears of a stationary chondrocyte. Note the rounded cell shape and centrally located cell nucleus.

As ECM is damaged, its ability to retain aggrecan in the collagen network is reduced, and the local concentration of aggrecan decreases below its normal concentration. When this occurs the chondrocyte is first observed to increase in size (which is consistent with reduced osmolality of the local ECM) (Urban et al. 1993;

Bush and Hall 2003). Then it begins to exhibit small projections from its surface (Bush and Hall 2003).

Initially, these projections are very small, attaching to nearby collagen network [see Fig. 1B (Poole et al. 1992)]. These projections become larger and more numerous the further the aggrecan concentration falls. In other words, there is a graded response in cell morphology (see for example Fig. 2.41). The cells begin to de-differentiate toward a migratory mode, exhibiting a more prominent smooth muscle α actin cytoskeleton (McNary et al. 2014). We observe that similar observations are made for fibroblastic cells in low density collagen networks (Iordan et al. 2010).

We mention here that there are three main isoforms of actin: α, β, and γ. α actin is associated with muscles and with cell migration, β actin is found around the edge of cells and is involved in the formation of cell projections, while γ actin is found as cytoskeletal tensile stress elements http://en.wikipedia.org/wiki/Actin.

In some cases, damaged collagen network is so loosened and enzymatically degraded that the chondrocyte extensions are many 'cell bodies' in length (Lyman et al. 2012). In the limit, some (artificially constructed) tissue scaffolds have no aggrecan and are of an 'open structure.' Under these conditions, chondrocyte extensions may become very long. At this end of the spectrum, chondrocyte migration can be very rapid (Pabbruwe et al. 2009; Ng et al. 2012)—cell migration rates are of the order of 1 μm per hour (Chen et al. 2001a)).

Again we note that as a chondrocyte's phenotype changes as ECM osmolality falls, so a chondrocyte's propensity to migrate increases. The population of migrating cells probably all express smooth muscle alpha actin. As the osmolality decreases, mechanical stimulation in the transitional zone induces chondrocytes to produce more PTHrP, stimulating the rate of chondrocyte proliferation. Mechanical stimulation also increases Wnt concentrations (both canonical and non-canonical).

As production of PTHrP and Wnts increases at the cartilage surface, cells at increasing depth from the cartilage surface may be induced to proliferate (via the 'canonical' (pathway) Wnts, e.g. Wnts 3 and 7) and to de-differentiate (via the 'non-canonical' (pathway) Wnts, e.g. Wnts 5 and 11) into a 'migratory mode' (Lories et al. 2013). Migratory cells usually migrate along aggrecan gradients in the direction of decreasing aggrecan (Pabbruwe et al. 2009). It is this process that gives rise to vertically oriented columns of chondrocytes in the deep zone, and a greater variety of directions for migration in the transitional zone of cartilage.

We note these vertical columns in the deep zone probably develop prior to adulthood and then become quiescent during much of adulthood, but they may be reactivated in adulthood if ECM damage is sufficiently large. As the concentration of aggrecan falls and becomes more heterogeneous in the upper deep and transitional zones, and as PTHrP and Wnt concentrations increase in the upper deep and transitional zones, the chondrocyte proliferation in the deep zone may be reactivated and chondrocytes proliferate and 'move upward' in (vertical) alignment seen in the deep zone, and then they 'spread out' (i.e. following a more variable course) through the transitional and superficial zones, as they make their way to where they are needed most (i.e. where ECM damage is greatest).

As migratory chondrocytes reach their final destination, and aggrecan concentrations begin to rise again in the ECM around chondrocytes, SOX9 concentrations rise inside chondrocytes. High aggrecan concentrations (both directly and indirectly via changing signaling molecule concentrations) then suppress chondrocyte migration (Davies 2006; Davies et al. 2008), inducing chondrocytes to once again become stationary. Having achieved chondrocyte (number) density homeostasis, and having restored aggrecan concentrations and tissue stiffness, chondrocyte proliferation is again suppressed and the repaired tissue becomes 'quiescent.' But if the ECM is again damaged, and aggrecan concentration falls, chondrocytes may again become 'reactivated,' and again begin to proliferate and migrate. By this means, the cartilage tissue maintains homeostasis in the face of its environmental challenges experienced throughout its life.

Of course the direction and speed of chondrocyte migration may be modified by various factors in addition to the aggrecan gradient, particularly the local density and quality of the collagen network. Chemotactic factors, including IGF-1, bFGF, PDGF, periostin (McGregor et al. 2011; Chang et al. 2003; Chinzei et al. 2018; Qu et al. 2017) will modify the chemotactic response of the migrating chondrocytes. Further, a number of signaling molecules have been shown to suppress cell migration, including IGF-1 and nitric oxide (Nadzir et al. 2013; Frenkel et al. 1996). IGF-1 (and other signaling molecules) can increase or decrease the rate of migration depending on microenvironmental conditions.

In normal cartilage ECM subject to normal usage, aggrecan concentrations are higher more distant from the cartilage surface. Therefore, most chondrocytes in the deep zone where the ECM is dense are stationary. Chondrocytes most likely to be migrating in normal adult cartilage are those chondrocytes in lower aggrecan environments found closer to the cartilage surface, in a comparatively de-differentiated state. The most likely route of migration in normal adult cartilage is from the upper transitional zone into the superficial zone, replacing apoptotic cells in the superficial zone. In healthy cartilage, chondrocyte migration occurs very slowly (i.e. migration rates are probably of the order of less than 1 μm per day).

Because chondrocyte migration rates in healthy cartilage are so slow, there is debate as to whether or not 'slow migration' actually occurs at all (in contrast, fast migration is well documented). However, there is experimental evidence that comparatively well-differentiated chondrocytes are in fact capable of migrating, as shown in Fig. 2.44 (Pabbruwe et al. 2009). Note in this particular case, chondrocytes are migrating from the edge of bovine nasal cartilage (which almost certainly has mechanically disrupted collagen network near the cut surface, due to the cutting procedure employed to recover the cartilage specimen), into a relatively loose collagen network in the collagen scaffold, and migration from a collagen scaffold pre-seeded with chondrocytes into the natural cartilage (see Fig. 2.44). Importantly, in the context of slow migration and collagen network repair, we note that: '*The cartilage–implant–cartilage sandwich appeared macroscopically as one continuous piece of tissue at the end of 40 day cultures*' (Pabbruwe et al. 2009). Further, slow migration of chondrocytes across the cartilage-scaffold interface resulted in

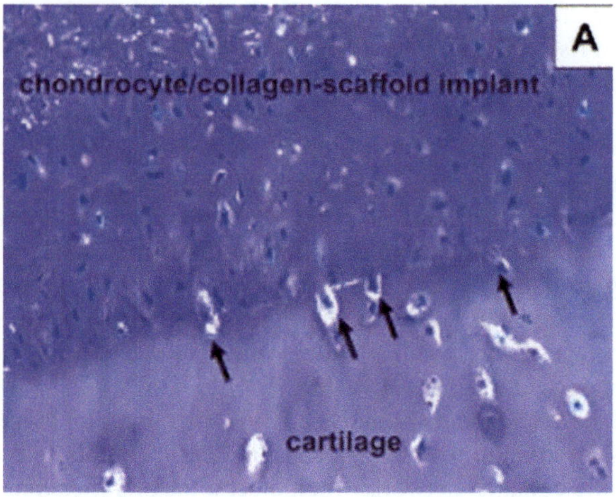

Fig. 2.44 Cell migration in [24–30-month bovine nasal] cartilage integration constructs. Cell migration from the chondrocyte/collagen-scaffold implant to the surrounding nasal cartilage was determined by pre-labeling the bovine nasal chondrocytes with fluorescent dye PKH26. In **a** the pre-labeled chondrocyte/collagen-scaffold implant was inserted between two pieces of living bovine nasal cartilage and one of the two interfaces is shown. **a** Stained with Toluidine blue and viewed under white light (200× magnification). Arrows indicate cells in the process of migrating across the border between the cartilage and the membrane (Pabbruwe et al. 2009). Importantly, note that normal well-differentiated mature bovine chondrocytes begin to migrate from normal nasal cartilage (high aggrecan concentration) into chondrocyte-collagen-scaffold implant (with low aggrecan concentration), down the aggrecan concentration gradient. This cell migration leads to 'integration' between scaffold and cartilage (almost certainly indicating significant collagen formation at the cartilage–construct interface), confirmed by tensile testing that reveals a significant strength that can only be conferred by collagen fibrils. See Pabbruwe et al. (2009) for further details

collagen network formation across the interface, and this collagen formation following chondrocytes migration gave the interface considerable mechanical strength: 'we have provided new evidence to support the hypothesis that cell migration between cartilage surfaces is critical for integration of those tissues' (Pabbruwe et al. 2009). How can such mechanical strength arise at the interface following slow chondrocyte migration across it?

Further evidence of migration by relatively rounded chondrocytes (i.e. morphologically well-differentiated chondrocytes in bovine (18–24 months) cartilage explant test samples) is provided in another experiment exploring repopulation of a zone of collagen network depleted of chondrocytes (i.e. the so-called zone chondrocyte death (ZCD)) (see Fig. 2.45) (McGregor et al. 2011). The ZCD is the region of cell death next to the 'cut surface' (the cut being necessary to obtain the cartilage explant test samples).

Importantly, for chondrocyte migration to be readily detected in the experimental time frame of 4 weeks, bovine explants were exposed to collagenase for 10 min.

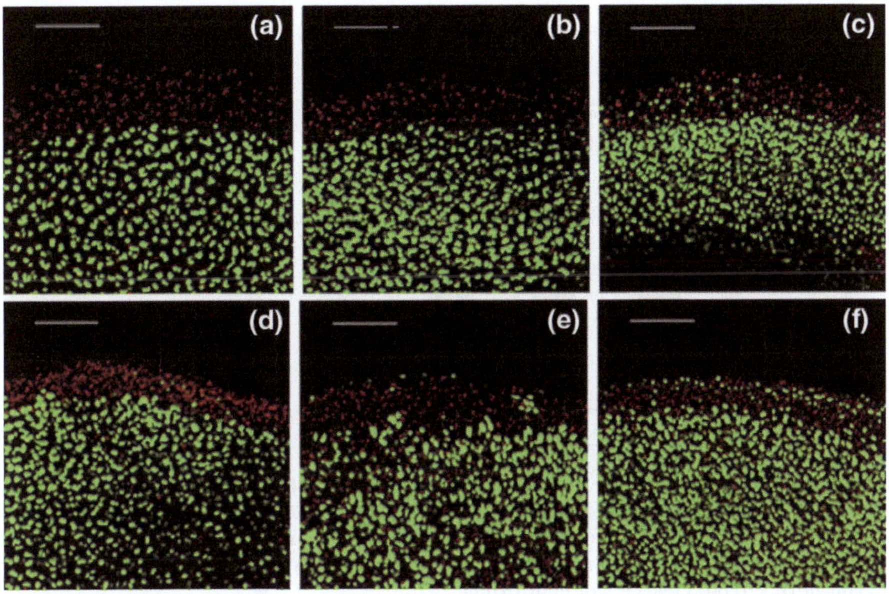

Fig. 2.45 Confocal images of the effect of various treatments on the repopulation of the ZCD. **a** Initial ZCD after 10 days of culture compared to different treatments after 4 weeks of culture: **b** untreated (control), **c** collagenase treatment only (0.6% for 10 min), **d** collagenase treatment and PDGF-bb (25 ng/mL), **e** collagenase treatment and bFGF (25 ng/mL), and **f** collagenase treatment and IGF-1 (25 ng/mL). Magnification of 40×; scale bar = 200 μm (McGregor et al. 2011)

During this procedure, it is reported that the collagenase penetrated about 175 μm into the cartilage explant, which is approximately the thickness of the ZCD. This collagenase procedure is designed to speed chondrocyte migration. The collagenase has the effect of further degrading the collagen network, thereby presumably facilitating more rapid migration of nearby chondrocytes to repopulate the ZCD. Equally importantly, we are suggesting that chondrocyte migration in vivo is stimulated by significant damage to the collagen network (i.e. by damage sufficient to cause local aggrecan depletion and the creation of local aggrecan gradients that chondrocytes migrate down—see discussion Sect. 2.9.4), so this experiment is analogous to what might be expected to happen in vivo.

As evidenced by Figs. 2.45 and 2.46, for the case of collagenase treatment followed by exposure to 25 ng/ml IGF in the support media, the fastest moving adult bovine chondrocytes migrated at a rate of (180–50 μm)/28 days = 4.6 μm per day over a 4-week period, all the while maintaining a relatively well-differentiated chondrocyte morphology (McGregor et al. 2011).

It seems plausible that mechanically damaged collagen network (see for example Fig. 3.5b, c, e, f) is loosened to such an extent by repeated cyclic loadings that the damage is equivalent to cutting and collagenase treatment (as described above). We

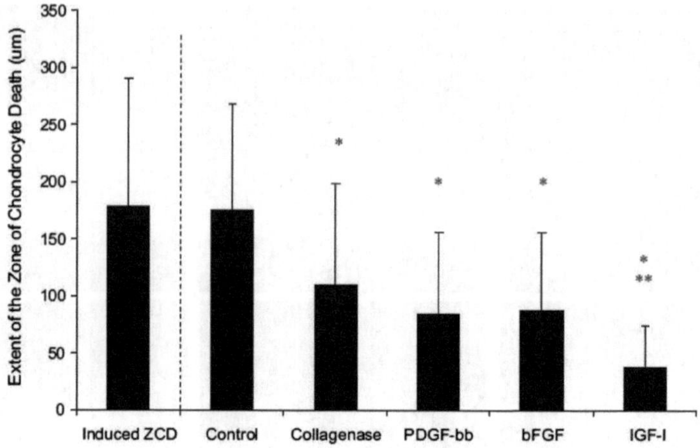

Fig. 2.46 ZCD measured from the cut surface of bovine articular cartilage explants following a 10-min treatment with 0.6% collagenase, or collagenase plus a chemotactic growth factor (PDGF-bb, bFGF, or IGF-I) at 25 ng/mL. Data expressed as mean with 95% CIs ($n = 5$). *Significantly different than control ($P < 0.02$); **Significantly different than collagenase-only treatment ($P < 0.001$) (McGregor et al. 2011)

also note in passing that it has been reported that soluble bFGF is released by cutting cartilage (probably from the damaged ECM chondrocytes) which induces upregulation of MMP1 and MMP3 (Vincent et al. 2002). In other words, following mechanical damage, in the normal course of events this in turn upregulates collagenases, which helps degrade the mechanically damaged collagen network. Given the experimental conditions in vitro are analogous to significant collagen damage in vivo, one can plausibly assume that the observed in vitro chondrocyte behavior can be extrapolated to in vivo chondrocyte behavior. On this basis, it is estimated that well-differentiated chondrocytes in vivo may be induced by aggrecan depletion together with very significant collagen network damage to migrate at speeds up to about 4–5 μm per day. At this rate of migration, it would take about 200–250 days for chondrocytes to migrate through 1 mm of cartilage. In other words, chondrocytes could theoretically migrate virtually the full thickness of human tibial cartilage (i.e. about 2.5 mm) over a period of about 2 years. This estimate is useful, for it gives us a 'lower-bound estimate' of the time required for repair of extensively damaged collagen network in human tibial cartilage. The actual time for repair may be longer if migration is slower or if multiple passes of successive chondrocytes is needed to repair a section of damaged tissue, or if further damage is occurring.

There is some in vivo evidence for slow mode migration. Once again, this behavior becomes prominent only when the collagen network has been 'mechanically loosened,' for example, by ongoing cyclic loading following ACL transections in mature dogs (3–8 years old). The magnitude and type of this cyclic loading on the surface of the articular cartilage following ACL transection is abnormal, and this abnormal loading causes the collagen network to be 'loosened.' Examining articular cartilage six months after ACL transection, Vignon et al. (1983) report:

> Such a repair process is suggested by the observation that dogs killed more than six months after operation do not show any degenerative alterations but a marked thickening of the superficial membrane due to apposition of new tangential fibers. This is also suggested by the proliferation of the remaining hypertrophic viable cells, which may be able to migrate in the lacerated matrix and to achieve its repair Such a possibility of repair might be relevant to the problem of chondromalacia patella [i.e. patellofemoral pain syndrome], which may undergo spontaneous recovery.

In the report by Vignon et al. (1983), chondrocyte migration in vivo is inferred from the data on the basis that: (i) the loosened, and initially much thicker cartilage [mean increase in articular cartilage thickness is 30% more than the normal thickness] following the ACL transection, displays a marked reduction in chondrocyte number density, (ii) yet the thicker cartilage later displays marked proliferation [indicated by chondrocyte proliferation and increased uptake of tritiated thymidine] over the following months, and finally, (iii) the thicker articular cartilage has a normal chondrocyte density at 6 months. At six moths the cartilage is thicker, but is otherwise normal looking cartilage. In short, the initial reduction in chondrocyte density occurs at early times, followed by proliferation and cloning, becoming a normal chondrocyte density at 7 months. The transition from proliferation and cloning to a normal appearance 7 months later implies chondrocyte proliferation followed by chondrocyte migration (otherwise chondrocyte cloning would persist, but in fact it is greatly reduced—for example, compare Fig. 1c (at 3 months) with Fig. 2b (at 7 months) in Vignon et al. 1983). As explained by Vignon et al. (1983):

> Cell hyperactivity is probably associated with division of middle and deep zone chondrocytes. This is suggested by the normal or high total cell number despite the reduction of the superficial zone cell density and by the normal cell density of the middle and deep zones in spite of the increased thickness of the tissue. Cell division following section of the anterior cruciate ligament is also suggested by histological studies showing clones of 2 or more cells and by metabolic studies showing an increased incorporation of tritiated thymidine.
>
> Cartilage hypertrophy due to cell division and hyperactivity thus appears to be an initial event preceding the appearance of fibrillation in joint-laxity induced osteoarthritis. These changes indicate an unexpected remodeling capacity of the tissue in a reactive adaptation to induced abnormal mechanical stresses.

As discussed in Sect. 3.5, similar reports have been made for cats by Clark et al. (2005) and for dogs by Adams and Brandt (1991) and Panula et al. (1998). This means the report of an 'unexpected remodeling capacity' by Vignon et al. (1983) is not an isolated report, but rather these studies show that articular cartilage repair is a well-documented tissue response to suddenly increased (and abnormal) surface loadings. Taken together, these three reports (Vignon et al. 1983; Pabbruwe et al. 2009; McGregor et al. 2011) provide substantial evidence that slow chondrocyte migration does in fact occur in vivo, while maintaining a well-differentiated chondrocytic phenotype, providing the collagen network is significantly damaged (i.e. where collagen network 'loosening' is sufficient to lead to aggrecan depletion and the formation of significant local aggrecan gradients). We also mention that these reports are consistent with the abovementioned experiments that show well-differentiated chondrocytes may, under suitably favorable conditions, migrate through damaged collagen network at speeds up to 5 μm per day.

What does this mean for the repair of damaged articular cartilage? If cartilage is damaged to a depth of 1 mm, we may expect that nearby healthy chondrocytes could migrate through the significantly damaged region of cartilage over a period as short as 200 days (though as the amount of damage decreases it would take longer than 200 days). Assuming less than optimal migration rates through the damaged collagen network, this slower migration will 'push out' the estimated timescale for collagen network repair from around 6 months to one year, or perhaps even two plus years. At this point, we observe that these estimated rates of collagen network repair are approximately the timescales reported for some types of in vivo cartilage repair following cartilage injury (Chu et al. 2014).

2.9.3 The Relationship Between Fast and Slow Migration Modes

In practice, a 'slow mode' and 'fast mode' of chondrocyte migration may be viewed as two extremes on a continuum of states. For the slow mode, secretion of aggrecan from the trailing surface of the chondrocyte is prominent, generating the majority of the net force that pushes the chondrocyte into the collagen network along the leading surface of the chondrocyte. The resistance offered by the collagen network to chondrocyte migration is reduced by enzymatic cleavage of the collagen network at or nearby the leading surface. Membrane-bound MMP14 and various other soluble MMPs cut the collagen network, opening up a passageway through the ECM for the migrating chondrocyte.

However, if the combined stimuli for chondrocyte migration are substantial, the actual rate of chondrocyte migration relative to the combined stimulus rate of chondrocyte migration may be too slow. In this case, the migrating chondrocyte has a choice. It can either generate more pushing from behind by secreting more aggrecan, or it can generate less resistance ahead by increasing the amount of collagen network cutting at the leading surface of the chondrocyte. In dense collagen networks, most speed gain can be achieved by increasing the amount of collagen network cutting at the leading surface, rather than increasing the amount of aggrecan secreted along the trailing surface. Indeed for very dense collagen networks, aggrecan secretion along the trailing surface may at some level have negligible further effect on migration speed.

Because MMP14 is probably rate limiting in cutting the collagen network, and because MMP14 is membrane bound and limited by its membrane 'density,' the most effective way to increase the rate of cutting at the leading surface is to increase the cell membrane surface area. To increase the surface area at the leading surface (and so the number of MMPs cutting the collagen network), chondrocytes then extend more and larger processes in the direction of migration. The size and extent of the lamellipodia and filopodia then determine the maximum rate of collagen cutting possible, and so the rate of chondrocyte migration through the ECM (see Fig. 2.41).

At some point, the polarization of the cell associated with extending processes from the leading surface of the chondrocyte begins to dominate the polarization associated with the secretion along the trailing surface of the chondrocyte. In this state, the chondrocyte may derive more of its motive force from pulling on the collagen network ahead of the chondrocyte's cell body (using the contraction machinery within chondrocyte processes), than by the secretion of aggrecan along the trailing surface of the chondrocyte. At this point, the chondrocyte may be said somewhat arbitrarily, to have swung from the slow mode of migration to the fast mode of migration.

We simply note here that different modes of migration are probably chosen by the chondrocyte based on particular environmental circumstances the chondrocyte experiences. For example, it seems likely that the slow mode of chondrocyte migration is particularly effective in maintaining the normal homeostasis of the collagen network, while the fast migration mode may be more concerned with getting the chondrocytes to the site of substantial damage in very challenging circumstances following traumatic injury. Having reached their destination, chondrocytes that have migrated in a fast mode can then proliferate and effect local repair. As this repair of the collagen network begins to be successful and aggrecan concentrations begin to rise, the chondrocytes again can more fully differentiate. Finally, a more permanent collagen network repair is effected by chondrocytes migrating into the damaged collagen network in the slow migration mode.

2.9.4 A Mechanism that Explains Slow Mode Chondrocyte Migration

It is all well and good identifying that chondrocytes can slowly migrate while maintaining a pronounced chondrocytic phenotype, but exactly how does migration occur? Here we would like to propose a new cell migration mechanism—a migration mechanism that may be unique to tissues with high proteoglycan content, of which articular cartilage is the prime example. To get to grips with this mechanism, we would like to detail the physical steps necessary that the proposed mechanism of slow migration mode implies, and then search the literature to see if there is any evidence to support the proposed steps. Here we find the evidence is both diverse and scattered, and that each piece of evidence by itself is not very convincing. However, when all the evidence is considered together, the proposed mechanism is much more plausible. We also suspect that it is likely that the proposed mechanism for slow chondrocyte migration has remained so well 'hidden,' precisely because slow migration is just that, slow. The slowness, and so long timescales required for observable change, creates all sorts of experimental difficulties in actually witnessing slow chondrocyte migration.

In the following, we work through the steps for slow chondrocyte migration, and the evidence supporting these steps. Like all new theories, we are really interested in the predictive power of the conceptual model. We find that the proposed model for slow chondrocyte migration does have explanatory power, and leads to the conclusion (also reached by others), that articular cartilage is capable of healing itself.

Turning now to the mechanistic steps for slow migration, and what evidence exists, it is proposed that slowly migrating chondrocytes first 'polarize' (as do fast migrating chondrocytes), and then differentially express proteases at the leading edge of a cell to 'clear the way,' while aggrecan is secreted along the trailing edge of the cell to provide the 'push.'

Indeed, probably all migrating cells are polarized, so this is not such a surprising starting point. Cell polarization arises due to localization of specific proteins to defined portions of the cell membrane, so polarized cells lose their symmetry. This also usually involves a redistribution of cell organelles within the cytoplasm, and so the cell nucleus moves to one 'side,' and the cell is morphologically noticeably asymmetric. We observe that activation of the non-canonical 'planar cell polarity pathway' by Wnt signaling is known to result in pronounced epithelial cell polarization, and it is plausible this (or a similar) signaling system may play a role in polarizing articular cartilage chondrocytes. Explains Prydz (2015):

> In eukaryotic cells, the secretory pathway transports secretory and plasma membrane proteins, PGs, and lipids from the ER [endoplasmic reticulum] to the cell surface. In addition, most endosomal and lysosomal proteins follow the secretory pathway from the ER through the Golgi apparatus, from where there are several routes directly to the cell surface, and a number of routes to compartments along the endocytic pathway. In polarized cells, like epithelial cells and neurons, the plasma membrane is divided into different regions, a situation that requires additional pathways for targeting of proteins, PGs and lipids to their respective acceptor membrane domains.

So as well as reorganizing the cell cytoplasm (e.g. the cell nucleus moves to 'one pole' of the cell and organelles in the cytoplasm reorganize), chondrocyte polarization results in asymmetric distribution of proteins at the cell membrane.

Specifically, more MMP14 is expressed on the cell membrane in the direction of migration (i.e. at the cell's 'apical pole,' referred to here as the cell's 'leading edge' in a 2D section or 'leading surface' in 3D). Nagase et al. report that such MMP14 activities are often found at the 'invasive front' when cells migrate through their local ECM (Nagase et al. 2006). Further they state exactly how MMP14 is localized to a particular section of the cell surface remains to be investigated, but that a unique cysteine in the cytoplasmic domain is post-translationally palmitoylated, and this modification is essential for the function of MMP14 in promoting cell migration.

MMP14 has been shown to be localized to the leading edge of migrating (and simultaneously resorbing) osteoclasts, as shown in Fig. 5a of Sato et al. (1997). Specifically as stated in Sato et al. (1997): *Double-stain fluorescence pictures of an osteoclast with the typical phenotype of a 'walking' cell. The pictures were taken adjusting the focus to the level of the contact between the cell and the glass-slide. Phalloidin staining shows lamellipodiae rich in actin (B) at the leading edge and retraction fibers at the rear. The MT1-MMP antibody shows immunoreactivity at the leading edge of the same cell (A).*

Though not a migrating chondrocyte, this at least demonstrates that membrane localization of MMP14 is not uncommon in migrating cells. An example in the literature illustrating MMP14 has a non-uniform localization of MMP14 on chondroclast cell membrane (possibly at 'ruffled borders') is shown in Fig. 5 h of Sakura et al. (2007). This image shows chondroclasts during resorption of Meckel's cartilage in embryonic mice (Sakakura et al. 2007). MMP14 is often found in resorbing or 'invasive cells' (with MMP14 often located on specific parts of cell membranes in osteoclasts (Sato et al. 1997) and odontoclasts).

In this context, we note MMP14 is found to be the key regulator of pericellular collagenolysis by pulmonary fibroblasts (Rowe et al. 2011) and is also involved in 'pannus invasion' of cartilage by synoviocytes in rheumatoid arthritis (Miller et al. 2009). MMP14 expression and localization on the leading edge of the cell (possibly directed by integrins) may be required for all cells migrating through their ECM (Itoh et al. 2006). Indeed, Van Doren (2015) remarks:

> Association of collagen fibrils with cell surfaces and MMPs was hypothesized to allow cells to move on collagen [e.g. keratinocyte migration on collagen]. Also fulfilling this hypothesis are the collagenolytic activities of: (i) MMP-8 supporting neutrophil migration and (ii) MT1-MMP [MMP14] in developing the full force of cells migrating through 3D collagen-based tissue models.

Collier et al. provide more details about the mechanism(s) for MMP14 activity at the cell surface (Collier et al. 2011):

> We have previously demonstrated that secreted MMP-2 is tethered to the cell surface and activated by $(MT1\text{-}MMP)_2$/TIMP-2-dependent mechanism. The resulting cell-surface collagenolytic complex $(MT1\text{-}MMP)_2$/TIMP-2/MMP-2 can initiate (MT1-MMP) and complete (MMP-2) degradation of an underlying collagen fibril... Most instructive is the finding that the inactivation of the enzymatic activity of MT1-MMP has a detectable negative effect on the cell force developed in miniaturized 3D tissue constructs. We propose that the collagenolytic complex $(MT1\text{-}MMP)_2$/TIMP-2/MMP-2 represents a Mobile Cell Surface – Collagen Substratum Interface. The biological implications of MT1-MMP acting as a molecular ratchet tethered to the cell surface in complex with MMP-2 suggest a new mechanism for the role of spatially regulated peri-cellular proteolysis in cell-matrix interactions.....Activated MT1-MMP can initiate the digestion of collagen fibril and is a Brown Ratchet capable of biased diffusion similar to the secreted collagenase MMP-1. MMP-2/TIMP-2 complex is capable of random lateral diffusion on the same substrate and can assist MT1-MMP dependent degradation of the fibril to small peptides. Integrin, a collagen receptor, can bind to collagen fibril and potentially interact with the MT1-MMP complex. The exact mechanisms of that interactions remains to be elucidated.

See schematic of MMP14–TIMP2–MMP-2 complex shown in Fig. 9 of Collier et al. (2011).

MMP14 can cleave a wide variety of ECM components including collagen Types 1, 2, and 3, as well as degraded collagen, fibronectin, laminin, vitronectin, perlecan, tenascin, and proteoglycans including aggrecan. It is an impressive list, and it appears nothing much in the ECM can resist this 'Swiss-army knife' of MMPs (only MMP1 and MMP13 have a comparable range of activities). Undoubtedly any remaining ECM components can be degraded with the assistance of MMP3 (which degrades collagen III–V and collagen IX), MMP2 (which degrades collagen VI), and membrane-bound MMP15, MMP16, and MMP17, which are all expressed in cartilage, although these are expressed in relatively small quantities compared to that of MMP14—see Fig. 2.61.

Most importantly, it has been established by Wagenaar-Miller et al. that MMP14 is *essential for chondrocytes* to degrade pericellular collagen type II network (see Fig. 3 in Wagenaar-Miller et al. 2007). We also note that Wagenaar-Miller et al. report that an intracellular pathway for collagen degradation that may operate alongside pericellular degradation. This intracellular collagen degradation pathway involves collagen fibrils or fragments binding to cell surface receptors (e.g. β1 integrins or UPARAP/Endo180 receptors) followed by endocytosis and lysosomal degradation of collagen by cathepsins.

While the limited existing experimental data has found that MMP14 mRNA expression does not appear to increase in osteoarthritis (see MMP mRNA expression shown in Fig. 2.61), nevertheless the amount of MMP14 detected by immunohistochemistry does increase significantly (see Fig. 2.47). This may mean that MMP14 is regulated elsewhere [e.g. the rate of membrane turnover changes (Kimura et al. 2010)]. However, an increase in MMP14 mRNA expression would be consistent with increases in transcription factor SAF-1 expression (Bal et al. 2006), which is known to promote MMP14 expression. The transcription factor SAF-1 induces the mRNA expression of other MMPs too (e.g. MMP1, MMP9) (Kumar et al. 2009; Ray et al. 2005).

In fact, there is a good correlation between MMP14 activity by chondrocytes and the 'Mankin Score' (Fig. 2.48), which indicates the severity of osteoarthritis as assessed on multiple histological criteria. This is almost certainly important, because part from its own ECM degradation ability, MMP14 plays an important role in activating MMP2 (along with TIMP2), MMP9, and MMP13, and consequently MMP14 activity is correlated with MMP2, MMP9, and MMP13 activity (Bal et al. 2006; Imai et al. 1997).

We now turn to evidence for polarized secretion of molecules around chondrocytes. We first note there is some evidence for polarized secretion of hyaluronic acid and ADAMTS5 around chondrocytes in vivo (see for example Fig. 2.49) (Plaas et al. 2007). This is relatively weak evidence in the sense that it only documents a specific instance of hyaluronic acid and ADAMTS5 membrane segmentation, but it nevertheless demonstrates that it does occur. There is also some evidence for polarized secretion of aggrecan (see for example Fig. 6 in Poole et al.

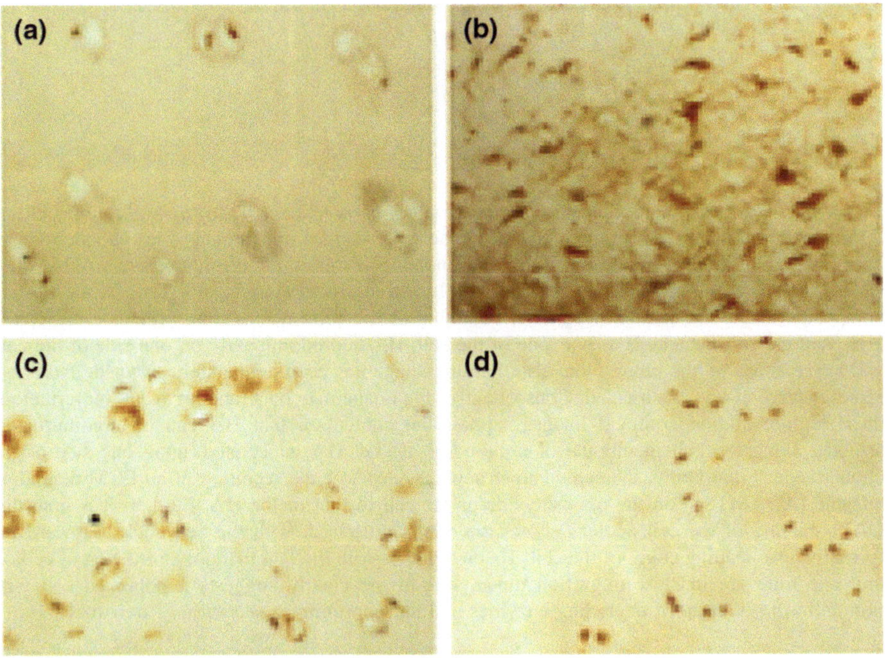

Fig. 2.47 Immunohistochemical analysis of MMP-14 in human joint cartilage. Cartilage from normal cadaver (panel **a**) and those of three representative OA patients collected from the medial femoral condyle or tibial plateau or patella were subjected to IHC. Cells positive for MMP-14 show brown color (Bal et al. 2006). SAF-1 expression is also significantly increased in osteoarthritis compared to controls, suggesting that in osteoarthritis MMP14 expression is also increased (we observe that the promoter of MMP14 is said to contain a 'SAF-1 binding element' for the transcription factor SAF-1)

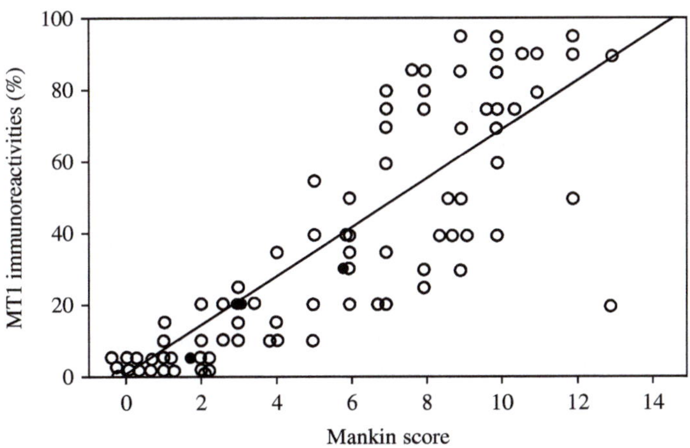

Fig. 2.48 Correlation of MMP14 (MT1-MMP1) immunostaining with histological grade. A positive, direct correlation between MMP14 immunoreactivity and Mankin score is found by Spearman's rank correlation ($R2 = 0.849$; $P = 0.0001$; $n = 79$) (Imai et al. 1997)

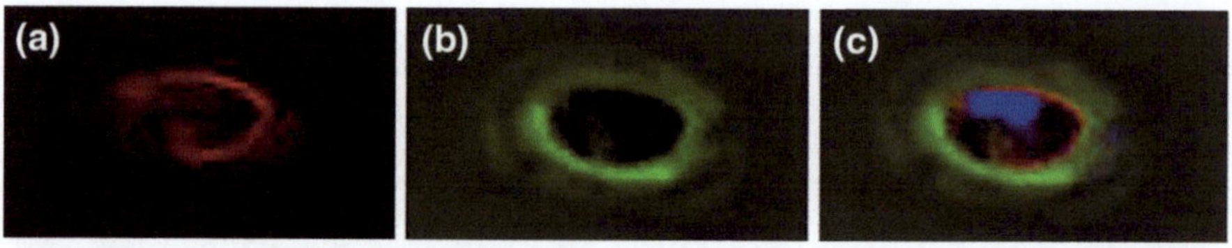

Fig. 2.49 (This particular single cell image has been extracted from) a high-resolution, fluorescent, immunohistochemical images of normal, human transitional zone chondrocyte in 'slow migration mode.' Figure 6a shows immunohistochemistry for the aggrecanase ADAMTS5 (in red), 6b shows immunohistochemistry for hyaluronic acid (HA is in green), while 6c is a composite image and also showing DAPI staining of the cell nucleus (blue) (Plaas et al. 2007). Note that the cell nucleus is to one side of the cell, HA is predominently on one side of the cell, and aggrecanase on the other. Note also that this image has been deliberately selected, and is not representative of all cells imaged. Consequently, this image may be a random event. Nevertheless, these high-resolution confocal images support the contention that HA can be non-uniformly secreted along the cell membrane of a chondrocyte, i.e. HA is secreted along one side of the chondrocyte. Importantly, the same chondrocyte is seen to have secretion of an ECM degrading enzyme (ADAMTS5) on the opposite side of the cell to that of the HA. Further, it is note that DAPI staining of the cell nucleus shows the cell is polarized, with the cell nucleus toward the (presumably) 'leading edge' of the cell. Taken together with the fact that only a fraction of cells at any one time are in slow migration mode, this image clearly suggests a polarized cell with non-uniform secretion of degrading enzymes and the non-uniform secretion of new mers

1992), polarized secretion of collagen VI (see Fig. 2.52 and Fig. 1 in Poole et al. 1992), and polarized secretion of fibronectin (see Fig. 2 in Chang et al. 1997). Taken together, this provides reasonably strong evidence that builds the case that chondrocytes can polarize, which provides the foundation for the proposed mechanism of slow chondrocyte migration.

We now detail our proposed mechanism for slow chondrocyte migration. Integrins expressed on the chondrocyte membrane (Aszodi et al. 2003) bind adjacent collagen via short processes extending from the cell (see for example Fig. 1B in Poole et al. 1992), and pull the cell membrane and collagen network closer together. We note Collier et al. also suggests that the MMP14-TIMP2-MMP-2 complex may also pull the cell and collagen network together (Collier et al. 2011). Cells generate force using an internal traction apparatus, probably involving myosin and smooth muscle α-actin (Qiu et al. 2000; Povysil et al. 2008; Hinz et al. 2001). Indeed, we note that Collier et al. report that the full traction force cannot be generated without the MMP14–TIMP2–MMP-2 complex operating (Collier et al. 2011), which suggests they are functionally coordinated, and this force is generated where the MMP14 is located (at the leading (invasive) edge (surface) of the cell).

Importantly, we observe that cells moving in a slow migration mode cannot pull themselves forward by their attachment to the collagen network, as there is no action–reaction force generated other than that to pull the cell membrane and collagen network toward one another at the leading edge of the cell.

Short processes extending from the cell to the side of the cell may also serve to hinder soluble secreted cell products freely flowing (in the space immediately adjacent to the cell) from the front of the cell to the back of the cell (or vice versa). It is proposed that the spatial separation of integral proteins in the cell membrane and perimeter concentration gradients in secretion products around the cell effectively generates two 'chemical compartments,' one at the 'front' and one at the 'back' of the chondrocyte, which enable slow chondrocyte migration.

We propose it is the differences in concentration of GAG in these two chemical 'compartments' that enable cell migration in the direction of its aggrecan gradient. In this context, we draw attention to reports by Poole et al. (1991a) about differences in keratin and chondroitin sulfate staining intensity (see Figs. 3, 4, and 5 in this publication) within canine chondrons maintained in agarose gel:

> This study has focused on identifying the light microscopic distribution of several proteoglycan/glycosaminoglycan epitopes within the microstructure of the isolated chondron. Antibodies to keratin sulfate (5-D-4, EFG-11, BCD-4, KPC-190) showed the greatest variability in localization and staining reaction. In general, the thinnest portion of the capsule that surrounds the chondrocytes (and which was easily recognized in ruptured capsular ghosts) showed only a weak to moderate staining reaction, whereas the thicker portions of the capsule formed by the tail and interconnecting segments of chondron columns stained intensely....Antibodies to chondroitin-4-sulfate (2-B-6, BE-123) and chondroitin-6-sulfate (MK-302) showed a more consistent and intense staining pattern, with a clearly defined boundary between the pericellular capsule and the collagen fibers that adhere to the isolated chondron. These staining reactions were found throughout the pericellular capsule but were usually stronger in the tail and interconnecting segments between adjacent chondrons...Similarly, the polyclonal antibody against hyaluronate binding region showed a strong reaction in the chondron. The persistence of a strong reaction after chondroitinase ABC digestion and its reduction after testicular hyaluronidase extraction are consistent with previous studies reporting high pericellular concentrations of hyaluronate.

In vivo, the local aggrecan gradient occurs in the context of the regional aggrecan gradient, so the two gradients (local and regional) are summed. It is the total aggrecan gradient around the migrating chondrocyte that enables the forward movement of the cell, as explained further below.

We note in passing there are some computational models that attempt to explain cell migration through ECM involving MMP secretion (Harjanto and Zaman 2010) and some studies describe various modes of migration (Zhu and Mogilner 2016). But we see that the modes of cell migration described by Zhu and Mogilner (2016) (e.g. see Fig. 3 in that publication) do not really include the slow migration mechanism described here. However, the migration modes summarized in Fig. 3 of Zhu and Mogilner (2016) probably do play an increasing role for chondrocytes in intermediate and fast chondrocyte modes.

Short processes containing contraction apparatus at the leading edge of the chondrocyte pull the cell and collagen together. MMP14s expressed on the membrane directly cut the collagen (Harjanto and Zaman 2010; Iordan et al. 2010; Wagenaar-Miller et al. 2007). MMP14s perform further important tasks at the membrane surface such activating cascades of soluble MMPs. MMP14 activates proMMP2, which in turn activates proMMP9 and proMMP13 (Hembry et al. 2001; Sakakura et al. 2007). It is likely there are local signaling molecules (e.g. possible nitric oxide and/or OPG) acting in the pericellular and territorial matrix help orchestrate these cell activities. Together these ECM degradation processes facilitate fragmentation of the ECM immediately ahead of the migrating cell (Page-McCaw et al. 2007).

Simultaneously, the chondrocyte secretes aggrecan along its trailing edge (i.e. at the 'basal pole' of the chondrocyte). It is the difference in osmotic pressure between the trailing and leading edge (i.e. across the chondrocyte between leading and trailing surfaces) that provides most of the motive force to push the cell forward (see Fig. 2.50, showing presumed stationary (top) and presumed migrating chondrocytes (bottom)). Of course, this local aggrecan gradient generated by the migrating chondrocyte is additive to the 'regional' (or regional average) aggrecan gradient. This aggrecan gradient generating a motive force on the chondrocyte is balanced by the resistance offered by collagen network in the direction of movement.

Figure 3 in Buschmann et al. (1996b) shows representative examples of images taken of groups of articular cartilage specimens labeled with different concentrations of radioactive sulfate 35. These images show it is possible to find chondrocytes with asymmetric distributed sulfate 35 grain densities around these chondrocytes, which is indicative of asymmetric distributions of aggrecan molecules. This could happen by chance, but it could be due to polarization of some chondrocytes. In this context, we also note asymmetric sulfate 35 grain density distributions are shown in Fig. 6 of Poole et al. (1991b).

See also image in Chap. 3 of textbook by Cross and Mercer (1993), showing a transmission electron microscope image of a polarized chondrocyte. This image shows a nucleus toward the leading edge of the chondrocyte, and an extensive endoplasmic reticulum and Golgi favouring the trailing edge of the chondrocyte. The chondrocyte has a ruffled border, as vesicles merge with the cell membrane, suggesting vigorous secretory activity along the 'trailing edge' of the chondrocyte.

The collagen network immediately ahead of the migrating chondrocyte is reported to be relatively more densely packed, as predicted by the slow migration model (see Fig. 2.50), while the collagen network is looser behind the cell (see the very helpful illustration of a chondron shown in Fig. 3 of Poole 1997).

Figure 3 in Poole (1997): *Three-dimensional cutaway representing of a single chondron showing dense compaction of pericellular collagen at the apical pole (Ap) [i.e. in direction of migration], and loosely woven tail tapering to the basal pole (Bp) [remnants of the digested collagen network].*

See also Fig. 1b in Poole et al. (1984): *A pair of middle layer chondrocytes each surrounded by pericellular matrix with pericellular capsule.* This transmission electron microscope image shows a 'chondrocyte couplet' following cell division (Poole et al. 1984). Secretion of aggrecan and hyaluronan by the topmost

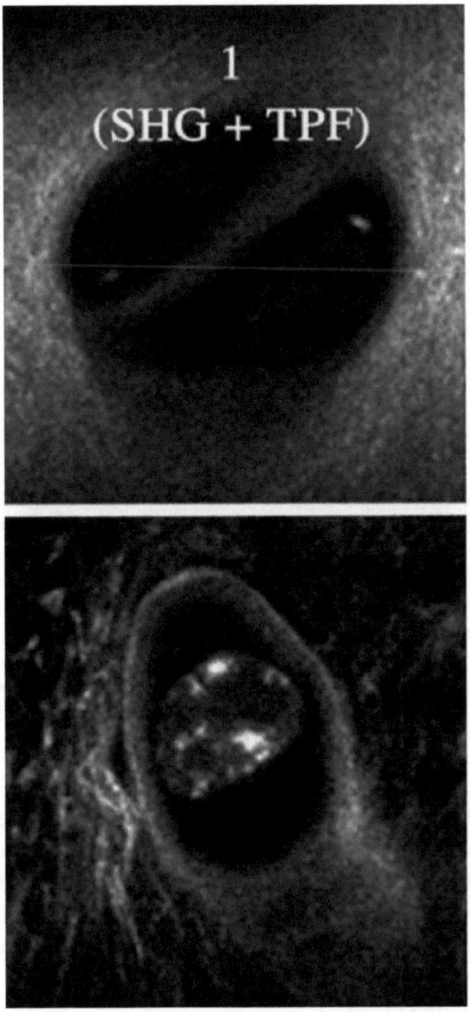

Fig. 2.50 Nonlinear optical microscopy images of chondrocytes and surrounding ECM (Yeh et al. 2005). Top image, normal cartilage—note the amorphous appearance of ECM around two stationary chondrocytes (image presumably shows a stationary chondrocyte 'couplet,' i.e. two chondrocytes in close apposition following cytokinesis). Bottom image, early fibrillar matrical degeneration cartilage—note oriented appearance of 'bullet-shaped' (presumably) migrating chondrocyte, within a 'bullet-shaped' chondron (see Fig. 2.51), with distinct fibrillar appearance of ECM immediately around chondron (notice similarity of this image to that shown in Fig. 1c in Poole 1997). Laser light detects both collagen (SHG 'second hormonic generation' signal at 400 nm) and proteoglycan (TPF 'two photon (auto)fluorescence' signal at 520 nm). Scale bar 8 μm. Notice that in the normal cartilage, proteoglycan and collagen surround the chondrocyte couplet uniformly, while in degenerating cartilage, the aggrecan is non-uniformly distributed and the collagen network is 'dense' in the (presumed) direction of migration, fibrillated along the side and 'loose' at the rear of the chondrocyte

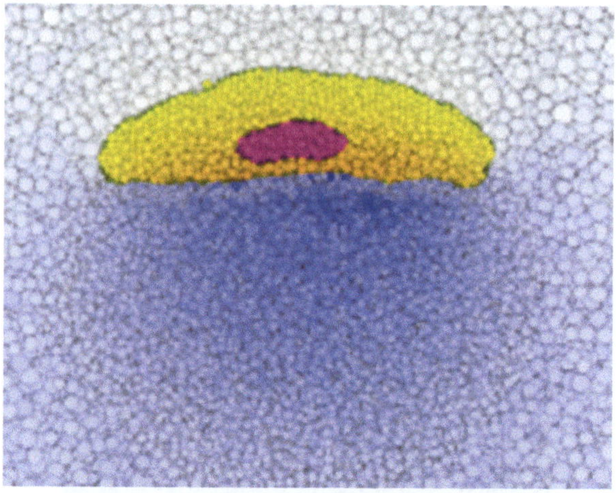

Fig. 2.51 This is the solution output from a discrete computational model simulating asymmetric ECM secretion to one 'side' of a single chondrocyte, similar to the autoradiograph images shown in Fig. 3 in Buschmann et al. (1996b). Notice flattening of chondrocyte, due to the cell being sandwiched between the higher concentration of aggrecan below (indicated by darker color) and resistance to chondrocyte migration offered by the ECM in the direction of migration above. The asymmetric distribution of aggrecan shown in this figure is broadly consistent with epitope distributions shown in Figs. 3, 4, 5 in Poole et al. (1991a), demonstrating asymmetric distributions of keratin sulfate and chondroitin sulfate epitopes around canine chondrocytes in agarose gel. In addition, chondrocyte flattening is frequently observed in migrating chondrocytes or proliferating chondrocytes in a single lacuna (see Figs. 2.50 and 2.49)

chondrocyte into the space between the two chondrocytes pushes it into the existing ECM in its migration path. This directed cell movement is soon resisted by the ECM ahead of it. The osmotic force is envisaged as pushing the topmost cell from 'behind' (at the trailing surface), and with ECM resistance ahead (at the leading (invasive) surface). This force distribution often results in flattening on the side of the chondrocyte, sometimes resulting in a 'bullet-shaped cell' (see Figs. 2.50 and 2.51 and compare shapes) (Fig. 2.52).

As the collagen network is gradually degraded by proteolysis at the leading (invasive) surface, the resistance in the direction of cell migration is reduced, and the migrating cell can then slowly 'creep' forward. A fibroblast migrating through a loose collagen network leaves a 'resorption tunnel' immediately behind it, in the cell's 'wake' so to speak (e.g. see for example, the collagen resorption 'tunnels' made by fibroblasts in collagen gels shown in Fig. 5D of Iordan et al. (2010)).

However, because the collagen network in articular cartilage is densified as it is compressed against a slowly migrating chondrocyte (due to the significant proteoglycan gradient across the cell), a 'clean-cut resorption tunnel' is much more difficult to form completely than the resorption tunnels shown in Fig. 5D of Iordan et al. (2010). In the case of slow migration through articular cartilage, any

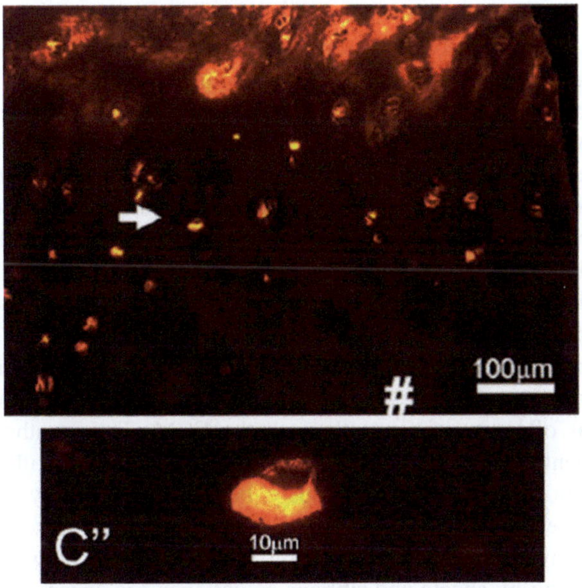

Fig. 2.52 Collagen VI changes distribution from the pericellular region to the interstitial matrix and there is an overall upregulation of collagen VI in articular cartilage with advanced OA (top). Magnified region (bottom) showing single cell identified by white arrow (top) (Nugent et al. 2009). Again we note the more extensive distribution of type VI collagen relative to normal cartilage (top) (which is probably indicative of ECM damage associated with OA) and in the bottom image we observe again an asymmetric distribution of labeled cell product, this time of collagen type IV in pericellular region

remaining untrimmed collagen network 'springs back' into the void space as the leading edge passes (see for example Figs. 2C and 4A in Poole et al. 1988a). Indeed, it appears likely that chondrocytes migrating through the collagen network create what has been named 'chondrons' [see Figs. 1B, C (Poole et al. 1992) and Fig. 3 (Poole 1997)]. 'Tapering chondrons' have dense packed collagen network at the 'apical pole' and a partially degraded collagen network chondron 'tail,' forming the 'basal pole' of the chondron. Reports (Poole et al. 1988a):

> Tapered tails common to single chondrons appear continuous with the pericellular capsule and were often associated with frayed collagen fibers…The persistence of a tail in many ghosts [i.e. chondrons devoid of a chondrocyte] indicates that they form an integral component of the capsular structure.

We propose that a specific type of chondron, the so-named tapered chondron (see Fig. 3 in Poole 1997), probably originates from partially digested collagen network forming around slowly migrating chondrocytes in adult articular cartilage. The tapered chondrons are not exclusively collagen, but are intermingled with ECM products secreted into the pericellular space (e.g. aggrecan, collagen VI, fibronectin). We note the existence of 'capsular ghosts' (i.e. chondrons devoid of a chondrocyte) clearly indicates that upon chondrocyte apoptosis, the chondron persists in the collagen network (as might be expected), presumably they remain until another migrating chondrocyte intersects (and remodels) the ghost chondron. The fact that 'capsular ghosts' and chondrons containing 'non-viable chondrocytes' exist as a substantial fraction of the total number of chondrons extracted from articular cartilage (Poole et al. 1991b) is consistent with chondrocyte death and chondrocyte proliferation being ongoing processes in normal cartilage tissue.

The chondrons in articular cartilage can be clearly imaged, as shown in Fig. 1 (Poole et al. 1992). See Fig. 1 in Poole et al. (1992): *Image B shows intensely stained canine chondrons, (labeled by streptavidin biotinylated horseradish peroxidase and collagen type VI antibody), which are aligned parallel with the radial collagen network (arrowheads show head to tail chondrons).* Note in this image the 'aligned' chondrons actually 'face' in slightly different directions, presumably following paths of least resistance through the ECM. These paths are guided by aggrecan gradients and/or regions of loose collagen network (offering less resistance to migration), while simultaneously being 'corralled' by dense sheets or ribbons of collagen we know exist in interterritorial zones. In those cartilage regions with generally low aggrecan concentration and/or collagen network damage, and less tight corralling by sheets and ribbons of collagen (as occurs in the transitional zone), chondrons directions 'spread out' and become more 'randomly' distributed, thereby maximizing their beneficial impact on repairing damaged ECM.

See Fig. 1 in Poole et al. (1992): *Light microscopy. (A) In chondrons labeled with FITC, each negatively contrasted chondrocyte (c) was surrounded by an intense staining reaction with obvious divisions between adjacent chondrons (arrowheads).* Note that the schematic shown in Fig. 3 in Poole (1997) can help interpret images shown in Fig. 1 (Poole et al. 1992).

After the chondrocyte has 'swept past' ('swept' in the sense of projecting events in real time onto a compressed timescale), the remaining partially degraded collagen network is broken up by proteases, until after several cell widths behind the trailing surface, it finally disappears altogether and becomes indistinguishable from the background collagen network. Meanwhile, it is proposed here that the region behind the cell is filled with ECM secretion products emanating from the trailing surface of the chondrocyte, including GAGs, linker protein, aggrecan, hyaluronic acid (HA), and tropocollagen mers formed from procollagen. Tropocollagen mers (with a diameter of about 1.5 nm about 300 nm long) are guided into assembling into new collagen network by a plethora of scaffold molecules (e.g. fibronectin and vitronectin), nucleating fibrils (e.g. type XI) and regulatory collagens secreted into the pericellular region (e.g. collagen IX). See distribution of fibronectin throughout chondrons (shown in Fig. 2 of Chang et al. 1997). In addition, there may be helpful roles played in collagen network formation by small leucine-rich proteoglycans (particularly lumican, decorin, bigylcan, and fibromodulin) and integrins (Aszodi et al. 2003; Chang et al. 1997).

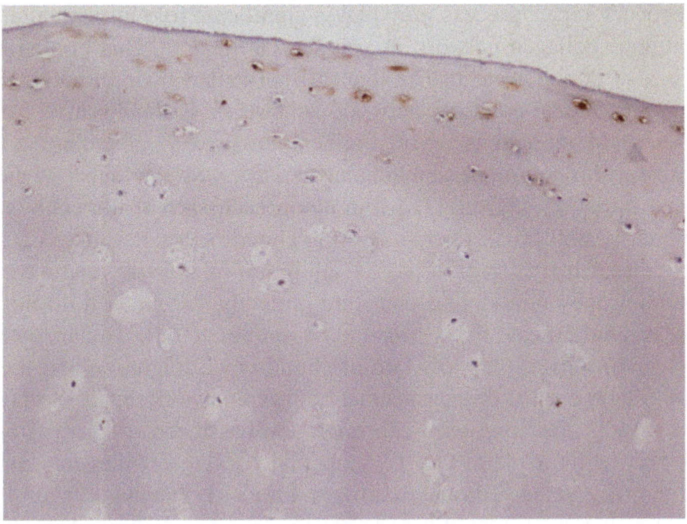

Fig. 2.53 α-actin protein-positive chondrocytes are predominantly in the superficial layer of the cartilage from normal femoral head (magnification times 200) (Povysil et al. 2008). See also Figs. 1 and 2 in Kim and Spector (2000), and images of the normal cartilage surface shown in Rolauffs et al. (2008)

See Fig. 2 in Chang et al. (1997): *Isolated chondrons from dogs are labeled with antifibronectin antibodies. Fibronectin staining was concentrated in the pericellular capsule and matrix around each chondrocyte, but in contrast to type VI collagen, remained relatively strong in the tail and interconnecting segments (white arrow).*

Acting together, the ECM molecules in the 'wake' of a migrating chondrocyte guide the tropocollagen mers to polymerize and form the new collagen network. As this new collagen network forms, we say the collagen has 'remodeled.' This intense polymerase activity in the 'wake' of a migrating chondrocyte explains the concentration of molecules known to be involved in collagen polymerization in this region. It also probably explains the plumes of type II collagen neoepitopes that appear to emanate from some (presumably migrating) chondrocytes (e.g. see Fig. 2.31).

Importantly, it has been reported that migrating α actin-positive chondrocytes can in fact secrete type II collagen while they are migrating (Kinner and Spector 2001; Chang et al. 2003). Collagen synthesis and secretion of collagen type II mers into the wake of a slowly migrating chondrocyte is clearly consistent with the polymerization of new collagen network in the wake of migrating chondrocytes. The newly formed collagen network enables the retention of aggrecan and the 'rebuilding' of the ECM. It is possible that under normal operating conditions, this collagen network repair process may play a significant role in maintaining normal transitional zone collagen network in articular cartilage (Aigner et al. 1993).

But it is also important to realize that this particular mode of collagen network repair can only repair collagen network as fast as a chondrocyte can migrate through the ECM—which is usually rather slowly (e.g. at around a couple of microns per day through moderately damaged tissue). So if a 'cube' of the collagen network that measures say only 0.1 mm along each side of the cube is damaged (side face area 10,000 μm^2), many migrating chondrocytes need to pass through to fully repair this damaged region (e.g. a minimum of 50–100 migrating chondrocytes (assuming they remodeling paths are perfectly straight and do not overlap), and almost certainly many, many more, need to migrate through damaged collagen network in vivo). Given the slow rate of chondrocyte migration and the need for multiple migration events, this means that collagen network repair by chondrocytes may take years. If this 'long-term cell migration-based repair' process operates as described, this helps explain why C telopeptide collagen fragments may remain elevated for many years following cartilage injury (Lohmander et al. 1996), and even remain elevated decades (for example see Fig. 4 in Lohmander et al. 2003a).

It is interesting to note that α actin-positive cells can after a period of time, revert to α actin negative phenotype (Chang et al. 2003). This suggests that chondrocyte migration through the articular cartilage may be a stop–start process, as is required to carefully 'titrate' the cell-based collagen network repair response to the degree of local collagen network damage.

McNary et al. notes that chondrocytes develop a more prominent (α, β, γ) actin cytoskeleton as they de-differentiate in the superficial zone, and assume a more fibroblastic phenotype (McNary et al. 2014). At this point, we remind ourselves about the three main types of actin (α, β, and γ). α actin (of which there are about four isoforms) is associated with (muscle and) cell migration, β actin is found around the edge of cells and is involved in the formation of cell projections, while γ actin is found as cytoskeletal tensile stress elements (http://en.wikipedia.org/wiki/Actin). Normally, a cell expresses a range of actin isoforms simultaneously depending on the functional needs of the cell (i.e. the isoform expression changes depending on the cell type and depending on what the cell is doing at the time, e.g. dividing, migrating, stationary).

Importantly, we observe that Povysil et al. (2008) identify two populations of chondrocytes: (i) those not expressing α smooth muscle actin and (ii) those expressing α smooth muscle actin (which these authors refer to as 'myochondrocytes') (Povysil et al. 2008). We here note that Wnt 5a is reported to lead to both cell polarization and expression of α smooth muscle actin (Tock et al. 2003).

Kinner and Spector report that in the upper-half of human articular cartilage, about 60% are α actin positive, while in the lower half, about 28% are α actin positive (Kinner and Spector 2001). Povysil et al. (2008) report that α actin smooth muscle-positive cells in normal cartilage were predominantly in the upper zone of normal articular cartilage (18% of cells are α actin positive), while only 10% of cells in the deep zone are α actin positive (Povysil et al. 2008). Variability in the reported proportion of articular chondrocytes that are α actin positive is to be expected, as each joint in each animal is likely to have different requirements of their resident chondrocytes based on the animal's age, the degree of cell proliferation, the collagen network damage history, disease states, and the animal's current environment.

Povysil et al. (2008) says:

> Actin-positive chondrocytes [myochondrocytes] are formed predominantly in response to injury to the osteoarthrotic cartilage, at sites of defective healing, and in newly form cartilage after autologous chondrocyte transplantation.

They go on to say:

> It seems that actin-positive chondrocytes in normal hyaline cartilage, with the exception of the surface layer, occur in association with arthrotic changes and predominantly during the process of reparation….Our study confirmed that human articular cartilage has some potential for spontaneous regeneration….This finding suggests that chondrocytes may have an adaptation mechanism enabling such transformation of the immunophenotype of cartilage cells depending on different situations. The chondrocytes probably took on smooth muscle features during the healing process.

Note that in normal cartilage, most of the α smooth muscle actin chondrocytes is expressed in the upper transitional and superficial zones (see Fig. 2.53), while in osteoarthritic cartilage all cartilage zones may display α smooth muscle actin-positive chondrocytes (see Fig. 2.54). Further, we mention that surgically created full depth defects in canine articular cartilage results in a significantly increased number of α actin-positive chondrocytes in the cartilage tissue surrounding the defect (including chondrocytes in the deep zone) (Wang et al. 2000).

These observations reported by Povysil et al. (2008) appear to be corroborated by observations made on arthroplasty explant tissue by Qiu et al. (2000):

> The objectives of this study were to investigate the effect of various enzymatic treatments on the outgrowth of chondrocytes from explants of adult human articular cartilage and the expression of a specific contractile protein isoform, α-smooth muscle actin, known to facilitate wound closure in other connective tissues. Explants of articular cartilage were prepared from specimens obtained from patients undergoing total joint arthroplasty. The time to cell outgrowth [i.e. the time for 'release of round' chondrocytes (i.e. the time for chondrocytes with round morphology to exit the collagen network onto the cartilage explant surface), which then spread within a few days] in vitro was determined and the expression of α-smooth muscle actin shown by immunohistochemistry. Treatment of the explants with collagenase for 15 min reduced the time to outgrowth from more than 30 days to 3 days….We conclude that human articular chondrocytes have the capability to migrate through enzymatically degraded matrix and express a contractile actin isoform.

It is very interesting to note that the tenfold reduction in time for 'outgrowth' of chondrocytes following exposure to collagenase (nothing else, including hyaluronidase, chondroitinase or trypsin, has a similar effect on 'outgrowth' to that of

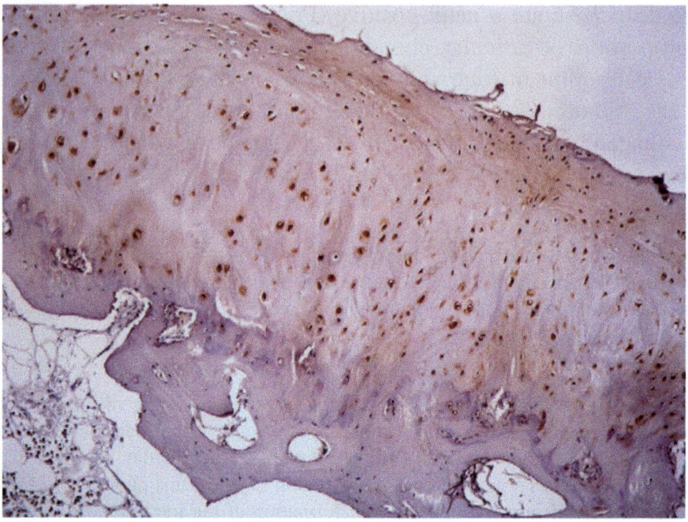

Fig. 2.54 The fibrocartilaginous repair tissue from ostearthritic cartilage predominantly contain smooth muscle α actin-positive chondrocytes (magnification times 200) (Povysil et al. 2008)

collagenase). We can appreciate that because 'outgrowth' (i.e. chondrocyte migration while maintaining a round morphology) takes so long to observe in untreated explants, the explant tissue itself may change over the long period in its support media. This makes in explant experiments difficult, while in vivo experiments are exceptionally difficult to perform.

At this point, we propose that for normal articular cartilage, negative α smooth muscle actin chondrocytes overlap primarily with the population of stationary chondrocytes, while the positive chondrocytes expressing α smooth muscle actin overlap primarily with the population of migratory chondrocytes. This is consistent with the observations of Qiu et al. (2000), Kim and Spector (2000), Kinner and Spector (2001), and Povysil et al. (2008). In normal articular cartilage, the largest fraction of α smooth muscle actin chondrocytes is found in the upper transitional and superficial zones (Povysil et al. 2008), where ongoing damage to the articular cartilage is greatest.

It is not uncommon following injury and subsequent healing for normal fibroblasts in various tissues to become so-called activated fibroblasts, and then become 'myofibroblasts' (expressing smooth muscle α actin and large quantities of collagen) (Hinz et al. 2007). For example, when the corneal epithelial cells are damaged, corneal stromal cells are 'activated' and begin expressing smooth muscle α actin and TGFβ (Barbosa et al. 2010; Wilson et al. 2014) (Interestingly we note in passing that activated corneal stromal cells also secrete IL-1 (Barbosa et al. 2010)]. Similarly in the lung and kidneys, activated fibroblasts become myofibroblasts (expressing smooth muscle α actin) and secrete large quantities of collagen causing fibrosis (Scotton and Chambers 2007; Hinz et al. 2007). We see that it appears to be 'standard practice' in multiple tissues for 'activated cells' to be involved in the repair of damaged tissues and to express smooth muscle α actin while doing so.

We also observe that the region occupied by the population of chondrocytes expressing the CD36 receptor (which is involved in debris uptake and intracellular degradation) is coincident with the region occupied by the population of chondrocytes expressing smooth muscle α actin (Pfander et al. 2000). This is a functionally consistent co-location of expression, for it is reported that CD36 expression increases focally at locations of cartilage damage (Cecil et al. 2009).

At this point, we mention that Simkin also proposes that chondrocytes migrate through cartilage (Simkin 2008). Without specifying the mechanism of migration, Simkin proposes chondrocytes migrate long distances from the junction of the synovial membrane and edge (or lateral border) of articular cartilage (i.e. this junction is identified as the 'groove of Ranvier' by Karlsson et al. (2009), which is the usual location of osteophyte growth). See also Brisby et al. (2013), who follows up with some further experimental data. Chondrocytes are postulated to migrate from the groove of Ranvier, across the superficial layer, and then to migrate from the superficial layer downwards into the middle and deep layers of cartilage. This theory is primarily motivated by the not unreasonable starting proposition that chondrocytes in articular cartilage move through the same cycle of differentiation as chondrocytes at the growth plate bone junction.

These authors believe it is possible that this happens in early postnatal development, while the synovial joint is still forming. Indeed, until adult zonation of the cartilage is complete, it is likely that the cartilage continues to grow at or near its surface, thereby ensuring congruence between the opposing cartilage surfaces. And surely chondrocytes do migrate laterally to some extent (indeed, this explains the 'rows' or 'strings' of chondrocytes that are observed in the superficial zone). Perhaps during the neonatal and postnatal period of growth, chondrocytes do have the appearance of migrating 'downward,' as new cartilage continues to be formed through interstitial growth in the superficial zone (Hunziker et al. 2007).

It appears to us that adult cartilage tissue operates differently. It seems much more likely chondrocytes migrate toward the surface of the cartilage rather than the reverse, for a variety of reasons. One compelling reason is that it has been established that while there are similarities between the Ihh-PTHrP signaling system at the growth plate and articular cartilage, there are important differences in chondrocyte behavior related to the relative importance of mechanical loading in stimulating PTHrP expression, rather than Ihh (see discussion above).

A second reason is the distance from the groove of Ranvier to the position experiencing high contact loads is measured in centimeters, a journey that would take too long and involve far too much risk to a cell's survival, having to migrate so close to the cartilage surface in a high strain environment.

A third reason is that well-formed columns of chondrocytes formed in the deep zone are most plausibly explained by proliferation of chondrocytes in the deep zone, rather than chondrocytes migrating randomly from the top of the cartilage and then somehow aligning themselves and 'coming together' into the highly regular columns of cells observed in the deep zone. It seems much more likely that locally proliferating chondrocytes give rise to the columns of cells in the deep zone, much as proliferating chondrocytes give rise to rows of cells observed in the superficial zone.

A fourth reason is that cells migrating away from the surface would have to be migrating in the opposite direction to the general movement of aggrecan toward the cartilage surface. It is more plausible that they are partially 'advected' through the matrix toward the cartilage surface along with aggrecan.

Finally, it is likely that there are chemotactic gradients for several different signaling molecules that induce cell migration from the deep zone to the superficial zone in the cartilage, e.g. the IGF gradient that exists in normal articular cartilage (Zhang et al. 2013a).

Taken together, while the idea that chondrocytes migrate through cartilage ECM seems to us plausible, the idea that they migrate many centimeters laterally through the cartilage and then downward to organize into regular columns of chondrocytes in adult tissue seems to us implausible. However, the fact that this has to be debated only highlights how little we actually know about the evolution of articular cartilage, and how little we know about the normal function of articular cartilage in adulthood.

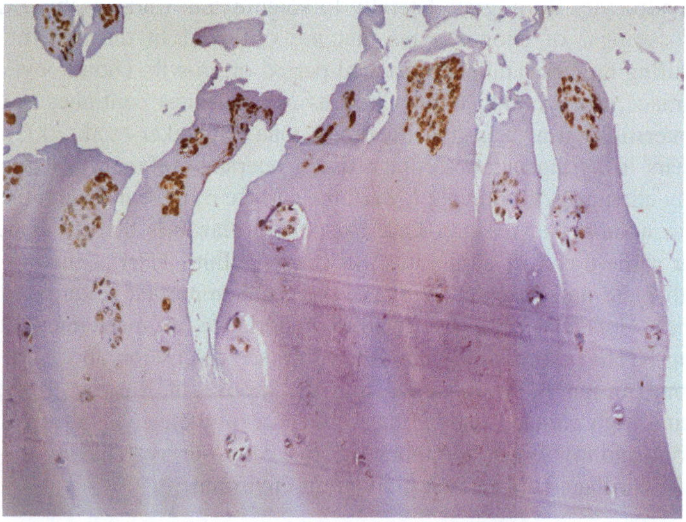

Fig. 2.55 Femoral head articular cartilage from an osteoarthritic patient. The chondrocytes comprising clonal grouping stained for α smooth muscle actin chondrocytes (magnification times 200) (Povysil et al. 2008)

2.9.5 Imbalances Between Chondrocyte Proliferation and Chondrocyte Migration

As has been reported above a 'chondroprogenitor population' of proliferating cells appears to normally reside in the upper transition and superficial zones of articular cartilage (Hayes et al. 2008; Seol et al. 2012; Jadin et al. 2006). Chondrocyte cloning (or chondrocyte cluster formation caused by proliferation within a single lacunae) is a feature of osteoarthritis (see Fig. 2.55 and Rolauffs et al. 2010). Chondrocyte cloning is a feature of rapid proliferative states in articular cartilage following severe tissue trauma (see for example Fig. 6 in Hembry et al. 2001).

It seems reasonable to suppose that persistent chondrocyte cloning represents an imbalance between normal chondrocyte proliferation and normal chondrocyte migration. Clearly 'clones' form when the rate of chondrocyte proliferation is significantly greater than the rate at which cells can migrate away from the site of proliferation. This may be a normal temporary state of affairs (Vignon et al. 1983), but it becomes abnormal if it becomes long-standing, because the chondrocytes cannot migrate and disperse throughout the tissue.

Importantly, it has been shown that 'reversion-inducing cysteine-rich protein with Kazal' motifs (RECK) is upregulated in osteoarthritic cartilage when persistent cloning is present (see Figs. 2.56 and 2.57) (Kimura et al. 2010). Increasing RECK has the effect of decreasing MMP14 on the surface of chondrocytes (so decreasing these chondrocytes' ability to degrade ECM at their invasive surfaces), while simultaneously increasing the propensity of chondrocytes to proliferate. Together, the combined effects probably drive abnormal chondrocyte cloning in osteoarthritic cartilage.

Studies have shown that RECK downregulates the expression of MMPs, including MMP2, MMP7, and MM14 (see for example Fig. 6 in Kondo et al. 2007). However, RECK's effect is very different to that of TIMPS, in that a RECK knockout is lethal while a TIMP KO causes minor abnormalities. Unlike TIMPS, RECK is (glycosylphosphatidylinositols) anchored to lipid rafts in the cell membrane (Sangiorgio et al. 2004). It is known that increasing RECK speeds the rate of endocytic pathways for MMP14 and CD13 at the cell membrane, which decreases the amount of MMP14 on the surface of chondrocytes. In light of this evidence, it is apparent that decreasing MMP14 on the cell membrane decreases the chondrocytes' ability to migrate through the ECM.

It is noted that IGF-BP5 is degraded by MMPs (MMP expression is usually increased in osteoarthritis), increasing the amount of free IGF-1 available for interaction with IGF-1R on chondrocytes (Visse and Nagase 2003). See Fig. 4 in Zhang et al. (2013b), which calculates IGF concentrations in articular cartilage. Importantly, increasing free IGF upregulates RECK expression (Fig. 2.58), and so if free IGF-1 concentrations are too high in cartilage, IGF-1 will contribute to chondrocyte cloning via RECK (Kimura et al. 2010).

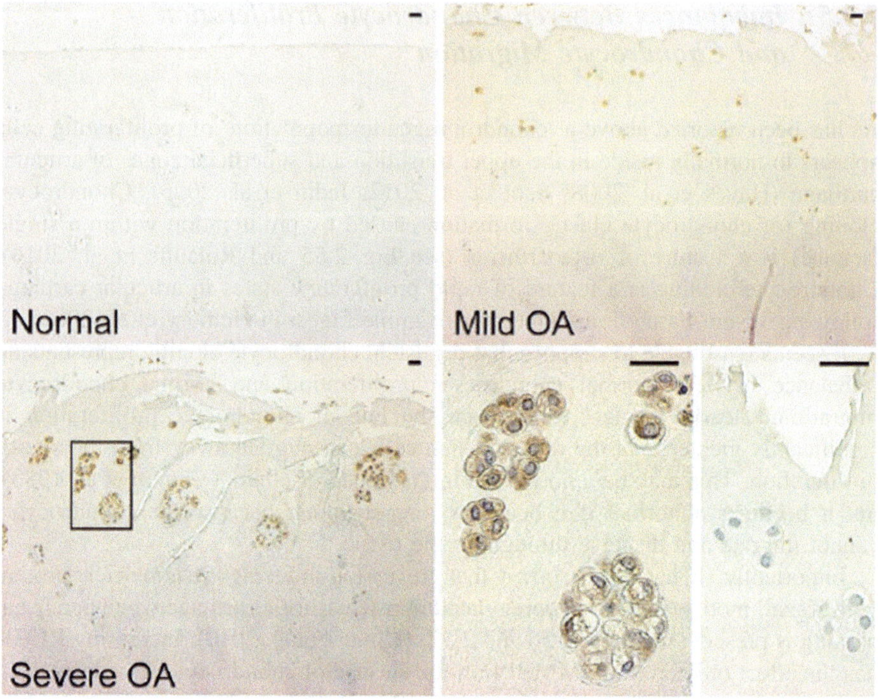

Fig. 2.56 Enhanced expression of RECK in human OA cartilage and correlation between RECK immunoreactivity and cartilage parameters. Immunohistochemical analysis of RECK in normal and OA cartilage samples. Paraffin sections of normal cartilage (Mankin score 1), mild OA cartilage (Mankin score 4), and severe OA cartilage (Mankin score 8) were immunostained with anti-RECK antibody (Kimura et al. 2010)

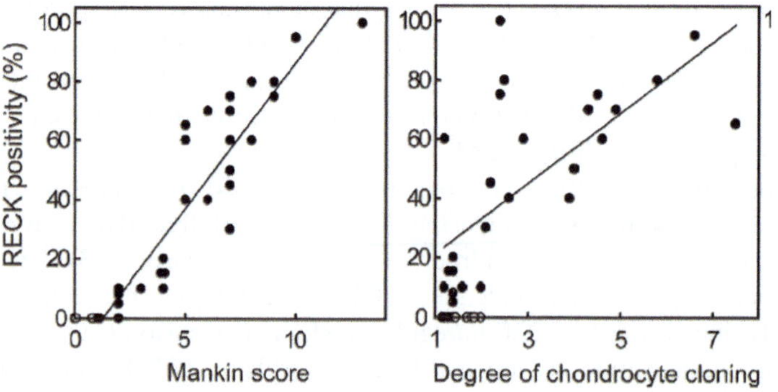

Fig. 2.57 Enhanced expression of RECK in human OA cartilage and correlation between RECK immunoreactivity and cartilage parameters. Correlations of RECK immunoreactivity with Mankin score (left), degree of chondrocyte cloning (right) (Kimura et al. 2010)

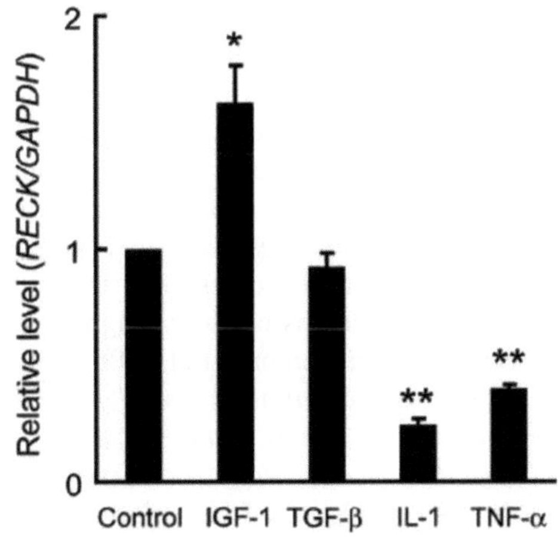

Fig. 2.58 Effects of cytokines and growth factors on RECK expression in OA chondrocytes. **a** Effects of IGF-1, TGF-β1, IL-1α, and TNFα on RECK expression in OA chondrocytes. RECK expression was examined by RT-PCR and immunoblotting (Kimura et al. 2010)

In addition, we also observe that inflammatory cytokines TNFα and IL-1β downregulate RECK (see Fig. 2.58). Downregulation of RECK has the effect of increasing a chondrocyte's propensity to successfully migrate, while simultaneously decreasing its propensity to proliferate. This raises the possibility that an osteoarthritic state driven by inflammatory cytokines is less likely to display chondrocyte cloning, than a non-inflammatory osteoarthritis (i.e. osteoarthrotic) state. In other words, it is possible that chondrocyte cloning is indicative of a distinctive osteoarthritic phenotype, induced by factors other than inflammatory cytokines (i.e. osteoarthroses disease states).

But in addition to RECK, other signaling systems may contribute to imbalances between chondrocyte proliferation and migration too (e.g. Wnts, TGFβ, Ihh-PTHrP). For example, Lories et al. summarized the functional role of the canonical pathway as being proliferative, while the functional role of non-canonical pathways as stimulating de-differentiation. Both pathways are inhibitory of each other (Lories et al. 2013). This signaling interaction also helps to explain the balance between cell proliferation and cell migration—for imbalances between proliferation and migration may also be driven by imbalances in canonical and non-canonical Wnt signaling. Too much canonical Wnt signaling may promote proliferation while suppresses de-differentiation of the chondrocytes, while too much Wnt signaling may promote de-differentiation of the chondrocytes and promote migration while suppressing chondrocyte proliferation (see Sect. 5.6.3). Clearly, it is possible for several signaling pathways to be altered and for them all to contribute toward an imbalance between chondrocyte proliferation and migration (see discussion Sect. 2.9.5).

Similar imbalances between chondrocyte proliferation and chondrocyte migration appear to occur following severe trauma to the cartilage. For example, for a partial thickness defect in pig articular cartilage (see Fig. 6 in Hembry et al. 2001;

Muldrew et al. 2001). Indeed within eight days of defect creation (Hembry et al. 2001), increased numbers of cell nuclei were evident. At three, six, and eight weeks, the expansion of the cell clusters continued, and the clones can sometimes 'bleb' through the necrotic zone adjacent to and into the defect cavity itself, helping to seed fibrous ECM formation in the defect. This imbalance may be a normal and desirable response (the experiment ran for only eight weeks), as this behavior could represent an initial response to severe cartilage damage. It would be deemed abnormal only if it becomes long-standing [e.g. so clonal clusters remains after say, six to twelve months or more (Vignon et al. 1983)].

See Fig. 6 in Hembry et al. (2001): *A summary of the major features following formation of partial thickness defects in pig articular cartilage. Macrophages infiltrating the defects at days 2 and 8 synthesize and deposit MMP-9 onto the damaged cartilage matrix, the zone of cartilage necrosis. The MMP-derived DIPEN341 neoepitope (from cleaved aggrecan) and the loss of matrix metachromasia (i.e. differences in staining) are co-localized to the damaged cartilage matrix at eight days and at six weeks, indicating aggrecan degradation. Strong induction of MMP-13 synthesis occurs in chondrocyte clusters that form at the interface between the zone of cartilage necrosis and the normal cartilage* (Hembry et al. 2001).

2.10 Radial Zonation Around Chondrocytes

2.10.1 How Can ECM Synthesis and Degradation Occur Simultaneously?

Based on a range of environmental cues including cell–cell signaling, composition of the ECM, and mechanical strain (ECM-cell signaling), the cells 'decide' what they need to be doing. Chondrocytes may proliferate, de-differentiate, daughter cells may migrate to new locations and redifferentiate again. Chondrocytes may also secrete molecules that either remove damaged ECM or create new ECM. Sometimes chondrocytes do both simultaneously.

The processes involved in matrix synthesis and degradation are described collectively as 'ECM turnover.' ECM turnover involves two basic processes: (i) ECM synthesis (primarily of structural collagen and aggrecan) and (ii) ECM degradation, in normal cartilage primarily driven by mechanical deformations, and to a much lesser extent, but the action of proteases. When the rate of synthesis is equal to the rate of degradation (including loss of ECM components from the tissue), there is 'steady-state' tissue turnover. A fundamental problem for understanding the radial structure of cartilage is explaining how both synthesis and degradation of the ECM can occur simultaneously in the same spatial domain, for it is immediately apparent that degrading enzymes may potentially degrade newly synthesized mers. If too much degradation happens relative to mer production, it could potentially prevent the formation of new extracellular matrix. How then, are these potentially conflicting processes resolved into useful functional outcomes for the tissue?

Theoretically, one could image a system whereby no particular action is taken, and so some fraction of all newly synthesized molecules are simply degraded soon after they are produced, while the remaining fraction is incorporated in the ECM. This is a feasible strategy, becoming more 'efficient' in normal tissues as the rate of degradation is reduced relative to the rate of production. As discussed for aggrecan turnover in normal articular cartilage, this appears to be the case—that is, aggrecanase degradation is a small fraction of total aggrecan loss (e.g. 5–20% of the total loss), and so also a small fraction of aggrecan production in a homeostatic state (see Sect. 2.3.3). Of course this is essentially what happens during development (i.e. remodeling associated degradation is initially very small relative to ECM molecule production). Clearly, the hormonal and cytokine environment of the organism plays an important role in determining general long-term strategies promoting tissue development versus promoting tissue maintenance.

Theoretically, one can imagine another management system whereby damaged molecules, by virtue of their damage, possess a 'molecular signature' that can be detected by degrading enzymes, thereby allowing newly synthesized molecules without this signature to evade destruction. This system is in fact implemented. For example, MMPs known as gelatinases only degrade damaged collagen fragments leaving newly synthesized collagen relatively unscathed (Rosenblum et al. 2010a). Such a system should not be too surprising, given that molecular shape recognition is fundamental to the operation of all chemical systems in the body (see Appendix A for discussion).

Theoretically, one can also imagine another management system where ECM degradation and formation share the same spatial domain, but there is temporal separation of these two processes—i.e. a time is allocated for ECM removal and another time for ECM repair. Though this management system has not yet been demonstrated experimentally, the evidence for circadian gene expression is now strong and growing. Almost certainly temporal separation of gene expression occurs to some extent in all tissues and is prominent skeletal tissues. From a functional standpoint, it appears to be very likely that skeletal tissues are damaged through the day, as part of their normal daily usage, and repaired through the night (i.e. during sleep). This cycle is probably driven by a combination of variations in activity levels throughout the day and night, by 'peripheral (tissue) circadian clocks' and finally by 'central circadian clocks.'

For example, it has been demonstrated that damage and repair of the human cornea show a circadian pattern, with marked temporal variations in the ratio of MMPs to TIMPs present on the eyeball surface (Markoulli et al. 2012). Further it has been shown mouse chondrocytes contain a circadian clock (see Fig. 2.46). In mammals, the circadian clock is driven by the transcriptional/translational feedback loops consisting of the transcriptional activator complex CLOCK and BMAL1 (bone and muscle ARNT-like 1), and repressor complex PERIOD1/2 and CRY1/2. PERs/CRYs periodically inhibit the CLOCK/BMAL1-mediated transactivation of the Pers and Crys through Ebox elements in their promoters (Yeung et al. 2014).

It has also been shown that 615 genes in mouse chondrocytes (i.e. 3.9% of the total number of genes) display a circadian rhythm (Gossan et al. 2013). The 615 genes included transcriptional activators and repressors, as well as nuclear receptors. Consequently, it appears likely that some of the 'circadian genes' will inevitably involve genes regulating cartilage homeostasis. Gossan et al. (2013) notes that the amplitude of PER2 in the circadian oscillation decreased with age. Further it is found that loss of BMAL1 leads to a progressive arthropathy (Bunger et al. 2005). Says Bunger et al. (2005):

> …BMAL1 null animals develop joint ankylosis due to flowing ossification of ligaments and tendons and almost complete immobilization of weight-bearing and non-weight-bearing joints.

This suggests that circadian expression of genes is important in musculoskeletal health, as clearly circadian rhythms are associated with at least some tissue critical processes. Despite the practical experimental difficulties, it seems very likely that understanding circadian variations in gene expression will become a crucially important area of future musculoskeletal research (Yeung et al. 2014), particularly for understanding temporal tissue damage and disruption of normal in vivo tissue

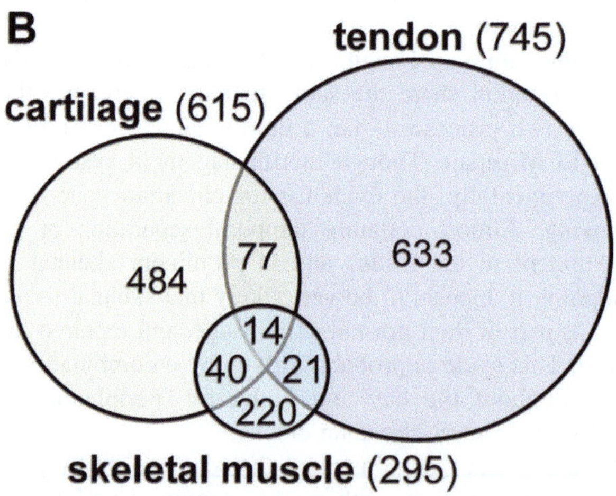

Fig. 2.59 Specificity of tendon clock genes. Venn diagram comparing the number of circadian genes of tendon, cartilage, and skeletal muscle. The total number of genes identified as circadian in each tissue is represented in brackets; areas of overlap indicate common genes (Yeung et al. 2014). *Note* Yeung et al. (2014) state: 'We have identified cell-autonomous oscillations of clock genes in murine and human tenocytes and shown that the circadian clock is a master regulator of BMP signaling.' Almost certainly diurnal variations (such as gremlin2 in tendon) will play an important role in cartilage physiology too

repair cycles. It has been reported that increasing 'internal desynchronization' of the circadian clock occurs in aging chondrocytes, and that this is associated with the development of OA (van der Kraan 2017) (Fig. 2.59).

Theoretically, one can also imagine a management system that enables efficient simultaneously synthesis and degradation, whereby newly synthesized molecules bind to a 'carrier molecule', which then shepherds these them through a 'hostile' chemical environment into an environment more conducive to the molecules functional purpose. Some of the SLRPS are obvious candidates to bind various newly synthesized mers to affect this kind of strategy, but nothing of this kind has really been reported to date. Despite this, it is seems highly likely that such a strategy is operational in normal articular cartilage tissue.

In practice, to overcome the fundamental problem of newly synthesized molecules immediately being degraded, it appears as though chondrocytes employ multiple management strategies including: (i) accept some degradation of newly synthesized and ECM molecules, (ii) only target particular molecular signatures (damage-associated molecular patterns or DAMPs) rather than molecular signatures of newly synthesized and intact molecules, (iii) temporal separation of gene expression via circadian gene expression, and (iv) temporary protective chemical binding and transport interactions. The particular mix of potential strategies adopted by chondrocytes at any one time to manage simultaneously synthesis and degradation depend on the functional demands placed on the tissue. Doubtless other strategies are employed as well, and no doubt will be uncovered in the future.

2.10.2 *Uncovering the Chemical Principles for Self-organization of the ECM*

We already know matrix turnover leads to a depth-dependent zonation of articular cartilage, and further, that there is a local radial zonation around chondrocytes. That is, cartilage ECM is typically described histologically as having pericellular, territorial, and interterritorial zones (see Fig. 1c in Poole 1997). This radial zonation is particularly apparent in the deep zone, where individual chondrocytes are responsible for maintaining a huge volume of matrix relative to the cell volume (Hunziker et al. 2002). But radial zonation is also apparent around chondrocytes in the transitional and superficial zones too.

The basic question we wish to begin to address here is this—how does radial zonation of the ECM around chondrocytes arise? It is clearly apparent that it is not the chondrocytes directing assembly of the territorial and interterritorial matrix, as by definition there is no chondrocyte (or part thereof, e.g. filamentous chondrocyte processes) present in the territorial or interterritorial zones. The assembly of extracellular matrix has to happen via chemical processes alone. So we are compelled to consider 'physio-chemical processes' that will facilitate the 'self-organization' of the ECM around chondrocytes. This leads to ask: what might these physio-chemical processes be?

Unfortunately, the details of the processes and the processes themselves responsible for the development of radial zonation around chondrocytes remain largely undefined in the literature. There is only a fragmentary integrated discussion of synthetic and degradative processes, and only few attempts (Heinegard and Oldberg 1989) have been made to organize this information into a meaningful whole.

It is known there are a number of known 'high-level interactions,' which couple ECM degradation with synthesis. For example, active TGFβ (larger quantities of latent TGFβ are known to be stored in ECM) and free IGF are released from the ECM when the ECM is damaged, which then directly stimulates a biosynthetic response from chondrocytes (see discussion Sect. 5.6.1). Clearly 'latent molecules' involving signaling mechanisms such as TGFβ's can couple ECM degradation with chondrocyte synthesis.

Another 'high-level coupling' occurs when ECM fragments are resorbed via receptors on the surface of chondrocytes. Indeed, molecular fragments from degradative processes can provide highly specific information to chondrocytes about what damage is occurring in the ECM. For example, many different types of molecules may be detected in the pericellular matrix via CD36 receptors on the cell surface (Pfander et al. 2000; Gonzalez and Ortiz-Navarrete 2013), or via $\alpha 2\beta 1$ receptor on the cell surface (which interacts with vimentin within the chondrocyte or the DGEA sequence found on collagen), or through the $\alpha 5\beta 1$ (or $\alpha V\beta 1$ and $\alpha 4\beta 7$) receptors for fibronectin, or the $\alpha 1\beta 1$ and $\alpha 3\beta 1$ receptors for laminin (Knudson and Loeser 2002; Loeser 2014). Once attached, the ligand and receptor are recycled [e.g. via lysosomal degradation (Bromme and Wilson 2011)]. This not only provides specific information to the chondrocyte about the degradative processes going on in the ECM, the degradation products also provide some of the raw materials required for the synthesis of new ECM. By these means, the process of degradation is coupled to the process of synthesis.

But the fraction of the degraded ECM internalized by chondrocytes in this way has not been quantified. There appears to be an underlying implied consensus assumption in the literature that a good fraction of the degraded ECM escapes to the synovial fluid, with only a small fraction being recycled, but exactly how much of various ECM components are recycled within articular cartilage is an open research question. In tendon, it is likely that most procollagen synthesized is recycled, indicating that the recycling of newly synthesized molecules can be significant in some tissues. However, while these general high-level couplings between ECM and chondrocytes are both interesting and potentially important, they do not really provide much insight into how zonation develops around chondrocytes.

To gain insight into our basic question (i.e. what are the physio-chemical processes responsible for self-organization of the ECM?), we would really need some insight into the processes regulating chemical reactions in the extracellular matrix.

To gain such insight, we need to allow ourselves to extrapolate from known in vitro chemistry to in vivo biochemistry. Here we rely on chemical and biochemical principles, together with our usual informed speculation and what literature there is to guide us.

Based on in vitro experiments, it is apparent that self-organization of the extracellular matrix must be regulated by the following local variables:

(i) The general environmental conditions of the cartilage in which chemical reactions occur (including the local temperature, local redox potential, local pH, local fixed charge density (which varies with aggrecan concentration) and the ionic strength of the synovial fluid (which is generally fixed at 'isotonic' conditions by the kidney);

(ii) The specific local chemical milieu, e.g. the local concentrations of structural mers in solution, within each radial zone of the extracellular matrix, and the local ionic composition of the cartilage pore fluid, e.g. the local concentration of calcium (or copper or zinc ions in solution required during the manufacture or operationalization of various enzymes);

(iii) The local concentration of reacting molecules. Most importantly, reactant A may have several competing partners, and so reactant A may potentially end up as several alternate reaction products. For example, a collagen fibril may grow in diameter by aggregation of additional tropocollagen molecules, or its growth may be arrested by binding of collagen type IX, or by binding of molecules such as SLRPs (e.g. decorin and biglycan). Which reaction predominates in situ (i.e. growth of the fibril diameter or growth arrest of fibril diameter) depends of the local concentrations of reactants and their local affinity constants. For example, the affinity of decorin for a collagen fibril is low for small fibril diameters but becomes high as the fibril diameter grows (i.e. molecular affinity is a function of the surface stereochemistry. For further explanation, see Eq. (A.5) and associated discussion in Appendix A, as well as Sect. 2.10.5).

(iv) The local concentration of various *regulatory* molecules. Regulatory molecules control enzymatic or nucleation reactions in the ECM. For example, the concentration of molecules that activate other molecules (e.g. TIMP concentrations may allow one protease to activate another protease (Nagase et al. 2006), which then allows procollagen to become tropocollagen upon the removal of C- and N-terminal procollagen telopeptides). Nucleators lower the 'activation energy' to initiate the formation of polymers [e.g. type XI collagen is a nucleator for type II collagen formation (Kadler et al. 2008)]. One molecular may protect another molecule from degradation by binding to it (Gardiner et al. 2011; Zhang et al. 2010b).

(v) The interplay between the rates of chemical reactions and the rates of chemical transport through the ECM. The so-called Damkohler number (https://en.wikipedia.org/wiki/Damk%C3%B6hler_numbers) for various chemical reactions is of crucial importance in determining the spatial distribution of chemical reactions around chondrocytes.

Let us now explore these general statements through examples, with a view to stating our postulated Principles for Self-Organization of the ECM in the next section.

One crucially important example reaction is the ECM polymerization reaction of aggrecan and HA with link protein to form aggregate. Aggregate is first formed through the binding of the G1 domain on aggrecan with sites on the HA polymer → aggregate. This bond is subsequently 'stabilized' by link protein (Roughley and Mort 2014; Day and Prestwich 2002).

Experimentally, it is found the fraction of G1 binding with HA is pH dependent. As the pH is reduced from pH 7.4 to 6.0, the bound fraction increases somewhat (see Fig. 4 in Heng et al. 2008), and then as the pH is reduced further to pH 5.0 the bound fraction rapidly reduces (see Fig. 1 in Parkar et al. 1998). This pH dependence of aggrecan binding to HA is also illustrated in Fig. 13 of http://glycoforum.gr.jp/science/hyaluronan/HA16/HA16E.pdf. From this, we infer that the binding affinity between aggrecan and HA will vary spatially with the spatial variation of pH. Practically, this means the spatial binding affinity variation favors the formation of aggrecan in the territorial and interterritorial zones, where the pH is higher, rather than in the pericellular matrix where the pH is relatively lower.

TSG-6 is a small molecule that is secreted during inflammatory states [e.g. in osteoarthritis (Bayliss et al. 2001)]. TSG-6 is believed to compete with the G1 domain on aggrecan for binding to HA (Heng et al. 2008). Like the G1 domain of aggrecan, TSG-6 also exhibits a pH-dependent binding affinity, but TSG-6 only really competes with the G1 domain for binding to HA if the pH is less than 6. As the pH rises from pH 6 to pH 8, there is a dramatic reduction in binding affinity of TSG-6 for HA (Heng et al. 2008) and (Fig. 13 of http://glycoforum.gr.jp/science/hyaluronan/HA16/HA16E.pdf). Intriguingly, we see that binding of G1 and TSG-6 to HA is 'differentially regulated' by pH (Heng et al. 2008). By this example, two important principles in radial zonation become apparent: (i) The spatial distribution of pH in the cartilage ECM has a strong influence on the spatial distribution of biochemical reactions, and (ii) that the competition between reactants for binding sites is also pH dependent.

But we can see further, if now we consider time. It is apparent the binding affinity of G1 with HA decreases with pH, meaning K_d increases with decreasing pH [as shown in Fig. 2 (Heng et al. 2008)]. Now $K_d = \frac{k_r}{k_f} = \frac{[L][P]}{[LP]}$—see Eq. (A.2). This equation means the fraction of time that aggrecan and HA are unbound (i.e. the reactants are free in solution) increases with decreasing pH, and so the free concentration rises. A greater free concentration gives the free mers greater opportunity to be transported away as mers, as opposed to being transported as a complex. But aggrecan and HA take longer to react as pH decreases (i.e. the rate of the forward reaction, k_f, decreases). Experiments on cartilage explants have shown fifty percent conversion of HA and aggrecan to aggregate requires about 5.7 h at pH 7.45, but this increases to 11 h at pH 6.99 (Sah et al. 1990).

Interestingly at physiological chondroitin sulfate concentrations in solution, the principal driving force for the migration of GAG becomes independent of the size of the GAG chains, depending only on the total GAG concentration (see Sect. 7.3.3). While the similar considerations hold for aggrecan and aggregate regarding their respective driving forces (i.e. the driving force for transport depends on the total GAG concentration, rather than details about the molecules containing GAG), the ease with which molecules can make their way through cartilage ECM greatly decreases with increasing size of the molecules. In other words, the effective diffusion coefficient greatly reduces for larger and longer molecules, meaning aggrecan will move faster through the ECM than aggregate (i.e. for aggrecan molecules attached to HA) (see discussion Sect. 7.3.3).

Clearly, the slower reaction rate at lower pH, and the lower K_d at higher pH, both favor aggregate formation in the territorial and interterritorial zones. But how much of this potential aggregate formation is realized depends on the capacity of the mers to be transported to the territorial and interterritorial zones and assembled into aggregate. It is possible that transport will rate limit the aggregate formation rate in the territorial and interterritorial zones.

From this example, another important principle in radial zonation becomes apparent: that the timescale for polymerization and the timescale for mer transport need to be appropriately matched to get an optimal outcome. For if transport is too fast, there will be little polymer formation and the mers will be deposited beyond where they are required, while if transport is too slow polymers will be deposited too close to the chondrocyte. At this point, we mention this type of problem, involving a transport rate and a rate of chemical reaction is common in chemical engineering. For these types of problems, it is usual to define a non-dimensional ratio: the timescale for chemical reaction (k_f) over the timescale for diffusive transport (L^2D). This non-dimensional time is known as the Damkohler number https://en.wikipedia.org/wiki/Damk%C3%B6hler_numbers. Unfortunately, there is little (if any) reference to Damkohler numbers in the cartilage literature to date. Clearly, this is a research area ripe for growth (maybe Damkohler number contour maps would be helpful in understanding ECM dynamics?).

From this example of aggregate formation, a key organizing principle for radial zonation around chondrocytes becomes evident: A 'field' of small ions or molecules around a chondrocyte (in this case, the hydrogen ion concentration) can act as a (chemical) 'indicator field,' specifying where other chemical reactions should or should not take place (e.g. the binding of aggrecan to hyaluronan). In the example of aggregate formation, the pH field around the chondrocyte is the 'indicator field' that directs the rate and extent of the chemical binding between the mers. In other words, this indicator field is employed to localize the polymerization reactions at specific radial distances from the chondrocyte, in 'sympathy' with the radial distribution of pH (which may of course itself vary over time). As we shall see from further examples, pH may be the most important chemical indicator field for radial zonation around chondrocytes, but there may well be other indicator fields we are not yet aware of [e.g. it is possible that fibronectin provides another indicator field, at least within the pericellular matrix. Fibronectin is a multifunctional glycoprotein that has multiple binding sites for a number of macromolecules and cell surface receptors (interestingly, fibronectin has been shown to be an absolute requirement for fibroblast migration into fibrin clots) (Makogonenko et al. 2002)].

In humans, the interterritorial zone in articular cartilage only develops in the late teens. After reaching adulthood and with normal loadings, collagen matrix turnover in the deep zone slows very significantly. In fact, the collagen fibrils in the radial zone of articular cartilage gradually grow in size as they mature. Then cross-linking of the collagen mers proceeds over weeks, months, and many years, which greatly slow the reversibility of the chemical reactions that led to the formation of the collagen fibrils in the first place. Of course cross-linking of collagen mers within collagen fibrils serves the useful purpose of increasing the strength and stiffness of the collagen fibrils, as well as limiting reversibility of formation reactions.

The enzymes that do the cross-linking of collagen fibrils are typically found to be small. It is not unreasonable to suppose that their small size enables them to diffuse through small pore spaces in the ECM to their 'site of action' at the collagen fibrils. Indeed when reviewing molecular transport through articular cartilage tissue, DiDomenico et al. conclude: '*the most crucial factor when designing new therapeutic molecules is solute size*' (DiDomenico et al. 2018). For example, a particularly important copper-containing cross-linking enzyme is lysyl oxidase https://en.wikipedia.org/wiki/Lysyl_oxidase. Lysyl oxidase is about 46–48 kDa in its intracellular 'pro-form' but in its secreted active form it is 30–32 kDa (Gacheru et al. 1990; Rucker et al. 1998).

If we assume a 'globular' protein is spherical in shape, we can calculate its radius (Erickson 2009), viz.,

$$r_p^{\min} = \left[\frac{3V}{4\pi}\right]^{1/3} = \left[\frac{3\bar{v}M}{4\pi}\right]^{1/3} = 0.066M^{1/3} \quad (2.1)$$

where $\bar{v}$ is the molar volume for protein and M is the molecular weight in Daltons. For a 30 kDa protein such lysyl oxidase, we can assume it is globular and estimate the radius as $r_{\text{lysyl}}^{\min}$ equals 2.05 nm, while for a globular 100 kDa protein $r_p^{\min}$ equals 3.06 nm. Because real molecules are continually jostled by thermal agitation, the actual radii based on measurements of the radius of gyration of the protein or somewhat larger—but not by much, probably about 5% (Smilgies and Folta-Stogniew 2015). In this context, it is interesting to note that molecular dynamic simulations of the average distance between GAG molecules at 20 mg/ml and 80 mg/ml have shown that they are separated by about 4.0 and 2.0 nm (Bathe et al. 2005). This is an average separation, and the GAG chains wiggle (although they are quite stiff), so they wiggle stiffly. In addition, it is known that lysyl oxidase it not in fact spherical, but is about 50% longer than it is wide [so the minimum length across the molecule ($l_{\text{lysyl}}^{\text{min}}$) is about 3.0 nm (Bhuvanasundar et al. 2014)]. Lysyl oxidase also wiggles and deforms somewhat with thermal agitation [so amino acids in the backbone fluctuates in position by about 2.5 Å, while the 'carboxyl tail' fluctuates in position by about 5.0 Å (Bhuvanasundar et al. 2014)]. With the wiggles of the GAG chains and with the wiggles of lysyl oxidase, lysyl oxidase can diffuse through GAG at about 80 mg/ml concentration. However, neutral globular proteins with a MW of 100 kDa (i.e. about 6.0 nm diameter) are known to be excluded from cartilage, which helps gives some idea of the maximum for the distribution of transient pore size openings in normal cartilage. See the schematic diagram of showing hydrodynamic radii for molecules of various sizes shown in Fig. 1 of DiDomenico et al. (2018).

From this example, it begins to become evident that a molecule's size is very important if the molecule is to travel a considerable distance from the chondrocyte through the small pore openings in the ECM. Indeed as the ECM becomes denser with molecules in solution and their ECM 'precipitates,' the median pore size opening of the ECM (multiphase material) decreases. This may well be one critically important physical factor regulating tissue maturity, by 'signaling' tissue maturity. In any case following secretion, the ECM matrix clearly acts as a 'size-selective barrier' to the transport of molecules away from the chondrocyte. It is therefore desirable for the ECM with the smallest pore sizes to be most distant from the chondrocyte (i.e. the interterritorial zone), otherwise this may limit the volume of ECM that can be serviced by a chondrocyte.

It is apparent that a large molecular size will hinder molecular transport through the ECM, but a variety of effects come into play. Consider the blood protein albumin, which is crucially important for retaining plasma within the vascular space, and is probably the main protein in synovial fluid. As a first approximation albumin may be thought of as a 'globular protein,' but a more detailed analysis suggests that it changes its conformal shape with pH, being described as 'heart-shaped' at pH 7.4, while at pH 3.5 it is an elongated 'cigar-shape,' and at pH 2.7 expands into a fully 'elongated shape' (Baler et al. 2014). At pH 7.4 albumin only penetrates cartilage in small quantities. In this case, the extracellular matrix in the superficial zone is apparently acting as a reasonably effective size exclusion barrier. However, at pH 7.4 albumin is also negatively charged (with net charge about −16 at this pH). The negative charge also influences its 'partitioning' cartilage (see discussion on fixed charge and Donnan partitioning below). The molecular weight (MW) of albumin is 65 kDa unglycosylated and about 67 kDa glycosylated. We can use Eq. (2.1) to estimate the radius for albumin at $r_{\text{albumin}}^{\text{min}}$ equals 2.7 nm [compare radius of gyration 2.74 nm (Kiselev et al. 2001)]. With the abovementioned limitations in mind, this provides another guide to the limits to the MW of globular proteins that can be transported through articular cartilage ECM.

Physical chemistry tells us larger molecules have lower diffusion coefficients (so for example, full-sized antibodies) cannot usually enter cartilage ECM. But this is not invariably the case [e.g. see data on measured diffusion coefficients for dextran moving through cartilage (Leddy and Guilak 2003)]. We have already seen previously that hyaluronan can penetrate into cartilage about 300 μm in 48 h (Forsey et al. 2006). Clearly the shape of the molecule is also important to its transport—for if it is 'elongated,' it may travel more easily through the ECM than if it is 'globular.' Remarkably, large linear molecules can move through the ECM more easily (e.g. hyaluronan or aggregate) than many smaller globular-shaped proteins.

The process by which long polymer chains are believed to 'wiggle their way' through small pore spaces is referred to as 'reptation' (analogous to 'slithering' or 'reptilian movement'). The diffusive movement of a long polymer can be thought as migration through a 'tube' (in space) within the matrix containing the polymer. In a similar vein, the movement of long relatively stiff polymers may be conceived as analogous to a somewhat-like an articulated 'spear.' In a polymer gradient, the spear is pushed unequally at both ends to generate a net force in the direction of the polymer gradient. This net driving force is then primarily resisted by viscous drag along the sides of the spear. It turns out that thermal movements of linear molecules can be large, as is illustrated by molecular dynamics simulations and NMR measurements of hyaluronan (see for example http://www.glycoforum.gr.jp/science/hyaluronan/HA31/HA31E.pdf). It can be shown that the time required for the polymer to move the length of a tube is approximately proportional to the length of the tube squared, and inversely proportional to the viscous drag between the tube wall and the matrix (see http://en.wikipedia.org/wiki/Reptation). Note that this means if the molecule is very long its diffusivity will be small, while if the polymer and matrix have the same charge then viscous drag while be reduced and so diffusivity increases.

To have a larger diffusion coefficient, the molecules need to be smaller, more linear, and if linear then shorter, and finally have the same charge as the matrix. Now this helps to explain why component mers (rather than even larger polymers) are more easily transported from the site of their production (i.e. secreted from the surface of chondrocytes) to their more distant assembly site in the ECM. For example, aggrecan and hyaluronan are transported to and through the territorial matrix as mers, where they are assembled into super-sized molecule known as aggregate. Once in place, aggregate molecules are further bound together by 'linker molecules' such a fibulin and tenascin to form even larger 'super-aggregates.' In other words, these super-aggregates are constructed sequentially in situ from many small molecules that are transported to the site of assembly. As these super-aggregates form and grow larger, they become more and more 'fixed' to the nearby collagen network, and so the ECM becomes more solid-like as the solid phase density increases.

Because aggregate is very large, it has a very small diffusion coefficient in a solution of aggregate. Aggrecan has a very small diffusion coefficient that decreases with increasing solution concentration (Gribbon and Hardingham 1998), so aggregate has an even smaller diffusion coefficient. The collagen network then reduces the diffusion coefficient several orders of magnitude again. The very slow rate of transport through normal cartilage extracellular matrix helps ensure large molecules like aggregate are retained in the collagen network for physiologically useful lengths of time.

For fast reptation through cartilage, elongated polymers need to have the same negative charge as thc aggregate—as does aggrecan and hyaluronan. But it is still difficult to conceive exactly how a large 'bottlebrush' molecule like aggrecan (MW 2.0 to 2.5 MDa) is transported into and through the territorial matrix. Aggrecan is about 350 nm long, and 60 nm in diameter, so in terms of usual molecular dimensions (e.g. of globular proteins) the dimensions of aggrecan are enormous. The GAG chains attached to the core protein are relatively stiff, much like rods that have limited articulation along their length. The repulsion between the closely spaced GAG molecules (G1 domains on the aggrecan core protein attached along hyaluronan are spaced at 15–40 nm intervals) stiffens the aggrecan core protein.

Given this, it seems plausible to suppose that aggrecan is transported with either the G1 or G3 domain pointing in the direction of movement. Given the asymmetry of the keratin sulfate GAG chains at one end, and chondroitin sulfate GAG chains at the other end, it seems probably that one or other domain is more likely to 'lead.' The GAG chains forming bottlebrush shape would be 'swept back,' confined to a smaller diameter 'tube,' as the GAG chains push their way into and through the surrounding GAG chains that define the 'tube' through which an aggrecan molecules moves. Clearly, the osmotic pressure gradient across the aggrecan molecules during transport must be considerable to effect this movement, but we know that unusually large osmotic pressures are generated by chondroitin sulfate (see Appendix D).

The 'secret' of aggrecan transport through the ECM is probably that the large molecule is flexible, so individual components (i.e. the individual GAG chains) making up the aggrecan molecule deform while maintaining a constant pressure on the surrounding ECM that is resisting its movement. So the individual parts of the aggrecan molecule are then 'primed,' and ready to take advantage of local thermal fluctuations in ECM pore size openings, and advance that part of the molecule when a local pore size increase allows. In other words, the whole molecule does not advance *en bloc*, but each of the molecule's many component parts advance individually as opportunities present themselves (for if the molecules was rigid, the probability of pore size openings being simultaneously large enough everywhere to permit advance of the whole molecule would be vanishingly small). In this way, the diffusion coefficient for the aggrecan molecule becomes the diffusion coefficients for its component parts. This provides a way of explaining how the diffusion coefficient for some large molecules (e.g. aggrecan) becomes independent of their molecular weight. For example, it is known that the diffusion coefficient for chondroitin sulfate polymers become independent of their molecular weight (e.g. see for example Fig. 2 in Comper 1991).

Significant ‘random’ aggrecan transport is easily verified by labeling newly synthesized aggrecan with radioactive sulfate 35 (Buschmann et al. 1996b). It is conceivable that over longer periods of time, ‘holes’ (a hole being defined here as a region with at least one missing aggrecan molecule in an otherwise uniform aggrecan concentration field) may diffuse toward the surface of the aggrecan concentration field (probably by the detachment of the G1 domain from HA followed by the reattachment of the same G1 domain further along the HA molecule). It seems plausible to suppose that aggrecan molecules would be in some cases transported ‘single file’ and in others transported ‘on mass’ to some location (Bayliss et al. 2000), where they then disperse by diffusion (see further discussion Sect. 3.4). The relative importance of the transport mode appears to change with age-dependent maturation of the cartilage, and with each zone in adult cartilage.

If the macromolecule to be transported has a local, exposed positive charge on the outside of the molecule (as it may even when the net charge is negative), then the positively charged region(s) of the molecule will be tend to bind to the negatively charged aggregate by electrostatic forces, and either ‘slow down’ or perhaps ‘temporarily immobilize’ the molecule’s movement. This observation highlights the fact that reversible (or temporary) ‘binding reactions’ limit the rate at which molecules can be transported through the ECM.

Partitioning is the preferential distribution of a molecule between two phases or two compartments. The quantitative effect of electrostatic interactions on the partitioning of charged molecules between a ‘bathing fluid’ (e.g. in vivo, the bathing fluid is synovial fluid while ex vivo the bathing fluid is the ‘support media’) and the cartilage extracellular can be estimated by analyzing the Donnan (electrochemical) equilibrium of the molecule (Grodzinsky 2011). Cartilage ECM cartilage has a negative fixed-charge density (FCD). For 6% GAG by weight, the FCD has been estimated to about 0.2 mEq/gram for cartilage (Maroudas 1976). Another estimate of FCD at a GAG concentration of 102 mg/ml is 0.4 mEq/ml (using the equation described in the Methods section of Chahine et al. 2005). Because of the negative FCD, cations are attracted into the cartilage ECM while anions are repelled.

If a cartilage is bathed in ‘reference solution’ of isotonic saline (i.e. about 0.15 M NaCl solution) and the fixed charge is 0.2 M (i.e. 0.2 mEq/ml), then the Donnan equilibrium sodium ion concentration in the cartilage ECM is 0.28 M (almost double the 0.15 M concentration of Na in the (reference) bathing fluid), while the equilibrium chloride ions concentration is 0.08 M (almost half the 0.15 M concentration of Cl in the bathing fluid). The difference between the Na and Cl concentrations in the cartilage is accounted for by the fixed charge density, so overall charge balance is maintained at all times (0.2 + 0.08 = 0.28 mEq/ml). This ‘Donnan analysis’ implies that small highly negatively charged molecules will partition strongly toward the synovial fluid, i.e. they are ‘excluded’ from the cartilage ECM, while small highly positively charged molecules will partition strongly toward the cartilage, i.e. they are ‘attracted’ to the cartilage ECM. We mention in passing that this is an important consideration when designing a drug to enter or leave cartilage ECM (Kar et al. 2016).

From the foregoing, we may infer that 'largish' molecules such as COMP (COMP is a pentamer, with each mer comprising the pentamer about 100 kDa in size) and tenascin [tenascin is a hexamer about 1.5 MDa in size (Patel et al. 2011)], both of which are found in the territorial and interterritorial matrix, are probably transported as component mers deep into the territorial and interterritorial matrix, where they are assembled into the pentamers and the hexamers (as seen in electron micrographs and as illustrated in Fig. 4 in Aspberg (2012).

2.10.3 Chemical Principles for Self-organization of the ECM

Given the discussion and examples in the previous section, let us now try and summarize some of the physicochemical Principles for Self-Organization of the ECM:

(i) Local environmental conditions determine which reactions are fast and which reactions are slow (or next to impossible). As for most biochemical reactions, the rate of production of reaction product is generally increased by: (i) increasing the concentration of reactants, (ii) by having more enzyme present (which promotes the reaction), and (iii) by having the enzyme in favorable 'matrix conditions,' which increases the enzyme's efficiency of operation (e.g. having certain macromolecules or ions in solution, e.g. calcium, and up to a certain point, increasing the temperature).

(ii) To form the ECM around the chondrocyte, molecules must be transported away from the chondrocyte to where they are functionally operative in the ECM. In this case, the rate of the chemical reaction relative to the rates of transport of reactants and reaction products is important, as this determines where the reactant products form relative to the chondrocyte.

(iii) The transport properties are generally characterized by the 'diffusivity' (random thermally induced movement through a static matrix) or the 'permeability' of the porous medium in the case of bulk advective transport (i.e. plug flow) of molecules. Bulk transport may be the advective flow of the solvent (and any dissolved solutes, which may be transported at the same velocity as the solvent or at a reduced rate, i.e. they are hindered), or it may be advective transport of macromolecules themselves (e.g. aggrecan) through a notionally static solvent). Both modes of transport are generally related to the pore size distribution and their connectivity in the porous medium. It is noted that for charged molecules being transported through the charged porous material, the apparent pore size distribution is influenced by the Debye double-layer thickness (e.g. so diffusing anions are excluded from at least some of the pore space if the surface charge on the solids is negative, and vice versa).

(iv) The concentration of reactants and/or reaction products alters the diffusivity or permeability of the porous medium. The concentration can do this in a number of different ways. An increased concentration may increase the apparent viscosity of the solution (i.e. so diffusion is hindered relative to a free solution). This influences both diffusion and permeability. There are also 'volume exclusion effects,' which can influence osmotic pressure gradients (see Chap. 6). Irreversible reactions that form solid precipitates alter the porosity of the ECM. Reactants and reaction products may reversibly bind to solids, altering the apparent porosity of the porous material (and so the net cross-section area available for transport through the porous media), or the apparent diffusivity (the interpretation of diffusivity depending on the mathematical formulation employed). Generally, these effects become more important as the ECM becomes more and more crowded and the solution becomes less 'ideal' (see Chap. 6).

(v) The size and shape of the molecule are important determinants of its transport properties through the cartilage ECM. The shape of the molecule is important to its transport properties—globular molecules have more difficulty moving through cartilage ECM than linear polymers of the same MW. A large size may lead to size exclusion, either partial or complete, depending on the size and connectivity of the pore spaces in the ECM relative to the size and shape of the molecule. We note in passing that size separation is the principle by which 'gel electrophoresis' operates (molecules of different sizes can move more or less easily through the gel, so they become separated over time as they are transported through the gel). One 'strategy' enabling a molecule to penetrate further from the chondrocyte is by progressively cleaving the molecule into smaller and smaller pieces, while retaining its biological function. So for example, some proteases and aggrecanases are progressively cut into smaller and smaller pieces, allowing them to travel through smaller pore openings, while still retaining their bioactivity;

(vi) The charge on the molecule is important to its transport properties. Charge has a variety of effects on macromolecules. For example, macromolecules may change their shape as the charge on their surface changes, which may in turn changes their apparent viscosity, diffusivity, and their reaction kinetics. So the shape of the molecules may change with pH. A charged molecule may bind to fixed charge on the ECM, delaying its transport through the ECM. Alternatively, one molecule may reversibly bind to another molecule (perhaps electrostatically) and then 'piggy-back' to elsewhere in the ECM, whereupon it 'unhitches' itself as it enters a different chemical microenvironment. A 'transporter molecule' thereby enhances another molecules transport properties. Further if the molecule is an enzyme, a change in the 'background matrix' conditions may change its charge distribution, which in turn may change its 'rate constant,' and so its bioactivity.

(vii) pH is an important determinant of the electrostatic charge on a molecule, because carboxyl (acidic COOH groups), the amino group (basic NH_2 groups) and sulfate containing groups ionize at different pHs (as quantified by the various pKa of these groups). As the different groups ionize, the overall charge on the molecule changes (Uquillas and Akkus 2012). Importantly, the isoelectric point (pI) is defined as the pH at which the net charge on the molecule is zero. For a pH below the pI, the net charge on the molecule is positive, while for a pH above the pI, the net charge on the molecule is negative https://en.wikipedia.org/wiki/Isoelectric_point. Therefore, the key 'indicator' for net electric charge on a molecule in the ECM is the pI of the molecule relative to the local pH. Given that electrostatic charge is an important determinant of both the molecular shapes and reaction kinetics of the molecules involved matrix reactions, this explains why pH is such a critically important 'indicator function' for structuring radial zonation around chondrocytes.

(viii) For any reasonable normal physiological state, the ECM of cartilage is always negatively charged due to the low pKas of sulfate, carboxyl, and uronic groups found in GAG molecules. Most importantly, molecules that have the same charge as the (negative) 'fixed charge' on the ECM move more easily through the ECM. This is because like electrostatic charges repel each other, which reduces the likelihood of binding interactions between the molecule and the fixed charge on the ECM. However while electrostatic binding is minimized, non-specific interactions such as van der Waals forces or hydrophobic binding interactions are still present. Consequently, broadly speaking the repulsion between like-charged molecules being transported through the ECM reduces the 'apparent resistance' of that molecule to transport through the ECM, though there remains some resistance due to non-specific binding interactions (i.e. the resistance to transport does not become zero for any real molecule through any real solution).

(ix) Conversely, molecules that have opposite charge to the fixed charge on the ECM (the ECM is negatively charged at physiological pH) will generally not move easily through the ECM, but rather partition to the fixed charge. Partitioning slows the rate advance of molecules through the tissue, approximately in inverse proportion to the fraction of molecules bound to immobile molecules. Of course, if all the ECM binding sites have all been occupied by the molecule being transported, then transport may once again 'speed up' (provided there is adequate space remaining to allow transport through the solution). However for many molecules, in practice the number of ECM binding sites greatly outnumbers the number of molecules to be transported (e.g. as is the case for MMPs, aggrecanases, TIMPs, procollagens), and so occupying all the binding sites to increase the rate of transport through the ECM may not be a feasible strategy to speed transport. It is for this reason that electrostatic charge on the molecule to be transported becomes a particularly important issue when the molecule is to be transported a significant distance from the pericellular matrix into the negatively charged ECM found in articular cartilage.

Given its central importance in regulating the net electrostatic charge distribution on proteins and glycans, let us now look at the pH distribution around chondrocytes in more detail.

We first note that pH is the negative $\log_{10}$ of the average hydrogen ion concentration in a volume of solution, viz.,

$$pH = -\log_{10}[H^+] \tag{2.2}$$

This is a straightforward concept in simple homogeneous solutions, but a meaningful definition depends on the existence of a representative volume element (RVE), that is, a sufficiently large volume over which the hydrogen ion concentration can be suitably averaged (see Grodzinsky 2011 and Sect. 3.4.1). If the volume is too small then time-dependent fluctuations in the concentration of hydrogen ions will occur, particularly when hydrogen ion concentrations are small. Wide variations in local pH may become relevant to molecular kinetics when the volumes being considered are very small and/or the time-period considered is very short.

For example, in the presence of a high density fixed charge (as occurs in cartilage because of dissociation of carboxyl, sulfate, and other groups on GAGs attached to aggrecan), the local hydrogen ion concentration may vary significantly through the thickness of the diffuse double layer around the fixed charge. This means that a molecule introduced into a high-density fixed-charged environment may experience a range of pHs over a very short distance (i.e. over a Debye length, which is only about 0.8 nm for isotonic physiological conditions). It is noted that the Debye length is a function of the ionic strength of the solution. While the ionic strength of the bathing solution is usually fixed (because the ionic strength of the blood is held nearly constant by the kidneys), the ionic strength can change locally as the cartilage experiences volume strain during deformations. The relevant 'averaging volume' increases with the size of the molecule. Despite local variations in the spatially heterogeneous fixed charge density and variations in the thickness of the Debye length, it is probably still reasonable to define an average hydrogen ion concentration based on a volume somewhat larger than the molecules being considered here, and therefore accept that pH is a definable function of position within articular cartilage.

We know the pH of normal synovial fluid is basic (pH 7.3–7.7) and so we can assume that the surface of the cartilage is fixed at a pH around 7.4, because synovial fluid is derived from blood plasma which normally has a pH of 7.4. The pH of normal cartilage varies predictably from its surface, generally decreasing with increasing distance from the cartilage surface, while pH also decreases locally with increasingly proximity to a chondrocyte (Konttinen et al. 2002; Parkar et al. 1998).

This decrease in pH with depth is driven by two processes: (i) passive hydrogen ion partitioning [i.e. (a) due to the increasing fixed charge density that is associated with an increasing aggrecan concentration with depth from the cartilage surface e.g. this probably explains about 1 unit of decrease in pH with depth (Parkar et al. 1998), and (b) by the increase in aggrecan concentration radially to a maximum within the pericellular region (e.g. which may explain a further 0.25–0.5 unit decrease in pH locally)], and (ii) by active production of hydrogen ions [i.e. because of the production of lactic acid (Milner et al. 2012) during anaerobic metabolism by chondrocytes (Parkar et al. 1998)] leads to a depth-dependent lactic acid profile in the cartilage. Because both anaerobic metabolism increases with depth and aggrecan concentration increases with depth, together these processes inevitably lead to a decrease in pH with depth (Fig. 2.60).

Because of high local concentration of aggrecan in the pericellular matrix and because of the anaerobic metabolism taking place within the chondrocytes

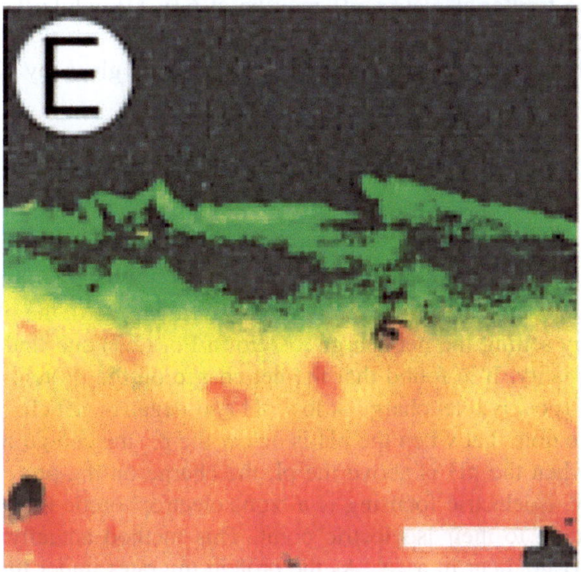

Fig. 2.60 Acridine orange staining of femoral head articular cartilage taken from patients with primary hip OA that underwent total hip replacement. Figure shows apparently clinically normal cartilage retrieved from an OA patient and stained (scale bar is 100 μm). Based on the calibration reported in the paper (see Fig. 2 in the paper), acridine orange stains approximately green for pH > 7.0, yellow for pH 6.0–7.0, light orange pH 5.0–6.0, and dark orange less than pH 5.0 (Konttinen et al. 2002). Note that the tissue pHs suggested by this staining are indicative only. Actual tissue pHs may be higher or lower, and the actual pHs for this specimen are in any case probably different to normal articular cartilage, as this specimen was taken from OA cartilage, even though it was graded as 'normal.' Nevertheless, it is clear there is a general decrease in pH with depth from the cartilage surface. It is also apparent that around individual chondrocytes (or around chondrocyte clusters), there is a local decrease in pH (as indicated by orange 'spotting' about the same size as chondrocytes within the ECM)

(producing a lactic acid gradient from the chondrocyte into the surrounding ECM), the pericellular matrix of chondrocytes in the deep zone is relatively acidic (e.g. probably in the range pH 5.0–6.0 in normal cartilage). Close to the cartilage surface, say within the superficial zone, the pericellular pH may be considerably higher (e.g. pH 6.5–7.5). Because of anaerobic metabolism, the more distant interterritorial matrix may be one pH unit higher (and possibly as much as one and a half pH units higher) than the pericellular matrix next to the chondrocyte. In other words, the pH rises with increasing distance from chondrocytes. Note that this increase in pH with distance from the chondrocyte may not rise as a smooth curve, but may have a pH jump at the interface between the pericellular matrix and territorial matrix (and quite possibly another jump at the interface between the territorial matrix and interterritorial matrix), primarily because of differences in aggrecan concentration.

Despite the inevitable uncertainties, to focus our arguments we attempt to put some numbers to the local and regional variations of pH for normal articular (human knee) cartilage. For example in the deep zone, the territorial matrix may be at pH 5.5–6.5, while the interterritorial zone may be at pH 6.0–7.0. In the transitional zone, the territorial matrix pH may be at 6.0–7.0, while the interterritorial matrix pH may be at 6.5–7.25.

It can be shown that the net charge on a protein is given by Kirkwood et al. (2015),

$$\text{net charge} = \sum_{i=1}^{n^-} \frac{-1}{1 + 10^{pK_n - pH}} + \sum_{j=1}^{n^+} \frac{-1}{1 + 10^{pH - pK_p}} \tag{2.3}$$

where pK_n and pK_p are the pK values of negatively and positively charged amino acids, respectively. When the net charge on the protein is zero, the pI = pH. This pI can be found by setting the net charge to zero in Eq. (2.3) and finding the pH that satisfies the equality. It is found that the net charge on the protein is a monotonic function of pH (positive net charge at low pH and negative net charge at high pH). Importantly, we note that when the pH is below the pI, the charge on the protein is positive, and when the pH is above the pI, the charge on the protein is negative.

Interestingly, isoelectric focusing is a 'zone electrophoretic method' to separate proteins according to their isoelectric point. This method actually relies upon an immobilized pH gradient in a gel. As the protein migrates under the influence of a direct electric current, it stops migrating at the location where pI = pH, as the net protein charge is then zero and the electric field has no effect on the protein's migration. Sharp focusing can be obtained using this method, with separation of proteins that have a pI that differs by as little as 0.01 units https://en.wikipedia.org/wiki/Isoelectric_focusing.

Interestingly, Baskin et al. (2006) show that for short distances, say up to 10 μm or so, a protein evenly distributed over an immobilized pH gradient will, given some time, concentrate itself at the location on the gel where pI = pH. This occurs because there is a chemical potential gradient in the protein distributed along the gel, and the protein diffuses down its chemical potential gradient to where the chemical potential is minimized. The chemical potential is defined to be equal to the energy required to add a single molecule at any point along the pH gradient. Clearly the amount of energy required to be added changes as a result of the electrostatic repulsion between proteins that have the same charge. The higher the charge, the greater energy that is required to be added with the addition of each molecule, so this means the chemical potential is higher (see discussion of potentials in Appendix K). Baskin et al. (2006) demonstrate pH gradients exist inside cells and then demonstrate isoelectric focusing of GFP inside a HaLe cell attached to a bead of gel with fixed pH. The same considerations doubtless play some role in defining where molecules are transported within the ECM, and so the spatial distribution of ECM reactions.

2.10.4 MMPs, Aggrecanases, TIMPs, Cathepsins, and Cystatins

Let us now try to better understand proteases and their inhibitors in articular cartilage, while simultaneously growing our qualitative understanding of the processes responsible for radial zonation around chondrocytes. We attempt to do this by trying to deepen our understanding of the proteases operating in the ECM by applying our new-found physicochemical Principles for Self-Organization of the ECM described in Sects. 2.10.2 and 2.10.3, to aid our understanding the behaviour of proteases and their inhibitors in articular cartilage.

To do this, we first need to 'get to grips' with proteases. But there are so many to choose from! There are, for example, some 607 proteases expressed by the human genome, including 31% that are serine proteases, 25% cysteine proteases, 33% metalloproteases, and 4% are aspartic proteases (Bromme and Wilson 2011). Which to focus on?

MMP activity was discovered by Gross and Lapiere in 1962 while studying tail resorption during the metamorphosis of frogs (Gross and Lapiere 1962; Murphy and Nagase 2008). Traditionally, the MMPs have been considered the main proteases in extracellular degradation because of their membrane or extracellular localization, their optimal activity at neutral pH and their ability to degrade collagen, elastins and proteoglycans (see Fig. 2.63) (Bromme and Wilson 2011). Similar to studies on other tissues, the primary focus for articular cartilage has been on MMPs. However, it is important to keep in mind that MMPs are not the only proteases in the extracellular matrix of articular cartilage (Struglics and Hansson 2010) and at certain times they may not be the most important protease system operating (e.g. aggrecanases, cathepsins). But first let us focus on MMPs.

We now provide a basic outline on what is known about MMPs. It turns out that pre-proMMPs are synthesized within the cell. Some ten proMMPs have a

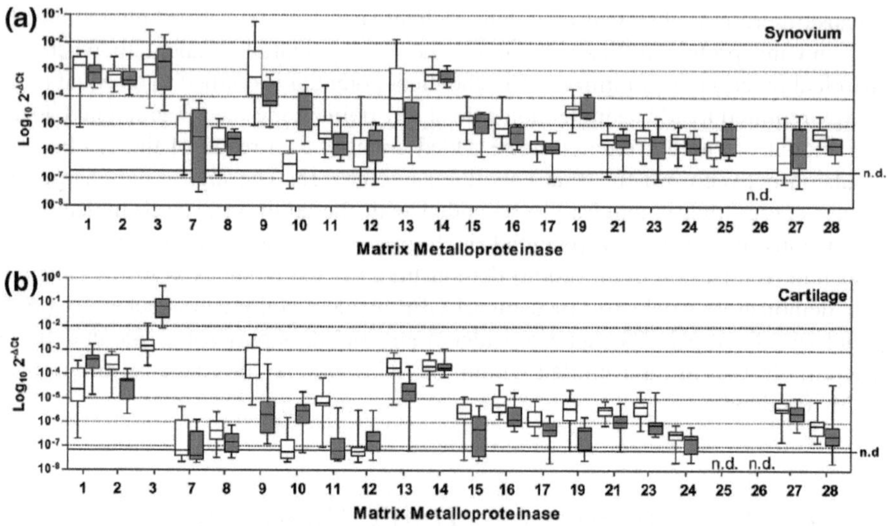

Fig. 2.61 Comparative expression of MMP genes in the fractured NOF (i.e. used as normal reference control cartilage) (filled boxes) compared with osteoarthritis (open boxes) in hip synovium and cartilage. The expression level of each gene was determined and normalized to the level of *18S* rRNA gene expression using ΔC_T [C_T(target gene) − C_T(18S)]. The box-and-whisker plot shows median values, and each section represents a quartile of the data. Samples on or below the dotted line have C_T(target gene) = 40 and are therefore not detectable (n.d.). (Davidson et al. 2006). Note that the top three MMPs (normalized median expression level above 10^{-4}) in normal cartilage are MMP1, MMP3, and MMP14, while in osteoarthritic cartilage the top five MMPs (normalized median expression level above 10^{-4}) are MMP2, MMP9, MMP13 and again MMP3 and MMP14. Note MMP mRNA expressions in the synovium generally remain fairly constant, though MMP2 and MMP13 are upregulated in osteoarthritis, suggesting the primary pathology is cartilage related. Perhaps synoviocytes are responding to the same signaling molecules and ECM damage fragments in the synovial fluid that the chondrocytes are responding to within the cartilage?

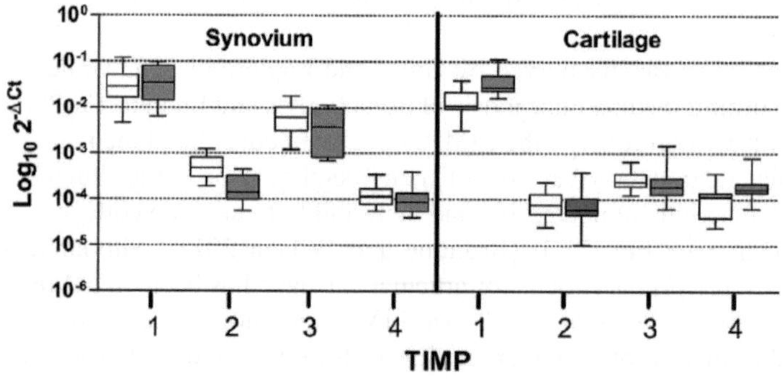

Fig. 2.62 Comparative expression of TIMP genes in the fractured NOF (i.e. used as normal reference control cartilage) (filled boxes) compared with osteoarthritis [cartilage] (open boxes) in hip synovium and cartilage. The expression level of each gene was determined and normalized to the level of 18S rRNA gene expression using ΔC_T [C_T(target gene) − C_T(18S)]. The box-and-whisker plot shows median values, and each section represents a quartile of the data (Davidson et al. 2006)

'furin-like proprotein convertase recognition sequence' (RX[K/R]R) at the end of the propeptide. They are likely to be activated intracellularly and then either are bound to the cell surface or secreted. Thirteen MMPs are secreted from the cell as proMMPs and activated extracellularly (Nagase et al. 2006). For example, MMP-3 and MMP-10 digest a number of ECM molecules and participate in proMMP activation. Membrane-bound MMP14 (sometimes referred to as MT1-MMP) plays a key role in activating proMMP-2 (while requiring participation of TIMP2), and MMP14 also plays a role in activating proMMP-13 at the cell surface (Nagase et al. 2006). Secreted proMMPs can be either activated directly or in a step-wise fashion (i.e. part of propeptide is first removed, while the other part of the propeptide is removed by either the intermediate MMP or by another activated MMP). This provides one explanation for a variety of reported molecular weights for the same MMP. The activities of all these MMPs are regulated by the tissue-specific location of activating enzymes, competitive inhibition, and inactivation by endogenous inhibitors and by the proteolysis of activated MMPs.

MMP14 is a crucially important collagenase that degrades interstitial collagens as well as a large range of other ECM molecules. Importantly, Nagase et al. (2006) note these MMP14 activities are usually found at the 'invasive front' as cells migrate through ECM (Nagase et al. 2006) (see discussion Sect. 2.9.4). mRNA expression levels for MMP1 through to MMP28 in both normal and osteoarthritic synovium and cartilage are shown in Fig. 2.61.

The local concentration of bioactive MMPs is regulated by TIMPs (tissue inhibitors of metalloproteases). TIMPs inhibit active MMPs by binding one to one at the active site of the MMP (Murphy 2011), so the ratio of the two is often viewed as an important variable that is related to the rate of ECM degradation. Binding constants between TIMPs and MMPs are typically in the low picomolar range (Murphy and Lee 2005), so TIMPs bind very tightly to MMPs. In addition to local strain and the concentration of molecules that attach to collagen fibrils, which provides 'competitive inhibition' for activated MMPs, it is the local TIMP to enzyme ratio (i.e. either the TIMP/MMP ratio or the inverse TIMP/ADAMTS ratio) (Kashiwagi et al. 2001) that provides the final regulatory control over the local rate of tissue degradation.

mRNA expression levels in normal and osteoarthritic cartilage for TIMP1 through to TIMP4 are shown in Fig. 2.62. All the TIMPs inhibit all the (soluble) MMPs. However, some TIMPs are specialized and have specific roles to play in the ECM. For example, TIMP1 does not inhibit membrane-bound MMPs (e.g. MMP14), while TIMP2 facilitates the cell surface activation of MMP2 by MMP14 (see Fig. 5.14). This arrangement presumably ensures that TIMP2 is present when MMP2 is activated.

Says Murphy (2008): '*Like many 'housekeeping genes'—that is, genes that are always expressed because they encode proteins that are constantly required—TIMP1, TIMP2, and TIMP4 lack obvious TATA boxes.*'

TIMP1 inhibits ADAM10 (which binds fibrillin-1), while TIMP2 inhibits ADAM12 (which degrades COMP). TIMP3 has a much broader spectrum of ADAM and ADAMTS inhibition than the other TIMPs (i.e. it inhibits ADAM10, 12,17, 19, 33 and ADAMTS1, 2, 4, 5), while TIMP4 inhibits ADAM17 and 28 (Murphy 2011). Importantly, TIMP3 inhibits ADAMTS4 and 5 (which are the main ADAMTSs in articular cartilage). Once again we begin to get multiple hints about the very specific roles within specific chemical systems being played by each molecule in the ECM. We observe that the specific (and often multiple) roles played by each molecule undoubtedly explain the 'bewildering variety' of molecules in the ECM—but on the flip side it is this bewildering variety that permits exquisite control of the chemical systems operating in the ECM. Doubtless the variety of TIMPs offer more precise process controls over several protease systems, which help enable functionally relevant outcomes for the tissue.

Says Murphy (2011): '*Knowledge of the precise functions and spatio-temporal expression patterns of metalloproteinases are still somewhat limited.*' At this point, we observe this appears to be a recurring theme when it comes to our understanding of articular cartilage.

Let us now try to expand our knowledge about MMPs and TIMPs by applying our Principles for Self-Organization of the ECM. In the interterritorial and territorial zones, the TIMP to MMP ratio and the TIMP to ADAMTS ratio are probably normally (functionally) close to one, as supported by the observation of a very slow rate of proteolytic degradation of ECM in normal cartilage (see Sect. 2.3.3 for discussion). This raises the question as to how the TIMP/MMP ratio is usually kept higher than one in the interterritorial zone, which probably has a significantly slower rate of turnover than the average for the whole ECM. In addition to the relative rates of production of TIMPs and MMPs (which may also vary temporally depending on the local activation of signaling pathways), another potentially important mechanism contributing to zonation around chondrocytes is the relative rates of transport of TIMPs and MMPs through the ECM. This gives us an opportunity to test the theoretical predictions of our Principles for Self-Organization of the ECM against experimental findings. If we can repeatedly confirm the theoretical predictions made by applying the Principles for Self-Organization of the ECM, the probability of them being correct increases, and our confidence in them will begin to grow.

If we wish to have the TIMP/MMP ratio increase with increasing distance from a chondrocyte (thereby ensuring greater resistance to proteolytic degradation with distance from a chondrocyte), then based on the Principles for Self-Organization of the ECM, one would theoretically *a priori* predict that the TIMPs should be generally transported more easily through the ECM than the MMPs. Assuming that both TIMPs and MMPs are degraded at similar rates, and assuming that TIMPs and MMPs are both approximately globular in shape at physiological pH, and knowing that size is important to their transport rates through ECM—we then might predict a priori, that the TIMPs will be smaller molecules than the MMPs. For other things being equal, the (smaller) TIMPs will have larger diffusion coefficients then the (larger) MMPs. Larger TIMP diffusion coefficients relative to the MMP diffusion coefficients would enable the TIMPs to penetrate further and faster than MMPs. This set of conditions will result in the TIMP/MMP ratio increasing with distance from a chondrocyte. This outcome depends on the TIMPs being smaller in size than MMPs, which we now test.

It is found that all the TIMPs are expressed in articular cartilage. And as predicted on the basis of The Principles for Self-Organization of the ECM describes above, it is found experimentally that the TIMPs are in fact generally smaller MW molecules relative to the MMPs they regulate. For example, TIMP1 is 28 kDa, TIMP2 is 22 kDa, TIMP3 is 24 or 27 kDa, while TIMP4 is 22 kDa (i.e. they are all less than 200 amino acids in length) (Kanyama et al. 2000; Gomez et al. 1997). In contrast to the TIMPs, the MMPs are larger (generally the MMPs are more than twice the size of the TIMPs). Let us confirm that the MMPs are generally larger than TIMPs.

Often MMPs are secreted as proMMPs and then activated. For example, proMMP1 is 52–55 kDa, while bioactive MMP1 is 42 or 48 kDa (http://herkules.oulu.fi/isbn9514270789/html/x561.html). MMP1 is a collagenase, capable of digesting collagen types I, II, and III [this MMP is one of only four proteases capable of doing this (the four collagenases are cathepsin K, MMP1, MMP8 and MMP13) (Bromme and Wilson 2011)]. It is noted in passing that MMP1 is deactivated in vivo by meprin proteases (Broder et al. 2013). Meprins very likely help to provide a 'switch' between debridement of damaged collagen by MMPs and initiation of repair via collagen assembly (Murphy and Nagase 2008), for meprins not only degrade the collagenase MMP1 but also play a role the removal of telopeptides from procollagen.

MMP2 is 72 kDa (MMP2 digests collagen types IV, V, VI, VII, X, XI, as well as proteoglycans fibronectin and denatured collagen). ProMMP3 is 55 kDa while bioactive MMP3 is either 45 kDa or 28 kDa (http://herkules.oulu.fi/isbn9514270789/html/x561.html) (MMP3 digests type III procollagen, aggrecan, decorin, fibrinogen, and activities proTGFβ, proMMP1, 3, 8, and 9, substance P, T kinnogen) (Nagase 2001). ProMMP8 is 55 kDa while bioactive MMP8 is 42–46 kDa (http://herkules.oulu.fi/isbn9514270789/html/x561.html) (MMP8 digests collagen types I, II, and IIIa). MMP9 is 83–95 kDa (MMP9 digests collagen types I, II, III, IV, V, and VI as well as proteoglycans, fibronectin, and denatured collagen). ProMMP13 is 60–65 kDa, while bioactive MMP13 is 50–55 kDa (http://herkules.oulu.fi/isbn9514270789/html/x561.html) (MMP13 digests collagen types I, II, and III, aggrecan, fibronectin, tenascin) (Kanyama et al. 2000; Ishiguro et al. 2002; Veidal et al. 2011; Jaiswal et al. 2011). In each case, we see that the active MMPs are indeed significantly larger than the TIMPs.

With a larger diffusion coefficient, in the same amount of time the TIMPs can be transported further into the surrounding extracellular matrix than MMPs. So if equal amounts of TIMP and MMP are secreted by a chondrocyte and they are degraded at similar rates, other things being equal, then their relative rates of transport through

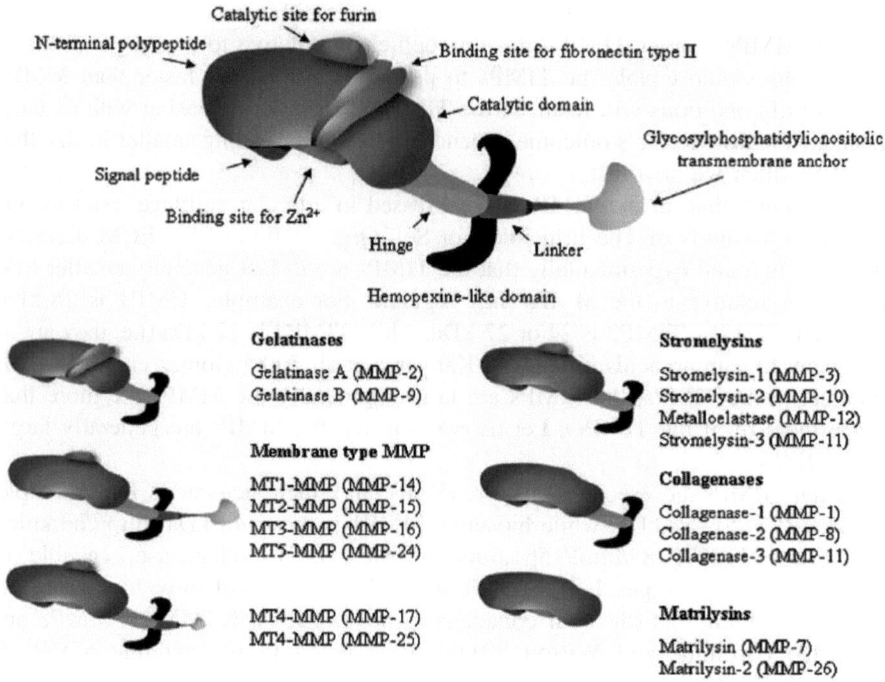

Fig. 2.63 Characterization of single MMPs according to their structural differences (Huska et al. 2009). Note in the figure collagenase-3 should be labeled MMP13

the cartilage ECM will 'swing' the TIMP/MMP ratio higher at greater distances from the chondrocyte. By this reasoning, ECM close to the chondrocyte would tend to be degraded more readily by MMPs, while more distant ECM would tend to be preserved from MMP degradation. This provides a simple but valuable example of how molecular size and transport rates may influence the local concentration distribution around a chondrocyte (in this case the TIMP/MMP ratio). Though this example gives us only a small insight into one mechanism contributing to the radial zonation ECM around chondrocytes, a number of small insights may sum to much explanatory power.

So let us now consider other opportunities to the test the Principles of Self-Organization of the ECM, even though they are simple examples yielding small incremental insights. Collagen type II is usually highly resistant to proteolytic degradation. Indeed, only a few proteases can actually hydrolyze the peptide bonds in the triple helix region of collagen (types I and II). The collagenases are cathepsin K, MMP1, MMP8, and MMP13 (Bromme and Wilson 2011). Most importantly, we note that MMP13 cleaves collagen type II with 40 times greater efficiency than either MMP1 or MMP8 (http://herkules.oulu.fi/isbn9514270789/html/x561.html). MMP1 is the collagenase active in normal cartilage, while MMP13 is often strongly upregulated when cartilage is injured or diseased. MMP13's efficiency in cleaving type II collagen helps explain its destructive potential in disease processes such as osteoarthritis (Goldring et al. 2011).

In addition to the soluble MMPs, there are various membrane-bound MMPs including MMP14, 15, 16, and 17, which are all expressed in cartilage. However, MMP14 is the key 'broad spectrum' membrane-bound MMP, as MMP14 digests almost the full range of ECM components including collagens types I, II, III, proteoglycan, fibronectin, vitronectin, and laminin. It also plays a role in activating other proMMPs (Ishiguro et al. 2002).

Importantly, we note that MMP14 is well suited to coordinating the degradation of pericellular type II collagen, via direct cleavage of any type II collagen that comes in contact with it. MMP14 also directs a proteolytic cascade through activation of proMMPs, e.g. activation of MMP2 and MMP13 (Tam et al. 2002). By regulating this proteolytic cascade, MMP14 makes itself a central player in the process of chondrocyte migration, degrading ECM in the path of a migrating chondrocyte. MMP14 is regulated at the cell surface by various processes including induction of MMP14 through integrin binding of collagens, palmitoylation of MMP14, MMP14 shedding, MMP14 endocytosis (Kimura et al. 2010) and TIMP binding (Tam et al. 2002).

MMP2 and MMP9 are known as 'gelatinases' (Rosenblum et al. 2010a). Gelatinases contain three 'fibronectin type II modules' (see Fig. 2.63) that efficiently bind 'denatured collagen,' known as gelatin. Denatured collagen may be produced enzymatically (leaving so-called ¾ and ¼ fragments), or thermally (tropocollagen is thermally unstable at body temperature and begin to slowly unwind) or mechanically (collagen may have been stretched, plastically deformed or fractured, changing molecular recognition sites on its surface) (An et al. 2014; Veres et al. 2014). Gelatinases are usually confined to the *pericellular matrix* and are particularly effective in binding and finally degrading denatured collagen [e.g. MMP9 binds to the 'tail' of collagen fragments (Rosenblum et al. 2010a)]. We note that MMP1, MMP8, and MMP13 do not normally bind to the surface of denatured collagen, as the usual surface MMP molecular recognition sites for these collagenases have been altered by the denaturation process, requiring 'fibronectin type II modules' that target and can bind to the denatured collagen fragments.

The MMPs normally operating in the territorial and interterritorial regions of articular cartilage (e.g. MMP1, MMP8, and MMP13) only partially degrade normal collagen, solubilizing collagen by cutting the collagen mer to create ¼ and ¾ fragments, which are then transported away from the collagen fibril. The partially degraded collagen is then either degraded further in the pericellular matrix by the gelatinases (where the end products may potentially be resorbed by a chondrocyte) or the partially degraded collagen fragments escape the cartilage and enter the synovial fluid.

Some MMPs may themselves be 'degraded' into smaller bioactive molecules. For example, MMP1 has a 28 kDa bioactive fragment while MMP3 has a 45 kDa bioactive fragment. Importantly, the aggrecanases may also be 'degraded' while retaining their bioactivity. For example, ADAMTS5 is active in a range of sizes from 150, 60, 40, down to 30 kDa (Plaas et al. 2007), while ADAMTS4 (which has a predicted MW of 90 kDa in the pro-form) has active forms of 68 and 53 kDa ((Gao, Westling et al. 2002; Gao, Plaas et al. 2004) and http://www.sigmaaldrich.com/etc/medialib/docs/Sigma/Datasheet/6/a4726dat.Par.0001.File.tmp/a4726dat.pdf). It is noted that the ADAMTS4 (which cleaves the interglobular domain (between G1 and G2) of aggrecan) is activated in the pericellular matrix by MMP17 (a membrane-bound MMP) (Gao et al. 2004; Patwari et al. 2005; Clements et al. 2011).

By now it may be apparent that again the size of MMPs and ADAMTSs plays an important role in determining the concentrations of these ECM degrading enzymes in the zones around a chondrocyte. Now using our Principles for Self-Organization of the ECM, we can predict which 'territories' around a chondrocyte MMP molecules are likely to occupy. We note that the gelatinase MMPs are large—MMP2 is 72 kDa while MMP9 is 95 kDa. ADAMTSs (e.g. ADAMTS5—the active sizes ranging 150–60 kDa, while active ADAMTS4 is around 68–53 kDa), so they are usually restricted by virtue of their large size to the immediate environs of the chondrocyte. In other words, MMP2 and MMP9, the larger forms of ADAMTS5 and the larger active forms of ADAMTS4, are normally restricted to the *pericellular matrix*.

When these molecules are required to move beyond the pericellular matrix, the cascade of reactions regulating the decreasing enzyme size of MMP1 (the main collagenase in ECM of normal cartilage) and of ADAMTS5 [probably the main aggrecanase found throughout the ECM of normal cartilage (Naito et al. 2007)] enables these molecules to shrink in size, and by this means expand their range of influence. We see that regulation of 'sequential cleavage while retaining bioactivity' is an important means for controlling how much degradation occurs at distance from the chondrocyte in normal cartilage. But size alone is not enough to explain the distribution of molecules in the ECM.

Can we use our Principles for Self-Organization of the ECM to gain further insights into the structuring of molecular distributions around chondrocytes? Electrostatic binding of TIMPs, MMPs, and ADAMTSs to solids can effectively remove molecules from solution in specific regions by sequestration. Immobilized on fixed solids, such molecules clearly cannot move deeper into the surrounding ECM. So, for example, if a molecule (e.g. a TIMP) is sequestered to say aggregate molecules, it may effectively be rendered immobile, and quite possibly inactive.

For this reason, it is considerable interest to consider the isoelectric points of TIMPs, MMPs, and ADAMTSs in relation to the pH profiles around chondrocytes (as described in the Principles for Self-Organization of the ECM), for the local pH determines the net electrostatic charge on a molecule. Importantly, we note that the negative charge on the aggrecan is virtually independent of pH over the physiological range of pH in normal cartilage tissue—so it is only the charge on the TIMPs, MMPs, ADAMTSs, and other ECM component molecules that are likely to vary significantly in the pH field around chondrocytes.

Clearly, charge variation with pH may alter the sequestration status of all mobile molecules with a pI in the range of pH variation in cartilage ECM. Importantly, we first mention that there is often a range of pIs for any one molecule, as the molecule may be 'post-translationally modified' by the addition of charged groups (e.g. increasing sulfation and phosphorylation typically decrease the pI of the molecule). In some cases, the change in pI due to post-translational modification may be substantial (e.g. the pI may change up to 3 pH units with post-translational attachment of about 10 phosphate groups).

The isoelectric points of the TIMPs are given by Gomez et al. (1997) as: TIMP1 pI = 8.5; TIMP2 pI = 6.5, TIMP3 pI = 9.1, and TIMP4 pI = 7.2 (Gomez et al. 1997). However, the pI for each molecule depends on its phosphorylation status. We give several examples: (i) the 'basal' pI of TIMP1 is 8.5, but the pI may vary from 8.5 down to 7.0 with the addition of three phosphate groups; see http://www.phosphosite.org/proteinAction.do?id=3565715&showAllSites=true, (ii) the 'basal' pI of TIMP2 is 7.5, but the pI may vary from 7.5 to 6.3 with the addition of two phosphate groups, (iii) the 'basal' pI of TIMP3 is 9.0, but the pI may vary from 9.0 to 8.1 with the addition of 4 phosphate groups, and (iv) the 'basal' pI of TIMP4 is 8.0, but the pI may vary from 8.0 to 6.6 with the addition of three phosphate groups.

TIMP2 may be the most mobile in the pericellular region (as it is closest to having a near neutral charge at a pH = 6.0 to 6.5 in the pericellular region). The other TIMPs have a net positive charge in the normal pericellular region (i.e. pH 6.0 to 6.5), though this depends on the phosphorylation state). In the 'base state' (without additional phosphorylation), these molecules are probably mainly electrostatically bound (i.e. they are sequestered in various amounts depending on pH and pI and the quantity of negatively charged aggrecan molecules present). If aggregate is immobilized, then any bound molecule also becomes immobile. However if a newly synthesized aggrecan is mobile, then a sequestered TIMP molecule may transported along with aggrecan into the territorial region (and beyond). In other words, the smaller (e.g. TIMP) molecule may potentially 'piggy-back' on the larger (e.g. aggrecan) molecule, and so be transported into the ECM at the rate of transport for the larger molecule. Though aggrecan is a larger molecule than the molecules it carries, it may be transported surprisingly quickly for a large molecule along its osmotic/concentration gradient around chondrocytes, and more generally along its osmotic/concentration gradient toward the surface of the cartilage. Significant radial aggrecan transport is easily verified by labeling it with radioactive sulfate 35.

It is noted that TIMP3 (pI 9.1) is likely to be the TIMP most strongly sequestered to aggrecan (Yu et al. 2000). If sequestration does indeed occur in vivo, then TIMP3 would inevitably travel with the aggrecan/GAG as it is transported along its concentration gradient to the territorial and interterritorial regions. Interestingly, we recall that TIMP3 is the broadest spectrum TIMP, inhibiting MMP1 and ADAMTS5, which are present in the territorial and interterritorial zones. But the other TIMPs may also bind to aggrecan/GAG to varying fractions depending on pH.

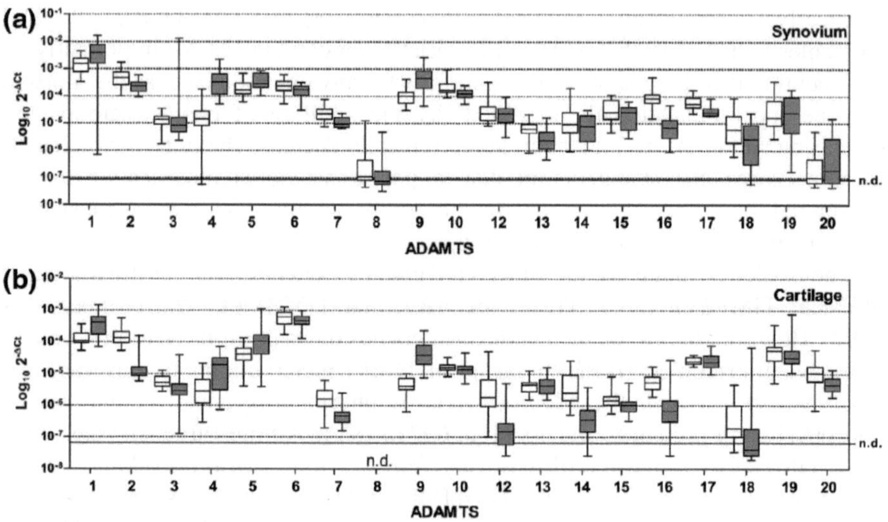

Fig. 2.64 Comparative expression of ADAMTS genes in the fractured NOF (i.e. used as normal reference control cartilage) (filled boxes) compared with osteoarthritis (open boxes) in hip synovium and cartilage. The expression level of each gene was determined and normalized to the level of *18S* rRNA gene expression using ΔC_T [C_T(target gene) − C_T(18S)]. The box-and-whisker plot shows median values, and each section represents a quartile of the data. Samples on or below the dotted line have C_T(target gene) = 40 and are therefore not detectable (n.d.) (Davidson et al. 2006). *Note* ADAMTS4 and ADAMTS5 are both downregulated in OA. This is most likely due to the cartilage being end-stage osteoarthritis (cartilage was obtained from patients undergoing total hip replacement), rather than early-stage osteoarthritis. Note also that ADAMTS2 mRNA expression increases significantly in osteoarthritis, probably reflecting the need to remove more N telopeptides from procollagen type II, which is required for collagen fibril repair. At the same time, ADAMTS12 mRNA expression significantly increases, which degrades COMP

If TIMPs do bind to aggrecan/GAG molecules, and the whole complex is transported away from the chondrocyte, as the pH rises with increasing distance from the chondrocyte, so the TIMPs 2, 4, 1, and 3 would be sequentially released (in that order, at least according to the pIs reported by Gomez et al 1997) from the aggrecan/GAG/TIMP complex. As they are released, the TIMPs locally would inhibit MMPs at that radius from the chondrocyte. Again, this analysis points to TIMP3 regulating the activity of MMP1 and ADAMTS5 in the distant interterritorial region. Though these insights are small and hard won, our Principles for Self-Organization of the ECM appear to be beginning to uncover a new perspective on the ways for structuring ECM around chondrocytes.

Something interesting happens for aggrecanases, which is also consistent with our Principles for Self-Organization of the ECM. ADAMTS5 has a predicted isoelectric point 9.13–8.27 depending on phosphorylation state (http://www.phosphosite.org/isoelectricCalcAction.do?id=1284982&residues=14), while ADAMTS4 is predicted to have a pI 8.2 though to 7.24 depending on phosphorylation state. This suggests

ADAMTS5 will be strongly electrostatically sequestered to aggrecan/GAG in the pericellular matrix, and just like TIMP3, probably travels with the aggrecan/GAG to the distant territorial and interterritorial matrix. This suggests that TIMP3 and ADAMTS5 can travel to the same region of ECM together. However, it has been shown that the sequestration of TIMP3 on aggrecan enhances TIMP3's affinity for ADAMTS4 (in other words a trimer is formed). This trimer so formed strongly enhances ADAMTS4 inhibition by TIMP3 (i.e. inhibition of ADAMTS4 by TIMP3 is increased some 20-fold using as little as 10 μg/ml of aggrecan) (Wayne et al. 2007). This means that if TIMP3 is present in the pericellular matrix, it will strongly inhibit the activity of ADAMTS4. Trimer formation may happen for ADAMTS5 too (Troeberg et al. 2012), though this has not been experimentally confirmed. This demonstrates that binding may change the shape of the molecule, and increase its functional activity, rather than diminish it as usually assumed.

In addition to the aggrecanases ADAMTS4 and ADAMTS5, there are many other ADAMTSs expressed in synovium and cartilage tissue. mRNA expression levels in normal and osteoarthritic cartilage for ADAMTS1 through to ADAMTS20 are shown in Fig. 2.64. All the ADAMTSs would doubtless have specific roles to play in ECM biology. For example, ADAMTS2 removes N telopeptides from procollagen type II, while ADAMTS12 degrades COMP (Liu 2009).

The comparatively low pI of MMP2 relative to other MMPs catches our attention (5.3–6.1 according to Sariahmetoglu et al. 2007) and 5.3–4.8 with up to 12 phosphorylation sites according to http://www.phosphosite.org/proteinAction.do?id=10002&showAllSites=true). This suggests that electrostatic interactions between MMP2 (which in normal cartilage has a pH in the range 5.0–7.5 depending on proximity to the cartilage surface—see previous discussion) and aggrecan are comparatively weak in the pericellular matrix, and that this will enhance MMP2's mobility in the pericellular matrix. The greater comparative mobility of TIMP2 relative to the other TIMPs in the pericellular matrix probably also enables it to 'partner' MMP14 to activate MMP2 (see Nagase et al. 2006). Again we see that Principles for Self-Organization of the ECM are beginning to help us to make functional sense of what we perceive (in our previous ignorance) as apparently 'random properties' of molecules. The molecular properties are not at all random, but in fact confer highly specific functional attributes on particular molecules, which we can begin to appreciate.

Continuing in this vein, we again note that the gelatinase MMP2 has a MW of 72 kDa. So MMP2 is also relatively large, which helps to confine it to the pericellular matrix. It is noted that MMP2 digests collagen types IV, V, VI, VII, X, XI, as well as proteoglycans and fibronectin. The other gelatinase MMP 9 (MW 83–95 kDa) is also a large MMP, and it too has a low isoelectric point and similar substrate profile to MMP2 (pI = 5.7 to 5.5 with 2 phosphorylation sites http://www.phosphosite.org/proteinAction.do?id=4675&showAllSites=true). It becomes more clearly apparent that it is MMP2, MMP9 and MMP14 and TIMP2, working together, that degrade all the damaged ECM collagen, lectican and other ECM fragments that find their way back to the pericellular matrix.

MMP1 has a pI of 6.5 (ranging down to 5.5 with increasing phosphorylation) (http://www.phosphosite.org/proteinAction.do?id=15877&showAllSites=true). We note that MMP1 is usually 52–55 kDa, but also has a bioactive fragment of 28 kDa. As the pH rises with distance from the chondrocyte, this small fragment will become progressively more negatively charged, which facilitates its mobility at distance from a chondrocyte. With its small size and a strong negative charge, it is maximally mobile through dense aggrecan and so can reach the distant interterritorial region (much like lysyl oxidase, discussed in Sect. 2.10.2). Once there, providing TIMP3 is in sufficiently low concentrations, providing there is not too much competitive inhibition, and providing the collagen type II fibrils are not too stretched, MMP1 can bind to the neutral or positively charged sites on the collagen type II fibril, and once suitably positioned, begin to degrade it.

The cathepsins are regulated by cystatins (MW around 10–13 kDa) in the picomolar range (cystatin C regulates cathepsins in the extracellular matrix) (Konttinen et al. 2002; Morko et al. 2004; Bromme and Wilson 2011). Clearly, the family of cathepsins has properties that enable them to be highly mobile and effectively remodel cartilage ECM, just as the MMPs do. It is noted that in normal cartilage, cathepsins may play a role in activating MMP2 and MMP3 (Fosang and Beier 2011).

In disease states such as OA, the pH of cartilage may be one to two pH units lower than that normally found throughout the cartilage (Konttinen et al. 2002). At these lower pHs, many MMPs operating in the territorial and interterritorial regions may become less effective (possibly due to their electrostatic charge becoming the same as that of their substrate, and increasing repulsive forces leading to reductions in binding affinities). In such disease states, ECM degradation may be accomplished by another group of proteases, such as the cathepsins, as they often operate optimally in an acidic environment (Dejica et al. 2012b).

The family of cathepsins are typically around 320 amino acids in the pro-form (i.e. MW about 35 kDa) and 220 amino acids in the active form (i.e. MW about 25 kDa). The isoelectric points (pIs) of the cathepsins vary widely: cathepsins B, C, H, L, and K have pIs 4.3–5.4, 5.3 and 5.9, 5.2–5.7, 7.0–8.7, respectively (Lynn and Labow 1984; Lecaille et al. 2002). However, the cathepsins are generally active at an acidic pH (Bromme and Wilson 2011), e.g. the optimum activity of cathepsin B is 4.5–5.5 (Giusti et al. 2008). Some report that cathepsin K operates in the acidic range (indeed it is generally accepted that cathepsins K is primarily responsible for collagen I degradation during osteoclast resorption) (Bromme et al. 1996; Wilson et al. 2009). Cathepsin K clearly has a very high pI, which should see it bound to aggrecan. How then, do we explain its activity in cartilage?

The first thing to note is that cathepsins have multiple substrates, including proteins, aggrecans, and elastins. Cathepsin K is one of the few proteases that can cleave collagen in the triple helical domain (Bromme and Wilson 2011; Dejica et al. 2012a). Cathepsin K (and S and L) is also aggrecanases. For cathepsin K, the aggrecanase site is found on the 'back side' of the protein, away from the reactive site for the protease, close to the N- and C-termini (Bromme and Wilson 2011). This means that cathepsin K can bind to another molecule and still be an active collagenase. For example, cathepsin K forms collagenolytic complexes with glycosaminoglycan fragments (Hou et al. 2003; Panwar et al. 2013). Now it has been shown that cathepsin K can exist in two distinct functional conformations with chondroitin sulfate at pH 7 (see http://www.fkkt.uni-lj.si/attachments/3110/novinec_ysf09.pdf). In other words, cathepsin K has multiple substrates and behaves in rather subtle ways forming intermediate complexes that are not easily predicted on the basis of the simple considerations of pI and pH of enzyme and substrate that we have been using above.

We have learned a great deal about MMPs and their inhibitors, and our Principles for Self-Organization of the ECM have proven their worth by giving us some small but valuable insights in how MMP and TIMP operate to maintain ECM homeostasis. However, the final example serves to highlight limitations in predicting molecular behaviors using our Principles for Self-Organization of the ECM. We have come to appreciate that there seems to be no end to biological inventiveness and complexity in the ECM.

2.10.5 Aggregate

Let us now try to better understand the assembly of aggrecan and HA into aggregate, while trying to grow our qualitative understanding of the processes responsible for radial zonation around chondrocytes. Again we will try applying our Principles Self-Organization of the ECM to see how simple theoretical predictions line up with experimental findings, and see what insights into mechanisms might be uncovered using this approach.

In vitro hyaluronic acid and the G1 domain of aggrecan interact strongly (though non-covalently) with each other over the pH range 6–8, to form complexes. As mentioned in Sect. 2.10.2, these interactions are pH dependent, and the binding affinity is greatly reduced at pH 5.0 (becoming negligible at lower pH) (Parkar et al. 1998). Closer to in vivo conditions, experiments on cartilage explants have shown fifty percent conversion of HA and aggrecan to aggregate requires around 5.7 h at pH 7.45, but this increases to 11 h at pH 6.99 (Sah et al. 1990). This data suggests that pH 6 to 8 is probably the target in vivo extracellular pH range for the assembly of extracellular matrix aggregates. We note that the time for reaction products to form becomes longer the lower the pH—indeed the reaction rate for aggregate formation is very slow around pH 5.0, and negligible at pHs below 5.0.

Importantly, we note that the pH ranges for aggregate assembly, overlap with the indicative pH ranges discussed above for normal articular cartilage. Interpreting the physiological significance of this, we infer that in the deep zone, aggregate formation in the pericellular matrix may be suppressed (e.g. where a normal

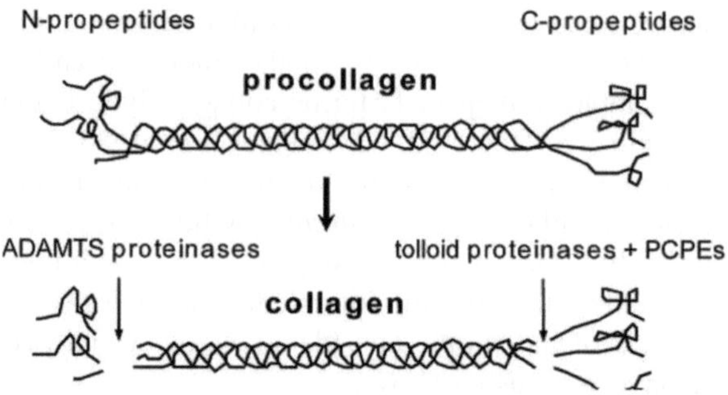

Fig. 2.65 Fibrillar procollagens and cleavage. N- and C-terminal processing leads to spontaneous assembly of collagen fibrils, which are subsequently stabilized by the formation of covalent cross-links (via the lysyl oxidase family of enzymes) (Hulmes 2008). See also Fig. 1 in Bekhouche and Colige (2015). It is noted that meprins cleave both the C- and N-terminal telopeptides (Broder et al. 2013)

pericellular matrix around a chondrocyte may be as low as pH 5.0). In this case, newly synthesized aggrecan and hyaluronan will not have time to assemble into aggregate in the pericellular matrix before they are transported away to the preferred aggregate assembly regions in the territorial (e.g. pH 5.5–6.0) and (particularly) in the interterritorial zones (e.g. pH 6.5). It is possible the comparatively fast reaction kinetics of aggregate formation at pH 6.5 in the interterritorial zone may contribute to a denser interterritorial ECM relative to the territorial zone.

Further, the relatively slow reaction kinetics of the aggregate in the pericellular and territorial zones may also facilitate the transport of recently synthesized aggrecan from the deep zone into the transitional (intermediate) and even superficial zones, along 'transport corridors' (i.e. along confluent territorial zones that form around adjacent chondrocytes—see discussion Sect. 3.4). We can appreciate that the pH-dependent reaction kinetics between aggrecan and HA provides a mechanism enabling chondrocytes in the deep zone to be an important source of aggrecan, HA and link protein that forms aggregate in the transitional and superficial zones. The line of reasoning is indeed consistent with the experimental data in Fig. 2.9, which suggests that most aggrecan is normally produced in the deep zone.

Taken together, these important experimental findings on the pH dependence of reaction rates for aggregate formation help us understand why the territorial and interterritorial zones develop in the deep zone, and how aggrecan can be transported from the deep zone to the transitional zone and superficial zone before it polymerizes. Clearly, this data strongly suggests that the locations of aggregate assembly in relation to chondrocytes may vary both radially and with depth in the cartilage (see discussion Sect. 3.4).

2.10.6 Collagens

Let us now try to better understand collagens in articular cartilage, while growing our qualitative understanding of the processes responsible for radial zonation around chondrocytes. Let us again do this by applying our physicochemical Principles for Self-Organization of the ECM to further our understanding the transport of procollagens through articular cartilage ECM.

Collagens are trimeric molecules in which each of the three chains comprises a repeating Gly-X-Y triplet, where X and Y can be any amino acid residue but are usually proline and hydroxyproline (Canty and Kadler 2005). Each procollagen consists of a triple helical domain about 300+ nm in length, but they also have N- and C-terminal domains (telopeptides) at each end that do not conform to the Gly-X-Y triplet (Fig. 2.65). We note that the lysine in the telopeptide region may be hydroxylated, and that these hydroxylysine amino acid resides may be glycosylated (to form galactosyl-hydroxylysine and glucosyl-galactosyl-hydroxylysine). Other post-translational modifications of the amino acids may occur (e.g. sulfation of tyrosine and 'glycation reactions,' i.e. meaning some amino acids along the triple helix domain are glycated). These alterations can change the pI of the procollagen.

There are five structural collagens (I, II, III, V, and XI), and many more 'regulatory collagens.' For collagen type II, each (left handed) chain (or alpha-1(II) helix) in the (right handed) triple helix, is the same. For collagen type III, each chain (or alpha-1(III) helix) in the triple helix is also the same (though different from collagen type II helix). However, the collagen type I triple helix is different to types II and III, consisting of two alpha-1(I) chains that are the same and another different alpha-2(I) chain.

Procollagen is 'activated'—here meaning it becomes tropocollagen upon removal to telopeptides and so can assemble into a collagen fibril. We note that removal of just one telopeptide promotes collagen assembly, while removal of both telopeptides promotes much stronger collagen assembly. ADAMTS2 typically removes 'N-terminal telopeptides' from collagens type II (and collagen type III, while ADAMTS2, 3, and 14 typically removal the telopeptides from collagen Type I), while 'C terminal telopeptides' are removed by BMP1, tolloid-like proteases and furins (Hulmes 2008; Bekhouche and Colige 2015). The absence of ADAMTS2 can lead to weak collagen assemblies that are misshapen (e.g. see Fig. 2 in Bekhouche and Colige 2015). It has been demonstrated that the meprins (α and β) can also remove both C and N-terminal telopeptides from procollagen (Broder et al. 2013).

Interestingly, procollagen type II comes in two isoforms: A and B (Ryan and Sandell 1990). The B isoform is expressed in adult human tissue while the A isoform is expressed in maturing articular cartilage, and is re-expressed in osteoarthritis. The A isoform has an additional 69 amino acids (which are 'cysteine rich') expressed in the N-terminal propeptide region, but these 69 amino acids have been deleted in the B isoform. We will return to this later when considering transport of procollagen through cartilage ECM.

Procollagen is very soluble, but upon removal of the telopeptides the remaining triple helix (i.e. the tropocollagen mer) becomes comparatively insoluble. For example, intact procollagen type I will assemble into sheets at concentrations greater than 1 mg/ml. However, fully processed procollagen will assemble into collagen fibrils at the critical concentration of 1 μg/ml (i.e. polymerization occurs at a critical concentration 1000 times lower than that for procollagen) (Hulmes 2008). When the C-terminal telopeptide is removed, but the N-terminal telopeptide remains intact, sheet-like collagen type I assembly begins at 100 μg/ml (i.e. ten times lower critical concentration than procollagen). It is likely that the N-propeptide polymer forms 8 nm sheets of collagen because the N-terminal telopeptides folds back again onto the main triple helical region, preventing aggregation of mers in one direction (Hulmes 2008). See also Bekhouche and Colige (2015).

Following fibril rupture, collagen repair assembly proceeds from the rupture surfaces. But fibril assembly has to compete with alternate potential reactions at the fibril surfaces, including the binding of molecules other than tropocollagen to the fibril surfaces (e.g. the binding of COMP, decorin, biglycan, collagen type IX and matrilins, to name a few contenders). The geometry of the surface is of principal importance in this competition. For example, it has been shown that the spatial distribution of decorin along a tapering collagen fibril depends on the fibril diameter, i.e. the frequency of binding is less toward the thin 'pointy ends' of a growing collagen fibril (Hulmes 2008). This observation points to the stereochemistry of the collagen surface changing so that the binding constant for decorin increases with increasing fibril diameter. This allows a new collagen fibril to grow lengthwise unhindered by decorin, however as the fibril grows in diameter, lateral growth of the fibril becomes more and more difficult and eventually is arrested by decorin (providing decorin concentrations are sufficiently high). Biglycan and other SLRPs probably plays similar roles, so deletion of any one SLRP is compensated by the other, while an insufficient total SLRP concentration tends to lead to similar fibril abnormalities. We note that COMP may form a 'protective shield' around the growing fibril tip, preventing immediate removal of newly formed collagen by any local proteases. It becomes apparent that through the coordinated effects of several regulatory macromolecules, long collagen fibrils can develop in vivo, rather than ribbons or sheets of collagen.

When collagen is forming in articular cartilage, initially two groups of fibrils of uniform diameter develop—a small diameter group of fibrils made of collagen type XI, and then a group of 30 nm diameter type II fibrils appear—perhaps regulated by decorin and other SLRPs). Over time fibrils grow in length, while fibril diameter may increase tenfold or more (Canty and Kadler 2005). This increase in diameter may be partly due to lateral fusion of small 30 nm fibrils (Chen and Birk 2013)—but presumably fusion of fibrils would first require the removal of any molecules attached to the fibril surface, e.g. SLRPs, matrilins, COMP, collagen IX. It is possible that fibril fusion may be partly mechanically driven? In this case, direct contact pressure between fibrils may be responsible for forcing (or squeezing) molecules off the collagen surface.

Exactly how type II collagen fibril diameter is regulated as it matures is not known. Typically collagen type IX is attached to the surface of mature type II fibrils. Eyre has suggested that cleavage by MMP3 of the NC4 domain at the ends of collagen type IX may permit further fibril growth, obviating any need for removal collagen type IX before a fibril can grow (Eyre et al. 2006). Excessive tensile, squeezing, or bending strain may change the stereochemistry of the fibril surface sufficiently that some of the attached molecules spontaneously detached from the fibril surface, which may help permit further growth in fibril diameter. Ongoing cross-linking between any new mers and the existing fibril (by the lysyl oxidase family of enzymes) increases fibril strength and stiffness many times and stabilizes the mers in a growing collagen fibril.

Interestingly, it has been recognized for some time that 'basement membrane defining proteins,' required for attaching epithelial cells to their basement membrane, are actually found in the pericellular matrix around chondrocytes. For example, collagen type IV, laminin, nidogen, matrilin, and perlecan are found in the pericellular matrix. This suggests that the pericellular matrix is the functional equivalent of a basement membrane for chondrocytes (Kvist et al. 2008). Collagen type VI appears to act like a basement membrane collagen next to chondrocytes in articular cartilage (though collagen IV is primarily in basement membrane secreted epithelial cells, there is a small amount of collagen VI adjoining the epithelial cells).

The collagen VI mer is a molecule with globular ends (which have homology with von Willebrand factor A-domains). These mers first polymerize intracellularly, associating side to side to form dimers [the dimers are 'antiparallel' with N-terminal globular domains at each end of the dimer (Knupp and Squire 2001)] and tetramers (Ball et al. 2003). When secreted to the extracellular environment, these dimers and tetramers associate end to end (Lamande et al. 2006), to finally form a collagen VI meshwork around the chondrocyte. Fibronectin forms a cell anchoring meshwork that co-localizes with the collagen type VI (Poole 1997; Horikawa et al. 2004; Poole et al. 1992). This meshwork is found between the territorial matrix and the chondrocyte, i.e. throughout the pericellular matrix (Groulx et al. 2011; Poole 1997). We mention that keratinocytes in corneal stroma also secrete collagen VI, which forms a meshwork next to keratinocytes at interfaces between stromal collagen lamella.

It has been shown than von Willebrand factor A-domain-related protein (WARP) also co-localizes with the collagen type VI meshwork (Hansen et al. 2012). It appears quite likely that with an adequate supply of collagens IX and XI, the fibronectin meshwork in place, with co-localized WARP, collagen type VI, decorin, biglycan, matrilin, and perlecan (Melrose et al. 2008a, b), a chondrocyte can presumably initiate new fibril formation. Once initiated, the network continues to organize and structure the collagen type II fibril meshwork assembly (Kadler et al.

2008). For example, it is known that decorin, biglycan, and matrilin attach collagen VI fibrils to collagen II fibrils (Wiberg et al. 2003). Presumably this happens in the pericellular matrix, which with maturation, eventually becomes the territorial matrix around a stationary chondrocyte. Similarly, the 'pericellular matrix' left behind a slowly migrating chondrocyte (with an adequate supply of collagens IX and XI, the fibronectin meshwork in place, with co-localized WARP, collagen type VI, decorin, biglycan, matrilin, and perlecan), presumably forms a collagen network that matures to become territorial matrix.

Details of this collagen network initiation and maturation process are sketchy. However, the importance of the fibronectin meshwork is illustrated by the experimental finding that for fibronectin null-fibroblasts collagen fibril formation will not occur until the addition of exogenous fibronectin to the cell culture medium (Canty and Kadler 2005). It is not clear if chondrocytes are essential too, but presumably they are essential for initiation of fibril formation in vivo (conditions in vivo are quite different to a simplified experimental media typically used in vitro).

(Kadler et al. 2008) offers a model of how collagen types I and II fibril assembly may occur in vivo—first fibronectin binds to the cell membrane via $\alpha5\beta1$ integrins causing fibronectin to 'open out' and to bind additional receptors on the cell membrane. Fibronectin complexes then engage with collagen mers (types I and V in the case of collagen type I, and types II and XI in case of collagen type II) and collagen fibril polymerization is nucleated and the fibril grows—see Fig. 2 (Kadler et al. 2008). Kadler et al. (2008) point out that while collagen fibrils can assemble in vitro, mystifyingly it requires the presence of cells to do this in vivo. Kadler et al. (2008) explains:

> The answer to this riddle most probably lies in the fact that fibrillar collagens have $\sim$50 known binding partners in vivo. This number of binding partners is presumably required to generate the diversity of fibril patterns, which range from parallel bundles in tendon and ligament, to orthogonal lattices in cornea, and interlocking weaves in blood vessels, skin, and bone. On the contrary, having a large number of binding partners can spell disaster for a protein self-assembly/polymerization process. Purified collagen spontaneously assembles into fibrils in vitro because collagen molecules are free to bind to other collagen molecules, and only collagen molecules. In vivo, however, the situation is very different. Faced with so many potential-binding partners, collagen molecules might easily be sequestered into dead-end molecular interactions, which would lower the effective concentration of collagen monomers available to form fibrils. In the remainder of the review, we explore the possibility that cells use collagen V and XI to nucleate collagen fibrils, and fibronectin (FN) and integrins to specify their site of assembly.

This suggests that small amounts of collagen network damage (e.g. breakages of isolated fibrils) may be repaired by tropocollagen polymerization distant from cells, while more extensive collagen network damage (e.g. see Fig. 3.5) may require the involvement of chondrocytes for collagen network restitution. The involvement of chondrocytes to form new collagen network may be required due to the level of regulation over potential unwanted chemical reactions, thereby ensuring the desired end product (a new collagen network) is actually formed from the chemical milieu in the pericellular matrix. This observation lends additional support to the idea that slowly migrating chondrocytes may play a crucially important role in maintaining collagen network homeostasis over the life span of long-lived animals, nucleating and creating new collagen network as they migrate through extensively damaged ECM (see Sect. 2.9.2). On the other hand, fast migrating chondrocytes play a key role in creating new collagen network (albeit usually fibrocartilage) after their arrival at the site of damaged ECM following traumatic injury (see Sect. 2.9.1).

Rather surprisingly, it has been observed both in vitro and in vivo that chondrocytes are found encapsulated in physical structures dubbed 'chondrons,' and their function has been the source of speculation (Poole et al. 1988b; Poole 1997). The structure of chondrons formed in vitro usually comprises a fine collagen VI network immediately surrounding the pericellular matrix (Nagai et al. 2008). But in adult cartilage, in vivo chondrons may become much more substantial structures—the so-named tapering chondrons [see Fig. 3 (Poole 1997)]. We propose that tapering chondrons arise due to slow mode of chondrocyte migration (see discussion Sect. 2.9.4). In our view, the tapering chondrons represent compressed collagen network ahead of the leading edge of a slowly migrating chondrocyte, while partial digested and newly forming collagen network forms at the chondrons trailing edge. In other words, slow chondrocyte migration leads to the reported 'tapered structured' of adult cartilage chondrons.

While collagen network assembly in vivo may have to be initiated at a cell surface (see Kadler et al. 2008 discussion above), fibril growth and repair may nevertheless occur at any distance from the chondrocyte given suitable conditions. Once secreted by the chondrocyte, procollagen needs to be transported through the ECM to the site of fibril growth or repair. It seems likely that if both structural and regulatory collagens are to be transported through the cartilage extracellular matrix to sites of fibril growth and repair, then based on the pI–pH analysis described in Sects. 2.10.1 and 2.10.4, these molecules are expected to have an pI < tissue pH. In this way, they will carry a net negative charge (as does aggrecan) and so rather than binding to the ECM, the procollagen could be transported through it. Using our Principles of Self-Organization of the ECM as this conceptual starting point, how might procollagen achieve a negative charge suitable for transport?

Telopeptides may confer a net negative charge on the procollagen, facilitating its transport though the negatively charged aggrecan. We note in passing that even if the hydrophilic telopeptides are not themselves negatively charged, post-translation modifications of the amino acids (e.g. glycosylation, sulfation) may result in a net negative charge. For example, *O*-glycosylations, which include galactosyl-hydroxylysine and glucosyl-galactosyl-hydroxylysine, are common in telopeptide regions of procollagens (Taga et al. 2013).

When the procollagens have reached their location for incorporation into a growing collagen fibril, the telopeptides are removed, and the pI of the tropocollagen increases close to that of the collagen fibril. With the change in pI, and possibly with the assistance of regulatory molecules (e.g. regulatory collagens and SLRPs such as fibromodulin, lumican, biglycan, decorin (Martinek et al. 2007)), the tropocollagen can then be maneuvered into the correct position and correct alignment relative to the collagen fibril. The tropocollagen mer is actually experiencing continuous collisions in solution with solute and solvent molecules, and so it appears to rapidly and randomly vibrating in the pore fluid. At some point, when the tropocollagen molecule becomes close enough to the fibril, hydrophobic and other non-specific entropic interactions ensure that the collagen mer 'snaps' into position at the surface of the developing collagen fibril, where it is held fast until cross-linking secures the mer more permanently in its final position.

Collagen VI also binds (near its N-terminal end) to molecules decorin and biglycan, and may form dimers and tetramers of collagen VI (Wiberg et al. 2001). Given that biglycan is mainly pericellular, one may propose that biglycan is the usually binding partner of collagen VI (Wiberg et al. 2001). So assuming that decorin and biglycan facilitate transport of collagen VI, one might then suppose that collagen type VI needs no further assistance to be transported through the acidic pH environment within the pericellular matrix, and consequently collagen VI has no need for an acidic pI. This proves not to be the case, as collagen VI has a pI between 5.0 and 6.0 (Wegrowski et al. 1990). In this case, collagen VI transport appears to be adequately facilitated by its own pI in the acidic range. Collagen VI is in any case about 5 times larger than biglycan, so it is likely biglycan and decorin may perform other roles when they bind to collagen VI (e.g. perhaps shepherding and facilitating collagen VI in its interactions with the other pericellular components such as fibronectin, and together with matrilin, helping collagen VI bind to collagen II).

Some of these interactions are summarized in the schematic depiction of molecules associating in the pericellular, territorial, and interterritorial matrix (Heinegard 2009).

See Fig. 1 in Heinegard (2009): *Schematic illustration of molecular constituents in cartilage and their arrangement into large multimolecular assemblies. The different compositions and organizations at the cell surface with a number of receptors interacting with specific matrix molecules, at the interterritorial matrix closer to the cells and the interterritorial matrix at a distance are indicated.* Or see similar illustration in Fig. 3C of Han et al. (2011c).

Another useful example to consider is the transport of procollagen type III through articular cartilage, as collagen type III is an important structural collagen in cartilage, comprising up to 10% of the total collagen in articular cartilage, being particularly prominent in the transitional and superficial zones. Collagen III attaches to and cross-links collagen II, forming a network of collagen III fibrils (Wu et al. 2010). Collagen III is also reported to bind to molecules like decorin (Thieszen and Rosenquist 1995). However, if procollagen type III is to migrate to distant locations in the ECM, it appears necessary for procollagen type III to have an acidic pI too (6.2, but can be as low as 5.6), http://www.signaling-gateway.org/molecule/query;jsessionid=f4099259c475ef82f46c4aa44642cad379c0668817cb4ed7557eda1623bfd75c?afcsid=A005368&type=sequence&adv=latest. Again transport appears to be adequately facilitated by the acidic pI, at least toward the surface of articular

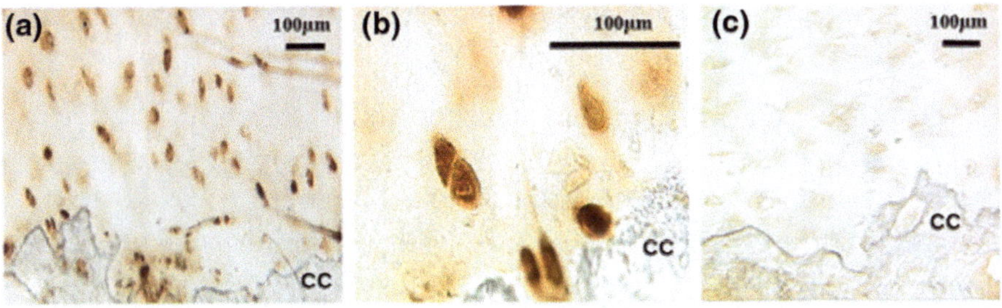

Fig. 2.66 Immunostaining for type IIA procollagen was frequently strong in the cells closest to the calcified cartilage (CC) and bone. **a**, **b** type IIA procollagen and **c** normal rabbit control; 46-year-old female 14 months post-ACI (autologous chondrocyte implantation) (Roberts et al. 2009). It was noted by the authors that there is no collagen type IIA staining reported in normal cartilage from individuals age over 40. When present, immunostaining for type IIA procollagen is strongest closest to the bone (i.e. where the pH is lowest), suggesting that type IIA expression is related to reductions in cartilage pH

cartilage where the pH is normally close to 7.0 and collagen type III formation most prominent. From these two examples, an impression is created that collagen transport is more dependent on an acidic pI than binding to a 'transport molecule' (e.g. potentially like decorin or biglycan).

Indeed, it turns out that the N-terminal region of the collagen III has a very low pI (3.1) (Niemela et al. 1985). It is noted that this N-terminal region has a pI somewhat lower than its theoretical pI (3.8), probably due to amino acid modifications (the amino acid sequence of collagen III and post-translation modifications can be found at http://www.uniprot.org/uniprot/P02461). It is appears likely this particularly low pI N telopeptide region (which incidentally, is rich in cysteine amino acids) allows collagen type III to move easily through normal and damaged ECM, which may be otherwise impenetrable to most other procollagens. This may help explain the importance of collagen type III in early stages of tissue repair.

We now consider the transport of procollagen type II. It is likely that the pI of individual collagen type II alpha chains is in the range pI = 8.0 to 8.5, (see autoradiograph Figs. 2 and 3 (Hermansson et al. 2004)). The precise pI of 'collagen type II' shown in the autoradiograph probably depends on its form, i.e. it is type II procollagen or 'partially processed' procollagen or tropocollagen. The theoretical pI of collagen type II isoform B (http://isoelectric.ovh.org/) containing the amino acids 26 to 1487 (i.e. the whole propeptide) is 6.34 (http://www.uniprot.org/uniprot/P02458), while the pI of amino acids 32–90, 201–1241, and 1253–1487 is 7.2, and for amino acids 201 to 1241 the pI = 9.75. From this, we conclude that the pI increases as various domains of amino acids are progressively removed from an intact procollagen type II molecule.

Type II alpha chain collagen probably has a pI around 8.2 (Hermansson et al. 2004), which is lower than the theoretically predicted pI for amino acids 201 to

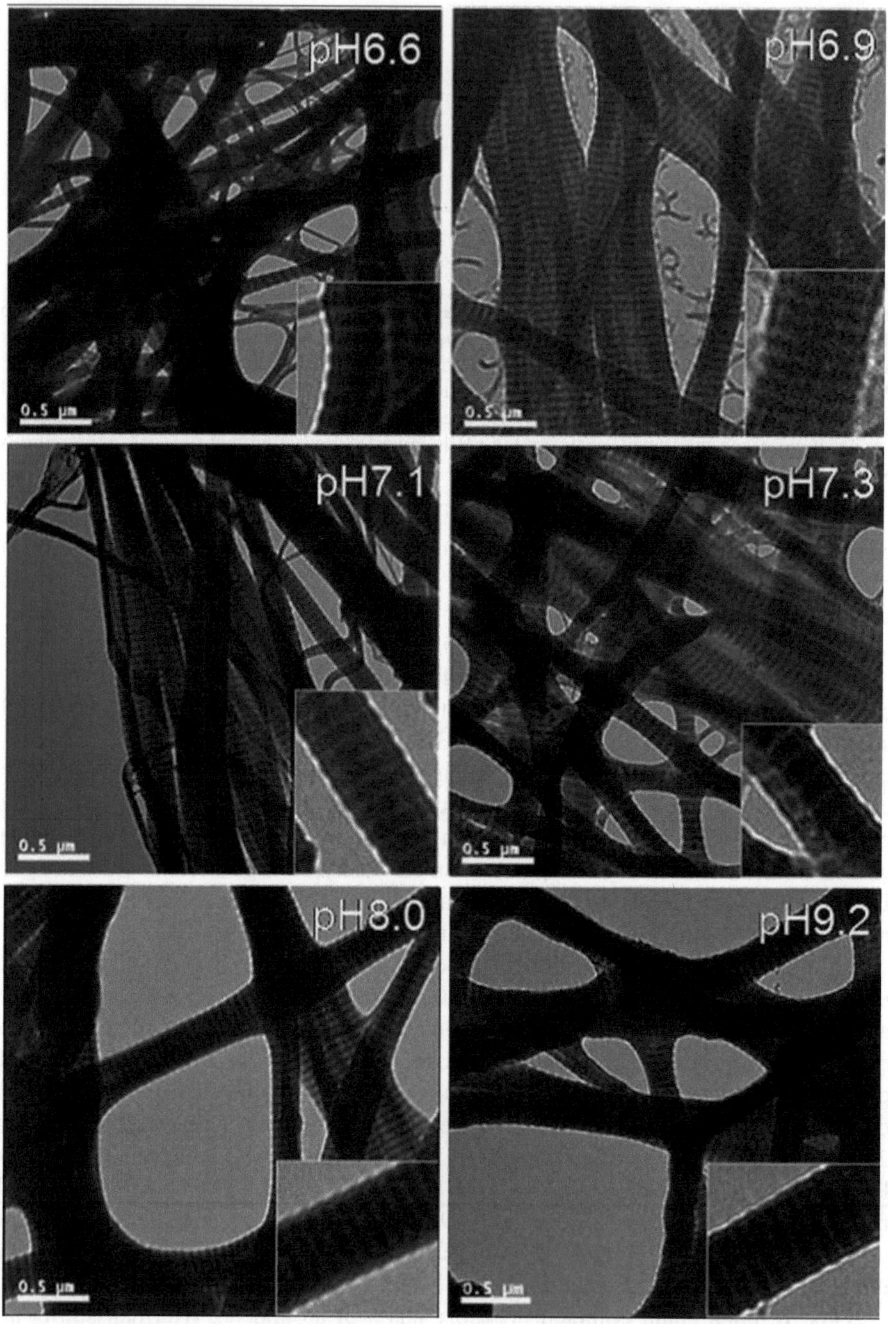
pH6.6
0.5 μm
pH6.9
0.5 μm
pH7.1
0.5 μm
pH7.3
0.5 μm
pH8.0
0.5 μm
pH9.2
0.5 μm

◂**Fig. 2.67** SEM images of in vitro collagen type I fibrils obtained in different pH conditions. The fibrils form branched networks with non-fibrillar collagen. Scale bar is 10 μm for all the images (Li 2009). Note pH influences density of the collagen network, and degree of collagen fiber entanglement. It is suggested that collagen microfibrils begin to repel each at a pH close to or just above the pI of the collagen, leading to separation of the fibers (ref Assembly of collagen into microribbons: effects of pH and electrolytes). This leads to a change from sheet-like collagen structures at low pH to the appearance of individual collagen fibrils that have a 'woven' appearance at higher pH. This observation is consistent with the change in collagen structure observed between the radial zone and the transitional zone

1241 pI 9.75. There is extensive glycosylation along amino acids 201 to 1241 (http://www.uniprot.org/uniprot/P02458), which probably responsible for reducing the pI of the tropocollagen to around 8.2. It appears that the final pI of collagen type II depends on amino acid post-translational modifications, e.g. hydroxylation, glycation, and sulfation (and possible attachment of collagen type IX—collagen type IX has a very high isoelectric point globular domain (i.e. pI = 9.8)).

See autoradiograph Figs. 2 and 3 (Hermansson et al. 2004): *Autoradiographs of 2D-gels (MW versus pH) of radiolabeled proteins secreted by osteoarthritic cartilage explants [The 2D-gels are created by isoelectric focusing (IEF) followed by and SDS-PAGE electrophoresis]. Approximately 3 g of cartilage was used for each 2D gel. Proteins indicated are (a) TIMP-1, (b) YKL-39, (c) collagen type II C-propeptide, (d) inhibin βA, (e) procollagen C-proteinase enhancer, and (f) collagen type VI* (Hermansson et al. 2004). Note that pH location corresponds to the pI for each molecule identified.

Interestingly, there are two splice variants of COL2A1 gene (Ryan and Sandell 1990). Type II A, which has an additional 69 aa cysteine-rich region in the N-terminal propeptide compared to the type II B isoform, is expressed in young animals with developing cartilage. Collagen type II B (which has no cysteine-rich N-terminal propeptide) is expressed in adult tissues. Presumably importantly, it is noted that the type II A isoform is re-expressed in adults with osteoarthritis (Aigner et al. 1999).

Theoretical pIs for the two isoforms of collagen type II (http://www.uniprot.org/uniprot/P02458) can be calculated (http://isoelectric.ovh.org/). This reveals that the procollagen type IIB isoform has a pI of 6.34, while the type IIA isoform (i.e. with the additional 69 aa) has a pI of 5.5 (note that post-translations modifications of amino acids may increase this difference further, but there is probably at least 0.84 pH units difference). In other words, it seems likely that the presence of the cysteine-rich domain, present in the A isoform, lowers the pI of the procollagen type II by around 1 pH unit.

From the perspective of transport through aggrecan and our Principles for Self-Organization of the ECM, the type IIA isoform is likely to be more easily transported through the negatively charged aggrecan, particularly at lower tissue

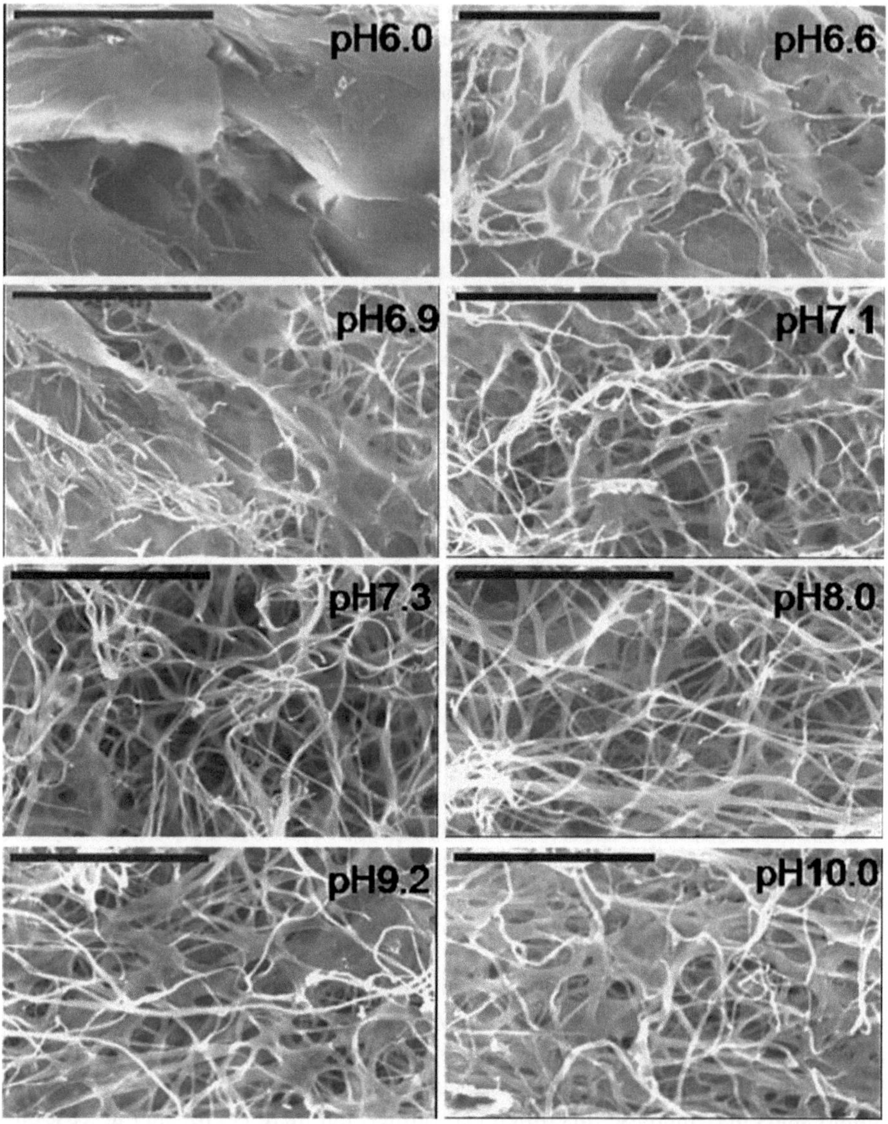

Fig. 2.68 TEM images of in vitro self-assembled collagen type I fibrils after three days of fibrillogenesis. Scale bar 500 nm (Li 2009). Note that the pH influences the thickness of the collagen fibrils. Further the ionic strength may lead to a general alignment of collagen fibers that are somehow woven and twisted around one another (Jiang et al. 2004)

pHs, than the type IIB isoform. Lower tissue pHs are more likely during active remodeling of cartilage during development [and again in old age with the onset of early OA (Konttinen et al. 2012)]. Indeed in normal cartilage, the type IIA isoform shows an age-dependent distribution—normal cartilage from individuals less than

13 years of age showed weak positive immunohistochemical distribution of type II A procollagen throughout the ECM, with more concentrated staining in the pericellular region. Weaker staining again was see in a 19 year old, but there was little type II A procollagen staining in any samples from older individuals, and no type IIA procollagen staining from individuals greater than 40 years of age (see Fig. 2.66 from Roberts et al. 2009).

Further, we observe that when procollagen type II A is expressed, it is more frequently strong expression closer to the osteochondral junction, where pH is known to be lower (Fig. 2.66). In addition, the re-expression of the collagen type IIA isoform in osteoarthritis [and following osteoarthritic repair procedures, such as autologous chondrocyte implantation (Roberts et al. 2009)], which generally has a pH one to two pH units lower than normal cartilage, is consistent with the idea that the additional amino acids present in the type IIA isoform are an adaption to low pH environments. The procollagen type II A isoform may well be related to its transport through cartilage ECM at lower pHs, which would be consistent with the discussion above.

Interestingly but perhaps unsurprisingly, it has been shown that collagen type II isoforms are under the control of transcription factors that control the RNA splicing machinery. In the case of collagen type II, it has been shown that the A and B isoform expression is under the control of splicing factor SRSF10 (or TASR-1 or SRp38) (Matsushita et al. 2007). When chondrocytes are metabolically 'stressed' (e.g. in rapidly maturing cartilage or osteoarthritic cartilage) SRSF10 is downregulated and the collagen type IIA isoform is expressed (along with the many other cell pathways linked to 'stress' and cell survival, such as 'endoplasmic reticulum linked apoptosis'). When cartilage is under less stressed conditions as in normal adult cartilage tissue, SRSF10 is upregulated and the collagen type IIB isoform is then expressed (Zhou et al. 2014b).

At this point, we note that the local pH can also have a profound effect on the assembly of the collagen mers in vitro. Depending on pH, tropocollagen may assemble into sheets or fibril networks of various degrees of connectedness and fibril diameter. Figures 2.67 and 2.68 illustrate the variety of fibril assemblies of collagen type I as the pH and ionic strength are varied. While it is to be expected that collagen type II and III will create sheets and fibril networks somewhat differently to collagen type I, the responsiveness of fibril assembly to changing pH and ionic strength is likely to be a shared characteristic. Again we note that collagen network formation in the much more complex solution matrix present in vivo may require the presence of chondrocytes to initiate collagen network formation as do collagen networks formed by fibroblasts (Kadler et al. 2008).

In the extracellular matrix of articular cartilage, collagen is the most important substrate for lysyl oxidase, as it promotes cross-linking of collagen mers, stiffening, and strengthening collagen fibrils. We also mention that the isoelectric point (pI) of lysyl oxidase is approximately 6.0 (Li et al. 2011). It is also reported that lysyl oxidase can oxidize lysine residues in various globular proteins with basic isoelectric points (pI > 8). Taken together, this data suggests that electrostatic interactions between the (anionic) lysyl oxidase and (cationic) protein substrates are important for enzyme–substrate interaction. Again this provides further general evidence supporting a working range of pH in the territorial and interterritorial zones of about 6.0–8.0.

2.11 What Are the Implications for Osteoarthritis?

Having identified key processes leading to ECM formation and tissue homeostasis, what insights does this give us into osteoarthritis?

The first impression created by the information covered in this chapter is the incredible molecular complexity behind the apparent simplicity of the glistening white cartilage tissue covering the ends of long bones. The myriad of molecular reactions and molecular components means that a defect in any one tissue process may lead to a failure of ECM component homeostasis, which then leads to a cascading failure, ending in articular cartilage disease states (Krishnan and Grodzinsky 2018). In one conception, osteoarthritis may be broadly characterized as a 'slowly evolving cascading failure of cartilage' within a 'slowly evolving cascading failure of the joint,' arising from the failure of homeostatic mechanisms in the joint to cope with the local environmental conditions that inflict damage on joint tissues. The normal operation of homeostatic processes within cartilage is extremely complex, with so many component parts it is difficult to comprehend it all.

Indeed like many diseases of complex etiology—twin studies demonstrate that somewhere between 40 and 60% of osteoarthritis is genetic in origin (Spector and MacGregor 2004; Valdes and Spector 2008), rather than environmental in origin. The uncertainty in the estimate is partly related to the fact that there is interplay between an individual's genome and their environment, both in real time and historically. Epigenetics provides a mechanism for the environment one's recent ancestors experienced to become part of your genetic inheritance (Heijmans et al. 2008) (see discussion Appendix A). The circular interplay between multiple genes and the environment makes it often difficult to ascribe causation to one or the other, but at least theoretically (if not quantitatively) we may define a genetic susceptibility to OA based on one's genetic inheritance. In another words, 'family history' is known to be a significant risk factor for OA (though clearly this includes confounding environmental factors, e.g. similar occupations among family members).

Given the complexity of normal homeostatic mechanisms in cartilage, it is unlikely any two cartilages from different people would be the same. There are hundreds of well-defined 'point abnormalities' and probably many tens of billions of 'cell system abnormalities' possible. Multiple (possibly hundreds or thousands of), subtle genetic abnormalities can interact cumulatively to make one individual more susceptible to OA than another. New gene suspects are being turned up regularly using GWAS (gene-wide association studies). For example, there is

GLIS3, another nuclear gene activator and repressor (Casalone et al. 2018). Unfortunately, there is almost no end to what may theoretically be defective, due to 'gene mutations' (i.e. SNPs or otherwise), abnormal gene interactions 'epigenetic silencing' and then the operation of multiple homeostatic mechanisms in a challenging and changing environment.

For example, consider aggrecan, a key structural component of cartilage. Table 1 in Aspberg (2012) details 16 known genetic defects pertaining to the aggrecan molecule alone. We see from Table 1 in Aspberg (2012) that an absence of aggrecan is lethal, but there are many non-lethal defects of the aggrecan molecule itself. Perhaps unsurprisingly, defects in enzymes leading to glycosylation of sulfation of aggrecan lead to dwarfism and chondrodysplasias. New genetic defects or mechanisms are being added to this list all the time. For example, it has been shown that osteochondritis dissecans arises when a genetic defect in the C-type lectin domain of collagen IX no longer suppresses complement activation within the cartilage (Furst et al. 2013). And for every frank clinical disease entity, there are multiple sub-clinical disease entities that may potentially be uncovered by genetic and/or cell testing.

There is another very complex story around the genetic defects associated with the 28 different collagen molecules (Tsipouras and Ramirez 1987; Markova 2007; Murphy-Ryan et al. 2010; Nose 2011; Sahlman et al. 2004). Reginato and Olsen (2002) list some 30+ collagen type II diseases that lead to osteochondrodysplasias (see Table 2). Undoubtedly, there are other genetic defects that plague the hundreds of different molecules making up cartilage ECM. Indeed, we expect that at least a similar number of defects (10–20) are likely for every molecule expressed in articular cartilage. The number of possible genetic defects of cartilage ECM is itself truly staggering. SNPs alone are numbered in the tens of millions, while network interactions multiply possible permutations to very, very large numbers of possibilities, explaining why no two individuals are likely to have exactly the same articular cartilage. Indeed, it this diversity that provides the 'raw material' on which selective pressures acts to drive evolution of the species.

Epigenetic changes that also influence chondrocyte behavior in vivo are only beginning to be uncovered. Iliopoulos et al. found that epigenetic mechanisms regulate leptin's expression in chondrocytes, influencing expression of its downstream target MMP-13 (Iliopoulos et al. 2007). Hashimoto et al. (2009a, b) report that inflammatory cytokines can change the DNA methylation status at key CpG sites, resulting in long-term induction of IL-1β in human articular chondrocytes (Hashimoto et al. 2009b). Imagawa et al. have shown that hyper-methylation of CpG sites occur on the collagen IX gene in osteoarthritis, which significantly downregulates the mRNA expression of collagen IX (Imagawa et al. 2014). Indeed, when six methylation CpG sites were significantly hyper-methylated in collagen type IX (COL9A1 gene), they reduced collagen type IX mRNA levels 6000-fold, effectively silencing this vitally important cartilage gene. Clearly, epigenetic change can significantly influence cartilage's capacity to maintain homeostatic control of the tissue. We mention that one's epigenetic profile changes throughout life (Raddatz et al. 2013).

The great challenge for making practical sense of human genome is to understand in any one individual the implications of many minor genetic changes acting in concert (i.e. minor alleles are defined to be those alleles that change the odds ratio for disease by less than 10%). Many genes with minor changes acting in concert lead to 'polygenic diseases' (Valdes and Spector 2008). Most of the major disease processes afflicting the developed world are in fact polygenic in origin (e.g. most cancers and most cardiovascular diseases are polygenic). Unfortunately, no one yet knows how to interpret SNPs in terms of functional outcomes in tissues, though efforts to predict the functional effects of SNPs are being made (see for example http://genetics.bwh.harvard.edu/pph2/).

Ideally, we would like to be able to integrate data from various sources into a physiological model of cartilage and use the model to then make predictions about possible future pathologies. In recent years, great excitement in the OA research community has centered on the predictions made possible by the very strong associations between genetic factors (i.e. combinations of gene alleles), environmental factors (e.g. body mass index, activity levels), and the risk of developing OA. This initial enthusiasm has been dampened somewhat by the lack of reproducibility of the genetic studies to date, but the cost of genome sequencing is falling rapidly, and high-quality whole genome data will soon be widely available. So people will keep searching for a solution using ever better databases. Clearly if effective use can be made of this information, this has the potential to transform patient specific disease risk prediction. But the fundamentally important question is how to effectively integrate genetic data with environmental data to extract useful biological information (and for example, predict a tissue response in an individual?). This is an open research question of the highest importance (Gardiner et al. 2016).

Some have sensibly proposed polygene models of, for example, collagen diseases [e.g. see Fig. 10 (Nose 2011)]. This is undoubtedly part of the answer. More generally, we believe the answer to this question is to put 'risk relevant' genetic information in the correct context with environment experienced by the cell in tissues. This means we need to interpret both environmental and genetic sources of information for their effect on cellular microenvironments within a tissue. Through computational modeling at various length and timescales, we can at least in theory, link data on genetic and environmental 'risk factors' to the changes in cellular microenvironments, to changes in regulation of cartilage matrix turnover, and finally to deterioration in cartilage ECM integrity (Gardiner et al. 2016). Such an integrative approach bridges the 'information gap' between population level genetic and environmental risk studies (which provide statistical associations between risk factors and OA), organ/tissue physiology, and detailed cell and molecular biology studies that take a 'reductionist approach' to understanding chondrocyte mediated cartilage matrix turnover.

While the number of models at different length scales required to capture changes across a broad range of polygenic diseases is enormous, in principle most of the models are quite straightforward to develop. How the models are connected together can be challenging. But progress is being made on models at all length scales. For example, at the molecular scale, consider two interacting proteins and one is mutated or contains a snSNPs. Very importantly, it is feasible to predict the likely change in binding interaction based on the known mutation/snSNP using several different programs. For example, the program PolyPhen (Polymorphism Phenotyping) estimates the likely effect of a change in the amino acid sequence on protein function http://genetics.bwh.harvard.edu/pph/pph_help_text.html. This web reference provides details of how this molecular-level prediction uses databases and models to obtain its prediction. Crucially, programs such as PolyPhen working together with mechanistic models of cell and tissue function (represented by systems of ODE and PDE equations) promise to link genome with tissue phenotype under different environmental stressors.

There can be little doubt that bridging the 'information gaps' between genomic, cell, tissue and environmental data, to predict their combined effect on cartilage tissue (and leading for example, to the onset of OA) needs to be (and will be) a key focus of future research. This work has begun, through consortia like the physiome project http://physiomeproject.org/, and at some private companies such as Entelos http://www.entelos.com/. Our own 'engineering approach' to this problem is to first develop a computational framework that incorporates mechanistic (causal) models of cartilage physiology, and then employ reliability theory to incorporate uncertainty and population variability to predict the probability of cartilage failure (e.g. for osteoarthritis) (Gardiner et al. 2016). Our multiscale model aims to integrate data from diverse sources (e.g. genetic, environmental, mechanical, and statistical) so that each can be put in their proper 'system level' context.

Once the source of a disease process has been accurately diagnosed, one looks to treatment. While the treatment of genetic disease has been a 'pipe dream' until quite recently, so-called gene therapy (often involving the delivery of defect free genes via viruses, though some are attempting other therapies including the development of artificial chromosomes) is now rapidly evolving and moving closer to reality (see http://ghr.nlm.nih.gov/handbook/therapy/genetherapy and http://www.nature.com/gt/index.html).

But genome engineering is evolving rapidly in new directions—one revolutionary example is so-called CRISPR-Cas9 technology (clustered regularly interspaced short palindromic repeats—CRISPR associated genes) (Baltimore et al. 2015) https://www.technologyreview.com/s/609722/crispr-in-2018-coming-to-a-human-near-you/. The gene editing system is a natural defense system in bacteria, which is part of the acquired eukaryotic immune system protecting against phages (viruses) that invade the hosts DNA. Operationally, this involves small fragments of viral DNA being incorporated in the CRISPR system of the bacteria, which then acts as a memory of a previous viral invasion (see further discussion Appendix Genes, SNPs, and reaction binding constants).

Amazingly, the CRISPR system can be introduced into a prokaryotic (e.g. animal) system using lentiviruses, where it works successfully. The technology is inexpensive, incredibly simple to use (allowing anyone with molecular biology training to modify genes) and incredibly powerful (allowing precise modification of specific genes, e.g. to correct genetic defects in any animal) (Ma et al. 2017). But the existence of this technology will only intensify the search for knowledge about how to use it effectively to cure diseases that have a genetic basis (Brunger et al. 2017). Like all treatments though, there are potentially 'off-target' gene modifications that may result in 'side effects' (Kosicki et al. 2018), so this technology needs to be carefully evaluated.

The second important issue arising from the almost overwhelming molecular complexity of the normal homeostatic processes described above is how to develop measurements that provide a practically useful window into this world. It is a clinical bugbear that more useful biomarkers of osteoarthritic disease onset and progression are not available, despite some determined efforts to change this (Gobezie et al. 2007; Larsson et al. 2009; Neuman et al. 2017). Indeed, the limitations of current biomarkers available are lamented as being an important barrier to more effective and rapid research and clinical progress in developing new therapies (Hunter et al. 2014). Given this, the identification of new biomarkers is clearly a clinical priority. Barring specific disease causing abnormalities, it is less likely that any one biomarker by itself will be clinically useful. Biomarkers are also more likely to be clinically useful if they can be focused on identifying normal and abnormal processes within tissues, rather than end states of tissues (which represent the integrated effects of tissue processes over time). It is also clear that sensitivity and specificity of biomarkers need to be greatly improved if biomarkers are capable of identifying tissue-specific processes (just as there is a search for tissue-specific therapeutic targets).

While the analysis of blood and urine samples are commonplace (and are in some cases valuable in relation to bone and cartilage), they suffer because often many tissues contribute to the composition of blood and urine, confounding its interpretation. Much more tissue-specific information can be obtain by sampling local transudates (e.g. as is the case of sweat, saliva, cerebrospinal fluid, and aqueous humor). In the case of the knee joint, the obvious fluid to sample is synovial fluid, as it is in direct contact with joint tissues. Though synovial fluid turnover is relatively short, clearly the composition of synovial fluid provides a 'window' on local processes occurring within the (knee) joint.

Larsson et al. (2009) advocate strongly for the analysis of synovial fluid for aggrecan epitopes to better understand disease progression. These authors point out that measuring total GAG concentration in synovial fluid is not a good predictor of disease or progression (as confirmed by Neuman et al. 2017). However, the concentration of the so-called ARGS epitope offers a much more sensitive assessment of disease processes, with ARGS fragments in synovial fluid increasing between 34 and 177 times following acute injury or following acute onset inflammatory arthritis. In contrast, GAG concentrations in synovial fluid only increase about twofold in these injury or disease states. We note in passing that the chondroitin sulfate epitope 846, found on newly synthesized GAG, has also been shown to be elevated two- to threefold in synovial fluid for up to twenty years following knee injury relative to uninjured knees (Lohmander et al. 1999). This suggests that a tissue repair response is ongoing over many years. Indeed, the decline in ARGS concentration generally occurs over many years too, which again suggests that repair processes within cartilage may take many years to run their course (Lohmander et al. 1999; Larsson et al. 2009).

Cartilage repair takes many years to resolve (Neuman et al. 2017; Struglics et al. 2015; Chu et al. 2014), which is consistent with healing via slow migration of chondrocytes through the cartilage matrix, renewing the collagen network. At this point, we note that both slow and fast migrations of chondrocytes are only likely to be effective when repairing partial thickness cartilage defects, though repair as a result of slow migration results in hyaline cartilage, while repair as a result of fast migration lead to fibrocartilage formation.

How well do ARGS fragments track radiological progression of knee OA? By themselves, not very well. Perhaps unexpectedly Larsson et al. (2009) report a weak negative association of ARGs fragments and OA progression. The 'unexpected' weak negative association may possibly be explained by ongoing repair within the cartilage, which may temporarily reduce the rate of osteoarthritis progression, even as OA worsens. The most promising single indicator of progression to OA appears to be TIMP1 concentrations in synovial fluid (aggrecan, COMP and MMP1 were ineffective biomarkers) (Neuman et al. 2017).

Despite the limited research at present investigating molecular profiling of synovial fluid (Struglics et al. 2009; Gobezie et al. 2007; Hermansson et al. 2004), and some setbacks that may demand a rethink of experimental designs (Larsson et al. 2009; Neuman et al. 2017) this is clearly a promising area for future investigations. Analysis of synovial fluid appears to be more promising than the analysis of (blood) serum (Mickiewicz et al. 2016; Honsawek et al. 2015) (Remarks Micjiewicz et al. (2016): '*Our study indicates that systemic pharmacological and genetic factors have a greater effect on serum metabolic profiles than site-specific factors such as surgery*'). Driven by new technologies in proteomics and new understanding of cartilage processes and responses, synovial fluid profiling will doubtless experience considerable growth in the future. Proteomics is currently employed for the diagnosis of renal diseases (Bramham et al. 2009). An analysis of a larger suite of molecules in the synovial fluid will almost certainly reveal molecular profiles that are predictive of knee joint healing, disease progression, and the likely effectiveness of therapeutics. An important step may be following patient-specific profiles (clearly averaging data across groups can hide important relationships occurring in individuals).

It has been shown that the proteome of synovial fluid has distinctly different profiles for different tissue states. In normal synovial fluid, high concentrations of cystatin A, serine protease inhibitors (e.g. ATIII, C1 inhibitor, PEDF, α_1-anti-trypsin, α_1-anti-chymotrypsin, kinninogen, and chondroitin sulfate) are found in contrast to osteoarthritic synovial fluid (Gobezie et al. 2007). Gobezie et al. even found that two (molecular) 'phenotypes' of osteoarthritis revealed by distinct clustering of proteins in the synovial fluid. Digestion of cartilage explants by MMPs reveals breakdown products of collagen types II and III, decorin, biglycan, prolargen, fibromodulin, clusterin, fibronectin, COMP, CILP, aggrecan, and lumican (Zhen et al. 2008).

It has also been shown that groups with rheumatoid arthritis can be distinguished from osteoarthritis and normal cartilage on the basis of interleukin and LIF concentrations in the synovial fluid (i.e. on the basis of IL-1, IL-6, IL-8, IL-11, and LIF concentrations) (Kokebie et al. 2011). Patel et al. (2011) found that tenascin C concentrations in synovial fluid correlate positively with GAG release in osteoarthritis (Patel et al. 2011). Tsuchida et al. found that IL-6, IL-13, oncostatin M, and interferon (IFNγ) were elevated in synovial fluid obtained from patients with cartilage pathology (Tsuchida et al. 2014). Lohmander et al. and Reijman et al. report that CTX-II is associated with OA onset and progression (Lohmander et al. 2003a; Reijman et al. 2004), while later studies have been largely encouraging (Sharif et al. 2007; Sowers et al. 2009; Kaneko et al. 2014). Various COMP fragments in synovial fluid have been associated with different joint diseases (Ahrman et al. 2014). MMP concentrations in synovial fluid become elevated upon exposure to bacterial LPS in the blood (Luo et al. 2017). Periostin concentrations in synovial fluid have been shown to be positively correlated with radiographic knee arthritis (Honsawek et al. 2015). Chinzei et al. have shown that there is cross-talk between remnants of damaged ligament tissue and articular cartilage via periostin, which suggests there should be a 'cross-talk signature' present in the synovial fluid following traumatic injury of the joint (Chinzei et al. 2018). Further discussion of proteomics and osteoarthritis can be found in the number of reviews (Hsueh et al. 2014; Lotz et al. 2014; Nguyen et al. 2017).

Clearly, differential diagnosis of individual patients via profiling of synovial fluid needs to be extended to a wider suite of macromolecules including growth factors, their inhibitors and binding proteins, hormones and their binding proteins, proteases and their inhibitors, inflammatory cytokines, chemokines, metabolites, as well as ECM molecules and their degradation products. It is the complexity, and the meaning we can attach to this huge molecular 'ecosystem' in cartilage ECM, that awaits discovery. Almost certainly, these profiles will contain molecular 'fingerprints' to help reconstruct events that would otherwise be unrecognized. Of course 'event reconstruction' will require a deep understanding of normal and pathogenic tissue processes taking place in articular cartilage (indeed, this chapter and the next chapter are aimed at promoting just such an understanding).

Based on this discussion in this chapter, one can easily imagine that finding telopeptides indicating re-expression of the collagen type II A isoform in the synovial fluid would not be a good prognostic sign (because it indicates the tissue is metabolically 'stressed,' and (optimistically?) suggesting a prolonged repair process is required, if indeed repair is possible). Similarly for increasing production of MMP13 over time, both indicate a tissue under significant damage–repair stress, and likely to be indicative of progression toward early osteoarthritis. Tracking trajectories for lubricin concentration, zinc and copper concentrations, inflammatory molecule concentrations, and procollagen concentrations are all promising targets for synovial fluid investigation, but doubtless there are many more.

There are many promising joint imaging modalities [including the various types of weighted MRI (Matzat et al. 2013) along with those using contrast-enhancing agents such a gadolinium (Wedig et al. 2005; Matzat et al. 2013)], sodium MRI

(Zbyn et al. 2014), X-ray CT (which is particularly suited to characterizing bone and 'bone marrow lesions'—see for example Fig. 3.17), ultrasound (which is particularly useful for characterizing soft tissues including cartilage, tendon, and ligament) and newer technologies such as optical coherence imaging (OCT) (Chu et al. 2010), Fourier Transform Infrared Imaging Spectroscopy FT-IRIS) (Bi et al. 2007), second harmonic generation (SHG) (Brown et al. 2014), and Raman spectroscopy (Parkes et al. 2017). It seems likely that combining imaging with molecular profiling of synovial fluid will offer the clearest diagnostic and prognostic window on a joint over time—excluding taking repeated cartilage biopsies, which is almost universally avoided in clinical practice [but very occasionally has been adopted for research purposes only, e.g. Roberts et al. (2009)].

Hormones levels play an important role in every tissue of the body throughout life. It is apparent that systemic hormones (e.g. the IGFs, adiponectin, leptins, osteocalcin, glucocorticoids, growth hormone, thyroxine, parathyroid hormone, insulin, glucagon, melatonin, and the sex hormones) all have significant effects on the response of chondrocytes to local signaling factors and mechanical load. Hormonal levels often decrease with age (e.g. median IGF-1 concentration decrease at about 1% per year, while growth hormone secretion decreases exponentially registering a sixfold reduction between puberty and age 55). Estrogen decreases abruptly at menopause in women, and median testosterone concentrations in males decrease at about 1% per year from age 40. Sometimes, hormonal levels may increase with age, parathyroid hormone being an example that may directly play a role in the signaling systems operating within cartilage. Clearly, all hormonal changes will influence the 'set points' within chondrocytes, affecting ECM production and chondrocyte proliferation and migration, and so the intensity of cartilage tissue maintenance (see Sect. 5.4).

We also observe that the functional integrity of the osteochondral junction is essential for normal cartilage and subchondral bone. It is apparent that water and then small signaling molecules such as PTHrP will be the first to cross this tissue junction when the integrity of the junction is compromised (e.g. due to microfractures of the subchondral bone (due to excessive joint loading or bone 'thinning') or when excessive bone remodeling occurs). Transport may lead to loss of fluid pressure in the deep cartilage, and gain or loss of PTHrP from articular cartilage, depending on the direction of transport. As the osteochondral junction becomes more compromised, so larger molecules such as TGFβ, sclerostin, RANKL, and OPG to name a few, may make their way into cartilage tissue from bone, disrupting local signaling systems in both cartilage and bone. Ways of suppressing inappropriate interactions between cartilage and bone are being explored (Cui et al. 2016).

Clearly, disease states in any of the major endocrine systems can have significant effects on articular cartilage and may lead to secondary osteoarthritis. For example, hyperthyroidism or hypothyroidism can have significant effects on most tissues in the body including cartilage (Lewinson et al. 1994; Rosenthal et al. 2003). There are numerous other examples of secondary osteoarthritis arising from endocrine disease states including hypoparathyroidism, diabetes, acromegaly, and Cushing's syndrome (Mahajan et al. 2005).

Even in the absence of disease states, normal age-related changes in hormonal concentrations may influence cartilage indirectly, via declining muscle strength and activity levels with age (Hersch and Merriam 2008). Declining muscle strength and activity levels reduces joint loading. In this way, declining muscle strength and activity levels are associated with reduced aggrecan production, which is presumably one cause of age-related 'cartilage thinning,' reduced collagen production, and reduced the proliferative drive associated with mechanical loading of articular cartilage. Exactly how mechanical stimulus interacts with hormonal stimuli to influence cartilage tissue integrity is largely unexplored.

In addition to system-level hormonal changes, there are many tissue-specific hormonal systems, including TGFβ signaling, IHH-PTH signaling, RANKL-OPG signaling, and Wnt signaling to name a few (see Sect. 5.6). Further, it is becoming more clear epigenetic changes occur in cartilage over time [as they do in most if not all tissues of the body (Raddatz et al. 2013)]. While cartilage-specific epigenetic changes are only starting to be uncovered, it appears as though they may significantly influence the capacity of the tissue to repair itself (Imagawa et al. 2014) and may accelerate tissue 'aging' (Gonzalez et al. 2016). It seems likely that just as blood tests are required to assess systemic hormonal status, so molecular profiling of the synovial fluid is needed to infer the operating state of the cartilage, so in the future genetic and epigenetic testing of chondrocytes will be needed to assess the capacity of the tissue to repair itself (perhaps fragments of DNA can be recovered from the synovial fluid to facilitate such testing?).

Healthy adult cartilage represents a homeostatic steady state, with aggrecan, collagen, and cell damage being matched by aggrecan, collagen, and cell repair (or replacement). In the case of aggrecan, it appears as though for normal adult cartilage the primary damage mechanism is mechanical rather than enzymatic. In disease states including osteoarthritis, the rate of enzymatic degradation increases many times and then can exceed the rate of mechanically induced damage. It is very likely that similar statements can be made about rates of collagen network damage.

It is also likely that in vivo the primary means of chondrocyte destruction is excessive mechanical loading, with the threshold for cell death being lowered by inflammatory cytokines and raised by anabolic molecules. One important excessive mechanical load type is traumatic injury of a joint, as exemplified by application of a single large impact load. Mechanical testing of cartilage explants, both bovine and human, has shown that chondrocyte apoptosis is 'dose-dependent' [i.e. beginning at about 4.5 MPa (Loening et al. 2000), increasing to 30% at 14 MPa (D'Lima et al. 2001) and to 50% at around 20+ MPa (Loening et al. 2000)].

Importantly following traumatic injury apoptosis evolves over about 7 days (at least in young bovine tissue), which provides a window of opportunity to influence the time course of chondrocyte apoptosis. Importantly it has been shown that some cyclic mechanical loading post-injury can reduce total chondrocyte apoptosis. However, above a threshold of mechanical load, rates of apoptosis increase (Li et al. 2013). This suggests that a 'therapeutic dose' of mechanical loading post-injury needs to be carefully titrated. Despite tantalizing experiments such as these to date, unfortunately far too little is known quantitatively about the damage and repair functions in cartilage to intervene confidently. Despite perhaps a hundred years of clinical observations and data (Benedek 2006), this remains a crucially important area of future basic and clinical research.

Finally, to realize the potential of the discipline known as biomechanics (with an emphasis on bio) the mechanical understanding of cartilage really needs to be further integrated with the biology of cartilage. This will necessarily involve: (i) the systematic study of cartilage mechanics alongside hormonal data, (ii) molecular profiling of the synovial fluid to assess the operating state of the cartilage, (iii) genetic and epigenetic profiling to help establish the capabilities of the cartilage system for repairing itself, together with more traditional approaches such as (iv) sophisticated imaging over a range of length scales, and of course, (v) clinical data. It seems almost certain that only when such a deep integration of data occurs will translation from 'bench-top to clinic' really begin to gather pace.

Chapter 3
Cartilage Tissue Dynamics

3.1 Introduction

Now we want to further develop the view that articular cartilage is a dynamic tissue that adapts to its environmental demands. As described previously, if we rely solely on the experimental literature, this is not an easy task as much of reported experimental literature relates to the behavior of cartilage at 'snapshots' in time, or the experimental data is over relatively short times compared to the timescales over which articular cartilage adapts itself. Adaption can be measured in months and years rather than days and weeks, which is a fairly usual duration for many cartilage experiments. This 'separation of timescales' leads to a significant hiatus in data because of the practical, ethical, and cost considerations involved in gathering the longitudinal data required to develop and test concepts. But despite the difficulties, many ingenious experiments have been carried out that give us starting points and insights into what is happening within articular cartilage over time. This allows us to use our knowledge developed above, to interpolate between the findings of various experiments, and extrapolate when necessary, to create a more complete picture of articular cartilage over a timespan long enough to begin to understand cartilage as a dynamic tissue, adapting to its changed circumstances in both health and disease.

As noted in the foreword, our broad aim is to develop a new dynamic view of cartilage, based on our background in systems modeling of cartilage and related tissues, and grounded in physics, chemistry, engineering, and the biological disciplines. Our hypothetico-deductive approach is developed in the context of a wide re-reading of the cartilage literature with 'systems frameworks' in mind. However, we again mention that we do not claim that our proposed systems view of cartilage is the 'final word.' It is in fact just a beginning. We do believe a sustained attempt to develop a detailed view of cartilage as a dynamic, adaptable tissue is worthwhile, because such a model can potentially serve as a valuable basis for a more connected, functionally adaptive basis for understanding cartilage tissue physiology.

D. W. Smith et al., *Articular Cartilage Dynamics*,
https://doi.org/10.1007/978-981-13-1474-2_3

When developed, such a dynamic model of cartilage can be used to help reinterpret existing experiments, interpret new experiments, suggest new hypotheses, and guide the development of new studies directed at critical missing information. The potential rewards justify the considerable effort and the risks involved.

The adaption of cartilage tissue to new environmental conditions stands in contrast to the view in the literature that cartilage is a 'static tissue.' In this view of cartilage, cartilage cannot repair itself. See, for example, Newman (1998) and Sect. 4 by Henry Mankin at http://cal.vet.upenn.edu/projects/saortho/chapter_05/05mast.htm—says Makin:

> The second issue that must be discussed in relation to the possible response of articular cartilage to mechanical trauma involves the pathophysiology of trauma. Since cartilage is an avascular tissue, significant modifications must occur in the normal response of the body's tissues to trauma, modifications that are likely to affect the healing process. The general response to injury in vascularized mammalian tissues is a phasic one, so similar for most organs and structures as to be almost stereotypic….In considering the application of this scheme to injuries to the hyaline articular cartilages, it is apparent that the response is likely to vary in an important way in view of the avascular state of the cartilage…
>
> The experiment in which a superficial (not extending to the underlying bone) laceration is produced in the articular cartilage of an animal's joint and the tissue then evaluated at regular intervals for evidence of healing has been repeated often. The earliest recorded observation was that of Hunter in 1743 who stated, 'from Hippocrates to the present age it is universally allowed that ulcerated cartilage is a troublesome thing and that, once destroyed, it is not repaired.' Since then, numerous studies have defined the relatively meager and, indeed, ineffectual response of articular cartilage to lacerative injury. …Longer term follow-up of superficial lacerative injuries (slices) has demonstrated no further healing, but oddly, neither has there been evidence of progression to osteoarthritis.
>
> CONCLUSIONS: The clinical implications of the observations described above should be quite apparent. A single lacerative injury to the articular surface should cause little concern. Although the lesion is unlikely to heal, the likelihood of a progression to osteoarthritis is small. Multiple lacerative injuries or even lesions resembling chondromalacia that do not violate the underlying bone can be treated by 'shaving' of the articular surfaces, but unless the underlying bone is violated in the process of shaving, there is little likelihood that one will stimulate repair of the articular cartilage or that new cartilage will grow at the local site. A tangential slice of articular cartilage removed as a result of trauma or the surgeon's knife will remain as a defective area in the cartilaginous surface without evidence of repair.

In this view, adult chondrocytes are in a 'post-mitotic' state and so non-proliferative (Muir 1995). Chondrocytes are viewed as being immobile, 'entombed' within an unchanging collagen network (Pfeiffer et al. 1995; Langelier et al. 2000; Buschmann et al. 1995). Further, it has been reported that the collagen half-life of cartilage is possibly measured in hundreds of years (Eyre et al. 2006), which suggests a level of permanence not afforded to even large animals. Indeed, some hold that if the collagen network is damaged, it is irreversibly damaged (Brandt 2003).

How could such a tissue be dynamically adaptive? In short, it cannot be dynamically adaptive if all these attributes are in fact true. In contrast, we present a 'natural history' of cartilage that shows it to be a complex, dynamic, responsive tissue, that is capable of continually remodeling both collagen network and aggrecan in response to changing environmental loads and mechanical damage throughout the life of the animal (Nelson et al. 1998; Responte et al. 2012).

3.2 Overview of Cartilage as a Dynamic Tissue

Here, we identify and propose novel homeostatic processes operating in articular cartilage and find that the processes involved in maintaining homeostasis of each cartilage tissue component are different, sometimes involving a variety of complex interdependent processes that require temporal coordination. For example, both aggrecan and collagen are lost through the cartilage surface. We propose that the shedding of aggrecan from the cartilage surface is a form of 'sacrificial wear.' This sacrificial wear helps protect the collagen network from abrasive wear. We explain how sacrificial wear of the cartilage tissue contributes, along with other molecules such as PRG4, aggrecan, and hyaluronan, to effective 'mixed mode lubrication' at the cartilage surface.

We find there are complex homeostatic interactions between the chemical and mechanical environments within the cartilage, which together shape its adult structure. For example, mechanical 'segregation failure' of the collagen network and aggrecan may result in new structural features in the collagen network that enhance the collagen network strength and durability, making the collagen network less likely to fail in the future. We also find that failure of one material component in the ECM can lead to a cascading failure of the whole tissue. For example, damage to the collagen network can lead to excessive aggrecan loss, while damage to aggrecan can lead to excessive strains that damage the collagen network. Both scenarios lead to excessive strains that damage chondrocytes faster than they can be replaced, which ultimately leads to failure of the cartilage tissue. Cascading type failures can sometimes lead to very rapid tissue failure [e.g., cartilage may fail within 12 months following joint injury (Driban et al. 2014)].

The interactions discussed here are admittedly complex and are often difficult to study experimentally, but hopefully, this chapter suggests real ways in which cartilage can be viewed as a much more subtle and dynamically adjustable tissue than many traditionally have believed possible.

3.3 Material Failure Modes Within the Cartilage Biocomposite

There is a class of cartilage failures associated with limitations of the structural materials comprising the cartilage biocomposite. Indeed, these material limitations may come to the fore when the materials themselves are 'immature,' the materials in the biocomposite are present in the wrong proportions, the materials have been damaged, or the materials themselves are defective in some way. For example, if the collagen network is genetically defective (Williams and Jimenez 1995), the collagen network is more compliant and cannot sustain normal tensile loads [e.g., a genetic defect in the type II collagen molecule, e.g., Stickler syndrome (Liberfarb et al. 2003), or one of numerous chondrodysplasias (Arnold and Fertala 2013)]. Or perhaps the collagen network is insufficiently cross-linked, perhaps because the animal is young and there has been insufficient time for cross-linking to occur, or because there is a defect in cross-linking of collagen, e.g., malnutrition and/or a deprivation of vitamin C, which is essential for normal collagen cross-linking (Eyre et al. 1988, 2010; Eyre 2002). Of course, if the collagen network is excessively stretched, it may be damaged and not return to its normal shape.

In each case, the whole cartilage biocomposite becomes more compliant. The problem here is that repeated loading involving large strains may quickly cause further damage to the collagen network, to the aggrecan molecules and finally to the chondrocytes. If the chondrocytes are damaged at a rate faster than they can be replaced, this clearly compromises the functional integrity of the whole tissue. We see that an apparently minor molecular defect may lead to a cascading failure with the end result being the capacity of the homeostatic mechanisms to restore equilibrium of the tissue is exceeded.

The capacity of normal homeostatic mechanisms to restore equilibrium may be exceed in a variety of other ways. The chondrocyte response may be compromised. So, for example, aggrecan may be produced in insufficient quantities (decreased synthesis is correlated with aging (Bayliss et al. 2000) and probably at least in part related to a changing systemic hormonal environment). Or if aggrecan core protein is produced in adequate quantities, the final aggrecan molecule itself may be of inferior quality [i.e. shorter aggrecan core protein with fewer glycosaminoglycan side-chains is correlated with aging (Lee et al. 2013b)].

Aggregate suffers fatigue damage because of the incessant squeezing and relaxing of the tissue. On occasions, the aggregate molecule is literally torn apart by infrequent but violent pushing and/or pulling within the tissue, which ruptures the internal attachments of aggrecan molecules to HA, as well as breaking covalent bonds within large molecules. As aggregate is fragmented, aggrecan is lost and the repulsive forces between the remaining aggrecan molecules are reduced. The stiffness of the cartilage to compression is then compromised. Indeed, softened articular cartilage may be said quite literally to have had the ‘stuffing knocked out of it.’ The cartilage tissue then experiences strains that are too large, leading to further damage of the collagen network, aggregate, and the chondrocytes. When cartilage reaches a point where it can no longer be repaired adequately and so damage accumulates, then by a definition of failure as ongoing damage accumulation that degrades tissue function monotonically, the biocomposite may be deemed to have failed.

Let us now consider stiffness and damage to the collagen network. A collagen network may develop a permanent stretch [see Figs. 5 and 6 (Tang et al. 2010)]. The collagen fibrils are said to have been ‘plastically’ deformed (Tang et al. 2010; Svensson et al. 2010). By definition, a plastic deformation means the collagen fibrils do not return to its original conformation after being stretched. Repeated loadings lead to more stretching and collagen fibril damage, and over time, the ultimate tensile strength and stiffness of the collagen are compromised as collagen damage accumulates.

Let us now try to put some numbers on stiffness, fatigue damage, and ultimate tensile strength of the collagen network in cartilage. At a slow rate of loading (5 mm/min), Kempson (1982) finds the tensile strength of human superficial zone of collagen (tested in the direction of the split lines) is at a maximum around 25–30 years of age, with a peak ultimate tensile strength of about 35–40 MPa, declining thereafter, being less than half the maximum value at around 60–80 years of age (see Fig. 3.1) (Kempson 1982).

The tensile stiffness at 5 MPa preloading also peaks at about the same age range at around 100 MPa and then declines, again approximately halving by 60–80 years of age (see Fig. 3.2). We note in passing that the tensile stiffness at 10 MPa

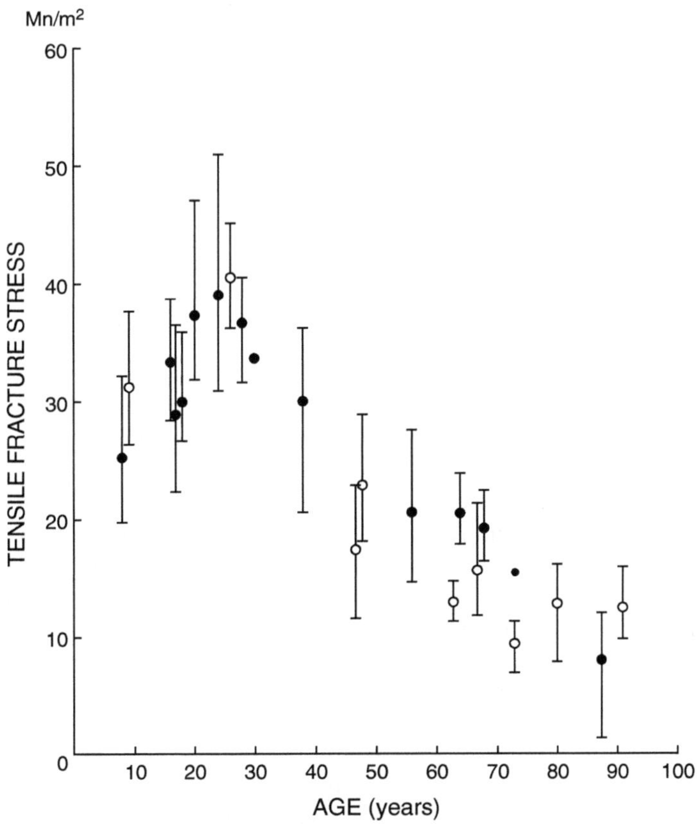

Fig. 3.1 Tensile fracture stress versus age for [human] superficial zone specimens from the medial and lateral femoral condyles. Males -●-. Females -o-. Rate of sample loading 5 mm per minute (Kempson 1982)

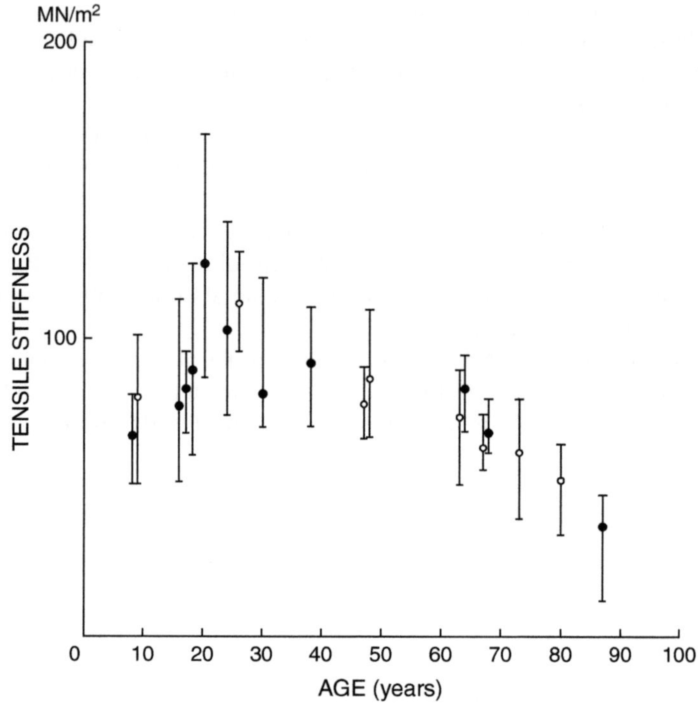

Fig. 3.2 Tensile stiffness at 5 MPa versus age for [human] superficial zone specimens from both femoral condyles. Males -●-. Females -o-. Rate of sample loading 5 mm per minute (Kempson 1982)

preloading increases to around 125 MPa at 20 years of age and then again declines with increasing age (Kempson 1982). In other words, in the superficial zone cartilage, there is nonlinearity in the collagen stiffness with tensile stress magnitude and age.

Testing has also been carried out on bovine articular cartilage. Two to three year old bovine patella experienced no rupture (i.e. splitting of the surface collagen) even for impact stress of 28 MPa, but the same tissue ruptured at 15 MPa when loaded statically (Flachsmann et al. 2001). Two-year-old bovine humeral head cartilage was found to have a transient tensile collagen stiffness of around 28 MPa, while the equilibrium tensile stiffness was around 15 MPa. A strong increase in stiffness is noted with age, and a strong decrease in stiffness with depth was reported by the authors (Charlebois et al. 2004). Ahsanizadeh and Li report that rate stiffening of the collagen network in adult bovine cartilage (16–24 months old) can result in an increase of measured collagen Young's modulus from 35 MPa at 0.1% strain rate to 55 MPa at 80% strain rate at 5% strain (Ahsanizadeh and Li 2015). Taken together this data suggests that while the ultimate tensile strength of adult bovine cartilage is similar to the ultimate tensile strength of adult human cartilage (as described in the previous paragraph), the tensile

Fig. 3.3 [Ligament] tissue stretched to 5.8% strain displaying [type I] collagen fibril rupture (the [collagen] fiber containing this region appeared intact). Shortly after [the stretch] injury, some fibers are intact but contain ruptured collagen fibrils that contain a 'gap' between ruptured fibril ends (magnification 20,000×) (Provenzano et al. 2005)

stiffness of bovine cartilage appears to be about two to three times less. This may be an age effect, as measured human adult cartilage is much older than measured adult bovine cartilage.

We note that young-adult cartilage data on humans implies that the (engineering) strain at ultimate tensile strength is likely to be around 30%. However, it is very likely that damage to collagen fibrils within the cartilage network start at much lower strains, as it does in other tissues. For example, tensile straining of about 6% results in clear evidence of breakage of some collagen fibrils in rat ligament (Provenzano et al. 2005) (Fig. 3.3).

Tensile fatigue tests on femoral head cartilage show that collagen network in cartilage is highly susceptible to fatigue damage (Weightman et al. 1978). Weightman et al. tested 20 femoral heads loaded parallel to the predominant collagen fiber orientation. Each specimen was loaded for 1 s in 20 s with essentially a 'square waveform' with a rise time of 0.1 s until fatigue failure (see data from four fatigue tests shown in Fig. 3.4).

Weightman et al. (1978) found an average fatigue behavior, based on 20 tests (average age 47), could be well approximated by the fatigue curve:

$$\sigma = 25.4 - 0.15a - 1.65\log(n) \tag{3.1}$$

where σ is the (nominal) cartilage (tensile) stress (MPa), 'a' is the age in years, and n is the number of cycles to fatigue failure. We note here that this data is consistent with the ultimate tensile strength measurements (i.e. when $N = 1$) of Kempson (1982) (see Fig. 3.1), despite Kempson's data being for knee cartilage rather than femoral head cartilage. Says Weightman et al. (1978):

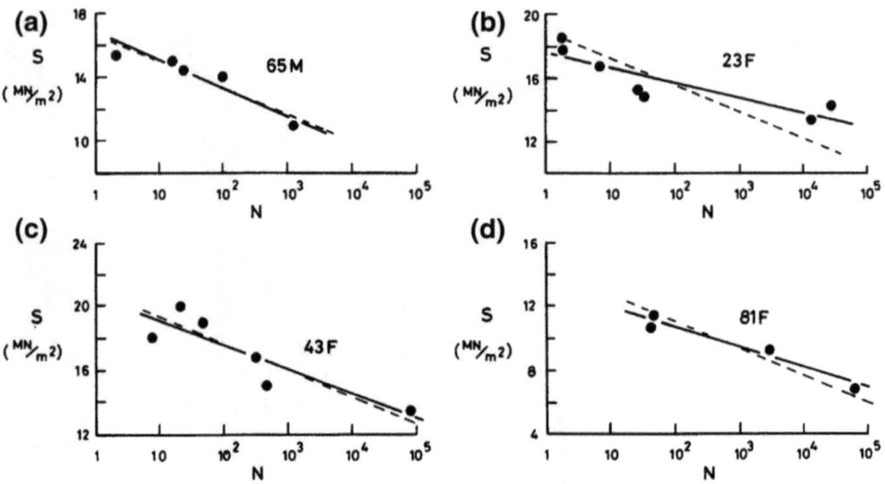

Fig. 3.4 Individual fatigue curves for cartilage from different femoral heads. S = stress magnitude; N = no. of load cycles to fracture; M = male; F = female (Weightman et al. 1978)

> The fact that increasing osteoarthrotic fibrillation and decreasing fatigue resistance both occur with advancing age suggests the possibility of a causal connection between these phenomena. However, it may be that the stresses experienced in vivo could not produce fatigue failure in even old cartilage within a sensible period of time, and this possibility requires investigation.

Weightman et al. (1978) note a reasonable estimate for the total number of load cycles collagen that articular cartilage may be expected to endure before being replaced is two million cycles per year for 5 years (i.e. this equivalent to walking about 5.5 km per day, with 0.5 loadings per step and two stress peaks per loading), i.e. 10^7 load cycles total. Then, setting the fatigue life at 5 years of loading cycles (and so n equal to 10 million loading cycles) and then choosing an adult age, Eq. (3.1) yields alarming low levels of tensile stress predicted to cause failure within 5 years (i.e. at 10 million cycles, assuming no collagen repair). For example, using the previous equation 10 million cycles at age 40 yields a maximum allowable collagen tensile strength of only 8 MPa, and only 2 MPa at age 80. From this, Weightman et al. (1978) conclude:

> In fact the results suggest that a stress of 8 MPa will produce fatigue fracture in 39-year-old cartilage after 5 years, and that the stress would have to be lower than, say, 2 MPa (i.e. a factor of more than 10 less than the maximum tensile stress) for fatigue to be physiologically irrelevant. It follows that the tensile fatigue properties of the collagen fiber meshwork in adult human articular cartilage are such that fatigue failure in life is a possibility.

In this context, it is also relevant to note that similar findings are made in relation to the fatigue loading of tendon tissue. For example, Shepherd and Screen (2013) summarize:

> However, even with appropriate test methods, in vitro tests [on tendon tissue] have, without fail, provided values of fatigue life vastly unrealistic for the necessary lifetime of tendons in situ. (Schechtman and Bader 1997), analysing the human extensor digitorum longus, estimated that in vitro data predicted in vivo failure after about 4 months of normal walking activity.

Weightman's analysis and observations only serve to highlight the crucial dependence of cartilage longevity on ongoing collagen repair (see discussion Sect. 3.5). Indeed, Weightman's analysis assumes the collagen will be 'replaced' after five years (Weightman et al. 1978), which appears to be exceptionally rapid in light of estimates for the half-life of cartilage collagen ranging between 100 and 400 years (Eyre et al. 2006; Verzijl et al. 2000). However, a detailed analysis of cartilage wear and formation rates suggest the collagen is in fact replaced, and it is replaced at a rate approximately that needed to avoid fatigue failure of the collagen network (see discussion and analysis Sect. 3.5).

It has been found after performing 72 fatigue tests on cartilage obtained from a single 48-year-old knee (which had visually fibrillated cartilage) that the permanent stretch in the collagen network of articular cartilage in vivo was highly variable, ranging between 2 and 140% of its initial collagen network length (Bellucci and Seedhom 2001). It is reported that usually an obvious permanent stretch develops in the collagen network after 10,000–100,000 cycles of loading, depending on the load magnitude and the permanent stretch at failure (Bellucci and Seedhom 2001). Plastic deformation and collagen fibril fracture allow greater shear straining in damaged collagen network compared to normal cartilage. For example, Nguyen et al. (2010) report that: '*normal adult femoral condyle cartilage tested with normal synovial fluid (SF), shear strain reached 5.6% near the articular surface and 2.8% overall, increasing markedly with degeneration to 28 and 4.1%, respectively*' (Nguyen et al. 2010).

Clearly, this process of collagen network 'loosening' with mechanical loading, and so the continual expansion of the collagen network with repeated loadings, either due to plastic deformation or fibril breakage combined with inflexibly persistent aggrecan 'inflation,' is an intrinsic feature of articular cartilage in vivo. Expansion of the collagen network is most likely where collagen network stresses are largest, as this is where collagen fibril damage will be greatest. In adult cartilage experiencing everyday normal loadings, this is most likely to be in the superficial and transitional zones of the cartilage beneath the area of cartilage-to-cartilage contact. Note that under extreme repeated surface shear stress, the collagen network at the osteochondral junction may also fail as the collagen fibers are unfavorably aligned for this type of extreme load state (Levy et al. 1996).

To effect adequate repair of any attendant cell death due to excessive cyclic straining, it is expected that chondrocytes in the superficial and transitional zones will also be most active proliferatively (which appears to be the case—see discussions in Sects. 2.7, 2.8 and 2.9.4). However, perhaps somewhat surprisingly, we have seen that synthetically the superficial zone is not the most active. Indeed, highly synthetic chondrocytes generally have an extensive endoplasmic reticulum, numerous Golgi and are generally 'plump' with a 'rounded' morphology, rather than have the 'spindle-shaped' morphology observed in the superficial zone. But we note the ECM in the superficial cartilage is 'supported' deep zone chondrocytes producing ECM components. That is, the majority of the aggrecan, link protein, hyaluronan, and procollagen (the structural molecules) produced in the deep zone are surplus to the needs of the deep zone. In fact, the majority of molecules secreted in the deep zone continually migrate through the ECM toward the cartilage surface (see discussions in Sects. 2.3.2 and 3.4). However, the superficial (and transitional) zone does normally produce more of some specialized molecules, such as lubricin (which helps reduce shear stress in the cartilage surface) being a prime example, and collagen type III another structural collagen. We observe that a primary ECM damage centre, and the primary source of ECM repair molecules, are spatially separated in normal articular cartilage.

Given that cartilage is a biocomposite that depends on both aggregate and the collagen network for its structural integrity, clearly an important issue is the *differential rate* of fatigue damage for each of the material components in the biocomposite. Damage to the collagen fibers is likely to increase significantly when there is substantial deformation in the direction of the collagen fibers (rotation into the load direction by collagen fibers in a collagen network reduces strain in those fibers that rotate, so that average collagen fiber strain is normally less than average tissue strain). In the superficial zone, tensile forces in the collagen fibrils are usually maximal where the curvature of the cartilage surface is greatest. Indeed, the important role of the meniscus in distributing surface loadings and reducing contact stress thereby reducing tensile forces in the tibiofemoral cartilage surface collagen is apparent. Increased collagen curvature and collagen stress in the superficial zone following total meniscectomy promote fatigue damage of superficial zone collagen, which explains the often rapid onset of fibrillation of the tibiofemoral cartilage surface (which almost inevitably progresses to osteoarthritis).

The transitional zone has a well-developed collagen network, which is comparatively more isotropic than either the deep or superficial zones. Shear stress is likely to cause increasing collagen network damage with depth, as the fraction of collagen fibers oriented optimally to resist shear stress decreases with depth from the cartilage surface. It is reported that shear damage in the transitional may be rather localized, as the collagen network is surprisingly heterogeneous in its tensile fatigue properties (Bellucci and Seedhom 2001). Collagen network heterogeneity may be amplified by interactions that accelerate local damage, leading to 'strain localization' in the biocomposite material. For example, collagen damage leads to aggrecan loss, which then amplifies collagen damage, for the magnitude of the shear strain required to cause collagen network damage partly depends on the amount of prestress in the collagen network. With high aggrecan concentrations in the deep zone cartilage and large aggrecan gradients in the transitional and superficial zones, the collagen network normally has significant prestress.

When the cartilage is very stiff in compression (i.e. when true aggrecan concentrations are high), initially, comparatively small shear deformations may lead to significant collagen network damage, as loading becomes focused on particular collagen fibers in the network in the direction of loading. Such small shear strains are much less likely to damage aggrecan, as small strain deformations (alone) maintain nearly constant volume of the cartilage. Indeed, constant volume deformations (i.e. undrained conditions) permit aggregate molecules to relatively easily adjust their shape and accommodate such deformation. For shear loadings of cartilage with significant prestress, it is probably the three-dimensional collagen network that provides most of the resistance to the shear load. In other words, it is possible that repeated small shear deformations of the cartilage may bias damage toward collagen network and not of aggregate. Of course, as the collagen network is damaged by repeated loadings and the collagen network then loosens (Chen et al. 1999), cartilage shear deformations then grow. In this case, the aggregate suffers larger shears strains, and it too may then be damaged more easily. Finally, as deformations grow, the chondrocytes themselves are damaged, and if they cannot be replaced quickly enough, the tissue loses its capacity for some types of repair (see discussion Sect. 2.9.4).

Alternatively, cartilage may suffer predominantly compressive volumetric strains, which unloads and preserves most of the collagen network from tensile mechanical damage (tensile loadings are mainly confined to the superficial zone collagen). In this case, repeated volumetric straining may bias damage toward the aggregate molecules at depth, sparing the collagen network. However, as the aggregate is damaged by repeated loadings and escapes from the cartilage faster than it is replaced, the volumetric strains become larger, and this strain may eventually impinge on the collagen network causing increased rates of damage. Again, approaching an end stage, chondrocytes themselves are damaged, and the tissue again loses capacity to repair itself.

Collagen damage to eight-month-old bovine cartilage, due to five applications of a 4-mm-diameter round indenter, has been reported by (Wilson et al. 2006b). Wilson et al. (2006a) conclude that: '*for our load conditions, cartilage damage starts with damage due to excessive shear strain along the collagen fibrils, and that damage due to excessive fibril strain occurs at more excessive loads.*' In a similar study, Hosseini et al. (2013) explain: '*Damage to the non-fibrillar PG network is less likely, because PGs are expected to be more resilient than collagen fibers.... The remaining explanation is that the collagen network was damaged, yet that we were unable to capture this damage with the Col2-3/4M staining. However, this method has proven very sensitive in the past, and also to stain collagen that was damaged by both excessive tensile and shear loading.* (Hosseini et al. 2013)' It seems likely that collagen damage could not be detected because either samples tested were from 'young animals', the antibody did not work, or there were unexplained problems with the experimental method.

Of course, damage may also be initiated enzymatically, rather than mechanically. For example, for a rat model of osteoarthritis, papain is injected leading to sGAG depleted articular cartilage, which is then subject to mechanical loading via normal exercise (Siebelt et al. 2014). It becomes apparent that the expression of various MMPs, cathepsins, and aggrecanases in cartilage can, when coupled with mechanical loading associated with normal exercise, lead to permanent damage of the collagen network and to a 'downward spiral' into osteoarthritis. Clearly, considerably more research is required to better define the quantitative relationships between the load type, frequency and duration and the relative and absolute damage to aggregate and the collagen network, with and without inflammatory cytokines. Fortunately, early numerical experiments exploring cartilage ECM component damage have commenced (Quiroga et al. 2017).

However, even simple mechanical damage models can help explain why transection of cruciate ligaments often leads to osteoarthritis. Transection of the ACL results in increased reciprocal sliding motion between tibia and femur (Imhauser et al. 2016). This leads to significantly increased shear loading on the articular cartilage surface. If the cartilage is normal prior to transection (i.e. the cartilage adequately inflated by aggrecan), by the abovementioned processes of mechanical damage, this will result in a greatly increased rate of collagen network damage. It has been observed that ACL transection often leads to early onset osteoarthritis—there are many animal ACL models of osteoarthritis (Hasler et al. 1998, Dargel et al. 2007). The damaging effect of ACL transection leading to collagen network expansion is discussed in some detail in Sects. 2.9.2 and 3.5.

Another potential 'failure mode' for the cartilage biocomposite is 'segregation' of the two main structural components—aggrecan and collagen. Rapid segregation of structural components is most likely to be a failure mode when surface loadings are high and stress states are biaxial—that is, when there is a large tensile force in one direction and a compressive force in the normal direction (i.e. a 'squeeze and pull' combination), generating large shear forces in the biocomposite. In this material failure mode, the aggrecan is forced to move through the pore openings in the collagen network, whereupon it collects in distinct 'pools' of aggrecan (i.e. regions with collagen density less than average and aggrecan density above average). Meanwhile, the collagen network collapses on itself forming compact 'bands' of collagen (Kobayashi et al. 1996; Bevill et al. 2010; Nickien et al. 2013). Band formation is probably then aided by fusion of adjacent collagen fibrils when there is little or no aggrecan to separate them, much as occurs in formation of the superficial zone collagen network. For a more extreme but closely related process leading to collagen fibril aggregation into bands, see the discussion of collagen 'destructuring,' and collagen fibril 'aggregation,' motivated by physical models of the collagen network (these physical models are supported by their similarity to observed transmission electron microscope images of collagen in cartilage (Nickien et al. 2013): See, for example, Fig. 3.5).

Fig. 3.5 Series of images of an analog of the fibrillar network in cartilage and illustrating its ▶ potential to undergo destructuring and aggregation and created from an initial radial array of parallel strings incorporating repeating sites of interconnectivity along their lengths. In all images, the radial direction representing the 'native' orientation of the array elements is vertical. The fibrillar model represented in images (**a**–**c**) contains radial elements that have 100% entwinement-based interconnections: **a** is the intact array, **b** and **c** showing partial and advanced destructuring, respectively. Note the increasing amount of near-transverse tangling that is always associated with any localized destructuring of the fibrillar network into near parallel, radial aggregates. The fibrillar model in images (**d**–**f**) incorporates only 30% entwinement-based interconnections (identified by solid circles), the remaining 70% being no entwinement based and thus able to be removed without residual tangling; **d** is the intact array, **e** and **f** illustrate partial and advanced destructuring, respectively. Note that this advanced degree of 'degradation' leads to large-scale rearrangement of array into aggregated parallel bundles aligned in the radial direction (Nickien et al. 2013)

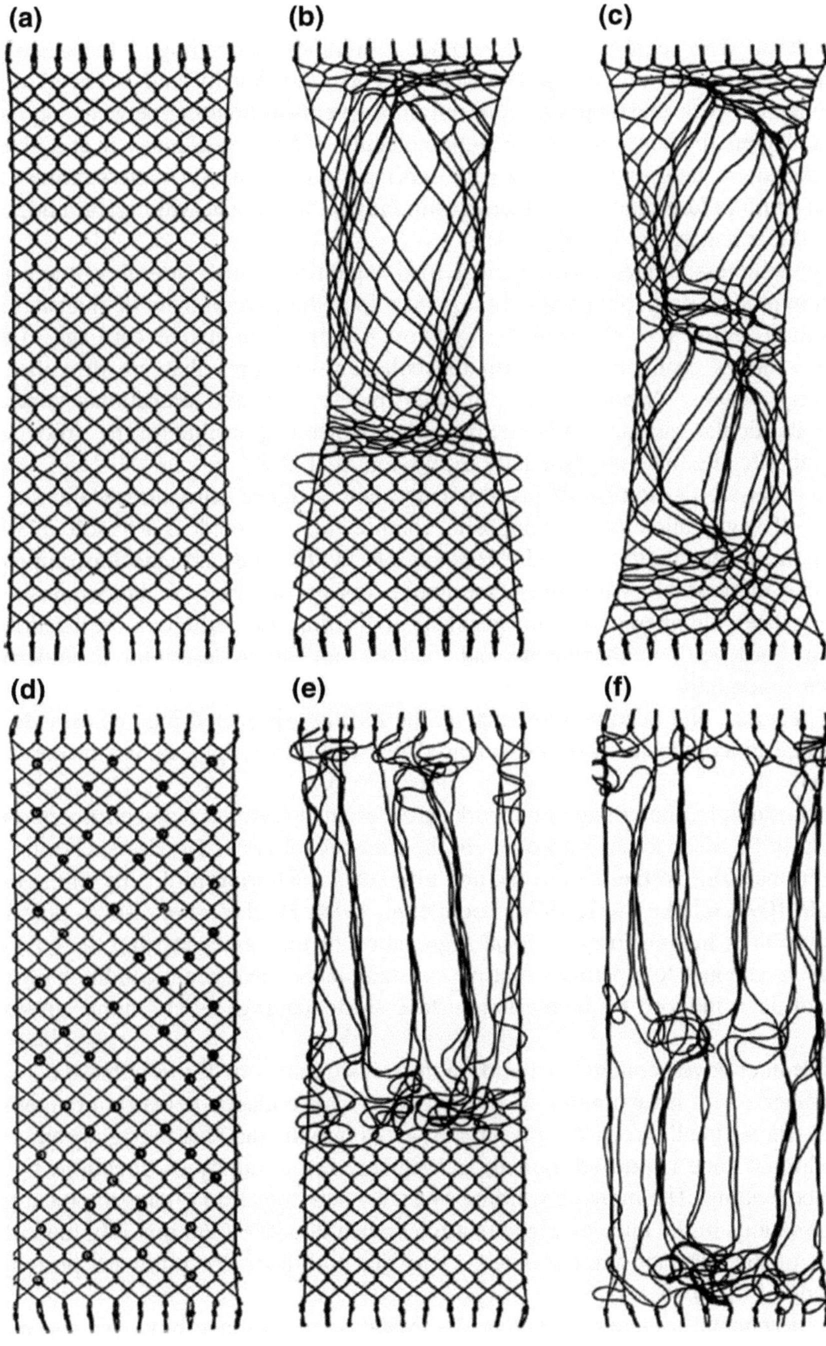
(a)
(b)
(c)
(d)
(e)
(f)

The mechanics of this segregation failure mode may also occur over longer time periods. A 'slow segregation' failure mode is complex, depending on the size of the effective pore opening in the collagen network, the density and strength of the collagen network, the aggrecan concentration, chemical bonding between aggrecan and collagen, the concentration of 'link molecules' like such as tenascin and fibulin, and the strain states experienced by the cartilage. To complicate matters further, this segregation failure mode may interact with fatigue failure and with repair modes for both the collagen and the aggregate.

Nevertheless, under physiological load conditions, such complex interaction 'failure modes' probably play an important role in the development of characteristic structural features of the collagen network observed in native cartilage. These failure modes probably help form the collagen structures that are described as 'sheets,' 'leaves,' 'ribbons,' and 'arcades' in the deep and transitional zones of normal articular cartilage and quite possibly are also responsible for the structure of the superficial zone itself (see later discussion Sect. 3.5).

In other words, somewhat paradoxically, newly 'failed' structures (failed in the sense of segregation of biocomposite components), formed as a result of focal component segregation in high-stress regions within the cartilage biocomposite, become more stable with time. One could loosely describe the biocomposite as having been 'tempered' by the mechanical loading, or perhaps more accurately 'strain hardened' by the mechanical loading, but either descriptor is imperfect. When segregation occurs under higher loadings arising as the animal increases in size, matures, and become more active, the locally segregated regions provide the cartilage biocomposite with an increased resistance to fatigue failure in the adult animal.

For example, the collagen network 'arcades' that form as collagen fibers originating in the deep zone 'turn over' in the transitional zone, thereby anchoring the transitional zone to the deep zone, and ultimately anchoring it directly to the bone (Poole 1997; Clarke 1971, 1974; Poole et al. 1984; Hughes et al. 2005), plausibly develop their mature form by focal segregation of the aggrecan from the collagen fibers in response to repetitive, relative constant, large shear strain conditions within the cartilage induced by increasing surface shear loadings as an animal grows in size.

The interwoven collagen network in the transitional zone forms a dense network of collagen with fiber running in all directions. This collagen network arrangement resists forces in all directions, presumably reflecting the fact that the aggrecan in the transitional zone is pushed, pulled, and squeezed in a variety of directions by the surface loading. The more demanding everyday activities lead to physiological load deformations in the knee cartilage that may reach 20–30% (and possibly more than 30%) of the cartilage thickness (e.g., walking up-hill or down-hill for prolonged periods of time).

Experimental systems to explore the evolution of cartilage biocomposite materials under cyclic loading have yet to be developed, but clearly, these experimental systems promise exciting future discoveries.

3.4 Aggrecan Shedding to the Synovial Fluid

While the collagen network in cartilage resists the dispersion of aggregate and aggrecan by both mechanical entanglement and chemical interactions with the collagen network, aggregate/aggrecan does nevertheless slowly migrate along its chemical potential gradient from higher concentrations to lower concentrations (Bayliss et al. 2000). This tendency to minimize 'strain energy' (or more generally, minimize 'chemical potential energy,' as strain energy can be regarded as a component of chemical potential energy) leads to a general movement of aggregate and aggrecan from the deep zone of the cartilage toward the superficial zone (Comper and Williams 1987), whereupon it eventually joins the synovial fluid. Superposed on this global aggregate/aggrecan movement is the local movement of newly synthesized ECM molecules produced by chondrocytes 'dotted' throughout the cartilage tissue. Generally, this superposed newly synthesized aggrecan movement is away from locally high concentrations of aggrecan in the chondrocyte's pericellular region, toward the relatively lower concentrations in the more distant territorial and interterritorial regions (Poole 1997; Leddy et al. 2008). With this initial description, let us now consider this more closely.

In normal cartilage (and in early stage osteoarthritis when aggrecan production is increased), the predominant direction of aggregate/aggrecan movement is from the osteochondral junction, where true average aggrecan concentrations are normally highest in an unloaded state, i.e. up to 80–100 mg/mL of articular cartilage [and possibly 125 mg/mL (Wedig et al. 2005)] (beneath the regions in articular cartilage that experience the highest regional loads), toward the superficial zone (where average aggrecan concentrations are considerably lower in the unloaded state, e.g., around 40–60 mg/mL in high load areas of knee cartilage). Under these conditions, chondrocytes producing newly synthesized aggrecan molecules contribute 'aggrecan plumes,' which can be seen extending from the chondrocytes. Compelling evidence for the existence of these plumes of newly synthesized aggrecan (and presumably other ECM components) is shown in Figs. 3.6, 3.7 and 3.9 and Fig. 4 of (Bayliss et al. 2000). The aggrecan plumes from individual chondrocytes are made visible by the labeling newly synthesized GAGs attached to aggrecan. This newly synthesized aggrecan joins the general flux of (unlabeled) aggregate/ aggrecan already 'streaming' (in the sense of moving 'on mass') toward the cartilage surface (see Fig. 4 of Bayliss et al. (2000), Figs. 3.6, 3.7 and 3.9).

Figure 4 of Bayliss et al. (2000) show images of human articular cartilage obtained from 9 years old, using a protocol of a 4 h pulse with radiolabeled sulfate 35 [Fig. 4a of Bayliss et al. (2000)], followed by an 18 h chase (Fig. 4b). After fixing the tissue, seven micron slices were taken and exposed to auto-radiographic emulsion for 4 weeks (Bayliss et al. 2000). Note that newly synthesized aggrecan labels chondrocytes and nearby pericellular matrix largely by diffusive transport during the pulse period [Fig. 4a of Bayliss et al. (2000)], but 18 h later advective-diffusive 'plumes' of newly synthesized aggrecan clearly stream from individual chondrocytes toward the cartilage surface [Fig. 4b of Bayliss et al.

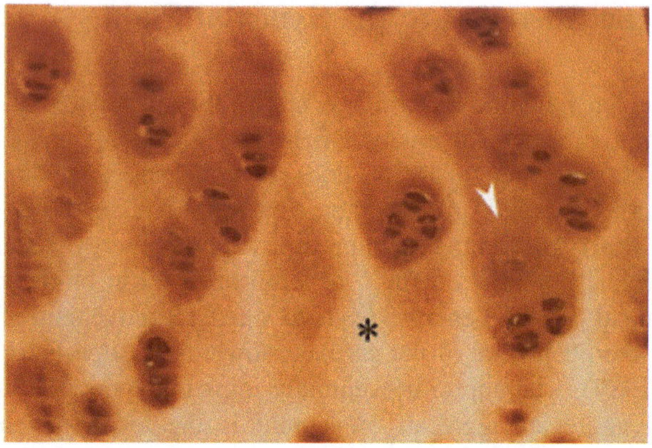

Fig. 3.6 Newly synthesized aggrecan in 18-month-old bovine cartilage, identified by an antibody targeting aggrecan's C-type lectin domain. This newly synthesized aggrecan closely surrounds cells, but showed a distinctly territorial localization. Similar immunohistochemical staining patterns were observed in normal human articular cartilage (Kvist et al. 2008). Kvist et al. identify the region of aggrecan advective plumes, territorial matrix, while they identify the 'asterisk' region as interterritorial matrix. This suggests that the aggrecan in the territorial zone around chondrocytes has a comparatively rapid turnover relative to the aggrecan in the interterritorial zone

(2000)]. This clearly indicates aggrecan and aggrecan fragments are continuously advected toward the cartilage surface, supporting the availability of aggrecan availability in more superficial zones, but in addition providing a 'shedding defense' of the cartilage tissue against articular surface wear.

Based on these images, it appears as though the newly synthesized aggrecan moves along paths in the ECM defined by contiguous territorial zones, around chondrocyte columns in the deep zone cartilage [see schematic depiction Fig. 21 in (Poole et al. 1984)], and into contiguous territorial zones in the superficial zone. It appears that aggrecan molecules are transported 'on mass' from the site of production at a chondrocyte to some distant location. This 'on mass' aggrecan transport rate through the territorial zones is relatively rapid. Based on the rate of newly synthesized aggrecan transport in young animals, the transport rate varies between 2 to more than 10 μ *per hour*. This suggests newly synthesized aggrecan could theoretically traverse the average depth of human femoral knee cartilage (about 2.2 mm) in as little as 10–50 days. This provides strong evidence for there being (at least) two pools of aggrecan in cartilage (i.e. one with a faster turnover (e.g., newly synthesized (unattached) aggrecan) and one with a slower turnover (e.g., aggrecan attached to hyaluronan becoming aggregate)—the average turnover rate for the two pools being estimated to be around 1.75 years in a young adult—see Appendix E).

Newly synthesized aggrecan (presumably along with aggrecan link protein, hyaluronan, procollagen, COMP, and many other molecules) is transported from the deep zone (where damage rates are usually very low in healthy cartilage) toward the transitional and superficial zones (where damage rates are usually considerably higher). Repeated mechanical straining of aggregate in the transitional and

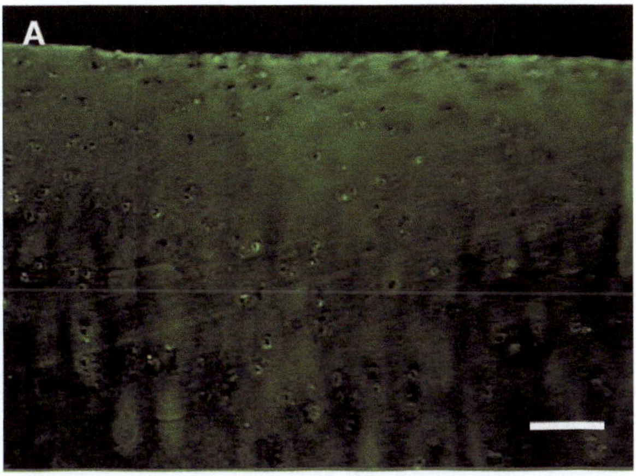

Fig. 3.7 7D4 (green) epitope staining of native chondroitin sulfate in bovine cartilage. All cell nuclei were counterstained with propidium iodide (red fluorescence). A strong (green) staining was seen in the territorial and pericellular regions of normal (non-loaded) adult articular cartilage scale bar 200 μ (Lin et al. 2004). Evidence supporting the concept of aggrecan advection from the deep zone toward the superficial zone is the advective 'plumes' of native chondroitin sulfate emanating from cells (see above). Note that the aggrecan concentration is not zero at the cartilage surface. This is because the small densely packed collagen fibrils, together with abundant type III collagen in the superficial zone of normal cartilage, offers substantially more resistance to the migration of large aggrecan and aggregate molecules through the superficial zone than collagen in the deep zone of the cartilage (Smith et al. 2016). The low diffusivity of aggrecan is partly due to the high density of the collagen fibers in the superficial zone and partly due to the abundance of type III collagen that is covalently linked to the type II collagen or part of type IX collagen. The positively charged N-terminal domain on the type III collagen can bind (and sequester) negatively charged molecules like tenascin, COMP, and aggrecan (see Fig. 3 for the distribution of *N*-terminal domain of type III collagen in cartilage (Eyre et al. 2006) or http://www.biology-online.org/articles/articular_cartilage_collagen_irreplaceable/figures.html). If this resistance to aggrecan migration afforded by the superficial zone is lost [e.g., because of fraying of the surface collagen or splitting of the cartilage surface (Workman et al. 2017)], then aggrecan loss from the cartilage is greatly increased (see Fig. 2.27)

superficial zones may result in aggrecan molecules becoming detached from the hyaluronan molecule, creating fresh sites for attachment of newly synthesized aggrecan and link protein (see discussion on aggrecan turnover Sect. 2.3.3). In this way, aggregate molecules in the transitional and superficial zones that have been damaged by mechanical loading may be continuously repaired by the aggrecan and link protein transported from the deep zone.

Indeed, it begins to become apparent that aggrecan has 'evolved' (via 'trial and error' selection through evolution) to develop a damage avoidance mechanism whereby it detaches and then reattaches to hyaluronan, thereby minimizing mechanical damage to the aggrecan molecule itself. If damage to the aggrecan molecule is minimized by controlling the magnitude of tensile forces carried by an aggrecan molecule by controlling the hyaluronan detachment force, then aggrecan can be detached and reattach multiple times on its journey to the articular cartilage

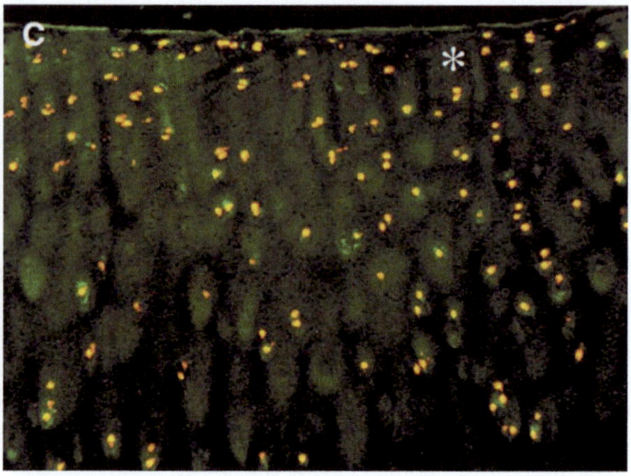

Fig. 3.8 [Compare this figure with Fig. 3.7 shows] a loss of 7D4 staining in the superficial zone of articular cartilage loaded for 1 h at 1 MPa (Lin et al. 2004). This clearly demonstrates that static loading suppresses aggrecan synthesis by chondrocytes. Suppression of aggrecan synthesis moved deeper into the cartilage tissue as the duration of loading increased (Lin et al. 2004)

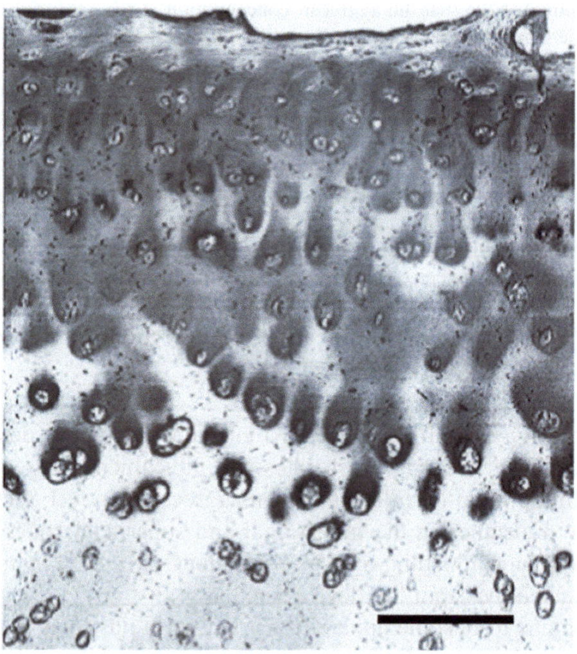

Fig. 3.9 Immunostaining of femoral condylar cartilage from an osteoarthritis patient (Dejica et al. 2012a). Scale bar 200 μ. Evidence supporting the concept of aggrecan advection from the deep zone toward the superficial zone is the advection of collagen fragments streaming from individual chondrocytes, as evidenced by the immunostaining of collagen type II neoepitopes (the collagen fragments are clearly visible because of collagen type II production, and collagenases production are both upregulated in osteoarthritis, resulting in prominent plumes of type II collagen fragments) (Dejica et al. 2012)

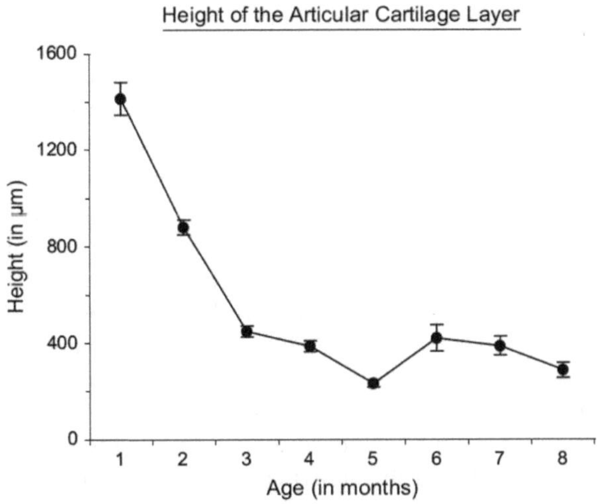

Fig. 3.10 Graph depicting the overall height of the articular cartilage layer as a function of the postnatal age of the New Zealand white rabbits. The height of the articular cartilage layer decreases precipitously and almost linearly up to the third postnatal month, at which juncture the mature structural organization of the tissue is achieved and the animals attain sexual maturity. Thereafter (between 3 and 8 months), the height of the articular cartilage layer does not change significantly. Mean values S.E.M. are represented (Hunziker et al. 2007)

surface. Then, we realize that aggrecan link protein further titrates the 'strength of the linkage' between hyaluronan and aggrecan, providing a simple mechanism to regulate the speed of aggrecan progress through interterritorial zones toward the cartilage surface, by controlling the concentration of aggrecan link protein in the ECM by regulating link protein production and secretion.

In other words, the chondrocytes in the deep zone are supporting the chondrocytes in the transitional and superficial zones by producing excess aggrecan, and helping to preserve the cartilage surface by sacrificing some of that excess aggrecan to surface wear. Given this spatial separation of production to functional usage, logically, we may then expect signaling molecules to be exchanged between the zones over the depth of the cartilage to coordinate chondrocyte synthetic activities in the deep zone with damage in more superficial zones. Indeed, this appears to be this case (see discussion of various signaling systems operating in articular cartilage in Sect. 5.6).

Comparing Figs. 3.7 and 3.8, we have a direct visualization of the profound suppression of synthesis caused by static loading of articular cartilage. In as little as one hour, significant suppression of aggrecan secretion is evident, and it is reported that this suppression moved deeper into the cartilage over time (Lin et al. 2004). This visual impression appears to be consistent with the data presented in Fig. 2.6, which shows very significant suppression of aggrecan (and collagen) production after 12 h static compression at various strain levels (Sah et al. 1989).

Upon reaching the cartilage surface, both intact and partially separated and/or degraded aggregate-HA-aggrecan finally enter the synovial fluid, whereupon it is removed via lymph vessels located in the sub-intima of the synovial membrane (Page-Thomas et al. 1987). Due to the continual removal of ECM debris from the synovial fluid, the synovial fluid has a relatively low concentration of aggrecan (and other ECM) molecules. The half-life of molecules in the synovial fluid depends on molecular size, but is probably measured in hours. This low concentration means the synovial fluid acts as an effective 'sink' for both intact and partially degraded aggrecan molecules exiting the cartilage tissue (Belcher et al. 1997).

Importantly though, as the aggrecan and hyaluronan molecules transition from the superficial cartilage to the synovial fluid they make important contribution to the mixed mode of joint lubrication at the surface of the cartilage (see schematic depictions of this in Figs. 3.11 and 4.7) (Seror et al. 2011; Lee et al. 2013a). Alongside the specialized boundary lubricating molecule PRG4 [or lubricin, which is secreted by chondrocytes in the superficial and upper transitional zones (Rhee et al. 2005b; Jay and Waller 2014)], this flux of aggrecan and hyaluronan into the synovial fluid helps to maintain the fraction of 'hydrodynamic' lubrication when in the 'mixed-mode' lubrication regime (see more discussion Chap. 4) and thereby reduce friction at the cartilage surface. Lee et al. (2013a) suggest:

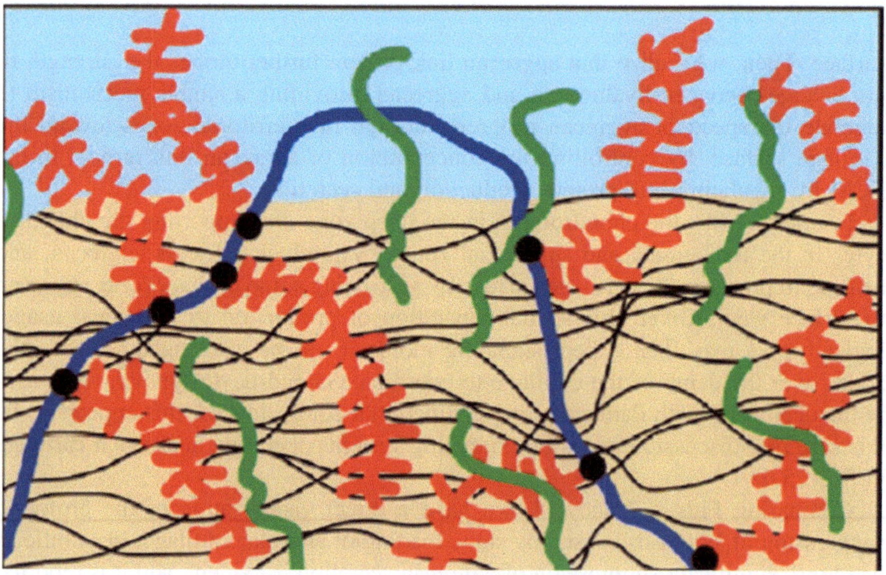

Fig. 3.11 Detailed structure at the outer cartilage surface is thought to comprise charged macromolecules as shown in this schematic. The major charged molecules are hyaluronan (darker, thick curves, blue online), bottlebrush-like aggrecans (red online), and lubricins (lighter, thick curves, green online), extending from the surface to form a brush-like layer. Reprinted with copyright permission (Seror et al. 2011) and Copyright 2011 American Chemical Society. See also Fig. 4.7 for another schematic depiction of the cartilage–cartilage interface

> ....the main function of HA is to act as a surface anchored protective layer rather than a bulk viscosity lubricant. To act as a well-functioning protective layer, HA should have a high enough molecular mass to both be trapped in the porous structure of the cartilage and entangle lubricin (LUB) for efficient boundary lubrication....

Aggrecan improves joint lubrication by increasing the duration of the mixed-mode of joint lubrication. As explained in Lee et al. (2013a):

> The failure to maintain an osmotic balance within the cartilage and preserve the hydration layer at the interface explains the significant increase in the friction force after GAGs digestion. Furthermore, interpenetration and entanglements of HA chains between opposing surfaces is facilitated because of the absence of GAGs, further leading to higher friction forces....

The main functional purpose of maintaining low friction surfaces in synovial joints is to reduce the rate of cartilage surface wear (Jay et al. 2007; Lee et al. 2013a). It is something of a paradox to observe that to avoid significant rates of cartilage wear, the cartilage tissue in effect accepts a controlled rate of wear, via continually shedding its ECM to the synovial fluid. It is noted here that a corollary is that if the rate of aggrecan and hyaluronan shedding to the joint space is reduced (e.g., with age), then the rate of wear of the collagen fibers at the surface would be expected to increase. Let us now look into collagen shedding at the articular surface.

3.5 Collagen Network Shedding to the Synovial Fluid

In addition to the shedding of aggrecan (and hyaluronan and other molecules) from the cartilage, it is worthwhile first postulating and then working through the proposition that the superficial zone (which has collagen fibers oriented parallel to the cartilage surface) in fact represents a zone for gradual shedding of the collagen network into the synovial fluid (Kobayashi et al. 1997; Lohmander et al. 2003b).

This postulate is based on a key deduction—the collagen network continually expands toward the joint space. In fact, articular cartilage expansion is an intrinsic property of working articular cartilage. This can be readily deduced using the following facts:

(i) aggrecan/aggregate swelling prestresses the collagen network (so the collagen network intrinsically wants to expand) and;
(ii) the collagen network is continually 'loosened' (meaning collagen fibrils are plastically deformed and/or fractured) through repeated mechanical cyclic loading. Cyclic mechanical loading induces microfatigue damage on the collagen fibrils in the collagen network, which leads to collagen network loosening via fibril plastic deformation and fibril fracture (Bellucci and Seedhom 2001; Oungoulian et al. 2015) and;
(iii) some collagen fibrils are also loosened via proteolytic debridement, and sometimes collagen fibrils are severed enzymatically.

As a consequence of (i) to (iii), the collagen network in the cartilage will necessarily expand toward the joint space, with the 'outward expansion' driven by expansion of the aggrecan confined by the collagen network. This means that because the collagen network is continually damaged as part of its everyday usage, in vivo there must be continual growth (or expansion) of the collagen network toward the joint space. In other words, continual growth of the collagen network is an intrinsic property of normal, working articular cartilage in vivo.

As the articular cartilage biocomposite incrementally expands, repair of the collagen network can still be effected within the cartilage by fibril repair and by slow chondrocyte migration (see discussion of collagen network repair processes in Sect. 2.6). However, it is difficult to conceive how any of the collagen network repair process can restore the remaining collagen network to precisely the original state of the collagen network prior to it being 'loosened' (i.e. once the collage network is plastically deformed, or fractured, or enzymatically debrided or cut, the collagen network irreversibly adopts a new configuration). Consistent with this analysis, it has been shown experimentally that damage to the collagen network results in extracellular matrix expansion in cartilage (Broom and Flachsmann 2003; Oungoulian et al. 2015). Such collagen network damage is often reported as 'softening' of the cartilage or reported as a 'swelling ratio' [i.e. ratio of weight weights 'after' to 'before' testing (Oungoulian et al. 2015)]. But shear strain is also increased in cartilage with damaged collagen network. For example, Nguyen et al. (2010) report: '*normal adult femoral condyle cartilage tested with normal synovial fluid (SF), shear strain reached 5.6% near the articular surface and 2.8% overall, increasing markedly with degeneration to 28 and 4.1%, respectively*' (Nguyen et al. 2010).

Accepting that repaired collagen network cannot be restored precisely to its original configuration (i.e. geometry), it follows that the collagen network will expand with each damage event, and by this means, articular cartilage gradually expands into the synovial space. Importantly, Oungoulian et al. report that for their 'wear experiments,' internal damage to the ECM mainly occurred just below the interface between the transition and superficial zones ["...*a very notable finding of this study because it indicated that wear damage was primarily initiated by fatigue failure of the extracellular matrix in a subsurface region [below the superficial zone], thus leading to delamination, rather than abrasive wear of the articular surface*" (Oungoulian et al. 2015)]. For an image showing delamination at the cartilage surface, see Fig. 4b of Mow et al. (1992).

So if the collagen network in the transition zone is damaged and expands (see 'swelling ratio' data shown in Fig. 4 of Oungoulian et al. (2015), where the swelling ratio is defined to be the ratio of wet weights after and before testing), it will expand against the overlying superficial zone collagen network (or abnormally, start to 'delaminate' if the damage is very extensive relative to the areal extent of the cartilage). Consequently, when loaded fibrils at this damaged interface between the transitional and superficial zones are compressed, they buckle and fold. A very narrow, exaggerated form of such buckling and folding in the collagen network is the so-named ‘chevron discontinuity’ reported by some authors (see Fig. 2.25) (Bevill et al. 2010).

By this means, most of the random networks of collagen fibers that are damaged expand toward and into the zonal interface, where they are coaxed into bending to become parallel to the cartilage surface, as they press against the bottom of the superficial zone (Glaser and Putz 2002). This re-orientation of collagen fibers is aided by both the loss of aggrecan from the collagen network (i.e. the reduction in aggrecan concentration toward the cartilage surface somewhat deflates the collagen network, facilitating densification of the more superficial collagen network under loading).

Kiraly et al. (1998) report that many hours of in vitro cyclic loading (up to 20 h at 4.1 MPa and 0.5 Hz in unconfined compression) cause ‘*packing or condensation of the tissue.*’ Kiraly et al. say the thickness and birefringence of both superficial, transitional, and deep zones changes with loading, namely both thickness and birefringence of the superficial zone increased [*e.g., data suggests that the loaded superficial cartilage showed higher orientation of collagen fibrils or that the fibrils were packed more in a more parallel manner and closer to each other’* (Kiraly et al. 1998)], while the converse holds for the deep zone. In vivo, rolling, sliding, kneading, along with other mechanical deformations of the cartilage surface complete the collagen fiber re-alignment (Paetzold et al. 2012; Hughes et al. 2005) [collagen fiber reorientation by compressive loading is illustrated schematically in Fig. 11 of Clark et al. (2005)]. By this means, the fibrils at the zonal interface buckle and re-orientate, realigning collagen fibers in the upper transitional zone into collagen fibers parallel with the cartilage surface (Kobayashi et al. 1996; Paetzold et al. 2012; Hughes et al. 2005).

The mechanical effect of this folding and fiber re-orientation process at the interface between the transitional and superficial zones has been demonstrated experimentally. Upon application of a cyclic shear loading to the cartilage surface, it can be shown that shear stiffness of the collagen network over the depth of articular cartilage reaches a minimum just below the interface between the transitional and superficial zones (Buckley et al. 2008). This nadir in shear stiffness is explained by the greater preponderance of collagen fibers preferentially buckling (and folding) at this junction under compressive loading, leading to a plane of low shear stiffness at this location [see Fig. 9 of Buckley et al. (2008)]. This experimental data fits nicely with the findings of Oungoulian et al. (2015), by helping us interpret the functional significance of a zone with reduced shear wave velocity.

For static compressive loadings of long duration (for example, standing for long periods), or for ongoing repeated compressive-shear loadings in vivo (for example, cyclic loading experienced during normal activities), which time dependently densifies the cartilage ECM, and with ongoing damage and expansion of the collagen network and loss of aggrecan at superficial zone due to aggrecan loss through the cartilage surface, eventually ‘cross-linking’ between adjacent collagen fibrils becomes possible in the deformed (buckled or shear) position. Cross-linking between collagen fibrils may involve molecules like COMP, CILP, tenascin, SLRPs, and asporin, as well as polymerization of new collagen (from types II and

III procollagen or possibly even type VI dimers or tetramers). Which molecule or molecules are the most 'sticky' is not known. Together, these molecules bridge between two initially separated collagen type II fibrils, holding the newly recruited fibrils together in a permanent new alignment. As cross-linking takes place, the collagen network is changed from the open, somewhat randomly oriented network found in the transition zone, to the more densely packed and aligned collagen fibers in the superficial zone, all largely oriented parallel to the cartilage surface. The density of collagen fibril network with depth is revealed by Fourier transform infrared imaging, and is typically is denser at the surface, decreases to a minimum around the transitional zone, and then increases again with depth to a maximum in the deep zone (Nieminen et al. 2015; Yin et al. 2012; Zhang et al. 2015).

A more detailed description of the mechanical and chemical conditions enabling buckling and cross-linking processes that work hand-in-hand to create new superficial zone is unfortunately unknown at the present time and so requires further research. However, plausible estimates of the deformation and time necessary for these processes to operate effectively can be guesstimated based on observed or calculated strains found in normal cartilage at the top of the transitional zone, and likely duration fibers spend each day in a buckled state. From the observational evidence available, it appears likely that cartilage strains of around 15–20% for at least several hours per day (i.e. standing or any repetitive activity for relatively long periods) over a period of weeks to months are sufficient to lead to the formation of new superficial zone in cartilage. In this context, we note that for mouse cartilage, collagen fibers oriented parallel to the cartilage surface developed within 5 days of birth, and within 18 days, mouse cartilage was fully zonate and could not be distinguished from adult mouse articular cartilage (Hughes et al. 2005). For domestic pigs at 4 months of age, femoral groove cartilage had most collagen fibers oriented parallel with the surface over the full depth of the cartilage, but by 21 months, the cartilage was fully zonated and had developed a clear Benninghoff structure (Rieppo et al. 2009). For cartilage constructs implanted into Gottingen minipigs, collagen fibers parallel to the cartilage surface were present after 2 weeks to a depth of 10–20% of the cartilage thickness, while at greater depths collagen fiber orientation was 'random' (Paetzold et al. 2012). However, by 12 months a fully zonated collagen network had developed in the cartilage constructs, which was very similar to the native articular cartilage. This data confirms that normal loading of articular cartilage creates zonation cartilage and gives some idea of the likely time necessary for collagen network in articular cartilage to be re-oriented at the surface.

This proposed mechanism implies that a relatively uniform 'lighter-duty' collagen network is initially present at some depth in an articular cartilage (e.g., collagen in a cartilage construct (Paetzold et al. 2012), or neonatal or juvenile articular cartilage (Hughes et al. 2005; Rieppo et al. 2009), or adult transitional zone cartilage (Oungoulian et al. 2015)—it does not really matter which type of collagen network is initially present, it just needs to be there).

In this context, we note that the density of articular cartilage collagen with depth is revealed by Fourier transform infrared imaging, and typically, the collagen network is denser at the surface, decreases to a minimum at the transitional zone, and then increases again with depth to a maximum in the deep zone (Nieminen et al. 2015; Yin et al. 2012; Zhang et al. 2015). The lower collagen density and the lower birefringence [suggesting greater disorder, i.e. a more random structure (Matcher 2015)] in the transitional zone suggest it is a lighter duty collagen network, which would be preferentially 'loosened' by surface shear loading.

If the cartilage surface subsequently experiences higher environmental surface loads, it would be predicted that significant thickening of the superficial zone cartilage would develop over time. We note that the formation of the superficial zone is evidenced as a newborn animal becomes more active and then grows in size. Initially, there is little superficial zone evident. But as the animal matures and becomes more active and increases in weight, the superficial zone concurrently increases in density and thickness. In an adult animal, preconditions for rapid development collagen fiber reorientation and thickening of the superficial zone are reported in an ACL transection model with dogs. Following significant, ongoing collagen network damage as a result of abnormally large surface loadings following ACL transection, the formation of new tangential collagen fibers in the superficial zone femoral trochlea cartilage is speeded up, as reported by Vignon et al. (1983). Vignon et al. remark on their experimental findings: '*Such a repair process is suggested by the observation that the cartilage of dogs killed more than 6 months after operation does not show any degenerative alterations but a marked thickening of the superficial membrane due to apposition of new tangential fibers.*' These observations are as predicted by our proposed model for superficial zone formation.

But thickening of the superficial zone sets in train events that then limit further thickening of the superficial zone. Importantly, thickening of the superficial zone leads to increased resistance to aggrecan flux through the cartilage surface. The retention of aggrecan leads to increased aggrecan concentrations deeper in the cartilage and so to increased compressive and shear resistance. The increased compressive resistance of the cartilage leads to a decrease in (compressive and shear) straining in the collagen network and so a reduction in damage and buckling of the collagen fibers (which are primary events for initiation of superficial zone formation). This self-limiting feedback mechanism is almost certainly an important process in the homeostatic mechanisms for maintaining the thickness of the superficial zone and so the integrity of the entire articular cartilage.

Through the abovementioned collagen network 'aggregation process,' new superficial zone continually forms 'from below.' But what happens 'above' (i.e. at the top of the superficial zone)? Collagen network reaching the cartilage surface is either enzymatically degraded or it is abraded away (i.e. it is mechanically 'worn away' by frictional contact) (Zimmerman et al. 1988; Drewniak et al. 2012; Stachowiak et al. 2006; Kaleva et al. 2009; Li et al. 2011; Lee et al. 2013a). Consequently, new collagen fibers will over time gradually progress through the superficial zone, toward the joint space, as additional superficial zone is formed from below, and the old superficial zone is removed from above.

Clearly, the superficial zone cannot be too thin, for then excessive amounts of aggrecan will be lost through the cartilage surface, which temporarily increases protection of the cartilage surface while increasing collagen network deformation below—both processes speed up growth in thickness of the superficial zone. These processes tend to thicken the superficial zone when it is too thin.

So on one hand, if the superficial zone grows too thick, too much aggrecan is retained leading to less aggrecan production and so less aggrecan is lost through the cartilage surface, which increases wear rates at the cartilage surface (Lee et al. 2013a), while simultaneously decreasing superficial zone formation below. On the other hand, if the superficial zone is too thin, both these processes operate in reverse, thickening the superficial zone. Because too thick and too thin superficial zones are 'self-correcting,' the superficial zone reaches a steady-state thickness (or a 'balanced' thickness state). These self-correcting processes may be properly called homeostatic processes, but they are driven simply by the physics of the biocomposite material as it serves its intended function in vivo, rather than driven by chondrocytes (though the chondrocytes participate by actively sensing their environment and secreting molecules that facilitate superficial zone homeostasis).

It is then a small step to suppose that these processes leading to the formation of the superficial zone have a role to play in shaping the extent (or geometry) of the collagen network (and so the geometry of the articular cartilage itself) and in doing so promote congruence of the articulating cartilage surfaces. By way of example, New Zealand white rabbits have articular cartilage that at birth is very thick. But the articular cartilage thickness decreases precipitously up to puberty (i.e. at between 3 and 4 months). This decrease in thickness is caused by new (interstitial) growth at the top of the cartilage being exceeded by cartilage resorption and bone formation at the base of the cartilage (Hunziker et al. 2007). Together, these two processes ensure coarse-scale congruence of the maturing joint.

The articular cartilage thickness then changes at a more modest rate, driven by various processes including growth, resorption, straining and buckling of the collagen network and bone formation, which together with an active mechanical environment, shape the final form of the adult cartilage at around 8 months (see Fig. 3.10). See also descriptions of cartilage development for the mouse (Hughes et al. 2005) and domestic pigs (Rieppo et al. 2009).

How is a congruence of the opposing cartilage surface promoted during joint development by an active mechanical environment? As the superficial zone forms and the joint is mechanically loaded, local superficial zone processes help promote fine-scaled congruence of the cartilage surfaces. For if the cartilage surface is too prominent at one location, it attracts additional loading and suffers more shear and compressive strain. The transitional zone collagen network immediately below the area of increased loading experiences relatively more damage and fiber buckling at that location, and so fiber accretion increases at this location. With the fiber accretion process in operation (Paetzold et al. 2012; Hughes et al. 2005), over time, this leads to a slightly thicker superficial zone and a less prominent surface at that location. By this process, joint contact stresses converge toward some preferred distribution over time, and the cartilage surfaces are shaped into optimally congruent (or mating) surfaces. Doubtless, this particular cartilage congruence promoting process acts alongside other cartilage congruence promoting processes (see, for example, the analysis proposed by (Plochocki et al. 2009; Hamrick 1999) and also the discussion in Appendix C).

As mentioned above, it is important that abrasive wear on the surface of the collagen network not be excessive and thereby exceed the repair capacity of the cartilage. An interesting case study of excessive wear following ACL transection is reported in (Clark et al. 2005). The most obvious means for reducing surface wear is lubrication (see discussion next Chapter). However, wear can be reduced by a number of other means. A rather subtle (and probably overlooked) way to reduce surface wear of the superficial zone collagen is to have ‘compliant anchorage’ of the superficial zone collagen sheet when it abuts to bone. A compliant sheet anchorage system reduces stresses in the superficial zone collagen sheet itself, shifting shear loading through the articular cartilage and into subchondral bone, rather than retaining the shear loading in the superficial zone collagen sheet. This reduces damage to the collagen sheet.

Such desirable ‘compliant attachment’ to the bone is achieved by the collagen in the superficial zone ‘wrapping over’ the subchondral bone, before it inserts into the bone [see, for example, Fig. 2 in (Goodwin et al. 2004)]. This clever ‘wrap-over attachment,’ which traps aggregate between bone and the collagen sheet, creates a forgiving, compliant attachment of the collagen sheet to the bone. This compliant attachment reduces tensile damage in the ‘tension resisting diaphragm’ (i.e. the superficial zone) by reducing in-plane ‘sheet loading’ [as the superficial zone has been described in Meachim and Stiockwell (1979) according to Poole (1997)].

The compliant attachment allows ‘load sharing’ by enabling sufficient vertical deformation of the collagen sheet to occur without excessive tension, so the collagen network beneath (in the transitional and deep zones) picks up some of the surface shear load, while the vertical deformation and concentration of aggrecan pick up most of the normal load (apart from a small component in the tension diaphragm). In other words, the risk of tensile damage to the superficial zone is reduced by having an attachment system that enables load sharing. It is noted here that depending on cartilage curvature, the superficial zone collagen may be in tension even at rest. The residual tensile stress state in the superficial zone is revealed by the cartilage ‘curling’ (i.e. reversing its in situ curvature), when full thickness cartilage is sliced away from the bone (Poole 1997).

Clearly, abrasive wear is also reduced by extending the duration of (mixed-mode) lubrication at the contact surfaces of cartilage (see Chap. 4). So protection against abrasive wear is assisted by good congruence between the articulating cartilage surfaces. Better congruence helps preserve fluid pressure at the contact interface, which reduces wear. Protection against surface wear is also offered by aggrecan and HA molecules, which continuously emerge from the cartilage surface, assisting boundary lubrication at the cartilage surface. And finally, surface lubrication is assisted by shear stress stimulating the production of PRG4 (or lubricin, which tightly attaches itself to collagen fibrils). But all these lubrication and load management strategies cannot and do not completely stop abrasive wear at the surface of the collagen network. Wear is ever present at contacting surfaces.

In the knee, the cartilage surface is abraded by the sliding action of menisci on the tibial cartilage surface and by the combined sliding (and rolling) action of the femoral cartilage on the menisci and the tibial surface (Kaleva et al. 2009; Neu et al. 2008; McCann et al. 2009). Indeed, large sliding displacement under load during walking may explain the propensity of cartilage on the femoral condyles, as well as the tibial cartilage exposed to large shear sliding (in the 'central' area not in contact with menisci), to be often the first regions of articular cartilage to show evidence of 'focal lesions' that sometimes grow and progress to OA.

For example, for an group of patients with osteoarthritis (Ding et al. 2007) report: '*Progression of cartilage defects occurred frequently in the central regions of the femur and tibia as well as the posterior femur region which is consistent with previous* ex vivo *studies,*' while for another sports-related injury group, the same authors report: '*Lesions located in the central region of the medial compartment were more likely to progress to more advanced cartilage pathology than lesions in the anterior and posterior regions or lesions located in the lateral compartment.*'

We note that exactly what drives growth and atrophy of focal lesions is not known, though they appear to grow from a surface defect. Their initiation is attributed to 'trauma' (the most common cause), osteochondritis dissecans, and osteonecrosis (Falah et al. 2010). It is reported that 'excessive shear' causes fracture of the cartilage (Falah et al. 2010), though presumably excessive normal impact loading or fatigue damage may do the same thing. Indeed, (Liukkonen et al. 2017) found that using a simple collagen fibril fracture rule (i.e. collagen fibrils in the superficial zone fracture when tensile stresses in the collagen sheet are >7 MPa) results in initiation of what appears to be 'focal lesions' on the cartilage surface. Focal lesions doubtless grow due to the altered mechanical conditions (Venalainen et al. 2016; Men et al. 2017) and chemical conditions around the defect, which creates a deleterious feedback loop, which may be self-limiting.

Presumably, focal defects diminish in size due to repair processes (Ding et al. 2007). Repair of large focal defects is undoubtedly cell driven and is probably due to a combination of collagen network repair and proliferation following by fast (and later slow) chondrocyte migration into the defect. However, for small focal lesions, if the collagen network is shed at a rate greater than the focal lesion can grow, then the focal lesion will simply be 'washed out' of the superficial zone cartilage as it is renewed from below. This is probably how most small surface defects are in fact repaired in vivo. But clearly understanding focal defect dynamics requires further research.

Unfortunately, the rate of abrasive wear of the articular cartilage surface has not yet been quantitatively estimated for normal joints in vivo, and so this is yet another area requiring further research. However, wear experiments have been performed on very small cartilage samples using the atomic force microscope (lubricated with either equine synovial fluid or phosphate buffer solution). It is found that moderate to higher loading combined with slow sliding velocity results in 'stick-slip sliding' motion. Lee et al. report that 10 h of continuous smooth sliding (normal stress < 0.08 MPa with velocities greater than 1 μ per second) saw no change on roughness parameters (a proxy for surface wear) (Lee et al. 2013a). However,

Lee et al. (2013) report that 10 h of continuous stick-slip shearing (normal stress > 0.08 MPa with velocities less than 1 μ per second) sees 'medium level' wear on normal cartilage (the measured wear depth is about 1.0 μ), but much shorter times (1.5 h) are required to see similar wear effects for abnormal cartilage or cartilage with abnormal lubrication. This study clearly demonstrates that high loads at very slow velocities may cause substantial cartilage wear and damage to the cartilage surface, but unfortunately, this does not really provide us with a lot of insight into cartilage wear rates in vivo, just that wear can be rapid under certain conditions.

Surface wear in whole joints can be measured using a 'pendulum friction simulator.' Such a simulator was employed to measure friction and wear on adult bovine knee joints ($n = 6$, 18–4 months old). The joints were tested intact, for peak contact normal stress of 4.9 MPa (without meniscectomy), and for peak contact normal stress of 17 MPa (with meniscectomy), at 1 Hz for 3,600 cycles (McCann et al. 2009). Surfaces were measured using a 'stylus profilometer' and a 9.4 T MRI. Limitations of the test include a 23° swing angle (rather than 60° during gait), the use of a mixture of serum and saline as a lubricant (rather than synovial fluid), and the small number of test cycles (only 3,600). Intact joints had a measured friction coefficient between 0.06 and 0.09, while those joints without meniscus had a friction coefficient of 0.08 and 0.12. There was no detected wear volume for the intact joint with average contact shear stress 0.5 MPa (without meniscectomy), but the wear volume increased dramatically to 70 mm^3 for an average contact shear stress of 1.8 MPa (with meniscectomy). This demonstrates that wear rates can increase dramatically with increasing contact shear stress, even for relatively few load cycles (i.e. measured in the several thousands of cycles, which is a tiny fraction of the load cycles experienced in vivo over one year of normal activity).

Perhaps the best evidence for cartilage wear rates in the human knee is a single cadaver wear test in a knee simulator using a single patellarfemoral joint, reported by Li et al. to have: '*with no visible signs of articular degeneration in the region of contact*' (Li et al. 2011). The specimen was wear tested for 375,000 cycles of simulated gait (Li et al. 2011). Unfortunately, the specimen was bathed in phosphate buffer solution (rather than synovial fluid), which is known to have a friction coefficient two to three times larger than synovial fluid (Gleghorn and Bonassar 2008), which will presumably accelerate wear. Li et al. found the wear rate of patella cartilage was higher (maximum wear depth 0.63 mm ± 0.13 mm) than that for femoral cartilage in contact with the patella (maximum wear depth 0.29 mm ± 0.13 mm) (Li et al. 2011). Based on data shown in the paper (see Fig. 5 (Li et al. 2011) showing progressive wear geometry), average wear rates over the contact area are estimated to be about one-third of the reported maximum values. This data clearly shows cartilage wear rates can be very large, however, in a normal in vivo joint, the joint is bathed in synovial fluid, with lubricin on the surface of the cartilage, which can be replenished over an extended period of time, and in addition, collagen repair processes can also repair the collagen network and so reduce wear-induced changes in articular cartilage geometry.

Despite these limitations, based on an estimated *average* femoral cartilage wear rate per million cycles of 0.26 mm found in the experiment, and then assuming the presence of synovial fluid and lubricin (adjust by dividing by 3), gives an estimate wear rate of 65 μ per million cycles (i.e. about 65 μ per year). Clearly, this estimate is subject to significant uncertainty, and it is expected the actual wear rate in a normal adult human knee joint may range anywhere between 20 and 130 μ per year. Assuming the dry volume of collagen at the cartilage surface is 21% [somewhat larger than the average of about 18% (Maroudas et al. 1992)], and that the specific weight of collagen is 1.4 (Basser et al. 1998), and noting the wear area [based on the cadaver experiment is 435 mm^2 (Li et al. 2011)], this rate of surface wear is equivalent to a rate of collagen loss of $0.21 \times 1.4 \times (0.02–0.13\ mm) \times 435\ mm^2 = 2.6$ mg to 17 mg/year. Noting that given the contact area of the medial and lateral menisci is around [$400\ mm^2 \times$ (2 condyles)] = $900\ mm^2$, then we might expect a total rate of collagen fiber loss for a whole human adult knee joint to be some 3 times larger, i.e. in the range 8–70 mg/year.

In contrast to the paucity of data for cartilage wear rates, the rate of abrasion in hip and knee replacement joints has been carefully measured over time in vivo and in the laboratory. Lundberg et al. report that for typical polyethylene and metal prostheses in vivo: '*The per-duty-cycle computed wear depth [is] typically tens or hundreds of picometers per cycle*' (Lundberg et al. 2005). For their 'base case' (i.e. typical) analysis, Lundberg et al. assume a polyethylene wear rate of 80 μm per million cycles (i.e. 80 pm per cycle) resulting in a wear volume of 20.4 mm^3 (Lundberg et al. 2005). When 10% of the surface roughened (so the wear coefficient increases 50 times), they report the wear volume increases 7.5 times).

With improvements in the formulation of cross-linked ultra-high molecular weight (UHMW) polyethylene in a '*conforming fixed-bearing, posterior cruciate ligament retaining*' prosthesis from DuPey, Kretzer et al. (2014) report (GUR1020) polyethylene wear rates were 7.3 mm^3 per million cycles. Assuming a specific gravity for polyethylene of one, the abrasion weight is about 7.3 mg of polyethylene. Based on Lundberg et al. (2005) report, presumably the wear rate is then is about 30 pm per cycle (or 30 μ per year).

In another study, Zhang et al. (2017) test a '*Zimmer NK-II cruciate retaining*' prosthesis over 5 million cycles, reporting wear rates around 11 mm^3 per million cycles, which is somewhat higher than that reported by Kretzer et al. (2014). Neu et al. point out that cartilage wear coefficients range over many orders of magnitude, depending on the mode of joint lubrication [see Fig. 5 in (Neu et al. 2008)], so clearly the rate of cartilage wear will be strongly dependent on the lubrication mode, which in turn depends on both knee loading (e.g., magnitude and timing), tissue strain, and how closely the knee joint approximates normal physiological conditions (some critical parameters being normal menisci, lubricin, aggrecan production, and surface roughness).

Nevertheless, based on the wear estimates above, it seems reasonable to expect that the in vivo cartilage wear rate is of the order of magnitudes for polyethylene prostheses reported by Lundberg et al. (2005), Kretzer et al. (2014) and Zhang et al. (2017) (i.e. 30–80 μ per year), assuming for cartilage both a normal joint and normal activity. Indeed, these prosthetic wear rates appear to be consistent with our range of adjusted wear rates based on the cadaver test as described above (i.e. 20–130 μ per year). The estimated wear mass for the polyethylene from the prosthesis is about 7–20 mg/year, compared to the wear mass of knee collagen estimated to be the range of 8–70 mg/year.

We can check this estimated articular cartilage wear rate for the normal human adult knee in the following very approximate way. Let us first assume that most of the collagen network damage is in the 'high load' region (say 20% of the total cartilage area), and in this region, new collagen network formation each year is one percent of the total network collagen in the high load region (note that procollagen production by articular is likely to be considerably larger than the fraction incorporated into new collagen network). Let us then assume nearly all of the new collagen fibril formation is a result of collagen network 'loosening' occurring predominantly in the transitional zone [for normal articular cartilage, the collagen network in the deep zone is very stable and less easily damaged (Nguyen et al. 2010)]. The transitional zone accounts for 20% of normal knee cartilage thickness (Changoor et al. 2011), and let us estimate it accounts for 15% of the total collagen over the cartilage depth. Let the average cartilage thickness for the whole knee be 2 mm (Blewis et al. 2007; Eckstein et al. 1998) and the average thickness in the high load region be 3 mm. Then, assuming the transitional zone collagen network expands about 7% per year (i.e. adding 1% total collagen due to repair of the damage to the 15% of total collagen in the transition zone), this is 3 mm $\times$ 0.2 $\times$ 0.07 = 42 μ per year (i.e. somewhat less than one micron per week). This expansion of the transition zone reorientates the collagen fibrils and accretes to form say, about 30 μ of new superficial zone. Then, assuming steady-state conditions for the superficial zone thickness (that is, the rate of new superficial zone formation equals the rate of removal of superficial zone surface), and assuming about one million steps per year for each knee, the rate of wear is about 30 μ per million load cycles (i.e. somewhat more than half a micron per week). The estimated wear rate for the polyethylene from the prosthesis is about 30–80 μ/year, compared to the wear rate of articular knee collagen estimated to be the range of 20–130 μ/year. We note that the 30 μ estimate based on assuming a collagen expansion involving replacement of 1% of total collagen (over the depth), which is replaced in the intermediate zone per year (i.e. 7% of the intermediate zone collagen) is consistent with the estimated range of experimental wear rates described above.

We can check if this rate of wear makes sense, at least in an approximate way, by starting from the above estimate and looking at the expected rate of (existing) collagen fibril removal through the synovial fluid. The total volume of the human knee is about 23 mls (Eckstein et al. 1998), and the average total dry volume of collagen is 14% of this volume (Basser et al. 1998). Then, since the specific gravity of dry collagen is about 1.4 (Basser et al. 1998), the total mass of knee collagen is 23 * 0.14 * 1.4 = 4.5 g. Assuming 33% of the total collagen is in the high load region (i.e. 20% of total area and average 3 mm thickness), then if one percent of total collagen is removed from the cartilage surface each year, then there would be a flux of 15 mg through the synovial fluid per year, which is a flux of about 40 ng through the synovial fluid per day. This mass flux of collagen is somewhat less than mass flux of polyethylene reported by Lundberg et al. (2005) (about 20 mg/year), but somewhat more than that reported by Kretzer et al. (2014) and Zhang et al. (2017) (around 10 mg/year) and toward the lower end of the estimated range of articular collagen wear rates (8–70 mg/year) based on the cadaver experiment described above.

Abraded collagen fibrils would probably accumulate in the synovial fluid because of their large size, but the rate of accumulation and degradation of the collagen fibrils would reach a steady state. Degradation by synovial fluid proteases would probably cut the fibrils into small fragments, which would then have a reasonably rapid turnover through the synovial fluid. Assuming a half-life for the small degraded collagen fragments in the synovial fluid of about 4 h, 40 ng of collagen/day equates to a small fragment concentration of degraded collagen in the synovial fluid at any one time of about 10 ng per joint/5 mL per joint = 2.0 ng/mL. Though CartiLaps is not a quantitative assay, this appears to be approximately consistent with measurements of CTX-II concentrations in normal synovial fluid (about 1 to 2 ng/mL) (Lohmander et al. 2003a). We also mention that urinary excretion of (ng CTX-II)/(mmol creatinine) = 159 ng/mmol) (Sowers et al. 2009) (note the average 70 kg person excretes about 11 mmol of creatinine per day, so the total CTX-II excreted per day is 1,750 ng). This appears to be somewhat high compared with the estimated 40 ng/day produced from a single knee joint (but we note it is not precisely clear what actually comprises CTX-II fragments in vivo, and so CartiLaps is not a quantitative measurement).

Finally, we can check the rate of procollagen II synthesis by adult articular cartilage and compare the rate of synthesis with the rate of incorporation of tropocollagen into the collagen fiber network (e.g. average estimate 15 mg/year; range 8–70 mg/year). We would expect that procollagen synthesis is greater than the rate of incorporation of procollagen into the collagen fiber network, perhaps many times greater (based on previous experience (e.g. with tendon tissue), perhaps in the range 2–10 times greater, but at times much more). Certainly, it cannot be less than the rate of incorporation of procollagen into the collagen fiber network. Nelson et al. (1998) performed extensive testing on the rate of procollagen synthesis by articular cartilage using several different methods that enabled cross-checking of measurements for both bovine and human cartilage explants. They correlate the rate of procollagen type II production with the rate of production of C-peptide from type II procollagen (CPII). Together with several other authors, they also find the half-life of CPII in cartilage tissue is relatively short (at around 16 h) and that most of the procollagen produced has its C-peptide removed ['*The analysis of* ^{35}*S-methionine–labeled CPII revealed that the radiolabel was predominantly in free CPII removed from procollagen*' (Nelson et al. 1998)]. This means that CPII concentrations correlate closely with the total rate of production of procollagen type II by chondrocytes within the tissue. Using ^{35}S-methionine to label newly synthesized procollagen and a pulse-chase method, they found that the concentration of CPII in normal adult tissue (obtained at autopsy within 16 h of death (age range 47–81 years, $n = 6$)) was 0.45 ng/mg wet weight of tissue (range $\pm$ 0.27 ng/mg) [see Table 2 (Nelson et al. 1998)]. Given that procollagen has a molecular weight of

141 kDa, while CPII has a MW of 35 kDa (Nelson et al. 1998), this means the mass of tropocollagen produced in a 2 h period is about 3 times larger than the mass of CPII, i.e. about 1.35 ng/mg wet weight of tissue. From this, we can estimate the yearly production of tropocollagen by an adult knee joint, that is, 1.35 ng/mg * 23 g cartilage tissue per knee * 12 (2 h periods in a day) * 365 days in a year = 180 mg/year (range 90–290 mg/year).

We compare this with the estimated rate of tropocollagen incorporated into the collagen fibril network, i.e. 15 mg/year, range 8–70 mg/year. Comparing the ratio of mean estimates of synthesis and mean rate of incorporation of collagen gives 180/15, a factor of 12, with a range of 1.25–35. We see that this approximate estimate sits comfortably with all our other approximate estimates above. We note in passing that Nelson et al. (1998) reports that the rate of CPII production in different joints is approximately proportional to the amount of collagen network present in the cartilage, but is strongly age dependent (Nelson et al. 1998). For example, bovine fetal production is about 60 times greater than normal adult cartilage, newborn bovine cartilage about 20 times greater than normal adult cartilage, and OA adult cartilage about 8 times greater than normal adult cartilage) (Nelson et al. 1998).

Taking into account all the estimates made above, all these (very) approximate quantitative analyses suggest that the wear rate of collagen in the high load regions of normal articular cartilage could be in the range 20–60 μ per year. In other words, without collagen replacement, under normal physiological conditions about 0.5–1.5 mm of articular cartilage would be removed every 25 years. In other words, it is again clearly apparent that without ongoing collagen repair, normal articular cartilage would completely wear away prematurely in most people. That is, we reach a similar conclusion to (Weightman et al. 1978) about the likelihood of cartilage surviving throughout adulthood without repair (i.e. Weightman (1978) concluded this was very unlikely).

But most importantly, wear can be advantageous if new collagen network is moving progressively upwards, for it can resolve Weightman's dilemma—namely that collagen in the superficial zone needs to be replaced within about 5 years; otherwise, it will fail (i.e. break) due to fatigue damage (Weightman et al. 1978). For if the rate of collagen replacement amounts to 0.13–0.4 mm per 5 years (see the previous paragraph), this is just the rate required to replace the collagen network in the superficial zone and avoid the fatigue failure that (Weightman et al. 1978) feared. This ties together quite pleasingly. And given the wear rates estimated above (i.e. 0.5–1.5 mm of articular cartilage is removed every 25 years), it is very unlikely the collagen network will survive for five or more decades of adulthood in the absence of collagen network repair (see discussion Sect. 3.3). But with collagen repair mainly in the intermediate zone, this is no longer problematic. Based on this analysis, it seems highly likely that collagen is repaired (and grows in amount) at approximately the same rate it is worn away. This repair of collagen mainly occurs in the intermediate zone, which moves progressively upward to become superficial zone collagen, which is mainly worn away (rather than mainly repaired) before it exceeds its fatigue life. Further, the progressive upward movement of collagen in the superficial zone may help remove small focal lesions in the superficial zone of articular cartilage. So there appears to be a number of significant benefits that accrue to articular cartilage if such a system is in fact operational in articular cartilage.

At this point, it is instructive to consider the possibility that substantial long-term changes in the total thickness of adult cartilage can occur in response to changes in environmental loadings (e.g. perhaps as a result of sustained activity levels increasing significantly (e.g. seasonal sports activity), or as a result of abnormalities in joint biomechanics, e.g. as a result of rupture of the ligament, or as a result of valgus or varus). We have seen that it is possible for cartilage thickness to change over time as the very young animal matures by balancing interstitial growth toward the articular surface and osteochondral interface progression into the cartilage. But in the adult, it is possible for cartilage thickness to change by rebalancing the rate at which articular cartilage continually expands and the rate at which the collagen network is shed to the joint space. It is clear that a step increase in exercise levels, or a sudden abnormality in joint biomechanics, could result in more extensive mechanical damage to the collagen network than usual, potentially causing the collagen network to expand significantly more quickly than it usually does (Broom and Flachsmann 2003; Oungoulian et al. 2015; Clark et al. 2005).

This is demonstrated most clearly in animal experiments. For example, this is demonstrated by an anterior cruciate ligament (ACL) transection experiment (n = 11) in the mature dog [i.e. compare cartilage images shown in Fig. 1a, b in Vignon et al. (1983)]. This experiment demonstrates the effects of ACL transection on trochlea cartilage model. One remarkable feature of this experiment is the marked thickening of femoral cartilage with an intact surface three months after ACL transection: *'An early and progressive decrease in superficial cell density and a later progressive increase in cartilage thickness without any change in the cell density of the middle and deep cartilage layers was found'* [see Fig. 1b in Vignon et al. (1983)].

Similar results have been reported in a similar ACL transection experiments in dogs. For example, Adams and Brandt (1991) report: '*Thus, in these animals, knee instability did not lead to loss of cartilage but to an active synthetic response by the chondrocytes resulting in hypertrophic cartilage repair which was sustained for up to 64 weeks of observation.*'

Similar results are reported in dogs with a right-side knee osteotomy that led to a 30° valgus (i.e. a bow-legged deformity) (Panula et al. 1998). Two groups of sixteen dogs (eight controls) were sacrificed at 7 and 18 months. Panula et al. report: '*Thus, in general, the thickness of the superficial zone decreased and that of the deep zone and of the total uncalcified articular cartilage increased. It is noteworthy that these results were from sites where cartilage surface was macroscopically intact.*' They go on to say: '*In the light of earlier reports, the increase of cartilage thickness can be interpreted to represent a hypertrophic repair phase of canine OA*' (Panula et al. 1998).

Similar results have been observed in an anterior cruciate ligament (ACL) transection cat model (n = 6) at 54 months (Clark et al. 2005). Clark et al. point out that the two cartilages in the patellafemoral joint displayed quite different responses: The patella cartilage is grossly thickened with a 'feathered surface,' while for four or six animals the femoral cartilage appeared slightly thickened but otherwise apparently normal with an intact surface. In an exceptional and excellent discussion of the differences in patella and femoral cartilage that may explain their different histological appearances, (Clark 2008) consider three hypotheses: (i) load duration, (ii) material properties of articular cartilage, and (iii) cell responses. Clark remarks:

> The differences between patellar and femoral groove cartilages and their corresponding chondrocytes are numerous. Further consideration of the potential interactions between these properties multiplies the complexity of this 'simple' joint.... It is clear that even in the simple patellofemoral joint, where cartilage interacts directly with cartilage, the mechanical and metabolic interactions are abundant.", [and notes] "Ankle cartilage is more resistant to progressive degeneration and osteoarthritis than knee cartilage. It is also thinner, stiffer, and less permeable and has a higher proteoglycan content compared with knee cartilage. In addition, ankle chondrocytes synthesize more glycosaminoglycan and protein than knee chondrocytes and up-regulate matrix synthesis as opposed to collagen degradation in response to degeneration. It is interesting that a number of these differences between ankle and knee cartilage match those between femoral groove and patellar cartilage."

These interesting observations on femoral versus patella cartilage, and the similarity of femoral cartilage to ankle joint cartilage, suggest that for a similar chondrocyte synthetic capacity, thick cartilage may be more vulnerable to shear damage than thin cartilage (i.e. large shear forces at the contact surface may be a particular vulnerability for thick cartilage relative to thin cartilage). We also note that a vulnerability of a tissue is amplified by 'interactions' within the tissue, potentially becoming a cascading tissue failure, which may explain the striking difference in outcomes for patella and femoral cartilage.

Clearly, abnormal joint loading arising from the ACL transection damages the transitional and superficial zone collagen network in femoral cartilage, while leaving the cartilage surface largely intact (see abovementioned experiments). The femoral collagen network then expands, as predicted, due to 'inflation' of the collagen network by aggrecan within. If this newly damaged and expanded network is now repaired in situ (see discussion of collagen network repair mechanisms Sect. 2.6), then this repaired collagen network will become the new, thicker articular cartilage. Remarkably, such repair appears to be have occurred in vivo in these experiments [see, for example, (Vignon et al. 1983; Clark et al. 2005; Adams and Brandt 1991; Panula et al. 1998)], where it is dubbed 'hypertrophic repair.' Of course, if the rate of mechanical damage happens to exceed the rate of repair, then given enough time the cartilage tissue will eventually degrade (Clark et al. 2005; Brandt et al. 1991b).

See Figs. 1a, b and 2a, b (Vignon et al. 1983): *LHS image: Normal appearance of femoral trochlea dog cartilage. Middle image: Femoral trochlea dog cartilage, 3 months after transection of anterior cruciate ligament. Notice abnormal shear loading results in focal damage to collagen network (indicated by translucent region) beneath the cartilage surface. RHS image: Femoral trochlea dog cartilage, 9 months after transection of anterior cruciate ligament. It is reported that the cartilage grew progressively thicker (about 50% thicker over 9 months). Nine months post-operatively the cartilage appeared normal, showing normal or higher aggrecan content and normal cell density (but greater total cell number, indicating cell proliferation)* All slides magnification × 100 (Vignon et al. 1983).

The experimental evidence provided by Vignon et al. (1983), Clark et al. (2005), Adams and Brandt (1991), Panula et al. (1998) supports the notion of collagen network expansion, followed by repair, and chondrocyte proliferation (and possibly migration with additional repair). Notice that in this injury model the superficial zone of the cartilage remains intact, but thickens with the formation of new tangential collagen fibers in the superficial zone, while the collagen network in the intermediate zone is damaged and loosened [see figures in Vignon et al. (1983)]. To orchestrate this sophisticated repair outcome, see the later discussion on cell–cell signaling between chondrocytes in different zones of the articular cartilage.

There is evidence for human cartilage thickening following ACL injury too (Eckstein et al. 2015). One hundred and twenty-one adults (ages 18–35 years) had an MRI imaging within 4 weeks of ACL rupture and follow-up assessments two and five years later. Some 62 patients had early ACL reconstruction, 30 chose delayed ACL reconstruction, and 29 chose no ACL reconstruction. Total femorotibial cartilage thickness was the focus on the study. It would found that overall, femorotibial cartilage thickness increased by 31 μ per year over a five year period (95% confidence interval 18–44 μ), particularly in the medial compartment and in younger subjects (Eckstein et al. 2015). Importantly, we note that some regions gained in thickness more than average (e.g. central tibia cartilage, medial femoral cartilage), while others thinned (e.g. posterior region of lateral compartment). The region of articular cartilage with a maximum rate of thickness gain was 55 μ per year, and the articular cartilage region with the maximum rate of thinning was −48 μ per year. The rate of change of cartilage thickness did not differ significantly between the first two years and the later 3 years, while younger subjects had significantly greater thickening.

We would not expect such thickening to continue indefinitely; rather, cartilage would eventually stabilize or progress to OA. Importantly, there is some evidence for cartilage healing over a 2 year period, following ACL injury in humans (Chu et al. 2014). Having used ultra-short time echo of T2 signaling [UTE-T2*] to image cartilage (Robson et al. 2003) before anterior cruciate ligament surgical repair and 2 years later, Chu et al. (2014) conclude: '*Through cross-sectional and longitudinal evaluations, this study shows both a potential for intact articular surfaces of cartilage to heal deep tissue injuries and the diagnostic potential of UTE-T2* mapping, a new non-invasive quantitative MRI metric, to track clinical disease states.*' Note that in both these examples of cartilage healing (Vignon et al. 1983; Chu et al. 2014), the articular surface remained intact.

Clearly, the spatial distribution of rates of cartilage thickening is consistent with most internal articular damage being focused on high contact load regions [see Fig. 2 in Eckstein et al. (2015)]. As cartilage thickening is a result of an imbalance between the rate of internal cartilage damage (generating new superficial zone cartilage) and the rate of surface wear, it is not possible to independently estimate the rates of expansion and wear, only their difference. But we do observe that the average expansion rate of 31 μ per year is consistent with the magnitudes of our previously estimated normal cartilage thickness rates of expansion and wear (i.e. normally 30 μ per year collagen expansion, based on estimating one percent of total collagen replaced each year in the transitional zone of the high load region, while estimated normal wear rates are 20–60 μ per year).

Based on the expansion and wear model for normal articular cartilage described above, the data from Eckstein et al. (2015) suggests a *lower bound estimate* of a twofold increase in collagen network damage following ACL injury (e.g. a minimum twofold increase in damage is driven by a doubling in the rate of damage while the wear rate remains unchanged). This too, appears to be a reasonable estimate. We would not expect cartilage thickening to continue indefinitely. As the cartilage thickens, eventually wear rates at the articular surface will increase. Indeed, many of these injured joints will progress to osteoarthritis in ten to fifteen years, and for many of these cases, the first signs will be 'roughening' of the cartilage surface (indicating increased surface friction with its attendant increase in wear). This suggests that roughening of the articular cartilage surface may be a (later) sign that internal collagen network damage rates within articular cartilage have increased.

In this context, we note that when for a large group of OA patients (Kellgren Lawrence (KL) grades 1, 2, or 3 and average age 61), Wirth et al. (2016) report 'thinning rates' (driven by joint narrowing) for the central medial tibiofemoral compartment are around 75 μ per year (i.e. thinning cartilage at this rate would require a lower bound surface wear rate about two to four times greater than our estimated normal wear rate, i.e. range 20–60 μ per year). For patients with OA that progresses in KL grade over a two year period, rates of cartilage thinning increase to about 150 μ per year (Wirth et al. 2016) (i.e. thinning cartilage at this rate would require a lower bound surface wear rate about four to eight times greater than our estimated normal wear rate i.e. range 20–60 μ per year). Our estimates of normal cartilage wear rates appear to sit comfortably with the thinning rates observed in OA reported by Wirth et al., as the lower bound estimates of fourfold to eightfold increases in rates of wear required to obtain these observable changes associated with OA are reasonable. We also mention in passing these estimates agree with the computational modeling estimates of a fourfold and eightfold increase in wear volumes between normal medial tibial cartilage and OA cartilage grades KL2 (fourfold increase) and KL3 (eightfold increase) reported in Liukkonen et al. (2017) [which builds on the modeling work of Mononen et al. (2016)].

Taken together, this extensive set of consist experimental evidence build some confidence in the notion that if there is a large, prolonged step change in environmental loading on a joint, then articular cartilage may respond over a period of months to years, growing or shrinking its thickness (of the collagen network) at a comparatively rapid rate. This is likely to be particularly relevant for younger adults who have high rates of tissue metabolism as a consequence of their hormonal status, as well as leading very active physical lifestyles (e.g. participating in occupations, sports of all kinds, and other activities that place very demanding loads on joints).

Finally, having discovered the intrinsic property of the collagen network in articular cartilage is to expand as it is damaged, it is interesting to consider if the same thing might occur in other types of cartilage, such as auricular and nasal cartilages. These tissues are subject to large deformations, and it seems likely that the collagen network in these tissues are also both mechanically and enzymatically damaged, and so they too are expected to expand over time. However, unlike articular cartilage, these tissues have no clear mechanism for removing collagen network via surface wear mechanisms, as does articular cartilage. By this reasoning, one would expect in auricular and nasal cartilage, having cartilage volume production in the absence of cartilage volume reduction, will result in a net enlargement of auricular and nasal tissues over time.

The proposed cartilage damage-expansion mechanism may explain the well-documented fact that cartilaginous ears and noses grow in size with age. For example, Sforza et al. (2009) report that the vertical ear length increases about 0.3 mm per year (Sforza et al. 2009). Sforza et al. (2011) also find the volume of the nose continues to increase with age (with about a 20–25% increase in volume between age 25 and 65 years) (Sforza et al. 2011), while Ito et al. (2001) report the all the dimensions of human auricle increased with age (length, thickness, lobe length), suggesting a volumetric expansion of the ear (Ito et al. 2001).

Unfortunately to date, there is no conclusive evidence available to support or refute this proposed explanation for progressive auricular and nasal cartilage growth (i.e. the growth of the ear may possibly be due to continued interstitial or appositional cartilage growth). But we do note that (Ito et al. 2001) also suggest disruption of elastic fibers as one cause for ear growth, having observed histologically elastic fiber fragmentation that increases with age. We also note that auricular cartilage contains a majority of chondrocytes expressing α-actin (Kana et al. 2006), which is consistent with auricular cartilage also being a dynamic tissue, subject to active collagen network remodeling by chondrocytes as the auricular collagen network is damaged (even though such ECM remodeling does not actively reduce ear or nose volume, ECM repair is ongoing in these tissues—see further discussion of α-actin positive cells and remodeling of the articular cartilage collagen network via cell migration in Sect. 2.9).

3.6 The Cartilage Surface

By themselves, shedding of aggrecan and hyaluronan is probably insufficient to maintain the integrity of the cartilage surface, as demonstrated by observations on PRG4 knock-out mice (Jay et al. 2007). Knockout of the molecule PRG4 (also known as 'lubricin' or 'superficial zone protein') shows it to be required for the maintenance of stable, functional articular contact surfaces in mice, and one would image this to be true for larger sized animals too. PRG4 aids joint lubrication in the 'boundary lubrication regime' (see Chap. 4), reducing mechanical interlocking of the collagen fibers and abrasive wear of the cartilage surface. In PRG4's absence, the cartilage surface roughens quickly and develops shallow, regularly spaced splits normal to the cartilage surface. Cartilage surface degradation rapidly progresses to osteoarthritis (Jay et al. 2007; Coles et al. 2010). These findings indicate PRG4 plays an important role in the boundary lubrication regime for mice, and almost certainly, it does for other animals too. We mention that PRG4 concentrations decrease in the synovial fluid of OA patients and in the synovial fluid of temporomandibular joint derangements (Leonardi et al. 2016). What special properties does PRG4 possess that enables it to fulfill its function as boundary lubricant for articular cartilage?

PRG4 is produced by chondrocytes in the superficial zone of the cartilage (Ogawa et al. 2014) and by cells in the synovial membrane (PRG4 is also expressed in tendon). The molecular weight of PRG4 is around 206 kDa, making it a reasonably large molecule. The normal concentration of PRG4 in synovial fluid is in the range 0.02 mg/mL (Swann et al. 1977) to 0.35 mg/mL (Neu et al. 2010). It has a comparatively long half-life in the joint cavity of around six days (Jay et al. 2010). The size of the molecule suggests a shorter half-life (e.g. albumin (67 kDa) has a half-life of about 1 day), but this does not take into account binding of PRG4 to collagen fibrils—its stickiness probably explains it slows turnover in the joint cavity [indeed, radiolabeled PRG4 is detectable up to 28 days following intraarticular injection (Jay and Waller 2014)].

What is the structure of PRG4 that enables it to reduce boundary friction so effectively? PRG4 has a molecular weight of around 206 kDa and is approximately 'half protein' and 'half oligosaccharides.' The 1404 amino acid protein has four main domains: a PEX domain at one end. It is likely the so-called four-bladed propeller comprising the PEX domain binds to the type II collagen in the superficial zone, as well as binding to type II collagen exposed on the cartilage surface. The PEX domain serves much the same functional purpose they do for MMPs. Next to the PEX domain there is a large mucin-binding domain along PRG4's mid-region (attached mucins are extensively O-glycosylated), and finally heparin binding and somatomedin B domains at the other end of the molecule. It is noted that the binding of PRG4 to the cartilage surface is prevented at very high salt concentrations, suggesting the interactions between the PEX domain and collagen are electrostatic in origin (Gleghorn et al. 2009). It is also noted that the amino acid sequence of PRG4 is a paralogue of vitronectin, while the amino acid sequence in the four-bladed propeller region of the PEX domain is related to those four-bladed propeller regions found in vitronectin and MMPs (Rhee et al. 2005a). The PRG4 is about 220 nm in length (Zappone et al. 2008), which in the mouse is approximately the thickness of the so-named 'lamina splendens' (Jay et al. 2007). The lamina splendens is a thin, acellular layer (of probably glycoproteins) found on the surface of articular cartilage [see Fig. 2.26 and Figs. 3, 4, 5, 6, 7, 8, and 9 in Kobayashi et al. (1996)].

It is plausible that the lamina splendens of articular cartilage form around a layer of PRG4 molecules attached to the cartilage surface. The firmly attached PRG4 molecules are joined by a variety of other molecules including aggrecan (either free in the joint space or partially embedded in the cartilage surface), hyaluronan (either free in joint space or partially embedded in the cartilage surface) as well as a variety of other molecules including albumin, nectins, macroglobulins and ECM debris. Kneaded together by contact forces, they form a protective, hydrated boundary layer on the surface of the cartilage (see the schematic depiction Fig. 3.11 (and also Fig. 4.7) of the cartilage–cartilage interface).

A detailed understanding of joint lubrication for bearings made of compliant, saturated porous materials such as cartilage is contentious (for example see http://blacklistedbiomechanics.com/biphasiclubrication.htm), for there is no close analogue for these types of bearings employed in industry, and so this bearing type has not become the subject of intense industrially sponsored research, as have most other types of bearing (Hamrock et al. 2004).

It is likely that in normal operating conditions in animals, lubrication of the contacting surfaces in diarthrodial joints is mainly operating in a so-called mixed mode of lubrication. Mixed mode lubrication means there is a mixture of hydrodynamic lubrication (i.e. two surfaces are held apart by viscous forces in the fluid) and boundary lubrication (i.e. lubrication at asperity contacts) (Gleghorn and Bonassar 2008). See further discussion Chap. 4.

McCutcheon suggested in the late 1950s that so-called weeping lubrication may make an important contribution to mixed-mode lubrication. 'Weeping' refers to the load-driven exudation of pore fluid from the cartilage tissue into the interfacial region between contacting cartilage surfaces (i.e. fluid exudation into the 'contact interface'). Once there, the fluid exudate helps to support the load between the contacting surfaces, extending the duration of hydrodynamic lubrication and so the duration of mixed-mode lubrication (Lewis and McCutchen 1959; McCutchen 1983, 1998, 2011).

Weeping of pore fluid to the contact interface results from an interaction between the boundary and hydrodynamics modes of lubrication that only becomes more prominent as porous bearing surfaces are relatively rough. It turns out that by the standards of manufactured bearings, the cartilage surface is relatively rough (Smyth et al. 2012). Walker et al. suggest that normal cartilage is 3–15 rougher than engineering bearing surfaces, and osteoarthrosic cartilage was found to be between 20 and 200 times as rough (Walker et al. 1968). There is some evidence that surface roughness varies between joints (Smyth et al. 2012). It is this surface roughness that enables weeping lubrication to play a significant role in the lubrication of diarthrodial joints (see discussion Chap. 4).

While on the subject of joint lubrication, it is noted that aggrecan and hyaluronan molecules 'poke' through the surface of the cartilage (see Figs. 3.11 and 4.7), and they also help obstruct the removal of pore fluid from the contact interface region. Aggrecan (and to a lesser extent hyaluronan) does this by reducing the lateral transmissivity of the pore fluid along the contact interface. This lateral resistance to fluid flow between the rough contacting surfaces also lengthens the duration of hydrodynamic lubrication. This is one means by which the aggrecan flux through the cartilage surface helps extend the duration of hydrodynamic lubrication, so maintaining the fraction of mixed-mode lubrication that may be attributed to hydrodynamic lubrication. In other words, the presence of aggrecan (and hyaluronan) at the cartilage surface changes the shape of the 'Stribeck curves' (Lee et al. 2013a). This is discussed in more detail in Chap. 4.

For cartilage tissue to be sustainable in vivo, additional 'surface issues' need to be managed. Importantly, chondrocytes close to the surface of the cartilage are in a rather vigorous 'mixing zone,' in the sense that this region of cartilage tissue is being variably influenced by events taking place deep within the cartilage (via diffusion or advective transport toward the surface of extracellular matrix components produced or degraded deep within the cartilage, which exit the cartilage surface) *and* by events taking place in the synovial membrane and synovial fluid (via diffusion or advective transport of molecules from the synovial fluid into the cartilage). While this 'mixing zone' is an unavoidable consequence of enabling transport of much-needed resources into the cartilage tissue so as to sustain it, clearly unwanted or damaging molecules may also enter the cartilage, placing the cartilage tissue at considerable risk. For once molecules have entered the mixing zone, they influence the local environment experienced by the chondrocytes for better or worse (Carlsen et al. 2008; de Grauw et al. 2009; Fell and Jubb 1977).

It is clear there are many possible molecules in the synovial fluid that are likely to cause damage to the cartilage tissue itself (e.g. there are a variety of inflammatory-related molecules, while concentrations of key signaling molecules within the tissue may also be disrupted). One important example of potentially damaging group of inflammatory molecules is those associated with the powerful complement system.

The complement system is employed by the innate immune system, which aims to identify inflammatory 'targets' deemed potentially harmful to the organism. Once identified, complement amplifies the immunological defense responses to the identified target. It is noted that the complement system is activated by some fragments of aggrecan degradation (Furst et al. 2013) and by collagen type II (Hanauskeabel et al. 1982). However, other molecules produced throughout the cartilage ECM, such as SLRPs and the C-lectin domain of collagen IX, help suppress complement activation (Sjoberg et al. 2009). Some of the molecules that inhibit the complement system are produced specifically by chondrocytes in the superficial zone. Notable examples of complement inhibitors are vitronectin and clusterin (Khan et al. 2001; Jenne and Tschopp 1992).

We see that is a competition between activation and inhibition of the complement system in the superficial layers of articular cartilage. Failure to inhibit the complement system can lead to cartilage destruction. For example, a missense mutation in the C-type lectin domain found on the ends of the collagen IX molecules leads to osteochondritis dissecans (Stattin et al. 2010). It has been shown that osteochondritis dissecans arises in these patients because the C-type lectin domain of collagen IX no longer suppresses complement activation within the cartilage (Furst et al. 2013).

Further, inflammatory processes elsewhere in the body, or (bacterial) sepsis (Bao et al. 2012) or inflammation of the synovial membrane itself, can be disastrous for the integrity of the articular cartilage. Clearly, articular cartilage is susceptible to damage, because the synovial fluid is formed as an exudate of blood plasma. So though not yet experimentally reported, it is conceivable that an excess of signaling molecules generated from say bone tissue (as occurs in diseases such as Paget's disease) may travel via the blood to the synovial fluid (just as inflammatory molecules from distant sites of inflammation do), where they may influence the coordination of processes within articular cartilage. In recent years, it has been proposed that rheumatoid synovial fibroblasts can enter the bloodstream and migrate to a new joint, spreading the rheumatic arthritic disease from the joint to joint (Hillen et al. 2017). Undoubtedly, much remains to be discovered about long distance interactions between organ systems and articular joints in health and disease.

However, using 'first principles,' it is clear that the greater the flux of aggrecan and hyaluronan out of the cartilage, the smaller the extent (depth) of the mixing zone next to the cartilage surface (as there is greater 'flushing' of the mixing zone). In other words, an increased flux of aggrecan from the cartilage enables the chondrocytes close to the surface to experience a more normal cartilage microenvironment. By this reasoning, we predict that aggrecan and hyaluronan 'flushing' of the mixing zone is protective of articular cartilage.

It therefore follows that young animals with high rates of aggrecan and hyaluronan synthesis would be expected to be less prone to damage arising from an abnormal composition of the synovial fluid. In this context, we note that the most frequent age of onset for rheumatoid arthritis is middle age, which correlates with a decline in both the rates of aggrecan synthesis and the quality of aggrecan produced by chondrocytes (Quinn et al. 2006; Bizzaro et al. 2013).

3.7 Dynamics of the Calcified Cartilage Zone

The cartilage calcified zone (CCZ) lies between the so-named 'tidemark' (a thin region a few microns thick at the interface between the CCZ and the hyaline cartilage) and the 'cement line' (another thin region a few microns thick at the interface between the cartilage and the subchondral bone).

The tidemark is so named because '*Fawns and Landells likened this landmark to the irregular line of debris that is left on a sandy beach by the high point of a now-receding tide*' (Simkin 2012). Simkin suggests the 'debris' is the remains of apoptotic cells (Simkin 2012). At the cement line, there is a sharp demarcation between type II collagen present in articular cartilage and type I collagen present in the subchondral bone.

The thickness of the CCZ tends to decrease with increasing age. In articular cartilage of the femoral head and head of humerus, the CCZ is about 200 μ thick at age 20, but this decreases linearly with age to about 100 μ at age 80 (Lane and Bullough 1980). However, the number of 'tidemarks' in the CCZ increases with age [see Fig. 1 in Simkin (2012)], averaging one at age 20, but increases to an average of two at age 80. It is noted that the CCZ thickens in early OA, but decreases in late stage OA (Hwang et al. 2008). Further data on CCZ thickness is provided by Hoemann et al. (2012). There are various possible roles of the CCZ in articular cartilage, which we now consider. First, it is possible that the CCZ allows the thickness of the cartilage to be dynamically adjusted in response to environmental loadings. This may partly explain the presence of a CCZ in normal cartilage (e.g. see, for example, mouse CCZ shown in Fig. 1 in Hughes et al. (2005), for the CCZ permits fine adjustments to be made in the functional thickness of deep zone cartilage, just as adjustments are made to the superficial zone thickness (though adjustments are made by completely different mechanisms). Such reversible fine adjustments to function cartilage thickness offered by the CCZ have the additional advantage of being dynamic. For if additional cartilage thickness is required again (perhaps due to excessive wear at the cartilage surface), 'reserve cartilage' can be recruited and pressed into service via 'retreat' of the CCZ zone via decalcification.

See Fig. 1 in Hughes et al. (2005): *Adult mouse articular cartilage fixed either in formaldehyde (figures A–D) decalcified, embedded, sectioned, stained in picrosirius red and imaged by LM (figures A and B) and PLM (figures C and D), or cryofixed (figures E and F), dried, fractured, and imaged by SEM to create red(L)-cyan(R) anaglyph stereo-pair images (6° tilt). The chondrocyte (C) orientations indicate the zonation of the tissue in the LM images (figures A and B), and there are also clear zonal differences in the collagen matrix shown in the PLM images (figures C and D). I, II, III, and IV indicate the positions of the superficial, intermediate, deep, and calcified zones, respectively, and SB indicates the position of the subchondral bone. The tidemark boundary between the deep and calcified zones (open arrows) by each of the microscopic methods is clearly distinguished. More intensely stained (light microscope) and birefringent (polarized light microscope) regions of the collagen matrix can also be observed, located adjacent to the chondrocytes and their surrounding pericellular capsules and representing the regions of territorial matrix (TM) (closed arrows). The TM surrounds columns of cells, apparent in all the images. The SEM images (figures E and F) show both the matrix and chondrocyte orientations with the superficial zone arching over the deeper zones of the tissue (figure E) and lifted away, exposing the underlying chondrocytes (figure F).*

Practically, decalcification of the CCZ can be easily effected by decreasing cartilage pH. For example, pH in the deep zone decreases as anaerobic metabolism is 'ramped up.' When deep zone chondrocytes are required to produce more aggrecan and collagen (e.g. in the event of even more severe synthetic demands, they may even be required to proliferate), these chondrocytes will require more metabolic energy for increased synthesis. Because chondrocytes obtain at least part of its energy from anaerobic metabolism of glucose, producing more lactic acid will decrease local pH. A decrease in pH dissolves calcium deposits, which thins the CCZ. This 'metabolic mechanism' for adjusting the thickness of the CCZ and deep zones fits nicely with helping to match the functional capability of the cartilage tissue to the environmental demands placed upon it.

It is likely that such dynamic adjustments of cartilage metabolism (and so CCZ thickness) occur over adulthood in response to environmental stressors. Variations in rates of deep zone chondrocyte metabolism and synthesis may explain the presence of multiple tide-marks in the cartilage (presumably due to repeated recession and advancement of the tidemark), which tend to increase in number with age. Unfortunately, tidemarks do not provide a complete record of stress events because while tidemarkings will faithfully record a series of stressors of decreasing severity, doubtless they provide a much less reliable record of stress events of increasing severity. The tidemark record is therefore probably an incomplete historical record of past events experienced by articular cartilage, which lessens its potential prognostic value.

Secondly, it is important to realize that it has been shown that the CCZ increases the interfacial shear resistance of cartilage to bone (St-Pierre et al. 2012). Presumably, tying the collagen type II network more firmly to its base is advantageous by reducing the likelihood of (collagen) 'pull-out failure.' Improvement in resistance to shear failure is probably better explained by more gradually matching the compliance of collagen in shear, to the compliance of bone at collagen's insertional support, via a gradient in compliance through the CCZ, thereby reducing the likelihood of shear failure of collagen fibrils. For if the CCZ was not present, collagen fibrils would inserted directly into bone, which is several orders of magnitude harder than collagen. Such an abrupt, large change in material stiffnesses between cartilage collagen and bone reduces resistance to shear failure of the collagen fibers. This reduction is shear resistance arises because strain concentrations in the collagen fibers at the osteochondral interface may become large enough to mechanically disrupt covalent bonds in the collagen fibrils. In other words, when the cartilage is subject to shear without a CCZ, the bone at the osteochondral junction acts much like a knife blade on the softer (collagen network and) cartilage, leading to structural failure at the osteochondral interface. Once such a failure begins, shear stresses on the remaining intact interface are amplified, which accelerates further failure, and this positive feedback damage process quickly grows to cause extensive failure. This is detected as 'osteochondral delamination' (see Fig. 1 in Levy et al. (1996), which demonstrates delamination at the osteochondral interface may be surprisingly extensive, due to the deleterious positive feedback damage process). Says Levy et al: '*The incidence of chondral injuries in these soccer players may be consistent with the concept that these high velocities and repetitive pivoting decelerations that the knee experiences are capable of placing extreme stress on the articular cartilage*'). In simple terms, the presence of the CCZ, with its intermediate shear stiffness, helps to blunt bone sharpness when during extreme shear states, it acts as a 'knife blade' that delaminates the overlying cartilage.

Third, the CCZ has been shown to reduce solute transport between hyaline cartilage and the subchondral bone, and vice versa. It is apparent that the CCZ normally completely prevents the transport of very large molecules like aggrecan from cartilage to bone. In other words, the CCZ helps create an impermeable boundary for large molecules at the interface between cartilage and bone. In theory at least, calcium precipitates may also help 'heal' any bone microfractures at the bone surface, again helping to create an impermeable osteochondral boundary. Though not reported in the literature, mineral precipitation may be an important mechanism for maintaining functional separation of normal cartilage from subchondral bone tissue (and vice versa).

However, it has been shown experimentally that calcified cartilage is permeable to low molecular weight solutes such as fluorescein and rhodamine (M.W. around 400), starting from either the superficial or subchondral sides. The effective diffusion coefficient of these two solutes in equine CCZ was of the order of 9×10^{-9} $cm^2\ s^{-1}$, about fivefold less than for uncalcified hyaline cartilage (measured with a diffusion coefficient of 4.5×10^{-8} $cm^2\ s^{-1}$) (Arkill and Winlove 2008). It is noted that dilute concentrations of small ions in water have (free) diffusion coefficients of the order of 1×10^{-5} $cm^2\ s^{-1}$, while the (free) diffusivity of fluorescein and rhodamine in water or phosphate buffer solution (PBS) is of the order of 2.5-4.2 $\times\ 10^{-6}$ $cm^2\ s^{-1}$ [e.g. Lee et al. (2011)]. So the diffusivity of fluorescein and rhodamine is about 100 times less in equine hyaline cartilage compared to its diffusion coefficient in pure water. In a murine study of diffusion in the CCZ (Pan et al. 2009), fluorescein diffusivity in the CCZ was measured as $0.7 \pm 0.3 \times 10^{-9}$ $cm^2\ s^{-1}$ (i.e. about 10 times less than that reported for equine cartilage), while the fluorescein diffusion coefficient in murine hyaline cartilage was measured as $2.6 \pm 2.2 \times 10^{-9}$ $cm^2\ s^{-1}$ (i.e. about four times larger than the diffusion coefficient for the mouse CCZ). So this evidence suggests the CCZ does help reduce any potential transport of small molecules through the osteochondral junction.

It is also of considerable interest to identify the permeability of water at the osteochondral junction complex. Testing articular cartilage tissue obtained at arthroplasty (mean age 71 years, $n = 16$) and from a tissue bank (mean age 24 years, $n = 12$) shows that normal subchondral bone plate (which is this particular study is defined to include the calcified cartilage zone together with the underlying subchondral bone (see Fig. 3.12)) has a measured fluid permeability of 90 mm^2/MPa s, about 350–600 times greater than the hyaline cartilage itself (which is reported to be 0.15 mm^2/MPa s) [see Table 1 (Hwang et al. 2008)]. What this data implies is that the usual modeling assumption of an 'impermeable

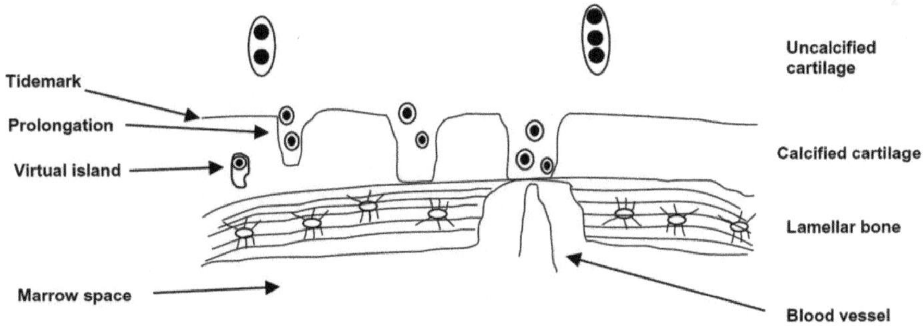

Fig. 3.12 Revised representation of the components of the chondro-osseous junctional region of articular cartilage (based on the serial sections of knee cartilage from a 27-year-old male). This shows a virtual island, one prolongation of the uncalcified cartilage into the calcified cartilage layer, one prolongation that abuts the bone, and one prolongation that abuts the marrow space containing a blood vessel (Lyons et al. 2006). This model of the chondro-osseous junction appears to be supported by other studies [see, for example, Figs. 2 and 5 (Hwang et al. 2008)]

boundary' at the osteochondral junction is probably incorrect, opening up the possibility of significant fluid flow across the osteochondral junction. If true, this has very significant implications for the time to complete consolidation of articular cartilage when loaded (i.e. we note that halving the length of the drainage path leads to a fourfold decrease in the time for consolidation).

Analysis of the CCZ shows that it contains an organic phase that is mainly composed of collagen type II and an inorganic phase that is a mixture of hydroxyapatite, calcium phosphate and calcium carbonate (Zhang et al. 2012c). It is reported that the mineral crystals in the CCZ are typically small. Small crystals have a large surface area so they are more readily dissolved, or grow, conditions permitting. In other words, the small mineral crystals in the CCZ are labile. This fits with the CCZ being dynamically adjusted depending on local (chemical) conditions around the mineral crystals.

Tetracycline can be used to identify newly mineralized tissues. The binding of tetracycline does not depend on cells in the tissue—rather, there appears to be a direct reaction between tetracycline and one or more inorganic components of the mineralized tissue. The most likely mineral species binding to tetracycline in bone is hydroxyapatite, $Ca_{10}(PO_4)_6(OH)_2$ (Perrin 1965), but there is also direct labeling of calcium phosphate and calcium carbonate. Tetracycline labeling shows the tidemark in cartilage slowly advances in the direction of the non-calcified cartilage. But analysis of double-tetracycline labeled cartilage shows that this is not an appositional phenomenon like that occurring in bone, for when double-labeling occurs, several tidemarks can be labeled at the same time. (see Fig. 3.13). Each tidemark may therefore be depositing hydroxyapatite, calcium phosphate, and

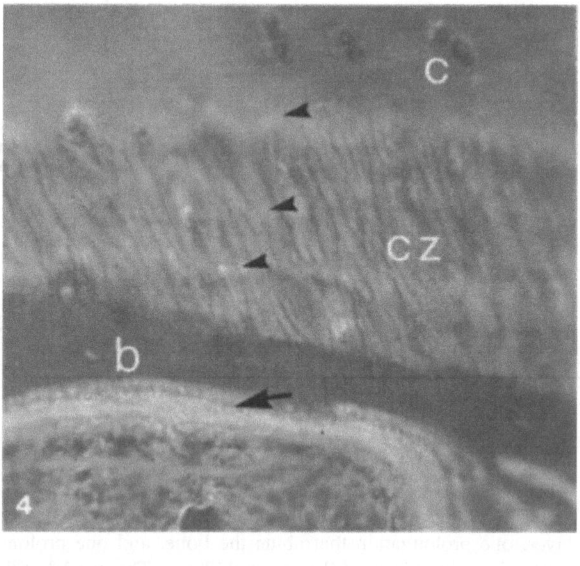

Fig. 3.13 Medium power view of tetracycline labeling of multiple tidemarks (arrowheads) in osteoarthritic cartilage (C = hyaline cartilage, CZ = calcified zone, and b = bone) (Revell et al. 1990)

calcium carbonate. In other words, it need not be just the tidemark adjacent to hyaline cartilage that incorporates calcium, but several tidemarks can do so simultaneously (Lane and Bullough 1980).

This makes sense because the density of crystals in the calcified cartilage zone does not preclude penetration of ions that can form new crystals at any level within the CCZ. It is therefore possible for the same or different minerals to be deposited at different depths in the cartilage. The mineral phases are labile (e.g. calcium carbonate and hydroxyapatite are strongly influenced by the pH in the deep zone cartilage). So the mineral phases may also redissolve and go into solution and then reprecipitate at a new location within the CCZ depending on local factors influencing precipitation (e.g. existing crystal size and composition, local pH, local ionic composition, and presence or absence of crystal 'nucleating factors').

It is apparent that the CCZ is a three-dimensional structure that dynamically evolves over time (Hoemann et al. 2012). Normally, there are no blood vessels (or the very occasional one) beyond the CCZ penetrating into the deep zone cartilage (Bonde et al. 2005). In some cases, the distance between the bone marrow and BMUs, and articular cartilage can become very small, permitting the movement of signaling molecules between the tissues (along preferential transport pathways including Haversian canals and canaliculi), and even permit the loss of mers and aggrecan from the cartilage to the bone marrow (see Fig. 3.14). Clearly, the flow could be bi-directional, allowing the bone marrow ECM and signaling molecules to enter articular cartilage. Such transport would only be exacerbated if random heavy loadings lead to microfractures through the remaining CCZ and subchondral bone (microfractures in the cortical bone are probably invariably associated with bone marrow lesions in the trabecular bone, though this remains to be confirmed), permitting fluid flow between the cartilage and bone compartments (van Dijk et al. 2010). Once fractured, this would trigger a remodeling response in healthy bone, activating basic multicellular units (BMUs), which then set about repairing the damaged bone, but this repair process (which involves opening up substantial resorption cavities) may themselves temporarily open up connections between bone and cartilage further.

Of course, once contiguous pathways exist (e.g. along Haversian canals and canaliculi and along microfractures), signaling molecules like PTHrP may be transported along the pathways through the fluid. PTHrP from cartilage may then lead to bone resorption in bone and/or promote the formation of bone marrow lesions (BMLs). But the exact role of the osteochondral junction plays in health and disease is still being evaluated. Says Yuan et al. (2014),

> The perspective that calcified cartilage and bone is an insurmountable and impermeable barrier to soluble molecules and functional interaction has changed, and new evidence indicates the important role played by the bone–cartilage biomechanical unit in the development and progression of OA.

It is important to note though because the CCZ is labile, upon the cessation of damaging loading, any microfractures appearing between cartilage and bone may over time 'self-heal,' as new minerals are deposited within the microfracture. We

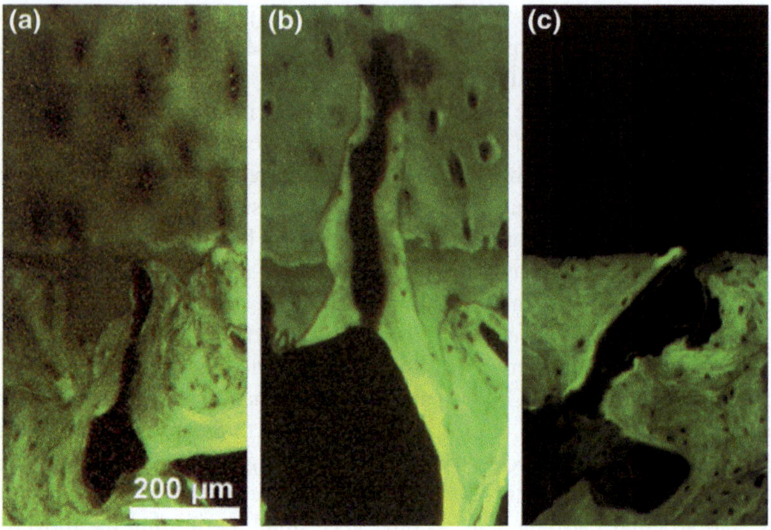

Fig. 3.14 Typical vascular canals penetrating **a** into calcified cartilage in normal chondro-osseous junction, **b** into deep zone cartilage in partially eroded chondro-osseous junction, and **c** to the surface of fully eroded chondro-osseous junction (Hwang et al. 2008). Note in **a** the blood vessel within a BMU penetrates to the CCZ and then terminates (termination is driven by exposure to a high OPG/RANKL ratio in the cartilage). But in OA, BMUs from the subchondral bone tunnel through the CCZ and into the hyaline cartilage in the deep zone (progression is facilitated by low OPG/RANKL ratio in the cartilage), forming a bony canal with direct communication from the bone marrow to the hyaline cartilage (Bonde et al. 2005)

begin to perceive that the mineral dynamics of the CCZ may serve a variety of useful functions for maintaining long-term cartilage homeostasis and so long-term cartilage integrity.

It is noted here that low concentrations of PTHrP within the cartilage lead to mineralization (calcification) of articular cartilage (Chen et al. 2008), while elevated levels of systemic PTH (as occurs in hyperparathyroidism) may also lead to calcification of the cartilage (Pritchard and Jessop 1977). This is yet another example of signaling molecules concentrations, either too low or too high, leading to tissue pathology. In short, all signaling molecules need to be within their 'normal operating range' for articular cartilage to function normally.

3.8 Dynamics of Subchondral Bone

Bone is distinguished by its highly mineralized extracellular matrix. But it is a biocomposite material comprised of mineral and collagen, just as cartilage is a composite material of aggrecan and collagen. When bone is demineralized (e.g. via mineral resorption, e.g. via exposure to weak acids such as acetic acid or citric acid), it is found that the remaining 'protein skeleton' can no longer carry significant compressive loads. On the other hand, if the collagen protein in bone is destroyed (for example by heating bone), the bone can no longer carry tensile loads, and in this case, the bone simply crumbles when loaded.

These experimental findings demonstrate that bone (like cartilage) only has its useful functional properties when it is a complete biocomposite material. Much like reinforcing steel in 'reinforced concrete,' or the collagen network in cartilage, the collagen network in bone primarily carries the tensile loads in the bone tissue. And much like cement in reinforced concrete, or aggrecan in cartilage, the mineral crystals of hydroxyapatite and calcium carbonate in bone primarily carry the compressive loads. So each component of bone has a specialized mechanical role. Of course, the compressive deformations in bone are much less than the compressive deformations in cartilage, because the small strain stiffness of mineralized cortical bone is around 10–20 GPa, which is much greater than the large strain stiffness of aggrecan in cartilage at around 10–20 MPa. In other words, bone is at least 1,000 times stiffer than cartilage.

Bone is normally classified as being either 'dense' bone (sometimes dense bone is referred to as 'cortical' bone) or 'trabecular' bone (sometimes trabecular bone is referred to as 'cancellous' or 'spongy' bone). In long bones, cortical bone normally encases the trabecular bone (which in turn contains (red or yellow) marrow—the yellow marrow is mainly adipose tissue, while the red marrow is mainly myeloid tissue). As an alternative to dense bone, trabecular bone has less strength than dense bone, but has the advantage of being considerably lighter. A reduction in bone weight provides metabolic efficiencies in locomotion, as well as gains in speed and agility. Such gains in speed and agility become very significant in light of the evolutionary 'fitness competition' between species. However, if a bone becomes too light, the risk of bone fracture increases dramatically. So there is a clear risk-cost trade-off between bone strength and lightness that can be realized by each bone having different a fraction of dense and trabecular bone (e.g. the ossicles bones in the ear are solid, while the femur and tibia are mainly trabecular bone to reduce weight as much a practically possible without running too great a risk of bone fracture).

Between the 'plate-like sheets' and 'strut-like rods' of the trabecular bone resides bone marrow tissue. Bone marrow comprises a large number of hemopoietic cells types producing circulating blood cells including erythrocytes and platelets, granulocyte cells (including neutrophils, basophils and eosinophils), as well as containing lymphoid tissues, including B and T cells. With advancing years, the haemopoietic cells are gradually replaced by increasing amounts of fat cells. All these cells in marrow are mechanically protected by the cortical bone and supported by the trabecular bone.

Within both dense and trabecular bone is a network of bone cells called osteocytes. The osteocytes communicate with each other and the bone marrow space via processes extending through small canals known as canaliculi (which are about 0.1–0.2 μ in diameter). The osteocytes obtain nutrients and dispose of waste products via these canaliculi. Bathed within the fluid within the canaliculi are the osteocyte processes, so the osteocytes can directly communicate with each other, and they form a dense communication network throughout the entire bone. The network of osteocytes almost certainly detects mechanical loading and damage to the bone (e.g. the osteocyte processes may be stretched by (shear forces upon) fluid flow and are severed as the bone is damaged and fractured, which allows the osteocytes to indicate the present state of the bone matrix to adjoining cells via signaling molecules).

The osteocyte network can detect systemic hormonal signals (e.g. thyroxine, PTH, calcitonin, glucocorticoids, and the sex hormones) and respond appropriately to these signals by adjusting the ion concentrations in the bone compartment and in the blood compartment (e.g. increases in PTH cause bone resorption, thereby releasing Ca^{2+} ions into the bloodstream). Of course, there are many local signaling molecules in bone too. Some of the most important signaling systems in bone are the RANK-RANKL-OPG system, the PTHrP system, the Wnt systems, and of course the TGF-β and the IGF systems. It will be noticed that all these signaling systems are also present in articular cartilage, as they are in many tissues.

The osteocyte processes contact bone lining cells, which cover the surface of bone. The bone lining cells on the surface of the bone normally form a continuous sheet of cells over all bony surfaces. This cell sheet serves to separate both trabecular bone and dense bone compartments from the bone marrow compartment. This is important because there are many signaling systems within the bone marrow that serve different functions to those in bone. For example, the RANK-RANKL-OPG system is used for communication between lymphocytes in the bone marrow, while in bone it is being used for signaling between osteocytes and BMUs, and between osteoblasts and osteoclasts. In the absence of bone lining cells, signaling 'cross-talk' may result in dysregulation of the bone tissue signaling and immune systems. Just as for bone and cartilage, this example again highlights the crucial importance of maintaining compartmental integrity if there is to be normal organ/tissue homeostasis and function in both tissues.

For dense and trabecular bone in animals larger than about the size of a rabbit, the osteocyte network alone cannot provide sufficient nutrients through the canalicular system connected directly to the bone lining cells and marrow space. Despite increasing canaliculi number (which is one way of increasing transport fluxes between two regions), diffusion distances become too large as the bone thickens, which is necessary to cope with the increased forces as an animal grows larger. To overcome the nutrient flux limitation to osteocytes throughout the bone, blood vessels have to penetrate the dense and trabecular bone. So blood vessels form a network within the bone tissue, as they do in most tissues. The blood vessels reside in bony canals. The canals normal to the long axis of long bones are known as Volkmann canals, while canals parallel with the long axis of the bone are known as Haversian canals (typically Haversian canals are 20–40 μ in diameter, while Volkmann canals are somewhat larger). The vascular system is connected to periosteal and endosteal bone surfaces via the Volkmann canals, where arteries pierce the bone surface. Of course, to maintain compartment integrity, bone lining cells line the inside of both Volkmann and Haversian canals (except at BMUs where bone resorption is taking place—see below).

Repeated loading of bone inevitably leads to tissue damage (as it does in all tissues). Bone damage principally involves the formation of microfractures. If left unrepaired, the microfractures grow in number and length upon repeated loading, and eventually, the microfractures join together to form a 'macroscale fracture.' Bone is then said to have a 'stress fracture' or be overtly 'broken.' As is well known, broken bones lead to functional and structural deficits, and they are usually associated with significant pain, because bone is well innervated. We mention in passing that nerve signals normally play some role in maintaining bone homeostasis (Elefteriou 2005), as they do in most tissues.

To avert microfractures coalescing into a macrofracture, both cortical and trabecular bone is continuously repaired by coordinated groups of specialized bone cells called 'basic modeling units' (or BMUs) (Frost 1987b). Near the front of an advancing BMU, precursor osteoclasts exit a growing blood vessel and through a variety of signaling systems mature into osteoclasts. The osteoclasts move to the front of the BMU and begin to secrete hydrochloric acid (which dissolves bone minerals). Osteoclasts also secrete proteases that enzymatically cut collagen fibrils in acidic environments, primarily via the protease cathepsin K. Acting together, acid and collagenase entirely resorbed the bone matrix leaving a 'resorption cavity' (typically about 200–250 μ in diameter) (see Fig. 3.15). Further back in the advancing BMU, progenitor osteoblasts proliferate around the blood vessel, migrate radially/laterally toward the wall of the resorption cavity, and then attach themselves to the wall of the cavity and mature into osteoblasts (see Fig. 3.15). The osteoblasts then secrete osteoid in distinct layers known as bone lamella. Within days, the osteoid in each lamella starts to mineralize, and this mineralization process is usually 90% complete within a month, though further mineralization of the osteoid may occur over a number of years.

In long bones, the BMU usually migrates in the direction of the average maximum principal stress in the bone (this is usually along the axis of a long bone); however, BMUs may be deviated from this general course, under the influence of gradients in signaling molecules (such as RANKL and OPG), which are released by the osteocytes in damaged bone. BMUs also deviate in the direction of resorption by local stress fields—BMUs tend to migrate in the direction of the principal stresses—so if the local stress field changes, BMU direction changes. The speed of BMU migration is usually of the order of 10–40 μ per day (meaning they can traverse about 1 mm of bone every 1–4 months).

Bone repair also operates in trabecular bone, much as it does in dense bone. However, the details are somewhat different due to the local environment and nature of the repair required. In trabecular bone, bone lining cells first 'lift' from the bone surface adjacent to bone damage forming a 'canopy' (i.e. the bone lining cells remain contiguous so as to always maintain bone and marrow compartment separation as far as possible) (Wesseling-Perry 2014), and a blood vessel penetrates the canopy and enter the 'bone compartment.' Haemopoietic progenitor or 'monocyte' cells travel in the blood and exit the tip of a (leaky) blood vessel, where in the right

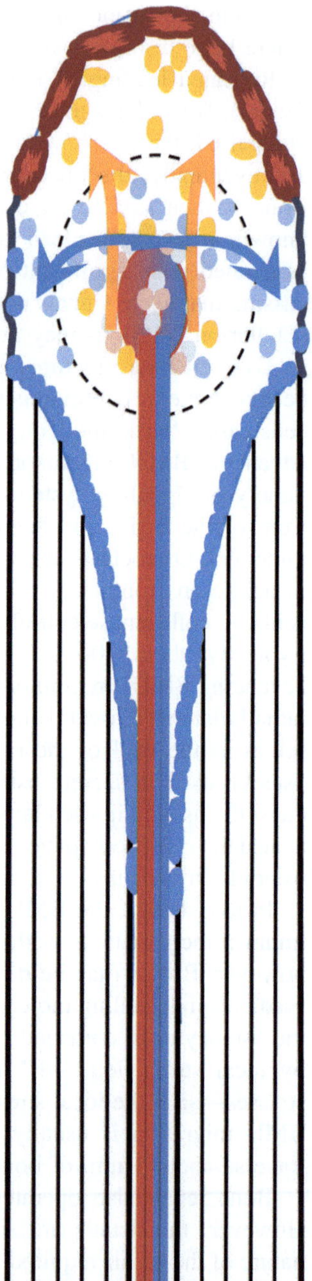

Fig. 3.15 Schematic of a longitudinal section through a 'bone modeling unit' (BMU). At the front of the advancing BMU are 'resorbing osteoclasts' (depicted as large red cells), which resorb bone matrix through the secretion of acid and cathepsins. Resorbing osteoclasts are renewed/maintained through repeated fusion with mature osteoclasts (depicted as pale orange cells next to resorbing osteoclasts). The mature osteoclasts originated from progenitor osteoclasts found around a central blood vessel, moving toward the front of the BMU as they differentiate and mature (orange arrows). RANKL-RANK signaling activation in progenitor osteoclasts is the crucial regulator of osteoclast maturation from progenitor osteoclasts. Progenitor osteoblasts (pale blue) differentiate and mature as they migrate toward the wall of the BMU (blue arrows), where they mature and become active osteoblasts. The active osteoblasts lay down osteoid in distinct layers, known as lamellae. The osteoid mineralizes to form hard bone within about a month. We note that the bone strength is temporarily reduced by the cavity opened in the bone as a result of BMU activity (Smith et al. 2012)

signaling environment, they differentiate into mature osteoclasts. The mature osteoclasts then resorb the damaged bone.

Mesenchymal 'progenitor cells' traveling in the blood also exit the tip of the blood vessel and differentiate into osteoblasts, which then secrete osteoid to replace the resorbed bone with new bone. The level of trabecular 'filling' depends on the local stress fields and local signaling molecule concentrations (e.g. TGF-β and IGF play important roles). Because of their similarity to 'osteonal' BMUs found in dense bone, the trabecular repair process has been described as a 'hemi-osteon' (i.e. functionally the hemi-osteon can be viewed as half of the osteonal BMU described above for dense bone, which is basically what one might expect to see if a BMU engaged in 'surface repair').

Over decades of research, it is clear that much like the view of cartilage described here, apparently 'static bone' is in fact in a vigorous dynamic equilibrium. As for articular cartilage, mechanical fatigue damage is continually inflicted on bone and in normal people, bone remodeling is ongoing, so bone repair keeps pace with the bone damage (as indeed occurs in most tissues). In fact, bone remodeling is so vigorous that in humans the entire skeleton is on average completely replaced every 7–10 years. However, individual parts of the skeleton may be subject to more or less damage, and so remodeling takes place more or less quickly relative to the average.

For example, the delicate ossicles of the middle ear and cochlear in the inner ear are essentially not remodeled at all, normally remaining unchanged throughout adult life (importantly, we note that these bone possess have very high levels of OPG, which very effectively blocks remodeling) (Zehnder et al. 2006). The cortical bone in the weight-bearing long bones is replaced on average about every ten to twelve years. However, the trabecular bone within the long bones is replaced on average about every three to five of years.

In high load areas of trabecular bone, such as subchondral trabecular bone beneath the subchondral bone plate, the turnover of trabecular bone may be even more rapid, perhaps averaging complete renewal every couple of years. We note that the trabecular bone is more 'labile' than dense bone for a variety of reasons, including higher rates of mechanical damage, the complex signaling environment in the adjacent bone marrow tissue (which may dysregulate bone signaling if compartment integrity is not maintained; e.g. lymphoid tissue also uses the RANKL–RANK–OPG signaling system) and of course the spongy bone's very large surface area (facilitating rapid resorption in the event of mechanical damage or signaling dysregulation).

We have seen that articular cartilage covering the ends of long bone is a dynamic tissue, continually adjusting to its environmental circumstances. So given this example, it is perhaps not surprising that bone is also a dynamic tissue, and so the overlying articular cartilage has to accommodate an evolving bone structure beneath it. Under the effect of constant loading-induced remodeling, on a timescale measured in decades, the bone making up the large weight-bearing joints very slowly and permanently deforms—i.e. it is gradually pushed out of its initial shape by the forces it has to resist. This is because the bone repair process cannot return damaged bone precisely to its original geometrical configuration (much as the collagen network in cartilage cannot be precisely returned to its original geometrical configuration). So if trabecular bone is damage and plastically deforms (meaning it does not recover its initial position upon unloading), trabecular bone is repaired in its deformed geometry, much like the collagen network in cartilage is repaired in its damaged configuration. And in an analogous way, while gradual 'progressive expansion' of the collagen network is an intrinsic property of working articular cartilage, so gradual 'progressive collapse' of subchondral bone is an intrinsic property of subchondral bone found beneath articular cartilage in joints. We mention in passing that identifying these similarities between bones and articular cartilage helps give context to our understanding of each tissue and helps to build our confidence in both tissue models.

For example, it has been shown that the lateral and medial compartments of the tibial plateau grow in surface area at the rate of about 1.5 and 2.2% (respectively) per year. The medial compartment normally experiences higher loads than the lateral compartment and consequently has a faster average rate of areal growth, while a faster rate of increase in tibial plateau area is also correlated with increasing BMI, which is a surrogate for increasing mechanical loading (Wang et al. 2005). Increased tibial areas have also been associated with meniscal tears (Davies-Tuck et al. 2008), which also tends to increase cartilage and bone stress and so damage rates in both tissues. The observed increases in tibial plateau areas over time supports the idea that this occurs because bone repair processes repair the bone in its damaged shape. Because of this, even though repair is effected, the damage and repair processes in bone result in a gradual 'creep' in bone geometry. This may be poetically viewed as an exceedingly slow structural collapse of bone.

It is possible that this bone 'slow-motion collapse' in highly loaded joints may adversely affect the overlaying cartilage. It is not too difficult to then infer that the 'spread' of the tibial plateau may cause curvature changes in articular cartilage, which may lead to separation and possibly splitting of the cartilage from below and perhaps separation/splitting from above too. 'Irregularities' in cartilage at the osteochondral junction have been identified in vivo (Ding et al. 2007). Splits in the cartilage would increase the ease with which aggrecan can find its way to the cartilage surface and so may compromise aggrecan concentrations in cartilage, potentially leading to articular cartilage damage.

While it is difficult to identify precisely the initial defect in the tissue and so causation—partly because the same load stimulus has multiple effects on cartilage and bone that are difficult to separate and partly because bone and cartilage repair themselves too—it has been shown that growth in the tibial plateau is correlated with an increase in cartilage defects (Dore et al. 2010). It certainly is plausible that if the rate of tibial spreading exceeds the rate at which defects in the osteochondral junction and cartilage can be repaired, then slow-motion collapse of subchondral bone tissue would provide a mechanism for initiating osteoarthritis (Ding et al. 2007).

Tissue compartmentalization is crucial to the functional integrity of both subchondral bone and articular cartilage tissues. While there may be some small molecule exchanges between the two tissues (e.g. oxygen and water molecules may be exchanged to some extent), we may then suppose that normally there is little signaling cross-talk between the tissues. In normal tissue, there are few if any blood vessels traversing the CCZ into cartilage, though in osteoarthritic states there are many (Bonde et al. 2005). Canaliculi contacting the osteochondral junction are too few in number to transport significant quantities of material, and those canaliculi that do contact the osteochondral junction are capped by a cement line, while CCZ precipitates provides an additional barrier that few (if any) blood vessels normally cross. For if there is an exchange of signaling molecules between tissues, the different roles played by signaling molecules in each tissue would almost certainly quickly lead to dysregulated signaling and disease states in the other tissue (see discussion Sect. 3.9).

The interface between collagen types I and II is usually well-defined, continuous, and stable. However, should a BMU remodeling in the subchondral bone plate 'stray' too close (or into) cartilage tissue, it wayward progress is 'shut down' as it approaches (or crosses) the cartilage-bone interface, because of the high OPG concentrations in articular cartilage. OPG will start to diffuse from the relatively high concentrations in cartilage through the canalicular network into the resorption cavity, where it will blocks RANKL–RANK signaling, preventing bone resorption progression toward (or into) cartilage. This is a critically important protective mechanism in cartilage, one that is required if cartilage integrity is to be maintained. If this protective mechanism breaks down for any reason [e.g. as in osteoarthritis (Bonde et al. 2005)], then the cartilage can be destroyed by bone remodeling (i.e. BMUs) penetrating into articular cartilage and turning articular cartilage into bone [(Bucay et al. 1998; Shimizu et al. 2007) and personal communication with Prof. Colin Dunstan].

Even with this OPG protective mechanism in cartilage in place, we reason that over time the number of 'close encounters' between active BMUs and the osteochondral junction will increase. As the rate of bone turnover increases, the statistical likelihood of a close encounter in any given time increases [which is of particular interest in disease states, where bone turnover may increase 20-fold (Bailey et al. 2004)]. Continuous pathways between cartilage and bone may then open up (see Fig. 3.14 and accompanying discussion). This will lead to increased transport of various molecules between the bone and cartilage, potentially disrupting normal signaling in both tissues (Zhang et al. 2012a; Sharma et al. 2013; Bonde et al. 2005).

It is known that thickened, radio-dense, sclerotic subchondral bone plate, composed of abnormal, poorly mineralized, porous bone (see Fig. 3.16) is a hallmark of osteoarthritis (Li et al. 1999; Bailey et al. 2004). While the sclerosis is probably initiated as a normal homeostatic response by the bone to increased mechanical loading (Hayami et al. 2006), at some point, it is likely that bone sclerosis is accelerated by the breakdown in the integrity of the osteochondral junction under significant bone remodeling stress. Many more blood vessels may then cross the CCZ into articular cartilage (Bonde et al. 2005). This may lead to the migration of Wnts (up-regulating osteoblastogenesis), GAGs (down-regulating

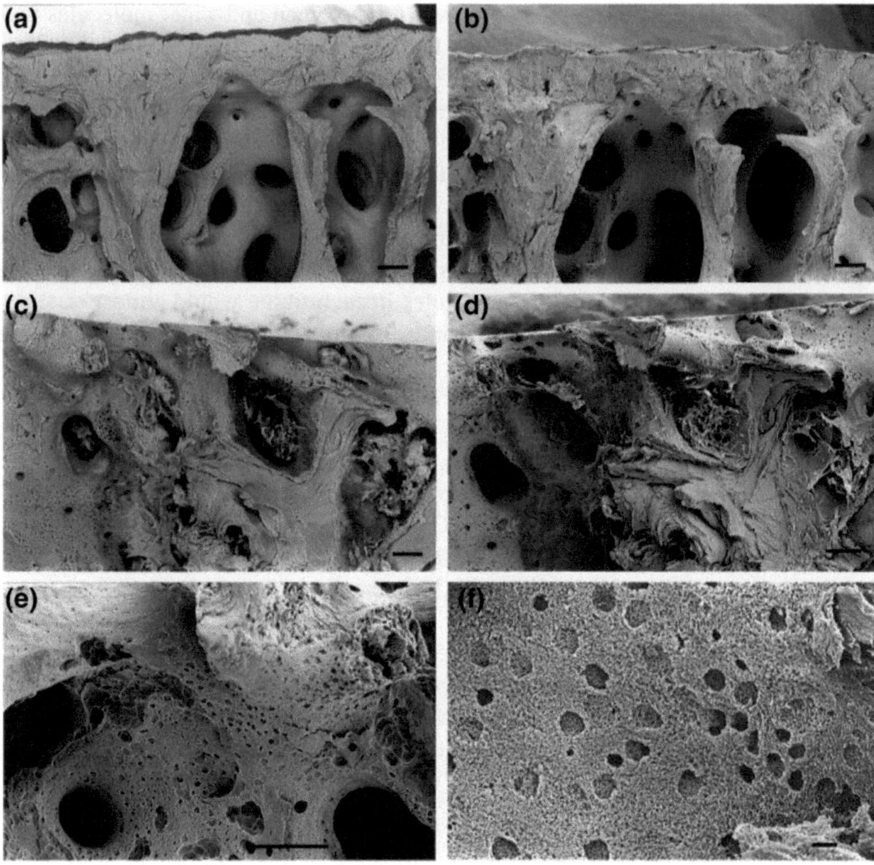

Fig. 3.16 Osteoarthritic bone appeared markedly different from normal in both backscatter and secondary emission images. **a** From a 75-year-old female and **c** from a 70-year-old female. Corresponding secondary emission images (**b**, **d**) with total or almost complete absence of the dark surface layer. Bars 200 μ. The bone surfaces are penetrated by numerous small holes which can also be seen on the fractured ends of trabeculae (**e**). Bar, 200 μ. At higher magnification (**f**) a finer texture is also apparent. Bar 20 μ (Li et al. 1999)

osteoclastogenesis), and other molecules across the interface, amplifying the sclerosis so much that it leads to poor quality bone formation (Ling et al. 2010; Sharma et al. 2013).

Clearly, repeated mechanical loadings may exceed the capacity of the bone to repair itself (Sormaala et al. 2006). Indeed, we have noted previously that the bone repair process itself may pose an increased risk of structural failure, for a BMU forms a reasonably sized 'resorption cavity' as part of a normal repair process, together with a substantial volume of partially mineralized osteoid, both of which last for a considerable period of time, and both temporarily reduce the load carrying capacity of the subchondral bone. This effect becomes more pronounced as the

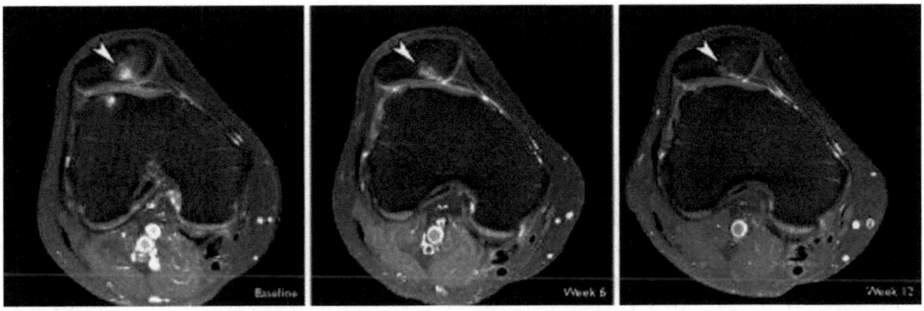

Fig. 3.17 Example of a patient with a patellar [bone marrow] lesion [BML] changing over 12 weeks (denoted by white arrow)—axial view (Felson et al. 2012). See also Fig. 2 (Felson et al. 2001)

numbers of remodeling BMUs increase, and so the resorption volume represents a more significant fraction of the total subchondral bone volume. It is known that BMU density is related to the rate of bone damage, which in turn is related to the intensity of the mechanical environment experienced by the bone. So when bone damage is substantial, BMU numbers are high. In this state, the repair process may then significantly weaken an already damaged bone.

Indeed, subchondral bone lesions are acknowledged as one of the hallmarks of osteoarthritis (see Fig. 3.17). Bone lesions are labile—they change in size in a matter of weeks (Kwoh 2013). The histopathology of bone marrow lesions reveals a variety of findings including damaged and abnormal trabeculae, bone marrow necrosis, fibrosis, oedema, and microfractures in various stages of healing. It is possible that microfractures of the subchondral bone following injurious loadings may in some cases initiate the bone loss and/or the formation of bone marrow lesions (van Dijk et al. 2010). The spatial distribution of bone marrow lesions has been correlated with areas of bone denuded of cartilage. Importantly, subchondral bone lesions are associated with more rapid progression to knee replacement (Tanamas et al. 2010). Furthermore, bone marrow lesions are associated with knee pain.

For example, Felson et al. found that about 75% of those persons with OA knee pain had bone marrow lesions, while only 30% of those with no knee pain had bone marrow lesions (Felson et al. 2001). We infer that bone marrow lesions are an important source of knee pain (e.g. perhaps explaining as much as 30–40% of knee pain). This appears to be supported by reported pain relief upon injection of bone cement into bone lesions (Bonadio et al. 2017), though pain is notoriously subject to the placebo effect [and it is well known that the placebo effect for OA-related symptoms is substantial (de Campos 2015)]. It is noted that other key sources of knee pain are fluid flow into the bone compartment (van Dijk et al. 2010), meniscal damage (e.g. meniscal extrusion due to a meniscal tear), and synovial membrane inflammation (Stoppiello et al. 2014) (e.g., synovitis is aggravated by stretching the knee, or ‘overgrowth’ of synovial membrane that is subsequently squeezed or crushed between contacting cartilages). Further sources of pain are the 'fat pads' around the knee joint and ligaments within the joint.

Importantly, we note that pain is a powerful negative feedback mechanism, normally helping to rebalance the damage rate to more closely match the repair rate. This is because pain leads to a reduction in usage of a joint and mechanical damage, which then gives repair processes within the tissue a better chance of exceeding the new, revised rate of damage. This is the conceptual basis for the traditional medical concept of treating an injured joint by 'resting' the joint, but it is now recognized that some use of the joint is beneficial (Quicke et al. 2014). But exactly how much joint usage is optimal is the key, moot point. Apart from a range of general health benefits arising from exercise, the general (unspoken) concept that for some modest amount of exercise, the rate of repair may initially increase more than the rate of damage, but at some higher threshold rate of exercise, this situation reverses. But unfortunately, how much exercise is too much exercise is simply not known. Given the central importance of physical therapy in nearly all musculoskeletal disease treatments, how much exercise is too much following knee injury is a question that needs much more intensive research evaluation. Of course, symptomatic pain relief may lead to undesirable overuse of the joint, so pain abatement needs to be accompanied with suitable patient education.

Structurally, the subchondral bone plate and the underlying trabecular bone are structurally like a 'propped beam' in a 2D section or a 'propped plate' in a 3D volume. For the dense subchondral bone to act structurally as a beam or plate, the beam or plate has to carry a large moment and a substantial shear force. When a beam or plate is partially supported by props, the moment and shear force in the plate are reduced. However, it remains likely that for a partially propped beam or plate, the shear will still be at a maximum at the edge of the beam/plate. Therefore, bone marrow lesions observed toward the edge of the cartilage are suggestive of a shear-type failure in the dense subchondral bone [see Fig. 2 in Felson et al. (2001)]. Bone marrow lesions toward the center of the beam or plate are suggestive of inadequate propping by the plate-like trabecula below, leading to excessive moments in the beam or plate toward its center [see Fig. 2 in Felson et al. (2001)]. An excessive moment may lead to fracture of subchondral bone plate, suddenly shifting load from the plate to the trabecular bone, which may then be followed by collapse of the underlying rods and/or the plate-like trabecula. Alternatively, it is possible that the subchondral trabecular bone fails first, and then, the subchondral dense bone fails, through an overload shear failure or an overload bending moment failure.

No matter which element fails first, ongoing cyclic loading as a result of continued activity leads to chronic refracturing of the trabecular bone undergoing repair. Chronic damage accompanied by an inflammatory response probably explains the formation of the bone marrow lesions beneath the subchondral plate.

Bone marrow lesions are highly dynamic. At least 50% of BMLs change their volume by more than 50% over a six to twelve week period (Felson et al. 2012). Importantly, from a practical viewpoint, we note the observation that patellofemoral knee bracing for 6 weeks has been shown to reduce bone marrow lesions and reduce knee pain, e.g. http://www.sciencedaily.com/releases/2013/10/131027122929.htm. In terms of known bone physiology, the effectiveness of this treatment makes sense conceptually in terms of bone homeostasis being a dynamic equilibrium between damage and repair (Frost 1987a). By reducing the rate of damage (through use of a knee brace), the dynamic balance can be tilted in favor of repair. This appears to be borne out in practice.

3.9 Transport of Signaling Molecules Across the Osteochondral Junction

Knowledge of the IHH-PTHrP signaling system in articular cartilage (see Sect. 5.6.2) is potentially important in aiding our understanding of some otherwise puzzling observations on bone–cartilage interactions across the osteochondral junction. In cartilage, PTHrP suppresses hypertrophic differentiation of chondrocytes. PTHrP concentrations in the deep articular cartilage are usually high. However, elevated concentrations of PTHrP in bone are a common cause of bone loss disorders. In bone, elevated PTHrP concentrations result in more RANKL expression, and less OPG expression, by osteoblasts and other cells of the osteoblastic lineage. An abnormally high RANKL/OPG ratio usually results in 'negative bone balances' during bone remodeling events and consequently is often associated with osteoporosis. Could cartilage health and osteoporosis be correlated, the strength of the correlation depending on the integrity of the barrier presented by the osteochondral junction between the two tissues? Alternatively, could a comprised osteochondral barrier allow PTHrP in the deep zone to diffuse into subchondral bone compartment, so promoting local osteoporosis?

When assessing the integrity of the osteochondral junction, the size of the molecule is crucially important. Clearly, small molecules are more likely to move across the osteochondral junction (Pan et al. 2009). Parathyroid-related protein (PTHrP) is highly variable in its size depending on 'cell processing' of PTHrP (e.g. estimates range from 3.5 to 11.7 kDa (Brandt et al. 1991a), to 10–18 kDa to 24–25 kDa). Often cells may secrete multiple variations on PTHR1 active molecules. Given the variability in its size, PTHrP is among the smaller of signaling proteins. By comparison, RANKL is a 31 kDa monomer, which is functional as a homotrimer (i.e. 91 kDa), while OPG can exist as a 60 kDa monomer, or a homodimer or as a homotrimer (i.e. 120 kDa or 180 kDa) (Kostenuik 2005). We note there are many additional molecules that modify signaling systems, e.g. DKKs (MW 24–29 kDa) modify Wnt signaling (which is important in bone and cartilage), and potentially, the transport of microRNAs will also have some influence on cell behaviors in an adjacent tissue.

But PTHrP is a critically important regulator of chondrocyte differentiation states in articular cartilage, preventing hypertrophic differentiation, and is normally present in relatively high concentrations in articular cartilage, so we focus on this molecule. Based on a size exclusion effect at the osteochondral interface, it is clear that PTHrP is generally much smaller than either OPG or RANKL, and so if microfractures exist between cartilage and subchondral bone, PTHrP will be among the first of the protein signaling molecules to cross to adjacent tissues.

If a person has long-standing osteoarthritis, usually their subchondral bone shows evidence of sclerosis (i.e. the bone is abnormally thick, but usually the bone is of poor quality) (see Fig. 3.16) (Mastbergen and Lafeber 2011; Zhang et al. 2012b). It then seems reasonable to suppose small molecules will be more easily transported across the more 'porous' subchondral bone. If PTHrP levels in cartilage are elevated in normal (and perhaps more elevated in OA), then transport of PTHrP molecules of cartilage origin into the bone would raise the concentration of PTHrP in nearby bone. As described above, it is well known that PTHrP elevates RANKL and reduces OPG concentrations in bone, thereby raising the RANKL/OPG ratio in bone. Elevated RANKL/OPG ratios usually result in increased rates of bone remodeling. In fact, a 20-fold increase in bone remodeling is reported in osteoarthritis (Bailey et al. 2004). Osteoid production is poor, and quality is low (e.g. the collagen quality is poor) resulting in an abnormal osteoid structure and mineralization of the newly formed bone. The poor quality bone weakens the subchondral bone plate, which potentially contributes to the formation of bone marrow lesions immediately beneath the subchondral bone plate.

But there may be further complications as the quality the osteochondral junction deteriorates. OPG and RANKL are produced in cartilage. Usually, the RANKL/OPG ratios are very low in cartilage. Indeed, it is the elevated OPG concentrations in cartilage that serves [among other functions (Shimizu et al. 2007)], to deter bone remodeling units (BRUs) from crossing the osteochondral interface and tunneling into and damaging articular cartilage (resulting in matrix loss, surface fibrillation and chondrocyte apoptosis) (Bolon et al. 2015). However, it is known that in osteoarthritic states, OPG concentrations fall and RANKL concentrations increase articular cartilage. In chronic OA states, subchondral bone loss is correlated with increased articular cartilage RANKL expression (Martinez-Calatrava et al. 2012). In other words, increased rates of bone remodeling, potentially driven by PTHrP, RANKL, and TGF-β leakage into the bone microenvironment, may first lead to subchondral bone loss in the bone microenvironment, and lower than normal OPG concentrations in cartilage also allow BMUs to intersect the osteochondral interface and tunnel into articular cartilage. Once BMUs cross into articular cartilage with some regularity, many more signaling molecules in much greater quantities can cross the osteochondral interface, and large molecules like glycosaminoglycans (GAGs) can leak from cartilage tissue into the bone tissue. It is known that many GAGs can result in dysregulation of OPG-RANK-RANKL signaling by binding tightly to OPG (Lamoureux et al. 2007). In this context, we note that for a mouse ACL transection OA model, 7 days after ACL transection, large quantities of active TGFβ1 are transported from cartilage to the bone marrow compartment during subchondral bone resorption, which then recruits $nestin^+$ MSCs to form marrow osteoid islets and angiogenesis (Zhen and Cao 2014).

3.10 What Are the Implications for Osteoarthritis?

It is clear that mechanical loading helps shape the collagen network fibril orientation, collagen fiber thickness, and zonation of cartilage during its development, which then enables articular cartilage to more effectively resist physiological loadings experienced in adulthood (Paetzold et al. 2012; Hunziker et al. 2007). For example, the thick bands and Benninghoff arcades that form in the collagen network of knee cartilage reflect the particular and demanding load conditions experienced by that particular joint.

The superficial zone forms as animal grows in size and matures, and it then dynamically adjusts its thickness as loads compress the collagen network, thereby promoting congruence of the contacting cartilage surface. Damage to the collagen network results in aggrecan-induced expansion of the collagen network, which is an intrinsic property of articular cartilage. Damage followed by expansion provides the preconditions for the collagen network in the upper transition zone to become newly formed superficial zone collagen network from 'below.' Collagen is continually removed through wear at the articular cartilage surface 'above.' Wear on cartilage is intermittent, as the friction coefficient at the cartilage surface depends on the duration and type of loading. Mechanical loading also drives increased rates of apoptosis in the superficial zone, which are replaced by cell division in the superficial zone and upper transitional zone. When required, cell division in the transitional zone is followed by cell migration across the transitional zone into the superficial zone. Chondrocyte proliferation is promoted by damage associated 'loosening' of the collagen network in the transition zone, and chondrocyte migration toward the cartilage surface along the local aggrecan gradient is assisted by the collagen network expansion. The collagen network is continually repaired by direct collagen assembly at debrided surfaces, and by slow chondrocyte migration (see Sect. 2.5). Slow chondrocyte migrations help to repair damage at the interface between the transitional and superficial zones. Where chondrocytes slowly cross a zonal interface, it is analogous to a 'spot-weld' in the sense that new collagen network forms in the path taken by a slowly migrating chondrocytes.

Every time we use our joints, aggregate is mechanically damaged and so it is either repaired (i.e. aggrecan reattaches to hyaluronan) or its replaced (i.e. a newly synthesized aggrecan molecule attaches to hyaluronan). Newly synthesized, mechanically detached and proteolytically degraded aggrecan all make their way toward the cartilage surface, and along with hyaluronan and PRG4, support mix-mode lubrication as they exit the articular cartilage surface (see further discussion Chap. 4).

All the above processes happen simultaneously in normal cartilage, as the tissue dynamically responds to its environmental loadings. Large step increments in environmental load conditions usually result in a 'step-up' in the activity level of normal tissue processes in an attempt to maintain homeostasis of the tissue. Most importantly, this includes increased rates of synthesis of ECM components and increased rate of cell proliferation (which increases the synthetic potential of the tissue). Slow chondrocyte migration effects permanent repair of the collagen network. Importantly, there is now some MRI evidence for cartilage healing (in the sense that MRI collagen signals return to normal) taking place in human cartilage over a two year period (Chu et al. 2014). We note this timescale is consistent with the long time periods for cartilage remodeling, and for repair of the collagen network by slow chondrocyte migration.

But cartilage tissue repair clearly has limitations, and it is not difficult to imagine that rates of tissue damage can in some circumstances exceed rates of repair in articular cartilage. Cyclic shear deformations are particularly damaging to articular cartilage when combined with an elevated friction coefficient at the sliding interface. It is also clear that failure to adequately repair any one of (i) aggregate/aggrecan, (ii) the collagen network, or (iii) chondrocytes will result in a 'cascading failure' of the other components and ultimately to failure of the whole tissue. This is because each one of these structural tissue components is necessary for the maintenance of the other in a challenging mechanical environment. A cascading failure may lead to tissue destruction surprisingly quickly (e.g. late-stage osteoarthritis may develop within 12 months of joint injury (Driban et al. 2014), though normally knee osteoarthritis develops over a decade or more following traumatic injury, or over many decades for age-related osteoarthritis.

Clearly, osteoarthritis may be theoretically classified (and potentially treated) on the basis of which structural component leads the cascading failure. However, practically, it is not usually possible to identify which tissue component fails first, and so conservative treatment usually retreats to managing environmental loading on the tissue. Encouraging experimental findings suggest that 'appropriate loadings' can reduce net damage to all the tissue components simultaneously by increasing the rate of repair more than the rate of damage, while reducing cell apoptosis, but excessive loadings can reverse these favorable relationships (Li et al. 2013). The crucially important practical problem becomes identifying 'appropriate loadings' that will maximize cartilage repair in a patient-specific context. This is only likely to be achieved with the aid of computational models to estimate actual tissue loading, which is likely to be an important area of future research (Gardiner et al. 2016). It seems likely that in some cases (e.g. following joint injury), appropriate load regimes (together with the avoidance of inappropriate joint loadings, e.g. fall-related injuries) may well have an important role to play in preventing or mitigating a cascading failure, and so at least delaying the development of severe osteoarthritis following injury. As part of a cascading joint failure (as opposed to a cascading articular cartilage tissue failure), it is noted that not only may joint injury lead to osteoarthritis, but osteoarthritis may lead to joint injury (Englund et al. 2009). This positive feedback cycle can rapidly lead to joint failure (Driban et al. 2014).

Clearly, cartilage health also depends on surrounding joint tissues. Vital to articular cartilage is the normal operation of the synovial membrane, as it nutritionally supports articular cartilage. Indeed, articular cartilage is totally dependent on the synovial membrane, so dysregulation of this tissue inevitably leads to problems for cartilage, which can be disastrous.

In practical mechanical terms, it appears that the 'factors of safety' on mechanical loading in the normal synovial joint are comparatively low by engineering standards [factors of safety are probably in the range of 2.0–4.0 (Biewener 1990), while normal engineering for durable structures requires a factor of safety of at least 3 (or 4 according to Biewener (1990) and sometimes more depending on estimated risk-cost)]. This means comparatively small changes in nearby joint tissues, (i) the ligaments which regulate the motion of the femoral cartilage on the tibial plateau), (ii) the meniscus (which disperses and so reduces contact loads to more manageable levels), (iii) the nerves, tendons, and muscles (which control the magnitude and timing of the loads applied across joints), and (iv) bone (which structurally supports the cartilage and maintains the primary shape of the articulating surfaces), may all have significant impacts on articular cartilage. Indeed, 'damage' to anyone of these tissues alters either the rate of articular cartilage damage or its repair capacity, thereby determining the fate of the cartilage tissue. Given this, it is perhaps unsurprising that about half of traumatic injuries to the knee joint progress to osteoarthritis within ten to fifteen years (Simon et al. 2015).

While the previous chapter made clear the intrinsic vulnerability of articular cartilage to genetic defects at the molecular level, this chapter helps make more clear the mechanical vulnerability of articular cartilage to abnormal loading or the unsatisfactory performance of the surrounding tissues. In the introduction to this book, we noted the paucity of research data on exactly how the tissues making up the whole joint come to their normal equilibrium state and how injury to one tissue may impact on others. However, some data exists.

One example is animal models of anterior cruciate ligament transection-induced osteoarthritis. For example, experiments on dogs reported in this chapter have helped to define the processes and capacity of articular cartilage to respond to sudden changes in environmental loadings over a period of months and years (Vignon et al. 1983; Clarke 1974). This experimental data gave us considerable insight into the intrinsic property of cartilage to expand and soften as it is damaged and then to apparently heal! It seems likely that something similar is happening in humans following ACL rupture, given the expansion of cartilage thickness at the rate of about 30 μ/year over five years following ACL rupture (Eckstein et al. 2015). The capacity of the tissue to go through a healing process is also consistent with MRI evidence for human cartilage healing taking place following anterior cruciate ligament injury over a two year period (Chu et al. 2014). Says Chu et al. (2014): '*Through cross-sectional and longitudinal evaluations, this study shows both a potential for intact articular surfaces of cartilage to heal deep tissue injuries and the diagnostic potential of UTE-T2* mapping, a new non-invasive quantitative MRI metric, to track clinical disease states.*'

Meniscal damage models of osteoarthritis may involve transection of one or more meniscal attachments and subsequent removal of meniscal tissue. This damage changes the loading on the cartilage surface to such an extent that the repair capacity of the cartilage is then exceeded and eventually the tissue fails. Remarks Papalia et al. (2011): '*The main finding of the present study is that the amount of meniscal tissue removed remains the strongest predictor of long-term onset of osteoarthritis.*' It is interesting to observe that in two such animal models, the release of proteases by the meniscal damage into the synovial fluid resulted in IGFBP5 degradation and consequently decreased concentrations of IGF in the synovial fluid and articular cartilage. Importantly, it was found that inhibiting proteases in the synovial fluid reduces the incidence and severity of osteoarthritic lesions (Clemmons et al. 2002; Yates et al. 2010). These experiments demonstrate that it is not possible to clearly separate mechanical signaling from chemical signaling (including ECM signaling), which always confounds the interpretation of mechanical load.

Nevertheless, experiments such as these suggest that short-term pharmacological interventions following trauma to a joint may offer considerable hope by altering the course of disease (i.e. using a 'window of opportunity') for patients at high risk of developing osteoarthritis (Lieberthal et al. 2015). After many disappointments in the search for an anti-OA drug, this has more recently led to a focus on investigations of drug usage that may be useful either just prior to, or soon after, traumatic joint injury (e.g. very low dose dexamethasone, IGF supplementation, protease inhibitors, or some combination of these), along with a 'dose' of mechanical loading (Li et al. 2013, 2015).

Given its importance to cartilage tissue, one would imagine TGFβ might be a suitable drug candidate (i.e. TGFβ is a powerful anabolic growth and differentiation factor in cartilage tissue—see Sect. 5.6.1). Unfortunately, addition of TGFβ to the synovial fluid results in fibrosis of the synovial membrane and joint tissues (Shen et al. 2014; van der Kraan 2017). However, we note that this fibrosis is abrogated in ADAMTS5 knockout mice (Plaas et al. 2011), which points to the importance of ADAMTS5 to TGFβ signaling (see Sect. 5.6.1).

The complexity of pharmacological interventions to modify cartilage tissue behavior is nicely illustrated by NSAIDs. NSAIDs (which inhibit prostaglandin production) '*fall into one of three categories: those exemplified by aceclofenac and tenidap that are capable of stimulating matrix synthesis, while others, which include piroxicam, aspirin, and nabumetone, do not appear to have a major effect on matrix synthesis, and a third group, which includes ibuprofen, indomethacin, nimesulide, and naproxen, that are markedly inhibitory in terms of matrix synthesis*' (Dingle and Parker 1997). Clearly, this suggests that for articular cartilage, not all 'COX inhibitors' are alike. The authors suggest that based on in vivo and in vitro data, aceclofenac (an analogue of diclofenac, which is a COX2 inhibitor that decreases PGE_2 synthesis by chondrocytes) appears promising, avoiding some of the drawbacks of alternative NSAIDs at the tissue level (Dingle and Parker 1997; Dooley et al. 2001). We note here in passing that a more recent study finds that NSAIDs help to block the resorption of cartilage and nearby bone by keeping the OPG/RANKL ratio elevated (Moreno-Rubio et al. 2010). For further discussion, the results of clinical trials, and therapeutic options with contraindications for the use NSAIDs in treating arthritis see Crofford (2013), Lories (2012), Losina et al. (2018).

Nevertheless, the behavior of any NSAID drug in vivo is almost certainly patient-specific, i.e. a drug's efficacy depends on the osteoarthritis phenotype: is the underlying pathology osteoarthrosis, or are inflammatory mediators such as

IL-1, TNF, and Il-6 involved? And what role is played by about Toll-like and RAGE receptors? This again highlights the importance of biomarkers (e.g. molecular profiling of the synovial fluid to establish drivers for disease progression), which presumably would help indicate the most appropriate conservative and pharmacological interventions for each subgroup of patients.

In the context of pharmacological treatments, one important issue is the druggability of targets within thc cartilage tissue itself. Because cartilage carries a high negative fixed charged (due to the abundant glycosaminoglycan content in the extracellular matrix), drugs that ionize and carry a net negative charge cannot easily enter cartilage because of Donnan exclusion, even if they are small enough for size exclusion to not be an issue. (Kar et al. 2016). This has led to the investigation of positively charged 'carrier molecules,' such as avidin, which are readily taken up by cartilage tissue (Bajpayee and Grodzinsky 2013).

Chapter 4
Lubrication, Friction, and Wear in Diarthrodial Joints

4.1 Introduction

Friction force is the resistance to relative motion between contacting bodies. Friction is the principal cause of energy dissipation and wear at contacting surfaces (Amiri and Khonsari 2010). Lubricants are employed to reduce friction and wear at sliding contacting surfaces. Reducing friction not only reduces the amount of mechanical work dissipated as heat (and so the amount of work required to power relative movement), but crucially, it also reduces the wear at contacting surfaces. Tribology is the discipline devoted to the study of friction, wear, and lubricants.

The main modes of wear of contacting solids identified by tribologists are adhesive wear (Neu et al. 2008) (which is often associated with stiction, i.e. increased 'static friction'), stick–slip wear (which is wear associated with stop–start sliding) (Lee et al. 2013), abrasive wear [often associated with plowing (Linn 1967) or gouging of the surface (Correro-Shahgaldian et al. 2011)], chemical wear [which may include enzymatic degradation coupled with mechanically induced wear (Lee et al. 2013)], surface fatigue damage (due to repeated surface contacts) and surface cracking (often due to surface impact loading). The different modes of surface wear may interact (Katta et al. 2008). For example, enzymatic degradation is often accelerated by surface abrasion (as fresh reactants are continually revealed by surface abrasion), while surface cracking is often nucleated by increased surface roughness associated with sliding. We also mention that when one material rolls on another, a rolling resistance is generated as contacting materials deform (Ateshian and Wang 1997). This deformation mainly dissipates energy within the material (by molecular deformation or fluid flow in a poroelastic material), but rolling may also loosen molecules close to or on contacting surfaces as the surface is deformed leading to fatigue damage of the surface.

Both surface friction and wear are reduced by lubrication. There are two main modes of lubrication—fluid-film lubrication (Hamrock et al. 2004) and boundary lubrication. If both lubricating mechanisms are operating simultaneously, it is

D. W. Smith et al., *Articular Cartilage Dynamics*,
https://doi.org/10.1007/978-981-13-1474-2_4

referred to as 'mixed-mode' lubrication. If the fluid shearing is mainly water, fluid-film lubrication is also referred to as hydrodynamic lubrication. Early investigators puzzled over whether friction in synovial joints was either hydrodynamic lubrication or boundary lubrication (Charnley 1960). Based on previously published data, Neu et al. (2008) estimated that the film thickness is less than 400 nm over 60% of the gait cycles and approaches zero at times, which is much less than the waviness (https://en.wikipedia.org/wiki/Waviness) or roughness of the cartilage surface. Now, it is generally accepted that synovial joints usually operate in a mixed-mode lubrication regime (Gleghorn and Bonassar 2008; Neu et al. 2008; Moore and Burris 2017).

In normal diarthrodial joints, both contacting surfaces are cartilage (a porous material), and the fluid is synovial fluid. Typical synovial fluid contains about 3–4 mg/mL of hyaluronic acid and around 20 mg/mL (10–30 mg/mL) of protein (mostly albumin) and a few grams of lipid (Faryna and Goldenberg 1990; Prekasan and Saju 2016). The glycan and protein content of synovial fluid, give it an 'egg-white' consistency, means its flow behavior is not like that of isotonic saline. Like all other types of contacting surface, it has been shown that friction between cartilage surfaces leads to wear particle formation (Kuster et al. 1998; Oungoulian et al. 2013; Katta et al. 2008). Importantly, it has been established that increasing friction and/or increasing applied normal loads while sliding, so increasing the friction force acting on the cartilage surface, leads to increased rates of wear at the cartilage surface (Jay et al. 2007; McCann et al. 2009; Lee et al. 2013; Katta et al. 2008; Jones et al. 2015).

Synovial fluid reduces friction between cartilage surfaces. For small fluid shear rates, synovial fluid is more viscous than water due to the presence of hyaluronan and protein in the synovial fluid. As a rough approximation, normal synovial fluid is about 50 times more viscous than water [note that this estimate is based on measuring the time required for 59.4 mm^3 of synovial fluid to flow through a 1 mm capillary tube of 10 cm length under a 9.81 kPa of pressure drop (Jebens and Monk-Jones 1959)]. But the measurement of synovial fluid viscosity is difficult because the actual viscosity of the synovial fluid depends on both the duration of fluid shear and the rate of fluid shear.

In fact, the synovial fluid is found to be 'rheopectic' due to the synovial fluid's protein content (Oates et al. 2006). A rheopectic fluid increases its viscosity over time as it is shaken (and undergoes shear). We note in passing that a 'thixotropic fluid' decreases its viscosity over time as it is shaken (and undergoes shear).

While synovial fluid is rheopectic, when synovial fluid viscosities are considered at the same time duration but at different rates of shear (e.g. one sample is shaken more vigorously than the other for the same amount of time), synovial fluid also displays shear thinning. A 'shear-thinning' fluid decreases its viscosity with increasing shear rate. And for completeness, we note that a 'dilatant fluid' increases its viscosity with increasing shear rate. These findings mean that synovial fluid is both rheopectic and shear thinning, so it should not be regarded as a simple fluid.

Given the complexities of both the lubricating fluid itself and the in vivo conditions under which the fluid operates [we also note this includes temperature variation, and that the viscosity of synovial fluid is also temperature dependent (Hasselbacher 1976)], it is difficult to say exactly what synovial fluid viscosities occur in vivo. Some have found that generally synovial fluid viscosity decreases with age and with disease states (such as osteoarthritis), as hyaluronan concentration is reported to decrease with age (Jebens and Monk-Jones 1959; Bridges et al. 2010; Hasselbacher 1976).

It has been established, using both knockout animal models and through observations on particular human disease states [e.g. observations of patients or animals with camptodactyly–arthropathy–coxa vara–pericarditis syndrome, who do not produce the lubricating molecule PRG4 (Jay and Waller 2014)], that PRG4 is an essential 'boundary lubricant' for normal diarthrodial joint function [data suggests that PRG4 reduces sustained sliding friction about two to fourfold (Gleghorn et al. 2009; Chang et al. 2014)]. During joint loading and motion, normal cartilage usually operates in a mixed-mode lubrication regime. This means hydrodynamic lubrication acts simultaneously with boundary lubrication during normal joint operation. Based on animal experiments and observation on humans, it is found that boundary lubricants such as PRG4 are critically important for joint health. This implies that a significant fraction of mixed-mode lubrication is normally boundary mode lubrication, and that without PRG4, 'boundary mode' friction, along with its attendant shear straining throughout the cartilage as well as surface wear, becomes too high for cartilage to survive under what might otherwise be regarded as normal day-to-day loading.

It has also been found that during mixed-mode lubrication, the two lubrication types (i.e. hydrodynamic and boundary lubrication) interact in an unusual way when the bearing surfaces are made of saturated porous materials, which includes cartilage. Indeed given the complexity of porous media mechanics, the complexity of the synovial fluid material properties, and the complexity of the possible interactions during the mixed-mode lubrication in synovial joints, it is perhaps unsurprising that the details of exactly how joint lubrication operates have been controversial (see, e.g. http://blacklistedbiomechanics.com/biphasiclubrication.htm). What is certain is that the friction coefficient can be as little as 0.005 up to more than 0.5 (Neu et al. 2008), which indicates that profound changes in lubrication can and do occur when cartilage surface contact one another. Undoubtedly, much remains to be explained, quantified, and discovered about the lubrication and wear of diarthrodial joints.

In the following, we outline both qualitatively and quantitatively our understandings of the essential relationships between friction in known lubrication regimes and applied loading, the fluid pressure between the contacting saturated porous surfaces, the pore fluid pressure within the cartilage tissue, directions of fluid flow and (the importance of) lengths of fluid flow and with cartilage consolidation. We finally conclude with a section on some of the relevant and important experimental observations on friction and wear in cartilage.

4.2 Friction

When one material is dragged across another, a 'friction force' is generated at the interface in a direction opposing the direction of motion. Inevitably some of the mechanical work generating the motion is dissipated as heat and surface damage. Mechanical work is dissipated at asperity contacts through a variety of processes including vibrations (generating phonons and heat), stiction, stick–slip sliding, 'ploughing' and abrasion of the surface. At the nanoscale, mechanical work may also be dissipated via mechanical damage to individual molecules within the bearing material, by any of dissipative processes at the material surface. For example, abrasion of the surface can simply tear molecules apart or pulls them from the surface. The friction force tangential to the cartilage surface generates shear within the contacting bodies, which in turn may generate additional dissipative processes within the material itself (e.g. shear disruption of molecules, plastic deformations or fracture of molecular assemblies).

If a fluid or solid lubricant is present, mechanical work may be dissipated as heat by shear forces between lubricant molecules rubbing past each other, as well as by molecular damage to the lubricant molecules themselves. Because the friction force and so frictional damage is much less in fluid lubricants and boundary lubricants than between dry, unlubricated surfaces, clearly this is the potential advantage afforded by lubricants (explaining the long-standing interest in lubricant technology). It has been found that generally fluid lubricants have a much lower friction force and so frictional damage than boundary lubricants [so the wear coefficient for fluids is at least 3 orders of magnitude smaller than for most boundary lubricants (Neu et al. 2008)]. So when operating in a mixed-mode lubrication regime, there is a substantial potential advantage afforded by delaying the onset of significant boundary lubrication when in mixed-mode lubrication, by delaying the growth in boundary lubrication, by sustaining fluid lubrication for as long as possible.

Asperities are microscopic peaks and valleys on a surface. When surfaces contact, asperities contact. Dissipative processes at sliding asperity contacts, including adhesion, stick–slip, and surface abrasion (microscale pulling, grinding, ploughing, and mechanical interlocking) lead to significant wear at contact asperities (e.g. material is scraped from the surface of contact asperities on the cartilage surface). Collectively, these processes cause the gradual loss of the collagen network at the cartilage surface. The loss of collagen network leads to changes in the cartilage surface geometry and to changes in the surface properties of the cartilage. One of the most important surface properties that change as the surface wears is the 'friction coefficient,' because if loading remains unchanged, a change in friction coefficient changes the rate of wear. It is not hard to imagine that a positive feedback process can be established at the cartilage surface, where wear causes a higher friction force, and higher friction force causes higher rates of wear, and so on. Once such a positive feedback process becomes established, this process can lead to rapid degradation of articular cartilage, ending in bone rubbing directly on bone (which is usually very painful). Unfortunately, for the study of articular cartilage, there is no simple relationship between surface roughness and the friction coefficient, or the friction coefficient and surface wear (Katta et al. 2008; Neu et al. 2008).

The friction coefficient for material 1 sliding on material 2 (denoted most generally as $\mu^{\mathrm{mat1-mat2}}$) is defined to be (Neu et al. 2008):

$$\tau_{\max}^{\mathrm{mat1-mat2}} = \mu^{\mathrm{mat1-mat2}} \sigma_{\mathrm{normal}}^{\mathrm{S}} \tag{4.1}$$

where τ is the interfacial shear stress between the two contacting materials and $\sigma_{\mathrm{normal}}^{\mathrm{S}}$ is the solid stress acting normal to the surface. Solid stress is the stress between solid particles (defined as those particles comprising the porous material that are capable of maintaining similar spatial relationships with one another, as the material deforms under load). Substituting the equation for the normal solid stress (i.e. the effective stress equation; see caption Fig. 4.6 or Eq. (7.41) in Sect. 7.2.1), we find that for saturated porous materials contacting (Miramini et al. 2017)

$$\tau_{\max}^{\mathrm{mat1-mat2}} = \mu^{\mathrm{mat1-mat2}} \left(\sigma_{\mathrm{normal}}^{T} + p^{\mathrm{interface}}\right) \tag{4.2}$$

Importantly, we note that this means that if the pressure in the fluid at the interface between the contacting surfaces is approximately equal in magnitude to the applied total normal stress (but opposite in sign), then the 'apparent friction coefficient' at the interface is very small. This is because the frictional force in the supporting fluid is very small, and the solid stress on the cartilage surface is very small (meaning there are few asperities contacting other asperities), relative to the total stress acting on the contact surface. This state of affairs occurs in vivo when a cartilage joint is initially loaded. The friction coefficient in this initial state is known as the 'initial coefficient of friction' and denoted μ_0.

We mention that if the solid stress on the surface of the cartilage were zero (i.e. so the total stress equals the fluid pressure), then the mixed-mode lubrication regime has transitioned to the hydrodynamic lubrication regime. On the other hand, when the pore water pressure is zero, then the mixed-mode lubrication has transitioned to boundary only lubrication. We note here that hydrodynamic lubrication regime does not normally occur in synovial joint alone (or if it does, it does so for only a very brief period following initial load application). Normally, hydrodynamic lubrication acts together with boundary lubrication in mixed-mode lubrication regime (see discussion Sect. 4.3.4). Normally, pure boundary lubrication is also relatively rare, which is fortunate because that is when surface friction and attendant wear is largest.

What happens to the friction coefficient when the applied normal loading is maintained? The pore fluid pressure at the contact interface begins to decrease or 'dissipate' as fluid exits the edges (or sides) of contact interface (see, e.g. Fig. 7 in Ahmad et al. (1998), which pertains to the analysis of ink flow following pressure increase in a printing press). As fluid pressure in the contact interface decreases over time, the apparent friction coefficient increases as the solid stress σ^{S}_{normal} increases. After some period of time, the friction coefficient plateaus as the so-called equilibrium friction coefficient (μ_{eq}) is approached. Under 'static load conditions' (i.e. cartilage not sliding on cartilage along the contact interface), the excess pore fluid pressure eventually decreases to zero, and at this equilibrium condition $\sigma^{S}_{normal} = \sigma^{total}_{normal}$ and so the mix-mode lubrication regime transitions to a completely boundary (only) lubrication regime (see further discussion Sect. 4.3.3).

Importantly, under some kinematic load conditions (e.g. cartilage sliding reciprocally on cartilage), the excess pore pressure may not fall monotonically to zero, but the pressure oscillates, staying at some elevated mean pressure as synovial fluid is 'dragged' between the contacting, sliding surfaces. Note that the synovial fluid or 'background' pressure is normally atmospheric pressure, which is taken as zero. If the pore water pressure between the contacting surfaces does not go to zero, the mixed-mode lubrication continues to operate indefinitely. But usually, this permanent residual pressure in the fluid at the contact interface due to sliding is much less than the initial pore water pressure at the contact interfaces experienced soon after initial loading (probably, this fluid pressure is normally in the range 0.1–0.5 of the initial pressure, but can be up to 0.9 under very favor conditions, depending on tissue strain, reciprocal sliding velocity, and other details of the loading) (Moore and Burris 2017; Linn 1967). Somewhat confusingly, both static and kinematic states lead to what is described as an equilibrium friction coefficient (μ_{eq}). Equilibrium static friction is normally higher than equilibrium dynamic friction for most materials and cartilage is no exception (e.g. see Fig. 2 in Bell et al. (2006) and/or Fig. 4.23).

Importantly, for saturated porous bearings, upon initial loading the initial friction coefficient is generally considerably less than the equilibrium friction coefficient (i.e. $\mu_0 < \mu_{eq}$). This unusual behavior of saturated porous bearings was first investigated by McCutcheon, where his early experiments involved rubbing various saturated porous materials on glass (McCutchen 1959). These initial findings have been supported by later studies [see, e.g. (Hodge et al. 1986)].

For normal cartilage rubbing normal cartilage, upon step loading, the initial friction coefficient is generally very small (e.g. typically in the range 0.005–0.02 (Merkher et al. 2006), though sometimes lower values are reported, e.g. 0.0012–0.008 (Jones et al. 2015; Linn 1968)). Indeed, it is often stated that the 'friction coefficient of cartilage' is less than that of Teflon rubbing on Teflon [which has a friction coefficient around 0.04–0.08 (e.g. Jay and Waller 2014)]. Of course, Teflon is famous for being a very slippery (manufactured) surface.

The equilibrium friction coefficient for normal cartilage reciprocally rubbing normal cartilage is found to be experimentally around 0.05–0.1 (i.e. about an order of magnitude larger than the initial friction coefficient), depending on sliding velocity, tissue strain, and other details of the loading (Gleghorn and Bonassar 2008). However, if PRG4 is absent at the interface, the equilibrium friction coefficient for cartilage rubbing on cartilage increases substantially to around 0.3 (which is fairly typical of unlubricated everyday materials rubbing together, e.g. like stone rubbing on concrete). It has been demonstrated experimentally that there is a PRG4 concentration dependent decrease of the equilibrium friction coefficient from 0.3 to about 0.1 as PRG4 concentration in the synovial fluid is increased (Gleghorn et al. 2009). The normal concentration of PRG4 in synovial fluid is around 200 μg per mL (Gobezie et al. 2007). We note that as surface shear stress increases for any reason, chondrocyte apoptosis increases (Jay and Waller 2014; Waller et al. 2013).

Clearly, adequate lubrication is critically important for sustainable, normal synovial joint operation. With these preliminaries, let us now try to grow our understanding of the various types of lubrication theoretically possible in synovial joints (i.e. hydrodynamic lubrication, boundary lubrication, and mix-mode lubrication), and how hydrodynamic and boundary lubrication interact during mixed-mode lubrication.

4.3 Soft Porous Bearings

4.3.1 Hydrodynamic Lubrication

Hydrodynamic lubrication is a type of fluid-film lubrication. During fluid-film lubrication, the load is entirely supported by viscous forces in the fluid between the opposing surfaces. Generally, the viscous dissipation is comparatively small when hydrodynamic lubrication acts alone, and the fluid layer is sufficiently thick that asperities do not contact. Mechanical work is then only dissipated by viscous shear of the fluid. The viscous shear stress applied to the surface of the tissue is generally low, and so surface wear is minimal. If the opposing surfaces are not sliding relative to one another, it is referred to as 'hydrostatic lubrication,' while if the surfaces are sliding relative to one another it is referred to as 'hydrodynamic lubrication.'

A simple model of hydrostatic or hydrodynamic lubrication is illustrated in Fig. 4.1.

Let us now consider the pressure distribution in the fluid between two cartilage plates under the assumption that hydrodynamic lubrication is acting alone (i.e. as depicted in Fig. 4.1). The time-dependent pressure in the fluid between the contacting surfaces is shown in Fig. 4.2.

At this point, we may wonder what the pressure distribution in the pore fluid is within the cartilage plates. It turns out that under relatively mild assumptions (e.g. the time for consolidation of the cartilage is very long relative to the duration of hydrostatic lubrication), the pore water pressure in the cartilage tissue (see Fig. 4.3) is exactly the same as the pressure at the interface (see Fig. 4.2).

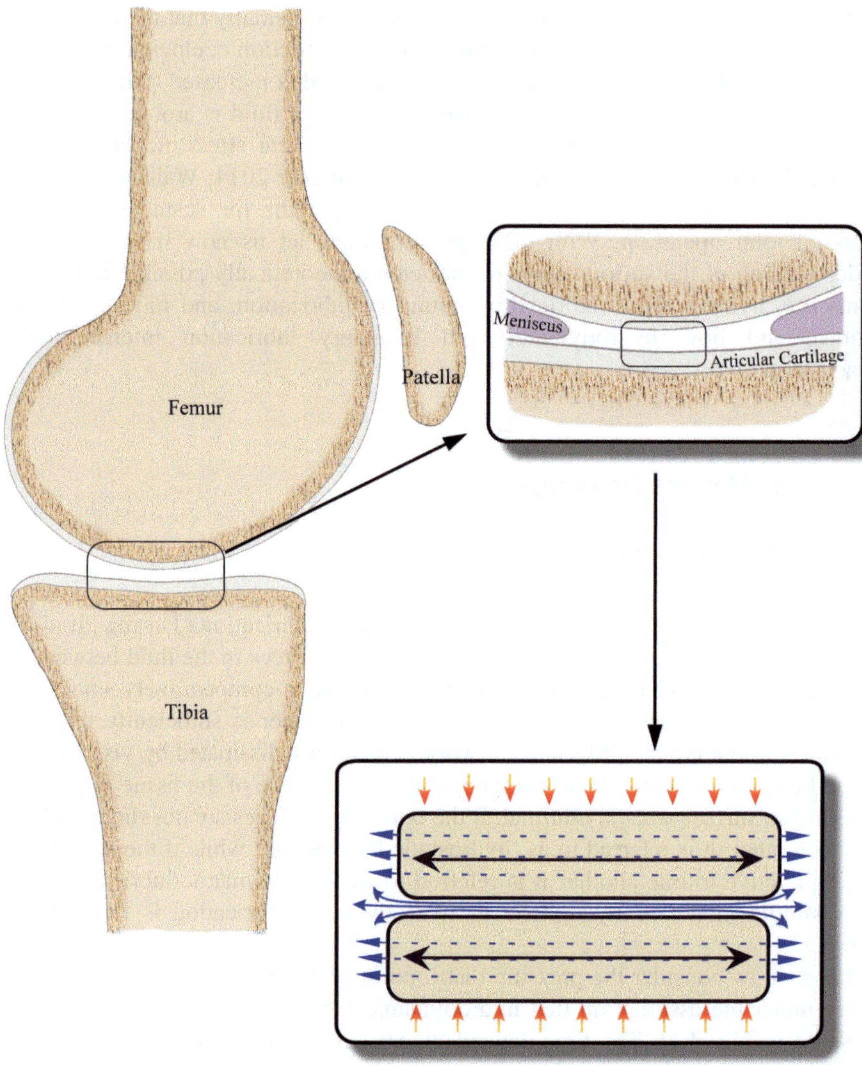

Fig. 4.1 Model 1—hydrodynamic lubrication. Viscous forces in the fluid carry the (vertically applied) compressive joint loading. The fluid is being squeezed from between loaded cartilage (schematic shows knee joint in cross section—e.g. femur condyle cartilage top, tibial cartilage below). Fluid flows along the pressure gradient toward the reference pressure within the joint cavity (i.e. background atmospheric pressure is normally taken as zero pressure). While hydrodynamic lubrication mode is probably only rarely found in synovial joints (Ateshian 2009; Neu et al. 2008), when it occurs (perhaps immediately following joint loading), it is present for only very brief periods. Consequently, this figure is only for illustrative and instructional purposes only. Nevertheless, hydrodynamic lubrication does occur frequently as part of mix-mode lubrication, which involves both hydrodynamic lubrication acting together with boundary lubrication. Figure drawn by Yvette Harrap

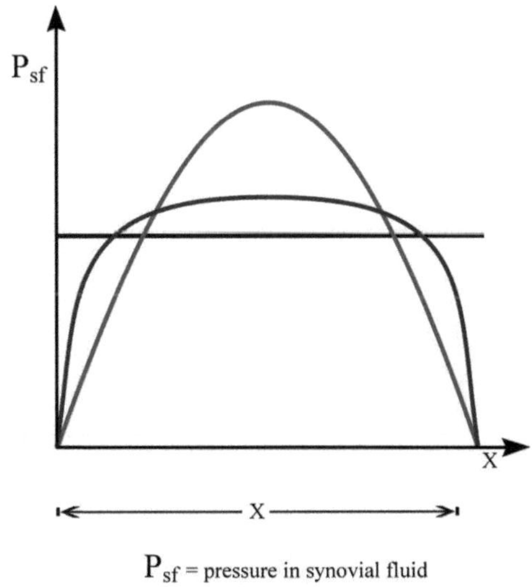

Fig. 4.2 Pressure distribution in the fluid between cartilage plates subject to a constant load. Upon application of a step loading at t_0, the initial pressure distribution is a constant (black line). This fluid pressure distribution is in equilibrium with the applied normal load. However, the joint space immediately begins to narrow as fluid exits the interface between the two opposing plates. Pressure in the fluid falls to zero at the edge of the contact zone (as the edge of the interface makes direct contact with the external synovial fluid, which is at zero pressure) (see blue and green lines). But the pressure actually rises toward the center of the interface (first blue line, then at a later time, the green line), so that force equilibrium with the load is maintained at all times. In other words, assuming a constant load is applied to the cartilage plates, the area under the fluid pressure curve is constant at all times during purely hydrodynamic lubrication (i.e. constant area under black, blue green lines curves). Figure drawn by Yvette Harrap

4.3.2 Consolidation of Two Loaded Cartilage Plates

Consider now a contrasting case where there is no lubricating fluid at all between two smooth cartilage plates, which are initially in 'perfect contact' (i.e. no contact interface detectable). In this case, the time-dependent behavior of the two stacked, saturated porous plates can be explained using Biot's theory of poroelasticity (Biot 1941).

More specifically, we consider consolidation of two smooth flat, saturated cartilage plates initially in perfect contact with one another. There is no preferential

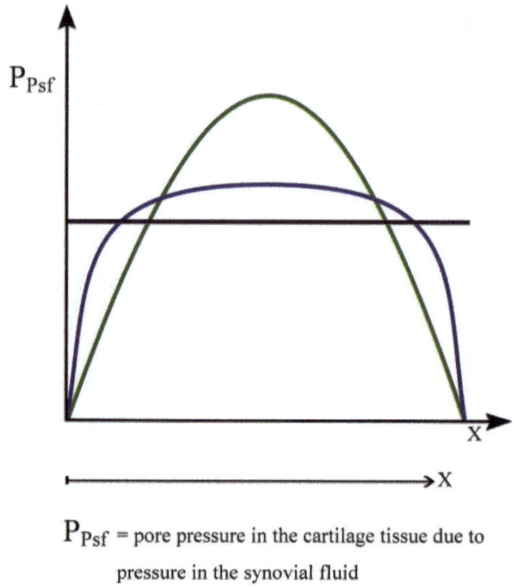

Fig. 4.3 Pressure distribution in the pore fluid within in the two cartilage plates during hydrodynamic lubrication assuming that the cartilage plates are impermeable at the osteochondral junction, while the sides of the cartilage plates are assumed to have a zero pressure boundary condition. Importantly, we note that for these example boundary conditions, the pressure distribution in the fluid between the opposing surfaces is exactly the same as the pore fluid pressure distribution in the cartilage tissue above and below. We note that this pore fluid pressure in the cartilage tissue here is not referred to as 'excess pore pressure,' as pore fluid pressure in the cartilage tissue during hydrodynamic lubrication is not caused by direct contact of asperities on the cartilage surfaces (see Fig. 4.10), and does not cause tissue deformation (i.e. assuming the flow has reached equilibrium and the solid and fluid phases are incompressible). Figure drawn by Yvette Harrap

flow path at the contact interface due to the perfect contact, so in this case the permeability along the contact interface is exactly the same as that of cartilage tissue itself (Fig. 4.4).

The time-dependent excess pore water pressure in the cartilage at various times is shown in Fig. 4.5, while the time-dependent solid stress in the cartilage is shown in Fig. 4.6. Note that the solid matrix (collagen network and aggrecan) of the cartilage only deforms in response to changes in solid stress. As collagen and aggrecan are incompressible hydrostatically, a hydrostatic change in water pressure has no effect on the solid matrix of the cartilage.

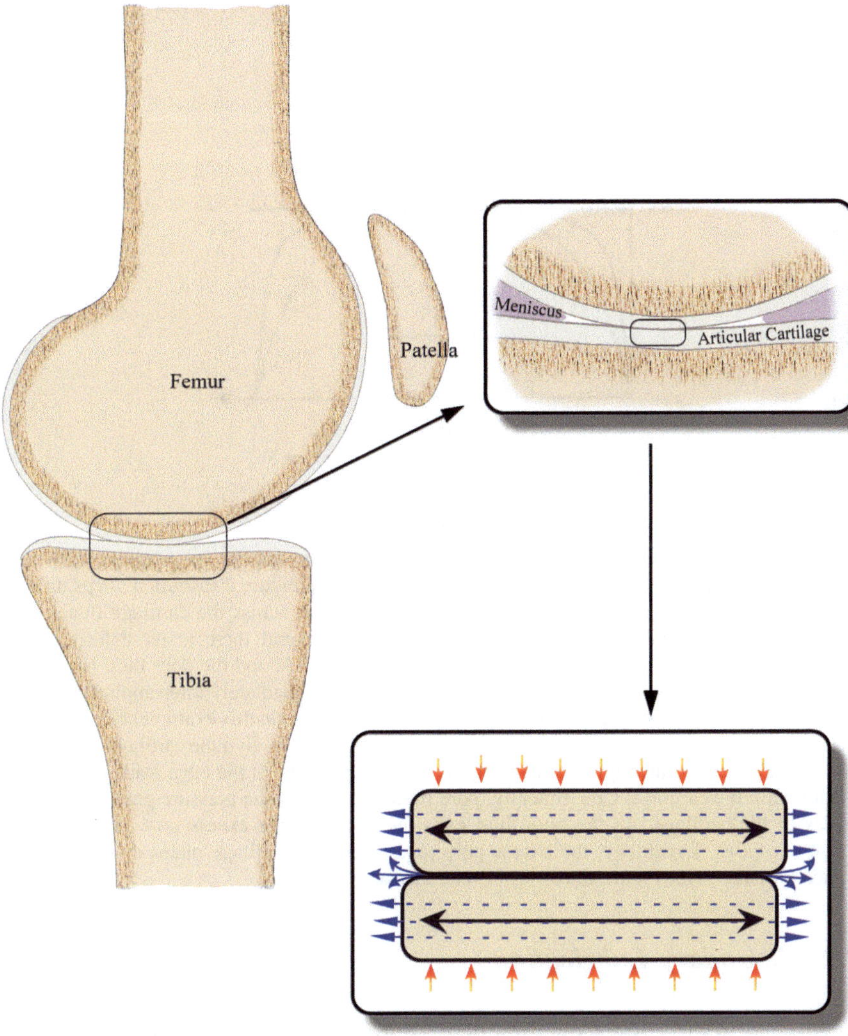

Fig. 4.4 Model 2—loading of two smooth saturated cartilage plates in perfect contact. There is no preferential flow path at the contact interface. Pore fluid can only escape laterally. Pore fluid flows laterally through the cartilage along the pressure gradient toward the zero pressure boundary condition on the sides of the cartilage plates (again, there is no flux boundary condition at the 'osteochondral junction' (top and bottom) is assumed). Figure drawn by Yvette Harrap

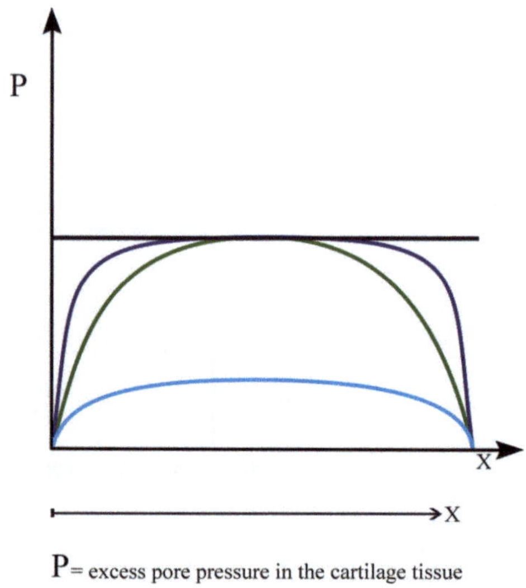

Fig. 4.5 Change in 'excess pore pressure' (p) distribution throughout the cartilage with time. Upon application of a step loading at t_0, the initial excess pore pressure distribution 'steps up' with the application of the step loading and is also initially constant across the cartilage (black line). This excess pressure distribution supports the load initially, and there is no deformation of cartilage. This is because the pore fluid is assumed incompressible and the pore fluid has not had time to exit the cartilage (which would allow it to deform). Immediately after application of the step loading, pressure falls to zero along the edges of the cartilage (in this example, the sides of the cartilage plates are at reference pressure zero, but we mention that in vivo the cartilage continues to extend laterally beyond the region of contact). This fall in pressure at the edge leads to a pressure gradient within the cartilage. Consequently, pore fluid flows down the pressure gradient and exits the sides of the cartilage. As the pore fluid exits the cartilage, the excess pore pressure in the cartilage decreases. Eventually, the excess pore pressure in the cartilage plates dissipates to the reference pressure zero. Figure drawn by Yvette Harrap

4.3.3 Boundary Lubrication

Real surfaces are not perfectly smooth. Any real surface has microscopic 'valleys and peaks' creating surface 'asperities.' When two real surfaces contact, they normally interact via 'asperity contacts' (see Fig. 4.8). Experimentally, it is found that asperity contacts are responsible for the large initial static equilibrium friction coefficient observed for most solid materials. A static equilibrium friction coefficient of 0.3 is common for many materials (e.g. dry wood on dry wood, concrete on concrete, steel on steel, and cartilage on cartilage (e.g. in phosphate buffer solution (PBS) without PRG4 present)).

But the static friction coefficient may be elevated above the equilibrium sliding coefficient by mechanical interlocking or by adhesive stiction. Indeed, if for example, mechanical interlocking is a function of contact stress, the initial and equilibrium friction coefficient under reciprocal cycling can vary with load magnitude (Katta et al. 2008). Heavy loads usually result in elevated rates of wear.

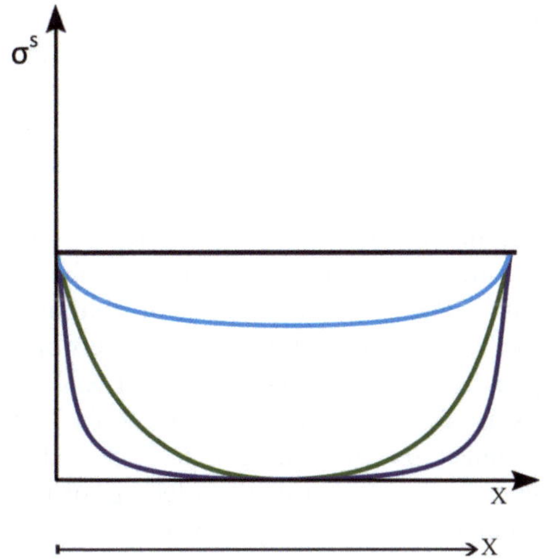

Fig. 4.6 As the excess pore pressure dissipates in the cartilage, the stress solid stress (σ^{S}) increases (see ruby, green and blue curves, at increasing times), causing the solid phase to deform. However the sum of the excess pore water and the solid stress always equals the applied compressive load, i.e. $\sigma^{total} = \sigma^{S} - p$ (i.e. the constant applied compressive total stress equals the sum of two curves, one in each of Figs. 4.5 and 4.6 both with the same time stamp). Note the usual solid mechanics stress convention applies, i.e. solid phase tension is taken to be positive, while fluid phase compression is taken to be positive. Figure drawn by Yvette Harrap

When $\mu_0 > \mu_{eq}$, and providing the sliding velocity is small enough, this may result in 'stick–slick' sliding behavior as one material slides across another. Stick–slip sliding usually results in elevated rates of surface wear.

On the other hand, the equilibrium sliding coefficient may be lowered by the addition of a lubricant (i.e. $\mu_0 < \mu_{eq}$). When a material is present between asperities contacts and the equilibrium friction coefficient is reduced, it is acting as a lubricant. If the lubricant is effective, in the sense of causing a significant reduction in the friction coefficient because the lubricant can stay attached to the sliding interface for a useful amount of time without being squeezed out or abraded away, it is then referred to as a 'boundary lubricant.'

As noted above, the boundary lubricant PRG4 (or lubricin), which is secreted by superficial zone chondrocytes, plays an important role in decreasing the equilibrium friction coefficient (Jay and Waller 2014). When cartilage rubs on cartilage in the presence of normal synovial fluid containing PRG4, the friction coefficient is reduced from about 0.3 to about 0.1 (Gleghorn et al. 2009). Jay et al. (2007) suggest PRG4 reduces stick–slip sliding, similar to the effect observed for industrial boundary lubricants such as molybdenum sulfide.

PRG4 binds strongly to denatured, amorphous, and fibrillar collagen. When PRG4 is attached at both contacting surfaces, PRG4 is reported to exert a strong repulsive force that increases with increasing concentration of PRG4 (Chang et al. 2014). Chang et al. (2014) found that repulsive force starts to rise exponentially when the distance between contacting surface is between 50 and 200 nm, the separation distance for initial interaction forces depending on concentration (i.e. higher concentration of PRG4 the longer range of repulsion), while the friction coefficient in the presence of high concentrations of PRG4 decreased around three or four times. But other molecules also play important roles in boundary lubrication too. See for example, the effects of hyaluronan and phospholipids on cartilage friction (Forsey et al. 2006), the effects of hyaluronan and aggregate on friction (Seror et al. 2011; Bell et al. 2006) and the effects of depleting hyaluronan, aggrecan, and collagen from the cartilage surface (Lee et al. 2013; Katta et al. 2008). Indeed, there is usually a well-developed gel some 7–10 μm thick, covering the surface of pig cartilage [see Figs. 3, 4, 5, 6, 7, 8, and 9 in Kobayashi et al. (1996)]. Simply wiping the natural cartilage surface prior to testing is reported to change the friction coefficient substantially (Naka et al. 2005).

Some have argued that hydration layers attached to 'polymer brushes' (e.g. HA and aggrecan) have special properties that prevent water being squeezed from between the asperity contacts, while retaining a fluid-like response to shear deformations (Murakami et al. 2009; Gaisinskaya et al. 2012) (see Fig. 4.7). Perhaps, the simplest way to think about so-called hydration lubrication is to again

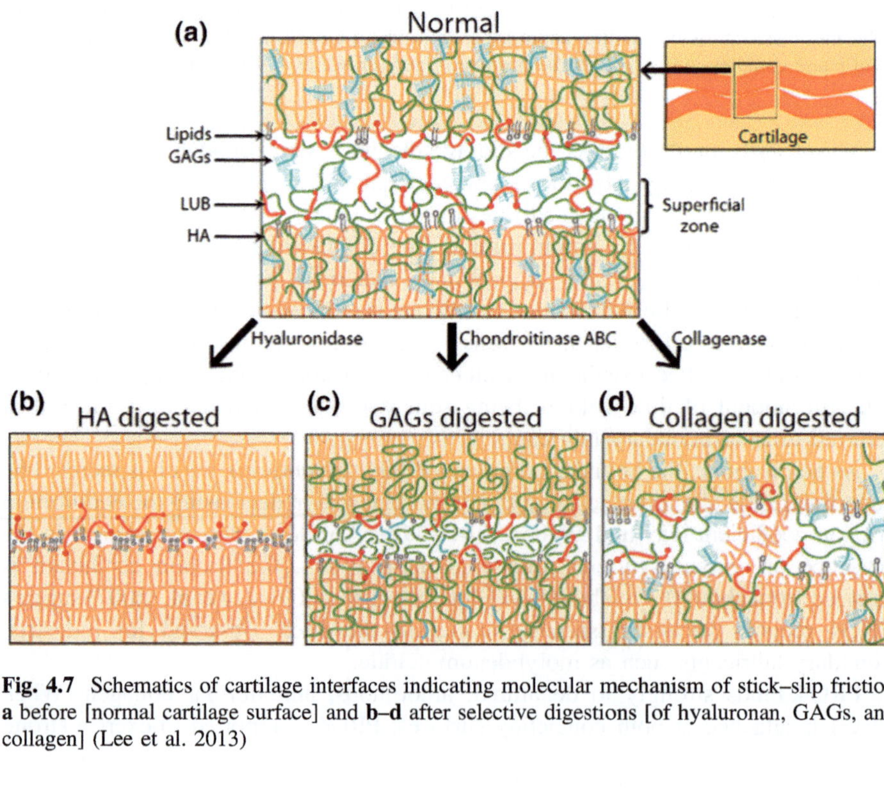

Fig. 4.7 Schematics of cartilage interfaces indicating molecular mechanism of stick–slip friction **a** before [normal cartilage surface] and **b–d** after selective digestions [of hyaluronan, GAGs, and collagen] (Lee et al. 2013)

use the concept of osmotic pressure (see discussion in Chap. 6). For as the collection of proteins, glycans and proteoglycans on the surface of the cartilage is compressed, the osmotic pressure in this gel increases, which enables the gel to retain its water. The retention of water by the gel results in a 'hydrated gel' on the surface of the cartilage [see Figs. 3, 4, 5, 6, 7, 8, and 9 in Kobayashi et al. (1996)]. This gel has the ability to shear without offering much shear resistance as it deforms, because somc of gel is beyond the extent of the collagen network (which largely is responsible for resisting shear forces). By this means, the gel acts as an effective boundary lubricant. From this perspective, it is perhaps unsurprising that the removal of hyaluronan, aggrecan, and collagen from the surface of the cartilage leads to functional deficits in the frictional performance of the surface (see, e.g. schematic depictions of these functional important changes in Fig. 4.7). Consequently, rates of surface wear are increased when these molecules are individually removed from the surface gel (Lee et al. 2013).

By definition, boundary lubrication can only operate when asperities are in contact and are transmitting normal forces. So the surfaces are usually close together when boundary lubrication begins. How close together depends on the roughness properties of the two contacting surfaces (together with the range of repulsion forces due to boundary lubricants, e.g. PRG4 at high concentrations has a repulsion range up to 200 nm, i.e. 0.2 μm).

Roughness is a measure of the vertical deviation of a real surface from an idealized smooth surface. Many different surface roughness measures have been defined (which differ from country to country and even from manufacturing company to manufacturing company). Two common roughness measures are (i) the average (of the absolute surface height deviation from the mean height) roughness denoted as R_a (Smyth et al. 2012; Tseng et al. 2016):

$$R_\mathrm{a} = \frac{1}{L}\int |z(s) - \bar{z}|\mathrm{d}s \tag{4.3}$$

and (ii) the root mean square (of the sum of the squares of surface height deviation from the mean height) roughness denoted as R_q :

$$R_\mathrm{q} = \sqrt{\frac{1}{L}\int |(z(s) - \bar{z})|^2\mathrm{d}s} \tag{4.4}$$

One may also define the skewness of the roughness, denoted S_k as

$$S_k = \frac{1}{LR_\mathrm{q}^3}\int |z(s) - \bar{z}|^3\mathrm{d}s \tag{4.5}$$

and the kurtosis, denoted K, as

$$K = \frac{1}{LR_q^4} \int |z(s) - \bar{z}|^4 \mathrm{d}s \tag{4.6}$$

Naturally, line integrals may be replaced by surface integrals as desired, while continuous measures of roughness may be replaced by discrete measurements at some predefined number of equally spaced sampling locations. Sometimes, maximum peak height (R_p), maximum valley depth (R_v) and maximum height of profile ($R_t = R_p - R_v$) are reported, along with skewness and kurtosis of the peaks, valleys, and maximum height profile. We note that a more complete description of the surface properties has to include frequency information, for the R_q of two surfaces can be the same, yet have very different numbers of asperities (i.e. the R_q of sin ($n\pi x$) over the interval zero to one is the same for all even n—similarly for R_a).

Clearly, from the definitions, we see roughness measures can be defined as a range of length scales that depend on the averaging length L. It is found that over length scales of 500 μm or so in normal cartilage, some asperities are relatively high and measured at ten or so microns (e.g. it is reported that for a scan square area with 250 μm side length for mature bovine cartilage, R_t = 12 μm (range 6–18 μm) was found (Graindorge et al. 2006). For equine joints, a similar magnitude of R_t roughness is shown in Fig. 4 of Smyth et al. (2012). A typical R_a for cartilage rubbing on cartilage is reported to be little over 1 μm (this varies with rubbing surface) (Katta et al. 2008). Over smaller length scales of 1 or 2 μm, asperities are tiny and measured in tens or hundreds of nanometres (a collagen fibril has a diameter around 100–150 nm). Clearly, if the average height of asperities over a 250 μm length scale on two cartilage surfaces is say 10 μm, and if the length scale of the contact region is greater than the relevant roughness averaging length (i.e. contact length > L), as it usually is, then asperities are in contact with one another when the surfaces are about 20 μm apart.

Importantly, we note here that while increases in surface roughness are usually (but not always) indicative of wear (i.e. as pieces of cartilage extracellular matrix are detached from the cartilage surface, roughness usually increases). While friction usually increases with an increase in roughness associated with surface wear, unfortunately there is no simple relation between cartilage surface roughness and friction—or cartilage surface friction and surface wear (Lee et al. 2013; Katta et al. 2008). Further discussion of the many reasons behind the complex relationship between the roughness and friction coefficient for cartilage is discussed in the next section on mixed-mode lubrication.

4.3.4 Mixed-Mode Lubrication

In normal diarthrodial joints, two cartilage surfaces may reciprocally, roll and slide across each other, with a ‘mix’ of hydrodynamic lubrication and boundary lubrication reducing friction between the contacting surfaces. Rolling generally involves less frictional resistance to motion than sliding, but rolling resistance does dissipate energy just like sliding resistance. It is interesting and probably important to observe that for the knee joint, the tibial cartilage surface and menisci experience limited amounts of sliding during each gait cycle (typically measured in millimeters), while the femoral cartilage surface and menisci experience comparatively large amounts of sliding during each gait cycle. This difference in sliding rates on these two surfaces probably has important implications for optimal joint structure (e.g. differences in cartilage thickness and composition on the tibial and femoral bones), and so on operational performance.

As described above, in this so-called mixed mode of lubrication, the apparent initial friction coefficient of normal cartilage can be very low—around 0.002–0.02. But in steady-state operation, the kinematic equilibrium friction coefficient usually rises to about 0.05–0.1 (depending on sliding velocities and the strain in the cartilage tissue (Gleghorn and Bonassar 2008), as well as on load fractions attributable to rolling and sliding).

It is an empirical fact that joint lubrication is improved by relative motion between contacting surfaces. Classically, a shaft rotating in a bearing is analyzed using so-called Stribeck curves (Gleghorn and Bonassar 2008). When the load on the rotating shaft is large and its rotation slow, bearing friction is primarily related to the effectiveness of boundary lubrication. In this case, friction is relatively high. As shaft rotation speed pick up, hydrodynamic lubrication increases relative to boundary lubrication, and the friction between shaft and bearing decreases. This is because as shaft rotation speed increases, more fluid is ‘dragged’ into the gap between the shaft and the bearing, leading to a greater proportion of hydrodynamic lubrication relative to boundary lubrication. We then appreciate that viscosity is not only important for the rate of fluid being squeezed from the gap, but also for rate of fluid being entrained into the gap by relative movement between fluid and the contact interfaces. Fluid is ‘dragged’ into the gap at the ‘leading edge’ of a sliding contact, primarily due to the propensity of the fluid and solute molecules to attach to the sliding surfaces and interact with one another (e.g. macromolecules may deform, break, become entangled, segregate, etc.), and inertial effects influencing fluid motion (explaining why fluid entrainment increases with increasing speed of relative motion between the contacting surfaces).

So there is a trade-off between energy dissipation and lubrication effectiveness—greater energy dissipation with fluid shear occurs at higher fluid viscosity, but the benefit is that higher viscosity promotes hydrodynamic lubrication via entrainment. Entrainment of synovial fluid into the contact interface at slow entrainment velocities probably helps to explain why normal synovial fluid is initially about 40–50 times more viscous than water (Wright and Dowson 1976), as this viscosity increase promotes low-speed synovial fluid entrainment. At higher entrainment velocities, viscosity (and energy dissipation) can reduce while maintaining the same amount of fluid entrainment. In this context, we note that the aforementioned complexity of the synovial fluid shear properties actually promotes adequate entrainment of synovial fluid at any entrainment velocity. Of course, appropriate surface roughness probably aids fluid entrainment, but there is little known about this interaction.

The interplay between lubrication modes and friction is captured in so-called Stribeck curves (see Fig. 4.8) or Stribeck surfaces (Jay and Waller 2014; Gleghorn and Bonassar 2008).

In mixed-mode lubrication, soft porous media bearings with rough surfaces (e.g. cartilage) experience some interesting interactions. First as the asperities on two cartilage surface contact the cartilage deforms. This changes the shape of the contact surfaces. If loadings are maintained over a considerable period time (e.g. it usually takes about two to three hours or more for the pore fluid to completely exit a normal loaded cartilage and for it to reach static equilibrium), the cartilage deformation may significantly increase the (gross) surface area of contact. Any increase in contact area increases the congruence of the cartilage surfaces, and reduces nominal contact stresses.

An increase in congruence with an increase in cartilage contact area has another important effect. For the larger the contact area and more congruent the cartilage surfaces are within a joint, the longer the average pore fluid drainage path becomes (i.e. fluid flow paths within the cartilage tissue and along the contact interface both become longer). In this context, it is noted that as the meniscus in the knee joint increase contact area and congruence between the femoral condyles and the tibial surface. The meniscus not only distributes load over a wider area on the articular cartilage attached to subchondral bone (reducing nominal contact stress), but in

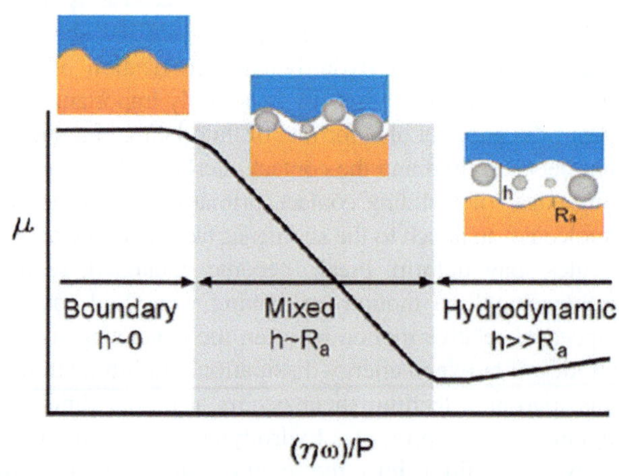

Fig. 4.8 Sketch of a Stribeck curve relating friction coefficient μ to the Hersey number $\eta\omega/P$, a non-dimensional expression of dynamic viscosity (η), shaft speed (ω), and mean contact pressure (P) for a typical journal bearing. Lubrication modes of a system are related to asperity–asperity interaction and fluid separation of two surfaces as suggested by the inset cartoons (Gleghorn and Bonassar 2008). See also Stribeck surface for the dog ankle joint shown in Fig. 2 of Linn (1968)

addition the presence of the meniscus greatly lengthens the paths for fluid to drain from between contacting surfaces. Importantly, the longer the fluid drainage pathway, the longer the duration of hydrodynamic lubrication. In other words, more congruent joints act to reduce both average nominal normal stresses, and they have a lower apparent friction coefficient for longer. This reduction in average normal nominal stress while extending the duration of hydrodynamic lubrication reduces the time-averaged friction coefficient at the surface, thereby reducing wear at synovial joint surfaces.

To model the contact interface, we could write a volume balance equation (i.e. a mass balance equation for the incompressible fluid in the gap) for the gap volume between the contact surfaces (Ahmad et al. 1998), viz.,

$$\frac{\partial \varepsilon_z}{\partial t} = \nabla \cdot v \tag{4.7}$$

where ε_z is the strain in the direction of gap closure (z direction) and v is the mean fluid velocity along the gap. One may use constitutive relationships of any desired complexity to convert this into an equation in terms of (excess) fluid pressure in the gap fluid. For example, one could assume a Stokes flow solution in the gap (Ahmad et al. 1998). Slip at the interface of porous media appears to not make much difference to the solution (Prakash and Vij 1974). Or perhaps, one could employ an exponential function to describe gap closure as a function of 'effective stress' in the gap, due to asperity contact (see Eq. (7.41) and/or Fig. 4.10), together with Darcy's flow law for the 'porous media' in the gap space. What becomes apparent is that the closure of the gap essentially becomes a 1-D consolidation equation (e.g. in axisymmetric coordinates), of the type described in Sect. 7.2.1. One can then use such an analysis to estimate the characteristic time (t_{90}) for gap closure (and so answer the question: how long does it take for 90% closure of the average gap height?).

In mixed-mode lubrication of soft porous media bearings with rough surfaces, a second interesting interaction occurs between the hydrodynamic and boundary lubrication modes. In synovial joints, pore fluid can move from the cartilage tissue into the cartilage contact interface. This process of fluid moving from the cartilage tissue toward the contact interface and exuding between the contact surfaces holding open the gap, and then being lost by sideways movement along the gap, was named 'weeping lubrication' by McCutchen (1983).

Weeping results from an interaction between hydrodynamic lubrication and boundary lubrication modes at a rough contact interface involving saturated porous media. There is much confusion as to how 'weeping' at an interface occurs. To develop an understanding of how 'weeping' to an interface occurs, it is helpful to first consider a simple, yet related 'thought model.' Our thought model is simply a porous, highly permeable material sandwiched between two plates of cartilage (see Fig. 4.9). To focus on the basic ideas and to simplify the explanation, we require that the 'sandwiched material' at the interface has the same stiffness as the two cartilage plates, the only material property difference being that the sandwiched

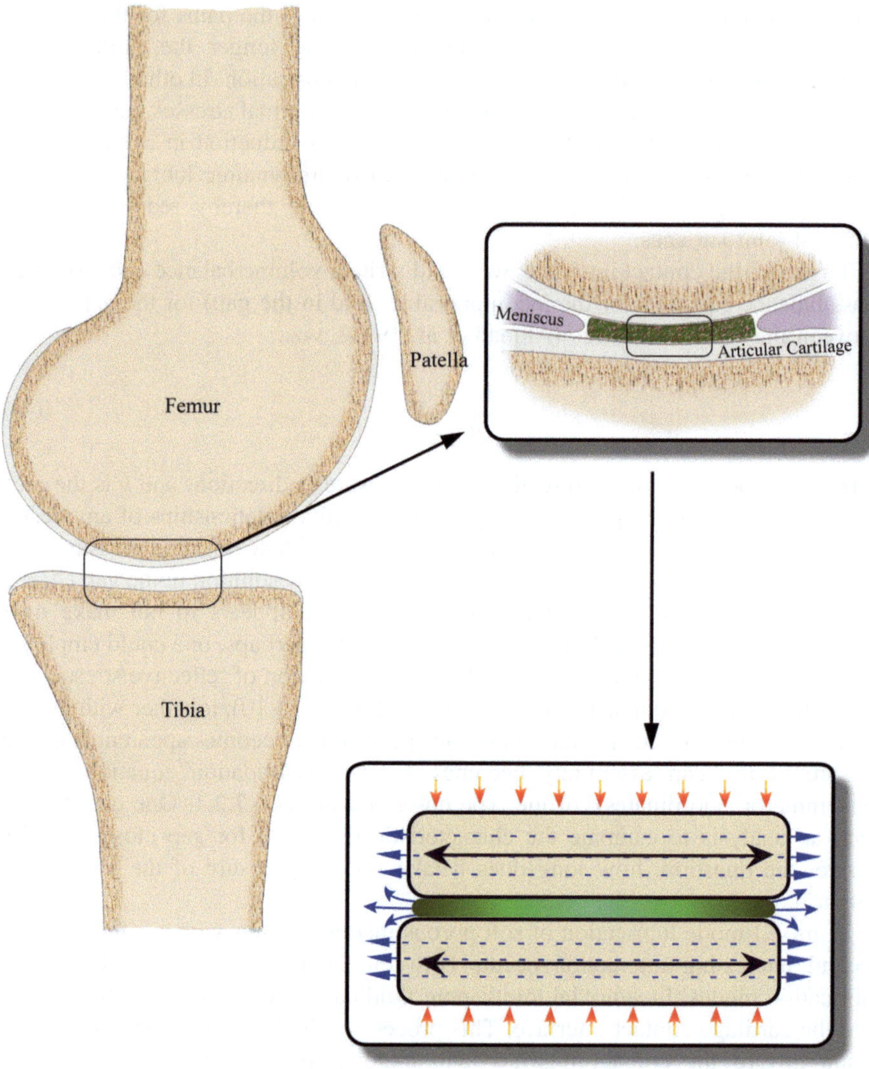
Femur
Patella
Tibia
Meniscus
Articular Cartilage

◄**Fig. 4.9** Change in 'excess pore fluid pressure' distribution throughout the cartilage with time. Upon application step loading at t_0, the initial excess pore pressure distribution also 'steps up' and is constant through the cartilage and the porous interface material (shown in green). The initial excess pressure distribution again supports the load entirely, and there is initially no deformation of cartilage, because the solids and pore fluid are incompressible and the pore fluid has not had time to exit either the cartilage or the cushion bearing at the interface. Pressure falls to zero at the edges of the cartilage and cushion bearing. This leads to a pressure gradient within the cartilage and cushion bearing. Fluid now exits the sides of the cartilage and the cushion bearing. However, because the porous cushion bearing has a higher permeability, pore water pressure falls more quickly in the cushion bearing (due to faster lateral drainage), than in the cartilage. This leads to a pressure gradient between the two materials, and consequently, there is a flow of pore fluid from the cartilage into the cushion bearing. This pore fluid flow into the cushion bearing helps maintain pore fluid pressure in the cushion bearing, at the expense of causing a faster decrease of pore fluid pressure in the cartilage. Importantly, we note that analogously to pore fluid flow from the cartilage into the cushion bearing, the fluid flow into the interfacial region between two rough contacting cartilage surfaces helps maintain mixed-mode lubrication in vivo. Eventually, the pore pressure dissipates to zero. Again in this example, there is no flux through osteochondral junction, while the cartilage plate sides are zero pressure conditions (we note that in vivo the cartilage continues to extend laterally beyond the region of contact). We also mention that the product of permeability and thickness of the porous interfacial material is known as the 'lateral transmissivity' of a porous material (e.g. geotechnical engineers use geotextiles with high lateral transmissivity sandwiched between two soils (each with relatively low permeability) to speed the consolidation of the low permeability soils). Finally, we note that if the permeability of the cushion bearing is lower than that of the cartilage, then fluid would flow from the cushion bearing into the cartilage (i.e. fluid will flow in the opposite direction to fluid flow in 'weeping lubrication'). This situation, with flow from the cushion bearing into the cartilage, has been called 'boosted lubrication' (Walker et al. 1968). Figure drawn by Yvette Harrap

material has a larger permeability than cartilage. This sandwiched material could be view as a porous 'cushion bearing.' It is interesting to note in passing that so-called cushion bearings, which fit between the contacting cartilage surfaces, have been investigated as a possible treatment of osteoarthritis (Stewart et al. 1997) (so this thought experiment may have practical application after all).

Clearly, pore fluid will drain to the side of the cartilage in the same way as in previous model. But inserting a porous cushion bearing with a higher relative permeability between the cartilage plates, results in something interesting happening. Excess pore water pressure falls more quickly within the porous interfacial material than within the cartilage itself, due to the higher permeability and so faster lateral drainage in the cushion bearing. Consequently, pore fluid in the cartilage (which maintains a higher relative excess pore pressure because it less permeable) drains toward and into the cushion bearing, whereupon it then drains laterally through the cushion bearing to exit at the side of the cushion bearing.

This new drainage route from the cartilage, through the more permeable cushion bearing, maintains a higher pore water pressure in the cushion bearing (slowing its closure), but speeds up consolidation and settlement of the cartilage. If the cushion bearing has a very large lateral transmissivity, fluid flow in the cartilage will become essentially normal to the surface, and the duration of consolidation will be short-lived relative to what it may have been without the cushion bearing. We note

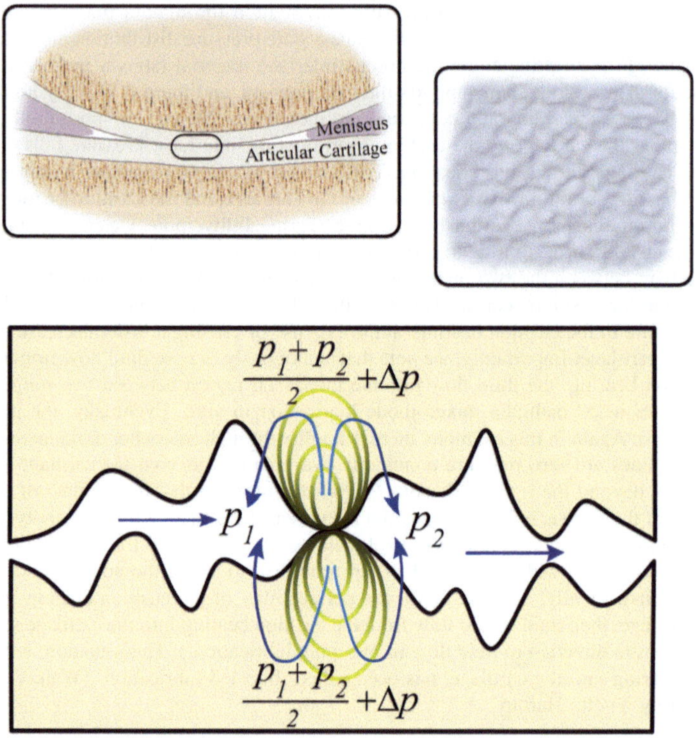

Fig. 4.10 Section through two asperities contacting in a saturated porous media in mixed-mode lubrication. In the gap between the contacting cartilage surfaces, fluid flows along the gap and down its pressure gradient, from p_1 to p_2 (in the direction of horizontal blue arrows). The mean pressure within the gap at the contact interface around the location of the asperity contact is $(p_1+p_2)/2$ (see Fig. 4.10), as was the case in model 1. However, the loading through the asperity contact causes an 'excess pore water pressure' to develop, as in model 2 (Δp) is the excess pore water pressure in the cartilage tissue immediately above and below the asperity contact (see contours show iso-excess pore water pressures). The total pore water pressure within the cartilage near the asperity contacts is given by $\frac{p_1+p_2}{2}+\Delta p$. Note that because $\frac{p_1+p_2}{2}+\Delta p>\frac{p_1+p_2}{2}$ pore fluid flows down the pressure gradient from the cartilage tissue to the contact interface. This 'weeping' of pore fluid to the contact interface (see flow paths of blue curved arrows) supports hydrodynamic lubrication in the mixed lubrication mode. But the weeping is driven by increasing asperity contact area. In other words, the cartilage below the asperity consolidates expelling fluid through the articular cartilage surface, helping to keep the contact interface. As this flow from the cartilage reduces, the excess pore water pressure in the gap gradually dissipates, and then solid phase stress in the asperity contact increases. As the effective stress in the asperity increases, the asperity gradually deforms, reduces its height, increases its contact area, and increases its friction coefficient. For both 'side' and 'plan' views, see also illustration Fig. 5(1) in Moore and Burris (2017). Figure drawn by Yvette Harrap

in passing that this same technique is employed by geotechnical engineers to speed up consolidation settlement of embankments constructed on soft clay ground. For example, vertically installed (so-named) 'wick drains' and horizontally installed 'drainage mats' are used to speed consolidation embankments on soft clay, which is a low permeability soil.

This thought experiment, which considers the behavior of a cushion bearing between two cartilage plates, is directly analogous to the gap space between contacting surfaces of normal cartilage within synovial joints. But if there is no cushion bearing normally present in synovial joints as described in the previous illustrative model, what is playing the role of a cushion bearing in vivo?

At least in theory, when two perfectly smooth cartilage plates moving together in a fluid, there is an abrupt transition from a complete reliance on hydrodynamic lubrication (which has a very low-friction coefficient and very low wear rates) to complete reliance on boundary lubrication (which has a much higher friction coefficient, and so higher wear rates) [see Fig. 5 (Neu et al. 2008)]. However, for a rough cartilage surface, there is now the possibility of a gradual transition from hydrodynamic lubrication to boundary lubrication through a mix-mode lubrication regime. Real contact surfaces are rough (at a range of length scales) and so asperities begin to contact well before all the fluid has been expelled from the contact interface. So it is the contact asperities protruding from the cartilage surface that are analogous to the cushion bearing material in the previous example, while the gap space is analogous to the void space within the cushion bearing. In exactly the same way as fluid flows (weeps) into the cushion bearing, so pore fluid flows (weeps) into the gap at the contact interface, and then move sideways in the gap between the cartilage plates. This flow is due to the relatively high lateral transmissivity of the contact interface gap relative to the cartilage plates (creating a lower pressures in the contact interface and so drawing fluid into it), while the load passing through the asperity contacts maintains an elevated (i.e. an excess) fluid pressure in the cartilage plates [see, e.g. illustration Fig. 5(1) in Moore and Burris (2017)]. This weeping flow into the gap at the contact interface prolongs the duration of mix-mode lubrication substantially, and because the gap is small, only a small amount of fluid is required from the cartilage to keep the pressure up in the gap, thereby minimizing the decrease in the time for consolidation of the cartilage plates.

To now see in more detail exactly how asperity contact leads to weeping of pore fluid from the cartilage tissue into the contact interface, consider the single asperity contact shown in Fig. 4.10. The (weeping) drainage of fluid toward the contact interface from the cartilage tissue provides additional fluid, which maintains pressure along the contact interface for longer, which in turn lengthens the duration of mixed-mode lubrication of rough articular cartilage well beyond what it would be if articular cartilage were not saturated with pore fluid. Indeed, 'weeping flow' that provides support for mix-mode lubrication does come at the cost of speeding up consolidation of the cartilage, which ultimately leads to higher friction coefficients. Faster weeping and so more fluid entering the gap at the contact interface may lead to a greater fraction of hydrodynamic lubrication initially, but faster weeping flow rates will hasten the consolidation time, and so hasten the onset of boundary lubrication [and raise friction coefficient; in the context, we note the consolidation analysis by Miramini et al. (2017)]. In other words, hydrodynamic lubrication will fade more quickly as flow rates from cartilage to contact interface increase and the cartilage drains more quickly.

But this trade-off can be partially offset by using a smaller gap height and by having a thicker cartilage, which extends the duration of consolidation, and so extends the duration of mixed-mode lubrication. Most importantly, we now appreciate that the duration of mixed-mode lubrication is now linked to the timescale for consolidation, which is much, much longer than the timescale for gap closure would have otherwise been. At this point, we also note that larger animals tend to have thicker cartilage (see discussion Sect. 1.6), which has potentially more fluid available for to support 'weeping' lubrication. Larger animals also have longer fluid drainage paths along longer congruent contact interfaces. These joint features help to extend the duration of mixed-mode lubrication of synovial joints in heavier, larger animals.

In mixed-mode lubrication of the cartilage contact gap coupled to cartilage consolidation, there is a long, gradual transition from mixed-mode lubrication dominated by hydrodynamic lubrication to one dominated by boundary lubrication. The pore fluid weeping from the cartilage tissue to the contact gap (see Fig. 4.10 for an explanation of why pore fluid in the tissue weeps to the contact interface) causes an increased pore fluid pressure in the gap, which slows the increase in effective stress within the asperities within the gap, which slows the deformation of the asperity (asperities have an excess pore water pressure similar to the excess pore water pressure in the gap). As fluid flow into the gap reduces as the degree of cartilage consolidation increases, so the excess pore water pressure in the gap slowly reduces, the effective stress in the asperity increases, the asperity contact area slowly increases, and the two contact surfaces move closer together. As the gap closes, more asperity contacts are recruited. The slow process of expelling the pore fluid from the cartilage via the gap space enables a prolonged, gentle transition between the hydrodynamic and boundary lubrication, ensuring a prolonged period of mixed-mode lubrication for the contacting surfaces.

To model weeping from the cartilage into the contact gap, we simply need to add a source term to the gap closure Eq. (4.7), viz.,

$$\frac{\partial \varepsilon_z}{\partial t} = \nabla \cdot v + s \tag{4.8}$$

where s describes the flow of fluid from the articular cartilage into the contact gap. What now becomes apparent is that while consolidation of the articular cartilage speeds up somewhat (due to the fluid escaping from the cartilage to the gap), the gap closure time (t_{90}) itself is greatly extended (because of the fluid flowing from the cartilage into the gap). Indeed, the time for gap closure is extended until the rate of fluid flow from the articular cartilage slows to a volume significantly less than the initial rate of flow from the gap. In other words, the great benefit of weeping lubrication is that gap closure time becomes coupled to the consolidation drainage time for the articular cartilage. Depending on the initial gap thickness, in theory the gap closure time without weeping lubrication may be extended from a time measured in seconds or minutes to one measure in at least several tens of minutes and possibly several hours.

When applied load increases, more fluid is induced to flow from the cartilage tissue to the contact interface, increasing the proportion of hydrodynamic lubrication in mixed-mode lubrication for a significant length of time. Of course, during reciprocal loading, the cartilage reaches a steady state where it imbibes fluid while unloaded (due to negative excess pore pressures in the cartilage), and expels fluid while loaded. If the contacting cartilage surfaces slide forward during the period of negative pressures and imbibition, this pressure gradient into the cartilage greatly assists the entrainment of synovial fluid into the contact gap, and the movement of pore fluid from the contact gap into the articular cartilage, thereby increasing the fraction of water in articular cartilage.

While the flow of fluid from tissue to contact gap increases the duration of the hydrodynamic contribution to the mixed-mode of lubrication, we see that more pore fluid movement into the contact interface means greater hydrodynamic support, but this necessarily means there is an inevitable leads to faster consolidation and greater reliance on boundary lubrication. If the load on the cartilage is maintained at a constant level until steady state is reached, after some time no more asperity contacts are recruited, fluid flow from the cartilage to the contact interface slows, and eventually it stops. When fluid flow stops the asperity contact area is steady, mixed-mode lubrication ends and in this state only boundary lubrication remains. Friction and wear rates are higher in a boundary only lubrication mode. This strongly suggests that the friction coefficient and wear is likely to be higher when cartilage is consolidated (Miramini et al. 2017).

This leads to the tantalizing suggestion that the degree of consolidation can be employed as an indicator of the friction coefficient (e.g. a high-resolution image estimating the degree of consolidation may provide information about the friction coefficient at in vivo cartilage surfaces (Miramini et al. 2017), but further research is required to confirm this). From a clinical viewpoint, the take-away message is that synovial joints are most durable when lightly loaded for short periods of time, and least durable when heavily loaded for long periods of time. In other words, synovial joints have optimal performance under intermittent, light load conditions (e.g. short walk and sit down to rest is probably optimal).

Of course if the joint is cyclically loaded and unloaded (as occurs during walking or running), the two contact surfaces reciprocally slide over one another repeatedly, the pore fluid pressures at the contact interface and within the cartilage reach an oscillating steady state about a zero mean excess pore water pressure. Synovial fluid can be re-imbibed by the negative excess pore water pressures in the contact gap and cartilage tissue during unloading, as more synovial fluid is dragged and 'sucked' into the contact gap (by the reversal the excess pore pressure gradient in the contact gap and within articular cartilage due to the aggrecan wanting to swell) during the sliding associated with the swing phase of the gait cycle.

The viscosity of the synovial fluid and the cartilage surface roughness also aid entrainment of the synovial fluid between the reciprocating contacting surfaces (indeed, synovial fluid entrainment during a reciprocating load may be an important determinant of optimal asperity height and spacing). For such dynamic loading situations, the fluid pressures at the contact interface are continually fluctuating (with negative fluid pressures (suction) associated with the swing phase of the gait cycle, and positive pressures with the stance phase of the gait cycle). By this means we see that repetitive intermittent loading and interleaved with reciprocal sliding may cause the mixed-mode lubrication to be extended indefinitely, as there remains a significant mean positive fluid pressure at the contact interface during stance phase while walking/running etc.

The relationship between the magnitude of the excess pore pressure increase in the contact gap and the relative sliding speed of the contacting surfaces has been experimentally investigated. For example, Moore and Burris (2017) report that for mature bovine knee cartilage sliding on glass in phosphate buffer solution, with a stroke length of 20 mm and a sliding speed of 60 mm/s, and for a contact diameter of 5 mm and contact pressure of 0.25 MPa, then fluid support is 90% of the applied normal load, and the initial friction coefficient reduces by an order of magnitude (from 0.2 to 0.02). We also note that when lubricated with phosphate buffer solution, the film thickness during sliding is in the range 0.5–1 μm. When the sliding speed is reduced to 10 mm/s, the fluid support is about 40–50% of the applied load, and the friction coefficient was approximately half its initial value.

Clearly, entrainment of synovial fluid at a sliding contact, driving an increase in gap pore fluid pressure, is almost certain to be one important mechanism contributing to femoral–meniscal gap function (as well as influencing the effective stress in femoral articular cartilage itself), for the femoral–meniscal contact gap normally experiences a large shear sliding speed during the normal gait cycle (unlike the tibial–meniscal gap in the same joint, which comparatively does not slide very much and so move very far). But exactly how this sliding contact entrainment mechanism translates to in vivo conditions remains uncertain. In this context, we mention that (Moore and Burris 2017) obtained qualitatively similar behavior to that of reported by Linn (1967).

Importantly, the experiments by Linn (1967) were on whole dog ankle joints, which were loaded to 1 MPa and cycled at 40 times per min (Linn 1968; Linn and Radin 1968). Presumably, this approximates fairly closely the in vivo conditions experienced by real synovial joints. For Linn's experiment, it took approximately 5 min of reciprocal sliding to reach a new equilibrium and a lower kinematic equilibrium friction coefficient [reported by Moore and Burris (2017)].

Extrapolating from this experiment on a real joint, it seems plausible that if a person stands until their cartilage is nearly consolidated, and then commences walking, for the first 100 cycles or so the friction coefficient will, during this period, be much higher than usual. During this 'start-up' period, we hypothesize that significantly more cartilage internal damage and surface wear may occur. In other words, it may be that articular cartilage damage is concentrated into a relatively brief period of time associated with certain cartilage states. This may be analogous to machine wear being concentrated during certain periods, when most of the machine wear occurs (e.g. for internal combustion engines, the great majority of engine wear occurs soon after starting a cold engine, in the time it takes to reach its normal operating temperature). Avoidance of these high wear states may lead to a dramatic reduction in wear rates (e.g. car engines that are nearly always warm (e.g. taxis) last about five times as many kilometres compared to normal usage cars). This is a highly relevant and probably important research that remains to be explored.

Of course if the cartilage first consolidates under a static loading (so it initially has a large degree of consolidated state), and if reciprocal sliding then commences while loading remains unchanged, and providing the sliding between the cartilage surfaces is fast enough, then cartilage tissue will expand (or swell), as the increased (positive) excess pore fluid pressure in the contact gap due to the sliding will cause fluid to flow along the gap and into the cartilage tissue, reducing the solid (or effective) stress in the cartilage tissue (Moore and Burris 2017). In other words, instead of 's' being a source term in Eq. (4.8) (i.e. articular cartilage feeding fluid into the gap), 's' becomes a sink term (i.e. the gap feeds fluid into the articular cartilage tissue).

Moore and Burns (2017) report on such an experimental finding, which occurs at high sliding velocities (i.e. 60–80 mm/s), and they choose to call this tissue response 'tribological rehydration' [see also (Graham et al. 2018)]. But this tissue response clearly sits within the known poroelastic framework described in Sects. 4.2 and 4.3.2 above, and it seems unlikely that such additional esoteric nomenclature (i.e. 'tribological rehydration') will progress an understanding of cartilage behavior. Alternatively, much as 'unload swell' complements 'load consolidation,' so 'slide swell' complements 'slide consolidation' (the difference in tissue responses depending not on the loading, but on the initial consolidation state of the tissue). However, even for laboratory test conditions using simplified contact conditions, the specific details of how these factors are actually interacting in real synovial joints is only just beginning to be elucidated, providing many research opportunities. In this context, the experimental findings of Linn (1967), Nguyen et al. (2010), Moore and Burns (2017) and Graham et al. (2018) make for very welcome additions to the cartilage literature.

It is clear is there is a trade-off in the frictional response at a cartilage contact interface with asperity height. For one could decrease the height of the asperities, and in the limit there be little discernible synovial fluid film at the contact interface, as synovial fluid would probably not be entrained as well by completely smooth surfaces. In this case, the cartilage is largely relying on the articular cartilage ECM to provide the mixed-mode lubrication. This may work satisfactorily for a while (depending on the quality and amounts of the boundary lubricant at the sliding surfaces e.g. the amount of lubricin in the articular tissues, the amounts of aggrecan and hyaluronan etc.), but collagen fibers would be more easily pulled from the smooth surface due to the lower quality of boundary lubrication (mechanical interlocking increases with decrease in the quality of boundary lubricant). The friction coefficient would then rapidly increase, and the surface would probably not retain its smoothness for very long (i.e. it would become fibrillated).

Or one could increase the height of asperity contacts, so load transmission between the contacting surfaces first occurs when the surfaces are further apart. But in the limit, asperities become too high (and the surface too rough), and mechanical interlocking of the asperities then becomes much more prominent. Once again, the friction coefficient increases. Indeed, mechanical interlocking (on either a very small scale or a larger scale) is probably what happens when cartilage surfaces become 'fibrillated' [e.g. in OA; for example, see increase in surface roughness shown Fig. 3 in Wright and Dowson (1976)]. We appreciate there is probably some optimal asperity height, somewhere between these two extremes (i.e. neither very smooth nor very rough).

In the case of soft porous bearings like articular cartilage, asperities prove to be useful because the asperities help to ensure the highest quality mixed-mode lubrication, enabling optimal synovial fluid entrainment by two reciprocating surfaces and an optimal boundary lubricant composition at the contacting asperities. Optimization of lubrication conditions ensures friction and wear are minimized. Perhaps the 'orange peel' appearance (see Fig. 4.25), observed on the surface of normal bovine cartilage (Shekhawat et al. 2009), provides the basis for an optimal surface configuration for cartilage contact? It is not known if the surface roughness changes appreciably with compressive deformations. More research is really needed to clarify the issues.

At this point, we mention that some argue that synovial fluid flows from the interfacial region into the cartilage, creating an alternate theory to weeping lubrication known as 'boosted lubrication' (Walker et al. 1968; Ateshian 2009). Now we have already met a type of 'boosted lubrication' in the discussion above—it will occur during the swing phase of gait. That is, as cartilage is unloaded it develops negative pore water pressures (i.e. suctions), and it imbibes water from the overlying gel at the articular cartilage surface (e.g. the gel being synovial fluid mixed with molecules adhered to or exiting the articular cartilage surface). As synovial fluid is drawn into the contact gap by negative pressures in the contact gap during the swing phase, water is drawn into the articular cartilage from the contact gap. This water moving into the cartilage partial dehydrates the water content of the gel on the surface of the cartilage. When loading comes on the joint again during the stance phase, these flows are reversed, and the cartilage weeps fluid into the contact gap. At steady state, the flow into the cartilage during the swing phase exactly equals the flow out of the cartilage during the stance phase [see, e.g. the analysis in (Zhang et al. 2015)].

However as originally conceived by Maroudas (1967) and discussed by Walker et al. (1968) and later by Dowson et al. (1970), boosted lubrication refers to the existence of '*trapped pools of synovial fluid*' in the contact gap, the pools being confined by confluent asperity contacts. This behavior results from assumptions about the lateral transmissivity of the interfacial region. For if the lateral transmissivity of the interfacial region is lower than that of the cartilage, then fluid may 'pool' within any surface depressions as the two cartilage surfaces contact (rather than escaping to the side) [for schematic illustration, see Fig. 6 in Walker et al. (1968)], and when the pressure in the cartilage drops below the pressure in the interfacial region (due to migration of the pore fluid to the free surface to the side of the contact region, or possibly through a defective osteochondral junction), the pressure gradient is then reversed relative to that driving weeping lubrication, inducing fluid to flow from the interfacial region into the cartilage (see Fig. 4.10). Dowson et al. (1970) also contemplates a 'molecular straining' effect, where fluid continues to move laterally, albeit it more slowly, yet hyaluronan molecules are retained in the contact gap.

Clearly, the type of behavior the fluid exhibits depends on the lateral transmissivity of the interfacial region and the integrity of the osteochondral junction. We note that for normal cartilage with normal surfaces, it seems most likely that weeping lubrication normally dominates during transient states (i.e. during walking before steady-state conditions arc achieved), while at steady state, both weeping lubrication (stance phase during gait) and a type of boosted lubrication (swing phase during gait) operates together, with equal intensity to maintain steady state (dynamic) conditions.

But the period of weeping lubrication dominating the flow back and forth in articular cartilage driven by a load cycle may be curtailed if the lateral transmissivity of the interfacial region falls too low (i.e. if the permeability in the contact gap falls below the permeability of an equivalent thickness of cartilage). If any pooled fluid remains at the interface at this time, then one may expect a switch in net flow direction, leading to so-named 'boosted lubrication' being dominant (which 'thickens' the gel layer on the surface), as conceived by Maroudas (1967) and discussed by Walker et al. (1968) and later by Dowson et al. (1970).

While further research is required to say definitely what happens at various cartilage surfaces during activities such as walking, it seems most likely that normally net flow into the contact gap from articular cartilage (i.e. 'weeping lubrication') is transiently sustained, and this net flow just fades gradually to zero. Then a type of boosted lubrication (where boosted lubrication is defined to mean fluid flow reversal (and imbibition of fluid from the contact gap)—but in this type of boosted lubrication, the fluid flow from the contact gap is driven primarily by negative excess pore water pressure in the articular cartilage itself during the swing phase, rather by positive pore water pressures in the contact gap during the stance phase) gradually increases. The boosted lubrication grows until at steady state for each complete gait cycle, the 'boosted lubrication' (i.e. imbibition of pore fluid by articular cartilage) occurring during the swing phase is exactly equal to the 'weeping lubrication' (i.e. expulsion of pore fluid from articular cartilage) occurring during the stance phase of each cycle.

A complete understanding of the costs and benefits involved in the various wear mitigation strategies (e.g. synovial fluid composition, surface roughness, weeping lubrication, boosted lubrication, and boundary lubrication, joint operating conditions) depends on many factors that are yet to be determined. But the factors influencing the trade-offs are likely to include joint motion (both the speed of motion and the magnitudes and relative proportions of rolling and sliding at the contact surfaces), the concavity/convexity of the cartilage surface, the congruence of the contacting surfaces (including the length of fluid drainage length at the contact interface), the magnitude and duration of the repetitive loadings to which the joint is subjected, the properties of the surface (including its roughness and slipperiness), the properties of the cartilage close to the surface (including the collagen and aggrecan contents, its stiffness and its permeability) and at depth (including cartilage thickness, stiffness and permeability), the concentration of lubricating materials (including aggrecan, hyaluronan, PRG4 and other molecules at the cartilage surface) and the lateral transmissivity of synovial fluid at the contact interface when two contacting cartilage surfaces are close together (which partly depends on the roughness and presence of hyaluronan and aggrecan attached to the cartilage surface, and the fraction of these molecules protruding from the contact interface). Undoubtedly there are many other nonlinear interactions, which may lead to unexpected in vivo behaviors [e.g. the measured friction may decrease somewhat at small loadings and then increase again at higher loadings, or with a change with reciprocal 'stroke length,' surface roughness and/or lubricant properties (Katta et al. 2008; Nguyen et al. 2010; Bell et al. 2006)].

Taking all these and other factors into account, presumably there is a (weighted) global optimal proportion of hydrodynamic and boundary lubrication that minimizes the mixed-mode friction coefficient over a range of environmental loading stressors. It also appears likely that this friction and wear optimum will change with age, disease states and with therapeutic interventions. Clearly, much remains to be discovered through both theoretical and experimental approaches.

4.4 Experimental Data Relating to Cartilage Friction

Given the importance of asperities, it is helpful to first consider the geometry of asperities on the surface of normal cartilage. We first need to recognize there are 'asperities' at a range of length scales. Indeed it has been pointed out that there is some confusion over what is 'cartilage geometry' and what are 'cartilage asperities' (Smyth et al. 2012).

Longmore and Gardner (1975) identify four length scales to define the contours of the cartilage surface, namely:

(i) 'Primary contours' (measured in mm of height and centimetres in extent) (these contours presumably relate to 'cartilage geometry' rather than to asperities);
(ii) 'Secondary undulations' (or 'surface waviness') are asperities defined on a length scale of about 0.2–0.5 mm (i.e. length scale about 200–500 μm). The height of secondary undulations at this length scale is not specified by Longmore and Gardner (1975), but secondary undulations are identified by Smyth et al. (2012) (in equine carpal joint cartilage, they appear to be about 10–15 μm in height [see, e.g. Fig. 4 (Smyth et al. 2012)], and they are specified in Graindorge et al. (2006), where asperities are reported to have an average height of 12 μm in mature bovine cartilage (range 6–18 μm), with an R_a of 1.4 μm (see, e.g. Fig. 4.25 showing MRI images of 'orange peel' appearance of the surface waviness in bovine cartilage);
(iii) The next sized asperities are 'tertiary hollows' on a length scale 20–45 μm. The height of asperities at this length scale is 0.5–2 μm (see Figs. 4.11, 4.12 and 4.13) and

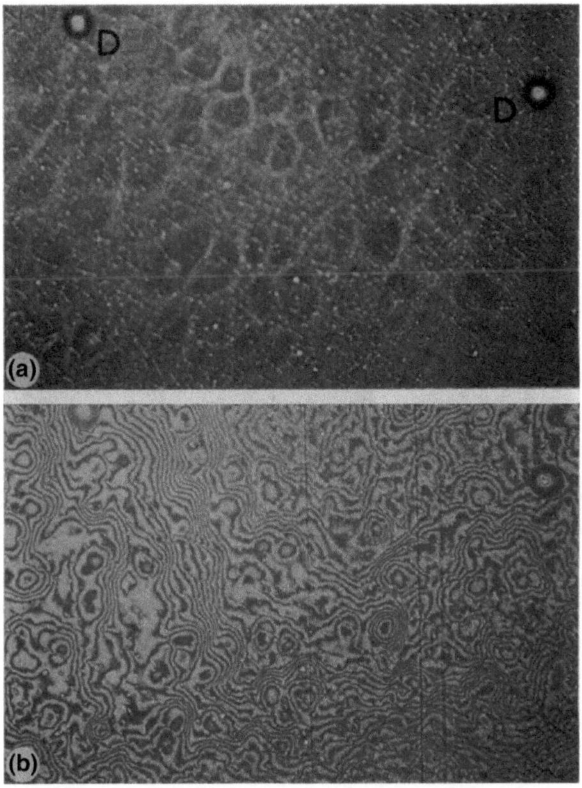

Fig. 4.11 Articular surface of the left lateral condyle of a 10-year-old male. 'Tertiary hollows' as paired figure-of-eight hollows are present. Labels D represent saline droplets **a** image magnification 170, and **b** interference pattern produces a contour map of same area (magnification 170) (Longmore and Gardner 1975)

(iv) Finally there are 'quaternary ridges' on a length scale of 1–4 μm. Asperities at this length scale are 130–275 nm in height (Longmore and Gardner 1975; Crockett et al. 2005). Similar asperity heights at this length scale are reported for human femoral cartilage by Crockett et al. (2005).

For fresh calf bovine cartilage surfaces, Shekhawat et al. (2009) report that

> *To the naked eye, the cartilage surfaces appeared smooth with slight "orange peel" texture, and their differentiation based on location and aspect was not evident.*

The 'orange peel' texture reported above is presumably related to asperities that are classified as 'secondary undulations' or 'surface waviness' above. The 'orange-peel' surface waviness is visualized by microMRI imaging of mature bovine cartilage (see Fig. 4.25). Based on Fig. 4.25, for normal cartilage the

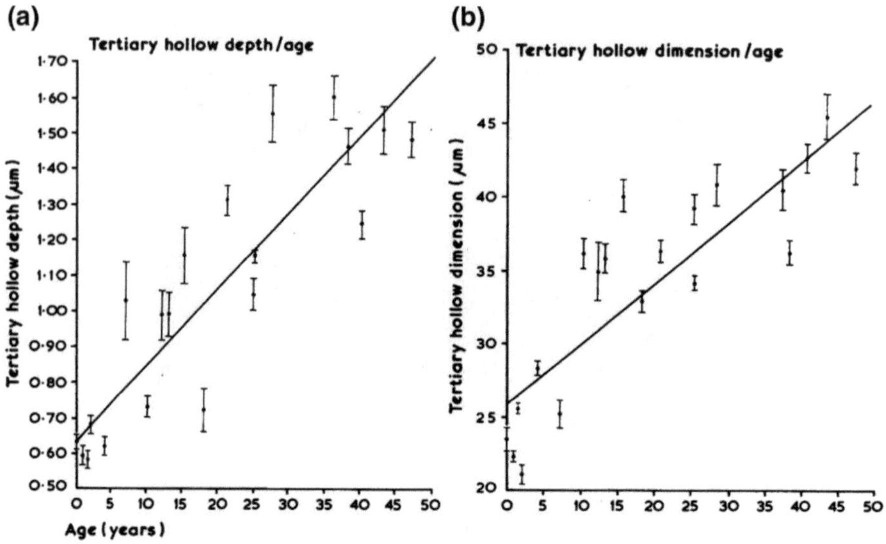

Fig. 4.12 **a** Graph of tertiary hollow depth against age. Regression of depth on age is given by y (depth in microns) = 0.65 + (0.021) × (age in years); $P < 0.001$; **b** graph of tertiary hollow diameter against age. Regression of diameter on age is given by y (tertiary hollow dimension in microns) = 26.3 + (0.41) × (age in years); $P < 0.001$ (Longmore and Gardner 1975). In a later publication, they confirmed the previous findings and added relationships that show the number of tertiary hollows per mm squared decreased with age (i.e. from about 300 per mm at age 10 years to 100 per mm at age 50 years) (Longmore and Gardner 1978)

secondary undulations or surface waviness have the appearance of sand ripples or ridges formed by waves in shallow water on a sandy beach. When normal cartilage is subject to loadings three times larger than normal for one hour and 1 Hz, the ridged appearance is similar but the height and spacing of the ridges has been strongly influenced by the magnitude of repetitive applied loading. Given the test conditions for this particular experiment reasonably closely approximate in vivo conditions, it seems likely that something similar probably occurs in vivo. We note that if there is traumatic injury to the cartilage, the surface changes (see Fig. 4.25), and this initial surface change probably progresses to surface fibrillation (Wright and Dowson 1976), and if the loading continues, it probably progresses to early stage osteoarthritis.

Each of the length scales to define cartilage surface contours can be investigated using a variety of measurement techniques. For example, 'primary contours' are revealed effectively by magnetic resonance imaging, while 'secondary undulations' can be revealed by laser scanning or profilometry or high-resolution MRI. 'Tertiary hollows' are characterized using light microscope interferometry, atomic force microscope or by scanning electron microscope imaging, while 'quaternary ridges' are characterized by scanning electron microscope imaging or an atomic force microscope.

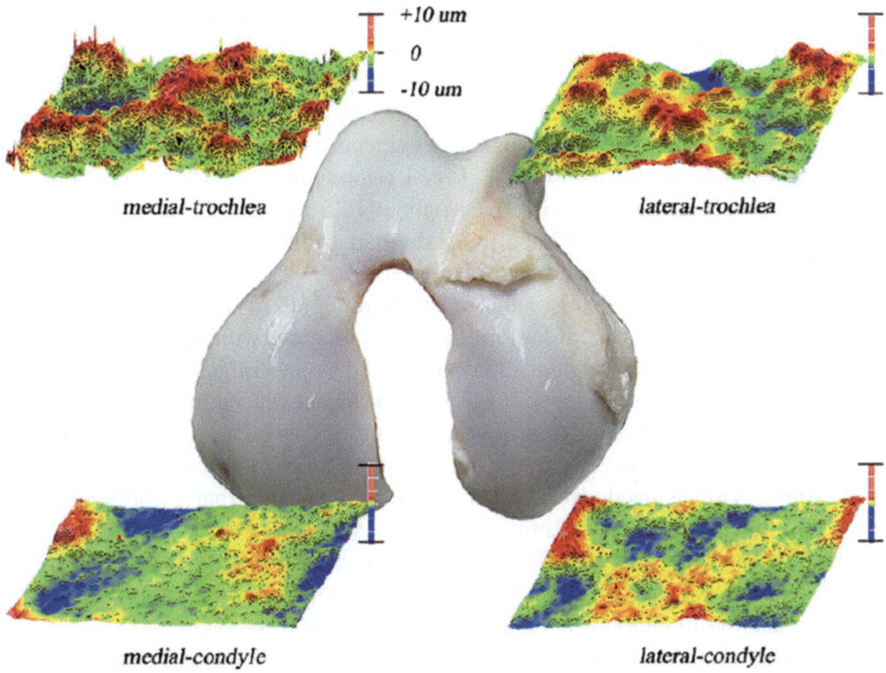

Fig. 4.13 3-D plots of the cartilage surface topography at four locations on bovine knee from 6- to 8-month-old animals (Shekhawat et al. 2009). Scan area is 0.70 mm by 0.52 mm. With very careful attention to sample preparation, differences are apparent in the roughness of the cartilage surface at various locations on femoral cartilage. Shekhawat et al. (2009) compare average trochlea roughness (R_a = 1.15 μm) with average condyle roughness (R_a = 0.6 μm) shown above. We note that Graindorge et al. (2006) reported similar data but provided more details on surface roughness for mature bovine articular cartilage (Graindorge et al. 2006). These authors found that upon scanning square areas, side length of 250 μm, with the 'surface adherent layer' removed: R_a = 1.39 μm, R_q = 1.75 μm. However they also reported, R_p = 5.36 μm (range 2–11 μm), R_v = −5.56 μm (range −3 to −7 μm) and R_t = 12 μm (range 6–18 μm). All these measurements of surface roughness were reduced somewhat when the surface adherent layer is intact, or reconstituted (i.e. the surface adherent layer smoothed the actual surface roughness) (Graindorge et al. 2006)

See Fig. 4 (Smyth et al. 2012): *Sample surface leveled with second-order polynomial*. Note that secondary undulations revealed by profilometry of equine carpel cartilage, have a height of about 10–15 μm above and below the mean. See also Fig. 2 in Walker et al. (1968) for measurements of secondary undulations on normal femoral heads: fetus at 8 months (asperities heights above and below the mean about 2–4 μm), and adults at 26 years (asperity heights above and below the mean about 1–2 μm) and 67 years (asperity heights above and below the mean about 2–5 μm) and osteoarthritic cartilage (asperity heights above and below the mean about 15 μm). See also Fig. 3 in Wright and Dowson (1976) for some measurements on human cartilage from normal (secondary undulations about 2–5 μm) to osteoarthritic (secondary undulations about 20 μm).

In this context, we note that when the knee joint is 'unloaded' in the supine position, and imaged in a standard clinical 1.5T MRI, the contact 'gap' between the two cartilage contacting surfaces is clearly evidenced as a continuous dark line [of greater or lesser distinctness—see, e.g. Fig. 1 (Goodwin et al. 2004)]. That this gap is evident in an image with a pixel size of about 1.25 mm (i.e. for a 1.5T machine), suggests that for an unloaded knee joint, the gap width can be initially large, i.e. potentially of the order of hundreds of microns.

It is interesting to note that Longmore and Gardner (1975) report that the number of so-named 'ridges' on lateral femoral condyles increase with age (Longmore and Gardner 1975, 1978). These ridges may be related to the 'ridges' observed in MRI images of the surface (see Fig. 4.25). Ridges probably represent accumulations of 'displaced material' adjacent to grooves, but ridges can be an artifact of cartilage surface drying (e.g. in an 'environmental scanning electron microscope' (Crockett et al. 2005), so care needs to be taken). It seems possible that compression of a convex cartilage surface will initially lead to slackness in the superficial zone cartilage, and when coupled with the large shear displacements experience by the femoral condyles rubbing on menisci during knee rotation, this environment may well increase the likelihood of ridge formation in this tissue.

It is likely that the surface roughness of the cartilage is tuned to the functional requirements of the joint. Indeed it has been shown that even within a joint, the surface roughness can vary with position over the cartilage contact surface (see Fig. 4.13). How this 'tuning' occurs requires further research, but clearly local surface wear is a function of local environmental conditions.

Based on scanning electron micrograph imaging of glutaraldehyde (1.5%) fixed cartilage tissues that appeared normal on dissection, it can be observed that convex, flat and concave cartilage surfaces may have very different roughnesses. For example, the talus bone forms a complex series of joint surfaces with the tibia, calcaneus, navicular and cuboid bones. It has been shown that there are deep surface depressions form on the concave aspect of the talar–calcaneus joint when unloaded, while the flat calcano–cuboid surface is much less rough and had rows of fibers parallel to the gliding direction (Myers et al. 1983) (Fig. 4.14).

Indeed, it seems likely that cartilage surface roughness of cartilage would be partly a function of the deformation of the cartilage, which changes the tensile stress within the superficial zone (i.e. convex surfaces are in tension when unloaded, compression when initially loaded, and finally in tension as deformation progresses, while for concave surfaces, this behavior is reversed (i.e. in compression when unloaded and in tension when loaded)). This may influence initial asperity heights, which are generally measured in an unloaded state.

In certain load and velocity regimes cartilage displays 'stick–slip' behavior. Stick–slip sliding occurs when the static friction is larger than the dynamic friction, resulting in jerky, stop–start sliding. Stick–slip most often occurs at high loads and slow sliding velocities. Lee et al. (2013) examined stick–slick behavior with and without various pre-treatments (pre-treatments include digestion of GAG, HA and

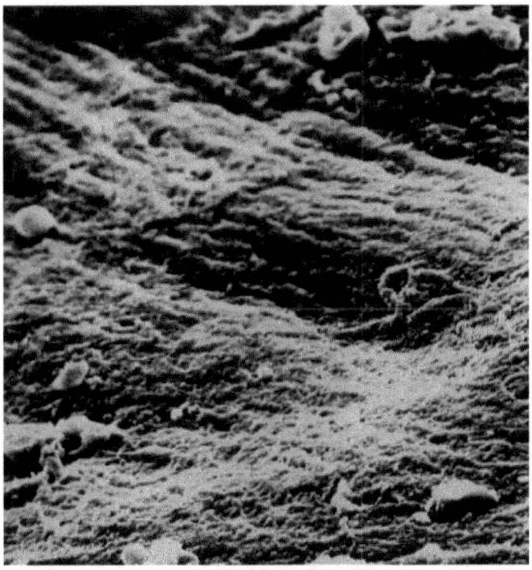

Fig. 4.14 Surface depressions [approximately 100 by 200 μm in extent] observed using SEM on concave surfaces [i.e. on the concave aspect of the talar–calcaneus joint]. Talar facet from 54-year-old male (magnification 1800) (Myers et al. 1983)

collagen from the cartilage). Lee et al. examined 'porcine leg cartilage' (full thickness cartilage was separated from bone and glued to glass slides). Within a micro-tribometer, the glass slides slid reciprocally with a stroke length of 100 μm in each direction at various velocities (up to 100 μm per s). Untreated cartilage was tested with equine synovial fluid as lubricant, while all other tests used phosphate buffer solution. Specimens were imaged under a white light interferometer and tertiary surface roughness evaluated (R_q) over a square with side length 80 μm (see Figs. 4.15, 4.16, 4.17 and 4.18). The roughness magnitudes measures reported appears to correlate within the previously mentioned classification as 'tertiary hollows.'

It is clear from these experiments that ineffective boundary lubrication (in the regime of high loads and slow sliding velocities) or tests conducted with 'abnormal composition' of the superficial zone cartilage, result in stick–slip sliding and friction (Lee et al. 2013). When stick–slip sliding occurs, it can result in substantial rates of wear (Jay et al. 2007; Lee et al. 2013). In these experiments, it is also apparent that hyaluronic acid and GAG plays a very important role in minimizing stick–slip regimes, and so total wear on the surface of the cartilage. Lee et al. (2013) report that 10 h of continuous stick–slip shearing (average normal stress > 0.08 MPa with velocities less than 1 μm per s) is required to see 'medium level' wear effects on normal cartilage, but much shorter times (only 1.5 h) is required to see effects on abnormal cartilage or cartilage with abnormal lubrication. It is also reported that 10 h of continuous smooth sliding (normal stress < 0.08 MPa with velocities greater than 1 μm per s) saw no change on roughness parameters. This data suggests that mild loading combined with slow movements may cause significant cartilage wear damage.

PRG4 plays a very important role in reducing the equilibrium friction coefficient —from around 0.3 to about 0.1 (see Fig. 4.19) (Gleghorn et al. 2009; Waller et al.

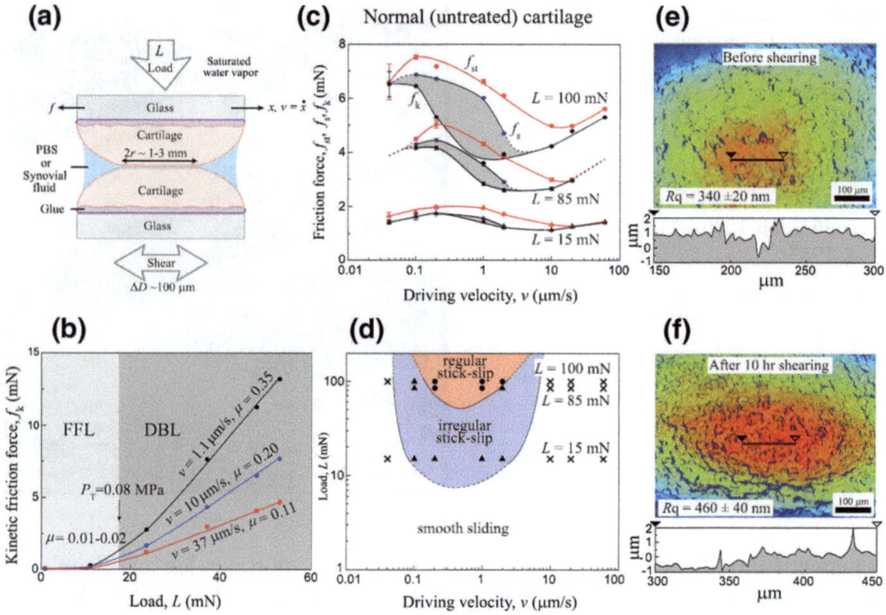

Fig. 4.15 **a** Schematic of the experimental setup used with measured and calculated variables. **b** Kinetic friction force (f_k) versus load (L) curve under different driving velocities (v = 1.1, 10, and 37 μm/s) showing low-friction coefficient (μ = 0.01–0.2) in the FFL regime (L < 16 mN) and high-friction coefficient (μ = 0.11–0.35) in the DBL regime (L > 16 mN). **c** Friction forces (f_{st}, f_s, and f_k) versus driving velocity (v) curve measured at three different loads (L = 100, 85, and 15 mN). The shaded regions indicate the stick–slip sliding regime. **d** Friction map showing representation of cartilage lubrication profiles. The dotted lines indicate the observed and measured trends based on the experiments and theories (15). **e** and **f** Topographic images (top view) of the contact zone of normal (non-digested) cartilage (**e**) before and (**f**) after 10 h shearing in stick–slip conditions. Red and blue colors indicate higher and lower heights, respectively. Single height profiles are also shown below each image (Lee et al. 2013). *Notes* (i) equine synovial fluid used for **c** and **d** above, and (ii) the surface roughness R_q increased from 370 to 460 nm after 10 h stick–slip shearing, but no change was observed after 10 h of smooth sliding

2013). It also reduces the likelihood of stick–slip sliding, and in doing this reduces articular cartilage wear (Jay et al. 2007). PRG4 has also been shown to reduce cartilage damage in an ACL transection model of osteoarthritis (an ACL transection tends to shift knee motion away from 'cartilage rolling on cartilage' toward much more 'cartilage sliding on cartilage,' i.e. shear straining increases) (Jay et al. 2010).

Shear straining of cartilage modulates chondrocyte secretional and metabolic responses (Nguyen et al. 2010; Ogawa et al. 2014; McNary et al. 2012). And importantly, it has been shown experimentally that a direct correlation exists between increased friction coefficient and increased levels of chondrocyte

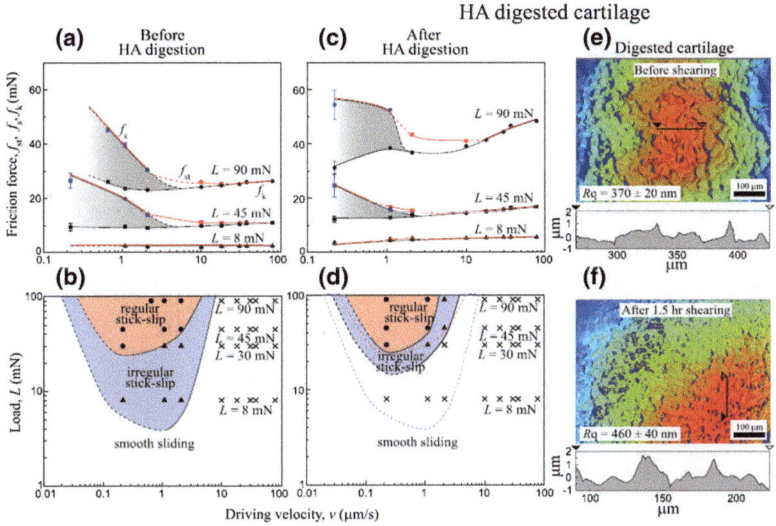

Fig. 4.16 **a** and **c** Friction forces versus sliding velocity curve; **b** and **d** friction maps (**a** and **b**) before and (**c** and **d**) after GAGs digestion. **e** and **f** 3D images (top view) of HAs-digested cartilage (**e**) before and (**f**) after 1.5 h shearing (1 h stick–slip and 0.5 h smooth sliding conditions) (Lee et al. 2013). *Note* PBS used as lubricant

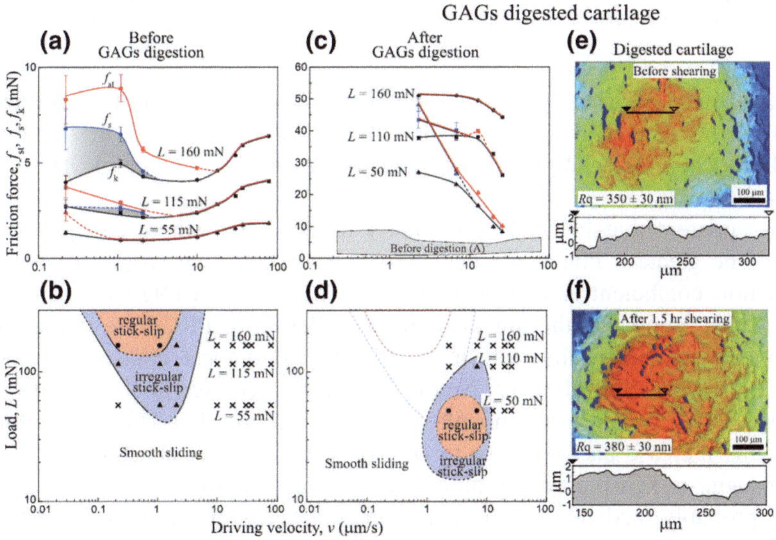

Fig. 4.17 **a** and **c** Friction forces versus sliding velocity curve; **b** and **d** friction maps (**a** and **b**) before and (**c** and **d**) after GAGs digestion. **e** and **f** 3D images (top view) after GAG digestion (**e**) before and (**f**) after 1.5 h shearing (1 h stick–slip and 0.5 h smooth sliding conditions) (Lee et al. 2013). *Note* PBS used as lubricant

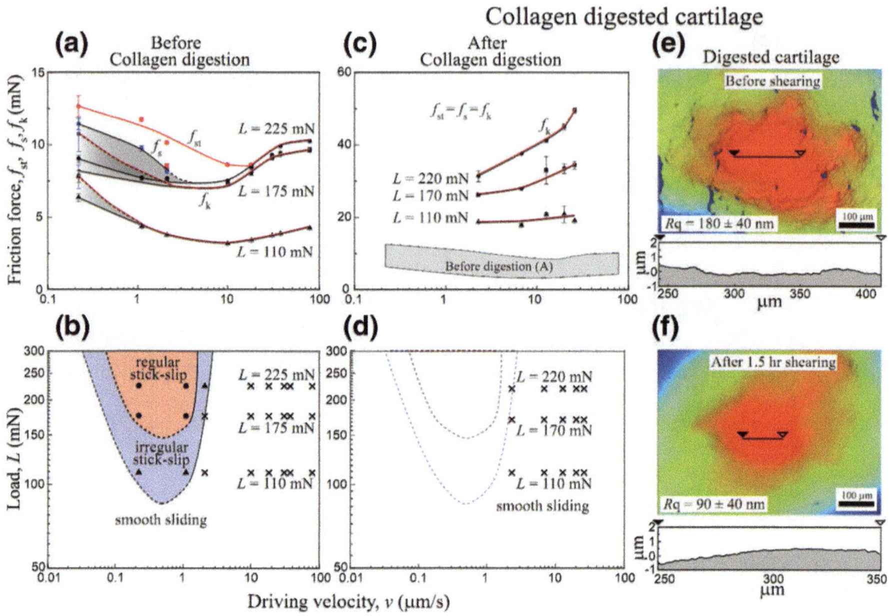

Fig. 4.18 **a** and **c** Friction forces versus sliding velocity curve; **b** and **d** friction maps (**a** and **b**) before and (**c** and **d**) after GAGs digestion. **e** and **f** 3D images (top view) after collagen digestion (**e**) before and (**f**) after 1.5 h shearing (1 h stick–slip and 0.5 h smooth sliding conditions) (Lee et al. 2013). *Note* PBS used as lubricant

apoptosis, with increasing amounts of apoptosis towards the cartilage surface (see Fig. 4.20) (Waller et al. 2013; Jay and Waller 2014). We note that addition of PRG4 reduces the friction coefficient and so the degree of chondrocyte apoptosis (Fig. 4.20).

It is apparent that PRG4 is particularly needed most where contact loads are likely to be greatest. Potentially destructive shear stress will vary in proportion to the friction coefficient and the magnitude of normal load (Nguyen et al. 2010). Indeed, it is found that the highest expression levels of PRG4 are in the anterior regions bovine stifle joint, which experience the highest contact stresses (see Fig. 4.21) (McNary et al. 2012; Neu et al. 2007). Increases in shear stress on the cartilage result in increased amounts of PGR4 secretion, but TGFβ inhibitors can return this increase back to close to control levels (Neu et al. 2007). This observation suggests TGFβ plays an important signaling role for PRG4 production. The link is probably mediated by shear deformation mechanically inducing proteases, which that activate latent TGFβ in the ECM (we note that mechanical deformation of devitalised (i.e. cell free) tissue does not activate latent TGFβ) (Albro et al. 2013).

It has been shown that PGR4 production is CREB dependent, while CREB is activated by ATP, PGE2 and PTHrP signaling (Ogawa et al. 2014). Interestingly we note that ATP, PGE2 and PTHrP are all produced in response to fluid flow shear

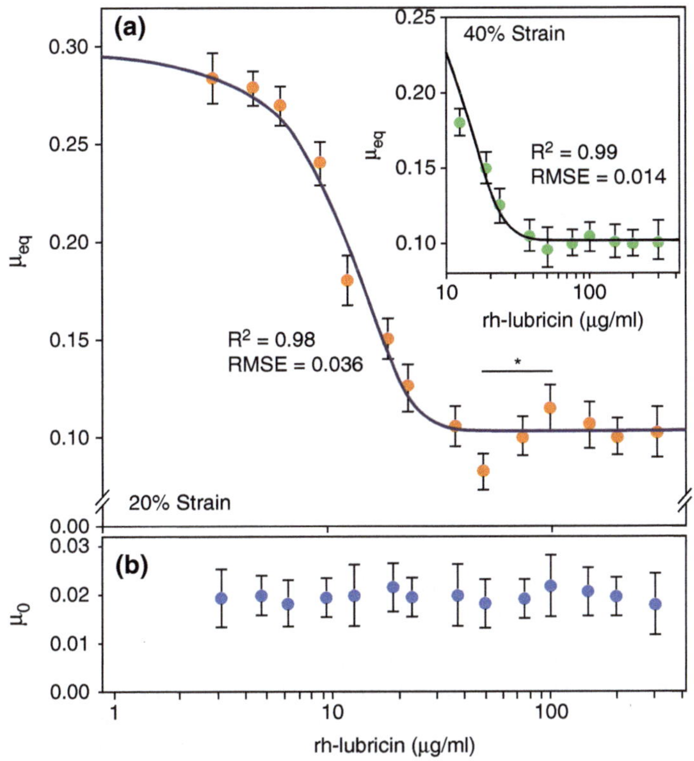

Fig. 4.19 Concentration-dependent response on equilibrium friction coefficient (**a**) and initial friction coefficient (**b**) for a range of rh-lubricin concentrations at velocity = 0.33 mm/s and 20% normal strain. VSCR model fit to data ± SD for 20% strain (**a**) and 40% strain (**a**, inset). Data represented as mean ± SD with $n = 6$ per rh-lubricin concentration ($^*p < 0.05$) (Gleghorn et al. 2009)

stress arising from mechanical loading (Ogawa et al. 2014). Presumably, PGE2 and PTHrP contain shear–stress response elements (SSRE) in their gene promoter regions, though this remains to be confirmed.

It is an important practical observation that bearing lubrication is usually improved by relative motion between contacting surfaces, as fluid lubricant is dragged between the sliding surfaces (by drag (viscous) forces in the fluid). Classically, a rotating shaft in a bearing is analysed using so-called Stribeck surfaces (Gleghorn and Bonassar 2008). When the load on the rotating shaft is large and its rotation slow, bearing friction is primarily related to the efficacy of boundary lubrication. As shaft rotation speed pick up, hydrodynamic lubrication increases relative to boundary lubrication, and the friction between shaft and bearing decreases. This interplay between lubrication modes and friction is presented as a so-called Stribeck curve or Stribeck surface. Stribeck surfaces for cartilage are shown in Figs. 4.22 and 4.23.

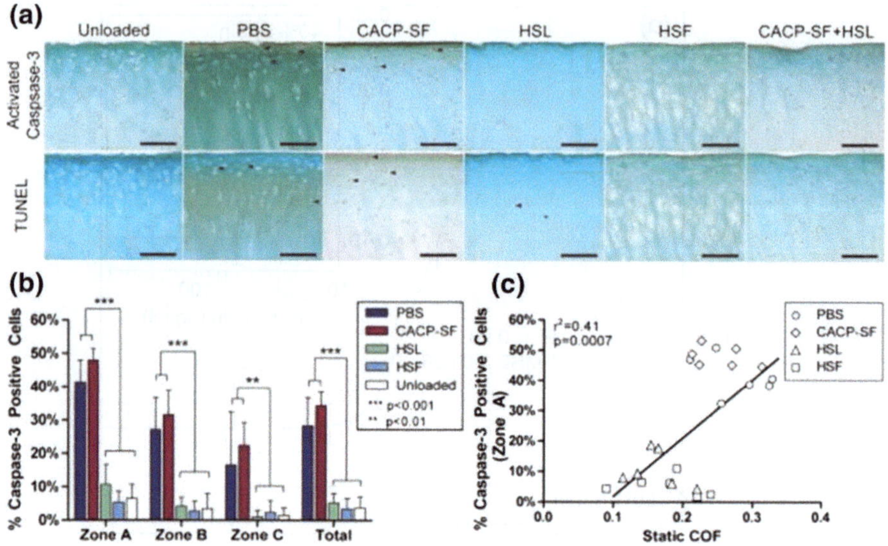

Fig. 4.20 **a** Apoptosis in lower bovine cartilage bearings analyzed by zone. Cartilage explant bearings tested with PBS and CACP-SF [i.e. autosomal recessive disease camptodactyly–arthropathy–coxa vara–pericarditis syndrome—synovial fluid (which lacks lubricin (PRG4))] had a greater number of cells positive for activated caspase-3 (brown, arrowheads) than bearings lubricated with HSL (human lubricin (PRG4) in distilled water), HSF (normal human synovial fluid), or CACP-SF+HSL and unloaded bearings. Cells negative for activated caspase-3 are stained blue. TUNEL staining (brown) confirmed apoptosis in bearings lubricated with PBS and CACP-SF (arrowheads), but few cells were TUNEL positive in the bearings lubricated with HSL, HSF, or CACP-SF+HSL or in unloaded bearings. Cells negative for TUNEL are stained blue. (Scale bars, 100 μm.) **b** Percentage of activated caspase-3 in lower bovine cartilage bearings. Bearings tested with PBS and CACP-SF had significantly higher percentages of apoptotic cells in the superficial and upper middle zones compared with unloaded bearings and bearings lubricated with HSL and HSF. Error bars indicate SD. **c** Correlation of static COF and activated caspase-3 in zone **a**. A significant correlation (r^2 = 0.41) between static COF and activated caspase-3—positive cells in zone **a** (articular surface, 100 μm) was observed for the mechanically tested bearings, across the different lubricants (Waller et al. 2013)

The repetitive cyclic load-unload cycles enable the cartilage tissue to re-imbibe fluid from the bathing synovial fluid during the swing phase while walking and running (Linn 1967; Moore and Burris 2017). This recharges the fluid on the contact surface and recharges the underlying cartilage tissue with fluid, so it swells. The fluid can then again weep into the contact region during the next load cycle. This repetitive loading prevents the fluid in the cartilage ever reaching its static equilibrium state. In normal joint usage, load–unload cycles are combined with sliding of cartilage on cartilage. Unfortunately, not a lot is known about the details of lubrication during normal walking and running (Hodge et al. 1986; Moore and Burris 2017), or during abnormal load regimes associated with pathologies such as ACL, meniscectomy and OA, or when different types of interventions are considered e.g. insertion of a ‘cushion bearing’ into the joint space (Stewart et al. 1997).

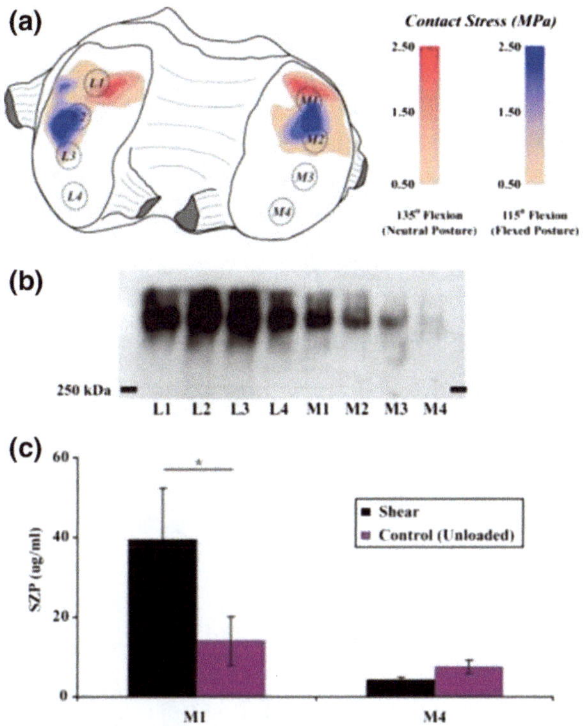

Fig. 4.21 Expression of SZP [PRG4] in (**a**) regions of high in vivo loading (M1) in the femoral condyles of bovine stifle joints is (**b**) significantly greater than in regions experiencing lower contact stresses (M4). L, lateral condyle; M, medial condyle; 1–4, locations on condyle numbered from anterior to posterior. SZP expression is (**c**) increased by sliding shear loading in cartilage explants obtained from M1. SZP synthesis in M4 explants is unaffected by shear (Mean–SEM, *$p = 0.021$) (Neu et al. 2007; McNary et al. 2012)

Given the low-friction coefficient associated with hydrodynamic lubrication relative to boundary lubrication, it is intuitively plausible to suppose that it is beneficial to keep hydrodynamic lubrication happening for as long as possible. Articular cartilage prolongs the period of hydrodynamic lubrication by coupling contact gap fluid flow with fluid flow from articular cartilage itself—effectively changing the duration of gap fluid flow from a time measured in seconds to potentially hours—the timescale for consolidation of the articular cartilage (see discussion Sect. 4.3.4). Indeed, based on the foregoing discussion and experimental testing, it is not unexpected that the cartilage favors a mixed-mode of lubrication rather than a boundary mode of lubrication, as the mixed-mode friction coefficient is around 3–5 times smaller than the boundary mode friction coefficient.

Consequently anything that reduces the duration of mixed-mode lubrication is likely to increase the time-averaged friction coefficient, increase the time averaged shear forces within the cartilage, and this will in turn lead to increased rates of damage both within the cartilage as well at the cartilage surface. Clearly, in disease states, changes occur that influence factors regulating both relative proportions and absolute values of hydrodynamic and boundary lubrication. When such changes are made to normal synovial joint operation, the time-averaged friction coefficient of the joint invariably rises, along with joint damage, as the synovial joint will move away from its normally optimal operating conditions.

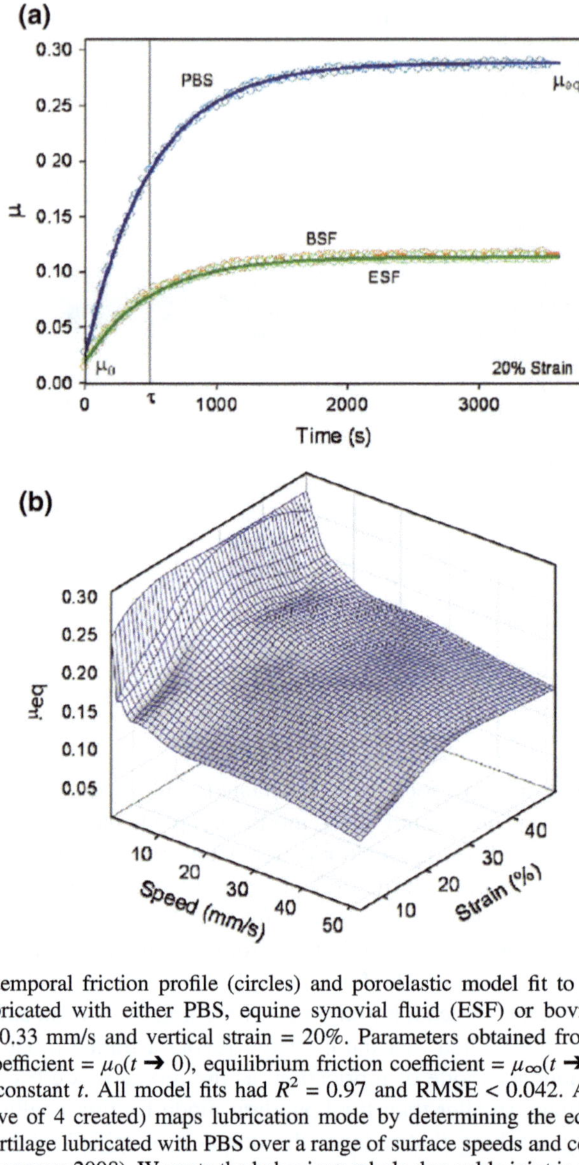

Fig. 4.22 The temporal friction profile (circles) and poroelastic model fit to the data (line) of cartilage (**a**) lubricated with either PBS, equine synovial fluid (ESF) or bovine synovial fluid (BSF) with v = 0.33 mm/s and vertical strain = 20%. Parameters obtained from the models are initial friction coefficient = $\mu_0(t \rightarrow 0)$, equilibrium friction coefficient = $\mu_\infty(t \rightarrow \text{infinity})$, and the relaxation time constant t. All model fits had R^2 = 0.97 and RMSE < 0.042. A Stribeck surface (**b**) (representative of 4 created) maps lubrication mode by determining the equilibrium friction coefficient for cartilage lubricated with PBS over a range of surface speeds and compressive strains (Gleghorn and Bonassar 2008). We note the behavior a whole dog ankle joint is shown in Fig. 2 of Linn (1968)

For example, testing on mature bovine knee cartilage (animals 18–24 months old, joints stored for 24–48 h prior to testing in pendulum friction simulator with articular cartilage contacting articular cartilage with 25% bovine serum in physiological saline as lubricant) with and without a meniscus. Without a meniscus results in much larger friction coefficients (see Fig. 4.24) (McCann et al. 2009). No wear

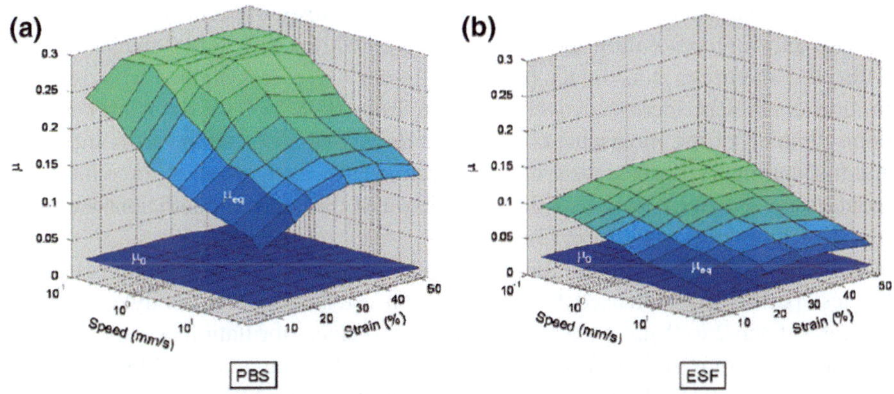

Fig. 4.23 Stribeck surfaces for the initial and equilibrium friction coefficient for cartilage lubricated with either PBS (left) or equine synovial fluid (right) (Gleghorn and Bonassar 2008). Notice the friction coefficients are much lower when lubricated with animal synovial fluid, compared to phosphate buffer solution (PBS). Notice also that as the cartilage contact velocity increases, the equilibrium friction coefficient decreases, particularly above 1 mm per second (while at around 10 mm/s the equilibrium friction coefficient is about 0.025 for small strains, but increases to about 0.05 for large strains)

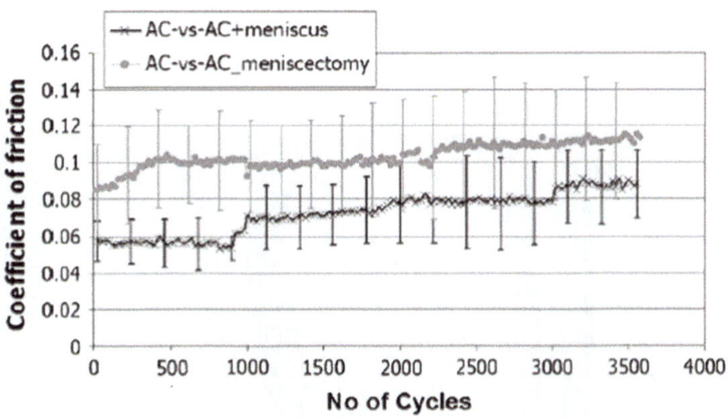

Fig. 4.24 Coefficient of friction over 3600 cycles at 1 Hz for AC-versus-AC [articular cartilage] with meniscus and AC-versus-AC_meniscectomy at a peak load of 1000 N [equivalent to peak contact stress around 5 MPa load with intact meniscus, and peak contact stresses around 17 MPa with meniscus removed (stress measured using Fuji film)]. Data presented as mean (n = 6) ± 95% confidence intervals (CI). There was a statistical difference ($P < 0.05$; ANOVA (analysis of variance) single factor + T-method) between the mean coefficients of friction (averaged over the 3600 cycles) (McCann et al. 2009)

was detectable using a 9.4T microMRI (normal resolution in-plane pixel size 78 μm, and high-resolution in-plane pixel size 20 μm) over 3,600 cycles at 1 Hz while the meniscus was intact (contact stresses were less than 5 MPa and the friction coefficient increased over one hour from 0.06 to 0.09). But immediately upon removal of the meniscus, contact area decreases and contact stress increases, and fibrillation and surface wear increase dramatically as the friction force increased (contact stresses were then less than 17 MPa and the friction coefficient increased over one hour from 0.08 to 0.12).

Meniscectomy resulted in a four to fivefold increase in surface roughness during the test period (R_a increased from 1 to 5 μm on the femoral condyle, while R_a increased from 2 to 9 μm on the tibial plateau), surface fibrillation became apparent and some 70 mm^3 of surface wear was detected using a microMRI after just 3,600 cycles at 1 Hz. Importantly, we mention that at 17 MPa compressive stress, surface damage was visible on the condylar surfaces after less than one hour of testing at 1 Hz, and 70 ± 18.6 mm^3 ($n = 6$, 95% confidence interval) of surface wear occurs. This clearly suggests that 17 MPa is not a viable stress level for maintaining normal cartilage.

Typical MRI images of the bovine cartilage surface are shown in Fig. 4.25. Not that the distinctive surface undulations (i.e. ridges and waviness) on the surface of cartilage. Normal cartilage was found to have a R_a of 1–2 μm (before and after testing at 5 MPa for 1 h at 1 Hz), while the meniscectomy group had a R_a of 5.3 and 8.7 μm after testing at 17 MPa for 1 h at 1 Hz (see Fig. 4.25).

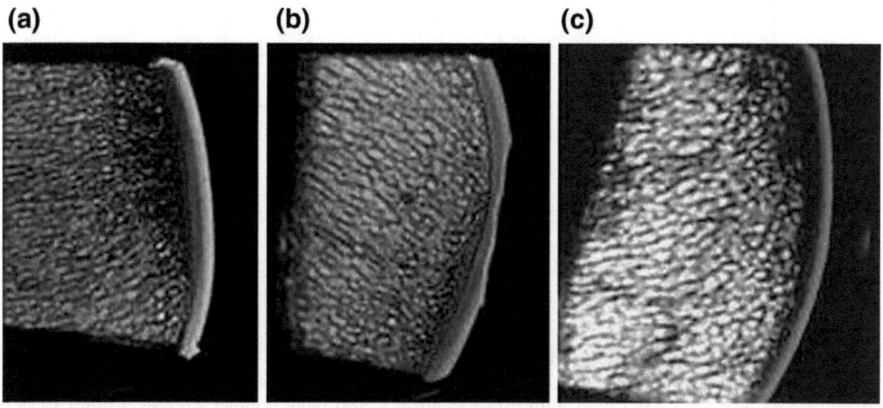

Fig. 4.25 Typical MRI images. **a** From the 'articular cartilage contacting articular cartilage with meniscus' group where there was no detectable change in wear volume at 78 mm × 78 mm in-plane pixel resolution [*Note* R_a = 1 or 2 μm, which is similar to the R_a of 1.4 μm reported by Graindorge et al. (2006) for bovine cartilage]. **b** The 'articular cartilage contacting articular cartilage with meniscectomy' group where surface damage was detectable at 78 mm × 78 mm in-plane pixel resolution [*Note* R_a = 5.3 or 8.7 μm] and **c** a high-resolution scan 20 mm × 20 mm in-plane pixel resolution of an undamaged sample [*Note* R_a = 1 or 2 μm] (McCann et al. 2009)

4.5 Cartilage Inspired Artificial Porous Bearing

In 2013, researchers developed 'cartilage inspired porous bearings' from a combination of polymers in saturated a solution (Greene et al. 2013). Interestingly, the properties of this artificial porous material, based on cellulose (i.e. wood) fibers (a polysaccharide of β1 → 4 linked D-glucose molecules) appear to have properties surprisingly similar to those of cartilage tested under similar conditions. For example, a friction coefficient of 0.2 is obtained, compared to a friction coefficient for cartilage in the range of 0.1. With an improvement on PEG as a 'boundary lubricant' (e.g. an equivalent of hrPRG4 employed in cartilage), the performance of synthetic biocomposite may well approach even more closely that of cartilage. While not there yet (Paetzold et al. 2012; Moutos et al. 2016), advances in tissue engineering may lead to the development of cellular hyaline cartilage that can replace damaged cartilage.

4.6 Fundamental Hypotheses for Clinical Biomechanics and the Relevance to Osteoarthritis

In the introduction to Chap. 1, we wrote:

> *Taking into account that stress distribution over the contact area of femoral cartilage and menisci on the tibial cartilage is not uniform [i.e. contours of stress magnitude can be defined over the cartilage-on-cartilage contact area over the tibial plateau; see for example the experimental data presented by Marzo and Gurske-DePerio (*2009*)], then it is not difficult to estimate that peak contact stresses could be as high as 20 MPa during such activity and walking up or down stairs, or up and down a hill. However experiments with normal cartilage rubbed back and forth on normal cartilage continuously for 1 h, under 17 MPa of compressive stress, have shown visible damage to the cartilage surface (McCann et al.* 2009*). Something extraordinary must be going on within the knee joint in vivo to prevent damage rapidly accumulating, as it is obviously apparent that people often walk downhill over long distances repeatedly loading their cartilage, and yet do not develop diseased joints (at least rapidly).*

To try to resolve these apparently disparate findings, it is of crucial importance to consider the normal operating conditions for the knee joint relative to experimental conditions. Joints are normally used intermittently by people and animals. During usage, joints normally operate reciprocally; i.e., there is repetitive oscillating relative motion between the cartilage surfaces. Further, normally joint loading accompanies reciprocal stroke motion in one direction (e.g. during the stance phase while walking or running), while unloading accompanies stroke motion in the other direction (e.g. during the swing phase while walking or running). Intermittent, reciprocal load–unload cycles mean that normal joints are capable of very low-friction force and so associated surface wear and internal cartilage tissue damage (Bell et al. 2006; McCann et al. 2009). Together, this adds up to a set of conditions that normally reduce joint tissue damage to low levels, despite the significant load magnitudes. Let us attempt to explore and explain this further, and define conditions that are both favorable and unfavorable in terms of friction force and cartilage damage.

By definition, 'intermittent joint usage' means that joints are loaded for short periods of time, and so the normal operating state of cartilage in vivo is toward the 'undrained' end of the consolidation spectrum (the extremes of the consolidation spectrum are fully 'undrained' to fully 'drained,' the spectrum being quantified by the 'degree of consolidation'—see discussion Sect. 1.6). So 'intermittent joint usage' implies that the joint is initially unloaded and in a relaxed state, and that upon loading, it is initially in a fully undrained (i.e. unconsolidated) state. This is because the *joints* are usually only used (and so loaded continuously) for short periods of time. Humans normally do something for 'a while' and then stop and have a rest, where 'a while' is usually some relatively small fraction of time (say 10%, i.e. 15 min or so) compared to the time required for complete consolidation of the cartilage [which is around one hour to three plus hours, depending on the particulars of a joint (Herberhold et al. 1999)]. Both animals and humans normally exercise strenuously in short bursts of activity (because they suffer muscle fatigue), and then they have a rest (often they sit or 'lie down' to recover).

Maintaining a 'separation of timescales' for periods of intermittent joint loading and the characteristic time for consolidation of articular cartilage, ensures that (net) fluid weeping to the contact interface makes an important contribution to hydrodynamic lubrication over the period of joint usage. We note that when weeping flow is significant, then the relative proportion of hydrodynamic to boundary lubrication increases, and friction is low when hydrodynamic lubrication dominates mixed-mode lubrication (see, e.g. (Miramini et al. 2017) for estimates of net flow rates of fluid exudation during cartilage consolidation). In the initial transient state of consolidation, when fluid flow is substantial (and weeping flow to the contact interface is most significant), the friction coefficient is very low. When the friction coefficient is very low, the friction force at the cartilage surface and the shear force within the cartilage tissue are minimized, thereby minimizing damage to the cartilage tissue and the contacting cartilage surfaces.

By definition, 'prolonged joint usage' means that joints are loaded for long periods of time relative to the characteristic time for consolidation. In this case, the operating state of cartilage in vivo is toward the 'drained' end of the consolidation spectrum (the extremes of consolidation spectrum are 'undrained' to 'drained'—see discussion Sect. 1.6). So 'prolonged joint usage' implies that the joint is initially consolidated (as opposed to unconsolidated), and that it is operating in this drained (consolidated) state. Weeping lubrication due to consolidation is spent, at least for static loading. Humans reasonably often operate in an 'endurance mode' (ignoring muscle fatigue, or training to build resilience against muscle fatigue). This operational mode could also mean a static loading for long periods of time (e.g. stalking), as well as prolonged reciprocal loading (e.g. chasing). It is likely that this endurance capability offers an evolutionary competitive advantage (Bramble and Lieberman 2004; Raichlen et al. 2011). In the modern workplace, 'prolonged joint usage' can become 'institutionalized,' as it may be largely unavoidable or uneconomic with current technology to do otherwise in many manual occupations (e.g. skilled or unskilled manual laborers).

Having identified these two broad modes of joint loading (i.e. intermittent and prolonged, relative to the characteristic time for cartilage consolidation), we can consider their implications for friction coefficient and cartilage wear. Stribeck surfaces for cartilage show that rapid reciprocating movement between the cartilage surfaces results in a much lower kinetic friction coefficient than static friction, and wear of the cartilage surface is thereby minimized. However, we also note that even for reciprocating conditions, the friction coefficient increases substantially with tissue consolidation (from zero up to about 30% tissue strain the friction coefficient increases two to three times; see Fig. 4.23).

In the worst-case scenario, the joint endures static loading until it is fully consolidated, and then begins a reciprocal motion while load is kept constant. In this case, the friction coefficient may be initially very large (e.g. 0.2–0.5) and only decrease over a period of approximately 5 min as 'slide swelling' occurs (Moore and Burris 2017; Linn 1967). During this 5 min period (particularly, in the first couple of minutes), the friction forces are very large and wear could be far more significant than it normally is. Though not yet experimentally validated, presumably if this scenario is repeated multiple times (as often occurs in manual occupations), this may well result in accelerated surface wear and accelerated internal articular cartilage damage. It is postulated that repeated unfavorable situations for articular cartilage, such as described in this paragraph, may contribute significantly to the development of osteoarthritis in many people. If this could be demonstrated experimentally, it may form the basis to begin developing general health advice on how to maintain healthy articular cartilage. Today, health advice on how to maintain healthy synovial joints is notably absent.

Taken together, practically this means cartilage tissue and surface damage is minimized by light loadings, applied to initially undrained (i.e. *no consolidation*) cartilage tissue, for short durations, while heavy loads carried for long periods of time maximize chances of significant cartilage damage. For joints that experience significant shear at the contact interface, fast reciprocal loadings are preferred to slow reciprocal loadings, to maximize entrainment of synovial fluid into the contact gap. A worst-case wear scenario is deduced to be the commencement of slow reciprocal loading for a fully consolidated cartilage under heavy constant load, as the friction coefficient may then be orders of magnitude greater than under the most favorable circumstances (Moore and Burris 2017; Linn 1967; Miramini et al. 2017). Practically, this unfavorable outcome could be realized by an overweight/obese person standing for a period of time so their cartilage tissue has a high degree of consolidation, and then the person begins to slowly walk/squat/ascend stairs/descend a hill. Such a worst-case scenario is often realized during the course of routine tasks in many occupations, particularly those involving heavy manual labor (e.g. plumbing, carpentry, roof tiling, mining). A detailed task-time analysis, computational analysis, and experimental data together may help explain the correlation reported between manual occupations and osteoarthritis (Hunter and Eckstein 2009). Of course, another worst-case scenario is repeated impact loading [e.g. dropping from as little as 1 m in height is reported to induce knee joint loads some 25 times body weight upon landing (Wright and Dowson 1976)]. Such very large loads are again experienced in many manual occupations (e.g. booting a shovel into resistant earth), not to mention contact sports (e.g. which involve collisions of all kinds, and so extreme loads on joint tissues), and even non-contact sports such as soccer (e.g. which involves abruptly changing speed or direction to avoid contact, and so extreme loads on joint tissues) (Levy et al. 1996).

While too much loading is deleterious to cartilage, too little loading may also be as deleterious, as normal joint loading stimulates chondrocytes to have the potential to repair the tissue at a rate greater than the rate of damage, and so too little loading may lead to an inadequate repair response. In the aged, too little loading is often combined with periodic (impact) overloading [e.g. too little exercise leads to poor balance, which leads to more falls (Gschwind et al. 2013), while stiffness or pain may lead to abnormal gait and joint loadings]. Of course, falls can be damaging to both joints and long bones, as can repeated abnormal joint load. Both the rate of cartilage repair and the rate of damage are functions of cartilage contact stresses (i.e. joint loading), and because both functions have different properties and characteristics, both functions need to be taken into account when attempting to optimize tissue repair. Optimal joint health (or an optimal recovery) following joint damage almost certainly requires an optimal 'dosage' of mechanical loading, which depends on the shapes of the repair and damage curves. An optimal dosage of mechanical loading presumably not only reduces the risk of developing osteoarthritis, an optimal dosage of mechanical loading presumably can reduce the rate of osteoarthritis progression.

From these observations, we may formulate the 'fundamental hypotheses for clinical biomechanics': (i) *A healthy synovial joint has an optimal load dosage that will maximize the functional lifespan of that joint* and (ii) *an injured joint has an optimal load dosage that maximizes the rate of repair, and an optimal load dosage that minimizes the rate of damage—these two optimums may be different. At the optimal load dosage for healing (defined as joint repair minus joint damage), the net rate of joint healing is maximized.* Similar statements may be made for each tissue in a synovial joint.

Titrating or estimating the optimal load dosage for each joint in each person is a fundamentally important research question for both basic and clinical biomechanics. The central challenge is to identify those loading patterns that increase repair of each tissue while reducing damage to each tissue, so that the rate of healing for that tissue is optimal. Currently, how to achieve this on a patient-specific basis is an open research question, but some progress on this front, which necessarily involves combining experimental and clinical data with computational modeling because many critical variables cannot be directly experimentally accessed, appears to be commencing (Liukkonen et al. 2017).

We note here that on the basis of the foregoing discussion, to minimize cartilage damage, joint loading should be intermittent and light, i.e. light loads of short duration with frequent rest periods to allow cartilage to swell to its equilibrium state. How long this should be continued is an open research question. Such a load regime may provide adequate stimulus for an adequate repair response, but again this is an open question.

Because the optimal loading for each defined outcome is also tissue specific, this leads to the question of identifying differences between tissues, together with any trade-offs in the context of the whole joint. While it seems plausible that tissues would have evolved to 'work together' while in a healthy state, differences in optimal conditions are more likely when one or more tissues are injured or diseased. And given the range of genetic and behavioral differences between people, it is inevitable that optimal loading patterns will be person specific. The practical question then becomes identifying the most appropriate load patterns for each person's circumstances. At present, advice on appropriate joint loading is very general—keep exercising but avoid tissue trauma (Bosomworth 2009) and stop if there is pain. But it is well known that pain is not well correlated with structural tissue damage (Goldring and Goldring 2004; Hunter et al. 2013), so the practical usefulness of such advice for those wanting to minimize progression of their osteoarthritis is questionable.

We mention that emerging measurement techniques such as recording acoustic emissions from synovial joints (which may indicate stick–slip sliding) (Shark et al. 2010; Abbott and Cole 2013), molecular profiling of synovial fluid (Gobezie et al. 2007) or urine, and the many imaging technologies now available, offer many practical ways to track the relationship between load dosage and the evolution of joint damage. Finally, we note that to maximize the information derived from such measurement techniques, computational modeling almost invariably has something to offer (e.g. the mechanical environment within cartilage, as experienced by chondrocytes, cannot be measured but can be computed using an appropriate model).

Chapter 5
A Systems Approach to Articular Cartilage

5.1 Introduction

A cell obeys all known physical and chemical laws. But the way cells 'evolved through natural selection' means that cellular systems have come to function in ways that are more complex, and often surprisingly so, than our more familiar 'engineering systems' (see discussion and explanation in Appendix B how this came to pass). The structure of engineering systems tend to reflect the way they were conceived in a human mind, while cellular systems have been created using a 'trial and error' approach in ever changing environments. Evolutionary processes have (and are) constantly testing new things. So just about any and every potentially useful connection between any two systems you care to name, across the full range of length scales from molecules to cells to organisms and groups, has been explored by evolutionary processes, a bewildering number of times over the course of life on earth.

When an engineer, who designs everyday objects like cars, buildings, computers looks at cells, tissues, and organisms after evolution has done its work, there is much that is familiar. An engineer can see, for example, that factors of safety against structural failure are of more or less familiar magnitudes (most often on the low side). It is apparent that some tissues have been optimized for maximum strength with minimum weight that feedback control systems are operating to maintain a constant internal state (homeostasis of the internal environment is a foundational physiological principle), and that an organism burns fuel to power the chemical and mechanical systems within it. Engineers and physicists can identify many familiar non-dimensional numbers regulating the system operation (Reynolds number, Peclet number, Damkohler number, etc.), and that all the known laws of chemistry and physics are scrupulously obeyed. So much 'feels' familiar.

But there is much that is unfamiliar too. For example, the 'deep integration' of systems (both within and between systems), the redundancy in and between system functions, the surprising/perplexing/flabbergasting/'weird' connections within and

D. W. Smith et al., *Articular Cartilage Dynamics*,
https://doi.org/10.1007/978-981-13-1474-2_5

between the systems, the way that cell or organism resists being categorized by knowledge from a single discipline (i.e. nearly all problems involve both chemistry and physics and/or knowledge from multiple engineering disciplines). The question is how should we think about organisms to gain insights into how they work?

5.2 Networks and Basic Functional Units in Tissues

It is helpful and useful to conceptualize organisms as 'bundles of (chemical and neuronal) networks,' within 'bundles of (chemical and neuronal) networks'—and so on, branching tree-like into more bundles, only with many surprising bundle 'cross-interactions' that lie outside any systematic classification system that might be proposed. Such systems are often described as 'complex.' But apart from being difficult to keep all the variables in mind simultaneously, what do we mean by complex? Carlson and Doyle (2002) offer this explanation of complex systems:

> What distinguishes the internal configurations of systems as complex? It is not the mere number of component parts. Any macroscopic material has a huge number of molecules. It is the extreme heterogeneity of the parts and their organization into intricate and highly structured networks, with hierarchies and multiple scales. (Some researchers have suggested that "complicated" be used to describe this feature.)

'*It is the extreme heterogeneity of the parts and their organization into intricate and highly structured networks, with hierarchies and multiple part*s' certainly describes biological systems. There seems to be no end to the heterogeneity of the parts (in addition to gene alleles, the splice variants of genes are themselves complicated). Indeed, simply defining 'hierarchies' and 'networks' pose significant ongoing challenges in their own right.

But one useful way of categorizing complicated networks is based on the observed functional role of bundles and networks (Roux 2014). For example, the traditional way of classifying networks in humans is based on 'organ system' networks, where each organ system is identified by its function in the organism. There are twelve organ systems usually identified—name and function are described as follows:

(i) Circulatory (heart and vessels). Function: convective transport of materials via the blood;
(ii) Lymphatic (vessels). Function: removal of proteins (e.g. albumin but including toxins) from interstitial compartment, infection control via the immune system;
(iii) Respiratory (nasal passages, trachea, lung). Function: provide a large gas–liquid interface for rapid exchange of gases with the atmosphere;
(iv) Digestive (the gut is tube from mouth to anus). Function: provide a large selective, liquid–liquid interface for the rapid extraction of molecules (providing energy and nutrients) from the environment;

(v) Urinary (kidney, ureters, bladder, and urethra). Function: retention of salt and water to adequately fill the cardiovascular system, control of chemical composition of the plasma by removal of excess or waste products;
(vi) Muscular (skeletal, smooth, and heart). Function: provides a motive force for blood flow, locomotion, speech, etc
(vii) Skeletal (bones, joints). Function: enables locomotion by providing structural support and protection for bone marrow;
(viii) Sensory systems (eyes, ears, taste, other sensory inputs such as pain, temperature, and joint position). Function: provide information about the external environment;
(ix) Integumentive (external skin, mucosal surfaces). Function: provides tough mechanical barrier and a (selective) chemical barrier with small surface area that separates self from the environment;
(x) Reproduction system (ovaries, uterus, testis). Function: reproduction;
(xi) Nervous (brain, spinal cord, and peripheral nerves). Function: cultural adaptions and higher-level decision making, low-level reflexes, unconscious neuro(-endocrine) organ coordination;
(xii) Endocrine (hormones). Function: coordination of organs.

The point to be made is that each organ systems is coordinated by a complex network of signals. There are many chemical signals [via endocrine, regional and local (paracrine or autocrine) signaling systems], and there is a complex of electrical–chemical signals (via the nervous system). Understanding the function of the organs themselves, untangling the functional networks within the organ that enables the organ systems, and identifying the regulatory signals coordinating the organ systems, is the traditional discipline known as 'physiology.'

In this sense, physiology (of the organ, tissue, and cell) is central to biological and clinical science, yet it has in some ways been pushed aside in modern university courses. The sub-disciplines of physiology—biochemistry, molecular biology, genetics, epigenetics, etc.—which have enjoyed greater prominence at the expense of physiology in recent decades, are given context and functional significance by the discipline of physiology. Sometimes, the push for reductionism in science causes us to lose sight of how the parts identified are integrated with one another to achieve functional outcomes. Physiology is about how the various parts and systems in the organism are integrated to achieve functional outcomes.

As technology has advanced, smaller-scale processes have been better resolved. Indeed, one of the fundamental technical obstacles to better understanding of biological systems is their tiny size (which explains the obsession of many experimentalists have with new high resolution—and preferably 'high throughput'—technologies). Because of the insights possible using these new technologies, we have a much better idea of how each cell is different (Trapnell 2015), how cells organize themselves and communicate with one another within organs.

In understanding organ systems, it is very helpful to realize that most organ systems are made up of groups of cells that form 'basic functional units' within the organ. These functional units of cells then repeat a large number of times to make up the tissue. So our 'gray matter' in the brain has a basic structural unit of about 10,000 neurons known as a 'cortical column' (believed to be responsible for basic thought processes) (see http://bluebrain.epfl.ch/). In the kidney, the functional group of cells is organized around a 'nephron' (which is a tubule responsible for controlling chemical composition of the body by resorption of what the body requires), the liver is made up of lobules (for processing various chemicals traveling in the blood), while in bone a functional grouping of several thousand cells is known as the 'basic modeling unit' [or BMU—which is responsible for remodeling bone (Frost 1987)].

What is the basic functional unit in cartilage tissue? It is not usual to identify any basic functional unit comprising many cells in articular cartilage [though some have argued it is the multicellular chondron (Poole 1997)]. Rather it is more usual to identify tissue 'zones,' containing either single or clusters of chondrocytes distributed throughout the ECM, that are in similar differentiation states, and so presumably have a similar function. Does this mean cartilage tissue is different to other tissues? In the preceding chapters, we have argued that the chondrocytes in different zones cooperate with one another, and coordinate their function by cell–cell signaling between and within the different zones over the depth of the cartilage. So functionally, the basic functional unit in cartilage could be imaged as a 'rectangular prism of tissue, with height the thickness of articular cartilage.' So we may say that the 'functional unit' in articular cartilage are the cells in an arbitrary volume of articular cartilage over the depth of the tissue (e.g. say an arbitrary rectangular prism 1 mm by 1 mm by *D* mm, where *D* is the depth or 'thickness' of the articular cartilage). In normal human adult tibial cartilage, such a rectangular prism of normal articular tissue would involve approximately 100,000 to 200,000 chondrocytes.

The actions of cells within a functional unit have to be coordinated to achieve the desired functional outcome. At the macro-level level, general hormones and neuronal signals help coordinate the cells in tissues, but within the function unit itself, cells secrete chemical signals that are sensed either by the cell itself (autocrine signaling) or by nearby cells in the functional unit (paracrine signaling), to fully coordinate the functional units' activities. As a general principle, cells always integrate both local, regional *and* whole organism-level information before deciding what to do. This means that in vivo chondrocytes integrate local information (both signals from themselves and neighboring cells) and systemic information (e.g. thyroxine from the thyroid, growth hormone from the pituitary, IGF-I from the liver, IGF-II from muscle, steroids from the adrenal, sex hormones from the gonads, melatonin from the pineal, adipokines and leptin signals from adipose tissue, insulin from the beta pancreatic cells, etc.). Additional hormones may also play roles e.g. EPO from the kidney, ghrelin from the gut, vitamin D, retinoic acid, and so on. A chondrocyte set of normal chemical signaling inputs about the state of the organism is extensive.

The extracellular matrix itself (which provides many types of signals to chondrocytes, and is a record of past signals and events), together with mechanical and chemical signals around each cell, provides the chondrocytes in a functional unit with critically important information about what is going on in its regional and local environments, and within the organism (e.g. the mechanical load environment). This information is used by the cell to decide if it should survive. What state it should be in (i.e. the cell's differentiation state), and so what it should be doing at that particular time (e.g. what metabolic state, what secretion profile, proliferating and dividing, migrating, resting), and when it should die (e.g. programmed necrosis or apoptosis). The study and understanding of these communication pathways between cells is an important part of 'tissue physiology.'

In the mid-twentieth century, great advances were made in understanding the 'metabolic networks' within cells. So the aerobic glycolysis (Krebs) cycle and anaerobic glycolysis (lactate) networks were discovered, as were a huge number of networks that ensure the production of basic chemicals required for cell maintenance. The complexity of these chemical networks is nicely conveyed by the Boehringer Metabolic Map http://web.expasy.org/cgi-bin/pathways/show_thumbnails.pl. These maps are worth a look if you have not already seen them, as they help give a visual impression of the complexity of what is going on inside cells.

As the twentieth century advanced, more and more cellular signaling and molecular processes were uncovered, as conveyed by the Boehringer Cellular and Molecular Processes Map http://biochemical-pathways.com/#/map/2. Those discovering the 'molecular' networks (i.e. networks of chemical reactions taking place in the cell nucleus) chose to call themselves 'molecular biologists,' to distinguish themselves from (old-fashioned) 'biochemists' (who delved into the workings of the cell's cytoplasm using the 'traditional tools' of metabolic research). But such distinctions hold little intrinsic value. Indeed, probably the main driver of such distinctions is the competition between researchers in various disciplines for limited available research funding. What discipline is perceived as 'hot' tends to get more research funding. In other words, 'marketing' is important even to basic researchers —how else to get the Human Genome Project commenced (Collins et al. 2003)? It has been found that most of the basic chemical network control principles uncovered for metabolic networks, apply equally well to signaling networks and networks for gene regulation. Collectively, understanding the chemical networks internal to the cell is referred to as 'cell physiology.'

So we see that networks may be grouped based on the type of signaling e.g. chemical or neuronal or mechanical signaling. Or networks may be grouped on the length scale over which they operate—we have the whole organism, an organ system, the tissue scale and cell scale physiological networks. But we may also group these networks based on function, e.g. metabolic pathways, or genetic networks or immune network signaling.

Networks conveying information to, within and from cells, belong to a particularly important group of networks, as defects in these 'signal-transduction systems' mean that crucial information may either not be received by the cell, or perhaps processed incorrectly by the cell (a lot of mutations constitutively activate a pathway, or they inactivate a pathway). Clearly, if this occurs, the normal physiology of the organ or tissue will be disrupted. Indeed growing attention is focused on understanding both communication systems and 'decision systems' between and within cells. For better or worse, those studying such signaling system have decided to call themselves 'systems biologists' (though physiologists such as Arthur Guyton for example, have always been systems biologists too). Wonderful examples of the basic signaling and decision networks are being uncovered in many species, including the nucleus of the urchin (red sea), fish (zebra), worms (C. Elegans), flies (Drosophila) and of course humans. See, for example, the sea urchin networks at http://www.biotapestry.org/#newWay and http://sugp.caltech.edu/endomes/index.html. We look more closely at so-called systems biology and signal-transduction pathways in the context of cartilage physiology in Sect. 5.6.

5.3 Growing to Understand Networks in Organisms

While this discussion of biological networks so far is all well and good, it really does not increase our predictive capability that much, apart from telling us that the networks are operating in tissues and are obviously important. What we would really like to know are the 'design rules' behind chemical networks, i.e. the range of ways that specific functional needs can be met by different combinations of network elements. For if we know the design rules underlying networks, our analytic and predictive capabilities should be greatly increased.

Graph theory is sometimes offered as a way forward [see, for example, Fig. 1 (Neph et al. 2012)]. There has been a useful identification of regulatory 'motifs' (Milo et al. 2002), which represent simple regulatory sub-networks (see Fig. 5.1a), which are building blocks for more complex networks. An example is a 'feedforward motif' (i.e. where one TF interacts directly with another TF, but also via an intermediate TF) (see Fig. 5.1b).

Recognizing similarities between biological networks and electronic networks, synthetic biologists have set about making sub-networks such as oscillators, pulse generators, and signaling filters. While these elements have led to many useful applications in genetic engineering, synthetic biological networks have not yet reached the 'plug and play' status of electronic networks (Rekhi and Qutub 2013), but impressive results are being achieved toward this outcome (Sun et al. 2017).

Graphs can be a useful way of presenting large amounts of gene interaction data generated by high throughput technologies (e.g. ChIP sequencing), and so have been occupying a lot of research effort. For example, graphs can be used to get a sense of the transcription factors that a single transcription factor (TF) interacts within different cell types. What comes out of these observations is that functionally similar cells share similar TF networks (Neph et al. 2012). Usually, there are a small number of TFs networks that are specific to each cell type (reflecting the specific specializations of that cell type), but many more TF networks are common (see Fig. 5.2).

Perhaps most importantly, networks have been found to usually have an identifiable and quite specific functional purpose. Indeed, it is found that by combining functional module information with raw transcription factor (TF) interaction data,

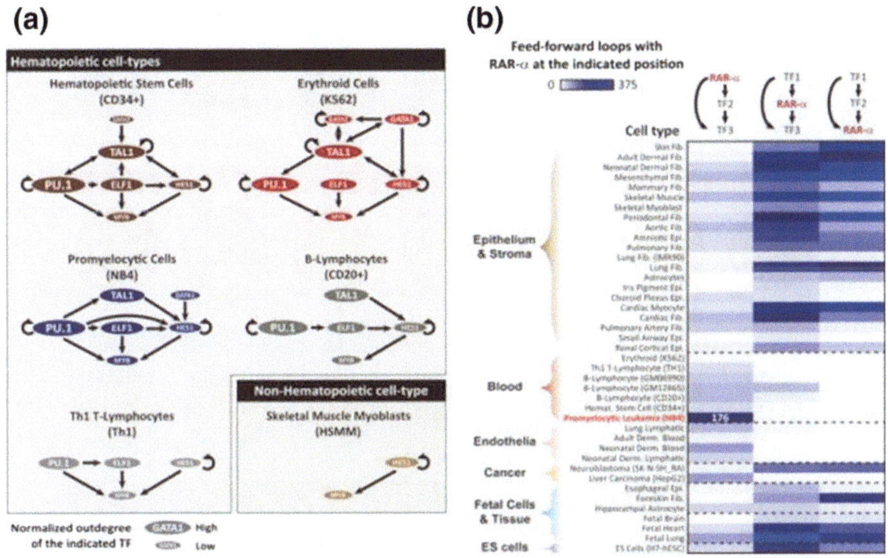

Fig. 5.1 **a** Shown are regulatory sub-networks comprising edges (arrows) between seven major hematopoietic regulators in five hematopoietic and one non-hematopoietic **cell types**. For each TF, the size of the corresponding colored oval is proportional to the normalized out-degree (i.e. out-going regulatory interactions) of that factor within the complete network of each cell type. The early hematopoietic fate decision factor PU.1 appears to play the largest role in **hematopoietic stem cells** (CD34+) and in promyelocytic **leukemia** (NB4) cells. The erythroidspecific regulator GATA1 appears as a strong driver of the core TAL1/PU.1/HES1/MYB network specifically within erythroid cells. In both B cells and T cells, the sub-network takes on a directional character, with PU.1 in a superior position. By contrast, the network is largely absent in non-hematopoietic cells (muscle, HSMM, bottom right). **b** Heatmap showing the frequency with which the retinoic acid receptor-alpha (**RAR-α**) is positioned as a driver (top) or passenger (middle or bottom) within feedforward loops (FFLs) mapped in 41 cell type regulatory networks. Note that in most cell types, RAR-α participates in feedforward loops at 'passenger' positions 2 and 3. However, within blood and endothelial cells, RAR-α switches from being a passenger of FFLs to being a driver (top position) of FFLs. In acute promyelocytic leukemia cells (NB4), RAR-α acts exclusively as a potent driver of feedforward loops (Neph et al. 2012)

network identification is much more efficient and productive (in other words, raw TF interaction data from various cells does not reveal very much that is useful, but including functional information does) (Karczewski et al. 2014) (Fig. 5.3).

While these efforts are yielding steadily increasing knowledge about the structure, functional purposes of networks and their relationship to disease states [see, for example, Fig. 1 in Kerkhofs et al. (2016)], a set of network design rules remain elusive. This is because even writing down design specifications and optimization criteria for biological networks are difficult, partly because all their interactions and functions are not known, and partly because of unpredictable 'emergent behaviors' (i.e. even simple rules given to many agents can lead to unpredictable states) (Rekhi and Qutub 2013). Emergent behaviors are particularly likely in dysregulated cell

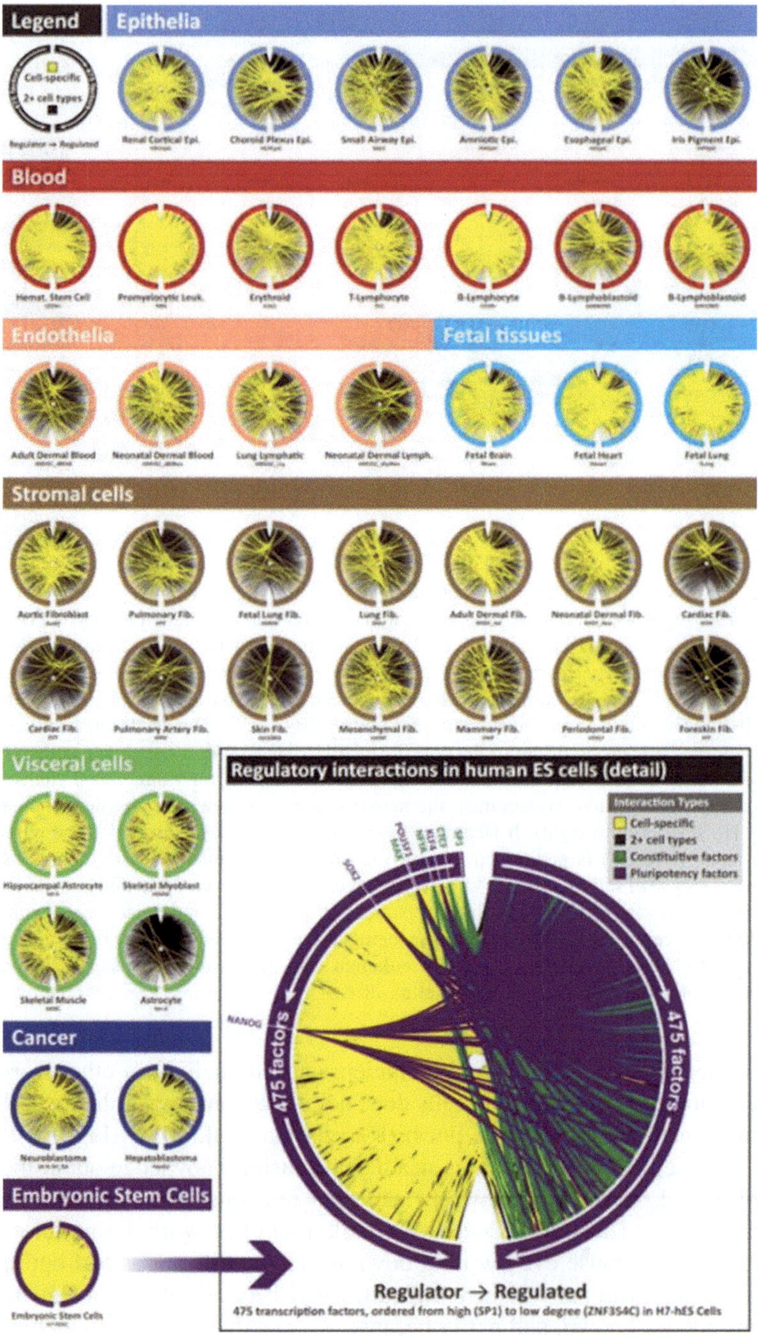
Legend
Cell-specific
2+ cell types
Regulator → Regulated
Epithelia
Renal Cortical Epi.
Choroid Plexus Epi.
Small Airway Epi.
Amniotic Epi.
Esophageal Epi.
Iris Pigment Epi.
Blood
Hemat. Stem Cell
Promyelocytic Leuk.
Erythroid
T-Lymphocyte
B-Lymphocyte
B-Lymphoblastoid
B-Lymphoblastoid
Endothelia
Adult Dermal Blood
Neonatal Dermal Blood
Lung Lymphatic
Neonatal Dermal Lymph.
Fetal tissues
Fetal Brain
Fetal Heart
Fetal Lung
Stromal cells
Aortic Fibroblast
Pulmonary Fib.
Fetal Lung Fib.
Lung Fib.
Adult Dermal Fib.
Neonatal Dermal Fib.
Cardiac Fib.
Cardiac Fib.
Pulmonary Artery Fib.
Skin Fib.
Mesenchymal Fib.
Mammary Fib.
Periodontal Fib.
Foreskin Fib.
Visceral cells
Hippocampal Astrocyte
Skeletal Myoblast
Skeletal Muscle
Astrocyte
Cancer
Neuroblastoma
Hepatoblastoma
Embryonic Stem Cells
Embryonic Stem Cells
Regulatory interactions in human ES cells (detail)
Interaction Types
Cell-specific
2+ cell types
Constitutive factors
Pluripotency factors
SP1
CTCF
KLF4
NFYA
POU5F1
MAX
SOX2
NANOG
475 factors
475 factors
Regulator → Regulated
475 transcription factors, ordered from high (SP1) to low degree (ZNF354C) in H7-hES Cells

◄**Fig. 5.2** Cell-specific versus shared regulatory interactions in TF networks of 41 diverse cell types. Shown for each of 41 cell types are schematics of cell type-specific versus non-specific (black) regulatory interactions between 475 TFs. Each half of each circular plot is divided into 475 points (not visible at this scale), one for each transcription factor. Lines connecting the left and right half-circles represent regulatory interactions between each factor and any other factors with which it interacts in the given cell type. *Yellow lines* represent TF-to-TF connections that are specific to the indicated cell type. *Black lines* represent TF-to-TF connections that are seen in two or more cell types. The order of TFs along each half-circular axis, and represents a sorted list (descending order) of their degree (i.e. number of connections to other TFs) in the ES cell network, from highest degree on top (SP1) to lowest degree on bottom (ZNF354C). Cell types are grouped based on their developmental and functional properties. Insert on bottom right shows a detailed view of the human ES cell network, and highlights the interactions of four pluripotent (KLF4, NANOG, POU5F1, SOX2) and four constitutive factors (SP1, CTCF, NFYA, MAX) with purple and green edges, respectively (Neph et al. 2012)

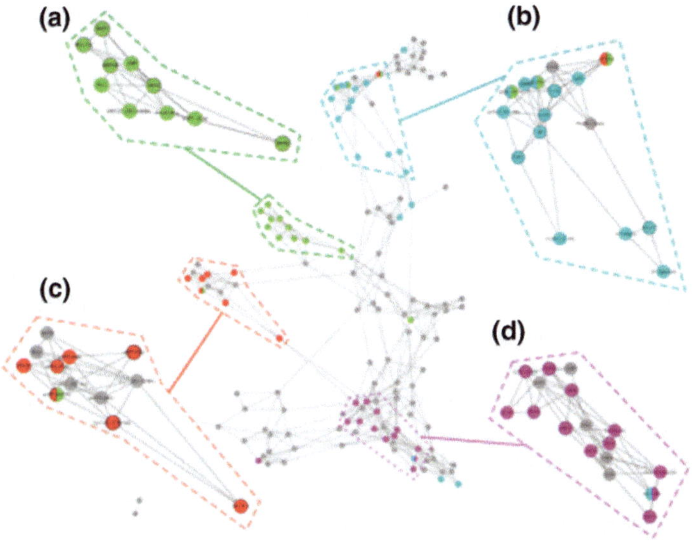

Fig. 5.3 Transcription factor interaction network reveals functional and disease sub-networks. Transcription factors are connected solely on the basis of the similarity of the modules that they regulate. Transcription factors are colored according to a selection of diseases; (**a**, green): AIDS; (**b**, blue): arrhythmia; (**c**, pink): breast cancer; (**d**, red): hemorrhage. Nodes are annotated with strong (dashed black borders) and weak (solid gray borders) (Karczewski et al. 2014)

states like cancer, which involve interactions that quickly become impossibly complex to comprehend without computational methods, but presumably this may be an important element in the behavior of normal cells too. Says Kerkhofs et al. (2016):

> Great strides in our understanding of cartilage biology have been made in the past 2 decades, supported by tried techniques like immunohistochemistry, ever more specific mouse models such as cre-lox systems and high-throughput in vitro assays alike. As this

> data accumulates in the wake of the field's progress, it becomes increasingly harder to glean insight by human intuition alone, making computational approaches that can smartly reuse this data all the more indispensable. Therefore we applied a qualitative approach that employs this readily available data, allowing for a first cursory exploration of the regulatory network in chondrocyte differentiation as a whole…Several observations correspond with a limited stability of the SOX9 attractor. First, the chondrocytic phenotype is notoriously difficult to keep intact in vitro as the cells rapidly dedifferentiate and lose expression of vital cartilage components.

We note that it may turn out that some aspects of cell behavior (e.g. emergent behaviors of cell populations), may only be understood using computational models of them (e.g. the relationship between the operation of individual neurons and brain function)—but we do not know yet.

Where complex objective functions can be defined, a most promising approach to finding optimal solutions are so-called genetic (including evolutionary) algorithms (McCall 2005). Evolutionary algorithms are inspired by natural evolutionary processes that gave rise to in vivo networks, and in essence attempt to mimic the evolutionary processes with a user-specified set of transformation and selection rules. Using this approach, some promising progress has been made in understanding the origin of non-intuitive configurations of signaling networks.

Metabolic networks usually aim to supply adequate energy or molecules that are basic building blocks for cell maintenance. When reserves are low, production is ramped up, and when there are more than adequate amounts present, the reaction products usually shut-down further production. This is a classical positive feed-forward and negative feedback found in system control theory. Indeed (distributed) control theory seems an appropriate framework in which to begin to quantify the behavior of these networks (see Sect. 5.5). A controller system requires a sensor (that detects changes in the controlled variable), an effector (which can adjust the controlled variable up or down) and a feedback controller. Control theory focuses on differential, proportional and integral controllers, and their respective merits in different situations http://en.wikipedia.org/wiki/Controller_%28control_theory%29. How these different types of controller are realized in practice is an open question. Control theory again raises questions about an 'objective function' (what is being minimized or maximized http://en.wikipedia.org/wiki/Mathematical_optimization) for the system, and raises questions about the 'set points' for the system.

Some argue that robustness means something more general than homeostasis. However, it appears that robustness is defined in essentially the same way as homeostasis [e.g. *robustness is a property that allows a system to maintain its functions despite external and internal perturbations* (Kitano 2004, 2007)], when the function of interest is defined to be the homeostatically controlled variable. Carlson and Doyle (2002) and Kitano (2004) make the point that robust systems also evolve, which links homeostasis with evolution (Carlson and Doyle 2002; Kitano 2004). In other words, effective homeostatic systems are selected, and once established, may be stable for hundreds of millions of years. Carlson and Doyle (2002) also make the point that complexity drives robustness. Contrary to what might first be supposed, more complex systems are generally more reliable. Carlson and Doyle (2002) offer the example of modern cars (which have ride control, engine control, cruise control, anti-lock braking, anti-skid braking, collision avoidance systems, satellite navigation, automatic temperature control, driverless cars, etc.) versus the first cars that had none of these features.

While complex systems are generally more reliable, they often have 'weak points' that can lead to catastrophic failure. For example, if the engine controller fails, the car goes nowhere. Similarly, there may be 'cascading failures,' e.g. where the on-board computer malfunctions, which leads to inappropriate anti-skid responses, which leads to a crash. Often, it is an apparently inconsequential part that causes very serious problems. For example, a simple part costing 57 cents in the ignition switch of GM cars lead to multiple deaths and the recall of 2.6 million vehicles http://money.cnn.com/2014/04/02/news/companies/gm-recall-part/index.html?iid=HP_River.

Similar weak points and cascading failures are seen in biological systems. One critically important weak point is at signal transduction cascades inside cells—if one of the proteins in the cascade is defective, the pathway may not operate to specifications, and normal cell–cell communications are then disrupted. For example, the Wnt signaling pathway is constitutively activated by a defect in APC, which accounts for 85% of all spontaneous colorectal cancers. Another critical 'weak point' is the fidelity of DNA replication, and the repair responses to DNA damage. Any DNA damage that is not repaired may lead to cascading failures, as is seen in many types of cancers (after multiple failures, eventually the whole genome is badly damaged, and no recovery is possible). Carlson and Doyle (2002) call this feature of complexity systems fragility (Carlson and Doyle 2002). So while complexity can bring reliability, it also brings fragility.

5.4 Objective Functions, Set-Points and Regulating Chondrocyte Behavior in Articular Cartilage

Regulation of articular cartilage is complicated because the tissue is an integrated system of interacting components that respond to specific needs in changing environments. For example, a chondrocyte population within the tissue needs to mitigate or respond to damage in order to maintain normal tissue function. Feedback control processes are required to maintain each tissue component (e.g. to protect and maintain aggrecan concentrations throughout the tissue, protect and maintain the collagen network and protect, and maintain chondrocyte populations in each zone of the cartilage). In other words, the networks operating at any one time in a tissue are context specific, reflecting tissue-specific needs at that time.

We remind ourselves that cells in tissues make decisions based on information available to them. Cells gather information from systemic, regional and local chemical signaling systems, from the extracellular matrix, as well as from mechanical inputs. They then integrate this information into cell behavior in a distributed way. In other words, there is no 'centralized processing unit' that makes decisions in a cell. Rather decision making is distributed throughout a cell, and into the extracellular matrix. Because decisions are distributed and emergent, this makes it difficult to find a familiar analogy about the way a cell 'perceives' information presented to it.

Perhaps 'a regulated market-based economy' is somewhat analogous to a cell, in that there are countless interactions between people (analogous to molecules in a living system), some of which result in transactions of goods or services (analogous to chemical reactions forming reactants). At any one time, it is difficult to know just how all the independent interactions by people across a country making up the economy will ensure that shops will be stocked with the stuff people require over the next week. How does bread say, appear on the shelves every day of the week, as if by 'magic' (as far as the average customer is concerned). Somehow the economy manages to do this, each and day for every week of the year. Adam Smith described the unintended social benefit of self-interested individual actions an 'invisible hand.' The invisible hand is an analogy for distributed decision making (we cannot actually measure the decisions themselves), that results in social benefits (we can measure the results of decisions), such as bread on shelves in shops.

How does a 'market economy' make decisions that result in bread on shelves? A market economy is regulated by continuously changing, simultaneous 'price signals' in multiple markets, to which people respond. The prospect of profit (simply income minus costs) incentivizes people's behavior, guiding them toward the 'unintended social benefit' of bread on the shelf. We cannot actually see or measure the decision, just the behaviors of people and the outcomes of their labor (in this case bread). Much like a cell, we appreciate that 'decision making' in a market economy is distributed. Much like a cell, energy is crucial to both cell and market economy. But within a cell supply and demand is regulated not by price, but by regulatory chemical interactions occurring at multiple levels (so the analogy breaks down).

But we get the idea that 'decision making' is distributed throughout the whole cell with nothing happening completely independently of everything else, just as happens in a market economy. We note that it for both a cell and a market economy, it is very difficult to measure what is going on in a cell or economy at any one time, leaving many unanswered questions (how efficient or how reliable or how responsive is this system?). It is these types of unanswered questions about these systems that explains why people looking for answers turn to computational models.

Some proxy for this 'distributed decision apparatus' is useful when trying to developing predictive computational models. Because a cell cannot yet be practically represented by a complete set of chemical equations (which in principle should capture a cell's decision making, along with everything else it does), we seek to model an alternative system, one that is inevitably simplified. At a high level, we perceive the information presented to us about cells through a variety of measurement methods (e.g. tissue morphology, aggrecan production, cell division and migration rate, apoptosis, etc.), which we can interpret as cells accomplishing these tasks. So it appears to us as though there are a variety of functional objectives being pursued by cells (Roux 2014), even though it is certain a cell holds no such conception of what it is doing.

From this perspective, we can infer a 'cell decisions' based on functional outcomes (i.e. cellular decisions are tracked by how much of some functional outcome is achieved—or not achieved). More formally, we simply define that cellular decision making can be measured by how much of some predefined outcome is achieved. For example, the speed that a cell migrates (a measurable functional outcome enabled by the internal state of a cell), reflects the speed the cell 'decides is appropriate' for those particular environmental conditions.

Control theory defines functional objectives quantitatively using a so-called objective function. The objective function may be a very complex, reflecting the fact that more than one objective may be pursued simultaneously by a cell (e.g. repair damaged tissue, where tissue repair could involve multiple simultaneous actions). Clearly, the objective function may change over time, reflecting changing (cell) objectives. So one may think of a single broad objective function (e.g. tissue homeostasis or tissue repair), as made up of many separate (functional) parts, each having a time-dependent weighting. In this way the objective function (and so cell behavior) can evolve over time. Often an objective function is framed as a minimization of a 'least squares sum' [or 'least products sum' (Ludbrook 2012)]. 'Least squares sum' usually means the sum of the squares of the difference between a desired state and the actual state. The cell changes its behavior so as to minimize this least squares sum (i.e. minimize the objective function).

For example, part of a complete tissue level homeostatic objective function may involve an objective function module controlling 'tissue strain.' This tissue strain module may be further broken down into strain control of various tissue components (e.g. aggrecan and collagen network, with volumetric strain and shear strains in both). Component strains may be controlled through objectives functions that attempt to avoid tissue failures (Gardiner et al. 2016; Mononen et al. 2016). There may be a variety of failure modes in the tissue—some of the more important may be avoiding loads exceeding the ultimate tensile strength of the tissue, avoiding fatigue failure of the tissue, avoiding a wear failure at the cartilage surface, and avoiding chondrocyte cell death in each cartilage zone. And when tissue failure inevitably does occur through usage, additional objective functions are required to minimize the damaged volume through processes representing specialized repair processes.

Another part of the homeostatic objective function may involve an objective function that uses strain to control the state of cell differentiation and the cell's secretome. Chondrocytes in the superficial zone of articular cartilage are in tension in the plane of the superficial zone collagen fibrils, and in compression normal to the plane of the superficial zone (i.e. in a biaxial strain state), which probably contributes to de-differentiation of chondrocytes toward a more fibroblastic/ tenocyte-like phenotype, and so to decreased aggrecan production.

We mention that biaxial strain/stress states are not uncommon in tissues. For example, when tendon is compressed as it is bent around a bone (i.e. in so-called 'wrap around' tendons), tenocytes are simultaneously placed in biaxial tension and compression, while tendon fibrils are also compressed when twisted fibers are stretched, and at tendon entheses (i.e. insertional points into bone). Mirroring the behavior of chondrocytes, in a biaxial stress state, tenocytes tend to de-differentiate toward a more chondrocytic morphology (i.e. they become more rounded rather than 'spindle shaped'). These rounded tenocytes increase their production of aggrecan (Kaplan 1980). Interestingly we note that aggrecan (and TGFβ) production increases very significantly when the tendon collagen fibers break (Maeda et al. 2011). Aggrecan production probably protects, and indeed saves tenocytes in a shear zone (a local shear zone occurs between intact and damaged collagen fibrils, as the tenocytes straddle both). Later during repair, aggrecan probably facilitates the movement of repair molecules to the site of collagen fibril damage from the tissue between tendon fascicles. As another example of cell behavior in a biaxial stress state, the pressure in the aorta produces what may be summarized as 'hoop tension and radial compression.' In this case, the smooth muscle cells between the elastic lamina secrete the molecule versican to resist radial compression (Jacob 2003), much as chondrocytes and tenocytes secrete aggrecan to resist compression. Versican is a close relative or aggrecan but it has fewer GAG side chains, and so tends to be used in low to medium compressive stress applications, in contrast to medium to high compressive stress applications where aggrecan is deployed.

Other components of a homeostatic objective function would include maintaining metabolic homeostasis of tissue chondrocytes (i.e. ensuring the right balance between aerobic and anaerobic metabolism, availability of molecules required to construct the molecules in chondrocyte secretions) and maintaining homeostasis of 'supporting tissues' in the joint (e.g. synovial tissue, bone, tendon and ligament).

From the above discussion, one can gather that a very, very complicated objective function is required to completely describe joint behavior over time. Though this objective function is not known (researchers are a long way from formulating comprehensive objective functions, and we need to keep in mind this may even be an inappropriate formalism in which to express a cell's behavior e.g. deep neural networks may yield better outcomes, but as yet, we do not really understand how these function—much like our brain). However at least in principle, we know what the general form this complex objective function must take. The key ideas are that there are 'actual states' and 'desired states' required for tissue integrity, while tissue and cellular processes attempt to minimize the difference between these two states. In the parlance of negative feedback theory https://en.wikipedia.org/wiki/Control_theory (Modell et al. 2015) the difference between the two states is the 'error,' the actual output states are monitored by 'sensors,' the references (or desired) states are simply tissue reference or 'set points,' while the 'controller' adjusts the inputs to the (tissue) processes to generate the desired (tissue) output. When the two states cannot be reconciled despite the controller's best efforts, then the limits of homeostatic control have been reached, and then the tissue has in some sense failed.

Interestingly, we know that these set points need to adequately reduce the risk of failure to an acceptable level. One simple way to manage the risk of failure employed by engineers is to introduce the concept of a 'factor of safety' (or 'partial factors of safety' in a refinement of the method). A factor of safety is a measure of the amount of overload required for the system to fail (however failure may be defined), relative to some normal loading (typically involving the 95th centile of normal loading). Obviously, the factor of safety chosen depends on the statistical distribution of the variability of loading, the intended lifespan of the system and the importance of failure states (which is assessed by 'risk-cost'). It appears that the factors of safety on mechanical loading in the normal synovial joint are comparatively low by engineering standards. Factors of safety are probably in the range of 2.0–4.0 (Biewener 1990), while normal engineering for durable structures requires a factor of safety of at least 3 or 4 according to Biewener (1990), and sometimes more depending on estimated risk-cost.

We note that ultimate load states causing bone fracture are about 2 times greater than typical maximum working loads (e.g. experienced when running, hopping or galloping), so the factor of safety is about 2 for long bones in most animals, independent of their size [see Fig. 1 (Biewener 1990)]. While larger factors of safety reduce the likelihood of breakage due to unusual loading, higher factors of safety inevitably mean heavier bones, which would compromise acceleration and/or running speed. Importantly, such trade-offs are the norm for complex multiobjective functions used to describe any tissue. This highlights the fact that cells, tissues and animals are subject to the same types of constraints and trade-offs as are found in any engineered system.

Rather than using factors of safety to manage risk, an alternative engineering method is to directly estimate the 'probability of failure.' This type of design approach is usually adopted for machines or infrastructure that imposes a very high cost on the community if it fails (e.g. airplane design and nuclear power station design). Typically a design methodology known as 'reliability analysis' is employed.

Reliability analysis involves defining (in a multidimensional space) generalized loadings, failure modes and generalized structural resistances associated with the identified failure modes. In its simplest form, the generalized loading could be a mechanical loading (e.g. variable cyclic mechanical loading), the failure mode may be material fatigue, while the structural resistance is obtained from various experiments on the material. Integrating this data, for any time a probability of failure (i.e. 1—reliability) can be estimated. In principle this simple example can be expanded to take into account 'generalized loadings' (e.g. enzymatic degradation) as well as taking into account repair of the tissue (e.g. aggrecan replacement through de novo synthesis of aggrecan by chondrocytes). This appears to be a promising avenue for future research (Gardiner et al. 2016).

5.5 Negative Feedback and Set-Points

Physical systems themselves are often stable in the sense that any perturbation (i.e. change in the environment) sets in motion events that tend to counteract the imposed change, and so limit the system moving to far from its former state. Indeed at a fundamental level, Le Chatelier's principle itself [or the 'equilibrium law for chemical systems' https://en.wikipedia.org/wiki/Le_Chatelier%27s_principle, and the Principle of Least Action for the free energy of chemical systems (Karl 2012)], aid the stability of chemical systems within organisms. But these are passive responses that do not restore the system close to its original state—the passive systems aid homeostasis, but don't ensure homeostasis. To restore the system to its original state requires some form of active control system, usually referred to as a negative feedback control system (though often positive feedback often occurs as part of the initial homeostatic process, which then is balanced by a negative feedback response that becomes dominant at the steady state, close to the set point).

One of the central principles of physiology is homeostasis http://en.wikipedia.org/wiki/Homeostasis (Modell et al. 2015). Homeostasis means the maintenance of the *status quo* within the organism, despite variations in the 'outside environment.' In other words, homeostasis means that the biological system is controlled to maintain a comparatively constant 'internal state.' For example, core body temperature is maintained at close to a pre-determined set point for all warm-blooded animals, despite wide changes in ambient temperature. Indeed homeostasis has been used to explain behavioral responses of all the major organ systems.

For the haemopoietic system, the concept of homeostasis helps us understand physiological responses to everything from blood temperature, blood pH, blood pressure and blood protein control, as well as the fraction of red blood cells in blood (i.e. the hematocrit), to controlling the oxygen, glucose and ATP levels in cells in the marrow. Indeed, it appears likely that the chemical reactions constituting life are bundles of homeostatic processes within bundles of homeostatic processes. Homeostasis is one of the most potent and important organizing principles in physiology (sitting alongside natural selection in importance, which applies at the population level).

The general method of operation of a 'negative feedback control system' is to have a sensor element [a sensor may be distributed throughout the tissue (e.g. see discussion of TGFβ), which generates a signal]. A comparison of this signal is made to some reference state or 'set point.' This step is done by 'information processing networks' within the cell, but what that exactly involves can be something of a mystery. Modell says of 'set points' that physiological systems behave as though there are set points to regulate physiological variables, though we do not necessarily understand the mechanisms and the set points may change over time or be cyclic (Modell et al. 2015).

We see that we define a simple conceptual model of homeostasis, but when we apply it to a cell, the simple model quickly becomes multilayered and much more complex, which reflects the complexity of the cell system being modeled. A cell 'decides' upon a response, which aims to restore any detected deviation in the system from normal back toward the desired state, but a cell's information processing networks and set-points may themselves be modified by the cells differentiation state and by its local microenvironment. In other words, we understand that the details of feedback processes are context specific, and so a sensor system, 'set-point' and effectors are often transient.

Rather than negative feedback, a cell may exhibit positive feedback. Positive feedback tends to amplify a cell response. While this is often useful, a positive feedback process always tapers off. It has too, as something always limits the positive feedback. Sometimes both positive feedback and negative feedback processes both oppose one another, creating a stable dynamic state. When negative feedback processes containing positive feedback processes fail, the system becomes dysregulated and often the tissue is destroyed (e.g. many chronic inflammatory states result from dysregulated feedback signaling systems).

Finally, an interesting type of negative feedback systems arises if time delays are introduced into the process. A time delay can mean that the system drifts further from a set point than might otherwise be the case. When the system does correct, and starts to move toward the set point, then the correction lasts longer than might otherwise be the case. This leads to periodic undershooting and overshooting of the set point, and so the system can develop oscillations. These oscillating systems are sometimes deliberate. For example, diurnal (or circadian) rhythms involving the expression (with time delays in negative feedback) of two transcription factors may regulate the periodic expression of genes (see discussion Sect. 2.10.1).

Signaling molecules are some of the more easily detectable parts of the positive and negative feedback loops operating in tissues. For this reason, often more is known about signaling molecules than about the other (important) details of the positive and negative feedback loops. Accepting this experimental bias in data collection, we focus on signaling molecules. Cells produce many signaling molecules that are perceived by 'receptors,' which are sensors. If the receptors for chemical signaling molecules are on nearby cells, it is known as paracrine signaling, while if they are on more distant cells, it is known as endocrine signaling. Sometimes cells signal to themselves, which is referred to as autocrine signaling.

On first impression, autocrine signaling is perplexing—why would a cell want to signal to itself? It is eventually realized that looping a 'chain of events' out into the extracellular environment (through the autocrine signal pathway in the ECM) and then returning the signal to the cell via a cell-surface receptor and its associated signal-transduction processes inside the cell, has the advantage of the signaling cascade interacting with, and so in some sense 'sampling,' the extracellular environment. In effect, the cell is sending out a 'test signal' into the external environment to gather information about its environment. Based on the signal that comes back to the cell, presumably the cell can infer information about the cell's outside environment (in much the same way as sonar puts a mechanical signal into the surrounding environment). In other words, autocrine signaling enables a cell to 'sample' or 'test' their local microenvironment, using the information that returns to the cell to help it choose how to respond most appropriately in that environment.

For example, Wnt molecules are often secreted by a cell, which are detected by the same cell's Frizzled receptors (see Sect. 5.6.3). However if the Wnt molecule meets Secreted Frizzled-Related Protein (SFRP) in its external environment, then the Wnt molecules may never return to the cell's Frizzled receptor, which gives a very clear signal to the cell indicating that it should not undergo cell division.

We note that for cartilage tissue, mechanical loading is normally a critically important environmental variable that determines how the tissue feedback networks respond. Mechanical loading is sensed by chondrocytes throughout the depth of the tissue as a combination of local osmotic pressure change (Urban et al. 1993; Negoro et al. 2008), collagen network strain, hydrostatic fluid pressure (Lee et al. 2003) and fluid flow (Zhang et al. 2008a). The chondrocytes in the deep zone can perceive loading throughout the chondrocyte via changes in hydrostatic pressure, even though collagen strains in their immediate locale may be negligible. By the pressure signal, chondrocytes throughout the whole tissue receive important information about the surface mechanical loading that the tissue is experiencing (see discussion Sect. 2.3.2).

Long-range regulation of chondrocytes over the depth of the cartilage tissue also occurs by cell–cell signaling between populations of chondrocytes in each cartilage zone. The chondrocytes in the superficial zone communicate with cells in the transitional zone, while chondrocytes in the transitional zone communicate with cells in the deep zone. Of course, even longer range chemical signaling may enable chondrocytes immediately beneath the contact load to communicate directly with more distant chondrocytes in the deep zone or in the lateral cartilage. While all the cell–cell signals have yet to be fully identified and elucidated (there appears to be dozens of them), it is known that some of the major soluble signaling systems involved in communication between chondrocyte cell populations in articular cartilage are TGFβ signaling, Ihh-PTPrP signaling and Wnt signaling.

Other signaling systems play more limited, but often crucially important roles. More appear to be discovered all the time. For example, it is known that RANKL expression in the long-bone cartilage template is responsible for signaling its resorption by so-called chondroclasts (which are derived from monocytes) (Nakashima et al. 2011). Articular cartilage represents a cartilage 'remnant' of the original long-bone template (Poole 1997). This cartilage remnant's existence is due to its resorption being prevented by the presence of OPG (which binds and blocks the action RANKL, so free RANKL levels are not high enough to signal chondroclast resorption). In this case, the OPG-RANKL signaling from the cartilage to monocytes determines whether it is resorbed or preserved. It is also known that in adult articular cartilage OPG remains a crucially important signal preventing bone remodeling units tunneling from the subchondral bone tissue into the cartilage tissue (Shimizu et al. 2007).

Examples of other signaling systems include nitric oxide (NO) signaling (Amin and Abramson 1998; Lee et al. 2003) and prostaglandin (PGE2) signaling (Attur et al. 2006; Otsuka et al. 2009), which appear to be involved in regulating processes in the pericellular space. It is noted in passing that the NO signal has a very short half-life of only a few seconds, and so by necessity it has to be a local signaling molecule. FGF signaling (Sahni et al. 1999), BMP signaling (Rountree et al. 2004) and activin signaling (Plaas et al. 2011) modify chondrocyte proliferation and differentiation states. Clearly, the composition of the ECM may influence all the chemical signaling systems (Otsuki et al. 2010), either by changing the transport or binding of the chemical.

While there are additional signaling systems (e.g. cartilage derived morphogenic proteins—CDMPs), here we focus on a more detailed description for a subset of the better-known system including TGFβ, IHH-PTHrP, Wnt and IGF signaling systems, and give shorter descriptions for many other signaling systems: TNFα, IL-1, OPG-RANKL-RANK signaling, nitric oxide signaling, prostaglandin signaling, FGF signaling HIF1-alpha, BMP/activin, integrin, calcium, TOLL and RAGE signaling etc.

5.6 Extracellular Signaling Systems in Cartilage

5.6.1 TGFβ Signaling

The TGFβ signaling family of molecules has at least 35 members [including the three isoforms of TGFβ: TGFβ1, TGFβ2 and TGFβ3 and activin, together with the (bone morphogenic protein) BMP and (growth and differentiation factor) GDF groups (van der Kraan 2017)]. TGFβ acts in a large number of different ways, influencing most cell decisions including differentiation states, e.g. sometimes it stimulates proliferation, while also blocking terminal differentiation of chondrocytes (Zhen and Cao 2014). Because of this variety of reported effects, TGFβ's effects are said to be 'pleiotropic' (i.e. it has multiple phenotypic effects).

Most importantly, TGFβ is synthesized as an inactive precursor, consisting of an association between TGFβ, latency associated protein (LAP) and latent TGFβ binding protein (LTBP) (Zhen and Cao 2014), which stick together until active TGFβ is released (Annes et al. 2003; van der Kraan 2017) (Fig. 5.4). Unlike most signaling systems, high concentration of latent signaling molecules, attached to the ECM, enables spatial and temporal regulation of TGFβ (and chondrocytes) in response to changes in articular cartilage ECM. Connective tissue growth factor (CTGF) appears to play a role in one TGFβ activation pathway at the cell membrane (Tang et al. 2017), and in mice removing GTGF has been shown to increase SMAD2 signaling, production of latent TGFβ, and increase cartilage thickness (Tang et al. 2017). Premature activation of TGFβ can lead to skeletal disorders such as Camurati-Engelmann disease, Marfan syndrome, Loeys-Dietz syndrome and Shprintzen-Goldberg syndrome, while constitutive activation of TGFβ leads to fibrosis (Zhen and Cao 2014).

High concentrations of active TGFβ are essential to maintain the chondrocyte and the articular cartilage tissue phenotype (van der Kraan 2017), and for tissue homeostasis (Zhen and Cao 2014). Says Zhen and Cao (2014): '[TGFβ] *inhibits terminal differentiation of chondrocytes, thereby blocking cartilage matrix calcification and vascular[ization] to maintain extracellular matrix (ECM) integrity. Interruption of TGFβ signaling in the articular cartilage results in loss of proteoglycans and cartilage degeneration... Dysregulation of any factor involved in TGFβ signaling transduction may affect cartilage integrity.*'

Concentrations of latent TGFβ in articular cartilage tissue is around 300 ng/ml (Albro et al. 2013). Joint loading stimulates TGFβ production and activates TGFβ signaling (which can be shut-down by molecules blocking ALK5, ALK6 and ALK7 activity) (van der Kraan 2017). In contrast to cartilage, where active TGFβ concentrations are normally high and ADAMTS5 concentrations are normally low, active TGFβ concentrations are normally low in tendon while ADAMTS5 is high (Mehdizadeh et al. 2017). Within bone tissues, active TGFβ concentrations are reasonably high within BMUs. In all three tissues (cartilage, tendon and bone) TGFβ concentrations increase during tissue repair, helping to coordinate tissue repair processes.

In cartilage (and bone and tendon) TGFβ signaling is usually 'anabolic,' helping to regulate cell responses while building new cartilage ECM, and so restoring tissue to its former state (Lu et al. 2014). In cartilage (and bone and tendon) latent TGFβ attaches itself to the ECM, so that when the ECM is degraded by proteases, active TGFβ is released, thereby stimulating tissue renewal and repair. This is one important way in which tissue damage is coupled to tissue repair. It is noted that in all of these tissues, TGFβ normally works cooperatively with IGF signaling while building new ECM (i.e. where one growth factor is found, the other is generally not very far away). TGFβ, IGF-1, FGF-2, and BMP7 are considered the main anabolic molecules in cartilage (Mariani et al. 2014).

As well as enhancing ECM production, TGFβ signaling promotes cell proliferation (Grimaud et al. 2002). Elevated levels of all three isoforms of TGFβ activate Smad2 and 3 signaling [and suppresses terminal (runt related) RUNX2 signaling] via ALK5, which actives the transcription factor SOX9. SOX9 in turn is responsible for chondrogenesis and expression of the type II collagen gene Col2A1 (Bell et al. 1997; Furumatsu et al. 2005, 2009). This strong linkage between TGFβ and the SOX9 transcription factor largely explains TGFβ's central role in maintaining the cartilage tissue phenotype. Activation of the SMAD1, 5, 8 signaling via the ALK1 induces RUNX2 and leads to hypertrophic differentiation (and expression of Collagen X, MMP13, osteopontin, osteocalcin, alkaline phosphatase and VEGF by chondrocytes) (Mariani et al. 2014).

We note in passing that Rock/Rho signaling downregulates SOX9 expression, so Rho signaling needs to be downregulated to promote a stable chondrocyte phenotype (Woods et al. 2005). RhoA/ROCK signaling suppresses cortical actin organization, a hallmark of differentiated chondrocytes. We also note that for mice, TGFβRII knock-down leads to upregulation of MMP13 and ADAMTS5, which progresses to an OA phenotype (Shen et al. 2013).

Importantly some research is pointing to age-dependent differences in TGFβ signaling within human chondrocytes relating to changes in Smad signaling (i.e. age-related changes are associated with changes in the ratio of ALK1/ALK5 receptors at the chondrocyte cell membrane) (van der Kraan et al. 2012). van der Kraan et al. suggest an increase in ALK1/ALK5 ratio on the cell surface induces a switch in the TGFβ signaling pathway from 'canonical' SMAD2/3 activation toward 'non-canonical' SMAD1/5/8 activation (Mariani et al. 2014; Wang et al. 2014), which increases expression of RUNX2 and MMP13 (which could contribute

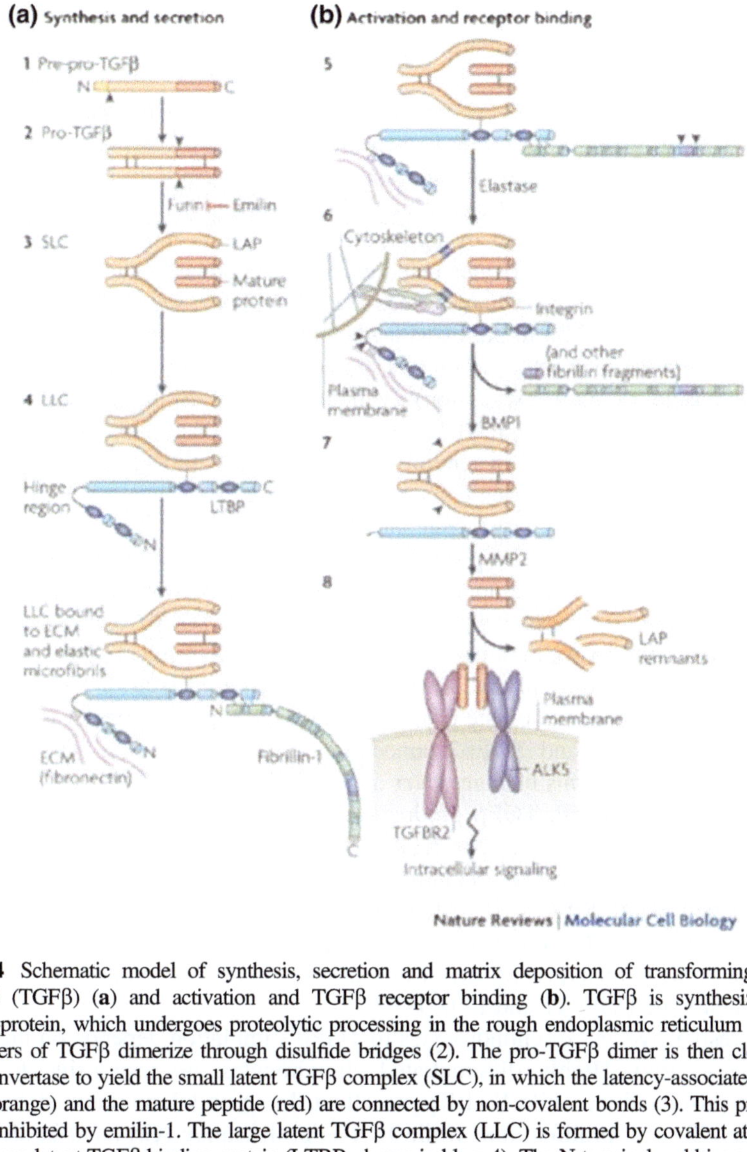

Fig. 5.4 Schematic model of synthesis, secretion and matrix deposition of transforming growth factor-β (TGFβ) (**a**) and activation and TGFβ receptor binding (**b**). TGFβ is synthesized as a pre-pro-protein, which undergoes proteolytic processing in the rough endoplasmic reticulum (1). Two monomers of TGFβ dimerize through disulfide bridges (2). The pro-TGFβ dimer is then cleaved by furin convertase to yield the small latent TGFβ complex (SLC), in which the latency-associated peptide (LAP; orange) and the mature peptide (red) are connected by non-covalent bonds (3). This processing step is inhibited by emilin-1. The large latent TGFβ complex (LLC) is formed by covalent attachment of the large latent TGFβ binding protein (LTBP, shown in blue; 4). The N-terminal and hinge region of LTBP interact with extracellular matrix (ECM) components such as fibronectin; this interaction can be covalent owing to crosslinking by transglutaminase. The C-terminal region of LTBP (blue) interacts non-covalently with the N-terminal region of fibrillin-1 (green; 4). As part of TGFβ activation and receptor binding (**b**), an internal fragment of fibrillin-1 (indicated in purple in 5) can be released by proteolysis (mediated by elastases at sites indicated by black arrowheads in 5) and interacts with N-terminal region of fibrillin-1 to displace LTBP and release LLC (6). The LLC can be targeted to the cell surface by binding to integrins via RGD sequences (blue regions) in LAP (6). Bone morphogenetic protein-1 (BMP1) can cleave two sites in the hinge region of LTBP (arrowheads in 6), which results in the release of LLC (7). Matrix metalloprotease-2 (MMP2) (and other proteases) can cleave LAP (black arrowheads in 7) to release the mature TGFβ (red). Mature TGFβ can then bind to its cognate receptors, TGFBR2 and ALK5 (8) (ten Dijke and Arthur 2007)

to the development of early osteoarthritis). From this we see that TGFβ may not be exclusively anabolic, but can have a so-called 'dark-side' that drives catabolic processes (van den Berg 2011).

Following a brilliant review of injury responses in various tissues, together with their own data, Plaas et al. come to a position on TGFβ1 signaling in cartilage diametrically opposed to the conventional one, as described in the papers above (Plaas et al. 2011). Plaas et al. argue that SMAD1/5/8 is the normal chondrogenic signaling pathway (modulated by aggrecan and HA concentration in pericellular matrix), arguing that chondrocytes respond to tissue damage more like soft tissues (e.g. like dermis) rather than hard tissues (e.g. like bone) (see Fig. 5.4). They find an increase in ADAMTS5 in the ECM leads to more SMAD2/3 activation, and this induces reversion to a more primitive fibrogenic phenotype (as evidenced by smooth muscle α actin expression), which can lead to tissue destruction in osteoarthritic states (Plaas et al. 2011). In a later paper, Plaas and colleagues report that ADAMTS5 is required in tendon to shift from a pathological 'chondrogenic response' to a normal fibrogenic (healing) response (Bell et al. 2013), which is consistent with the signaling schema shown in Fig. 5.4.

All the TGFβ isoforms are produced by chondrocytes. In pig mandibular cartilage, there is a long-term temporal variation in the expression of TGFβ isoforms. TGFβ3 expression is constant from newborn onwards, TGFβ2 expression significantly increased from six months of age until 24 months, while TGFβ1 expression significantly increased at 24 and 36 months (Moroco et al. 1997). TGFβ isoforms are expressed in all zones of the cartilage, while the type II collagen increased in parallel with TGFβ1 and TGFβ2 concentrations. So apparently, the expression profile of TGFβ isoforms changes over an animal's lifetime.

Crucially like PTHrP, TGFβ production is driven by compressive mechanical loading (Li et al. 2010; van der Kraan 2017), while tension suppresses TGFβ production [which helps to explain the low TGFβ concentrations in tendon (Mehdizadeh et al. 2017, 2018)]. It is also noted that TGFβ1 and TGFβ3 show auto-induction (i.e. TGFβ induces its own expression in a positive feedback loop), which is perhaps unsurprising given this growth factor is central to the chondrogenic phenotype (Villiger and Lotz 1992).

The responsiveness of chondrocytes to TGFβ stimulation depends on their state of differentiation. Highly differentiated chondrocytes are responsive to TGFβ induced proliferation, while IL-1 has no effect. In an undifferentiated state, TGFβ has little effect on proliferation, while IL-1 is proliferative (Grimaud et al. 2002). This clearly demonstrates that growth factors have different effects depending on differentiation state.

Most importantly, we note that the total concentration of TGFβ in human cartilage tissue is around 70 ng/ml (i.e. about 5 nM), with 98% being in the latent form, implying about 100 pM is in the active form (Albro et al. 2013). The total concentration of TGFβ in normal human synovial fluid is about 1.8 ng/ml (i.e. about 70 pM), with less than 0.05 ng/ml in the active form (i.e. about 2.0 pM) (Albro et al. 2012). We also note that synovial fluid shear can activate TGFβ,

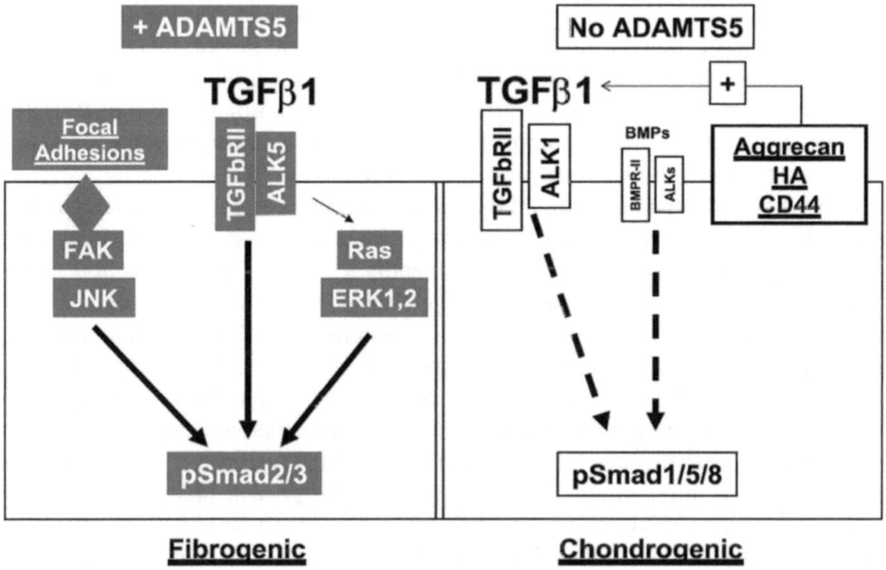

Fig. 5.5 Schematic of ADAMTS5-mediated control of pro-fibrotic/prochondrogenic TGFβ1 signaling in mesenchymal cells. The schematic describes the proposed modulation of TGFβ1 signal transduction through the ALK5-fibrogenic pathway and the ALK1-chondrogenic pathway. ALK5/Smad2, 3 signaling is shown to require ADAMTS5 and can be further supported by pFAK at focal adhesion and pERK1, 2 generated via the non-canonical TGFβ1 pathway. ALK1/Smad1, 5, 8 occurs in the absence of ADAMTS5 when it is enhanced by the presence of HA-aggrecan bound near the cell surface by CD44 (Plaas et al. 2011)

increasing the active concentration of TGFβ in the synovial fluid to about 0.6 ng/ml (i.e. about 25 pM) (Albro et al. 2012). We also mention that the mean concentrations of active TGFβ in synovial fluid from OA joints is reported to be about 4 ng/ml (Fava et al. 1989). Rising active TGFβ concentrations in OA joints may promote osteophyte formation (van der Kraan 2017) (Fig. 5.5).

5.6.2 Ihh-PTHrP Signaling

Parathyroid hormone (PTH) is a hormone produced by the parathyroid gland, while parathyroid hormone-related peptide (PTHrP) is a paracrine/autocrine signaling molecule produced within a tissue. Despite their limited homology, they both bind and have high affinity for the same PTH/PTHrP receptor (Schipani and Provot 2003). This means the total concentration of PTH/PTHrP in a tissue is a weighted sum of the two (though calculating a weight sum appears to be uncommon in experimental practice).

Indian hedgehog (Ihh) and parathyroid hormone-related peptide (PTHrP) signaling is of critical importance in regulating cartilage growth plates and bone formation (see Fig. 5.6 and Appendix C Fig. C.2) at the growth plate. So we first consider growth plate dynamics, and consider articular cartilage second. PTHrP is reported to be an essential component of the mechanotransduction complex necessary to stimulate chondrocyte proliferation (Wu et al. 2001).

For the growth plate system, PTHrP is produced by cells in the perichondrium, while Ihh is produced by distant prehypertrophic cells, setting up two opposing signaling gradients (see Fig. 5.6 and Appendix C Fig. C.2). A reduction in one signaling partner from one cell population (Ihh or PTHrP) sets up compensatory processes that boost the other signal (and so restores the cell population producing the signal), so a new equilibrium is re-established. This mutually dependent signaling system maintains two chondrocyte populations, with cells between these two populations progressing through a sequence of differentiation states (see Fig. C.2).

Actual concentrations of Ihh and PTHrP control the rate of flux of cells through the sequence of differentiation states supporting the growth plate (and bone formation). Ihh drives production of PTHrP and cell proliferation in the perichondrium and 'adjacent articular cartilage' [i.e. the higher the Ihh concentration the more 'round cells' next to the perichondrium (see round cells Fig. 5.6)]. Ihh also induces BMP production, which increases Ihh secretion in a positive feedback loop [i.e. BMP7 induces Ihh (Seki and Hata 2004)], while FGFs repress Ihh production [see Fig. 6 (Schipani and Provot 2003)]. As round cell numbers and PTHrP production in increases, this suppresses Ihh production, setting up a negative feedback loop between round cells and pre-hypertrophic Ihh producing cells. Round cells produce

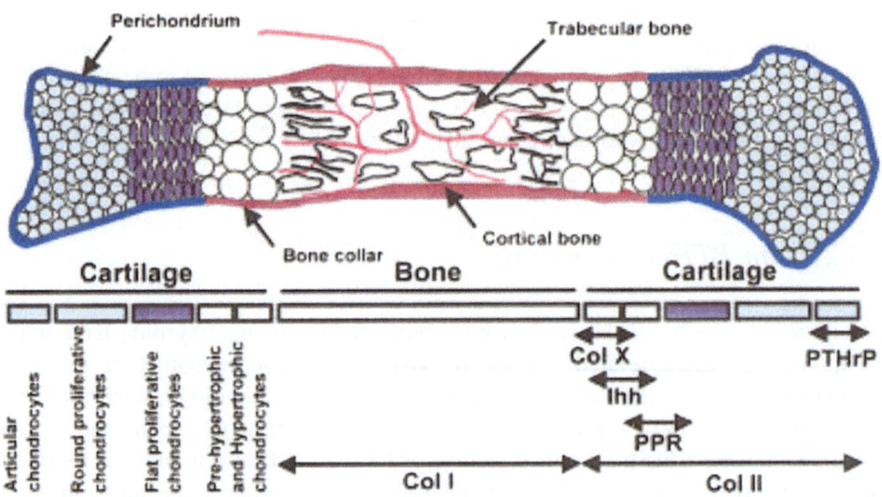

Fig. 5.6 Endochondral bone formation. Schematic representation of a mouse tibia at late stage of fetal development. Characteristic markers for bone, periarticular, flat, prehypertrophic, and hypertrophic chondrocytes are noted (Schipani and Provot 2003)

columns of chondrocytes and collagen type II. Round cells mature in a series of intermediate states into pre-hypertrophic chondrocytes (which produce Ihh). The pre-hypertrophic cells that are most distant from the PTHrP source 'escape' the influence of Ihh suppression by PTHrP, and under the influence of more elevated Ihh concentrations, differentiate into post-mitotic hypertrophic chondrocytes, which then produce collagen type X and mineralized matrix. Finally, hypertrophic chondrocytes undergo cell death, blood vessels invade, and the calcified cartilage is resorbed by osteoclasts. Bone is then formed (see Fig. 5.6 and Appendix C Fig. C.2), and by this means the bone lengthens. We mention that the spatial extent of PTHrP and Ihh signaling is lengthened by higher concentrations of (sulfated) chondroitin sulfate (Cortes et al. 2009).

In simple terms, at the growth plate, the stronger the Ihh drive for perichondrium and round cells to produce PTHrP and increase round cell numbers through proliferation, the more strongly the perichondrium and round cells produce a PTHrP signal that directly reduces Ihh production and the number of prehypertrophic Ihh producing cells, down-regulating Ihh production at its source. So when Ihh achieves its intended effect (a supply of chondrocytes to form bone), the drive to produce more Ihh is suppressed by a negative feedback. The two cell populations reach a steady state, such that if one cell population changes either up or down, it sets in train a series of events restoring both populations toward a new equilibrium point. As might be expected the equilibrium point is itself variable (e.g. it depends primarily on growth and sex hormonal concentrations), which controls the rate of flux of chondrocytes to form bone.

What does developmental growth of long bones have to do with articular cartilage? The importance of this growth signaling system for articular cartilage is that during development, the formation of a secondary ossification center in the epiphysis, splits the original single Ihh-PTPrP signaling system for the growth plate bone into two independent Ihh-PTHrP signaling systems. There is one signaling system to one-side of the secondary ossification center that remains with the growth plate, while the second signaling system is on the other side of the second ossification center, in what is to become the articular cartilage (Chau et al. 2011; Chen et al. 2008; Vortkamp et al. 1996). The evidence for splitting of the Ihh-PTHrP system by the secondary ossification center is shown in Fig. 5.7.

In articular cartilage, the region equivalent to the 'perichondrium' are chondrocytes located in what will become the transitional/superficial zone in adult articular cartilage, while the equivalent region to the prehypertrophic zone contains the articular chondrocytes that will be located in the future deep/radial zone in adult articular cartilage. However, Iwamoto et al. note that the cells forming the superficial and transitional zone also experience a different growth factor environment as the joint forms (being exposed to Wnts, GDF5, Erg and PTHLG) (Iwamoto et al. 2013) (see further discussion Appendix C). In this context of articular cartilage formation during joint development, it appears relevant to also mention that during maturation of embryonic stem cells that '*…specification with GDF5, together with inhibition of hedgehog and BMP signaling pathways, generated a population of*

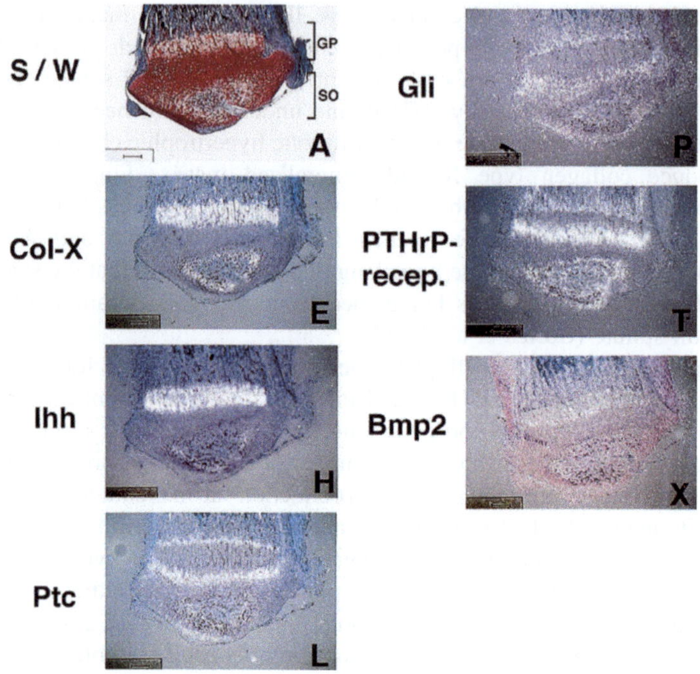

Fig. 5.7 Gene expression in the growth plate of a P13 (post-natal day 13) mouse tibia. Parallel sections of the growth plate of a P13 mouse tibia were stained with Safranin–Weigert (A) or hybridized with 33P-UTP labeled RNA probes for *Col-X* (E), *Ihh* (H), *Ptc* (L), *Gli* (P), *PTH/PTHrP receptor* (T) and *Bmp2* (X). Safranin–Weigert stains bone blue, cartilage red and the large hybertrophic chondrocytes appear white. *Col-X* is expressed in the hypertrophic chondrocytes in the growth plate and in the secondary ossification center. The expression domain of *Col-X* is overlapped and flanked by the successive expression of *Ihh* and the *PTH/PTHrP receptor* in less differentiated chondrocytes, respectively. *PTH/PTHrP receptor* is also expressed in osteoblasts in the periosteum flanking the bone and in the osteogenic cells replacing the hypertrophic chondrocytes. *Ptc* and *Gli* are expressed in the less differentiated chondrocytes distal to the *Ihh* expressing chondrocytes in both the primary and secondary ossification centers. In addition both genes are also expressed in the osteogenic cells replacing the hypertrophic cartilage. The expression of *Bmp2* is found in the early differentiating chondrocytes flanking the distal expression domain of *Ihh*. The same sections are shown at low magnification to show the relative expression of each gene (A, E, H, L, P, T, X); scale bar: 200 mm, while higher magnifications are shown in Fig. 4 of Vortkamp et al. (1998). GP = growth plate; SO = secondary ossification center (Vortkamp et al. 1998)

non-hypertrophic chondrocytes that displayed properties of articular chondrocytes' (and formed *hyaline* cartilage) (Craft et al. 2013).

Further, Ray et al. explain the role of noggin in suppressing BMP expression [i.e. BMP2 and 4 (Sandell and Adler 1999)] in the chondrocyte destined to become articular cartilage (Ray et al. 2015):

> We show that nascent cartilage cells are capable of differentiating as articular or transient [bone forming] cartilage, depending on exposure to Wnt or BMP signaling, respectively.

The spatial organization of the articular cartilage results from a band of Nog-expressing cells, which insulates these proliferating chondrocytes from BMP signaling and allows them to differentiate as articular cartilage under the influence of Wnt signaling emanating from the interzone.

In chick and mouse embryos, a band of cells expressing noggin suppress the action of bone morphogenic proteins (BMPs), and so the ossification of articular cartilage. However in adult cartilage, rather than noggin suppressing BMP2 and BMP4 (Sandell and Adler 1999), it is more likely gremlin and follistatin do this (Tardif et al. 2004). For example, Leijten et al. (2012) reports: '*Taken together, our study identified Gremlin 1, FRP, and Dkk-1 as natural brakes on hypertrophic differentiation in articular cartilage.*' We mention in passing that all BMPs are not the same in developing articular cartilage, as BMP2 and 4 appear to induce chondrocyte hypertrophy, while BMP7 blocks chondrocyte hypertrophy (Zhong et al. 2015).

In addition to gremlin [which blocks BMPs 2 and 4, and to 'some extent' BMP7 (Myllarniemi et al. 2008)], FRP (frizzled-related protein) and Dkk-1 (which blocks Wnt signaling), a high concentration of PTHrP in the deep zone directly hinders the differentiation of deep zone chondrocytes into hypertrophic cells (with its attendant bone formation) (Chen et al. 2015; Jiang et al. 2008; Provot et al. 2006; Zhong et al. 2015), and indirectly hinders further differentiation by lowering Ihh concentrations (Jiang et al. 2008; Zhou et al. 2014).

Clearly, this second independent Ihh-PTHrP signaling system that has 'split away' from original single system experiences a very different chemical and mechanical environment during articular cartilage formation to that at the growth plate. It is apparent that subchondral bone forms up to a certain depth from the articular cartilage surface, at which point the PTHrP produced by the transitional zone chondrocytes has sufficient concentration to maintain very low concentrations of Ihh, and together high PTHrP and low Ihh keep deep zone chondrocytes in a prehypertrophic state.

We presume this implies that higher concentrations of PTHrP [together with higher concentrations of TGFβ (Tang et al. 2017)] in the articular cartilage during development will support the development of thicker articular cartilage in the adult. Importantly, mechanical loading strongly induces PTHrP expression, so this suggests that a more active lifestyle during development may potentially lead to thicker cartilage in the adult. Indeed it is conceivable that variation in environmental loadings during growth and maturation is primarily responsible for the variation in cartilage thickness observed in adults. It appears plausible that a more active lifestyle during development may not only thicken cartilage, but also increase the fraction of cartilage thickness designated 'deep zone' (see discussion Sect. 5.10).

If PTHrP concentrations within the deep zone of articular cartilage were to fall and Ihh concentrations rise for any reason, bone formation may resume. In this context, we note that PTHrP concentrations in osteophytes are some 5 times lower than in cartilage (Zhang et al. 2012). Further transgenic mice with induced Ihh expression develop cartilage damage resembling OA (Zhou et al. 2014), and that blocking hedgehog signaling (via Ihh) prevents osteophyte formation (Ruiz-Heiland et al. 2011, 2012). To prevent mineralization of articular cartilage due to low levels of PTHrP, PTHrP supplementation either directly or via a viral vector has been proposed (Zhang et al. 2012). We should keep in mind that PTHrP/Ihh signaling in not only influenced by local concentrations of PTHrP and Ihh, but also by receptor densities (for example PTHR-1).

In articular cartilage PTHrP producing cells are those chondrocytes in the superficial and transitional zones of the articular cartilage. The more differentiated Ihh producing cells remain in the deep zone of the articular cartilage, from where they drive proliferation of chondrocytes in the transitional zone. Importantly, the rudiments of signaling between the transitional and deep zones have been demonstrated experimentally in vitro using a co-culture system [e.g. http://www.ors.org/Transactions/56/0003.pdf (Jiang et al. 2008)]. As might be expected, it is Ihh production is upregulated in osteoarthritis (Wei et al. 2012). Wei et al. reports:

> Ihh expression was increased 2.6 fold in OA cartilage and 37% in OA synovial fluid when compared to normal control samples. Increased expression of Ihh was associated with the severity of OA and expression of markers of chondrocyte hypertrophy: type X collagen and MMP-13, and chondrocyte size. Chondrocytes were more spherical with increasing severity of OA. There was a significant correlation between Mankin score and cell size (r^2 = 0.80) and Ihh intensity (r^2 = 0.89).

Of course, elevated levels of Ihh will drive higher PTHrP levels, which will provide a strong proliferative drive to chondrocytes in the transitional and superficial zones, thereby increasing chondrocyte numbers to help repair damaged tissue. In articular cartilage, proliferating chondrocytes normally migrate toward the cartilage surface (down the aggrecan gradient toward a less well-differentiated chondrocyte phenotype), rather than into the deep zone (and a more differentiated state). We observe that chondrocyte migration in the articular zone in the opposite direction to chondrocyte migration in the growth plate in the same bone (which might be expected on the basis of splitting the growth plate chondrocyte signaling system). However in both cases chondrocyte migration is down the local aggrecan gradient [for example, see aggrecan/GAG gradients from bone and growth plate through to articular cartilage surface, for mice, as shown in Fig. 4 of Mertz et al. (2012) or qualitative data shown as intensity of alcian blue staining of growth plates in 12 week old rats, as shown in Fig. 3 (Bassett et al. 2006)]. Any significant reduction in PTHrP production by the transitional zone chondrocytes may result in mineralization of the articular cartilage in the deep zone (Chen et al. 2008) (Fig. 5.8).

In articular cartilage the dominant regulatory signal for PTHrP production by chondrocytes in the transitional zone appears to be mechanical loading (Chen et al. 2008; Ng et al. 2006; Ogawa et al. 2014). High concentrations of PTHrP support a less differentiated phenotype for the chondrocytes in the transitional and superficial zones, and PTHrP (and Ihh) supports cell proliferation in the transitional and superficial zones. As a consequence of mechanical loading is probably the dominant driver for PTHrP expression by chondrocytes in the transitional zone in normal cartilage, the production of Ihh in the deep zone is suppressed by mechanical loading below that which it would normally be, due to its contribution to PTHrP

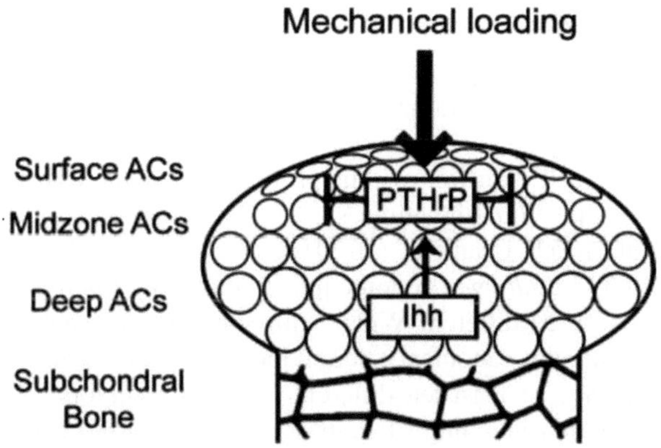

Fig. 5.8 Schematic representation of regulation by the parathyroid hormone–related protein (PTHrP)–Indian hedgehog (Ihh) axis in articular chondrocytes (ACs) (Chen et al. 2008) Ihh is produced by chondrocytes in the deep zone, while PTHrP is produced by cells in the transitional zone. PTHrP signaling is driven by Ihh signaling. However, it is likely that PTHrP production is dominated by mechanical loading [http://www.ors.org/Transactions/56/0003.pdf (Ng et al. 2006)]. This PTHrP suppresses Ihh production in the deep zone, together with suppression of Ihh production in the deep zone by PTHrP and hydrostatic pressure (generated by mechanical loading on the surface of the cartilage). PTHrP promotes proliferation of cells in the transitional zone, but prevents further differentiation of cells in the transitional zone to prehypertrophic cells in the deep zone. This prevents progression of the deep zone into the existing transitional zone

production by mechanical loading above that driven by Ihh alone, and by hydrostatic pressure in the deep zone. But Ihh concentration will 'rebound' when mechanical loading ceases. Clearly, this reversion to Ihh driven PTHrP expression ensures that transitional chondrocytes always have a minimum drive for PTHrP expression, which is essential to maintain healthy cartilage. In other words, this minimum drive for PTHrP expression maintains SOX9 expression, and prevents mineralization of the deep zone cartilage.

This novel arrangement, with two pathways driving PTHrP production by the transitional chondrocytes in articular cartilage (Ihh and mechanical loading), ensures that the tissue will remain cartilage even during periods of inactivity or disuse, when joint loading is minimal or absent. This Ihh proliferative drive is also important for maintaining the health of articular cartilage in the lateral regions of the joint, where loads are small (and possibly absent). There is probably a lot more to learn and understand about the crucially important Ihh-PTHrP signaling system in articular cartilage.

5.6.3 Wnt Signaling

Wnts are lipo-glycoproteins and are involved in most aspects of organism development, as well as playing essential roles in tissue maintenance. For example, they play a key role in cell turnover within proliferative adult tissues. Perhaps now needless to say, the Wnt signaling system is complex (Lories et al. 2013; MacDonald et al. 2009; Usami et al. 2016) (see also the Wnt signal pathways annotated at KEGG http://www.genome.jp/kegg/pathway/hsa/hsa04310.html).

There are at least 20 different Wnts (with subtypes for each), three signaling pathways (canonical, non-canonical calcium pathway, and the non-canonical planar cell polarity pathway), at least ten different 'frizzled receptors' (Killock 2012), a variety of co-receptors (including LRP5/6, Ror2, and kremin) and a large range of Wnt signaling antagonists (including Dkk1–4 and sclerostin) (Warde 2011) (which inhibit the LRP5/6 co-receptors), secreted frizzled-related proteins (SFRPs) 1–10 and WIF (which all bind to Wnts) not to mention Wnt agonists [e.g. R-spondin http://www.ncbi.nlm.nih.gov/pubmed/22439850 (Abed et al. 2011) and Norrin (Chang et al. 2015)]. Even biglycan plays a role in Wnt signaling, perhaps serving as a reservoir in the pericellular space (Berendsen et al. 2011). Wnts are typically hydrophobic due to palmitoylation (Gao and Hannoush 2014).

This signaling complexity is like an extensive 'palette of different colors' that can be mixed and matched to achieve the intended effect. For example, by: (i) allowing Wnt signaling to vary spatially throughout the tissue due to gradients in SFRPs, (ii) the same Wnt can have different effects because the cells express different frizzled receptors, or (iii) allowing several different Wnts to carry different information to the same location, allowing each signal to be interpreted in different ways by a different Frizzled receptor expressed on each cell type. The variety of combinations is enormous, indicating the capacity for a nuanced Wnt signal in environments with a lot of things happening simultaneously. To give one example, activation of non-canonical Wnt signaling through overexpression of the cell-surface Wnt receptor Fzd7 and receptor-associated protein Ror2, by addition of recombinant Wnt5a, promotes the initiation of the chondrocyte columnar architecture in growth plates (Randall et al. 2012).

In the post-natal growth plate of mice and rats (which are known to be somewhat different to human growth plates) six Wnts are expressed, three for canonical signaling and three for non-canonical signaling. Of these Wnts -2b, -4, and -10b signal through the canonical β-catenin pathway and Wnts -5a, -5b, and -11 signal through the non-canonical calcium pathway (Andrade et al. 2007). Summarizes Yang et al. (2016): '*Biologically, the canonical WNT/β-catenin signaling pathway usually has crucial roles in regulating cell fate, proliferation, and survival, whereas the non-canonical WNT signaling is more associated with differentiation, cell polarity, and migration*' (Yang et al. 2016).

In human articular cartilage, various Wnts have been identified including Wnt2b, Wnt3, Wnt5a, Wnt5b, Wnt10a, Wnt10b, and Wnt16—Wnt16 is strongly upregulated following injury (Dell'accio et al. 2008). We note that Wnt16 is also involved in joint space formation, inducing de-differentiation of chondrocytes (Blom et al. 2009), which may explain the rapid de-differentiation of chondrocytes reported by Lyman et al. (2012). Further, Chun et al. report involvement of Wnt3a, Wnt5a, Wnt5b and Wnt11 (Chun et al. 2008). Numerous Wnts, their receptors and their modulators are reported to be differentially regulated in both synovium and articular cartilage in OA models [see, for example, Table 1 in Blom et al. (2009)]. Blom et al. report that β catenin is upregulated in experimental OAs, while WISP-1 is upregulated in the synovium and superficial zone of the cartilage (Blom et al. 2009) (WISP-1 induces the upregulation of numerous MMPs, independent of IL-1 activation). Wnt signaling is reported to be modulated by TGFβ (Dong et al. 2005).

Wnt signaling appears to be essential for maintaining cartilage, as under-expression of Wnts can result in tissue atrophy, but excessive signaling can result in hypertrophy and bone formation (Chan and Little 2012). In this context, we observe that SOX9 expression occurs in epithelial cells in intestinal crypts, but its expression is dependent on Wnt signaling (and downstream, formation of the β-catenin-TCF4 signaling complex). We also mention that high Wnt expression suppresses SOX9 expression, the key transcription factor inducing the chondrocyte phenotype. Clearly, if a Wnt antagonist is suppressed (e.g. one of the SFRP concentrations decrease, or one of the Dkk concentrations decrease) and Wnt signaling becomes too high, this may push the tissue toward a bone phenotype. Fortunately, it appears as though SFRPs are tightly and robustly regulated, so that if one is knocked down, others are upregulated to ensure the cartilage phenotype is maintained (Lodewyckx et al. 2012).

An understanding of this delicate Wnt balance operating to maintain tissue homeostasis is only slowly becoming apparent. Lories et al. (2013) summarized the functional role of the canonical pathway as being proliferative, while the functional role of non-canonical pathways is to stimulate de-differentiation (Lories et al. 2013). According to Lories et al. apparently both signaling pathways are inhibitory of each other (see Fig. 5.9) (Lories et al. 2013). If true, this would help to explain the homeostatic balance between cell proliferation and cell migration—then imbalances between proliferation and migration may be driven by imbalances in canonical and non-canonical Wnt signaling. Too much canonical Wnt signaling may then promote proliferation while suppressing dedifferentiation of the chondrocytes, which promotes chondrocyte migration (see discussion on proliferation and migration imbalances in Sect. 2.9.5). The inhibitory effect between canonical and non-canonical Wnt signaling is also consistent with chondrocytes either proliferating or migrating, but (understandably) not doing both simultaneously.

Chun et al. provide a diagram summarizing the effects of non-canonical Wnt signaling (see Fig. 5.10) (Chun et al. 2008). It appears likely that in humans Wnt3a and Wnt7a result in a proliferation with a small amount of de-differentiation (Wnt3a modulates articular cartilage phenotype by activating both proliferation and some de-differentiation), while activation of the Wnt pathway by Wnt5a and Wnt11 result in strong de-differentiation toward a fibroblastic phenotype (Matzelle et al. 2012), together with cell polarization and expression of smooth muscle α-actin (Tock et al. 2003) (see previous discussions on cell polarization for migration in Sect. 2.9.4).

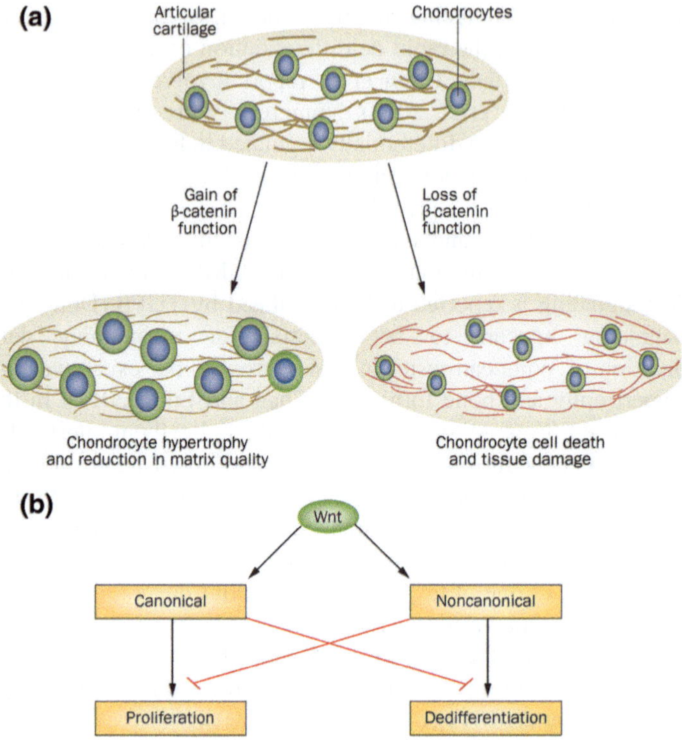

Fig. 5.9 Complex roles of Wnts in cartilage homeostasis and disease. **a** Both overexpression and loss of β-catenin in the articular cartilage lead to joint damage. Overexpression β-catenin results in chondrocytes hypertrophy and loss of matrix quality, whereas loss of β-catenin function results in tissue damage through chondrocyte death. **b** Canonical and noncanonical Wnt signaling pathways keep each other in check through reciprocal inhibition. The canonical pathway appears to stimulate proliferation. The noncanonical pathway stimulates de-differentiation (Lories et al. 2013)

We mention that it has been shown that periostin can work in conjunction with canonical Wnt signaling to promote the expression of MMP13 in cartilage (Attur et al. 2015). In addition, Wnt7a is human and mouse cartilage is pro-anabolic and blocks effects of inflammatory signaling (Gibson et al. 2017), though Wnt7a appears to simultaneously signal through both canonical and non-canonical signaling pathways, which is consistent with inducing a small amount of de-differentiation. Wnt5b signaling activates planar cell polarity, disrupts cell aggregation (by decreasing expression of cadherins) and upregulates cell migration through JNK signaling (Wnt5b regulates mesenchymal cell aggregation and chondrocyte differentiation through the planar cell polarity pathway). Wnt5b signaling is greater closer to the cartilage surface, suggesting chondrocytes in the superficial zone produce most of the Wnt5b (Fukui et al. 2008). It is likely JNK signaling via integrins is augmented by JNK signaling via Wnt5a (Pereira et al. 2011). It is not a simple system.

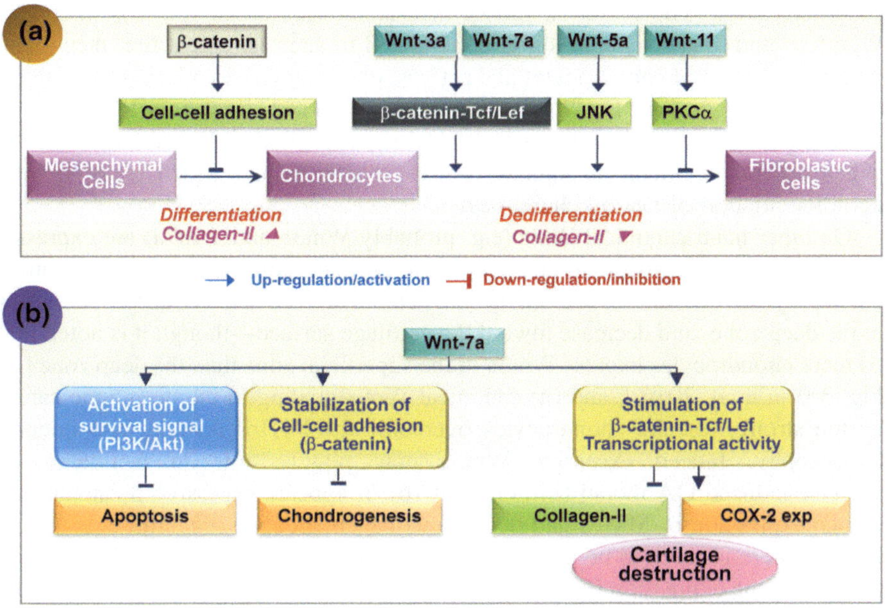

Fig. 5.10 Schematic representation of selected Wnt proteins in the regulation of articular chondrocytes. **a** In primary culture articular chondrocytes, Wnt-3a and -7a blocks type II collagen expression, whereas Wnt-5a inhibits this expression via the noncanonical pathway. By contrast, Wnt-11 enhances type II collagen expression via the noncanonical pathway. **b** Wnt-7a in chondrocytes inhibits type II collagen expression and induces cyclooxygenase (COX)-2 expression via the canonical pathway. Wnt-7a also causes accumulation of β-catenin and increased β-catenin concentrations stabilize cell–cell adhesion and inhibit chondrogenesis. Wnt-7a blocks nitric oxide (NO)-induced apoptosis by activation of survival signals (Wnt signaling in cartilage development and degeneration)

It is generally accepted that Wnt signaling is essential to tissue homeostasis, and given that Wnts5a,b and Wnt11 induce smooth muscle α actin expression, then based on the usual expression profile of smooth muscle α actin over the depth of the cartilage tissue, it is reasonable to suppose that Wnts5a,b and Wnt11 are normally expressed by the chondrocytes in the superficial (and possibly the upper transitional) zone of the cartilage. The superficial zone is subject to considerable tensile loadings as the tissue is loaded. It is likely that the level of Wnts5a,b and Wnt11 expression is related to the tension experienced by chondrocytes in the superficial zone of the cartilage, as these Wnts promote de-differentiation of the chondrocytes toward a more fibroblastic phenotype. Increased levels of Wnts5a,b and Wnt11 would then appropriately signal to the transitional zone chondrocytes, stimulating the chondrocytes in the transition zone to migrate into the superficial zone to replenish those cells that have undergone apoptosis—so maintaining the integrity of the superficial zone cartilage [see Povysil et al. (2008)].

Immunohistochemical staining suggests that Wnt5a is in fact expressed throughout normal articular cartilage, but somewhat more is found toward in the superficial zone (Thorfve et al. 2012). In osteoarthritis processes, it is likely that Wnts5a,b and Wnt11 expression is upregulated to such an extent that they signal de-differentiation of chondrocytes and activate chondrocytes to migrate over the full depth of cartilage [see Fig. 2.55 (Povysil et al. 2008)]. Real-time PCR analysis following IL-1 stimulation shows that Wnt5a is upregulated (Thorfve et al. 2012). When articular cartilage is injured, Wnt16 is upregulated which leads to rapid de-differentiation of nearby chondrocytes.

On other hand canonical Wnts (e.g. probably Wnt3a and Wnt7a) are expressed in the superficial, transitional and deep zones, which along with PTHrP, stimulate chondrocyte proliferation. It is likely that canonical Wnt concentrations are largest in the deep zone, and decrease toward the cartilage surface—though it is noted that: (i) more chondrocytes express Wnt3a in the superficial zone than the deep zone [see Fig. 3 (Liu et al. 2016)], and (ii) canonical Wnt signaling in the deep zone cannot be too strong, lest the chondrocytes overcome the PTHrP restraint and undergo hypertrophy. Indeed excessive Wnt3a may play a role in facilitating an exercise-induced OA model (Liu et al. 2016). It appears excessive β-catenin may induce expression of MMPs and decrease collagen II expression (Liu et al. 2016). Clearly, there is much to be learned, as illustrated by Wnt7a blocking IL-1 induced catabolic gene expression (Gibson et al. 2017). Wnt2b's expression is mainly in the superficial zone, so this signal readily finds its way into the synovial fluid. It seems probable canonical Wnts provides a steady signal for proliferation over the depth of articular cartilage.

This steady canonical Wnt signal for proliferation is normally balanced by a non-canonical Wnt signal for de-differentiation and a migratory drive. This Wnt balance helps determine the balance between cell proliferation and migration to sites of chondrocyte apoptosis. Clearly, one or other signals in this balance may become over-stimulated or under-stimulated, and so become unbalanced in injury and disease processes (e.g. perhaps leading to chondrocyte cloning). For example, should the superficial zone develop splits in its surface (which means it can no longer generate tensile strains) then the non-canonical Wnt migration stimulus may be substantially reduced. This is expected to then unbalance the canonical-non-canonical Wnt signaling balance, leading to excessive proliferation in relation to chondrocyte migration. This imbalance probably contributes to 'chondrocyte cloning,' which is characteristic finding in histological examination of OA cartilage (see discussion Sect. 2.9.5).

There are many different ways for chondrocyte proliferation and migration to become unbalanced. For example, if sFRP or Dkk signaling is reduced then canonical Wnt signaling upregulated. This may happen for example, simply by 'softening' of the cartilage in compression leading to more 'washout' of molecules close to the cartilage surface with intermittent loading. It has been noted that polymorphisms in genes involved in Wnt signaling—particularly in the gene encoding sFRP-3, are associated with an increased susceptibility to the development of OA (Corr 2008). This may reduce the capacity of chondrocytes to de-differentiate and migrate, and may indeed cause the deep chondrocytes to undergo hypertrophy, and for the cartilage to thin as bone is formed.

Further, there are many important details influencing homeostatic cell–cell signaling processes that may appear as 'side effects' relative to the 'big picture' signaling systems. For example, in addition to excessive Wnt3a provoking a catabolic expression profile (Liu et al. 2016), it is likely that an increased concentration of canonical Wnts in the synovial fluid leads to activation of progenitor cells in the groove of Ranvier (Karlsson et al. 2009), which may play a role in osteophyte formation in osteoarthritis (Funck-Brentano et al. 2014). Chondrocytes in osteophytes are probably driven to terminal differentiation and bone formation by Wnt2b (Blom et al. 2009). In another example, it is mentioned that surface shear force increases the expression of PRG4 in the superficial chondrocytes (see, for example, Fig. 4.21) (Grad et al. 2005), while Ihh increases PRG4 expression and canonical Wnt3a decreases PRG4 expression (Iwakura et al. 2013). Intriguingly we also mention that research on familial cardiomyopathy that shows Dkk3 can activate the canonical Wnt signaling system, while suppressing non-canonical signaling (Lu et al. 2016), suggesting we have only begun to understand the subtleties of Wnt signaling.

The Wnt system is clearly very complex, with diverse and critically important physiological influences on the development and maintenance of adult articular cartilage, both in health and disease states.

5.6.4 IGF Signaling

IGF has a profoundly anabolic effect on articular cartilage. IGF strongly upregulates the de novo synthesis of aggrecan and type II procollagen. IGF generally complements the anabolic actions of TGFβ (Davies et al. 2008). Importantly the effects of IGF are synergistic with cyclic mechanical loading, as experienced during everyday activities (Grodzinsky et al. 2000). IGF may well be the most powerful anabolic regulator of ECM homeostasis on a day to day basis.

Like all signaling systems, the insulin-like growth factor system is complex (Tahimic et al. 2013). It is comprised of two insulin-like growth factors (i.e. IGF-I and -II), types I and II IGF receptors (i.e. IGF-IR and IGF-IIR), the insulin receptor (IR), a family of IGF binding proteins (e.g. IGFBP1 through to IGFBP6) and IGFBP-degrading proteases (Zhang et al. 2008b, 2010a, 2013a). Growth hormone regulates the IGF-I production by the liver that is the source of the majority of IGF-I found in plasma (Velloso 2008). On the other hand, IGF-II and IGFBPs found in the serum are most likely sourced from a variety of tissues (e.g. liver, muscle, brain, kidney being the principal sources) (Mohan and Baylink 2002).

IGF also occurs in a number of splice variants, which may have different actions, e.g. in muscle, IGF-1Ea (produced by the liver) promotes muscle cell differentiation, while IGF-1Ec (produced locally in muscle and induced by exercise) promotes muscle cell growth (Goldspink 2005). The significance of IGF splice variants in cartilage is unknown.

IGF is not normally produced by healthy cartilage, though chondrocytes do produce IGF-1 in disease states such as osteoarthritis. IGF-1 from the liver normally enters articular cartilage via the synovial fluid. IGF-1 is usually bound to IGFBP3, which is also produced by the liver, or IGFBP5, which is produced by articular cartilage (Fukui et al. 2008). IGF concentrations in the blood plasma are usually constant throughout over a 24 h period. That is, there is no strong circadian variation in the supply of IGF-1, as observed for other hormones, so it is not immediately clear how articular cartilage (or any other tissue) regulates its supply of IGF to suite its requirements.

A series of papers by the authors have shown that the free concentration of IGF is regulated within the tissue by varying the rate of degradation of the IGF-IGFBP complexes (Gardiner et al. 2011; Zhang et al. 2010b, 2013b). In other words, the same proteases that cause tissue degradation increase the amount of free IGF and thereby stimulate the drive for ECM renewal. In this context, we see that IGF works similarly and cooperatively with TGFβ (TGFβ is also released from a 'latent form' by proteases). It has been shown computationally that because of the IGF-IGFBP complex degradation by proteases, free concentrations of IGF can rise well above those in the synovial fluid. We mention in passing that the same principle may be employed to target drugs of various kinds to specific tissues (Gardiner et al. 2011). Computational modeling of IGF in articular cartilage has shown that free IGF concentrations are highest in the superficial and upper transitional zone of articular cartilage, declining toward the synovial fluid and toward the deep cartilage [see for example, Fig. 4 in Zhang et al. (2013b)].

Some abnormalities of cartilage are attributed to IGF signaling being either too high or too low. For it is known that enzymatic degradation of IGF-BP5 *within the synovial fluid* (leaving free IGF to be degraded by proteases within the synovial fluid) can lead to low IGF levels in the cartilage. The cartilage then develops some features of osteoarthritis (Clemmons et al. 2002). It is reported that 'meniscal tear' leads to increased degradation of IGF-BP5 in the synovial fluid, which can lead to osteoarthritis. It has been shown that blocking the degradation of IGFBP5 degradation by a serine protease inhibitor following meniscal tear can prevent the development of cartilage lesions (Yates et al. 2010). It is likely that the NSAID aceclofenac exerts a positive effect on matrix synthesis via IGF signaling (Dingle and Parker 1997). Taken together this suggests that protecting cartilage from low IGF concentrations is beneficial to articular cartilage.

Further it is known that excessive tissue IGF levels arising by upregulation of binding protein degradation within articular cartilage (Okada et al. 2008), leads to a downregulation of MMP14 expression on the chondrocyte membrane [via reversion-inducing cysteine-rich protein with Kazal motifs (RECK) signaling]. Downregulation of MMP14 effectively blocks cell migration through the cartilage ECM. This results in an imbalance between chondrocyte proliferation (which is relatively too high) and chondrocyte migration (which is relatively too low) leading to what is referred to histomorphometrically as 'chondrocyte cloning' (Kimura et al. 2010) (see discussion Sect. 2.9.5).

5.6.5 OPG-RANKL-RANK Signaling

OPG-RANKL-RANK signaling is the principal regulator for the rate of bone remodeling (Boyce and Xing 2008). OPG is soluble 'decoy ligand' that binds to RANKL, preventing RANKL from activating RANK. This reduces osteoclast numbers and so bone resorption. A normal OPG/RANKL ratio is crucial for maintaining normal bone mass. Low OPG/RANKL ratios result in higher rates of bone remodeling by basic multicellular units (BMUs), which carries the risk of a 'negative bone balance' and so loss of bone mass (Smith et al. 2012). In bone, OPG and RANKL is produced by cells of the osteoblastic lineage (which includes osteocytes), while RANK is expressed by cells of the osteoclastic lineage (Tat et al. 2009).

Chondrocytes in articular cartilage also express both OPG and RANKL, primarily in the superficial and transitional zones (Moreno-Rubio et al. 2010). There are normally no chondrocytes expressing the RANKL receptor, RANK [though OPG does bind to membrane-bound RANKL (Tat et al. 2009)]. During development, RANKL induces the differentiation of chondroclasts—chondroclasts are slightly different to osteoclasts (Wlodarski et al. 2014), and resorb cartilage. Unless RANKL's action is blocked, cartilage will be resorbed. So it is important that the potential resorption of adult articular cartilage be blocked. It is noted that OPG is upregulated by Wnts (Dell'accio et al. 2008), which is consistent with OPG's role in preserving tissue, and Wnt's role in rebuilding tissue via canonical Wnt signaling.

A key role of OPG in adult cartilage is to block any stray signaling by RANKL from adjoining bone, as well as blocking any endogenous RANKL in cartilage. This is vital to preserve articular cartilage tissue from invasion by BMUs from adjoining bone. In chronic OA states, subchondral bone loss is correlated with increased articular cartilage RANKL expression (Martinez-Calatrava et al. 2012).

There is apparently no effect of RANKL on chondrocytes themselves in normal adult articular cartilage (Kwan Tat et al. 2009), so chondrocytes are probably not the target of RANKL (this is consistent with chondrocytes not normally expressing the receptor RANK). Though the role of RANKL in cartilage is uncertain, it is very likely that RANKL's role is to orchestrate the resorption of cartilage by chondroclasts, when normal OPG/RANKL ratios in cartilage fall. Indeed, this is RANKL's role in metaphyseal bone formation (i.e. bone formation from the cartilage template during skeletogenesis) (Wang et al. 2015a). If OPG/RANKL ratios fall sufficiently, even if only very small amounts of RANKL are expressed, then given sufficient time it is almost certain that cartilage will be resorbed. Indeed, it is likely that any tissue that expresses high levels of RANKL will be resorbed quickly, so potent is the effect on osteoclasts differentiation by RANKL. Conversely, tissues that must be preserved express very high levels of OPG to supress RANKL completely, e.g. as occurs in bone ossicles. It seems possible that RANKL plays a role in debridement of cartilage defects or debridement of the cartilage surface, by activating macrophages in the synovial fluid and stimulating them to differentiate into chondroclasts to remove damaged cartilage tissue.

High levels of OPG have been reported to be protective against osteoarthritis progression in mouse models (Shimizu et al. 2007), which is consistent with its protective role in bone and in cartilage (Ota et al. 2009). However, it appears as though IL-1β, TNFα, and PGE2 also upregulate OPG, while OPG has been shown to upregulate PAR2 and MMP13 synthesis and production (Kwan Tat et al. 2009; Moreno-Rubio et al. 2010). This data appears to contradict the studies that have shown OPG reduces rates of osteoarthritis progression, rather suggesting a catabolic role for OPG. However, it may well be that OPG plays a significant local catabolic role as part of tissue homeostasis, perhaps signaling in the pericellular and territorial matrix, perhaps involving MMP13 in the turnover of the collagen network. For example, OPG may signal locally during cell migration, so in this broader sense OPG is in fact involved in tissue renewal (as well as tissue preservation). Upregulation of MMP13 is associated with OA like pathologies, but this is hardly surprising (Neuhold et al. 2001).

It is noted that PGE_2 upregulates RANKL expression more than OPG expression, and COX2 inhibitors reduces the concentration of PGE_2. In other words, NSAIDs help block the resorption of cartilage and nearby bone by keeping the OPG/RANKL ratio elevated, as indicated in the study of Moreno-Rubio et al. (2010).

5.6.6 IL-1, TNFα, Nitric Oxide and Prostaglandins

When cartilage is injured, chondrocytes produce the inflammatory cytokines TNFα and IL-1 (α and β). IL-1 and TNFα induces powerful catabolic effects on cartilage by upregulating proteases, which degrade cartilage tissue. IL-1 and TNFα probably act via the NF-κβ pathway (Montaseri et al. 2011). IL-1β and TNFα also induce the production of nitric oxide (NO) and prostaglandin production (e.g. PGE2).

Nitric oxide (NO) is a very small molecule with a very short duration, while PGE_2 is a small molecule with intermediate duration. Both molecules help regulate the pericellular environment. It is known that local oxygen concentration can influence nitric oxide synthesis (Fermor et al. 2007). Nitric oxide probably increases the propensity of a chondrocyte to undergo apoptosis (Hashimoto et al. 1998). NO mediates some of the effects of proinflammatory cytokines, such as TNFα and IL-1β (Abramson 2008; Hsu et al. 2017). Administration of L-NAME, which blocks NO synthesis, reduces TNFα and IL-1 production in a rat meniscectomy model of OA (Hsu et al. 2017). iNOS knockout mice are found to be resistant to OA (Scher et al. 2007). Pre-incubation of chondrocytes with TGFβ resulted in an increased ability of IL-1 to produce nitric oxide, which suggests interactions between anabolic growth factors and catabolic cytokines.

PGE_2 may be catabolic or anabolic, depending on the local environment (Hashimoto et al. 1998). In osteoarthritic cartilage explant experiments, it has been shown that PGE_2 inhibits MMP1 secretion but promotes MMP13 secretion (Attur et al. 2008), which may explain its differential effects on articular cartilage in different states. PGE_2 promotes CREB dependent induction of PGR4, suggesting that it plays a role in normal cartilage homeostasis (Ogawa et al. 2014). NO appears to suppress PGE2 production (Fermor et al. 2007).

Of course, these molecular signals regulating cell behavior are always modulated by the local mechanical environment. For example, IL-1β production of nitric oxide may be downregulated by cyclic loading (Chowdhury et al. 2008). And while chondrocytes in the superficial zone produce PRG4 in response to shear stress, their production rate is influenced by the local level of TGFβ (Neu et al. 2007). It seems possible that IL-1 is part of the normal homeostatic control mechanisms operating in articular cartilage [see for example, the effect of IL-1 on growth plate dynamics (Simsa-Maziel and Monsonego-Ornan 2012)].

We mention here that for porcine synovial fluid, median concentrations of IL-1α were 0.043 ng/mL for mild OA, and 0.288 ng/mL for moderate OA, whereas IL-1β concentrations were 0.109 ng/mL for mild OA and 0.122 ng/mL for moderate OA (McNulty et al. 2013).

5.6.7 HIF Signaling

Hypoxia-inducible factor (HIFs) expression help chondrocytes to adapt and survive in hypoxic conditions (Fernandez-Torres et al. 2017). It has been found that HIF-1α supports cartilage homeostasis (promoting ECM synthesis and chondrocyte differentiation), while HIF-2α is catabolic (and functionally supports chondrocyte hypertrophy and bone formation) (Fernandez-Torres et al. 2017; Husa et al. 2010).

HIF-1α activation upregulates the expression of TGFβ, NOS2, VEGF, EPO, GLUT1, SOX9, IGF2, and COL2A1, supporting the chondrocyte phenotype (Fernandez-Torres et al. 2017; Grimshaw and Mason 2001; Robins et al. 2005), while HIF-2α activation upregulates the expression of COL10A1 and RUNX2 (Fernandez-Torres et al. 2017). HIF-1α and 'stress' (any chemical adversity or radiation experienced by the cell) induce the expression of local EPO, an anti-inflammatory signaling molecule, while inflammatory cytokines suppress EPO expression (Brines and Cerami 2008). We see that EPO is part of a mechanism that pushes cells into 'phenotypic attractors.' Importantly we note that EPO (erythropoietin) production is induced in chondrocytes by blunt impact trauma to the cartilage surface, causing a penumbra of EPO producing cells that presumably help to limit the extent of damage induced by IL-6, IL-1 and TNFα following the impact (Wang et al. 2015b).

5.6.8 IL-6 Signaling

In cartilage IL-6 is generally considered catabolic, suppressing aggrecan production and inducing the upregulation of MMP13, which can lead to OA phenotypes. IL-6 is usually activated when TNFα is activated (Stannus et al. 2010). It has been reported that IL-6 is the catabolic effector of HIF-2α activation (see discussion Sect. 5.6.7) and Notch activation (Ryu et al. 2011; Zanotti and Canalis 2013). However, for apparently healthy suspended human chondrocytes, adding IL-6 to the media is reported to increase GAG production (Tsuchida et al. 2012), which demonstrates that any experiment on cells not in their normal environment need to be interpreted with caution and additional context.

5.6.9 BMP Signaling

There are bone morphogenetic proteins (BMPs) 1–7, as well as BMP8a, 8b, 10 and 15, which signal through receptors BMPR1a and b and BMMPR2, to activate SMADs (typically SMADs 1/5/8) (see http://www.genome.jp/kegg-bin/show_pathway?hsa04350). For the growth plate, BMP2–7 are primarily expressed in the perichondrium, BMP7 is expressed by proliferating chondrocytes, while BMP6 is found exclusively in hypertrophic chondrocytes, together with BMP2 (Hidaka and Goldring 2008).

Overexpression of noggin in transgenic mice (which blocks BMP signaling) results in an absence of most cartilage components, which suggests BMP signaling is essential for cartilage formation (Tsumaki et al. 2002).

While a BMPR1 knockout is lethal, it is found that cre-floxed GDF5 controlled knock-down of BMPR1 results in an osteoarthritic phenotype—the surface of articular cartilage wearing away at an accelerated rate [partly due to a reduction in Col2 expression, but also suggesting inadequate PRG4 expression (Kim et al. 2015; Ogawa et al. 2014)]—which demonstrates that in mice, BMP signaling is essential for maintaining normal articular cartilage homeostasis (Rountree et al. 2004).

Interestingly, we mention that Gremlin2 (a BMP antagonist) shows a circadian expression in tendon. Gremlin2 is essential for tendon homeostasis (inhibiting ossification of tendon tissue by repressing BMP2 and BMP7 signaling), and it seems possible that it may play a similar role in articular cartilage.

5.6.10 FGF Signaling

There are at least 18 fibroblast growth factors with four tyrosine kinase receptors (FGFR1–4) (Ornitz and Itoh 2015), acting through AKT, PKC and Stat signaling pathways [see Fig. 3 (Ornitz and Itoh 2015)]. The effects of the FGFs during skeletal development are wide ranging. During development, FGFs play important roles in anlagen formation, joint formation, growth plate regulation (Minina et al. 2002), and they play a role in adult cartilage homeostasis. For examples, FGF-2 (ECM bound FGF-2 is released upon tissue damage, having a catabolic role), FGF-18 (a well-known anabolic agent) and FGF-8 (Ellman et al. 2013). FGFs also play a role in cartilage repair (Moore et al. 2005). Human mutations in FGFR3 result in dwarfism and achondroplasia (Minina et al. 2002).

Based on numerous, at times conflicting experimental data, it has been proposed that in adult cartilage FGF2 (Gavrilovic 2009; Vincent et al. 2002) and FGF 8 exert primarily catabolic effects, while FGF18 exerts primarily an anabolic effect (promoting cell differentiation and ECM formation, but inhibiting cell proliferation) (Ellman et al. 2013). Interestingly, we mention that in a mouse model, FGF2 appears to upregulate MMP1 and MMP3 expression, but downregulate ADAMTS5 expression (Gavrilovic 2009), offering preferential catabolism of the collagen network over aggrecan. Further, in a rat meniscal tear model, injections of FGF18 into the synovial joint are reported to result in increases of the cartilage thickness on the tibial plateau due to new cartilage formation at the articular surface and the joint periphery (Moore et al. 2005).

It is known that FGF23, produced by osteocytes, regulates serum phosphate and 1,25-dihydroxyvitamin D levels in the blood via the kidney—FGF23 results in increasing binding to the FGF receptor in kidney epithelial cells, and a reduction in phosphate reabsorption by the proximal tubules. Though FGF23 is mainly produced by bone osteocytes, it appears as though FGF23 also plays a potentially important role in cartilage. Orfanidou et al. (2009) report a proposed signaling system whereby normally SOX9 (the main TF inducing the chondrocyte phenotype) suppresses RUNX2 (the main TF inducing the osteoblastic differentiation and bone formation), however in osteoarthritis, increased FGF23 signaling overrides this suppression resulting in increased levels of RNUX2 expression, which in turn upregulates MMP13 expression [see Fig. 6 (Orfanidou et al. 2009)]. Again we see that FGF23 (like EPO) may be part of a mechanism that pushes cells into 'phenotypic attractors.'

5.6.11 Integrin Signaling

Integrins are heterodimers expressed on the surface of most cells, including chondrocytes. The α and β subunits are 'single-pass' transmembrane proteins, which non-covalently associate at the cell surface. In evolutionary terms, β 'tails' are more conserved than the diverse α tails. The dimers bind various molecules in the extracellular matrix. Integrins can transmit signals 'outside-in' [by binding to extracellular molecules, they induce clusters to form inside the cell (influencing the activity of FAK, ILK and Src), and attaches to actin]. Integrins can also transmit signals 'inside-out' (by events inside the cell changing the molecules they bind) (Harburger and Calderwood 2009).

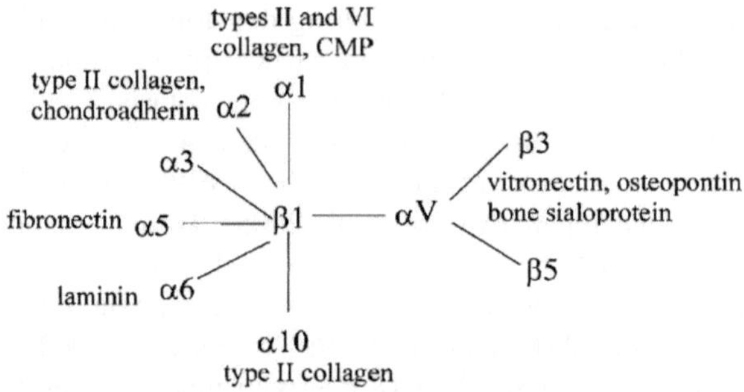

Fig. 5.11 Chondrocyte integrins and their ligands. The members of the αXβ1 and αVβY (where X = 1, 2, 3, 4, 5, 6 and 10, and Y = 3 and 5) integrin subfamilies reported to be expressed by chondrocytes are shown along with the extracellular matrix proteins which have been demonstrated to bind to the specific integrin subunits (CMP, cartilage matrix protein) (Knudson and Loeser 2002). See also Loeser (2014)

A range of integrins are expressed on the surface of normal adult tissue chondrocytes (including α1β1, α3β1, α5β1, α10β1, αVβ1, αVβ3 and αVβ5) and a number of additional integrins are expressed on the surface of chondrocytes from osteoarthritic tissue (including α2β1, α4β1 and α6β1) (Loeser 2014). The integrins bind to ECM molecules (see Fig. 5.11), regulating tissue homeostasis and chondrocyte survival. We note that an integrin β1 sub-unit knockout results in a severe chondrocyte dysplasia, and chondrocytes fail to organize into columns. This is apparently caused by a lack of chondrocyte motility [chondrocytes in the growth plate normally divide normal (i.e. at right angles) to the 'chondrocyte stack,' and then 'slide' over one another to join the chondrocyte stack alignment, but the sliding capacity is lost in the β1 sub-unit knockout], reportedly because the chondrocytes cannot bind ECM including collagen type II (Aszodi et al. 2003). Chondrocyte cytokinesis is also impaired.

5.6.12 *Purinergic Signaling*

Extracellular signaling is mediated by purine nucleotides and nucleosides such as adenosine and ATP. Says Baroja-Mazo et al. (2013):

> ATP-mediated signaling has been identified in virtually all tissues and cell types and appears to be the most widespread and omnipresent of all known extracellular signaling molecules.

Activation of P2Y receptors by ATP, ADP (and pyrimidines such as UTP, UDP) and activation of P1 receptors by adenosine result in activation of G protein-coupled receptors, while gating of the P2X receptor by ATP leads to opening of ligand-gated ion channels (e.g. calcium channels).

Though much remains to be discovered, it is known that purinergic signaling can play an important role in both bone and cartilage homeostasis, and in several pathologies (Strazzulla and Cronstein 2016). Mechanical loading of articular chondrocytes triggers ATP release through connexin hemichannels into the pericellular microenvironment, which activates purine receptors triggering autocrine and paracrine signaling involving intracellular calcium (Baroja-Mazo et al. 2013). Baroja-Mazo et al. also indicate that so-called primary cilia mechanosensitivity probably involves purinergic driven Ca^{2+} signaling, releasing ATP through connexin 43 hemichannels, with subsequent signaling through the P2 family (G coupled and Ca signaling) receptors. It is noted that the connexin 43 expression is confined to the upper 200 microns closest to the cartilage surface, i.e. in the superficial (upper transitional) zones (which oxygen concentrations are highest).

We mention that Clodronate (a first-generation bisphosphonate, down-regulating inflammatory markers such a IL-1 and TNFα) exerts a significant anabolic effect on chondrocytes, increasing calcium signaling transduced through purinergic signaling pathways (Rosa et al. 2014).

5.6.13 Calcium Signaling

Intracellular calcium concentration exerts a significant effect on signal-transduction pathways, because of its allosteric effects on many enzymes within the cell (i.e. binding of Ca^{2+} causes changes in the shape of that molecule, which changes the ligand binding constants for other binding sites on the same molecule), including protein kinases in signal-transduction pathways (see signaling network influenced by Ca^{2+} at http://www.genome.jp/kegg-bin/show_pathway?hsa04020). Calcium ions can also activate the cAMP (second messenger) pathway (see http://www.genome.jp/kegg-bin/show_pathway?ko04024+K04498).

Normally resting intracellular calcium concentrations are very low—in the range 10–100 nM. Extracellular concentrations are much higher (in the mM range). Various triggers can lead to calcium channels opening, resulting in calcium 'flooding' into the cell cytoplasm (e.g. leading to cytoplasmic Ca^{2+} concentrations of up to 500–1000 nM). Calcium stores (calcium storing vesicles within cells) may also open and flood the cytoplasm with additional calcium to ramp calcium concentrations further. Calcium signaling is particularly important in muscle cells (regulating actin-myosin interactions) and in neurons (regulating synaptic activity), but it is likely calcium signaling also plays an important role in chondrocytes. Chondrocytes contain calcium stored in intracellular vesicles.

Calcium ion channels are abundantly expressed by chondrocytes. Chondrocytes have various calcium channels in their cell and vesicle membranes, including transient receptor potential vanilloid 4 (TRV4) Ca^{2+} ion channels (Phan et al. 2009) and mechanosensitive ion channels that can be blocked by gadolinium (Erickson and Guilak 2000). These ion channels regulate the flux of calcium across membranes. It appears that calcium transient upon exposure to hyperosmotic stress is regulated by the gadolinium sensitive ion channels, while the calcium transient to hypo-osmotic stress is regulated by the TRV4 ion channels.

It has been demonstrated many times that calcium concentrations in chondrocytes fluctuate depending on mechanical deformation (Madden et al. 2014) and osmotic stress (Pritchard et al. 2008). So it is almost certain that at least some of the mechanical and osmotic stress responses by chondrocytes are mediated via calcium, but clearly there are other pathways for mechanical signals to be transducted (e.g. via integrins) and for osmotic signals to be transducted (e.g. by changing transcription factor and promoter binding site concentrations in the cell nucleus, as the nucleus changes volume in response to osmotic stress).

We note that strontium ranelate probably increases the concentration of calcium and strontium inside chondrocytes, by increasing the concentration of strontium in the extracellular fluid (Henrotin et al. 2001). This appears to have an anabolic effect on collagen and aggrecan production (increasing both quantity and quality). Part of strontium ranelate's mechanism of anabolic action may also involve increasing IGF levels in the tissue (Henrotin et al. 2001).

Strontium ranelate decreases osteoclast formation, increases osteoblast formation, and decreases bone turnover. Consequently, it has been used for many years in the treatment of osteoporosis. A clinical trial (Reginster et al. 2013) and an animal model (Yu et al. 2013) using strontium ranelate for the treatment of osteoarthritis appear promising (viz., strontium ranelate is found to both decrease subchondral bone turnover and stimulate cartilage ECM production), though osteophyte formation may potentially be an unwanted side-effect when treating osteoarthritis (Chu et al. 2017).

It appears likely these outcomes will stimulate more clinical trials with other anti-osteoporotic treatments for the treatment of osteoarthritis (e.g. some bisphosphonates). This should help separate the relative contributions of treating bone or cartilage in slowing the progression of osteoarthritis. For example, it has been shown that Clodronate exerts a significant anabolic effect on chondrocytes, increasing calcium signaling transduced through purinergic signaling pathways (Rosa et al. 2014).

5.6.14 Toll Signaling

Toll-like receptors (TLRs) have important roles in activating the innate (and specific) immune system during microbial infections. They do so by recognizing a limited set of specific molecular patterns (so-called PAMPs or MAMPs, such as lipopolysaccharide, lipoteichoic acid, flagellin, double-stranded RNA and unmethylated CpG sequences, that commonly arise from infections (see http://en.wikipedia.org/wiki/Pathogen-associated_molecular_pattern)). We mention that lumican may play an important role in TLR activation (Amjadi et al. 2013). So far about 10 TLRs have been identified in humans, and all are present on the surface of chondrocytes. Several low-molecular-weight hyaluronan, heparan sulfate, biglycan, tenascin C and fibronectin fragments [and aggrecan core protein fragments (Fosang 2015)]—have been identified as potential TLR ligands (Hodgkinson and Ye 2011).

On this basis, Konttinen et al. proposed that osteoarthritis was an 'autoinflammatory disease' (Konttinen et al. 2012), rather than '*wear and tear*' or '*the rate of damage simply exceeding the rate of repair.*' Konttinen et al. propose that it is both wear and tear interacting with (auto)inflammation, suggesting it is '*biomechanical wear and tear*' interacting with '*activation of inflammatory cytokines*' [which then goes on to interact with the synovial membrane (which contains macrophage like Type A lining cells as well as fibroblast-like Type B cells), causing a osteoarthritic synovitis] (Konttinen et al. 2012). Under this scenario, autoinflammatory molecules (e.g. fibronectin) activate TLRs on chondrocytes [which activate the p38 pathway (Bobacz et al. 2007) and NF-κβ pathway, leading to the release of TNFα, IL-1β, and IL-6, etc.]. These inflammatory cytokines kick-off the degradative process by provoking a synovitis in the synovium, causing cartilage damage that is amplified by enzymatic and mechanical damage of the cartilage.

The only difficulty with this thesis is that TLR's are primarily activated by molecular pattern associated with microbes. That is not to say that TLR's cannot be activated by ECM or other potentially microbial-like fragments (Fosang 2015). Naturally, there is going to be some overlaps between the multitude of molecule patterns in ECM fragments and microbial fragments, and indeed, some of these ECM fragment patterns may well help guide cells such as neutrophils to the source in the infected tissue.

It is undoubtedly true that avoiding or minimizing infections of all kinds are prime concerns of all tissues. Cartilage is no exception. The innate immune system is complicated, and of course things go can wrong, and this may explain some osteoarthritis. However this does not necessarily mean that this innate defense is the primary driver for the majority of osteoarthritis. If it is, this remains to be confirmed.

In the context of the innate immune system being complicated, it is interesting to consider MMP3 further. We are familiar with it as a protease. We first note that in normal cartilage the amount of MMP3 mRNA produced is about two orders of magnitude greater than any other MMP (see Fig. 2.62). MMP3 has many substrates (e.g. proTGFβ, collagen III, and IX, and activating other MMPs such as MMP1 and MMP9), some of which are present in abundance in the pericellular environment (e.g. fibronectin). This may explain the abundance of MMP3, but could it also be doing other things? Is it possible that a pool of MMP3 mRNA is maintained to ensure a rapid cellular response to an environmental stressor? What might such a stressor be?

MMP3 is reported to play an important role in the cellular response to the Dengue virus in the Raw264.7 murine macrophage cell line, a murine embryonic fibroblast cell line, epithelial cell lines and cell other lines (Zuo et al. 2014). It is reported that MMP3 co-localizes with NFκβ, and is transported into the cell nucleus. Zuo et al. say (2014):

> Confocal microscopy showed that MMP3 co-localized with NFκB RelA (p65) in DENV infected cell nucleus...Our data suggested that MMP3 is translocated into nucleus upon DENV infection, interacts with and activates NFκB, thereby promotes the production of anti-viral cytokines... However, very little is known about the distinct function of MMP3 in nucleus.

This suggests that MMP3 is involved in intracellular signaling, potentially acting with a transcription factor-like function, and plays a role in immunomodulation. A similar role is reported for MMP3 acting on the gene for 'connective tissue growth factor' (CTGF) (Eguchi et al. 2008). It has been found that CTGF is latent TGFβ binding protein, helping to determine cartilage thickness (Tang et al. 2017). Though chondrocyte cell lines were not tested in the study by Zuo et al. (2014), such (expected) additional functional roles for MMP3 may help to explain the abundance of mRNA in chondrocytes, by providing a rapid responsiveness by MMP3 to environmental stressors. All this serves to remind us that the innate immune system is complicated (things can go wrong and explain some osteoarthritis), most biological molecules are multifunctional (which explains some of the most surprising interactions when investigating biological systems), and that much remains to be discovered (e.g. MMP3 may help regulate CTGF in chondrocytes).

5.6.15 RAGE Signaling

Another pattern recognition receptor in tissue is the RAGE receptor (an acronym for Receptor for Advanced Glycation Endproducts). Apparently, the RAGE receptor is only found in mammals, and apparently has evolved from the cell adhesion molecules (CAMs) (Sessa et al. 2014). Many proteins are glycated over time by saccharides via non-enzymatic 'Maillard reactions,' forming Amadori products that then react to form 'advanced glycation endproducts' (AGE) on proteins, lipids and nucleic acids. These glycation products constitute a set of molecular patterns that are recognized by the receptor RAGE. RAGE also appears to be also activated by HMGB1—a DNA binding protein that is released by necrotic cells as well as some immune cells (e.g. macrophages), together with some other proteins such as S100 proteins [which are Damage-Associated Molecular Pattern molecules (DAMPs) released during tissue damage] (Sessa et al. 2014). RAGE is believed to up-regulate inflammatory pathways (see, for example, http://www.wikipathways.org/index.php/Pathway:WP2324), which is thought to be responsible for the inflammatory diseases associated with diabetes, Alzheimer's and some cancers.

It has been shown that chondrocytes express the RAGE receptor (Loeser et al. 2005). It has also been shown that RAGE activation inhibits the differentiation step between prehypertrophic and hypertrophic chondrocytes (Kosaka et al. 2014).

It is reasonable to suppose that reaction products from unregulated reactions between proteins and saccharides result in altered proteins that do not possess normal functional activity, and so it is desirable that these reaction products be removed from the ECM (lest they cause further abnormal structural or enzymatic behaviors in the ECM). And it reasonable to suppose that if these unwanted reaction products are detected in high concentrations, then inflammatory pathways would be activated to speed their removal, as occurs in disease states such as diabetes. It appears that type II diabetes is a significant risk factor for joint replacement (Schett et al. 2013), which suggests that 'metabolic syndrome' (including RAGE activation) may play a significant role in some types of (inflammatory driven) osteoarthritis. It is noted that type II diabetes remained a significant risk factor even after adjusted for BMI, age and other risk factors.

5.6.16 Heparin Sulfate Proteoglycans

Heparin sulfate is now recognized to function as a potent cofactor for canonical FGF signaling, together with a wide range of other signaling pathways including bone morphogenetic proteins (BMPs), Wnts, and Hedgehog signaling (Ornitz and Itoh 2015). The heparin sulfate proteoglycans (HSPs) are either membrane bound [e.g. syndecan (Pap and Bertrand 2013) and glypican] or secreted (e.g. perlecan, agrin and type XVIII collagen). Membrane-bound HSPs play important roles in binding or acting as co-receptors for regulatory molecules, while the secreted HSPs play an important role in organizing the basement membrane of cells (e.g. the pericellular matrix around chondrocytes). It is noted that in cartilage syndecan IV plays a role in activating ADAMTS5 (Echtermeyer et al. 2009; Husa et al. 2010).

See Fig. 2 (Bishop et al. 2007): *HSPGs function as co-receptors for growth factors and their receptor tyrosine kinases, which are present either on the same cell (**a**) or on adjacent cells (**b**). They transport chemokines across cells (**c**) and present them at the cell surface (**d**). Proteolytic processing leads to the shedding of syndecans and glypicans from the cell surface (**e**), and heparanase cleaves the HS chains (**f**), liberating bound ligands (such as growth factors). Cell-surface HSPGs are actively taken up by endocytosis (**g**) and can recycle back to the surface or be degraded in lysosomes (**h**). HSPGs also facilitate cell adhesion to the extracellular matrix (**i**) and form bridges to the cytoskeleton (**j**). Secreted HSPGs are involved in the formation of organized extracellular matrices that form physiological barriers (**k**) and sequester growth factors and morphogens for later release (**l**). Serglycin carrying highly sulphated heparin chains is packaged into secretory granules of haematopoetic cells (**m**). Finally, some experiments suggest that HS chains exist in the nucleus (**n**), although their function in this location is unknown.*

See also Sarrazin et al. (2011) for further information on heparin sulfate proteoglycans.

5.6.17 *Periostin Signaling*

Periostin is a secreted ECM protein involved in cell adhesion and cell recruitment. Periostin downregulates sclerostin (sclerostin is a negative regulator of Wnt) and so periostin is anabolic in bone tissue (Bonnet et al. 2009). Periostin is normally expressed in cartilage and is upregulated in osteoarthritis (Lourido et al. 2014). Periostin has been shown to have a complex set of interactions in cartilage, changing the expression of TGFβ, IL-1β, MMP13, collagens and other molecules in a context-specific way. Periostin has been shown to work in conjunction with canonical Wnt signaling in cartilage to promote the expression of MMP13 (Attur et al. 2015). Periostin has been shown to be involved in cross-talk between damaged ligament and articular cartilage (Chinzei et al. 2018).

Periostin has a number of interesting properties in other tissues that may in the future be shown to be relevant to articular cartilage. For example, periostin expression may be involved in expression of smooth muscle α-actin (Elliott et al. 2012). In addition, periostin has also been shown to contribute to the acquisition of stem cell-like qualities and a mesenchymal phenotype in epithelial tissues (Wang et al. 2013), and to enhance the regeneration of pancreatic tissue (Smid et al. 2015). Periostin may prove very relevant to processes regulating chondrocyte proliferation and in tissue repair processes in vivo.

5.7 Intracellular Signaling Pathways

There are 40 or more intracellular signaling pathways http://en.wikipedia.org/wiki/Signal_transduction. These pathways are annotated in detail at such sites as the Kyoto Encylopedia of Genes and Genomes (http://www.genome.jp/kegg/), Netpath (http://www.netpath.org/) and Reactome (http://www.reactome.org/PathwayBrowser/#DIAGRAM=2028269&PATH=162582Paging%20and%20Bottom%20Toolbar%20%20%20Previous%20Item%20Next%20ItemConnected%20to%20Microsoft%20Exchange). These databases are very valuable for exploring overlaps and connections between signal-transduction pathways, though it needs to be kept in mind that not all connections shown in these databases are relevant to a particular cell type (e.g. relevant to cartilage tissue). Rather the databases show generic pathways based on data from many different of cell types using the full range of experimental methods.

One of the most researched signaling pathway is the MAP kinase pathway (see http://www.genome.jp/kegg-bin/show_pathway?hsa04010). There are three main arms of the MAP kinase pathway. The classical MAPK pathway which stimulates: (i) the ERK (extracellular signal-regulated kinases) pathway, which is generally anabolic. This arm is usually activated by the GCPRs (G-couple protein receptors) and RTKs (receptor tyrosine kinases); (ii) the JNK (Jun amino-terminal kinases) pathway. This arm is mainly activated by inflammatory cytokines, heat, oxidative stress and DNA damage; and (iii) the p38 (stress-activated protein kinases) pathway (this arm is also activated by inflammatory cytokines and environmental stressors), which stimulates IKK, NF-κB, and NFAT, which are mainly catabolic (Morrison 2012).

Another key anabolic pathway is the PI3K-AKT signaling pathway (http://www.genome.jp/kegg-bin/show_pathway?hsa04151). This pathway is activated by various receptor tyrosine kinases (RTKs), including the crucially important cartilage growth factors IGF-I and IGF-II.

Interestingly the ERBb3 signaling pathway also signals through the AKT-ERK pathways http://www.genome.jp/kegg-bin/show_pathway?hsa04012 as does the insulin signalling pathway http://www.genome.jp/kegg-bin/show_pathway?hsa04910. Various studies have shown that if one of these pathways is blocked (e.g. by a drug or a mutation), signaling may occur through a different pathway to similar effect (e.g. similar effects may include pro-survival, growth, proliferation and differentiation). This illustrates redundancy in signaling pathways, which is one reason that many single drug treatments that influence a single molecule in a signaling pathway are usually either ineffective or only effective transiently.

Of course, the cAMP-PKA-CREB pathway is also important in cartilage (see http://www.genome.jp/kegg-bin/show_pathway?ko04024+K04498). CREB facilitates chondrocyte proliferation and helps maintain the chondrocyte phenotype by upregulating SOX9.

The so-called Jak-Stat signaling is highly conserved throughout evolution, and ubiquitous throughout cells (http://www.genome.jp/kegg-bin/show_pathway?hsa04630). Jak denotes the receptor Janus Kinase, while Stat denotes the Signal Transducer and Activator of Transcription. Jak-Stat signaling is activated by the interleukins, interferons and various cytokines (or immuno-modulators). The pathway is negatively regulated by SOCS (suppressors of cytokine signaling).

The NF-κβ signaling pathway is normally activated in stress and inflammation related environments. The role played by this pathway in early osteoarthritis has been reported as 'unknown' (Goldring et al. 2011). However, throughout the body, NF-κβ signaling is generally activated by inflammatory cytokines, and chondrocytes appear to be no different (Montaseri et al. 2011).

Rho signals through guanine-nucleotide-exchange factors (GEFs) and GTPase-activating proteins (GAPs) to control the activation state of the small GTPases Rho, Rac and Cdc42. Once activated, the GTPases bind to a spectrum of effectors to stimulate downstream signaling pathways (Schwartz 2004). ROCK/Rho signaling downregulates SOX9 expression, so Rho signaling needs to be downregulated to promote the chondrocyte phenotype (Woods et al. 2005). Downregulation of ROCK/Rho signaling is accompanied by increased GAG synthesis, increased SOX9 expression and reorganization of the cortical actin cytoskeleton, a hallmark of differentiated chondrocytes.

A central cell proliferation pathway is the so-called canonical Wnt signaling pathway (http://www.genome.jp/kegg/pathway/hsa/hsa04310.html). Interestingly there is interaction between the Wnt and AKT pathways through GSK-3β (which is an integral phosphorylating protein in the destruction complex of the Wnt signaling pathway). Some researchers even refer to this linkage as the 'AKT-Wnt axis'

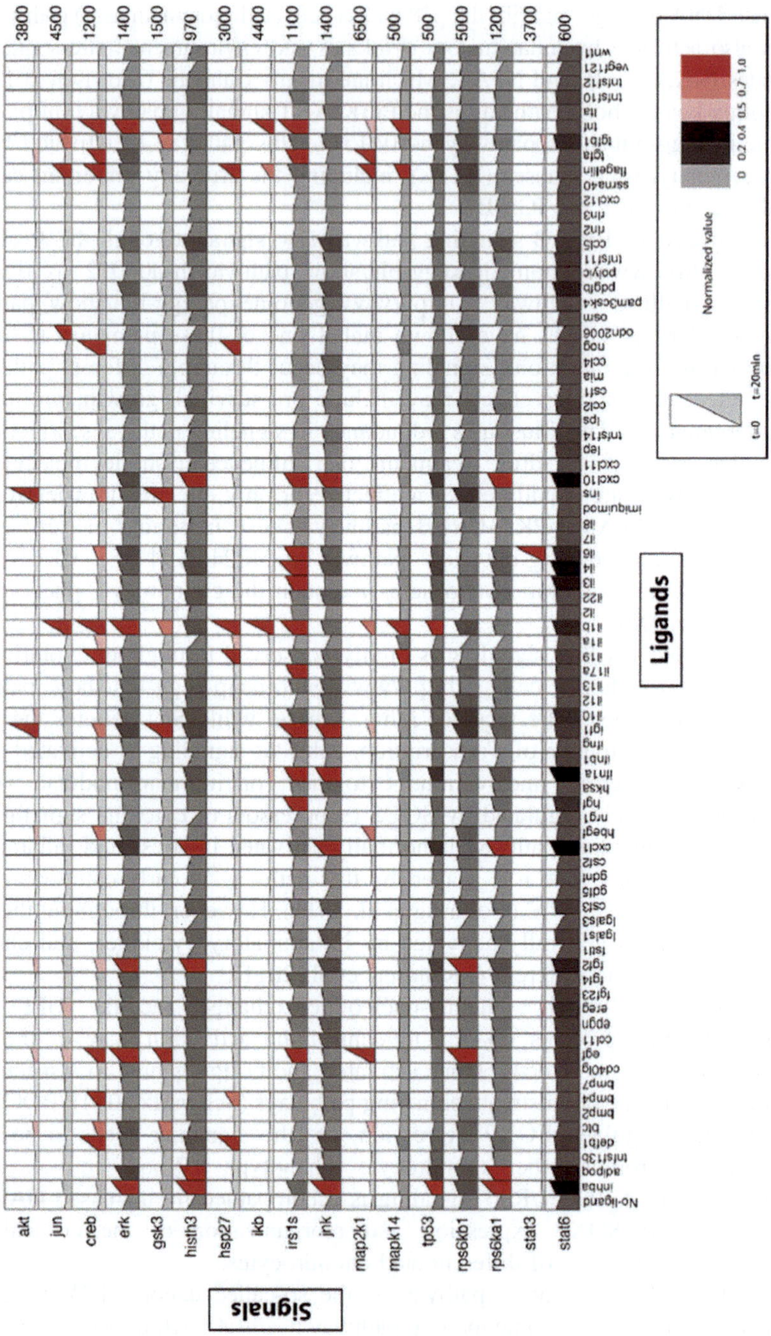
Signals
akt
jun
creb
erk
gsk3
histh3
hsp27
ikb
irs1s
jnk
map2k1
mapk14
tp53
rps6kb1
rps6ka1
stat3
stat6
3800
4500
1200
1400
1500
970
3000
4400
1100
1400
6500
500
500
5000
1200
3700
600
Ligands
Normalized value
t=0
t=20min
0 0.2 0.4 0.5 0.7 1.0

◂**Fig. 5.12** Phosphoproteomic data: The time course of the phosphoprotein signals from the unstimulated state to the average early response is illustrated. The rows correspond to the 17 phosphoproteins measured and the columns to the 79 ligand treatments (including the No-ligand treatment). In each subplot, the first point shows the unstimulated activity of the respective signal (zero time point); the second point shows the raw measurement of the signal (in fluorescent units) 5 + 25 min after stimulation; while the color code corresponds to the normalized value (between 0 and 1) of the signal. The numbers on the right hand side of the figure show the maximum phosphorylation value of each signal in fluorescent units (Melas et al. 2014)

(Fukumoto et al. 2001). Non-canonical Wnt signaling has a variety of effects other than proliferation, e.g. induces cell polarisation, de-differentiation and cell migration (Skurk et al. 2005).

Interactions between signal-transduction pathways are common (Majumder et al. 2016; Manoranjan et al. 2017), adding considerably to the complexity in understanding these signal-transduction pathways (see, for example, http://www.systemscenters.org/wordpress/wp-content/uploads/2011/01/ncsb-2011_cdp_morris.pdf or http://christophe.dessimoz.org/revcompbiol/_media/2013/t2-slides.pdf).

Intracellular pathway interactions in chondrocytes have been investigated. In one impressive paper (Melas et al. 2014), chondrocytes obtained from osteoarthritis patients were first isolated, 78 receptors were activated, 17 intracellular phosphoproteins were measured (AKT, JUN, CREB, ERK, GSK3β, HISTH3, HSP27, IKB, IRS1S, JNK, MAP2K1, MAPK14, TP53, RPS6KB1, RPS6KA1, STAT3, STAT6 using the Luminex assay) following binding of 78 ligands (e.g. IL1β, TNF, EGF, TGFA, INS, IGF1 and IL6) to the 78 receptors, and 55 secreted cytokines were measured in the supernatant (CCL27, CCL11, FGF2, CSF3, CSF2, CXCL1, HGF, ICAM, IFNA2, IFNG, IL10, IL12, IL12P40, IL13, IL15, IL16, IL17, IL18, IL1B, IL1RA, IL2, IL2R, IL3, IL4, IL5, IL6, IL7, IL8, IL9, CXCL10, LIF, CCL2, CCL7, CSF1, MIF, CXCL9, CCL4L1, CCL4L2, NGF, PDGFB, CCL5, KITLG, SCGFB, CXCL 12, TNF, LTA, TNFSF 10, VCAM1, VEGF121, MMP1, MMP10, MMP13, MMP3, MMP8, MMP9) (Melas et al. 2014) (Fig. 5.12). It seems plausible that these larger data sets describing cell activity (advocated by 'systems biologists') may well help us understand homeostatic imbalances that arises under environmental stress (e.g. growth/maturation and ageing, both with their associated hormonal profiles and activity levels) (Kerkhofs et al. 2016).

5.8 Interactions Between Endocrine, Paracrine and Autocrine Factors, Signaling Pathways, Transcription Factors and MicroRNAs

5.8.1 General Observations on Interactions Between Systemic and Local Signaling

What gradually becomes apparent from a general reading of the research literature on cell signaling systems are several general principles, namely:

(i) Several systemic and local cell signaling systems usually act cooperatively to achieve the required functional outcome (note: 'endocrine signaling' and 'systemic signaling' are used interchangeably in this section). Cooperativity between signaling systems enables checks on important cell decisions, and together with variability between individual cells (meaning no two cells in a tissue are in exactly the same state at the same time), builds robustness into the signaling systems for achieving desired functional outcomes.

(ii) But the robustness of signaling systems has their limitations. Importantly too much or too little of any of the major signaling molecules will lead to communication abnormalities between cells. Sometimes communication abnormalities are minor, but at other times they are lethal. In many cases they lead to disease states, which may only become evident as the organism ages.

(iii) While the full complement of information transfer capacity through cell–cell signaling systems is employed to coordinate cell behaviors during development, maintaining adult tissue does not require the same intensity of signaling. Adult tissue probably only employs a subset of the original signaling systems required for development, and of those employed in adulthood, only a fraction of the whole signaling system's capability is employed to maintain to homeostasis. However, in disease states, 'dormant' signaling systems may come into play and parts of signaling systems in play may be reactivated.

We try to identify these 'principles' based on examples. From the discussion in Sect. 5.5 it becomes apparent that TGFβ, Ihh-PTHrP, Wnt and other signaling systems, interact with one another in subtle ways, though many of the interactions are either not known or fully understood (Kerkhofs et al. 2016). Interactions between signal-transduction systems presumably allowed a nuanced functional response by the cell to its local tissue conditions. This is achieved by nuanced signaling interactions generating cell-specific profiles of transcription factors within the cell nucleus, and so nuanced gene expression and cell secretion profiles (Dong et al. 2005).

For example, it is known that growth hormone promotes IGF-1 production by the liver. IGF-1 concentration in the blood provides a signal to all the cells in the body about the nutritional status of the liver, and so the liver's ability to support cell activities (e.g. cell metabolism, growth and proliferation) (Yu and Rohan 2000). IGF-1 from the liver diffuses from the blood to synovial fluid and into articular cartilage (Zhang et al. 2013a). Thyroid hormones (which affect the basal metabolic rate of cells) increase expression of the IGF-1R receptor within articular cartilage (see Fig. 5.13) (Wang et al. 2010).

The activation of IGF-1R by IGF-1 via AKT signaling is known to interact with Wnt signaling through GSK3β (Wu and Pan 2010). Participation of GSK3β enables the 'destruction complex' to remove free beta-catenin from the cytosol, reducing

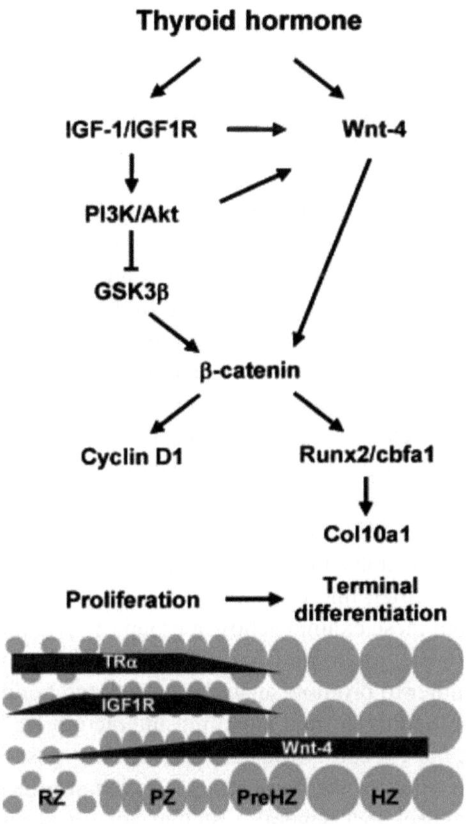

Fig. 5.13 Schematic diagram of the proposed interactions between thyroid hormone, IGF-1/IGF1R, and β-catenin signaling pathways in regulating cell proliferation and terminal differentiation of growth plate chondrocytes. RZ-resting zone; PZ-proliferating zone; preHZ-prehyperprehypertrophic zone; HZ-hypertrophic zone (Wang et al. 2010). See also Fig. 1 in Williams (2013)

the available beta-catenin in the Wnt signaling pathway. Beta-catenin is the key effector protein in the Wnt signaling pathway (beta-catenin binds to TCF or LEF and translocates to the cell nucleus, where a large number of genes are activated) (Wang et al. 2007, 2010). Increased IGF-1 signaling through AKT withdraws GSK3β from the destruction complex, and so upregulates Wnt signaling (Skurk et al. 2005). By this means, we see that IGF-1 is not only pro-survival, but also pro-proliferative. Such an interaction between signaling systems is not unusual in cells, but in fact commonplace. These kinds of interactions are usefully kept in mind when reading the experimental literature.

In another example, we mention that a clinical trial of a TRβ-selective agonist (TRβ is one of several thyroxine receptors) for the treatment of hypercholesterolemia, was terminated due to pre-clinical toxicology studies that identified dose-related articular cartilage damage in dogs treated for 12 months (Williams

2013). Increased DIO2 enzyme expression increases active thyroxine concentration in tissues, and high active thyroxine concentrations has been identified as an independent genetic risk factor for OA (Meulenbelt et al. 2008). It is known that thyroxine promotes chondrocyte mitosis (Henricson et al. 1982), but clearly too much thyroxine signaling appears to be deleterious to cartilage. How?

Increased thyroxine signaling is known to increases calreticulin (CALR) expression in articular chondrocytes (Bomer et al. 2016). Calreticulin is a multi-function, multicompartmental protein found in the endoplasmic reticulum, where it can bind calcium ions and act as a quality control protein chaperone and modulator of steroid receptor expression, but is also found on the cell surface and in the ECM (Gold et al. 2010). While it has been proposed that increasing CALR concentration may increase the likelihood of chondrocyte apoptosis, it has been demonstrated that increasing CALR expression leads to cartilage destruction (Bomer et al. 2016). We mention in passing that chondrocytes (and most other cells) express CD46, CD55 and CD59 [CD59 is otherwise known as 'protectin,' which protects the cells from complement-mediated damage (Davies et al. 1994)]. CALR competitively binds to CD59 (Kimberley et al. 2007), possibly increasing the probability of complement activation. Again this illustrates interactions between different signaling systems are commonplace, and are often unpredictable.

Here we see in both these examples how a group of systemic and local hormones (growth hormone, IGF-1, thyroxine and Wnt signaling) normally cooperatively interact with one another to promote anabolic growth of chondrocytes and cartilage tissue, but disruption of this normal signaling interaction leads to cartilage abnormalities. This is not the exception, but the rule in most tissues.

How do we make sense of the complexity? As a first principle, we propose that one can generally expect that several systemic and local signaling systems will 'test' the environment to see if they can cooperatively act so as to achieve a functional outcome for the tissue. Activated cellular signaling systems invariably involve both system level signaling ligands and local signaling ligands. For example, IGF-I carries information about the nutritional status of the liver, providing information about whether it can offer the required support for the local tissue to achieve a functional outcome. Local level signaling ligands such as Wnt are secreted locally, testing the local environment about the need for cell proliferation. If free canonical Wnt signals gather in sufficient concentration in the ECM (i.e. they are not bound by SFRPs or Dkks or other antagonists), this indicates to local cells that they should consider proliferating. When both system level and local level signals align, the likelihood of cell proliferation is then much stronger. Though there are many additional checks that may either permit or abort cell proliferation depending on their status, it is likely that (canonical) Wnt signaling is a fundamental signaling system, indicating to cells they need to consider proliferating. In normal adult articular cartilage this Wnt signal needs reinforcement through IGF-1 signaling, PTHrP signaling, Ihh signaling, and thyroxine signaling (not to mention the sex hormones), before cell division occurs.

Growth hormone, thyroxine, IGFs TGFβs, Wnts, PTHrPs and BMP signaling all activate signaling pathways that evolved hundreds of millions of years, and so they have been performing essential cellular and tissue coordination functions for very, very long periods of time in many different species. Because of this consistency, is it is noted that if the intended functional task can be divined by the researcher (Roux 2014), it is usually possible to predict which signaling systems need to be activated, as the same signaling systems are linked to more or less predictable functions or outcomes across species and time. This provides a valuable opening for future model development.

The cooperativity between multiple signaling systems not only enables complex functions to be coordinated, but it also builds in system ‘checks.’ For example, growth hormone signals directly via the growth hormone receptor, and it also signals indirectly via IGF-1 and its receptor IGF-1R, probably because doing so checks on the resource availability at the liver—for if growth hormone merely signalled alone, a cell may be induced by the growth hormone signal to divide without the necessary resources being available to do so (this would be a bit like starting a project without having the resources to complete the project).

Cooperativity builds in redundancy (e.g. IGF, ERBb3 and insulin all activate the ERK and AKT pathways). So if one signaling system becomes inoperable, another system often ‘picks up’ at least some of the signaling role (albeit the information provided by the new signaling system may provide less specific information than the signaling ligand being replaced).

So-called cellular checkpoints help ascertain whether or not it is in fact possible to achieve the desired functional outcome (e.g. cell-cycle checkpoints ascertain if the various stages of mitosis are in fact feasible). These internal checks, along with signaling cooperativity, bestow robustness on the cell’s activities. By ‘robustness’ we mean achieving consistent functional outcomes in the face of variability in the cell and its environment. Or in other words, signaling systems do not drive a cell into impossible situations that may ultimately prove fatal.

Interestingly an emerging theme in biology is uncovering unexpected variability in the states of cells in an apparently homogenous cell population. At any one time, it turns out that some cells are very ‘active,’ while others are comparatively ‘quiet,’ even in apparently similar environmental conditions (Trapnell 2015). Explains Trapnell (2015):

> Single-cell genomics experiments will revolutionize our understanding of gene regulation during development and disease because they access a level of information that has simply never been available. Fundamental questions regarding how cell identity is defined and maintained will be, for the first time, answerable. Robust unsupervised clustering analysis will help catalogue cell types and states. Trajectory analysis will track cells as they travel across the landscape of possible states. In time, we may even be able to construct complete regulatory networks and genome-scale dynamical systems of the cell from these measurements. However, much work remains to be done, both in the experimental and computational single-cell domains.

It is likely that this variability in cell responses, subject to the same environmental stimuli, can lead to robustness in the cell population’s response to variable environmental stimuli, because some cell states are better suited to respond appropriately to the particular environmental circumstances. For example, if cells are exposed to an environmental toxin, a cell that is quick to proliferate will perish, while a cell that is slow to proliferate may survive as the toxin dissipates.

A good example of robustness in cell behavior is provided by cell-cycle progression checkpoints during cell growth and proliferation (Barnum and O'Connell 2014). But the concept of checkpoints is wider than just cell growth and cycling, and as we have tried to explain, can be applied to 'sensing' general environmental conditions such as growth factor availability, nutrient availability, O_2 supply and stress and indeed to all critical cell operations required to achieve functional outcomes (e.g. cell maintenance, cell migration, cell secretion and cell apoptosis) (Beier et al. 2001; Levine et al. 2006).

Despite all the efforts by evolution to ensure robustness, robustness has it limits. A second general principle is that too much or too little of any of the major signaling molecules involved in a functional outcome (see, for example, Hedgehog signaling http://en.wikipedia.org/wiki/Hedgehog_signaling_pathway) (or regulators of these molecules, e.g. microRNAs or gene splicing regulators) will lead to abnormalities and/or disease. Indeed, some 15% of protein 'knockouts' are lethal, indicating they are critically important. We note in passing that the lethality of KO can be overcome by 'time-shifting' when the KO occurs using so-named 'conditional (or inducible) knockouts' (van Amerongen and Berns 2006), or using siRNA https://www.labome.com/method/siRNAs-and-shRNAs-Tools-for-Protein-Knockdown-by-Gene-Silencing.html or CRISPR-Cas9 http://med.stanford.edu/tktc/service/transgenesis-animal-services/CMKMS.html or using one of a number of other mechanisms https://en.wikipedia.org/wiki/Regulation_of_gene_expression.

Genetic and epigenetic abnormalities and disease are more likely to occur when cooperation is needed between several functional systems within a cell, or between groups of cells, as risks of errors mount not only because of the multiple systems involved, but because molecules involved in coordinating the functional systems also present a risk. Risks do not scale linearly with system size. The usual outcome is that the cell or developmental or homeostatic process is 'dysregulated,' meaning that the various elements required to achieve a functional outcome (e.g. proliferation followed by cell migration) are not properly operational or not properly coordinated in time (see, for example, the discussion on the complex processes involved in cell proliferation and cell migration in cartilage that may be dysregulated—Sect. 2.9.5). If dysregulated states become apparent in early life, these states are referred to as 'genetic abnormalities' (e.g. chondrodysplasias), but if they become appear later in life, they contribute to the 'incidence of age-related diseases' (e.g. osteoarthritis). Indeed, twin studies have revealed that osteoarthritis is about 50% environmental and 50% genetic in origin (Spector and MacGregor 2004), so this 'polygenic disease' has a significant genetic component, even though it is usually hidden from clinical view (though at least it is traditionally accessible in general terms through a 'family history,' which is a very important part of a clinical history).

Standard tools used by biologists today are to either upregulate or downregulate a gene (http://en.wikipedia.org/wiki/Regulation_of_gene_expression). What is clear is that up and down regulation of some genes does result in abnormalities or conditions that resemble disease states, and so these signaling molecules are then deemed 'most important' (Loughlin 2015; Ramos and Meulenbelt 2015; Reynard et al. 2016). Sometimes, the same signaling system is knocked down in several different ways (i.e. by reducing expression single proteins in the same signaling pathway, or by using 'kinase inhibitors'). In the early stages of interrogating an unknown biological system, this often leads to considerable repetition of essentially the same research (e.g. 'knockdown' of various components of the Wnt signaling pathway) result in pretty much the same system outputs—a Wnt signaling deficiency (van Amerongen and Berns 2006). We now know that too little or too much signaling is likely to result in abnormal cartilage tissue or disease states.

The key point is the information transfer capacity of signaling systems, which has not really been quantitated, other than via phenotypic outcomes observable in whole organism, which is a useful but blunt tool. Construction of an organism requires many highly specific instructions be carried out in environments with a lot of signaling traffic that are variable. This need for highly specific instructions in a confusing and noisy signaling environment drives complexity in the signaling systems. So for example, there are over 20 different Wnts signaling to seven frizzled receptors, and these interactions can be modified be a very large number of molecules in the extracellular environment (e.g. SFRPs) and co-receptors (LRP5/6) (see discussion on Wnt in Sect. 5.6.3). This complexity is necessary during development, but in less challenging environments, signaling complexity may be wound right down. So in adult tissue, normal homeostasis may be maintained by just a small fraction of the signaling intensity that the whole signaling system can muster.

For example, in adult articular cartilage tissue, it is probable that most of homeostatic Wnt signaling in articular cartilage is maintained by just four Wnts [two canonical Wnts (Wnt3a and Wnt7a) and two non-canonical Wnts (Wnt5a and Wnt11)]. However, when the tissue is injured or disease states arise, the signaling intensity often rises. For example, in addition to the four Wnts mentioned above for tissue homeostasis, during development Wnts2b, 5b, 10a, 10b, and 16 are also deployed. For example, Wnt16 plays a key role in specifying the location of joint development (see discussion Appendix C) and Wnt16 is also strongly upregulated when cartilage tissue is injured.

The final point relates to mRNAs. It is now known that microRNAs play important roles in regulating skeletal growth and maturation, and cartilage is no exception. For example, miR-93 regulates MMP3 and collagen loss in the nucleus pulposus of a spinal disk (Jing and Jiang 2015). It has been shown a large number of microRNAs are differentially expressed over time and space in cartilage during development (McAlinden et al. 2013). The most highly expressed miRNAs in cartilage are PC, DC and HYP, miR-140-5p. Interestingly, miR-140-5p is most specific to cartilage. It has also been shown that miRNAs are important in regulating cartilage homeostasis (Gibson and Asahara 2013).

For example, it has been found that miRNA-140 knock out mice have a short stature—but do survive. At one month of age, the knee is grossly normal, but close inspection shows abnormalities including surface fibrillation and proteoglycan loss.

Overexpression of miRNA results in resistance to developing osteoarthritis. The propensity to develop osteoarthritis when miRNA levels are low has in part been attributed to elevated expression levels of ADAMTS5 (Miyaki et al. 2010).

Importantly, it has been found that many of the miRNAs exert their effects through regulating growth factor signaling pathways. Some of the growth factor signaling pathways influenced by miRNAs including IGF-1, TGFβ, BMP (including activin A), Wnt (including Wnt5a), FGF, VEGF, and Hedgehog signaling (McAlinden et al. 2013). This is consistent with the role of miRNAs is regulating whole signaling pathways. miRNAs usually take down several key proteins in a pathway, rendering it almost completely silent, or the reverse—they promote the production of key phosphokinases, upregulating signaling through a specific pathway. Undoubtedly, much remains to be discovered about the ways miRNAs can influence cartilage behavior in both health and disease.

5.8.2 Interactions and Growth Factors

There is increasing evidence that pro-anabolic growth factors (e.g. IGFs, PTHrP, TGFβ, PDGFbb, canonical Wnts) and appropriate mechanical loading can offset the catabolic effects of inflammatory mediators such as TNFα and IL-1β (Bader et al. 2011; Gibson et al. 2017; Li et al. 2013; Montaseri et al. 2011; Torzilli et al. 2010). It has been shown that growth factor activation of Src, PI3K, and AKT signaling can downregulate NF-κβ activation (Montaseri et al. 2011). Exploiting opposing actions mediated by intracellular pathway interactions is likely to prove clinically useful.

Perhaps the most important set of experiments relating to signaling interactions between growth factors in cartilage are reported by Shi et al. (2013). They used a viral gene transfer technology to transfect adult bovine chondrocytes in a cell culture system with additional copies of several cell–cell signaling molecules known to be important anabolic and mitogenic molecules in cartilage. The genes transfected were FGF2, IGF-1, TGFβ, BMP2, and BMP7 (they were transfected one, two, three, four and five at a time), and they interpreted the effects of transfection by looking at cell proliferation, GAG production, and collagen production. Shi et al. explain

> Gene transfer is a promising approach to delivering such factors. No single growth factor gene is likely to optimize these cell functions, but multiple growth factor gene transfer remains unexplored. We tested the hypothesis that multiple growth factor gene transfer selectively modulates articular chondrocyte proliferation and matrix synthesis....The transgene pair encoding IGF-I and FGF-2 maximized cell proliferation. The three-transgene group encoding IGF-I, BMP-2 and BMP-7 maximized matrix production and also optimized the balance between cell proliferation and matrix production. These data demonstrate a potentially tunable approach to articular chondrocyte regulation and suggest that certain growth factor gene combinations have potential value for cell-based articular cartilage repair.

They found that IGF-1 with FGF-2 had the most powerful effect on cell proliferation (ninefold increase in cell DNA over controls in 6 days), while each of the cell signaling molecules acting individually had less than a twofold increase in cell DNA [see Fig. 1 in Shi et al. (2013)]. Regarding GAG production normalized to DNA content, they found that IGF-1 and either BMP2 or BMP7 were most effective (3.5 fold increase over controls in 6 days), while each of the cell signaling molecules had about a maximum twofold increase (IGF-1, BMP2, or BMP7) [see Fig. 4 in Shi et al. (2013)]. Regarding collagen production normalized to DNA content, they found that IGF-1 was most effective, inducing a 1.7 fold increase in 6 days [see Fig. 4 in Shi et al. (2013)]. Shi et al. found that FGF2 alone or in combination invariably decreased GAG and collagen production. It seems possible in these experiments FGF2 is redirecting metabolic energy into cell proliferation, rather than ECM production.

We note that the same viral technology has been employed to introduce BMP-2 and BMP-6 into osteochondral defect repair model using ponies. This technology using two BMPs met with modest success in the short-term, but the osteochondral repair was not adequate in the long-term (Menendez et al. 2011). While it seems unlikely that the addition of two these BMPs by themselves would make the difference between a successful long-term repair and not, this gene technology appears very promising. This is evidenced by the method of elucidating Wnt7a's role in preventing articular damage in an experimental OA animal model (Gibson et al. 2017). Presumably, similar results may soon be achieved more simply using CRISPR-cas9 technology (Mout et al. 2017).

5.8.3 Interactions Between Signaling Pathways

Interactions between pathways are very common, adding considerably to the complexity in understanding how these signal-transduction pathways operate. See for example, the large number of interactions shown between the signal-transduction pathways illustrated in Kerkhofs et al. (2016) and http://christophe.dessimoz.org/revcompbiol/_media/2013/t2-slides.pdf.

As a general principle interactions between signal-transduction signaling system help facilitate some functional outcome. It is noted that often interactions occur at multiple sites to reinforce their effect, particularly when they share many common proteins. For example, there is interaction between the (canonical) Wnt and AKT pathways through GSK-3beta. GSK3beta is an integral phosphorylating protein in the 'destruction complex' of the canonical Wnt signaling pathway. This has been called the 'GSK3beta/beta-catenin axis' (Skurk et al. 2005). When AKT signaling increases, GSK3beta concentration reduces. This leads to an upregulation of beta-catenin concentrations (as the concentration of the destruction complex is reduces as GSK3beta reduces). The increase in beta-catenin concentrations makes the cell more sensitive to Wnt signaling, i.e. smaller increments of extracellular Wnt signaling molecules is likely to trigger the cell to divide. There is also interaction between canonical Wnt signaling and IL-1 catabolic signaling pathways (Gibson et al. 2017).

The issue becomes not whether or not interactions between signal-transduction pathways are useful or not (they are), but what the interactions achieve, how are regulated and when they are operational. There are basically two ways that one signaling pathway may dominate other pathways. The first way involves the two signaling pathways sharing a common molecule in the two signaling cascades. In this case there is competition between the pathways for the same molecule. So for example, receptor tyrosine kinases (RTKs), toll receptor signaling pathway, JAK/Stat and focal adhesion signaling all pass through the AKT signaling molecule (see http://www.genome.jp/kegg-bin/show_pathway?hsa04151). The upstream pathway that predominates depends on which pathway claims the greatest role in activating AKT. AKT then signals to multiple downstream pathways (e.g. eNOS, CREB, IKK, MDM2). The downstream signaling pathway(s) that predominates depends on the relative concentrations and binding affinities of eNOS, CREB, IKK, and MDM2 etc. for activated AKT.

The competition for binding (through reaction rates, affinities and concentrations) occurs right up to the transcription factors themselves. One transcription factor in abundance may bind to and inhibit another transcription factor. For example, FOXO (i.e. forkhead box-O transcription factor) binds to β-catenin, blocking β-catenin/TCF-LEF signaling and so cell proliferation (Hoogeboom et al. 2008).

The second less well-researched way of selecting one pathway over another is frequency encoding of the signaling information (Utzny et al. 2005). This is approach is exploiting differences not only in concentration and binding affinity, but the rates of (one or more) chemical reactions in the forward or reverse directions, and the rates of transport processes (e.g. the rates of translocation of various transcription factors from the cytosol to the nuclear compartment and back again).

5.8.4 Transcription Factors in Cartilage

SOX9 appears to be an essential transcription factor (TF) the chondrocyte phenotype, causing the differentiation of mesenchymal progenitor cells into chondrocytes (Bi et al. 1999). L-SOX5 and SOX 6 are reported to be essential TFs too (Smits et al. 2001). The expression of SOX5 and SOX6 depend on the expression of SOX9 (Akiyama et al. 2004).

The transcription factor Erg (Ets-related gene) and its splice variant C-1-1 appear to be important in normal articular cartilage development, and the formation of the growth plate (suppressing Runx2 and bone formation via chondrocyte hypertrophy) (Hidaka and Goldring 2008). Erg appears to be under the control of transcription factor GDF5, which is involved in joint formation (Iwamoto et al. 2007). Exposure to GDF5 may be essential for hyaline cartilage formation (Yamashita et al. 2015).

While SOX9 has been associated with aggrecan, link protein and collagen type II gene expression, it has been mentioned that SOX9 and collagen type II gene expression is tightly coupled. In osteoarthritis, SOX9 expression reduces (Haag et al. 2008) (as PTHrP levels decrease), but collagen type II expression increases (Aigner et al. 2003). This can be explained by taking into account a very important interaction between SOX9 and β-catenin.

It can be shown that overexpression of SOX9 or inactivation of β-catenin result in dwarfism. Inactivation of SOX9 and stabilization of β-catenin produces a similar phenotype of chondrodysplasia. This evidence, together with molecular evidence, builds a case supporting the proposition that SOX9 functionally blocks the action of β-catenin, directly binding to β-catenin, blocking the proliferative activity of chondrocytes (Akiyama et al. 2004).

It is possible that this provides an explanation for the proliferative quiescence of chondrocytes observed in adult cartilage, and the re-emergence of proliferative activity in osteoarthritis (SOX9 levels fall as PTHrP levels fall). As SOX9 levels fall, Wnt activity increases and chondrocyte proliferation increases. Migration also increases with proliferation, particularly as aggrecan concentration also fall in osteoarthritis. Chondrocyte migration and falling aggrecan concentrations lead to an increase in collagen type II expression. The fall in PTHrP levels also stimulate Ihh levels to increase in osteoarthritis, which promote the production of hyaluronic acid. The production of collagen type II and hyaluronic acid are essential steps in cartilage ECM repair. Clearly, there is much to explore.

5.9 How Can a Systems Approach Progress Cartilage Research?

Rather unusually for a body tissue, articular cartilage contains no blood vessels, lymphatics or nerves, is composed of predominantly 'extracellular matrix,' and contains only one cell type. What tissue is simpler? What better place to start understanding the biology of a tissue than 'rubber-like' cartilage?

A moment's reflection on the description in previous sections on signaling molecules in cartilage would suggest a lot could be actually be going on in this 'simple, rubber-like tissue.' Indeed, given the staggering complexity of the networks within the cell nucleus and cell cytoplasm, anytime a cell is involved in a response, we know a priori it is a going to be a sophisticated response. For example, we might immediately expect an extraordinary range of chemical reactions to be going on in that apparently 'simple, rubber-like' extracellular matrix. The extracellular matrix 'self-organizes,' so we expect many feedback control systems to be operating in the extracellular matrix (see Fig. 5.14). In other words, we may well expect there would be powerful interlocking networks regulating homeostasis of the tissue, as well as networks facilitating adaption of the tissue if environmental circumstances change for a significant length of time.

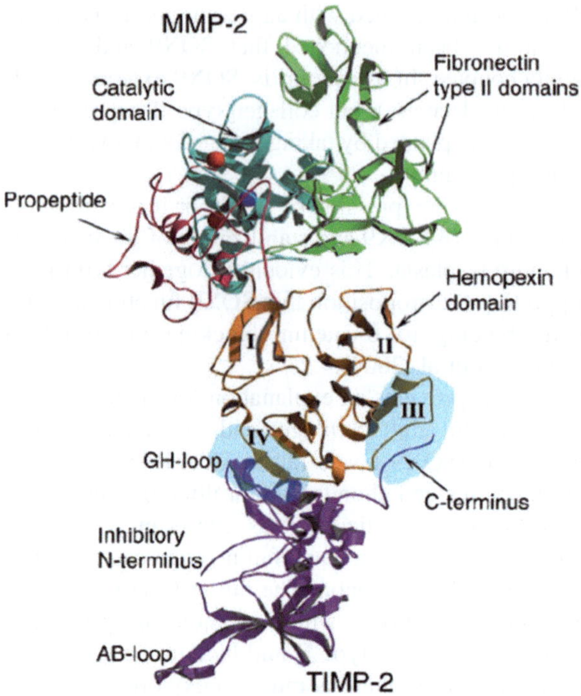

Fig. 5.14 There are many feedback control loops in both the cell and the extracellular matrix of cartilage. For example, MMP is secreted as a propeptide that has to be activated. Once activated, it is inhibited by TIMPs. The structure of the proMMP-2_TIMP-2 complex is shown in the accompanying figure. Overall conformation: the proteinase and inhibitor interact via their C-terminal domains. The catalytic site of MMP-2 and the inhibitory active site of TIMP-2 are turned away from each other. This topology excludes an inhibitory interaction between the proteinase and inhibitor and implies that both proteins remain fully functional in the complex. Catalytic and structural Zn2_ ions are colored red and Ca2_ ion purple. The _propeller blades of the hemopexin domain are numbered from I to IV. Two light blue ellipsoids in blades III and IV indicate two areas of interaction between proMMP-2 and TIMP-2 molecules (Morgunova et al. 2002)

After a moment's reflection, we see the foregoing suggests immediately that cartilage is not a 'simple, rubber-like tissue' at all. Indeed, this line of reasoning argues that cartilage is probably not simple at all, exhibiting complex behaviors that are not easily explained.

Despite the elementary nature of the arguments above, the notion that cartilage is a 'simple, rubber-like tissue' proved irresistibly attractive for many researchers, and took a firm hold on the collective imagination of researchers. For optimistic researchers, cartilage seemed the most likely tissue to yield up its secrets.

Diseases of cartilage including osteoarthritis appeared to offer an attractive business case for drug development, for osteoarthritis is a chronic condition, so any successful drug treatment is likely to require ongoing use of a commercial drug. The optimistic research case was that just a little research investment was required to unlock the next 'blockbuster drug' for this 'simple tissue'! This set of perceptions provided an attractive 'drug development environment' for 'big pharma,' and so a lot of drug companies established drug development programs to identify so-called disease-modifying osteoarthritis drugs (DMODs). A common development strategy for drug companies was to focus on the degradation of the extracellular matrix. A common drug development research strategy appeared to be finding an enzyme that broke down the extracellular matrix (an MMP or aggrecanase), and then find the drug that blocked its action. This was a 'simple plan' for a 'simple tissue.' What could go wrong? Part of the organizational attractiveness of this approach is that it is readily comprehended by everyone involved, everyone knew what needed to be done, and clear milestones could be devised thereby enabling drug companies to manage their risk. What could be simpler?

Unfortunately, development plan failure upon development plan failure has gradually made it clear to drug companies and researchers alike that cartilage is not the simple, rubber-like tissue it was originally conceived to be. Unfortunately, the experience to date is that the simple strategies to develop DMODs tried so far by drug companies have invariably failed. After spending many billions of dollars on cartilage research, big pharma have pulled back from cartilage DMOD development—they now look elsewhere to develop 'blockbuster drugs' (Hunter et al. 2014). Today the language of big pharma has changed, and most are now talking about doing 'systems biology,' and tackling a chosen disease through 'multiple drug targets' simultaneously. Most drug companies are now outsourcing risk to university research groups, only becoming involved after something interesting turns up. For those researchers participating in cartilage research over decades, it is 'back to the biological basics' so to speak.

The fundamental question remains: What do we really know about the physiology of cartilage? Frustratingly, the answer is that we know a lot of specific facts about specific situations, but there is not really a coherent narrative that links of the specific facts about specific situations together into concepts, and then turns concepts into a predictive system model of cartilage. While undoubtedly a lot remains to be discovered about cartilage, a great deal remains to be explained based on what we already have discovered about cartilage. Reducing the very obvious gap between the 'mountains' of experimental data and using concepts to explain how cartilage functions is what this book attempts to do.

First, we need the concepts, then data can be employed to build mathematical models. We believe that one promising way to provide a coherent, quantitative narrative is through analytic and computational modeling of cartilage. A mathematical framework can in principle address all the abovementioned systems issues in cartilage. In other words, in principle all the abovementioned information can be packaged in sophisticated multiscale models of cartilage (which would probably need to run on a supercomputer). While development of such a model seems daunting and maybe even impossible today, a reasonably firm foundation has been developed by mathematicians for such analytic modeling over the centuries, and more recently, a relatively firm foundation has been developed by physicists and engineers for the numerical modeling of complex biological systems.

Indeed, the so-called physiome project sets out an impressive agenda, and has some inspiring achievements to date http://physiomeproject.org/. Work has begun on the virtual human http://physiomeproject.org/, on the development of virtual organs (e.g. the heart http://heart.physiomeproject.org/ and liver http://www.virtual-liver.de/wordpress/en/), and incredibly detailed models of how the brain operates, i.e. modeling the functional units of cortical columns in the brain http://bluebrain.epfl.ch/. Based on known physics, chemistry and physiology of these organs, computational models are leading the discovery of new insights into how these organ systems operate (e.g. http://actu.epfl.ch/news/neural-simulations-hint-at-the-origin-of-brain-w-4/). We believe breakthroughs can be achieved with the appropriate use of computational modeling of cartilage combined with other ways of knowing (e.g. epidemiological, biomedical engineering, biomedical science, and clinical knowledge).

5.10 What Are the Implications for Osteoarthritis?

Zhong et al. (2015) helpfully explain:

> Chondrocyte hypertrophy is not under the strict control of a single pathway but appears to be regulated by an intricately regulated network of multiple signaling pathways, such as Wnt, Bone morphogenetic protein (BMP)/Transforming growth factor-β (TGFβ), Parathyroid hormone-related peptide (PTHrP), Indian hedgehog (Ihh), Fibroblast growth factor (FGF), Insulin like growth factor (IGF) and Hypoxia-inducible factor (HIF).

Agreed. Indeed, it is complex, and dysregulation of the various signaling systems and their interactions doubtless lead to a variety of osteoarthritic cartilage phenotypes. However, it appears as though the PTHrP/Ihh is a foundational cartilage signaling system in the sense that if the joint has little mechanical stimulation, the PTHrP/Ihh negative feedback loop continues to operate, preventing complete involution of articular cartilage. With the addition of normal loading, PTHrP production in the transitional and superficial zones is upregulated, while in the deep zone Ihh is downregulated. It seems likely that this reduction in production of Ihh leads to less stimulation of transitional zone chondrocytes, for normal cartilage physiology this would mean the deep zone slowly 'advances' on the transitional zone (i.e., the deep zone expands and the transitional zone contracts), with articular cartilage deep zone taking up a greater fraction of the total cartilage thickness. Various chemical signals (aggrecan feeding back on the chondrocyte, EPO, FGF23, etc.) serve as 'phenotype attractors' that stabilize full differentiation of chondrocytes in normal deep zone articular cartilage.

As noted previously, for humans, the proportions of each articular cartilage zone relative to the total thickness for normal human femur hip and knee articular cartilage is reported as being $7 \pm 1\%$ for the superficial zone, $19 \pm 7\%$ for the transitional (middle) zone and 75 ± 9% for the deep (radial) zone (Changoor et al. 2011). However, the fractions of cartilage thickness in each zone change dramatically for ‘degraded’ cartilage [note: the meaning of ‘degraded’ in this context is defined in Fig. 1 of Changoor et al. (2011), but largely it pertains to tissue obtained during arthroplasties (i.e., joint replacement)]. It is found that the relative proportions of each zone in cartilage classified as ‘degraded’ are 10 ± 3% for the superficial zone, 47 ± 10% for the transitional zone, and 46 ± 10% for the deep (i.e. radial) zone. As OA develops, the transitional zone actually ‘advances’ on the deep zone. It is clearly apparent that there is a substantial expansion of the transitional and superficial zones at the expense of the deep zone for ‘degraded’ (i.e. OA) cartilage. This expansion in the transitional and superficial zone is simply the tissue’s attempt to maintain tissue homeostasis. An increased rate of damage to the collagen network in the transitional zone provides a mechanism for cartilage tissue expansion, with newly expanded cartilage tissue being repaired and pushing upwards to form new superficial layer from below (see discussion Sect. 3.5).

As one ages, normal activity level declines, which decreases the amount of mechanical joint loading and so the mechanical stimulation of cartilage tissue. With the onset of stiffness and pain that occurs in a majority of OA suffers, activity levels decrease further. On the basis of decreasing joint loading and knowing, the behavior of PTHrP/Ihh system, we would expect the PTHrP concentrations to decrease and Ihh concentrations to increase in articular cartilage, and so we expect the transitional zone to expand in thickness, while the deep zone decreases in thickness until a new equilibrium is found. The predicted increase in Ihh concentration by this model appears to occur in OA. For example, Wei et al. (2012) reports:

> Ihh expression was increased 2.6 fold in OA cartilage and 37% in OA synovial fluid when compared to normal control samples. Increased expression of Ihh was associated with the severity of OA and expression of markers of chondrocyte hypertrophy: type X collagen and MMP-13, and chondrocyte size.

With increasing age and decreasing activity, the PTHrP/Ihh signaling system promotes cell proliferation and growth of the transitional zone.

Increasing collagen network damage can be compatible and co-exist with decreasing activity levels. One way this can occur is for collagen network repair capability to decline faster than the rate of collagen network damage, leading to net increase in collagen network damage. As damage accumulates, increased fibrillation of the cartilage surface increases the surface friction coefficient and so shear stress experienced by cartilage tissue, which accelerates damage. In addition, decreased activity may be accompanied by increased number of large/injurious loadings as balance deteriorates (Graham et al. 2016), or joint loadings may become physiologically abnormal as joint stiffness and/or pain increases and gait patterns change. We observe that collagen network damage in the transitional zone ‘pushes’ in the same direction as an increase in Ihh signaling to the transition zone, thickening the transitional zone. Measuring the relative thicknesses of the deep and transitional zones (e.g. using MRI) may provide an index of articular cartilage healthiness or conversely, susceptibility to OA.

The theme of our system biology discussion is the interconnectedness of biological systems. At every level—from molecule, to cellular transduction, to cellular 'checkpoints'—from cell–cell communication systems, to regional autocrine signaling, to multiple hormonal systems through to mechanical signals—organ systems are communicating with one another, and so are tissues within organs, cells within tissues, and so is the ECM to a cell, and a cell to the ECM. Our bodies are composed of bundles of networks within bundles of networks.

Consequently, to say any disease impacts all systems in the body is certainly true, and to say all systems in the body have some influence on a disease process is also true. To draw a somewhat arbitrary boundary around certain systems in the body and associate these with a disease process is something we do as humans, due to our finite knowledge and finite power to make discriminating measurements. We also draw these boundaries to limit the scope of the disease to something more manageable and comprehensible, and then we limit the scope further when we think about treatments for a disease process. Our conception of a disease is in fact an (imperfect) conceptual model of that disease process within the body, which may 'more or less' overlap with how the actual disease interacts within the body, depending on our understanding of the disease process and the quality of our model.

Of course, our knowledge of any disease process is a work in progress. In some cases, our knowledge is exquisitely detailed (e.g. genetic abnormalities of metabolism), while in other cases our knowledge is primitive (e.g. diseases processes associated with brain function). As we develop the ability of evaluate new potential interactions, our understanding of a disease progresses, and sometimes this transforms our understanding (e.g. the discovery that most stomach ulcers are caused by the Helicobacter pylori bacterium). Indeed, there is much excitement around interactions of the microbiome with organ systems in the body, e.g. the gut microbiome and its influence on health of every organ system in the body (research activity in this area has increased largely due to new, cheaper technologies that bring genetic and proteomic analyses within practical reach). The remarkable thing about the body is it is connected in so many ways that new things are bound to be uncovered, some of which may transform our knowledge of disease (as for the case of stomach ulcers). In this context, perhaps OA too is influenced by the gut microbiome in ways we do not yet appreciate? Perhaps, interactions of the microbiome with the immune system results in a drift toward a generalized inflammatory state throughout the whole body, and synovial joint tissues are just one of the first group of tissues to respond?

Most diseases associated with ageing have a complex etiology, as they result from lifelong interactions between genes and the organism's environment. Over a lifetime, the organism is tested in many, many different ways—and as a consequence of that testing the organism changes throughout life as it is forced to respond to the stressors in its own unique way. Some of these changes do in fact affect its genome (e.g. epigenetic change may influence any cell, while the immune system manages a controlled system of gene mutation), and other changes result in damage accumulation in the organisms tissues, which then compromise the organism's ability to cope with stressors in the future more or less permanently [e.g. myocardial infarction (i.e. 'heart attack'), cerebrovascular events (i.e. 'strokes'), and 'scarring' of the ECM damage tissue (e.g. lung fibrosis, malaligned fractures)].

In this context, the suggestion that osteoarthritis may be a 'systemic musculoskeletal disease' (Malemud 2015) is a given. That osteoarthritis may be driven along by 'lifestyle/diet choice diseases' collectively referred to as 'metabolic diseases' is also a given (Courties et al. 2017; Sellam and Berenbaum 2013). Is any disease process not influenced by chronic low-grade inflammation, irrespective of its source (e.g. be it bacterial, viral, immune system dysregulation, dyslipidemias)? The advantage of 'labels' like 'systemic musculoskeletal disease' and 'metabolic disease' is that they remind us of the connections in the body. The disadvantage is that broad definitions of disease processes by definition do not really have that much explanatory power, because the connections we know about are weak. It is up to discovery-led research to either make the connections stronger (or not) or show how they interact synergistically with other known factors in driving an OA phenotype.

The truth of the matter is that we already know of plenty of factors that have strong connections in initiating and then driving OA progression (e.g. usually those factors are related to either driving abnormal or excessive mechanical loading of the joints, and/or the production of inflammatory cytokines within joint tissues). One can make a strong argument we know what most of the strong connections are that lead to OA; the problem is we are yet to find a way of harnessing that knowledge to either mitigate or prevent OA.

This suggests that perhaps we need a new type of discovery mechanism, A discovery mechanism that can answer different questions involving the integration of the knowledge we already have. We need a discovery mechanism that can increase the discriminating power of this knowledge to make predictions that are relevant to people's lives, either by preventing or mitigation of OA, or by the treatment of OA using one or more drugs simultaneously. This line of reasoning would suggest we need new skills and new methods to generate new insights that help unlock the knowledge we have already gained and will gain in the future through new experiments.

This is where mathematical modeling and computational modeling has something to offer—a systematic way for integrating the knowledge we already have and providing a means for knowledge to be systematically organized, and most importantly, become cumulative. For if knowledge is to systematically cumulative, then new knowledge is not tested against a 'random event model,' but rather it is tested against the reported findings of many past experiments. This is perhaps best implemented using a 'Bayesian approach' to updating computational models (and so updating our knowledge). By this means, we can pull ourselves out of a morass of qualitative and semi-quantitative observations, into a quantitative world where models gradually improve (and new findings are 'weighed' against our cumulative knowledge) and into a quantitative world where proposed strategies, individual therapies and the associated risks can be more rationally assessed. How to commence development of such a modeling approach is discussed in the next two chapters.

Further, we would suggest that the most cost-effective approach to better classification of OA is combining basic and clinical experimental research with computational modeling of normal and pathological joints, in a 'virtual (modeling) environment' that can be employed to quantitate the likelihood of causes and outcomes for a potentially unlimited number of scenarios. This approach can potentially help delineate the relative contributions of system disease processes on local tissue processes in joints. Perhaps, metabolic syndrome(s) does have an outsized effect on OA progression due to synergistic interactions with mechanical loads? Or perhaps metabolic syndrome(s) have an outsized effect on late stage OA progression, or perhaps metabolic syndrome(s) are responsible for initiating OA (Schott et al. 2018)? We don't really know. It seems plausible that these scenarios could be investigated cost effectively in a virtual environment using computational models that have been appropriately calibrated and validated (which means development of the computational model needs to occur together with appropriately designed experimentally based discovery research).

Unfortunately, a modeling approach takes some time to get used to, involving different skill sets, more related to engineering than to the traditional health professional skill sets typically practiced and honed by clinicians over many years. Given that there is not usually enough time to practically develop these two skills sets in a single person, this means there needs to be more and better cooperation between the two groups, so that different skills and viewpoints can be alloyed into stronger conceptualizations than either group are capable of acting alone. Many around the world are currently attempting to do just this—alloy two group's skills and viewpoints—but it has proved to be not an easy thing to do, for a whole host of interconnected reasons that are well-known to human resource and organizational specialists. These reasons are explored elsewhere, but clearly alloying of skills and viewpoints is a large work in progress, to which this book attempts to make a small but useful contribution.

Let us now turn to computational modeling of joint tissues, with our main focus on cartilage. We begin by developing our understanding of osmotic pressure, which plays an important role in the behavior of cartilage tissue.

Chapter 6
Osmotic Pressure, Solid Stress, and the Diffuse Double Layer

6.1 Introduction

Osmotic pressure is often mentioned as being fundamental to developing an understanding of articular cartilage. And indeed, we have mentioned osmotic pressure on several occasions as being responsible for the swelling pressure of articular cartilage. For cartilage in the unloaded state is pre-stressed. The tensile stress in the collagen network is balanced by the osmotic swelling pressure of the aggrecan molecules, bound within the collagen network. And when cartilage is loaded and consolidates, the aggrecan is compressed and the osmotic swelling pressure increases and provides resistance to the loads applied on the surface of articular cartilage. So osmotic pressure is fundamental to understanding the mechanical behavior of articular cartilage, and so worthy of close examination.

Despite being the subject of intensive research in many disciplines for over 100 years, the origin of osmotic pressure is still controversial. For this reason, we start at the beginning and work our way through the issues, deciding that osmotic pressure arises physically because of a boundary phenomenon. Any physical situation that blocks the movement of solute particles, but allows the free movement of solvent, between two solutions that have different numbers of osmotically active particles, gives rises to increased solvent pressure in the solution with more osmotically active particles. This rise in solvent pressure is referred to as osmotic pressure.

In the following, we first consider osmotic pressure in ideal solutions, for uncharged molecules and charged macromolecules. We then describe how osmotic pressure can arise between two charged plates or two charged macromolecules, and how osmotic pressure in a solution can be calculated in the presence of electric fields. We describe how the osmotic pressure in ideal solutions can be extended to non-ideal solutions, and finally we consider how osmotic pressure (and other) pressures can be practically measured.

D. W. Smith et al., *Articular Cartilage Dynamics*,
https://doi.org/10.1007/978-981-13-1474-2_6

6.2 Osmotic Pressure

'Osmotic pressure' is a property of a solution, much like 'freezing point depression' is a property of a solution. In both cases, the property is due to the presence of solute in a solvent. The osmotic pressure of a solution does not have to be realized, and so knowing the osmotic pressure of a solution does not tell you the mechanical fluid pressure in the solution, anymore than knowing the freezing point depression tells you the temperature of the solution (Denbigh 1981). Mechanical osmotic fluid pressure in a solution is realized in the presence of a 'semipermeable membrane'—an 'ideal semipermeable membrane' realizes all the osmotic pressure potential of the solution. Osmotic pressure is generated by a difference in concentration of osmotically active particles acting across a semipermeable membrane. The semipermeable membrane is permeable to the solvent (e.g. usually water in biological applications), but it is not permeable to the osmotically active particles (i.e. solutes) in solution (hence the name, semipermeable membrane). Osmotically active particles (atoms or molecules) are those that cannot cross the semipermeable membrane. As shown in Fig. 6.1, the osmotic fluid pressure in the presence of a semipermeable membrane may be measured under laboratory conditions by the height of fluid in a so-named Pitot tube.

It is important to realize that osmotic pressure is realized because of the semipermeable membrane. To say this another way, classical osmotic pressure is caused by the presence of the semipermeable membrane ['classical' is this context means osmotic pressure as studied by van 't Hoff and Lewis (1908), or as typically described under the heading 'colligative properties' to be found in most textbooks on Physical Chemistry (e.g. Atkin's Physical Chemistry)]. Without a semipermeable membrane, the colligative particles (the osmotically active particles) simply diffuse down their activity gradients until a uniform concentration in the solution is achieved at the equilibrium state.

Osmotic pressure is normally calculated assuming an 'ideal solution.' So we need to first define what is meant by an ideal solution. An ideal solution is one that is analogous to a mixture of ideal gases. An ideal solution has zero enthalpy of mixing. That is, there is no interaction energy between solute and solvent as they mix to generate a heat flow across the system boundary. A 'regular' solution has a nonzero enthalpy of mixing, but is otherwise ideal. The solvent and solute molecules in the mixture can be approximated as identically sized hard spheres, which collide with one another and so intermingle.

When solvent and solute are mixed, the volume of the ideal mixture is just the sum of the volumes of the individual components of the mixture. Importantly, mixing solute and solvent increases the entropy of the solution. The entropy of mixing for an ideal solution can be found as the sum of entropy changes due to an isothermal volume expansion of solute and solvent considered as two ideal gases, changing their volumes from pre- to post-mixing volumes [see Eq. (L.21)]. Since the solute undergoes the largest volume change, most of the entropy change associated with mixing is due to entropy change of the solute. For a regular

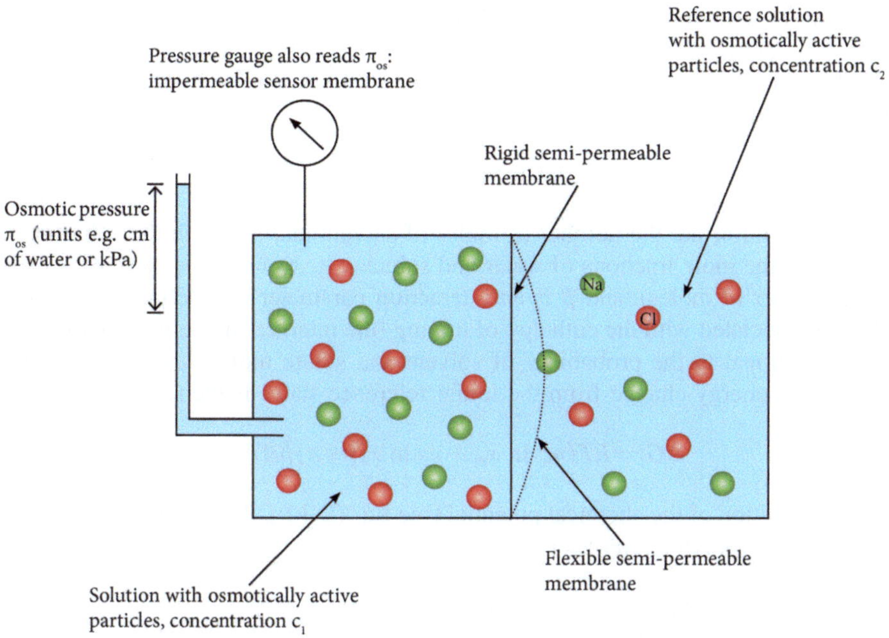

Fig. 6.1 Osmotic pressure arises when there is a difference in concentration of osmotically active particles across a semipermeable membrane. Osmotically active particles may be salts in solution [e.g. Na ions (green) and Cl ions (red) in solution], or any colloidal particles (e.g. uncharged molecules such as sugar molecules). In this example, the container is filled with solvent (e.g. water). The concentration of osmotically active particles on the left-hand side, c_1, is greater than concentration of osmotically active particles on the right-hand side, c_2 (which is our 'reference solution'). Our reference solution is taken to be at zero pressure (at the top of the container). The Pitot tube measures the mechanical fluid pressure in the left-hand side compartment. The pressure gauge has an impermeable sensor membrane, and so it measures the total stress [which is equal to (any) solid stress minus the fluid pressure, i.e. $\sigma_T = \sigma^s - p$]. For this example, the pressure gauge measures the fluid pressure (as solid stress is taken to be zero). The sign change arises because fluid compression by convention is taken as positive. The semipermeable membrane is so named because it is permeable to water, but it is not permeable to the colloidal molecules in solution. It is the semipermeable membrane that gives rise to an imbalance in momentum exchange between the molecules in the solutions, which results in a pressure jump at the entrance to the membrane pore. However, this pressure jump causes fluid to flow from right-hand side to the left-hand side of the membrane (down the chemical potential gradient of the solvent, water). This flow is transient and causes a mechanical pressure in the solution with more osmotically active particles. This pressure in the fluid is the osmotic pressure, i.e. $\pi_{os} = RT(c_2 - c_1)$. The transient fluid flow and resultant pressure difference across the membrane causes the fluid in the manometer (Pitot tube) to rise, the pressure gauge to read the fluid pressure, and the flexible semipermeable membrane to bend. In physiological experiments, the Pitot tube is often replaced by a more convenient 'wick needle' or 'glass micropipette.' All these devices measure the 'interstitial fluid pressure' (which is the net effect of osmotic and other mechanical pressures in the fluid between cells). Figure drawn by Yvette Harrap

solution, an additional term is included to capture the entropy change associated with heat energy crossing the system boundary.

The change in entropy associated with mixing for a regular solution is given by Benedek and Villars (2000),

$$\Delta S = R(n_w \ln x_w + n_s \ln x_s) + n_T \beta R x_w x_a \tag{6.1}$$

where n_w and n_s are the number of moles of solvent and solute in the solution, x_w and x_s are the mole fractions of water and solute (e.g. $x_w = n_w/n_T$, where n_T is the total number of moles), and β is an interaction parameter that defines the entropy change associated with the enthalpy of mixing (the interaction energy is assumed to be proportional to the probability of solvent and solute molecules colliding). The Gibbs free energy change from the initial reference states is then,

$$\Delta G = RT(n_w \ln x_w + n_s \ln x_s) + n_T \beta RT x_w x_a \tag{6.2}$$

By definition of the chemical potential [see Eq. (L.70)], the chemical potential of solvent in the mixture (μ_w^{mixing}) is

$$\mu_w^{\text{mixing}} = \frac{\partial \Delta G}{\partial n_w}|_{T,p,n_s} = RT \frac{\partial}{\partial n_w}((n_w \ln x_w + n_s \ln x_s) + n_T \beta x_w x_a) \tag{6.3}$$

Upon evaluating Eq. (6.3), we find

$$\mu_w^{\text{mixing}} = RT\left(\ln x_w + \beta(x_w^2 - x_w x_s)\right) \tag{6.4}$$

Now to find the osmotic pressure, we set μ_w^{mixing} equal to the change in chemical potential due to pressure [see Eq. (L.90)], so

$$p\bar{v} = RT(\ln x_w + \beta x_w(x_w - x_s)) \tag{6.5}$$

Note for small concentrations of solute $\ln x_w = \ln(1 - x_s) \to x_s$ and $\bar{v}n_T \to V$, then

$$p = RTc_s + \beta RT x_w(c_w - c_s) \tag{6.6}$$

We see that for exothermic reactions between solvent and solute (i.e. heat moves out of the isothermal solution and $\beta < 0$), the osmotic pressure is reduced, while for endothermic reactions between solvent and solute (i.e. heat moves into the isothermal solution and $\beta > 0$) osmotic pressure is increased. For an ideal solution $\beta = 0$, and Eq. (6.6) reduces to the well-known van ‘t Hoff equation.

$$p = RTc_s \tag{6.7}$$

Consequently, when analyzing a solution and a reference solution separated by a semipermeable membrane (see Fig. 6.1 for illustration), we find that

$$\pi = p_1 - p_2 = RT(c_2 - c_1) \tag{6.8}$$

where by definition π is the osmotic pressure with respect to the reference solution. The osmotic pressure of the solution is that osmotic pressure found in relation to the solvent at 'standard temperature and pressure' in the reference solution.

Having calculated the property of osmotic pressure, a natural question arises: What physical mechanism realizes this (osmotic) pressure as a mechanical pressure in the solvent? We know that we need a semipermeable membrane, for if the solution itself is an isolated thermodynamic system (in the sense of there being no exchange of matter), this system takes on the pressure exerted by its surroundings. Alternative if the solution can exchange matter with a reference solution without a semipermeable membrane being present, the solute would simply diffuse down its concentration gradient until the solute was dispersed throughout the whole volume. While there is momentum exchange between solute particles and solvent particles during diffusion (Benedek and Villars 2000), the solvent pressure is (for all practical purposes) constant, as the pressure surrounding the system is held constant (usually atmospheric pressure for laboratory conditions). So how does the presence of semipermeable membrane cause a mechanical pressure in the solvent?

One explanation says that water molecules simply diffuse down their chemical potential gradient (Denbigh 1981). That is, if osmotically active particles were to suddenly appear on one side of an ideal semipermeable membrane connected to a reference solution, the mixing entropy decreases the chemical potential ($\partial(\Delta G)/\partial n_{\mathrm{w}}|_{T,p,n_{\mathrm{s}}}$) in the compartment with the added solute. Then solvent molecules simply diffuse through the semipermeable membrane (which by definition of a semipermeable membrane they can do) from a higher chemical potential to a lower chemical potential. This continues until the pressure rises in the solution, which raises the chemical potential in the solution ($\bar{v}\Delta p$) until it is equal to the chemical potential in the reference solution.

Importantly, we note that net diffusion in one direction is not necessarily driven by solvent molecule number density (i.e. a difference in concentrations of water), rather it is driven by the chemical potential gradient. In other words, initially the increase in entropy upon mixing drives solvent diffusion preferentially in one direction. To say this in another way, more mechanistically, movement across the membrane in one direction becomes more probable than in the other direction, so the direction of the random walk of the solvent particles is biased. The increase in entropy due to mixing initially creates a 'rectified solvent diffusion' (meaning a biased random walk in one direction) close to or at the membrane, which leads to net movement of fluid toward the side with lower chemical potential. This solvent flow continues until the (osmotic) pressure in the solution rises, which raises the chemical potential of the solvent. Eventually the chemical potential in the solvent in the solution equals the chemical potential of the solvent in the reference solution and net fluid flow stops (though exchange of solvent molecules via membrane pores continues). A gradient in the number of osmotically active particles occurs very close to or at the semipermeable membrane, so it is apparent that this region of 'rectified solvent diffusion' is a boundary phenomenon or a 'membrane interfacial phenomena.'

Another explanation for osmotic pressure is based on the kinetic theory of solutions. For this theory, there is momentum exchange between the semipermeable membrane and solute particles (which clearly cannot happen in the bulk solution, as there is no semipermeable membrane present there). This momentum exchange from the semipermeable membrane to the solute particles results in a 'body force' in the solvent in the region of the membrane pore openings (Benedek and Villars 2000; Kramer and Myers 2012, 2013). Kramer and Myers (2012) explain that the semipermeable membrane may be conceived as a force field that repels solute, and then through viscous interactions between solute and solvent, the momentum in the solute is rapidly distributed to the solvent. In this way, the semipermeable membrane exerts a repulsive (body) force on the solution as a whole.

To explore this further, we now follow Benedek and Villars (2000) in their explanation for osmotic pressure, but in an effort to define as clearly as possible the interactions involved, alongside the equations describing the solute we show equations describing the solvent, and arrive at the same conclusion. By pairing solute and solvent momentum equations, every step of reasoning is clearly revealed. Benedek and Villars (2000) first develop a simple 1-D model of solute–solvent interaction, by deriving equations for mass and momentum balance for the solute. They begin by considering number density of solute particles moving from left to right (denoted c_{s}^{+}) and from right to left (denoted c_{s}^{-}). They define a representative elementary volume (REV) in the solution and write an equation for the time rate of change of momentum for all solute particles in the REV [i.e. Eq. (2.142) in Chap. 2 of Statistical Mechanics (Benedek and Villars 2000)]. Note that we choose to write two equations, describing the time rate of change of momentum for solute and solvent molecules, viz.

$$\begin{aligned} \frac{1}{V_{\mathrm{REV}}}\frac{\mathrm{d}p^{\mathrm{solute}}}{\mathrm{d}t} &= -mv^2(\partial c_{\mathrm{s}}/\partial x) - 2m\gamma j \\ \frac{1}{V_{\mathrm{REV}}}\frac{\mathrm{d}p^{\mathrm{solvent}}}{\mathrm{d}t} &= -(1-\chi)m_{\mathrm{w}}v_{\mathrm{w}}^2(\partial c_{\mathrm{w}}/\partial x) - \chi\frac{\partial p}{\partial x} + 2m\gamma j \end{aligned} \tag{6.9}$$

where V_{REV} is the REV volume and P^{solute} is the solute particle momentum in the REV. The time rate of change of solute momentum per unit volume (i.e. $\frac{1}{V_{\mathrm{REV}}}\frac{\mathrm{d}p^{\mathrm{solute}}}{\mathrm{d}t}$) is made up of two terms: (i) the net flow of momentum into the REV carried by all solute particles (i.e. $-mv^2(\partial c_{\mathrm{s}}/\partial x)$, where m is the particle mass, v is the mean particle velocity and $c_{\mathrm{s}} = c_{\mathrm{s}}^{+} + c_{\mathrm{s}}^{-}$), and a second term; (ii) which represents the net force the solvent molecules exert on the solute as a result of collisions between solute and solvent.

Benedek and Villars (2000) employ a simple model of complete momentum exchange between solvent and solute at each periodic collision. The rate of momentum exchange creates a (body) force ($F^{\mathrm{solvent-solute}}$ where superscript solvent–solute means solvent acting on solute) acting on the solute particles from the solvent in the REV, equal to $F^{\text{solvent-solute}}/V_{\text{REV}} = -2m\gamma j = -(jkT/D_{\text{s}})$, where j is the 'particle current' $v(c_{\text{s}}^{+} - c_{\text{s}}^{-})$ (or particle flux per unit time), γ is the collision rate between solute particles and solvent per unit volume (so $1/\gamma = t_{\text{c}}$, where t_{c} is the time interval between collisions). Clearly by Newton's third law, the body force acting on the solvent particles $F^{\text{solute-solvent}}/V_{\text{REV}}$ in a transient concentration profile in the bulk solution is equal and opposite to that acting on the solute particles, that is, $-F^{\text{solvent-solute}}/V_{\text{REV}} = F^{\text{solute-solvent}}/V_{\text{REV}} = 2m\gamma v(c_{\text{s}}^{+} - c_{\text{s}}^{-}) = jkT/D_{\text{s}}$. We note that Benedek and Villars (2000) show that D_{s}, the diffusion coefficient for the solute in the solvent, is equal to $D_{\text{s}} = \frac{1}{2}\frac{kT}{m\gamma}$ and we know the translational kinetic energy for particles moving in one spatial dimension is $1/2mv^2 = 1/2kT$.

The solvent may also sustain a pressure gradient $(\partial p/\partial x)$ which can also oppose a body force in the fluid [e.g. $F^{\text{solute-solvent}}/V_{\text{REV}} = +2m\gamma j$ in Eq. (6.9)]. But the sum of the rates of change of momentum $\partial p/\partial x$ and $m_{\text{w}}v_{\text{w}}^{2}(\partial c_{\text{w}}/\partial x)$ must equal the body force in the fluid (i.e. the solvent cannot accelerate away—see Appendix J). This equality leads to $m_{\text{w}}v_{\text{w}}^{2}(\partial c_{\text{w}}/\partial x)$ decreasing as $\partial p/\partial x$ increases.

But whether or not a pressure gradient actually develops in the solution depends on the experimental conditions. For most diffusion experiments, the surface of the fluid is open to the atmosphere, so the pressure in the solution is fixed at atmospheric pressure. In this case, there is no gradient of pressure in the solution or solvent, and χ is zero. If the pressure gradient is zero, the only term opposing the body force in the solvent is the net flow of momentum into the REV carried by all solvent particles. If the pressure gradient is not zero in the solvent, then the net flow of momentum into the REV carried by all solvent particles reduces in proportion to the pressure gradient. Importantly, the pressure gradient can reduce net flow of momentum into the REV carried by all solvent particles to zero, providing the pressure gradient is large enough. So χ may range from zero to one (excluding the possibility that additional pressure may be applied to the solution, as occurs in reverse osmosis).

We are now in a position to introduce an external force field (e.g. gravity or a centrifugal force) into the analysis. The external force field compels all solute particles (c_{s}) to undergo a time-averaged translation in one direction. f is the magnitude of the force acting on a single solute particle, so the contribution this solute force per unit volume makes to the momentum balance of the solute is fc_{s}. Solute particles collide with solvent particles as they drift—if solute particles did not collide with solvent particles, they would accelerate away under the influence of the force field. But because solute particles do collide with solvent particles as they drift, the force field also contributes fc_{s} to the momentum balance of the solvent. The external force field generates a uniform solute current density $j = uc_{\text{s}}$, where u is the uniform drift velocity. It can be shown that $u = \frac{D}{kT}f$. The proper interpretation of the drift velocity is that the external force causes a bias in random walk of the particles, such that the bias results in a time-averaged translation in the direction of the force field (Benedek and Villars 2000).

Putting this new information together with Eq. (6.9), we now have for the time rate of change of momentum balance of solute and solvent,

$$\begin{aligned}\frac{1}{V_{\mathrm{REV}}}\frac{\mathrm{d}p^{\mathrm{solute}}}{\mathrm{d}t} &= -kT(\partial c_{\mathrm{s}}/\partial x) - \tfrac{kT}{D_{\mathrm{s}}}v(c_{\mathrm{s}}^{+} - c_{\mathrm{s}}^{-}) + fc_{\mathrm{s}} \\ \frac{1}{V_{\mathrm{REV}}}\frac{\mathrm{d}p^{\mathrm{solvent}}}{\mathrm{d}t} &= -(1-\chi)kT(\partial c_{\mathrm{w}}/\partial x) - \chi\tfrac{\partial p}{\partial x} + \tfrac{kT}{D_{\mathrm{s}}}v(c_{\mathrm{s}}^{+} - c_{\mathrm{s}}^{-}) + fc_{\mathrm{s}}\end{aligned} \tag{6.10}$$

The solute particle current density (or particle flux) is given by,

$$j = v(c_{\mathrm{s}}^{+} - c_{\mathrm{s}}^{-}) = -D_{\mathrm{s}}(\partial c_{\mathrm{s}}/\partial x) + \frac{D}{kT}fc_{\mathrm{s}} \tag{6.11}$$

Now we assume that the boundary conditions limit solute movement in the force field through the solution, and this results in *steady-state concentration profile* of solute particles [e.g. at the steady-state particle concentration profile in a solution within a spinning centrifuge, this concentration profile is known as 'sedimentation equilibrium,' which is in fact employed in the experiments of Comper and associates to estimate the properties of chondroitin sulfate, e.g. Comper (1991)]. Because steady-state conditions prevail, the particle current is zero,

$$v(c_{\mathrm{s}}^{+} - c_{\mathrm{s}}^{-}) = 0 \tag{6.12}$$

so $c_{\mathrm{s}}^{+} = c_{\mathrm{s}}^{-}$, and setting $j = 0$ in Eq. (6.11) reveals that for steady-state conditions,

$$kT(\partial c_{\mathrm{s}}/\partial x) = fc_{\mathrm{s}} \tag{6.13}$$

Further, we assume that the boundary conditions put a limit on solvent's movement too, resulting in a steady-state concentration profile of solvent particles, with zero solvent flux. In other words, $\chi \to 1$.

Substituting Eqs. (6.12), (6.13), and $\chi \to 1$ in Eq. (6.10) leads to steady-state momentum balance equations,

$$\begin{aligned}\frac{1}{V_{\mathrm{REV}}}\frac{\mathrm{d}p^{\mathrm{solute}}}{\mathrm{d}t} &= -kT(\partial c_{\mathrm{s}}/\partial x) + fc_{\mathrm{s}} = 0 \\ \frac{1}{V_{\mathrm{REV}}}\frac{\mathrm{d}p^{\mathrm{solvent}}}{\mathrm{d}t} &= -\tfrac{\partial p}{\partial x} + kT(\partial c_{\mathrm{s}}/\partial x) = 0\end{aligned} \tag{6.14}$$

The rate of change of momentum for both solute and solvent at every position in the solution is now zero because of the steady state. We see that for steady-state conditions, a gradient in solvent pressure develops equal to the body force distribution on the solute. Equation (6.14) confirms the statement by Benedek and Villars (2000) that: '*Since the solute molecules are part of each volume element in the fluid as a whole, the semipermeable membrane in effect, is exerting a body force on the volume element of fluid located directly in front of the pore.*'

Having considered the two previous problems (transient and steady-state solute concentration profiles in a solution), we now are in a position to consider a semipermeable membrane separating the reference solution from our solution of interest (as shown in Fig. 6.1). The semipermeable membrane can be modeled as a source of time rate of change of momentum to the solute. The semipermeable membrane is effectively the same as an external force field. The only substantive difference is rather than a continuous force field, the force field is a discrete distribution of δ functions fixed in space throughout the REV, which only acts on solute particles moving left to right in our solution of interest. By definition, a semipermeable membrane as no (direct) effect on solvent particles—they can pass unobstructed through the membrane. The contribution to the time rate of change of solute momentum made by the semipermeable membrane is then $fc_s = -2m\beta vc_s^+$, where β is the collision rate of c_s^+ particles in the solution of interest with the semipermeable membrane and we assume steady-state conditions, so $c_s^+ = c_s^-$, again Eqs. (6.12), (6.13), and $\chi \to 1$ all hold. Substituting into Eq. (6.10), the result is again Eq. (6.14).

Equation (6.14) implies,

$$\frac{\partial p}{\partial x} = kT(\partial c_s/\partial x) \tag{6.15}$$

Now integrating Eq. (6.15) w.r.t. distance from a point well within the solute solution up to the surface of the semipermeable membrane finally leads to [Eq. (2.203) in Chap. 2 of Statistical Mechanics (Benedek and Villars 2000)],

$$\Delta p = p_{\text{solvent}} - p_{\text{ref}} = kT\Delta c_s = kT(c_s - c_{\text{ref}}) = \pi \tag{6.16}$$

At time zero, when the solute has just been added to the solvent, and pressure in the solution of interest is still at the reference pressure, the solvent pressure at the mouth of a pore opening in the semipermeable membrane is negative. The flow of fluid through the semipermeable membrane is given by

$$j_w = L_p(\Delta p - \pi) \tag{6.17}$$

where L_p is the filtration coefficient for the membrane. As fluid flows into the solution of interest from the reference solution, the solvent pressure in the solution of interest rises, and the flow decreases until $\Delta p = \pi$, at which point the flow ceases and $j_w = 0$.

We observe this explanation also leads to a 'rectified solvent diffusion' (i.e. a biased random walk next to the semipermeable membrane in the solution of interest). But now the altered solvent diffusion is driven by a physical mechanism—momentum exchange between solute particles and the semipermeable membrane, which results in a body force on the solvent particles and a pressure in the solvent, rather driven by a chemical potential gradient (which includes an entropic contribution to the free energy) in the solvent across the membrane.

This 'body force' in the fluid arises right at the entrance to membrane pores, which generates a pressure jump at the mouth of the membrane pore (Benedek and Villars 2000; Kramer and Myers 2012, 2013). This pressure jump in the fluid is then transmitted throughout the solvent with the greater number of osmotically active particles (i.e. throughout the left-hand side of the container shown in Fig. 1). This makes clear that (classical) osmotic pressure arises because of a 'membrane interfacial phenomena,' and so again osmotic pressure has its origins in a 'boundary phenomena.'

Indeed, given an idealized semipermeable membrane can be conceived in the 'thought experiment' described above '*as a force field that repels solute but has no effect on the solvent,*' then any physical process that confines the solute particles, but not the solvent particles, becomes a 'virtual wall' (Luo and Roux 2010), and may give rise to an osmotic pressure. For example, electrostatic fields associated with overlapping double layers in a solution may also generate a suitable 'force field' that creates a 'virtual wall' in the solution. We discuss this further in Sect. 6.4.

We mention in passing that the origin of osmotic pressure has been argued for over century and continues to be argued [see, e.g. Bowler (2017), Nelson (2017)]. At the molecular level, osmotic pressure can also be calculated using molecular dynamics simulations [e.g. Kohns et al. (2016), Luo and Roux (2010), Ilani (2015), Taherian and Mousavi (2017), Luo et al. (2013) and for a visual impression see https://www.youtube.com/watch?v=OylJ58r0X8g]. We make the observation that some molecular dynamic simulations actually apply the concept of a 'virtual wall,' directly calculating the osmotic pressure in the solvent from the force required to hold a virtual wall in place, even for concentrated, non-ideal solutions. For example, Luo and Roux (2010) explains:

> The basic idea is to introduce 'virtual' walls to represent the effect of ideal semipermeable membranes, separating a high concentration region from a pure water region. The walls keep the ions confined, but the water molecules are allowed to pass freely, permitting an equalization of their chemical potential throughout the entire system. The mean force per unit area exerted on the ions by the virtual walls during the simulations can be directly related to the osmotic pressure… The osmotic pressure is calculated directly from F/A, where A is the (constant) cross-sectional area of the system and F wall is the instantaneous force exerted by one wall onto the ions.

We observe that the method of implementing a molecular dynamics model for solvent–solute interactions in the presence of a semipermeable membrane, as described by Luo and Roux (2010), Luo et al. (2013) and Kohns et al. (2016), with their method for calculating the osmotic pressure, again confirms the analysis presented by Benedek and Villars (2000).

The principle virtue of the analysis presented above is that one can follow the all the reasoning—from the entropy change due to mixing of an ideal solution, through to the thermodynamic concept of solvent chemical potential, to the momentum balance for solvent and solute particles for both transient and steady-state conditions, and finally the specific term for the momentum interaction between the solute particles and the semipermeable membrane. One can follow the argument back the other way from the solute particles interacting with the membrane through to the entropy of mixing solute and solvent. From this perspective, we can see that the explanations offered by Denbigh (1981) and Benedek and Villars (2000) are consistent with one another: Denbigh (1981) employs thermodynamic potentials and the concept of that solvent diffusion occurs, while Benedek and Villars (2000) uses the kinetic theory of solutions and momentum balance equations to arrive at the same conclusion, though it appears Benedek and Villars (2000) have the advantage of identifying the initiating event for osmotic pressure in the solvent, as the momentum interaction between solute particles and a semipermeable membrane [or more generally for momentum interaction between solute particles (but not solvent particles) and a suitable 'virtual wall'].

6.3 Donnan Osmotic Pressure

Donnan osmotic pressure is also a mechanical pressure in the solvent, arising when there is a difference in concentration of charged (ionizable) *macromolecules* across a semipermeable membrane. Again classical Donnan osmotic pressure is calculated assuming an ideal solution, and again this is an interfacial (or boundary) membrane phenomena. But in this case, the semipermeable membrane is permeable to both solvent molecules *and* ions in solution, but the membrane is impermeable to the macromolecules.

The macromolecules may be synthetic or biological molecules [e.g. intracellular proteins, or extracellular matrix molecules (proteins or glycoaminoglycans), or they may be macromolecules found in blood plasma (e.g. albumin)]. We note for biological systems, the macromolecules tend to have a net negative charge at physiological pH (i.e. pH = 7.4.) rather than a net positive charge (though there are exceptions). If the macromolecules all have the same charge (negative or positive), they are known as polyelectrolytes, while if they have a mix of negative and positive charges (as the majority of macromolecules do) they are known as polyampholytes—so most proteins in solution are polyampholytes at some pH.

It is also important to understand how an ideal solution is defined when a solute is charged. A charged macromolecule has counterions in the solution to maintain electroneutrality of the macromolecule. The counterions 'screen' the charge on the macromolecule (reducing the spatial extent of the macromolecules Coulombic interaction potentials in the solution), but the counterions in solution also increase the apparent size of the macromolecule. The spatial distribution of counterions in the solution around the macromolecule may be described using the so-called triple-layer model. Conceptually, the triple-layer model suggests that counterions that are very strongly attracted to the charged molecule are 'adsorbed' onto the charged surface, which are described as 'inner sphere' and then 'outer sphere' complexes, but when the counterions are less strongly attracted to the charged surface, they are found in a 'diffuse double layer' (DDL) (Bourg and Sposito 2011). The diffuse double layer is like a diffuse ion 'atmosphere' around the charged particle, with more counterions near the charged surface, and fewer further away (http://www.feynmanlectures.caltech.edu/II_07.html). The distribution of ions next to charged surface is analogous to the earth's atmosphere (assuming constant temperature), which is more dense closer to the earth's surface.

To understand how low the concentration of macromolecules needs to be before the solution of ionized macromolecules can be considered ideal, we need to understand more about ion distributions at charged surfaces, and so the 'effective size' of the molecule in a solution. For example, to estimate the size of the molecule, we need to know whether or not we can use a so-called mean field theory. Once we know the effective size of the ionized macromolecule, can use this size to estimate the average distance between molecules, and so define the range of concentrations where the distance between macromolecules is sufficiently large for the solution to be considered dilute. To this end, let us briefly review the necessary background to estimate the effective size of charged macromolecules.

How close an ion resides to the surface of a macromolecule depends on the magnitude of the Coulombic attraction potential between the charged molecule and counterions in solution, relative to the thermal energy of the molecules in the solution. We can quantify this ratio using the so-called Bjerrum length (defined in standard units as $l_B = e^2/4\pi\varepsilon\varepsilon_r kT$). The Bjerrum length is the distance over which two interacting elementary charges have an electrostatic energy (at least) equal to the background thermal energy. We note that with water as the solvent, at 298 °K, the Bjerrum length is about 0.71 nm.

Clearly if the distance between charges is less than the Bjerrum length, then Coulombic interactions between macromolecules and mobile counterions tend to dominate the background thermal energy, and so 'pairwise interactions' (between a point charge on the macroion and an ion in solution) become prominent. Because thermal collisions have comparatively less effect on such strong molecular interactions, counterions are more likely to be closely associated with the charged surface (or increasingly, associated with each other at the surface). Ions in this state are described as being 'adsorbed' onto the charged surface. Some polymer textbooks refer to this general state of close association of macromolecule and counterions as 'counterion condensation', while others textbooks refer more specifically to adsorption of counterions in a triple layer, as described above (i.e. forming inner and outer sphere complexes in the triple layer). In any case, when the distance between charges is much greater than the Bjerrum length, then thermal energy tends to dominate the Coulombic pair interactions, and so Coulombic pair interactions are strongly disrupted by the averaging effect of thermal collisions. This thermal agitation works against Coulombic-driven spatial correlations between ions developing in the solution, tending to average out ion concentrations, so they are more 'randomly' distributed. It is the dominance of thermal collisions that result in counterions and co-ions being in a 'diffuse double layer' (i.e. the third layer of a triple-layer model of ion adsorption).

When thermal agitation dominates, it turns out that the concentration of counterions in the diffuse double layer is not randomly distributed, but conforms to a 'Boltzmann distribution' (i.e. the probability of finding counterions in a particular energy state varies exponentially as the negative of the energy divided by $k_B T$). To understand practically what this distribution means in terms of conceptualization of ion adsorption, we now need to estimate the 'thickness' of a diffuse double layer. The thickness of the double layer is defined as a 'Debye length' (in standard units)

$\kappa^{-1} = 1/(\sqrt{8\pi\lambda_B N_A I})$, where λ_B is the Bjerrum length, N_A is Avogadro's number and 'I' is the ionic strength of the solution in moles ($I = 0.5\sum_i \bar{c}_i z_i^2$). The Debye length is the distance over which the electric potential in the solution decreases to $1/e$ (about 37%) of the electric potential at the charged surface. So for dilute solutions, we find that is about two-thirds (or more precisely 63%) of all the counterions are within one Debye length of the charged surface.

For a 1:1 electrolyte at 37 °C with water as solvent, the Debye length (in nanometers) equals $0.316/\sqrt{I}$. So for a solution of ionic strength 10^{-6} M, the Debye length (or the 'double-layer thickness') is about 300 nm. We see in this case $\kappa^{-1} \gg \lambda_B$ (i.e. the double-layer thickness is much greater than the Bjerrum length). This means nearly all the counterions will be in the DDL, rather than sorbed to the surface as inner and outer sphere complexes. However, we note that for a solution of ionic strength 0.15 M (isotonic saline), the Debye length (or the 'double-layer thickness') is about 0.82 nm.

When the DDL dominates ion adsorption, we can then use a 'mean field theory' to calculate the ion distributions around the charged surface on the macromolecule. Mean field theories express the statistically averaged behavior of a large number of discrete particles (an example of a mean field quantity are chemical potentials associated with a collection of particles) in terms of only their mean concentrations and their mean electrical potential in the solution. Classical Gouy-Chapman (1910–1913), Debye-Huckel (1923), and so-called DLVO theory (1941–1948), and many subsequent theories describing interactions between colloidal particles are all based on mean field theory—and all use the mean field Poisson–Boltzmann equation (Che et al. 2008) https://en.wikipedia.org/wiki/Poisson%E2%80%93Boltzmann_equation.

However, when the Debye length is less than the Bjerrum length, then mean field theory may no longer be sufficiently accurate (we note that for conditions of isotonic saline at 37 °C, the DDL length is about the same size as the Bjerrum length, so we might expect mean field theory to start to break down). To be more accurate, we then need to define individual chemical potentials for discrete molecules. We do this by summing the pairwise interactions between the molecule of interest and all the species with which they interact in the solution. To calculate electrochemical potentials of individual molecules, we need to take an integral over a volume, from the molecular boundary of the central molecule up to an infinite distance, with the integrand taking into account the interaction energy with distance between any two particles in the volume, and the statistical probability that the second particle is at a given radial distance from the central molecule. The probability of the second molecule being at some distance r from the central molecules is called the radial distribution function for the second molecule around the central molecule. We note that for an ideal gas, the radial distribution function is unity, representing complete mixing. For an introduction on how to do these calculations on discrete molecules, see Bazant http://ocw.mit.edu/courses/chemical-engineering/10-626-electrochemical-energy-systems-spring-2014/study-materials/MIT10_626S14_S11lec28.pdf.

Calculations to determine chemical potentials of individual molecules are often done using molecular dynamics simulations. Molecular dynamics simulations use spatial derivatives of interaction potentials to calculate forces between molecules at a distance, and Newtonian mechanics to update their positions in the solution. The power of molecular dynamics simulations is that they can approximate mean field theories, but in addition they can take into account statistical correlations that may develop in solution when Coulombic forces dominate thermal forces. The main limitations of molecular dynamics simulations are: (i) the interaction potentials proposed in the simulation may deviate from the actual interaction potentials (we note that proposed interaction potentials are typically 'adjusted' so that predicted radial distribution functions agree with experimental neutron scattering data on radial distribution functions), (ii) computational limitations associated with modeling large particle systems usually mean the number of particles in the simulation has to be small, and the duration of the simulations is very short to be computationally tractable and, (iii) numerical limitations—on some occasions the net forces acting on particles are the difference of two very large numbers (e.g. hydrogen bonding in water creates a very large attractive pressure while thermal energy creates a very large repulsive pressure), yet the properties of water depend on this net force (i.e. the difference of two very large numbers), which creates numerical errors. Nevertheless for small particle systems, this approach is both feasible and revealing. For example, in the paper by Bourg and Sposito (2011), one can visualize how molecular dynamics simulations can make clear the details of triple-layer adsorption of counterions near a charged surface (Bourg and Sposito 2011). We note that some recent papers have attempted to model GAGs in cartilage using full atomistic molecular dynamics simulations, but further work appears required.

So to define an ideal solution for macromolecules, we first need to make sure that mean field theory is appropriate to use (this implies our ideal solution must have a small salt concentration, ensuring that the DDL thickness of counterions surrounding macromolecules is large compared to the Bjerrum length). Then we can use a mean field theory to calculate the effective size of the macromolecules, taking into account the DDL thickness. The effective size may be the molecular radius plus some multiple of the Debye length (to ensuring double-layer overlap is minimal). We can then use this effective size to define the maximum molar concentration of a dilute solution, so that the macromolecules and associated counterions have plenty of space in which to diffuse unimpeded by neighboring molecules in the solvent (e.g. so the volume for each macroion is say, an order of magnitude larger than the macromolecule radius and DDL thickness). Under these conditions, a solution of charged macromolecules approaches the behavior of an ideal solution. In principle, we can check how close a solution approaches an ideal solution by comparing the measured osmotic pressure with the theoretical osmotic pressure $\pi_{os} = RT(N/V) = RT\sum c_i$, where N is the sum number of all osmotically active particles and V is the solution volume.

For such an ideal solution, the Donnan osmotic pressure is given by exactly the same expression as for osmotic pressure in the Section preceding (based on entropy related free energy changes associated with mixing of solvent and solute), providing all osmotically active particles on each side of the semipermeable membrane are taken into account (note that both ions and counterions in solution are taken into account, and they redistribute to minimize their Gibbs free energy while maintaining electroneutrality). The ideal Donnan osmotic pressure can also be expressed in terms of the macromolecule (e.g. protein) concentration in the solution (c_p) in moles and its valence z. For a charge protein in a 1:1 salt solution at concentration (c_s) in moles, the ideal Donnan osmotic pressure (in kPa) is then given by Philipse and Vrij (2011):

$$\pi_\mathrm{D} = RT\left(c_\mathrm{p} + 2c_\mathrm{s}\left[\left(1 + y^2\right)^{(1/2)} - 1\right]\right); \quad y = \frac{zc_\mathrm{p}}{2c_\mathrm{s}} \tag{6.18}$$

where π_D is the Donnan osmotic pressure and R = 8.31 (L kPa/K^{-1} mol^{-1}). We note this equation assumes the volume of salt reservoir is very large compared to the volume of the macromolecular solution on the other side of the semipermeable membrane.

One interesting thing is that the presence of the macromolecules causes 'salt depletion' in the solution containing macromolecules (Philipse and Vrij 2011). Salt depletion refers to the fact that upon addition of the macromolecules to one of the reference solutions, which is initially on both sides of the semipermeable membrane, many electroneutral pairs of ions cross from the solution now containing the macromolecules to the reference solution. This means the osmotic pressure exerted by the addition of the macromolecules to the solution is not simply the sum of all macroions and counterions added to the solution (the Donnan osmotic pressure is actually much less due to salt depletion).

Another interesting thing is that even though the membrane is completely permeable to the ions in solution, the ions still contribute to the Donnan osmotic pressure, as they are not uniformly distributed on both sides of the semipermeable membrane. The reason this occurs is that (positive) 'counterions' in solution are in fact electrostatically 'attached' to the (negatively) charge macromolecules, so even though the membrane is permeable to ions, they are still restrained in their movement by their interaction potential with the macromolecule, while the charged macromolecule itself cannot pass though the membrane and is confined to one side.

'Donnan osmotic pressure' like 'standard osmotic pressure' is due to the presence of pores in the membrane that allow solvent and some solutes to pass, but not larger solutes. Donnan osmotic pressure for an ideal solution is again clearly an interfacial membrane or boundary phenomena, for as stated by Ilani (2015):

> The numerical analysis of several Donnan systems showed that in the concentration range of a fraction of a millimolar and above, the total change in voltage, ion density and pressure occurs within the [membrane] pore, less than a few hundred Å from the interface.

In other words, the pressure jump in the fluid occurs at the 'mouth' of the pores, or very close to the semipermeable membrane.

Ilani (2015) goes on to say:

> The pressure that develops in a Donnan system is equal to the difference between the osmotic values of the two solutions [i.e. $\pi_\mathrm{D} = RT^*(C_1 - C_2)$] .... Thus, a Donnan system seemingly challenges the concept that for a membrane to be able to develop a pressure that

> matches the full difference in osmotic values between two solutions, it must be an ideal semipermeable membrane. Instead, for a membrane to develop the full difference between the osmotic values of the two solutions, it is necessary only that there is no net solute flow across it. This is the case when the membrane is ideally semipermeable and there is no flow of solutes due to impermeability, or in a Donnan system where no net flow of solute occurs due to the absence of a [electrochemical] driving force.

Perhaps the simplest way to think about Donnan osmotic pressure is again considering a 'virtual wall.' Because the pores in the membrane exclude charged macromolecules, the pores also exclude the ions responsible for maintaining electroneutrality of the charged macromolecules. In other words, there will be no net flux of counterions associated with macromolecules across the membrane, even though all ions are free to cross the membrane (in pairs, so as to maintain electroneutrality). The electrical attraction between macromolecules and their counterions result from the electrical 'force field' acting on the counterions, creating a virtual wall that is just as effective as if the physical semipermeable membrane acted on the macroion counterions.

Alternatively, we may conceive of an attraction between the counterions and the electrical field throughout the solution (i.e. the so-called Donnan potential), which is generated by the difference in 'smeared fixed charge' (representing the smeared effect of charged macromolecules in solution) on either side of the membrane. This Donnan potential results in counterions that try to cross the membrane transferring their momentum to the fixed charge in solution, via electric field local perturbations in the Donnan potential across the membrane (in this case, the 'virtual particles' or 'gauge bosons' transmitting electromagnetic interactions are photons https://en.wikipedia.org/wiki/Gauge_boson). This is interesting, and we again realize that abrupt changes in spatial gradients of electric fields in solution may physically act like an ideal semipermeable membrane. Either way, the 'virtual wall' causes a pressure jump at the entrance of the pore. This fluid pressure jump at the boundary is then transmitted throughout the solvent.

6.4 Solid Stress or Effective Stress

'Solid stress' is the stress between solid particles that is capable of maintaining similar spatial relationships between the particles as the material deforms under load. 'Solid stress' is a term synonymous with several other terms used in different disciplines to specify the same quantity, with the same idea recurring in the course of different experiments in different disciplines. The term 'effective stress' is used in soil mechanics and geotechnical engineering, 'swelling pressure' is used in soil science (and often in biology), and 'disjoining pressure' is used in colloid science. So how does Donnan osmotic pressure relate to 'solid stress'? The Donnan osmotic pressure arises when charged molecules in an ideal solution are trapped by a semipermeable membrane. When ideal assumptions are no longer valid, we clearly need more sophisticated theories to take into account solution non-ideality.

As the concentration of charged solute species increases, ideality is lost for a number of reasons. One important reason may be that concentration of macromolecules is sufficiently high so that diffuse double layers around the charged particles always overlap significantly, and so the molecules can no longer move in the solution independently of one another, as assumed for an ideal solution. At sufficiently high particle concentrations, instead of the macromolecules being freely mobile in the solution, they can become trapped in a lattice-like structure.

A lattice-like structure in the solution may also arise by decreasing the ionic strength of the solution, and in doing so 'unscreen' the fixed charge (i.e. increasing the range of electrostatic interactions). For example, Horkay et al. reports that chondroitin sulfate molecules in solution become more 'ordered' (and so more lattice-like) as the ionic strength of the solution decreases and the Debye length of the double layer increases (Horkay et al. 2012). This ordering is partly initially driven by the stiffening of the aggrecan core protein and GAG chains, as their fixed charged is unscreened. But at higher ionic strengths, which result in significant charge screening, the chondroitin sulfate GAG molecules are still highly extended and so still rod-like, due to the 'intrinsic stiffness' of covalent linkages in the backbone chain (Bathe et al. 2005). For a 0.15 M solution, based on the measured radius of gyration (Fig. 7b in Bathe et al. 2005), the chondroitin sulfate GAG chains are 70% of the way to becoming 'rigid rods' rather than 'random coils'—but based on the osmotic pressure generated by chondroitin sulfate, they are hardly different from rigid rods (Fig. 7a in Bathe et al. 2005), i.e. the conformational freedom they do have makes little contribution to the osmotic pressure.

Of course in a complex real extracellular matrix environment, additional positional constraints on macromolecules may also be imposed by other extracellular matrix molecules in solution. For example, chondroitin sulfate chains are partially structured by their attachment to aggrecan core protein. Likewise, binding of aggrecan to hyaluronic acid molecules imposes an additional structure. Overlap between aggrecan molecules imposes yet further positional constraints. In addition, aggregate molecules are embedded in a collagen network, which imposes significantly more structure, while some macromolecular molecules that link aggrecan molecules together impose further positional constraints. Through these mechanisms, we see that aggrecan molecules in articular cartilage tissue at physiological concentrations are confined in a more or less a lattice-like structure. Such structures are sometimes described as (extracellular matrix) gels https://en.wikipedia.org/wiki/Gel.

Even though these 'particles' in a lattice do not physically touch one another, they transmit a force from molecule to molecule, which gives the lattice an 'elasticity' and a 'finite shear strength' (Russel 1980; Benzing and Russel 1981; Russel and Benzing 1981). In their paper titled *Viscoelastic properties of ordered lattices*, Russel and Benzing explain (1981):

> The mechanics of ordered suspensions [e.g. latex spheres in FCC or BCC structures] are equally interesting because of the elasticity imparted by the screened electrostatic forces between particles. Disordered suspensions of spheres generally behave as non-Newtonian fluids with a Newtonian shear limit. With the onset of order, however, the low shear

viscosity becomes infinite, indicating that a finite stress must be applied to initiate steady flow. The suspension then behaves as a linear viscoelastic solid with a nonzero static shear modulus and a low dynamic viscosity, which allows the suspension to propagate low frequency shear waves, and high osmotic pressure.

The lattice ordering becomes more and more apparent as interaction potentials between macromolecules become more significant than thermally induced Brownian motion (Russel 1980). The fact that a finite stress must be applied to initiate steady (fluid) flow is diagnostic that the suspension is behaving, at least initially, as a linear viscoelastic solid (or often more accurately, as a linear poroelastic solid, which is a particular kind of viscoelastic solid). Similar behavior is reported for concentrated aggrecan-aggregate solutions (Meechai et al. 2001). Given these experimental facts, it becomes clear that for such non-ideal 'solutions,' we could define material properties for a 'poroelastic solid.' For example, we could define solid material properties like moduli [e.g. Young's modulus (*E*), shear modulus (*G*) and bulk modulus (*K*)] for the 'ordered lattice structure.' These material properties can then be used in constitutive equations to relate stresses with strains in the ordered lattice, as for any solid material. We have seen previously that persistently overlapping double layers in a non-ideal solution may lead to 'solid stress' within the solution, assuming 'fixity' of the macromolecules in space, and then like any solid, we can then relate this stress to the strain in the non-ideal 'solution,' and so define various moduli.

Again we note that different disciplines use different terminology. So for porous media, we relate 'effective stress' (denoted σ' where the prime denotes effective stress) to strain, while Russel (1980) relate osmotic pressure (π) to strain. Now as Barclay et al. (1972) point out clearly (e.g. in Fig. 2 and accompanying discussion in their paper), the so-called osmotic pressure *between* two charged parallel plates with counterions in the solution between two plates is defined by Barclay et al. (1972):

$$\pi_{\text{os}} = \frac{\partial A}{\partial \varepsilon_{ii}} = -\sigma'_{ii} \tag{6.19}$$

where A is the (Helmholtz) free energy of two plate system as they are brought together from a distance apart, and ε is the strain (in Barclays' case, strain in one spatial dimension). This 'osmotic pressure' is exactly the same as the 'effective stress' employed in porous media theory for clays (which are typically charged plate-like particles). Barclay et al. (1972) point out this same stress is named the 'disjoining pressure' by Derjaguin (who represents the 'D' in so-called DLVO theory) in two papers published in 1955 and 1964, and indeed Derjaguin continues to use this terminology in multiple later papers on the same topic, e.g. Derjaguin and Churaev (1977), as does the colloid science community. So osmotic pressure generated by overlapping double layers referred to here is known as the 'effective stress' or 'solid stress' in the solid mechanics and geomechanics literature (while

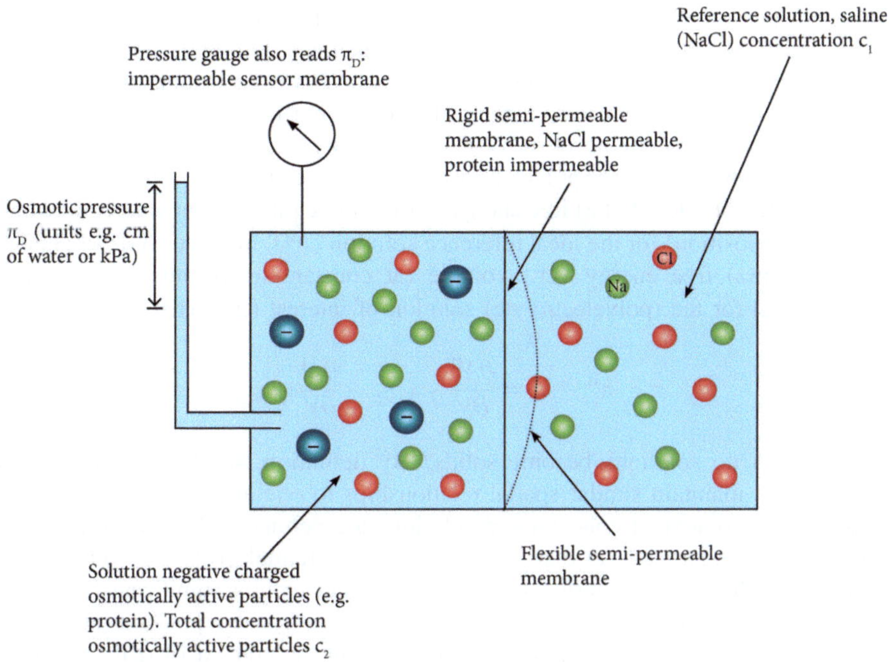

Fig. 6.2 Donnan osmotic pressure arises when there is a difference in osmotically active particles [i.e. macromolecules (blue), Na ions (green), and Cl ions (red) in solution] across a semipermeable membrane, but in this case, water and ions in solution are all free to move through the membrane pores—only the (usually negatively charged) macromolecule is not. In Fig. 6.2, the macromolecule is free to move in the solution on the left-hand side. For Donnan osmotic pressure, ions redistribute in response to the negatively charged protein on the left-hand side. At equilibrium, electroneutrality is maintained and so only pairs of ions can cross the membrane. AT equilibrium, an equal probability of ions pair crossing in either direction across the membrane, implies that the product of ions concentrations on each side of a 1:1 salt solution must be equal. This results in difference in concentration of osmotically active particles on either side of the membrane, and so at infinite dilution $\pi_D = RT^*(c_1 - c_2)$. The Pitot tube measures the mechanical fluid pressure in the left-hand side compartment. The pressure gauge has an impermeable sensor membrane, and so measures the total stress (which is equal to the solid stress minus the fluid pressure, i.e. $\sigma_T = \sigma^s - p$). Because of the infinite dilution of macromolecules in this example, the pressure gauge measures the fluid pressure (as solid stress is zero). Figure drawn by Yvette Harrap

the second derivative of the Helmholtz free energy with respect to strain is a deformation modulus).

Barclay et al. (1972) also point out that this 'osmotic pressure' (i.e. the effective stress) itself may be comprised of a number of different components including Born repulsion (hard-core repulsion occurs when two particle physically collide with one another and electron clouds begin to overlap), electrostatic repulsion, van der Waals attraction, steric, configurational, excluded volume, conformational contributions as well as mechanical pressure due to frictional contact forces (which may be one source of hysteresis upon repeated loading and unloading) (Barclay et al. 1972).

But often one or other of these interactions dominate a regime of behaviors. For many practical applications, electrostatic interactions dominate, and then it is commonplace as Bathe et al. (2005) and Kovach (1995) point out, to distinguish between charge dependent and charge independent portions of osmotic pressure.

We note here that Bathe et al. (2005), Kovach (1995), and others define the osmotic pressure (π) in the (polyelectrolyte) solution of interest to be equal to the rate of change of (Helmholtz) free energy w.r.t volume (at constant temperature and particle composition) for the ideal reference solution (π^{II}), minus the rate of change of (Helmholtz) free energy w.r.t volume (at constant temperature and particle composition) for the (polyelectrolyte) solution of interest (π^{I}). That is,

$$\pi = \pi^{\mathrm{II}} - \pi^{\mathrm{I}} = \frac{\partial A^{\mathrm{II}}}{\partial V}\Big|_{T,N,\mu_{\mathrm{w}}} - \frac{\partial A^{\mathrm{I}}}{\partial V}\Big|_{T,N,\mu_{\mathrm{w}}} \tag{6.20}$$

But when do solutions become solids? By definition, solids are made up of particles that maintain similar spatial relationships to one another as they deform. Solids in the context of concentrated solutions are defined to be (macro)particles sufficiently concentrated, so that their interaction potential energy is significantly greater than the thermal energy (*kT*), so they do not exhibit significant Brownian motion in the solution and may therefore be regarded as fixed in position (at least on the timescale of interest to the investigator). But in the context of real tissues, as mentioned above, the degree of fixity does not just depend on the macromolecule concentration, but also on other extracellular matrix molecules. For example, polyelectrolytes like aggrecan may be structured by surrounding collagen network. Clearly molecules making up the collagen network maintain similar spatial relationships with one another due to covalent linkages, and so are usually chosen as the frame of reference for 'solids.' Additional molecules in the extracellular matrix that remain in a defined volume attached to the collagen network may also then be regarded as fixed, as they will transfer load directly to the collagen network to maintain their position in a defined volume as it deforms. While large molecules like chondroitin sulfate and aggrecan may be more or less fixed within the defined volume, it is important to realize that small molecules in solution, both uncharged (e.g. sugar molecules) and charged (e.g. ions, small proteins), are free to exhibit unrestricted Brownian motion in the solvent, even while the Brownian motion of the macroparticles is significantly curtailed.

To define the relevant 'osmotic pressure' for a complex material like cartilage tissue, Eq. (6.20) says we need to take the difference in the rates of change free energy w.r.t volume of two solutions, while the solid particle number does not change, and the solvent chemical potential remains the same in both the solution and reference solution. Our interest in this seemingly odd set of conditions stems from the fact this is precisely what happens to real tissues in the body (e.g. cartilage). The reference solution for cartilage is synovial fluid. Cartilage extracellular matrix is initially a semidilute collection of macromolecules and salt in a collagen network, and as it becomes more concentrated, what limited motion the macromolecules have, becomes even further curtailed as more pore fluid drains from the articular cartilage tissue as the tissue is compressed and reduces its volume.

For an ideal solution, the free energy associated with mixing solute and solvent is relatively easily captured by a simple model, but in non-ideal solutions there are many more effects that change the system's free energy. For example, Nap and Szleifer (2008) reports a model that has sixteen interactions influencing the free energy of the system. Nevertheless for dilute and semidilute solutions of charged colloids (e.g. albumin) and charged polyelectrolytes (e.g. chondroitin sulfate solutions, gels, and cartilage tissue), there is frequently a very important region of practical interest where electrostatic interactions dominate the system osmotic pressure. When electrostatic repulsion dominates such colloidal systems, two particles that maintain the same spatial relationship need not physically touch one another, yet transmit a mechanical force between them. Indeed the particles are held apart by a mechanical (osmotic) pressure in the fluid *between* the particles. In other words, a (local) osmotic pressure is generated in the fluid *between* particles. It then becomes apparent that for real heterogeneous solution systems involving charged macromolecules with some degree of fixity, the osmotic pressure in the solvent is spatially variable. In contrast for a simple dilute solution, random fluctuations in the osmotic pressure in the solution help drive Brownian motion of the particles (i.e. higher particles concentrations are pushed apart as solvent moves toward regions with higher osmotic pressure), but these fluctuations in osmotic pressure can never depart very far from the mean solution osmotic pressure, in contrast to particles that are nearly fixed, where spatial variations in osmotic pressure in the solution may then become large.

Spatial variations in osmotic pressure do not only occur in non-ideal solutions of charged macromolecules. We mention in passing that spatial variability of osmotic pressure can also arise in dilute solutions of particles with different sizes. Such a variation in osmotic pressure explains the 'depletion force' (which tends to cause flocculation of some larger colloidal particles). The depletion force is caused by local variations in osmotic pressure, which arise from small scale spatial variations in the concentration of smaller particles due to an 'excluded volume' effect of the larger particles in the solution https://en.wikipedia.org/wiki/Depletion_force (Singh-Zocchi et al. 1999). In other words, the larger particle creates a barrier around a volume of fluid that the smaller particles cannot enter. This barrier or 'virtual wall' has the same effect as if a microscale semipermeable membrane surrounded the tiny excluded volumes.

For the case of a solution containing 'fixed charged macromolecules' (together with their very much 'unfixed' counterions in overlapping diffuse double layers), how can we calculate the spatial variation of the effective pressure in the solvent?

It turns out that when a linearized Poisson–Boltzmann theory applies (e.g. a particle surface voltage is not too large, and the particles are bathed in 1:1 electrolyte such as isotonic saline), the effective pressure in the solvent at any location in the double layer can be calculated from difference in the ideal osmotic pressure [the ideal osmotic pressure that is predicted by a van 't Hoff model in the absence of an electric field, and equal to $RT \sum c_i$, which can be more conveniently expressed in terms of the mean field (electric) potential, ϕ, as $\frac{1}{2}\varepsilon_0\varepsilon_r\kappa^2\phi^2$], and the 'Maxwell stress' in the diffuse double layer (Carnie et al. 1994), viz,

$$P_{ij} = \frac{1}{2}\varepsilon_0\varepsilon_r\kappa^2\phi^2 I - \sigma_{ij}^{el} \tag{6.21}$$

where P_{ij} is a stress tensor and σ_{ij}^{el} is a Maxwell stress tensor, which may be defined as $\sigma_{ij}^{el} = \varepsilon_0\varepsilon_r E_i E_j - 0.5\delta_{ij}\varepsilon_0\varepsilon_r E_k E_k$ (Grodzinsky 2011). I is the identity matrix and the electric field strength is $E_i = -\nabla\phi \cdot e_i$.

The stress tensor P_{ij} is referred to by various names in the double-layer literature, for example, it is called the 'total stress' in Stankovich and Carnie (1996), the 'total pressure' in Deserno and Holm (2001) and the 'full stress tensor' in Lu et al. (2005). The Maxwell stress tensor has units of momentum per unit of area per unit time (N/m^2) and gives the flux of momentum parallel to the ith axis crossing a surface normal to the jth axis (in the negative direction) per unit of time. In matrix notation, the components of the Maxwell stress can be expressed as,

$$\sigma_{ij}^{el} = \begin{bmatrix} \frac{1}{2}\varepsilon_0\varepsilon_r(E_x^2 - E_y^2 - E_z^2) & \varepsilon_0\varepsilon_r E_x E_y & \varepsilon_0\varepsilon_r E_x E_z \\ \varepsilon_0\varepsilon_r E_x E_y & \frac{1}{2}\varepsilon_0\varepsilon_r(E_y^2 - E_z^2 - E_x^2) & \varepsilon_0\varepsilon_r E_y E_z \\ \varepsilon_0\varepsilon_r E_x E_z & \varepsilon_0\varepsilon_r E_y E_z & \frac{1}{2}\varepsilon_0\varepsilon_r(E_z^2 - E_x^2 - E_y^2) \end{bmatrix} e_i \otimes e_j \tag{6.22}$$

It is apparent that the total electromagnetic forces (for static conditions) represented by the Maxwell stress tensor above have both normal pressure components and shear stress components, i.e. Maxwell stresses can sustain both normal and shear loadings.

We mention that the Maxwell stress in the solution representing the '*the electrical forces 'on the fluid,' are actually exerted on the solute ions; the latter transfer their force to the fluid*' (Grodzinsky 2011). See further discussion in Sect. 6.2, which explains how the presence of a solute particles 'reflected' from a semipermeable membrane realize an osmotic pressure in the solvent, again by the reflected solute particles transferring 'their force to the fluid.'

To calculate the total force on a charged particle (f_{particle}) (e.g. a macroion, polyelectrolyte, or other solid particle), one can evaluate the integral,

$$f_{\text{particle}} = \int_A P_{ij} \cdot \hat{n}\, dA \tag{6.23}$$

where the integral is taken over the particle's surface. We mention that if a group of particles are on the surface of a body, and the only forces within the body are those associated with repulsive double layers, then the sum of the particle forces calculated using Eq. (6.23) (i.e. $\sum f_{\text{particle}}$) divided by their (gross) surface area, is the surface traction acting on the body. The surface traction is the applied mechanical load on the surface of the body, which is resisted by the repulsive forces between fixed-charged particles bathed in an ionic solution [see Eq. (7.34) for a definition of surface traction].

To calculate the osmotic pressure change between any two points (p_1, p_2) in the solution, one can evaluate the line integral,

$$\Delta\pi = \int_{p_1}^{p_2} (\nabla \cdot P_{ij}) \cdot \hat{t} \mathrm{d}S \tag{6.24}$$

where $\hat{t}$ is the unit vector tangent to the line curve S.

Interestingly, we can now interpret the concept of a 'virtual wall' (as mentioned in Sect. 6.2) as it relates to double layers associated with polyelectrolytes in a solution. The 'virtual wall' is created by a nonzero divergence of P, as this term generates a gradient in the osmotic pressure. We see the divergence of P acts like a (net) body force acting on the charged particles in a solvent (much as the semipermeable membrane acts on particles in a solvent). For the linearized Poisson–Boltzmann theory, the divergence of P can be calculated from Eq. (6.21) as,

$$\nabla \cdot P_{ij} = \frac{1}{2}\varepsilon_0\varepsilon_r\kappa^2\nabla\phi^2 I - \nabla \cdot \sigma_{ij}^{el} \tag{6.25}$$

We can now define a 'virtual wall' around a spatial region caused by overlapping double layers as simply an (arbitrary) magnitude in the maximum local gradient of the osmotic pressure calculated using the osmotic pressure, defined by Eq. (6.24). We note a semipermeable membrane may create an osmotic pressure gradient on a length scale measured in nanometers, and indeed overlapping diffuse double layers associated with pairs of charged macromolecules in close proximity can result in equally large variations in osmotic pressure over a similar length scale.

For a 'free diffuse double layer' (i.e. a double layer that does not overlap with another double layer), the (osmotic) solvent pressure in the double layer does not rise [see the example in Sect. 4.7 of Grodzinsky (2011) for an illustration of a free diffuse double layer]. The (osmotic solvent) pressure does not rise because the gradient in $RT\sum c_i$ (i.e. the gradient in the ideal osmotic pressure) and the gradient in the divergence of σ_{ij}^{el}, exactly cancel one another, and so the solvent pressure in a free diffuse double layer is equal to the background solvent pressure. In other words, for an isolated charged plane surface in a solvent (which is a good approximation of most surfaces given sufficient proximity to the particle surface), the body force in the fluid arising from the divergence of the Maxwell stress component acting in the direction normal to, and pointing away from, the fixed charged particle surface, exactly counters the (hydrostatic) body force in the fluid due to the increasing (total) ion concentration acting in a direction normal to, and pointing toward the fixed charge particle surface. So an isolated fixed charge with counterions in solution (necessary to maintain electroneutrality of the particle with fixed charge) has no influence on the spatial distribution of osmotic pressure in the solvent in a solution. This follows from the linearized Poisson–Boltzmann theory. We can see this by evaluating Eq. (6.25), which shows

$$\nabla \cdot P_{ij} = \kappa^2 \phi - \nabla \cdot \nabla \phi = 0 \tag{6.26}$$

everywhere within a free double layer, i.e. so $\pi =$ constant [e.g. say the value of π at p_1 in Eq. (6.24)].

However when double layers on fixed charge overlap, and where the divergence of the Maxwell stress is zero, the ideal osmotic pressure $RT\sum_i c_i$ is realized in the solvent. The divergence of the Maxwell stress is usually zero somewhere between two charged particles—for example, along the mid-plane between two identical long parallel charged plates, where E equal zero [this case is considered in Sect. 4.8 of Grodzinsky (2011)]. It is this osmotic pressure in the solvent that supports load transfer throughout the solution between fixed charged particles. For the case of particle fixity (which is defined to occur when the interaction potential energy between particles is significantly greater than the thermal energy (kT), the osmotic pressure in the solution between the fixed charge macromolecules may be regarded as a 'solid stress,' even though the solid stress passes through the solvent in the solution between the solid particles (with their fixed charge). See further discussion on solid (or effective) stress in Sect. 7.2.1.

We also mention that for the case of long parallel charged plates that have an overlapping double-layer [e.g. the example shown in Sect. 4.8 of Grodzinsky (2011)], the 'excess' ion concentration in this overlapping diffuse double-layer can be defined to be that ion concentration over and above the ion concentration found when the Maxwell stress is zero. These 'excess ions' behave like ions in a free diffuse double layer (i.e. these ions are rendered unable to create an ideal osmotic pressure in the solution as they are prevented from doing so by the presence of the body force arising from divergence of the Maxwell stress in the solution), and so these 'excess ions' do not contribute to any change in effective osmotic pressure with position throughout the region between the flat plates. In other words, the effective osmotic pressure π is constant between (infinitely wide) parallel charged plates that have an overlapping double-layer [see the example shown in Sect. 4.8 of Grodzinsky (2011), or see Eq. (2) in Adar and Andelman (2018)]. Employing Eq. (6.24), we can check this by evaluating the change in osmotic pressure between any two points between the charged plates, which again yields the result shown in Eq. (6.26).

However if the charged parallel plates are of finite width (rather than of infinite width), the only place the osmotic pressure varies significantly is around the 'edge' (or 'sides') of the (nearly) constant pressure region between the finite width plates. In this 'edge region,' the osmotic pressure changes from its (nearly) constant (elevated) level in the region between the plates, back to the background solvent pressure in the surrounding solution. The rate of change of osmotic pressure in this edge region is determined by the magnitude of the body forces [see Eq. (6.25)] that appears in the solvent around the edge region of the (nearly) constant pressure region between the plates. In this edge region,

$$\nabla \cdot P_{ij} = \kappa^2\phi - \nabla \cdot \nabla\phi = f_i^{\nabla P} \neq 0 \tag{6.27}$$

The (net) body forces in the solution within the edge region $\left(f_i^{\nabla P}\right)$, point toward the (nearly) constant pressure region between the plates. Integrating $f_i^{\nabla P}$ along a line from a point in the background solution (or reference solution) to a point between the plates gives the osmotic pressure difference between the background solution and the point between the plates [see Eq. (6.24)]. The value of this line integral is independent of the path taken between the two points in the solution.

If the width of the charge parallel plates is narrow in relation to their separation distance, then there may be no constant pressure region across the width, and then the 'edge region' extends across the width of the plates, i.e. the osmotic pressure varies continuously across in the direction of the plate width. And finally, if the two plates have a heterogeneous charge distribution on their surfaces, or they are inclined at an angle to one another, or are 'rolled up' to form cylinders, or some combination of these, then the osmotic pressure between the plates/cylinders will vary from point to point between the plates/cylinders, as $f_i^{\nabla P}$ will be a vector field defined over the whole solution domain. Again, the variation in osmotic pressure between any two points in the solution can be evaluated using Eq. (6.24), while the repulsive force between any two like-charged particles can be found by integrating the osmotic pressure along a line normal to the line joining the centroids of the particles.

Now given that the spatial distribution of fixed charge on macromolecules in real systems is heterogeneous, our virtual wall (which could be imagined to be like a 'nanoscale semipermeable membrane') forms a complex 3-D surface in space enclosing a volume (or forms a closed curve along any cut-plane through the region). It is interesting to consider how the size of the volume enclosed by the 3-D surface may vary. Clearly in dilute solutions that has heterogeneously distributed fixed charge with variable diffuse double-layer overlaps, a closed surface may contain only a single pair of fixed charges that are close together, and so the enclosed volume would be relatively small. But as the spatial density of the fixed charge grows, so more double layers may overlap contiguously, and the enclosed volume grows. In real tissues at rest, with heterogeneous distributions of fixed charge at low density (e.g. in such tissues of the body as skin, sclera, arteries), there may be no or only tenuous continuity of the virtual walls enclosing volumes. But as the tissue is compressed and the density of fixed charge increases, the virtual walls merge and size of the volumes bounded by a continuous virtual walls increases, along with the compressive load carrying capacity of the material.

While GAG molecules in solution can sustain an all-round pressure, they cannot sustain a shear stress (when subject to torque or shear in a solution, the GAG molecules simply move to the side out of the way). In articular cartilage, GAG molecules can sustain an all-round pressure, but a torque or shear force on the GAG molecules can only be sustained if the GAG molecules are prevented from moving to the side out of the way. Usually, such restricted movement occurs because of the collagen network, and it is to the collagen network the GAG molecules transfer their torque or shear force. In general as the GAG density increases, so the degree of rotation and the magnitude of the shear displacement required before GAG molecules transfer significant torque or shear force to the collagen network reduces. This is because as GAG concentration increases, GAG molecules have less space to move to the side and out of the way. So a reduction in space available increases the GAG molecules capacity to directly transfer torque or shear force to the collagen network, probably through a combination of mechanical interlocking and the transfer of forces along the direction of the covalent linkages between saccharides making up the GAG molecules, and/or along the core aggrecan protein.

Unfortunately, we have only a very primitive understanding of interactions between GAG molecules, aggrecan molecules, aggregate molecules, COMP, tenascin, etc., and the type II collagen network in articular cartilage.

6.5 Osmotic Pressure for Non-ideal Solutions

As the solids fraction of a solution increases, it is apparent that we need more sophisticated solution models than that provided by an ideal solution model, i.e. we need an appropriate model that can be generalized to non-ideal solutions.

To guide us, we first note that to extend the theory of ideal gases to non-ideal gases, van der Waals developed a theory that took into account: (i) the finite size of the gas molecules and their excluded volume and (ii) included interaction forces between the gas molecules. In Appendix K, we find that for a van der Waal's gas, the excluded volume leads to an increase in pressure above that for an ideal gas, due to increased frequency of collisions with the container wall, while the attractive force between gas particles reduced gas pressure at the container wall. A similar approach can be used to extend this theory for ideal solutions (which is based on mixing two ideal gases) to non-ideal van der Waal's solutions (Cruz et al. 2010) (which is also based on mixing two solutions). When solute particles have a finite size and large, the entropy of mixing is increased, and the solute particles collide with the container wall more frequently, increasing osmotic pressure of the solution. It is found that osmotic pressure predictions, based on the comparatively simple van der Waals equation of state for non-ideal solutions, are quite satisfactory for PEG, polystyrene, hydroxylpropyl cellulose, and many other uncharged polymers in concentrated solutions (Cruz et al. 2010).

However, charged macromolecules require account be taken of electrostatic interactions as well as excluded volume. So-named DLVO theory includes two forces between the macromolecules, namely attractive van der Waals interactions (including dipole–dipole, dipole-induced dipole, and dispersion interactions) and a repulsive diffuse double-layer (DDL) interaction https://en.wikipedia.org/wiki/

DLVO_theory. When modeling the DDL, one can upgrade the linear Poisson–Boltzmann to a nonlinear Poisson–Boltzmann theory. One can also use a number of more sophisticated continuum models, e.g. the Korteweg–Helmholtz model that takes into account dipoles and electrostriction (Melcher 1981; Grodzinsky 2011). To take account of excluded volume, which Vilker et al. (1981) reports provided the major contribution to the second and third virial coefficients, one can also include short range repulsive interactions, such as a 'hard' sphere repulsion [this approach is taken by Wu and Prausnitz (1999)]. Of course, one also has recourse to molecular dynamics simulations of varying complexity [e.g. Nap and Szleifer (2008)].

Examination of the macromolecule albumin is of interest here, as this is an important biological molecule. It is the principal oncotic protein in (blood) plasma. Albumin is negatively charged at pH 7.4 (the normal pH of blood). Vilker et al. (1981) used the following semiempirical equation to predict osmotic pressure for albumin solutions,

$$\Pi = RT\left[2\left((zc_\mathrm{p}/2M_\mathrm{p})^2 + c_\mathrm{s}^2\right)^{(1/2)} - 2c_\mathrm{s}\right] + \frac{RT}{M_\mathrm{p}}\left(c_\mathrm{p} + A_2 c_\mathrm{p}^2 + A_3 c_\mathrm{p}^3\right) \tag{6.28}$$

where the first term on the right-hand side relates to the ideal Donnan osmotic pressure (again c_p is the protein concentration, c_s the salt concentration, M_p is the molecular weight of albumin and z is the protein valence), while the second term involving a virial expansion takes account of excluded volume and other effects. The 'excluded volume' of a molecule is the volume that is inaccessible to other molecules in the system as a result of the presence of the first molecule, due to the finite size of one molecule (see further discussion in Appendix K).

The data shown in Fig. 10 of Vilker et al. (1981) [and confirmed by Wu and Prausnitz (1999)] shows that the osmotic pressure increases strongly with increasing albumin concentration. As mentioned above, this strong increase in osmotic pressure at very high concentrations is largely driven by 'excluded volume' effects.

Using Eq. (6.28), Vilker et al. (1981) found that the 'reduced osmotic pressure' (defined as π/c_p) for (bovine) albumin is about 0.4 (pH 7.4, 0.15 M NaCl) at 40 g/l albumin (which is approximately the concentration in human plasma). This suggests that the osmotic pressure of albumin is about 16 mm Hg at a concentration of 40 g/l. The Donnan osmotic pressure predicted using Eq. (6.18) is 15 mm Hg, suggesting that at 40 g/l albumin is behaving largely as an ideal macromolecule. We note that in vivo albumin is one of a number of proteins found in blood plasma, and these proteins would increase the in vivo osmotic pressure of plasma albumin above these predictions due to excluded volume effects. Weisberg (1978) estimates that the total osmotic pressure exerted by plasma colloids is about 25 mm Hg—about 0.5% of the total osmotic pressure of plasma. We note that at 120 g/l, albumin has as osmotic pressure of 24 mm Hg, which gives some idea about the possible magnitude of excluded volume effect due to the presence of other proteins in blood plasma (i.e. the effective excluded volume due to the other proteins in plasma is equivalent to an additional 80 mg/ml of albumin).

Indeed, neutron scattering shows that at 40 g of albumin per liter, in 1:1 salt solution at 0.15 M and pH 7.4, the albumin is not in a lattice structure [in fact, it has been suggested this lattice structure only appears at concentrations above 200 g/l (Sharma et al. 2011) as osmotic pressure approaches 200 mm Hg]. Simple calculations can show for a 40 g/l solution of albumin (MW = 69 kDa), the albumin protein molecules are confined to an average volume equal to a cube of side length 14 nm. How does this compare to the size of albumin? The size of albumin is variously reported in the literature, but Kiselev et al. (2001) reports that:

> The size of a human serum albumin molecule in aqueous solution containing 150 mM NaCl was studied using small-angle neutron scattering. The molecular radius of gyration was estimated to be 27.4 ± 0.35 A.

which is also consistent with reported radii of gyration in Smilgies and Folta-Stogniew (2015). For a discussion of 'radius of gyration' of macromolecules, see Benedek and Villars (2000). This data suggests that albumin at 40 g/l has considerable freedom in its movement, and so could be reasonably approximated using an ideal Donnan model [as discussed above—see Vilker et al. (1981)].

For dilute and semidense polymer solutions, the Flory–Huggins theory is usually employed to calculate entropy changes in the solution https://en.wikipedia.org/wiki/Flory%E2%80%93Huggins_solution_theory. For a charged polymer (e.g. GAGs) attached to a backbone (e.g. aggrecan core protein), covalent bonds between GAG and protein put structural constraints on polymer configurations in solution, which are different to those for the standard Flory–Huggins theory.

Kovach (1995) has used a modified Flory–Huggins theory to calculate the entropy change associated with excluded volume effects (i.e. volume dependent entropy change) and configurational entropy https://en.wikipedia.org/wiki/Configuration_entropy [i.e. entropy dependent on position, which may include volume dependent position entropy (e.g. entropy of mixing) and conformational dependent position entropy (often applied to polymers in solution https://en.wikipedia.org/wiki/Conformational_entropy)]. Kovach (1995) finds that the configurational entropy of the GAG dominates the mixing entropy of the aggrecan molecules, for all but very dilute solutions of chondroitin sulfate.

Kovach (1995) assumes the free energy of the polymer solution is the sum of the free energies due to mixing entropy of the aggrecan, configurational entropy of the GAG molecules and the Donnan free energy contributions of the GAG molecules. When the salt concentration is 1.5 M (note: isotonic saline is 0.15 M), which greatly suppresses double-layer thickness below the approximately 1 nm for 0.15 M salt solution, Kovach (1995) reports he found a good correlation between his model of configurational entropy and electrostatic potential and the osmotic pressure data reported by Urban et al. (1979).

However, more sophisticated molecular dynamics models [which include updated aspects of the Kovach (1995) model] are reported by Bathe et al. (2005) and Nap and Szleifer (2008), which suggest that for reference salt solutions of 0.15 M and GAG concentrations less than 80 mg/ml (i.e. 80 mg/ml aggrecan results in about an average of 2 nm separation distance between GAG chains), that electrostatic contributions to osmotic pressure dominate. At higher aggrecan concentrations, configurational entropy of the GAG chains can make a more important contribution to osmotic pressure than electrostatic interactions. This highlights the fact that for charged polymer solutions, the dominant effect responsible for osmotic pressure itself changes with polymer concentration (i.e. with strain magnitude in articular cartilage). And as also discussed further in Appendix K, the effects responsible for the dominant contribution to a material's stiffness also depend on strain magnitude and rate of deformation.

6.6 Cell Models for Polyelectrolyte Solutions

Another simplifying approach to calculate osmotic pressure, applicable to charged macromolecules that do not have long range interactions, but are dominated by electrostatic interactions, is to focus on the behavior of an individual macromolecule, rather than analyze the whole solution. One approach is to use a van der Waals type of equation to calculate the osmotic pressure in the solution based on the properties of a single repeating 'unit cell.' For this 'cell model' approach, each macroion occupies a cell volume that is directly related to macroion concentration. The cell contains one macroion, counterions, and co-ions. The behavior of a single cell then represents the behavior of the whole polyelectrolyte solution [see, e.g. Fig. 1 in Deserno and Holm (2001)]. Explains Deserno and Holm (2001):

> The cell model approximation consists in restricting the theoretical description of the total system to just one cell. While the interactions between the small ions with "their" macroion as well as with small ions in the same cell are explicitly taken into account, all interactions across the cell boundary are neglected. Note that the existence of cells, all of which have essentially the same size, requires correlations to be present between the macroions. However, these correlations are no longer the subject of study. The cell model can thus be viewed as an approximate attempt to factorize the partition function in the macroion coordinates, i.e., replacing the many polyelectrolyte problem by a one polyelectrolyte problem.... in computer simulations the cell could be identified with the replicating unit of periodic boundary conditions. This requires space-filling cells, for instance, cubes. In an analytical treatment one usually tries to maximize the symmetry of the problem. Hence, spherical colloids are centered in spherical cells. The main advantage of this strategy is that a density functional approach neglecting symmetry-breaking fluctuations becomes a one-dimensional problem.

A 'partition function' (Z) describes the statistical properties of a thermodynamic system at equilibrium https://web.stanford.edu/~peastman/statmech/statisticaldescription.html. The partition function is a measure of the number of quantum states the system can occupy at a given temperature. In principle, if the statistical mechanics partition function is known, it can tell us everything we wish to know about the thermodynamic system under equilibrium conditions. Being able to determine the partition function is one of the most important achievements of statistical mechanics, at least from the perspective of physical chemistry. The problem with trying to use a partition function is that in most cases it is intractably complex. What makes it so complex is that all molecules in a system can interact with one another, implying all their positions are correlated with each other in some unknown way [so it is not possible to factorize (and so simplify) the molecular correlations in terms of 'molecular coordinates'].

However in a few situations, simplifications of the partition function are possible, and these can result in useful results. One example simplification is provided by the N-particle distribution function in the partition function is replaced by a product of N identical one-particle distribution functions (which leads to an N-dimensional integral being replaced by N one-dimensional integrals). By ignoring interactions, each one-dimensional integral is much like the approach taken for the 'cell model,' except it is now factorized in the counterion–co-ion coordinates. Such an approximate partition function can lead to the 'mean field' theory known as the Poisson–Boltzmann equation. Another approach to problem simplification is calculating the derivative of the partition function only at the boundaries of a cell model (Wennerstrom et al. 1982).

Here we consider this second case further, and we follow the explanation for calculating the derivative of the partition function at a cell boundary, as described by Deserno and Holm (2001). Let the Helmholtz free energy of the cell A be defined as $A = kT \ln Z$ where Z is the partition function [$Z = \mathrm{Tr}(e^{-\beta H})$, where the 'trace' represents the integrals over phase space], where $\beta = 1/kT$ and H is the Hamiltonian of the unit cell [i.e. the total energy of the unit cell, i.e. $\beta \sum_i U_i$, including kinetic energy and potential energies (e.g. Coulombic energy)]. Let an individual model cell be approximated as two concentric 'spheres' [where r_0 is the radius of the charged macroion, and R the radius of the outer (insulated) boundary of the cell] with domain between. The osmotic pressure of the cell is defined as the rate of change of the Helmholtz free energy w.r.t. volume, that is,

$$\pi = -\frac{\partial A}{\partial V} = \frac{kT}{A_R Z}\frac{\partial Z}{\partial R} \tag{6.29}$$

As the outer boundary of the cell carries no charge and is insulating, the (Coulombic) energy of the system is independent of R (Deserno and Holm 2001), i.e. the Coulombic energy only depends on the distances between any two microions and the distance of any microion from the macroion, and so Coulombic terms do not involve R [see Wennerstrom et al. (1982) for further detail]. In this case, R only enters the partition function via the upper limit in the configuration integrals for the partition function. That is, R enters when evaluating the partition function as the trace over the phase space is an N-fold volume integral over all particle coordinates, each containing a radial integration from r_0 up to R. But by the fundamental theorem of calculus (which says a derivative of Z reverses an integral of Z), the radial coordinate in the integrand for the first ion is just set to R (the position where we are to evaluate the derivative of the free energy). The integral of the two remaining spatial coordinates results in the area A_R. Since all ions are identical, the N volume integrals for each individual ion are identical. In a last step, we used the fact that the trace over all particles but the first one (which is evaluated at R) is equal to Z times the probability distribution of the first particle (i.e. the partition function is defined so that $p_1(R)Z(R) = e^{-\beta E_1(R)}$), so multiplication by N gives the ion density $c(R)$ [see details Deserno and Holm (2001) and Wennerstrom et al. (1982)]. The outcome is

$$\frac{\partial Z(R)}{\partial R} = A_R Z(R) c(R) \tag{6.30}$$

Combining the previous two equations shows

$$\pi(R) = -\frac{\partial A}{\partial V} = kTc(R) \tag{6.31}$$

This means the osmotic pressure at the cell boundary R is equal to the kT times the ion density at R (where the electric field is zero). If there is more than one ion type present, we replace a single ion concentration by a sum of ion concentrations. No assumption has been made about an ideal gas—in other words, this relation also holds for dense concentrations of ions (i.e. non-ideal solutions) (Deserno and Holm 2001).

One can also calculate the derivative of the free energy with respect to the inner boundary at r_0 too. This results in an additional term due to the surface being charged—for a planer surface of infinite extent this leads to (Deserno and Holm 2001),

$$\pi(r_0) = -\frac{\partial A}{\partial V} = kTc(r_0) - 2\pi kTl_B\sigma^2 \tag{6.32}$$

where $\pi(r_0)$ is the osmotic pressure at the inner boundary, $\sigma(r_0)$ is the surface charge density and l_B is the Bjerrum constant [see similar expression in Wennerstrom et al. (1982)]. This equation is known as the 'contact value theorem,' since it gives the contact density of ions at the surface of the macroion as a function of its surface charge density and the osmotic pressure at r_0. The occurrence of the second term is related to the presence of an electric field (generated by the surface charge on the macroion and all microions in solution), which vanishes at the outer cell boundary, but contributes its share to the total pressure at other locations via the Maxwell stress tensor.

By subtraction of the two previous equations, for the plane-wall case we have:

$$\pi(r_0) - \pi(R) = kT\sum_i c_i(r_0) - 2\pi kTl_B\sigma^2 - kT\sum_i c_i(R) \tag{6.33}$$

If we assume $\pi(r)$ is constant between r_0 and R, then

$$kT\sum_i c_i(r_0) - kT\sum_i c_i(R) = 2\pi kTl_B\sigma^2 \tag{6.34}$$

The derivation of this equation was first found in the context of mean field (Poisson–Boltzmann) theory by Graeme in 1947 (Deserno and Holm 2001). For planer surfaces and using the relation,

$$\sigma = -\varepsilon\frac{\partial\phi(r)}{\partial r} \tag{6.35}$$

and then assuming mean field theory is applicable, we may generalize the previous equation to any distance, viz.

$$\pi(r) = kT\sum_i c_i(r) - \frac{1}{2}\varepsilon E^2(r) \tag{6.36}$$

This equation can again be generalized to the osmotic pressure at any location in a double layer with electric field of any complexity as the difference between the local osmotic pressure and Maxwell stress, viz.

$$\pi(r) = kT\sum_i c_i(r) - \sigma_{ij}^{el}(r) \tag{6.37}$$

where $\sigma_{ij}^{el} = \varepsilon E_iE_j - (1/2)\delta_{ij}\varepsilon E_kE_k$, as given previously [see Eq. (6.21), or see Appendix in Stankovich and Carnie (1996)]. If there is a background salt solution present, then we may write

$$\pi(r) = kT\sum_i c_i(r) - kT\sum_i c_i^{\mathrm{ref}}(r) - \sigma_{ij}^{el}(r) \tag{6.38}$$

where c_i^{ref} are the ion concentrations in the background or reference salt solution.

In the next chapter, we describe the basics for developing mathematical and computational models of cartilage using so-called continuum models. While current modeling efforts fall well short of addressing all the issues covered so far in this book, a good start on many aspects has been made. In the decades to come, we expect to see extraordinarily sophisticated models of cartilage being developed that offer new insights, which are predictive of cartilage behavior. Ultimately, we expect such models guide decisions about general strategies and the selection of specific therapies in the treatment of cartilage disease processes, including OA.

6.7 Pressure Measurement for Solutions and Tissues

This brings us to important practical point—how to measure osmotic (and other) pressures in a solution and tissues that contain macroparticles with varying degrees of fixity?

We could use a so-named osmometer, which has a semipermeable membrane between the reference solution and solution of interest. An osmometer is the 'gold standard' to measure the osmotic pressure of solutions, including those with any degree of non-ideality (Vilker et al. 1981; Wu and Prausnitz 1999). The solution or tissue just needs to be a saturated porous media or a solution. If the test is on ex vivo tissue, we can the wrap the tissue in a semipermeable membrane and place it in a solution of known osmotic pressure, and by this means achieve a known osmotic pressure in the tissue (Maroudas et al. 1992; Chen et al. 2001). We could imagine doing the reverse experiment to obtained data on in vivo tissue—that is, place the solution in an osmotic membrane, then place this package in vivo (via surgery), wait until it equilibrates, and then remove the package and measure the osmotic pressure of the solution. But while the first two approaches are practical, the third approach is impractical to scale up.

But an osmometer cannot measure spatial variability in osmotic pressure, as occurs in some macroparticle solutions (e.g. solutions of particles with different sizes) and tissues (e.g. cartilage). To measure spatial variability in osmotic pressure, we could imagine an ideal sub-nanoscale pressure measuring device that has a pressure sensor that was impermeable to fluid—an ideal nanopipette. We could imagine inserting this ideal nanopipette into the solution of macroparticles or tissue that is surrounded by a reference solution. Providing the macroparticles or tissue was heterogeneous and the macroparticles had some fixity (or excluded volume effect), this nanopipette could be used to measure nanoscale spatial variations in solvent pressure throughout the domain. We expect that it would measure a high fluid pressure between the plate-like particles as shown in Fig. 2 of Barclay et al. (1972), but well away from the charged particle, beyond the influence of double-layer overlap, it would measure the background or reference solvent pressure.

The nanopipette could measure spatial variations in osmotic pressure associated with the 'depletion force.' This nanopipette could also be inserted in vivo, and the spatial variation of osmotic pressure measured in tissues in vivo. But if the particles in solution have no fixity, as is the case for a simple solution, the ideal nanopipette would measure essentially the same osmotic pressure everywhere throughout the solution and obtain the same value for osmotic pressure as an osmometer. Presumably if the nanosensor had an incredibly fast response time, it may be able to detect slight fluctuations in the osmotic pressure within these solutions, but these fluctuations would be small (or the order of kT).

But such a device does not exist, and we are left to use and interpret readings from macroscale solvent pressure devices. Practical macroscale devices to measure solvent pressure include Pitot tubes, servo-controlled glass micropipettes (about

0.5–3.0 μ in diameter) (Wiederhielm et al. 1964; Morgan et al. 1998), and 'wick-in-needle' probes (which are the thickness of the needle in which the wick is inserted) (Fadnes et al. 1977). These last two methods measure interstitial mechanical solvent pressure in vivo with a high degree of correlation and accuracy (Wiig et al. 1981).

Now if such macroscale pressure measuring devices are inserted into a solution, where the macromolecules free to undergo Brownian motion, what will they measure? If a Pitot tube or micropipette or wick-in-needle contains a 'reference solution' that has solvent at higher chemical potential than the device, and the micropipette or wick-in-needle is inserted into a solution, the reference fluid in the devices will simply empty into the solution. Small negative pressures can be sustained by the servo-control device in the micropipette as it empties of fluid. But such devices can only sustain negative pressures up to the 'cavitation pressure.' If a micropipette or wick-in-needle is inserted into an ideal solution surrounded semipermeable membrane, that is in turn surrounded by a reference solution at atmospheric pressure, and the devices contain reference solution at the same chemical potential as the solution of interest, then the solvent osmotic pressure will be measured (as shown in Fig. 6.1 using a Pitot tube).

What would such a macroscale device read if inserted into a non-ideal polymer solution, where the polymer molecules are free to move in solution (such as for example, a chondroitin sulfate solution), if surrounded by semipermeable membrane that is in turn surrounded by a reference solution at atmospheric pressure? Again if the measuring device contained reference solution, the solvent osmotic pressure will be measured, the same as it was measured as shown in Fig. 6.1 using a Pitot tube for an ideal solution. The same pressure is measured using a micropipette or using a wick-in-needle device.

What pressure would be measured if the same micropipette is now inserted to some depth into a piece of healthy cartilage tissue, which is at equilibrium with a surrounding isotonic saline solution at atmospheric pressure (much like the reference solution is exposed to in vivo)? We may assume that the micropipette is filled with the same reference solution as bathes the tissue. For a micropipette or wick-in-needle, there is no physical semipermeable membrane present between the tissue and the reference solution. We mention that for human cartilage tissue, the average collagen fibril diameter increases with depth from about 56 ± 9 nm at the surface of the cartilage to 108 ± 2 nm at depth (Changoor et al. 2011), but in addition to these larger fibrils, fine collagen fibrils, some 4–10 nm in diameter, fill the interfibrillar space between the larger fibrils [see, e.g. Fig. 5 in Lane and Weiss (1975)]. We may also assume that the collagen network (together with molecules such as hyaluronan) restrains the movement of the aggregate, the aggrecan, and the GAG within the cartilage tissue.

Aggrecan and GAG cannot enter the end of the micropipette as they are restrained by the (coarse and fine) collagen network spanning across the micropipette tip. The chemical potential of the solvent in the tissue is identical to that in the micropipette (as this is the same as the reference solution surrounding the tissue), so initially there is no tendency for the solvent to enter or leave the micropipette, and consequently no pressure change would be recorded in the micropipette (Wiederhielm et al. 1964; Morgan et al. 1998).

There is a difference in cation and anion concentration in the two solutions, so both cation and anion would begin to diffuse down their respective concentration gradients, only to create a diffusion (Donnan) potential that immediately stops further ion diffusion, thereby preventing further change to the composition of the reference solution in the micropipette. As the chemical potential of the solvent in the tissue then remains identical to that in the micropipette, there is no tendency for the solvent to enter or leave the micropipette, and consequently no pressure change would be recorded in the micropipette long term. Similarly for readings made using a wick-in-needle device inserted into cartilage tissue, or a Pitot tube placed against the surface of the cartilage tissue.

Importantly, we recognize that the heterogeneous distribution of fixed charge with the cartilage tissue creates a Donnan potential (and a 'virtual wall') at the interface with the micropipette and cartilage tissue, as surely as if the micropipette had a semipermeable membrane covering its tip. Realizing this, it is unsurprising that the micropipette reads zero fluid pressure, as the solution in the micropipette is the same as the reference solution, which is also at zero pressure. Because the heterogeneous fixed charge in the cartilage tissue creates a 'virtual wall' throughout the substance of the cartilage tissue itself (as describe above), even though osmotic pressures may vary dramatically over very short distances within the tissue, these variations in osmotic pressure cannot be measured by a macroscale micropipette.

But importantly, we also note that both a micropipette and a wick-in-needle measurement devices can measure 'excess pore water pressure' [i.e. transient fluid pressures due to loading on the surface of the cartilage (or on the surface of a gel). And excess pore water pressure in cartilage (or gel) can be measured very accurately using a micropipette (Wiig et al. 1981). For example, Wiig et al. report '*hydrostatic pressures supposed on the HA gel or saline were accurately reflected by the micropipette in the range −10 to +10 mm Hg.*']. Excess pore water pressures are typically measured during consolidation testing of cartilage (Oloyede and Broom 1991).

What pressures do we measure if we use a different type of macroscale pressure measuring device, one that has a solid metal impermeable/impervious contact surface, rather than a fluid–fluid contact interface, as for the micropipette or wick-in-needle? Such a pressure measuring device may be side-mounted in a needle, as described by McNally et al. (1992). This device can measure solvent pressure in an ideal solution at equilibrium with its reference solution, and provide it presses firmly against the cartilage tissue (with just enough pressure to relieve any restraint to the aggrecan offered by a say a collagen network), it can measure solid (osmotic) pressure in cartilage tissue at equilibrium with isotonic saline. Both pressures can be measured with a very short equilibration time.

The device can do this because the impermeable/impervious metal surface actually measures total stress, i.e. the sum of solid stress in the cartilage (or gel) and any pore fluid pressure (either transient or constant). This device cannot distinguish between these two stresses, or distinguish various contributions to these two stresses that may make up the same total stress measurement. For an ideal solution, the device will be measuring solvent pressure (which is the osmotic fluid pressure). The device will measure weight pressure (or gravitational hydrostatic pressure) of a fluid. If the probe is inserted into cartilage tissue (that is at equilibrium with isotonic saline at atmospheric pressure), this device will measure the 'solid stress' in the cartilage (sometimes referred to as 'swelling pressure' or 'osmotic pressure'). However, it may measure different pressures depending on how much the tissue deforms next to the pressure sensor (e.g. the pressure will rise if nearby tissue deforms). As explained above, this effective or solid stress is the osmotic fluid pressure between charged aggrecan and GAG molecules. This solid stress arises for non-ideal conditions and may include osmotic pressure contributions from van der Waals attraction, steric, configurational, conformational and excluded volume effects, as well as electrostatic interactions as discussed above. This pressure measuring device is measuring simultaneously, the sum of all the contributions to osmotic pressure, and cannot distinguish between them.

We know that the solid (osmotic) pressure in cartilage varies continuously throughout the tissue, with 'depth-dependent zonation' (superficial, intermediate, and deep zones), with 'radial zonation' around chondrocytes (pericellular, territorial, and interterritorial zones), and with local variations in composition of aggrecan and collagen in each zone (e.g. thick collagen bundles have large excluded volume effect, and interdigitating/overlapping aggrecan molecules may both increase local fixed charge density). Indeed it is possible for the osmotic pressure to change dramatically over very short distances in cartilage tissue, given the diffuse double-layer thickness $(1/\kappa)$ on GAG molecules in isotonic salt solution is less than 1 nm (see discussion Sect. 6.2). Providing that there are 'GAG free' regions (possible about 5–10 nm wide) exist in cartilage tissue (e.g. within bundles of collagen fibers), it is quite possible that these regions may contain essentially isotonic saline at reference pressure. It is possible such regions are contiguous throughout cartilage tissue, at least in the undeformed state. But even if such low osmotic pressure regions exist in cartilage tissue (and this is currently unknown), at 10 nm across these regions cannot be resolved using any kind macroscale device that typically average pressures over distances of hundreds of microns in each direction.

If cartilage tissue is loaded externally, a McNally et al. (1992) type pressure measuring device will in addition to the effective stress measure the excess pore water pressures generated in the cartilage tissue. This type of measuring device will follow simultaneously the decrease in excess pore water pressure and rise in effective stress as the tissue consolidates. Interestingly, we see that if the initial effective stress is zero, and the tissue is loaded, the McNally et al. (1992) type measuring device would measure a constant stress, for as the effective stress rises the excess pore water pressure decreases in proportion, leaving a constant reading equal to the total applied stress. When consolidation is complete, this pressure measuring device will measure the new equilibrium osmotic pressure (effective stress) within the tissue. Given the above, it is apparent that a McNally et al. (1992) type measuring device measures the total stress in the cartilage tissue.

The fact that a McNally et al. (1992) type pressure measuring device measures total stress is nicely illustrated by the pressure data presented for (caprine) intervertebral disk tissue (i.e. the measuring probe is inserted into the nucleus pulposus), as shown in Figs. 4 and 5 of Emanuel et al. (2018). In this paper, a load is applied to an intervertebral disk that has an initial osmotic pressure (i.e. an initial effective stress) of about 400 kPa immediately prior to loading (more accurately, probably about 370 kPa) (400 kPa is approximately equivalent to a aggrecan concentration of about 80 mg/ml based on the approach described to estimating osmotic pressure of aggrecan solutions described in Appendix D).

Upon application of the external stress of about 700 kPa (more accurately, probably about 650 kPa), the initial total stress read by the McNally et al. (1992) type measuring device is about 1100 kPa (equal to the initial effective stress of 400 kPa plus excess pore water pressure 700 kPa induced by the external loading). The pressure measurement device records a decrease in the sum of the effective stress and excess pore water pressure of 400 kPa (as excess pore water pressures dissipate, while simultaneously the effective stress increases), to a final osmotic pressure reading (i.e. a final effective stress) of about 700 kPa (equal to the external applied load). Only about 300 kPa (more accurately, probably about 280 kPa) of excess pore water pressure dissipation is associated with the effective stress rising from an initial value of 400 kPa to its final value of 700 kPa. Of course, some 700 kPa of excess pore water pressure actually dissipates (i.e. initially the excess pore water pressure is 700 kPa, while at equilibrium the excess pore water pressure is zero), but the (measured) effective stress only rises 300 kPa (as measured in the nucleus pulposus). This is due to the initial effective stress being about 400 kPa in the intervertebral disk in the direction of loading (i.e. equivalent to 400 kPa tension in the direction of loading in the annular fibrosis, which is not measured).

To an external observer without any information about the initial internal stress state of the system [i.e. that the compressive (osmotic) effective (pre)stress in the nuclear pulposus is in equilibrium with the tensile stress in the ring of collagen making up the annular fibrosis], on applying an external load of 700 kPa, based on poroelastic theory, they would conclude correctly that the excess pore water pressure would rise 700 kPa. But they would incorrectly conclude that the incremental effective stress in the aggrecan of the nucleus pulposus has also risen 700 kPa, as they were unaware of the equal and opposite initial prestress in the aggrecan and collagen. However, based on poroelastic theory, they would correctly conclude the total change in effective stress in the intervertebral disk is 700 kPa. To see this, we have to add together the 300 kPa change in effective stress in the aggrecan (which is measured) and the 400 kPa change in the effective stress in the collagen in the direction of loading (i.e. in this case the 400 kPa stress change in the direction of loading occurs in the annular fibrosis, and this stress change is not measured). We also mention that if the stiffness of the measured material in the nucleus pulposus were calculated without knowledge of the prestress, the stiffness estimated would be a combination of initial collagen stiffness and aggrecan stiffness (see Fig. 1.8). However, if they were aware of the collagen stiffness and the amount of prestress, it would be possible to calculate the stiffness of the aggrecan acting alone (see Fig. 1.8).

Finally given the previous example, which had tissue pressure measured in an intervertebral disk using a pressure measuring device that has an impermeable/imperious pressure sensing surface, what pressure would be measured if a micropipette (with a fluid interface) were now inserted to some depth into the nucleosis pulposus? We assume that the intervertebral disk is surrounded by isotonic saline, as described in Emanuel et al. (2018). We also assume there is no restraining collagen network in the nucleosis pulposus, and that the aggrecan present is confined above and below by impervious bone and by a thick collagen annular ring (annular fibrosis) around the 'sides.' For this situation, it is as if we have a charged polymer confined in a stiff semipermeable porous container. While the aggrecan can very slowly migrate through the thick annular collagen ring, for practical purposes it cannot migrate, at least on a timescale of interest here. The aggrecan will swell to completely occupy the space within the container, as it can imbibe water through the thick annular ring of collagen. This situation is much like measuring the osmotic pressure of GAG surrounded by a semipermeable membrane. Consequently, when a micropipette filled with reference solution is inserted into the nucleosis pulposus, it would measure the osmotic pressure of the aggrecan/charged polymer in the nucleosis pulposus, just as if the aggrecan in the nuclear pulposus were surrounded by a semipermeable membrane and set in a reference solution (as described in the example above).

To formalize these observations, the momentum balance of the solution says that the incremental total stress (σ_{ij}^T) in the solution is given by:

$$\sigma_{ij}^T = \sigma_{ij}^s - pI \tag{6.39}$$

where p is the excess pore fluid pressure (i.e. the transient background fluid pressure acting throughout the fluid phase due to applied loading) and σ_{ij}^s is the incremental average (osmotic/solid/effective) stress between the (fixed) solid particles/macromolecules in solution (e.g. as in the case of cartilage tissue).

For completeness of our discussion on pressure measurement, we mention that many biological tissues have comparatively low concentrations of (fixed) proteoglycans, which contribute to the solid stress in the tissue [see, e.g. the illustrative diagram Fig. 5 Attili and Richter (2013)], which shows aggrecan at low concentration (e.g. <4 mg/ml) interacting with hyaluronan, but with large free spaces between the aggrecan. This is likely to be similar to conditions found in many normal tissues. In some circumstances (e.g. inflammatory states) simultaneously, another group of mobile molecules (e.g. proteins such as albumin, which is common in the interstitial environment following injury and/or inflammation) may reside within a semipermeable membrane (the role of semipermeable membrane being played by the surrounding tissues) creating a mechanical (osmotic) fluid pressure [which is measureable using either a micropipette (or similar device) or using a pressure sensor with an impermeable surface]. The difference between the pressure measured using an impermeable pressure sensor and the pressure measured using a wick-in-needle (or micropipette) has led to controversies (Chauhan et al. 2014).

By making observations using a combination of an impermeable (total) pressure sensor, a wick-in-needle (fluid) pressure sensor, and observing fluid pressures with and without a contribution due to vascular pressure (by performing measurements in vivo and at postmortem), we note that Chauhan et al. (2014) appear to be analyzing their particular problem, relating to pancreatic ductal adenocarcinoma, in the main correctly. Their particular problem is to identify the components of equilibrium total stress, which potentially involves a solid stress contribution to total stress (where the solid stress may be transferred through solids by covalent bonds and other chemical interactions including Born repulsion, and by electrical and osmotic stress in over-lapping double-layers), and a fluid pressure contribution to the total stress (where the fluid pressure contribution may be due to either an osmotic fluid pressure (generated by the presence of a semipermeable membrane), or it may be driven by vascular pressure (generated by the heart contracting), or some combination of the two). In this particular case, equilibrium conditions prevail, and so there is no transient excess pore fluid stress due to external loading (which would lead to consolidation of the tissue). Since the interstitial fluid pressure is more or less permanent (only reducing to zero as blood pressure went to zero), this pressure may be regarded as a background pressure. To obtain the solid stress in the tissue (in this case, generated by cell growth), one should add the background interstitial fluid pressure (p_{IFP}) to the total stress (σ_T) to obtain the solid stress $\left(\sigma'_{\mathrm{cells}}\right)$, i.e. $\sigma'_{\mathrm{cells}} = \sigma_T + p_{\mathrm{IFP}}$.

In geomechanics, the stress balance shown in Eq. (6.39) is known as the 'Principle of Effective Stress' and was first being published by Karl Terzaghi in his book Erdbaumechanik—'Soil Mechanics' (Terzaghi 1925). We mention in passing that this Principle of Effective Stress applies to both sand [where the effective stress is largely due to Born repulsion between (fixed) particles that are so large they do not exhibit Brownian motion in a solution] and fixed charged clay soils (where the effective stress is mainly due to electrostatic repulsion between fixed charged particles, much as they are in articular cartilage). The Principle of Effective Stress was later employed by Biot in his 2D and 3D analyses of soil consolidation (Biot 1941), which has been successfully applied to tissues. This is further discussed in detail in Sect. 7.2.1.

Chapter 7
Theory for Modeling Articular Cartilage

7.1 Introductory Remarks on Solid Mechanics and Poromechanics

Cartilage is usually modeled as a 'poroelastic continuum.' A continuum model is a spatially averaged model of a material. Developing a continuum model for porous media entails choosing a length scale for 'representative elementary volumes' (REVs) (Bear and Bachmat 1991), over which the spatial averaging of physical quantities takes place. The REV length scale needs to be large enough to average out the effect of discrete components that make up a porous media on quantities of interest (e.g. porosity), but still small enough to provide detailed information that is useful for the intended purpose. At smaller REV length scales, the discrete components making up the material become apparent (e.g. the distribution of porosity), so details about interactions between the components may be investigated using so-named 'multiscale models' (Halloran et al. 2012). Multiscale models are essentially a hierarchy of continuum models that have different averaging length scales. These continuum models are joined together by handing information back and forth between the models for different length scales. Often, the output from a lower (length) scale is used as a parameter in the higher (length) scale model, while the higher scale model provides appropriate boundary conditions for the lower-scale model. Rather than use a hierarchical modeling approach, some choose to model cartilage by including information from a somewhat smaller length scale than that length scale used for the REV length scale. For example, Pierce et al. explicitly include some directional collagen fiber reinforcement, embedding the collagen fibres within a porous continuum (Pierce et al. 2013).

Poromechanics continuum models analyze materials containing solids that have interconnected pore spaces throughout the solid. For biological materials, the interconnected pore spaces are always filled with water. A saturated poroelastic continuum model approximates the real material as two continua, one continuum representing the solid phase and one continuum representing the fluid phase

D. W. Smith et al., *Articular Cartilage Dynamics*,
https://doi.org/10.1007/978-981-13-1474-2_7

(so these models are sometimes referred to as 'biphasic'). The two continuum models interact with one another, so the two continuum models are said to be 'coupled.' Because water in the pore spaces can reasonably be regarded as incompressible at physiologic loadings, the porous solid cannot change its volume until the pore fluid flows through the pore spaces, allowing the porous solid to change its volume. Volume changes by the solid-phase couples to the fluid-phase continuum model as a fluid-phase source or sink term in the fluid mass balance equation. The so-named 'Darcy fluid flow' in the porous material is driven by pressure gradients in the pore fluid, which develop upon loading the poroelastic material. The fluid flow through the porous solid causes drag forces to be exerted on the solid phase (which appear as transient body forces in the solid mechanics analysis), so the fluid flow also deforms the solid. Because the solid deformation influences the fluid flow, and the fluid flow influences the solid deformation, the governing equations are said to be 'fully coupled' (there is mathematical modeling analogy between fully coupled heat flow in a solid and a continuum model of saturated porous media). Volume changes take time to occur because it takes time for fluid to migrate through saturated porous materials when it is loaded due to the drag forces on the fluid, so saturated porous media have a natural timescale associated with the 'consolidation' process (i.e. the 'settlement' process) that follows load application.

The continuum theory of poroelasticity was developed over the twentieth century, commencing with the geotechnical engineer Karl Terzaghi (http://en.wikipedia.org/wiki/Karl_von_Terzaghi, *Erdbaumechanik auf bodenphysikalischer Grundlage 1925).* Important later contributions were made by Maurice Biot [http://en.wikipedia.org/wiki/Maurice_Anthony_Biot and (Biot 1941)]. Rice and Cleary clarified the meaning of some linear model parameters in the theory (Rice and Cleary 1976), while many different extensions of this basic theory have been made over the following decades—a few extensions associated with the authors are partial saturated materials (Sheng et al. 2003), thermal consolidation (Smith and Booker 1993), and to electro-osmotic consolidation, streaming potentials, etc.). Poroelastic theories developed from these early approaches are known collectively as 'phenomenological theories,' because the governing equations were inferred directly from experimental findings, rather than deduced from more general (physics and chemistry) theories. Unsurprisingly, later researchers have sought to deduce the governing equations for porous media from more general physics and chemistry theories.

In the 1950s and 1960s, various mechanicians such as Clifford Truesdell re-thought the theoretical basis for mechanics (http://en.wikipedia.org/wiki/Clifford_Truesdell), and in doing so developed a set of assumptions, methods, and analysis that came to be known as 'rational mechanics.' From rational mechanics developed 'mixture theory.' Mixture theory is a more general theory of 'multiphase materials' than 'phenomenological poromechanics' (http://en.wikipedia.org/wiki/Mixture_theory) (Bowen 1980, 1982).

Phenomenological theories and mixture theories have run alongside one another ever since, with some researchers preferring one approach over the other. Mixture theory is definitely more theoretical (so the more mathematically inclined often prefer models based on mixture theories), but one can easily get 'lost' in the theory, in the sense it is easy to lose touch with how the theoretical variables relate to experimentally measured quantities. The main advantages of mixture theories are: (i) They include thermodynamics of the material, in contrast to most phenomenological poromechanics (phenomenological theories do not usually specifically require the second law of thermodynamics not be contravened, but instead rely on the phenomenological observations of the actual physical system to ensure that the thermodynamics is correct), and (ii) if some new physical phenomenon is to be included in mixture theory, it is generally clear where it 'belongs' in the theoretical framework (Frijns et al. 1997) (see also http://mate.tue.nl/mate/pdfs/467.pdf).

The disadvantage of mixture theory is that for all the theoretical considerations of mixture theories (not to mention the considerable heat generated by academic controversy over the years), it turns out there is essentially no important difference between the predictions made using either mixture theories or phenomenological theories (Verruijt 2013). For example, including thermodynamic considerations of mixture theory makes very little measurable difference (if any) to the standard phenomenological theories. The fact that there is essentially no difference between the phenomenological and mixture theories is perhaps best explained by the fact that phenomenological theories are based on experimental data, while the mixture theories ultimately have to agree with the experimental data too. If either theory does not agree with experiment, then their predictive ability is compromised, and depending on the application, they may be judged unsuitable for that application. We mention that when including 'coupling effects' in both phenomenological and mixture theories for porous media (e.g. streaming potentials, electro-osmosis), early scientific investigators (e.g. Onsager) developed useful criteria (e.g. the 'Onsager reciprocal relations') that when observed ensure the second law of thermodynamics is not contravened for these processes.

For these reasons, we explain here the more easily accessible phenomenological theory of poroelasticity and observe that once the phenomenological theories are understood, models based on mixture theory are more easily understood too. In any case, the phenomenological theory of poroelasticity is complicated itself. So to make this description of phenomenological theories more accessible, we also take a historical approach, explaining Terzaghi's 1-D consolidation theory before Biot's multidimensional consolidation theory.

The second important issue to consider in this introduction is the way the mathematical problem of poroelasticity is formulated. In fluid mechanics, a 'spatial' (or Eulerian) formulation is used. For this approach, the coordinate system for the domain is fixed while fluid, particles move through the domain. For usual fluid analysis, one keeps track of what is happening to measurable quantities at each point in the domain, rather than keeping track of the particles making up the fluid, and these displacements can be very large. The quantities (such as pressure, density, flux, shear rate) usually do not depend on identifying particular particles. These quantities satisfy mass balance equations (which are material independent) or are related to one another through 'constitutive relations' (which are material dependent). Constitutive relations describe material behavior. For example, stress–strain relations (or stress–strain rate relations in the case of fluids), or driving force and material flux relationships in the porous media material. Since constitutive relationships depend on material behavior, relationships between variables are (usually) determined by experiment (though some advanced (lower scale) models of the materials themselves may be employed in some cases to deduce constitutive relationships).

For a solid, the situation is somewhat different. Each particle of a continuum solid is attached to specific neighboring particles, and so material particles cannot mix freely. The 'solid stress' in the material is the stress between solid particles that is capable of maintaining similar spatial relationships between the particles as the material deforms under load. The displacements of individual particles are deemed to be important because based on the amount of stretch between neighboring particles, the stresses (force per unit area) in the material can be deduced with the help of the material's stress–strain relationships. The way particles stretch when loaded determines the way the material deforms, if the material is likely to be internally damaged by the loading, or even if it will break apart. These are clearly important practical concerns for all solid materials in applications, and cartilage is no exception. Because neighboring particles making up the material usually stay close together in continuum solid mechanics, a so-called material formulation is normally adopted. For this type of analysis, the coordinate base vectors are usually attached to material particles and deform with the material as it deforms. This material coordinate system is used to identify relevant physical quantities within the material (e.g. density, stress, fluid flow).

Now the so-named 'material formulation' and 'spatial formulation' are related, for in principle one could use either coordinate base vectors to describe the solid or the fluid, though one or other coordinate system may be more mathematically convenient. Indeed, for small deformation problems the material and (small strain) spatial formulations of the problem are identical. But we mention that the 'small strain analysis' is not 'rotationally invariant.' This means that upon loading of a solid, if the rotations of REVs in the material are significant, a small strain analysis will give an incorrect solution (see effect of rotation on small strain tensor http://www.continuummechanics.org/smallstrain.html; we note that the small strain tensor is defined as $\varepsilon = 0.5(F + F^{\mathrm{T}}) - I$, where the deformation gradient tensor F is defined below and I is the identity tensor). We mention that a material formulation has the computational advantage of eliminating moving boundaries, since the boundaries are the material particles at the domain surface.

Indeed, every modeling choice brings with it advantages and disadvantages. Achieving mathematical convenience may involve a coordinate system or analytic method that involves non-physical variables that are naturally difficult to understand. Mathematical convenience may be gained, but intuition is lost. Because the material formulation involved in solid mechanics results in non-physical variables (i.e. intuition is traded for mathematical convenience—remarks (MacMinn et al. 2016) '*Although powerful, these tools can be cumbersome from the perspective of [developing] physical insight,*' the approach that we adopt here is to first write the governing equations for the solid in a spatial coordinate formulation, because this formulation involves physical variables that are more familiar and intuitive. Physical variables in a spatial coordinate system are more familiar and intuitive because our perceptions of the physical world, and our experimental results, are normally reported with respect to an inertial Cartesian (i.e. spatial) coordinate system, i.e. laboratory frame of reference with the material in its current configuration (i.e. the laboratory frame of reference at time t), rather than for the material in its initial configuration (when the experiment began, i.e. at t equals zero). Once the problem is formulated, we transform the model described in spatial coordinates into a model described in material coordinates in initial configuration, using geometrical transformation rules of differential geometry ('differential geometry' studies coordinate transformations in curvilinear coordinates). We do not need to keep track of individual fluid particles, as we do for the solid. Fluid movement in a saturated porous solid is defined relative to the material, as this definition allows us to link volume changes in the fluid phase with volume changes in the solid phase, and because it is relative movement between fluid and solid that results in drag forces between the two phases.

By definition, tensors remain the same under coordinate transformations, so this explains an inevitable modeling preoccupation with defining and using tensors as descriptors of physical quantities in governing equation for problems requiring coordinate transformations. We now describe tensors and coordinate transformations that we require for the development of a large deformation continuum mechanics model.

A zero-order tensor is a Scaler (e.g. fluid pressure, temperature, density), and a first-order tensor is a vector (e.g. temperature gradients and pressure gradients, a displacement vector), while a second-order tensor is often called a 'tensor' (e.g. stress, strain, permeability). In fact the word 'tensor' arose in the consideration of how to describe stress at a point in a material (Bonet and Wood 2008). There are higher-order tensors too (e.g. the elasticity tensor, which relates stress to strain, is a fourth-order tensor).

Higher-order tensors are formed as 'tensor products' (sometimes called 'outer' or 'dyad products') of lower-order tensors. For example, the dyad 'tensor product' of two vectors $u = a_1e_1 + a_2e_2 + a_3e_3$ and $v = b_1e_1 + b_2e_2 + b_3e_3$ (both vectors defined on an orthogonal basis vector set e_i) is a second-order tensor T. A second-order tensor T is found from the 'outer product' of two vectors $\left(a_ib_j^{\mathrm{T}}\right)$ http://en.wikipedia.org/wiki/Dyadics. For example, a dyad product tensor is

$$\begin{aligned} T = u \otimes v &= \begin{bmatrix} a_1e_1 \\ a_2e_2 \\ a_3e_3 \end{bmatrix} \begin{bmatrix} b_1e_1 & b_2e_2 & b_3e_3 \end{bmatrix} \\ &= \begin{bmatrix} a_1b_1\, e_1 \otimes e_1 & a_1b_2\, e_1 \otimes e_2 & a_1b_3\, e_1 \otimes e_3 \\ a_2b_1\, e_2 \otimes e_1 & a_2b_2\, e_2 \otimes e_2 & a_2b_3\, e_2 \otimes e_3 \\ a_3b_1\, e_3 \otimes e_1 & a_3b_2\, e_3 \otimes e_2 & a_3b_3\, e_3 \otimes e_3 \end{bmatrix} \end{aligned} \tag{7.1}$$

This tensor product matrix is too long to write out all the time, so this is abbreviated using several notations. For example, 'matrix notation' may be used (e.g. T is a matrix of tensor coefficients) or 'indicial notation' may be used (e.g. T_{ij}, which identifies the individual coefficients in the matrix of tensor coefficients). So

$$T = u \otimes v = u_i \otimes v_j = \begin{bmatrix} a_1b_1 & a_1b_2 & a_1b_3 \\ a_2b_1 & a_2b_2 & a_2b_3 \\ a_3b_1 & a_3b_2 & a_3b_3 \end{bmatrix} e_i \otimes e_j = a_ib_je_i \otimes e_j = T_{ij} \quad (7.2)$$

where i and j take the values 1, 2, and 3 depending on the number of spatial dimensions. We see that the components of the tensor product (a_ib_j) are just the ordinary products of the vector components (see http://en.wikipedia.org/wiki/Tensor_product). This observation holds generally for higher-order tensors and for non-orthogonal basis sets too (see http://en.wikipedia.org/wiki/Tensor_product https://en.wikipedia.org/wiki/Tensors_in_curvilinear_coordinates.

A linear combination of dyads with Scaler coefficients is called a dyadic. Every second-order tensor can be represented as a dyadic (but not all tensors can be written as dyads—so these tensors are called simple tensors).

By definition, a second-order tensor transforms (or maps) one vector into another vector (the vector transformation must be independent of the coordinate system; otherwise, it is not a tensor—it is just a matrix of numbers). For example, a second-order Cauchy stress tensor takes a direction vector as input and as an output produces a traction on a surface normal to the direction vector. A second-order permeability tensor takes a vector of pressure gradient components as input and as output produces a vector of Darcy flux components.

More specifically, a simple second-order tensor $u \otimes v$ linearly transforms a vector w into a vector $u(v \cdot w)$. This can be visualized as the magnitude of the projection of vector v onto the vector w multiplied by the magnitude of w, multiplied by the magnitude of vector u, pointing in the direction of the vector u. It is this unfamiliar operation of a simple (dyad) tensor on a vector, which involves three vectors, that tends to make the effect of second-order tensors on vectors difficult to grasp by intuition, but a detailed intuition becomes nearly impossible when a general dyadic tensor (i.e. which is a linear combination of simple tensors) acts on a vector. This means the evaluation of the effect of second-order tensors on vectors (and other tensor operations, e.g. product of two tensors, differentiation of tensors) is reduced to rule-based mathematical manipulations. Unfortunately there are a lot of identities, 'tricks,' and methods (i.e. a lot of rules) to remember (see, for example, http://homepages.engineering.auckland.ac.nz/~pkel015/SolidMechanicsBooks/Part_III/).

Having defined tensors, we now need to look at coordinate transformations. Using an orthogonal coordinate system, a vector can be described in only one way, but in curvilinear (non-orthogonal) coordinates, the same vector can be described in two ways (for a helpful explanation see https://www.youtube.com/watch?v=AKPZkHvqTao). For curvilinear coordinates, one description uses covariant coordinates and contravariant vector components (by convention, this description is usually adopted), while the second way involves contravariant coordinates and covariant vector components. In curvilinear coordinates it is conventional to identify G_i as the covariant base vectors in the reference configuration (and contravariant base vectors as G^i in the reference configuration) and g_i as the covariant base vectors for the current configuration (and contravariant base vectors as g^i in the current configuration). We note that covariant quantities use a subscript index (the third letter in covariant, '*v*', points downward as a reminder the index is subscript), while contravariant quantities use a superscript index (the third letter in contravariant '*n*' points upward, as a reminder the index is superscript).

However, when base vectors are orthogonal, the two sets of base vectors become the same, avoiding the complication of covariant and contravariant bases and coordinates, so we usually prefer to work with orthogonal (Cartesian) coordinates where possible. For the moment, let one coordinate system in the reference configuration be denoted by an orthogonal triad of basis vectors E_I (the undeformed or reference configuration is denoted by uppercase letters), and let another coordinate system in the current configuration be denoted by an orthogonal triad of basis vectors e_i (the deformed or current configuration is denoted by lowercase letters). We can represent second-order tensors in the current e_i coordinate system using $e_i \otimes e_j$, and the same second-order tensor in the reference E_I coordinate system using $E_I \otimes E_J$. Naturally, the matrix components representing the tensors change with the coordinate system, though there are 'tensor invariants' found from combinations of the matrix components that are independent of the coordinate system (e.g. the trace of the tensor matrix is the first invariant $(\mathrm{tr}(T))$, and the determinant of the tensor matrix is the third invariant $\det(T)$, while the second invariant is more complicated—it is one-half, the trace of the tensor matrix, squared, minus half of the trace of the square of the tensor matrix, i.e. $0.5\left((\mathrm{tr}(T))^2 - \mathrm{tr}(T^2)\right)$. There are well-known rules of differential geometry for performing coordinate transformations that explain how to compute the matrix components and base vectors in any coordinate system (e.g. (Arfken et al. 2013) or http://homepages.engineering.auckland.ac.nz/~pkel015/SolidMechanicsBooks/Part_III/).

In solid mechanics, it turns out that the deformation gradient in the material is fundamental to coordinate transformations. The reason coordinate transformations are useful for understanding how the material deforms is that coordinate transformations deal with how the vector components of a line element in one coordinate system change when one attempts to describe it in another rotated or stretched coordinate system, which is mathematically analogous to how the Cartesian vector components of a short line in a solid change as the solid deforms. Explains (Rice 1994) http://esag.harvard.edu/rice/e0_Solid_Mechanics_94_10.pdf:

> It is sometimes convenient to think in terms of [initial] lines in the material, parallels to the 1, 2 and 3 axes, being carried along, or convected, with the deformation, so that the labels X_1, X_2, X_3 correspond to the labeling of a system of curvilinear coordinates, convected with the deformation. This viewpoint was introduced by Hencky in 1925. When the square of the length of an infinitesimal line element is written as a quadratic form in the

corresponding increments of curvilinear reference coordinates, the coefficients in the quadratic form are called the metric tensor.

For an illustration of how the initial lines in the material become curved as they are '*carried along, or convected*' as the material is deformed, see Sect. 2.10, Fig. 2.10.6 at http://homepages.engineering.auckland.ac.nz/~pkel015/SolidMechanicsBooks/Part_III/

Unfortunately, the coordinate transformation we require to solve a particular solid mechanics problem is usually unknown. It is not known (and generally not at all obvious) because the coordinate transformation matrix that maps our solid from its initial configuration to its current (deformed) configuration depends nonlinearly on the loadings applied to the solid (e.g. surface loads and body forces), the boundary conditions of the surface of the material (tractions, displacements and fluid pressures, fluxes), and on the material properties of the solid itself (the constitutive relationships). In other words, to find the mapping function for the coordinate transformation, we have to solve the mechanics problem. From a coordinate transformation viewpoint, one could imagine our solid mechanics problem as finding a suitable coordinate transformation matrix subject to the constraints imposed by solid mechanics equations. But the interesting thing is the solution of the solid mechanics problem can be simplified by solving the mechanics problem in the most advantageous coordinate system. For solid mechanics problems, normally the most advantageous coordinate system is an orthogonal material coordinate system in the undeformed (reference) geometry.

Why are the orthogonal material coordinates in the undeformed (reference) geometry most advantageous for this type of problem? We could try and solve the stress equilibrium equations in the current configuration, but to do this, we need to know the deformed geometry. But to know the deformed geometry, we need to know the stresses. We could try and guess the deformed geometry (or the deformation gradient); however, for all but the simplest problems, this proves to be an impossible task. We could try and compute the Cauchy stress tensor in the undeformed geometry, but then we would be analysing different volume elements to those in the deformed geometry, so that is incorrect. The solution of the solid mechanics problem is unusually awkward, because unlike most problems we encounter, the problem domain is not fixed but part of the solution we seek. Clearly, we need to solve for the deformed geometry and the stresses together in the same coordinate system.

This dilemma forces us back to the known reference configuration, which has the virtue of being known, and once there, we find it advantageous to adopt orthogonal material coordinates. We then transform (or 'map' depending on how you look at it) all the governing equations and boundary conditions appropriate for the deformed geometry in a spatial coordinate system into the undeformed geometry with a material coordinate system. We do this transformation (or mapping) by writing the governing equations in the undeformed material coordinate system in such a way each variable in the governing equations involves the (as yet unknown) deformation gradient. The deformation gradient transforms (or maps) the reference geometry into the deformed geometry and can transform (or map) other quantities from the undeformed to the deformed geometry (e.g. the second Piola–Kirchhoff stress transforms into Cauchy stress), while the inverse of the deformation gradient transforms (or maps) all quantities from the deformed geometry back to the reference configuration. It is the deformation gradient that provides the connection between the (known) undeformed and the (as yet unknown) deformed material. By introducing this unknown deformation gradient, we can solve for the unknown stresses and the unknown deformed geometry simultaneously in the undeformed (or reference) configuration (we solve for the deformation gradient that satisfies all the governing equations and boundary conditions in the deformed configuration, and do so in the undeformed (reference) configuration). Once we have the solution in the reference geometry, we can then 'push-forward' to find the deformed geometry and the solution in the deformed geometry (e.g. find the Cauchy stresses in the spatial coordinate system from the second Piola–Kirchhoff stresses in the material coordinate system).

To better understand this process, we mention here that it is instructive to look at a simple one-dimensional problem, because for these mechanics problems the second-order tensor F reduces to a Scaler quantity (i.e. the ratio of a material volume element in the current configuration to the volume of the same material element in the initial configuration), which simplifies the analysis. So we will consider the one-dimensional consolidation problem later, but first we formulate the general problem in two (or three) spatial dimensions.

As is now apparent, the central quantity in the solution of a solid mechanics problem is the deformation gradient (denoted F). In general the deformation gradient F is a second-order (mixed) tensor. In two spatial dimensions, the material deformation gradient tensor F at point P, identified by coordinates X_P in the orthogonal material coordinate system, can be represented as (see, for example, http://en.wikipedia.org/wiki/Deformation_gradient#Deformation_gradient_tensor),

$$\begin{aligned} Fe_i \otimes E_J = F_{iJ} e_i \otimes E_J = \left[\frac{\partial \phi_i}{\partial X_J}\right] = x_{i,J} \\ = \begin{bmatrix} \frac{\partial \phi_1(X_P,t)}{\partial X_1} & \frac{\partial \phi_1(X_P,t)}{\partial X_2} \\ \frac{\partial \phi_2(X_P,t)}{\partial X_1} & \frac{\partial \phi_2(X_P,t)}{\partial X_2} \end{bmatrix} = \begin{bmatrix} \frac{\partial x_1}{\partial X_1} & \frac{\partial x_1}{\partial X_2} \\ \frac{\partial x_2}{\partial X_1} & \frac{\partial x_2}{\partial X_2} \end{bmatrix} \end{aligned} \tag{7.3}$$

where $\phi(X,t)$ [or $x(X, t)$] is an (unknown) function that maps material particles in the initial configuration of the solid directly into any current configuration at time t, described in spatial coordinates. Lowercase x denotes coordinates in the spatial (Cartesian) coordinate system (with orthogonal base vectors e_i), while uppercase X denotes coordinates in the material (Cartesian) coordinate system (with orthogonal base vectors E_J) [note: the initial geometry is usually described using orthogonal basis vectors rather than general curvilinear coordinates (so the $G_J \rightarrow E_J$)]. Using 'matrix notation,' the deformation gradient tensor is denoted simply as F, but more explicitly F may be denoted using 'indicial notation' as F_{iJ}. We mention that F and F_{iJ} are also sometimes denoted as $x_{i,J} = \text{Grad}x$. The (material) deformation gradient F_{iJ} is an example of a (second order) 'two-point' tensor because it straddles the material and spatial coordinate systems. Example deformation gradients can be found at http://www.continuummechanics.org/cm/deformationgradient.html.

Note that the continuous function $\phi(X,t)$ [or $x(X, t)$] is invertible (i.e. as an inverse function $\phi(X,t)$ exists so F is also invertible). The deformation and gradient may be nonlinear, so the deformation gradient may in general change from material point to material point. However, the local deformation gradient for a small region around each point may be linearly approximated by F_{iJ}, the second-order deformation gradient tensor. This explains why we only consider short (differential) vectors (dX) and small (differential) volumes dV for the examples below, but more generally we allow F_{iJ} to be a continuous function of position. We can 'stitch' the linear approximations for adjacent small regions together into lines or larger regions through integration. For example, let the displacement of the material particles be the vector $u(X,t)$ so that

$$\phi(X,t) = X + u(X,t) \tag{7.4}$$

The displacement vector at some point P in the material can then be calculated from the deformation gradient by

$$u(X_P,t) = \int_{X_{\text{BFD}}}^{X_P} (F - I)\text{d}S \tag{7.5}$$

where I is the identity matrix, and the integral is along an arbitrary line path beginning at a known displacement (X_{BFD}), usually on the material surface, and ending at the material point P. Equation (7.5) explains why every well-posed 'boundary value problem' in solid mechanics must have at least one point that has a known displacement, for if this is not known, the displacements are indeterminate (as the lower limit for the integral above is unknown).

Let us now consider how the deformation gradient tensor at a point transforms vector components between coordinate systems. Let dX and dx define the components of the same vector described in either the material or spatial coordinate systems, respectively. Then, the deformation gradient relates the two vector representations of the same vector in each coordinate system by,

$$\text{d}x = \text{d}x_i = F_{iJ}\text{d}X_J = F\text{d}X \tag{7.6}$$

The vector dx in the spatial coordinate system is equivalent to and can be replaced by the same vector FdX in the material coordinate system.

Now taking an alternative viewpoint, consider the physical interpretation of a vector mapping by the second-order (deformation gradient) tensor. We consider what happens to a short line in the solid as it deforms (the coordinate systems e_i and

E_J have the same origin, so both effectively become the same Cartesian coordinate systems). If $\mathrm{d}\hat{X}$ and $\mathrm{d}\hat{x}$ are both defined as unit vectors, then Eq. (7.6) may be written as

$$\lambda \mathrm{d}\hat{x} = F\mathrm{d}\hat{X} \tag{7.7}$$

where λ is the stretch ratio, i.e. the ratio of the current vector length to thc original vector $\mathrm{d}\hat{X}$. Physically this can be interpreted as the original (unit) vector $\mathrm{d}\hat{X}$ in the initial configuration being stretched by λ and rotated into a new vector $\lambda\mathrm{d}\hat{x}$ in the current configuration. The displacement from the undeformed to deformed configuration can then found using Eq. (7.5), and the stretch and rotated vector displaced to its correct position in the deformed configuration. Based on this physical interpretation, the transformation of some tensor quantity described in the (initial/undeformed) material coordinate system to the (current/deformed) spatial coordinate system is described in solid mechanics as a 'push-forward.' For example, Eq. (7.6) is described a push-forward of the vector $\mathrm{d}X$ in the initial undeformed material configuration to the vector $\mathrm{d}x$ in the current deformed configuration. The reverse process (describing a tensor quantity in the (current) spatial configuration in the initial undeformed material coordinates) is referred to as 'pull-back.'

Consider again how the deformation gradient tensor at a point transforms vector components between coordinate systems. The vector $F^{-1}\mathrm{d}x$ in the current deformed configuration described in a spatial coordinate system represents the same vector $\mathrm{d}X$ in the undeformed configuration described in a material coordinate system, viz.,

$$\mathrm{d}X = \mathrm{d}X_J = F_{Ji}^{-1}\mathrm{d}x_i = F^{-1}\mathrm{d}x \tag{7.8}$$

Or taking an alternate viewpoint, this may be interpreted physically as a vector mapping between the deformed and undeformed material. Assuming $\mathrm{d}\hat{X}$ and $\mathrm{d}\hat{x}$ are both defined as unit vectors, then Eq. (7.8) may be written as

$$\frac{1}{\lambda}\mathrm{d}\hat{X} = F^{-1}\mathrm{d}\hat{x} \tag{7.9}$$

where $1/\lambda$ is the stretch (in length) of the original unit vector $\mathrm{d}\hat{x}$.

Now, switching to curvilinear coordinates for a moment, the square of the vector length (i.e. $\mathrm{d}s^2$) in a forward convected curvilinear coordinate system (with base vectors g_J and material coordinates X^J) is related to the same vector in the reference configuration set (with G^J base vectors in the (undeformed) equal to orthogonal base vectors in the (undeformed) reference configuration E^J) through the (covariant) metric tensor, g_{IJ}. That is,

$$\mathrm{d}s^2 = \mathrm{d}x_i\mathrm{d}x_i = g_{IJ}\mathrm{d}X^I\mathrm{d}X^J = F_{iJ}^{\mathrm{T}}F_{iJ}\mathrm{d}X^I\mathrm{d}X^J = (\delta_{IJ} + 2E_{IJ})\mathrm{d}X^I\mathrm{d}X^J = \mathrm{d}X_J\mathrm{d}X^J \tag{7.10}$$

where $F = F_{i,j} = g_J \otimes G^J = g_J \otimes E^J = \frac{\partial x^i}{\partial X^J} e_i \otimes E^J$ and dX^J are known as the contravariant curvilinear vector components for the vector dx_i described in the current deformed configuration. The X^J vector components are paired with the covariant curvilinear coordinate system base vectors $g_J = F^{\mathrm{T}}_{iJ} e_i$, which do change with the coordinate convection. Similarly, dX_J are the covariant vector components of the vector dx_i described in the curvilinear coordinate system. The X_J vector components are paired with the contravariant material coordinate system base vectors $g^J = F^{-1}_{iJ} e^i$. We see $2E_{IJ}$ is the change in metric tensor of the convected coordinates in going from the undeformed reference configuration to the deformed configuration (Rice 1994) http://esag.harvard.edu/rice/e0_Solid_Mechanics_94_10.pdf), where E_{IJ} is the Green–Lagrange finite strain tensor [see Eq. (7.14)]. For schematic depictions of the coordinate transformations (along with a lot of additional explanation and information, as the reader may require), see Sect. 2.10, Figs. 2.10.6 and 2.10.7 at http://homepages.engineering.auckland.ac.nz/~pkel015/SolidMechanicsBooks/Part_III/.

When we have ‘forward convected’ the curvilinear coordinates, we could then transform from the covariant coordinate system g_J to the Cartesian e_i coordinate system. This forward convection and transformation to a Cartesian coordinate system turns out to be the same as transforming a vector (or tensor) directly from an orthogonal material coordinate systems for the undeformed reference geometry to an orthogonal spatial coordinate system for the deformed configuration. In this case, the vector components transform as we have shown in Eq. (7.6). This direct transformation between orthogonal coordinate systems enables intermediate curvilinear coordinate transformations (along with covariant and contravariant coordinate systems) to fade into the background for most solid mechanics applications.

Push-forward and pull-back are also descriptors for coordinate transformations used in differential geometry, and it is apparent that the rules for geometrical transformations of tensors between coordinate systems (e.g. using metric tensor g_{IJ} to raise or lower indices) are intertwined with the physics of the material deformation itself (i.e. $g_{IJ} = F^{\mathrm{T}}_{iJ} F_{iJ}$). Naturally, this is potentially confusing, as it is in other applications where the physics gets ‘mixed up’ with the metric tensor, such as in general relativity (where a mass deforms space-time around it). In solid mechanics, it is normally the physics of the material deformation that is being captured by the expressions ‘push-forward’ (from the initial state to the deformed state) and ‘pull-back’ (from the deformed state to the initial state), as they refer to the material, rather than coordinate transformations in differential geometry, but the coordinate transformations are lurking in the background. The coordinate transformation rules described above have also helped identify an ‘objective’ strain (meaning a coordinate invariant strain e.g. we could use $g_{IJ} = F^{\mathrm{T}}_{iJ} F_{iJ}$ or E_{IJ}), so adopting E_{IJ} as a strain measure allows large deformation solid mechanics predictions of stress to accord more closely with experimental observations. And taking the viewpoint of coordinate transformations has let us see that the solution of the solid mechanics problem in the current deformed configuration can be achieved by means of a coordinate transformation to material coordinates in the undeformed configuration and that the deformation gradient makes this coordinate transformation possible.

Now, the deformation gradient tensor F can be decomposed into a product of two tensors

$$F = RU \tag{7.11}$$

where U is a tensor representing stretching of a vector line element (we note in passing that $U - I$ is known as the Biot strain tensor, which can be employed as a large deformation strain in place of E) and R represents a tensor that rotates the vector line element. The important thing here is that a rotation does not contribute to stress in the material, while a stretch does contribute to stress in a material. Because $R^{\mathrm{T}}R = I$ we see that

$$F^{\mathrm{T}}F = U^{\mathrm{T}}R^{\mathrm{T}}RU = U^{\mathrm{T}}R^{-1}RU = U^2 = C \tag{7.12}$$

The fact that $R^{\mathrm{T}}R = I$ is the 'secret' to the success of the tensor C being 'rotationally invariant' and so being an 'objective' strain measure (note the rotational invariance of C contrasts with the small strain tensor, defined as $\varepsilon = 0.5(F + F^{\mathrm{T}}) - I$, which is not rotationally invariant).

C is known as the right Cauchy–Green tensor, which is the metric tensor g_{IJ}, which ensured lengths are preserved in coordinate transformations [see Eq. (7.8)]. The tensor C can be decomposed into the sum of three tensors representing the 'square of the stretches' (eigenvalues of C) in each of the three principal strain directions (the eigenvector directions of C are defined in the undeformed state), viz.,

$$U^2 = \sum_{\alpha=1}^{3} U_\alpha^2 = \sum_{\alpha=1}^{3} \lambda_\alpha^2 \hat{N}_\alpha \otimes \hat{N}_\alpha \tag{7.13}$$

where λ_α^2 represents the square of the stretch ratio and $\hat{N}_\alpha$ represents the unit vector in a local orthogonal coordinate system aligned with the α principal strain direction.

We can form a range of strain tensors from C (which will all be 'objective' too, meaning that rotation of the body will not influence the calculated strain). The most commonly used strain tensor is the Green–Lagrangian (or Lagrangian) finite strain tensor

$$E = \frac{1}{2}(C - I) \tag{7.14}$$

E measures how much C differs from the identity tensor I and halves the amount (so this normalized strain measure conveniently reduces to small deformation strain measures at small strains). Being able to describe the stress using a strain measure that is a rotational invariant is a crucial advantage of introducing a material coordinate system. E is particularly useful for problems that involve large rotations with modest strains, but a disadvantage of using E as a strain measure is that its strain remains finite as compressions become infinitely large (Neff et al. 2016). This disadvantage is ameliorated somewhat by use of the Biot strain $(U - I)$, but this is more difficult to calculate than E, so it is not widely used. Some advocate using the Hencky strain (ln U) as the most suitable strain measure, which is probably the most useful strain measure for problems that involve large rotations and large strains (Neff et al. 2016), but its use is not widespread.

The deformation gradient F proves useful at every step of the way to finding a solution to the solid mechanics problem. For example, the determinant of F (denoted J) turns out to transform a volume element in the material configuration $(\mathrm{d}V)$ to a volume element in the current configuration $(\mathrm{d}v)$. So,

$$\mathrm{d}v = J\mathrm{d}V \tag{7.15}$$

We mention that $J = \prod_{\alpha=1}^{3} \lambda_\alpha$, i.e. the product of the three principal stretches. The volume relationship shown in Eq. (7.15) is helpful for transforming integration variables (i.e. $\int \mathrm{d}v \to \int J\mathrm{d}V$). For example, rates of production per unit volume in spatial coordinates $s(x,t)$ can be transformed to rates of production $S^{\mathrm{m}}(X,t)$ in material coordinates, viz.,

$$\int_v s(x,t)\mathrm{d}v = \int_V Js^{\mathrm{sm}}(\phi(X),t)\mathrm{d}V = \int_V S^{\mathrm{m}}(X,t)\mathrm{d}V \tag{7.16}$$

We note that Scaler field quantities can be mapped from the deformed geometry to the undeformed geometry by $s(x,t) = s(\phi(X,t),t) = s(X+u(X,t),t) \to s^{\mathrm{sm}}(x-u(X,t),t) = s^{\mathrm{sm}}(\phi^{-1}(x,t),t) = s^{\mathrm{sm}}(X,t)$, where the superscript 'sm' denotes a mapping of the quantity from the deformed geometry to the undeformed geometry. In other words, $s(x,t)$, which is the time-dependent source term in the deformed geometry described in spatial coordinates at time t, is mapped onto the domain of undeformed geometry described in material coordinates in X and is a time-dependent source term in the undeformed geometry at time t, denoted $s^{\mathrm{sm}}(X,t)$. J arises from a change in integration variable associated with the change in coordinate systems associated with the undeformed (reference) and deformed (current) configurations. We can view the J term as adjusting the source term strength in the undeformed geometry, so in the undeformed geometry, it is of equivalent strength to that in the deformed geometry. Then, when the two source terms in either domain are integrated over space and time, the equivalence of source strength is confirmed by the same mass change occurring in the undeformed configuration as in the deformed configuration.

We also realize that we can transform other Scalers between deformed and undeformed coordinates in a similar way, so that for example a concentration of solute in the fluid fraction can be mapped from the deformed to the undeformed geometry $c^{\mathrm{s}}(x,t) \to c^{\mathrm{sm}}(X,t)$. It turns out that fluid porosity maps in a similar way to a source term, so that $n_{\mathrm{f}}^{\mathrm{s}}(x,t) \to n_{\mathrm{f}}^{\mathrm{sm}}(X,t)$, and we define $Jn_{\mathrm{f}}^{\mathrm{sm}}(X,t) = N_{\mathrm{f}}^{\mathrm{m}}(X,t)$, where $N_{\mathrm{f}}^{\mathrm{m}}(X,t)$ has been referred to as the 'fluid Lagrangian porosity' (Coussy

2004). The fluid pressure is interesting as it may be regarded as a second-order tensor or a Scaler, depending on context. When a Scaler, $p(x,t)\rightarrow p^{sm}(X,t)$ in the usual way of Scalers, but when regarded as a second-order tensor, it transforms differently (see discussion below).

A volume $(\mathrm{d}v)$ may be created by dotting an area $\mathrm{d}a$ (with the unit normal $\hat{n}$) and an arbitrary vector $\mathrm{d}l$ in the spatial coordinate system. This volume in the spatial coordinate system transforms to the volume $\mathrm{d}V$ in the material coordinate system, created by dotting an area $\mathrm{d}A$, (with the unit normal $\hat{N}$ in the material coordinate system) with the same vector $\mathrm{d}L$ in the material coordinate system. Then, employing Eq. (7.15), it is apparent these two volumes are related via

$$\mathrm{d}v = \mathrm{d}l \cdot \hat{n}\mathrm{d}a = J\mathrm{d}V = J\mathrm{d}L \cdot \hat{N}\mathrm{d}A \tag{7.17}$$

Now realizing that $\mathrm{d}l_i = F_{iJ}\mathrm{d}L_J$ leads to

$$F_{iJ}\mathrm{d}L_J\hat{n}_i\mathrm{d}a = \mathrm{d}L \cdot (F^{\mathrm{T}} \cdot \hat{n}\mathrm{d}a) = \mathrm{d}L \cdot (\hat{N}J\mathrm{d}A) \tag{7.18}$$

Finally, because $\mathrm{d}L$ is arbitrary the terms in brackets must be equal, which leads to the relation between areas in spatial and material coordinates as

$$\hat{n}\mathrm{d}a = JF^{-\mathrm{T}} \cdot \hat{N}\mathrm{d}A \tag{7.19}$$

Nanson's relation as it is known is important when transforming quantities involving areas (e.g. the flux or divergence of some quantity) from one coordinate system to another, as it enables the surface integral to be performed over the correct area with unit normal in each coordinate system.

We note in passing that $F^{-1} = F_{Ji}^{-1} e_J \otimes e_i$ and $F^{-\mathrm{T}} = F_{Ji}^{-1} e_i \otimes e_J$. In both cases, the indices of the inverse deformation gradient matrix are the same, while the transpose of the matrix implies that the order of the base vectors is reversed.

It is helpful to realize that the transformation of a velocity vector (e.g. ms^{-1}) between coordinate systems is different to the transformation of a flux (e.g. $\mathrm{m}^3\mathrm{m}^{-2}\mathrm{s}^{-1}$), which is a flow rate normal to a unit area. For example, consider the Darcy velocity vector in spatial coordinates $v_{\mathrm{d}}^{\mathrm{s}}$ (ms^{-1}), which transforms as an ordinary vector, viz.,

$$v_{\mathrm{d}}^{\mathrm{m}} = F^{-1} v_{\mathrm{d}}^{\mathrm{s}} \tag{7.20}$$

However, multiplying both sides of Eq. (7.19) by the Darcy velocity vector to obtain the Darcy flow rate $\mathrm{d}q_{\mathrm{d}}^{\mathrm{s}}$ ($\mathrm{m}^3\mathrm{s}^{-1}$) in the spatial coordinate system (a Scaler quantity defined as the fluid volume discharge rate across an area $\mathrm{d}a$ with associated unit normal n) shows the Darcy flow rate transforms to the material coordinate system as

$$\mathrm{d}q_{\mathrm{d}}^{\mathrm{s}} = v_{\mathrm{d}}^{\mathrm{s}} \cdot \hat{n}\mathrm{d}a = Jv_{\mathrm{d}}^{\mathrm{s}}F^{-\mathrm{T}} \cdot \hat{N}\mathrm{d}A = JF^{-1}v_{\mathrm{d}}^{\mathrm{s}} \cdot \hat{N}\mathrm{d}A = Jv_{\mathrm{d}}^{\mathrm{m}} \cdot \hat{N}\mathrm{d}A = \mathrm{d}q_{\mathrm{d}}^{\mathrm{m}} \quad (7.21)$$

Note that the Darcy flux (discharge flow rate per unit area, $\mathrm{m}^3\mathrm{m}^{-2}\mathrm{s}^{-1}$) in the spatial coordinate system (deformed configuration) is defined as $v_{\mathrm{d}}^{\mathrm{s}} \cdot \hat{n}(= \mathrm{d}q_{\mathrm{d}}^{\mathrm{s}}/\mathrm{d}a)$, while the equivalent Darcy flux (discharge flow rate per unit area, $\mathrm{m}^3\mathrm{m}^{-2}\mathrm{s}^{-1}$) in the material coordinate system (undeformed configuration) is $Jv_{\mathrm{d}}^{\mathrm{m}} \cdot \hat{N} = (\mathrm{d}q_{\mathrm{d}}^{\mathrm{m}}/\mathrm{d}A)$. Again, the J can be viewed as adjusting the Darcy velocity vector in the material coordinate system due to the change in area between deformed and undeformed configuration. By this means, the Darcy flux in the undeformed configuration is the same as that the Darcy flux in the deformed configuration.

It is very helpful to know how to transform a gradient operator (∇^{s}) and a divergence operator ($\nabla^{\mathrm{s}}\cdot$) in spatial coordinates, into a gradient operator (∇^{m}) and a divergence operator ($\nabla^{\mathrm{m}}\cdot$) in material coordinates. When a derivative is taken in curvilinear (material) coordinates, the chain rule suggests it is not only the components of the quantity of interest that change, but also the base vectors themselves change. This raises the prospect of rather complicated derivatives when using material coordinates in solid mechanics problems. For example, the covariant gradient of Darcy velocity vector $v_{\mathrm{d}}^{\mathrm{s}}$ in spatial and a general curvilinear coordinate system is

$$\nabla^{\mathrm{s}} v_{\mathrm{d}}^{\mathrm{s}} = \left(\frac{\partial v_{\mathrm{d}}^{L}}{\partial X^{J}} + v_{\mathrm{d}}^{K} \Gamma^{L}_{.\mathrm{JK}} \right) F^{-1} \quad (7.22)$$

where $v_{\mathrm{d}}^{\mathrm{s}}$, v_{d}^{L}, and v_{d}^{K} are the same vector field in the spatial coordinate system and $\Gamma^{L}_{.\mathrm{JK}}$ is a Riemann–Christoffel symbol of the second kind. The first term on the right-hand side relates to the differentiation of the vector components, while the second term relates to the differentiation of the base vectors (the Riemann–Christoffel symbol manages the 'connection' between the basis functions and their derivative).

Fortunately though, a significant simplification occurs due to the so-named 'compatibility equations' in solid mechanics, which ensure no gaps develop between material particles, and a theorem due to Riemann that the Riemann–Christoffel symbol formed from a symmetric metric tensor should be zero if the tensor is a metric tensor in Euclidean space. It appears that (Kozak 2000) has worked this through, concluding that the compatibility equations obtained from the principle of complementary virtual work are equivalent to the Riemann–Christoffel symbols always being zero.

So for this reason, in solid mechanics the gradient operator transforms between material and spatial coordinate systems as,

$$\nabla^{\mathrm{s}} = \nabla^{\mathrm{m}} F^{-1} \quad (7.23)$$

We note the difference between a 'vector transformation' [e.g. Eqs. (7.6) and (7.20)] and a 'gradient operator transformation' [i.e. Equation (7.23)]. The vector transformation is said to be contravariant, while the gradient transformation is covariant http://en.wikipedia.org/wiki/Covariance_and_contravariance_of_vectors. So

$$\nabla^{\mathrm{s}} v_{\mathrm{d}}^{\mathrm{s}} = \nabla^{\mathrm{m}} F^{-1} F v_{\mathrm{d}}^{\mathrm{m}} = \nabla^{\mathrm{m}} v_{\mathrm{d}}^{\mathrm{m}} \tag{7.24}$$

It is apparent that the vector components change in the opposite direction to the basis function, so if the basis function rotates in one direction, the vector components rotate in the opposite direction, and if the basis function is stretched, the vector components proportionally shrink. By this means, the vector itself is independent of the coordinate system employed to represent it.

Using Eqs. (7.21) and (7.23), we can, for example, transform Darcy's constitutive equation in spatial coordinates to Darcy's equation in material coordinates using the above transformation rules, viz.,

$$\begin{gathered} v_{\mathrm{d}_i}^{\mathrm{s}} = -k_{ij} \nabla_j^{\mathrm{s}} p \\ JF^{-1} v_{\mathrm{d}}^{\mathrm{s}} = JF_{Ii}^{-1} v_{\mathrm{d}_i}^{\mathrm{s}} = -JF_{Ii}^{-1} k_{ij} F_{jJ}^{-\mathrm{T}} \nabla_J^{\mathrm{m}} p^{\mathrm{sm}} = -K_{IJ}^{\mathrm{m}} \nabla_J^{\mathrm{m}} p^{\mathrm{sm}} = q_{\mathrm{d}_I}^{m} \end{gathered} \tag{7.25}$$

where $p(x,t) \rightarrow p^{\mathrm{sm}}(X,t)$ and $K_{IJ}^{\mathrm{m}} = JF_{Ii}^{-1} k_{ij} F_{jJ}^{-\mathrm{T}}$. The fluid pressure transforms as a Scaler here, while the fluid flow permeability transforms as a second-order tensor. Because a second-order tensor involves two vectors (a vector quantity and the unit normal vector to the area associated with that quantity), it is perhaps not too surprising that the (inverse) deformation gradient appears twice in the transformation of a second-order tensor (though the normal 'area vectors' are both unit vectors, the areas associated with these units vectors change between deformed and undeformed configurations). It is the inverse deformation gradient that appears in the transformation because it is from the deformed configuration to the undeformed configuration. J arises from a change in integration variable [i.e. $\mathrm{d}v \rightarrow J\mathrm{d}V$; see Eq. (7.15)] associated with the maintaining equivalent divergences in each of coordinate systems associated with the deformed (current) and undeformed (reference) configurations.

We need to include the change in material properties associated with the deformation in the analysis on the undeformed configuration, i.e. the terms JF_{Ii}^{-1} and $F_{jJ}^{-\mathrm{T}}$ in K_{IJ}^{m} are to do with coordinate transformations, not the change in material properties associated with the deformation. So to use K_{IJ}^{m} in an analysis in the undeformed configuration, we need to take into account that permeability (k_{ij}) is changing in the deformed configuration, and express this change in the undeformed configuration using material coordinates. We can do this in different ways, but if the permeability is a Scaler (i.e. the permeability is independent of direction), then for particulate systems. it is commonplace to express the permeability as a function of the porosity $n_{\mathrm{f}}^{\mathrm{s}}(x,t)$. So, for example, there is the Kozeny–Carmen equation—one form being (Xu and Yu 2008),

$$k_{ij}^{\mathrm{s}} = \frac{d^2}{180} \frac{(n_{\mathrm{f}}^{\mathrm{s}})^3}{(1 - n_{\mathrm{f}}^{\mathrm{s}})^2} \tag{7.26}$$

where d is the average particle size that has the same specific surface area as the particles in the porous media. See Xu and Yu (2008) for numerous other forms of the Kozeny–Carmen equation, which may be more or less suitable depending on the application. We can then express the spatial porosity in terms of porosity in material coordinates and J (this is shown later to be $n_{\mathrm{f}}^{\mathrm{s}}(x,t) = 1 + \frac{(n_0(X,0)-1)}{J}$, where n_0 is the original fluid porosity in the undeformed configuration). So we see that the permeability k_{ij}^{s} can be expressed as a function of J; i.e. the change in permeability in the spatial coordinate system is mapped across to the same change in material coordinate system, viz., $k_{ij}^{\mathrm{s}}(n_{\mathrm{f}}^{\mathrm{s}}) \rightarrow k_{ij}^{\mathrm{sm}}(J)$. Another approach, more appropriate for cartilage, is to express the cartilage permeability as a function of the aggrecan concentration. Expressing the current concentration in terms of initial concentration again turns out to be a function of J (this is shown later to be $c^{\mathrm{sm}}(X,t) = \frac{c^{\mathrm{m}}(X,0)}{1+(J-1)/n_0}$). When k_{ij}^{s} is a second-order tensor rather than a Scaler, each component of the second-order tensor has its own functional relationship in terms of porosity (or other relevant variables).

Integration of Nanson's relation [Eq. (7.19)] over the surfaces of the differential volumes in the undeformed and deformed configuration (e.g. of the velocity $\int v_{\mathrm{d}}^{\mathrm{s}} n da = \int JF^{-\mathrm{T}} v_{\mathrm{d}}^{\mathrm{s}} \cdot N dA$), applying the divergence theorem, then equating the integrand multiplied by volume differential for the deformed and undeformed configurations, and using Eq. (7.15), leads to the divergence operator transformation rule

$$\nabla^{\mathrm{s}} \cdot = \frac{1}{J} \nabla^{\mathrm{m}} \cdot JF^{-1} \tag{7.27}$$

For the divergence of the Darcy velocity, vector we find

$$\nabla^{\mathrm{s}} \cdot v_{\mathrm{d}}^{\mathrm{s}} = \frac{1}{J} \nabla^{\mathrm{m}} \cdot JF^{-1} v_{\mathrm{d}}^{\mathrm{s}} = \frac{1}{J} \nabla^{\mathrm{m}} \cdot J v_{\mathrm{d}}^{\mathrm{m}} = \frac{1}{J} \nabla^{\mathrm{m}} \cdot q_{\mathrm{d}}^{\mathrm{m}} \tag{7.28}$$

and substituting the Darcy flux vector from Eq. (7.25) find,

$$\nabla^{\mathrm{s}} \cdot v_{\mathrm{d}}^{\mathrm{s}} = -\frac{1}{J} \nabla^{\mathrm{m}} \cdot K_{IJ}^{\mathrm{m}} \nabla_J^{\mathrm{m}} p^{\mathrm{m}} \tag{7.29}$$

When we have solved the governing equations in material coordinates, we would like to express the solution in spatial coordinates, so we need to know how to transform tensors from the undeformed configuration to the deformed configuration. First, we note that a second-order tensor (say the Cauchy stress tensor $\sigma_{ij} = \sigma$) in the spatial coordinate system can be 'pulled back' to form the so-called second Piola–Kirchhoff stress tensor (S) in the material coordinate system using

$$S = JF^{-1}\sigma F^{-\mathrm{T}} \tag{7.30}$$

just as the permeability tensor was pulled back (i.e. $k_{IJ}^{\mathrm{m}} = JF_{Ii}^{-1}k_{ij}^{\mathrm{sm}}F_{jJ}^{-\mathrm{T}}$). And of course the second Piola–Kirchhoff stress tensor S in material coordinates can be 'pushed forward' to the Cauchy stress in spatial coordinates using

$$\sigma = \frac{1}{J}FSF^{\mathrm{T}} \tag{7.31}$$

But deformation of saturated porous media solid phase is not controlled by the total stress σ_{T}, rather it is controlled by the solid stress σ^{s}. The 'principle of effective stress' states that the total Cauchy stress equals the solid stress minus the pore water pressure, viz., $\sigma_{\mathrm{T}} = \sigma^{\mathrm{s}} - p$ in the spatial coordinate system. So the total Cauchy stress tensor is in two parts, and the total Cauchy stress in the spatial coordinate system can be 'pulled back' to the material coordinate system by

$$S_{\mathrm{T}} = S^{\mathrm{s}} - P^{\mathrm{m}} = JF^{-1}\sigma_{T}^{\mathrm{sm}}F^{-\mathrm{T}} = JF^{-1}\sigma^{\mathrm{sm}}F^{-\mathrm{T}} - JF^{-1}p^{\mathrm{sm}}IF^{-\mathrm{T}} \tag{7.32}$$

where S_{T} is the total second Piola–Kirchhoff stress in the material coordinate system, S^{s} is the second Piola–Kirchhoff solid stress in the material coordinate system, $P^{\mathrm{m}} = JF^{-1}F^{-\mathrm{T}}p^{\mathrm{sm}} = JHp^{\mathrm{sm}}$ is the second Piola–Kirchhoff pore water pressure in the material coordinate system (Armstrong et al. 2016), H is the inverse of Finger's (left Cauchy–Green) strain tensor ($B = FF^{\mathrm{T}}$), and p^{sm} is the fluid pressure in the spatial coordinate system (deformed configuration) mapped across the material coordinate system (undeformed configuration). Using Eqs. (7.27) and (7.31), we see the divergence of the total Cauchy stress $\nabla^{\mathrm{s}} \cdot \sigma_{\mathrm{T}}^{\mathrm{sm}}$ expressed in material coordinates is given by

$$\begin{aligned} &\tfrac{1}{J}\nabla^{\mathrm{m}} \cdot JF^{-\mathrm{T}}[\sigma^{\mathrm{sm}} - p^{\mathrm{sm}}I] \\ &\tfrac{1}{J}\nabla^{\mathrm{m}} \cdot [J\sigma^{\mathrm{sm}}F^{-\mathrm{T}} - Jp^{\mathrm{sm}}IF^{-\mathrm{T}}] \\ &\tfrac{1}{J}\nabla^{\mathrm{m}} \cdot [P^{\mathrm{s}} - P^{*}] \end{aligned} \tag{7.33}$$

where P^{s} is a two-point tensor known as the first Piola–Kirchhoff stress tensor ($P^{s} = J\sigma^{s}F^{-\mathrm{T}}$) and P^{*} is the first Piola–Kirchhoff pore water pressure in a mixed coordinate system. Comparing Eqs. (7.32) and (7.33), we see P^{*} and P^{m} are related by $P^{*} = FP^{\mathrm{m}}$. We observe in passing from Eq. (7.33) that the gradient of the pore water pressures act like body forces in the equations describing deformation of the solid phase in terms of effective stresses.

If we multiply the Cauchy stress in the deformed configuration by an area ($\mathrm{d}a$ with unit normal n), we find the force $\mathrm{d}f$ in the spatial coordinate system acting on that area. Then, employing Nanson's relation [Eq. (7.19)], we find

$$\mathrm{d}f = t\mathrm{d}a = \sigma^{\mathrm{s}} \cdot \hat{n}\mathrm{d}a = J\left(\frac{1}{J}FSF^{\mathrm{T}}\right)F^{-\mathrm{T}} \cdot \hat{N}\mathrm{d}A = P^{\mathrm{s}} \cdot \hat{N}\mathrm{d}A = T\mathrm{d}A = \mathrm{d}f_0 \quad (7.34)$$

where $\mathrm{d}f_0$ in the material coordinate system is the same force vector $\mathrm{d}f$ in the spatial coordinate system. Note that the first Piola–Kirchhoff stress tensor ($P^{\mathrm{s}} = FS$) dotted with the unit vector normal to the surface in the undeformed body (i.e. the surface traction T in material coordinates in the reference configuration) is the actual traction on the surface of the deformed body (t) in spatial coordinates. Consequently, the second Piola–Kirchhoff stress dotted with the unit vector normal to the surface of the undeformed body is not the actual traction on the surface of the deformed body.

Similarly, for the traction force due to the fluid pressure, i.e. $FP^{\mathrm{m}} \cdot \hat{N} = P^* \cdot \hat{N} = p^{\mathrm{sm}}JF^{-\mathrm{T}} \cdot \hat{N} = pI \cdot \hat{n}$. It is apparent that the first and second Piola–Kirchhoff pore water pressures vary with position and with direction in the material coordinate system (because stretching is direction specific), unlike the fluid pressure counterpart in the spatial coordinate system, where the fluid pressure is independent of direction. Note that for the fluid balance equation in material coordinates [Eq. (7.29)], we can treat pressure as a Scaler and use the mapping transformation $p(x,t) \rightarrow p^{\mathrm{sm}}(X,t)$, because the necessary tensor transformation to account for directional dependence of the fluid pressure in the material coordinate system is 'hidden' in the transformation associated with the permeability tensor, i.e. in $K^{\mathrm{m}}_{IJ} = JF^{-1}_{Ii}k^{\mathrm{sm}}_{ij}F^{-\mathrm{T}}_{jJ}$.

Correct 'conjugate variables' need to be chosen so that their product has units of work per unit volume. But this statement is ambiguous because there are two unit volumes, namely the unit volume in the reference configuration and the unit volume in the current configuration. Either work rate may be of interest, but when comparing work rates in the two configurations, we need to make sure that the local work rates are for the same material element. This means that the '*spatial stress power must be equal to its Lagrangian counterpart*' (Neff et al. 2016). The work rate conjugate of the second Piola–Kirchhoff stress S is the Green–Lagrange strain tensor rate $\frac{\mathrm{d}E}{\mathrm{d}t}$ (which are both in the reference configuration), i.e. $P = S : \dot{E}$ (where P is the power, i.e. work per unit time), while the work conjugate of σ^{s} is d^{s} (i.e. $P = \sigma^{s} : d^{s}$), the rate of deformation in spatial coordinates ($d^{\mathrm{s}} = \nabla^{\mathrm{s}}v^{\mathrm{s}}_{\mathrm{s}}$ where $v^{\mathrm{s}}_{\mathrm{s}}$ is the velocity of the solids in the deformed configuration—we note that $d^{\mathrm{s}} = F^{-\mathrm{T}} \cdot \dot{E} \cdot F^{-1}$). So to compare the work rates over the same material element in undeformed and deformed configurations, we have

$$S : \frac{\mathrm{d}E}{\mathrm{d}t}(\mathrm{d}V) = \sigma^{\mathrm{s}} : d^{\mathrm{s}}(\mathrm{d}v) \quad (7.35)$$

That is the work rate per unit reference (undeformed) volume is (Neff et al. 2016),

$$S : \frac{\mathrm{d}E}{\mathrm{d}t} = J\sigma^{\mathrm{s}} : d^{\mathrm{s}} \tag{7.36}$$

To solve the equations of solid mechanics, we need a constitutive relationship between stress and strain. If the strain energy is 'path independent' (meaning that the order of applying load increments has no effect on the final deformed state), then strain energy is a thermodynamic potential (e.g. a Helmholtz potential). Path-independent constitutive relations are known as 'hyperelastic' models and have a strain energy density function. Path-dependent models are known as 'hypoelastic' models and do not have a strain energy density function https://en.wikipedia.org/wiki/Hypoelastic_material. Suitable (Helmholtz) strain energy density functions (denoted Ψ^e) that are often employed in practice are so-named neo-Hookean, polynomial, or Mooney–Rivlin strain energy density functions https://en.wikipedia.org/wiki/Hyperelastic_material (there are many other models). We note that the strain energy density function is defined per unit volume of the initial configuration. By definition of the strain energy density function, stress and strain are related by

$$\mathrm{d}S^{m}_{IJ} = \frac{\partial \Psi^{e}}{\partial E_{IJ} \partial E_{\mathrm{KL}}} \mathrm{d}E_{\mathrm{KL}} \tag{7.37}$$

It is necessary to take the time derivative in the reference (undeformed) configuration (e.g. $d^{s} = F^{-\mathrm{T}} \cdot \dot{E} \cdot F^{-1}$), for if the time derivative is taken in the deformed configuration, the time derivative is a partial derivative with respect to time because this time derivative (in the spatial coordinate system) does not take account of the time rate of change in the deformation gradient. It is well known that the material (or total) derivative of a Scaler quantity with respect to time (d/dt) in the material coordinate system is related to the time derivative in the spatial coordinate system by

$$\frac{\mathrm{d}}{\mathrm{d}t} = \frac{\partial}{\partial t} + v^{\mathrm{s}}_{\mathrm{s}} \nabla^{\mathrm{s}} \tag{7.38}$$

while for a tensor quantity of order one (i.e. a vector) and larger, the gradient operator is replaced by a divergence operator, so the material time derivative is related to the spatial time derivative by

$$\frac{\mathrm{d}}{\mathrm{d}t} = \frac{\partial}{\partial t} + v^{s}_{s} \nabla^{s} \cdot \tag{7.39}$$

The advantage of using the material coordinate system is the 'time invariance' of tensors. Sometimes, the stress–strain relations are written in a rate form (Ji et al. 2013), e.g. when considering inelastic or hypoelastic materials. Unfortunately, the stress rate for the Cauchy stress does not transform as a tensor should (which means that if we use this matrix, we will get different answers depending on the coordinate system chosen). This is clearly unsatisfactory, as in reality, there is only one solution. The solution is to take the total or material derivative of the second-order tensor. To do this, we can pull-back the Cauchy stress to the material coordinate system, take the time derivative of the stress in the material coordinate system, and then push the stress rate forward to the spatial coordinate system. This process for finding a stress rate is an example of a so-called Lie derivative. For example, the Truesdell stress rate tensor $\tilde{\sigma}$ is Bonet and Wood (2008),

$$\tilde{\sigma} = \frac{1}{J} F \left[\frac{\mathrm{d}}{\mathrm{d}t} (J F^{-1} \sigma F^{-\mathrm{T}}) \right] F^{\mathrm{T}} \tag{7.40}$$

There are many other objective stress rate tensors, the common ones being the Oldroyd and Green–Naghdi stress rate tensors, as well as the time rate of change of the second Piola–Kirchhoff stress tensor $\frac{\mathrm{d}S}{\mathrm{d}t}$ (and its associated strain rate $\frac{\mathrm{d}E}{\mathrm{d}t}$).

The fact that the material coordinate system enables stress and strain tensors to be formulated that are independent of material or frame rotations, enabling 'objective' stress rate tensors to be defined, is the fundamental reason the material coordinate system is needed with all its attendant complications for large deformation problems, as described above (unfortunately, alternative approaches are generally even more complicated). This independence of material or frame rotations is sometimes referred to as the principle of material objectivity, though now it is simply referred to as 'frame invariance.' See http://en.wikipedia.org/wiki/Objectivity_%28frame_invariance%29 for further discussion.

With this introduction, we begin by considering Karl Terzaghi's contribution to one-dimensional 'consolidation theory.' It turns out that this theory is very helpful for understanding the basics of poroelastic consolidation. Though this theory was developed for soils, it turns out that the same arguments and theory can be directly applied to articular cartilage virtually unchanged.

7.2 Poroelasticity

7.2.1 *Terzaghi's 'Principle of Effective Stress' and 1-D Consolidation Theory*

In this section we develop Terzaghi's theory for consolidation of saturated soil. The theory was originally developed for sand or clay soils (which can transmit load between soil particles in quite different ways). But it turns out that as far as porous media theory is concerned, mathematically there is no significant difference between soil materials and soft biological tissues—including cartilage extracellular matrix. Both these materials are porous mixtures of small-sized solids (small, charged, or uncharged soil particles for clay soils, or small, charged, or uncharged macromolecules for biological materials), which are saturated (or bathed) in a pore fluid that almost (or does) entirely surround the clay particles or does entirely surround the biological macromolecules in solution.

Soil particles are the weathering product of rock. A saturated soil is a composite material made up of a solid phase and a fluid phase, that is: (i) the solid particles weathered from rock and (ii) water in the interconnected pore openings. There is a theory of partially saturated soils that has both water and gas in the pore openings, but that does not concern us here, as clearly all biological materials are saturated with salty water (referred to as isotonic saline solution).

When a compressive load is applied to the surface of soil, the saturated soil is compressed by the load and water is squeezed from the soil. Exactly, the same things happen for biological materials—water is squeezed from a tissue when it is loaded. For example, water is squeezed from your gluteal muscles when you sit, water is squeezed from epidermal tissues covering the soles of your feet when you stand, (and even a very small amount of water is squeezed from your weight-bearing bones when you stand—though bone is a hard tissue rather than a soft tissue).

In pathophysiological states, edematous tissue may exhibit 'pitting edema' (i.e. thumb pressure on the edematous region of soft tissue for 15 s or so leaves a distinct thumb impression in the tissue, which gradually resumes its previous state upon release of thumb pressure). To analyze the water being squeezed from soil or from tissue, we first need a theory that transforms the physical problem into a mathematical problem, which can hopefully tell us something useful about the time-dependent 'consolidation settlements.' Prior to 1925, no published mathematical theory existed about time-dependent consolidation settlements.

In 1925, Karl Terzaghi published a book describing how to estimate the settlements of buildings placed on the surface of water-saturated soils (*Erdbaumechanik auf bodenphysikalischer Grundlage 1925*). This book is important because it contained two critically important advances: (i) the use of 'partial stresses,' and (ii) it showed how the partial stresses, a conservation equation, and constitutive equations can be assembled, to form a well-posed mathematical problem that has a unique solution. The solutions of this mathematical problem can then be employed to predict how the soil will deform when loaded. For example, Terzaghi applied his theory to predict future settlements of the Great Dome at MIT campus in Massachusetts, which soon after construction finished, began to noticeably settle into the 'Boston Blue' clay on which it is founded.

Terzaghi's key idea for the analysis of this (and other problems) is that each phase making up the 'soil composite' carries some portion of the total load. That is, the incremental total applied load is carried partly by the soil particles and partly by the water, viz.,

$$\sigma_{\mathrm{T}} = \sigma^{\mathrm{s}} - pI \tag{7.41}$$

where σ_{T} is the (incremental) 'total stress' carried by the soil composite (i.e. the porous media), σ^{s} is the (incremental) 'solid stress' carried by the solid phase of the porous media, p is the (incremental) fluid stress carried by the fluid phase of the porous media, and I is the identity matrix (often, the identity matrix is not formally written down, as its context in Eq. (7.41) means the identity matrix is implied).

When referring to solid stress, we said above that the stress is 'carried by' the solid phase. We can be more specific and say the solid stress is the stress between solid particles, and the solid particles are capable of maintaining similar spatial relationships between the particles as the material deforms under load. This is in contrast to a fluid, where the particles form new particle associations as the fluid deforms. In Chap. 6, it became apparent that this solid stress could actually pass from solid particle to adjacent solid particle through the intervening fluid in a diffuse double-layer between the particles. So solid stress does not necessarily only pass through solid particles, it can on some occasions pass through the intervening fluid.

One could write Eq. (7.41) more precisely as

$$\Delta\sigma_{\mathrm{T}} = \Delta\sigma^{\mathrm{s}} - \Delta pI \tag{7.42}$$

to be explicit about the fact that all quantities are incremental, but it conventional to write Eq. (7.42) as shown in Eq. (7.41). By convention, tensions are taken as positive for the stress tensors, while a negative pore fluid pressure is taken to be positive. The same fluid pressure convention is adopted in solid mechanics (and cartilage) mechanics, but unlike cartilage mechanics, for soil mechanics, the convention is compressions are taken as positive for the stress tensors (i.e. so we note for the soil mechanics literature, Eq. (7.41) is normally written $\sigma^{\mathrm{s}} = \sigma_{\mathrm{T}} - p$, where the identity matrix is implied). We also mention that in the soil mechanics literature, the solid stress (σ^{s}) is denoted σ', where the prime denotes the 'effective stress.' Terzaghi's discovery of the principle of effective stress is described in detail by Skempton (1960).

The incremental fluid pressure p is called the 'excess pore water pressure,' to emphasize that it is the incremental fluid pressure arising from mechanical loading, separating this pressure from any other fluid pressure, whatever its origins may be (see further discussion on fluid pressures below). We also mention that if there is an initial self-equilibrating prestress within the material (as there is in cartilage), the change in effective stress is the sum of the changes in effective stress of the material components making up the material. So, for example, if collagen is initially in tension while is in aggrecan in compression, the change in effective stress due to the change in total stress σ_{T} caused by surface loading of the material is the sum of the change in effective stresses for collagen and aggrecan. The changes in effective stress in each component of the biocomposite can only be uncovered with additional information about the behavior and internal arrangement of the component materials.

The solid stress carried by the solid phase is called the 'effective stress' σ^{s} because it is the solid stress that is 'effective' in causing the assembly of soil particles to deform. Here, we use effective stress or solid stress interchangeably. It is important to realize that a hydrostatic fluid pressure p does not cause the porous material to deform; i.e. there is no deformation of the particles caused by hydrostatic pressure p (the particles are so stiff that usual physiological fluid pressures do not cause them to deform significantly, and this deformation can be ignored). The incremental fluid pressure p is called the 'excess pore pressure' because this distinguishes fluid pore pressure due to the loading from any long-term fluid pore water pressure. The long-term fluid pressure is clearly not involved in carrying an incremental applied load, as this fluid pressure was there prior to the application or removal of any load.

As an important aside, there may be existing pore fluid 'driving pressure' (p_d) that causes fluid to flow through the porous material or tissue; that is, there is a pre-existing Darcy velocity in the tissue given by $v_d^s = -k_w^s \nabla^s p_d$, where $\nabla^s p_d$ is the 'driving pressure' gradient. A driving pressure gradient may be pre-existing in a porous material and continue long after any consolidation event is complete. For example, there is a mean arterial pressure gradient across the arterial wall that causes a very small amount of fluid to flow continuously through the arterial wall and similarly for capillaries.

More specifically, it is the gradient in 'driving pressure' (p_d) that causes fluid to flow through a saturated porous material, not a gradient in the measured fluid pressure (p_m). The driving pressure can be calculated from the measured pressure using $p_d = p_m - \rho_f g z$, where z is aligned with the direction of gravity. For hydrostatic conditions (meaning no flow conditions), p_d = constant. The position for the origin of z can be adjusted so that p_d = constant = 0. When $p_d = 0$, the measured pressure (p_m) equals the gravitational hydrostatic pressure (due to the weight of the overlying fluid) at every position in a static fluid. There is no gradient in driving pressure, and so there is no fluid flow.

A nonzero driving fluid pressure gradient ($\nabla p_d \neq 0$) may be generated by pumping of the heart. But often the driving pressure, the measured pressure, and the gravitational hydrostatic pressure are not clearly identified as specific descriptors, and only 'pressure' is discussed. This can lead to confusion. To avoid measuring the wrong pressure, it is for this reason that 'blood pressure' (which normally refers to measured arterial pressure at the level of the aortic valve) needs to be consistently measured in the upright position at the same level as the heart (or more specifically, at the same level as the aortic valve). In this way, the gravitational hydrostatic component ($\rho_f g z$) of the measured pressure is then the same at both the heart and at the position of blood pressure measurement. Because the column of blood in the arterial vasculature above the level of the heart is approximately the same at the level of the aortic valve as it is at the level of a person's mid-arm, the mid-arm position is the clinically preferred point of blood pressure measurement. At this location, most of the recorded blood pressure is due to pressure generated by the heart acting as a pump. Any blood pressure measurement error is related to the vertical position of the aortic valve relative to the blood pressure cuff at mid-arm (which should be small), and any pressure loss as the blood flows from the aortic valve to the brachial artery in the mid-arm (which is also small).

The gravitational hydrostatic component of measured blood pressure at the mid-arm is likely to change with body size and shape. This potential confounding effect is comparatively small relative to other confounding effects on blood pressure (e.g. misusing a blood pressure measuring device, or 'white coat hypertension'), so size and shape of the person are normally neglected when comparing blood pressure measurements in adults. But we mention that if we expect that the tissue perfusion pressure at the top of the brain to be constant in different people, then we should expect there will be a variation in blood pressure with a person's height (blood pressure being on average higher in a taller person). There is evidence to support this, but height is not usually considered in adult blood pressure tables.

We also mention that when in the upright position, it is the gravitational hydrostatic pressure in the large veins at the level of the heart that drives filling of the heart with blood. As the right atrium expands, the pressure inside an expanding right atrium is normally taken to be zero, so in the upright position, the gravitational hydrostatic pressure at the level of the heart (measured from the atrial opening to the vena cava, to the height of zero pressure in the jugular vein) is essentially the 'filling pressure' for the heart. When in the supine position, the filling pressure for the heart is referred to as the 'central venous pressure.'

For cartilage at equilibrium, the total fluid pressure in the cartilage minus the gravitational hydrostatic fluid pressure in the cartilage is constant (i.e. the driving pressure is constant). So the driving pressure gradient is zero, and there is no fluid flow through cartilage tissue at equilibrium. The only way there could be permanent driving pressure gradient and so long-term flow through articular cartilage is if there is a driving pressure difference between the fluid in the interstitial space of the bone marrow and the synovial fluid.

At rest, fluid pressure in synovial fluid is actually slightly negative (e.g. Roth et al. (1984) reports minus 3.8 mm Hg in the pig temporomandibular joint at rest, but this pressure increases with mastication, while Wood et al. (1988) report pressures of about minus 5 ± 1.5 mm Hg in the human knee joint. The synovial membranes containing the synovial fluid in the knee joint are quite distensible, so significant excess fluid is required to elevate synovial fluid pressure above atmospheric pressure [e.g. approximately 5 mls of additional fluid raises the fluid pressure to positive values (Wood et al. 1988)]. Pathological states (e.g. trauma, bursitis, or synovitis) may result in very high synovial fluid pressures (2–65 mm Hg), while osteoarthritis may result in significantly elevated synovial pressures of 7–21 mm Hg.

Bone marrow interstitial pressure in rats is found to be about 10 mm Hg (Iversen et al. 2001). This interstitial pressure is mainly driven by leakage of fluid from bone marrow capillaries. Assuming human bone marrow interstitial fluid pressures are similar, the maximum normal driving pressure gradient between marrow and synovial fluid is probably about 15 mm Hg (i.e. 2 kPa). Normally, the osteochondral junction is considered impermeable, but with increasing age and disease (e.g. OA) the osteochondral junction may become permeable. In this case, there may up to about 2 kPa driving pressure from the marrow to the synovial fluid through articular cartilage tissue. If synovial fluid pressures are elevated (as they often are in OA), then this pressure gradient may reverse, and there may be up to a 2 kPa driving pressure gradient from the synovial fluid to bone marrow. Because the permeability of articular cartilage is so low, this small driving pressure results in Darcy velocities through the tissue of the order of 1 nm/s (i.e. at this velocity, it would take of the order of 20 days for a water molecule to pass through tibial cartilage).

Gravitational hydrostatic pressures in cartilage tissue at rest are normally tiny (of the order of only plus or minus 20 Pa or so, i.e. the pressure exerted by a couple of mm of water across the thickness of articular cartilage). Typically, interstitial fluid pressures are zero to slightly negative for most tissues of the body (Hall 2015), even in the normal leg after standing for normal lengths of time. This is somewhat counter-intuitive given the large hydrostatic pressure arising from the column of blood in the large vessels that extend from the knee to the top of the head, while a similar 'column' of interstitial fluid extends from the knee to the top of the head. But a slightly negative pressure is maintained in the tissue by the removal of proteins and interstitial fluid via the lymphatic system. Indeed, for normal tissues it is ongoing fluid removal by the lymphatic system $\left(s_{\text{lymphatic drainage}}\right)$ that prevents hydrostatic interstitial fluid pressures becoming significantly elevated. The interstitial fluid pressure is controlled by,

$$C_{\text{interstitial compliance}} \frac{\mathrm{d}p_{\text{interstitial}}}{\mathrm{d}t} = s_{\text{vertical drainage}} + s_{\text{from plasma}} - s_{\text{lymphatic drainage}} \tag{7.43}$$

where the terms in this equation are self-explanatory. The hydrostatic gravitational interstitial fluid pressure is dissipated by the downward movement (i.e. by the vertical drainage) of interstitial fluid, so it never gets a chance to develop significant fluid pressure in the interstitial space given the magnitude of the lymphatic sink term. And insufficient fluid enters from the blood plasma across the endothelial cells lining the vessels, so this source too does not get a chance to develop significant fluid pressure in the interstitial space given the magnitude of the lymphatic sink term.

While hydrostatic pressures in the interstitial spaces of normal leg tissues do increase somewhat upon standing, the increase in hydrostatic interstitial pressure is limited by the myogenic reflex that increases flow resistance at the level of arterioles (which reduces capillary pressure), the presence of valves in veins (which reduces the height of the hydrostatic pressure column on the venous side, at least when there are muscle contractions), and there is increased fluid flow in the lymphatics (which can actively pump fluid along small vessels, driven by (lining) cell contractions. In addition, the lymphatics have valves to prevent backflow. As a bonus, the increased flow through the interstitial space (from capillary to lymphatics) causes a reduction in interstitial protein content (which helps to retain fluid in the capillaries) (Hall 2015).

In contrast to the small rise in interstitial fluid pressures in leg tissues, the pressure in capillaries can rise by 50 mm Hg upon standing ['*…capillary pressure in the feet increases by more than 50 mm Hg relative to values measured when supine, owing to the large hydrostatic column in arteries and veins.*' (Scallan et al. 2010)], so it is important to realize that all capillaries are not the same throughout the body. Some capillaries are fenestrated and fluid easily passes through the endothelial lining cells, while other capillaries are very impermeable and so have little fluid flow. Indeed, it is a general principle in the body that tissues and cells with the same anatomical name can be very highly specialized for their particular location, and so functionally, they can be very different in nominally similar tissues.

The capillaries in the feet are a case in point and must have a very low filtration coefficient relative to other capillaries that are not subject to high hydrostatic pressure. This low filtration coefficient reduces the rate of fluid flow into tissues to a more manageable rate (i.e. the flow rate is so low that the lymphatics can easily manage this rate). We again mention that the synovial tissues are well supplied with lymphatics. We also observe that fluid redistribution in the interstitial compartment (i.e. vertical drainage) is very slow because interstitial fluid permeability is very low (e.g. so fluid in muscle tissue does not drain quickly into the knee joint) and the small amount of fluid that does move due to gravity is normally easily removed by the lymphatic system.

When a tissue is mechanically loaded, it generates 'excess pore water pressures' in the tissue, which drive fluid flow through the tissue. The excess pore water pressures are the transient fluid pressures driving a transient fluid flow associated with consolidation of the tissue. For this reason, excess pore pressures are defined as incremental fluid pressures associated with the mechanical loading, and they are calculated by subtracting the current driving fluid pressure from any enduring, long-term fluid driving pressure. By this means, enduring, long-term driving fluid pressures do not enter into a consolidation analysis.

In any case, the enduring, long-term driving fluid pressure in articular cartilage is small. Articular cartilage tissue fluid driving pressure is totally dominated by 'excess pore fluid pressure' (denoted here p, where the subscript 'd' has been dropped for notational simplicity) generated by mechanical loading at the surface of cartilage tissue. Mechanical loading of the knee is due to both body weight above the level of the knee, and due to muscle forces acting across the knee to maintain the upright position when standing. Typically, these two contributions to mechanical loading are of similar magnitude when standing upright. Because of this mechanical loading, the driving fluid pressure in normally loaded knee cartilage upon assuming a standing position relative to the resting supine condition is normally around several MPa. If the long-term driving pressure between synovial fluid and marrow is a maximum of 2 kPa, we see that there is approximately three orders of magnitude difference in driving pressures (driving pressure gradients are often larger again). So nearly always, the 'fluid pressure' that drives fluid flow in cartilage refers to the 'excess pore fluid pressure' arising from mechanical loading at the cartilage surface. We also note in passing that loading of small joints of the hand during normal hand activities results in stresses of magnitude measured in hundreds of kilopascals, so again any long-term driving pressure through articular cartilage of the hand is minimal compared to the driving pressures generated by mechanically loading hand joints.

When articular cartilage is mechanically loaded, the so-called apparent (or nominal) partial stresses in each phase of a porous material are calculated by dividing that fraction of the total incremental force acting on each phase, by the total (or gross) cross-sectional area. These (apparent) partial stresses are not the actual stress experienced by that phase, because a phase usually only occupies a fraction of the total cross-sectional area. For example, the solid particles only occupy a fraction of the cross-sectional area n_s (where n_s is solid fraction), so the 'true average stress' in the solid phase is actually σ^s/n_s. This true average stress is always higher than the apparent stress (σ^s), because the solid fraction n_s cannot be larger than one.

We also recognize that the 'actual stress' experienced by the solid phase at a specific location may be higher or lower than the true average stress. For example, in soils like sand soils, the particles only make (a small, 'point-like') Hertzian contact with one another, so the actual stresses at contact 'points' may be higher or lower than the true average stress, depending on the details of the contacting materials and the load being carried through particles of different shapes. Of course, it is this actual stress in the material that changes and potentially damages materials.

It is clear we use apparent partial stresses knowing that they are averaged quantities, which have very specific meanings. If we wish to estimate true average stress or actual local stress in soil particles or biological materials, then we need to do some additional calculations (e.g. we can use the apparent stresses to estimate the true average stresses). But it turns out that the apparent stresses prove to be useful in practice, and helpfully, they are easy to measure. For example, we typically record experimental stress–strain curves presented in terms of apparent (sometimes called 'nominal') stresses because it is relatively easy to divide a force by the gross cross-sectional area—it is much more difficult to estimate by measurement the true cross-sectional area of each component making up the material. We therefore formulate constitutive equations in terms of apparent stresses, so it is customary to work with these stresses (e.g. the Cauchy stress σ_{ij} in a poroelastic material is an apparent (or nominal) stress acting over the gross cross-sectional area of the porous material.

From his experiments on sand soils, Terzaghi also noticed something else very interesting and very important. This fact is not immediately obvious given our discussion immediately above about the relationship between apparent stresses and true average stresses in the solid phase. Terzaghi found experimentally that the apparent stress in the fluid phase is equal to the true average stress in the fluid phase, which is equal to the measured fluid pressure in the fluid phase. This is strange—(it is as if $n_w = 1$ in the formula p/n_w to calculate the true average fluid stress from the partial fluid stress), when clearly n_w is less than one in a saturated soil. In fact, we know that the sum of the solid and fluid porosities is one (i.e. $n_s + n_w = 1$; fluid porosity is defined as the volume of 'void space' divide by the total volume, while the solid porosity is the volume of the solids divided by the total volume. We note that 'areal porosity' is taken to be equal to 'volumetric porosity').

The reason for the difference between 'theory' and experiment is that in soils the contact areas between the particles are very small (i.e. 'point-like' Hertzian contacts)—the fluid pressure can effectively exert a pressure all around each solid particle. So when you 'cut out' an (imaginary) small representative elementary volume (REV) of porous media and draw a 'free body diagram,' you would see that the forces acting on the solid cross section includes forces being transmitted through the solid particles along 'stress chains,' plus the pressure from the fluid just inside the small section acting on the solid particles. Because the fluid acts all around the particle (the actual solid-to-solid contact areas that water does not access are only a very small fraction of the total cross-sectional area), the fluid pressure appears as a stress acting through both the solid phase and through the fluid phase. In this way, to a close approximation the fluid pressure acts over the whole surface (i.e. the nominal surface area is very close to the true surface area for the fluid). A similar conclusion can be drawn by using a 'wavy' surface (rather than planes) to define the surfaces of a representative elementary volume, that is, using an REV whose surfaces only passes through the particle contacts (which are small). For a REV defined using a wavy-surface, (nearly) the whole surface of the representative volume element cuts through fluid, and so the fluid pressure p is again seen to have an effective cross-sectional porosity of (very close) to one.

We note that when the fluid does *not* exert a pressure all over the surface of the solid-phase particles (e.g. the solid is continuous so the fluid is confined within pockets within a solid), then the apparent pressure does approach the true average fluid as calculated using p/n_w. So for soils, when it comes to calculating true average pressure in the fluid, we should use $n_w = 1$ (as then the calculated true average pressure accords closely with experimental measurements). Importantly, charged macromolecules in cartilage only have exceedingly small physical contact surfaces with one another; i.e. water usually intercedes between the macromolecular contacts, so the same reasoning applies to cartilage as to porous materials such as soil. Indeed, clay soils closely approximate the contact behavior of charged macromolecules. That consolidation theory developed for particulate soils applies to cartilage has been confirmed experimentally. For example, see the experimental and theoretical results of Oloyede and Broom (1991, 1994), Armstrong et al. (1984), Edwards (1966), Wellard et al. (2014), Bursac et al. (1999), Miramini et al. (2017), Zhang et al. (2015), Fick et al. (2010). These papers show plots of excess pore water pressures and 1-D or 3-D settlements that are essentially the same as experimental results for soils. We mention in passing that consolidation theory also holds for the nucleosis pulposus of intervertebral disk (Emanuel et al. 2018).

The reason for the experimental and theoretical agreement in the poromechanical description of soft tissues and soils is that just like aggregates of charged clay particles, the water in cartilage can exert a pressure (nearly) all around the structural macromolecules in cartilage. In cartilage, water exerts a pressure all around aggrecan macromolecules, and water can exert a pressure (almost) all around the collagen tropocollagen mers making up collagen fibers. Self-evidently, water can exert a pressure all around the aggrecan molecules as these molecules are in solution and exert forces between adjacent molecules via osmotic pressure between the macromolecules (see detailed discussion in Chap. 6). But like the soil particles, the water can also exert a pressure nearly all around the collagen fibers too, for two reasons. First, the collagen fibers themselves have water-filled (interconnected) 'nanoporosity' (i.e. the tropocollagen molecules are made of three amino acid polymers ('triple helices') with water molecules between the polymers and between the triple helices). When these tropocollagen molecules are assembled end to end into collagen fibrils, they are 'staggered' [see, for example, Fig. 6 (Tang et al. 2010)] with water in the gaps between the end-to-end tropocollagen molecules, at least when they are unmineralized. The end result is nearly all the porosity in a collagen fibril is filled with water—the only regions that water cannot access in collagen fibrils are at collagen 'cross-links' (cross-links are either molecules that have covalent molecular bonds between the atoms (usually linking adjacent tropocollagen molecules), or they are multivalent cations bridging negative charge centers, e.g. calcium ions, which link tropocollagen molecules along their side or end to end (Nudelman et al. 2013; Pang et al. 2017)].

And second, water forms an integral part of the collagen fiber itself. The collagen fiber is in fact hydrated (Palfi and Perczel 2010). The water molecules are involved in hydrogen bonding between the tropocollagen molecule and water molecules coating much of its primary amino acid polymer surface, and these water molecules transmit fluid pressure into the collagen fibril from nearby water molecules. Indeed, hydrogen bonding between CO–HN groups on adjacent amino acid helices and water molecules intervening between the adjacent helices play a key role in determining the mechanical properties of tropocollagen and so the collagen fibril [e.g. see Figs. 1, 2 ,5, 6, and 10 (Palfi and Perczel 2010): '*To sum it up, in total there are five threads of water spirals around a tropocollagen fiber that wrap it like a net*' (Palfi and Perczel 2010)]. For these reasons, the water pressure p acts (nearly) all around the collagen fibers and all around the aggrecan molecules. So when it comes to drawing an REV in cartilage, we can also draw a 'wavy-surface' that mainly passes through water, just as we could for particulate soils. As aggrecan and collagen together account for the principal structural solid materials in cartilage extracellular matrix, these molecules essentially define its structural response. So we do not really need to consider additional macromolecules in cartilage from a structural viewpoint (even though these macromolecules behave similarly to aggrecan and collagen in the sense that these large polymer macromolecules often contain porosity filled with water or have water between the mers). For further discussion on 'effective stress,' see Appendix I Terzaghi's principle of effective stress.

Of course, theoreticians do not like the apparent arbitrariness of modifying the theory in light of experimental results—quite reasonably they would prefer more consistency. But the fact is the experimental finding tells us that when interpreting apparent, true average and local actual stresses, we need to be aware of the internal arrangement of the phases with each other, as this influences how we should interpret the partial stresses if they are to make practical sense and line up with measurements. Of course, these kinds of conceptual 'fiddles' create theoretical uncertainties, which have led to long-running debates and controversies in the cartilage community, e.g. how lubrication occurs in vivo (McCutchen 2011) and http://blacklistedbiomechanics.com/index.htm. We note in passing that sometimes theoreticians have stuck with their theoretical analysis and finished up with predictions that do not line up with experimental measurements. But the upshot is that experiments show that cartilage behaves for all practical purposes the same as particulate soil, and so to a high degree of accuracy, one can apply essentially the same methodologies for cartilage mechanics, and more generally to soft tissue poromechanics, as that for soil mechanics.

The second thing Terzaghi did was assemble the principle of effective stress, the momentum balance equation, the constitutive equation, and boundary conditions into a coherent whole, so that they presented a well-posed mathematical problem that can be solved. One way to see how the equations can be assembled is to do so by example. Here, we first write down the one-dimensional mass conservation equations for each of phase of saturated soil in spatial coordinates. We do this because these equations are comparatively simple to understand from first principles and because most experimental data is gathered in spatial coordinates (i.e. in the current configuration). For example, 'Cauchy stresses' are normally measured experimentally in the current configuration, not equivalent 'second Piola–Kirchhoff stresses' that are referenced to the initial configuration of the material, while Darcy fluid flows are measured in the current configuration, not equivalent nominal Darcy fluid flows in the original configuration. We can write down the mass balance equations and then introduce constitutive equations, which relate pressure gradients to fluid flow and solid stress to volumetric deformation of the porous material. Once, we have the governing equations in spatial coordinates, and we can then transform the equation into a material coordinate system using coordinate transformations mentioned in the introduction to solid mechanics.

We begin by writing down the 1-D mass conservation equation for the solid particles. Considering a representative elementary volume (REV) with volume dv in a fixed spatial coordinate system (where $\mathrm{d}v = \mathrm{d}z \times 1$ in one spatial dimension). For the mass balance of the solids, we have

$$-\frac{\partial[n_{\mathrm{s}}^{\mathrm{s}} \rho_{\mathrm{s}} v_{\mathrm{s}}^{\mathrm{s}}]}{\partial z} = \frac{\partial[n_{\mathrm{s}}^{\mathrm{s}} \rho_{\mathrm{s}}]}{\partial t} \tag{7.44}$$

and if we assume ρ_{s} is constant (the particles are very stiff and so their deformation is negligible) and employ our relation $n_{\mathrm{s}}^{\mathrm{s}} + n_{\mathrm{f}}^{\mathrm{s}} = 1$, we note for later reference that this equation reduces to

$$-\frac{\partial[(1 - n_{\mathrm{f}}^{\mathrm{s}}) v_{\mathrm{s}}^{\mathrm{s}}]}{\partial z} = \frac{\partial n_{\mathrm{s}}^{\mathrm{s}}}{\partial t} \tag{7.45}$$

where $n_{\mathrm{s}}^{\mathrm{s}}$ is the volume fraction of solid particles in the REV, $n_{\mathrm{f}}^{\mathrm{s}}$ is the volume fraction of fluid in the REV, $v_{\mathrm{s}}^{\mathrm{s}}$ is the velocity of the solid particles, and ρ_{s} is the actual density of the solid particles themselves, i.e. the mass of solid in a particle divided by the volume of that particle.

We note in passing that unlike soil particles, which retain the same density irrespective of the soil particles around them, for macromolecules making up biological materials, we note that component densities in a mixture are the inverse of the partial specific volumes of each component. The partial specific volume is the change in volume of the mixture with the addition of a unit mass of a component to the mixture. We note that this partial specific volume (or its inverse, density) may change with the composition of the mixture. For example, if aggrecan is added to hypotonic saline, the measured partial specific volume of aggrecan is different to that measured when aggrecan is added to isotonic saline solution (i.e. the measured density of the macromolecule depends on its matrix).

We expand Eq. (7.44) using the chain rule and then regroup terms according to Eq. (7.38) to find the material derivative with respect to time. Adopting material coordinates has the advantage that the mass of solids in the REV is constant in time. Material coordinates imply that the REV volume to be employed when estimating spatial porosity at time t is not arbitrary, but connected back to the initial REV volume used to estimate initial porosity. The volume relationship is $J = \mathrm{d}v/\mathrm{d}V$. Constant mass of solids in the REV implies $n_s^s\rho_s \mathrm{d}v = \rho_0 \mathrm{d}V$, i.e. $n_s^s\rho_s = \rho_0/J$, where ρ_0 is the initial apparent solids density (i.e. the mass of solids divided by the initial REV volume). So using the chain rule and regrouping, we find from Eq. (7.44),

$$-\frac{\partial v_s^s}{\partial z} = \frac{1}{n_s^s\rho_s}\frac{\mathrm{d}(n_s^s\rho_s)}{\mathrm{d}t} = \frac{1}{\rho_0/J}\frac{\mathrm{d}(\rho_0/J)}{\mathrm{d}t} = -\frac{1}{J}\frac{\mathrm{d}J}{\mathrm{d}t} \tag{7.46}$$

where $\frac{\mathrm{d}}{\mathrm{d}t}$ is the material (or total) derivative with respect to time. We notice that a negative sign has 'crept' in front of the time derivative. This is because time has a direction, and in this case, time ran 'backward' from the spatial coordinate system to the material coordinate system (this happened when we used the chain rule and had to evaluate $\mathrm{d}J^{-1}/\mathrm{d}J$). Normally, we want time to increase from the reference (undeformed) configuration in material coordinates to the current (deformed) configuration in spatial coordinates. To fix this, we simply change the sign on the right-hand side of Eq. (7.46).

The soil particle assembly is saturated with water, so we now write down the mass conservation equation for the fluid phase. The 1-D conservation for the fluid phase is given by

$$-\frac{\partial[n_f^s\rho_f v_f^s]}{\partial z} = \frac{\partial[n_f^s\rho_f]}{\partial t} \tag{7.47}$$

where ρ_f is the fluid density and v_f^s is true velocity of the fluid relative to the (fixed) spatial coordinate system. If the fluid is considered incompressible—which is a good assumption for typical pressures encountered in both soil and cartilage mechanics—then fluid densities may be eliminated from Eq. (7.47) and employing partial time derivatives on the identity $n_s^s + n_f^s = 1$ leads to,

$$-\frac{\partial(n_f^s v_f^s)}{\partial z} = \frac{\partial n_f^s}{\partial t} = -\frac{\partial n_s^s}{\partial t} \tag{7.48}$$

Now, adding Eqs. (7.45) and (7.48) reveals

$$\frac{\partial[n_{\mathrm{f}}^{\mathrm{s}}(v_{\mathrm{f}}^{\mathrm{s}}-v_{\mathrm{s}}^{\mathrm{s}})]}{\partial z}+\frac{\partial v_{\mathrm{s}}^{\mathrm{s}}}{\partial z}=\nabla^{\mathrm{s}}\cdot v_{\mathrm{d}}^{\mathrm{s}}+\nabla^{\mathrm{s}}\cdot v_{\mathrm{s}}^{\mathrm{s}}=0 \tag{7.49}$$

where $v_{\mathrm{d}}^{\mathrm{s}}$ is the Darcy velocity $v_{\mathrm{d}}^{\mathrm{s}}=n_{\mathrm{f}}^{\mathrm{s}}(v_{\mathrm{f}}^{\mathrm{s}}-v_{\mathrm{s}}^{\mathrm{s}})$. This means the Darcy velocity divided by the porosity is the true average fluid velocity relative to the solids velocity. It is this relative velocity between solid and fluid that generates a drag force interaction between fluid and solid phases. This drag force causes the solid phase to deform. But because the drag force is of equal magnitude and opposite direction on each phase, it has no effect on overall momentum balance for the porous media.

Employing Eq. (7.46) (with sign change) in Eq. (7.49) leads to

$$\nabla^{\mathrm{s}}\cdot v_{\mathrm{d}}^{\mathrm{s}}=\frac{1}{J}\frac{\mathrm{d}J}{\mathrm{d}t} \tag{7.50}$$

We note that the volume strain is given by $e_v=J-1$, so we could write this as

$$\nabla^{\mathrm{s}}\cdot v_{\mathrm{d}}^{\mathrm{s}}=\frac{1}{(1+e_v)}\frac{\mathrm{d}e_v}{\mathrm{d}t} \tag{7.51}$$

This equation is known as the 'storage equation' in consolidation theory.

We notice that the storage equation above is written in a mixed form in the sense that it involves a spatial coordinate system on the LHS, and a material coordinate system on the RHS. To remedy this, we introduce an (unknown) mapping function $z=\phi(a,t)$, which maps particles described in the material coordinate a to particles described in the spatial coordinate z. At time zero, the material is in its reference state, and the same material particle is then referenced by the material coordinate a at all future times. In this one-dimensional example, there is no rotation and deformation can only involve stretch. Because we wish to formulate the analysis in material coordinates (i.e. using the undeformed reference state), this has the benefit of removing issues to do with moving boundaries, one of the practical benefits of the so-named 'Lagrangian formulation.'

If we want to transform a quantity in the reference state to the actual state, we need to 'push-forward' from the initial position of the particles to the actual positions of material particles in the spatial coordinate system. The relationship between vectors described in the reference coordinate system and vectors described in the spatial coordinate system is revealed by the deformation gradient F. The deformation gradient in one dimension is given by $F=\frac{\partial z}{\partial a}$, where $z=z(a,t)$ is the position in the spatial coordinate system expressed as a function of the material coordinate a at time t.

The Jacobian determinant of F, denoted (J), is $\det[F]$. For one spatial dimension, the Jacobian determinant is given by $J=\det[\frac{\partial z}{\partial a}]=F$. In other words, for one spatial dimension, the Jacobian determinant is the Jacobian itself. We again note that in general the Jacobian determinant of the deformation gradient relates the current volume of a differential volume element dv to its initial differential volume $\mathrm{d}V$, that is $\mathrm{d}v = J\mathrm{d}V$. The inverse of the deformation gradient is given by $F^{-1} = \frac{\partial a}{\partial z} = \frac{1}{J}$, which for a one-dimensional analysis is the same as the transpose of the inverse of the deformation gradient, i.e. $(F^{-1})^{\mathrm{T}} = F^{-1}$.

Darcy's law of fluid flow in porous media says the velocity of the fluid relative to the solid particles in the deformed state is given by

$$v_{\mathrm{d}}^{\mathrm{s}} = -k_w^{\mathrm{s}} \nabla^{\mathrm{s}} p \tag{7.52}$$

where p is the excess pore water pressure. We recognize that generally the permeability may be a second-order tensor rather than a Scaler, as in this 1-D example. Starting with Eq. (7.52), we can first write gradient operator in material coordinates, viz.,

$$v_{\mathrm{d}}^{\mathrm{s}} = -k_w^{\mathrm{s}} \nabla^{\mathrm{s}} p = -k_w^{\mathrm{sm}} F^{-1} \nabla^{\mathrm{m}} p^{\mathrm{sm}} \tag{7.53}$$

We can use treat pressure as a Scaler and use the transformation $p(x,t) \rightarrow p^{\mathrm{sm}}(X,t)$, because the necessary tensor transformation to account for stretch or directional dependence of the fluid pressure in the material coordinate system is 'hidden' in F^{-1}.

The divergence is defined as the net exit flux per unit volume per unit time at a particular location. But if we are to use material coordinates, we need to recognize that the actual area over which the flux takes place may be stretched or contracted when described in material coordinates, so the unit volume at time zero may no longer be a unit volume at a later time. For this reason, the divergence transforms as

$$\nabla^{\mathrm{s}} \cdot = \frac{1}{J} \nabla^{\mathrm{m}} \cdot JF^{-1} = \frac{1}{J} \frac{\partial}{\partial a} \tag{7.54}$$

Now, using Eqs. (7.53) and (7.54) in Eq. (7.50) leads to

$$\nabla^{\mathrm{m}} \cdot JF^{-1} k_w^{\mathrm{sm}} F^{-\mathrm{T}} \nabla^{\mathrm{m}} p^{\mathrm{sm}} = \frac{\partial}{\partial a} \frac{k_w^{\mathrm{sm}}}{J} \frac{\partial p^{\mathrm{sm}}}{\partial a} = \frac{\mathrm{d}J}{\mathrm{d}t} \tag{7.55}$$

In this equation, the term JF^{-1} can be viewed as converting a true flux in the spatial coordinate system into a nominal flux in the undeformed material coordinate system, while $F^{-\mathrm{T}}$ connects the spatial to the material coordinate systems. Alternatively, we can view F^{-1} and $F^{-\mathrm{T}}$ as two, two-point tensors connecting the diffusivity tensor in the spatial coordinate system with the material coordinate system, while J is a 'two-point' Scaler that ensures the divergence is evaluated over the correct volume (i.e. evaluated over the actual volume rather than evaluated over the nominal volume).

Recognising the coordinate transformation rule for a second-order tensor is $K_w^{\mathrm{m}} = JF^{-1} k_w^{\mathrm{sm}} F^{-\mathrm{T}}$ Eq. (7.55) could be written as

$$\nabla^m \cdot K_w^m \nabla p^{\mathrm{sm}} = \frac{\partial}{\partial a} K_w^m \frac{\partial p^{\mathrm{sm}}}{\partial a} = \frac{\mathrm{d}J}{\mathrm{d}t} \tag{7.56}$$

However, to practically use K_w^{m} in material coordinates, we need to express the permeability in spatial coordinates $(k_w^{\mathrm{sm}}(n_{\mathrm{f}}^{\mathrm{s}}(x,t)))$ in material coordinates, so we can capture the change in this material property as the material deforms. As describe in the introduction to this chapter, we can do this by expressing the spatial porosity in terms of material coordinates (it is shown later to be $n_{\mathrm{f}}^{\mathrm{s}}(x,t) = 1 + \frac{(n_0(X,0)-1)}{J}$, where n_0 is the original fluid porosity in the undeformed configuration). In this way, the permeability in the spatial coordinate system k_w^{s}, which changes with deformation, can be mapped to the material coordinate system to become $k_w^{\mathrm{sm}}(J)$ and become a function of J.

The 1-D storage equation may be written as

$$\nabla^{\mathrm{m}} \cdot JF^{-1} k_w^{\mathrm{sm}}(J) F^{-\mathrm{T}} \nabla^{\mathrm{m}} p^{\mathrm{sm}} = \frac{\partial}{\partial a} \frac{k_w^{\mathrm{sm}}(J)}{J} \frac{\partial p^{\mathrm{sm}}}{\partial a} = \frac{\mathrm{d}J}{\mathrm{d}t} \tag{7.57}$$

We note in passing that when consolidation is finished, unless volume (i.e. mass of material) is injected into the soil (which may occur if water and/or solid particles are pumped into the soil), $\frac{\mathrm{d}J}{\mathrm{d}t}$ is zero.

We could have derived Eq. (7.56) in material coordinates. We do this by starting with the equation for mass balance of the solid phase in material coordinates, viz.,

$$\rho_{\mathrm{s}}(1 - n_{\mathrm{f}}^{\mathrm{m}}(a,0))\mathrm{d}V = \rho_{\mathrm{s}}(1 - n_{\mathrm{f}}^{\mathrm{sm}}(a,t))\mathrm{d}v = \rho_{\mathrm{s}}(1 - n_{\mathrm{f}}^{\mathrm{s}}(x,t))\mathrm{d}v \tag{7.58}$$

where $n_{\mathrm{f}}^{\mathrm{s}}(x,t) = n_{\mathrm{f}}^{\mathrm{s}}(a+u(a,t),t) \rightarrow n_{\mathrm{f}}^{\mathrm{sm}}(x-u(a,t),t) = n_{\mathrm{f}}^{\mathrm{sm}}(a,t)$. Then, for an incompressible solid,

$$\frac{\mathrm{d}[(1 - n_{\mathrm{f}}^{\mathrm{sm}}(a,t)J]}{\mathrm{d}t} = 0 \tag{7.59}$$

That is,

$$\frac{\mathrm{d}J}{\mathrm{d}t} = \frac{\mathrm{d}[Jn_{\mathrm{f}}^{\mathrm{sm}}(a,t)]}{\mathrm{d}t} = \frac{\mathrm{d}N_{\mathrm{f}}^{m}}{\mathrm{d}t} \tag{7.60}$$

where the 'Lagrangian fluid porosity' $N_{\mathrm{f}}^{\mathrm{m}}(a\ ,t) = Jn_{\mathrm{f}}^{\mathrm{sm}}(a,t) = (v_{\mathrm{s}}^{\mathrm{s}}(t)/V(0)) \times (v_{\mathrm{f}}^{\mathrm{s}}(t)/v_{\mathrm{s}}^{\mathrm{s}}(t)) = v_{\mathrm{f}}^{\mathrm{s}}(t)/V(0)$. Integrating Eq. (7.60) w.r.t time can show that $N_{\mathrm{f}}^{\mathrm{m}}(a\ ,t) = n_0 + J - 1$ e.g. see discussion following Eq. (7.134). We observe that the Lagrangian porosity in the material coordinate system keeps track of the change of fluid volume in the undeformed configuration.

The continuity equation for an incompressible fluid in material coordinates is

$$\frac{dN_f^m}{dt} = -\nabla^m \cdot Jv_d^m \quad (7.61)$$

Substituting Eqs. (7.60) in (7.61) leads to

$$\frac{dJ}{dt} = -\nabla^m \cdot Jv_d^m \quad (7.62)$$

Now, substitution of Darcy's law in material coordinates [e.g. see Eq. (7.25)] leads directly to the derivation of Eq. (7.57).

With the help of Appendix G (the rate of volume change in material coordinates), we convert Eq. (7.57) to,

$$\nabla^m \cdot JF^{-1}k_w^m F^{-T}\nabla^m p^m = \frac{dJ}{dt} = JF^{-1}F^{-T} : \frac{dE}{dt} = \frac{1}{J}\frac{dE_{zz}}{dt} \quad (7.63)$$

For one spatial dimension, $E_{zz} = \frac{1}{2}(C-1) = \frac{1}{2}(F^T F - 1) = \frac{1}{2}(J^2 - 1)$. Upon substituting E_{zz} in Eq. (7.63), we can verify that in one spatial dimension $\frac{1}{J}\frac{dE_{zz}}{dt} = \frac{dJ}{dt}$ as expected.

To progress our analysis further, we now need a constitutive equation to describe how the solid skeleton deforms when loaded [this constitutive equation describes the relationship between stress and strain under equilibrium (or drained) conditions; i.e. this constitutive relationship describes the deformation response to loading when the excess pore water pressure is zero (that is, at thermodynamic equilibrium)]. This immediately raises the issue of choosing appropriate conjugate variables for the analysis. Given the above discussions, we could use the second Piola–Kirchhoff stress tensor for the solid phase S_s^m and the Green–Lagrangian strain tensor E as conjugate variables and then choose a suitable relationship (i.e. constitutive equation) between stress and strain in the reference configuration.

If we follow this path, we first 'pull-back' Terzaghi's effective stress, i.e. Eq. (7.41), and find second Piola–Kirchhoff stress tensor for the solid phase as $S_T = S^s - P^m = JF^{-1}\sigma_T^{sm}F^{-T} = JF^{-1}\sigma^{sm}F^{-T} - p^{sm}JF^{-1}IF^{-T}$, where P^m is the fluid pressure tensor in the material coordinate system. We know that the work conjugate for the second Piola–Kirchhoff stress is dE. We could now choose the simplest constitutive relation possible in the material coordinate system, and assume that the second Piola–Kirchhoff stress is linearly related to the Lagrange–Green strain. A material with this constitutive relationship is known as a St. Venant material.

When we use a St. Venant constitutive relation to analyze the one-dimensional extension or compression of an unconstrained square bar, the (bar) force in the spatial coordinate system (f) can be shown to be a cubic function of the engineering strain. By differentiating the force in the spatial coordinate system with respect to the engineering strain, and setting this to zero (to find the minimum stiffness), it can be shown that for one dimensional compression, the stiffness of bar goes to zero at a (compressive) strain of minus 42% https://www.comsol.com/blogs/modeling-linear-elastic-materials-how-difficult-can-it-be/. In other words, for a simple St. Venant constitutive relationship, unfortunately the geometric nonlinearity of the Green–Lagrange strain results in the stiffness of the St. Venant material becoming zero in the spatial coordinate system, and so the solid mechanics analysis fails (due to division by zero).

The usual approach taken to address this problem is choosing another constitutive relationship in the reference configuration. Rather than a linear relationship between the second Piola–Kirchhoff stress and Green–Lagrange strain, which results in the force in the spatial coordinate system for the example bar problem described above becoming a cubic function of engineering strain, we chose a constitutive relationship that results in the force in the spatial coordinate system becoming a more complicated function of the engineering strain that does not result in zero stiffness at some finite strain. An example of such a stress–strain relationship is the so-named neo-Hookean model material. For a compressible neo-Hookean material, the (Helmholtz) strain energy density function Ψ^e, (where the strain energy density function is defined per unit volume of the reference configuration) is given by

$$\Psi^{\mathrm{e}} = C_1(\bar{I}_1 - 3) + D_1(J - 1)^2 \tag{7.64}$$

where $\bar{I}_1 = I_1 J^{-(2/3)}$ (is the first invariant of the isochoric part of the distortional strain tensor, i.e. $CJ^{-(2/3)}$), $I_1 = \mathrm{Tr}(C)$ is the first strain invariant of C, and C_1 and D_1 are material constants. Then, by definition of the strain energy density function,

$$\mathrm{d}S_{ij}^{\mathrm{m}} = \frac{\partial \Psi^e}{\partial E_{ij} \partial E_{\mathrm{kl}}} \mathrm{d}E_{\mathrm{kl}} \tag{7.65}$$

It can be shown that for our unconstrained square bar example with a neo-Hookean strain energy density function, the force (f) in the bar in the spatial coordinate system (area times the Cauchy stress) is given by https://en.wikipedia.org/wiki/Neo-Hookean_solid,

$$f = 2C_1 J^{-\left(\frac{2}{3}\right)} \left(\lambda - \frac{J}{\lambda^2}\right) \tag{7.66}$$

where λ and J are related to each other by the equation,

$$D_1 J^{\left(\frac{8}{3}\right)} - D_1 J^{\left(\frac{5}{3}\right)} + \frac{C_1}{3\lambda} J - \frac{C_1 \lambda^2}{3} = 0 \tag{7.67}$$

which can be solved numerically via iteration.

The main problem with the above approaches to finding a suitable strain energy function is it becomes an exercise in ‘curve fitting,’ because the nonlinearity in the strain energy density function is primarily geometric, rather than being related to the behavior of the material itself. In fact, the composition of cartilage evolves over time with the deformation, which implies that the material parameters should be a function of its time-dependent composition, rather than the deformation itself. But material parameters are constant in the neo-Hookean model, as they are for all strain energy density functions derived from the generalized Rivlin model (also known as 'polynomial hyperelastic models'). In these models, the 'bulk material parameters' that appear in the stress–strain relationship do not have any physically measurable relationship to material behavior at finite strains (because it is 'mixed up' in a complex way with the geometric parameters describing the deformation). For small strain, when the geometric nonlinearity inherit in the (nonlinear) strain measure is linearized by the small strain (and so the material behavior as we measure it in the laboratory can be revealed), it can be arranged so that the neo-Hookean material parameters are related to the familiar shear modulus $(\mu = 2C_1)$ and bulk modulus $(\kappa = 2D_1)$.

It is odd and frustrating that the geometric nonlinearity dominates the constitutive behavior rather than nonlinearity of the material itself in response to straining. For cartilage tissue, the preferred approach is to develop a suitable constitutive relationship based on the known material composition of the cartilage. Halloran et al. (2012) explains:

> Hence, with knowledge of the properties of cartilage constituents, biochemically and histologically measurable data, rather than a set of fitted parameters, could ultimately be used as inputs for the computational model. Microscale composition could then be used to inform the tissue level response, alleviating the need to obtain site- or specimen-specific bulk mechanical parameters (commonly performed using inverse analysis).

For an example of how a constitutive model might be driven for cartilage loaded in compression, the key constitutive relationship would relate the local aggrecan concentration to the local volumetric compressive stiffness, which evolves over time as the cartilage consolidates.

While we could proceed along the lines of using a strain energy density function, it is apparent that any polynomial hyperelastic model we may choose is less than ideal for the reasons mentioned above, and even for the simplest neo-Hookean material, the analysis becomes rather complicated. Indeed, a neo-Hookean model turns out to be unnecessarily complicated for our one-dimensional analysis of consolidation considered here.

A 1-D analysis of consolidation has no rotation and $F = J$, which enable us to choose a more suitable stress and conjugate strain. Rather than use the second Piola–Kirchhoff stress and its conjugate strain rate, we can choose the first Piola–Kirchhoff stress $(P^{\mathrm{s}} = J\sigma^{\mathrm{s}}F^{-\mathrm{T}})$ and its conjugate strain rate $\left(\frac{\mathrm{d}F}{\mathrm{d}t}\right)$. For our 1-D conditions, the first Piola–Kirchhoff stress reduces to $P^{\mathrm{s}} = \sigma^{\mathrm{s}}$, and the strain rate becomes $\frac{\mathrm{d}F}{\mathrm{d}t} = \frac{\mathrm{d}J}{\mathrm{d}t}$. In other words, an engineering strain measure $(\varepsilon_z^{\mathrm{m}} = J - 1 = U - I$, i.e. the Biot strain measure) and a Cauchy stress can be adopted as strain energy conjugates in the material coordinate system—all we need to do is map the Cauchy stress and fluid pressure in the current configuration to the material coordinate system (i.e. $\sigma^{s}(x,t)\rightarrow\sigma^{sm}(X,t)$ and $p(x,t)\rightarrow p^{sm}(X,t)$), and similarly for strain $\left(\varepsilon_{z}^{m}=J-1=U-I\right)$.

Now, for the one-dimensional problem, we can proceed by choosing a suitable constitutive relationship between our 'new' (and familiar) conjugate variables, say $\sigma_{z}^{sm}=\beta_{1}E_{z}^{sm}\varepsilon_{z}^{sm}$, where E^{sm} is a stiffness of the soil skeleton in the z direction (i.e. the familiar Young's modulus) and β_{1} is a constant that may be found from elasticity theory (at least for small strains) and equilibrium conditions.

At this point, we mention that for the stiffness of a biocomposite material, like cartilage, E^{sm} may be nonlinear. Indeed, for compression of a prestressed material such as articular cartilage, the initial stiffening effect of the prestressed collagen network (see Fig. 1.8) can be taken into account by have a larger E^{sm}, which then reduces (as the collagen is fully unloaded) to the stiffness of aggrecan in compression. As a first approximation, a strain-dependent bilinear stiffness $\left(E^{sm}\left(\varepsilon_{z}^{sm},\varepsilon_{z\text{-threshold}}^{sm}\right)\right.$, see Fig. 1.8) may be incorporated into a consolidation model of a prestressed material.

Taking stresses as positive in tension (unlike the convention in soil mechanics), we insert our constitutive equation into Eq. (7.57), which leads to the 1-D storage equation as,

$$\frac{\partial}{\partial a}k_{w}^{sm}F^{-1}\frac{\partial p^{sm}}{\partial a}=-\frac{1}{\beta_{1}E^{sm}}\frac{d\sigma_{z}^{sm}}{dt} \tag{7.68}$$

Now, a rate form of Terzaghi's principle of effective stress is

$$\frac{d\sigma_{z}^{sm}}{dt}=\frac{d\sigma_{Tz}^{sm}}{dt}-\frac{dp^{sm}}{dt} \tag{7.69}$$

and employing this rate form in the previous equation, we find the storage equation in terms of total stress and pore water pressure

$$\frac{\partial}{\partial a}k_{w}^{sm}F^{-1}\frac{\partial p^{sm}}{\partial a}=\frac{1}{\beta_{1}E^{s}}\left(\frac{dp^{sm}}{dt}-\frac{d\sigma_{T}^{sm}}{dt}\right) \tag{7.70}$$

This is our 1-D consolidation equation in material coordinates. The surface loading may be a step loading, in which case $\frac{d\sigma_{Tz}^{sm}}{dt}=0$. Then Eq. (7.70) reduces to a diffusion equation in material coordinates with non-linear parameters, viz.,

$$\frac{\partial}{\partial a}k_{w}^{sm}F^{-1}\frac{\partial p^{sm}}{\partial a}=\frac{1}{\beta_{1}E^{sm}}\frac{dp^{sm}}{dt} \tag{7.71}$$

Because of the hyperbolic character induced by the nonlinear material parameters, this equation is sometime said to be 'self-sharpening' (i.e. the diffusion front narrows over time, rather than spreading over time, which diffusion equations usually do). If the soil deformation is small (so that $\frac{\partial z}{\partial a}\approx 1$), Eq. (7.70) simplifies to

$$\frac{\partial}{\partial z} k_w^s \frac{\partial p}{\partial z} = \frac{1}{\beta_1 E^s} \left(\frac{\mathrm{d}p}{\mathrm{d}t} - \frac{\mathrm{d}\sigma_T}{\mathrm{d}t} \right) \tag{7.72}$$

Noting the definition of a material time derivative Eq. (7.72) may then be written as

$$\frac{\partial}{\partial z} k_w^s \frac{\partial p}{\partial z} - \frac{v_s^s}{\beta_1 E^s} \frac{\partial (p - \sigma_T)}{\partial z} = \frac{1}{\beta_1 E^s} \left(\frac{\partial p}{\partial t} - \frac{\partial \sigma_T}{\partial t} \right) \tag{7.73}$$

which demonstrates that there is a hyperbolic character to the governing equation. However, this character only becomes significant at large deformations in soft materials (e.g. which we mention is reasonably common in many biological applications). However, for materials that do not deform too much (e.g. hard tissues such as bone and hard soils), a suitable small deformation consolidation equation is

$$\frac{\partial}{\partial z} k_w^s \frac{\partial p}{\partial z} = \frac{1}{\beta_1 E^s} \left(\frac{\partial p}{\partial t} - \frac{\partial \sigma_T}{\partial t} \right) \tag{7.74}$$

If we the assume the permeability of the pore fluid and the stiffness of the porous material assembly are constant in space and time (which is approximately true if the deformations are small) and that we have step load applied to the soil surface, then Eq. (7.74) reduces to the classical diffusion equation,

$$c_v \frac{\partial^2 p}{\partial z^2} = \frac{\partial p}{\partial t} \tag{7.75}$$

where $c_v = \beta_1 k_w^s E^s$ is known as the coefficient of consolidation. This linear partial differential equation may be non-dimensionalized, viz.,

$$\frac{\partial^2 p}{\partial L^2} = \frac{\partial p}{\partial T_v} \tag{7.76}$$

where $L = z/l$ is the non-dimensional depth of the soil layer (and l is length of the longest drainage path) and $T_v = \frac{c_v t}{l^2}$ is the non-dimensional time. If the non-dimensional time is taken as one (i.e. $T_v = 1$), then the characteristic time t for consolidation is l^2/c_v.

In relation to the characteristic time for consolidation, we note that if a dynamic loading frequency has about the same period as the characteristic time for consolidation, then viscous dissipation in a saturated porous material is maximized. However, if the loading frequency is much higher (and loading period much shorter than the characteristic time), then the saturated porous material behaves as if it is undrained (the pore fluid does not have time to move in and out of the porous material) and the viscous loss is small. On the other hand, if the loading frequency is much lower (and loading period much longer than the characteristic time), then there is plenty of time for dissipation of excess pore water pressures, so velocity gradients are small, and energy dissipation again becomes small. Interestingly, a suddenly applied load, sustained to equilibrium and then suddenly released, can lead to complete dissipation of mechanical work done on the porous material by the loading within the porous material—as discussed below.

We have seen that the traditional 1-D consolidation equation is, under simplifying approximations, equivalent to the 1-D classical diffusion equation with constant coefficients. This equation can be solved analytically using either separation of variables and Fourier series (see any soil mechanics textbook) or using Laplace transform methods (Verruijt 2013). Suitable boundary conditions are: (i) a Dirichlet boundary condition (e.g. zero excess pore water pressure at a 'drained' surface; $p = 0$) or (ii) a Neumann boundary condition (e.g. no flux boundary condition at an impermeable surface; $v_d = 0$) or (iii) a Robin boundary condition, which describes a resistance to flow at the surface of the material (e.g. $v_d = h(p - p_{ref})$ where h and p_{ref} are arbitrary constants).

For pore fluid drainage at the top of the consolidating porous material and no flux across the bottom of the porous material, it turns out that when $T_v = 0.01$, the consolidation process is about 10% complete, when $T_v = 0.5$ the consolidation process is about 75% complete, and when T_v equals 1.0, the consolidation process is about 95% complete.

It is informative to now consider the mapping function ϕ for the 1-D consolidation problem discussed above. When consolidation is complete, we assume that a constant load and constant material parameters over the domain lead to a linear mapping function (even if the stiffness is nonlinear). That is,

$$\phi = a(1 + m) \tag{7.77}$$

where m is a constant. The deformation gradient is then,

$$F = \frac{\partial \phi}{\partial a} = (1 + m) \tag{7.78}$$

Noting Eq. (7.5), we find the displacement u_z is given by

$$u_z = \int_0^L m da \tag{7.79}$$

and it becomes clear to those familiar with these types of problems that m is in fact the engineering strain throughout the material (i.e. $m = \varepsilon_z = \frac{\partial u_z}{\partial a} = J - 1 = F - 1$).

In light of later discussion, it is helpful to also estimate the energy lost as viscous dissipation during 1-D consolidation. It is first noted that it is the 'unbalanced stress' (i.e. the difference between the total stress and the internal 'effective' or solid stress) that results in dissipation of energy. A time-dependent 'unbalanced stress' is a typical feature of viscous dissipative systems. Flow of pore fluid through a porous media is a particular type of viscous dissipation. Viscous dissipation may arise from a number of microscale processes occurring within a material as it deforms. Often, 'internal microscale processes' within a material (e.g. associated with 'relaxation' of molecules from a higher thermodynamic state into a lower thermodynamic state) occur on a timescale that may be longer than the time scale of loading. In this case, there is a 'creep process' that continues due to relaxation of higher energy states after loading has ceased.

In a 'quasi-static' system (one that only departs infinitesimally from equilibrium), there is no unbalanced stress and excess pore water pressures are zero at all time (i.e. quasi-static loading means the timescale for loading is much longer than the timescale required for the internal relaxation process to reach thermodynamic equilibrium). So in this view, the degree of excess pore water pressure is a measure of the degree of thermodynamic nonequilibrium of the system. The gradient of the excess pore water pressure plays a key role in the dissipation function for the porous material [i.e. ∇p is the so-called thermodynamic force (which in this case is a gradient of the mechanical pressure in the fluid) and v_{d} is the 'thermodynamic flux' (which is the fluid flow through the porous medium)]. The product of the thermodynamic force and thermodynamic flux gives the rate of energy dissipation in the porous medium (Hackl and Fischer 2008; Coussy and Scherer 2004; Haase 1990).

Hence, the rate of energy dissipation as heat (in Jm^{-3} using SI units) at some location P in the poroelastic material (Ψ^{df}) is given by

$$\Psi^{\mathrm{df}} = v_{\mathrm{d}}^{\mathrm{s}} \cdot \nabla p = k_{ij}^{-1} v_{\mathrm{d}}^{\mathrm{s}} \cdot v_{\mathrm{d}}^{\mathrm{s}} = (K_{IJ}^{\mathrm{m}})^{-1} (J v_{\mathrm{d}}^{\mathrm{m}}) \cdot (J v_{\mathrm{d}}^{\mathrm{s}}) \tag{7.80}$$

where Ψ^{df} is known as the 'dissipation function' and we have included the equivalent quantity in both spatial and material coordinates. We note in passing that a quadratic (Rayleigh) 'dissipation potential' is defined as Coussy and Scherer (2004),

$$D = \frac{\Psi^{\mathrm{df}}}{2} = \frac{k_{ij}^{-1}}{2} v_{\mathrm{d}} \cdot v_{\mathrm{d}} \tag{7.81}$$

so naturally the derivative of the dissipation potential with respect to the flux gives the driving force. The total energy dissipated as heat over the whole domain by the consolidation process up to some time T is given by the integral of the dissipation function, viz.,

$$\Delta Q = \int_0^T \int_V \Psi^{\mathrm{df}} \mathrm{d}V \mathrm{d}t \tag{7.82}$$

where ΔQ is the heat generated.

Differentiation of Ψ^{df} with respect to the thermodynamic flux (i.e. in this case, w.r.t. to the Darcy flux) gives the 'thermodynamic force,' or more helpfully in this case, the gradient of thermodynamic pressure, ∇p, which has units of Nm^{-3} in the current configuration. Now, integration of the gradient of thermodynamic pressure (acting on a 'unit volume of fluid') with respect to distance yields another potential, known as the pressure potential Ψ^{f} http://en.wikipedia.org/wiki/Water_potential. In fluid mechanics, this pressure potential is the potential that arises from the so-called buoyancy force in the fluid. This potential has units of Jm^{-3}. The pressure potential represents the mechanical work required to expand fluid under pressure, so as to make room for a unit volume of new fluid at that pressure.

This pressure potential is sometimes confused with the thermodynamic pressure itself, because the units of the pressure potential $\left(\mathrm{Jm}^{-3}\right)$ and of thermodynamic pressure $\left(\mathrm{Nm}^{-2}\right)$ are the same. For this reason, the difference between the pressure potential and pressure often goes unnoticed, yet they represent different quantities.

Now, the pressure potential Ψ^{f} is closely related to the Gibbs free energy potential of the fluid per unit mass (μ^{f}) due to mechanical pressure on the fluid. Assuming the fluid is water, the Gibbs free energy per unit mass due to mechanical pressure is $\mu^{\mathrm{w}} = \int_0^p \bar{v}^{\mathrm{w}} \mathrm{d}p$, where $\bar{v}^{\mathrm{w}}$ is the specific volume of water (i.e. $\bar{v}^{\mathrm{w}} = \frac{1}{\rho_{\mathrm{w}}}$). The Gibbs free energy represents the mechanical work required to expand fluid under pressure, so as to make room for a unit mass of new fluid.

If we now consider that water is incompressible (which is a good approximation for our purposes here), then

$$\Psi^{\mathrm{w}}\left(\mathrm{Jm}^{-3}\right) = \frac{\mu^{\mathrm{w}}}{\bar{v}^{\mathrm{w}}} = p\left(\mathrm{Nm}^{-2}\right) \tag{7.83}$$

where for an incompressible fluid $\Psi^{\mathrm{w}}\left(\mathrm{Jm}^{-3}\right) = p\left(N\mathrm{m}^{-2}\right)$. Because the Gibbs free energy results from a mechanical pressure, it is clear that the gradient of 'thermodynamic pressure' described above is in fact a gradient of the mechanical pressure in the fluid. This is confirmed experimentally by being able to measure this mechanical pressure in the fluid of the porous medium.

Now the pressure potential per unit volume of porous medium in the material (undeformed) configuration is given by

$$\Psi^{\mathrm{wp}} = N_{\mathrm{f}}^{\mathrm{m}} \Psi^{\mathrm{w}} = N_{\mathrm{f}}^{\mathrm{m}} p^{\mathrm{sm}} \tag{7.84}$$

and so,

$$\frac{\Psi^{\mathrm{wp}}}{\mathrm{d}t} = N_{\mathrm{f}}^{m} \frac{\mathrm{d}p^{\mathrm{sm}}}{\mathrm{d}t} + p^{\mathrm{sm}} \frac{\mathrm{d}N_{\mathrm{f}}^{m}}{\mathrm{d}t} = \frac{\mathrm{d}\Psi^{\mathrm{lw}}}{\mathrm{d}t} + \Psi^{\mathrm{dw}} \tag{7.85}$$

where p^{sm} is the pressure in spatial coordinates in Nm^{-2} mapped to the undeformed configuration, Ψ^{lm} is the rate of change of pressure potential due to the rate of change of applied loading and Ψ^{dw} is the dissipative function expressed in terms of fluid pressure in material coordinates, $N_{\mathrm{f}}^{\mathrm{m}}(X,t) = Jn_{\mathrm{f}}^{\mathrm{sm}} = v_{\mathrm{f}}^{\mathrm{s}}(t)/V(0) = n_0(X,0) + J - 1$ [see discussion following Eq. (7.134)] is the fluid porosity in material coordinates (hence Ψ^{wp} is per unit volume in the material configuration), and $\frac{\mathrm{d}N_{\mathrm{f}}^{\mathrm{m}}}{\mathrm{d}t} = \frac{\mathrm{d}J}{\mathrm{d}t} = JF^{-1}F^{-\mathrm{T}} : \frac{\mathrm{d}E}{\mathrm{d}t}$ (see Appendix G, the rate of volume change in material coordinates). We note that

$$\frac{d\Psi^{\mathrm{dw}}}{d\dot{N}_{\mathrm{f}}^{\mathrm{m}}} = p^{\mathrm{sm}}(\mathrm{Nm}^{-2}) \tag{7.86}$$

Having defined Ψ^{dw} w.r.t a unit volume in material coordinates, this enables it to be added to the strain energy density Ψ^{e}, as it is also defined with respect to a unit volume in material coordinates.

However, we observe that we could have just as easily evaluated the actual mechanical pressure at some location in the fluid by integrating the pressure gradient in the spatial coordinate system along a line starting at some point where the pressure is known up to the pressure found at some arbitrary point within the domain. That is integration along a line element dS leads to

$$p = \int_{p_{\mathrm{ref}}}^{p} \frac{\partial \Psi^{\mathrm{df}}}{\partial v_d} \mathrm{d}S + p_{\mathrm{ref}} = \int_{p_{\mathrm{ref}}}^{p} \nabla p \mathrm{d}S + p_{\mathrm{ref}}(\mathrm{Nm}^{-2}) \tag{7.87}$$

where p is the pressure at some arbitrary point in the deformed configuration.

We note that the same dissipation potential may be formed using different conjugate variables, but only a combination of conjugate variables in the material coordinate system follows the same group of material particles. For examples, v_{d} and $\nabla^{\mathrm{s}}p$ (give a work rate per unit volume in spatial coordinates), or $Jv_{\mathrm{d}}^{\mathrm{m}}$ and $F^{-\mathrm{T}}\nabla^{\mathrm{m}}p^{\mathrm{sm}}$ (give a work rate per unit volume in material coordinates), or $\dot{N}_{\mathrm{f}}^{\mathrm{m}}$ and p^{sm}(give a work rate per unit volume in material coordinates), or $\dot{E}$ and P^{m} (give a work rate per unit volume in material coordinates). In each case, multiplication of the two conjugate variables produces units of $T\dot{\vartheta}$ that represents heat dissipation, where T is the absolute temperature and $\dot{\vartheta}$ is the rate of entropy production $(\mathrm{Jm}^{-3}\mathrm{s}^{-1})$.

Clearly, the choice of conjugate variables is to some extent arbitrary; the conjugate pair chosen just needs to be consistent with other pairs, in the sense that they refer to the same unit volume of material. If this occurs, they may be meaningfully summed together. But it is most convenient to define potentials with respect to material coordinates, as then the same particles are being considered over time (if a spatial coordinate system is chosen, the particles in a fixed REV change over time). The addition of the elastic potential for the porous solid in the material coordinate system and the dissipation function for the pore fluid in the material coordinate system provide an important example, as they can be summed (as they track the same particles) to find, after some manipulation, Terzaghi's 'principle of effective stress' (see Appendix I Terzaghi's principle of effective stress). We note that

Terzaghi's principle of effective stress was our starting point above, so it is reassuring to be able to recover that statement from the sum of elastic potential and the dissipation function in material coordinates.

Consider now a 1-D consolidation process with constant stiffness and permeability, subject to a constant load at a drained boundary (the other boundary being fixed and impermeable). For a small strain analysis (material and spatial descriptions are then identical), the strain energy stored at the end of consolidation is found to be one-half of the total work done by the load acting on the surface, i.e.

$$\int_V \sigma_{sz} \mathrm{d}\varepsilon_z = \frac{E_z}{2}\varepsilon_z^2 = \frac{1}{2}\sigma_z \Delta H \tag{7.88}$$

where ΔH is the settlement of the soil surface. We note in passing that the so-called elastic potential is defined as

$$\Psi^e = \frac{1}{2}E_z\varepsilon_z^2 \tag{7.89}$$

Differentiation of Ψ^e w.r.t. the stress gives the strain (or vice versa).

The total work done by the load is $\sigma_z \Delta H$, and so half of the work done at the soil surface has been stored as elastic strain energy within the material (Ψ^e). The other half of the work done at the soil surface has been dissipated by the viscous flow of the fluid through the porous material. That is,

$$\frac{1}{2}\sigma_z \Delta H = \int_0^\infty \int_V \Psi^{dp} \mathrm{d}V \mathrm{d}t \tag{7.90}$$

Removal of the load results in water being re-imbibed as the porous material resumes its former shape. In this case, all the elastic strain energy is dissipated by the viscous flow. That is,

$$\frac{1}{2}E_z\varepsilon_{\mathrm{zf}}^2 = \int_0^\infty \int_V \Psi^{\mathrm{dp}} \mathrm{d}V \mathrm{d}t = \frac{1}{2}\sigma_z \Delta H \tag{7.91}$$

In other words, during a step load–unload cycle, 100% of the work done on the soil surface during loading is viscously dissipated by fluid flow within the material.

For a step load–unload cycle, the porous medium could be thought of as a 'spring' and a 'hydraulic damper' (much like some car suspension systems, e.g. the MacPherson strut; see http://en.wikipedia.org/wiki/MacPherson_strut). This analogy to viscoelastic dampers motivates some engineers with experience of viscoelastic models to use viscoelastic models to model the poroelastic consolidation of soils. The difficulty with this approach is that the viscoelastic time constants are not material properties of the porous medium. Rather, the viscoelastic time constants for a problem depend on the problem geometry, the boundary conditions, and the material. In other words, to solve the poroelastic problem properly using viscoelasticity one needs to know what the solution is, which defeats the purpose of the viscoelastic modeling. To say this another way, it means parameters in the viscous dissipation model change for every problem (i.e. the parameters are simply curve fitting parameters). That is not to say that visco-poroelastic models cannot be employed to capture additional dissipative processing occurring within the material itself (in the case of soil, for example, such a process may be time dependent particle breakage, while in the case of cartilage, the collagen fibrils may break or they may be view as viscoelastic (Li et al. 2009) (probably because collagen fibrils have significant water-filled porosity themselves (which leads to the idea of having double porosity models of cartilage), so there is viscous dissipation as water leaves the collagen fibril when it is stretched). This is taken up again in the section titled nonequilibrium equations for large deformation of a porous medium.

It is interesting to reconsider the dissipative response if a 'step loading' applied to the surface of the porous medium is replaced by a 'quasi-static' loading applied to the surface of the porous medium. By quasi-static loading, we mean the loading is at a rate that is sufficiently slow such that the system conditions are always close to thermodynamic equilibrium. In this case, the loading is so slow that no excess pore water pressure develops in the porous medium, and so there is no viscous fluid dissipation induced by the loading. In fact, the presence of excess pore water pressures means that the loading process is thermodynamically, a nonequilibrium process (this is discussed in more detail section nonequilibrium equations for large deformation of a porous medium). Similarly, quasi-static unloading results in complete recovery of the elastic strain energy. So for a complete load cycle under quasi-static conditions, no energy is dissipated as heat throughout a load cycle.

We see that energy dissipation in the poroelastic material is connected to the rate of loading and unloading. More generally, if the loading frequency has a period similar to the characteristic time for consolidation, dissipation is a significant fraction of the work done by the external load, while if the loading frequency is either much smaller or much larger than the characteristic time for consolidation, then dissipation is a small fraction of the work done by the external load.

When the period of a load–unload cycle is much smaller than the characteristic time for consolidation, the response of the poroelastic material is said to be 'undrained,' as the pore water does not have time to escape the porous medium, while if the period of loading is much longer than the characteristic time for consolidation, the response of the poroelastic material is said to be 'drained,' as the pore water has plenty of time to escape the porous medium (and pore pressure gradients are small). The analysis of both drained and undrained responses of porous media is discussed later.

As mentioned in the introduction to this section, of crucial importance to this discussion is the consolidation equation holds for very fine particulate soils known as clays. Saturated clays are actually colloidal particles in water of different ionic strengths. In very soft clays, the colloidal particles may be completely suspended when in fresh water, so the particles do not contact one another directly at all. Some clay soils carry a very high negative charge (e.g. montmorillonite has a high negative 'fixed charge density'), and so there are counter-ions around each soil particle to maintain electroneutrality.

DLVO theory (named after Derjaguin, Landau, Verwey, and Overbeek) was developed in the 1940s to understand this distribution of counterions around colloidal particles in solution. The DLVO theory is discussed at some length in Chap. 6, and from this theory, it is possible to calculate the electro-osmotic force (and other nonspecific forces) acting between colloidal particles under various conditions (e.g. with varying ionic strength of the solution) (http://en.wikipedia.org/wiki/DLVO_theory). The Helmholtz-Smoluchowski theory may be employed to understand 'electrokinetic phenomena' of colloids; e.g. the so-called streaming potential may be calculated (http://en.wikipedia.org/wiki/Streaming_current). Various forces and fluxes of fluid, electricity, heat, and chemicals are coupled through Onsager's reciprocal relations in colloidal systems (http://en.wikipedia.org/wiki/Onsager_reciprocal_relations). For the application of these (colloidal) theories and principles to biological systems, see Forces, Fields and Flows in Biological Systems (Grodzinsky 2011).

The key point to made above is that consolidation theory applies to clay soil, and more generally, it applies to solutions of colloidal particles (which includes solutions of charged macromolecules) http://en.wikipedia.org/wiki/Colloid. In this way, we see that consolidation theory, which was developed for soils, can be employed to explain the deformation behavior of concentrations of highly charge macromolecules (e.g. aggrecan found in cartilage).

Due to the electro-osmotic forces between colloidal particles, the colloidal particles do not have to directly touch one another at all in order for there to be a mechanical force between them. This mechanical force between the particles is actually the 'effective stress' within the colloidal suspension. So the concept of an 'effective stress,' which controls the (nonlinear) deformation of a colloidal particle assembly, can also be applied to colloidal suspensions. This includes the deformation of aggrecan found in cartilage. That consolidation theory developed for soils applies to cartilage has been confirmed experimentally—as mentioned above, see the experimental and theoretical results of Oloyede and Broom (1991, 1994, Armstrong et al. (1984), Edwards (1966), Wellard et al. (2014), Bursac et al. (1999), Miramini et al. (2017), Zhang et al. (2015), Fick et al. (2010).

All this is very interesting and indeed brings us considerably closer to being able to analyze the time-dependent deformation of cartilage. But we still need to solve some additional practical issues. For example, we need to be able to generalize the 1-D Terzaghi analysis to three spatial dimensions (so we can do a 2-D or a 3-D analysis of cartilage if we so choose). Maurice Biot succeeded in formulating these equations in the 1940s (Biot 1941). We derive these equations in the next section.

In addition, we also need to develop a 'large deformation' theory, as the settlements found in cartilage may be beyond the range where small deformation theory remains sufficiently accurate. McNabb (1960) and Gibson et al. (1967, 1981) succeeded in doing this in one spatial dimension in the 1960–1980s. But we also need to generalize this 1-D large deformation analysis to 3-D.

And finally, we need to consider the transport of growth factors and other solutes through cartilage. It is desirable if we are able to couple solute transport within the mechanical deformation of the porous medium, as there can be interesting interactions between the two. Smith and Peters (Peters and Smith 2002; Smith 2000) coupled solute transport through (diffusion-dispersion) with (advection) of the pore fluid (due to consolidation), with a large deformation consolidation analysis for one spatial dimension. Clearly, it is desirable to consider how to do this in two or three spatial dimensions. This set of equations would enable us to analyze the transport of growth factors through cartilage as it is cyclically loaded with large deformations.

We now turn to developing the equations for Biot's general, small deformation, consolidation theory in three spatial dimensions.

7.2.2 Biot Poroelasticity Theory

An exact and powerfully simplifying assumption employed in the Terzaghi 1-D consolidation analysis in the previous section is that the rate of change of the total stress within the poroelastic material is always known; i.e. it is always proportional to the surface loading. For example, the total stress is a constant throughout thickness the poroelastic material for a step load in the 1-D consolidation analysis. This knowledge of the total stress at every point in the domain at all times made it possible to directly substitute Terzaghi's principle of effective stress relationship, and obtain Eq. (7.32). Unfortunately, for two- or three-dimensional consolidation problems, the total stress at a location in the material is not known a priori, so one is obliged to solve the storage equation together with the equations of elasticity in their entirety. Relative to the one-dimensional analysis, having to solve the equations of elasticity increases the complexity of two- and three-dimensional analyses greatly. Initially, it was not even clear how to write down the equations in two and three dimensions, let alone do such an analysis.

However, Maurice Biot managed to develop a three-dimensional phenomenological theory (Biot 1941), which generalized Terzaghi's one-dimensional theory of consolidation (see *General Theory of Three Dimensional Consolidation* (1941) http://www.pmi.ou.edu/Biot2005/papers/FILES/040.PDF). Here, we develop the Biot equations in three spatial dimensions, writing the equations out in full for an isotropic poroelastic material, both to follow the derivation and to more fully appreciate their meaning and import.

The analysis is based on small deformations, which means the deformed shape should not be too different to the initial shape (e.g. the model is often used if the expected rotations are not more than about 10°). This is often the case if loads are not too large and concentrated, and if material is stiff enough. If inertial loadings are included in the momentum equations, the model is also suited to the analysis of sound wave propagation through porous materials (i.e. the analysis of ultrasound).

Consider a poroelastic material, consisting of solid particles (including macromolecules in solution) with interconnected pore spaces. Following Professor Arnold Verruijt (1995), we begin with the mass conservation equations for fluid and solid phases in a spatial (i.e. fixed) coordinate system. For mass conservation of the fluid, we have

$$\frac{\partial(n_f \rho_f)}{\partial t} + \nabla \cdot (n_f \rho_f v_f) = 0 \tag{7.92}$$

and for mass conservation of the solid we have

$$\frac{\partial[(1 - n_f)\rho_s]}{\partial t} + \nabla \cdot [(1 - n_f)\rho_s v_s] = 0 \tag{7.93}$$

where n_f, ρ_f are the fluid porosity and fluid density, ρ_s, v_s are the solid density and velocity of the solid particles (w.r.t the fixed coordinate system), and v_f is the velocity of the fluid (w.r.t the fixed coordinate system). The 'solid deformation' is supposed to be caused by the rearrangement of the solid particles (or macromolecules in solution) in the porous medium. However, the particles themselves are taken to be incompressible (i.e. ρ_s = constant) as is the fluid (i.e. ρ_f = constant). In this case,

$$\begin{aligned} \tfrac{\partial n_f}{\partial t} + \nabla \cdot (n_f v_f) = 0 \\ -\tfrac{\partial n_f}{\partial t} + \nabla \cdot [(1 - n_f)v_s] = 0 \end{aligned} \tag{7.94}$$

Adding these equations reveals

$$\nabla \cdot [n_f(v_f - v_s)] + \nabla \cdot v_s = 0 \tag{7.95}$$

Now, as in the previous section, the Darcy velocity $v_d = n_f(v_f - v_s)$ is defined as the average of the fluid velocity across the gross cross-sectional area, which is proportional to the difference in the actual fluid velocity relative to the velocity of the solid-phase particles. It is the relative velocity between the phases that leads to momentum exchange between the phases. And using Eq. (7.51), we have

$$-\nabla \cdot v_s = \frac{1}{(1 + e_v)} \frac{\mathrm{d}e_v}{\mathrm{d}t} \approx \frac{\partial e_v}{\partial t} \tag{7.96}$$

for small deformations. Then, Eqs. (7.95) and (7.96) lead to

$$\nabla \cdot v_d = \frac{\partial e_v}{\partial t} \tag{7.97}$$

This is exactly the same storage equation, assuming small deformations, that we found previously for the large deformation 1-D consolidation analysis [see Eq. (7.51)].

Now, the Darcy velocity is given by the constitutive equation

$$v_{\mathrm{d}} = -\frac{K}{\mu_{\mathrm{f}}}\nabla p \tag{7.98}$$

where K is the intrinsic permeability (which depends only on the geometry of the interconnected pore space, i.e. the size and shape of the interconnected pore space), μ_{f} is the dynamic viscosity of the pore fluid, and p is the excess fluid pore pressure (as previously defined). Then, our storage equation (which is actually an amalgam of mass conservation equations for solid and fluid, assumptions about the compressibility of the phases, and a constitutive relation) becomes

$$\nabla \cdot k_{\mathrm{f}}\nabla p = \frac{\partial e_v}{\partial t} \tag{7.99}$$

This is all unsurprising, as we arrive at the same conclusions as for the 1-D analysis. We now turn to the momentum balance equations. We again adopt the solid mechanics convention for tensions in the solids being taken as positive, while compressive stress in the pore fluid is taken as positive. The basic momentum conservation equation says the divergence of the stresses is the body forces, viz.,

$$\begin{aligned}
\frac{\partial\sigma_{xx}}{\partial x} + \frac{\partial\sigma_{yx}}{\partial y} + \frac{\partial\sigma_{zx}}{\partial z} + f_x = 0\\
\frac{\partial\sigma_{xy}}{\partial x} + \frac{\partial\sigma_{yy}}{\partial y} + \frac{\partial\sigma_{zy}}{\partial z} + f_y = 0\\
\frac{\partial\sigma_{xz}}{\partial x} + \frac{\partial\sigma_{yz}}{\partial y} + \frac{\partial\sigma_{zz}}{\partial z} + f_z = 0
\end{aligned} \tag{7.100}$$

where σ_{ij} are total Cauchy stresses and f_i are body forces (and i, j take values 1–3 for the coordinate axes x, y, z). Now, we define the solid (or effective) stresses $(\sigma_{\mathrm{s}ij})$ as we previously for the 1-D analysis, i.e.

$$\begin{aligned}
\sigma_{xx} = \sigma_{\mathrm{s}xx} - p, \sigma_{xy} = \sigma_{\mathrm{s}xy}, \sigma_{xz} = \sigma_{\mathrm{s}xz}\\
\sigma_{yy} = \sigma_{\mathrm{s}yy} - p, \sigma_{yx} = \sigma_{\mathrm{s}yx}, \sigma_{yz} = \sigma_{\mathrm{s}yz}\\
\sigma_{zz} = \sigma_{\mathrm{s}zz} - p, \sigma_{xz} = \sigma_{\mathrm{s}xz}, \sigma_{yz} = \sigma_{\mathrm{s}yz}
\end{aligned} \tag{7.101}$$

The effective stresses are related to the strains for the particle assembly by the constitutive equation. A generalized form of nonlinear elasticity (i.e. the material 'constants' may be function of strain) is adopted as a first approximation. For an isotropic material this becomes

$$\begin{aligned}
\sigma_{\mathrm{s}xx} = \lambda e_v + 2\mu e_{xx}, \sigma_{\mathrm{s}xy} = 2\mu e_{xy}, \sigma_{\mathrm{s}xz} = 2\mu e_{xz}\\
\sigma_{\mathrm{s}yy} = \lambda e_v + 2\mu e_{yy}, \sigma_{\mathrm{s}yz} = 2\mu e_{yz}, \sigma_{\mathrm{s}yx} = 2\mu e_{yx}\\
\sigma_{\mathrm{s}zz} = -\lambda e_v + 2\mu e_{zz}, \sigma_{\mathrm{s}zx} = 2\mu e_{zx}, \sigma_{\mathrm{s}zy} = 2\mu e_{zy}
\end{aligned} \tag{7.102}$$

where $\lambda = K - \frac{2}{3}G, \mu = G$ are elastic coefficients such that where K is the bulk (or Lame) modulus and G is the shear modulus. These moduli are related to Young's modulus (E) and Poisson's ratio (ν) by $K = \frac{E}{3(1-2\nu)}$, $G = \frac{E}{2(1+\nu)}$(see http://en.wikipedia.org/wiki/Bulk_modulus for further relations).

The strain components are related to the displacements by

$$\begin{aligned}
\varepsilon_{xx} &= \frac{\partial u_x}{\partial x}, \varepsilon_{xy} = \frac{1}{2}\left(\frac{\partial u_x}{\partial y} + \frac{\partial u_y}{\partial x}\right), \varepsilon_{xz} = \frac{1}{2}\left(\frac{\partial u_x}{\partial z} + \frac{\partial u_z}{\partial x}\right) \\
\varepsilon_{yy} &= \frac{\partial u_y}{\partial y}, \varepsilon_{yz} = \frac{1}{2}\left(\frac{\partial u_y}{\partial z} + \frac{\partial u_z}{\partial y}\right), \varepsilon_{xz} = \frac{1}{2}\left(\frac{\partial u_y}{\partial x} + \frac{\partial u_x}{\partial y}\right) \\
\varepsilon_{zz} &= \frac{\partial u_z}{\partial z}, \varepsilon_{zx} = \frac{1}{2}\left(\frac{\partial u_z}{\partial x} + \frac{\partial u_x}{\partial z}\right), \varepsilon_{zy} = \frac{1}{2}\left(\frac{\partial u_z}{\partial y} + \frac{\partial u_y}{\partial z}\right)
\end{aligned} \quad (7.103)$$

The volumetric strain is given by

$$e_v = \varepsilon_{xx} + \varepsilon_{yy} + \varepsilon_{zz} \quad (7.104)$$

The completes the system of field equations. There are 22 unknowns (9 stresses, 9 strains, 3 displacements, and the pore water pressure). The system is well posed, because there are 22 independent equations (6 equilibrium equations, 9 compatibility equations, 6 independent stress–strain relations, and the storage equation). This system of equations can be considerably simplified by eliminating the stresses and strains and expressing the equilibrium equations in terms of displacements. For a homogeneous material (with constant λ, μ), these equations are:

$$\begin{aligned}
(\lambda+\mu)\frac{\partial \varepsilon_v}{\partial x} + \mu\nabla^2 u_x + f_x - \frac{\partial p}{\partial x} &= 0, \\
(\lambda+\mu)\frac{\partial \varepsilon_v}{\partial y} + \mu\nabla^2 u_y + f_y - \frac{\partial p}{\partial y} &= 0, \\
(\lambda+\mu)\frac{\partial \varepsilon_v}{\partial z} + \mu\nabla^2 u_z + f_z - \frac{\partial p}{\partial z} &= 0,
\end{aligned} \quad (7.105)$$

where ∇^2 is defined as

$$\nabla^2 = \frac{\partial^2}{\partial x^2} + \frac{\partial^2}{\partial y^2} + \frac{\partial^2}{\partial z^2} \quad (7.106)$$

where it is clear the gradients of the pore water pressures are acting like transient body forces on the solid phase. We also note that it is not possible to reduce 2-D and 3-D problems to a single equation as we could for the 1-D problem, because we do not know a priori in the 2-D and 3-D problems what the total stress actually is at each point in the solid skeleton, as we do for the 1-D problem. For the 1-D problem, the total stress at each time point is given by the boundary condition to the problem, but it is not obvious from the boundary conditions in 2-D and 3-D problems, so we need to solve the equations of elasticity together with the storage equation.

Now there are four unknowns (p, u_x, u_y, u_z) in four independent equations [i.e. the three displacements in the three equilibrium equations (7.105) and the excess pore water pressure in the storage equation (7.106)]. Solving these equations simultaneously, using say a finite element formulation in space combined with a 'forward marching' finite difference method in time, leads to the numerical solution of small deformation multidimensional consolidation problems.

7.2.3 *Drained and Undrained Poroelastic Responses*

When the period for a load–unload cycle is much smaller than the characteristic time for consolidation (that is, the rate of loading is comparatively fast compared to the characteristic time), the response of the poroelastic material is said to be 'undrained,' as the pore water does not have time to escape the porous medium. On the other hand, if the period of loading is much longer than the characteristic time for consolidation, the response of the poroelastic material is said to be 'drained' (the rate of loading is comparatively slow compared to the characteristic time), as the pore water has plenty of time to escape the porous medium (and so pore pressure gradients are small).

If during the period of interest (Δt) the divergence of the pore fluid velocity is essentially zero (which usually occurs when the permeability is very low and/or the loading fast), then the volume strain rate may be approximated as zero. In this case, the poroelastic material is behaving volumetrically as if it is infinitely stiff in 'hydrostatic compression,' though it is free to undergo shear deformation, as (small deformation) shear does not involve a volume change. This volumetric behavior is consistent with the pore water being assumed to be incompressible. From the definition of bulk modulus,

$$K = \frac{E}{3(1 - 2v)}$$

We see that the poroelastic material becomes infinitely stiff in hydrostatic compression when the Poisson's ratio approaches 0.5. A Poisson's ratio of 0.5 is consistent with no change in volume under hydrostatic compression [for example, the initial Poisson's ratio in cartilage has been experimental measured as 0.49 ± 0.08 (Wong et al. 2000)]. In practice, it is generally considered sufficiently accurate during finite element analyses of consolidation to approximate the undrained condition by using a Poisson's ratio that is 0.45 or larger. In other words, the undrained response of a poroelastic material may be approximated by a solid mechanics analysis using the drained Young's modulus with Poisson's ratio 'approaching 0.5.'

The (drained) Young's modulus is the measured stiffness of the material when excess pore water pressures are zero (i.e. at thermodynamic equilibrium). Similarly, the drained Poisson's ratio may be defined as the measured Poisson's ratio when pore water pressures are zero. Using the drained Young's modulus and the drained Poisson's ratio in a solid mechanics analysis results in the same stresses, strains, and displacements as for a consolidation analysis at equilibrium, i.e. after all the pore water pressure have dissipated.

7.2.4 Equilibrium Equations for Large Deformation of a Porous Medium

When a system is loaded 'quasi-statically,' the system is by definition always in thermodynamic equilibrium. For reversible, path-independent systems, all the state variables of the system may be determined from a 'state equation,' which is often referred to as the 'free energy function' for that material. For isothermal mechanical systems, we can reduce the number of state variables by choosing a Helmholtz free energy function Ψ^e(which is defined per unit volume in the material coordinate system) with $\mathrm{d}T = 0$). The isothermal condition is often a reasonable approximation to reality. As the material is loaded, it elastically deforms, and energy is stored within it. The energy stored in the material is independent of the path taken from the initial state to the final state, so the change in Helmholtz free energy per unit volume becomes a function of the final elastic material deformation (under isothermal conditions), not of the path taken to the final state.

Large deformations generate stretching and rotations in the material. As described above, we require a strain measure that varies with stretching, but is invariant to any rotation. Unfortunately, engineering strain—as used for the 'small deformation' Biot analysis (i.e. where rotations are assumed small)—proves to be unsuitable as a rigid body rotation leads to a finite strain, which is clearly incorrect (the small strain tensor is defined as $\varepsilon = 0.5(F + F^{\mathrm{T}}) - I$, where the deformation gradient tensor F is defined below and I is the identity tensor). We therefore need to find a suitable strain measure that is invariant with rotation. As mentioned in the introduction to this chapter, the strain measure usually chosen is the Green–Lagrangian strain tensor (Bonet and Wood 2008),

$$E = \frac{1}{2}(C - I) \tag{7.107}$$

where $C = F^{\mathrm{T}}F$ is the right Cauchy–Green tensor. Both C and E are invariant with respect to rigid body motion, so either is a suitable strain measure. We also note that $U - I$, known as the Biot strain tensor, can be used a large deformation strain measure, as can the logarithm of U (Neff et al. 2016).

The Green–Lagrangian strain tensor can be expressed in terms of the familiar displacement gradients in the spatial coordinate system (http://en.wikipedia.org/wiki/Finite_strain_theory#Finite_strain_tensors). For three spatial dimensions in material coordinates, this is,

$$E_{ij} = \frac{1}{2}\left(\frac{\partial U_i}{\partial X_j} + \frac{\partial U_j}{\partial X_i} + \frac{\partial U_m}{\partial X_i}\frac{\partial U_m}{\partial X_j}\right) \tag{7.108}$$

where U_i are the displacements in material coordinates. The Eulerian strain tensor or Almansi strain tensor in spatial coordinates (ε_{ij}) is the push-forward of the

Green–Lagrangian strain tensor, i.e. $\varepsilon = F^{-\mathrm{T}}EF^{-1}$. The Eulerian strain tensor can be defined in terms of the left Cauchy–Green deformation tensor as

$$\varepsilon = \frac{1}{2}(I - H) \tag{7.109}$$

where H is the inverse of the left Cauchy–Green (Finger's) strain tensor ($B = FF^{\mathrm{T}}$). Then,

$$\varepsilon_{ij} = \frac{1}{2}\left(\frac{\partial u_i}{\partial x_j} + \frac{\partial u_j}{\partial x_i} - \frac{\partial u_{\mathrm{m}}}{\partial x_i}\frac{\partial u_{\mathrm{m}}}{\partial x_j}\right) \tag{7.110}$$

For two spatial dimensions, this can be expanded to

$$\begin{aligned}
\varepsilon_{xx} &= \tfrac{1}{2}\left(\tfrac{\partial u_x}{\partial x} + \tfrac{\partial u_x}{\partial x} - \tfrac{\partial u_x}{\partial x}\tfrac{\partial u_x}{\partial x} - \tfrac{\partial u_y}{\partial x}\tfrac{\partial u_y}{\partial x}\right), \varepsilon_{xy} = \tfrac{1}{2}\left(\tfrac{\partial u_x}{\partial y} + \tfrac{\partial u_y}{\partial x} - \tfrac{\partial u_x}{\partial x}\tfrac{\partial u_x}{\partial y} - \tfrac{\partial u_y}{\partial x}\tfrac{\partial u_y}{\partial y}\right),\\
\varepsilon_{yx} &= \tfrac{1}{2}\left(\tfrac{\partial u_y}{\partial x} + \tfrac{\partial u_x}{\partial y} - \tfrac{\partial u_x}{\partial y}\tfrac{\partial u_x}{\partial x} - \tfrac{\partial u_y}{\partial y}\tfrac{\partial u_y}{\partial x}\right), \varepsilon_{yy} = \tfrac{1}{2}\left(\tfrac{\partial u_y}{\partial y} + \tfrac{\partial u_y}{\partial y} - \tfrac{\partial u_x}{\partial y}\tfrac{\partial u_x}{\partial y} - \tfrac{\partial u_y}{\partial x}\tfrac{\partial u_y}{\partial x}\right)
\end{aligned} \tag{7.111}$$

If the displacements are small, second-order terms can be neglected. In this case, the strains reduce to engineering strains employed for a small deformation analysis. We note that in the subsequent discussion of large deformation solid mechanics, we largely follow Bonet and Wood—Non-linear Continuum Analysis for Finite Element Analysis (Bonet and Wood 2008).

For small deformation theory, the work conjugate of the engineering strain tensor is the Cauchy stress tensor. It turns out that the work conjugate of the Green–Lagrangian strain tensor E_{kl} is the second Piola–Kirchhoff stress tensor S_{ij}. These two material coordinate tensors are related by the fourth-order elasticity tensor $\mathbb{C}_{ij\mathrm{kl}}$, also expressed in material coordinates. Because S is related to the strain by the elasticity tensor, we could signify this elasticity relationship using S^e, but for equilibrium thermodynamics, this does not add much, so we choose here just to use S. In the discussion on nonequilibrium thermodynamics, it proves helpful to distinguish between various second Piola–Kirchhoff stress tensors that arise from elastic deformations and dissipative processes separately.

For large deformation analysis, it is usual for the fourth-order elasticity tensor to be found by differentiating a strain energy density function $\Psi^e(E)$ w.r.t E or w.r.t the right Cauchy–Green tensor C (see the excellent explanation of how to take the derivative Ψ^e w.r.t. C, and the subsequent discussion of symmetry relations for $\mathbb{C}_{ij\mathrm{kl}}$ at http://www.umich.edu/~bme332/ch6consteqelasticity/bme332consteqelasticity.htm). The strain energy is the energy per unit volume of solid in the reference configuration. The stress–strain relation is

$$S_{ij} = \mathbb{C}_{ijkl} : E_{kl} \tag{7.112}$$

where

$$\begin{aligned} S_{ij} &= \frac{\partial \Psi^e}{\partial E_{ij}} = 2\frac{\partial \Psi^e}{\partial C_{ij}} \\ \mathbb{C}_{ijkl} &= \frac{\partial^2 \Psi^e}{\partial E_{ij} \partial E_{kl}} = 4\frac{\partial^2 \Psi^e}{\partial C_{ij} \partial C_{kl}} \end{aligned} \tag{7.113}$$

All second-order tensors, including the stress and strain tensors, can be decomposed as a sum of a 'hydrostatic tensor' and a 'deviatoric tensor.' For example, for the stress tensor, S can be decomposed as

$$S = \hat{S} + S^{m} \tag{7.114}$$

where $S^{m} = \frac{\mathrm{Tr}(S)}{3}\delta_{ij}$ is the hydrostatic tensor and $\hat{S}$ is the deviatoric tensor, $\hat{S} = S - S^{m}$.

Using this decomposition, the elastic strain energy density function Ψ^e is split into two parts that are functions of invariants of the volumetric strain $\left(\Psi^{evol}\right)$ and 'distortional strain' tensor $\left(\Psi^{edis}\right)$. The 'volumetric strain' is related to the Jacobian of the deformation gradient, i.e. $J - 1 = \det[F] - 1$, while the 'distortional strain' is related to $\hat{C} = J^{-(2/3)}C$.

Most often, strain energy density functions employed in large deformation analysis are for isotropic materials http://en.wikipedia.org/wiki/Hyperelastic_material. Further, most of the common strain energy density functions are simply curve fits to experimental data, meaning they are not derived from a deeper theory based on material behavior. An example of a popular strain energy density function employed for curve fitting to experimental data is the Mooney–Rivlin strain energy density function, which is a polynomial of the principal stretches or the strain invariants I_1, I_2 and I_3. For the so-called two parameter compressible Mooney–Rivlin potential http://en.wikipedia.org/wiki/Mooney%E2%80%93Rivlin_solid,

$$\Psi = \mathrm{C}_{10}(\bar{I}_1 - 3) + \mathrm{C}_{01}(\bar{I}_2 - 3) + \mathrm{D}_1(\mathrm{J} - 1)^2 \tag{7.115}$$

where $\bar{I}_1 = I_1 J^{-(2/3)}$ (which is the first invariant of the distortional strain tensor), $\bar{I}_2 = I_2 J^{-(4/3)}$(which is the second invariant of the distortional strain tensor), and C_{10}, C_{01} are material constants. $I_1 = \mathrm{Tr}(C)$ and $I_2 = C : C$ are the first and second strain invariants of C. We note that if a Mooney–Rivlin potential is formulated in terms of the left Cauchy–Green deformation tensor ($B = FF^{\mathrm{T}}$), rather than the right Cauchy–Green deformation tensor ($C = F^{\mathrm{T}}F$), then differentiation w.r.t B yields the Cauchy stresses directly, rather than the second Piola–Kirchhoff stresses that result from a Mooney–Rivlin potential formulated in terms in the right Cauchy–Green deformation tensor, $C = F^{\mathrm{T}}F$. Of course, either approach is fine, as the second Piola–Kirchhoff stress can be 'pushed forward' to yield the (same) Cauchy stress.

The first and second terms in Eq. (7.115) account for the distortional strain energy, while the third term in Eq. (7.115) accounts for the volumetric strain energy. It is noted that for pure shear $\bar{I}_1 = \bar{I}_2$. A generalized Rivlin model can be found at http://en.wikipedia.org/wiki/Mooney%E2%80%93Rivlin_solid. A helpful discussion for incompressible Mooney–Rivlin materials ($J = 1$) can be found at http://www.continuummechanics.org/cm/mooneyrivlin.html.

The Mooney–Rivlin material constants are related to the more familiar initial pure shear modulus μ, such that $\mu = 2(C_{10} + C_{01})$ and to κ, the initial bulk modulus, where $D_1 = \kappa/2$ for an isotropic material. When $C_{01} = 0$, the two-parameter compressible Mooney–Rivlin constitutive equation reduces to the compressible Neo-Hookean constitutive equation.

From the foregoing, it is clear that the initial Young's modulus and Poisson's ratio are related to C_{10} and κ. For example, when the Poisson's ratio is chosen as zero, then $E_0 = 4C_{10}$, while if Poisson's ratio is chosen to be 0.5, then $E_0 = 6C_{10}$. These relationships may be helpful when starting to model the elastic behavior of a poroelastic material using a hyperelastic strain energy density functions. We note here that a hyperelastic material is simply a perfectly elastic material that has a path-independent Helmholtz free energy density function. And previously, we have seen that all poroelastic materials are purely elastic when deformations take place in the quasi-static way, such that no excess pore water pressures are developed. If quasi-static deformation occurs, there is no entropy production, and the deformation can be considered to be in thermodynamic equilibrium. More usually, a real deformation occurs under nonequilibrium conditions, excess pore water pressures develop, and the dissipative viscous flow ensures there is entropy production (which occurs in all real spontaneous processes).

If one wishes to include fiber reinforcement in the continuum model, then it can be shown that the strain energy function for the single collagen fiber be defined simply as

$$\Psi(\hat{n}^r, I_n) = \xi(\hat{n}^r)(I_n - 1)^{\alpha(\hat{n}^r)} \tag{7.116}$$

where all parameters are defined at equation (O.3). The second derivative with respect to the stretch defines the collagen fiber stiffness, which can be added into the stiffness matrix of a finite element analysis.

Because the second Piola–Kirchhoff stress tensor is the pull-back of the Cauchy stress tensor, we can first write the momentum balance equation in spatial coordinates and then pull-back the momentum balance equation back to material coordinates. Implementing this approach to derive the momentum balance in material coordinates begins with the momentum balance equation in spatial coordinates (see Appendix J The momentum balance equation and strain work rate),

$$\nabla^{s} \cdot (\sigma_s + p) + f^{s} = 0 \tag{7.117}$$

Now if 'pull-back' to the material coordinate system, we find

$$\nabla^{\mathrm{m}} \cdot (FS) + \nabla^{\mathrm{m}} \cdot JF^{-\mathrm{T}} p^{\mathrm{sm}} + F^{\mathrm{m}} = 0 \tag{7.118}$$

where $F^{\mathrm{m}} = Jf^{\mathrm{s}}$. It is noted that $P = FS$, where P is the first Piola–Kirchhoff stress tensor. The work conjugate of P is the deformation gradient F.

We have previously developed the 'storage equation' in a suitable form for a large deformation analysis [see Eq. (7.63)], viz.,

$$\nabla^{\mathrm{m}} \cdot JF^{-1} k_{\mathrm{w}}^{\mathrm{sm}} F^{-\mathrm{T}} \nabla^{\mathrm{m}} p^{\mathrm{sm}} = JF^{-1} F^{-\mathrm{T}} : \frac{\mathrm{d}E}{\mathrm{d}t} = \frac{\mathrm{d}J}{\mathrm{d}t} \tag{7.119}$$

where the primary independent variable in Eq. (7.119) is the pore water pressure in the deformed spatial coordinate system mapped across to the material coordinate system, p^{sm}. This completes the governing equations for an elastic large deformation analysis of a poroelastic solid. The complete set of equations is (7.108), (7.112), (7.113), (7.118) and (7.119).

$$\begin{gathered}
\nabla^{\mathrm{m}} \cdot (FS) + \nabla^{\mathrm{m}} \cdot JF^{-\mathrm{T}} p^{\mathrm{sm}} + F^{\mathrm{m}} = 0 \\
S_{ij} = \mathbb{C}_{ij\mathrm{kl}} : E_{\mathrm{kl}} \\
S_{ij} = \frac{\partial \Psi^{e}}{\partial E_{ij}} \\
\mathbb{C}_{ij\mathrm{kl}} = \frac{\partial^2 \Psi^{e}}{\partial E_{ij} \partial E_{\mathrm{kl}}} \\
E_{ij} = \frac{1}{2}\left(\frac{\partial U_i}{\partial X_j} + \frac{\partial U_j}{\partial X_i} + \frac{\partial U_m}{\partial X_i}\frac{\partial U_m}{\partial X_j}\right) \\
\nabla^{\mathrm{m}} \cdot JF^{-1} k_{\mathrm{w}}^{\mathrm{sm}} F^{-T} \nabla^{\mathrm{m}} p^{\mathrm{sm}} = JF^{-1} F^{-\mathrm{T}} : \frac{\mathrm{d}E}{\mathrm{d}t} = \frac{\mathrm{dJ}}{\mathrm{dt}}
\end{gathered} \tag{7.120}$$

Finally, we note that if the poroelastic material consists of more than one type of solid (as the case for cartilage), then we may develop strain energy density functions for each component and sum them together. So the strain energy density function for aggrecan $\Psi^{e-\mathrm{agg}}$ may be summed with the strain energy density for collagen $\Psi^{e-\mathrm{col}}$ to give the elastic strain energy density for cartilage $\Psi^{e-\mathrm{cart}}$. That is,

$$\Psi^{e-\mathrm{cart}} = \Psi^{e-\mathrm{agg}} + \Psi^{e-\mathrm{col}} + \Psi^{e-\mathrm{agg-col}}$$

where $\Psi^{e-\mathrm{agg-col}}$ represents any elastic potential energy arising from interactions between the component materials. It is noted that if there are two or more collagen networks present with different characteristics (as can occur in cartilage), then each network could have its own elastic potential. As a first approximation, the elastic interaction potential energy term is set to zero. Unfortunately, we do not know that much about the interactions between aggrecan and collagen as cartilage is deformed. We know the change in osmotic pressure associated with an increase in aggrecan concentration does lead to a transfer of water from the intrafibrillar compartment to the extrafibrillar compartment (see discussion Appendix D). If there is an exchange of water (which is particularly evident at low osmotic pressures, i.e. less than about 1 MPa), one might expect some interaction between component materials, but this interaction has not been mechanically quantified.

Now as mentioned above, each strain energy density function may be split in a volumetric component Ψ^{evol} and a distortional component Ψ^{edis}. So setting $\Psi^{e-\text{agg}-\text{col}} = 0$, we may then further write,

$$\Psi^{e-\text{cart}} = \Psi^{evol-\text{agg}} + \Psi^{edis-\text{agg}} + \Psi^{evol-\text{col}} + \Psi^{edis-\text{col}}$$

Again, various approximations can be adopted depending on the measured properties of cartilage. For example, as a first approximation, one may argue that aggrecan is primarily responsible for the change in compressive volumetric strain energy density, while collagen is primarily responsible for the change in tensile volumetric strain energy density. We have derived theoretically a strain energy density function for the volumetric strain of aggrecan (see Appendix L).

Further, one may attribute proportions of the deviatoric strain energy density to either aggrecan or collagen. For example, one could assume that the component carrying most of the deviatoric free energy varies with the volumetric strain. So when the volumetric strain is zero, collagen may store most of the deviatoric elastic strain energy. On the other hand, when the compressive volumetric strain is large, aggrecan may store most of the deviatoric elastic strain energy. Unfortunately, the truth is we do not know very much about the interactions between aggrecan and collagen in cartilage as it is mechanically deformed particularly at large deformations.

7.2.5 Nonequilibrium Equations for Large Deformation of a Porous Medium

The rate of entropy change w.r.t. time, $\dot{\vartheta}$, can be decomposed into two parts, viz.,

$$\dot{\vartheta} = \dot{\vartheta}^{a} + \dot{\vartheta}^{\text{irr}} \tag{7.121}$$

where $\dot{\vartheta}^{a}$ accounts for the rate of entropy change in the system due to exchange of heat and matter with the system's surroundings, while $\dot{\vartheta}^{\text{irr}}$ accounts for the rate of entropy production within the system. $\dot{\vartheta}^{a}$ can be positive, zero, or negative, depending on the direction of exchange of heat and matter with the surroundings. In contrast, $\dot{\vartheta}^{\text{irr}}$ is never negative and becomes zero only in systems that are at thermodynamic equilibrium. In other words, for all nonequilibrium systems (with spontaneous, non-equilibrium processes within those systems), $\dot{\vartheta}^{\text{irr}}$ is positive.

When porous media deform quickly, non-equilibrium thermodynamic conditions are generated, and so dissipative processes occur within the porous medium. For non-quasi-static deformations, elastic processes occurring in the porous medium are represented by an isothermal Helmholtz 'free energy potential' Ψ^e (as described above), while the 'dissipation function' $\Psi^d = \dot{\vartheta}^{\text{irr}} T$ (where T is the absolute temperature) represents the free energy that is dissipated by processes internal to the material (i.e. the energy is no longer recoverable as mechanical work).

The dissipation function Ψ_i^{d} for the ith process is usually represented as products of so-named 'thermodynamic forces' and 'thermodynamic fluxes.' See, for example, Haase (1990). That is,

$$\Psi^{\text{d}} = \sum_{i=1}^{I} \Psi_i^{\text{d}} = \sum_{i=1}^{N} \chi_i \dot{\alpha}_i \tag{7.122}$$

where Ψ^{d} is the known as the dissipation function for the material, χ_i is the 'thermodynamic force' (which may be a tensor when thermodynamic stresses χ_{ij}^i are involved), and $\dot{\alpha}^i$ is 'thermodynamic flux' (which may be a tensor if thermodynamic strains rates $\dot{\alpha}_{ij}^i$ are involved) for the ith process. The dissipation functions are convex, so second derivatives of Ψ_i^d w.r.t. $\dot{\alpha}_i$ are positive, implying entropy is always produced (rather than removed—it would contravene the second law of thermodynamics if entropy decreases, i.e. entropy always increases in a spontaneous process, so $\Psi_i^{\text{d}} > 0$).

At least for processes not too far from equilibrium, the dissipation functions generally obey the 'normality condition,' that is the thermodynamic fluxes are normal to iso-contours of Ψ_i^{d} (Coussy and Scherer 2004). This is consistent with the principle that dissipative processes generally maximize the rate of entropy production (and so maximize the decrease in the potential). We note that there is no law of thermodynamics that says anything about the rate of entropy generation by a process, so the 'principle of maximum rate of entropy production' has the status of being a principle (Dyke and Kleidon 2010)), though some appear to consider this an additional law of thermodynamics (i.e. a law about the path taken by a system when approaching equilibrium). We mention in passing that individual processes may have thermodynamic fluxes non-normal to contours of isovalues of Ψ_i^{d} when there are constraints in the system.

Now, the rate of energy dissipation $\left(\text{Jm}^{-3}\text{s}^{-1}\right)$ is,

$$\Psi_i^{\text{d}} = T\dot{\vartheta}_i^{\text{irr}} = \sum_{i=1}^{N} \chi_i \dot{\alpha}_i \tag{7.123}$$

where $\dot{\vartheta}_i$ is the local rate of entropy formation $\left(\text{Jm}^{-3}\text{s}^{-1}\text{K}^{-1}\right)$ and T is the absolute temperature (in degrees Kelvin; K). For many biological systems, isothermal conditions may be reasonably assumed (as is the case for cartilage). For example, a process of isothermal diffusion of solute in a solvent generates entropy in the solution at the rate,

$$\Psi^{\text{ddif}} = T\dot{\vartheta}_i^{\text{irr}} = \nabla\mu_i \cdot f_{c_i} = \nabla\mu_i \cdot \frac{D_i c_i}{RT}(\nabla\mu_i) \tag{7.124}$$

where the conjugate variables are $\nabla\mu_i$ (i.e. the gradient of the chemical potential of the ith species) and the '(volume average) velocity of the solution minus the (volume averaged) velocity of the ith species in the solution' (but conjugate variables may be chosen to suit the problem variables). We notice that the thermodynamic flux is normal to contours of isovalues of chemical potential, so the rate of entropy production is maximized for a diffusive process.

For diffusion of a solute within an ideal solution ($\mu_i = RT \ln c_i$) under isotropic, isothermal conditions, we find

$$\Psi^{\text{ddif}} = T\dot{\vartheta}_i^{\text{irr}} = \frac{RT}{c_i}\nabla c_i \cdot D_i \nabla c_i = \frac{RTD_i}{c_i}(\nabla c_i)^2 \tag{7.125}$$

We note in passing that if multiple species are diffusing together there may be diffusion couplings between the species (Cussler 1984).

There are four main dissipative processes in cartilage. The first dissipative process is due to the diffusion of solutes through the aqueous solution. Even though the rate of diffusive transport may be relatively fast over short distances, the amounts transported are small or occur over long time periods, and consequently, the rates of dissipation due to diffusion are small for all solutes relative to other rates of dissipation in the porous medium. So are not considered further here.

The rate of free energy dissipation for aggrecan is very slow and so not important from an entropy production viewpoint, but given its structural important in cartilage, it is worthy of further comment. Aggrecan diffusion is very slow in cartilage because it is hindered by the collagen network, both by mechanical entanglement and by chemical interactions. Consequently, because of the collagen network acting as a constraint on aggrecan diffusion, dissipation of the free energy embodied in the aggrecan concentration gradient in cartilage typically takes place over months. In contrast, if the aggrecan concentration gradient in cartilage was in a free aqueous solution (i.e. one with no collagen network to restrain aggrecan movement), concentration gradients in cartilage would vanish in minutes, and the chemical potential energy of the aggrecan would be dissipated quickly. However, it is this delay in the rate of chemical potential energy dissipation that gives rise to an 'osmotic pressure' in cartilage, because water migrates and reaches equilibrium much more quickly than the aggrecan possibly can migrate. The gradient of osmotic pressure in cartilage generates a real mechanical pressure gradient that expands the collagen network and puts the collagen network in tension in an unloaded state. How this osmotic pressure is generated is explained in Chap. 6.

The second dissipative process is due to the flow of water through the porous medium due to consolidation upon loading. This process typically produces most of the heat being generated in cartilage. The third dissipative process is due to viscous dissipation in the collagen fibers and in the aggrecan during cyclic loading (e.g. due to sliding or partial unwinding of the helices comprising collagen mers, or distortion of the aggrecan molecules). We note that water movement between intrafibrillar and extrafibrillar collagen compartments, which contribute to energy dissipation, may be grouped with consolidation processes (this fluid flow could be viewed as simply a microscale porous media flow process via local nanoporosity occurring within and contributing to macroscale fluid flow through the porous media), or it could be lumped in with other viscous dissipative processes taking place in collagen and aggrecan.

And the fourth and final process is due to molecular damage processes in the cartilage extracellular matrix materials. Possible damage processes include elasto-plastic deformation of the collagen network, as well as outright breakage of collagen fibrils. Depending on the amount of cross-linking, collagen fibril fracture may occur as soon as the ultimate tensile stress is reached under a stress-controlled condition, with stored energy being dissipated as elastic recoil of the collagen fibril. Similarly, there may be breakage of aggregate (i.e. separation of aggrecan molecules attached hyaluronic acid) and/or breakage of aggrecan (composed of a proteoglycan core with many attached GAGs), into smaller component molecules.

While much of the necessary understanding has already been developed in previous discussion, we consider now these dissipative processes in turn, drawing the various descriptions together into a more complete picture.

We have found previously that the rate of viscous dissipation per unit volume of the initial configuration due to the pore fluid is given by [see derivation Eq. (7.85)],

$$\Psi^{\mathrm{dw}} = p^{\mathrm{sm}} \dot{N}_{\mathrm{w}}^{\mathrm{m}} \tag{7.126}$$

where p^{sm} is the thermodynamic pressure in mapped from spatial to material coordinates (which is in fact a mechanical pressure) and $\dot{N}_{\mathrm{f}}^{\mathrm{m}}$ is the rate of change of fluid porosity in material coordinates. The viscous dissipative stress due to flow of water through the porous medium is given by

$$S^{\mathrm{dw}} = p^{\mathrm{sm}} = \frac{\partial \Psi^{\mathrm{dw}}}{\partial \dot{N}_{\mathrm{w}}^{\mathrm{m}}} \tag{7.127}$$

The viscous dissipative function for a single collagen fiber is given by [see derivation equation (O.6)]

$$\Psi^{\mathrm{d}}(n^r, \dot{I}_n, k_i, k_0, \lambda_i, \tau) = \left[\frac{\xi(n^r)}{k_0} \sum_{i=1}^{N} k_i e^{-\lambda_i \tau}\right] (\dot{I}_n(t) - 1)^{\alpha(n^r)} \tag{7.128}$$

while the viscous dissipative stress due to all the collagen fibers at a given location in the cartilage is given by

$$S^{\mathrm{dcol}} = 2 \int_0^{2\pi} \int_0^{\pi} H(I_n - 1) \frac{\partial \Psi^{\mathrm{dcol}}}{\partial \dot{C}} \sin \varphi \mathrm{d}\varphi \mathrm{d}\theta \tag{7.129}$$

The viscous dissipation due to elastoplastic strain does not cause inelastic (viscous) stresses in the cartilage, as elasto-plastic strain is rate independent. However the elastoplastic strain causes energy dissipation (due to frictional sliding at constant stress) and a redistribution of elastic stress within the porous medium. When the stress state in the loaded cartilage reaches conditions for 'yield,' plastic strains develop, probably in the collagen network, but possibly in the conformation of the aggrecan molecules too. Unloading of the porous medium reveals the plastic (or 'permanent set') strains in the collagen network.

Unfortunately, the separation of strains into elastic and plastic strain is not straightforward in large deformation plasticity [typically involving strains defined with respect to intermediate deformed configurations, either plastic or elastic (Nematnasser 1979)], though it is straightforward in small deformation theory (the total strain is simply the sum of the elastic and plastic strains).

Noting that in the absence of plastic straining S^e is given by (see section on equilibrium stresses in the porous medium)

$$S^e = 2\frac{\partial \Psi^e}{\partial C} \tag{7.130}$$

and then, we find that the total second Piola–Kirchhoff stress in the cartilage is

$$S = S^e + S^{\mathrm{d}} = S^e + S^{\mathrm{dw}} + S^{\mathrm{dcol}} \tag{7.131}$$

The elastic second Piola–Kirchhoff stress is summed with the dissipative second Piola–Kirchhoff stresses in the porous medium (due to flow of water and viscous effects in the collagen), so that the total second Piola–Kirchhoff stress provides momentum balance to any applied loading. Over time, the dissipative stresses 'relax' and the elastic second Piola–Kirchhoff stress eventually balance the load applied to the surface. Nonequilibrium stress is zero in the limiting case of quasi-static loading, and in this case elastic stresses provide momentum balance to the applied loading at all times.

7.2.6 Transport Equations for Large Deformations of Porous Media

The transport of ions, growth factors, and other solutes (e.g. aggrecan) through porous media is of fundamental importance to the growth and maintenance of cartilage and, indeed, all tissues. For this reason, we now focus on the transport equations in a deforming porous media. Clearly, deformations lead to advective fluxes of pore fluid within the porous medium due to consolidation. The consolidation fluxes may be of relatively long duration as occurs with sustained load cycles (e.g. standing), or short duration as they follow individual cyclic loadings events (e.g. brief walk), or a combination of the two (e.g. sustained walking). In addition, the volumetric changes may lead to changes in local concentrations of fixed chemical species (i.e. those chemicals that cannot be transported with the pore fluid by virtue of their size or chemical attachment to the deforming porous media). In general, such changes in local concentrations may speed up or slow down chemical reactions within the porous medium (i.e. local rates of product formation depend on local chemical concentrations). Indeed, we note in passing that chemical reactions that repair molecules in damaged rubber have been reported to have been promoted as a result of large cyclic deformations (Ahagon and Kirino 2007). But as far as we are aware, this potentially important relationship between rates of chemical reaction and deformation has not been explored for cartilage.

Again, we adopt the same strategy as employed above; namely, we first write the equations in the deformed configuration using a spatial coordinate system and then use our coordinate transformation rules to write the transport equations in the undeformed configuration in material coordinates. A transport analysis may be based on ‘apparent concentrations’ or ‘volume average concentrations’ $(\hat{c})$, but the difficulty with this approach is that it is the actual concentrations in the pore fluid solution that determine the chemical potential of the solutes, and so the rates of product formation or removal of chemical species by chemically interacting with one another. In other words, the actual concentrations of the solutes are required for modeling chemical reactions in the porous media, not the volume-averaged concentrations. So rather than using apparent concentrations when formulating the transport equations in spatial coordinates, we generally use

$$\hat{c} = n_{\mathrm{f}} c \tag{7.132}$$

By introducing the fluid porosity n_{f}, we can then identify the actual concentrations c as an independent problem variable in the spatial coordinate system. One can transform concentrations between material coordinates (undeformed configuration) and spatial coordinates (deformed configuration) systems using the relations

$$Jn_{\mathrm{f}}^{\mathrm{sm}}(X,t)c^{\mathrm{sm}}(X,t)\mathrm{d}V = n_{\mathrm{f}}^{\mathrm{s}}(x,t)c^{\mathrm{s}}(x,t)\mathrm{d}v \tag{7.133}$$

which we may write as,

$$N_{\mathrm{f}}^{\mathrm{m}} c^{\mathrm{sm}} = Jn_{\mathrm{f}}^{\mathrm{s}} c^{\mathrm{s}} \tag{7.134}$$

where the superscript s denotes spatial coordinates, m denotes spatial coordinates and the superscript sm denotes a mapping of quantities in the deformed configuration (and spatial coordinates) into same quantities in the undeformed configuration (and material coordinates). Like the second Piola–Kirchhoff stress, the variable $N_{\mathrm{f}}^{\mathrm{m}}(X,t) = Jn_{\mathrm{f}}^{\mathrm{sm}}(X,t) = Jn_{\mathrm{f}}^{\mathrm{s}}(x,t) = (v_{\mathrm{s}}^{\mathrm{s}}(t)/V(0) \times (v_{\mathrm{f}}^{\mathrm{s}}(t)/v_{\mathrm{s}}^{\mathrm{s}}(t)) = v_{\mathrm{f}}^{\mathrm{s}}(t)/V(0)$ has no experimentally observable physical meaning. $N_{\mathrm{f}}^{\mathrm{m}}$, which has been referred to as the 'Lagrangian porosity' (Coussy 2004), is simply the ratio $N_{\mathrm{f}}^{\mathrm{m}} = Jn_{\mathrm{f}}^{\mathrm{sm}} = v_{\mathrm{f}}^{\mathrm{s}}(t)/V(0)$, where $v_{\mathrm{f}}^{\mathrm{s}}(t)$ is the volume of fluid in the deformed state at time t (i.e. the spatial coordinate system), while $V(0)$ is the undeformed volume at time zero.

We note that $N_{\mathrm{f}}^{\mathrm{m}}(X,t)$, the fluid porosity in material coordinates at time t, is given by $N_{\mathrm{f}}^{\mathrm{m}}(X,t) = N_{\mathrm{f}}^{\mathrm{m}}(X,0) + \int_0^t \frac{\mathrm{d}N_{\mathrm{f}}^{\mathrm{m}}}{\mathrm{d}_t}\mathrm{d}t = N_{\mathrm{f}}^{\mathrm{m}}(X,0) + \int_0^t \frac{\mathrm{d}J}{\mathrm{d}_t}\mathrm{d}t = N_{\mathrm{f}}^{\mathrm{m}}(X,0) + \int_0^t (JF^{-1}F^{-\mathrm{T}} : \frac{\mathrm{d}E}{\mathrm{d}_t})\mathrm{d}t$, which may be evaluated as $N_{\mathrm{f}}^{\mathrm{m}}(X,t) = n_0 + J(X,t) - 1 = n_{\mathrm{f}}^{\mathrm{sm}}(X,0) + J(X,t) - 1$ (see Appendix G, the rate of volume change in material coordinates for the origin these equalities). The fluid porosity in spatial coordinates $n_{\mathrm{f}}^{\mathrm{s}}(x,t)$ mapped across to the undeformed configuration $\left(n_{\mathrm{f}}^{\mathrm{sm}}(X,t)\right)$ can be evaluated using the relation $N_{\mathrm{f}}^{\mathrm{m}}(X,t) = Jn_{\mathrm{f}}^{\mathrm{sm}}(X,t)$, i.e. so $n_{\mathrm{f}}^{\mathrm{sm}}(X,t) = \frac{N_{\mathrm{f}}^{\mathrm{m}}(X,t)}{J} = 1 + \frac{(n_0 - 1)}{J}$. Both E and F are known from the large deformation solid poromechanics analysis.

Like the porosity, concentration is a Scaler, and so we can use the transformation $c^{\mathrm{s}}(x,t) \rightarrow c^{\mathrm{sm}}(X,t)$. Note that the concentration $c^{\mathrm{sm}}(X,t)$ is an actual concentration in the undeformed configuration that can be measured experimentally in the deformed configuration at the material point occupying the position x in the spatial coordinate system at time t, so it is suitable to use in equations representing chemical reactions.

Mass flux of the ith species results from both diffusive transport of the solute through the solvent and the convective transport of the solute with the solvent. Adding these two transport processes together results in the so-called generalized flux. For an orthogonal spatial coordinate system, we have

$$f_{M_i}^{\mathrm{s}} = -\frac{D_{i,\mathrm{eff}}^{*} c_i^{\mathrm{s}}}{RT} \nabla^s \mu_i^{\mathrm{s}} + \nabla^{\mathrm{s}} n_{\mathrm{f}}^{\mathrm{s}} c_i^{\mathrm{s}} v_{\mathrm{t}}^{\mathrm{s}} \tag{7.135}$$

where $D_{i,\mathrm{eff}}^{*}$ is the effective diffusion coefficient (which includes both porosity and tortuosity—see below), R the gas constant, T the absolute temperature, ∇^{s} is the gradient operator in spatial coordinates, μ_i^s is the electrochemical potential (normally a Gibbs electrochemical potential is employed https://en.wikipedia.org/wiki/Chemical_potential), $n_{\mathrm{f}}^{\mathrm{s}}$ is the fluid porosity, and $v_{\mathrm{t}}^{\mathrm{s}}$ is the true velocity of the pore fluid (all w.r.t. the deformed configuration and spatial coordinate system). We note that the moles flux $\left(\mathrm{Mm}^{-2}\,\mathrm{s}^{-1}\right)$ may be multiplied by the gram-molecular weight of the ith species to obtain the mass flux of the ith species $(\mathrm{gm}^{-2}\,\mathrm{s}^{-1})$, and the mass flux may be multiplied by the specific volume $(\bar{v})$ to obtain the volume flux of the ith species $(\mathrm{m}^3\mathrm{m}^{-2}\mathrm{s}^{-1})$.

For a porous medium, the effective diffusion coefficient (D_{eff}) is usually expressed as

$$D_{i,\mathrm{eff}}^{*} = n_{\mathrm{f}}^{\mathrm{s}} \tau_i^{\mathrm{s}} D_i^{*} \tag{7.136}$$

where D_i^* is the diffusion coefficient of the species at infinite dilution in the solvent and τ_i^s is the tortuosity of the connected pore space in spatial coordinates within the porous medium.

We note that where there is advective transport in a porous media, there is also 'hydrodynamic dispersion' (D_c) (Rowe et al. 2004), which is often represented mathematically as a diffusive process. So $D_c = D_{\text{eff}} + \alpha_d v_d$ where α_d is known as the hydrodynamic dispersion coefficient. Hydrodynamic dispersion accounts for the spreading of an advected solute observed in porous medium due to the tortuosity of the fluid flow paths through the porous medium, modeling this spreading as a diffusion process. We note that the hydrodynamic dispersion of the solute is dominated by the diffusion coefficient at low advection speeds, and this often occurs in porous media with very low permeability (e.g. in cartilage, so in this case the hydrodynamic dispersion is normally very small).

It is important to realize here that the gradient of the *chemical* potential $\nabla^s \mu_i^s$—which is the 'thermodynamic force' acting on the solute—is not a mechanical pressure acting on the solute particles. Diffusion occurs because of random thermal motion of the solute in the solvent. The solute particles in a chemical gradient are on average moving in minus the gradient of the concentration because there are more particles randomly moving in one direction (down the chemical gradient) than moving in the other (i.e. up the chemical gradient). At the same time as net solute is moving in one direction, there is an average movement of solvent in the direction opposite to the average movement of the solute. As this change in average position of solute and solvent occurs, there is a momentum exchange between solute and solvent. In a solution in which diffusion is occurring, the momentum exchange from solute to solvent is exactly equal to the momentum exchange from solvent to solute, so no mechanical pressure is generated in the solution as a result of diffusion. The free energy of the solution decreases as diffusional mixing progresses, while the entropy of the solution increases, as expected for any 'spontaneous process' (total entropy always increases for any spontaneous process).

While there is no mechanical force pushing the solute down its concentration gradient, this isothermal concentration gradient nevertheless has free energy, meaning it is capable of doing mechanical work, provided a suitable machine is devised to harness the free energy embodied in the chemical potential. In practice, a suitable machine can be realized using a semipermeable membrane attached to a piston, with solutions of different composition on either side of the semipermeable membrane. The piston arrangement can exploit the osmotic pressure difference on either side of the piston to realize mechanical work. If the piston expands very slowly—sufficiently slowly that no gradients in chemical potential are realized on either side of the piston (in other words, conditions of thermodynamic equilibrium exists at all times in solution)—then for isothermal conditions heat flows into the solution from surroundings (i.e. from the environment outside the system) to replace the kinetic energy that is extracted from solute molecules as mechanical energy by the slow expansion of the piston (much like mechanical work can be extracted from a isothermal gas using a piston impermeable to the gas).

Indeed, such a mechanical arrangement was employed by Albert Einstein (1908) in a 'thought experiment' that appeared in a paper (titled 'The Elementary Theory of the Brownian Motion[1]' appearing in *zeit. fur elektronchemie*, **14**, 1908, pp. 235–239.) to derive an estimate of the diffusion coefficient for the solute moving through a solvent. His estimate for the diffusion coefficient is,

$$D_i^{*\mathrm{m}} = \frac{k_{\mathrm{B}}T}{f} = \frac{k_{\mathrm{B}}T}{6\pi\mu_{\mathrm{f}}r_0} \tag{7.137}$$

where $D_i^{*\mathrm{m}}$ the diffusion coefficient for a single molecule in the solution, k_{B} is Boltzmann's constant, T is the absolute temperature, f is the 'friction factor' of the molecule with the solvent, μ_{f} is the (macroscopic) viscosity of the solution and r_0 is the 'effective radius' of the molecule. We note that in practice all diffusive processes are dissipative because they involve gradients of chemical potential (i.e. the system is in thermodynamic disequilibrium when diffusion is occurring).

It is also important to realise that the mechanical pressure acting on the piston in the machine described above is known as the 'osmotic pressure' of the solution. In Chap. 6 it was explained that the osmotic pressure is caused by an imbalance in the momentum transfer between the solute and the solvent molecules at the interface of the solution with a semipermeable, due to the presence of the semipermeable membrane. In other words, the solute molecules exchange part of their momentum with the osmotic membrane, creating and imbalance in the momentum transfer between solute and solvent. This momentum imbalance acts like a body force in the fluid close to the membrane, leading to the osmotic pressure in the fluid (Benedek and Villars 2000; Kramer and Myers 2012).

We note that in the case of a gradient of distributed fixed charge (which is unlike a semi-permeable membrane, in that both ions and water are freely transported throughout the heterogeneous fixed charge domain in the reference solution; i.e. usually, the reference solution is close to isotonic saline, e.g. blood or synovial fluid), the expansive 'swelling pressure' (originating from a site-specific 'osmotic pressure' between macromolecules in close proximity) is generated at locations where 'diffuse double-layer' counterions in the solvent overlap. The expansive swelling pressure is constant throughout the region between say two charged parallel plates carrying an applied load, even though the ion concentrations increase closer to the charge plates. This is because Maxwell stress generates a 'negative pressure' that opposes the increase in osmotic pressure (see Chap. 6 for further discussion). The counterions are heterogeneously distributed throughout the solution (i.e. the concentration of ions varies continuously through the domain), being associated with the heterogeneously distributed fixed charge. Higher concentrations of counterions are found in regions close to the fixed charge, while 'several' Debye lengths away ion concentrations are essentially back to their reference

[1]Einstein has a footnote: We mean by Brownian motion that irregular movement which small particles of microscopic size carry out when suspended in a liquid.

concentrations, i.e. isotonic saline. It is for this reason that these osmotic pressures are essentially beyond even 'microscopic' measuring techniques (e.g. micron-sized micropipettes cannot measure the behavior of the fluid between two fixed charges separated by several nanometres). Only the 'swelling pressure' (or 'osmotic pressure') is measurable macroscopically.

At certain locations in overlapping diffuse double layers, the gradient in the electric field is zero (hence, the Maxwell stress is zero at these locations). At these locations, the Maxwell stress is zero, and the ions in solution generate an osmotic expansion pressure in the solvent (i.e. in the water), which carries the mechanical load between fixed charges. It is this mechanical pressure in the water between the overlapping double layers than leads to the expansive (osmotic/swelling) pressure of aggrecan. It is the osmotic pressure of the aggrecan that is either resisted by the collagen network, or by the load applied to the cartilage. Indeed, if the applied load is larger than the osmotic pressure within the cartilage, the aggrecan is compressed into a smaller 'region' or volume, and so the fixed charge density and the osmotic pressure increase until the advance of the compressive load is stopped by the osmotic pressure generated between fixed charge on the densely packed (largely GAG/aggrecan) molecules. See further discussion in Chap. 6.

Now, Eq. (7.135) may be expressed in terms of the concentration gradient, the voltage gradient, and the Darcy velocity, giving our generalized electrochemical flux equation as (Cussler 1984)

$$f_{M_i}^{\mathrm{s}} = -\left[\frac{D_{i,\mathrm{eff}}^{*} c_i^{\mathrm{s}}}{RT}\frac{\partial \mu_i^{\mathrm{s}}}{\partial c_i^{\mathrm{s}}}\nabla^{s} c_i^{s} + \frac{D_{i,\mathrm{eff}}^{*} c_i^{\mathrm{s}}}{RT}\frac{\partial \mu_i^{\mathrm{s}}}{\partial \hat{\phi}}\nabla^{\mathrm{s}}\hat{\phi}\right] + \nabla^{\mathrm{s}} c_i v_{\mathrm{d}}^{\mathrm{s}} \tag{7.138}$$

For an ideal solute, the electrochemical potential μ_i^{s} is given by

$$\mu_i^{\mathrm{s}} = \mu_i^0 + RT \ln c_i^{\mathrm{s}} + z_i F\hat{\phi} \tag{7.139}$$

where μ_i^0 is the reference potential $\left(\mathrm{JM}^{-1}\right)$, R is the gas constant $\left(\mathrm{JM}^{-1}\mathrm{K}^{-1}\right)$, T is the absolute temperature (K), F is Faraday's constant $\left(\mathrm{CM}^{-1}\right)$, $\hat{\phi}$ $\left(\mathrm{JC}^{-1}\right)$ is the macroscopic average voltage, and z_i is the charge on the ith species (which is $z_i = 0$ for no charge, $+n$ for a positive charge, and $-n$ for a negative charge, where n is a positive integer equal to the valence of the charged species $|z_i|$). M denotes moles, J denotes Joules (note: not det $[F]$, as in large deformation analysis), and C denotes coulombs of charge (note: not $F^{\mathrm{T}}F$ as in large deformation analysis). The concentration c_i^{s} may be in a variety of units depending on the purpose, but once the other units are fixed, so the units of concentration $\left(\mathrm{Mm}^{-3}\right)$ are fixed. $z_i F\hat{\phi}$ $\left(\mathrm{JM}^{-1}\right)$ is simply the electrostatic potential of a charged species, which is added to the chemical potential RT ln c_i^{s} to obtain the electrochemical potential $\left(\mathrm{JM}^{-1}\right)$.

The voltage $\hat{\phi}$ may arise due to an externally imposed electric or time-varying magnetic field, or the voltage may arise internally due to co-transport of charged species [this is the so-named 'diffusion potential' that ensures two charged ions with different diffusion coefficients migrate as a pair of ions with diffusion coefficient that is the harmonic mean of the individual diffusion coefficients (Cussler 1984)], or the potential may be due to a spatial-varying fixed electric charge in the solution (this gives rise to a so-called Donnan potential).

Inserting the electrochemical potential for an ideal solution (7.139) into Eq. (7.138) leads to a generalized Nernst–Planck equation (Cussler 1984),

$$f^{\mathrm{s}}_{\mathrm{M}_i} = -[D^*_{i,\mathrm{eff}}(\nabla^{\mathrm{s}} c^{\mathrm{s}}_i + \frac{z_i F c^{\mathrm{s}}_i}{RT}\nabla^{\mathrm{s}}\hat{\phi})] + \nabla^{\mathrm{s}} c_i v^{\mathrm{s}}_{\mathrm{d}} \tag{7.140}$$

where $f^{\mathrm{s}}_{\mathrm{M}_i}$ is the moles flux $\left(\mathrm{Mm}^{-2}\mathrm{s}^{-1}\right)$.

Assuming that the species is uncharged and that the Darcy velocity v^s_{d} is zero, Eq. (7.140) recovers Fick's first law,

$$f^{\mathrm{s}}_{c_i} = -D^*_{i,\mathrm{eff}}\nabla^{\mathrm{s}} c^{\mathrm{s}}_i \tag{7.141}$$

If we set the mole flux and Darcy velocity equal to zero in Eq. (7.140) and integrate between points 1 and 2, we find can calculate the potential difference between the two points from equation as (7.140),

$$\hat{\phi}_2 - \hat{\phi}_1 = \frac{RT}{z_i F}\ln\frac{c_2}{c_1} \tag{7.142}$$

If a Donnan potential exists between a region of fixed charge and the surrounding solution, and assuming this difference in concentration is known, the Donnan potential can be calculated using Eq. (7.142). Alternatively, if the Donnan potential is known experimentally, then the distribution of charged species between the region of fixed charge and the surrounding solution can be calculated using Eq. (7.142). More usually, the difference in electrical potential is known and the concentration difference calculated, as potential difference is more easily measured experimentally. Once all the concentrations of the mobile ions are known, the fixed charge density may be calculated from the electroneutrality condition for the region of fixed charge, viz.,

$$z_{\mathrm{F}} c_{\mathrm{F}} = -\sum_{i=1}^{M} z_i c_i \tag{7.143}$$

where $z_{\mathrm{F}} c_{\mathrm{F}}$ is the fixed charge density.

The Donnan potential $\hat{\phi}$ refers to a macroscopic average potential, rather than the local potential ϕ at a given point within the region of fixed charge (Smith et al. 2004). We note here that the distribution of fixed charge is usually non-uniform (as the charge usually arises from dissociation of chemical species in solution), and so the distributions of counter-ions and co-ions within the region of fixed charge are generally non-uniform, as is the electric potential (Smith et al. 2004). This non-uniform distribution of charge is important when calculating the osmotic forces and electrical forces within the region of fixed charge (see Grodzinsky (2011) and Chap. 6 for more details).

Now, let the mass balance equation in spatial coordinates be,

$$\frac{\partial n_{\mathrm{f}}^{s} c^{s}}{\partial t} = -\nabla^{s} \cdot f_{c^{s}} + s_{\mathrm{m}}^{s} \tag{7.144}$$

where c^{s} is the concentration of solute in the pore fluid in the spatial coordinate system, $n_{\mathrm{f}}^{\mathrm{s}}$ is the (fluid) porosity in the spatial system, $f_{c^{s}}$ is the flux of solute in the spatial coordinate system, and $s_{\mathrm{m}}^{\mathrm{s}}$ is a source–sink term representing the rate of production or removal of solute mass per unit volume in the spatial coordinate system. We note that this source–sink term represents degradation or production of the chemical species, and so this is how chemical reactions enter into the transport analysis. We will return to this later.

Neglecting electrical forces for simplicity, the mass flux equation is given by

$$f_{c^{s}} = -n_{\mathrm{f}}^{\mathrm{s}} D_{c} \nabla^{\mathrm{s}} c^{\mathrm{s}} + n_{\mathrm{f}}^{\mathrm{s}} v_{\mathrm{f}}^{\mathrm{s}} c^{\mathrm{s}} \tag{7.145}$$

where D_c is 'hydrodynamic dispersion' (see discussion above). Substituting the flux equation in the mass balance Eq. (7.144) leads to

$$n_{\mathrm{f}}^{\mathrm{s}} \frac{\partial c^{\mathrm{s}}}{\partial t} + c^{\mathrm{s}} \frac{\partial n_{\mathrm{f}}^{\mathrm{s}}}{\partial t} = \nabla^{\mathrm{s}} \cdot n_{\mathrm{f}}^{\mathrm{s}} D_{c} \nabla^{\mathrm{s}} c^{\mathrm{s}} - \nabla^{\mathrm{s}} \cdot (n_{\mathrm{f}}^{\mathrm{s}} v_{\mathrm{f}}^{\mathrm{s}} c^{\mathrm{s}}) + s_{\mathrm{m}}^{\mathrm{s}} \tag{7.146}$$

Noting that

$$-\nabla^{\mathrm{s}} \cdot (n_{\mathrm{f}}^{\mathrm{s}} v_{\mathrm{f}}^{\mathrm{s}}) = \frac{\partial n_{\mathrm{f}}^{\mathrm{s}}}{\partial t} \tag{7.147}$$

leads to

$$n_{\mathrm{f}}^{\mathrm{s}} \frac{\partial c^{\mathrm{s}}}{\partial t} = \nabla^{\mathrm{s}} \cdot n_{\mathrm{f}}^{\mathrm{s}} D_{c} \nabla^{\mathrm{s}} c^{\mathrm{s}} - n_{\mathrm{f}}^{\mathrm{s}} v_{\mathrm{f}}^{\mathrm{s}} \cdot \nabla^{\mathrm{s}} c^{\mathrm{s}} + s_{\mathrm{m}}^{\mathrm{s}} \tag{7.148}$$

Now, employing a rearrangement of the definition of Darcy velocity (i.e. $v_{\mathrm{f}}^{\mathrm{s}} = \left[\frac{v_{\mathrm{d}}^{\mathrm{s}}}{n_{\mathrm{f}}^{\mathrm{s}}} + v_{\mathrm{s}}^{\mathrm{s}}\right]$) shows,

$$n_{\mathrm{f}}^{\mathrm{s}} \left(\frac{\partial c^{\mathrm{s}}}{\partial t} + v_{\mathrm{s}}^{\mathrm{s}} \cdot \nabla^{\mathrm{s}} c^{\mathrm{s}} \right) = n_{\mathrm{f}}^{\mathrm{s}} \frac{\mathrm{d} c^{\mathrm{s}}}{\mathrm{d} t} = \nabla^{\mathrm{s}} \cdot n_{\mathrm{f}}^{\mathrm{s}} D_{c} \nabla^{\mathrm{s}} c^{\mathrm{s}} - v_{\mathrm{d}}^{\mathrm{s}} \cdot \nabla^{\mathrm{s}} c + s_{\mathrm{m}}^{\mathrm{s}} \tag{7.149}$$

which is the required transport equation in spatial coordinates.

Using coordinate transformation rules for divergence, Scaler mapping and second-order tensor transformation described in the introduction to this chapter [see

Eqs. (7.15), (7.116), (7.121), (7.125), and (7.129)], we can now rewrite Eq. (7.149) in material coordinates, viz.,

$$Jn_{\mathrm{f}}^{\mathrm{sm}}\frac{\mathrm{d}c^{\mathrm{sm}}}{\mathrm{d}t}=\nabla^{\mathrm{m}}\cdot n_{\mathrm{f}}^{\mathrm{sm}}JF^{-1}D_{c}F^{-\mathrm{T}}\nabla^{\mathrm{m}}c^{\mathrm{sm}}-Jv_{\mathrm{d}}^{\mathrm{m}}\nabla^{\mathrm{m}}c^{\mathrm{sm}}+S_{\mathrm{m}}^{\mathrm{m}} \tag{7.150}$$

Using notation developed previously Eq. (7.150) may be written as

$$N_{\mathrm{f}}^{\mathrm{m}}\frac{\mathrm{d}c^{\mathrm{sm}}}{\mathrm{d}t}=\nabla^{\mathrm{m}}\cdot N_{\mathrm{f}}^{\mathrm{m}}F^{-1}D_{c}F^{-\mathrm{T}}\nabla^{\mathrm{m}}c^{\mathrm{sm}}-Jv_{\mathrm{d}}^{\mathrm{m}}\nabla^{\mathrm{m}}c^{\mathrm{sm}}+S_{\mathrm{m}}^{\mathrm{m}} \tag{7.151}$$

where $N_{\mathrm{f}}^{\mathrm{m}}(X,t)=Jn_{\mathrm{f}}^{\mathrm{sm}}(x,t)$ is the adjusted fluid porosity in undeformed material coordinates, $JF^{-1}v_{\mathrm{d}}^{\mathrm{s}}=Jv_{\mathrm{d}}^{\mathrm{m}}$ is the Darcy flux vector in undeformed material coordinates, and $S_{\mathrm{m}}^{\mathrm{m}}(X,t)=Js_{\mathrm{m}}^{\mathrm{m}}(X,t)$ is the source–sink term in undeformed material coordinates. This is our transport equation in (undeformed) material coordinates.

We can check this by deriving the same equation in material coordinates. First the mass conservation equation in material coordinates is

$$\nabla^{\mathrm{m}}\cdot f_{\mathrm{f}}^{\mathrm{m}}=\frac{\mathrm{d}[Jn_{\mathrm{f}}^{\mathrm{sm}}(X,t)c^{\mathrm{sm}}(X,t)]}{\mathrm{d}t} \tag{7.152}$$

and the divergence of the solute flux in the fluid phase is

$$\nabla^{\mathrm{m}}\cdot f_{\mathrm{f}}^{\mathrm{m}}=\nabla^{\mathrm{m}}\cdot JF^{-1}f_{c^{s}}=\nabla^{\mathrm{m}}\cdot[-JF^{-1}(n_{\mathrm{f}}^{\mathrm{sm}}D_{c}^{\mathrm{sm}})F^{-\mathrm{T}}\nabla^{\mathrm{m}}c^{\mathrm{sm}}+JF^{-1}v_{\mathrm{d}}^{\mathrm{s}}c^{\mathrm{sm}}] \tag{7.153}$$

At this point, we mention that if there is no transport of a chemical species, then Eq. (7.152) reduces to

$$\frac{\mathrm{d}[Jn_{\mathrm{f}}^{\mathrm{sm}}(X,t)c^{\mathrm{sm}}]}{\mathrm{d}t}=0 \tag{7.154}$$

Then, evaluating Eq. (7.154) by taking its integral with respect to time, from time zero up to time t reveals

$$n_{0}c^{\mathrm{sm}}(X,0)=Jn_{\mathrm{f}}^{\mathrm{sm}}(X,t)c^{\mathrm{sm}}(X,t) \tag{7.155}$$

Using the relation $N_{\mathrm{f}}^{\mathrm{m}}(X,t)=Jn_{\mathrm{f}}^{\mathrm{sm}}(x,t)=n_{0}(X,0)+J(X,t)-1$ in Eq. (7.155) leads to the useful relation for the concentration of a chemical species that is not transported changes as the material deforms, which is

$$c^{\mathrm{sm}}(X,t)=\frac{c^{\mathrm{sm}}(X,0)}{\frac{Jn_{\mathrm{f}}^{\mathrm{sm}}}{n_{0}}}=\frac{c^{\mathrm{sm}}(X,0)}{1+\frac{(J-1)}{n_{0}}} \tag{7.156}$$

Now, continuing the development of the governing transport equation in material coordinates, we substitute Eqs. (7.153) in (7.152), which leads to

$$\frac{\mathrm{d}[Jn_\mathrm{f}^\mathrm{sm}c^\mathrm{sm}]}{\mathrm{d}t} = \nabla^\mathrm{m} \cdot Jn_\mathrm{f}^\mathrm{sm}F^{-1}D_c^\mathrm{sm}F^{-\mathrm{T}}\nabla^\mathrm{m}c^\mathrm{sm} - \nabla^\mathrm{m} \cdot JF^{-1}v_\mathrm{d}^\mathrm{s}c^\mathrm{sm} \tag{7.157}$$

Then, expanding the first term in Eq. (7.157) using the chain rule, and noting the fluid continuity equation, viz., $\frac{\mathrm{d}[Jn_\mathrm{f}^\mathrm{sm}]}{\mathrm{d}t} = \nabla^\mathrm{m} \cdot JF^{-1}v_\mathrm{d}^\mathrm{s}$, we find

$$\frac{\mathrm{d}[Jn_\mathrm{f}^\mathrm{sm}c^\mathrm{sm}]}{\mathrm{d}t} = Jn_\mathrm{f}^\mathrm{sm}\frac{\mathrm{d}c^\mathrm{sm}}{\mathrm{d}t} + c^\mathrm{sm}\frac{\mathrm{d}[Jn_\mathrm{f}^\mathrm{sm}]}{\mathrm{d}t} = Jn_\mathrm{f}^\mathrm{sm}\frac{\mathrm{d}c^\mathrm{sm}}{\mathrm{d}t} + c^\mathrm{m}\nabla^\mathrm{m} \cdot JF^{-1}v_\mathrm{d}^\mathrm{s} \tag{7.158}$$

Then, combining Eqs. (7.157) and (7.158), and using the relation $\nabla^\mathrm{m} \cdot JF^{-1}v_\mathrm{d}^\mathrm{s}c^\mathrm{sm} - c^\mathrm{sm}\nabla^\mathrm{m} \cdot JF^{-1}v_\mathrm{d}^\mathrm{s} = v_\mathrm{d}^\mathrm{s}JF^{-1}\nabla^\mathrm{m}c^\mathrm{sm}$ and finally including a source/sink term S_m^m shows

$$Jn_\mathrm{f}^\mathrm{sm}\frac{\mathrm{d}c^\mathrm{sm}}{\mathrm{d}t} = \nabla^\mathrm{m} \cdot Jn_\mathrm{f}^\mathrm{sm}F^{-1}D_\mathrm{c}^\mathrm{sm}F^{-\mathrm{T}}\nabla^\mathrm{m}c^\mathrm{sm} - Jv_\mathrm{d}^\mathrm{m}\nabla^\mathrm{m}c^\mathrm{sm} + S_\mathrm{m}^\mathrm{m} \tag{7.159}$$

which is the same as Eq. (7.150). These equations are consistent with the one-dimensional contaminant transport equation in material coordinates reported in (Peters and Smith 2002) (note that for one dimensional problems $F^{-1} = F^{-\mathrm{T}} = \frac{1}{J}$).

We note that for Eqs. (7.150), (7.151), and (7.159), the term JF^{-1} converts a true flux in the spatial coordinate system into a nominal flux in the undeformed material coordinate system, while F^{-1} converts from the spatial to the material coordinate systems. Alternatively, we can view F^{-1} and $F^{-\mathrm{T}}$ as two, two-point tensors connecting the material coordinate system with the diffusivity tensor in the spatial coordinate system, while J is a two-point Scaler that ensures the divergence is evaluated over the correct volume (i.e. evaluated over the actual volume rather than the nominal volume). The porosity and concentration are simply mapped across from porosity and concentration in the spatial coordinate system (and expressed in material coordinates). Note that for the porosity mapping, the J in $Jn_\mathrm{f}^\mathrm{sm}\frac{\mathrm{d}c^\mathrm{sm}}{\mathrm{d}t}$ simply scales the REV so the nominal accumulated total mass of the transported species in the REV in the material coordinate system agrees with the actual accumulated mass of the transported species in the REV in the spatial coordinate system.

Let us now consider introducing a chemical reaction, which acts as source/sink terms. Let the nonlinear chemical reaction be:

$$a + b \rightleftarrows c \tag{7.160}$$

where chemical species a and b associate in the forward direction at rate k_f to produce complex c, and the complex c dissociates at rate k_r to produce a and b. If c_i is the concentration of the ith species, then we can write this chemical reaction as,

$$\frac{\mathrm{d}c_{\mathrm{a}}^{\mathrm{sm}}}{\mathrm{d}t} = k_r c_{\mathrm{c}}^{\mathrm{sm}} - k_{\mathrm{f}} c_{\mathrm{a}}^{\mathrm{sm}} c_{\mathrm{b}}^{\mathrm{sm}} \tag{7.161}$$

where superscript sm denotes actual concentrations mapping from the deformed configuration to the undeformed configuration. Constraints can be included now if so desired (e.g. $c_{\mathrm{a}}^{\mathrm{sm}} + c_{\mathrm{c}}^{\mathrm{sm}} = c_{a-\mathrm{total}}^{\mathrm{sm}}$). Then, Eq. (7.159) becomes

$$Jn_{\mathrm{f}}^{\mathrm{sm}} \frac{\mathrm{d}c_{\mathrm{a}}^{\mathrm{sm}}}{\mathrm{d}t} = \nabla^{\mathrm{m}} \cdot Jn_{\mathrm{f}}^{\mathrm{sm}} F^{-1} D_{\mathrm{c}}^{\mathrm{sm}} F^{-\mathrm{T}} \nabla^{\mathrm{m}} c_{\mathrm{a}}^{\mathrm{sm}} - Jv_{\mathrm{d}}^{\mathrm{m}} \nabla^{\mathrm{m}} c_{\mathrm{a}}^{\mathrm{sm}} + Jn_{\mathrm{f}}^{\mathrm{sm}} (k_r c_{\mathrm{c}}^{\mathrm{sm}} - k_{\mathrm{f}} c_{\mathrm{a}}^{\mathrm{sm}} c_{\mathrm{b}}^{\mathrm{sm}}) \tag{7.162}$$

This completes our derivation of the transport equation coupled to large deformations.

It is noted that when solving reaction-transport equation in a porous medium undergoing large deformations, it may well prove computationally advantageous to uncouple the transport equations from the chemical reaction equations, and so solve these sets of equations independently of each other. This approach considerably simplifies the reaction-transport analysis and can result in accurate solution providing the time steps are small enough to capture the reactions of interest.

7.2.7 Boundary and Initial Conditions

For a system of partial differential equations to have a unique solution, the governing equations require so-called initial conditions and boundary conditions. The initial conditions relate to the initial state of the system, while the boundary conditions relate to what is happening at the surface of the system. At the surface, the system interacts with the surroundings, so the boundary conditions generally specify how the system is 'seeing' its environment. When the governing equations and the initial and boundary conditions are specified correctly, the system is said to be 'well posed' and there exists a unique solution.

For a deformation analysis, the initial conditions relate to any initial strains or stresses in the body prior to the current analysis commencing. In the case of cartilage, the key initial condition for the poromechanical analysis is the self-equilibrating stresses in the collagen and aggrecan. The aggrecan stress causes an initial strain in the collagen network, and a complete analysis would represent the expansion of the collagen network by the aggrecan and subsequent deformation of the articular cartilage. Due to ongoing damage and repair processes within cartilage, there is almost certainly no stress free state for the collagen network, even in the absence of aggrecan.

Either because the initial strain in the collagen is assumed to be small, or because the focus is on an incremental analysis, the initial collagen strain is often neglected. In this case, the mechanical analysis confines itself to incremental stresses and strains from the equilibrium initial state. While the analysis starts with initial stiffnesses of collagen and aggrecan, the stiffness may change with deformation, and so is then nonlinear. Most analyses of cartilage to date confine themselves to incremental analyses.

Standard boundary conditions are Dirichlet, Neumann, and Robin boundary conditions. For a transport analysis, the initial conditions relate to initial concentrations over the domain.

7.3 Poroelastic Modeling of Cartilage

7.3.1 Introduction

For normal operating conditions, cartilage clearly behaves as a poroelastic biocomposite material, made up of three principal structural components: water (approximately 65–80% total cartilage weight is water), aggrecan (approximately 3–10% of the total cartilage weight is dry aggrecan), and collagen (approximately 15–20% of the total cartilage weight is dry collagen) (Basser et al. 1998; Maroudas et al. 1991). About 50–60% of the dry cartilage weight is collagen (Basser et al. 1998; Sophia Fox et al. 2009).

More specifically, if the partial specific volume of the collagen molecules is between 0.70 and 0.75 cm^3 per g (i.e. partial specific density is 1.33–1.43) (Basser et al. 1998; Levick 1987), and the intrafibrillar water content of the collagen fibrils ranges from 0.7 to 1.14 g of water per gram of collagen (Levick 1987; Maroudas et al. 1991), then the effective specific volume of collagen fibrils are 1.45–1.89 cm^3/g of collagen. For data shown in Table 1 (Basser et al. 1998), this means 17.3% collagen dry weight fraction is equivalent to a volume fraction of about 27–36% (assuming a partial specific volume of 0.7), as the intrafibrillar water content of the collagen fibrils ranges from 0.7 to 1.14 g of water per gram of collagen (the water content of the collagen depends on the local osmotic pressure, primarily exerted by the local GAG concentration) (Levick 1987; Maroudas et al. 1991). At normal concentrations of aggrecan, we can infer that the normal adult collagen volume fraction is in the region of 27–30% (see further discussion Appendix D).

Clearly, to model the structural behavior of cartilage as a poroelastic biocomposite material reasonably well, we need to reasonably represent the behavior of each component of the cartilage, together with their interactions. Next, we first consider the definition of effective stress in cartilage. Clearly, the definition of effective stress in a poroelastic material such as cartilage needs to be revised, as there are now two solids present (i.e. aggrecan and collagen) rather than one. We then discuss the behavior of each component in turn, identifying its transport properties, stiffnesses, and other relevant properties.

We then need to reconsider the storage equation. Clearly, in spatial coordinates, there are three fluxes to consider (i.e. the flux of water, aggrecan and collagen), but this can reduced to two if we take the collagen network as our reference material coordinates (i.e. so fluxes of water and aggrecan are referenced to the volume strain of the collagen network in the storage equation). The collagen network is the most suitable reference material coordinate, primarily because the collagen network is the most durable structural component (though it evolves over time too).

Finally, we can write down a set of governing equations that are reasonably useful for the analysis of cartilage as a structural material.

7.3.2 *The Definition of Effective Stress and Water Flow in Cartilage*

As we have seen in the derivation of the governing equations for poroelasticity, a simple and effective way to represent the transport of water through the solids in cartilage is using Darcy's law, viz.,

$$v_{\mathrm{d}} = -k\nabla p \tag{7.163}$$

where v_{d} is Darcy's velocity, k is the permeability tensor, and p is the excess pore water pressure. While this equation still holds in cartilage, at this point, we need to carefully define what is meant by 'excess pore water pressure' for cartilage. As always, we use the principle of effective stress, viz.,

$$\sigma_{\mathrm{T}} = \sigma^{\mathrm{s}} - pI \tag{7.164}$$

i.e. the total stress is the effective (solid) stress minus the excess pore water pressure. But in fact, there are two solid stresses in articular cartilage—one solid stress due to the collagen, and the other solid stress due to the aggrecan (see discussion Sect. 6.7). The excess pore water pressure is the mechanical pressure in the pore fluid. This excess pore fluid pressure is quite separate from the equilibrium 'osmotic pressure in fluid between the overlapping diffuse double layers,' which is generated by the fixed charges with diffusive double layers between aggrecan molecules in solution (see discussion Sect. 6.4). This osmotic pressure between GAG molecules on aggrecan should be properly regarded as a 'solid stress' component in the cartilage (acting through the GAG molecules), denoted here σ^{agg} after adopting the convention for solid stresses (i.e. tension positive), so that $\sigma^{\mathrm{agg}} = -\pi^{\mathrm{agg}}$. The excess pore water pressure is *not* simply $p = -\frac{\mathrm{tr}(\sigma^{\mathrm{total}} - \sigma^{\mathrm{col}})}{3}$ (i.e. the pore water pressure equals minus the mean total stress plus the mean solid stress in the collagen), but rather it is

$$p = -\frac{\mathrm{tr}(\sigma^{\mathrm{total}} - \sigma^{\mathrm{col}})}{3} + \sigma^{\mathrm{agg}} \tag{7.165}$$

i.e. the excess pore water pressure is equal to minus the hydrostatic total stress, plus the hydrostatic solid stress in the collagen, plus the solid stress in the aggrecan. Or in other words, Eq. (7.165) effectively says that any unbalanced hydrostatic stress in the fluid (p) arises from the residual of an incremental total stress due to applied load, incremental collagen stress, and the incremental osmotic pressure between the GAG molecules in the material.

We mention in passing that it is this ‘unbalanced (hydrostatic) stress’ that results in dissipation of energy in a poroelastic material (i.e. in a quasi-static system there is no unbalanced stress and excess pore water pressures are zero).

If the total stress is zero [i.e. $\sigma^{\text{total}} = 0$ due to sudden removal of an applied load from the surface of the articular cartilage, or if there is solids volume production within the cartilage and the cartilage tissue expands with no change in surface loading (Smith et al. 2016)], then Eq. (7.165) becomes

$$p = \frac{\text{tr}\sigma^{\text{col}}}{3} + \sigma^{\text{agg}} \tag{7.166}$$

For the case of sudden removal of an applied loading so then $\sigma^{\text{total}} = 0$, the pore water pressure becomes negative, and then the tensile force in the water minus the hydrostatic tensile force in the collagen is balanced by the compressive (swelling) force between the charged particles. If the hydrostatic collagen tensile force is also zero, the tensile force in the water balances the (repulsive) swelling force between the charged particles (usually the tensile force is generated either by surface tension at the air–water interface, or when saturated, by the finite time it takes for water to move into the porous media).

For Eq. (7.165), when the excess pore water pressure has reduced to zero, then consolidation is complete. Setting $p = 0$ in Eq. (7.165) we have

$$\frac{\text{tr}(\sigma^{\text{total}} - \sigma^{\text{col}})}{3} = \sigma^{\text{agg}} \tag{7.167}$$

This shows that at equilibrium, $\text{tr}(\sigma^{\text{total}} - \sigma^{\text{col}})/3$ becomes the stress in the aggrecan. For the special case of no applied load (Smith et al. 2016), and no excess pore water pressure, we see that

$$-\frac{\text{tr}\sigma^{\text{col}}}{3} = \sigma^{\text{agg}} \tag{7.168}$$

In other words, at equilibrium under no load, the initial mean tensile (pre)stress in the collagen is equal to the initial compressive (pre)stress in the aggrecan.

From Eq. (7.165), we see that the revised incremental ‘effective stress principle’ for cartilage becomes either

$$\frac{\text{tr}\sigma^{\text{total}}}{3} = \frac{\text{tr}\sigma^{\text{col}}}{3} + \sigma^{\text{agg}} - p = \frac{\text{tr}\sigma^{s}}{3} - p \tag{7.169}$$

or

$$\sigma^{\text{total}} = \sigma^{\text{col}} + \sigma^{\text{agg}}I - pI = \sigma^{\text{s}} - pI \tag{7.170}$$

We can compare the effective stress principle for cartilage with the previous effective stress principle for soil (i.e. Equation (7.41) is $\sigma_T = \sigma^s - p$). We see that for cartilage, it is necessary to take into account the additional partial pressure due to the aggrecan. We also see that irrespective of the prestress in collagen and aggrecan, at equilibrium, the sum of the change in effective stress in the collagen and the change in effective stress in the aggrecan (which due to sign convention is minus the change in osmotic pressure) equals the incremental total stress due to the incremental applied loading stress (i.e. $\sigma^s = \sigma^{total}$). See the discussion in Sect. 6.7 on the analysis of pressure data upon loading (caprine) intervertebral disk for an applied example of this principle.

With our new understanding of the meaning of p in cartilage, we can now use the definition of Darcy velocity $v_d = n_f(v_f - v_s)$ as the relative velocity between the pore fluid and solids to find

$$\nabla^s p = -k^{-1} n_f (v_f - v_s) \tag{7.171}$$

where $k^{-1} n_f (v_f - v_s)$ represents the momentum interaction between the fluid and solids (the term $k^{-1} n_f (v_f - v_s)$ is effectively acting like a body force in the fluid, which results in a pressure gradient in the fluid). Because there are two solids (collagen and aggrecan) in cartilage, it is reasonable to ask at this point which solid is responsible for most of the momentum interaction with water [or should the solid be some combination of two resistances to fluid flow (in which case, the correct value of v_s would be some weighted average of the two solid velocities)].

Because aggrecan is a long 'open' molecule with very large surface area, in addition to having a high charge density (with large numbers of counterions maintaining electroneutrality), the resistance aggrecan presents to fluid flow is very high. On the other hand, collagen is generally packed into fibrils that present a small interaction surface with water, so the relative resistance the collagen presents to fluid flow is comparatively small. Indeed, the pore openings in the collagen network have been measured in the range of 10 s of nanometres (the collagen fibrils themselves have exceedingly small pore openings, and so to a first approximation may be taken as impermeable relative to aggrecan), while the spacing between the aggrecan molecules is in the range 2–4 nm. Consequently, most of the resistance to fluid flow is due to the aggrecan within collagen.

However, some disagree with this assessment. For example, Levick et al. shows in Fig. 4 that the measured permeability for different tissues (including skin, cornea and cartilage) lies about two orders of magnitude below the line for the permeability of hyaluronate in synovial fluid, while Fig. 6 shows tissue permeability compared to calculated permeability of hyaluronate taking into account the volume occupied by collagen fibrils, the area available for flow, and tortuosity—again the theoretical estimate comes up more permeable than the experimental permeability (Levick 1987). However, taking account of the proteoglycan core protein as well as the abovementioned factors (GAGs, excluded collagen volume, area available for flow), theoretical and experimental permeabilities then agree quite closely (see Fig. 10 in Levick (1987). However, we note that for cartilage permeability (Holmes 1985), the measured permeability (based on centrifuge testing of chondroitin sulfate) is fairly well explained by the actual concentration of chondroitin sulfate alone (Zamparo and Comper 1989), taking into account the volume occupied by collagen and the area available for flow.

For this reason, it is a reasonable to assume that all the resistance to the fluid flow in the deep and transitional zones is due to the aggrecan. In the superficial zone, the aggrecan is at its lowest concentration and the collagen is most densely packed. In this zone, pore spaces between the collagen fibrils are smallest (and made smaller by fine collagen type III fibrils). While the collagen plays a more significant role, the aggrecan is probably still the major contributor to resistance of fluid flow through the pore openings of the collagen network. The simplest modeling approximation for this situation is to model a boundary impedance to flow (e.g. using a Robin boundary condition), to represent a possible increased resistance to the flow of water in the superficial zone. However, because this effect is relatively small, this boundary resistance can be neglected with little effect on the model predictions (note that this increased boundary resistance cannot be reasonably neglected for aggrecan transport—see later discussion of aggrecan transport).

Importantly, aggrecan offers resistance to water flow equally in all directions. This means the permeability tensor is isotropic, and so using a Scaler permeability is a reasonable approximation. We can now revisit Eq. (7.171). Assuming our material coordinate system is the collagen network, we find

$$\nabla^{s} p = -k^{-1} n_{\mathrm{f}} (v_{\mathrm{f}} - v_{\mathrm{agg}}) \tag{7.172}$$

where v_{f} is the true water velocity relative to a spatial coordinate system and v_{agg} is the true velocity of aggrecan relative to a spatial coordinate system.

The permeability of aggrecan to water flow at physiological salt concentrations (i.e. 0.15 M NaCl) has been measured in the laboratory. The permeability of aggrecan is clearly related to the concentration of aggrecan. For example, a curve fit of the data presented by Comper and Lyons (1993) leads to the estimate

$$k_{\mathrm{w}} = 1.02 \times 10^{-12} \times c^{-1.6} (\mathrm{m^2 Pa^{-1} s^{-1}}) \tag{7.173}$$

where c is the concentration in mg/ml of cartilage and the dynamic viscosity of water is taken to be 0.001 Pa s at 20 °C (note: the viscosity of water decreases about 35% at 40 °C, while the protein fraction decreases the cross-sectional area available for flow by about 35%, and consequently these two adjustments to permeability largely cancel one another, so using a viscosity of water for 20 °C appears to be reasonable).

7.3.3 Aggrecan Transport Through Cartilage

The transport of aggrecan through cartilage can be modeled using different constitutive equations. The usual way of modeling aggrecan transport is as a diffusive process. For example, aggrecan flux through isotonic saline/aggrecan solution may be represented as

$$f^{agg} = -D_{eff}\nabla^s c^{agg}_{eff} \tag{7.174}$$

where

$$D_{eff} = D^*_{eff}\frac{\partial \pi^{agg}}{\partial c^{agg}} = D^*_{eff}(\alpha_1 + 2\alpha_2 c^{agg} + 3\alpha_3 c^{agg^2}) \tag{7.175}$$

and D_{eff} is the effective diffusion coefficient for aggrecan. The effective diffusion coefficient may be viewed as a diffusion coefficient $D^*_{eff} = RT/f_{12}$ (where f_{12} is the frictional interaction coefficient between 1 mol of aggrecan and water), modified by the rate of change of osmotic pressure w.r.t. concentration (Comper 1991) (see Appendix M Diffusive transport of aggrecan for details).

It is important to note that this diffusion coefficient (D_{eff}) is not the actual diffusion coefficient (D^*_{eff}) for aggrecan diffusing through a solution of aggrecan. The diffusion coefficient shown in Eq. (7.175) is only to be used as a means for representing the mass transport of aggrecan in an aggrecan concentration gradient. If there is no concentration gradient—there normally is a gradient in cartilage—then the diffusion coefficient is greatly reduced [see Eq. (7.175)].

Estimates of *actual* diffusion coefficients of aggrecan in aggrecan solution (D^*_{eff}) are provided by (Gribbon and Hardingham 1998). For a concentration of aggrecan less than 1.6 mg/ml, they found $D^*_{eff} = 4.25 \pm 0.6 \times 10^{-12}$ m^{2}s^{-1}, while for concentrations above that (their data extended to 20 mg/ml—see their Fig. 5), they fitted an exponential function, viz.,

$$D^*_{eff} = 4.93 \times 10^{-12} e^{-0.18(\text{mL/mg})\times c(\text{mg/mL})}\ \text{m}^2\text{s}^{-1} \tag{7.176}$$

In Appendix M, it is also shown that the effective diffusivity in terms of the chemical potential of aggrecan is given by Comper (1991)

$$D_{eff} = \frac{D^*_{eff} c}{RT}\frac{\partial \mu^{agg}}{\partial c^{agg}} = D^*_{eff}\frac{\partial \pi^{agg}}{\partial c^{agg}} \tag{7.177}$$

Surprisingly, D_{eff} is similar for both aggrecan and chondroitin sulfate, at least at physiological concentrations found in cartilage, despite there enormous size difference (Comper 1991; Comper and Williams 1987). That is, based on centrifuge experiments on solutions of macromolecules, D_{eff} has been shown to be independent of the molecular weight of the molecule above concentrations of about 5–10 mg/mL [see Fig. 2 in Comper (1991)]. Comper (1991) states this is due to the dominance of the osmotic gradient term $\left(\frac{\partial \pi^{agg}}{\partial c^{agg}}\right)$ over the frictional term (D^*_{eff}) in Eq. (7.175). In other words at physiological concentrations of aggrecan in cartilage (say 10 (low) to 125 (high) mg/mL), the diffusion coefficient of aggrecan and chondroitin sulfate for a transport analysis is independent of molecular weight. So if, for example, aggrecan of different lengths (e.g. including aggrecan fragments) are diffusing together, they all have similar diffusion coefficients because their osmotic gradients are similar.

Comper and Williams (1987) and Comper (1991) performed centrifuge experiments from which effective diffusion coefficients were computed that are approximately represented by the following equation (see Fig. 2 in Comper (1991) for details),

$$D_{eff}(\mathrm{m^2 s^{-1}}) = 3.3 \times 10^{-12} \times c(\mathrm{mg/mL}) \tag{7.178}$$

Equation (7.178) is broadly consistent with theoretical Eq. (7.175) above, as D_{eff} in Eq. (7.178) increases with concentration as predicted by Eq. (7.175), but we note that the effective diffusion coefficient for aggrecan in aggrecan solutions D^*_{eff} decreases from 4.93×10^{-12} as the concentration increases Eq. (7.176), while equation (7.178) shows D_{eff} increasing from 3.3×10^{-12}. Bringing the experimental Eq. (7.176) together with the theoretical Eq. (7.177), one might predict that for aggrecan diffusing in aggrecan, the diffusion coefficient in cartilage could be estimated by

$$D_{eff} = D^*_{eff} \frac{\partial \pi^{agg}}{\partial c^{agg}} = 4.93 \times 10^{-12} e^{-0.18(\mathrm{mL/mg}) \times c(\mathrm{mg/mL})} \frac{\partial \pi^{agg}}{\partial c^{agg}} \ \mathrm{m^2\, s^{-1}} \tag{7.179}$$

Nevertheless, for the context-specific tests in a centrifuge involving a concentration gradient (and so $D_{eff} = D^*_{eff} \frac{\partial \pi^{agg}}{\partial c^{agg}}$, where the non-linear $\frac{\partial \pi^{agg}}{\partial c^{agg}}$ term), the experimental tests show an increase in effective diffusion coefficient. We note that for articular cartilage in vivo, there normally is a concentration gradient, much as in the Comper experiments. However, an aggrecan solution is not articular cartilage. Clearly, the extracellular matrix in cartilage includes both the collagen network that may impede collagen transport (e.g. mechanical entanglement of large molecules), and the binding of other 'bridging' or 'linking' molecules that may also hinder aggrecan transport. Cartilage is clearly not just other aggrecan molecules in solution. Consequently, D_{eff} of aggrecan through cartilage would be expected to have a much smaller effective diffusion coefficient than suggested by the laboratory experiments described by Comper and colleagues above; i.e. the D_{eff} described above clearly represent an 'upper bound' on effective diffusion coefficient, as no account is taken of the hindered transport offered by the collagen network and other aggregate molecules within cartilage. In short, the relationship between Comper's experiment and articular cartilage is unknown.

Alternatively, aggrecan may be conceptualized as a 'viscous fluid' moving through the collagen network in cartilage (rather than as molecules diffusing through an aqueous solution). In this case, a simple model approximating a constitutive equation for the aggrecan is Darcy's law (as employed to represent the transport of water), viz,

$$k^{\mathrm{agg}} \nabla^{\mathrm{s}} \sigma^{\mathrm{agg}} = n_{\mathrm{agg}} \left(v^{\mathrm{agg}} - v^{\mathrm{col}} \right) = v_{\mathrm{d}}^{\mathrm{agg}} \tag{7.180}$$

where n_{agg} is the porosity of the aggrecan, $V_{\mathrm{d}}^{\mathrm{agg}}$ is the 'Darcy velocity of aggrecan', and v^{agg} is the actual velocity of the aggrecan relative to the actual velocity of the collagen network v^{col}. The driving force for the aggrecan flow σ^{agg} (with sign convention as for solid stress, i.e. $\sigma^{\mathrm{agg}} = -\pi^{\mathrm{agg}}$), can be found from Eq. (7.169),

$$\nabla \sigma^{\mathrm{agg}} = \nabla \frac{\mathrm{tr} \sigma^{\mathrm{total}}}{3} - \nabla \frac{\mathrm{tr} \sigma^{\mathrm{col}}}{3} + \nabla p \tag{7.181}$$

We see that the driving force (i.e. gradient of σ^{agg}) is in general a function of the gradient of the hydrostatic total stress, the gradient of the hydrostatic collagen stress, and the gradient of pore water pressure. This makes sense, because in addition to momentum transfer from the collagen network to aggrecan molecules, there may be momentum transfer from the water to aggrecan molecules. In other words, if there is pore fluid flow in the cartilage, the drag force of the water on the aggrecan may help to either move the aggrecan in the (minus) direction of the osmotic gradient, or hinder the movement of aggrecan when pore water flow is in the opposite direction to the flow of aggrecan.

If there is no cyclic loading ($\sigma^{\mathrm{total}} = 0$), the flux of water is minimal ($\nabla p \approx 0$) and in this case $\sigma^{\mathrm{agg}} \approx -\sigma^{\mathrm{col}}$. This approach has been used to successfully model the development of an aggrecan profile similar to that observed in young bovine cartilage (Smith et al. 2016). A suitable equation to describe aggrecan permeability for calf cartilage was found to be Smith et al. (2016),

$$k^{\mathrm{agg}} (\mathrm{m}^2 \mathrm{Pa}^{-1} \mathrm{s}^{-1}) = \frac{2.7 \times 10^{-17}}{c^{\mathrm{agg}} (\mathrm{mg/mL})} \tag{7.182}$$

Interestingly, we can estimate an equivalent diffusivity based on this permeability estimate by equating the two fluxes (i.e. diffusive flux with advective Darcy flux). The diffusive flux of aggrecan is

$$f^{\mathrm{agg}} = -D_{\mathrm{eff}} \nabla^{s} c_{\mathrm{eff}}^{\mathrm{agg}} \tag{7.183}$$

where D_{eff} is given by Eq. (7.175).

We now put the Darcy flux of aggrecan in the same form as the diffusive flux of aggrecan to compare terms. To do this, we note that $v_{\mathrm{d}}^{\mathrm{agg}}$ represents the 'volume flux of aggrecan' (see Appendix O for details). To convert the flux of aggrecan volume into a mass flux, $f^{\mathrm{agg}} (\mathrm{kgm}^{-2} \mathrm{s}^{-1})$, we need to use the relation $\mathrm{mass}^{\mathrm{agg}} / (\mathrm{m}^2 \mathrm{s}) = v_{\mathrm{d}}^{\mathrm{agg}} / \bar{v}^{\mathrm{agg}}$, where $\bar{v}^{\mathrm{agg}}$ is the (partial) specific volume of aggrecan. If the specific volume of aggrecan is constant, then

$$f^{agg} = \frac{v_d^{agg}}{\bar{v}^{agg}} = -\frac{k^{agg}}{\bar{v}^{agg}}\frac{\partial \pi^{agg}}{\partial c^{agg}}\nabla c^{agg} \tag{7.184}$$

Comparing Eq. (7.174) and (7.184) leads to

$$D_{\mathrm{eff}} = \frac{k^{\mathrm{agg}}}{\bar{v}^{\mathrm{agg}}}\frac{\partial \pi^{\mathrm{agg}}}{\partial c^{\mathrm{agg}}} = \frac{k^{\mathrm{agg}}}{\bar{v}^{\mathrm{agg}}} RT(\alpha_1 + 2\alpha_2 c^{\mathrm{agg}} + 3\alpha_3 c^{\mathrm{agg}^2}) \tag{7.185}$$

For the young bovine cartilage reported in Eq. (7.182), we find

$$D_{\mathrm{eff}} = \frac{k^{\mathrm{agg}}}{\bar{v}^{\mathrm{agg}}}\frac{\partial \pi^{\mathrm{agg}}}{\partial c^{\mathrm{agg}}} = 2.7 \times 10^{-17}\frac{RT}{\bar{v}^{\mathrm{agg}}}\left(\frac{\alpha_1}{c^{\mathrm{agg}}} + 2\alpha_2 + 3\alpha_3 c^{\mathrm{agg}}\right) \tag{7.186}$$

Assuming the partial specific volume $\bar{v}^{\mathrm{agg}} = 5.5 \times 10^{-4}$ mL/mg [see Table 1 (Comper and Williams 1987)],$R = 8.31 \times 10^3 \mathrm{PaLK^{-1}mol^{-1}}$, $T = 300$ K and α_i(as shown in Appendix D), then

$$D_{\mathrm{eff}} = 1.25 \times 10^{-7}\left(\frac{1.4 \times 10^{-4}}{c^{\mathrm{agg}}} + 8.8 \times 10^{-6} + 1.7 \times 10^{-7} c^{\mathrm{agg}}\right) \tag{7.187}$$

We note the predicted form of D_{eff} for young calf bovine cartilage using Eq. (7.187) is somewhat more complex than the experimental Eq. (7.178) and about the same complexity as the combined experimental-theoretical Eq. (7.179), describing aggrecan diffusing through aggrecan. How do they compare quantitatively?

Let us now examine the various predictions. For aggrecan diffusing through aggrecan in centrifuge experiments, we estimate using Eq. (7.178), at $c^{\mathrm{agg}} = 10$ mg/mL, $D_{\mathrm{eff}} = 3.3 \times 10^{-11}$, while at $c^{\mathrm{agg}} = 100$ mg/mL, we find $D_{\mathrm{eff}} = 3.3 \times 10^{-10}$. Using Eq. (7.179), a combined experimental-theoretical equation (i.e. aggrecan diffusing in aggrecan with no osmotic pressure gradient, then multiplied by the osmotic pressure gradient), we estimate at $c^{\mathrm{agg}} = 10$ mg/mL, $D_{\mathrm{eff}} = 5.0 \times 10^{-10}$, while at $c^{\mathrm{agg}} = 100$ mg/mL we find $D_{\mathrm{eff}} = 5.0 \times 10^{-16}$, suggesting very large variation in diffusion coefficient with concentration. And finally, using Eq. (7.187) for calf cartilage at $c^{\mathrm{agg}} = 10$ mg/mL we find $D_{\mathrm{eff}} = 3.0 \times 10^{-12}$, while at $c^{\mathrm{agg}} = 100$ mg/mL we find $D_{\mathrm{eff}} = 3.3 \times 10^{-12}$, suggesting a very nearly constant value for the diffusion constant through calf cartilage.

In other words, we find that the estimated diffusion coefficient for calf cartilage [using Eq. (7.187)] is about 10 times smaller at low concentrations aggrecan (around 10 mg/mL), and about 100 times smaller at high concentrations of aggrecan (about 100 mg/mL), relative to the experimentally measured diffusion coefficient measured for aggrecan diffusing through aggrecan Eq. (7.178). This is perhaps not surprising, as we would expect the collagen network in calf cartilage to impede aggrecan transport.

On the other hand, we see that the combined experimental-theoretical equation [i.e. equation (7.179)] varies greatly across six orders of magnitude, first over-estimating then under-estimating the likely diffusion coefficient, due to the exponential expression to represent D^*_{eff}. Using Eq. (7.179) and assuming a gradient of aggrecan from 50 mg/mL aggrecan to zero over a distance of 100 μ, we estimate that D^*_{eff} is about 6.1×10^{-16} m^2 s^{-1}, while D_{eff} is about 1.5×10^{-12} m^2 s^{-1}. Happily, this results in an estimated diffusion flux about the same as that observed in calf cartilage, but estimates also differ by orders of magnitude above and below 50 mg/mL aggrecan. Though theoretically appealing, Eq. (7.179) does not appear to produce reasonable estimates for the diffusion coefficient of aggrecan through cartilage, apparently due to the exponential expression to represent D^*_{eff} dominating D_{eff} (but we note that Comper (1991) notes that the reverse happens for the centrifuge experiments; i.e. the nonlinear term $\frac{\partial \pi^{\text{agg}}}{\partial c^{\text{agg}}}$ dominates D_{eff}).

It seems likely that mean diffusion coefficients for adult cartilage are smaller than that predicted by Eq. (7.187), which is based on calf cartilage. For it is apparent that using plausible parameters for osmotic pressure gradient, a diffusion coefficient D_{eff} around 1.0×10^{-14} m^2 s^{-1} is required to obtain a flux through an adult knee joint of about 2.5 mg/day (see Appendix E). We note that this suggests that for adult cartilage, the permeability of aggrecan would be two orders of magnitude less than that suggested by Eq. (7.182), which was developed to describe calf cartilage. We note that this estimate of D_{eff} around 1.0×10^{-14} m^2 s^{-1} is a spatially weighted average diffusion coefficient, and so we expect there to be regional variations in the diffusion coefficient (e.g. the diffusion coefficient for the territorial zone will be considerably higher than for the interterritorial zone).

From this follows that the diffusion coefficient for aggrecan transport through cartilage will be uniform if the collagen network is uniform (as the distribution of collagen in young bovine articular cartilage tissue appears to be), and it will be non-uniform when the collagen network is non-uniform (as the distribution of collagen is in adult articular cartilage appears to be). This is very important because it supports the idea that the characteristics of the collagen network determine the transport of aggrecan through cartilage. In adult cartilage, zones appear that have distinct collagen networks, and so it is expected that the diffusion coefficient in each zone will be different. So D_{eff} varies spatially with collagen density and fabric. Such non-uniform transport is exemplified by the experimental findings discussed in Sect. 3.4, which shows in histological slides ‘plumes’ of aggrecan emanating from chondrocytes and moving toward the cartilage surface in ‘ribbons.’ The collagen network in the superficial zone is quite dense with a fabric unfavorable to aggrecan transport through the cartilage surface and so may be expected to have (perhaps) the lowest diffusion coefficient in the direction normal to the surface.

It is also of interest to compare the permeability of aggrecan moving through (calf) cartilage matrix with that of water moving through cartilage matrix. For the aggrecan moving through calf cartilage matrix, we have

$$k^{\mathrm{agg}}(\mathrm{m^2Pa^{-1}s^{-1}}) = \frac{2.7 \times 10^{-17}}{c^{\mathrm{agg}}(\mathrm{mg/mL})} \tag{7.188}$$

And for the permeability of water moving through aggrecan,

$$k_{\mathrm{w}}(\mathrm{m^2Pa^{-1}s^{-1}}) = \frac{1.02 \times 10^{-12}}{c^{1.6}} \tag{7.189}$$

Substituting values for concentration shows that at 10 mg/ml the ratio $k_{\mathrm{w}}/k_{\mathrm{agg}}$ is about 10^4, while at 100 mg/mL this ratio decreases fourfold to about 2.3×10^3. It is apparent that the permeability of water through calf cartilage is between 1,000 and 10,000 times greater than the permeability of aggrecan through calf cartilage. This large difference in timescales explains the common assumption that the proteoglycan is 'fixed' in the cartilage collagen network. We mention that in adult human cartilage, we would expect the difference to be about two orders of magnitude greater, further reinforcing the differences in transport times scales.

Finally, it seems likely that for some spatial domains of articular cartilage, the dominant transport mechanism is diffusion, while in for other spatial domains advective transport may dominate. For example, advective transport appears to dominant diffusive transport in the territorial zones in adult articular cartilage (see Sect. 3.4).

7.3.4 The Compressibility of Aggrecan in Cartilage

The change in (Helmholtz) free energy potential of the aggrecan can be found by integrating the osmotic pressure function with respect to volume, while the osmotic modulus can be found by differentiating the osmotic pressure function with respect to volume. The osmotic pressure function for aggrecan can be represented as a virial equation (Comper and Williams 1987; Comper 1991; Bathe et al. 2005), viz.,

$$\pi^{\mathrm{agg}} = RT(\alpha_1 c^{\mathrm{agg}} + \alpha_2 c^{\mathrm{agg}^2} + \alpha_3 c^{\mathrm{agg}^3}) \tag{7.190}$$

where R is the universal gas constant (8.3e3 Pa L/K mol), T is the absolute (e.g. 300 K) and α_1 = 1.4e−4, α_2 = 4.4e−6, α_3 = 5.6e−8, and c^{agg} is the concentration of aggrecan in mg/ml. It is noted that the first term captures ideal (van 't Hoff) behavior, while the second and third terms capture non-ideal behavior. Bathe et al. explain: '*second virial coefficient represents the contribution of two-body, intermolecular GAG interactions to the osmotic pressure, and together with [the third virial coefficient] quantifies the deviation of the osmotic pressure from ideality*' within the polymer solution (Bathe et al. 2005). The osmotic (bulk) modulus due to aggrecan is given by (see Appendix L),

$$\mathrm{OM} = \frac{\mathrm{d}\pi^{\mathrm{agg}}}{\mathrm{d}V/V} = c^{\mathrm{agg}} \frac{\partial \pi^{\mathrm{agg}}}{\partial c^{\mathrm{agg}}} \tag{7.191}$$

We note here that taking into account the temperature and the volume occupied by collagen, the osmotic modulus becomes

$$\mathrm{OM} = \frac{\mathrm{d}\pi^{\mathrm{agg}}}{\mathrm{d}V/V} = \frac{c^{\mathrm{agg}}}{n^{\mathrm{sm}}} \frac{\partial \pi^{\mathrm{agg}}}{\partial c^{\mathrm{agg}}} = RT(\alpha_1 c + 2\alpha_2 c^2 + 3\alpha_3 c^3)/n^{\mathrm{sm}} \tag{7.192}$$

Assuming the collagen volume fraction is constant, the effect of deformation changing the fluid porosity has a significant effect on increasing the apparent stiffness of the aggrecan in the cartilage ECM, both by changing the actual concentration of aggrecan and via its direct influence on the osmotic stiffness (see further discussion at equation (L.27)). Finally, the Young's modulus for aggrecan can be calculated from the bulk modulus as discussed in Appendix D.

The physico-chemical basis for the resistance to compression can be dissected into its component parts through molecular dynamics simulations, e.g. (Nap and Szleifer 2008). Nap and Szleifer show that at pH 5.6 the main contributor to the osmotic force between aggrecans is the counterions and co-ions in solution (i.e. over separation distances between aggrecan molecules of 2–10 nm); however, at very low separation distances (<1 nm), the 'polymer' contributions (or 'entropic' contributions: See discussion Appendix XI State variables, entropy and free energy potentials) become increasingly important and may dominate [see, for example, Fig. 6 in Nap and Szleifer (2008)]. It is noted that the average separation between the GAG chains on the aggrecan molecules is around 2–4 nm (Bajpayee and Grodzinsky 2013) corresponding to (true) GAG concentrations of 80–20 mg/mL (Bathe et al. 2005), so the counter and co-ions in the aggrecan solution are usually the principal source of osmotic pressure.

It is shown theoretically in Appendix D that if the aggrecan concentration is high, and the volume fraction of the (hydrated) collagen network is in the range 20–30%, the compressive modulus of aggrecan alone can explain most of the compressive resistance of cartilage in normal healthy load carrying joints like the knee. For example, if the aggrecan concentrations are normal (e.g. 80–120 mg/mL in high contact regions of cartilage in load-bearing joints, e.g. for tibial cartilage), it is found that it is very difficult to compress healthy cartilage more than 50% of its original height because the compressive stiffness increases very rapidly (see Tables shown in Appendix D). This rapid increase in stiffness is driven by the actual concentration of aggrecan in the cartilage varying inversely with the volumetric strain, i.e.

$$c^{\mathrm{sm}}(X,t) = \frac{c^{\mathrm{sm}}(X,0)}{\frac{Jn_{\mathrm{f}}^{\mathrm{sm}}}{n_0}} = \frac{c^{\mathrm{sm}}(X,0)}{1 + \frac{(J-1)}{n_0}} \tag{7.193}$$

where $c^{sm}(X,0)$ is the initial (true) aggrecan concentration in the pore fluid and J is the Jacobian of the deformation gradient tensor and n_0 is the initial (fluid) porosity. In other words, if the average strain is 50%, the local concentrations are well-above the initial (true) concentration of aggrecan and these concentrations result in very large compressive stiffness (see Tables Appendix D).

For example, if the initial (true) concentration is around 100 mg/mL and the collagen volume fraction is zero, then the Young's initial modulus (YM) in compression is about 2 MPa, but upon 50% compression the concentration is raised to 200 mg/mL, and the YM in compression is raised to about 13 MPa. But if the initial apparent concentration is around 100 mg/mL and the collagen volume fraction is 20%, then the actual initial aggrecan concentration is 125 mg/mL and the Young's initial modulus (YM) in compression is about 3.7 MPa, but upon 50% compression the concentration is raised to 333 mg/mL, and the YM in compression is raised to about 55 MPa. It is noted that many laboratory measurements (e.g. ultrasound) only test for small strain stiffnesses, which are usually considerable different than the working stiffness of cartilage (which can involve heavy loads for long times and so large strains).

Experimentally, it has been reported that nearly all the compressive stiffness is attributable to aggrecan (at least over the range 0–15% strain) for bovine and young human cartilage, but the aggrecan content could explain only about one half the compressive stiffness for older human cartilage (Han et al. 2011). Older human cartilage is known to have defective aggrecan in the sense that the GAG content of the aggrecan molecules is greatly reduced. Consequently, it is certainly possible that in cartilage without much aggrecan, collagen fibrils may be compressed directly onto collagen fibrils, and from experiments, the collagen may carry about 50% of the applied load(Han et al. 2011). However, it is noted that these conclusions are highly dependent on the collagen fraction in the cartilage, as well as the apparent aggrecan concentration. It also seems likely that when collagen fibrils are compressed on collagen fibrils, it is likely to quickly damage the fibrils, leading to aggrecan loss.

It is useful here to approximately fit a polynomial curve to the variation of Young's modulus with aggrecan concentration. Assuming zero collagen volume fraction (which represents a lower bound), over the range 25–200 mg/mL of aggrecan, the Young's modulus can be approximated by

$$E_{\text{comp}} = 0.5151 + 0.319c^{\text{agg}} + 0.0005c^{\text{agg}^2} \tag{7.194}$$

where c is the sGAG concentration in mg/ml and the estimated Young's modulus is in MPa (agreeing with the analysis in Appendix D) (r^2 = 0.998). Less accurately, we could use a power law approximation, e.g.

$$E_{\text{comp}} = 1.26 \times 10^{-3} \times c^{\text{agg}^{1.6}} \tag{7.195}$$

We note that this equation leads up to discrepancy of about 100% between the approximated and theoretical estimates of aggrecan stiffness over the range 25–200 mg/mL. Nevertheless, given that the stiffness varies over orders of magnitude, for our purpose here—considering the effect of stiffness and permeability on the consolidation coefficient—this equation has acceptably small relative errors.

For example, the one-dimensional consolidation coefficient is $c_z = \beta_1 k_e E_{sz}$ (see Sect. 7.2.1). Now, substituting Eqs. (7.173) and (7.194) into the consolidation coefficient, we find to a reasonable approximation that

$$c_z = \beta_1 (1.02 \times 10^{-12} \times c^{\mathrm{agg}^{-1.6}})(1.26 \times 10^{-3} \times c^{\mathrm{agg}^{1.6}}) = \text{constant} \qquad (7.196)$$

over a wide range of aggrecan physiological concentrations.

This result is a remarkable property of aggrecan, because it enables the cartilage to maintain its characteristic consolidation rate, despite wide variations in cartilage aggrecan content. To express this another way, the materials comprising cartilage enable the coefficient of consolidation of cartilage to be self-regulating, in the sense that the coefficient of consolidation remains approximately constant as the aggrecan concentration varies as the articular cartilage compresses.

Equation (7.196) also means that consolidation behavior of cartilage can be remarkably stable even as articular cartilage tissue degenerates. For example, Fick et al. tested bovine patella classified as normal and degraded (e.g. degraded cartilage had evident surface fibrillation) (Fick et al. 2010). Despite the degradation, the consolidation behavior of the normal and 'degraded' articular cartilage is remarkably similar [see Fig. 6 in (Fick et al. 2010)].

This 'invariance property' for the coefficient of consolidation has important implications for the duration of time-dependent settlement of loaded cartilage. For example, the time for the cartilage to reach its consolidated 'working state' for some activity (e.g. walking) is then approximately constant, despite variations in the initial aggrecan concentration with depth. It is also important because it means the duration of weeping lubrication occurring at cartilage–cartilage or cartilage–meniscus interfaces is also invariant (approximately of similar the characteristic time for articular cartilage consolidation), and so the duration of mixed mode lubrication at the cartilage–meniscus or cartilage–cartilage interface is resistant to change of aggrecan concentration, though the articular cartilage deformations change (see discussion Chap. 4).

Given the physiological importance of the constant coefficient of consolidation to normal joint homeostasis, other things being held constant, presumably this invariance of the characteristic time for consolidation with aggrecan concentration, bestowed through the fixed charge density, stiffness, and permeability properties of aggrecan, has been selected by evolution.

7.3.5 Representing the Tensile Strength of Collagen in Cartilage

Collagen fibrils are dispersed throughout the cartilage matrix. In some regions, the fibrils are isolated in the extracellular matrix, while in other regions, the fibrils aggregate into fibers that connect with one another and so carry force over long distances (e.g. from the cartilage surface to the bone–cartilage interface). The collagen fibers may be in a variety of geometries. For example, the fibers may be described as forming a 'collagen network' (which may be isotropic or anisotropic depending on cartilage zone), including fibrils that aggregate to form bundles or sheets of collagen (which are clearly anisotropic in the stiffness properties). Both the quantity and 'fabric' of the collagen vary in each zone of the cartilage, undoubtedly in response to the aggrecan-induced stress in the collagen network and to the load-induced stresses in the collagen network. These two primary sources of cartilage stress, shape, over time, the fiber arrangement of the collagen network throughout the cartilage.

It is generally accepted that collagen carries only tensile loads. Because collagen fibrils and fibers are usually long and thin, they generally buckle at a low load when in compression. In most cartilage models, they are usually represented as carrying no load when in compression. Because collagen only carries tensile loads, we clearly need to check if the strain in the collagen fiber is positive, in which case it carries some load, or negative, in which case it carries no load. In other words, the stiffness tensor is conditional on the strain state of the cartilage.

How to adequately represent the geometry and properties of collagen in a continuum model has been a puzzle for some time. At a basic level, one needs to take into account the measured anisotropic stiffness of the continuum due to the presence of collagen fibers. The first option is to use one of the well-established anisotropic constitutive models. For example, one could use a transverse isotropic or an orthotropic constitutive model to represent the influence of fiber directions in cartilage. As a first approximation, anisotropic nonlinear elasticity models can work well because in many cases the collagen fiber directions are aligned orthogonally and so with an orthogonal coordinate system. For example, 'radial' collagen fibers in the deep zone are predominantly vertical in orientation, while collagen fibers in the superficial zone are predominantly horizontal in orientation (these fiber directions coincide with the coordinate directions as usually chosen in either material or spatial coordinates). As a first approximation, the collagen direction in the intermediate zone could be modeled as random (i.e. isotopic).

A large strain analysis of fiber-reinforced matrixes (with the fibers made of many individual collagen fibers or collagen bundles) has the merit that it can capture the progressive reorientation of collagen fibers into the directions of principal stress. Indeed, the progressive recruitment of collagen fibers as the deformation progresses to the principal stress directions is an important structural feature of the collagen network behavior in tension (it is noted in passing that this behavior may be approximated in a conventional anisotropic model by using a nonlinear stress–strain relation in tension). This behavior may be important physiologically in cartilage under certain circumstances, e.g. large tensile strain or shear loads on the surface of the cartilage.

However, if we wish to do a large deformation analysis using anisotropic material properties, we usually need to find a suitable strain energy density function based on the material anisotropy. Unfortunately, suitable strain energy density function has proven difficult to identify, and so this difficulty has driven the movement towards using a strain energy density function for an isotropic material, and developing anisotropy in this basic isotropic model by adding to it collagen fibre reinforcement.

In addition to the merits of large deformation models of fiber-reinforced matrixes, some point to the difficulties and shortcomings of standard anisotropic constitutive models. One is the fact that collagen fibers are not necessarily oriented with the coordinate axes.

To address fiber orientation in any direction, it is desirable to introduce a texture (or fabric) tensor. To address this concern, we follow the approach of Ateshian et al. (2009), as this approach has the considerable merit of simply constructing an anisotropic continuum stiffness matrix by adding in the stiffness of collagen fibers one at a time, based on a strain energy density function for an individual collagen fiber. This approach not only solves the fabric issue, it also gets around identifying beforehand the strain energy density function at each point across the whole continuum (replacing it with the strain energy density function for a single collagen fiber, which is experimentally much more easily accessible, e.g. from an image of the collagen fibers and their orientation). Details of how to do this are shown in Appendix N (stress and elasticity tensors in a fiber-reinforced composite).

The strain energy density functions for the aggrecan and collagen are usually taken to be independent, and so the second Piola–Kirchhoff stresses for the cartilage biocomposite (S), and the stiffness tensor for cartilage biocomposite ($\mathbb{C}$) may be simply estimated as the sum of the collagen and aggrecan component stiffnesses to find

$$S = S^{agg} + S^{col}; \mathbb{C} = \mathbb{C}^{agg} + \mathbb{C}^{col} \tag{7.197}$$

For example, (Pierce et al. 2012) define the strain energy density function as

$$\Psi = \Psi^{agg}(J, I_1) + \Psi^{col}(I_1, I_4) \tag{7.198}$$

where

$$\begin{aligned} \Psi^{agg}(J, I_1) &= \tfrac{1}{\rho^s_{os}}\left(U(J) + \tfrac{1}{2}\mu^s(I_1 - 3)\right) \\ U &= \lambda^s_{cp}[\tfrac{1}{2}(\log J)^2 + \xi^s] - \mu^s \log J \end{aligned} \tag{7.199}$$

and

$$\Psi^{\mathrm{col}} = \frac{1}{\rho_{\mathrm{os}}^{\mathrm{s}}}\left\{\frac{k_1}{2k_2}[\exp\{k_2[(1-\rho)(I_1-3)+\rho(I_4-1)^2]-1\right\} \tag{7.200}$$

such that $I_1 = \mathrm{tr}(C), I_4 = C : M_0$ where $M_0 = a_0 \otimes a_0$ is a (Lagrangian) structural tensor, and the local principal collagen fibers is characterized by a (reference) vector a_0, where $|a_0| = 1$ (anisotropy only arises through the modified invariant I_4). We note that the collagen strain energy density function is modeled as an exponential function, while the shear stress in the cartilage is modeled as being carried by the aggrecan, and by any collagen fibres that are stretched by the shear deformation. Further, explanation of parameters can be found in (Pierce et al. 2012). This strain energy density function for cartilage shown above displays several similarities to the density functions derived previously for aggrecan and collagen separately, however, they also involve a considerable number of parameters (about 10), and the physical meaning of these parameters is not always clear.

The analysis may be extended to include viscoelasticity of the collagen network, e.g. (Li et al. 1999; Seifzadeh et al. 2011), and to remodeling of the collagen network to better align with the directions of principal stress [e.g. Grytz and Meschke (2010)].

7.3.6 The Storage Equation for Cartilage

The storage equation for a multicomponent mixture experiencing small deformation is given by (see Appendix XV the storage equation for cartilage)

$$\frac{1}{1+e_v}\frac{\partial e_v}{\partial t} = -\sum_{i \neq j} \nabla^s \cdot [v_{\mathrm{d}}^i] + \sum s_{n_i} \tag{7.201}$$

where e_v is the volume strain of the jth phase, v_{d}^i is the ith Darcy velocity w.r.t the jth phase, and $s_{n_i}(x,t)$ is the rate of production or removal of porosity per unit volume in spatial coordinates of the ith phase. It is noted that the porosity of the ith phase is

$$n_i = c_i \bar{v}_i \tag{7.202}$$

where $\bar{v}_i$ is the partial specific volume of the ith phase, and so the rate of production or removal of porosity can be expressed in terms of mass production or removal per unit volume if required. The corresponding equation for large deformation analysis is

$$\frac{\mathrm{d}J}{\mathrm{d}t} = -\sum_{i \neq j} \nabla^{\mathrm{m}} \cdot JF^{-\mathrm{T}}[v_{\mathrm{d}}^{i}] + \sum S_{n_i} \tag{7.203}$$

where $S_{n_i}(X,t)$ are now expressed in terms of material coordinates.

The main structural components of the cartilage biocomposite are water, aggrecan, and collagen so the storage equation needs to take account several components. Because both water and aggrecan can be transported through the collagen network, we take the collagen network as our reference material (j phase above). Therefore, our material coordinates for the analysis of cartilage coincide with the collagen network.

Darcy velocities for the remaining components (i.e. water and aggrecan) in our storage equation are now referenced to collagen. This is important to our interpretation of Darcy velocity because previously we defined the Darcy velocity of water $v_{\mathrm{d}}^{\mathrm{w}}$ w.r.t. aggrecan movement (which is the primary determinant of the measured permeability of cartilage), not to collagen $v_{\mathrm{d-col}}^{\mathrm{w}}$. However, when we use the 'Darcy flow equation' for water in the storage equation, we need to use Darcy flow relative to collagen (i.e. $v_{\mathrm{d-col}}^{\mathrm{w}}$). We can see how to do this from

$$v_{\mathrm{d-col}}^{\mathrm{w}} = n_{\mathrm{w}}(v^{\mathrm{w}} - v^{\mathrm{col}}) = n_{\mathrm{w}}[(v^{\mathrm{w}} - v^{\mathrm{agg}}) + (v^{\mathrm{agg}} - v^{\mathrm{col}})] \tag{7.204}$$

that is,

$$v_{\mathrm{d-col}}^{\mathrm{w}} = v_{\mathrm{d}}^{\mathrm{w}} + \frac{n_{\mathrm{w}}}{n_{\mathrm{agg}}} v_{\mathrm{d}}^{\mathrm{agg}} \tag{7.205}$$

Assuming a small deformation analysis and that aggrecan is transported according to Darcy's and using the expression for $v_{\mathrm{d-col}}^{\mathrm{w}}$, the storage equation for cartilage becomes

$$\frac{1}{1+e_v}\frac{\partial e_v}{\partial t} = -\nabla^{\mathrm{s}} \cdot v_{\mathrm{d}}^{\mathrm{w}} - \nabla^{\mathrm{s}} \cdot \left[1 + \frac{n_{\mathrm{w}}}{n_{\mathrm{agg}}}\right] v_{\mathrm{d}}^{\mathrm{agg}} + s_{\mathrm{n-col}} + s_{\mathrm{n-agg}} + s_{n-\mathrm{w}} \tag{7.206}$$

Employing the constitutive equations for water and aggrecan transport leads to

$$\begin{aligned} \frac{\partial e_v}{\partial t} &\approx \frac{1}{1+e_v}\frac{\partial e_v}{\partial t} \\ &= \nabla^{\mathrm{s}} \cdot k_{\mathrm{w}} \nabla^{\mathrm{s}} p + \nabla^{\mathrm{s}} \cdot \left[1 + \frac{n_{\mathrm{w}}}{n_{\mathrm{agg}}}\right] k_{\mathrm{agg}} \nabla^{\mathrm{s}} \sigma^{\mathrm{agg}} + s_{\mathrm{n-col}} + s_{\mathrm{n-agg}} + s_{n-\mathrm{w}} \end{aligned} \tag{7.207}$$

and using our transformation rules to accommodate a large deformation analysis, the storage equation is

$$\frac{\mathrm{d}J}{\mathrm{d}t} = \nabla^{\mathrm{m}} \cdot JF^{-1} k_{\mathrm{w}} F^{-\mathrm{T}} \nabla^{\mathrm{m}} p + \nabla^{\mathrm{m}} \cdot JF^{-1} \left[1 + \frac{n_{\mathrm{w}}^{\mathrm{sm}}}{n_{\mathrm{agg}}^{\mathrm{sm}}} \right] k_{\mathrm{agg}} F^{-\mathrm{T}} \nabla^{\mathrm{m}} \sigma^{\mathrm{agg}} + S_{\mathrm{n-col}} + S_{\mathrm{n-agg}} + S_{n-\mathrm{w}} \tag{7.208}$$

where $S_{n-i}(X, t)$ are now expressed in terms of material coordinates. One may carry out a similar analysis for the diffusive transport of aggrecan.

Appendix A
Key Concepts for Biology

Water, Biochemistry, and Cells

Liquid water provides the solvent and 'background matrix' in which the set of chemical reactions that constitute life takes place. Water molecules are 'polar,' which means that they interact strongly with each other through electrical interactions arising from small imbalances in the distribution of electrons around oxygen and hydrogen atoms. This leads to 'hydrogen bonding' between water molecules. Other polar molecules (e.g. alcohol, amino acids, glucose) usually dissolve readily in water, and so they are called 'water-loving' or 'hydrophilic.' The most hydrophilic water molecules also engage in hydrogen bonding with water molecules. Nonpolar molecules only interact weakly with water molecules, and so they are called 'hydrophobic.' Hydrophobic molecules tend to dissolve sparingly in water, so, for example, 'oily molecules' cluster together in droplets rather than dissolving in water. The body uses several different mechanisms to transport fats in the blood, e.g. fats may be coated in hydrophilic proteins, to control droplet size and help target them to certain cells (https://en.wikipedia.org/wiki/Blood_lipids).

Some macromolecules including some proteins and phospholipids are partly hydrophilic and partly hydrophobic, in which case a flexible, linear macromolecule will twist and turn in water so that the hydrophilic parts of the macromolecule tend to be found at the water–molecule interface, while the hydrophobic parts of the macromolecule tend to face inside and cluster within the macromolecule. In other words, the relative affinities of macromolecules for water orientate the macromolecules in solution. For example, phospholipids are used to create 'cell membranes' (https://en.wikipedia.org/wiki/Cell_membrane). In cell membranes, the phospholipids orientate themselves so the hydrophilic (the polar 'phosphate group') is in contact with the water molecules, while the hydrophobic, 'oily tails' cluster together away from the water. If the folding of protein is complex, other proteins known as chaperones may temporarily interact with a protein in the process of folding to make sure it interacts with the water in just the right way, and so that it

D. W. Smith et al., *Articular Cartilage Dynamics*,
https://doi.org/10.1007/978-981-13-1474-2

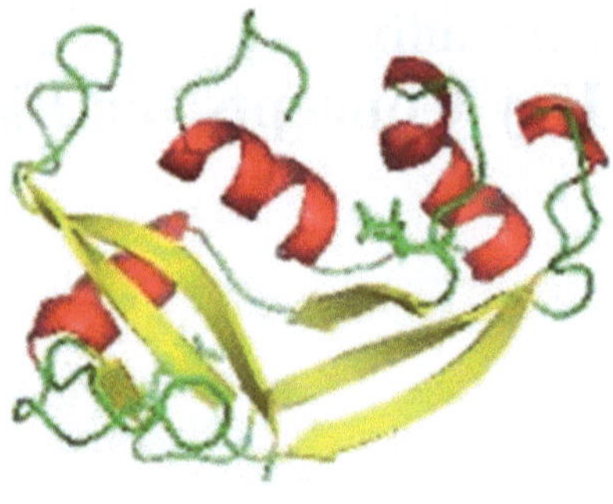

Fig. A.1 Tertiary structure of a protein, showing α helices (red) and β-pleated sheets (gold), from Wikimedia Commons

folds in the intended way (proteins that do not fold in the intended way are generally not functional, or even deleterious—so chaperones are important molecules). Because proteins are more likely to misfold at higher temperatures, chaperones are upregulated in heat shock (https://en.wikipedia.org/wiki/Heat_shock_protein), which serves to highlight the importance of thermal regulation of body temperature, as it is much harder to exercise control over the types and rates of chemical reactions if temperature is highly variable.

Proteins have said to have a primary structure (the amino acid sequence), a secondary structure (e.g. a so-called alpha helix https://en.wikipedia.org/wiki/Alpha_helix or a 'beta-pleated sheet' https://en.wikipedia.org/wiki/Beta_sheet), and a tertiary structure (the 3-D structure of the protein (Fig. A.1)—proper folding ensures a functional 3-D shape for the protein) (https://en.wikipedia.org/wiki/Protein). Often, structures form in proteins, which are known as domains or motifs. There are many fewer domains and motifs in proteins than there are proteins (https://en.wikipedia.org/wiki/Structural_motif). Sometimes, proteins aggregate in a 'quaternary structure' too, e.g. hemoglobin (see https://en.wikipedia.org/wiki/Hemoglobin).

A 'cell membrane' presents a 'selective barrier' to the molecular transport of proteins, ions, and other molecules in solution. The cell membrane acts to partition space, defining both the cell itself and cell structures within the cell (e.g. nucleus, endoplasmic reticulum, Golgi apparatus, vesicles are all cell structures within a cell).

The key function of membranes is to confine solutes and their attendant chemical reactions to certain compartments, and to regulate interactions with other compartments. It becomes apparent that each compartment gives the cell much more control over the chemical reactions that take place within that compartment (much the same as different saucepans are used to control separately the chemical reactions on different food types that go to make up a meal. So, for example, proteoglycans are glycosylated in the Golgi apparatus (https://www.ncbi.nlm.nih.gov/pmc/articles/PMC3062213/).

The 'permeability' of the cell membrane to selected solutes may be modified by the insertion of 'membrane proteins' (https://en.wikipedia.org/wiki/Membrane_protein), such as aquaporins (which facilitate the transport of water molecules in those tissues that require a higher transport capacity than provided by normal cell membrane, https://www.ncbi.nlm.nih.gov/pmc/articles/PMC3590904/), ions channels (which facilitate the transport of specific ions, e.g. sodium, potassium, calcium, chloride ion channels, https://www.ncbi.nlm.nih.gov/pmc/articles/PMC3935107/), and proteins that promote passage of certain molecules through 'facilitated diffusion' (e.g. molecules such as glucose, https://en.wikipedia.org/wiki/Glucose_transporter, and amino acids have specific facilitation transporters in the cell membrane, as do vitamins such as vitamin C, https://www.ncbi.nlm.nih.gov/pubmed/19391462). It is noted that molecular oxygen just diffuses through the cell membrane and has no special transporter, much as it diffuses through water. Lipid soluble hormones (e.g. the steroid hormones and vitamin D) also easily diffuse through cell membranes and then bind to 'receptors' in the cytosol or nucleus. A 'receptor' is usually a protein that receives a chemical message from a messenger molecule, which is generically referred to as a 'receptor ligand.'

Typical solutes both inside and outside the cell membrane are salts and organic molecules, including lipids and proteins. Because proteins are produced inside cells in order to regulate the chemical reactions taking place there, often the protein concentration within the cell is higher than the protein concentration outside the cell. The difference in protein concentration across the cell membrane results in an osmotic pressure difference across the cell membrane (often called oncotic pressure when the osmotic pressure is generated by proteins), which could result in cell lysis (i.e. bursting of the cell membrane due to movement of water into the cell). To prevent cell lysis, cells use metabolic energy to pump soluble salts (i.e. ions in solution) from inside the cell to outside the cell, until the total osmotic pressure on both sides of the cell is approximately the same. This ensures the cell will not lyze.

The pumping of ions across a cell membrane, together with differential permeability of the cell membrane to various ions, creates a 'membrane voltage' across the cell membrane. This adds a new functionality to cells. In the case of neurons and muscle cells, this membrane voltage is large (e.g. −100 mV, i.e. the inner surface of the cell membrane is negative w.r.t. the outside of the cell membrane in the resting state), which allows for electrical communication between cells via an 'action potential,' which is the name given to a rapid (cell) membrane depolarization followed by re-polarization. The membrane potential depends on the concentrations and membrane permeabilities of all the ions in solution (Fig. A.2). This membrane potential can be calculated using the so-called Goldman equation (http://www.physiologyweb.com/calculators/ghk_equation_calculator.html). An action potential is caused by a sequential time-varying change in the membrane permeability to various ions in solution.

It is clear that partitioning solutes and their chemical reactions by membranes into compartments is crucial to the functioning of cells as we know them. However, we also note that cells require metabolic energy in order to maintain themselves in a dynamic steady state.

Large amounts of solutes can be dissolved in water with surprisingly little change in the volume of the water. Molecular dynamics simulations of liquid water at room temperature and pressure show that water is a 'fluffy' material, with considerable open space between the water molecules (e.g. http://pubs.acs.org/doi/abs/10.1021/jp301299a, http://www.youtube.com/watch?v=B3cXuisH8PI). The amount of space between the water molecules depends on the balance between the attraction of water molecules for each other (e.g. by so-called hydrogen bonding between hydrogen and

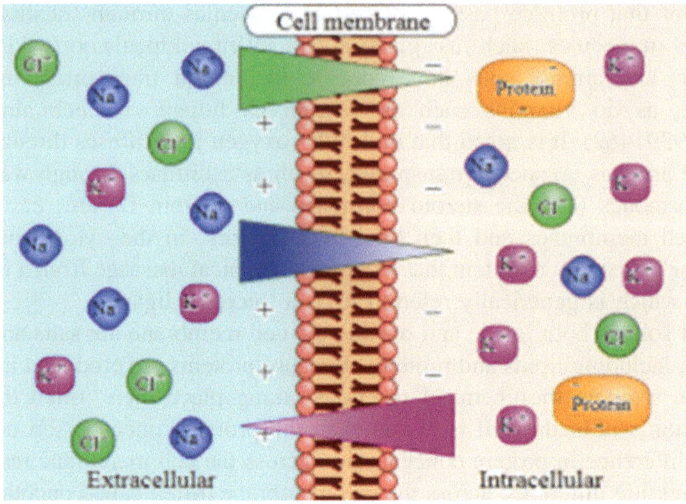

Fig. A.2 Illustration of the way that differences in ion concentration on opposite sides of a cell membrane produce a voltage difference (Wiki Commons). Only a minute fraction of particles needs to diffuse through the membrane to set up the voltage difference (i.e. a diffusion voltage), so the concentrations on either side of the membrane do not really change

oxygen atoms located on different water molecules), which draws the water molecules together, and thermal agitation of the water, which tends to push the water molecules apart. As the temperatures of the water increase, the thermal agitation increases and the water molecules are pushed apart with greater force making more space between the water molecules. Liquid water is 'fluffy' (there is considerable open space between the water molecules), and it becomes even 'fluffier' as temperature increases.

Thermal agitation gives the water molecules both 'kinetic energy' and 'position energy.' This energy is 'equi-partitioned' between molecular rotations, vibrations (which have positional and kinetic components), and translation motions of the water molecules (https://en.wikipedia.org/wiki/Equipartition_theorem). It is this thermally induced translation of solvent molecules and solutes in a solution that is responsible for concentration-dependent 'diffusion' of solutes through water.

Advection of a solution transports molecules long distances (e.g. along blood vessels), but diffusion is a crucially important transport mechanism at the length scale of cells (about 10 μm). We mention in passing that it can take decades (or hundreds of years) for a protein to diffuse an average distance of about one meter (i.e. mean square displacement from the start position is given by $6Dt$, where t is time), so advection or 'molecular motors' (such as kinesin) are required for transport at length scales longer than about 200 μm. Generally, the smaller the molecule, the larger its diffusion coefficient. For example, a water (molecular weight (MW) = 18) has a diffusion coefficient $D_{\text{water}} = 2.3 \times 10^{-5}$ cm^2 s^{-1} at 25 °C. This is a comparatively large diffusion coefficient. Nearly, all solutes in solution have a smaller diffusion coefficient than water (molecular oxygen in water has a diffusion efficient about the same value as water) and generally the larger the solute the smaller the diffusion coefficient. For example, compare glucose (MW = 180) which has a diffusion coefficient $D_{glucose} = 5.7 \times 10^{-7}$ cm^2 s^{-1}, with albumin (MW = 67 kDa) which has a diffusion coefficient $D_{albumin} = 6.1 \times 10^{-8}$ cm^2 s^{-1}. However, the temperature, shape, and charge on the molecule, as well as the presence of other molecules in solution, also can play significant roles in determining the actual diffusion coefficient for a particular molecule in a solution.

This same thermal agitation of the water molecules causing diffusion of solutes also promotes chemical reactions between solutes. According to 'collision theory' (https://en.wikipedia.org/wiki/Collision_theory), chemical reactions can only take place when molecules collide/interact with one another. The electron clouds of two or more solutes overlap temporarily during a collision, and this collision may provide sufficient 'activation energy' for reorganization of the electron cloud arrangements and so a chemical reaction takes place (or when several molecules interact/collide and one molecule is a catalyst, the catalyst reduces the activation energy for the chemical reaction and so promotes it without being changed itself). In cells, most collisions/interactions usually have insufficient kinetic/thermal energy or incorrect molecular orientation to promote a chemical reaction, but some collisions/interactions do have sufficient energy to overcome the activation energy. Clearly, these molecular collisions/interactions are of great importance because cells are fundamentally 'chemical machines' that rely upon chemical reactions to operate.

Chemical reactions may be divided into two types—reversible and irreversible chemical reactions. Irreversible reactions are a limiting case of reversible reactions, where the chemical equilibrium lies almost entirely toward reaction products (irreversible reactions typically proceed in one direction and have an equilibrium association constant (K_a) that is very large, e.g. greater than several hundred or more) (http://bip.cnrs-mrs.fr/bip10/stellen.htm). An example of irreversible cellular reactions is the initial hexokinase reaction involved in 'glycolysis' (in the first step of this pathway, glucose is transformed into glucose-6-phosphate by hexokinase, and the equilibrium constant for this reaction is about 250) (https://en.wikipedia.org/wiki/Glycolysis). The glycolytic pathway transforms glucose into two pyruvate molecules and produces two ATP molecules. Under anaerobic conditions, the pyruvate molecules are converted into lactate, which is enabled by the oxidation (oxidation reactions involve a loss of electrons) of NADH to NAD+.

'Aerobic glycolysis' involving the so-called Krebs (or citric acid) cycle is a series of chemical reactions that reduce (reduction reactions involve a gain of electrons) NAD+ to NADH (https://en.wikipedia.org/wiki/Citric_acid_cycle). This takes place in the cell cytosol. NADH then feeds into the oxidative phosphorylation (electron transport) pathway in mitochondria (which are redox reactions that end with oxygen as the electron acceptor). The energy released in the electron transport chain is used to pump hydrogen ions into a confined space between the two walls of mitochondria (https://www.ncbi.nlm.nih.gov/books/NBK21208/). This potential energy is then realized as mechanical work that is used to drive a rotary machine

(ATP synthase) that produces ATP from ADP (https://en.wikipedia.org/wiki/ATP_synthase). ATP is the 'energy currency' within the cell. Aerobic glycolysis produces about 36 or so ATP molecules per glucose molecule, but we note that chondrocytes in cartilage rely heavily on anaerobic glycolysis, which only produces two ATP molecules per glucose molecule.

In the case of reversible chemical reactions, these reactions do not run to completion. At chemical equilibrium, there is a mixture of both reactants and products. It turns out that the entropy change ($T\Delta S$) of the chemical reaction is 'balanced' by enthalpy change (ΔH) of the reaction, so the reaction does not run to completion, but reaches a dynamic equilibrium ($\Delta G = \Delta H - T\Delta S = 0$ at equilibrium). Most chemical reactions within cells are reversible. For example, polymerization reactions within a cell often reach equilibrium with a mixture of reactants and products remaining. Then (in accordance with Le Chatelier's principle, https://en.wikipedia.org/wiki/Le_Chatelier%27s_principle), the removal of mers from solution results in de-polymerization of the polymer, while the addition of mers to the solution results in additional polymerization, and so the formation of new polymer. An example is provided by microtubule polymerization and de-polymerization within a cell (http://en.wikipedia.org/wiki/Microtubule) (Fig. A.3).

Cells also have to make 'decisions' about what sets of chemical reactions are required in response to the cell's local circumstances. For example, should a cell secrete protein, extend a cell process by polymerizing microtubules, or should it divide and migrate? The 'decision apparatus' within cells is itself encoded through further sets of reversible chemical reactions. Usually, decision pathways are encoded in groups of proteins arranged in so-called signal transduction pathways

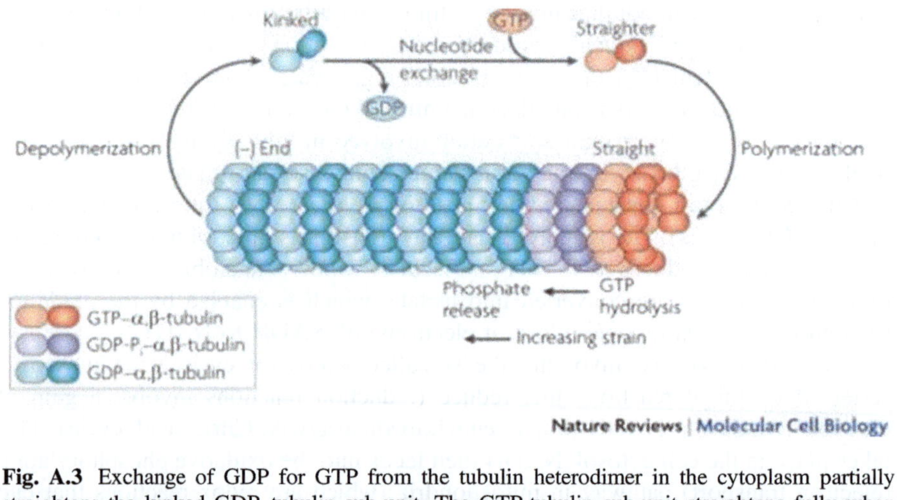

Fig. A.3 Exchange of GDP for GTP from the tubulin heterodimer in the cytoplasm partially straightens the kinked GDP–tubulin sub-unit. The GTP-tubulin sub-unit straightens fully when constrained in the wall of the microtubule after polymerization. Following hydrolysis, the GDP-tubulin dimer is even more strained, and this strain is the driving force for de-polymerization, thus completing the cycle (from growth, fluctuations, and switching at microtubule plus ends, http://www.nature.com/nrm/journal/v10/n8/fig_tab/nrm2713_F3.html)

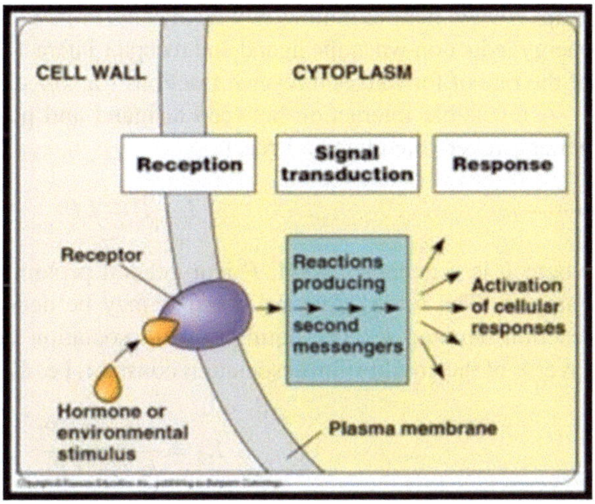

Fig. A.4 Signal transduction converts are stimulus into a response within a cell. http://simple.wikipedia.org/wiki/Signal_transduction

(Fig. A.4), or they are arranged in decision pathways encoded in groups of proteins called 'cell-cycle proteins').

A signal is transferred along a (chemical) decision pathway by one ligand binding reversibly (often with the help of catalysts) to the next molecule in the 'signaling pathway.' The signal transduction pathways are often long and convoluted. This length and convolution of the signaling pathway is a way of ensuring that all the relevant environmental information is taken into account by the chemical reactions in the signaling pathway involved in the decision process. To this end, many cells secrete signaling molecules, which then act on the cell itself (this is called 'autocrine signaling'). This contrasts to 'paracrine signaling,' where cells secrete signals that influence nearby cells. What is the purpose of autocrine and paracrine signaling? By this means, relevant environm ent conditions can be taken into account, as well as inputs from other signaling pathways within the cell, when making a decision about how the cell should respond. The cell can then make 'informed decisions'. Clever!.

Molecules conveying messages between cells are known collectively as 'cell–cell signaling molecules' [e.g. are paracrine (short-range) signals such as 'growth factors' (e.g. IGF, TGF-beta, BMPs) or (long-range) hormonal signals (e.g. thyroxine, growth hormone, and the steroidal hormones, which travel throughout the body)], while messages conveyed along a chemical pathway between the cell surface and the cell nucleus are called 'signal transduction molecules' (e.g. the molecules in the 'Wnt signaling pathway' or the 'Gli signaling pathway' to name two), and finally there is messaging within the cell nucleus (e.g. movement of transcription factors from cytoplasm into cell nucleus, regulation of transcription, splicing, and production of mRNA for export from the cell nucleus).

A receptor is a specialized protein that interacts with signaling ligands. It is the specificity of this interaction, often described as being like a 'lock and key' (or an 'induced lock and key'), that gives precision to relaying chemical messages (https://www.ncbi.nlm.nih.gov/pmc/articles/PMC5764188/). It is the interfacial energy reduction when the ligand and receptor interact that is usually a key determinant of the rate of forward and reverse reactions (or 'on–off' rate of ligand and receptor).

A reversible interaction between a ligand and protein may be represented by a general reversible chemical reaction,

$$L + P \rightleftarrows LP \tag{A.1}$$

where L is a general ligand, P is a general protein, and LP is the ligand–protein complex. The rate of forward reaction may be denoted k_{f} and the rate of reverse reaction denoted k_{r}. The equilibrium dissociation constant K_{d} (which equals the inverse of the equilibrium association constant, i.e. $K_{\mathrm{d}} = 1/K_{\mathrm{a}}$) is then defined to be:

$$K_{\mathrm{d}} = \frac{k_{\mathrm{r}}}{k_{\mathrm{f}}} = \frac{[L][P]}{[LP]} \tag{A.2}$$

If the ligand and protein bind strongly, the rate of forward reaction is relatively large and K_{d} is relatively small (e.g. a very small K_{d} = 0.1 pM), while if the ligand binds weakly the rate of reverse reaction is relatively large and K_{d} is large (e.g. a large K_{d} = 0.1 mM).

We note in passing that the binding coefficient can be translated into the Gibbs free energy (ΔG_{d}), which for the forward reaction to occur must be positive (Kastritis and Bonvin 2013):

$$\Delta G_{\mathrm{d}} = -RT \ln \frac{K_{\mathrm{d}}}{c_0} = \Delta H - T\Delta S_{\mathrm{d}} \tag{A.3}$$

where c_0 is the concentration of complex in the standard state (i.e. one molar) by convention, R is the universal gas constant (8.31 J K^{-1} mol^{-1} = 1.99 cal K^{-1} mol^{-1}), T is the absolute temperature (in Kelvin), ΔH is the change in enthalpy, and ΔS_{d} is the change in entropy. For example, if the K_{d} is one nanomole, then $\Delta G_{\mathrm{d}} =$ 54.4 kJ/mole (positive ΔG_{d} indicates the complexation reaction is spontaneous).

Now, let reactant A react with reactant x_1, which results in the complexation of A by x_1 (i.e. the formation of Ax_1). If the total amount of A_{T} is fixed in number, the fraction of the total available A_{T} that is complexed by x_1 is given by:

$$\frac{[Ax_1]}{[A_{\mathrm{T}}]} = \frac{x_1}{x_1 + K_{\mathrm{d1}}} \tag{A.4}$$

We now see that when $K_{\mathrm{d1}} = x_1$, the fraction of $[A_{\mathrm{T}}]$ that has reacted with $[x_1]$ is one half. This gives a readily understood physical meaning to K_{d}. So, for example, if decorin reacts with sites on collagen fibrils, and the $K_{\mathrm{d}} = 10^{-8}$ M, then half the collagen sites will be occupied with decorin when the concentration of decorin is 10^{-8} M.

However in most biological systems, reactant A has more than one possible reaction partner, i.e. there may be 'I' additional reactants, denoted x_i. In this case, we may wonder how a variety of chemical reactions pathways influence the fraction of $[A_T]$ bound by x_1. Clearly, competition for partnering with A will reduce the fraction of x_1 bound, but by how much? Taking into account all the competing reacting products, the total concentration of $[A_T]$ is given by

$$A_T = [A] + [Ax_1] + \sum_{i=2}^{N} Ax_i$$

Then, substituting equations like Eq. (A.2) in the RHS of the previous equation leads to:

$$\frac{[Ax_1]}{[A_T]} = \frac{x_1}{x_1 + K_{d1}\left(1 + \sum_{i=2}^{N} \frac{x_i}{K_{di}}\right)} \tag{A.5}$$

So, if there were two reactants competing for complexation with A, if both reactants had the same concentration equal to K_{d1}, and if $K_{d1}/K_{d2} = 10$, then the bound fraction Ax_1/A_T would be equal to 0.0825. In this example, the second reaction has had a significant effect on the bound fraction Ax_1/A_T, reducing it from 0.5 for the single reaction to 0.0825 in the presence of the second reaction. On the other hand, if x_1 and x_2 are reversed but the concentration remains at the same concentration equal to K_{d1}, then Ax_2/A_T reduces from a bound fraction 0.909 for a single reaction to 0.833 in the presence of the second reaction.

In words, this means if a molecule with strong binding affinity to a protein competes with molecules with weak binding affinity to a protein, and they have equal concentrations, this will have a big effect on bound fraction of weakly binding molecules, but if weak binding molecules compete with a molecule that has strong binding affinity, and they have the same concentrations, it causes only a small reduction in the bound fraction of the strongly binding molecule. Of course, the change in bound fraction is also modified by the relative concentrations according to Eq. (A.5).

It would be difficult to overemphasize the importance of Eq. (A.5), for it applies to all metabolic processes in the cell (e.g. distribution of oxygen, ATP), to all signaling pathways in the cell (e.g. competition for binding determines precisely what pathways are activated), to all reactions in the cell nucleus (competition for binding determines which genes are activated), and to all reactions in the extracellular matrix (e.g. collagen fibrils have approximately 50 extracellular reactants that may bind to surface sites, and these competing reactions regulate collagen fiber formation and so the collagen network morphology). In other words, this equation is operational everywhere all the time.

Since each of the variables in Eq. (A.5) is relatively difficult to measure in a real biological experimental system, particularly the concentration, and since not all competing reactions are known, most biologists do not really have any clear grasp of what is going on the cell (or ECM) in terms of Eq. (A.5), and so much if not most of the time, experimentalists are simply guessing what is happening in their system. Not being able to successfully apply Eq. (A.5) is a root cause of a great deal of uncertainty about the status of biological systems during experiments or clinical trials, and so uncertainty about the interpretation put on the experimental data. However, we mention in passing that this is rarely identified as an important source of uncertainty about the conclusions drawn from experiments.

As mentioned previously, while collision theory says that chemical reactions can only take place when molecules collide/interact, a collision/interaction does not necessarily mean a chemical reaction will proceed. This will only occur if the kinetic/thermal energy of the colliding/interacting molecules provides sufficient chemical potential energy to form an 'activated complex' and overcome the so-called activation energy required for the chemical reaction to proceed to completion. The change in number of molecular collisions with sufficient energy to induce a reaction leads to the often quoted approximation that biochemical reaction rates in many cells approximately double for each 10 °C increase in temperature (see http://en.wikipedia.org/wiki/Arrhenius_equation).

This basic chemical fact means that 'warm-blooded' mammals, with a body temperature of 37 °C, have a distinct advantage over 'cold-blooded' reptiles (e.g. they are usually speedier in cold weather), which have their body temperature and so metabolic reaction rates dictated by the environmental temperature. Even some pelagic fish (e.g. tuna) are described as being warm blooded (https://en.wikipedia.org/wiki/Pacific_bluefin_tuna). Of course, with the advantages come disadvantages; for example the increased rate of metabolism in warm-blooded animals and fish means they require a higher basal food intake to maintain their body temperature than do cold-blooded animals and fish.

In living cells, temperature is not the only important factor determining whether a chemical reaction occurs or not. Cells contain protein catalysts known as enzymes. Catalysts lower the activation energy for a chemical reaction (without being produced or consumed by the chemical reaction), thereby enabling *more* molecular collisions to result in the progression of a chemical reaction to completion (see, for example, http://chemwiki.ucdavis.edu/Physical_Chemistry/Kinetics/Complex_Reactions/Catalysis/Catalyst).

Fundamentally, it is the specificity of enzymatic proteins for their ligands that gives specificity to the chemical reactions taking place within the cell. These chemical reactions are regulated either by changing the amount of enzyme present or by changing the protein's binding constant for its ligand (e.g. usually by either positive or negative feedback via 'allosteric regulation' of enzymes, http://en.wikipedia.org/wiki/Allosteric_regulation, or by the addition or removal of some chemical group, e.g. a phosphate group).

Solutes and enzymes in solution are usually in relatively low concentrations, and so it is not immediately clear how molecules can actually find each other in a solution, especially given the vast number of intervening water molecules. For example, a liter of water at standard temperature and pressure contains about 55.5 mol of water molecules (55.5 mol of water is a huge number of molecules, i.e. about 3.3×10^{25} molecules). We can use a 'thought experiment' to convey how enormous the number 3.3×10^{25} actually is. Given that there is about 1.4 billion cubic kilometers of seawater in all the earth's oceans, this means that if one liter of labeled water molecules was stirred into and evenly distributed through the earth's oceans, at equilibrium, each liter of water scooped out of the sea would contain about 24,000 of the labeled water molecules!

Signaling molecules such as hormones may operate in vivo at very low concentrations, e.g. in the picomolar range. One picomolar concentration of signaling molecule means that there is about one signaling molecule per 3.3×10^{13} (33,000 billion) water molecules. This raises the question as to how small molecules (say less than a couple of hundred Daltons) can find their 'target' in solution. Again, the answer lies in the thermal agitation of water and solute molecules. The collisions between the water molecules in say 10^{-12} l of water (about the amount in an average 10 μm animal cell) at room temperature occur at such a phenomenal rate that essentially all of the small molecules in the cell get to 'meet' one another in a collision within a relatively short period of time. Does the same apply for large ligands and macromolecules? To give another example of the phenomenal rate of collisions and exchange of water molecules by diffusion across a boundary, it has been shown that the great majority of tritiated water injected into the central region of the vitreous of a rabbit eye (diameter greater than 1 cm) is removed from that eye by diffusion into choroidal veins within about 60 min (Moseley et al. 1984).

Water and solutes are said to move in so-called Brownian motion (http://en.wikipedia.org/wiki/Brownian_motion). Brownian motion is a collision induced random walk that involves frequent changes in direction of molecule. But large solutes move relatively slowly (i.e. diffusion coefficient is relatively small), and changes in direction are less abrupt than those of small solutes and water molecules. Consequently, the collision frequency between macromolecules in solution is considerably reduced compared to collision frequency between small molecules in solution. Some have estimated that the theoretical collision frequency for two protein macromolecules is too slow relative to that observed in practice [e.g. binding in the growth hormone receptor complex (Wells 1996)]. However, Brownian motion simulations show that when two proteins do meet there is 'diffusive entrapment,' which means that following the initial collision, the two proteins then collide repeatedly (due to the entrapment) with 'rotational re-orientation' during each encounter (Northrup and Erickson 1992). After taking diffusive entrapment into account, predicted and observed rates of collision between macromolecules can be brought into agreement. This 'Brownian molecular world' actually is very 'jittery' at the molecular scale. How jittery is conveyed visually by this animation of DNA binding to a protein (see http://www.wehi.edu.au/education/wehitv/restriction_enzyme_ecor1/).

In summary, it is apparent that thermal agitation of the molecules in the 'background matrix' of liquid water is fundamental to the operation of the chemical machine known as a cell. Thermal agitation of the molecules in solution drives diffusion within a cell, ensuring that all the molecules meet with one another in repeated collisions. The collisions promote chemical reactions within a cell, including the reversible chemical reactions concerned primarily with information exchange. It is this set of chemical reactions that collectively we refer to as life. Increase the temperature of the solution or provide catalytic proteins, and the number of molecular collisions with sufficient energy to result in a chemical reaction proceeding to conclusion is increased. This has given a speed advantage to warm-blooded animals, but comes with the disadvantage of increased energy (i.e. food) requirements.

Phospholipid molecules have a hydrophilic 'head' and hydrophobic 'tail,' and can associate to form membranes. Membranes are essentially containers that separate groups of solutes, thereby providing control over the types of chemical reactions taking place in each container. The container may be a cell. Cells with different chemical reactions are distinguished as characteristic of different tissues. The container may also be partitions within a cell, which distinguish organelles within the cell (e.g. cell nucleus, vesicles, Golgi apparatus, endoplasmic reticulum). Additional controls on the chemical taking place within the cell are provided by altering the concentration of enzyme in solution and by conformation changes in the enzymes active site (e.g. by allosteric regulation, or binding of some molecule to the protein, e.g. phosphorylation of the protein). By carefully controlling the set of chemical reactions taking place within cells enables the organism to respond to its external environment successfully.

The chemical reactions taking place within cells and in tissues all rely on proteins manufactured within the cell. The manufacture of proteins is ultimately regulated by the expression of genes. We consider the genetic complement within the cell nucleus next.

Genes, SNPs, and Reaction-Binding Constants

Human 'somatic' cells (i.e. all cells in the body other than 'gamete' cells) each have two strands of double helix DNA—one strand from each parent (https://geneed.nlm.nih.gov/topic_subtopic.php?tid=15&sid=16). Note that each strand of double helix contains two lengths of DNA wound in a spiral, so there are a total of four complete lengths of DNA in somatic cells. Because there are two strands of double helix DNA in each somatic cell, humans are said to have a diploid genome (the gamete cells are haploid, meaning they have only one strand of double helix in each gamete cell).

Each double helix strand has a total length of about 3 billion base pairs. Even though this number is large, the amount of matter involved is small. For example, one picogram is equivalent to about 978 million bases, so the DNA in human somatic cells has a mass of about 6 picograms.

The 3 billion base pairs are divided up into 23 parts called chromosomes. The sectioning of this very long strand of 3 billion base pairs into 23 parts presumably makes the DNA strand more manageable from a variety of viewpoints. But why it is sectioned into 23 parts of varying lengths is not clear. Two chromosomes are joined together at a structure called a centromere (remember somatic cells are diploid (one haploid gamete from each parent, so there are two strands of double helix)), so there are a total of 46 chromosomes and 23 pairs of chromosomes (http://ghr.nlm.nih.gov/handbook/illustrations/normalkaryotype).

Each of 22 sections of DNA double helix parts is called autosomes. An autosome is simply a chromosome that is not a sex-determining chromosome. Each pair of joined chromosomes contains equivalent sections of DNA (one from each parent), and for this reason the joined chromosomes are said to be 'homologous' chromosomes.

The remaining 23rd section of the 3 billion base pair strand is called an allosome (which is simply a name for the sex-determining chromosomes). Again, two allosomal chromosomes are also joined together by a centromere. Either two X chromosomes are joined to produce a female or X and Y chromosomes are joined to produce a male. We note that males have only one copy of the X chromosome and only one copy of the Y chromosome, and so males tend to be more susceptible than females to some genetic diseases; e.g. Duchenne muscular dystrophy, haemophilia, and Hunter syndrome are male genetic diseases (http://www.sciencedaily.com/releases/2006/03/060322184446.htm). The Y chromosome has about 59 million base pairs and 70 genes, while the X chromosome contains about 153 million base pairs and 800 genes (total number of protein-coding genes about 20,000–25,000).

On the other hand, females have two copies of X chromosome in each cell, which gives them some measure of safety against genetic defects on the X chromosome. It is noted though that this measure of safety is realized in a rather unexpected way (as is often the way in biology). For in each somatic cell, one X chromosome is randomly 'silenced' (by DNA methylation). This means females are a mosaic of cells, and so if there is a defect in an X chromosome, it will only affect half the cells in her body. This mosaic of cells gives some measure of protection against genetic disorders linked to the X chromosome.

In total, there are 46 chromosomes (http://ghr.nlm.nih.gov/handbook/illustrations/normalkaryotype). Chromosome 1 contains the largest section of DNA and so is the largest chromosome, while chromosome 22 contains the smallest section of DNA and so is the smallest chromosome. However, we note that the number of genes on a chromosome is not well correlated with the length of the DNA sections (and so size of the chromosome (http://en.wikipedia.org/wiki/Chromosome). Indeed, somewhat surprisingly the length of DNA in various organisms is not correlated with the number of genes in that organism or the apparent complexity of the organism (http://www.ncbi.nlm.nih.gov/pubmed/17295292). It turns out that there is a lot of so-called noncoding DNA in organisms.

If the DNA double helix were extended in a straight line, the total length of the 3 billion base pairs is about 1.8 m. This long string of base pairs has to fit into an average human cell with a diameter of about 10 μm, containing a cell nucleus diameter a fraction of that 10 μm. So, there is a 'packaging problem'—how to fit the two double helix strands of DNA into the cell nucleus?

In eukaryotic cells (which includes humans), the DNA double helix is wound around proteins called histones, much as thread is wound on a spool. Histones are positively charged (due to the presence of lysine, which has a low isoelectric point and so carries a positive charge at pH 7.4), while DNA is negatively charged at pH

7.4 (due to the low isoelectric point of phosphate in the phosphate deoxyribose backbone of the DNA molecule). Consequently, DNA and histone bind tightly because of electrostatic interactions.

When the DNA thread is wound on the histone spool, the structure is called a nucleosome. The nucleosomes then wrap into a 30-nm fiber called heterochromatins, which is then looped, coiled again, and then folded to form chromosomes (http://en.wikipedia.org/wiki/Chromatin). This packaging reduces the DNA length in the 46 chromosomes to about 90 μm (which is about 20–40 thousand times shorter than the DNA stretched out in a straight line). When packaged, the DNA fits easily inside the cell nucleus.

The histones on which the DNA is wrapped are not passive spools, say like cotton thread wrapped on a wooden spool. Rather, the histone spool can be modified by changing the affinity of the DNA double helix for the histone spool. This variable affinity facilitates the necessary packing or unpacking of the DNA double helix from the spool. Unpacking the DNA from the histone spool enables the transcriptional machinery to access the DNA strand. In turns out that controlling the affinity of the histones for the DNA provides one way for 'activation' or 'repression' of sections of the DNA to be controlled within the cell nucleus.

It has been found empirically that the affinity of the histone is controlled by the density of nucleosomes, local salt concentration, and temperature (http://www.ncbi.nlm.nih.gov/pubmed/11551187), as well as by attaching various functional groups to the amino acids comprising the histones (e.g. methylation (mono, di, tri-methylation), acetylation, phosphorylation, sumolyation, citrullination and ADP-ribosylation) (http://en.wikipedia.org/wiki/Histone). For example, acetylation of the lysine residues making up histone (by the enzyme histone acetyltransferase) reduces the positive charge on the lysine residues making up histone, providing an important mechanism that enables the DNA to unwind from the histone spool.

Interestingly, we note in passing that if super-tight packing of DNA is required, as occurs in the head of sperm, then histones are replaced by protamines (https://en.wikipedia.org/wiki/Protamine).

Somatic cells divide in a process called mitosis. In mitosis, the entire diploid DNA is first copied, and then using a structure called the 'mitotic spindle' (http://en.wikipedia.org/wiki/Spindle_apparatus), one complete set of diploid DNA is dragged into each daughter cell.

DNA is first copied in a so-called 'replication fork' by a molecule known as DNA polymerase (Fig. A.5). During the DNA copying process, a lot of copying errors occur. To minimize errors, the DNA copy is 'proofread' (which reduces the error rate 100–1000-fold), and then it is proofread again (reducing the error rate a further 100-fold) (http://www.ncbi.nlm.nih.gov/books/NBK21751/, http://jonlieffmd.com/blog/dna-proofreading-correcting-mutations-during-replication-cellullar-self-directed-engineering and http://dnareplication.cshl.edu/content/free/chapters/07_roberts.pdf). It is clearly very important that possible errors in DNA replication are minimized. Too many errors and the whole system will collapse.

DNA may also possess further mechanisms to reduce the number of mutations in the DNA of some cells. In this context, it is noted here that for so-called stem cells,

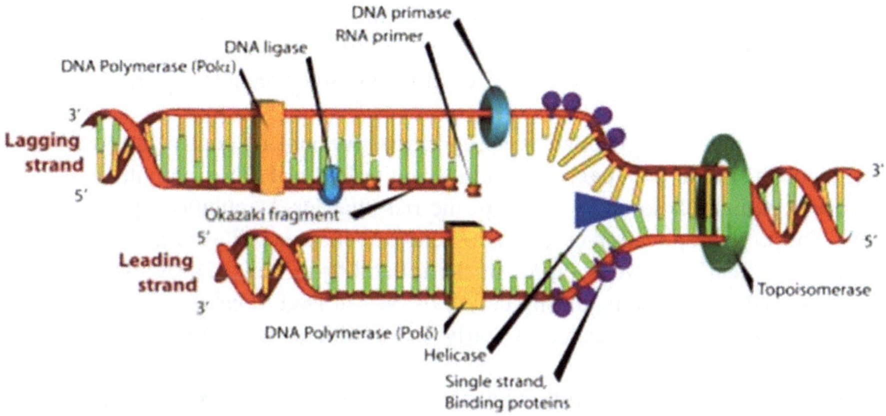

Fig. A.5 A depiction of a so-called replication fork, which involves various processes that produce two strands of double helix from a single initial strand of double helix. http://www.vanderbilt.edu/vicb/DiscoveriesArchives/dnareplication_ipond.html

the 'immortal strand' hypothesis postulates that the original (stem) cell keeps its original DNA, while the daughter cell obtains the newly synthesized DNA (along with any replication errors). The immortal strand hypothesis was formulated to provide a way to prevent transcription errors accumulating in the stem cell. Some evidence for the immortal strand hypothesis exists: see http://en.wikipedia.org/wiki/Immortal_DNA_strand_hypothesis.

Replication or not, DNA is continually damaged by exposure to high energy radiation, exposure to reactive oxygen species, and exposure to various other 'chemical toxins.' Interestingly, DNA damage is detected in situ by changes to the shape of the DNA double helix, which is repaired by a complex of proteins (http://en.wikipedia.org/wiki/DNA_repair). As noted earlier, the shape of molecules is very important in determining the functional behavior. Clearly, the cell makes a very large investment in ensuring cellular DNA damage and errors are minimized.

The so-called cell cycle is a process that results in a cell division known as mitosis (https://www2.le.ac.uk/projects/vgec/schoolsandcolleges/topics/cellcycle-mitosis-meiosis). A quiescent cell is said to be in G_0 phase of the cell cycle. If a cell is in the G_0 phase, it first has to 'decide' whether to stay in G_0 or enter the so-called G_1 phase (https://en.wikipedia.org/wiki/Cell_cycle). Cell decisions are made based on the state of the chemical system and the resources available to the cell for cell division. During the G_1 phase, there are so-called checkpoints (http://gleesonbiology.pbworks.com/w/page/7537859/C9), in which a cell 'assesses' transcription factor concentrations from both growth signaling pathways and apoptotic pathways. If a decision is reached to continue with cell division, then the cell enters the S phase. The cell's DNA is then replicated. The cell then enters the G_2 phase and another checkpoint. The cell has to be prepared for cell division (e.g. mitotic spindles form), and a decision has made that the cell cycle can be completed. If the 'all clear' is given, the cell then enters the M phase, and a 'mitotic spindle checkpoint' is then passed (confirming that all the chromosomes are aligned correctly ready for division). Finally, the cell divides. The daughter cells then have to make a decision as to whether they enter G_0 phase or reenter the G_1 phase ready for another cell division.

It is clear that the cell devotes an enormous amount of control information to 'checkpoints' in the 'cell cycle' and to the multiplicity of DNA copying and repair processes, which all serve to highlight the risk that DNA mutations present to the cell (http://dnareplication.cshl.edu/content/free/chapters/07_roberts.pdf). DNA mutation is clearly a 'weak point' in the reliability of cells. DNA mutations and chromosomal malfunctions during cell division can lead to cascading 'catastrophic failures' that we know as cancer. Clearly, extraordinary measures are taken by the cell to avoid this fate.

Gametes (i.e. egg and sperms cells) are formed in a different mitotic cell division process, known as meiosis. Egg and sperm cells are haploid, meaning they contain only one strand of double helix DNA. During meiosis, the two strands of double helix found in a normal somatic cell merge in a process called 'recombination' to form a single strand of DNA double helix in each cell (http://en.wikipedia.org/wiki/Genetic_recombination).

Many thousands of human genomes have now been completely sequenced, and this number is likely to expand exponentially with the advent of '$1000 genome sequence' in early 2014 (see http://www.illumina.com/systems/hiseq-x-sequencing-system.ilmn). A surprising early finding is that the 'coding DNA' in humans (i.e. that DNA containing genes that result in the production of proteins) is only about 1% of the total DNA. The remaining 99% of DNA is 'noncoding.' Suggested functions of noncoding DNA are regulation of the coding portions of DNA, controlling chromosome architecture (perhaps to facilitate expression of certain portions of the DNA) and epigenetic roles, but much remains to be discovered about the actual roles of noncoding DNA (much of the DNA may be 'junk').

Prior to sequencing of the first human genome (announced in April 2003), it was supposed that there is about 100,000 genes in humans. But sequencing the human genome showed there is a surprisingly small number—only about 20,000–25,000 genes. It has since become clear that many genes form several related proteins from the one gene (i.e. so-called exon processing can vary, so there are several splice variants of mRNA and so several variants of the protein). Of Course, sequencing the human genome, and cataloguing all the proteins and their variations, has given rise to database upon database upon database (example of databases are http://www.ncbi.nlm.nih.gov/ and http://www.uniprot.org/). Database formation is the modern equivalent of biological 'stamp-collecting' (http://en.wikiquote.org/wiki/Ernest_Rutherford) to borrow a put-down from Rutherford.

However, database formation is necessary to organize known information and then aid pattern seeking in this information (which gives rise to 'big data' bioinformatics). These databases are becoming very powerful (see, for example, 'GeneCards' from the Weizmann Institute of Science, http://www.genecards.org/).

The discipline of bioinformatics is concerned with identifying patterns by comparing data within and between databases, including databases of human genomes.

Potential differences in DNA patterns include frameshifts, single nucleotide polymorphisms (SNPs), multiple nucleotide polymorphisms (MSPs), indels (insertions and deletions in the DNA), microsatellites [or short sequence repeats (SSRs) or short tandem repeats (STRs)], and mutations. Gene differences are classified as mutations if they are rare (e.g. they occur with a frequency of less than 1%, but the actual cutoff varies with applications). All the other gene variations are deemed to be 'relatively common' (e.g. have a frequency of greater than 1–5% in the general population), and these variations are called 'common alleles.' The combination of alleles that a person has is known as their 'genotype.' There is great interest in a person's genotype.

Variations in DNA may lead to differences in phenotype (the organism's observable characteristics, http://en.wikipedia.org/wiki/Phenotype). Comparisons of human genomes have shown there is only about 0.1% of difference in the DNA between different humans http://www.genomenewsnetwork.org/resources/whats_a_genome/Chp4_1.shtml, and only approximately 4% difference between us and chimpanzees (the 4% difference made up of approximately 35 million single nucleotide differences and ~90 Mb of insertions and deletions, http://genome.cshlp.org/content/15/12/1746.long).

The variations in DNA between humans that lead to differences in body size and shape, skeletal size and shape, and skeletal tissue size and shape are all related to the specific gene variants within that person. The same applies to all tissues of the body (including the brain). Further, it is widely believed that the differences or variants in DNA lead to different susceptibility to disease—the so-called common disease–common variant hypothesis (http://en.wikipedia.org/wiki/Common_disease-common_variant).

We noted again that a common gene variation is called an 'allele' (http://en.wikipedia.org/wiki/Allele). If two alleles on homologous chromosomes are the same, then the person is said to be homozygous for that gene, while if the two alleles are different, the person is said to be heterozygous for that gene. An allele variant may cause a difference in phenotype (these variants are known as polymorphisms—meaning literally many shapes), but many genetic variations do not result in a change of phenotype.

If two alleles at the same gene locus are expressed differently, one allele is said to dominant, while the other is recessive. If the allele is recessive, then a person needs two copies of the allele (one from each parent) to express that (recessive) trait (i.e. an observable characteristic in the person). An example of a recessive allele is the polymorphisms for red hair http://en.wikipedia.org/wiki/Red_hair. We note that: 'A "dominant" allele is dominant to a particular allele of the same gene that can be inferred from the context, but it may be recessive to a third allele and codominant to a fourth [allele],' https://en.wikipedia.org/wiki/Dominance_(genetics).

Much early bioinformatics research effort has fallen on single nucleotide polymorphisms (SNPs), as these variants are by definition common and relatively easy to identify and measure (though undoubtedly other forms of DNA variation will be found to be important for explaining phenotypes and disease susceptibility). For example, STRs are prominent in many affective disorders (http://www.ncbi.nlm.nih.gov/pubmed/15744031 and https://www.ncbi.nlm.nih.gov/pmc/articles/PMC5054066/) and it now clear that defects in introns, including SNPs, may influence gene expression (http://www.genengnews.com/insight-and-intelligence/what-junk-dna-it-s-an-operating-system/77899872/).

SNPs are gene variants that usually have a frequency in the population of greater than 1–5% (http://compbio.pbworks.com/w/page/16252923/SNPs,%20CNVs,%20and%20SNP%20chip). In 2005, the number of SNPs was estimated to be 10 million, but by 2012 this estimate had grown to 45 million SNPs (Bryzgalov et al. 2013). Dividing 3 billion base pairs by 45 million SNPs provides an estimate of the SNP frequency: viz., about one SNP per 70 base pairs. Given that three base pairs specify an amino acid, this means there is a common variant in the amino acid sequence of a protein about every 20–30 amino acids proteins.

Because only 1% of the genome is protein-coding, only a small percentage of SNPs occurs in the 'coding' regions of the DNA (estimated to be about 1–3% of SNPs). Of this 'coding' subset of SNPs, 86% of the SNPs are actually (amino acid) coding SNPs (cSNPs—or more specifically non-synonymous cSNPs are known as nsSNPs—i.e. affect the amino acid sequence in expressed proteins), about 10% are splicing SNPs (i.e. sSNPs affect the intron–exon processing of the mRNA, which influences the final form of nearly all proteins), while the remaining SNPs are regulatory SNPs (i.e. rSNPs affect the promoter region of genes) (Bryzgalov et al. 2013) (http://www.hgmd.org/).

During meiosis, DNA in the two strands of double helix is swapped (or mixed) with one another, producing a new combination of genes that is different from the parent (see http://en.wikipedia.org/wiki/Genetic_recombination for details of the several mechanisms by which this occurs). The net result is two offspring from the same parents only share about 50% of their alleles. But the main point here is that during chromosomal recombination, blocks of DNA get swapped with each other, rather than small random pieces of DNA. Because blocks of DNA are swapped, all the genes in those blocks are swapped together. This leads to an association between genes known as 'linkage.' If blocks of DNA are swapped together, then the SNPs that occur in these blocks are also moved together, and so there is expected to be a linkage between SNPs. These linkages (or associations) are being sought in the so-called International HapMap Project, http://en.wikipedia.org/wiki/International_HapMap_Project. Perhaps unsurprisingly but usefully, it is found that if the genes are close together, linkage is more likely than if the genes are far apart (i.e. frequency of crossovers is related to the distance genes are apart on a chromosome).

How are some SNPs related to phenotype variation or to disease susceptibility? Proteins are macromolecules that have specific functions [e.g. that form signaling molecules (e.g. IGF), structural molecules (e.g. collagen), and molecules to protect against bacteria (e.g. defensins)]. To carry out their specific function, a molecule usually has identifiable domains that enable it to perform various functions. For example, an enzyme has an 'active site' where catalysis takes place, but it may also have regions on the protein that enable it to be located in a favorable orientation, say close to other enzymes with complementary functions. SNPs may be divided into two types—synonymous and non-synonymous SNPs. Synonymous SNPs have no effect on the amino acid sequencing in a protein, as the code for amino acids is redundant (i.e. more than one triplet of base pairs codes for the same amino acid), while non-synonymous SNPs do effect a change in the amino acid sequence of a protein. About 50% of SNPs are synonymous and 50% are non-synonymous.

Clearly, if a SNP occurs in the coding region of the active site of a protein, then it may change the amino acid sequence and alter protein function (http://genetics.bwh.harvard.edu/pph2/dokuwiki/overview). For example, when an amino acid is changed in the active site of the protein, it may change the binding affinity of the ligand for protein, or it may change catalytic function of the protein. If this occurs, the protein function may be either enhanced or degraded. As most active sites of protein catalysts are optimized by 'natural selection' for their catalytic function, most changes in amino acid sequences result in a degradation of function. The magnitude of the decrease in function depends on how closely related the functional groups on the two substituted amino acids are. For example, if an amino acid in the active site was a strongly acidic amino acid, and it is replaced by a neutral or basic amino acid, then it is likely that there would be a significant deleterious change in function of that enzyme. The similarity of amino acids is clearly of great interest, and so indexed by many different measures (https://en.wikipedia.org/wiki/Amino_acid_replacement).

The same arguments apply for ligand binding sites on the protein. If amino acids with different properties are substituted at the ligand binding site, this may have a significant effect on the binding constant of the ligand for the protein. Clearly, if the binding constant is either much larger or much smaller, it is likely that the signal from a signaling molecule will then be 'misunderstood' (at least in terms of the language of the chemical reactions transferring a message).

Clearly, changes in the binding constants of protein interactions in control networks would be expected to have a significant effect on the development of the organism. Indeed, it often does, and these differences are what we can see in various phenotypes. Of course, it is changes in the control networks that lead to differences in phenotype that is the raw material on which the evolution of organisms primarily operates. This explains why animals look different but are basically the same; e.g. they have similar body plan, similar metabolic pathways, similar organs, similar physiology.

It is to be expected that the functional effects of some SNPs will contribute to the incidence of genetic diseases, and many of these will be apparent by early adulthood. More often than not, the functional effects will be more subtle, and in this case disease may only become apparent later in life, following more prolonged exposure to a variety of environmental challenges (every day presents a new biological challenge, even if the person is unaware of this). Most of the important age-related diseases are believed to be 'polygenic,' meaning that the disease susceptibility is increased by the combined effect of many small gene changes (which on their own may have little effect on the risk of disease). Indeed, the changes in odds ratio between those with SNPs associated with disease and disease-free controls for many of the common age-related diseases (e.g. cardiovascular disease, cancer, and osteoarthritis) by themselves typically cause only very small changes in odds ratios. The most significant SNPs in polygenic disease might cause a 10% change in the odds ratio of realizing a disease compared to controls.

However, it is likely that when there is a large number of SNPs, all associated with the same disease, the odds ratio for disease may rise to be substantially more than a simple sum of the risks associated with each SNP individually. In other words, there is what may be termed a 'network effect' that increases the total risk more than expected from the sum of the individual risks. There is some evidence for this. For example, family history of a disease is a very significant risk factor clinically. This may be because the effect of lots of small differences in odds ratio is 'integrated' within a person and expressed as an increased likelihood for that disease. In the context of SNPs, the thought would be that groups of SNPs are linked into 'disease states,' causing groupings that are transferred between generations (http://en.wikipedia.org/wiki/International_HapMap_Project).

Rapid Change in Our Knowledge of and Treatments for Gene Function

The collective understanding we have of genetic networks and gene regulation is changing rapidly. Initially, the dogma was one gene → one mRNA → one protein (http://en.wikipedia.org/wiki/One_gene-one_enzyme_hypothesis). However in 1977, the Sharp and Roberts laboratories showed that genes were fragmented along the DNA double helix, with regions of noncoding DNA interspersed with the coding regions of DNA. They showed that when pre-mRNA was hybridized with DNA, regions of pre-mRNA looped out from the DNA, and these loops (later called 'introns') were removed in a process called splicing. Sharp and Roberts received the 1993 Nobel Prize in Physiology or Medicine for this work (https://www.nobelprize.org/nobel_prizes/medicine/laureates/1993/).

We now know that there is a 'spliceosome' that processes pre-mRNA (http://en.wikipedia.org/wiki/Spliceosome). The processing removes the introns, leaving the 'exons,' which are joined together to form mRNA, which is exported from the cell nucleus ready for transcription. It turns out that there are many ways the exons can be joined back together, meaning that there are many related mRNAs and many related proteins that can be produced from a single gene. Indeed, it is now known that most eukaryotic genes have at least several 'splice variants' (and sometimes thousands of splice variants).

In 1999, the Baulcombe laboratory discovered in plants small interfering RNA (siRNA) molecules (only about 24 base pairs long) that could bind to mRNA, thereby causing the mRNA to be broken down after transcription (so there can be no translation of the mRNA into protein) (http://en.wikipedia.org/wiki/Small_interfering_RNA). It is now known that this happens in eukaryotic cells too and that mRNA from any gene can be 'knocked down' (meaning rendered non-functional and degraded) (http://en.wikipedia.org/wiki/RNA_interference). Fire and Mello shared the 2006 Nobel Prize in Physiology and Medicine for their work on RNA interference (RNAi) (https://www.nobelprize.org/nobel_prizes/medicine/laureates/2006/).

siRNA is now used routinely to cause robust suppression of any gene of interest. It has been subsequently found that siRNA can also cause upregulation of some genes when it binds to the promoter region of the upregulated genes. In normal human cells, there are hundreds of micro-RNAs (miRNAs, https://en.wikipedia.org/wiki/MicroRNA), and their functions are only now becoming clear. What is also now apparent is that some miRNAs can block the transcription of multiple mRNAs simultaneously. By this means, whole signaling pathways can be effectively regulated (taken down or put up) within cells (http://www.ncbi.nlm.nih.gov/pubmed/20216554). For example, if several proteins in a signal transduction pathway are knocked out, then the pathway is rendered inoperable. Computational models of miRNA regulation of signal transduction pathways are being created (http://www.plosone.org/article/info%3Adoi%2F10.1371%2Fjournal.pone.0030140). It is likely that epithelial–mesenchymal cell transition (EMT) (https://en.wikipedia.org/wiki/Epithelial%E2%80%93mesenchymal_transition) (polarized epithelial cells change into migratory mesenchymal cells) involves micro-RNA expression (http://journals.sagepub.com/doi/full/10.1177/1010428317691682). The EMT transition is particularly important in metastasis of cancer.

It is interesting to observe in passing that when a cell wants to knock out a signaling pathway, it usually takes out multiple proteins in that pathway. The cell does not rely on a knockdown of a single protein, as do nearly all single purpose drugs, which may well explain why single purpose drugs are often not that effective in achieving their intended effect (http://www.nature.com/nbt/journal/v30/n10/full/nbt1012-906c.html).

It is generally recounted that Darwin and Lamarck held different views about inheritance of traits. Darwin argued that natural selection acts on heritable variations in fitness (http://plato.stanford.edu/entries/heredity/) (it was later discovered the variable heritable units are in fact genes, http://en.wikipedia.org/wiki/Mendelian_inheritance), while Lamarck argued that organisms can pass on to offspring features that they have acquired during their lifetime (http://en.wikipedia.org/wiki/Lamarckism). Until recently, it was held that Darwin was completely right and Lamarck completely wrong. A lesson that biology gradually teaches us is that there are no absolutes in biology—e.g. if there is potential merit/advantage in say two concepts/ideas, then evolution will find a way to exploit both to its advantage. And so, the field of 'epigenetics' has come to pass.

Epigenetics is the study of heritable changes in gene activity that are not caused by changes in the DNA sequence. Examples of mechanisms that produce such changes are DNA methylation and histone modification, each of which alters how genes are expressed without altering the underlying DNA sequence (see http://

commonfund.nih.gov/epigenomics/figure). At this point, we note that it has not yet been shown that histone modification (discussed above) can be inherited, but it has been shown that DNA methylation is inherited and DNA methylation alters gene function in the offspring (Heijmans et al. 2008).

Methylation is a biochemical reaction that adds a methyl group to adenosine and cytosine. The DNA methylation pattern generally changes (and increases) throughout a person's life (called de novo methylation). While a cell has a particular DNA methylation pattern divides, it generally stably expresses the same methylation pattern in daughter cells (i.e. called maintenance methylation). However, much of the methylation is removed during egg and sperm formation, though some of the methylation pattern remains. It is this remaining methylation that is inherited by the next generation. Indeed, DNA methylation is required for normal embryo development, being associated with gene imprinting (i.e. silencing of genes from either the mother or father) and X chromosome inactivation in females (as discussed previously). In cancers, hypomethylation is associated with loss of gene imprinting and hypermethylation is associated with transcriptional silencing.

The treatment of genetic disease has been a 'pipe dream' until quite recently. So-called gene therapy (e.g. involving the delivery of defect-free genes via viruses or by attempting other therapies including the development of 'artificial chromosomes') is moving closer to reality (see http://ghr.nlm.nih.gov/handbook/therapy/genetherapy and http://www.nature.com/gt/index.html).

But from about 2007, a series of papers have developed new 'gene-editing technologies' that are revolutionizing the field. (http://en.wikipedia.org/wiki/Genome_editing). Genome engineering is now evolving rapidly in new directions—one revolutionary example of gene editing is the so-called CRISPR-Cas9 technology (clustered regularly interspaced short palindromic repeats—CRISPR associated genes) (Baltimore et al. 2015, https://en.wikipedia.org/wiki/CRISPR).

This gene-editing system is a natural system in bacteria that is part of the acquired eukaryotic immune system protecting against phages (viruses) that invade the host bacterial DNA. This involves small fragments of viral DNA being incorporated in the CRISPR system of bacteria, which acts as a memory of previous viral invasions. When a new viral invasion occurs by a virus that has invaded previously, inserting itself in the bacterial DNA (as viruses do), the bacterial CRISPR system can identify and remove the newly invaded virus; i.e. the bacteria have a system to effectively remove abnormal viral DNA from its own DNA.

This natural system occurring in bacteria was slightly modified by researchers (e.g. the 'guide RNA' was joined to so-called 'tracr RNA') so that the system is now programmable (i.e. any short RNA sequence can be inserted into this programmable CRISPR system, which will in turn form an RNA template that is incorporated into Cas9—the DNA nuclease). This RNA template attached to Cas9 then directs precisely where the Cas9 nuclease will make the two-stranded cut in the target DNA—this relies on RNA–DNA base pairing recognition (between the guide RNA and the target DNA), which locates the target DNA site and cleaves both DNA strands, https://www.youtube.com/watch?v=SuAxDVBt7kQ). The new system can be applied to any DNA, meaning it can be applied to eukaryotic cells, not just bacterial cells from which it was developed.

Following the cut at a programmable, specific site in the genome by Cas9, the DNA is then repaired. DNA repair systems had been previously discovered (https://en.wikipedia.org/wiki/DNA_repair). Repair of double-stranded break is done either by 'non-homologous recombination' (which is error prone and leads to targeted 'gene silencing') or by 'homologous recombination' (which is not error pronc and leads to precise gene editing/repair). If donor DNA is available that has sequences the same as sequencing flanking the site of the DNA break (the similar sequencing are said to be 'homologous'), then the donor DNA is incorporated error free. This type of repair is therefore described as homologous directed, and the process is called homologous recombination.

The CRISPR system can be introduced into a prokaryotic (e.g. animal) system using lentiviruses. The technology is inexpensive, incredibly simple to use (allowing anyone with molecular biology training to modify animal genes), and incredibly powerful (allowing precise modification of specific genes, e.g. to correct genetic mutations in animals).

In 2015, calls were made to halt research of the technology on embryos until safety and ethics of the new technology could be worked out (http://www.nature.com/news/scientists-sound-alarm-over-dna-editing-of-human-embryos-1.17110). Society appears to be moving closer more quickly to a future portended in the movie Gattaca than many imagined possible (http://en.wikipedia.org/wiki/Gattaca). Genome editing of human embryos is pressing ahead, as might be inevitably expected (https://www.nature.com/news/take-stock-of-research-ethics-in-human-genome-editing-1.22632 and http://blogs.plos.org/thestudentblog/2017/08/18/molecular-ethics-crispr-genome-editing/).

Appendix B
Why Are Biological Systems so Complex?

Why are biological systems so complex? It is a common question. There seems to be no end to the complexity—the complexity of cell–cell signaling systems, interactions in the extracellular matrix, signal transduction pathways, transcription factors (about 10% of the genes in the genome code for transcription factors), gene isoforms, gene activation (with promoters, enhancers, silencers, etc. both local and remote), intron–exon processing, normally multiple proteins expressed by each gene, posttranslational RNA modifications, the fascinating details of epigenetics and 'interfering RNAs,' protein production, posttranslational modifications, protein folding, protein localization, protein function (with most proteins have more than one function), regulation of protein function(s), and decommissioning of protein. Why such mind-boggling complexity?

It is likely that such complexity is unavoidable for biological systems based on 'wet' chemical systems, for it is next to impossible to separate and so individually control all the possible chemical interactions. Imagine one wants to create a new functional capability in a living cell that involves creating a new protein. Imagine now creating this new protein and releasing it into the cell's cytoplasm, or releasing it to the cell's membrane or secreting it into the extracellular matrix. In each case, one would immediately find that in addition to the desired function, the new protein would interact with many other molecules in many other chemical systems, some interactions being quite unexpected.

Or alternatively rather than imaging a completely new protein being released, imagine an existing protein is 'somewhat changed' (e.g. one or more amino acids are substituted for other amino acids in the protein). Again, this protein would interact with most of its former molecules, but in somewhat different ways (i.e. the binding coefficients will be modified somewhat by the changes to the amino acid sequence, so changing the strength of chemical interactions and how long they are bound), but some interactions will be lost altogether and some new ones gained. Now, the chemical system will clearly 'shift' its behavior in response to this

D. W. Smith et al., *Articular Cartilage Dynamics*,
https://doi.org/10.1007/978-981-13-1474-2

change. This astounding complexity and the evolving nature of the chemical system as molecules (and control systems) are modified are the key driver of evolution.

The majority of unexpected new interactions would likely be deleterious to a functional existing chemical system, because a random change to an ordered system is very rarely more advantageously ordered and so more useful. But rarely a new chemical interaction may simultaneously have minor deleterious or no change in interactions, and it may be a chance to enhance the function of existing chemical systems (or very rarely, enable a new function altogether). Such a successful modification is more likely if the change is very small. Such rare beneficial interactions would be retained, as it would enhance the competitive advantage of the organism. This advantage would favor further evolution of that organism, building on this advantageous 'innovation.' Small changes accumulate. Over many, many generations, divergence between two or more organisms becomes more and more obvious, and eventually phenotype change is obvious to even the 'untrained eye'. We mention in passing that humans are naturally very good detectors of even small phenotypic changes in people, even if they cannot relate this phenotypic change to the causative change or changes in DNA (e.g. collections of small phenotypic changes are often identified by physicians as 'syndromes'). This provides a molecular perspective on evolution of organisms (which may or may not lead to external observable changes in the organism's phenotype).

Carroll (2008) says that he has presented the case for the genetic theory of morphological evolution of organisms that can be condensed into two statements: '*(1) form evolves largely by altering the expression of functionally conserved proteins, and (2) such changes largely occur by altering the expression of functionally conserved proteins and that such changes occur through mutation in the cis-regulatory sequences of pleiotropic developmental regulatory loci and of the target genes within the vast networks they control,*' commenting '*It is important to underscore that this theory is a statement about tendencies not absolutes, asserting that most, not all, pertinent mutations are regulatory*' (Carroll 2008).

So functional molecular evolution (i.e. developing new protein functions via many small changes) is very difficult when the organism has been making changes for billions of years and so most of the favorable possibilities have been explored already. Morphological evolution, which is driven by small changes in regulatory molecules (which governs the amount and timing of gene expression), may progress at comparatively much faster rates. Indeed, the theory of punctuated equilibrium holds that most morphological evolution occurs over relatively short timescales (relative to geological time), when there are significant changes in the earth's environment. Carrol (2008) is saying these evolutionary changes are mainly driven by changes in the control/regulatory systems, rather than by changes to the underlying chemical system itself (where change usually results in a decrease in functional fitness).

Importantly, we note that when an advantageous functional capability for an altered molecule is achieved, given the complexity of the chemical system in which it operates, it is almost inevitable that one (or more) additional unexpected functions for this altered molecule will also appear. This helps to explain why unexpected roles 'emerge' for molecules, and so why molecules can have completely unrelated functional roles in a cell. For example, beta-catenin acts in a structural role at cell membrane junctional complexes. But beta-catenin also acts in a signal transduction/information-carrying role, facilitating transcription within the cell nucleus by binding to the TCF-LEF proteins, which then bind to the promoter region of various Wnt target molecules. So, beta-catenin has both a structural role and a signaling role. In contrast, for most engineering applications different functions have separate systems using functionally specific materials.

Given this, the wonder is not that molecules have multiple functions (most molecules have several functions), but that they have relatively few functions. That proteins have relatively few functions points to the difficulty of finding new proteins that can meet the demanding criteria that the modifications are not deleterious to one or more systems. As the complexity grows, it becomes more and more difficult to meet the 'no deleterious interactions' requirement as new functions are added and the complexity of the chemical systems in the cell increases. This difficulty pushes evolutionary change away from modifying chemical systems, and toward modifications of the control of these functional systems. This explains why morphologically different animals are so similar at the cellular level, all running with basically the same functional chemical systems.

It is this 'evolutionary bind' in the further development of functional systems that drives partitioning of cells into compartments, for compartmentalization of chemical systems by definition separates them. From this perspective, we see that membrane compartments within cells allow specialized reactions to take place within that compartment that could not take place otherwise. For if other chemical systems were included, their interactions would effectively prevent the specialized reactions taking place in that compartment. One major division is between prokaryotic cells (without a nuclear membrane and no intracellular organelles) and eukaryotic cells (with a nuclear membrane and with many different intracellular organelles). For this reason, within so-called eukaryotic cells there is the nuclear membrane, the cell membrane, endoplasmic reticulum, mitochondria (or chloroplasts), Golgi apparatus, lysosomes, and many other different vesicles. And for multicellular organisms, additional compartmentalization of different sets of chemical systems occurs in what we choose to refer to as 'cell types.' An epithelial cell, a cardiac cell, a neural cell all have functionally specific systems of chemical reactions that facilitate their specialized function.

Classifying how many cell types there are in the body is a fraught process, as there are many functional differences between cells that have been given the same generic name (e.g. within the lymphocyte system, there are many cell types (e.g. T cells, B cells, and dendritic cells. Then, there are many different types of T cells, B cells, dendritic cells, etc.). In fact, no two cells in an animal are completely alike, even when in the same organ. However, with our current naming conventions there are at least 400 identifiable cell types (although each cell has a unique gene expression profile). Each of these 400 cell types contains their own sets of specialized chemical systems, which give them functional specificity. Indeed, when we say a cell has 'differentiated,' we mean that a significantly different set of chemical reactions is now taking place inside the cell relative to those taking place in the cell's initial state, and these changed chemical reactions confer functional specialization on that cell.

In addition to membrane compartments (which present physical barriers to mixing), partial (or effective) compartmentalization may be achieved using a variety of additional devices. One way is using so-called scaffold proteins. Several proteins bind to scaffold proteins, and by doing so, they change the probability of certain chemical interactions taking place. By means of scaffold proteins, deleterious interactions may be minimized (as the molecules are sequestered to a particular region of the cell compartment), while the desired reactions are more probable, as the molecules to react are in close proximity to one another. One can easily imagine how several scaffold proteins may interact to further enhance the probability of the desirable chemical reactions (e.g. the formation of the so-called destruction complex (between axin, APC, beta-catenin, GSK3beta, and associated kinases) found in the Wnt signaling pathway).

From this perspective, one can also see that a second way of compartmentalization may be achieved by having the chemical system in a linear tube network. Such an extensive tube network exists in the cell and is called the 'endoplasmic reticulum.' The endoplasmic reticulum tube can increase the probability of certain reactions occurring (while minimizing the probability of deleterious reactions) by confining a chemical system to certain regions of the tube, even if the tube is 'open ended.' In other words, mass transport requirements in a confined space can 'rate limit' unwanted chemical reactions while promoting others.

Indeed, the endoplasmic reticulum tube can be viewed as a process assembly line for chemical reactions, which operates much like the more familiar mechanical assembly line (e.g. for radios or cars). So-called rough endoplasmic reticulum ('rough' describing the presence of protein manufacturing ribosomes attached to tube wall) represents locations where new proteins are added to the evolving chemical system within the tube. Junctions of endoplasmic reticular tubes represent locations where the chemical systems within each tube are mixed to create more complex chemical environments. In other words, a system of tubes offers control over what chemical reactions are taking place in each part of the tubular network.

Endoplasmic reticulum runs into sheetlike spaces, known as Golgi apparatus, which provides new spaces within which to regulate chemical reactions by mass transport rates. From Golgi, the chemical system may be packaged in a vesicle, which provides a new environment for a new set of chemical reactions. Vesicles may even undergo retrograde transport and return to and fuse with the Golgi apparatus, thereby repeating a cycle of chemical environments in order to achieve the final desired reaction product mix.

Appendix C
The Formation of Diarthrodial Joints and Articular Cartilage

For the discussion on articular cartilage, it proves useful to have some understanding of how diarthrodial joints form. Luyten et al describe a model for joint formation (see schematic Fig. C.1, Luyten et al. 2009).

A cartilage template for a bone is first established. This cartilage template is known as the anlagen. The cartilage template develops first by 'appositional growth' (i.e. by secretion of ECM by perichondrial cells located on the surface of the cartilage template, which is supported by 'interstitial growth' within the thickness of perichondrium to replace those chondrocytes left behind in the ECM as the cartilage template grows. Second, the cartilage template grows internally by further 'interstitial growth' (i.e. by proliferation of chondrocytes left behind in the ECM, followed by secretion of ECM, which further expands the cartilage template). The great bulk of cartilage growth is interstitial growth, unlike bone, where bone is formed through appositional growth. Bone growth by apposition is necessary because interstitial growth is not possible in bone due to mineralization of the osteoid.

For endochondral ossification of long bones, the primary ossification center forms in the 'diaphysis' (i.e. at about the middle of the future long bone). A secondary ossification center forms at about the middle of each 'epiphysis' (epiphyses are found at each end of the long bone). The so-called growth plate forms between these two ossification centers. It is the growth plate that enables continued growth in length of the long bone throughout adolescence, driven by interstitial growth of the ECM and proliferating chondrocytes in the growth plate, which mineralizes to form bone. At some locations in the cartilage template, which is continuous and uninterrupted in the limbs (Decker et al. 2014), diarthrodial joints form (which enable mobility between the segments of long bones: see Fig. C.1). For example, the glenohumeral (shoulder) joint forms from a '*chondrogenic mass... [that] resolves into the humerous and scapula*' (Hita-Contreras et al. 2018). A similar thing happens in knee formation (Decker et al. 2014).

D. W. Smith et al., *Articular Cartilage Dynamics*,
https://doi.org/10.1007/978-981-13-1474-2

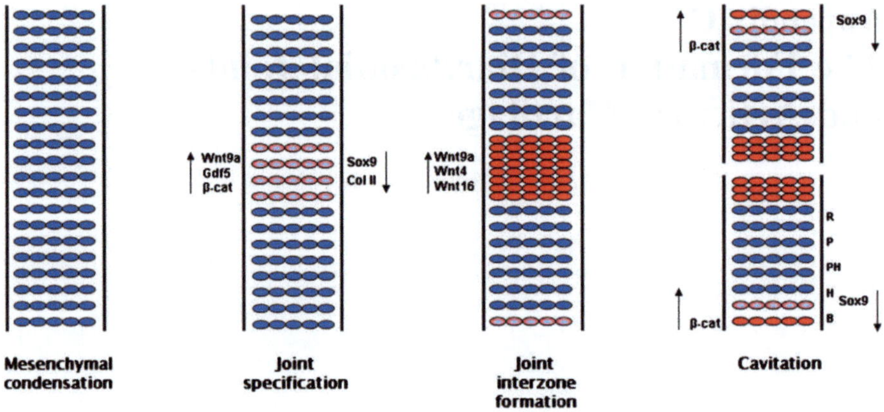

Fig. C.1 A model of joint formation. A mesenchymal condensation [cartilage template] expresses SOX9 (blue ovals). Upon Wnt9a signaling, transcription of β-catenin and Gdf5 is induced while SOX9 and collagen II repressed (red/blue ovals). Next, cells in the joint interzone cease expressing SOX9 and maintain the expression of β-catenin (red ovals). Finally, the cavitation process takes place with physical separation of the cartilaginous element. Concomitant with that process, SOX9 is repressed and β-catenin induced in hypotrophic chondrocytes: arrows up—induction, arrows down—repression, blue ovals—SOX9 positive cells, blue/red ovals—SOX9 and β-catenin positive cells, red ovals—β-catenin positive cells. R, P, PH, and H indicate resting, proliferating, pre-hypotrophic, and hypotrophic chondrocytes, respectively, and B indicates bone (Luyten et al. 2009). See also discussion of cell-tracing experiments (Iwamoto et al. 2013). Note that miR-140 micro-RNA appears to enhance SOX9 dimerization and so helps promote the normal chondrocyte phenotype (Yamashita et al. 2012)

The articular cartilage itself forms from a thin layer of 'remnant cartilage' adjoining the (diarthrodial) joint cavity. These cells are exposed to Wnts (Wnt 9a, Wnt 4, and Wnt 16), GDF5 (we note that loss of GDF5 results in joint fusion), erg (which possesses an anti-maturation activity on chondrocytes), PTH-LH during joint formation (Iwamoto et al. 2013). Rather than being a remnant cartilage, some suggest that 'interzone cells' located at the site of future joint formation (and invade the anlagen from surrounding tissues), which are exposed to the abovementioned growth factors, are the progenitors of the joint tissues (i.e. progenitors of articular cartilage, synovium, and ligaments) (Decker et al. 2014). The future articular cartilage avoids ossification, presumably by maintaining high PTHrP concentrations throughout the cartilage, suppressing Ihh (Koyama et al. 2008; Chen et al. 2008). The expression of noggin to block BMP (2 and 4) signaling in the future articular cartilage is also held to be critically important (Ray et al. 2015). Ray et al. (2015) says:

> We show that nascent cartilage cells are capable of differentiating as articular or transient [bone forming] cartilage, depending on exposure to Wnt or BMP signaling, respectively. The spatial organization of the articular cartilage results from a band of *Nog*-expressing cells, which insulates these proliferating chondrocytes from BMP signaling and allows them to differentiate as articular cartilage under the influence of Wnt signaling emanating from the interzone.

Hopefully, PTHrP concentrations remain sufficiently elevated and the cartilage remains unossified throughout life, as chondrocytes maintain their capacity to progress to a hypertrophic state (Iwamoto et al. 2013; Jiang et al. 2008). In adult cartilage, rather than noggin suppressing BMP2 and 4, gremlin and follistatin probably do so (Tardif et al. 2004). For example, Leijten et al. (2012) reports: '*Taken together, our study identified Gremlin 1, FRP, and Dkk-1 as natural brakes on hypertrophic differentiation in articular cartilage.*' We note in passing that BMP2 and 4 induce chondrocyte hypertrophy, while BMP7 blocks chondrocyte hypertrophy (Zhong et al. 2015).

The SOX9 transcription factor is characteristic of cartilage tissue and so unsurprisingly is highly expressed in the cartilage template (Takahashi et al. 1998). However, at the location of the future joint space SOX9 expression decreases, while β-catenin concentration increases. The joint cavity actually forms where Wnt expression and β-catenin levels are highest in the cartilage template (Fig. C.1). We also note that normal formation of joint also requires muscle activity across the joint during development (Schwartz et al. 2013), so the mechanics of the joint play an important role in defining the joint geometry and cell differentiation.

In addition to elevated levels of Ihh and BMPs 2 and 4, some Wnts also help to form bone from the cartilage template (Dao et al. 2012), by driving the 'pre-hypertrophic cartilage cells' to terminally differentiate into 'hypertrophic cartilage cells' (which then undergo apoptosis) Andrade et al. (2007). Following apoptosis of the chondrocytes, osteoblasts secrete osteoid, which then mineralizes to form bone.

At the growth plate, Indian hedgehog (Ihh) and parathyroid hormone-related peptide (PTHrP) together regulate the flux of cells through the growth plate itself (Sandell and Adler 1999; Kronenberg 2003; Schipani and Provot 2003). This feedback loop and associated signaling molecules are schematically illustrated by Sandell and Adler (1999). We note the feedback signals are strengthened by the presence of sulfated chondroitin sulfate (Cortes et al. 2009).

See Fig. 5 (Sandell and Adler 1999): *Embryological Development of Cartilage. IGF-1 = insulin-like growth factor 1, PTHrp = parathyroid hormone-related protein, Col IIA = collagen IIA, Col IIB = collagen IIB, BMPr = bone morphogenetic protein receptor, FGFr = fibroblast growth factor receptor, Ihh = Indian hedgehog, BMP = bone morphogenetic protein (Sandell and Adler 1999)* (https://www.bioscience.org/1999/v4/d/sandell/fulltext.htm).

See also an actual photomicrograph of growth plate zonation Fig. C.2.

These signaling systems are in operation prior to formation of the secondary ossification center. When the secondary ossification center forms, it actually splits the Ihh-PTHrP signaling system (Chen et al. 2008), forming two signaling Ihh-PTHrP systems (operating on either side of the secondary ossification center). This developmental sequence is helpful for understanding how Ihh-PTHrP signaling becomes established in articular cartilage (see discussion Ihh-PTHrP signaling).

We note that Ihh is produced by cells in the pre-hypertrophic region, while postnatally PTHrP is produced by cells in the resting zone (see Fig. C.2) (Chau et al. 2011), which shows the four stages of chondrocyte differentiation: from cells

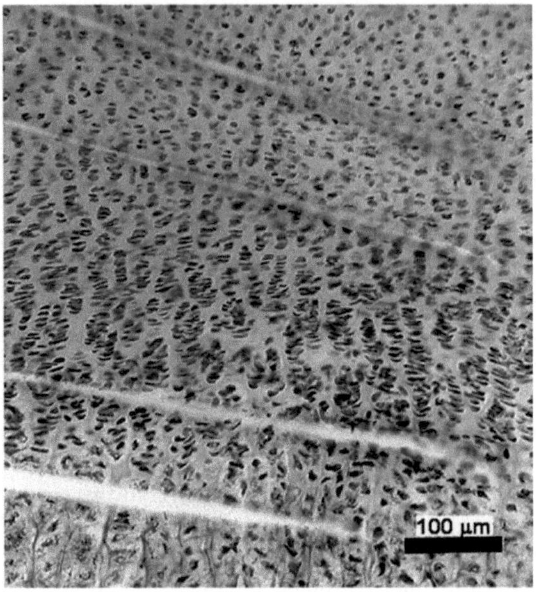

Fig. C.2 Zonation of the growth plate; 1-week-old mouse (Andrade et al. 2007)

in the resting zone through cells in the hypertrophic zone. It is of interest to note here that while chondrocytes are described generically as chondrocytes, they in fact exist in various states of differentiation. This proves relevant to our understanding of how articular cartilage functions, as the chondrocytes in articular cartilage alter their functional behaviors depending on their state of differentiation (or dedifferentiation).

Some of the key signaling molecules secreted by cells in each zone of cartilage adjoining the growth plate are shown in Fig. 5 (Sandell and Adler 1999) and Fig. 4 (Mackie et al. 2011). This also proves relevant to our understanding of articular cartilage dynamics, as the signaling systems found within the cartilage next to the growth plate (upper two panels of Fig. 4, Mackie et al. 2011) are the same signaling systems found to be operating within articular cartilage.

See also Fig. 4 (Mackie et al. 2011): *Overview of the effects of factors secreted by chondrocytes on growth plate cell function and invasion. Under the control of circulating hormones, chondrocytes secrete growth factors that act on chondrocytes to regulate their proliferation and hypertrophy (upper two panels) and on cells of the ossification front (lower panel), to regulate their invasion of the growth plate cartilage. Arrows indicate stimulatory pathways, and crossed lines indicate inhibitory pathways* (Mackie et al. 2011).

As the animal grows, the growth plate needs to expand laterally in proportion to the growing animal, as does the articular cartilage covering the bone of diarthrodial joints. To aid this growth, a population of progenitor cells is located at the junction of the growth plate and the articular cartilage (see Fig. 7, Karlsson et al. 2009). This population of cells 'feeds' newly formed cells into the growth plate promoting lateral expansion of the growth plate, and these cells also feed into the articular cartilage so it can expand laterally and cover the growing epiphyseal bone (Karlsson et al. 2009). It is this population of progenitor cells (in the groove of Ranvier: Fig. 7, Karlsson et al. 2009) that is (re)-activated in some disease conditions. For example, re-activation of this progenitor population at the cartilage–periosteal synovial membrane interface leads to the formation of osteophytes in some types of osteoarthritis (van der Kraan and van den Berg 2007).

See Fig. 7 (Karlsson et al. 2009): *Schematic overview of the localization of the BrdU-positive cells in the knee. (A) Perichondrial groove of Ranvier, (B) growth plate, (C) articular cartilage. Label-retaining cells were localized in the perichondrial groove of Ranvier as well as the articular cartilage. Of special interest are the cells in the perichondrial groove of Ranvier, in which label-retaining cells were detected as well as cells positive for progenitor markers. Cells above the perichondrial groove of Ranvier further proved to be a distinct population of cells compared to the surrounding tissue and are thus likely transient amplifying cells that later will mature into terminally differentiated cells.*

Karlsson et al. (2009) also say that they propose that it is these progenitor cells that have the ability to migrate [arrows shown diverging from (A)]; however, we argue that while migration occurs, it does so only over distances of the thickness of the cartilage, rather than centimeters as perhaps indicated. It is most likely the cells from the groove of Ranvier supply chondrocytes to enable lateral expansion of the growth plate and the lateral expansion of the articular cartilage covering the end of bones that grow by apposition of new bone.

Understanding joint development is helpful in understanding adult articular tissue homeostasis because most of the signaling systems that operate in the growth plate apparently continue to operate in articular cartilage, albeit in a modified way. For example, in articular cartilage the Ihh-PTHrP signaling system continues to operate, but in this case the system is modified so that mechanical loading at the cartilage surface becomes the primary driver of PTHrP production by chondrocytes within the transitional zone of articular cartilage—not IHH production by the pre-hypertrophic chondrocytes (see further discussion: Ihh-PTHrP signaling). The IGF and Wnt signaling systems remain key regulatory processes in articular cartilage (see discussion: extracellular signaling systems in cartilage).

Further, understanding the developmental process is relevant to understanding disease processes in articular cartilage. For example, many developmental signaling systems may be re-activated by the disease process (e.g. re-activation of BMP signaling system in osteoarthritis, or the re-activation of the progenitor cell population in the groove of Ranvier leading to the development of osteophytes in osteoarthritis).

At this point, it is interesting to note that following his success in developing a bone 'mechanostat theory,' Harold Frost formulated a less well-known theory that might explain joint congruence (Frost 1979). The theory Frost proposed said simply this: '*a growing joint surface tend[s] toward that shape which eliminates incongruous fit under the patterns of load and motion imposed by neuromotor anatomy and function*.' By way of general mechanism, Frost assumed that the cartilage growth would cease in regions bearing excessive load and increase in adjacent areas to increase bearing area (Hamrick 1999). While this appears to be perfectly reasonable proposition, the mechanism by which this may be achieved was not specified.

A couple of authors have subsequently proposed hydrostatic pressure as the variable regulating tissue growth. For example, Hamrick (1999) proposed that cartilage growth was zero at zero MPa, reaches a maximum at about 3.0 MPa hydrostatic pressure, and ceases at 10 MPa. Plochocki et al. (2009) attempted to optimize congruence between opposing cartilages based on maximizing hydrostatic pressure in the cartilage tissue (Hamrick 1999; Plochocki et al. 2009).

It seems most likely that joint growth and remodeling for (coarse) joint congruence involve both bone and cartilage tissue, and in light of Fig. 2.9, that growth and remodeling of the cartilage are a function not only of hydrostatic pressure but also of fluid flow and tissue strains. It is very likely that cartilage thickness is determined during this growth and remodeling phase and that cartilage thickness depends on the applied loads up to sexual maturity. It is likely that fine congruence adjustments are made by the surface loading shaping the collagen network within each of the cartilage zones based on strain distributions throughout the tissue (Paetzold et al. 2012) (see further discussion in Sect. 3.5).

Importantly though, time-weighted hydrostatic pressure in the cartilage tissue is maximized by improvements in joint congruence, which has the effect of minimizing the escape of pore fluid through the cartilage surface during loading. Minimizing the escape of pore fluid from the cartilage tissue maximizes hydrostatic pressure in the cartilage tissue, reduces tissue strains, and minimizes surface wear by maximizing the fraction of hydrodynamic lubrication in the contact interface (see discussion in Chap. 4).

Appendix D
The Compressive Stiffness of Glycosaminoglycans

Chondroitin sulfate is the main sulfated glycosaminoglycan (sGAG) of aggrecan, and so when discussing stiffness, aggrecan is at times used interchangeably with sGAG. We also note that the compressive resistance of sGAG is related to the equilibrium stiffness of cartilage. The equilibrium stiffness of cartilage is associated with zero excess pore water pressures in the cartilage (i.e. at equilibrium, all the compressive loads are being carried by the solid phase in cartilage). We note that the dynamic stiffness of cartilage can be considerably larger than the equilibrium stiffness, as during rapid loading part of the load is carried by the pore fluid (which at physiological loadings is well approximated as incompressible, so cartilage tissue appears transiently to be stiffer).

The equilibrium compressive resistance of sulfated glycosaminoglycans has been estimated experimentally and theoretically. For example, it was found that the osmotic pressure generated by chondroitin sulfate of 16 disaccharide repeats in length (7.3 kDa) in isotonic saline (0.15 M) could be approximated using a virial equation (Bathe et al. 2005):

$$\prod = RT(\alpha_1 c + \alpha_2 c^2 + \alpha_3 c^3) \tag{D.1}$$

where R is the universal gas constant (8.3e3 Pa L/K mol), T is the absolute (e.g. 300 K) and (for polymers of 16 disaccharides and MW 7.3 kDa) $\alpha_1 = 1.4\text{e-}4\,(\text{mol/g})$, $\alpha_2 = 4.4\text{e-}6(\text{mol}\,\text{L/g}^2)$, $\alpha_3 = 5.6\text{e-}8(\text{mol}\,\text{L}^2/\text{g}^3)$, and c is the concentration of aggrecan in (g/L or) mg/ml. It is noted that the second and third terms capture 'non-ideal' behavior of the polymer solution (Bathe et al. 2005). We have chosen the parameters for 16 disaccharides because Bathe et al. (2005) note that '*the intrinsic backbone stiffness characteristic of polysaccharides such as CS… is similar to that of a solution of charged rods for the 20-disaccharide chains*

D. W. Smith et al., *Articular Cartilage Dynamics*,
https://doi.org/10.1007/978-981-13-1474-2

considered.' Note also that this virial equation is only calibrated for solutions up to 80 mg/ml, so beyond 80 mg/ml GAG concentration is extrapolation (and is probably conservative for higher concentrations because (configurational) entropic stiffness becomes more important).

We observe that the experimentally measured (using centrifuge sedimentation) osmotic pressures for aggrecan (PGS), obtained from Swarm rat chondrosarcomas grown in Wistar rats, are about 15% larger at 60 mg/ml (and by extrapolation, 30% larger at 80 mg/ml) than the osmotic pressures predicted by Eq. (D.1) (see Fig. 4, Williams and Comper 1990). This suggests that Eq. (D.1) is a conservative estimate of actual aggrecan osmotic pressures. But we note also that the osmotic data (Chahine et al. 2005) is obtained for processed shark GAG and bovine trachea GAG. They found that at a GAG concentration of 102 mg/ml (fixed charge density, FCD) estimated as 0.4 mEq/ml using the equation described in the methods section of Chahine et al. (2005), the osmotic pressure was the nearly same as predicted by Eq. (D.1) (measured value 0.3 MPa at 0.15 M salt concentration (see Fig. 5, Chahine et al. 2005), while the predicted value using Eq. (D.1) is also 0.3 MPa). However, we note that at lower GAG concentrations the measured osmotic pressure is higher than predicted by Eq. (D.1), while at higher concentrations the measured osmotic pressure is lower than predicted by Eq. (D.1) (at 170 mg/ml, measured osmotic pressure is 0.6 MPa while predicted osmotic pressure is 1.1 MPa). We also mention the osmotic pressures reported for a very large number of human cartilage specimen by Maroudas et al. (1992). Referring to Table 1 (Maroudas et al. 1992), if we assume the concentration of GAG is 154 mg/ml at a fixed charge density of 0.6 mEq/ml (calculated from Chahine et al. (2005), i.e. 100 mg/ml being equivalent to FCD of 0.39 mEq/ml), this gives an predicted osmotic pressure of 0.83 MPa compared to a measured osmotic pressure of 1.01 MPa, suggesting Eq. (D.1) is conservative at up to 150 mg/ml. Taken together, this evidence suggests that Eq. (D.1) is probably somewhat conservative in its estimate of osmotic pressure.

The osmotic modulus (OS, that is, the bulk modulus for aggrecan) is estimated as (Smith et al. 2013) (see also Appendix L):

$$\mathrm{OM} = c \times \frac{d\prod}{dc} \tag{D.2}$$

Note that here we have taken a conservative estimate of the osmotic modulus, as shown in Eq. (D.2), rather than taking porosity into account (and using Eq. 7.192). We can compare the predicted modulus using Eq. (D.1) with the modulus inferred on the basis of dynamic light scattering of a chondroitin sulfate solution at 10% w/w (i.e. 104.5 mg/ml) and 20% w/w (i.e. 220 mg/ml) in 0.1 M salt solution (Horkay et al. 2012). At 104.5 mg/ml, we predict an osmotic modulus of 0.77 MPa (0.15 M salt solution) while Horkey et al. (2012) predict by measuring dynamic light scattering an osmotic modulus of about 0.6 MPa (note: Eq. (D.1) predicts an OM of 0.6 MPa at 95 mg/ml), while at 220 mg/ml we predict an osmotic modulus of 5.7 MPa (0.15 M salt solution) while Horkay et al. (2012) predict by dynamic light scattering a modulus of about 3.0 MPa (note: Eq. (D.1) predicts an OM of 3 MPa at 175 mg/ml). This suggests Eq. (D.1) may be somewhat unconservative at high concentrations.

From the definition of bulk modulus, it follows that the (small deformation) 'Young's modulus' is given by:

$$E_c = 3(1 - 2v)\mathrm{OM} \tag{D.3}$$

where E_c is the compressive modulus and v is Poisson's ratio for aggrecan in cartilage. The Poisson ratio for cartilage is typically measured as being small (often estimated in the range 0–0.1, e.g. Schinagl et al. (1996), Chen et al. (2001b), Chahine et al. (2004)), and here taken to be zero. We compare the calculated compressive modulus with measured equilibrium 'aggregate modulus' (H_A) reported by Treppo et al. (2000) later in this section. Note that H_A is usually defined as the initial equilibrium modulus measured in a one-dimensional confined compression test (Kwan et al. 1992). For confined compression test conditions and using small deformation theory, it is found that $H_A = E_c(1 - v)/[(1 + v)(1 - 2v)]$, so the aggregate modulus is equal to the equilibrium compressive modulus E_c when Poisson's ratio is zero.

Using the equations above, one can calculate the properties for different concentrations of aggrecan [the osmotic pressure, the osmotic modulus, the compressive Young's modulus (assuming a Poisson's ratio of zero), and the compressive Young's modulus at 50% strain (assuming no additional production of aggrecan and no transport of aggrecan during the compression)]. This is shown in the table below.

Concentration of aggrecan (mg/ml)	25	50	100	150	200
Π (MPa)	0.018	0.063	0.29	0.78	1.7
OM (MPa)	0.03	0.13	0.67	2	4.3
E_comp (MPa) at zero strain	0.09	0.38	2	6	13
E_comp (MPa) at 50% strain	0.38	2	13	41	93

Working backward, if the peak contact pressure in the knee joint on tibial cartilage are of the order of 4 MPa when walking (probably the upper end of the range), and the (average) compressive strain in the tissue beneath region of maximum stress is of the order of 30% when the tissue is fully consolidated, then both the dynamic and (eventually) the equilibrium average E_comp in the vertical direction for the cartilage have to be around 15 MPa. The table above suggests that if all of the compressive load is to be carried by the aggrecan, then the (average) aggrecan content of the cartilage over the thickness of the cartilage in this region would over 200 mg/ml in the compressed state (i.e. an average of about 140 mg/ml through the thickness in the uncompressed state). We note that this concentration seems very high relative to measured average GAG concentrations in the knee joint, with best estimates of in vivo tibial plateau GAG concentration of around 80 mg/ml (Wedig et al. 2005). Consequently, we are very interested in understanding how the measured GAG (apparent) concentrations in cartilage can be reconciled with the calculations in the table above for osmotic pressure and compressive stiffness.

Importantly, we first note that there is a significant proportion of the total volume in adult cartilage occupied by collagen fibrils, so aggrecan is excluded from this volume (Maroudas et al. 1991, 1992). For example, consider the case of 50% of the total volume is occupied by collagen (based on the discussion below, the total protein exclusion volume may be up to about 50% in adult cartilage)—see diagrams below.

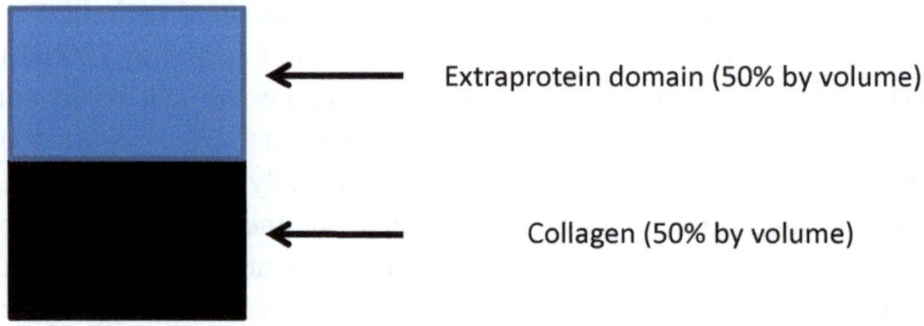

This diagram shows the volume fraction occupied by the proteins (black) and the volume fraction occupied by water (blue). Aggrecan and GAG molecules can only access the volume occupied by water, i.e. the extrafibrillar or more accurately, the extraprotein domain. In this case, if the apparent concentration of aggrecan (mass of aggrecan/total volume) in the superficial zone is 40 mg/ml, the actual concentration (mass of aggrecan/water volume) in the 'extraprotein' volume is 80 mg/ml.

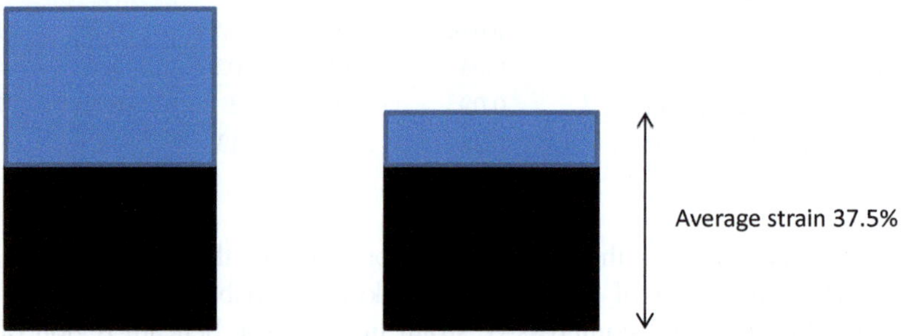

Now, consider a compressive strain in the superficial zone of 37.5%, and the collagen fibrils and non-collagenous proteins are considered incompressible (relative to the GAG). In this case, the concentration of GAG in the extrafibrillar space increases from the initial actual concentration of 80–320 mg/ml. The general formulae for the calculation of the true aggrecan concentration are given by:

$$c^{\text{true}}(t) = c_0^{\text{apparent}}/(1 + e_v(t) - n_{col}) = c_0^{\text{apparent}}/(J(t) - n_{\text{col}}) \quad \text{(D.4)}$$

where n_{col} is the initial volume fraction occupied by collagen (or more accurately, protein). For the case described above, the stiffness of aggrecan in compression would theoretically be 100+ MPa. Note that this E_comp is more than enough compressive stiffness to manage additional large static loadings back to reasonable deformation magnitudes. It is apparent that taking into account the volume occupied by the collagen and other proteins does have a very significant influence on equilibrium tissue stiffness. For this reason, we calculate the osmotic pressures and equilibrium compressive stiffnesses, assuming varying volume fractions (20, 30, 40, and 50%), which are occupied by collagen fibrils and other proteins.

For 20% collagen volume fraction, we find:

Apparent concentration of aggrecan (mg/ml)	25	50	100	150	200
Actual aggrecan concentration (20% volume fraction collagen)	31.25	62.5	125	187.5	250
$\prod$ (MPa)	0.026	0.10	0.50	1.4	3.0
OM (MPa)	0.045	0.21	1.2	3.6	8.1
E_comp (MPa) at zero strain	0.14	0.64	3.7	10.9	24.4
E_comp (MPa) at 50% total strain	1.3	8.0	55	176	400

For 30% collagen volume fraction:

Apparent concentration of aggrecan (mg/ml)	25	50	100	150	200
Actual aggrecan concentration (30% volume fraction collagen)	35.7	71.4	142.8	214.3	285.7
$\prod$ (MPa)	0.033	0.13	0.67	2.0	4.31
OM (MPa)	0.06	0.29	1.7	3.6	11.8
E_comp (MPa) at zero strain	0.18	0.88	5.2	15.8	35.5
E_comp (MPa) at 50% total strain	3.6	24.5	178	582	1360

For 40% volume fraction of collagen:

Apparent concentration of aggrecan (mg/ml)	25	50	100	150	200
Actual aggrecan concentration (40% volume fraction collagen)	41.7	83.3	166.6	250	333.3
$\prod$ (MPa)	0.044	0.19	1.0	2.0	6.6
OM (MPa)	0.08	0.43	2.6	3.6	18.3
E_comp (MPa) at zero strain	0.25	1.3	7.9	15.8	55
E_comp (MPa) at 50% total strain	14.7	104	786	582	6100

For 50% volume fraction of collagen:

Apparent concentration of aggrecan (mg/ml)	25	50	100	150	200
Actual aggrecan concentration (50% volume fraction collagen)	50	100	200	300	400
$\prod$(MPa)	0.063	0.29	1.7	4.9	11.0
OM (MPa)	0.13	0.67	4.3	13.6	18.3
E_comp (MPa) at zero strain	0.38	2.0	13	40.8	92.8
E_comp (MPa) at 50% total strain	∞	∞	∞	∞	∞

What becomes clear from these tables is that at a constant strain magnitude, the equilibrium stiffness of aggrecan becomes extraordinarily large as the volume fraction of collagen increases. Experimentally the equilibrium 'aggregate modulus'(H_A) is defined Treppo et al. to be the 'linear portion' of the equilibrium confined compression stress-strain curve (when strains were applied in 5–10% increments, to a maximum of 25% strain) (Treppo et al. 2000). Equilibrium stress was measured on distal femur, tibial plateau, and talar cartilage, using 3-mm-diameter specimens and cartilage obtained from the top 1 mm of cartilage for each cartilage sample (Treppo et al. 2000)]. At 5% of sGAG per wet weight of cartilage tissue, the aggregate modulus is about 1 MPa (see Fig. D.1).

Looking at the sGAG percentage in Fig. D.1, and checking the theoretical compressive Young's modulus estimates for various collagen volume fractions shown in the tables above, reveals that a collagen volume fraction of between 20 and 30% seems to offer reasonable approximation to the measured aggregate modulus for all these articular surfaces.

While testing normal human femoral cartilage, Chen et al. (2001b) found a depth-dependent compressive stiffness (H_A, confined compression), ranging from of 1.1 ± 0.2 MPa for the surface layer to 7.75 ± 1.45 MPa for the cartilage deepest layer (1.25–1.5 mm from the cartilage surface). This data suggests a collagen volume fraction somewhat larger than that estimated for the Treppo et al. data, Fig. D.1, perhaps in the range of 40% excluded volume (e.g. a stiffness of 7.9 MPa is estimated to be generated by an apparent aggrecan concentration of 100 mg/ml with 40% excluded volume—see tables above).

We also mention that the Chen et al. data on the ratio of H_A to the bulk modulus (bulk modulus, which is the OM, is denoted K_0 in Chen et al.) is about 3, which is in line with assuming the Poisson ratio is about zero (e.g. see Eq. 9 and Fig. 9 and Fig. 10 in Chen et al. (2001b), and Eq. D.3 above). Chen et al. also observe that H_A and K_0 do not vary linearly with the fixed charge density (FCD) and appear to be

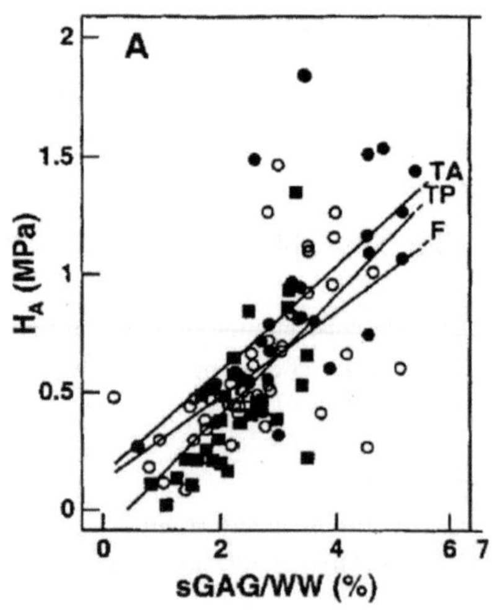

Fig. D.1 Scatter plots showing the variation in equilibrium modulus (H_A) with sulfated glycosaminoglycans (sGAG)/wet weight (WW) (human donors were aged 14, 22, 31, 38, and 45 years of age). The data is shown for the different joint surfaces: filled circles = talar (TA), open circles = distal femur (F), and filled squares = tibial plateau (TP) (Treppo et al. 2000)

surprised by this, commenting: '*It seems possible, though less likely, that the observed [depth-dependent] trends are due to the non-linear dependence of* H_A *and* K_0 *on the FCD.*'), yet a nonlinear dependence is to be expected on the basis of results of molecular dynamics simulations (Bathe et al. 2005; Nap and Szleifer 2008), and so the most likely explanation is that the fixed charge on GAG is primarily responsible for the observed material behaviour of cartilage under compression loadings.

We can calculate collagen volume fractions. Using the data shown in Table I (Basser et al. 1998), the data in the table implies that 17.3% collagen dry total weight fraction is equivalent to a hydrated collagen volume fraction of about 27–36% (assuming a partial specific volume of 0.7), because the intrafibrillar water content of the collagen fibrils ranges from 0.7 to 1.14 g of water per gram of collagen (as the water content of the collagen depends on the local osmotic pressure, primarily exerted by the local GAG concentration) (Levick 1987; Maroudas et al. 1991).

Interestingly, according to the data presented on a large number of human articular cartilage specimens (see Table 2, Maroudas et al. 1992), we note that the dry collagen weight fraction per gram of uncompressed cartilage tissue is remarkably stable with age (at 17–20 years, dry collagen weight fraction of cartilage is 18%, while at 70–81 years it is 17%), and mention in passing that the fixed charge density also increases with age (at 17–20 years, FCD is 0.2 mEq/ml of extrafibrillar water, while at 70–81 years FCD is 0.3 mEq/ml of extrafibrillar water).

But we also note that the non-collagen proteins probably account for up to another 10% of the volume of the cartilage (this is likely a lower limit, as this estimate is based on the dry weight and a specific gravity of 1.36 with no hydration) (see Table I in Basser et al. 1998), so it seems possible that protein exclusion volume fractions in cartilage could be around 35–45%.

We see that these protein volume fraction estimates are now somewhat above the initial estimate above of between 20 and 30% (based on theory and stiffness measurements). However, an estimated protein volume fraction of 35 to 45% appears to be consistent with the data presented by Chen et al. (2001b).

We also mention in passing that it is reported that convex articular surfaces associated with the talus joints (47.4% dry weight fraction collagen) contain significantly more collagen than concave articular surfaces (40.8% dry weight fraction), while flat surfaces have a collagen content in between these two averages (Myers et al. 1983). We finally mention that for femoral head cartilage reported in Basser et al. (1998), the dry weight fraction of collagen was estimated at about 52%.

Given that the importance of the volume fraction occupied by collagen for the estimation of the properties of the cartilage tissue. Given this, it becomes important to know how the volume fraction of collagen changes with changes in the osmotic pressure of GAG, for there exists a competition between the GAG and the collagen for the available water within cartilage. At equilibrium, the free energy potential of the water through the intrafibrillar and extrafibrillar phases must be equal. It is apparent from Fig. D.2 that the intrafibrillar water content can decrease by about 40% as the osmotic pressure increases from zero to 1 MPa, but thereafter intrafibrular water content approaches a constant value at higher osmotic pressures (Basser et al. 1998). Figure 7 in Basser et al. (1998) suggests this loss of water with

Fig. D.2 Intrafibrillar water as a function of applied osmotic stress, the osmotic pressure of PEG representing the extrafibrillar osmotic pressure (π_{EF}) (Basser et al. 1998)

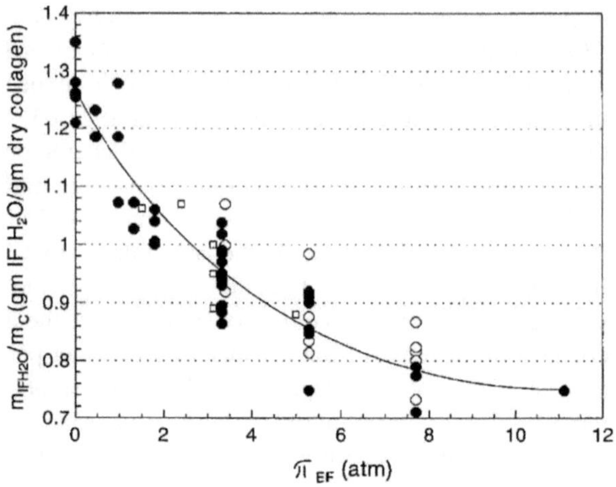

osmotic pressure changing from values approaching zero to 1 MPa amounts to a change in collagen volume of somewhere between 15 and 20%.

In other words, if the average aggrecan content of the cartilage is 25 mg/ml and collagen volume fraction of the cartilage is 30%, the collagen will be almost fully hydrated, as the osmotic pressure is only 33 kPa (see table above for 30% collagen volume fraction). However with the application of 50% strain to the cartilage, the osmotic pressure increases to 500 kPa. This implies that the collagen volume decreases about 15% due to the increase in osmotic pressure (i.e. there is a shift in 4.5% of the water fraction from the collagen to the extrafibrillar space). What effect does this have on the calculations shown in the table above for 30% volume fraction? The osmotic pressure reduces from 500 MPa to about 350 kPa, and the compressive stiffness decreases from about 3.6–2.5 MPa. So while there is some change, it is noted that if the strain increases slightly, i.e. say to 55%, then the previous state is recovered (i.e. 500 kPa osmotic pressure and a compressive stiffness of 3.6 MPa). Therefore, in light of all the uncertainties in many other (crucially) important parameters (e.g. GAG content, collagen volume fraction, non-collagen protein volume fraction, strain magnitude, and the virial equation itself—Eq. D.1), this represents another uncertainty, one among many.

We finally observe that after careful testing of articular cartilage specimens, it has been concluded that aggrecan alone can explain and so is responsible for, the measured osmotic pressure in cartilage (Maroudas et al. 1992), at least up to osmotic pressures of 1 MPa. Says (Maroudas et al. 1992): '*This indicates that in the final equilibrium state the applied pressure is indeed resisted only by the osmotic pressure of the proteoglycans in the extrafibrillar space*' (Maroudas et al. 1992).

Similarly, we observe here that taking the aggrecan content of cartilage and the volume fraction of collagen into account, experimentally it has been reported that nearly all the compressive stiffnesses are attributable to aggrecan, over the range 0–15% strain, for bovine and young human cartilage (Han et al. 2011b). But Han et al also report that aggrecan content could only explain about one half of the compressive stiffness for older human cartilage (which is known to have defective aggrecan, in the sense that the GAG content of the aggrecan molecules is greatly reduced (Han et al. 2011b), which we mention, sits at odds with the reported increase in fixed charged density with age reported in Table 2, Maroudas et al. 1992). It seems likely that for diseased and damaged cartilage without much aggrecan, collagen fibers may well be compressed onto collagen fibers and carry about 50% of the applied load. But in this case, it is expected that the collagen in such cartilage would degrade comparatively quickly when subjected to heavy loads, due to damage of collagen type II fibrils.

Appendix E
Estimating the Aggrecan Production Rate, Aggrecan Turnover and Aggrecan Flux Through the Surface of Knee Articular Cartilage

The aggrecan production rate of knee articular cartilage can be estimated in different ways. One way is to measure the rate of sulfate incorporation into articular cartilage from the knee. Maroudas (1975) did so for 16 people (6 with normal cartilage and 10 with osteoarthritis). Cartilage samples were obtained from a group with an average age of 64. After taking into account the Donnan equilibrium and so the actual inorganic concentration of sulfate in the fluid of the cartilage tissue, and taking into account the ratio of the sulfate radioactive decay counts (in mmol) in the interstitial fluid and the ECM, the measured sulfate 35 uptake rate incorporated into aggrecan for femoral condyle was 3.3×10^{-6} mmol/g of cartilage tissue (see Table 4, Maroudas 1975).

Now assuming 5000 sulfates per aggrecan molecule (i.e. 50 pairs of disaccharides of uronic acid and singly sulfated galactosamine, on 100 GAG side chains) with a molecular weight of 2.5×10^{6} leads to a production rate of aggrecan of 2.4×10^{-5} mol per day per gram of tissue. Given that the average total volume of knee cartilage is 23 ml (Eckstein et al. 1998), and assuming a weight of 1 g per ml, then the rate of aggrecan production is about 0.6 mg/day. Assuming that the rate of production of aggrecan is equal to the rate of loss of aggrecan to the synovial fluid, the rate of loss of aggrecan from knee cartilage is 0.6 mg/day. We note that this is for a group age 64 years, and so based on previous discussion of the age-related decrease in aggrecan production, we may suppose that the rate of production in a group of young adults would be 2–4 times larger (i.e. 1.2–2.4 mg/day) and larger again in the newborn and very young.

We note that the aggrecan turnover rate is found by simply dividing the total aggrecan in knee cartilage by the rate of production (or loss). Assuming an average knee cartilage aggrecan concentration of 50 mg in young adults, this leads to an estimated aggrecan turnover rate of (50 mg/ml $\times$ 23 ml/1.8 mg/day) = 1.75 years.

D. W. Smith et al., *Articular Cartilage Dynamics*,
https://doi.org/10.1007/978-981-13-1474-2

This is about the same as the 600 days for the rate of turnover estimated by Maroudas (1975) for the femoral condyle cartilage for the 64-year-old age group (Maroudas 1975). We note that the rate of aggrecan turnover in intervertebral disk has been estimated using racemization of aspartic acid as about 11–12 years in healthy nucleosis pulposus and 8–9 years in degenerate tissue (Sivan et al. 2006).

A second way is to estimate the aggrecan turnover rate is to first estimate the rate of removal of GAGs from the synovial joint space itself. We now estimate the rate of loss of aggrecan from adult knee cartilage each day by estimating the rate of aggrecan clearance from synovial fluid in an adult knee joint. Let c represent the concentration of aggrecan in the synovial fluid. The rate of loss of aggrecan from the joint space may be estimated using:

$$c = c_0 \mathrm{e}^{-\lambda t}$$

The half-life of labeled GAGs in rabbit knee joint = 12 h (Page-Thomas et al. 1987), while for the pig knee joint it was 8 h (Jensen et al. 1993). Using 8 h as an approximation for human knee joints:

$$\lambda = -\ln(0.5)/8 \approx 0.087\,\mathrm{h}$$

Using a differential equation for the mass flux that is consistent with the first equation above, at time zero the mass flux from joint is given by:

$$\frac{\mathrm{d}m^{\mathrm{agg}}}{\mathrm{d}t} = -\lambda c_0 V^{\mathrm{SF}}$$

From this equation, the rate of loss of GAG may be estimated. Since $c_0 = 0.1$ mg/ml (Larsson et al. 2009), $V^{SF} = 5$ ml (Kraus et al. 2007; Heilmann et al. 1996), the rate of mass loss from normal adult knee joint in a human is estimated as 0.043 mg/h = 1.04 mg/day. This estimate of the aggrecan removal rate from the synovial fluid accords reasonably well with the estimates based on the rate of sulfate incorporation into the cartilage ECM (i.e. 2.4 mg/day for the young and active and 0.6 mg for the aged and inactive).

Now, the surface area of cartilage in the normal human knee averages 122 cm^2 (Blewis et al. 2007). Let us work with our second estimate for the aggrecan production rate (i.e. 1.04 mg/day). With this assumption, the average flux per cm^2 of aggrecan through cartilage surface is 1.04 mg/day/122 = 0.0085 mg/day/cm^2.

Assuming the average concentration of aggrecan at the cartilage surface is 30 mg/ml, the velocity of the aggrecan flux through the cartilage surface per day is given by:

$$v = \frac{85\,\mathrm{mg/day/m^2}}{3 \times 10^7\,\mathrm{mg/m^3}} = 2.8 \times 10^{-6}\,\mathrm{m/day}$$

That is, about 2.8 μm per day is the average volume flux rate of aggrecan through the surface. However, it is noted that some (reasonably small) fraction of this aggrecan will be mechanically and proteolytically degraded into GAG and HA fragments within normal adult cartilage, so not all will be intact aggrecan that mainly exits the cartilage (though it appears that the majority of aggrecan will exit the cartilage surface intact: see discussion on aggrecan turnover in Sect. 2.3.3).

It is also reasonable to suppose that the thickness of cartilage on the tibial surface of the knee may vary two- or threefold, while the rate of aggrecan production under the contact load may be 2 or 3 times greater than at the edge of the cartilage surface. Therefore at the tibial plateau beneath the contact load, the rate of aggrecan flux through the surface of the cartilage in terms of local concentration at the surface may potentially be 10–25 μm per day.

It is known that the rate of aggrecan production varies with age, being much greater during early childhood, tapering in adolescence to reach the adult rate of flux through the cartilage surface. In this case, we may suppose that maximum rates of aggrecan flux through the cartilage surface of the knee may be higher again in the young. We further note it is known that the rates of mass flux from OA knees may be considerably higher than normal adult knee, e.g. twofold or more.

Finally, we make an observation on wear rates at the cartilage surface. For if we assume that for each step, on average about 1 nm of aggrecan is abraded away from the surface of the cartilage, we see that 1 μm of efflux of aggrecan affords protection for about 1000 steps (e.g. 1 kilometer of walking/running).

Appendix F
GAG and Aggrecan Flux Through Surface of Knee Cartilage

Consider the flux of molecules through the surface of cartilage in the human knee joint. A schematic of this flux of molecules is shown in Fig. 3.11 and Fig. 4.7.

Let the molecular weight of a GAG molecule (e.g. chondroitin sulfate) be 2×10^4. Assume 1 mg of GAGs is exiting the cartilage through the cartilage surface per day. Then, the number of moles of GAGs is:

$$1 \times 10^{-3}(\mathrm{g/day})/(2 \times 10^4(\mathrm{g}) = 5 \times 10^{-8}\,\text{moles per day.}$$

And the number of GAG molecules is:

$$5 \times 10^{-8}(\text{moles per day}) \times 6.02 \times 10^{23} = 3.1 \times 10^{16}\ \text{molecules per day.}$$

The average surface area of cartilage in the knee joint is 122 cm squared (Blewis et al. 2007), i.e. $1.22 \times 10^{16}\mathrm{nm}^2$. Then, the flux of GAG molecules through the cartilage surface is:

$$\begin{aligned} &3.1 \times 10^{16}\,\text{molecules per day}/1.22 \times 10^{16}\,\mathrm{nm}^2 \\ &= 2.5\ \text{GAG molecules}/\mathrm{nm}^2\ \text{per day.} \end{aligned}$$

This estimate assumes that an aggrecan molecule is completely disassembled into say 100 GAG chains. This provides a lower-bound estimate on the molecule flux. An upper-bound estimate can be obtained if we now assume the aggrecan molecule is intact.

The molecular weight of aggrecan is taken to be 2×10^6, and then the flux on aggrecan molecules through the cartilage surface is 100 times less than that for the GAG molecules, i.e. 2.5 aggrecan molecules/100 nm^2 per day.

D. W. Smith et al., *Articular Cartilage Dynamics*,
https://doi.org/10.1007/978-981-13-1474-2

The actual number of intact and partially degraded molecules fluxing through the surface would be somewhere between these upper- and lower-bound estimates, e.g. between 2.5 and 250 molecules/100 nm^2 per day.

This shows that there is a very substantial flux of molecules through the cartilage surface. The high-load contact area flux would be expected to be higher—possibly up to 10 times higher. So for each square with side-length of 10 nm, there could be between 25 and 2500 molecules per day, fluxing through the cartilage surface in high-contact areas of the knee.

The exiting molecules may be pulled from the knee cartilage surface during normal joint activities. This 'shedding' protects the cartilage surface from abrasive wear of the collagen network (through the aggrecan component being 'sacrificed'). In addition, partially exited but still anchored molecules may considerably reduce the flow of synovial fluid in the lateral direction of the joint space during loading, when the joint space is very narrow, so presumably these molecules help to support a mix-mode of joint lubrication (see discussion in Chap. 4).

Appendix G
The Rate of Volume Change in Material Coordinates

The ratio of the differential volume at time t in material coordinates to its initial volume (J) is the product of the three principal stretches (λ_α). That is,

$$J = \prod_{\alpha=1}^{3} \lambda_\alpha$$

The Green–Lagrange strain tensor is by definition

$$E = \tfrac{1}{2}(C^2 - 1)$$
$$E = \tfrac{1}{2}(U^T R^T R U - 1)$$
$$E = \sum_{\alpha=1}^{3} E_\alpha = \tfrac{1}{2}(U^2 - 1)$$

where $U = \sum_{\alpha=1}^{3} U_\alpha = \sum_{\alpha=1}^{3} \lambda_\alpha N_\alpha \otimes N_\alpha$where N_α are unit vectors in each of the three principal strain directions. Then, we can define the Green–Lagrange stretches in the principal directions as

$$\lambda_\alpha = (1 + 2E_\alpha)^{\frac{1}{2}} = U_\alpha$$

and so,

$$J = \prod_{\alpha=1}^{3} (1 + 2E_\alpha)^{\frac{1}{2}}$$

Now, differentiating J w.r.t. time leads to

D. W. Smith et al., *Articular Cartilage Dynamics*,
https://doi.org/10.1007/978-981-13-1474-2

$$\frac{dJ}{dt} = (1+2E_1)^{-\frac{1}{2}}\frac{dE_1}{dt}U_2U_3 + (1+2E_2)^{-\frac{1}{2}}\frac{dE_2}{dt}U_1U_3 + (1+2E_3)^{-\frac{1}{2}}\frac{dE_3}{dt}U_2U_3$$

and so,

$$\frac{1}{J}\frac{dJ}{dt} = \sum_{\alpha=1}^{3}\frac{1}{(1+2E_\alpha)}\frac{dE_\alpha}{dt}$$

That is,

$$\frac{1}{J}\frac{dJ}{dt} = \sum_{\alpha=1}^{3}\frac{1}{U_\alpha^2}\frac{dE_\alpha}{dt} = \sum_{\alpha=1}^{3}\frac{1}{U_\alpha^T R_\alpha^T R_\alpha U_\alpha}\frac{dE_\alpha}{dt} = F^{-1}F^{-T} : \frac{dE}{dt} \tag{G.1}$$

For deformation problems confined to one spatial dimension $F^{-1} = F^{-T} = J^{-1}$, and noting $E = 0.5(\lambda^2 - 1)$, this identity simplifies to

$$\frac{1}{J}\frac{dJ}{dt} = \frac{1}{2J^2}\frac{d\lambda^2}{dt}$$

For one-dimensional deformation, $\lambda = J$. Substitution in the previous equation and differentiation confirms the identity.

Appendix H
The Cauchy Stress Tensor

The Cauchy stress tensor is defined in spatial coordinate system as represented by σ_{ij}. The Cauchy stress is related to the first Piola–Kirchhoff stress by

$$P_{iK} = J\sigma_{ij}^{sm}F_{jK}^{-T}$$

where P_{iK} is the 'two-point tensor' derived from the Cauchy stress, F_{iJ} is the deformation gradient, and J is the determinant of F.

The second Piola–Kirchhoff stress is a tensor in the material coordinate system derived from the Cauchy stress by

$$S_{JK} = JF_{iJ}^{-1}\sigma_{ij}^{\mathrm{sm}}F_{jK}^{-\mathrm{T}}$$

The traction vector T_i, at any point in the spatial coordinate description of the material, on the plane normal to the direction of the n_j unit vector, is given by $T_i = \sigma_{ij}n_j$. This is known as Cauchy's formula.

Now because the Cauchy stress tensor is symmetric (see Appendix J The Momentum Balance Equation and Strain Work Rate), there exits three orthogonal principal stress directions. Our first aim is to find the directions n_i that make σ_{ij} diagonal. These directions can be found by an 'eigenvalue analysis' of the Cauchy stress tensor.

The traction vector T_i, on the plane normal to the direction of the n_j unit vector, is given by $T_i = \sigma_{ij}n_j$. For an eigenvalue analysis, we also require that $T_i = \lambda n_i$. Then in the usual way of eigenvalue analysis, we subtract the two and so

$$\sigma_{ij}n_j - \lambda n_i = 0$$

D. W. Smith et al., *Articular Cartilage Dynamics*,
https://doi.org/10.1007/978-981-13-1474-2

That is,

$$(\sigma_{ij} - \lambda\delta_{ij})n_j = 0$$

where δ_{ij} is the Kronecker delta. Then, the solution is given when $\det|\sigma_{ij} - \lambda\delta_{ij}| = 0$. The eigenvalues are the solution of the (cubic) 'characteristic equation' in λ, viz.,

$$-\lambda^3 + I_1\lambda^2 - I_2\lambda + I_3 = 0$$

where I_1, I_2, I_3 are the invariants of σ_{ij}. Because σ_{ij} is symmetric, the eigenvalues are real, as are the principal stresses. The invariants can be expressed in terms of the eigenvalues, so that (http://en.wikipedia.org/wiki/Cauchy_stress_tensor),

$$\begin{aligned} I_1 &= \lambda_1 + \lambda_2 + \lambda_3 \\ I_2 &= \lambda_1\lambda_2 + \lambda_2\lambda_3 + \lambda_3\lambda_1 \\ I_3 &= \lambda_1\lambda_2\lambda_3 \end{aligned}$$

Any stress tensor can be split into a deviatoric stress tensor and a hydrostatic stress tensor. For example, the Cauchy stress can be represented by

$$\sigma_{ij} = s_{ij} + \frac{(\sigma_{11} + \sigma_{22} + \sigma_{33})}{3}\delta_{ij} = s_{ij} + \frac{I_1}{3}\delta_{ij}$$

The characteristic equation for the deviator stress tensor is

$$|s_{ij} - \lambda\delta_{ij}| = -\lambda^3 - J_1\lambda^2 - J_2\lambda - J_3$$

where the invariants of the deviatoric stress tensor are

$$\begin{aligned} J_1 &= 0 \\ J_2 &= \tfrac{1}{3}I_1^2 - I_2 \\ J_3 &= \tfrac{2}{27}I_1^3 - \tfrac{1}{3}I_1 I_2 + I_3 \end{aligned}$$

The so-called von Mises stress σ_e is found from the second deviatoric stress invariant as

$$\sigma_e = \sqrt{3J_2} = \sqrt{\frac{1}{2}[(\sigma_1 - \sigma_2)^{\frac{1}{2}} + (\sigma_2 - \sigma_3)^{\frac{1}{2}} + (\sigma_3 - \sigma_1)^{\frac{1}{2}}]}$$

We note that the failure criterion σ_e equals a constant may be a suitable elasto-plastic failure criterion for the collagen network in cartilage.

Appendix I
Terzaghi's Principle of Effective Stress

One way to understand the meaning of a 'stress' in a material is through 'work rates.' The product of stress and deformation rate is work rate per unit volume, and so a stress can only sensibly be defined in relation to a conjugate variable that gives units of work. A stress may be defined in different coordinate systems (e.g. the Cauchy stress σ_{ij} transforms to the second Piola–Kirchhoff stress S_{ij}), so each stress representation needs to be combined with appropriate conjugate strain variable (rate of deformation in spatial coordinates d^{s}, and rate of deformation in material coordinates $\frac{\mathrm{d}E}{\mathrm{d}t}$, respectively), the product representing the work rates per unit volume in the deformed and undeformed configurations respectively.

With this in mind, let us consider a two-phase porous medium consisting of an elastic solid and viscous fluid, and let the total work rate per unit reference volume be denoted $\dot{\Psi}^{\mathrm{tot}}$. We add together the (solid) elastic work rate, $\dot{\Psi}^{\mathrm{e}}$, and the rate of (fluid) pressure work, $\dot{\Psi}^{p} = -p^{\mathrm{sm}}\frac{\mathrm{d}N_{\mathrm{f}}^{\mathrm{m}}}{\mathrm{dt}}$, where the negative sign accounts for the convention that a positive pressure is a negative (compressive) stress, and we define the Lagrangian fluid porosity $N_{\mathrm{f}}^{\mathrm{m}} = v_{\mathrm{f}}^{\mathrm{m}}(t)/V(0)$ (Coussy 2004), where $N_{\mathrm{f}}^{\mathrm{m}} = Jn_{\mathrm{f}}^{\mathrm{sm}}$ and $n_{\mathrm{f}}^{\mathrm{sm}}$ is the spatial (Eulerian) fluid porosity mapped across to the reference configuration. This gives the total work rate per unit reference volume $\dot{\Psi}^{\mathrm{tot}}$ (Coussy and Scherer 2004),

$$\frac{\mathrm{d}\Psi^{\mathrm{tot}}}{\mathrm{d}t} = \frac{\mathrm{d}\Psi^{\mathrm{e}}}{\mathrm{d}t} + \frac{\mathrm{d}\Psi^{\mathrm{p}}}{\mathrm{d}t} \tag{I.1}$$

that is,

$$S^{\mathrm{tot}} : \frac{\mathrm{d}E}{\mathrm{d}t} = S^{\mathrm{e}} : \frac{\mathrm{d}E}{\mathrm{d}t} - p^{\mathrm{sm}}\frac{\mathrm{d}N_{\mathrm{f}}^{\mathrm{m}}}{\mathrm{d}t} \tag{I.2}$$

D. W. Smith et al., *Articular Cartilage Dynamics*,
https://doi.org/10.1007/978-981-13-1474-2

Let us now identify rate of change of Lagrangian porosity $\frac{dN_f^m}{dt}$. Taking into account that both solid and fluid phase are incompressible, the rate of fluid volume change due to fluid flux leaving the material REV $\left(-\nabla^m \cdot Jv_d^m = dv_f^s/dt\right)$ must be same as the rate of change of the REV itself (dJ/dt). By definition of J (Eq. 7.15),

$$\frac{dJ}{dt} = \frac{d(v/V(0))}{dt} = \frac{d(v_f^s(t)/V(0))}{dt} = \frac{dN_f^m}{dt} \tag{I.3}$$

This relationship can also be derived based on incompressibility of the solid and fluid phases. Solid phase mass balance in material coordinates is

$$\rho_s(1 - n_f^m(X,0))dV = \rho_s(1 - n_f^{sm}(X,t))dv = \rho_s(1 - n_f^s(x,t))dv \tag{I.4}$$

where $n_f^{sm}(\xi(a),t)$ is $n_f^s(x,t)$ in spatial coordinates mapped across to material coordinates. Then for an incompressible solid,

$$\frac{d[(1 - n_f^{sm}(X,t))J]}{dt} = 0 \tag{I.5}$$

That is,

$$\frac{dJ}{dt} = \frac{d[J(n_f^{sm}(X,t))]}{dt} = \frac{dN_f^m}{dt} \tag{I.6}$$

where $N_f^m = Jn_f^{sm}(X,t) = v_f^s(t)/V(0)$.

And using Eq. (I.3) together with the derivation of Eq. (G.1) (shown in Appendix G titled The Rate of Volume Change in Material Coordinates), we have

$$\frac{dN_f^m}{dt} = \frac{dJ}{dt} = JF^{-1}F^{-T} : \frac{dE}{dt} \tag{I.7}$$

Substituting Eq. (I.7) into Eq. (I.2) leads to

$$S^{tot} : \frac{dE}{dt} = \left(S^e - Jp^{sm}F^{-1}F^{-T}\right) : \frac{dE}{dt} \tag{I.8}$$

which shows

$$S^{tot} = S^e - Jp^{sm}F^{-1}F^{-T} = S^e - P^m \tag{I.9}$$

This is Terzaghi's principle of effective stress in material coordinates (the work conjugate to the total second Piola–Kirchhoff stress is $\frac{dE}{dt}$). 'Pushing forward' Eq. (I.9) leads to

$$\sigma^{\mathrm{tot}} = \sigma^{\mathrm{e}} - p \tag{I.10}$$

which is widely known in small deformation soil mechanics as 'Terzaghi's principle of effective stress.' The same principle applies for the deformation analysis of cartilage.

The 'push forward' of the $\frac{\mathrm{d}E}{\mathrm{dt}}$ is

$$d_{\mathrm{s}} = F^{-\mathrm{T}} \frac{\mathrm{d}E}{\mathrm{d}t} F^{-1}$$

In spatial coordinates, the work conjugate of the Cauchy solid stress is $d_{\mathrm{s}} = \nabla^{\mathrm{s}} v_{\mathrm{s}}$where v_{s} is the velocity of the solids in the spatial coordinate system.

Appendix J
The Momentum Balance Equation and Strain Work Rate

Newton's law for a particle P is $\frac{\mathrm{d}}{\mathrm{d}t}(m_{\mathrm{P}} v_{\mathrm{P}}^{\mathrm{s}}) = F_{\mathrm{P}} = m_{\mathrm{P}} a_{\mathrm{P}}^{\mathrm{s}}$, where m_{P} is the mass of the particle, a_{P}^{s} is the acceleration of the particle in spatial coordinates, and F_{P} is the applied force (Newton's second law). The inertial force $m_{\mathrm{P}} a_{\mathrm{P}}^{\mathrm{s}}$ is equal and opposite in direction to the applied force (Newton's third law). Based on the integral form of Newton's second law for a particle P, the rate of change of linear momentum of the solid and fluid phases, making up continuous materials occupying the domain Ω_t, is equal to the integral of the body forces (f) over the domain plus the integral of the traction (T) over the surface of the material, that is,

$$\begin{aligned}&\frac{\mathrm{d}}{\mathrm{d}t}\int_{\Omega_t} \rho_{\mathrm{s}}(1 - n_{\mathrm{f}}^{\mathrm{s}}) V^{\mathrm{s}} \mathrm{d}\Omega_t + \frac{\mathrm{d}}{\mathrm{d}t}\int_{\Omega_t} \rho_{\mathrm{s}} n_{\mathrm{f}}^{\mathrm{s}} V^{f} \mathrm{d}\Omega_t \\ &\quad = \int_{\Omega_t} \rho f \mathrm{d}\Omega_t + \int_{\partial\Omega_t} T \cdot n \mathrm{d}s\end{aligned} \tag{J.1}$$

where $\rho(x) = (1 - n_{\mathrm{f}}^{\mathrm{s}})\rho_{\mathrm{s}} + n_{\mathrm{f}}^{\mathrm{s}}\rho_{\mathrm{f}}$, $V(x, t)$ is the velocity at spatial location x at time t, f is a general body force per unit mass, and T is the traction applied to the material surface at time t. The LHS of Eq. (J.1) represents inertial forces, while the RHS represents the applied body force and the applied surface loading. The rotational momentum balance is given by

$$\begin{aligned}&\frac{\mathrm{d}}{\mathrm{d}t}\int_{\Omega_t} x \times \rho_{\mathrm{s}}(1 - n_{\mathrm{f}}^{\mathrm{s}}) V^{\mathrm{s}} \mathrm{d}\Omega_t + \frac{\mathrm{d}}{\mathrm{d}t}\int_{\Omega_t} x \times \rho_{\mathrm{s}} n_{\mathrm{f}}^{\mathrm{s}} V^{\mathrm{f}} \mathrm{d}\Omega_t \\ &\quad = \int_{\Omega_t} x \times \rho f \mathrm{d}\Omega_t + \int_{\partial\Omega_t} x \times T \mathrm{d}s\end{aligned} \tag{J.2}$$

D. W. Smith et al., *Articular Cartilage Dynamics*,
https://doi.org/10.1007/978-981-13-1474-2

where x is the distance from the origin. The LHS of Eq. (J.2) is the rate of change of angular momentum about the origin, while the RHS represents the applied torque. Taking the time derivative of the above equations leads to

$$\begin{aligned}\int_{\Omega_t} \rho_s(1-n_f^s)a_s d\Omega_t + \int_{\Omega_t} \rho_s n_f^s a_f d\Omega_t \\ = \int_{\Omega_t} \rho f d\Omega_t + \int_{\partial\Omega_t} T ds\end{aligned} \tag{J.3}$$

and,

$$\begin{aligned}\int_{\Omega_t} x \times \rho_s(1-n_f^s)a_s d\Omega_t + \int_{\Omega_t} x \times \rho_s n_f^s a_f d\Omega_t \\ = \int_{\Omega_t} x \times \rho f d\Omega_t + \int_{\partial\Omega_t} x \times T ds\end{aligned} \tag{J.4}$$

where a_i is the acceleration of the ith phase, $[\rho_s(1-n_f^s)a_s + \rho_f n_f^s a_f]d\Omega_t$ are the inertial forces, and $[x \times \rho_s(1-n_f^s)a_s + x \times \rho_f n_f^s a_f]d\Omega_t$ are the inertial torques acting on the material.

Noting the surface traction T is equal to $\sigma \cdot n$ at the surface, where n is the unit normal to the tangent plane at the material surface, and making use of Gauss's divergence theorem (which says an integral of traction over the surface of the material is equal to an integral of the divergence of Cauchy stress over the volume of the material) leads to the linear momentum balance equation,

$$\begin{aligned}\int_{\Omega_t} \rho_s(1-n_f^s)a_s d\Omega_t + \int_{\Omega_t} \rho_s n_f^s a_f d\Omega_t \\ = \int_{\Omega_t} \rho f d\Omega_t + \int_{\Omega_t} \nabla^s \cdot \sigma d\Omega_t\end{aligned} \tag{J.5}$$

where $\nabla\cdot$ is the divergence operator. Collecting terms reveals

$$\int_{\Omega_t} (\nabla^s \cdot \sigma + \rho f - \rho_s(1-n_f^s)a_s - \rho_f n_f^s a_f) d\Omega_t \tag{J.6}$$

Since this equation holds for an arbitrary volume, this implies

$$\nabla^{\mathrm{s}} \cdot \sigma + \rho f - \rho_{\mathrm{s}}(1 - n_{\mathrm{f}}^{\mathrm{s}})a_{\mathrm{s}} - \rho_{\mathrm{f}} n_{\mathrm{f}}^{\mathrm{s}} a_{\mathrm{f}} = 0 \tag{J.7}$$

holds everywhere. This is the 'differential form' of the linear momentum equation.

If each phase in a porous medium is treated separately via an analysis similar to that above for the porous medium as a whole, then an additional term arises due to momentum transfer between (the fluid and solid) phases. For example, when there are two phases, the momentum balance equations for the solid and fluid phases are

$$\begin{aligned} \nabla^{\mathrm{s}} \cdot (1 - n_{\mathrm{f}}^{\mathrm{s}})\sigma_{\mathrm{s}} + (1 - n_{\mathrm{f}}^{\mathrm{s}})\rho_{\mathrm{s}} f_{\mathrm{s}} - (1 - n_{\mathrm{f}}^{\mathrm{s}})\rho_{\mathrm{s}} a_{\mathrm{s}} + f^{f-s} = 0 \\ -\nabla^{\mathrm{s}} \cdot n_{\mathrm{f}}^{\mathrm{s}} p + n_{\mathrm{f}}^{\mathrm{s}} \rho_{\mathrm{f}} f_{\mathrm{f}} - n_{\mathrm{f}}^{\mathrm{s}} \rho_{\mathrm{f}} a_{\mathrm{f}} + f^{\mathrm{s-f}} = 0 \end{aligned} \tag{J.8}$$

where $f^{\mathrm{f-s}}$ is the momentum transfer from the fluid phase to the solid phase and $f^{\mathrm{s-f}}$ is the momentum transfer from the solid phase to the fluid phase. We note that the transfer of momentum is equal and opposite requires

$$f^{\mathrm{f-s}} + f^{\mathrm{s-f}} = 0 \tag{J.9}$$

which means momentum is conserved during the exchange of momentum. For rotational momentum balance, we have

$$\begin{aligned} &\int_{\Omega_t} x \times \rho_{\mathrm{s}}(1 - n_{\mathrm{f}}^{\mathrm{s}})a^{s}\mathrm{d}\Omega_t + \int_{\Omega_t} x \times \rho_{\mathrm{s}} n_{\mathrm{f}}^{\mathrm{s}} a^{\mathrm{f}}\mathrm{d}\Omega_t \\ &= \int_{\Omega_t} x \times \rho f \mathrm{d}\Omega_t + \int_{\Omega_t} (x \times \nabla\sigma + \Lambda : \sigma^{T})\mathrm{d}\Omega_t \end{aligned} \tag{J.10}$$

where Λ is the third-order alternating tensor (i.e. ε_{ijk}). When Λ is double dotted with the second-order tensor σ^{T}, it becomes a vector, namely $2\begin{bmatrix} \sigma_{23} - \sigma_{32} \\ \sigma_{13} - \sigma_{31} \\ \sigma_{12} - \sigma_{21} \end{bmatrix}$.

Now, collecting terms leads to

$$\int_{\Omega_t} (x \times [\nabla\sigma + \rho f - \rho_s(1 - n_f^s)a^s - \rho_f n_f^s a^f] + \Lambda : \sigma^T)\mathrm{d}\Omega_t = 0 \tag{J.11}$$

Substituting the linear momentum balance Eq. (J.7) into Eq. (J.11) reveals

$$2\int_{\Omega_t} \begin{bmatrix} \sigma_{23} - \sigma_{32} \\ \sigma_{13} - \sigma_{31} \\ \sigma_{12} - \sigma_{21} \end{bmatrix} \mathrm{d}\Omega_t = 0 \tag{J.12}$$

Since this integral holds for an arbitrary volume, this implies $\sigma_{ij} = \sigma_{ji}$. In words, this means the Cauchy stress tensor is symmetric.

We can now calculate the strain deformation work rate for a single phase. We first multiply the linear momentum balance equation (Eq. (J.1) for a single phase) by the velocity (to obtain a work rate J m^{-3} s^{-1}) and integrate over the domain, viz.,

$$\int_{\Omega_t} [\nabla^s \cdot \sigma + \rho f - \rho a] \cdot v\mathrm{d}\Omega_t = 0 \tag{J.13}$$

where v is the spatial velocity field. We now note the identity,

$$\nabla \cdot (\sigma \cdot v) = (\nabla \cdot \sigma) \cdot v + \sigma : \nabla v \tag{J.14}$$

Inserting this equation reveals

$$\int_{\Omega_t} (\nabla \cdot (\sigma \cdot v) - \sigma : \nabla v + [\rho f - \rho a] \cdot v)\mathrm{d}\Omega_t = 0 \tag{J.15}$$

Now, using a symmetric rate of deformation tensor $\nabla v = d = \frac{1}{2}(l + l^T)$ along with Gauss's divergence theorem leads to

$$\int_{\partial\Omega_t} [T \cdot v]\mathrm{d}s + \int_{\Omega_t} [\rho f \cdot v - \rho\, a \cdot v]\mathrm{d}\Omega_t = \int_{\Omega_t} [\sigma : d]\mathrm{d}\Omega_t = \dot{\Psi}^{sr} \tag{J.16}$$

where $T = \sigma \cdot n$ and $\dot{\Psi}^{sr}$ is the deformation (stress) power. The stress power accounts for both energy stored by the internal stresses and energy dissipated by the internal stresses (so some mechanical work may appear as heat). Noting that $\int \rho a v \mathrm{d}\Omega_t = \frac{\mathrm{d}}{\mathrm{d}_t} \int [\frac{\varrho}{2} v^2] \mathrm{d}\Omega_t = \frac{\mathrm{d}}{\mathrm{d}_t} KE$, we find

$$\int_{\partial\Omega_t} [T \cdot v] \mathrm{d}s + \int_{\Omega_t} \rho f \cdot v \mathrm{d}\Omega_t = \int_{\Omega_t} [\sigma : d] \mathrm{d}\Omega_t + \frac{\mathrm{d}}{\mathrm{d}t} KE = P_T \tag{J.17}$$

where both sides represent the total power, P_T. The RHS says the total power may be decomposed into two parts: (a) the deformation of the material and (b) the change in KE of the material (i.e. bulk motion), while the LHS says the total power may be decomposed into two parts: (a) the work rate of the surface tractions and (b) the work rate of the body forces. To say this in other words, this equation says the rate of work done on the body by the sum of the work rates of applied surface tractions T plus the applied body forces f over the material is equal to the rate of strain work over the material $\int_{\Omega_t} [\sigma : d] \mathrm{d}\Omega_t$ plus the rate of change of kinetic energy. Equation (J.17) is known as the kinetic energy theorem.

Appendix K
State Variables, Entropy, and Free Energy Potentials

Entropy S is a 'state function' or 'potential,' which implies it has an 'exact differential.' Our strategy is to find this entropy potential by searching for a function that turns the inexact differential of the heat flow (Δq), into an exact differential equation (involving as new variable dS). The inexact differential for the heat flow across the system boundary is a result of the energy conservation equation for the thermodynamic system.

We begin with some background information on exact and inexact differentials. Let F be a function with variables x and y. The total increment of a function $F(x, y)$ is given by

$$\Delta F = \frac{\partial F}{\partial x}\Delta x + \frac{\partial F}{\partial y}\Delta y \tag{K.1}$$

The total increment of the function F may be classified as 'exact' or 'inexact.' Exact means that any line integral between two points will be the same, or in other words, the value of the line integral is independent of the path between the two points. Inexact means that the value of the line integral is path dependent, or in other words, that the 'function' F is a multivalued function. This multivalued nature of F means it has an inexact differential, which is signified by ΔF (rather than dF, which is reserved to denote exact differentials).

When a function F is exact, it can be shown that it will satisfy the following condition

$$\frac{\partial^2 F}{\partial y \partial x} = \frac{\partial^2 F}{\partial x \partial y} \tag{K.2}$$

D. W. Smith et al., *Articular Cartilage Dynamics*,
https://doi.org/10.1007/978-981-13-1474-2

In fact, this equivalence is a test for exactness. For example, let F have the properties,

$$\frac{\partial F}{\partial x} = 3x + 2y, \quad \frac{\partial F}{\partial y} = x \tag{K.3}$$

Then, we can use Eq. (K.2) to see that ΔF is inexact, as Eq. (K.2) is not satisfied. Now, it turns out that any two variable functions can be made into an exact function through an 'integrating factor.' For example, trial and error reveals that x is an integrating factor for function F in Eq. (K.3). That is,

$$\frac{\partial G}{\partial x} = 3x^2 + 2xy \quad , \quad \frac{\partial G}{\partial y} = x^2 \tag{K.4}$$

where $G = xF$ is exact, as can be verified using Eq. (K.2). The definition of the entropy potential reduces to finding a suitable integrating factor that makes the inexact differential form of the heat flow equation into an exact differential form using a new function called entropy.

Now, we move to the problem confronting physicists trying to understand the behavior of an ideal gas compressed by a piston. From the kinetic theory of gases https://en.wikipedia.org/wiki/Kinetic_theory_of_gases, they knew that the internal energy of the ideal gas was an exact differential (i.e. a state function or potential) because the internal energy only depends on the temperature T, viz.,

$$U = \frac{3NRT}{2} \tag{K.5}$$

However, the work required and the quantity of heat exchanged with its surroundings were different for each different path between two states in a P–V–T diagram. We can summarize this as a statement of energy conservation; namely, the change in internal energy U is equal to the work done on the system Δw plus the heat energy ΔQ that flows into the system. We note that Δw is positive when the force on the system boundary and the displacement of the system boundary both have the same sign, while ΔQ is positive when heat is added to the system. So, our energy conservation equation is

$$\mathrm{d}U = \Delta w + \Delta Q \tag{K.6}$$

where Δ signifies that the precise value of two increments in w and Q depends on the path between two system states—only the sum of increments is known—which is independent of the path and is equal to the exact differential $\mathrm{d}U$(which represents the change in (internal) energy of the system).

We have said that increments of w and Q depend on the path between two states. However, the measured values are also different again depending on the speed that the same path between the two states was traversed, i.e. depending on whether the path is traversed slowly or quickly. In this first part of the discussion, we are only focusing on the path dependence associated with (infinitely) slow movements along different paths—later we consider the effect of 'everyday movements' along paths between two states.

The amount of work done on the gas in the system is $\Delta w = -p\mathrm{d}V$ where p is the gas pressure (the minus sign follows from the convention that compressive pressure is taken as positive) and V is the volume of the thermodynamic system. The internal energy of an ideal gas is simply the translational kinetic energy of its constituent molecules, depending only on the temperature (Eq. K.5).

If there is no heat flow across the system boundary (which is known as an adiabatic condition), then the change in internal energy is exactly equal to pdV. And so for work done under adiabatic conditions, the temperature of the gas changes with work done on or by the system.

If no work is done (i.e. $\mathrm{d}V = 0$), then the internal energy can only be changed by heat flow ΔQ across the system boundary. Heat flow into the system increases its temperature T. The change in internal energy at constant volume is given by $\mathrm{d}U = c_v\mathrm{d}T$, where c_v is the heat capacity of the material at constant volume.

Therefore, we can write Eq. (K.6) as

$$\Delta Q = c_v\mathrm{d}T + p\mathrm{d}V \tag{K.7}$$

The 'equation of state' relates the 'state variables' for an ideal gas. This state equation tells us how this material responds when state variables are changed (there is a different constitutive equation for each material). For N moles of an ideal gas, the kinetic theory of gases leads to the constitutive equation describing how the state variables N, p, V, T are related theoretically (and experimentally), viz.,

$$pV = NRT \tag{K.8}$$

From Eq. (K.5), it is apparent theoretically (and confirmed experimentally) that $c_v = \frac{\partial U}{\partial T} = \frac{3NR}{2}$.

Given that potentials usually simplify the mathematical model of the physics (allowing, for example, a conservation statement about energy) and so improve our understanding of what is actually going on, one is motivated to find a function that expresses the heat exchange with the surroundings in terms of a potential. We now have the information required to do this.

To find our potential, we are searching for a function that turns the differential of the heat flow ΔQ (in the energy conservation equation) into an exact differential equation. Our task then is to find an integrating factor M that makes function $G = M\Delta Q$ exact. That is, find M such that

$$\mathrm{d}G = M\Delta Q = Mc_v\mathrm{d}T + Mp\mathrm{d}V \tag{K.9}$$

is exact. Because this is same mathematical problem as described earlier, we look to our exactness test to find a solution for an ideal gas. From the above exactness test and the equation of state for an ideal gas, it is clear that to find our potential we require the following equation be satisfied

$$\frac{\partial^2 G}{\partial V \partial T} = \frac{\partial M c_v}{\partial V} = \frac{\partial (\frac{3NMR}{2})}{\partial V} = \frac{\partial (\frac{MNRT}{V})}{\partial T} = \frac{\partial^2 G}{\partial T \partial V} \tag{K.10}$$

We note in passing that these 'exactness relations' are known as 'Maxwell relations' in the thermodynamics literature (http://en.wikipedia.org/wiki/Maxwell_relations). Now, if it is assumed as a 'trial guess' that M is independent of V, then it follows from Eq. (K.10) that

$$\frac{\partial (\frac{MNRT}{V})}{\partial T} = 0 \tag{K.11}$$

We see this relation is satisfied when $M = \frac{k}{T}$, which is our 'integrating factor.' The constant k is one if we adopt the 'thermodynamic temperature scale' for T. Therefore, we transform Eq. (K.7) to an exact differential using our integrating factor, viz.,

$$\mathrm{d}G = \frac{\Delta Q}{T} = \frac{c_v \mathrm{d}T}{T} + \frac{NR\mathrm{d}V}{V} \tag{K.12}$$

and so,

$$\mathrm{d}G = \frac{\Delta Q}{T} = c_v \mathrm{d}\ln T + NR\mathrm{d}\ln V \tag{K.13}$$

The function G is now exact and so is a 'state function' or 'potential' describing the energy conservation equation in terms of a new potential G. Integrating Eq. (K.13) leads to

$$G = c_v \ln T + NR \ln V + \text{constant} = S \tag{K.14}$$

Rudolf Clausius called the function G the 'entropy' of the material, which is usually now denoted by the symbol S, while Constantin Caratheodory later extended this type of analysis to systems with multiple variables using Pfaffian forms. We see that

$$\mathrm{d}S = \frac{\Delta Q}{T} \tag{K.15}$$

where ΔQ is the heat gained by the system from the surroundings. As did Clausius, we can now write our original energy balance equation as

$$TdS = dU + pdV \tag{K.16}$$

More generally, we can write this as

$$TdS = dU - \sum_{i=1}^{N} L_i dl_i \tag{K.17}$$

where L_i is a generalized work coefficient and dl_i are generalized work coordinates (which include pdV work). Equation (K.17) was known to the chemist Josiah Willard Gibbs and is now known as the 'generalized Gibbs equation' (Haase 1990).

Now in an attempt to familiarize ourselves with the physical significance of entropy, we ask does dS refer to the entropy change in the system dS_{sys} or to the entropy change in the surroundings dS_{sur}? dS refers to the change in entropy of the system (so $dS = dS_{sys}$), as it is derived from the energy conservation equation for the system. However for an reversible process, the conservation of energy also requires that the heat gained by the system is equal to the heat lost from the surroundings, so for an infinitely slow (i.e. reversible) interaction the entropy increase in the system is exactly equal to the entropy decrease in the surroundings (and vice versa), and so the total entropy of system and surroundings dS_{tot} is *unchanged*. In fact, this equality (i.e. $dS_{tot} = dS_{sys} + dS_{sur} = 0$, or $dS_{sys} = -dS_{sur}$) is the defining feature of any reversible (or equilibrium) processes in thermodynamics.

Very importantly, the state function S is defined not only for reversible (i.e. infinitely slow, equilibrium) processes but also for irreversible (i.e. 'fast,' non-equilibrium) processes too, providing the internal energy U and the work done on or by the system also remain well defined. This restriction on the use of S to irreversible processes with well-defined state variables is not too onerous, as it turns out that all irreversible processes in normal biological systems have well-defined internal energies and work functions, and consequently S is well defined. It turns out that S is only not well defined for 'violent processes' including explosions, but this is of no concern for most biological applications.

Now, because dS is well defined for reversible and irreversible process, we can now break dS into two parts (Haase 1990)

$$dS = dS_a + dS_{irr} \tag{K.18}$$

where dS_a is the entropy change due to the heat flow across the system boundary, while dS_{irr} are due to irreversible processes occurring within the system itself. When $dS_a = 0$, then the thermodynamic system is described as being 'adiabatic.' We particularly note that dS and dS_a may be negative, zero, or positive, but for dS_{irr} the following relations hold:

$$\begin{aligned} &\mathrm{d}S_{\mathrm{irr}} = 0 \quad \text{for reversible changes,} \\ &\mathrm{d}S_{\mathrm{irr}} > 0 \quad \text{for irreversible changes,} \\ &\mathrm{d}S_{\mathrm{irr}} < 0 \quad \text{is for all macroscopic systems, impossible.} \end{aligned} \tag{K.19}$$

These statements—(K.19)—are known as the 'second law of thermodynamics.' Uncovering this new law of physics is the 'payoff' for defining the entropy potential.

The second law means that $\mathrm{d}S_{\mathrm{tot}} = \mathrm{d}S_{\mathrm{sys}} + \mathrm{d}S_{\mathrm{sur}} = \mathrm{d}S_{\mathrm{a}} + \mathrm{d}S_{\mathrm{irr}} + \mathrm{d}S_{\mathrm{sur}} = \mathrm{d}S_{\mathrm{irr}} \geq 0$. The equality only holds for reversible processes, while all 'spontaneously occurring processes' (which is everything we see happening in the natural environment) result in an increase of total entropy (i.e. total entropy being the sum of the entropy in the system and the entropy in the surroundings). In fact, all actual processes are irreversible, though some are very slow processes that may approach reversible conditions (e.g. convection in the earth's mantle).

Let us have a look at an example involving an ideal gas. The entropy of an ideal gas is a function of both its volume and its temperature (which turns out to be true of all materials), and we know that a reversible process means that the total change in entropy is zero. When there is no heat flow across the system boundary ($\mathrm{d}Q = 0$), the process conditions are referred to as adiabatic conditions. When there is not heat flow across the system boundary, an adiabatic, reversible process is 'isentropic' (meaning $\mathrm{d}S = 0$ and so there is no change in either the entropy of the system or the entropy of the surroundings). So, we see from Eq. (K.14) that for an ideal gas, an isentropic process requires

$$\begin{gathered} \mathrm{d}S = c_v \mathrm{d}\ln T + NR\mathrm{d}\ln V = 0 \\ \ln\frac{T_1}{T_2} = \frac{NR}{c_v}\ln\frac{V_2}{V_1} = \frac{2}{3}\ln\frac{V_2}{V_1} \end{gathered} \tag{K.20}$$

This means that if work is extracted from a gas through volume expansion (and there is no heat flow across the boundary) then the temperature of the gas falls.

Turning to non-equilibrium thermodynamic processes, consider now an adiabatic expansion of the same gas, but this time very quickly, as opposed to 'quasi-statically' (i.e. infinitesimally slowly) as occurs in processes always in thermodynamic equilibrium. If the volume expands quickly enough, then $p\mathrm{d}V$ work approaches zero (the system boundary is moving away almost at the speed the gas molecules are moving toward it). In this limiting case, the temperature of the gas remains unchanged. So, we approach isothermal, adiabatic conditions. The change in entropy change is given by

$$\begin{gathered} \mathrm{d}S = c_v \mathrm{d}\ln T + NR\mathrm{d}\ln V \\ \Delta S_{\mathrm{irr}} = NR\ln\frac{V_2}{V_1} \end{gathered} \tag{K.21}$$

In this case, the rapid volume expansion resulted in irreversible processes within the rapidly expanding gas that produced entropy. It is clear that the total entropy of system and surroundings (i.e. change in entropy of the system $\Delta S_{\mathrm{irr}} = NR\ln\frac{V_2}{V_1}$ and change in entropy of the surroundings $\Delta S_{\mathrm{sur}} = 0$) has increased, i.e. $\Delta S_{\mathrm{tot}} = \Delta S_{\mathrm{irr}} > 0$, as it should according to the second law of thermodynamics for irreversible processes.

It is clear from this example that entropy in irreversible processes behaves quite unlike the 'conservation laws.' Entropy can simply 'appear' in non-equilibrium processes, but for conserved quantities like energy, this cannot and does not occur. When entropy simply appears during a process (which is the case for all practical situations), then the useful work that can be extracted from the thermodynamic system decreases (in the case of the very rapid gas expansion, $p\mathrm{d}V$ work approaches zero and little mechanical work is extracted). This helps to explain for example, why the efficiency of a petrol or diesel engine decreases at higher engine speeds.

We note that following an irreversible expansion (or compression) of the gas under adiabatic conditions, the system has been permanently changed by the irreversible process in the sense that states (i.e. positions in the P–V–T diagram) that were accessible previously under reversible, adiabatic conditions are no longer accessible. For example, if the temperature is unchanged but the system has a new volume (as in the previous example), then the isentropic compression line for future reversible processes is permanently shifted in the P–V–T diagram.

We also note that for this irreversible (adiabatic) process, the heat exchange with the surroundings ΔQ was zero. However, if the same change in states is done reversibly, say by an isothermal expansion, then $\Delta Q = NR\ln\frac{V_2}{V_1}$. We note that

$$\Delta Q_{\mathrm{irr}} < \Delta Q_{\mathrm{rev}} \tag{K.22}$$

We can also use Eq. (K.17) to gain greater insight into the origin of the 'thermodynamic pressure' p acting on the boundary of materials. To this end, we can re-arrange Eq. (K.16) as

$$-p = \frac{\mathrm{d}U}{\mathrm{d}V} - T\frac{\mathrm{d}S}{\mathrm{d}V} \tag{K.23}$$

and then find the rate of change of p w.r.t. volume. That is,

$$-\mathrm{d}p = \frac{\partial}{\partial V}\left(\frac{\mathrm{d}U}{\mathrm{d}V} - T\frac{\mathrm{d}S}{\mathrm{d}V}\right)\mathrm{d}V \tag{K.24}$$

The quantity $\mathbb{C} = \frac{\partial}{\partial V}\left(\frac{\mathrm{d}U}{\mathrm{d}V} - T\frac{\mathrm{d}S}{\mathrm{d}V}\right)$ is known as the 'bulk modulus', which is the stiffness of the material to volumetric compression.

Importantly, we see that the stiffness of a material is a function of rate of change of both the rate of change of internal energy and the rate of change of entropy w.r.t volume. Exactly how stiff a material is to compression depends on the material, but the relative sources of stiffness vary from material to material. For example, solids like metals and ceramics tend to have a stiffness that mainly arises from the rate of change of the rate of change of internal energy (due to stretching of chemical bonds), while in polymers like rubber and plastics the rate of change of entropy becomes much more important (conformational entropy https://en.wikipedia.org/wiki/Conformational_entropy). Indeed, the compression of many solids generates little reversible change in entropy of the solid material, and consequently there is little heat flow across the system boundary. This means most of the $p\mathrm{d}V$ work on the system surface is stored as internal (strain) energy in the material (e.g. stretching of chemical bonds). In contrast to ceramic and metal solids, where the rate of change of internal energy dominates its stiffness, for an ideal gas, the rate of change of entropy w.r.t volume completely dominates its stiffness.

For example, we can evaluate the stiffness of an ideal gas to an isothermal volumetric compression under reversible conditions. Evaluating Eq. (K.24), we find

$$\begin{aligned} \mathbb{C} &= \frac{\partial}{\partial V}\left[\frac{\mathrm{d}\left(\frac{3NRT}{2}\right)}{\mathrm{d}V} - T\frac{\mathrm{d}\left(\frac{3NR}{2}\ln T + NR\ln V + \text{constant}\right)}{\mathrm{d}V}\right] \\ \mathbb{C} &= \frac{\partial}{\partial V}\left[\frac{3NR}{2}\frac{\mathrm{d}T}{\mathrm{d}V} - \frac{3NR}{2}\frac{\mathrm{d}T}{\mathrm{d}V} - \frac{NRT}{V}\right] \\ \mathbb{C} &= \frac{NRT}{V^2} \end{aligned} \tag{K.25}$$

For isothermal, equilibrium conditions, we see that the stiffness of the gas is entirely entropic in origin. And so, a large amount of heat is exchanged with the surroundings to ensure that the change in total entropy is zero (as required for a reversible process). As mentioned above, this behavior is in contrast to many solids, where solid stiffness is dominated by the change in internal (strain) energy. Because the entropy change is small in many solids, the heat exchanged with the surroundings is small (e.g. isothermal compression of steel results in little heat flow between system and surroundings).

It is of interest here to consider 'excluded volume effects,' as excluded volume plays an important role in the behaviour of semidense solutions, and so in explaining the observed osmotic pressure of semidense solutions. The 'excluded volume' of a molecule is the volume that is inaccessible to other molecules in the system as a result of the presence of the first molecule, due to the finite size of one molecule. Consider the effect of excluded volume on the stiffness of a gas containing molecules that are a significant fraction of the total volume. A suitable model is a van der Waals gas, which is defined by the equation of state for N moles of gas as:

$$p = \frac{NRT}{V - Nb} - \frac{aN^2}{V^2} \tag{K.26}$$

where Nb represents the finite size of the molecules and a/V^2 represents the reduction in pressure due to interactions between molecules. Importantly, we see that if a is zero, the excluded volume (Nb) due to the finite size of the particles increases the pressure of the gas compared to an ideal gas with the same number of moles in the same volume. From the viewpoint of the kinetic theory of gases, the pressure is increased because the finite size of the molecules leads to more molecular collisions per unit time by the molecules with the walls of the container. And if there are more collisions with a semipermeable membrane by solute molecules in a solution, then this means the osmotic pressure will be increased because of the excluded volume of the finite sized molecules.

In passing, we also note that if a is greater than zero, then the pressure in the container is reduced. This is because the intermolecular attraction between the molecules results in the position of the molecules being not a uniform probability distribution throughout the volume, but instead the probability distribution varies with proximity to other molecules. In terms of the kinetic theory of gases, potential energy wells due to the interaction energy lead to a non-uniform radial distribution functions (a radial distribution function describes the proximity of other molecules about a single particle) and to spatial correlations in particle velocities. These spatial correlations in position and velocity lead to fewer collisions per unit time with the wall of the container and so a lower pressure.

For a van der Waals gas, Eq. (K.20) becomes

$$dS = c_v d \ln T + NR d \ln(V - Nb) \tag{K.27}$$

where c_v is independent of temperature and volume, as for an ideal gas. The total energy (U) of the van der Waals gas is

$$U = \frac{3NRT}{2} - \frac{\alpha N^2}{V} \tag{K.28}$$

Then using Eq. (K.24), the stiffness of the van der Waals gas is given by

$$\mathbb{C} = \frac{NRT}{(V - Nb)^2} - \frac{\alpha N^2}{V^3} \tag{K.29}$$

Setting α to zero, and comparing this Eq. (K.25), we see that for the same temperature, number of molecules and volume, the van der Waals gas is stiffer than an ideal gas. This is purely an excluded volume effect due to the finite size of the molecules. And importantly, the same stiffening effect happens in other materials that have a significant entropic contribution to their stiffness. For example, at high-concentration solutions of macromolecules like albumin and aggrecan, excluded volume effects lead to a very strong increase in osmotic pressure.

Reminding ourselves again of Eq. (K.24), we observe that for some materials the relative contributions to stiffness of strain energy and entropic effects vary with the strain level (i.e. with the concentrations of macromolecules). An example of this behavior is provided by polymer solutions like aggrecan in aqueous solution. When an aggrecan in aqueous solution is compressed, the work done on the mixture in the thermodynamic system is at first mainly stored as internal (osmotic) strain energy (due to double-layer overlap, e.g. over the strain ranges that result in polymer separation distances 2–10 nm, i.e. 80 mg/ml of chondroitin sulfate and less). But at very high aggrecan concentrations (e.g. resulting in polymer separation distances of

<1 nm), the work is mainly stored as entropic changes in the polymers [e.g. see the simulations of Nap and Szleifer (2008)]. Nap and Szleifer (2008) explains:

> Moreover, at very small separations of the cylinders, the loss of conformation entropy of the polymer becomes the leading contribution. This indicates that the increase of the repulsion cannot be solely attributed to the counterion confinement. Instead, the repulsions arise from a subtle interplay between the mixing entropy of the ions, the conformation entropy of the polymer chains, and the solvent osmotic pressure. Their relative contributions are also strongly distance dependent.

Let us now consider a non-equilibrium compression of an ideal gas. In this case, the volume of the gas decreases at a finite speed, which is the case for everyday compression of gases. We look to the equation of state for a general expression of the gas stiffness. Let us implicitly differentiate Eq. (K.8) to find

$$-p\mathrm{d}V - V\mathrm{d}p = nR\mathrm{d}T \tag{K.30}$$

Re-arranging this expression leads to

$$\mathrm{d}p = [\frac{NRT}{V^2} - \frac{nR}{V}\frac{\mathrm{d}T}{\mathrm{d}V}]\mathrm{d}V \tag{K.31}$$

and so the stiffness of the gas for a reversible or irreversible compression is given by

$$\mathbb{C} = \frac{nRT}{V^2} - \frac{nR}{V}\frac{\mathrm{d}T}{\mathrm{d}V} \tag{K.32}$$

Comparing Eq. (K.25) with (K.32), we see that the stiffness $\mathbb{C}$ in Eq. (K.25) is exactly the same as Eq. (K.32) provided $\mathrm{d}T = 0$, which of course it is for an equilibrium isothermal compression. However, if the compression of the gas has a finite speed, the stiffness increases proportionally with the rate of change of its temperature $(\mathrm{d}T)$ w.r.t. the volume change $(\mathrm{d}V)$; i.e. the stiffness is increased by the term $\frac{nR}{V}\frac{\mathrm{d}T}{\mathrm{d}V}$.

In other words, the pressure required to compress the gas at finite speed is increased relative to the pressure required for an infinitesimally slow speed, and so the work required to compress the gas is increased when compressed at a finite speed relative to the work required at an infinitesimally slow speed. Exactly how much extra work $\left(\Delta w_{\mathrm{dis}} = \int p_{\mathrm{dis}}\mathrm{d}V\right)$ is required to compress the gas depends on what controls the rate of rise of temperature with a finite deformation rate, namely the details of the system geometry, the arrangement of forces on the system boundary, the speed of compression, and the rate of thermal energy transfer from where thermal energy is generated, through the gas, to the boundary of the system where heat exchange takes place. Nevertheless, this extra work Δw_{dis} done on the system, above that work required for a reversible compression, is converted into 'new entropy' (which just 'appears' in an irreversible process), viz.,

$$\Delta S_{\mathrm{irr}} = \frac{\Delta w_{\mathrm{dis}}}{T} \tag{K.33}$$

This is consistent with the second law of thermodynamics as described by Eq. (K.19). For an irreversible process, we see that $S_{\mathrm{irr}} > 0$.

We note here that we have split work done on the system into reversible work associated with the work of an thermodynamic equilibrium process, $\mathrm{d}w_{\mathrm{rev},j} = \sum_j^J L_{\mathrm{rev},j}\mathrm{d}l_j$, and non-recoverable or dissipated work associated with a finite speed of compression, $\mathrm{d}w_{\mathrm{dis},j} = \sum_j^J L_{\mathrm{dis},j}\mathrm{d}l_j$. Here, generalized 'work coordinates' have been denoted l_j (e.g. one work coordinate is volume) and the generalized 'work coefficients' have been denoted L_j (e.g. one work coefficient is pressure). The work coordinates for reversible work and dissipated work are the same for a gas, but the work coefficients are different for reversible and irreversible processes. We note that for elasto-plastic deformation of solids, the reverse may be the case (i.e. the work coordinates may be different for the elastic and plastic deformations, but the work coefficients are the same; i.e. the same stresses in the material are driving both the elastic and plastic deformations of the material).

We know that the work done in compressing the gas at finite speed is greater than the work done compressing the gas reversibly (i.e. $\Delta w_{\mathrm{rev}} + \Delta w_{\mathrm{dis}} > \Delta w_{\mathrm{rev}}$). What is the relationship between the heat transferred across the system boundary for irreversible and reversible compression. We can write the energy balance equation for both irreversible and reversible processes occurring between two states. That is,

$$\begin{aligned} \Delta U &= \Delta w_{\mathrm{rev}} + \Delta w_{\mathrm{dis}} + \Delta Q_{\mathrm{irr}} \\ \Delta U &= \Delta w_{\mathrm{rev}} + \Delta Q_{\mathrm{rev}} \end{aligned} \tag{K.34}$$

Because ΔU is a state function and so only depends on the initial and end points, its change is the same for both reversible and irreversible processes (it is, after all, the law of conservation of energy). Subtracting both equations leads to the conclusion that

$$\Delta Q_{\mathrm{irr}} < \Delta Q_{\mathrm{rev}} \tag{K.35}$$

It is noted that $|\Delta Q_{\mathrm{irr}}| > |\Delta Q_{\mathrm{rev}}|$, but $|\Delta Q_{\mathrm{irr}}|$ has a negative sign (which explains relation (K.35)), meaning more heat is being lost from the system in the irreversible process than for a reversible process between the same two states. Clearly, more heat is being lost from the system in the irreversible process ($|\Delta Q_{\mathrm{irr}}| > |\Delta Q_{\mathrm{rev}}|$) because more work is done on the system (i.e. $\Delta w_{\mathrm{rev}} + \Delta w_{\mathrm{dis}} > \Delta w_{\mathrm{rev}}$).

We note that relation (K.35) is also true for the previous example of an irreversible, rapid expansion of a gas (see Eq. K.22), and indeed it turns out that this result is true for all irreversible processes (Mahan and Myers 1987). This means the thermodynamic behavior we have observed above for an ideal gas is similar to that for all material types in the sense that while the details of the irreversible processes taking place within the material may differ depending on material properties, the thermodynamic behavior can be represented in the same way. In other words, the thermodynamic framework is independent of the material, but different constitutive equations are required to make the various state quantities definite for a particular material and thermodynamic process.

For example, if a 'polymer material' is compressed at a finite speed, the work required to compress it is greater than if it were compressed slowly (just as we found for a gas). The additional work required Δw_{dis} is dissipated inside the material, producing entropy $\Delta S_{\text{irr}} = \frac{\Delta w_{\text{dis}}}{T}$ (just as we found for a gas). If the system is isothermal, the work coefficient (i.e. the stress in the material) is initially elevated as the material is compressed at finite speed above that for reversible compression $\Delta S_{\text{irr}} = \frac{\Delta w_{\text{dis}}}{T}$ (just as we found for a gas). When compression of the polymer at finite speed stops, the stress in the material relaxes back to its (reversible) equilibrium stress value (just as it did for isothermal compression of a gas at finite speed).

However, the magnitudes of the changes in state parameters vary with the material type. And further, the 'internal processes' dissipating work differ between materials. For example, in polymer solids the internal processes dissipating work and creating entropy are referred to as 'viscoelasticity.' However, gases are not (said to be) viscoelastic materials, as the internal (micro)processes responsible for the delay in each reaching the equilibrium stress are quite different (see the discussion for a gas loaded at a finite speed).

The brief explanation of the internal processes in polymers responsible for viscoelasticity is as follows (Grodzinsky 2011). As the polymer solid is deformed, it causes changes in the orientation and configuration of the polymer molecules. Whether a polymer displays viscoelasticity in an experiment depends on the relative rates of deformation that drive changes in polymer orientation and configuration, and the time constant for the polymer to relax back to its equilibrium configuration. On a short timescale, the polymer does not have time to relax, and the distortions in the polymer produce a new higher energy state for that material. The result is the material appears to be stiffer than it would have been had the compression been slow; consequently, it requires higher stress to deform it at the same rate, and so more work has to done to compress it at a finite speed. The time is taken for a polymer to relax back to its equilibrium position related to the material properties of the polymer (e.g. the stiffness of the polymer bonds and the degree of polymer branching), the density of the material (i.e. the spatial proximity of other polymers that may also be attempting to find their equilibrium position), the presence or absence of 'filler' materials (which may have to be squeezed out of the way to allow room for the polymer to relax).

At this point, we also note that the 'energy conservation equation' employed for our analysis of an ideal gas (Eq. K.6) naturally holds for any thermodynamic process on any system, so this equation can be modified to include 'dissipative work' done on the system in the energy balance for the system, that is,

$$\mathrm{d}Q = \mathrm{d}U + p\mathrm{d}V - \mathrm{d}w_{\mathrm{dis}} \tag{K.36}$$

where $\mathrm{d}Q$ is the heat crossing the boundary of the system. More generally, the heat transfer across the boundary $\mathrm{d}Q$ can be expressed using our generalized work coordinates, viz.,

$$\mathrm{d}Q = \mathrm{d}U - \sum_{j=1}^{J} L_{\mathrm{rev},j}\mathrm{d}l_j - \sum_{k=1}^{K} L_{\mathrm{dis},k}\mathrm{d}l_k \tag{K.37}$$

We mention that the generalized Gibbs equation (K.17), together with Eq. (K.37), provides the starting point for further developing the discussion of irreversible processes.

Now, we can generalize the previous analysis by allowing the addition or subtraction of material from the system. For example, N moles of gas may become $N + \mathrm{d}N$ moles of gas after injection of $\mathrm{d}N$ moles of gas into the system. If material can be added or subtracted from the system, the system is said to be an 'open' thermodynamic system. In this case, the increment of total internal energy is given by

$$\mathrm{d}U = T\mathrm{d}S - p\mathrm{d}V + \mu\mathrm{d}N \tag{K.38}$$

where μ is the 'chemical potential.' We note that internal energy is the total energy required to create the system and includes the kinetic energy of the constituent molecules (translational, rotational, and vibrational) U_{kinetic}, plus the external work done on the system $(-p\mathrm{d}V)$ and the potential energy of the system $U_{\mathrm{potential}}$ including positional potential energy, the chemical potential energy of chemical reactions and bonding between the atoms and molecules, mixing of solutions and plus any other potential energies such as internal strain energy stored at the interfaces between a mixture of phases (e.g. conformational entropy or electroosmotic forces arising from local charge heterogeneity and diffuse double layers comprising co-ions and counterions).

The chemical potential is defined to be

$$\mu = \frac{\partial U}{\partial N} \tag{K.39}$$

that is, μ is the rate of change of internal energy w.r.t. the moles in the system. For an ideal gas, we can find the chemical potential μ_{ig} from Eq. (K.5), viz.,

$$\mu_{\mathrm{ig}} = \frac{\partial U}{\partial N} = \frac{3RT}{2} \tag{K.40}$$

which is simply the translational kinetic energy of the a mole of gas. If there are several different species present in the system, then again we can have an open thermodynamic system, and various species may be added or subtracted from the thermodynamic system. In this case, we define the chemical potential to be

$$\mathrm{d}U = T\mathrm{d}S - p\mathrm{d}V + \sum_{i=1}^{N} \mu_i \mathrm{d}N_i \tag{K.41}$$

where $\mu_i = \frac{\partial U}{\partial N_i}$ (i.e. all other species are held constant, while $\mathrm{d}N_i$ is added to the system). μ_i is known as the 'chemical potential' or as the 'partial molar Gibbs free energy' (they are the same quantity).

We see that if the internal energy, entropy, and volume are all held constant, then

$$\sum_{i=1}^{N} \mu_i \mathrm{d}N_i = 0 \tag{K.42}$$

Under these conditions, the addition of $\mathrm{d}N_i$ will require adjustments in either masses or chemical potentials of the other species, so the species within the system are coupled to one another through a constraint created by Eq. K.42. However, usually it is very difficult for these variables to be held constant in an experiment, and so the internal energy, entropy are free to change with the addition of $\mathrm{d}N_i$. We see that if we wanted to get at such a constraint, we need to express it in terms of variables that are more experimentally accessible (see the later discussion on potentials other than internal energy and entropy, which have the advantage of using more experimentally accessible variables).

Because of the changes in internal energy and entropy, it is clear that the addition of new material to the system may change the internal pressure within the system (which balances the external applied pressure (see Eq. K.17)). If the volume is fixed, this additional material changes the stiffness of the system (see Eq. K.18). This is in fact what happens when cells synthesize and secrete new extracellular matrix—for example, the compressive stiffness of the ECM increases as newly synthesized aggrecan is added to cartilage tissue, while the tensile stiffness increases as collagen is polymerized.

Further, chemical reactions may occur between the various species. Importantly from the viewpoint of biological systems, some chemical reactions are coupled together through 'molecular machines' (which enable the construction of macromolecules like DNA, RNA, proteins, or the creation of many other smaller molecules via various 'metabolic pathways'). If these molecular machines are operating within our system boundaries, then they represent irreversible 'internal processes' and are responsible for the production of entropy within the system. This entropy production may not be strongly dependent on work and heat flow across the system boundary, but rather depend on the chemical potential energy transferred into the system along with mass $\mathrm{d}N_i$. For example, glucose may enter the thermodynamic system and be transformed by molecular machines into ATP within cells within the thermodynamic system. These machines produce entropy, no matter which way the machine runs—forward or backward. For example, the same molecular machine that generates ATP from ADP and P does so using the chemical energy stored in a proton gradient (between the inner and outer mitochondrial membranes) and can also generate a proton gradient from a chemical potential between ATP and ADP and P (i.e. the molecular motor can run forward or backward) (see http://en.wikipedia.org/wiki/ATP_synthase). However, the operation of this 'reversible machine' remains consistent with the second law of thermodynamics in that all irreversible processes produce entropy, i.e. $dS_{irr} > 0$.

The thermodynamic system may be open or closed. In either case, if there are chemical reactions in the solution, the dN_i may be related by the stoichiometry of the chemical reactions involved. At thermodynamic equilibrium, by definition reversible chemical reactions are also at thermodynamic equilibrium. However, if the state of the system is perturbed (e.g. if the system is compressed), consequent changes in the concentrations of species may lead to a new chemical equilibrium becoming established. In fact, this is an example of Le Chatelier's principle: See http://en.wikipedia.org/wiki/Le_Chatelier%27s_principle).

It is also apparent that reversible chemical reactions represent one way in which internal (chemical potential) energy may be stored within the system. As the thermodynamic system moves away from its initial equilibrium position, say through (drained) compression of the thermodynamic system, the chemical potential energy of the system increases. Entropic changes occur as the reaction proceeds. In this way, we see that chemical reactions within the system are coupled to the whole thermodynamic system, i.e. to the internal energy, to the mechanical work, to the heat flow, and to the entropy of the system. It is entirely possible, for example, that volumetric compression of the extracellular matrix promotes certain reactions while inhibiting others. In this way, we see that in biological system mechanical work and chemical equilibrium are intimately related [see also the Gibbs–Duhem relation, i.e. Eq. (K.91)]. There is little data on the way chemical reactions *in vivo* are perturbed by compression or tension (Yao et al. 2014).

As mentioned earlier, there may be other forms of work operating on the system in addition to pressure–volume (pdV) work. For example, pdV may be generalized to $\sigma_{ij}e_{ij}$ for solid materials (these conjugate variables include the possibility of work associated with shear stress and shear strain in solids as well as pressure–volume work), but there are other forms of work too, for example, moving a charge some distance in a magnetic field or electrochemical field, the work of creating a new surface for the system as it expands or contracts (e.g. $\sigma_{st}\Omega_{sa}$), or the work of a 'fiber' contracting and reducing the systems volume (e.g. the contraction of actin and myosin filaments within the system, or in a muscle fiber acting on the system).

Of course, the work conjugates may extract energy from the system and do useful work; i.e. work is done by the system on the surroundings. For example, pressure–volume work may become 'shaft work' (if coupled to a piston and crankshaft, or a proton gradient may become shaft work http://en.wikipedia.org/wiki/ATP_synthase), or pressure–volume work of the heart may pump blood around the body (https://en.wikipedia.org/wiki/Pressure%E2%80%93volume_loop_analysis_in_cardiology).

Including the additional forms of reversible work in our analysis, let the generalized 'work coordinates' be denoted l_j and the generalized 'work coefficients' be denoted L_j. Then, Eq. (K.41) may be modified to become

$$\begin{aligned} \mathrm{d}U &= T\mathrm{d}S - p\mathrm{d}V + \sum_{i=1}^{N} \mu_i \mathrm{d}N_i + \sum_{j}^{J} L_{\mathrm{rev},j}\mathrm{d}l_j \\ T\mathrm{d}S &= \mathrm{d}U + p\mathrm{d}V - \sum_{i=1}^{N} \mu_i \mathrm{d}N_i - \sum_{j}^{J} L_{\mathrm{rev},j}\mathrm{d}l_j \end{aligned} \tag{K.43}$$

State variables may be classified as 'intensive' or 'extensive.' Intensive variables do not depend on the size of the system. Examples of intensive variables are work coefficients L_j like surface tension (σ_{st}), pressure p and σ_{ij}, temperature T, and chemical potential μ, as well as material properties such as viscosity, density ρ, and specific volume $\bar{v}$.

Extensive variables are additive, so they increase as the size of the system increases. Examples of extensive variables include work coordinates l_j like surface area (Ω_{sa}), volume V, entropy S, and the number of moles in the system N, as well as others like length, electrical charge, and momentum.

Let some function F be a function of the intensive variables β_i and the extensive variables λ_j, i.e. $F(\beta_i, \lambda_j)$. If the function F itself is to be extensive, F then by definition must have the property,

$$F(\beta_i, k\lambda_j) = kF(\beta_i, \lambda_j) \tag{K.44}$$

But this is also the definition a homogeneous function of degree one. It then follows from Euler's theorem for homogenous functions (of degree one) that (https://en.wikipedia.org/wiki/Homogeneous_function),

$$F(\beta_i, \lambda_i) = \sum_{i=1}^{N} \lambda_i \frac{\partial F}{\partial \lambda_i} \tag{K.45}$$

where the partial derivative has the usual meaning (i.e. all λ_j's are held constant except λ_i). This is interesting, because if we let F be work, then because work is an extensive variable we find from Euler's theorem for homogenous functions that it must have the form

$$\Gamma(\beta_i, \lambda_i) = \sum_{i=1}^{N} \beta_i \lambda_i \tag{K.46}$$

where β_i is $\frac{\partial \Gamma}{\partial \lambda_i}$. In other words, work appears as the sum of the products of work producing 'conjugate pairs' (e.g. p and V or L_i and l_i or $\int \sigma_{ij} e_{ij} \mathrm{d}V$). We note that for each term $\beta_i \lambda_i$ to be additive they must have the same units, usually per unit volume or per unit mass or per mole. We note that if there are two volumes (e.g. deformed and undeformed volumes), we have to be careful that we adding the correct conjugate pairs so they have a meaningful sum (generally for large deformation solid mechanics, the preferred work per unit volume is per unit volume of the undeformed configuration is preferred and is used consistently for each conjugate pair).

We mention that division of one extensive quantity by another extensive quantity can transform extensive variables into intensive variables. Let $Z(T, L_i, N_k)$ be an extensive quantity (e.g. volume V, entropy S, internal energy U, enthalpy H, Gibbs free energy G, Helmholtz free energy A). Then, we note the total differential $\mathrm{d}Z$ is defined in the usual way,

$$\mathrm{d}Z = \frac{\partial Z}{\partial T}\mathrm{d}T + \sum_{j=1}^{J} \frac{\partial Z}{\partial L_j}\mathrm{d}L_j + \sum_{k=1}^{K} \frac{\partial Z}{\partial N_k}\mathrm{d}N_k \tag{K.47}$$

Now dividing one extensive quantity by another, the 'generalized density' $\hat{Z}$ is defined as

$$\hat{Z} = \frac{Z}{V} \tag{K.48}$$

where the density is an intensive variable. The 'mole fraction' is defined as

$$x_k = \frac{N_k}{\sum_{k=1}^{K} N_k} \tag{K.49}$$

where the mole fraction is an intensive variable. The 'molar concentration' is defined as

$$c_k = \frac{N_k}{V} \tag{K.50}$$

where the concentration is an intensive variable. The 'partial molar quantity' is defined to be

$$Z_k = \frac{\partial Z}{\partial N_k} \tag{K.51}$$

where Z_k is an intensive variable (e.g. partial molar volume V_k, partial molar entropy S_k, partial molar internal energy U_k, partial molar enthalpy H_k). As noted above, because Z is an extensive quantity (and so a homogenous function of first order), then Euler's theorem for homogenous functions must have the form

$$Z = \sum_{k=1}^{K} Z_k N_k \tag{K.52}$$

But in this case, the total derivative of Z must be

$$\mathrm{d}Z = \sum_{k=1}^{K} \frac{\partial Z}{\partial N_k}\mathrm{d}N_k + \sum_{k=1}^{K} \frac{\partial Z}{\partial Z_k}\mathrm{d}Z_k = \sum_{k=1}^{K} Z_k\mathrm{d}N_k + \sum_{k=1}^{K} N_k\mathrm{d}Z_k \tag{K.53}$$

Now, equating Eqs. (K.47) and (K.53) leads to

$$\frac{\partial Z}{\partial T}\mathrm{d}T + \sum_{j=1}^{J} \frac{\partial Z}{\partial L_j}\mathrm{d}L_j - \sum_{k=1}^{K} \frac{\partial Z}{\partial Z_k}\mathrm{d}Z_k = 0 \tag{K.54}$$

If $Z = G$, Eq. (K.54) reveals an important relationship known as the Gibbs–Duhem relation (see Eq. K.91).

$$S\mathrm{d}T + \sum_{j=1}^{J} l_j\mathrm{d}L_j - \sum_{k=1}^{K} N_k\mathrm{d}\mu_k = 0 \tag{K.55}$$

The internal energy U is a state function or potential, meaning it has an exact differential. However, we can find additional potentials by subtracting potentials to internal energy. Suitable potentials to subtract are pV and ST. Why would we want to find new potentials? As mentioned earlier, the key advantage is that new potentials facilitate a change of variables. The change of variables (referred to as 'Legendre transformations' in the thermodynamics literature) enables the various quantities in Eq. (K.41) to be 'dissected out' by removal (e.g. subtraction) of a potential from the original potential and the new potential employed theoretically. Or with the change of variables, thermodynamic quantities can be evaluated experimentally in much more experimentally feasible ways than would be possible otherwise. We will see that different thermodynamic potentials are best suited to particular purposes.

For example, subtracting ST from the internal energy results in the Helmholtz free energy potential A,

$$A = U - TS \tag{K.56}$$

We can see some logic in forming this new potential, as ST is in some sense the total amount of 'bound energy' (Slater 1939). So, subtracting this 'bound energy' from the 'internal energy' in some sense results in the 'free energy' that is available to do mechanical work.

Taking the differential of A and substituting Eq. (K.38) lead to

$$\begin{aligned} \mathrm{d}A &= T\mathrm{d}S - T\mathrm{d}S - S\mathrm{d}T - p\mathrm{d}V + \sum_{i=1}^{N} \mu_i \mathrm{d}N_i + \sum_{j}^{J} L_j \mathrm{d}l_j \\ \mathrm{d}A &= -S\mathrm{d}T - p\mathrm{d}V + \sum_{i=1}^{N} \mu_i \mathrm{d}N_i + \sum_{j}^{J} L_j \mathrm{d}l_j \end{aligned} \tag{K.57}$$

This says the (Helmholtz) free energy available to do work is equal to $-S\mathrm{d}T - p\mathrm{d}V + \sum_{i=1}^{N} \mu_i \mathrm{d}N_i + \sum_{j}^{J} L_j \mathrm{d}l_j$. At constant temperature and mass (meaning no mass transfer between system and surroundings and no chemical reactions), the incremental free energy available to do work is simply $-p\mathrm{d}V + \sum_{j}^{J} L_j \mathrm{d}l_j$. If the work coordinates l_j are also held constant, then

$$\mathrm{d}A = -p\mathrm{d}V \tag{K.58}$$

So we see the Helmholtz free energy of the material increases as it is reversibly compressed from a stress free state, and then decreases as work is extracted from the system as it is reversibly expanded again. This happens for gas, liquid, or solid materials (though the stiffness generally increases in the same order, and so the deformations are smaller in liquids than gases and smaller in solids than in liquids for the same confining pressure).

We know that loading a material at finite speed results in irreversible internal processes and dissipative work being done on the system. Then, because

$$0 < \sum_{j}^{J} L_{\mathrm{irr},j} \mathrm{d}l_j \tag{K.59}$$

it follows that

$$\mathrm{d}A < -S\mathrm{d}T - p\mathrm{d}V + \sum_{i=1}^{N} \mu_i \mathrm{d}N_i + \sum_{j}^{J} L_{\mathrm{rev},j} \mathrm{d}l_j + \sum_{j}^{J} L_{\mathrm{irr},j} \mathrm{d}l_j \tag{K.60}$$

We note that for materials with many internal (micro)processes happening on a variety of length and timescales, the reversible Helmholtz free energy is to some extent defined by the timescale of practical interest, as this defines what processes may be regarded as reversible and irreversible. Obviously, problems may arise if the timescale of laboratory tests is much smaller than the timescale for long-term loadings and attendant long-term internal microprocesses (which happens quite frequently in practice).

Examples of 'spontaneous internal processes' leading to changes in the Helmholtz free energy as the timescale of interest lengthens are viscoelasticity (which represents a large variety of processes within the material, which are often represented as a Prony series to reflect each (micro)process that is occurring on a different timescale), damage to the 'solid' material (e.g. the material irreversibly lengthens and cannot return to its former shape (this is referred to as a plastic deformation), or damage due to 'parting' of the material (resulting in 'crack formation'), or the 'solid' material may dissolve in solution (e.g. through 'dissolution reactions' or other chemical reactions occurring on the surface of the material).

Coupling of reactions between 'solid phases' and 'solution phases' (which can be a matter of semantics when polymers and mers are both suspended as 'colloids in solution') leads to the discipline of chemo-mechanics. Chemo-mechanics is an important part of engineering biology. In chemo-mechanics, the 'chemical potential' of the material in the solid phase is the sum of the 'strain energy potential' and the 'strain-free solid phase chemical potential' (as elastic strain energy is stored in chemical bonds between atoms). For example, the strain energy in bone contributes to any difference in chemical potential energy of the material in the solid phase relative to the chemical potential energy of the material in the solution phase, and so the strain energy may promote precipitation or dissolution of the solid phase. In the corrosion literature, this phenomenon is referred to as 'stress corrosion,' but it has hardly been explored in the literature on biomechanics.

Nevertheless, constant temperature and mass are often convenient conditions for experiments on 'solids' materials at room temperature. These conditions are common in many everyday situations (e.g. the analysis of engineering structures at 'ambient temperature' or biological structures at body temperature). And because experimentalists can measure $\sum_j^J L_j \mathrm{d}l_j$ under isothermal conditions easily in the laboratory, it is therefore not difficult to evaluate the Helmholtz free energy of a material. For this reason, Helmholtz potentials feature prominently in mechanical studies, including biomechanics. However, it is important to realize that just because the term $T\mathrm{d}S$ has been removed from the internal energy potential, the influence of entropic changes within the material has not been removed from the analysis. Indeed, entropic changes always take place in a material as it is compressed or stretched.

For example, we can show for an ideal gas that the Helmholtz free energy potential does not exclude the effects of changes in both entropy and internal energy on the deformation response for equilibrium conditions. For it is clear from Eq. (K.58) that

$$\frac{\mathrm{d}A}{\mathrm{d}V} = -p \tag{K.61}$$

Now using the constitutive equation for an ideal gas (K.8) and the definition of material stiffness (K.25), we find

$$\mathbb{C} = \frac{\partial}{\partial V}\frac{\mathrm{d}A}{\mathrm{d}V} = -\frac{\partial p}{\partial V} = \frac{nRT}{V^2} \tag{K.62}$$

This is exactly the result found previously in Eq. (K.25) for a reversible process. This shows that the same stiffness may be found by considering either the rate of change of internal energy and entropy (Eq. K.24) or the rate of change of the

Helmholtz free energy. Experimentalists choose the later approach as the internal energy and entropy of the material are usually difficult to measure or estimate, whereas the measurement of the load deformation response of the material is straightforward to find by comparison.

Something interesting happens here if we use a 'Maxwell relation' (i.e. our 'test for exactness'—see equation K.2). One Maxwell relationship of interest here is

$$\frac{\partial^2 A}{\partial V \partial T} = \frac{\partial S}{\partial V} = -\frac{\partial p}{\partial T} = \frac{\partial^2 A}{\partial T \partial V} \tag{K.63}$$

This says that rather than measuring the rate of change of entropy w.r.t. volume (while holding the temperature constant), we can do a much simpler experiment in its place—measure the rate of change of pressure w.r.t. temperature (while holding the volume constant). These examples begin to demonstrate the practical utility of forming new potentials from the internal energy potential.

Now, we may add another potential pV (or more generally add parts of $-\sum_j^J L_j \mathrm{d}l_j$ required to create the system) to the internal energy potential, which leads to a new potential known as the enthalpy H,

$$H = U + pV \tag{K.64}$$

We can see some logic in adding this potential, as pV is in some sense the negative of the total amount of mechanical energy required to make room for the system in its current state (we note this work assumes the current pressure is constant throughout the process of creating the system). Incrementally subtracting the 'mechanical energy' involved in creation of the system from the 'internal energy' of the system in some sense results in focusing on the remaining 'bound energy,' that is changes due to heat flow ($T\mathrm{d}S$) across the system boundary.

Taking the differential of H and substituting into Eq. (K.43) lead to

$$\begin{aligned} \mathrm{d}H &= T\mathrm{d}S - p\mathrm{d}V + p\mathrm{d}V + V\mathrm{d}p + \sum_{i=1}^{N} \mu_i \mathrm{d}N_i + \sum_{j}^{J} L_j \mathrm{d}l_j \\ \mathrm{d}H &= T\mathrm{d}S + V\mathrm{d}p + \sum_{i=1}^{N} \mu_i \mathrm{d}N_i + \sum_{j}^{J} L_j \mathrm{d}l_j \end{aligned} \tag{K.65}$$

This says the heat energy absorbed, after removing $p\mathrm{d}V$ work, is equal to $T\mathrm{d}S$ at constant pressure (and constant work coordinates and constant composition).

It is noted that many chemical experiments performed in a flask, open to the atmosphere in the laboratory, become much simpler to analyze using the enthalpy (as $\mathrm{d}p$ is zero) because it is very difficult to evaluate the pressure–volume work for an experiment that is open to the atmosphere, as reactants and products are consumed and formed and expand in volume against the atmosphere. However, heat produced by a slow chemical reaction is relatively easily measured using a calorimeter, and if the experiment is carried out at constant pressure, the change in entropy of the reaction can be calculated easily from the heat evolved from the reaction (the temperature is easily measured). In other words, the enthalpy enables experimental measurements of the entropy change resulting from chemical reaction. This makes the enthalpy of great interest to chemists.

It is interesting for us to first consider the enthalpy change due to an isothermal compression of an ideal gas. The relevant form of the enthalpy equation is,

$$\mathrm{d}H = T\mathrm{d}S + V\mathrm{d}p \tag{K.66}$$

From Eqs. (K.14) and (K.8), we find

$$\begin{aligned}\Delta H &= NRT\ln(V_\mathrm{f}/V_0) + NRT\ln(p_\mathrm{f}/p_0)\\ &= NRT\ln(V_\mathrm{f}/V_0) + NRT\ln(V_0/V_\mathrm{f}) = 0\end{aligned} \tag{K.67}$$

This shows that the entropy change in the gas due to pressure–volume work (and the resultant heat loss from the system to environment) does not influence enthalpy.

To show how chemists use enthalpy, consider the melting of ice. Ice is actually less dense than water (so ice floats in water). Melting ice results in the water molecules coming closer together, and so mechanical work is done on the water molecules by the atmosphere. However, the enthalpy of fusion ΔH_wf ignores the heat flow across the system boundary induced by this mechanical work (as seen in the previous example) and relates directly to the heat energy required to melt the ice itself. Since heat is absorbed, ΔH_wf is positive, and the entropy change for a mole of frozen water becoming a mole of liquid water is

$$\Delta S_\mathrm{wf} = \frac{\Delta H_\mathrm{wf}}{T_\mathrm{f}} = \frac{6.025\ \mathrm{kJ}}{273\ \mathrm{K}} = 22.07\ \mathrm{kJ/^\circ K} \tag{K.68}$$

We note that the heat energy transported with particles ($\mathrm{d}N_i$) at constant pressure for an open system is $H_i\mathrm{d}N_i$, where H_i is the partial molar enthalpy of the ith species.

We now turn to the Gibbs free energy. Subtracting the mechanical work (i.e. adding pV) and TS from the internal energy leads to a potential known as the Gibbs free energy, usually denoted by G, viz.,

$$G = H - TS = U + pV - TS \tag{K.69}$$

We can see some logic in subtracting the two potentials, as pV is in some sense the total amount of mechanical energy required to make room for the system (and so is involved in thermodynamic system's creation), while TS is in some sense the total amount of 'bound' energy required to create the entropy within the system at the current temperature (we note this assumes both the current pressure and current temperature are constant throughout the process of creating the system). Subtracting this 'mechanical energy' and 'bound energy' involved in the system's creation from the 'internal energy' of the system in some sense results in the Gibbs free energy remaining, which turns out to be the 'chemical potential energy' within the system. We note here that the chemical potential (μ_i) of the ith species is defined as the partial molar Gibbs free energy,

$$\mu_i = \frac{\partial G}{\partial n_i}|_{T,p,n_{j \neq i}} \tag{K.70}$$

or

$$\mu_i = H_i - TS_i \tag{K.71}$$

Taking the differential of G and substituting Eq. (K.43) lead to

$$\begin{aligned} \mathrm{d}G &= T\mathrm{d}S - T\mathrm{d}S - S\mathrm{d}T - p\mathrm{d}V + p\mathrm{d}V + V\mathrm{d}p + \sum_{i=1}^{N} \mu_i \mathrm{d}N_i + \sum_{j=1}^{J} L_j \mathrm{d}l_j \\ \mathrm{d}G &= -S\mathrm{d}T + V\mathrm{d}p + \sum_{i=1}^{N} \mu_i \mathrm{d}N_i + \sum_{j}^{J} L_j \mathrm{d}l_j \\ \mathrm{d}G &= \mathrm{d}H - T\mathrm{d}S - S\mathrm{d}T \end{aligned} \tag{K.72}$$

This says the (Gibbs) free energy available to do work, having excluded the (pressure–volume) $p\mathrm{d}V$ work and the (entropic) $T\mathrm{d}S$ work, is equal to $-S\mathrm{d}T + V\mathrm{d}p + \sum_{i=1}^{N} \mu_i \mathrm{d}N_i + \sum_{j}^{J} L_j \mathrm{d}l_j$.

The meaning of the Gibbs free energy is perhaps clearest when we consider the system to have $\mathrm{d}T = \mathrm{d}p = \mathrm{d}l_j = 0$, and we focus on the (irreversible) time evolution of a system of chemical equations $\sum_{i=1}^{N} \mu_i \mathrm{d}N_i$ (which usually involves heat transfers due to the chemical reactions in the system moving to an equilibrium state). In this case $\mathrm{d}H = \mathrm{d}Q$ (i.e. the enthalpy change $(\mathrm{d}H)$ as the chemical system evolves is the heat absorbed $(\mathrm{d}Q)$, while at constant pressure and temperature and the enthalpy is typically measured at standard temperature and pressure). The final entropy change between two chemical states is the same whether the change occurred reversibly or irreversibly, so $T\mathrm{d}S = Q_{\mathrm{rev}}$. Then, from Eq. (K.72)

$$\Delta G = \mathrm{d}H - T\mathrm{d}S = \Delta Q - \Delta Q_{\mathrm{rev}} \tag{K.73}$$

But by Eq. (K.35), $\Delta Q_{\mathrm{irr}} < \Delta Q_{\mathrm{rev}}$, so for spontaneous (irreversible) processes in chemical systems,

$$\mathrm{d}G < 0 \tag{K.74}$$

More generally, we can conclude that for any spontaneous process, the Gibbs free energy of the system decreases. Further, the maximum work that can be recovered at constant temperature and pressure is ΔG, but in all practical applications it is somewhat less, even when the reaction occurs slowly.

The Gibbs free energy also enables great insight into the equilibrium processes in chemical systems. Let a chemical reaction be denoted

$$aA + bB \rightleftharpoons cC + dD \tag{K.75}$$

where A, B, C, D are chemical species and a, b, c, d are the stoichiometric coefficients for the chemical reaction. According to Theophile de Donder's theory of 'affinities,' a system of chemical reactions reaches equilibrium when the weighted sum of the chemical potentials of the species is minimized. The weights are the stoichiometric coefficients for the chemical reaction.

First, consider the 'standard free energy of formation,' ΔG_f^0. ΔG_f^0 is the Gibbs free energy change associated with forming one mole of the chemical species in its standard state (e.g. formation of species A in Eq. (K.75)) from the elements in their standard states, viz.,

$$\Delta G_\mathrm{f}^0 = \Delta H_\mathrm{f}^0 - T\Delta S_\mathrm{f}^0 \tag{K.76}$$

where the formation takes place at a single temperature, often 298 K (https://en.wikipedia.org/wiki/Standard_state). We note that the standard free energy of formation of the elements is defined to be zero. The standard free energy of the elements is our fundamental reference point from which changes are measured. The standard state of a species is a convenient derived reference point from which subsequent changes are measured.

Second, consider the 'standard free energy change' ΔG^0 for the chemical reaction shown in Eq. (K.75). The 'standard free energy change' is defined to be the free energy change associated with conversion of the reactants in their standard states into products in their standard states via a chemical reaction, viz.,

$$\Delta G^0 = \sum \Delta G_\mathrm{f}^0(\text{products}) - \sum \Delta G_\mathrm{f}^0(\text{reactants}) \tag{K.77}$$

To calculate the standard free energy change, let us start with a chemical reaction involving ideal gases, as this is the simplest case. The standard state for an ideal gas is one atmosphere pressure and 298 K. For N_k moles of the kth ideal gas, the free energy change associated with a change in pressure p_k is given by Mahan and Myers (1987),

$$\begin{gathered}\int_{G_k^0}^{\Delta G_k} N_k dG = \int_{p^0}^{p_k} \frac{N_k RT}{p_k} \mathrm{d}p_k \\ N_k(\Delta G_k - G_k^0) = N_k RT \ln \frac{p_k}{p^0} = N_k RT \ln p_k\end{gathered} \tag{K.78}$$

where p_k is in atmospheres of pressure. Note that $p^0 = 1$ atmosphere, and so the pressure in $N_k RT$ ln p_k is to be understood in the sense that it is referenced to unit pressure (i.e. the standard state). In this way, the natural logarithm of the (normalized) pressure has no units (i.e. it is just a number).

Then, for Eq. (K.75) we have

$$\begin{aligned} \Delta G &= \sum \Delta G(\text{products}) - \sum \Delta G(\text{reactants}) \\ \Delta G &= c\Delta G(C) + d\Delta G(D) - a\Delta G(A) - b\Delta G(B) \end{aligned} \tag{K.79}$$

We can now define the free energy change for a chemical reaction w.r.t. the standard state. Now, substituting Eq. (K.78) into Eq. (K.79) leads to

$$\begin{aligned} \Delta G = {} & cG^0(C) + dG^0(D) - aG^0(A) - bG^0(B) \\ & + cRT\,\ln p_C + dRT\,\ln p_D - aRT\,\ln p_A - bRT\,\ln p_B \end{aligned} \tag{K.80}$$

where p_A is the partial pressure of species A. That is,

$$\Delta G = \Delta G^0 + RT\,\ln\left[\frac{(p_C)^c (p_D)^d}{(p_A)^a (p_B)^b}\right] \tag{K.81}$$

Let us now define the extent of the (forward) chemical reaction ξ, such that

$$\mathrm{d}N_k = v_k \mathrm{d}\xi \tag{K.82}$$

where v_k are stoichiometric coefficients (it is noted v_k for reactants are taken as negative, while v_k for products are taken as positive). Then, Eq. (K.81) may be represented by the extent of the chemical reaction, viz.,

$$\frac{\mathrm{d}G}{\mathrm{d}\xi} = \Delta G^0 + RT\ln\left[\frac{(p_{C_0+cd\xi})^c (p_{D_0+dd\xi})^d}{(p_{A_0-ad\xi})^a (p_{B_0-bd\xi})^b}\right] = \Delta G^0 + RT\ln\left[\frac{(p_C)^c (p_D)^d}{(p_A)^a (p_B)^b}\right] \tag{K.83}$$

where p_{A_0} is the initial partial pressure of the gas A. Let the chemical reaction now proceed to equilibrium. That is,

$$\frac{\mathrm{d}G}{\mathrm{d}\xi} = 0 \tag{K.84}$$

At this point, de Donder's chemical affinity for reaction $r(A_\mathrm{r})$ is zero, viz.,

$$A_\mathrm{r} = -\sum_{k=1}^{K} v_{rk}\mu_k = 0 \tag{K.85}$$

Then, Eq. (K.83) becomes

$$\Delta G^0 = -RT \ln K \tag{K.86}$$

where

$$K = \frac{(p_C)^c (p_D)^d}{(p_A)^a (p_B)^b}$$

where K is known as the 'equilibrium constant.'

It is difficult to overstate the importance of Eq. (K.86) in chemistry. For example, Eq. (K.86) relates the free energies of formation of a species, to the extent of chemical reactions involving that species. To see this, we re-arrange Eq. (K.86) to find

$$K = e^{\frac{-\Delta G^0}{RT}} \tag{K.87}$$

Now if $\Delta G^0 < 0$, then the exponent is positive and $K > 1$, and in this case the equilibrium favors the formation of products. Those reactions with the most negative ΔG^0 will proceed almost to completion, leaving little if any reactants. The reverse holds if $\Delta G^0 > 0$; i.e. the reaction favors formation of reactants.

We can gain more insight into how the free energies of formation are related to the extent of a equilibrium for the chemical reaction using Eq. (K.73). We see

$$K = e^{\frac{\Delta S^0}{R} + \frac{-\Delta H^0}{RT}} \tag{K.88}$$

This helps us to understand what is driving the equilibrium constant. The equilibrium constant is clearly subject to two competing influences. One influence is toward entropy formation in the system. If products formed from reactants result in a large, positive ΔS^0 (i.e. if 'the number of microstates' of the system increases), then this favors the formation of products. The other influence is minimization of enthalpy in the system (i.e. large negative ΔH^0 means energy is lost from the system to the surroundings, increasing entropy in the surroundings). So, the release of energy to the surroundings (by minimizing energy in the system as the reaction proceeds) also favors formation of products. The two influences may 'push' the chemical reaction in the same direction, or they may 'push' the reaction in opposing directions. Together, these two influences decide the extent of a chemical reaction proceeding to equilibrium.

We note here that the 'total reaction' may be complex and contain subsidiary 'coupled reactions.' It is important to realize that following a reaction the entropy of the system may be reduced (so ΔS^0 may be negative, meaning a more ordered molecular structure is created within the system—which must be common in biological systems), providing the enthalpy lost from the system is sufficiently large (so ΔH^0 is negative) ensures that the total entropy change (i.e. the sum of system and surroundings) increases. We note that for many reactions the necessary energy to be lost to make coupled reactions thermodynamically favorable is supplied by the hydrolysis of ATP. Indeed, this coupling of reactions enables living things to create 'highly structured' tissues within organisms and the cell so to be said to be 'alive' and creates ordered structures without contravening the second law of thermodynamics.

Non-ideality is introduced via correction factors known as 'fugacity coefficients' in gases and 'activity coefficients' in solutions, which account for non-ideal behavior of the gas or solution (http://en.wikipedia.org/wiki/Activity_coefficient).

We see that each thermodynamic potential has its advantages and so finds favor for particular applications. In this context, it is helpful to note here that the Gibbs free energy potential is a particularly useful quantity when analyzing fluids. For example, the Gibbs free energy of water at constant temperature is given by

$$\mu_{\mathrm{w}} = \int_{p_{\mathrm{ref}}}^{p} \bar{v}\mathrm{d}p \tag{K.89}$$

For the low pressures, as experienced by human body, water is well approximated as incompressible. So

$$\mu_{\mathrm{w}} = \bar{v}\int_{0}^{p} \mathrm{d}p = \bar{v}p \tag{K.90}$$

which is the work required to create by expansion the partial molar volume $\bar{v}$ against pressure p. This quantity represents the free energy due to the pressure on the fluid, that is, the energy available to do work. This free energy can be realized in practice if the pressure is held constant while the fluid is 'withdrawn' to a lower pressure environment. In fact, this practical situation is relatively common (e.g. the drawn down of water from the water reservoir to produce hydroelectric power, or the work required to inject a fluid into the blood stream) or the Gibbs free energy gradient driving fluid through a porous medium to minimize its Gibbs free energy of the water [i.e. for a spontaneous process $\mathrm{d}G < 0$: see Eq. (L.74)].

Next, we have the Gibbs–Duhem relationship, which is (L.55). We have

$$S\mathrm{d}T + \sum_{j=1}^{J} l_j \mathrm{d}L_j - \sum_{i=1}^{n} N_i \mathrm{d}\mu_i = 0 \tag{K.91}$$

For an isotropic material, this reduces to

$$S\mathrm{d}T - V\mathrm{d}p - \sum_{i=1}^{n} N_i \mathrm{d}\mu_i = 0 \tag{K.92}$$

Now dividing by the volume V to form intensive variables (see Eq. (L.48)), we find

$$\hat{S}\mathrm{d}T - \mathrm{d}p - \sum_{i=1}^{n} c_i \mathrm{d}\mu_i = 0 \tag{K.93}$$

and for isothermal conditions, we have

$$-\mathrm{d}p = \sum_{i=1}^{n} c_i \mathrm{d}\mu_i \tag{K.94}$$

Isothermal chemo-mechanics is an important part of engineering biology (but hardly researched at all). We see from Eq. (K.94) that the 'chemical potential' of the isotropic material is directly influenced by the applied hydrostatic stress. We note that the Gibbs–Duhem equation is very helpful in the analysis of phase transitions and for the analysis of biological membranes.

The five free energy potentials (internal energy, Helmholtz, enthalpy, Gibbs, and Gibbs–Duhem) normally complete the classical potential functions. However, there is also the so-called grand potential (or Landau potential) Φ_{G}, which is defined as

$$\Phi_{\mathrm{G}} = U - TS - \sum_{i=1}^{N} \mu_i N_i \tag{K.95}$$

Again, taking the differential of Φ_{G} and substituting Eq. (K.95) lead to

$$\begin{aligned} \mathrm{d}\Phi_{\mathrm{G}} &= T\mathrm{d}S - T\mathrm{d}S - S\mathrm{d}T - p\mathrm{d}V + \sum_{i=1}^{N} \mu_i \mathrm{d}N_i - \sum_{i=1}^{N} \mu_i \mathrm{d}N_i - \sum_{i=1}^{N} N_i \mathrm{d}\mu_i + \sum_{j}^{J} L_j \mathrm{d}l_j \\ \mathrm{d}\Phi_{\mathrm{G}} &= S\mathrm{d}T - p\mathrm{d}V - \sum_{i=1}^{N} N_i \mathrm{d}\mu_i + \sum_{j}^{J} L_j \mathrm{d}l_j \end{aligned} \tag{K.96}$$

We see that if $\mathrm{d}T = 0$, $\mathrm{d}V = 0$, and $\mathrm{d}l_j = 0$, then

$$\sum_{i=1}^{N} N_i \mathrm{d}\mu_i = 0 \tag{K.97}$$

This equation says if no mass is exchanged with the surroundings and the system has constant temperature and volume, it is not possible to change the chemical potential of one species without changing the chemical potential of at least one other species.

Appendix L
A Free Energy Potential for Aggrecan

The primary structural components of cartilage are water, collagen, and aggrecan. Here, we develop a free energy function for aggrecan based on an empirical virial expression for the osmotic pressure of chondroitin sulfate (the principal component that gives aggrecan its resistance to compression). It was found that the osmotic pressure generated by chondroitin sulfate of 16 disaccharide repeats in length (7.3 kDa) in isotonic saline (0.15 M) could be approximated using a virial equation (Bathe et al. 2005):

$$\pi^{agg} = RT(\alpha_1 c + \alpha_2 c^2 + \alpha_3 c^3) \tag{L.1}$$

where R is the universal gas constant (8.3e3), T is the absolute (e.g. 300 K) and α_1 = 1.4e−4, α_2 = 4.4e−6, α_3 = 5.6e−8, and c is the concentration of aggrecan in mg/ml. It is noted that the first term of the RHS captures ideal solute behavior, while the second and third terms of the RHS capture non-ideal behavior of the polymer solution (Bathe et al. 2005).

The compressive resistance of aggrecan in cartilage is a function of the concentration of the salt in the plasma exudate from the synovial membrane that bathes cartilage. However, we note that the salt concentration of plasma in the blood is powerfully regulated by the kidney, so that is effectively constant throughout life. It is therefore a very good approximation to assume that the salt solution bathing cartilage is an invariant isotonic solution (i.e. 0.15 M sodium chloride solution).

All multicomponent systems (including cartilage) have an internal energy. The internal energy of the system (U) is a function of extensive variables (S, V, N_i), that is,

$$U = f(S, V, N_i) \tag{L.2}$$

D. W. Smith et al., *Articular Cartilage Dynamics*,
https://doi.org/10.1007/978-981-13-1474-2

where S is the entropy, V is the volume, and N_i is the ith component particles (in moles) of the cartilage. Then, the total derivative of U is

$$\begin{aligned} dU &= \tfrac{\partial U}{\partial S} dS + \tfrac{\partial U}{\partial V} dV + \sum \tfrac{\partial U}{\partial N_i} dN_i, \\ dU &= TdS + \pi^{agg} dV + \sum \mu_i dN_i \end{aligned} \tag{L.3}$$

where it is assumed that aggrecan is the only component that contributes to the compressive resistance of the cartilage (π^{agg}) and μ^{agg} denotes the chemical potential of the aggrecan. It is mentioned that T, π^{agg}, μ^i are intensive variables that are conjugate to the extensive variables S, V, N^{agg} such that their product of the intensive and extensive variable has units of the free energy (i.e. J m^{-3} in SI units).

Let it now be assumed that all the components of the cartilage are held constant except the concentration of aggrecan, which is denoted N^{agg}. Then,

$$dU^{agg} = TdS + \pi^{agg} dV + \sum \mu^{agg} dN^{agg} \tag{L.4}$$

The internal energy U is the fundamental state function of thermodynamics (http://en.wikipedia.org/wiki/Internal_energy). U is a (thermodynamic) potential, meaning that its value is independent of the path taken between two states of the system. The fact that it is a potential leads to the well-known Maxwell relationships (http://en.wikipedia.org/wiki/Maxwell_relations).

Now, one can change the independent variables to produce a new potential function using so-called Legendre transformations (this amounts to adding or subtracting potential functions to the internal energy potential). For example, one can define the Helmholtz free energy (A) such that (http://en.wikipedia.org/wiki/Helmholtz_free_energy),

$$A = U - TS \tag{L.5}$$

so that,

$$dA = dU - TdS - SdT \tag{L.6}$$

Substituting for dU^{agg} above, we find

$$dA^{agg} = SdT + \pi^{agg} dV + \sum \mu^{agg} dN^{agg} \tag{L.7}$$

If all changes are isothermal (viz., $dT = 0$, which is usually a good assumption), then the Helmholtz free energy function simplifies to

$$dA^{agg} = \pi^{agg} dV + \sum \mu^{agg} dN^{agg} \tag{L.8}$$

If we now integrate dA^{agg} between a reference state and an end state, we can then find the free energy density function (Ψ^{agg}), which can be used in large deformation analysis (see previous discussion).

Given an empirical functional relation for π^{agg} (e.g. the first equation in this appendix), then Ψ^{agg} can be reconstructed in the following way. Our strategy is to use the Maxwell relationship,

$$\frac{\partial \pi^{agg}}{\partial N^{agg}} = \frac{\partial \mu^{agg}}{\partial V} \tag{L.9}$$

to find μ^{agg} and then integrate dA^{agg} between a reference state and the end state to find Ψ^{agg}. Let us proceed to do this.

First, we write π^{agg} as

$$\pi^{agg} = RT\left(\alpha_1 \frac{N^{agg}}{V} + \alpha_2 \frac{N^{agg^2}}{V^2} + \alpha_3 \frac{N^{agg^3}}{V^3}\right) \tag{L.10}$$

so as to clearly identify the extensive variables. Then using the Maxwell relation, we see

$$\frac{\partial \pi^{agg}}{\partial N^{agg}} = RT(\alpha_1 \frac{1}{V} + 2\alpha_2 \frac{N^{agg}}{V^2} + 3\alpha_3 \frac{N^{agg^2}}{V^3}) = \frac{\partial \mu^{agg}}{\partial V} \tag{L.11}$$

Integrating w.r.t. V from reference volume of one to V leads to

$$\begin{aligned} \mu^{agg} &= \int_1^V \frac{\partial \mu^{agg}}{\partial V} dV \\ &= RT\left[\alpha_1 \ln V + 2\alpha_2 N^{agg}\left(1 - \frac{1}{V}\right) + \frac{3}{2}\alpha_3 N^{agg^2}\left(1 - \frac{1}{V^2}\right)\right] \end{aligned} \tag{L.12}$$

Now, integrating μ^{agg} w.r.t. N^{agg} (from the reference zero up to N^{agg}) leads to the contribution of the chemical potential–aggrecan gain/loss work to the Helmholtz free energy potential $\left(A_\mu^{agg}\right)$, viz.,

$$\begin{aligned} A_\mu^{agg} &= \int_0^{N^{agg}} \mu^{agg} dN^{agg} \\ &= RT\left[\alpha_1 N^{agg} \ln V + \alpha_2 N^{agg^2}\left(1 - \frac{1}{V}\right) + \frac{1}{2}\alpha_3 N^{agg^3}\left(1 - \frac{1}{V^2}\right)\right] \end{aligned} \tag{L.13}$$

while integrating π^{agg} w.r.t. V (from the reference volume one up to V) leads to the contribution of the pressure–volume work to the Helmholtz free energy potential $\left(A_\pi^{\text{agg}}\right)$, viz.,

$$\begin{aligned} A_V^{\text{agg}} &= \int\limits_1^V \pi^{\text{agg}} dV \\ &= RT\left[\alpha_1 N^{\text{agg}} \ln V + \alpha_2 N^{\text{agg}^2}\left(1-\frac{1}{V}\right) + \frac{1}{2}\alpha_3 N^{\text{agg}^3}\left(1-\frac{1}{V^2}\right)\right] \end{aligned} \tag{L.14}$$

It is interesting to note that using $N^{\text{agg}} = 0$ and $V = 1$ as the reference state, our two free energy functions $A_\mu^{\text{agg}}, A_V^{\text{agg}}$ are identical. Adding the two contributions to the Helmholtz free energy leads our free energy density function for aggrecan in cartilage,

$$\Psi^{\text{agg}} = A_\mu^{\text{agg}} + A_V^{\text{agg}} \tag{L.15}$$

We can check that Ψ^{agg} is a potential by integrating the change from the reference state up to a final state along two different paths. For the first pathway, let the reference volume be filled with aggrecan and then let it expand to final volume V_f. In this case, the change in free energy from the reference state is

$$\begin{aligned} \Psi^{\text{agg}} &= A_\mu^{\text{agg}} + A_V^{\text{agg}} = RT\left[\alpha_1 N^{\text{agg}} \ln 1 + \alpha_2 N^{\text{agg}^2}\left(1-\frac{1}{1}\right) + \frac{1}{2}\alpha_3 N^{\text{agg}^3}\left(1-\frac{1}{1^2}\right)\right] \\ &\quad + RT\left[\alpha_1 N^{\text{agg}} \ln V_\text{f} + \alpha_2 N^{\text{agg}^2}\left(1-\frac{1}{V_\text{f}}\right) + \frac{1}{2}\alpha_3 N^{\text{agg}^3}\left(1-\frac{1}{V_\text{f}^2}\right)\right] \end{aligned} \tag{L.16}$$

We see all the changes in free energy are due to volume expansion. For the second pathway, let us expand the reference volume to the final volume with no aggrecan present and then fill this volume with aggrecan. In this case, all the changes in free energy are due to change in chemical potential of the aggrecan, viz.,

$$\begin{aligned} \Psi^{\text{agg}} &= A_\mu^{\text{agg}} + A_V^{\text{agg}} = RT\left[\alpha_1 N^{\text{agg}} \ln V_\text{f} + \alpha_2 N^{\text{agg}^2}\left(1-\frac{1}{V_\text{f}}\right) + \frac{1}{2}\alpha_3 N^{\text{agg}^3}\left(1-\frac{1}{V_\text{f}^2}\right)\right] \\ &\quad + RT\left[\alpha_1 0 \ln V_\text{f} + \alpha_2 0\left(1-\frac{1}{V_\text{f}}\right) + \frac{1}{2}\alpha_3 0\left(1-\frac{1}{V_\text{f}^2}\right)\right] \end{aligned} \tag{L.17}$$

The free energy change from the reference state to an arbitrary final state is identical. The path independence of the free energy Ψ^{agg} change from the reference state to an arbitrary final state demonstrates that Ψ^{agg} is a potential.

Let us now assume that the reference volume for the aggrecan $(V = 1)$ is the reference volume of the collagen network prior to aggrecan production by chondrocytes $\left(V_i^{\mathrm{col}} = V = 1\right)$. We can now fill the collagen network with aggrecan. The expansion pressure of the aggrecan will expand the collagen network to $V_{\mathrm{f}}^{\mathrm{col}}$. The free energy potential associated with the initial filling of the collagen network is

$$\Psi_i = RT\left[\alpha_1 N^{\mathrm{agg}} \ln V_{\mathrm{f}}^{\mathrm{col}} + \alpha_2 N^{\mathrm{agg}^2}\left(1 - \frac{1}{V_{\mathrm{f}}^{\mathrm{col}}}\right) + \frac{1}{2}\alpha_3 N^{\mathrm{agg}^3}\left(1 - \frac{1}{V_{\mathrm{f}}^{\mathrm{col}^2}}\right)\right] \tag{L.18}$$

If we now load the cartilage by compressing it to volume V_{f}, the final free energy of the aggrecan in the externally loaded state is

$$\begin{aligned}\Psi_{\mathrm{f}}^{\mathrm{agg}} = \Psi_i^{\mathrm{agg}} + \Delta\Psi_{\mathrm{load}}^{\mathrm{agg}} &= RT\left[\alpha_1 N^{\mathrm{agg}} \ln V_{\mathrm{f}}^{\mathrm{col}} + \alpha_2 N^{\mathrm{agg}^2}\left(1 - \frac{1}{V_{\mathrm{f}}^{\mathrm{col}}}\right) + \frac{1}{2}\alpha_3 N^{\mathrm{agg}^3}(1 - \frac{1}{V_{\mathrm{f}}^{\mathrm{col}^2}})\right] \\ &+ RT\left[\alpha_1 N^{\mathrm{agg}} \ln V_{\mathrm{f}}^{\mathrm{col}} + \alpha_2 N^{\mathrm{agg}^2}\left(\frac{1}{V_{\mathrm{f}}^{\mathrm{col}}} - \frac{1}{V_{\mathrm{f}}}\right) + \frac{1}{2}\alpha_3 N^{\mathrm{agg}^3}\left(\frac{1}{V_{\mathrm{f}}^{col^2}} - \frac{1}{V_{\mathrm{f}}^2}\right)\right]\end{aligned} \tag{L.19}$$

It is possible to decompose the osmotic pressure change due to a change in aggrecan concentration into two parts: one part due to aggrecan production and one part due to volume change.

Let $\pi^{\mathrm{agg}} = \pi^{\mathrm{agg}} = f(\alpha_i, c^{\mathrm{agg}})$ (where the α_i are internal (thermodynamics) variables like total ionic strength and pH of the bathing solution), then

$$\mathrm{d}\pi^{\mathrm{agg}} = \sum \frac{\partial \pi^{\mathrm{agg}}}{\partial \alpha_i}\mathrm{d}\alpha_i + \frac{\partial \pi^{\mathrm{agg}}}{\partial c^{\mathrm{agg}}}\mathrm{d}c^{\mathrm{agg}} \tag{L.20}$$

Let us assume all the internal variables are constant (i.e. $\mathrm{d}\alpha_i = 0$). Then,

$$\mathrm{d}\pi^{\mathrm{agg}} = \frac{\partial \pi^{\mathrm{agg}}}{\partial c^{\mathrm{agg}}}\mathrm{d}\left(\frac{N^{\mathrm{agg}}}{V}\right) = \frac{\partial \pi^{\mathrm{agg}}}{\partial c^{\mathrm{agg}}}\Big|_V \frac{\mathrm{d}N^{\mathrm{agg}}}{V} - \frac{\partial \pi^{\mathrm{agg}}}{\partial c^{\mathrm{agg}}}\Big|_{N^{\mathrm{agg}}} \frac{N^{\mathrm{agg}}}{V^2}\mathrm{d}V \tag{L.21}$$

Now, the rate of change of π^{agg} with $c^{\mathrm{agg}} = c$ (for convenience), at constant volume, is

$$\frac{\partial \pi^{\mathrm{agg}}}{\partial c}\Big|_V = RT(\alpha_1 + 2\alpha_2 c + 3\alpha_3 c^2) \tag{L.22}$$

and the rate of change of π^{agg} with $c^{\mathrm{agg}} = c$, at constant mass of aggrecan, is

$$\frac{\partial \pi^{\mathrm{agg}}}{\partial c}\Big|_{N^{\mathrm{agg}}} = -RT(\alpha_1 + 2\alpha_2 c + 3\alpha_3 c^2) \tag{L.23}$$

Substituting leads Eq. (L.18) to

$$\mathrm{d}\pi^{\text{agg}} = RT(\alpha_1 + 2\alpha_2 c + 3\alpha_3 c^2)\frac{\mathrm{d}N^{\text{agg}}}{V} - RT\left(\alpha_1 c + 2\alpha_2 c^2 + 3\alpha_3 c^3\right)\frac{\mathrm{d}V}{V} \tag{L.24}$$

Assuming no aggrecan is added or removed (i.e. $\mathrm{d}N^{\text{agg}}$ is zero), the term,

$$c^{\text{agg}}\frac{\partial \pi^{\text{agg}}}{\partial c^{\text{agg}}} = RT(\alpha_1 c + 2\alpha_2 c^2 + 3\alpha_3 c^3) = \frac{\mathrm{d}\pi^{\text{agg}}}{\mathrm{d}V/V} \tag{L.25}$$

is known as the osmotic modulus (which is equivalent to the bulk modulus of the aggrecan).

The first term on the RHS of Eq. (L.24) is the change in osmotic pressure due to an increase in aggrecan at constant volume, while the second term is the change in osmotic pressure due to a change in volume strain at constant aggrecan mass. This equation enables us to calculate the change in osmotic pressure due to aggrecan turnover by chondrocytes (first term RHS of Eq. L.24) and the change in osmotic pressure due to applied load (second term RHS of Eq. L.24). It is noted that often the timescales of the two changes are quite different; e.g. the timescale for loading is often seconds to hours, while the timescale for aggrecan turnover is usually weeks to months (and sometimes days to weeks).

We note that the estimate of osmotic modulus shown in Eq. (L.25) assumes that the whole fluid porosity is one, but in fact the volume fraction occupied by collagen in articular cartilage is significant (often around 30%), and so the fluid porosity is often around 70%. It has been shown that if the initial fluid porosity is denoted n_0, then assuming an incompressible fluid (so $\mathrm{d}V_\mathrm{f} = \mathrm{d}V$) the fluid porosity upon deformation is given by (see Sect. 7.2.6)

$$n_\mathrm{f}^{\text{sm}}(X,t) = \frac{N_\mathrm{f}^{m}(X,t)}{J} = 1 + \frac{(n_0 - 1)}{J} \tag{L.26}$$

To account for the fact that Eq. (L.25) relates to the fluid volume and not the whole volume (the two volumes are related by $n^{sm}(X,t) = V_\mathrm{f}/V$), we should then modify Eq. (L.25) to read

$$\frac{\mathrm{d}\pi^{\text{agg}}}{\mathrm{d}V/V} = \frac{c^{\text{agg}}}{n^{\text{sm}}}\frac{\partial \pi^{\text{agg}}}{\partial c^{\text{agg}}} = RT(\alpha_1 c + 2\alpha_2 c^2 + 3\alpha_3 c^3)/n^{\text{sm}} \tag{L.27}$$

Note that $c^{\text{agg}} = c$ is the actual concentration of the aggrecan in the representative elementary volume. Assuming the collagen volume is constant, the effect of deformation changing the fluid porosity has a significant effect on increasing the apparent stiffness of the aggrecan in the cartilage ECM, by changing both the actual concentration of aggrecan and directly via osmotic stiffness equation (L.27).

Appendix M
Diffusive Transport of Aggrecan

Aggrecan produced by chondrocytes continually moves down its concentration gradient toward the free surface of the cartilage, where intact and partially degraded aggrecan molecules exit the cartilage. This appendix is about representing aggrecan transport using a diffusion equation. The diffusion coefficients used in these equations are not actual diffusion coefficients for aggrecan diffusion in aggrecan (for estimates of actual diffusion coefficients, see Gribbon and Hardingham 1998b). The diffusion coefficients discussed here are for estimating mass fluxes of aggrecan through an aggrecan solution, in an aggrecan concentration gradient.

The transport of ideal solutes in solution is usually modeled using Fick's first law (where the flux is proportional to the concentration gradient); however, when the solute is non-ideal (e.g. at high concentrations), an alternative approach is required.

A way of calculating aggrecan flux through a solution with an aggrecan gradient using a diffusion model is suggested by Comper (1991). The data is collected in a centrifuge experiment in which the rate of sedimentation (due to centripetal acceleration) is balanced against by the rate of diffusion (due to minimization of chemical potential) along an aggrecan concentration gradient. Comper (1991) suggests way to calculate the mass transfer using the formalism of a diffusion of equation in the following way (Comper 1991), viz.,

$$f^{\mathrm{agg}} = -\frac{RT}{f_{12}}\nabla^{s} c_{\mathrm{eff}}^{\mathrm{agg}} = -\eta_{12} RT \nabla^{s} c_{\mathrm{eff}}^{\mathrm{agg}} = -D_{\mathrm{eff}}^{*} \nabla^{s} c_{\mathrm{eff}}^{\mathrm{agg}}$$

where D_{eff}^{*} is the effective diffusion coefficient. The effective diffusion coefficient may be viewed as the ratio of RT to the friction coefficient f_{12} between a mole of aggrecan molecules and a solvent (in this case water). The mobility of the species in solution η_{12} is inverse of the friction coefficient f_{12}. The 'effective concentration'

D. W. Smith et al., *Articular Cartilage Dynamics*,
https://doi.org/10.1007/978-981-13-1474-2

(which is taken to be an equivalent concentration of an ideal solute) is related to the osmotic pressure by $c_{\text{eff}}^{\text{agg}} = \frac{\pi^{\text{agg}}}{RT}$. Using the chain rule, Comper finds

$$f^{\text{agg}} = -D_{\text{eff}}^{*} \nabla^{s} \frac{\pi^{\text{agg}}}{RT} = -\frac{RT}{f_{12}} \frac{\partial \pi^{\text{agg}}}{\partial c^{\text{agg}}} \nabla^{s} c^{\text{agg}} = -D_{\text{eff}} \nabla^{s} c^{\text{agg}}$$

where

$$\begin{aligned} D_{\text{eff}} &= \frac{RT}{f_{12}} \frac{\partial \pi^{\text{agg}}}{\partial c^{\text{agg}}} = \frac{RT}{f_{12}} \left(\alpha_1 + 2\alpha_2 c^{\text{agg}} + 3\alpha_3 c^{\text{agg}^2}\right) \\ &= D_{\text{eff}}^{*} \left(\alpha_1 + 2\alpha_2 c^{\text{agg}} + 3\alpha_3 c^{\text{agg}^2}\right) \end{aligned}$$

An alternative derivation is shown below, based on the definition of chemical potential. The transport of ideal solutions is usually modeled using Fick's first law (where the flux is proportional to the concentration gradient). However for non-ideal solutions, the flux of the *i*th species is often modeled as

$$f_i = c_i v_i = -c_i \frac{D_{i,\text{eff}}^{*}}{RT} \nabla \mu_i$$

where μ_i is the chemical potential. In terms of a concentration gradient, this becomes

$$f_i = -\frac{D_{i,\text{eff}_i^*}}{RT} \frac{c_i \partial \mu_i}{\partial c_i} \nabla c_i = -D_{i,\text{eff}} \nabla c_i$$

where

$$D_{i,\text{eff}} = \frac{D_{i,\text{eff}}^{*}}{RT} \frac{c_i \partial \mu_i}{\partial c_i}$$

Here, we seek an expression for the diffusive flux based on the gradient of the chemical potential energy. To do this, we use the chemical potential for aggrecan derived previously. The chemical potential energy is given (see Eq. L.12),

$$\mu^{\text{agg}} = RT \left[\alpha_1 \left(\ln V - \ln N^{\text{agg}}\right) - 2\alpha_2 \frac{N^{\text{agg}}}{V} - \frac{3}{2} \alpha_3 \frac{N^{\text{agg}^2}}{V^2} \right]$$

where $\ln N^{\text{agg}}$ is an integration constant. That is,

$$\mu^{\text{agg}} = -RT \left[\alpha_1 \left(\ln c^{\text{agg}}\right) + 2\alpha_2 c^{\text{agg}} + \frac{3}{2} \alpha_3 c^{\text{agg}^2} \right]$$

Differentiating the chemical potential energy for aggrecan (μ^{agg}) w.r.t. c^{agg} at constant mass (i.e. w.r.t. V) of aggrecan leads to

$$\frac{\partial \mu^{agg}}{\partial c^{agg}} = RT\left(\frac{\alpha_1}{c^{agg}} + 2\alpha_2 + 3\alpha_3 c^{agg}\right)$$

and so,

$$D_{eff} = -D^*_{eff}(\alpha_1 + 2\alpha_2 c^{agg} + 3\alpha_3 c^{agg^2})$$

Differentiating the chemical potential energy for aggrecan (μ^{agg}) w.r.t. c^{agg} at constant volume (i.e. w.r.t. N^{agg}) leads to

$$\frac{\partial \mu^{agg}}{\partial c^{agg}} = \frac{\partial \mu^{agg}}{\partial N^{agg}} = -RT\left(\frac{\alpha_1}{N^{agg}} + 2\alpha_2 + 3\alpha_3 c^{agg}\right)$$

and so,

$$D_{eff} = -D^*_{eff}(\alpha_1 + 2\alpha_2 c^{agg} + 3\alpha_3 c^{agg^2})$$

These results agree with the Comper (1991).

Appendix N
Stress and Elasticity Tensors in a Fiber-Reinforced Composite

Following the approach of Ateshian et al. (2009), let the strain in the fiber-reinforced composite be represented by the familiar right Cauchy–Green tensor $C = F^{\mathrm{T}}F$. Let a collagen fiber of arbitrary direction be represented in spherical coordinates $\theta(0 \leq \theta \leq 2\pi)$ and $\varphi(0 \leq \varphi \leq \pi)$ by the unit vector $\hat{n}^r$ in the radial direction. This unit vector $\hat{n}^r$ in material coordinates is represented by

$$\hat{n}^r = \cos\,\theta\,\sin\,\varphi e_1 + \sin\,\theta\,\sin\,\varphi e_2 + \cos\,\varphi e_3 \tag{N.1}$$

The so-called texture tensor N^r is then defined as the tensor product $\hat{n}^r \otimes \hat{n}^r$. The components of the symmetric texture tensor N^r are:

$$\begin{aligned}
N^r_{11} &= \sin^2\varphi\,\cos^2\theta \\
N^r_{22} &= \sin^2\varphi\,\sin^2\theta \\
N^r_{33} &= \cos^2\theta \\
N^r_{12} &= \sin^2\varphi\,\cos\,\theta\,\sin\,\theta \\
N^r_{13} &= \sin\,\varphi\,\cos\,\varphi\,\cos\,\theta \\
N^r_{23} &= \sin\,\varphi\,\cos\,\varphi\,\sin\,\theta
\end{aligned}$$

The square of the fiber stretch $\left(\lambda_n^2\right)$ in the $\hat{n}^r$ direction is then given by Bonet and Wood (2008),

$$\lambda_n^2 = \hat{n}^r C \hat{n}^r = C : N^r \equiv I_n \tag{N.2}$$

Now, let the strain energy function for the fiber be defined simply as

D. W. Smith et al., *Articular Cartilage Dynamics*,
https://doi.org/10.1007/978-981-13-1474-2

$$\Psi(\hat{n}^r, I_n) = \xi(\hat{n}^r)(I_n - 1)^{\alpha(\hat{n}^r)} \tag{N.3}$$

where $\alpha(\hat{n}^r)$, which has no units, is ≥ 2 (note that if $\alpha(\hat{n}^r) > 2$, then at zero fiber stretch the contribution of the fibers to matrix stiffness is zero). The term $\xi(\hat{n}^r)$ is a scalar quantity that contributes to the stiffness of the collagen (i.e. related to $\alpha(\alpha - 1)\xi(\hat{n}^r)(I_n - 1)^{(\alpha(\hat{n}^r)-2)}$) oriented in the direction $\hat{n}^r$ within its surrounding extracellular matrix (which may be connected to more collagen fibers). $\xi(\hat{n}^r)$ is related to the stiffness properties of a collagen fiber and to the number of collagen fibers per unit gross cross-sectional area in the $\hat{n}^r$ direction. $\xi(\hat{n}^r)$ has units of force per unit area (e.g. MPa) and is a structural material property of the collagen in the extracellular matrix.

We note at this point that viscoelasticity of the collagen fiber may be introduced if so desired. For example, one could define the strain energy density function to be

$$\Psi(\hat{n}^r, I_n, \dot{I}_n) = \Psi^e(\hat{n}^r, I_n) + \Psi^d(\hat{n}^r, \dot{I}_n) \tag{N.4}$$

where Ψ^{d} is the so-called dissipation potential. One could choose a dissipation potential to be equivalent to a Maxwell (spring and damper in series) or Kelvin–Voigt model (spring and damper in parallel). Choosing a Kelvin–Voigt-like model, we may, for example, let

$$\Psi^{\mathrm{d}}(\hat{n}^r, \dot{I}_n, k_i, k_0, \lambda_i, \tau) = \left[\frac{\xi(\hat{n}^r)}{k_0}\sum_{i=1}^{N} k_i e^{-\lambda_i \tau}\right](\dot{I}_n(t) - 1)^{\alpha(\hat{n}^r)} \tag{N.5}$$

where $\dot{I}_n$ is the rate of change of stretch squared w.r.t. time, k_i, k_0, λ_i are constants chosen based on measured collagen responses to rates of stretching, and τ is $t - t'$where t is the current time and t' is a time in the past tied to the initiating event $\dot{I} = dI/dt'$. The type of material response requires an integral over time to evaluate its current state, as the material retains a 'memory' of previous applied loading rates. The sum $\sum_{i=1}^{N} k_i e^{-\lambda_i \tau}$ is known as a 'Prony series' (http://en.wikipedia.org/wiki/Viscoelasticity). The λ_i are adjusted so the model agrees with the time dependence of the experimental data on the material response to loading.

The contribution of the fiber associated with direction $\hat{n}^r$ to the second Piola–Kirchhoff stress tensor for the material is given by

$$S_n^t = S_n^e + S_n^d = 2\frac{\partial \Psi^e}{\partial C} + 2\frac{\partial \Psi^d}{\partial \dot{C}} = 2\frac{\partial \Psi^e}{\partial I_n}\frac{\partial I_n}{\partial C} + 2\frac{\partial \Psi^d}{\partial \dot{I}_n}\frac{\partial \dot{I}_n}{\partial \dot{C}} = 2\left(\frac{\partial \Psi^e}{\partial I_n} + \frac{\partial \Psi^d}{\partial \dot{I}_n}\right)N^r \tag{N.6}$$

We notice in passing the total stress tensor S_n^t is composed of two parts, namely the elastic stress tensor S_n^e derived from the elastic free energy function Ψ^e and the viscous stress tensor S_n^d derived from the dissipative function Ψ^d.

The contribution to the fourth-order stiffness tensor is given by (Curnier et al., 1995)

$$\mathbb{C}_n = 4\frac{\partial^2 \Psi}{\partial C^2} = 4\left(\frac{\partial^2 \Psi^e}{\partial I_n^2} + \frac{\partial^2 \Psi^d}{\partial \dot{I}_n^2}\right) N^r \otimes N^r \tag{N.7}$$

The contributions from each of the collagen fibers in n^r direction of the composite may be summed over a sphere of unit radius to give the total contribution of collagen fibers in the n^r direction to the second Piola–Kirchhoff stress at that location, viz.,

$$S_n^{\text{col}} = \int_0^{2\pi}\int_0^{\pi} H(I_n - 1) S_n \sin\varphi d\varphi d\theta \tag{N.8}$$

where H is an indicator function that is zero when the argument is negative and becomes one when the argument is positive (i.e. $H = 1$ when $I_n \geq 1$, that is, when the stretch is positive). The factor sin φ enables the integration to be performed over the surface of the sphere.

The elasticity tensor for all the collagen fibers in the n^r direction at that location may be found similarly, viz.,

$$\mathbb{C}_n^{\text{col}} = \int_0^{2\pi}\int_0^{\pi} H(I_n - 1)\mathbb{C}_n \sin\varphi \mathrm{d}\varphi \mathrm{d}\theta \tag{N.9}$$

Of course, the second Piola–Kirchhoff stress and the stiffness tensor can be pushed forward to find the Cauchy stress and the stiffness tensor for a single fiber,

$$\begin{aligned} \sigma_n &= 2J^{-1} I_n \frac{\partial \Psi^e}{\partial I_n}\left[\frac{1}{I_n} F N^r F^T\right] = 2J^{-1} I_n \frac{\partial \Psi^e}{\partial I_n} N \\ \mathbb{C}_n^s &= 4J^{-1} I_n^2 \frac{\partial^2 \Psi^e}{\partial I_n^2}\left[\frac{1}{I_n} F N^r F^T\right] \otimes [\frac{1}{I_n} F N^r F^T] = 4J^{-1} I_n^2 \frac{\partial^2 \Psi^e}{\partial I_n^2} N \otimes N \end{aligned} \tag{N.10}$$

where $N = \frac{1}{I_n} F N^r F^T$ is the texture tensor in the deformed state.

The Cauchy stress for all the collagen fibers in the matrix in the $\hat{n}^r$ direction at a location is given by the usual transformation of Eq. (N.8),

$$\begin{aligned} \sigma_n^{\text{s-col}} &= J^{-1} F S_n^{\text{col}} F^{\text{T}} \\ \mathbb{C}_n^{\text{s-col}} &= J^{-1} F \underline{\otimes} F : \mathbb{C}_n^{\text{col}} : F^T \underline{\otimes} F^{\text{T}} \end{aligned} \tag{N.11}$$

where $\underline{\otimes}$ is defined by $[A \underline{\otimes} B]X = AXB^{\text{T}}$ (Curnier et al. 1995).

Appendix O
The Storage Equation for Cartilage

We first develop the storage equation for a multicomponent material and then specialize this equation to cartilage. Let us define the ith apparent concentration (mass divided by total volume) of component i as c_i^{app} and the ith partial specific volume of component i as $\bar{v}_i$. Then,

$$c_i^{\text{app}} \bar{v}_i = \frac{v_i}{v} = n_i \tag{O.1}$$

where v_i is the volume of the mass of ith component and n_i is the ith porosity of component i. Then, the general conservation equation for the porosity of the ith component in spatial coordinates is

$$\frac{\partial n_i}{\partial t} = -\nabla^s \cdot f_{n_i} + s_{n_i} \tag{O.2}$$

where f_{n_i} is the porosity flux of the ith component and $s_{n_i}(x, t)$ is the rate of production of porosity per unit volume in spatial coordinates.

Let us assume that the partial specific volume of the ith component is constant, then

$$\nabla^s \cdot f_{n_i} = \nabla^s \cdot (\bar{v}_i f_{c_i}) = \bar{v}_i \nabla^s \cdot [-D_i \nabla^s (c_i^{\text{app}}/n_i) + v_i c_i^{\text{app}}] \tag{O.3}$$

where D_i is the ith effective diffusion coefficient, v_i is the ith velocity w.r.t. the spatial coordinate system, and c_i/n_i is the true concentration. Now, let the flux be due to advective transport alone (so all D_i are zero), then noting Eq. (O.1) leads to

$$\nabla^s \cdot f_{n_i} = \nabla^s \cdot [v_i n_i] \tag{O.4}$$

D. W. Smith et al., *Articular Cartilage Dynamics*,
https://doi.org/10.1007/978-981-13-1474-2

Now, let

$$v_i = v_i - v_{ref} + v_{\text{ref}} = \frac{v_d^i}{n_i} + v_{\text{ref}} \tag{O.5}$$

where v_d^i is the Darcy velocity of the ith component with respect to the reference velocity v_{ref}. Then,

$$\nabla \cdot f_{n_i} = \nabla^s \cdot \left[v_d^i\right] + \nabla^s \cdot \left[v_{ref} n_i\right] = \nabla^s \cdot \left[v_d^i\right] + v_{\text{ref}} \nabla^s \cdot [n_i] + n_i \nabla^s \cdot [v_{\text{ref}}] \tag{O.6}$$

Now, substituting Eq. (O.6) into the conservation equation for porosity leads to

$$\frac{\partial n_i}{\partial t} + v_{\text{ref}} \nabla^s \cdot [n_i] = -\nabla^s \cdot \left[v_d^i\right] - n_i \nabla^s \cdot [v_{\text{ref}}] + s_{n_i} \tag{O.7}$$

Using the definition of the material derivative w.r.t. time Eq. (O.7) can be written as

$$\frac{\mathrm{d}n_i}{\mathrm{d}t} = -\nabla^s \cdot \left[v_d^i\right] - n_i \nabla^s \cdot [v_{\text{ref}}] + s_{n_i} \tag{O.8}$$

Now, summing across all components shows

$$\frac{\mathrm{d}\sum n_i}{\mathrm{d}t} = -\sum \nabla^s \cdot [v_d^i] - \nabla^s \cdot [v_{\text{ref}}] \sum n_i + \sum s_{n_i} \tag{O.9}$$

and noting the identity,

$$\sum n_i = 1 \tag{O.10}$$

which holds at all times, leads to

$$0 = -\sum \nabla^s \cdot [v_d^i] - \nabla^s \cdot [v_{\text{ref}}] + \sum s_{n_i} \tag{O.11}$$

Neglecting source and sinks, we see that

$$\sum \nabla^s \cdot [v_d^i] = -\nabla^s \cdot [v_{\text{ref}}] \tag{O.12}$$

Now assuming that our reference velocity is the velocity of the jth component (where that the jth component defines our material coordinate at time zero), and employing Eq. (7.45) leads to

$$\frac{1}{J}\frac{\mathrm{d}(J-1)}{\mathrm{d}t} = \nabla^s \cdot [v_{\mathrm{ref}}] \approx \frac{1}{1+e_v}\frac{\partial e_v}{\partial t} \text{ (for small deformations)} \tag{O.13}$$

where $J - 1 = e_v$. Assuming small deformations, substituting Eq. (O.13) into Eq. (O.11) leads to

$$\frac{1}{1+e_v}\frac{\partial e_v}{\partial t} = -\sum_{i \neq j} \nabla^s \cdot [v_d^i] + \sum s_{n_i} \tag{O.14}$$

For large deformations, this small deformation approximation becomes

$$\frac{\mathrm{d}J}{\mathrm{d}t} = -\sum_{i \neq j} \nabla^m \cdot JF^{-\mathrm{T}}[v_d^i] + \sum S_{n_i} \tag{O.15}$$

where $S_{n_i}(X,t)$ are now expressed in terms of material coordinates (i.e. $Js_{n_i}^{sm}(x,t) = S_{n_i}(X,t)$7.16; see Eq.). We can now choose a constitutive equation for v_d^i such as

$$v_d^i = -k_i \nabla^s \eta_i \tag{O.16}$$

For small deformations, substituting this equation reveals

$$\frac{1}{1+e_v}\frac{\partial e_v}{\partial t} = \sum_{i \neq j} \nabla^s \cdot k_i \nabla^s \eta_i + \sum s_{n_i} \tag{O.17}$$

while for large deformations we find

$$\frac{\mathrm{d}J}{\mathrm{d}t} = \sum_{i \neq j} \nabla^m \cdot JF^{-1} k_i F^{-T} \nabla^m \eta_i^{sm} + \sum S_{n_i} \tag{O.18}$$

For cartilage, we take the reference material (j) as collagen, and so Darcy velocities for the remaining components (i.e. water and aggrecan) in our storage equation are referenced to collagen. This is important because previously we defined the Darcy velocity of water to aggrecan $v_{\mathrm{d}}^{\mathrm{w}}$, not to collagen $v_{\mathrm{d-col}}^{\mathrm{w}}$. However, when we use the 'Darcy flow equation' for water in the storage equation, we need to use Darcy flow relative to collagen (i.e. $v_{\mathrm{d-col}}^{\mathrm{w}}$). We can see how to do this from Smith et al. (2013), viz.,

$$v_{\mathrm{d-col}}^{\mathrm{w}} = n_{\mathrm{w}}(v^{\mathrm{w}} - v^{\mathrm{col}}) = n_{\mathrm{w}}[(v^{\mathrm{w}} - v^{\mathrm{agg}}) + (v^{\mathrm{agg}} - v^{\mathrm{col}})] \tag{O.19}$$

that is,

$$v_{\text{d-col}}^{\text{w}} = v_{\text{d}}^{\text{w}} + \frac{n_{\text{w}}}{n_{\text{agg}}} v_{\text{d}}^{\text{agg}} \tag{O.20}$$

Assuming a small deformation analysis that water and aggrecan are transported according to a Darcy law and using Eq. (7.163) the storage equation for cartilage becomes

$$\frac{1}{1+e_v}\frac{\partial e_v}{\partial t} = -\nabla^s \cdot v_d^w - \nabla^s \cdot \left[1 + \frac{n_{\text{w}}}{n_{\text{agg}}}\right] v_{\text{d}}^{\text{agg}} + s_{n-\text{col}} + s_{n-\text{agg}} + s_{n-w} \tag{O.21}$$

Employing the constitutive equations for water (Eq. 7.163) and aggrecan transport (7.180), with sign convention for σ^{agg} as a solid stress (i.e. $\sigma^{\text{agg}} = -\pi^{\text{agg}}$), leads to

$$\frac{1}{1+e_v}\frac{\partial e_v}{\partial t} = \nabla^s \cdot k_{\text{w}} \nabla^s p - \nabla^s \cdot \left[1 + \frac{n_{\text{w}}}{n_{\text{agg}}}\right] k_{\text{agg}} \nabla^s \sigma^{\text{agg}} + s_{n-\text{col}} + s_{n-\text{agg}} + s_{n-\text{w}} \tag{O.22}$$

and using our transformation rules to accommodate a large deformation analysis the storage equation is

$$\begin{aligned} \frac{\text{d}J}{\text{d}t} &= \nabla^m \cdot JF^{-1} k_w F^{-\text{T}} \nabla^m p^{sm} \\ &+ \nabla^m \cdot \left[1 + \frac{n_{\text{w}}^{sm}}{n_{\text{agg}}^{sm}}\right] JF^{-1} k_{\text{agg}} F^{-T} \nabla^m \pi^{\text{agg}-sm} + S_{n-\text{col}} + S_{n-\text{agg}} + S_{n-\text{w}} \end{aligned} \tag{O.23}$$

where $S_{n_i}(X, t)$ are expressed in terms of material coordinates.

References

Abbott, S. C., & Cole, M. D. (2013). Vibration arthrometry: A critical review. *Critical Reviews in Biomedical Engineering, 41,* 223–242.

Abed, E., Chan, T. F., Delalandre, A., Martel-Pelletier, J., Pelletier, J. P., & Lajeunesse, D. (2011). R-spondins are newly recognized players in osteoarthritis that regulate Wnt signaling in osteoblasts. *Arthritis and Rheumatism, 63,* 3865–3875.

Abella, V., Scotece, M., Conde, J., Lopez, V., Lazzaro, V., Pino, J., et al. (2014). Adipokines, metabolic syndrome and rheumatic diseases. *Journal of Immunology Research.*

Abramson, S. B. (2008). Osteoarthritis and nitric oxide. *Osteoarthritis Cartilage, 16*(Suppl 2), S15–S20.

Ackermann, B., & Steinmeyer, J. (2005). Collagen biosynthesis of mechanically loaded articular cartilage explants. *Osteoarthritis Cartilage, 13,* 906–914.

Adams, M. A. (2006). The mechanical environment of chondrocytes in articular cartilage. *Biorheology, 43,* 537–545.

Adams, M. E., & Brandt, K. D. (1991). Hypertrophic repair of canine articular-cartilage in osteoarthritis after anterior cruciate ligament transection. *Journal of Rheumatology, 18,* 428–435.

Adar, R. M., & Andelman, D. (2018). Osmotic pressure between arbitrarily charged planar surfaces: A revisited approach. *European Physical Journal E, 41,* 11.

Ahagon, A., & Kirino, Y. (2007). Mechanochemical reactions in black-filled Sbr vulcanizates under large deformation. *Rubber Chemistry and Technology, 80,* 265–278.

Ahmad, M. M. H. M., Gethin, D. T., Claypole, T. C., & Roylance, B. J. (1998). Numerical and experimental investigation into porous squeeze films. *Tribology International, 31,* 189–199.

Ahrman, E., Lorenzo, P., Holmgren, K., Grodzinsky, A. J., Dahlberg, L. E., Saxne, T., et al. (2014). Novel cartilage oligomeric matrix protein (COMP) neoepitopes identified in synovial fluids from patients with joint diseases using affinity chromatography and mass spectrometry. *Journal of Biological Chemistry, 289,* 20908–20916.

Ahsan, T., Harwood, F., McGowan, K. B., Amiel, D., & Sah, R. L. (2005). Kinetics of collagen crosslinking in adult bovine articular cartilage. *Osteoarthritis and Cartilage, 13,* 709–715.

Ahsanizadeh, S., & Li, L. P. (2015). Strain-rate-dependent non-linear tensile properties of the superficial zone of articular cartilage. *Connective Tissue Research, 56,* 469–476.

Aigner, T., Bertling, W., Stoss, H., Weseloh, G., & von der Mark, K. (1993). Independent expression of fibril-forming collagens I, II, and III in chondrocytes of human osteoarthritic cartilage. *Journal of Clinical Investigation, 91,* 829–837.

D. W. Smith et al., *Articular Cartilage Dynamics*,
https://doi.org/10.1007/978-981-13-1474-2

Aigner, T., & Dudhia, J. (1997). Phenotypic modulation of chondrocytes as a potential therapeutic target in osteoarthritis: A hypothesis. *Annals of the Rheumatic Diseases, 56,* 287–291.

Aigner, T., Gebhard, P. M., Schmid, E., Bau, B., Harley, V., & Poschl, E. (2003). SOX9 expression does not correlate with type II collagen expression in adult articular chondrocytes. *Matrix Biology, 22,* 363–372.

Aigner, T., Hemmel, M., Neureiter, D., Gebhard, P. M., Zeiler, G., Kirchner, T., et al. (2001). Apoptotic cell death is not a widespread phenomenon in normal aging and osteoarthritis human articular knee cartilage: A study of proliferation, programmed cell death (apoptosis), and viability of chondrocytes in normal and osteoarthritic human knee cartilage. *Arthritis and Rheumatism, 44,* 1304–1312.

Aigner, T., Zhu, Y., Chansky, H. H., Matsen, F. A., Maloney, W. J., & Sandell, L. J. (1999). Reexpression of type IIA procollagen by adult articular chondrocytes in osteoarthritic cartilage. *Arthritis and Rheumatism, 42,* 1443–1450.

Akiyama, H., Lyons, J. P., Mori-Akiyama, Y., Yang, X. H., Zhang, R., Zhang, Z. P., et al. (2004). Interactions between Sox9 and beta-catenin control chondrocyte differentiation. *Genes & Development, 18,* 1072–1087.

Akizuki, S., Mow, V. C., Muller, F., Pita, J. C., Howell, D. S., & Manicourt, D. H. (1986). Tensile properties of human knee joint cartilage: I. Influence of ionic conditions, weight bearing, and fibrillation on the tensile modulus. *Journal of Orthopaedic Research, 4,* 379–392.

Albro, M. B., Cigan, A. D., Nims, R. J., Yeroushalmi, K. J., Oungoulian, S. R., Hung, C. T., et al. (2012). Shearing of synovial fluid activates latent TGF-beta. *Osteoarthritis Cartilage, 20,* 1374–1382.

Albro, M. B., Nims, R. J., Cigan, A. D., Yeroushalmi, K. J., Shim, J. J., Hung, C. T., et al. (2013). Dynamic mechanical compression of devitalized articular cartilage does not activate latent TGF-beta. *Journal of Biomechanics, 46,* 1433–1439.

Alsalameh, S., Amin, R., Gemba, T., & Lotz, M. (2004). Identification of mesenchymal progenitor cells in normal and osteoarthritic human articular cartilage. *Arthritis and Rheumatism, 50,* 1522–1532.

Amin, A. R., & Abramson, S. B. (1998). The role of nitric oxide in articular cartilage breakdown in osteoarthritis. *Current Opinion in Rheumatology, 10,* 263–268.

Amiri, M., & Khonsari, M. M. (2010). On the thermodynamics of friction and wear—A review. *Entropy, 12,* 1021–1049.

Amjadi, S., Mai, K., McCluskey, P., & Wakefield, D. (2013). The role of lumican in ocular disease. *ISRN Ophthalmology, 2013,* 632302.

An, B., Abbonante, V., Yigit, S., Balduini, A., Kaplan, D. L., & Brodsky, B. (2014). Definition of the native and denatured type II collagen binding site for fibronectin using a recombinant collagen system. *Journal of Biological Chemistry, 289,* 4941–4951.

Anderson, A. E., Ellis, B. J., Maas, S. A., Peters, C. L., & Weiss, J. A. (2008). Validation of finite element predictions of cartilage contact pressure in the human hip joint. *Journal of Biomechanical Engineering-Transactions of the ASME, 130.*

Andersson, M. L., Petersson, I. F., Karlsson, K. E., Jonsson, E. N., Mansson, B., Heinegard, D., et al. (2006). Diurnal variation in serum levels of cartilage oligomeric matrix protein in patients with knee osteoarthritis or rheumatoid arthritis. *Annals of the Rheumatic Diseases, 65,* 1490–1494.

Andersson, M. L., Thorstensson, C. A., Roos, E. M., Petersson, I. F., Heinegard, D., & Saxne, T. (2006). Serum levels of cartilage oligomeric matrix protein (COMP) increase temporarily after physical exercise in patients with knee osteoarthritis. *BMC Musculoskelet Disord, 7,* 98.

Andrade, A. C., Nilsson, O., Barnes, K. M., & Baron, J. (2007). Wnt gene expression in the post-natal growth plate: Regulation with chondrocyte differentiation. *Bone, 40,* 1361–1369.

Annes, J. P., Munger, J. S., & Rifkin, D. B. (2003). Making sense of latent TGF beta activation. *Journal of Cell Science, 116,* 217–224.

Ap Gwynn, I., Wade, S., Ito, K. & Richards, R. G. (2002). Novel aspects to the structure of rabbit articular cartilage. *European Cells and Materials, 4,* 18–29.

Appleton, C. T. (2018). Osteoarthritis year in review 2017: Biology. *Osteoarthritis and Cartilage, 26,* 296–303.

Appleton, C. T., Hawker, G. A., Hill, C. L., & Pope, J. E. (Eds.). (2017). "Weighing in" on the Framingham osteoarthritis study: Measuring biomechanical and metabolic contributions to osteoarthritis. *Arthritis Rheumatology,* 69, 1127–1130.

Arfken, G. B., Weber, H. J., & Harris, F. E. (2013). *Mathematical methods for physicists.* Academic Press.

Arkill, K. P., & Winlove, C. P. (2008). Solute transport in the deep and calcified zones of articular cartilage. *Osteoarthritis and Cartilage, 16,* 708–714.

Armstrong, C. G., Lai, W. M., & Mow, V. C. (1984). An analysis of the unconfined compression of articular-cartilage. *Journal of Biomechanical Engineering-Transactions of the ASME, 106,* 165–173.

Armstrong, M. H., Tepole, A. B., Kuhl, E., Simon, B. R., & Vande Geest, J. P. (2016). A finite element model for mixed porohyperelasticity with transport, swelling, and growth. *PLoS ONE, 11.*

Arnold, W. V., & Fertala, A. (2013). Skeletal diseases caused by mutations that affect collagen structure and function. *International Journal of Biochemistry & Cell Biology, 45,* 1556–1567.

Arokoski, J. P. A., Hyttinen, M. M., Lapvetelainen, T., Takacs, P., Kosztbaczky, B., Modis, L., et al. (1996). Decreased birefringence of the superficial zone collagen network in the canine knee (stifle) articular cartilage after long distance running training, detected by quantitative polarised light microscopy. *Annals of the Rheumatic Diseases, 55,* 253–264.

Aspberg, A. (2012). The different roles of aggrecan interaction domains. *Journal of Histochemistry and Cytochemistry, 60,* 987–996.

Aspberg, A., Adam, S., Kostka, G., Timpl, R., & Heinegard, D. (1999). Fibulin-1 is a ligand for the C-type lectin domains of aggrecan and versican. *Journal of Biological Chemistry, 274,* 20444–20449.

Aspberg, A., Miura, R., Bourdoulous, S., Shimonaka, M., Heinegard, D., Schachner, M., et al. (1997). The C-type lectin domains of lecticans, a family of aggregating chondroitin sulfate proteoglycans, bind tenascin-R by protein-protein interactions independent of carbohydrate moiety. *Proceedings of the National Academy of Sciences of the United States of America, 94,* 10116–10121.

Aszodi, A., Hunziker, E. B., Brakebusch, C., & Fassler, R. (2003). Beta1 integrins regulate chondrocyte rotation, G1 progression, and cytokinesis. *Genes & Development, 17,* 2465–2479.

Atencia, L. J., McDevitt, C. A., Nile, W. B., & Sokoloff, L. (1989). Cartilage content of an immature dog. *Connective Tissue Research, 18,* 235–242.

Ateshian, G. A. (2009). The role of interstitial fluid pressurization in articular cartilage lubrication. *Journal of Biomechanics, 42,* 1163–1176.

Ateshian, G. A., Rajan, V., Chahine, N. O., Canal, C. E., & Hung, C. T. (2009). Modeling the matrix of articular cartilage using a continuous fiber angular distribution predicts many observed phenomena. *Journal of Biomechanical Engineering-Transactions of the ASME, 131.*

Ateshian, G. A., & Wang, H. (1997). Rolling resistance of articular cartilage due to interstitial fluid flow. *Proceedings of the Institution of Mechanical Engineers Part H—Journal of Engineering in Medicine, 211,* 419–424.

Attili, S., & Richter, R. P. (2013). Self-assembly and elasticity of hierarchical proteoglycan-hyaluronan brushes. *Soft Matter, 9,* 10473–10483.

Attur, M., Al-Mussawir, H. E., Patel, J., Kitay, A., Dave, M., Palmer, G., et al. (2008). Prostaglandin E2 exerts catabolic effects in osteoarthritis cartilage: Evidence for signaling via the EP4 receptor. *The Journal of Immunology, 181,* 5082–5088.

Attur, M., Dave, M., Patel, J., Al-Mussawir, H. E., Pillinger, M. H., & Abramson, S. B. (2006). Prostaglandin E2 induces mitochondrial dysfunction in osteoarthritic chondrocytes and exerts catabolic effects via the EP4 receptor. *Arthritis and Rheumatism, 54,* S93.

Attur, M., Yang, Q., Shimada, K., Tachida, Y., Nagase, H., Mignatti, P., et al. (2015). Elevated expression of periostin in human osteoarthritic cartilage and its potential role in matrix degradation via matrix metalloproteinase-13. *The FASEB Journal, 29,* 4107–4121.

Babyn, P. S., Kim, H. K., Lemaire, C., Gahunia, H. K., Cross, A., Denanassy, J., et al. (1996). High-resolution magnetic resonance imaging of normal porcine cartilaginous epiphyseal maturation. *Journal of Magnetic Resonance Imaging, 6,* 172–179.

Bader, D. L., Salter, D. M., & Chowdhury, T. T. (2011). Biomechanical influence of cartilage homeostasis in health and disease. *Arthritis, 2011,* 979032.

Bailey, A. J., Mansell, J. P., Sims, T. J., & Banse, X. (2004). Biochemical and mechanical properties of subchondral bone in osteoarthritis. *Biorheology, 41,* 349–358.

Bajpayee, A. G., & Grodzinsky, A. J. (2013). Avidin as a model for charge driven transport into cartilage: Relevance to osteoarthritis drug delivery. *Osteoarthritis and Cartilage, 21,* S290–S290.

Bajpayee, A. G., Scheu, M., Grodzinsky, A. J., & Porter, R. M. (2014). Electrostatic interactions enable rapid penetration, enhanced uptake and retention of intra-articular injected avidin in rat knee joints. *Journal of Orthopaedic Research.*

Baker, A. H., Edwards, D. R., & Murphy, G. (2002). Metalloproteinase inhibitors: Biological actions and therapeutic opportunities. *Journal of Cell Science, 115,* 3719–3727.

Bal, B. S., Ray, B. K., Shakya, A., & Ray, A. (2006). Overexpression of MMP-14 in human osteoarthritic joint is mediated by SAF-1. In *52nd Annual Meeting of the Orthopaedic Research Society.*

Baler, K., Martin, O. A., Carignano, M. A., Ameer, G. A., Vila, J. A., & Szleifer, I. (2014). Electrostatic unfolding and interactions of albumin driven by pH Changes: A molecular dynamics study. *Journal of Physical Chemistry B, 118,* 921–930.

Ball, S., Bella, J., Kielty, C., & Shuttleworth, A. (2003). Structural basis of type VI collagen dimer formation. *Journal of Biological Chemistry, 278,* 15326–15332.

Baltimore, D., Berg, P., Botchan, M., Carroll, D., Charo, R. A., Church, G., et al. (2015). Biotechnology. A prudent path forward for genomic engineering and germline gene modification. *Science, 348,* 36–38.

Bao, F., Wu, P., Xiao, N., Qiu, F., & Zeng, Q. P. (2012). Nitric oxide-driven hypoxia initiates synovial angiogenesis, hyperplasia and inflammatory lesions in mice. *PLoS ONE, 7.*

Barbero, A., Grogan, S., Schafer, D., Heberer, M., Mainil-Varlet, P., & Martin, I. (2004). Age related changes in human articular chondrocyte yield, proliferation and post-expansion chondrogenic capacity. *Osteoarthritis and Cartilage, 12,* 476–484.

Barbosa, F. L., Chaurasia, S. S., Kaur, H., de Medeiros, F. W., Agrawal, V., & Wilson, S. E. (2010). Stromal interleukin-1 expression in the cornea after haze-associated injury. *Experimental Eye Research, 91,* 456–461.

Barclay, L., Harrington, A., & Ottewill, R. H. (1972). The measurement of forces between particles in disperse systems. *Kolloid-Zeitschrift and Zeitschrift Fur Polymere, 250,* 655–666.

Barilla, M. L., & Carsons, S. E. (2000). Fibronectin fragments and their role in inflammatory arthritis. *Seminars in Arthritis and Rheumatism, 29,* 252–265.

Barker, H. E., Cox, T. R., & Erler, J. T. (2012). The rationale for targeting the LOX family in cancer. *Nature Reviews Cancer, 12,* 540–552.

Barnum, K. J., & O'Connell, M. J. (2014). Cell cycle regulation by checkpoints. *Methods in Molecular Biology, 1170,* 29–40.

Baroja-Mazo, A., Barbera-Cremades, M., & Pelegrin, P. (2013). The participation of plasma membrane hemichannels to purinergic signaling. *Biochimica et Biophysica Acta, 1828,* 79–93.

Baskin, E. M., Bukshpan, S., & Zilberstein, G. V. (2006). pH-induced intracellular protein transport. *Physical Biology, 3,* 101–106.

Basser, P. J., Schneiderman, R., Bank, R. A., Wachtel, E., & Maroudas, A. (1998). Mechanical properties of the collagen network in human articular cartilage as measured by osmotic stress technique. *Archives of Biochemistry and Biophysics, 351,* 207–219.

Bassett, J. H., Swinhoe, R., Chassande, O., Samarut, J., & Williams, G. R. (2006). Thyroid hormone regulates heparan sulfate proteoglycan expression in the growth plate. *Endocrinology, 147,* 295–305.

Bastow, E. R., Byers, S., Golub, S. B., Clarkin, C. E., Pitsillides, A. A., & Fosang, A. J. (2008). Hyaluronan synthesis and degradation in cartilage and bone. *Cellular and Molecular Life Sciences, 65,* 395–413.

Bathe, M., Rutledge, G. C., Grodzinsky, A. J., & Tidor, B. (2005). Osmotic pressure of aqueous chondroitin sulfate solution: A molecular modeling investigation. *Biophysical Journal, 89,* 2357–2371.

Bayliss, M. T., Howat, S., Davidson, C., & Dudhia, J. (2000). The organization of aggrecan in human articular cartilage. Evidence for age-related changes in the rate of aggregation of newly synthesized molecules. *Journal of Biological Chemistry, 275,* 6321–6327.

Bayliss, M. T., Howat, S. L. T., Dudhia, J., Murphy, J. M., Barry, F. P., Edwards, J. C. W., et al. (2001). Up-regulation and differential expression of the hyaluronan-binding protein TSG-6 in cartilage and synovium in rheumatoid arthritis and osteoarthritis. *Osteoarthritis and Cartilage, 9,* 42–48.

Bear, J., & Bachmat, Y. (1991). *Introduction to modeling of transport phenomena in porous media.* Dordrecht; Boston: Kluwer Academic Publishers.

Beekhuizen, M., Bastiaansen-Jenniskens, Y. M., Koevoet, W., Saris, D. B., Dhert, W. J., Creemers, L. B., et al. (2011). Osteoarthritic synovial tissue inhibition of proteoglycan production in human osteoarthritic knee cartilage: Establishment and characterization of a long-term cartilage-synovium coculture. *Arthritis and Rheumatism, 63,* 1918–1927.

Behrens, F., Kraft, E. L., & Oegema, T. R., Jr. (1989). Biochemical changes in articular cartilage after joint immobilization by casting or external fixation. *Journal of Orthopaedic Research, 7,* 335–343.

Beier, F., Ali, Z., Mok, D., Taylor, A. C., Leask, T., Albanese, C., et al. (2001). TGFbeta and PTHrP control chondrocyte proliferation by activating cyclin D1 expression. *Molecular Biology of the Cell, 12,* 3852–3863.

Bekhouche, M., & Colige, A. (2015). The procollagen N-proteinases ADAMTS2, 3 and 14 in pathophysiology. *Matrix Biology, 44–46,* 46–53.

Belcher, C., Yaqub, R., Fawthrop, F., Bayliss, M., & Doherty, M. (1997). Synovial fluid chondroitin and keratan sulphate epitopes, glycosaminoglycans, and hyaluronan in arthritic and normal knees. *Annals of the Rheumatic Diseases, 56,* 299–307.

Bell, C. J., Ingham, E., & Fisher, J. (2006). Influence of hyaluronic acid on the time-dependent friction response of articular cartilage under different conditions. *Proceedings of the Institution of Mechanical Engineers Part H—Journal of Engineering in Medicine, 220,* 23–31.

Bell, D. M., Leung, K. K., Wheatley, S. C., Ng, L. J., Zhou, S., Ling, K. W., et al. (1997). SOX9 directly regulates the type-II collagen gene. *Nature Genetics, 16,* 174–178.

Bell, R., Li, J., Shewman, E. F., Galante, J. O., Cole, B. J., Bach, B. R., et al. (2013). ADAMTS5 is required for biomechanically-stimulated healing of murine tendinopathy. *Journal of Orthopaedic Research, 31,* 1540–1548.

Bellucci, G., & Seedhom, B. B. (2001). Mechanical behaviour of articular cartilage under tensile cyclic load. *Rheumatology, 40,* 1337–1345.

Benaroch, T. E. (1990). *Biosynthetic response of young and adult human articular cartilage to growth factors.* Master of Science: McGill University.

Benedek, T. G. (2006). A history of the understanding of cartilage. *Osteoarthritis Cartilage, 14,* 203–209.

Benedek, G. B., & Villars, F. (2000). *Physics, with illustrative examples from medicine and biology.* New York: AIP Press.

Benninghoff, A. (1925). Form un Bau der Gelenkknorpel in ihren Beziehungen zur Funktion. II. Der Aufbau des Gelenkknorpels in seinen Bezeihungen zur Funktion. *Zeit Zellforsch und Mikroskop Anatomy, 2,* 783–862.

Benzing, D. W., & Russel, W. B. (1981). The viscoelastic properties of ordered lattices—Experiments. *Journal of Colloid and Interface Science, 83,* 178–190.

Berendsen, A. D., Fisher, L. W., Kilts, T. M., Owens, R. T., Robey, P. G., Gutkind, J. S., et al. (2011). Modulation of canonical Wnt signaling by the extracellular matrix component biglycan. *Proceedings of the National Academy of Sciences of the United States of America, 108,* 17022–17027.

Bevill, S. L., Briant, P. L., Levenston, M. E., & Andriacchi, T. P. (2009). Central and peripheral region tibial plateau chondrocytes respond differently to in vitro dynamic compression. *Osteoarthritis Cartilage, 17,* 980–987.

Bevill, S. L., Thambyah, A., & Broom, N. D. (2010). New insights into the role of the superficial tangential zone in influencing the microstructural response of articular cartilage to compression. *Osteoarthritis and Cartilage, 18,* 1310–1318.

Bhuvanasundar, R., John, A., Sulochana, K. N., Coral, K., Deepa, P. R., & Umashankar, V. (2014). A molecular model of human lysyl oxidase (LOX) with optimal copper orientation in the catalytic cavity for induced fit docking studies with potential modulators. *Bioinformation, 10,* 406–412.

Bi, W., Deng, J. M., Zhang, Z., Behringer, R. R., & de Crombrugghe, B. (1999). Sox9 is required for cartilage formation. *Nature Genetics, 22,* 85–89.

Bi, X. H., Yang, X., Bostrom, M. P. G., Bartusik, D., Ramaswamy, S., Fishbein, K. W., et al. (2007). Fourier transform infrared imaging and MR microscopy studies detect compositional and structural changes in cartilage in a rabbit model of osteoarthritis. *Analytical and Bioanalytical Chemistry, 387,* 1601–1612.

Biewener, A. A. (1990). Biomechanics of mammalian terrestrial locomotion. *Science, 250,* 1097–1103.

Bingham, J. T., Papannagari, R., van de Velde, S. K., Gross, C., Gill, T. J., Felson, D. T., et al. (2008). In vivo cartilage contact deformation in the healthy human tibiofemoral joint. *Rheumatology (Oxford), 47,* 1622–1627.

Biot, M. A. (1941). General theory of three-dimensional consolidation. *Journal of Applied Physics, 12,* 155–164.

Bishop, J. R., Schuksz, M., & Esko, J. D. (2007). Heparan sulphate proteoglycans fine-tune mammalian physiology. *Nature, 446,* 1030–1037.

Bizzaro, N., Bartoloni, E., Morozzi, G., Manganelli, S., Riccieri, V., Sabatini, P., et al. (2013). Anti-cyclic citrullinated peptide antibody titer predicts time to rheumatoid arthritis onset in patients with undifferentiated arthritis: Results from a 2-year prospective study. *Arthritis Research & Therapy, 15,* R16.

Blewis, M. E., Nugent-Derfus, G. E., Schmidt, T. A., Schumacher, B. L., & Sah, R. L. (2007). A model of synovial fluid lubricant composition in normal and injured joints. *European Cells and Materials, 13,* 39.

Blom, A. B., Brockbank, S. M., van Lent, P. L., van Beuningen, H. M., Geurts, J., Takahashi, N., et al. (2009). Involvement of the Wnt signaling pathway in experimental and human osteoarthritis: Prominent role of Wnt-induced signaling protein 1. *Arthritis and Rheumatism, 60,* 501–512.

Bobacz, K., Erlacher, L., Smolen, J., Soleiman, A., & Graninger, W. B. (2004). Chondrocyte number and proteoglycan synthesis in the aging and osteoarthritic human articular cartilage. *Annals of the Rheumatic Diseases, 63,* 1618–1622.

Bobacz, K., Sunk, I. G., Hofstaetter, J. G., Amoyo, L., Toma, C. D., Akira, S., et al. (2007). Toll-like receptors and chondrocytes: The lipopolysaccharide-induced decrease in cartilage matrix synthesis is dependent on the presence of toll-like receptor 4 and antagonized by bone morphogenetic protein 7. *Arthritis and Rheumatism, 56,* 1880–1893.

Boldrini, M., Fulmore, C. A., Tartt, A. N., Simeon, L. R., Pavlova, I., Poposka, V., et al. (2018). Human hippocampal neurogenesis persists throughout aging. *Cell Stem Cell, 22,* 589–.

Bolon, B., Grisanti, M., Villasenor, K., Morony, S., Feige, U., & Simonet, W. S. (2015). Generalized degenerative joint disease in osteoprotegerin (Opg) null mutant mice. *Veterinary Pathology, 52,* 873–882.

Bomer, N., Cornelis, F. M. F., Ramos, Y. F. M., den Hollander, W., Lakenberg, N., van der Breggen, R., et al. (2016). Aberrant calreticulin expression in articular cartilage of Dio2 deficient mice. *PLoS ONE, 11.*

Bonadio, M. B., Giglio, P. N., Helito, C. P., Pecora, J. R., Camanho, G. L., & Demange, M. K. (2017). Subchondroplasty for treating bone marrow lesions in the knee—Initial experience. *Revista Brasileira de Ortopedia, 52,* 325–330.

Bonassar, L. J., Grodzinsky, A. J., Frank, E. H., Davila, S. G., Bhaktav, N. R., & Trippel, S. B. (2001). The effect of dynamic compression on the response of articular cartilage to insulin-like growth factor-I. *Journal of Orthopaedic Research, 19,* 11–17.

Bonassar, L. J., Grodzinsky, A. J., Srinivasan, A., Davila, S. G., & Trippel, S. B. (2000). Mechanical and physicochemical regulation of the action of insulin-like growth factor-I on articular cartilage. *Archives of Biochemistry and Biophysics, 379,* 57–63.

Bonde, H. V., Talman, M. L. M., & Kofoed, H. (2005). The area of the tidemark in osteoarthritis—A three-dimensional stereological study in 21 patients. *APMIS, 113,* 349–352.

Bonet, J., & Wood, R. D. (2008). Nonlinear continuum mechanics for finite element analysis (2nd ed. preface). In *Nonlinear continuum mechanics for finite element analysis* (2nd Ed., Xv–+).

Bonet, J., & Wood, R. D. (2008). *Nonlinear continuum mechanics for finite element analysis* (2nd edn), Xv.

Bonnet, N., Standley, K. N., Bianchi, E. N., Stadelmann, V., Foti, M., Conway, S. J., et al. (2009). The matricellular protein Periostin is required for Sclerostin inhibition and the anabolic response to mechanical loading and physical activity. *Journal of Biological Chemistry, 284,* 35939–35950.

Bosomworth, N. J. (2009). Exercise and knee osteoarthritis: Benefit or hazard? *Canadian Family Physician, 55,* 871–878.

Bourg, I. C., & Sposito, G. (2011). Molecular dynamics simulations of the electrical double layer on smectite surfaces contacting concentrated mixed electrolyte (NaCl–$CaCl_2$) solutions. *Journal of Colloid and Interface Science, 360,* 701–715.

Bowen, R. M. (1980). Incompressible porous-media models by use of the theory of mixtures. *International Journal of Engineering Science, 18,* 1129–1148.

Bowen, R. M. (1982). Compressible porous-media models by use of the theory of mixtures. *International Journal of Engineering Science, 20,* 697–735.

Bowler, M. G. (2017). The physics of osmotic pressure. *European Journal of Physics, 38,* 055102.

Boyce, B. F., & Xing, L. (2008). Functions of RANKL/RANK/OPG in bone modeling and remodeling. *Archives of Biochemistry and Biophysics, 473,* 139–146.

Bramble, D. M., & Lieberman, D. E. (2004). Endurance running and the evolution of homo. *Nature, 432,* 345–352.

Bramham, K., Mistry, H. D., Poston, L., Chappell, L. C., & Thompson, A. J. (2009). The non-invasive biopsy-will urinary proteomics make the renal tissue biopsy redundant? *QJM—An International Journal of Medicine, 102,* 523–538.

Brandt, K. D. (2003). Response of joint structures to inactivity and to reloading after immobilization. *Arthritis and Rheumatism, 49,* 267–271.

Brandt, D. W., Burton, D. W., Gazdar, A. F., Oie, H. E., & Deftos, L. J. (1991). All major lung-cancer cell-types produce parathyroid hormone-like protein—Heterogeneity assessed by high-performance liquid-chromatography. *Endocrinology, 129,* 2466–2470.

Brandt, K. D., Myers, S. L., Burr, D., & Albrecht, M. (1991). Osteoarthritic changes in canine articular-cartilage, subchondral bone, and synovium 54 months after transection of the anterior cruciate ligament. *Arthritis and Rheumatism, 34,* 1560–1570.

Brandt, K. D., Radin, E. L., Dieppe, P. A., & van de Putte, L. (2006). Yet more evidence that osteoarthritis is not a cartilage disease. *Annals of the Rheumatic Diseases, 65,* 1261–1264.

Bridges, C., Karra, S., & Rajagopal, K. R. (2010). On modeling the response of the synovial fluid: Unsteady flow of a shear-thinning, chemically-reacting fluid mixture. *Computers & Mathematics with Applications, 60,* 2333–2349.

BRIET, M. & SCHIFFRIN, E. L. 2010. Aldosterone: effects on the kidney and cardiovascular system. *Nature Reviews Nephrology*, 6, 261-273.

Brines, M., & Cerami, A. (2008). Erythropoietin-mediated tissue protection: Reducing collateral damage from the primary injury response. *Journal of Internal Medicine, 264,* 405–432.

Brisby, H., Papadimitriou, N., Runesson, E., Sasaki, N., Lindahl, A., & Henriksson, H. B. (2013). Moderate physical exercise results in increased cell activity in articular cartilage of the knee joint in rats. *Cells Tissues Organs, 198,* 237–248.

Broder, C., Arnold, P., Vadon-Le Goff, S., Konerding, M. A., Bahr, K., Muller, S., et al. (2013). Metalloproteases meprin alpha and meprin beta are C- and N-procollagen proteinases important for collagen assembly and tensile strength. *Proceedings of the National Academy of Science USA*, 110, 14219–14224.

Bromme, D., Okamoto, K., Wang, B. B., & Biroc, S. (1996). Human cathepsin O2, a matrix protein-degrading cysteine protease expressed in osteoclasts. Functional expression of human cathepsin O2 in Spodoptera frugiperda and characterization of the enzyme. *Journal of Biological Chemistry, 271,* 2126–2132.

Bromme, D., & Wilson, S. (2011). Role of cysteine cathepsins in extracellular proteolysis. In W. C. Parks & R. P. Mecham (Eds.), *Extracellular matrix degradation, extracellular matrix degradation.* Berlin, Heidelberg: Springer.

Broom, N., Chen, M. H., & Hardy, A. (2001). A degeneration-based hypothesis for interpreting fibrillar changes in the osteoarthritic cartilage matrix. *Journal of Anatomy, 199,* 683–698.

Broom, N. D., & Flachsmann, R. (2003). Physical indicators of cartilage health: The relevance of compliance, thickness, swelling and fibrillar texture. *Journal of Anatomy, 202,* 481–494.

Brown, C. P., Houle, M. A., Popov, K., Nicklaus, M., Couture, C. A., Laliberte, M., et al. (2014). Imaging and modeling collagen architecture from the nano to micro scale. *Biomedical Optics Express, 5,* 233–243.

Brunger, J. M., Zutshi, A., Willard, V. P., Gersbach, C. A., & Guilak, F. (2017). Genome engineering of stem cells for autonomously regulated, closed-loop delivery of biologic drugs. *Stem Cell Reports, 8,* 1202–1213.

Bryan, N., Andrews, K. D., Loughran, M. J., Rhodes, N. P., & Hunt, J. A. (2011). Elucidating the contribution of the elemental composition of fetal calf serum to antigenic expression of primary human umbilical-vein endothelial cells in vitro. *Bioscience Reports, 31,* 199–210.

Bryzgalov, L. O., Antontseva, E. V., Matveeva, M. Y., Shilov, A. G., Kashina, E. V., Mordvinov, V. A., et al. (2013). Detection of regulatory SNPs in human genome using ChIP-seq ENCODE data. *PLoS ONE*, 8.

Bucay, N., Sarosi, I., Dunstan, C. R., Morony, S., Tarpley, J., Capparelli, C., et al. (1998). osteoprotegerin-deficient mice develop early onset osteoporosis and arterial calcification. *Genes & Development, 12,* 1260–1268.

Buckley, M. R., Gleghorn, J. P., Bonassar, L. J., & Cohen, I. (2008). Mapping the depth dependence of shear properties in articular cartilage. *Journal of Biomechanics, 41,* 2430–2437.

Buckwalter, J. A. (1995). Activity vs. rest in the treatment of bone, soft tissue and joint injuries. *Iowa Orthopaedic Journal, 15,* 29–42.

Buckwalter, J. A., Martin, J. A., & Brown, T. D. (2006). Perspectives on chondrocyte mechanobiology and osteoarthritis. *Biorheology, 43,* 603–609.

Bunger, M. K., Walisser, J. A., Sullivan, R., Manley, P. A., Moran, S. M., Kalscheur, V. L., et al. (2005). Progressive arthropathy in mice with a targeted disruption of the Mop3/Bmal-1 locus. *Genesis, 41,* 122–132.

Bursac, P. M., Obitz, T. W., Eisenberg, S. R., & Stamenovic, D. (1999). Confined and unconfined stress relaxation of cartilage: Appropriateness of a transversely isotropic analysis. *Journal of Biomechanics, 32,* 1125–1130.

Buschmann, M. D., Gluzband, Y. A., Grodzinsky, A. J., & Hunziker, E. B. (1995). Mechanical compression modulates matrix biosynthesis in chondrocyte/agarose culture. *Journal of Cell Science, 108*(Pt 4), 1497–1508.

Buschmann, M. D., Hunziker, E. B., Kim, Y. J., & Grodzinsky, A. J. (1996a). Altered aggrecan synthesis correlates with cell and nucleus structure in statically compressed cartilage. *Journal of Cell Science, 109*(Pt 2), 499–508.

Buschmann, M. D., Kim, Y. J., Wong, M., Frank, E., Hunziker, E. B., & Grodzinsky, A. J. (1999). Stimulation of aggrecan synthesis in cartilage explants by cyclic loading is localized to regions of high interstitial fluid flow. *Archives of Biochemistry and Biophysics, 366,* 1–7.

Buschmann, M. D., Maurer, A. M., Berger, E., & Hunziker, E. B. (1996b). A method of quantitative autoradiography for the spatial localization of proteoglycan synthesis rates in cartilage. *Journal of Histochemistry and Cytochemistry, 44,* 423–431.

Buschmann, M. D., Maurer, A. M., Berger, E., Perumbuli, P., & Hunziker, E. B. (2000). Ruthenium hexaammine trichloride chemography for aggrecan mapping in cartilage is a sensitive indicator of matrix degradation. *Journal of Histochemistry and Cytochemistry, 48,* 81–88.

Bush, P. G., & Hall, A. C. (2003). The volume and morphology of chondrocytes within non-degenerate and degenerate human articular cartilage. *Osteoarthritis Cartilage, 11,* 242–251.

Butz, K. D., Merrell, G., & Nauman, E. A. (2012). A biomechanical analysis of finger joint forces and stresses developed during common daily activities. *Computer Methods in Biomechanics and Biomedical Engineering, 15,* 131–140.

Calvet, J., Orellana, C., Gratacos, J., Berenguer-Llergo, A., Caixas, A., Chillaron, J. J., et al. (2016). Synovial fluid adipokines are associated with clinical severity in knee osteoarthritis: A cross-sectional study in female patients with joint effusion. *Arthritis Research & Therapy, 18.*

Camp, R. J., Liles, M., Beale, J., Saeidi, N., Flynn, B. P., Moore, E., et al. (2011). Molecular mechanochemistry: Low force switch slows enzymatic cleavage of human type I collagen monomer. *Journal of the American Chemical Society, 133,* 4073–4078.

Canty, E. G., & Kadler, K. E. (2005). Procollagen trafficking, processing and fibrillogenesis. *Journal of Cell Science, 118,* 1341–1353.

Capito, R. M., & Spector, M. (2007). Collagen scaffolds for nonviral IGF-1 gene delivery in articular cartilage tissue engineering. *Gene Therapy, 14,* 721–732.

Carlsen, S., Nandakumar, K. S., Backlund, J., Holmberg, J., Hultqvist, M., Vestberg, M., et al. (2008). Cartilage oligomeric matrix protein induction of chronic arthritis in mice. *Arthritis and Rheumatism, 58,* 2000–2011.

Carlson, J. M., & Doyle, J. (2002). Complexity and robustness. *Proceedings of the National Academy of Sciences of the United States of America, 99,* 2538–2545.

Carnie, S. L., Chan, D. Y. C., & Gunning, J. S. (1994). Electrical double-layer interaction between dissimilar spherical colloidal particles and between a sphere and a plate—The linearized Poisson-Boltzmann theory. *Langmuir, 10,* 2993–3009.

Carroll, S. B. (2008). Evo-devo and an expanding evolutionary synthesis: A genetic theory of morphological evolution. *Cell, 134,* 25–36.

Carter, D. R., Beaupre, G. S., Wong, M., Smith, R. L., Andriacchi, T. P., & Schurman, D. J. (2004). The mechanobiology of articular cartilage development and degeneration. *Clinical Orthopaedics and Related Research*, S69–S77.

Casalone, E., Tachmazidou, I., Zengini, E., Hatzikotoulas, K., Hackinger, S., Suveges, D., et al. (2018). A novel variant in GLIS3 is associated with osteoarthritis. *Annals of the Rheumatic Diseases, 77,* 620–623.

Cecil, D. L., Appleton, C. T. G., Polewski, M. D., Mort, J. S., Schmidt, A. M., Bendele, A., et al. (2009). The pattern recognition receptor CD36 is a chondrocyte hypertrophy marker associated with suppression of catabolic responses and promotion of repair responses to inflammatory stimuli. *Journal of Immunology, 182,* 5024–5031.

Chahine, N. O., Chen, F. H., Hung, C. T., & Ateshian, G. A. (2005). Direct measurement of osmotic pressure of glycosaminoglycan solutions by membrane osmometry at room temperature. *Biophysical Journal, 89,* 1543–1550.

Chahine, N. O., Wang, C. C. B., Hung, C. T., & Ateshian, G. A. (2004). Anisotropic strain-dependent material properties of bovine articular cartilage in the transitional range from tension to compression. *Journal of Biomechanics, 37,* 1251–1261.

Chai, D. H., Stevens, A. L., & Grodzinsky, A. (2007). Biomechanical aspects: Joint injury and osteoarthritis. *Bone and Osteoarthritis Topics in Bone Biology*, 165–179

Chan, B. Y., & Little, C. B. (2012). The interaction of canonical bone morphogenetic protein- and Wnt-signaling pathways may play an important role in regulating cartilage degradation in osteoarthritis. *Arthritis Research & Therapy, 14,* 119.

Chandran, P. L., & Horkay, F. (2012). Aggrecan, an unusual polyelectrolyte: Review of solution behavior and physiological implications. *Acta Biomaterialia, 8,* 3–12.

Chang, C., Lauffenburger, D. A., & Morales, T. I. (2003). Motile chondrocytes from newborn calf: Migration properties and synthesis of collagen II. *Osteoarthritis and Cartilage, 11,* 603–612.

Chang, D. P., Guilak, F., Jay, G. D., & Zauscher, S. (2014). Interaction of lubricin with type II collagen surfaces: Adsorption, friction, and normal forces. *Journal of Biomechanics, 47,* 659–666.

Chang, J., Nakajima, H., & Poole, C. A. (1997). Structural colocalisation of type VI collagen and fibronectin in agarose cultured chondrocytes and isolated chondrons extracted from adult canine tibial cartilage. *Journal of Anatomy, 190*(Pt 4), 523–532.

Chang, T. H., Hsieh, F. L., Zebisch, M., Harlos, K., Elegheert, J., & Jones, E. Y. (2015). Structure and functional properties of Norrin mimic Wnt for signalling with Frizzled4, Lrp5/6, and proteoglycan. *Elife, 4.*

Changoor, A., Nelea, M., Methot, S., Tran-Khanh, N., Chevrier, A., Restrepo, A., et al. (2011). Structural characteristics of the collagen network in human normal, degraded and repair articular cartilages observed in polarized light and scanning electron microscopies. *Osteoarthritis Cartilage, 19,* 1458–1468.

Charlebois, M., McKee, M. D., & Buschmann, M. D. (2004). Nonlinear tensile properties of bovine articular cartilage and their variation with age and depth. *Journal of Biomechanical Engineering, 126,* 129–137.

Charnley, J. (1960). The lubrication of animal joints in relation to surgical reconstruction by arthroplasty. *Annals of the Rheumatic Diseases, 19,* 10–19.

Chau, M., Forcinito, P., Andrade, A. C., Hegde, A., Ahn, S., Lui, J. C., et al. (2011). Organization of the Indian hedgehog–parathyroid hormone-related protein system in the postnatal growth plate. *Journal of Molecular Endocrinology, 47,* 99–107.

Chauhan, V. P., Boucher, Y., Ferrone, C. R., Roberge, S., Martin, J. D., Stylianopoulos, T., et al. (2014). Compression of pancreatic tumor blood vessels by hyaluronan is caused by solid stress and not interstitial fluid pressure. *Cancer Cell, 26,* 14–15.

Che, J., Dzubiella, J., Li, B., & McCammon, J. A. (2008). Electrostatic free energy and its variations in implicit solvent models. *Journal of Physical Chemistry B, 112,* 3058–3069.

Chen, C. T., Burton-Wurster, N., Borden, C., Hueffer, K., Bloom, S. E., & Lust, G. (2001). Chondrocyte necrosis and apoptosis in impact damaged articular cartilage. *Journal of Orthopaedic Research, 19,* 703–711.

Chen, C. T., Burton-Wurster, N., Lust, G., Bank, R. A., & Tekoppele, J. M. (1999). Compositional and metabolic changes in damaged cartilage are peak-stress, stress-rate, and loading-duration dependent. *Journal of Orthopaedic Research, 17,* 870–879.

Chen, S., & Birk, D. E. (2013). The regulatory roles of small leucine-rich proteoglycans in extracellular matrix assembly. *FEBS Journal, 280,* 2120–2137.

Chen, S., Fu, P., Cong, R., Wu, H., & Pei, M. (2015). Strategies to minimize hypertrophy in cartilage engineering and regeneration. *Genes & Diseases, 2,* 76–95.

Chen, S. S., Falcovitz, Y. H., Schneiderman, R., Maroudas, A., & Sah, R. L. (2001). Depth-dependent compressive properties of normal aged human femoral head articular cartilage: Relationship to fixed charge density. *Osteoarthritis and Cartilage, 9,* 561–569.

Chen, X., Macica, C. M., Nasiri, A., & Broadus, A. E. (2008). Regulation of articular chondrocyte proliferation and differentiation by Indian hedgehog and parathyroid hormone-related protein in mice. *Arthritis Rheum, 58,* 3788–3797.

Chen, X., Macica, C. M., Nasiri, A., & Broadus, A. E. (2008). Regulation of articular chondrocyte proliferation and differentiation by Indian hedgehog and parathyroid hormone-related protein in mice. *Arthritis and Rheumatism, 58,* 3788–3797.

Chinzei, N., Brophy, R. H., Duan, X., Cai, L., Nunley, R. M., Sandell, L. J., et al. (2018). Molecular influence of anterior cruciate ligament tear remnants on chondrocytes: A biologic connection between injury and osteoarthritis. *Osteoarthritis and Cartilage, 26,* 588–599.

Chowdhury, T. T., Arghandawi, S., Brand, J., Akanji, O. O., Bader, D. L., Salter, D. M., et al. (2008). Dynamic compression counteracts IL-1 beta induced inducible nitric oxide synthase and cyclo-oxygenase-2 expression in chondrocyte/agarose constructs. *Arthritis Research & Therapy, 10.*

Chu, C. R., Williams, A., Tolliver, D., Kwoh, C. K., Bruno, S., & Irrgang, J. J. (2010). Clinical optical coherence tomography of early articular cartilage degeneration in patients with degenerative meniscal tears. *Arthritis and Rheumatism, 62,* 1412–1420.

Chu, C. R., Williams, A. A., West, R. V., Qian, Y. X., Fu, F. H., Do, B. H., et al. (2014). Quantitative magnetic resonance imaging UTE-T2* mapping of cartilage and meniscus healing after anatomic anterior cruciate ligament reconstruction. *American Journal of Sports Medicine, 42,* 1847–1856.

Chu, J. G., Dai, M. W., Wang, Y., Tian, F. M., Song, H. P., Xiao, Y. P., et al. (2017). Strontium ranelate causes osteophytes overgrowth in a model of early phase osteoarthritis. *BMC Musculoskeletal Disorders, 18,* 78.

Chun, J. S., Oh, H., Yang, S., & Park, M. (2008). Wnt signaling in cartilage development and degeneration. *BMB Reports, 41,* 485–494.

Cicuttini, F. M., & Wluka, A. E. (2014). Osteoarthritis: Is OA a mechanical or systemic disease? *Nature Reviews Rheumatology, 10,* 515–516.

Clark, A. L. (2008). Osteoarthritis: What we have been missing in the patellofemoral joint. *Exercise and Sport Sciences Reviews, 36,* 30–37.

Clark, A. L., Leonard, T. R., Barclay, L. D., Matyas, J. R., & Herzog, W. (2005). Opposing cartilages in the patellofemoral joint adapt differently to long-term cruciate deficiency: Chondrocyte deformation and reorientation with compression. *Osteoarthritis Cartilage, 13,* 1100–1114.

Clarke, I. C. (1971). Articular cartilage: A review and scanning electron microscope study. 1. The interterritorial fibrillar architecture. *Journal of Bone and Joint Surgery. British Volume, 53,* 732–750.

Clarke, I. C. (1974). Articular cartilage: A review and scanning electron microscope study. II. The territorial fibrillar architecture. *Journal of Anatomy, 118,* 261–280.

Clemmons, D. R., Busby, W. H., Jr., Garmong, A., Schultz, D. R., Howell, D. S., Altman, R. D., et al. (2002). Inhibition of insulin-like growth factor binding protein 5 proteolysis in articular cartilage and joint fluid results in enhanced concentrations of insulin-like growth factor 1 and is associated with improved osteoarthritis. *Arthritis and Rheumatism, 46,* 694–703.

Coleman, P. J., Scott, D., Ray, J., Mason, R. M., & Levick, J. R. (1997). Hyaluronan secretion into the synovial cavity of rabbit knees and comparison with albumin turnover. *Journal of Physiology-London, 503,* 645–656.

Coles, J. M., Zhang, L., Blum, J. J., Warman, M. L., Jay, G. D., Guilak, F., et al. (2010). Loss of cartilage structure, stiffness, and frictional properties in mice lacking PRG4. *Arthritis and Rheumatism, 62,* 1666–1674.

Collier, I. E., Legant, W., Marmer, B., Lubman, O., Saffarian, S., Wakatsuki, T., et al. (2011). Diffusion of MMPs on the surface of collagen fibrils: The mobile cell surface-collagen substratum interface. *PLoS ONE, 6,* e24029.

Collins, F. S., Morgan, M., & Patrinos, A. (2003). The human genome project: Lessons from large-scale biology. *Science, 300,* 286–290.

Comper, W. D. (1991). Physicochemical aspects of cartilage extracellular matrix. In B. K. Hall & S. A. Newman (Eds.), *Cartilage: Molecular aspects*. Boca Raton: CRC Press.

Comper, W. D., & Lyons, K. C. (1993). Non-electrostatic factors govern the hydrodynamic properties of articular cartilage proteoglycan. *Biochemical Journal, 289*(Pt 2), 543–547.

Comper, W. D., & Williams, R. P. W. (1987). Hydrodynamics of concentrated proteoglycan solutions. *Journal of Biological Chemistry, 262,* 13464–13471.

Corr, M. (2008). Wnt-beta-catenin signaling in the pathogenesis of osteoarthritis. *Nature Clinical Practice Rheumatology, 4,* 550–556.

Correro-Shahgaldian, M. R., Colombo, V., Spencer, N. D., Weber, F. E., Imfeld, T., & Gallo, L. M. (2011). Coupling plowing of cartilage explants with gene expression in models for synovial joints. *Journal of Biomechanics, 44,* 2472–2476.

Cortes, M., Baria, A. T., & Schwartz, N. B. (2009). Sulfation of chondroitin sulfate proteoglycans is necessary for proper Indian hedgehog signaling in the developing growth plate. *Development, 136,* 1697–1706.

Courties, A., Sellam, J., & Berenbaum, F. (2017). Metabolic syndrome-associated osteoarthritis. *Current Opinion in Rheumatology, 29,* 214–222.

Coussy, O. (2004). *Poromechanics*. New York: Wiley.

Coussy, O., & Scherer, G. (2004). Special issue of concrete science and engineering on poromechanics of cement-based materials. *Materials and Structures, 37,* 2.

Craft, A. M., Ahmed, N., Rockel, J. S., Baht, G. S., Alman, B. A., Kandel, R. A., et al. (2013). Specification of chondrocytes and cartilage tissues from embryonic stem cells. *Development, 140,* 2597–2610.

Crockett, R., Roos, S., Rossbach, P., Dora, C., Born, W., & Troxler, H. (2005). Imaging of the surface of human and bovine articular cartilage with ESEM and AFM. *Tribology Letters, 19,* 311–317.

Crofford, L. J. (2013). Use of NSAIDs in treating patients with arthritis. *Arthritis Research & Therapy, 15*.

Cross, P. C., & Mercer, K. L. (1993). *Cell and tissue ultrastructure: A functional perspective* (2nd ed.). W.H: Freeman.

Cruz, R. D., Esteves, M. J. D., Teixeira, R. G. D., Cardoso, M. J. E. D., & Barcia, O. E. (2010). Calculation of the osmotic pressure and theta temperature of polymer solutions through cubic equations of state and the McMillan-Mayer solution theory framework. *Journal of Macromolecular Science Part B-Physics, 49,* 1083–1093.

Cui, Z., Crane, J., Xie, H., Jin, X., Zhen, G., Li, C., et al. (2016). Halofuginone attenuates osteoarthritis by inhibition of TGF-beta activity and H-type vessel formation in subchondral bone. *Annals of the Rheumatic Diseases, 75,* 1714–1721.

Curnier, A., He, Q. C., & Zysset, P. (1995). Conewise linear elastic-materials. *Journal of Elasticity, 37,* 1–38.

Cussler, E. L. (1984). *Diffusion, mass transfer in fluid systems*. Cambridge Cambridgeshire; New York: Cambridge University Press.

Czipri, M., Otto, J. M., Cs-Szabo, G., Kamath, R. V., Vermes, C., Firneisz, G., et al. (2003). Genetic rescue of chondrodysplasia and the perinatal lethal effect of cartilage link protein deficiency. *Journal of Biological Chemistry, 278,* 39214–39223.

D'Lima, D. D., Hashimoto, S., Chen, P. C., Colwell, C. W. Jr., & Lotz, M. K. (2001). Human chondrocyte apoptosis in response to mechanical injury. *Osteoarthritis Cartilage, 9,* 712–719.

Dao, D. Y., Jonason, J. H., Zhang, Y., Hsu, W., Chen, D., Hilton, M. J., et al. (2012). Cartilage-specific beta-catenin signaling regulates chondrocyte maturation, generation of ossification centers, and perichondrial bone formation during skeletal development. *Journal of Bone and Mineral Research, 27,* 1680–1694.

Dargel, J., Gotter, M., Mader, K., Pennig, D., Koebke, J., & Schmidt-Wiethoff, R. (2007). Biomechanics of the anterior cruciate ligament and implications for surgical reconstruction. *Strategies Trauma Limb Reconstr, 2,* 1–12.

Darling, E. M., Wilusz, R. E., Bolognesi, M. P., Zauscher, S., & Guilak, F. (2010). Spatial mapping of the biomechanical properties of the pericellular matrix of articular cartilage measured in situ via atomic force microscopy. *Biophysical Journal, 98,* 2848–2856.

Davidson, R. K., Waters, J. G., Kevorkian, L., Darrah, C., Cooper, A., Donell, S. T., & Clark, I. M. (2006). Expression profiling of metalloproteinases and their inhibitors in synovium and cartilage. *Arthritis Research & Therapy*, 8.

Davies-Tuck, M. L., Martel-Pelletier, J., Wluka, A. E., Pelletier, J. P., Ding, C., Jones, G., et al. (2008). Meniscal tear and increased tibial plateau bone area in healthy post-menopausal women. *Osteoarthritis Cartilage, 16,* 268–271.

Davies, L. C., Caterson, B., & Duance, V. (2006). Comparing the responsivenss of chondrocytes isolated from young and mature articular cartilage to IGF-1 and TGF-beta. *European Cells and Materials, 12,* 54.

Davies, L. C., Blain, E. J., Caterson, B., & Duance, V. C. (2008a). Chondroitin sulphate impedes the migration of a sub-population of articular cartilage chondrocytes. *Osteoarthritis Cartilage, 16,* 855–864.

Davies, L. C., Blain, E. J., Gilbert, S. J., Caterson, B., & Duance, V. C. (2008b). The potential of IGF-1 and TGFbeta1 for promoting "adult" articular cartilage repair: An in vitro study. *Tissue Engineering Part A, 14,* 1251–1261.

Davies, M. E., Horner, A., Loveland, B. E., & McKenzie, I. F. (1994). Upregulation of complement regulators MCP (CD46), DAF (CD55) and protectin (CD59) in arthritic joint disease. *Scandinavian Journal of Rheumatology, 23,* 316–321.

Dawes, C. (2003). What is the critical pH and why does a tooth dissolve in acid? *Journal/Canadian Dental Association (Journal de l'Association Dentaire Canadienne), 69,* 722–724.

Day, A. J., & Prestwich, G. D. (2002). Hyaluronan-binding proteins: Tying up the giant. *Journal of Biological Chemistry, 277,* 4585–4588.

de Campos, G. C. (2015). Placebo effect in osteoarthritis: Why not use it to our advantage? *World Journal of Orthopedics, 6,* 416–420.

de Grauw, J. C., van de Lest, C. H., & van Weeren, P. R. (2009). Inflammatory mediators and cartilage biomarkers in synovial fluid after a single inflammatory insult: A longitudinal experimental study. *Arthritis Research & Therapy, 11,* R35.

Decker, R. S., Koyama, E., & Pacifici, M. (2014). Genesis and morphogenesis of limb synovial joints and articular cartilage. *Matrix Biology, 39,* 5–10.

Dejica, V. M., Mort, J. S., Laverty, S., Antoniou, J., Zukor, D. J., Tanzer, M., et al. (2012). Increased type II collagen cleavage by cathepsin K and collagenase activities with aging and osteoarthritis in human articular cartilage. *Arthritis Research and Therapy, 14,* R113.

Dell'Accio, F., de Bari, C., Eltawil, N. M., Vanhummelen, P., & Pitzalis, C. (2008). Identification of the molecular response of articular cartilage to injury, by microarray screening: Wnt-16 expression and signaling after injury and in osteoarthritis. *Arthritis and Rheumatism, 58,* 1410–1421.

Denbigh, K. (1981). *The principles of chemical equilibrium.* Cambridge: Cambridge University Press.

Derjaguin, B. V., & Churaev, N. V. (1977). Disjoining pressure of thin-layers of binary-solutions. *Journal of Colloid and Interface Science, 62,* 369–380.

Deserno, M., & Holm, C. (2001). Cell model and Poisson-Boltzmann theory: A brief introduction. *Electrostatic Effects in Soft Matter and Biophysics, 46,* 27–50.

Dicesare, P. E., Morgelin, M., Carlson, C. S., Pasumarti, S., & Paulsson, M. (1995). Cartilage oligomeric matrix protein: Isolation and characterization from human articular cartilage. *Journal of Orthopaedic Research, 13,* 422–428.

Didomenico, C. D., Lintz, M., & Bonassar, L. J. (2018). Molecular transport in articular cartilage —What have we learned from the past 50 years? *Nature Reviews Rheumatology, 14*, 293–403.

Ding, C., Cicuttini, F., & Jones, G. (2007). Tibial subchondral bone size and knee cartilage defects: Relevance to knee osteoarthritis. *Osteoarthritis Cartilage, 15,* 479–486.

Ding, L., Guo, D., & Homandberg, G. A. (2008). The cartilage chondrolytic mechanism of fibronectin fragments involves MAP kinases: Comparison of three fragments and native fibronectin. *Osteoarthritis Cartilage, 16,* 1253–1262.

Ding, M., Dalstra, M., Linde, F., & Hvid, I. (1998a). Changes in the stiffness of the human tibial cartilage-bone complex in early-stage osteoarthrosis. *Acta Orthopaedica Scandinavica, 69,* 358–362.

Ding, M., Dalstra, M., Linde, F., & Hvid, I. (1998b). Mechanical properties of the normal human tibial cartilage-bone complex in relation to age. *Clinical Biomechanics, 13,* 351–358.

Dingle, J. T., & Parker, M. (1997). NSAID stimulation of human cartilage matrix synthesis—A study of the mechanism of action of aceclofenac. *Clinical Drug Investigation, 14,* 353–362.

Dong, Y., Drissi, H., Chen, M., Chen, D., Zuscik, M. J., Schwarz, E. M., et al. (2005). Wnt-mediated regulation of chondrocyte maturation: Modulation by TGF-beta. *Journal of Cellular Biochemistry, 95,* 1057–1068.

Dooley, M., Spencer, C. M., & Dunn, C. J. (2001). Aceclofenac—A reappraisal of its use in the management of pain and rheumatic disease. *Drugs, 61,* 1351–1378.

Dore, D., Quinn, S., Ding, C., Winzenberg, T., Cicuttini, F., & Jones, G. (2010). Subchondral bone and cartilage damage: A prospective study in older adults. *Arthritis and Rheumatism, 62,* 1967–1973.

Dowson, D., Unsworth, A., & Wright, V. (1970). Analysis of boosted-lubrication in human joints. *Journal of Mechanical Engineering Science, 12*, 364.

Dowthwaite, G. P., Bishop, J. C., Redman, S. N., Khan, I. M., Rooney, P., Evans, D. J., et al. (2004). The surface of articular cartilage contains a progenitor cell population. *Journal of Cell Science, 117,* 889–897.

Drewniak, E. I., Jay, G. D., Fleming, B. C., Zhang, L., Warman, M. L., & Crisco, J. J. (2012). Cyclic loading increases friction and changes cartilage surface integrity in lubricin-mutant mouse knees. *Arthritis and Rheumatism, 64,* 465–473.

Driban, J. B., Eaton, C. B., Lo, G. H., Ward, R. J., Lu, B., & McAlindon, T. E. (2014). Association of knee injuries with accelerated knee osteoarthritis progression: Data from the osteoarthritis initiative. *Arthritis Care Res (Hoboken), 66,* 1673–1679.

Dudhia, J., Davidson, C. M., Wells, T. M., Vynios, D. H., Hardingham, T. E., & Bayliss, M. T. (1996). Age-related changes in the content of the C-terminal region of aggrecan in human articular cartilage. *Biochemical Journal, 313*(Pt 3), 933–940.

Duer, M. J., Friscic, T., Murray, R. C., Reid, D. G., & Wise, E. R. (2009). The mineral phase of calcified cartilage: Its molecular structure and interface with the organic matrix. *Biophysical Journal, 96,* 3372–3378.

Dunlevy, J. R., & Rada, J. A. (2004). Interaction of lumican with aggrecan in the aging human sclera. *Investigative Ophthalmology & Visual Science, 45,* 3849–3856.

Dyke, J., & Kleidon, A. (2010). The maximum entropy production principle: Its theoretical foundations and applications to the earth system. *Entropy, 12,* 613–630.

Echtermeyer, F., Bertrand, J., Dreier, R., Meinecke, I., Neugebauer, K., Fuerst, M., et al. (2009). Syndecan-4 regulates ADAMTS-5 activation and cartilage breakdown in osteoarthritis. *Nature Medicine, 15,* 1072–1076.

Eckstein, F., Hudelmaier, M., & Putz, R. (2006). The effects of exercise on human articular cartilage. *Journal of Anatomy, 208,* 491–512.

Eckstein, F., Winzheimer, M., Westhoff, J., Schnier, M., Haubner, M., Englmeier, K. H., et al. (1998). Quantitative relationships of normal cartilage volumes of the human knee joint—Assessment by magnetic resonance imaging. *Anatomy and Embryology (Berlin), 197,* 383–390.

Eckstein, F., Wirth, W., Lohmander, L. S., Hudelmaier, M. I., & Frobell, R. B. (2015). Five-year followup of knee joint cartilage thickness changes after acute rupture of the anterior cruciate ligament. *Arthritis Rheumatol, 67,* 152–161.

Edwards, J. (1966). Physical characteristics of articular cartilage. *Proceedings of the Institution of Engineers, 181.*

Eguchi, T., Kubota, S., Kawata, K., Mukudai, Y., Uehara, J., Ohgawara, T., et al. (2008). Novel transcription factor-like function of human matrix metalloproteinase 3 regulating the CTGF/CCN2 gene. *Molecular and Cellular Biology, 28,* 2391–2413.

El-Tawil, S., Arendt, E., & Parker, D. (2016). Position statement: The epidemiology, pathogenesis and risk factors of osteoarthritis of the knee. *Journal of ISAKO: Joint Disorders and Orthopeadic Sports Medicine, 27*(2), 289–294

Elder, B. D., & Athanasiou, K. A. (2009). Hydrostatic pressure in articular cartilage tissue engineering: From chondrocytes to tissue regeneration. *Tissue Engineering Part B Reviews, 15,* 43–53.

Elefteriou, F. (2005). Neuronal signaling and the regulation of bone remodeling. *Cellular and Molecular Life Sciences, 62,* 2339–2349.

Elliott, C. G., Wang, J., Guo, X. L., Xu, S. W., Eastwood, M., Guan, J. J., et al. (2012). Periostin modulates myofibroblast differentiation during full-thickness cutaneous wound repair. *Journal of Cell Science, 125,* 121–132.

Ellman, M. B., Yan, D., Ahmadinia, K., Chen, D., An, H. S., & Im, H. J. (2013). Fibroblast growth factor control of cartilage homeostasis. *Journal of Cellular Biochemistry, 114,* 735–742.

Emanuel, K. S., van der Veen, A. J., Rustenburg, C. M. E., Smit, T. H., & Kingma, I. (2018). Osmosis and viscoelasticity both contribute to time-dependent behaviour of the intervertebral disc under compressive load: A caprine in vitro study. *Journal of Biomechanics, 70,* 10–15.

Englund, M., Guermazi, A., & Lohmander, S. L. (2009). The role of the meniscus in knee osteoarthritis: A cause or consequence? *Radiologic Clinics of North America, 47,* 703–712.

Erickson, G., & Guilak, F. (2000). Osmotic stress initiates intracellular calcium waves in chondrocytes through extracellular influx, the inositol phosphate and a G-protein mediated pathway. In *46th Annual Meeting of the Orthopaedic Research Society.*

Erickson, H. P. (2009). Size and shape of protein molecules at the nanometer level determined by sedimentation, gel filtration, and electron microscopy. *Biological Procedures Online, 11,* 32–51.

Evans, C. H., Kraus, V. B., & Setton, L. A. (2014). Progress in intra-articular therapy. *Nature Reviews Rheumatology, 10*(1), 11–22

Evans, R. G., & Bie, P. (2016). Role of the kidney in the pathogenesis of hypertension: Time for a neo-Guytonian paradigm or a paradigm shift? *American Journal of Physiology: Regulatory, Integrative and Comparative Physiology, 310,* R217–R229.

Ewers, B. J., Dvoracek-Driksna, D., Orth, M. W., & Haut, R. C. (2001). The extent of matrix damage and chondrocyte death in mechanically traumatized articular cartilage explants depends on rate of loading. *Journal of Orthopaedic Research, 19,* 779–784.

Eyre, D. (2002). Collagen of articular cartilage. *Arthritis Research & Therapy, 4,* 30–35.

Eyre, D. R., Dickson, I. R., & van Ness, K. (1988). Collagen cross-linking in human bone and articular cartilage. Age-related changes in the content of mature hydroxypyridinium residues. *Biochemical Journal, 252,* 495–500.

Eyre, D. R., McDevitt, C. A., Billingham, M. E., & Muir, H. (1980). Biosynthesis of collagen and other matrix proteins by articular cartilage in experimental osteoarthrosis. *Biochemical Journal, 188,* 823–837.

Eyre, D. R., Weis, M. A., & Wu, J. J. (2006). Articular cartilage collagen: An irreplaceable framework? *European Cells and Materials, 12,* 57–63.

Eyre, D. R., Weis, M. A., & Wu, J. J. (2008). Advances in collagen cross-link analysis. *Methods, 45,* 65–74.

Eyre, D. R., Weis, M. A., & Wu, J. J. (2010). Maturation of collagen Ketoimine cross-links by an alternative mechanism to pyridinoline formation in cartilage. *Journal of Biological Chemistry, 285,* 16675–16682.

Fadnes, H. O., Reed, R. K., & Aukland, K. (1977). Interstitial fluid pressure in rats measured with a modified wick technique. *Microvascular Research, 14,* 27–36.

Falah, M., Nierenberg, G., Soudry, M., Hayden, M., & Volpin, G. (2010). Treatment of articular cartilage lesions of the knee. *International Orthopaedics, 34,* 621–630.

Faryna, A., & Goldenberg, K. (1990). Joint fluid. In Walker, H. K., Hall, W. D., & Hurst, J. W. (Eds.), *Clinical methods: The history, physical, and laboratory examinations,* 3rd ed. Boston.

Fava, R., Olsen, N., Keskioja, J., Moses, H., & Pincus, T. (1989). Active and latent forms of transforming growth factor-beta activity in synovial effusions. *Journal of Experimental Medicine, 169,* 291–296.

Fell, H. B., & Jubb, R. W. (1977). The effect of synovial tissue on the breakdown of articular cartilage in organ culture. *Arthritis and Rheumatism, 20,* 1359–1371.

Felson, D. T., Chaisson, C. E., Hill, C. L., Totterman, S. M., Gale, M. E., Skinner, K. M., et al. (2001). The association of bone marrow lesions with pain in knee osteoarthritis. *Annals of Internal Medicine, 134,* 541–549.

Felson, D. T., Parkes, M. J., Marjanovic, E. J., Callaghan, M., Gait, A., Cootes, T., et al. (2012). Bone marrow lesions in knee osteoarthritis change in 6–12 weeks. *Osteoarthritis and Cartilage, 20,* 1514–1518.

Feng, J. Q., Guo, F. J., Jiang, B. C., Zhang, Y., Frenkel, S., Wang, D. W., et al. (2010). Granulin epithelin precursor: A bone morphogenic protein 2-inducible growth factor that activates Erk1/2 signaling and JunB transcription factor in chondrogenesis. *The FASEB Journal, 24,* 1879–1892.

Fermor, B., Christensen, S. E., Youn, I., Cernanec, J. M., Davies, C. M., & Weinberg, B. (2007). Oxygen, nitric oxide and articular cartilage. *European Cells and Materials, 13,* 56–65.

Fernandes, G. S., & Valdes, A. M. (2015). Cardiovascular disease and osteoarthritis: Common pathways and patient outcomes. *European Journal of Clinical Investigation, 45,* 405–414.

Fernandez-Torres, J., Zamudio-Cuevas, Y., Martinez-Nava, G. A., & Lopez-Reyes, A. G. (2017). Hypoxia-Inducible Factors (HIFs) in the articular cartilage: A systematic review. *European Review for Medical and Pharmacological Sciences, 21,* 2800–2810.

Fessler, J. H., & Fessler, L. I. (1978). Biosynthesis of procollagen. *Annual Review of Biochemistry, 47,* 129–162.

Fick, J. M., Thambyah, A., & Broom, N. D. (2010). Articular cartilage compression: How microstructural response influences pore pressure in relation to matrix health. *Connective Tissue Research, 51,* 132–149.

Fisk, G. R. (1952). Hyperplasia and metaplasia in synovial membrane. *Annals of the Royal College of Surgeons of England, 11,* 157–170.

Flachsmann, R., Broom, N. D., & Hardy, A. E. (2001). Deformation and rupture of the articular surface under dynamic and static compression. *Journal of Orthopaedic Research, 19,* 1131–1139.

Flannery, C. R., Little, C. B., Hughes, C. E., & Caterson, B. (1998). Expression and activity of articular cartilage hyaluronidases. *Biochemical and Biophysical Research Communications, 251,* 824–829.

Flynn, B. P., Bhole, A. P., Saeidi, N., Liles, M., Dimarzio, C. A., & Ruberti, J. W. (2010). Mechanical strain stabilizes reconstituted collagen fibrils against enzymatic degradation by mammalian collagenase matrix metalloproteinase 8 (MMP-8). *PLoS ONE, 5,* e12337.

Flynn, B. P., Tilburey, G. E., & Ruberti, J. W. (2013). Highly sensitive single-fibril erosion assay demonstrates mechanochemical switch in native collagen fibrils. *Biomechanics and Modeling in Mechanobiology, 12*, 291, 847–847.

Forsey, R. W., Fisher, J., Thompson, J., Stone, M. H., Bell, C., & Ingham, E. (2006). The effect of hyaluronic acid and phospholipid based lubricants on friction within a human cartilage damage model. *Biomaterials, 27,* 4581–4590.

Fosang, A. J. (2015). Bioactivity in an aggrecan 32mer fragment is mediated via Toll-like receptor 2. *Arthritis and Rheumatology, 67,* 10.

Fosang, A. J., & Beier, F. (2011). Emerging Frontiers in cartilage and chondrocyte biology. *Best Practice & Research in Clinical Rheumatology, 25,* 751–766.

Fraser, J. R., Laurent, T. C., & Laurent, U. B. (1997). Hyaluronan: Its nature, distribution, functions and turnover. *Journal of Internal Medicine, 242,* 27–33.

Frenkel, S. R., Clancy, R. M., Ricci, J. L., di Cesare, P. E., Rediske, J. J., & Abramson, S. B. (1996). Effects of nitric oxide on chondrocyte migration, adhesion, and cytoskeletal assembly. *Arthritis and Rheumatism, 39,* 1905–1912.

Frijns, A. J. H., Huyghe, J. M., & Janssen, J. D. (1997). A validation of the quadriphasic mixture theory for intervertebral disc tissue. *International Journal of Engineering Science, 35,* 1419–1429.

Frisbie, D. D., Cross, M. W., & McIlwraith, C. W. (2006). A comparative study of articular cartilage thickness in the stifle of animal species used in human pre-clinical studies compared to articular cartilage thickness in the human knee. *Veterinary and Comparative Orthopaedics and Traumatology, 19,* 142–146.

Frost, H. M. (1979). A chondral modeling theory. *Calcified Tissue International, 28,* 181–200.

Frost, H. M. (1987a). Bone "mass" and the "mechanostat": A proposal. *Anatomical Record, 219,* 1–9.

Frost, H. M. (1987b). The mechanostat: A proposed pathogenic mechanism of osteoporoses and the bone mass effects of mechanical and nonmechanical agents. *Bone Miner, 2,* 73–85.

Fukui, N., Miyamoto, Y., Nakajima, M., Ikeda, Y., Hikita, A., Furukawa, H., et al. (2008). Zonal gene expression of chondrocytes in osteoarthritic cartilage. *Arthritis and Rheumatism, 58,* 3843–3853.

Fukumoto, S., Hsieh, C. M., Maemura, K., Layne, R. D., Yet, S. F., Lee, K. H., et al. (2001). Akt participation in the Wnt signaling pathway through dishevelled. *Journal of Biological Chemistry, 276,* 17479–17483.

Funck-Brentano, T., Bouaziz, W., Marty, C., Geoffroy, V., Hay, E., & Cohen-Solal, M. (2014). Dkk-1-mediated inhibition of Wnt signaling in bone ameliorates osteoarthritis in mice. *Arthritis & Rheumatology, 66,* 3028–3039.

Furst, C. M., Morgelin, M., Vadstrup, K., Heinegard, D., Aspberg, A., & Blom, A. M. (2013). The C-type lectin of the aggrecan G3 domain activates complement. *PLoS ONE*, 8.

Furumatsu, T., Ozaki, T., & Asahara, H. (2009). Smad3 activates the Sox9-dependent transcription on chromatin. *International Journal of Biochemistry & Cell Biology, 41,* 1198–1204.

Furumatsu, T., Tsuda, M., Taniguchi, N., Tajima, Y., & Asahara, H. (2005). Smad3 induces chondrogenesis through the activation of SOX9 via CREB-binding protein/p300 recruitment. *Journal of Biological Chemistry, 280,* 8343–8350.

Gacheru, S. N., Trackman, P. C., Shah, M. A., O'Gara, C. Y., Spacciapoli, P., Greenaway, F. T., et al. (1990). Structural and catalytic properties of copper in lysyl oxidase. *Journal of Biological Chemistry, 265,* 19022–19027.

Gaisinskaya, A., Ma, L., Silbert, G., Sorkin, R., Tairy, O., Goldberg, R., et al. (2012). Hydration lubrication: Exploring a new paradigm. *Faraday Discuss, 156,* 217–233; *Discussion* 293–309.

Gao, X., & Hannoush, R. N. (2014). Single-cell imaging of Wnt palmitoylation by the acyltransferase porcupine. *Nature Chemical Biology, 10,* 61–68.

Gardiner, B., Smith, D., Pivonka, P., Grodzinsky, A., Frank, E., & Zhang, L. (2007). Solute transport in cartilage undergoing cyclic deformation. *Computer Methods in Biomechanics and Biomedical Engineering, 10,* 265–278.

Gardiner, B. S., Woodhouse, F. G., Besier, T. F., Grodzinsky, A. J., Lloyd, D. G., Zhang, L. H., et al. (2016). Predicting knee osteoarthritis. *Annals of Biomedical Engineering, 44,* 222–233.

Gardiner, B. S., Zhang, L., Smith, D. W., Pivonka, P., & Grodzinsky, A. J. (2011). A mathematical model for targeting chemicals to tissues by exploiting complex degradation. *Biology Direct, 6,* 46.

Gardner, K., Lavagnino, M., Egerbacher, M., & Arnoczky, S. P. (2012). Re-establishment of cytoskeletal tensional homeostasis in lax tendons occurs through an actin-mediated cellular contraction of the extracellular matrix. *Journal of Orthopaedic Research, 30,* 1695–1701.

Gavrilovic, J. (2009). Fibroblast growth factor 2: A new key player in osteoarthritis. *Arthritis and Rheumatism, 60,* 1869–1872.

Gelse, K., Poschl, E., & Aigner, T. (2003). Collagens—Structure, function, and biosynthesis. *Advanced Drug Delivery Reviews, 55,* 1531–1546.

Geng, H., Nandakumar, K. S., Pramhed, A., Aspberg, A., Mattsson, R., & Holmdahl, R. (2012). Cartilage oligomeric matrix protein specific antibodies are pathogenic. *Arthritis Research & Therapy, 14,* R191.

Gibson, G., & Asahara, H. (2013). MicroRNAs and cartilage. *Journal of Orthopaedic Research, 31,* 1333–1344.

Gibson, A. L., Hui Mingalone, C. K., Foote, A. T., Uchimura, T., Zhang, M., & Zeng, L. (2017). Wnt7a inhibits IL-1beta induced catabolic gene expression and prevents articular cartilage damage in experimental osteoarthritis. *Scientific Reports, 7,* 41823.

Gibson, R. E., England, G. l., & Hussey M. J. L. (1967). The theory of one-dimensional consolidation of saturated clays. I. Finite non-linear consolidation of thin homogeneous layers. *Géotechnique, 17*, 261–273.

Gibson, R. E., Schiffman, R. L., & Cargill, K. W. (1981). The theory of one-dimensional consolidation of saturated clays. 2. Finite non-linear consolidation of thick homogeneous Layers. *Canadian Geotechnical Journal, 18,* 280–293.

Gilmore, R. S., & Palfrey, A. J. (1988). Chondrocyte distribution in the articular cartilage of human femoral condyles. *Journal of anatomy, 157*, 23–31.

Giusti, I., D'Ascenzo, S., Millimaggi, D., Taraboletti, G., Carta, G., Franceschini, N., et al. (2008). Cathepsin B mediates the pH-dependent proinvasive activity of tumor-shed microvesicles. *Neoplasia, 10,* 481–488.

Glaser, C., & Putz, R. (2002). Functional anatomy of articular cartilage under compressive loading quantitative aspects of global, local and zonal reactions of the collagenous network with respect to the surface integrity. *Osteoarthritis Cartilage, 10,* 83–99.

Gleghorn, J. P., & Bonassar, L. J. (2008). Lubrication mode analysis of articular cartilage using Stribeck surfaces. *Journal of Biomechanics, 41,* 1910–1918.

Gleghorn, J. P., Jones, A. R., Flannery, C. R., & Bonassar, L. J. (2009). Boundary mode lubrication of articular cartilage by recombinant human lubricin. *Journal of Orthopaedic Research, 27,* 771–777.

Gobezie, R., Kho, A., Krastins, B., Sarracino, D. A., Thornhill, T. S., Chase, M., et al. (2007). High abundance synovial fluid proteome: Distinct profiles in health and osteoarthritis. *Arthritis Research & Therapy, 9,* R36.

Goggs, R., Carter, S. D., Schulze-Tanzil, G., Shakibaei, M., & Mobasheri, A. (2003). Apoptosis and the loss of chondrocyte survival signals contribute to articular cartilage degradation in osteoarthritis. *The Veterinary Journal, 166,* 140–158.

Gold, G. E., Chen, C. A., Koo, S., Hargreaves, B. A., & Bangerter, N. K. (2009). Recent advances in MRI of articular cartilage. *American Journal of Roentgenology, 193,* 628–638.

Gold, L. I., Eggleton, P., Sweetwyne, M. T., van Duyn, L. B., Greives, M. R., Naylor, S. M., et al. (2010). Calreticulin: Non-endoplasmic reticulum functions in physiology and disease. *The FASEB Journal, 24,* 665–683.

Goldring, M. B., Otero, M., Plumb, D. A., Dragomir, C., Favero, M., El Hachem, K., et al. (2011). Roles of inflammatory and anabolic cytokines in cartilage metabolism: Signals and multiple effectors converge upon MMP-13 regulation in osteoarthritis. *European Cells & Materials, 21,* 202–220.

Goldring, M. B., Otero, M., Plumb, D. A., Dragomir, C., Favero, M., el Hachem, K., et al. (2011). Roles of inflammatory and anabolic cytokines in cartilage metabolism: Signals and multiple effectors converge upon MMP-13 regulation in osteoarthritis. *European Cells and Materilas, 21,* 202–220.

Goldring, S. R., & Goldring, M. B. (2004). The role of cytokines in cartilage matrix degeneration in osteoarthritis. *Clinical Orthopaedics and Related Research,* S27–S36.

Goldspink, G. (2005). Mechanical signals, IGF-I gene splicing, and muscle adaptation. *Physiology (Bethesda), 20,* 232–238.

Gomez, D. E., Alonso, D. F., Yoshiji, H., & Thorgeirsson, U. P. (1997). Tissue inhibitors of metalloproteinases: structure, regulation and biological functions. *European Journal of Cell Biology, 74,* 111–122.

Gonzalez, S., Fragoso-Soriano, R. J., & Kouri, J. B. (2007). Chondrocytes interconnecting tracks and cytoplasmic projections observed within the superficial zone of normal human articular cartilage—A transmission electron microscopy, atomic force microscopy, and two-photon excitation microscopy studies. *Microscopy Research and Technique, 70,* 1072–1078.

Gonzalez, A., Vidal-Bralo, L., Lopez-Golan, Y., Mera-Varela, A., Rego-Perez, I., Horvath, S., et al. (2016). Specific premature epigenetic aging of cartilage in osteoarthritis. *Annals of the Rheumatic Diseases, 75,* 664.

Gonzalez, E., & Ortiz-Navarrete, V. (2013). CD36 expression in chondrocytes is involve in the clearance of apoptotic cells. *Journal of Immunology,* 190.

Goodwin, D. W., Wadghiri, Y. Z., Zhu, H. Q., Vinton, C. J., Smith, E. D., & Dunn, J. F. (2004). Macroscopic structure of articular cartilage of the tibial plateau: Influence of a characteristic matrix architecture on MRI appearance. *American Journal of Roentgenology, 182,* 311–318.

Gossan, N., Zeef, L., Hensman, J., Hughes, A., Bateman, J. F., Rowley, L., et al. (2013). The circadian clock in murine chondrocytes regulates genes controlling key aspects of cartilage homeostasis. *Arthritis and Rheumatism, 65,* 2334–2345.

Gottardi, R., Hansen, U., Raiteri, R., Loparic, M., Duggelin, M., Mathys, D., et al. (2016). Supramolecular organization of collagen fibrils in healthy and osteoarthritic human knee and hip joint cartilage. *PLoS ONE, 11,* e0163552.

Grad, S., Lee, C. R., Gorna, K., Gogolewski, S., Wimmer, M. A., & Alini, M. (2005). Surface motion upregulates superficial zone protein and hyaluronan production in chondrocyte-seeded three-dimensional scaffolds. *Tissue Engineering, 11,* 249–256.

Graham, D. F., Modenese, L., Trewartha, G., Carty, C. P., Constantinou, M., Lloyd, D. G., et al. (2016). Hip joint contact loads in older adults during recovery from forward loss of balance by stepping. *Journal of Biomechanics, 49,* 2619–2624.

Graham, B. T., Moore, A. C., Burris, D. L., & Price, C. (2018). Mapping the spatiotemporal evolution of solute transport in articular cartilage explants reveals how cartilage recovers fluid within the contact area during sliding. *Journal of Biomechanics, 71,* 271–276.

Graindorge, S., Ferrandez, W., Ingham, E., Jin, Z., Twigg, P., & Fisher, J. (2006). The role of the surface amorphous layer of articular cartilage in joint lubrication. *Proceedings of the Institution of Mechanical Engineers Part H, 220,* 597–607.

Greene, G. W., Olszewska, A., Osterberg, M., Zhu, H., & Horne, R. (2014). A cartilage inspired lubrication system. *Soft Matter, 10,* 374–382.

Gribbon, P., & Hardingham, T. E. (1998). Macromolecular diffusion of biological polymers measured by confocal fluorescence recovery after photobleaching. *Biophysical Journal, 75,* 1032–1039.

Grimaud, E., Heymann, D., & Redini, F. (2002). Recent advances in TGF-beta effects on chondrocyte metabolism. Potential therapeutic roles of TGF-beta in cartilage disorders. *Cytokine & Growth Factor Reviews, 13,* 241–257.

Grimmer, C., Balbus, N., Lang, U., Aigner, T., Cramer, T., Muller, L., et al. (2006). Regulation of type II collagen synthesis during osteoarthritis by prolyl-4-hydroxylases: Possible influence of low oxygen levels. *American Journal of Pathology, 169,* 491–502.

Grimshaw, M. J., & Mason, R. M. (2001). Modulation of bovine articular chondrocyte gene expression in vitro by oxygen tension. *Osteoarthritis Cartilage, 9,* 357–364.
Grodzinsky, A. J., Levenston, M. E., Jin, M., & Frank, E. H. (2000). Cartilage tissue remodeling in response to mechanical forces. *Annual Review of Biomedical Engineering, 2,* 691–713.
Grodzinsky, A. (2011). *Fields, forces, and flows in biological systems.* New York: Garland Publishing.
Gross, J., & Lapiere, C. M. (1962). Collagenolytic activity in amphibian tissues: A tissue culture assay. *Proceedings of the National Academy of Sciences, 48,* 1014–1022.
Groulx, J. F., Gagne, D., Benoit, Y. D., Martel, D., Basora, N., & Beaulieu, J. F. (2011). Collagen VI is a basement membrane component that regulates epithelial cell-fibronectin interactions. *Matrix Biology, 30,* 195–206.
Grytz, R., & Meschke, G. (2010). A computational remodeling approach to predict the physiological architecture of the collagen fibril network in corneo-scleral shells. *Biomechanics and Modeling in Mechanobiology, 9,* 225–235.
Gschwind, Y. J., Kressig, R. W., Lacroix, A., Muehlbauer, T., Pfenninger, B., & Granacher, U. (2013). A best practice fall prevention exercise program to improve balance, strength/power, and psychosocial health in older adults: Study protocol for a randomized controlled trial. *BMC Geriatrics, 13.*
Guo, R., Zhou, Q., Proulx, S. T., Wood, R., Ji, R. C., Ritchlin, C. T., et al. (2009). Inhibition of lymphangiogenesis and lymphatic drainage via vascular endothelial growth factor receptor 3 blockade increases the severity of inflammation in a mouse model of chronic inflammatory arthritis. *Arthritis and Rheumatism, 60,* 2666–2676.
Haag, J., Gebhard, P. M., & Aigner, T. (2008). SOX gene expression in human osteoarthritic cartilage. *Pathobiology, 75,* 195–199.
Haase, R. (1990). *Thermodynamics of irreversible processes.* New York: Dover.
Hackl, K., & Fischer, F. D. (2008). On the relation between the principle of maximum dissipation and inelastic evolution given by dissipation potentials. *Proceedings of the Royal Society A—Mathematical Physical and Engineering Sciences, 464,* 117–132.
Hadi, M. F., Sander, E. A., Ruberti, J. W., & Barocas, V. H. (2012). Simulated remodeling of loaded collagen networks via strain-dependent enzymatic degradation and constant-rate fiber growth. *Mechanics of Materials, 44,* 72–82.
Hall, J. E. (2015). *Guyton and hall textbook of medical physiology.* Amsterdam: Elsevier.
Halloran, J. P., Sibole, S., van Donkelaar, C. C., van Turnhout, M. C., Oomens, C. W. J., Weiss, J. A., et al. (2012). Multiscale mechanics of articular cartilage: Potentials and challenges of coupling musculoskeletal, joint, and microscale computational models. *Annals of Biomedical Engineering, 40,* 2456–2474.
Hamrick, M. W. (1999). A chondral modeling theory revisited. *Journal of Theoretical Biology, 201,* 201–208.
Hamrock, B. J., Schmid, S. R., & Jacobson, B. O. (2004). *Fundamentals of fluid film lubrication,* Marcel Dekker Inc.
Han, E., Chen, S. S., Klisch, S. M., & Sah, R. L. (2011a). Contribution of proteoglycan osmotic swelling pressure to the compressive properties of articular cartilage. *Biophysical Journal, 101,* 916–924.
Han, L., Grodzinsky, A. J., & Ortiz, C. (2011b). Nanomechanics of the cartilage extracellular matrix. *Annual Review of Materials Research, 41*(41), 133–168.
Hanauskeabel, H. M., Pontz, B. F., & Schorlemmer, H. U. (1982). Cartilage specific collagen activates macrophages and the alternative pathway of complement—Evidence for an immunopathogenic concept of rheumatoid-arthritis. *Annals of the Rheumatic Diseases, 41,* 168–176.
Hansen, P., Hassenkam, T., Svensson, R. B., Aagaard, P., Trappe, T., Haraldsson, B. T., et al. (2009). Glutaraldehyde cross-linking of tendon–mechanical effects at the level of the tendon fascicle and fibril. *Connective Tissue Research, 50,* 211–222.

Hansen, U., Allen, J. M., White, R., Moscibrocki, C., Bruckner, P., Bateman, J. F., & Fitzgerald, J. (2012). WARP interacts with collagen VI—Containing microfibrils in the pericellular matrix of human chondrocytes. *PLoS ONE*, 7.

Harburger, D. S., & Calderwood, D. A. (2009). Integrin signalling at a glance. *Journal of Cell Science, 122,* 159–163.

Harjanto, D., & Zaman, M. H. (2010). Computational study of proteolysis-driven single cell migration in a three-dimensional matrix. *Annals of Biomedical Engineering, 38,* 1815–1825.

Harms, S., Larson, R., Sahmoun, A. E., & Beal, J. R. (2007). Obesity increases the likelihood of total joint replacement surgery among younger adults. *International Orthopaedics, 31,* 23–26.

Harvey, L. A., & Herbert, R. D. (2002). Muscle stretching for treatment and prevention of contracture in people with spinal cord injury. *Spinal Cord, 40,* 1–9.

Hashimoto, K., Oreffo, R. O. C., Gibson, M. B., Goldring, M. B., & Roach, H. I. (2009a). DNA demethylation at specific CpG sites in the IL1B promoter in response to inflammatory cytokines in human articular chondrocytes. *Arthritis and Rheumatism, 60,* 3303–3313.

Hashimoto, S., Nishiyama, T., Hayashi, S., Fujishiro, T., Takebe, K., Kanzaki, N., et al. (2009b). Role of p53 in human chondrocyte apoptosis in response to shear strain. *Arthritis and Rheumatism, 60,* 2340–2349.

Hashimoto, S., Ochs, R. L., Komiya, S., & Lotz, M. (1998). Linkage of chondrocyte apoptosis and cartilage degradation in human osteoarthritis. *Arthritis and Rheumatism, 41,* 1632–1638.

Hashimoto, S., Takahashi, K., Amiel, D., Coutts, R. D., & Lotz, M. (1998b). Chondrocyte apoptosis and nitric oxide production during experimentally induced osteoarthritis. *Arthritis and Rheumatism, 41,* 1266–1274.

Hasler, E. M., Herzog, W., Leonard, T. R., Stano, A., & Nguyen, H. (1998). In vivo knee joint loading and kinematics before and after ACL transection in an animal model. *Journal of Biomechanics, 31,* 253–262.

Hasselbacher, P. (1976). Measuring synovial-fluid viscosity with a white blood-cell diluting pipette—Simple, rapid, and reproducible method. *Arthritis and Rheumatism, 19,* 1358–1362.

Havdrup, T., Henricson, A., & Telhag, H. (1982). Papain-induced mitosis of chondrocytes in adult joint cartilage. *Acta Orthopaedica Scandinavica, 53,* 119–124.

Hayami, T., Pickarski, M., Zhuo, Y., Wesolowski, G. A., Rodan, G. A., & Duong Le, T. (2006). Characterization of articular cartilage and subchondral bone changes in the rat anterior cruciate ligament transection and meniscectomized models of osteoarthritis. *Bone, 38*, 234–243.

Hayes, A. J., Tudor, D., Nowell, M. A., Caterson, B., & Hughes, C. E. (2008). Chondroitin sulfate sulfation motifs as putative biomarkers for isolation of articular cartilage progenitor cells. *Journal of Histochemistry and Cytochemistry, 56,* 125–138.

Heathfield, T. F., Onnerfjord, P., Dahlberg, L., & Heinegard, D. (2004). Cleavage of fibromodulin in cartilage explants involves removal of the N-terminal tyrosine sulfate-rich region by proteolysis at a site that is sensitive to matrix metalloproteinase-13. *Journal of Biological Chemistry, 279,* 6286–6295.

Hedlund, H., Hedbom, E., Heineg R. D. D., Mengarelli-Widholm, S., Reinholt, F. P. & Svensson, O. (1999). Association of the aggrecan keratan sulfate-rich region with collagen in bovine articular cartilage. *Journal of Biological Chemistry, 274*, 5777–5781.

Heijmans, B. T., Tobi, E. W., Stein, A. D., Putter, H., Blauw, G. J., Susser, E. S., et al. (2008). Persistent epigenetic differences associated with prenatal exposure to famine in humans. *Proceedings of the National Academy of Sciences of the United States of America, 105,* 17046–17049.

Heilmann, H. H., Lindenhayn, K., & Walther, H. U. (1996). Synovial volume of healthy and arthrotic human knee joints. *Zeitschrift fur Orthopadie und ihre Grenzgebiete, 134,* 144–148.

Heinegard, D. (2009). Proteoglycans and more—From molecules to biology. *International Journal of Experimental Pathology, 90,* 575–586.

Heinegard, D., Lohmander, S., & Thyberg, J. (1978). Cartilage proteoglycan aggregates—Electron-microscopic studies of native and fragmented molecules. *Biochemical Journal, 175,* 913–1000.

Heinegard, D., & Oldberg, A. (1989). Structure and biology of cartilage and bone matrix noncollagenous macromolecules. *The FASEB Journal, 3,* 2042–2051.
Heinegard, D., & Saxne, T. (2011). The role of the cartilage matrix in osteoarthritis. *Nature Reviews Rheumatology, 7,* 50–56.
Heinemeier, K. M., Schjerling, P., Heinemeier, J., Magnusson, S. P., & Kjaer, M. (2013). Lack of tissue renewal in human adult Achilles tendon is revealed by nuclear bomb (14)C. *The FASEB Journal, 27,* 2074–2079.
Heinemeier, K. M., Schjerling, P., Ohlenschlaeger, T. F., Eismark, C., Olsen, J., & Kjaer, M. (2018). Carbon-14 bomb pulse dating shows that tendinopathy is preceded by years of abnormally high collagen turnover. *The FASEB Journal*, fj201701569R.
Hembree, W. C., Ward, B. D., Furman, B. D., Zura, R. D., Nichols, L. A., Guilak, F., et al. (2007). Viability and apoptosis of human chondrocytes in osteochondral fragments following joint trauma. *Journal of Bone and Joint Surgery. British Volume, 89,* 1388–1395.
Hembry, R. M., Dyce, J., Driesang, I., Hunziker, E. B., Fosang, A. J., Tyler, J. A. & Murphy, G. (2001). Immunolocalization of matrix metalloproteinases in partial-thickness defects in pig articular cartilage. A preliminary report. *The Journal of Bone and Joint Surgery, American Volume, 83-A,* 826–838.
Heng, B. C., Gribbon, P. M., Day, A. J., & Hardingham, T. E. (2008). Hyaluronan binding to link module of TSG-6 and to G1 domain of aggrecan is differently regulated by pH. *Journal of Biological Chemistry, 283,* 32294–32301.
Henricson, A. S., Havdrup, T., & Telhag, H. (1982). The effect of growth-hormone and thyroxine on adult joint cartilage. *Clinical Orthopaedics and Related Research,* 270–275.
Henrotin, Y., Kurz, B., & Aigner, T. (2005). Oxygen and reactive oxygen species in cartilage degradation: Friends or foes? *Osteoarthritis Cartilage, 13,* 643–654.
Henrotin, Y., Labasse, A., Zheng, S. X., Galais, P., Tsouderos, Y., Crielaard, J. M., et al. (2001). Strontium ranelate increases cartilage matrix formation. *Journal of Bone and Mineral Research, 16,* 299–308.
Heraud, F., Heraud, A., & Harmand, M. F. (2000). Apoptosis in normal and osteoarthritic human articular cartilage. *Annals of the Rheumatic Diseases, 59,* 959–965.
Herberhold, C., Faber, S., Stammberger, T., Steinlechner, M., Putz, R., Englmeier, K. H., et al. (1999). In situ measurement of articular cartilage deformation in intact femoropatellar joints under static loading. *Journal of Biomechanics, 32,* 1287–1295.
Herbert, R. D., & Crosbie, J. (1997). Rest length and compliance of non-immobilised and immobilised rabbit soleus muscle and tendon. *European Journal of Applied Physiology and Occupational Physiology, 76,* 472–479.
Hermansson, M., Sawaji, Y., Bolton, M., Alexander, S., Wallace, A., Begum, S., et al. (2004). Proteomic analysis of articular cartilage shows increased type II collagen synthesis in osteoarthritis and expression of inhibin beta A (activin A), a regulatory molecule for chondrocytes. *Journal of Biological Chemistry, 279,* 43514–43521.
Hersch, E. C., & Merriam, G. R. (2008). Growth hormone (GH)-releasing hormone and GH secretagogues in normal aging: Fountain of youth or pool of tantalus? *Clinical Interventions in Aging, 3,* 121–129.
Hidaka, C., & Goldring, M. B. (2008). Regulatory mechanisms of chondrogenesis and implications for understanding articular cartilage homeostasis. *Current Rheumatology Reviews, 4,* 1573–3971.
Hillen, J., Geyer, C., Heitzmann, M., Beckmann, D., Krause, A., Winkler, I., et al. (2017). Structural cartilage damage attracts circulating rheumatoid arthritis synovial fibroblasts into affected joints. *Arthritis Research & Therapy, 19*, 40.
Hinz, B., Celetta, G., Tomasek, J. J., Gabbiani, G., & Chaponnier, C. (2001). Alpha-smooth muscle actin expression upregulates fibroblast contractile activity. *Molecular Biology of the Cell, 12,* 2730–2741.

Hinz, B., Phan, S. H., Thannickal, V. J., Galli, A., Bochaton-Piallat, M. L., & Gabbiani, G. (2007). The myofibroblast: One function, multiple origins. *American Journal of Pathology, 170,* 1807–1816.

Hirotani, H., & Ito, T. (1975). Chondrocyte mitosis in the articular cartilage of femoral heads with various diseases. *Acta Orthopaedica Scandinavica, 46,* 979–986.

Hirschmann, M. T., Zschäbitz, A., & Stofft, E. (2007). Immunohistochemical characterization of human synovial bursa cells by light and transmission electron microscopy. Where do these cells come from? *International Journal of Morphology, 25,* 9.

Hita-Contreras, F., Sanchez-Montesinos, I., Martinez-Amat, A., Cruz-Diaz, D., Barranco, R. J., & Roda, O. (2018). Development of the human shoulder joint during the embryonic and early fetal stages: Anatomical considerations for clinical practice. *Journal of Anatomy, 232,* 422–430.

Hockel, M., & Vaupel, P. (2001). Tumor hypoxia: Definitions and current clinical, biologic, and molecular aspects. *Journal of the National Cancer Institute, 93,* 266–276.

Hodge, W. A., Fijan, R. S., Carlson, K. L., Burgess, R. G., Harris, W. H., & Mann, R. W. (1986). Contact pressures in the human hip joint measured in vivo. *Proceedings of the National Academy of Science of the United States of America, 83,* 2879–2883.

Hodgkinson, C. P., & Ye, S. (2011). Toll-like receptors, their ligands, and atherosclerosis. *The Scientific World Journal, 11,* 437–453.

Hoemann, C. D., Lafantaisie-Favreau, C. H., Lascau-Coman, V., Chen, G. P., & Guzman-Morales, J. (2012). The cartilage-bone interface. *Journal of Knee Surgery, 25,* 85–97.

Hollander, A. P., Heathfield, T. F., Webber, C., Iwata, Y., Bourne, R., Rorabeck, C., et al. (1994). Increased damage to type II collagen in osteoarthritic articular cartilage detected by a new immunoassay. *Journal of Clinical Investigation, 93,* 1722–1732.

Holliday, C. M., Ridgely, R. C., Sedlmayr, J. C., & Witmer, L. M. (2010). Cartilaginous epiphyses in extant archosaurs and their implications for reconstructing limb function in dinosaurs. *PLoS ONE, 5.*

Holmes, M. H. (1985). A theoretical-analysis for determining the nonlinear hydraulic permeability of a soft-tissue from a permeation experiment. *Bulletin of Mathematical Biology, 47,* 669–683.

Homandberg, G. A., Wen, C., & Hui, F. (1998). Cartilage damaging activities of fibronectin fragments derived from cartilage and synovial fluid. *Osteoarthritis Cartilage, 6,* 231–244.

Honsawek, S., Wilairatana, V., Udomsinprasert, W., Sinlapavilawan, P., & Jirathanathornnukul, N. (2015). Association of plasma and synovial fluid periostin with radiographic knee osteoarthritis: Cross-sectional study. *Joint Bone Spine, 82,* 352–355.

Hoogeboom, D., Essers, M. A., Polderman, P. E., Voets, E., Smits, L. M., & Burgering, B. M. (2008). Interaction of FOXO with beta-catenin inhibits beta-catenin/T cell factor activity. *Journal of Biological Chemistry, 283,* 9224–9230.

Hopewell, B., & Urban, J. P. (2003). Adaptation of articular chondrocytes to changes in osmolality. *Biorheology, 40,* 73–77.

Horikawa, O., Nakajima, H., Kikuchi, T., Ichimura, S., Yamada, H., Fujikawa, K., et al. (2004). Distribution of type VI collagen in chondrocyte microenvironment: Study of chondrons isolated from human normal and degenerative articular cartilage and cultured chondrocytes. *Journal of Orthopaedic Science, 9,* 29–36.

Horkay, F., Basser, P. J., Hecht, A. M., & Geissler, E. (2012). Chondroitin sulfate in solution: Effects of mono- and divalent salts. *Macromolecules, 45,* 2882–2890.

Horky, D., & Tichy, F. (2004). Submicroscopic structure of canine articular cartilage. *Veterinarni Medicina, 49,* 207–216.

Hosseini, S. M., Veldink, M. B., Ito, K., & van Donkelaar, C. (2013). Is collagen fiber damage the cause of early softening in articular cartilage? *Osteoarthritis and Cartilage, 21,* 136–143.

Hotta, H., Yamada, H., Takaishi, H., Abe, T., Morioka, H., Kikuchi, T., et al. (2005). Type II collagen synthesis in the articular cartilage of a rabbit model of osteoarthritis: Expression of type II collagen C-propeptide and mRNA especially during early-stage osteoarthritis. *Journal of Orthopaedic Science, 10,* 595–607.

Hou, W. S., Li, Z. Q., Buttner, F. H., Bartnik, E., & Bromme, D. (2003). Cleavage site specificity of cathepsin K toward cartilage proteoglycans and protease complex formation. *Biological Chemistry, 384,* 891–897.

Howley, B. V., Hussey, G. S., Link, L. A., & Howe, P. H. (2015). Translational regulation of inhibin betaA by TGFbeta via the RNA-binding protein hnRNP E1 enhances the invasiveness of epithelial-to-mesenchymal transitioned cells. *Oncogene.*

Hsu, C. C., Lin, C. L., Jou, I. M., Wang, P. H., & Lee, J. S. (2017). The protective role of nitric oxide-dependent innate immunosuppression in the early stage of cartilage damage in rats. *Bone & Joint Research, 6,* 253–258.

Hsueh, M. F., Khabut, A., Kjellstrom, S., Onnefjord, P., & Kraus, V. B. (2016). Elucidating the molecular composition of cartilage by proteomics. *Journal of Proteome Research, 15,* 374–388.

Hsueh, M. F., Onnerfjord, P., & Kraus, V. B. (2014). Biomarkers and proteomic analysis of osteoarthritis. *Matrix Biology, 39,* 56–66.

Huang, C. Y., Stankiewicz, A., Ateshian, G. A., & Mow, V. C. (2005). Anisotropy, inhomogeneity, and tension-compression nonlinearity of human glenohumeral cartilage in finite deformation. *Journal of Biomechanics, 38,* 799–809.

Huey, D. J., Hu, J. C., & Athanasiou, K. A. (2012). Unlike bone, cartilage regeneration remains elusive. *Science, 338,* 917–921.

Hughes, L. C., Archer, C. W., & Ap Gwynn, I. (2005). The ultrastructure of mouse articular cartilage: Collagen orientation and implications for tissue functionality. A polarised light and scanning electron microscope study and review. *European Cells and Materials,* 9, 68–84.

Hui, A. Y., McCarty, W. J., Masuda, K., Firestein, G. S., & Sah, R. L. (2012). A systems biology approach to synovial joint lubrication in health, injury, and disease. *Wiley Interdisciplinary Reviews: Systems Biology and Medicine, 4,* 15–37.

Hulmes, D. J. (2008). Collagen diversity, synthesis and assembly. In P. Fratzl (Ed.), *Collagen structure and mechanics.* Springer.

Hunter, D. J., & Eckstein, F. (2009). Exercise and osteoarthritis. *Journal of Anatomy, 214,* 197–207.

Hunter, D. J., Guermazi, A., Roemer, F., Zhang, Y., & Neogi, T. (2013). Structural correlates of pain in joints with osteoarthritis. *Osteoarthritis and Cartilage, 21,* 1170–1178.

Hunter, D. J., Nevitt, M., Losina, E., & Kraus, V. (2014). Biomarkers for osteoarthritis: Current position and steps towards further validation. *Best Practice & Research in Clinical Rheumatology, 28,* 61–71.

Hunter, D. J., Zhang, W., Conaghan, P. G., Hirko, K., Menashe, L., Li, L., et al. (2011). Systematic review of the concurrent and predictive validity of MRI biomarkers in OA. *Osteoarthritis Cartilage, 19,* 557–588.

Hunziker, E. B., Kapfinger, E., & Geiss, J. (2007). The structural architecture of adult mammalian articular cartilage evolves by a synchronized process of tissue resorption and neoformation during postnatal development. *Osteoarthritis Cartilage, 15,* 403–413.

Hunziker, E. B., Quinn, T. M., & Hauselmann, H. J. (2002). Quantitative structural organization of normal adult human articular cartilage. *Osteoarthritis Cartilage, 10,* 564–572.

Husa, M., Liu-Bryan, R., & Terkeltaub, R. (2010). Shifting HIFs in osteoarthritis. *Nature Medicine, 16,* 641–644.

Huser, C. A., & Davies, M. E. (2006). Validation of an in vitro single-impact load model of the initiation of osteoarthritis-like changes in articular cartilage. *Journal of Orthopaedic Research, 24,* 725–732.

Huser, C. A., Peacock, M., & Davies, M. E. (2006). Inhibition of caspase-9 reduces chondrocyte apoptosis and proteoglycan loss following mechanical trauma. *Osteoarthritis Cartilage, 14,* 1002–1010.

Huska, D., Adam, V., Zitka, O., Kukacka, J., Prusa, R., & Kizek, R. (2009). Chronopotentiometric stripping analysis of gelatinase B, collagen and their interaction. *Electroanalysis, 21,* 536–541.

Hwang, J., Bae, W. C., Shieu, W., Lewis, C. W., Bugbee, W. D., & Sah, R. L. (2008). Increased hydraulic conductance of human articular cartilage and subchondral bone plate with progression of osteoarthritis. *Arthritis and Rheumatism, 58,* 3831–3842.

Hynes, R. O., & Naba, A. (2012). Overview of the matrisome—An inventory of extracellular matrix constituents and functions. *Cold Spring Harbor Perspectives in Biology, 4.*

Ilani, A. (2015). The development of pressure across membranes in Donnan systems. *Scientific Reports, 5,* 14695.

Ilic, M. Z., East, C. J., Rogerson, F. M., Fosang, A. J., & Handley, C. J. (2007). Distinguishing aggrecan loss from aggrecan proteolysis in ADAMTS-4 and ADAMTS-5 single and double deficient mice. *Journal of Biological Chemistry, 282,* 37420–37428.

Iliopoulos, D., Malizos, K. N., & Tsezou, A. (2007). Epigenetic regulation of leptin affects MMP-13 expression in osteoarthritic chondrocytes: Possible molecular target for osteoarthritis therapeutic intervention. *Annals of the Rheumatic Diseases, 66,* 1616–1621.

Imagawa, K., de Andres, M. C., Hashimoto, K., Itoi, E., Otero, M., Roach, H. I., et al. (2014). Association of reduced type IX collagen gene expression in human osteoarthritic chondrocytes with epigenetic silencing by DNA hypermethylation. *Arthritis & Rheumatology, 66,* 3040–3051.

Imai, K., Ohta, S., Matsumoto, T., Fujimoto, N., Sato, H., Seiki, M., et al. (1997). Expression of membrane-type 1 matrix metalloproteinase and activation of progelatinase A in human osteoarthritic cartilage. *American Journal of Pathology, 151,* 245–256.

Imhauser, C. W., Sheikh, S., Choi, D. S., Nguyen, J. T., Mauro, C. S., & Wickiewicz, T. L. (2016). Novel measure of articular instability based on contact stress confirms that the anterior cruciate ligament is a critical stabilizer of the lateral compartment. *Journal of Orthopaedic Research, 34,* 478–488.

Inamdar, S. R., Knight, D. P., Terril, N. J., Karunaratne, A., Cacho-Nerin, F., Knight, M. M., et al. (2017). The secret life of collagen: Temporal changes in nanoscale fibrillar pre-strain and molecular organization during physiological loading of cartilage. *ACS Nano, 11,* 9728–9737.

Intema, F., van Roermund, P. M., Marijnissen, A. C., Cotofana, S., Eckstein, F., Castelein, R. M., et al. (2011). Tissue structure modification in knee osteoarthritis by use of joint distraction: An open 1-year pilot study. *Annals of the Rheumatic Diseases, 70,* 1441–1446.

Ioannidis, J. P. (2016). Why most clinical research is not useful. *PLoS Medicine, 13,* e1002049.

Ioannidis, J. P. A. (2005). Why most published research findings are false. *PLoS Medicine, 2,* 696–701.

Iordan, A., Duperray, A., Gerard, A., Grichine, A., & Verdier, C. (2010). Breakdown of cell-collagen networks through collagen remodeling. *Biorheology, 47,* 277–295.

Iozzo, R. V. (1999). The biology of the small leucine-rich proteoglycans. Functional network of interactive proteins. *Journal of Biological Chemistry, 274,* 18843–18846.

Irie, T., Aizawa, T., & Kokubun, S. (2005). The role of sex hormones in the kinetics of chondrocytes in the growth plate. *Journal of Bone and Joint Surgery-British Volume, 87b,* 1278, 1582–1583.

Ishiguro, N., Kojima, T., & Poole, A. R. (2002). Mechanism of cartilage destruction in osteoarthritis. *Nagoya Journal of Medical Science, 65,* 73–84.

Ito, I., Imada, M., Ikeda, M., Sueno, K., Arikuni, T., & Kida, A. (2001). A morphological study of age changes in adult human auricular cartilage with special emphasis on elastic fibers. *Laryngoscope, 111,* 881–886.

Itoh, Y., Ito, N., Nagase, H., Evans, R. D., Bird, S. A., & Seiki, M. (2006). Cell surface collagenolysis requires homodimerization of the membrane-bound collagenase MT1-MMP. *Molecular Biology of the Cell, 17,* 5390–5399.

Iversen, P. O., Berggreen, E., Nicolaysen, G., & Heyeraas, K. (2001). Regulation of extracellular volume and interstitial fluid pressure in rat bone marrow. *American Journal of Physiology-Heart and Circulatory Physiology, 280,* H1807–H1813.

Iwakura, T., Inui, A., & Reddi, A. H. (2013). Stimulation of superficial zone protein accumulation by hedgehog and Wnt signaling in surface zone bovine articular chondrocytes. *Arthritis and Rheumatism, 65,* 408–417.

Iwamoto, M., Ohta, Y., Larmour, C., & Enomoto-Iwamoto, M. (2013). Toward regeneration of articular cartilage. *Birth Defects Research Part C-Embryo Today-Reviews, 99,* 192–202.

Iwamoto, M., Tamamura, Y., Koyama, E., Komori, T., Takeshita, N., Williams, J. A., et al. (2007). Transcription factor ERG and joint and articular cartilage formation during mouse limb and spine skeletogenesis. *Developmental Biology, 305,* 40–51.

Jablonski, C. L., Ferguson, S., Pozzi, A., & Clark, A. L. (2014). Integrin alpha1beta1 participates in chondrocyte transduction of osmotic stress. *Biochemical and Biophysical Research Communications, 445,* 184–190.

Jacob, M. P. (2003). Extracellular matrix remodeling and matrix metalloproteinases in the vascular wall during aging and in pathological conditions. *Biomedicine & Pharmacotherapy, 57,* 195–202.

Jadin, K. D., Schumacher, B. L., & Sah, R. L. (2006). Proliferation of chondrocytes within articular cartilage. In *52nd Annual Meeting of the Orthopaedic Research Society.*

Jadin, K. D., Wong, B. L., Bae, W. C., Li, K. W., Williamson, A. K., Schumacher, B. L., et al. (2005). Depth-varying density and organization of chondrocytes in immature and mature bovine articular cartilage assessed by 3D imaging and analysis. *Journal of Histochemistry and Cytochemistry, 53,* 1109–1119.

Jaiswal, A., Chhabra, A., Malhotra, U., Kohli, S., & Rani, V. (2011). Comparative analysis of human matrix metalloproteinases: Emerging therapeutic targets in diseases. *Bioinformation, 6,* 23–30.

Jay, G. D., Fleming, B. C., Watkins, B. A., McHugh, K. A., Anderson, S. C., Zhang, L. X., et al. (2010). Prevention of cartilage degeneration and restoration of chondroprotection by lubricin tribosupplementation in the rat following anterior cruciate ligament transection. *Arthritis and Rheumatism, 62,* 2382–2391.

Jay, G. D., Torres, J. R., Rhee, D. K., Helminen, H. J., Hytinnen, M. M., Cha, C. J., et al. (2007). Association between friction and wear in diarthrodial joints lacking lubricin. *Arthritis and Rheumatism, 56,* 3662–3669.

Jay, G. D., & Waller, K. A. (2014). The biology of Lubricin: Near frictionless joint motion. *Matrix Biology, 39,* 17–24.

Jebens, E. H., & Monk-Jones, M. E. (1959). On the viscosity and pH of synovial fluid and the pH of blood. *The Journal of Bone and Joint Surgery (The British Editorial Society of Bone & Joint Surgery), 41-B,* 388–400.

Jeffery, A. K., Blunn, G. W., Archer, C. W., & Bentley, G. (1991). Three-dimensional collagen architecture in bovine articular cartilage. *Journal of Bone and Joint Surgery. British Volume, 73,* 795–801.

Jeffrey, J. E., Thomson, L. A., & Aspden, R. M. (1997). Matrix loss and synthesis following a single impact load on articular cartilage in vitro. *Biochimica et Biophysica Acta, 1334,* 223–232.

Jenne, D. E., & Tschopp, J. (1992). Clusterin—The Intriguing guises of a widely expressed glycoprotein. *Trends in Biochemical Sciences, 17,* 154–159.

Jensen, L. T., Henriksen, J. H., Olesen, H. P., Risteli, J., & Lorenzen, I. (1993). Lymphatic clearance of synovial fluid in conscious pigs: The aminoterminal propeptide of type III procollagen. *European Journal of Clinical Investigation, 23,* 778–784.

Ji, W., Waas, A. M., & Bazant, Z. P. (2013). On the importance of work-conjugacy and objective stress rates in finite deformation incremental finite element analysis. *Journal of Applied Mechanics-Transactions of the ASME, 80,* 041024.

Jiang, F. Z., Horber, H., Howard, J., & Muller, D. J. (2004). Assembly of collagen into microribbons: Effects of pH and electrolytes. *Journal of Structural Biology, 148,* 268–278.

Jiang, J., Leong, N. L., Mung, J. C., Hidaka, C., & Lu, H. H. (2008). Interaction between zonal populations of articular chondrocytes suppresses chondrocyte mineralization and this process is mediated by PTHrP. *Osteoarthritis Cartilage, 16,* 70–82.

Jing, W., & Jiang, W. (2015). MicroRNA-93 regulates collagen loss by targeting MMP3 in human nucleus pulposus cells. *Cell Proliferation, 48,* 284–292.

Johnson, K., Farley, D., Hu, S. I., & Terkeltaub, R. (2003). One of two chondrocyte-expressed isoforms of cartilage intermediate-layer protein functions as an insulin-like growth factor 1 antagonist. *Arthritis and Rheumatism, 48,* 1302–1314.

Jones, B. K., Durney, K. M., Hung, C. T., & Ateshian, G. A. (2015). The friction coefficient of shoulder joints remains remarkably low over 24 h of loading. *Journal of Biomechanics, 48,* 3945–3949.

Juhasz, T., Matta, C., Somogyi, C., Katona, E., Takacs, R., Soha, R. F., et al. (2014). Mechanical loading stimulates chondrogenesis via the PKA/CREB-Sox9 and PP2A pathways in chicken micromass cultures. *Cellular Signalling, 26,* 468–482.

Julkunen, P., Wilson, W., Isaksson, H., Jurvelin, J. S., Herzog, W., & Korhonen, R. K. (2013). A review of the combination of experimental measurements and fibril-reinforced modeling for investigation of articular cartilage and chondrocyte response to loading. *Computational and Mathematical Methods in Medicine, 2013,* 326150.

Jungmann, O., Nikolovska, K., Stock, C., Schulz, J. N., Eckes, B., Riethmuller, C., et al. (2012). The dermatan sulfate proteoglycan decorin modulates alpha2beta1 integrin and the vimentin intermediate filament system during collagen synthesis. *PLoS ONE, 7,* e50809.

Kaab, M. J., Gwynn, I. A., & Notzli, H. P. (1998). Collagen fibre arrangement in the tibial plateau articular cartilage of man and other mammalian species. *Journal of Anatomy, 193*(Pt 1), 23–34.

Kadler, K. E., Hill, A., & Canty-Laird, E. G. (2008). Collagen fibrillogenesis: Fibronectin, integrins, and minor collagens as organizers and nucleators. *Current Opinion in Cell Biology, 20,* 495–501.

Kalamajski, S., Bihan, D., Bonna, A., Rubin, K., & Farndale, R. W. (2016). Fibromodulin interacts with collagen cross-linking sites and activates lysyl oxidase. *Journal of Biological Chemistry, 291,* 7951–7960.

Kalchishkova, N., Furst, C. M., Heinegard, D., & Blom, A. M. (2011). NC4 domain of cartilage-specific collagen IX inhibits complement directly due to attenuation of membrane attack formation and indirectly through binding and enhancing activity of complement inhibitors C4B-binding protein and factor H. *Journal of Biological Chemistry, 286,* 27915–27926.

Kaleva, E., Saarakkala, S., Jurvelin, J. S., Viren, T., & Toyras, J. (2009). Effects of ultrasound beam angle and surface roughness on the quantitative ultrasound parameters of articular cartilage. *Ultrasound in Medicine and Biology, 35,* 1344–1351.

Kamisan, N., Naveen, S. V., Ahmad, R. E., & Tunku, K. (2013). Chondrocyte density, proteoglycan content and gene expressions from native cartilage are species specific and not dependent on cartilage thickness: A comparative analysis between rat, rabbit and goat. *BMC Veterinary Research, 9,* 62.

Kana, R., Dundr, P., Tvrdik, D., Necas, E., & Povysil, C. (2006). Expression of actin isoforms in human auricular cartilage. *Folia Biologica (Praha), 52,* 167–172.

Kaneko, H., Liu, L., Kurosawa, H., Sadatsuki, R., Hada, S., Kinoshita, M., et al. (2014). Urine Crosslinked telopeptide of type II collagen predicted osteoarthritis progression. *Osteoarthritis and Cartilage, 22,* S76–S77.

Kanyama, M., Kuboki, T., Kojima, S., Fujisawa, T., Hattori, T., Takigawa, M., et al. (2000). Matrix metalloproteinases and tissue inhibitors of metalloproteinases in synovial fluids of patients with temporomandibular joint osteoarthritis. *Journal of Orofacial Pain, 14,* 20–30.

Kaplan, S. A. (1980). Factors affecting the relationship between rate of loading and measured compressive strength of concrete. *Magazine of Concrete Research, 32,* 79–88.

Kapyla, J., Jaalinoja, J., Tulla, M., Ylostalo, J., Nissinen, L., Viitasalo, T., et al. (2004). The fibril-associated collagen IX provides a novel mechanism for cell adhesion to cartilaginous matrix. *Journal of Biological Chemistry, 279,* 51677–51687.

Kar, S., Smith, D. W., Gardiner, B. S., & Grodzinsky, A. J. (2016). Systems based study of the therapeutic potential of small charged molecules for the inhibition of IL-1 mediated cartilage degradation. *PLoS ONE, 11*, e0168047.

Karczewski, K. J., Snyder, M., Altman, R. B., & Tatonetti, N. P. (2014). Coherent functional modules improve transcription factor target identification, cooperativity prediction, and disease association. *PLoS Genetics, 10,* e1004122.

Karl, F. (2012). A free energy principle for biological systems. *Entropy, 14,* 2100–2121.

Karlsson, C., Thornemo, M., Henriksson, H. B., & Lindahl, A. (2009). Identification of a stem cell niche in the zone of Ranvier within the knee joint. *Journal of Anatomy, 215,* 355–363.

Karsdal, M. A., Madsen, S. H., Christiansen, C., Henriksen, K., Fosang, A. J., & Sondergaard, B. C. (2008). Cartilage degradation is fully reversible in the presence of aggrecanase but not matrix metalloproteinase activity. *Arthritis Research & Therapy, 10,* R63.

Kashiwagi, M., Tortorella, M., Nagase, H., & Brew, K. (2001). TIMP-3 is a potent inhibitor of aggrecanase 1 (ADAM-TS4) and aggrecanase 2 (ADAM-TS5). *Journal of Biological Chemistry, 276,* 12501–12504.

Kastritis, P. L., & Bonvin, A. M. (2013). On the binding affinity of macromolecular interactions: Daring to ask why proteins interact. *Journal of the Royal Society Interface, 10,* 20120835.

Kataru, R. P., Jung, K., Jang, C., Yang, H., Schwendener, R. A., Baik, J. E., et al. (2009). Critical role of CD11b+ macrophages and VEGF in inflammatory lymphangiogenesis, antigen clearance, and inflammation resolution. *Blood, 113,* 5650–5659.

Katta, J., Jin, Z., Ingham, E., & Fisher, J. (2008). Biotribology of articular cartilage—A review of the recent advances. *Medical Engineering & Physics, 30,* 1349–1363.

Kempson, G. E. (1982). Relationship between the tensile properties of articular-cartilage from the human knee and age. *Annals of the Rheumatic Diseases, 41,* 508–511.

Kerkhofs, J., Leijten, J., Bolander, J., Luyten, F. P., Post, J. N., & Geris, L. (2016). A qualitative model of the differentiation network in chondrocyte maturation: A holistic view of chondrocyte hypertrophy. *PLoS ONE, 11.*

Khan, I. M., Salter, D. M., Bayliss, M. T., Thomson, G. M., & Archer, C. W. (2001). Expression of clusterin in the superficial zone of bovine articular cartilage. *Arthritis and Rheumatism, 44,* 1795–1799.

Kiani, C., Chen, L., Wu, Y. J., Yee, A. J., & Yang, B. B. (2002). Structure and function of aggrecan. *Cell Research, 12,* 19–32.

Killock, D. (2012). Osteoarthritis: Frzb knockout reveals the complexity of Wnt signaling in joint homeostasis. *Nature Reviews Rheumatology, 8,* 123.

Kim, A. C., & Spector, M. (2000). Distribution of chondrocytes containing alpha-smooth muscle actin in human articular cartilage. *Journal of Orthopaedic Research, 18,* 749–755.

Kim, P. G., Nakano, H., Das, P. P., Chen, M. J., Rowe, R. G., Chou, S. S., et al. (2015). Flow-induced protein kinase A-CREB pathway acts via BMP signaling to promote HSC emergence. *Journal of Experimental Medicine, 212,* 633–648.

Kimberley, F. C., Sivasankar, B., & Paul Morgan, B. (2007). Alternative roles for CD59. *Molecular Immunology, 44,* 73–81.

Kimura, T., Okada, A., Yatabe, T., Okubo, M., Toyama, Y., Noda, M., et al. (2010). RECK is up-regulated and involved in chondrocyte cloning in human osteoarthritic cartilage. *American Journal of Pathology, 176,* 2858–2867.

Kinner, B., & Spector, M. (2001). Smooth muscle actin expression by human articular chondrocytes and their contraction of a collagen-glycosaminoglycan matrix in vitro. *Journal of Orthopaedic Research, 19,* 233–241.

Kiraly, K., Hyttinen, M. M., Parkkinen, J. J., Arokoski, J. A., Lapvetelainen, T., Torronen, K., et al. (1998). Articular cartilage collagen birefringence is altered concurrent with changes in proteoglycan synthesis during dynamic in vitro loading. *Anatomical Record, 251,* 28–36.

Kirkwood, J., Hargreaves, D., O'Keefe, S., & Wilson, J. (2015). Using isoelectric point to determine the pH for initial protein crystallization trials. *Bioinformatics, 31,* 1444–1451.

Kiselev, M. A., Gryzunov I. U. A., Dobretsov, G. E. & Komarova, M. N. (2001). Size of a human serum albumin molecule in solution. *Biofizika, 46,* 423–427.

Kitamura, H. P., Yanase, H., Kitamura, H., & Iwanaga, T. (1999). Unique localization of protein gene product 9.5 in type B synoviocytes in the joints of the horse. *Journal of Histochemistry and Cytochemistry, 47,* 343–351.

Kitano, H. (2004). Biological robustness. *Nature Reviews Genetics, 5,* 826–837.

Kitano, H. (2007). Towards a theory of biological robustness. *Molecular Systems Biology, 3,* 137.

Knight, M. M., McGlashan, S. R., Garcia, M., Jensen, C. G., & Poole, C. A. (2009). Articular chondrocytes express connexin 43 hemichannels and P2 receptors—A putative mechanoreceptor complex involving the primary cilium? *Journal of Anatomy, 214,* 275–283.

Knudson, W., & Loeser, R. F. (2002). CD44 and integrin matrix receptors participate in cartilage homeostasis. *Cellular and Molecular Life Sciences, 59,* 36–44.

Knupp, C., & Squire, J. M. (2001). A new twist in the collagen story—The type VI segmented supercoil. *EMBO Journal, 20,* 372–376.

Ko, F. C., Dragomir, C., Plumb, D. A., Goldring, S. R., Wright, T. M., Goldring, M. B., & Van Der Meulen, M. C. (2013). In vivo cyclic compression causes cartilage degeneration and subchondral bone changes in mouse tibiae. *Arthritis Rheum, 65,* 1569–1578.

Kobayashi, K., Matsuzaka, S., Yoshida, Y., Miyauchi, S., Wada, Y., & Moriya, H. (2004). The effects of intraarticularly injected sodium hyaluronate on levels of intact aggrecan and nitric oxide in the joint fluid of patients with knee osteoarthritis. *Osteoarthritis Cartilage, 12,* 536–542.

Kobayashi, S., Yonekubo, S., & Kurogouchi, Y. (1996). Cryoscanning electron microscopy of loaded articular cartilage with special reference to the surface amorphous layer. *Journal of Anatomy, 188*(Pt 2), 311–322.

Kobayashi, T., Yoshihara, Y., Samura, A., Yamada, H., Shinmei, M., Roos, H., et al. (1997). Synovial fluid concentrations of the C-propeptide of type II collagen correlate with body mass index in primary knee osteoarthritis. *Annals of the Rheumatic Diseases, 56,* 500–503.

Kohns, M., Reiser, S., Horsch, M., & Hasse, H. (2016). Solvent activity in electrolyte solutions from molecular simulation of the osmotic pressure. *Journal of Chemical Physics, 144,* 084112.

Kokebie, R., Aggarwal, R., Lidder, S., Hakimiyan, A. A., Rueger, D. C., Block, J. A., et al. (2011). The role of synovial fluid markers of catabolism and anabolism in osteoarthritis, rheumatoid arthritis and asymptomatic organ donors. *Arthritis Research & Therapy, 13,* R50.

Kondo, S., Shukunami, C., Morioka, Y., Matsumoto, N., Takahashi, R., Oh, J., et al. (2007). Dual effects of the membrane-anchored MMP regulator RECK on chondrogenic differentiation of ATDC5 cells. *Journal of Cell Science, 120,* 849–857.

Konttinen, Y. T., Mandelin, J., Li, T. F., Salo, J., Lassus, J., Liljestrom, M., et al. (2002). Acidic cysteine endoproteinase cathepsin K in the degeneration of the superficial articular hyaline cartilage in osteoarthritis. *Arthritis and Rheumatism, 46,* 953–960.

Konttinen, Y. T., Sillat, T., Barreto, G., Ainola, M., & Nordstrom, D. C. E. (2012). Osteoarthritis as an autoinflammatory disease caused by chondrocyte-mediated inflammatory responses. *Arthritis and Rheumatism, 64,* 613–616.

Kosaka, T., Fukui, R., Matsui, M., Kurosaka, Y., Nishimura, H., Tanabe, M., et al. (2014). RAGE, receptor of advanced glycation endoproducts, negatively regulates chondrocytes differentiation. *PLoS ONE, 9,* e108819.

Kosicki, M., Tomberg, K., & Bradley, A. (2018). Repair of double-strand breaks induced by CRISPR–Cas9 leads to large deletions and complex rearrangements. *Nature Biotechnology, 36,* 765–771.

Kostenuik, P. J. (2005). Osteoprotegerin and RANKL regulate bone resorption, density, geometry and strength. *Current Opinion in Pharmacology, 5,* 618–625.

Kovach, I. S. (1995). The importance of polysaccharide configurational entropy in determining the osmotic swelling pressure of concentrated proteoglycan solution and the bulk compressive modulus of articular-cartilage. *Biophysical Chemistry, 53,* 181–187.

Koyama, E., Shibukawa, Y., Nagayama, M., Sugito, H., Young, B., Yuasa, T., et al. (2008). A distinct cohort of progenitor cells participates in synovial Joint and articular cartilage formation during mouse limb skeletogenesis. *Developmental Biology, 316,* 62–73.

Kozak, I. (2000). Principle of complementary virtual work and the Riemann-Christoffel curvature tensor as compatibility condition. *Journal of Computational and Applied Mechanics, 1,* 9.

Kramer, E. M., & Myers, D. R. (2012). Five popular misconceptions about osmosis. *American Journal of Physics, 80,* 694–699.

Kramer, E. M., & Myers, D. R. (2013). Osmosis is not driven by water dilution. *Trends in Plant Science, 18,* 195–197.

Kraus, V. B., Blanco, F. J., Englund, M., Karsdal, M. A., & Lohmander, L. S. (2015). Call for standardized definitions of osteoarthritis and risk stratification for clinical trials and clinical use. *Osteoarthritis and Cartilage, 23,* 1233–1241.

Kraus, V. B., Stabler, T. V., Kong, S. Y., Varju, G., & McDaniel, G. (2007). Measurement of synovial fluid volume using urea. *Osteoarthritis and Cartilage, 15,* 1217–1220.

Kretzer, J., Reinders, J., Sonntag, R., Hagmann, S., Streit, M., Jeager, S., et al. (2014). Wear in total knee arthroplasty-just a question of polyethylene? *International Orthopaedics, 38,* 335–340.

Krishnan, Y., & Grodzinsky, A. J. (2018). Cartilage diseases. *Matrix Biology, 71–72,* 51–69.

Kronenberg, H. M. (2003). Developmental regulation of the growth plate. *Nature, 423,* 332–336.

Kuhn, K., D'Lima, D. D., Hashimoto, S., & Lotz, M. (2004). Cell death in cartilage. *Osteoarthritis and Cartilage, 12,* 1–16.

Kumar, D., Ray, A., & Ray, B. K. (2009). Transcriptional synergy mediated by SAF-1 and AP-1: Critical role of N-terminal polyalanine and two zinc finger domains of SAF-1. *Journal of Biological Chemistry, 284,* 1853–1862.

Kurz, B., Lemke, A. K., Fay, J., Pufe, T., Grodzinsky, A. J., & Schunke, M. (2005). Pathomechanisms of cartilage destruction by mechanical injury. *Annals of Anatomy, 187,* 473–485.

Kuster, M. S., Podsiadlo, P., & Stachowiak, G. W. (1998). Shape of wear particles found in human knee joints and their relationship to osteoarthritis. *British Journal of Rheumatology, 37,* 978–984.

Kuster, M. S., Wood, G. A., Stachowiak, G. W., & Gachter, A. (1997). Joint load considerations in total knee replacement. *Journal of Bone and Joint Surgery. British Volume, 79,* 109–113.

Kvist, A. J., Nystrom, A., Hultenby, K., Sasaki, T., Talts, J. F., & Aspberg, A. (2008). The major basement membrane components localize to the chondrocyte pericellular matrix—A cartilage basement membrane equivalent? *Matrix Biology, 27,* 22–33.

Kwan, M. K., Hacker, S. A., Woo, S. L. Y., & Wayne, J. S. (1992). The effect of storage on the biomechanical behavior of articular-cartilage—A large strain study. *Journal of Biomechanical Engineering-Transactions of the ASME, 114,* 149–153.

Kwan Tat, S., Amiable, N., Pelletier, J. P., Boileau, C., Lajeunesse, D., Duval, N., et al. (2009). Modulation of OPG, RANK and RANKL by human chondrocytes and their implication during osteoarthritis. *Rheumatology (Oxford), 48,* 1482–1490.

Kwoh, C. K. (2013). OSTEOARTHRITIS Clinical relevance of bone marrow lesions in OA. *Nature Reviews Rheumatology, 9,* 7–8.

Lamande, S. R., Morgelin, M., Adams, N. E., Selan, C., & Allen, J. M. (2006). The C5 domain of the collagen VI alpha 3(VI) chain is critical for extracellular microfibril formation and is present in the extracellular matrix of cultured cells. *Journal of Biological Chemistry, 281,* 16607–16614.

Lamoureux, F., Baud'Huin, M., Duplomb, L., Heymann, D., & Redini, F. (2007). Proteoglycans: Key partners in bone cell biology. *BioEssays, 29,* 758–771.

Lane, N. E., Brandt, K., Hawker, G., Peeva, E., Schreyer, E., Tsuji, W., et al. (2011). OARSI-FDA initiative: Defining the disease state of osteoarthritis. *Osteoarthritis and Cartilage, 19,* 478–482.

Lane, L. B., & Bullough, P. G. (1980). Age-related-changes in the thickness of the calcified zone and the number of tidemarks in adult human articular-cartilage. *Journal of Bone and Joint Surgery-British Volume, 62,* 372–375.

Lane, J. M., & Weiss, C. (1975). Review of articular-cartilage collagen research. *Arthritis and Rheumatism, 18,* 553–562.

Langberg, H., Rosendal, L., & Kjaer, M. (2001). Training-induced changes in peritendinous type I collagen turnover determined by microdialysis in humans. *Journal of Physiology, 534,* 297–302.

Langelier, E., Suetterlin, R., Hoemann, C. D., Aebi, U., & Buschmann, M. D. (2000). The chondrocyte cytoskeleton in mature articular cartilage: Structure and distribution of actin, tubulin, and vimentin filaments. *Journal of Histochemistry and Cytochemistry, 48,* 1307–1320.

Langevin, H. M., Cornbrooks, C. J., & Taatjes, D. J. (2004). Fibroblasts form a body-wide cellular network. *Histochemistry and Cell Biology, 122,* 7–15.

Larsson, S., Lohmander, L. S., & Struglics, A. (2009). Synovial fluid level of aggrecan ARGS fragments is a more sensitive marker of joint disease than glycosaminoglycan or aggrecan levels: A cross-sectional study. *Arthritis Research & Therapy,* 11.

Last, R. J. (1948). Some anatomical details of the knee joint. *Journal of Bone and Joint Surgery-British Volume, 30,* 683–688.

Lauer-Fields, J. L., Broder, T., Sritharan, T., Chung, L., Nagase, H., & Fields, G. B. (2001). Kinetic analysis of matrix metalloproteinase activity using fluorogenic triple-helical substrates. *Biochemistry, 40,* 5795–5803.

Lawrence, R. C., Felson, D. T., Helmick, C. G., Arnold, L. M., Choi, H., Deyo, R. A., et al. (2008). Estimates of the prevalence of arthritis and other rheumatic conditions in the United States. *Arthritis and Rheumatism, 58,* 26–35.

Lecaille, F., Choe, Y., Brandt, W., Li, Z., Craik, C. S., & Bromme, D. (2002). Selective inhibition of the collagenolytic activity of human cathepsin K by altering its S2 subsite specificity. *Biochemistry, 41,* 8447–8454.

Leddy, H. A., Christensen, S. E., & Guilak, F. (2008). Microscale diffusion properties of the cartilage pericellular matrix measured using 3D scanning microphotolysis. *Journal of Biomechanical Engineering-Transactions of the ASME, 130.*

Leddy, H. A., & Guilak, F. (2003). Site-specific molecular diffusion in articular cartilage measured using fluorescence recovery after photobleaching. *Annals of Biomedical Engineering, 31,* 753–760.

Lee, D. W., Banquy, X., & Israelachvili, J. N. (2013). Stick-slip friction and wear of articular joints. *Proceedings of the National Academy of Sciences, 110,* E567–E574.

Lee, D. A., Bentley, G., & Archer, C. W. (1993). The control of cell division in articular chondrocytes. *Osteoarthritis Cartilage, 1,* 137–146.

Lee, C., Grad, S., Wimmer, M. Alini, M. (2006). The influence of mechanical stimuli on articular cartilage tissue engineering. In R. L. Reis & N. Ashammakhi (Eds.), *Topic in tissue engineering.* http://www.oulu.fi/spareparts/ebook_topics_in_t_e_vol2/list_of_contr.html.

Lee, H. Y., Han, L., Roughley, P. J., Grodzinsky, A. J., & Ortiz, C. (2013). Age-related nanostructural and nanomechanical changes of individual human cartilage aggrecan monomers and their glycosaminoglycan side chains. *Journal of Structural Biology, 181,* 264–273.

Lee, M. S., Trindade, M. C., Ikenoue, T., Schurman, D. J., Goodman, S. B., & Smith, R. L. (2003). Intermittent hydrostatic pressure inhibits shear stress-induced nitric oxide release in human osteoarthritic chondrocytes in vitro. *Journal of Rheumatology, 30,* 326–328.

Lee, J. I., Sato, M., Ushida, K., & Mochida, J. (2011). Measurement of diffusion in articular cartilage using fluorescence correlation spectroscopy. *Bmc Biotechnology, 11.*

Leighton, M. P., Nundlall, S., Starborg, T., Meadows, R. S., Suleman, F., Knowles, L., et al. (2007). Decreased chondrocyte proliferation and dysregulated apoptosis in the cartilage growth plate are key features of a murine model of epiphyseal dysplasia caused by a matn3 mutation. *Human Molecular Genetics, 16,* 1728–1741.

Leijten, J. C., Emons, J., Sticht, C., van Gool, S., Decker, E., Uitterlinden, A., et al. (2012). Gremlin 1, frizzled-related protein, and Dkk-1 are key regulators of human articular cartilage homeostasis. *Arthritis and Rheumatism, 64,* 3302–3312.

Leonardi, R., Perrotta, R. E., Almeida, L. E., Loreto, C., & Musumeci, G. (2016). Lubricin in synovial fluid of mild and severe temporomandibular joint internal derangements. *Medicina Oral Patologia Oral Y Cirugia Bucal, 21,* E793–E799.

Leppanen, V. M., Tossavainen, H., Permi, P., Lehtio, L., Ronnholm, G., Goldman, A., et al. (2007). Crystal structure of the N-terminal NC4 domain of collagen IX, a zinc binding member of the laminin-neurexin-sex hormone binding globulin (LNS) domain family. *Journal of Biological Chemistry, 282,* 23219–23230.

Levick, J. R. (1987). Flow through interstitium and other fibrous matrices. *Quarterly Journal of Experimental Physiology and Cognate Medical Sciences, 72,* 409–438.

Levick, J. R., & Michel, C. C. (2010). Microvascular fluid exchange and the revised Starling principle. *Cardiovascular Research, 87,* 198–210.

Levin, A. S., Chen, C. T., & Torzilli, P. A. (2005). Effect of tissue maturity on cell viability in load-injured articular cartilage explants. *Osteoarthritis Cartilage, 13,* 488–496.

Levine, A. J., Hu, W., & Feng, Z. (2006). The P53 pathway: What questions remain to be explored? *Cell Death and Differentiation, 13,* 1027–1036.

Levy, A. S., Lohnes, J., Sculley, S., Lecroy, M., & Garrett, W. (1996). Chondral delamination of the knee in soccer players. *American Journal of Sports Medicine, 24,* 634–639.

Lewinson, D., Bialik, G. M., & Hochberg, Z. (1994). Differential effects of hypothyroidism on the cartilage and the osteogenic process in the mandibular condyle: Recovery by growth hormone and thyroxine. *Endocrinology, 135,* 1504–1510.

Lewis, P. R., & McCutchen, C. W. (1959). Experimental evidence for weeping lubrication in mammalian joints. *Nature, 184,* 1285.

Lewis, G. N. (1908). The osmotic pressure of concentrated solutions, and the laws of the perfect solution. *Journal of the American Chemical Society, 30,* 668–683.

Li, B. H., Marshall, D., Roe, M., & Aspden, R. M. (1999a). The electron microscope appearance of the subchondral bone plate in the human femoral head in osteoarthritis and osteoporosis. *Journal of Anatomy, 195,* 101–110.

Li, L. M., Patil, S., Steklov, N., Bae, W., Temple-Wong, M., D'Lima, D. D., et al. (2011a). Computational wear simulation of patellofemoral articular cartilage during in vitro testing. *Journal of Biomechanics, 44,* 1507–1513.

Li, L. P., Cheung, J. T. M., & Herzog, W. (2009). Three-dimensional fibril-reinforced finite element model of articular cartilage. *Medical & Biological Engineering & Computing, 47,* 607–615.

Li, L. P., Soulhat, J., Buschmann, M. D., & Shirazi-Adl, A. (1999b). Nonlinear analysis of cartilage in unconfined ramp compression using a fibril reinforced poroelastic model. *Clinical Biomechanics (Bristol, Avon), 14,* 673–682.

Li, W. D., Zhou, J., Chen, L. J., Luo, Z. J., & Zhao, Y. Z. (2011b). Lysyl oxidase, a critical intra- and extra-cellular target in the lung for cigarette smoke pathogenesis. *International Journal of Environmental Research and Public Health, 8,* 161–184.

Li, Y. (2009). *The mechanism of collagen self-assembly: Hydrophobic and electrostatic interactions.* Ph.D., University of Florida.

Li, Y., Frank, E. H., Wang, Y., Chubinskaya, S., Huang, H. H., & Grodzinsky, A. J. (2013). Moderate dynamic compression inhibits pro-catabolic response of cartilage to mechanical injury, tumor necrosis factor-alpha and interleukin-6, but accentuates degradation above a strain threshold. *Osteoarthritis Cartilage, 21,* 1933–1941.

Li, Y., Wang, Y., Chubinskaya, S., Schoeberl, B., Florine, E., Kopesky, P., et al. (2015). Effects of insulin-like growth factor-1 and dexamethasone on cytokine-challenged cartilage: Relevance to post-traumatic osteoarthritis. *Osteoarthritis and Cartilage, 23,* 266–274.

Li, Z., Kupcsik, L., Yao, S. J., Alini, M., & Stoddart, M. J. (2010). Mechanical load modulates chondrogenesis of human mesenchymal stem cells through the TGF-beta pathway. *Journal of Cellular and Molecular Medicine, 14,* 1338–1346.

Liberfarb, R. M., Levy, H. P., Rose, P. S., Wilkin, D. J., Davis, J., Balog, J. Z., et al. (2003). The Stickler syndrome: Genotype/phenotype correlation in 10 families with Stickler syndrome resulting from seven mutations in the type II collagen gene locus COL2A1. *Genetics in Medicine, 5,* 21–27.

Lieberthal, J., Sambamurthy, N., & Scanzello, C. R. (2015). Inflammation in joint injury and post-traumatic osteoarthritis. *Osteoarthritis and Cartilage, 23,* 1825–1834.

Lin, P. M., Chen, C. T., & Torzilli, P. A. (2004). Increased stromelysin-1 (MMP-3), proteoglycan degradation (3B3- and 7D4) and collagen damage in cyclically load-injured articular cartilage. *Osteoarthritis Cartilage, 12,* 485–496.

Ling, L., Murali, S., Stein, G. S., van Wijnen, A. J., & Cool, S. M. (2010). Glycosaminoglycans modulate RANKL-induced osteoclastogenesis. *Journal of Cellular Biochemistry, 109,* 1222–1231.

Linn, F. C. (1967). Lubrication of animal joints. I. Arthrotripsometer. *Journal of Bone and Joint Surgery-American Volume, A 49,* 1079.

Linn, F. C. (1968). Lubrication of animal joints II the mechanism. *Journal of Biomechanics, 1,* 13.

Linn, F. C., & Radin, E. L. (1968). Lubrication of animal joints. 3. Effect of certain chemical alterations of cartilage and lubricant. *Arthritis and Rheumatism, 11,* 674.

Lipson, A., Nickoloff, E. L., Hsu, T. H., Kasecamp, W. R., Drew, H. M., Shakir, R. & Wagner, H. N., Jr. (1979). A study of age-dependent changes in thyroid function tests in adults. *Journal of Nuclear Medicine, 20,* 1124–1130.

Little, C. B., Hughes, C. E., Curtis, C. L., Janusz, M. J., Bohne, R., Wang-Weigand, S., et al. (2002). Matrix metalloproteinases are involved in C-terminal and interglobular domain processing of cartilage aggrecan in late stage cartilage degradation. *Matrix Biology, 21,* 271–288.

Little, C. B., Meeker, C. T., Golub, S. B., Lawlor, K. E., Farmer, P. J., Smith, S. M., et al. (2007). Blocking aggrecanase cleavage in the aggrecan interglobular domain abrogates cartilage erosion and promotes cartilage repair. *The Journal of Clinical Investigation, 117,* 1627–1636.

Liu, C. J. (2009). The role of ADAMTS-7 and ADAMTS-12 in the pathogenesis of arthritis. *Nature Clinical Practice Rheumatology, 5,* 38–45.

Liu, F., Kozanek, M., Hosseini, A., van de Velde, S. K., Gill, T. J., Rubash, H. E., et al. (2010). In vivo tibiofemoral cartilage deformation during the stance phase of gait. *Journal of Biomechanics, 43,* 658–665.

Liu, S. H., Yang, R. S., Al-Shaikh, R. & Lane, J. M. (1995). Collagen in tendon, ligament, and bone healing. A current review. *Clinical Orthopaedics and Related Research*, 265–278.

Liu, S. Q., & Fung, Y. C. (1996). Indicial functions of arterial remodeling in response to locally altered blood pressure. *American Journal of Physiology, 270,* H1323–H1333.

Liu, S. S., Zhou, P., & Zhang, Y. Q. (2016). Abnormal expression of key genes and proteins in the canonical Wnt/-catenin pathway of articular cartilage in a rat model of exercise-induced osteoarthritis. *Molecular Medicine Reports, 13,* 1999–2006.

Liu, X., Sun, J. Q., Heggeness, M. H., Yeh, M. L., & Luo, Z. P. (2006). Force-mediated dissociation of proteoglycan aggregate in articular cartilage. *Biorheology, 43,* 183–190.

Liukkonen, M. K., Mononen, M. E., Klets, O., Arokoski, J. P., Saarakkala, S., & Korhonen, R. K. (2017). Simulation of subject-specific progression of knee osteoarthritis and comparison to experimental follow-up data: Data from the osteoarthritis initiative. *Scientific Reports, 7,* 9177.

Lodewyckx, L., Cailotto, F., Thysen, S., Luyten, F. P., & Lories, R. J. (2012). Tight regulation of wingless-type signaling in the articular cartilage—Subchondral bone biomechanical unit: Transcriptomics in Frzb-knockout mice. *Arthritis Research & Therapy, 14.*

Loening, A. M., James, I. E., Levenston, M. E., Badger, A. M., Frank, E. H., Kurz, B., et al. (2000). Injurious mechanical compression of bovine articular cartilage induces chondrocyte apoptosis. *Archives of Biochemistry and Biophysics, 381,* 205–212.

Loeser, R. F. (2003). Systemic and local regulation of articular cartilage metabolism: Where does leptin fit in the puzzle? *Arthritis and Rheumatism, 48,* 3009–3012.

Loeser, R. F. (2009). Aging and osteoarthritis: The role of chondrocyte senescence and aging changes in the cartilage matrix. *Osteoarthritis and Cartilage, 17,* 971–979.

Loeser, R. F. (2014). Integrins and chondrocyte-matrix interactions in articular cartilage. *Matrix Biology, 39C,* 11–16.

Loeser, R. F., Goldring, S. R., Scanzello, C. R., & Goldring, M. B. (2012). Osteoarthritis: A disease of the joint as an organ. *Arthritis and Rheumatism, 64,* 1697–1707.

Loeser, R. F., Yammani, R. R., Carlson, C. S., Chen, H., Cole, A., Im, H. J., et al. (2005). Articular chondrocytes express the receptor for advanced glycation end products: Potential role in osteoarthritis. *Arthritis and Rheumatism, 52,* 2376–2385.

Lohmander, L. S., Atley, L. M., Pietka, T. A., & Eyre, D. R. (2003). The release of crosslinked peptides from type II collagen into human synovial fluid is increased soon after joint injury and in osteoarthritis. *Arthritis and Rheumatism, 48,* 3130–3139.

Lohmander, L. S., Ionescu, M., Jugessur, H., & Poole, A. R. (1999). Changes in joint cartilage aggrecan after knee injury and in osteoarthritis. *Arthritis and Rheumatism, 42,* 534–544.

Lohmander, L. S., Yoshihara, Y., Roos, H., Kobayashi, T., Yamada, H., & Shinmei, M. (1996). Procollagen II C-propeptide in joint fluid: Changes in concentration with age, time after knee injury, and osteoarthritis. *Journal of Rheumatology, 23,* 1765–1769.

Lohmander, S., Antonopoulos, C. A., & Friberg, U. (1973). Chemical and metabolic heterogeneity of chondroitin sulfate and keratin sulfate in guinea pig cartilage and nucleus pulposus. *Biochimica et Biophysica Acta, 304,* 430–448.

Long, F. X., Schipani, E., Asahara, H., Kronenberg, H., & Montminy, M. (2001). The CREB family of activators is required for endochondral bone development. *Development, 128,* 541–550.

Longmore, R. B., & Gardner, D. L. (1975). Development with age of human articular cartilage surface structure. A survey by interference microscopy of the lateral femoral condyle. *Annals of the Rheumatic Diseases, 34,* 26–37.

Longmore, R. B., & Gardner, D. L. (1978). The surface structure of ageing human articular cartilage: A study by reflected light interference microscopy (RLIM). *Journal of Anatomy, 126,* 353–365.

Lories, R. J., Corr, M., & Lane, N. E. (2013). To Wnt or not to Wnt: The bone and joint health dilemma. *Nature Reviews Rheumatology, 9,* 328–339.

Lories, R. J. U. (2012). Changing the outcome of osteoarthritis: Still a challenge for cyclooxygenase 2 inhibitors. *Arthritis and Rheumatism, 64,* 37–39.

Losina, E., Usiskin, I. M., Smith, S. R., Sullivan, J. K., Smith, K. C., Hunter, D. J., et al. (2018). Cost-effectiveness of generic celecoxib in knee osteoarthritis for average-risk patients: A model-based evaluation. *Osteoarthritis Cartilage, 26,* 641–650.

Lotz, M., Hashimoto, S., & Kuhn, K. (1999). Mechanisms of chondrocyte apoptosis. *Osteoarthritis Cartilage, 7,* 389–391.

Lotz, M., Martel-Pelletier, J., Christiansen, C., Brandi, M. L., Bruyere, O., Chapurlat, R., et al. (2014). Republished: Value of biomarkers in osteoarthritis: Current status and perspectives. *Postgraduate Medical Journal, 90,* 171–178.

Loughlin, J. (2015). Genetic contribution to osteoarthritis development: Current state of evidence. *Current Opinion in Rheumatology, 27,* 284–288.

Lourido, L., Calamia, V., Mateos, J., Fernandez-Puente, P., Fernandez-Tajes, J., Blanco, F. J., et al. (2014). Quantitative proteomic profiling of human articular cartilage degradation in osteoarthritis. *Journal of Proteome Research, 13,* 6096–6106.

Lu, B. Z., Cheng, X. L., Hou, T. J., & Mccammon, J. A. (2005). Calculation of the Maxwell stress tensor and the Poisson-Boltzmann force on a solvated molecular surface using hypersingular boundary integrals. *Journal of Chemical Physics, 123,* 084904.

Lu, C. H., Yeh, T. S., Yeh, C. L., Fang, Y. H. D., Sung, L. Y., Lin, S. Y., et al. (2014). Regenerating cartilages by engineered ASCs: Prolonged TGF-beta 3/BMP-6 expression improved articular cartilage formation and restored zonal structure. *Molecular Therapy, 22,* 186–195.

Lu, D., Bao, D., Dong, W., Liu, N., Zhang, X., Gao, S., et al. (2016). Dkk3 prevents familial dilated cardiomyopathy development through Wnt pathway. *Laboratory Investigation, 96,* 239–248.

Ludbrook, J. (2012). A primer for biomedical scientists on how to execute model II linear regression analysis. *Clinical and Experimental Pharmacology and Physiology, 39,* 329–335.

Lundberg, C., Asberg, I., Ionescu, M., Reiner, A., Smedegard, G., & Poole, A. R. (1996). Changes in cartilage proteoglycan aggrecan after intra-articular injection of interleukin-1 in rabbits: Studies of synovial fluid and articular cartilage. *Annals of the Rheumatic Diseases, 55,* 525–534.

Lundberg, H. J., Stewart, K. J., Callaghan, J. J., & Brown, T. A. (2005). Kinetically critical sites of femoral head roughening for wear rate acceleration in total hip arthroplasty. *Clinical Orthopaedics and Related Research*, 89–93.

Lundborg, G., & Rank, F. (1978). Experimental intrinsic healing of flexor tendons based upon synovial fluid nutrition. *Journal of Hand Surgery American, 3,* 21–31.

Luo, S., Shi, Q., Chen, J., Wang, H., Wu, W., & Zha, Z. (2017). Expression and significance of MMPs in synovial fluid, serum and PBMC culture supernatant stimulated by LPS in osteoarthritis patients with or without diabetes. *Experimental and Clinical Endocrinology & Diabetes.*

Luo, Y., Jiang, W., Yu, H. B., Mackerell, A. D., & Roux, B. (2013). Simulation study of ion pairing in concentrated aqueous salt solutions with a polarizable force field. *Faraday Discussions, 160,* 135–149.

Luo, Y., & Roux, B. (2010). Simulation of osmotic pressure in concentrated aqueous salt solutions. *Journal of Physical Chemistry Letters, 1,* 183–189.

Luyten, F. P., Tylzanowski, P., & Lories, R. J. (2009). Wnt signaling and osteoarthritis. *Bone, 44,* 522–527.

Lyman, J. R., Chappell, J. D., Morales, T. I., Kelley, S. S., & Lee, G. M. (2012). Response of chondrocytes to local mechanical injury in an ex vivo model. *Cartilage, 3,* 12.

Lynn, K. R., & Labow, R. S. (1984). A comparison of four sulfhydryl cathepsins (B, C, H, and L) from porcine spleen. *Canadian Journal of Biochemistry and Cell Biology, 62,* 1301–1308.

Lyons, T. J., Mcclure, S. F., Stoddart, R. W., & Mcclure, J. (2006). The normal human chondro-osseous junctional region: Evidence for contact of uncalcified cartilage with subchondral bone and marrow spaces. *BMC Musculoskeletal Disorders, 7.*

Ma, B., Landman, E. B., Miclea, R. L., Wit, J. M., Robanus-Maandag, E. C., Post, J. N., et al. (2013). WNT signaling and cartilage: Of mice and men. *Calcified Tissue International, 92,* 399–411.

Ma, H., Marti-Gutierrez, N., Park, S. W., Wu, J., Lee, Y., Suzuki, K., et al. (2017). Correction of a pathogenic gene mutation in human embryos. *Nature.*

Macdonald, B. T., Tamai, K., & He, X. (2009). Wnt/beta-catenin signaling: Components, mechanisms, and diseases. *Developmental Cell, 17,* 9–26.

Mackie, E. J., Tatarczuch, L., & Mirams, M. (2011). The skeleton: A multi-functional complex organ: The growth plate chondrocyte and endochondral ossification. *Journal of Endocrinology, 211,* 109–121.

Macminn, C. W., Dufresne, E. R., & Wettlaufer, J. S. (2016). Large deformations of a soft porous material. *Physical Review Applied, 5.*

Madden, R. M., Han, S. K., & Herzog, W. (2014). The effect of compressive loading magnitude on in situ chondrocyte calcium signaling. *Biomechanics and Modeling in Mechanobiology.*

Maeda, T., Sakabe, T., Sunaga, A., Sakai, K., Rivera, A. L., Keene, D. R., et al. (2011). Conversion of mechanical force into TGF-beta-mediated biochemical signals. *Current Biology, 21,* 933–941.

Mahajan, A., Verma, S., & Tandon, V. (2005). Osteoarthritis. *Journal of the Association of Physicians of India, 53,* 634–641.

Mahan, B., & Hmyers, R. J. (1987). *University chemistry*. Menlo Park, CA: Benjamin/Cummings Pub. Co.

Majumder, M., Xin, X. P., Liu, L., Tutunea-Fatan, E., Rodriguez-Torres, M., Vincent, K., et al. (2016). COX-2 induces breast cancer stem cells via EP4/PI3K/AKT/NOTCH/WNT axis. *Stem Cells, 34,* 2290–2305.

Makogonenko, E., Tsurupa, G., Ingham, K., & Medved, L. (2002). Interaction of fibrin (ogen) with fibronectin: Further characterization and localization of the fibronectin-binding site. *Biochemistry, 41,* 7907–7913.

Makris, E. A., Responte, D. J., Paschos, N. K., Hu, J. C., & Athanasiou, K. A. (2015). Developing functional musculoskeletal tissues through hypoxia and lysyl oxidase-induced collagen cross-linking (Vol. 111, pp. E4832, 2014). In *Proceedings of the National Academy of Sciences of the United States of America* (Vol. 12, pp. E2978–E2978).

Malda, J., Benders, K. E., Klein, T. J., de Grauw, J. C., Kik, M. J., Hutmacher, D. W., et al. (2012). Comparative study of depth-dependent characteristics of equine and human osteochondral tissue from the medial and lateral femoral condyles. *Osteoarthritis Cartilage, 20,* 1147–1151.

Malda, J., de Grauw, J. C., Benders, K. E., Kik, M. J., van de Lest, C. H., Creemers, L. B., et al. (2013). Of mice, men and elephants: The relation between articular cartilage thickness and body mass. *PLoS ONE, 8,* e57683.

Malemud, C. J. (2015). Biologic basis of osteoarthritis: State of the evidence. *Current Opinion in Rheumatology, 27,* 289–294.

Malpas, S. (2009). Editorial comment: Montani versus Osborn exchange of views. *Experimental Physiology, 94,* 381–382.

Mann, H. H., Ozbek, S., Engel, J., Paulsson, M., & Wagener, R. (2004). Interactions between the cartilage oligomeric matrix protein and matrilins—Implications for matrix assembly and the pathogenesis of chondrodysplasias. *Journal of Biological Chemistry, 279,* 25294–25298.

Manoranjan, B., Venugopal, C., Mahendram, S., Subapanditha, M., Savage, N., Qazi, M., et al. (2017). A CD133-Akt-Wnt signaling axis provides functional insight into the role of CD133 in glioblastoma brain tumor-initiating cells. *Neuro-Oncology, 19,* 54.

Mariani, E., Pulsatelli, L., & Facchini, A. (2014). Signaling pathways in cartilage repair. *International Journal of Molecular Sciences, 15,* 8667–8698.

Markoulli, M., Papas, E., Cole, N., & Holden, B. A. (2012). The diurnal variation of matrix metalloproteinase-9 and its associated factors in human tears. *Investigative Ophthalmology & Visual Science, 53,* 1479–1484.

Markova, T. G. (2007). Genetic characteristics of hearing disorders in changes in genes responsible for collagen synthesis. *Vestnik Otorinolaringologii*, 17–21.

Maroudas, A. (1967). Hyaluronic acid films. In *Lubrication and Wear in Living and Artificial Joints: A Symposium.* Institution of Mechanical Engineers Proceedings.

Maroudas, A. (1975). Glycosaminoglycan turn-over in articular cartilage. *Philosophical Transactions of the Royal Society of London. Series B, Biological Sciences, 271,* 293–313.

Maroudas, A. I. (1976). Balance between swelling pressure and collagen tension in normal and degenerate cartilage. *Nature, 260,* 808–809.

Maroudas, A., Bayliss, M. T., Uchitel-Kaushansky, N., Schneiderman, R., & Gilav, E. (1998). Aggrecan turnover in human articular cartilage: Use of aspartic acid racemization as a marker of molecular age. *Archives of Biochemistry and Biophysics, 350,* 61–71.

Maroudas, A., Mizrahi, J., Benaim, E., Scheiderman, R., & Grushko, G. (1992). Swelling pressure of cartilage: Roles played by proteoglycans and collagen. In T. K. Karalis (Ed.), *Mechanics of swelling—From clays to living cells and tissues*. Berlin: Springer.

Maroudas, A., Wachtel, E., Grushko, G., Katz, E. P., & Weinberg, P. (1991). The effect of osmotic and mechanical pressures on water partitioning in articular-cartilage. *Biochimica et Biophysica Acta, 1073,* 285–294.

Martin, J. A., & Buckwalter, J. A. (2001). Telomere erosion and senescence in human articular cartilage chondrocytes. *Journals of Gerontology Series A—Biological Sciences and Medical Sciences, 56,* B172–B179.

Martin, J. A., & Buckwalter, J. A. (2002). Aging, articular cartilage chondrocyte senescence and osteoarthritis. *Biogerontology, 3,* 257–264.

Martinek, N., Shahab, J., Sodek, J., & Ringuette, M. (2007). Is SPARC an evolutionarily conserved collagen chaperone? *Journal of Dental Research, 86,* 296–305.

Martinez-Calatrava, M. J., Prieto-Potin, I., Roman-Blas, J. A., Tardio, L., Largo, R., & Herrero-Beaumont, G. (2012). RANKL synthesized by articular chondrocytes contributes to juxta-articular bone loss in chronic arthritis. *Arthritis Research & Therapy, 14,* R149.

Marzo, J. M., & Gurske-Deperio, J. (2009). Effects of medial meniscus posterior horn avulsion and repair on tibiofemoral contact area and peak contact pressure with clinical implications. *American Journal of Sports Medicine, 37,* 124–129.

Mastbergen, S. C., & Lafeber, F. P. (2011). Changes in subchondral bone early in the development of osteoarthritis. *Arthritis and Rheumatism, 63,* 2561–2563.

Mastbergen, S. C., Saris, D. B., & Lafeber, F. P. (2013). Functional articular cartilage repair: Here, near, or is the best approach not yet clear? *Nature Reviews Rheumatology, 9,* 277–290.

Matcher, S. J. (2015). What can biophotonics tell us about the 3D microstructure of articular cartilage? *Quantitative Imaging in Medicine and Surgery, 5,* 143–158.

Matsumoto, K., Li, Y., Jakuba, C., Sugiyama, Y., Sayo, T., Okuno, M., et al. (2009). Conditional inactivation of Has2 reveals a crucial role for hyaluronan in skeletal growth, patterning, chondrocyte maturation and joint formation in the developing limb. *Development, 136,* 2825–2835.

Matsushita, H., Blackburn, M. L., Klineberg, E., Zielinska-Kwiatkowska, A., Bolander, M. E., Sarkar, G., et al. (2007). TASR-1 regulates alternative splicing of collagen genes in chondrogenic cells. *Biochemical and Biophysical Research Communications, 356,* 411–417.

Matzat, S. J., van Tiel, J., Gold, G. E., & Oei, E. H. (2013). Quantitative MRI techniques of cartilage composition. *Quantitative Imaging in Medicine and Surgery, 3,* 162–174.

Matzelle, M. M., Gallant, M. A., Condon, K. W., Walsh, N. C., Manning, C. A., Stein, G. S., et al. (2012). Resolution of inflammation induces osteoblast function and regulates the Wnt signaling pathway. *Arthritis and Rheumatism, 64,* 1540–1550.

McAlinden, A., Varghese, N., Wirthlin, L., & Chang, L. W. (2013). Differentially expressed microRNAs in chondrocytes from distinct regions of developing human cartilage. *PLoS ONE, 8,* e75012.

McCall, J. (2005). Genetic algorithms for modeling and optimisation. *Journal of Computational and Applied Mathematics, 184,* 205–222.

McCann, L., Ingham, E., Jin, Z., & Fisher, J. (2009). Influence of the meniscus on friction and degradation of cartilage in the natural knee joint. *Osteoarthritis Cartilage, 17,* 995–1000.

McCutchen, C. W. (1959). Sponge-hydrostatic and weeping bearings. *Nature, 184,* 1284–1285.

McCutchen, C. W. (1983). Joint lubrication. *Bulletin of the Hospital for Joint Diseases Orthopaedic Institute, 43,* 118–129.

McCutchen, C. W. (1998). A theoretical formulation for boundary friction in articular cartilage. *Journal of Biomechanical Engineering, 120,* 547.

McCutchen, C. W. (2011). Joints are not lubricated in the way Greene et al. say they are. *Proceedings of the National Academy of Sciences of the United States of America, 108,* E461; author reply E462.

McGill, S. M., & Cholewicki, J. (2001). Biomechanical basis for stability: An explanation to enhance clinical utility. *Journal of Orthopaedic and Sports Physical Therapy, 31,* 96–100.

McIntyre, M. E. (2017). On multi-level thinking and scientific understanding. *Advances in Atmospheric Sciences, 34,* 1150–1158.

McGregor, A. J., Amsden, B. G., & Waldman, S. D. (2011). Chondrocyte repopulation of the zone of death induced by osteochondral harvest. *Osteoarthritis Cartilage, 19,* 242–248.

McNabb, A. (1960). A mathematical treatment of one-dimensional soil consolidation. *Quarterly of Applied Mathematics, 17,* 337–347.

McNally, D. S., Adams, M. A., & Goodship, A. E. (1992). Development and validation of a new transducer for intradiscal pressure measurement. *Journal of Biomedical Engineering, 14,* 495–498.

McNary, S. M., Athanasiou, K. A., & Reddi, A. H. (2012). Engineering lubrication in articular cartilage. *Tissue Engineering Part B: Reviews, 18,* 88–100.

McNary, S. M., Athanasiou, K. A., & Reddi, A. H. (2014). Transforming growth factor beta-induced superficial zone protein accumulation in the surface zone of articular cartilage is dependent on the cytoskeleton. *Tissue Engineering Part A, 20,* 921–929.

McNulty, A. L., Rothfusz, N. E., Leddy, H. A., & Guilak, F. (2013). Synovial fluid concentrations and relative potency of interleukin-1 alpha and beta in cartilage and meniscus degradation. *Journal of Orthopaedic Research, 31,* 1039–1045.

Meachim, G., & Stiockwell, R. A. (1979). The matrix. In M. A. R. Freeman (Ed.), *Adult articular cartilage* (2nd ed.) Tunbridge Wells; United Kingdom: Pitman Medical.

Meechai, N., Jamieson, A. M., Blackwell, J., Carrino, D. A., & Bansal, R. (2001). Nonlinear viscoelasticity of concentrated solutions of aggrecan aggregate. *Biomacromolecules, 2,* 780–787.

Mehdizadeh, A., Gardiner, B. S., Lavagnino, M., & Smith, D. W. (2017). Predicting tenocyte expression profiles and average molecular concentrations in Achilles tendon ECM from tissue strain and fiber damage. *Biomechanics and Modeling in Mechanobiology, 16,* 1329–1348.

Mehdizadeh, A., Gardiner, B. S., Lavagnino, M., & Smith, D. W. (2018). Effect of collagen length distribution and timing for repair on the active TGF-beta concentration in tendon. *Connective Tissue Research, 59*(5), 396–409.

Melas, I. N., Chairakaki, A. D., Chatzopoulou, E. I., Messinis, D. E., Katopodi, T., Pliaka, V., et al. (2014). Modeling of signaling pathways in chondrocytes based on phosphoproteomic and cytokine release data. *Osteoarthritis Cartilage, 22,* 509–518.

Melcher, J. R. (1981). *Continuum electromechanics*. Cambridge: MIT Press.

Melrose, J., Fuller, E. S., Roughley, P. J., Smith, M. M., Kerr, B., Hughes, C. E., et al. (2008). Fragmentation of decorin, biglycan, lumican and keratocan is elevated in degenerate human meniscus, knee and hip articular cartilages compared with age-matched macroscopically normal and control tissues. *Arthritis Research & Therapy, 10,* R79.

Melrose, J., Hayes, A. J., Whitelock, J. M., & Little, C. B. (2008). Perlecan, the "jack of all trades" proteoglycan of cartilaginous weight-bearing connective tissues. *BioEssays, 30,* 457–469.

Men, Y. T., Jiang, Y. L., Chen, L., Zhang, C. Q., & Ye, J. D. (2017). On mechanical mechanism of damage evolution in articular cartilage. *Materials Science & Engineering C-Materials for Biological Applications, 78,* 79–87.

Menendez, M. I., Clark, D. J., Carlton, M., Flanigan, D. C., Jia, G., Sammet, S., et al. (2011). Direct delayed human adenoviral BMP-2 or BMP-6 gene therapy for bone and cartilage regeneration in a pony osteochondral model. *Osteoarthritis Cartilage, 19,* 1066–1075.

Merkher, Y., Sivan, S., Etsion, I., Maroudas, A., Halperin, G., & Yosef, A. (2006). A rational human joint friction test using a human cartilage-on-cartilage arrangement. *Tribology Letters, 22,* 29–36.

Merline, R., Schaefer, R. M., & Schaefer, L. (2009). The matricellular functions of small leucine-rich proteoglycans (SLRPs). *Journal of Cell Communication and Signaling, 3,* 323–335.

Mertz, E. L., Facchini, M., Pham, A. T., Gualeni, B., de Leonardis, F., Rossi, A., et al. (2012). Matrix disruptions, growth, and degradation of cartilage with impaired sulfation. *Journal of Biological Chemistry, 287,* 22030–22042.

Meulenbelt, I., Min, J. L., Bos, S., Riyazi, N., Houwing-Duistermaat, J. J., van der Wijk, H. J., et al. (2008). Identification of DIO2 as a new susceptibility locus for symptomatic osteoarthritis. *Human Molecular Genetics, 17,* 1867–1875.

Mickiewicz, B., Shin, S. Y., Pozzi, A., Vogel, H. J., & Clark, A. L. (2016). Serum metabolite profiles are altered by erlotinib treatment and the integrin alpha1-null genotype but not by post-traumatic osteoarthritis. *Journal of Proteome Research, 15,* 815–825.

Miller, M. C., Manning, H. B., Jain, A., Troeberg, L., Dudhia, J., Essex, D., et al. (2009). Membrane type 1 matrix metalloproteinase is a crucial promoter of synovial invasion in human rheumatoid arthritis. *Arthritis and Rheumatism, 60,* 686–697.

Milner, P. I., Wilkins R. J., & Gibson, J. S. (2012). Principles of osteoarthritis: Its definition, character, derivation and modality-related recognition. In B. M. Rothschild (Ed.), *Cellular physiology of articular cartilage in health and disease.* InTech.

Milo, R., Shen-Orr, S., Itzkovitz, S., Kashtan, N., Chklovskii, D., & Alon, U. (2002). Network motifs: Simple building blocks of complex networks. *Science, 298,* 824–827.

Minarelli, A. M., & Liberti, E. A. (1997). A microscopic survey of the human temporomandibular joint disc. *Journal of Oral Rehabilitation, 24,* 835–840.

Minina, E., Kreschel, C., Naski, M. C., Ornitz, D. M., & Vortkamp, A. (2002). Interaction of FGF, Ihh/Pthlh, and BMP signaling integrates chondrocyte proliferation and hypertrophic differentiation. *Developmental Cell, 3,* 439–449.

Miramini, S., Smith, D. W., Zhang, L., & Gardiner, B. S. (2017). The spatio-temporal mechanical environment of healthy and injured human cartilage during sustained activity and its role in cartilage damage. *Journal of the Mechanical Behavior of Biomedical Materials, 74,* 1–10.

Mitchell, N. S., & Cruess, R. L. (1977). Classification of degenerative arthritis. *Canadian Medical Association Journal, 117,* 763–765.

Miwa, M., Saura, R., Hirata, S., Hayashi, Y., Mizuno, K., & Itoh, H. (2000). Induction of apoptosis in bovine articular chondrocyte by prostaglandin E(2) through cAMP-dependent pathway. *Osteoarthritis Cartilage, 8,* 17–24.

Miyaki, S., Sato, T., Inoue, A., Otsuki, S., Ito, Y., Yokoyama, S., et al. (2010). MicroRNA-140 plays dual roles in both cartilage development and homeostasis. *Genes & Development, 24,* 1173–1185.

Mobasheri, A., Vannucci, S. J., Bondy, C. A., Carter, S. D., Innes, J. F., Arteaga, M. F., et al. (2002). Glucose transport and metabolism in chondrocytes: A key to understanding chondrogenesis, skeletal development and cartilage degradation in osteoarthritis. *Histology and Histopathology, 17,* 1239–1267.

Modell, H., Cliff, W., Michael, J., McFarland, J., Wenderoth, M. P., & Wright, A. (2015). A physiologist's view of homeostasis. *Advances in Physiology Education, 39,* 259–266.

Moffatt, P., Lee, E. R., St-Jacques, B., Matsumoto, K., Yamaguchi, Y., & Roughley, P. J. (2011). Hyaluronan production by means of Has2 gene expression in chondrocytes is essential for long bone development. *Developmental Dynamics, 240,* 404–412.

Mohan, S., & Baylink, D. J. (2002). IGF-binding proteins are multifunctional and act via IGF-dependent and -independent mechanisms. *Journal of Endocrinology, 175,* 19–31.

Mok, S. S., Masuda, K., Hauselmann, H. J., Aydelotte, M. B., & Thonar, E. J. (1994). Aggrecan synthesized by mature bovine chondrocytes suspended in alginate. Identification of two distinct metabolic matrix pools. *Journal of Biological Chemistry, 269,* 33021–33027.

Moller, H. J., Larsen, F. S., Ingemann-Hansen, T., & Poulsen, J. H. (1994). ELISA for the core protein of the cartilage large aggregating proteoglycan, aggrecan: Comparison with the concentrations of immunogenic keratan sulphate in synovial fluid, serum and urine. *Clinica Chimica Acta, 225,* 43–55.

Mononen, M. E., Tanska, P., Isaksson, H., & Korhonen, R. K. (2016). A Novel method to simulate the progression of collagen degeneration of cartilage in the knee: Data from the osteoarthritis initiative. *Scientific Reports, 6.*

Montaseri, A., Busch, F., Mobasheri, A., Buhrmann, C., Aldinger, C., Rad, J. S., et al. (2011). IGF-1 and PDGF-bb suppress IL-1beta-induced cartilage degradation through down-regulation of NF-kappaB signaling: Involvement of Src/PI-3K/AKT pathway. *PLoS ONE, 6,* e28663.

Moore, A. C., & Burris, D. L. (2017). Tribological rehydration of cartilage and its potential role in preserving joint health. *Osteoarthritis and Cartilage, 25,* 99–107.

Moore, E. E., Bendele, A. M., Thompson, D. L., Littau, A., Waggie, K. S., Reardon, B., et al. (2005). Fibroblast growth factor-18 stimulates chondrogenesis and cartilage repair in a rat model of injury-induced osteoarthritis. *Osteoarthritis Cartilage, 13,* 623–631.

Morales, T. I. (2007). Chondrocyte moves: Clever strategies? *Osteoarthritis and Cartilage, 15,* 861–871.

Moreno-Rubio, J., Herrero-Beaumont, G., Tardio, L., Alvarez-Soria, M. A., & Largo, R. (2010). Nonsteroidal antiinflammatory drugs and prostaglandin E(2) modulate the synthesis of osteoprotegerin and RANKL in the cartilage of patients with severe knee osteoarthritis. *Arthritis and Rheumatism, 62,* 478–488.

Morgan, W. H., Yu, D. Y., Alder, V. A., Cringle, S. J., Cooper, R. L., House, P. H., et al. (1998). The correlation between cerebrospinal fluid pressure and retrolaminar tissue pressure. *Investigative Ophthalmology & Visual Science, 39,* 1419–1428.

Morgunova, E., Tuuttila, A., Bergmann, U., & Tryggvason, K. (2002). Structural insight into the complex formation of latent matrix metalloproteinase 2 with tissue inhibitor of metalloproteinase 2. *Proceedings of the National Academy of Sciences of the United States of America, 99,* 7414–7419.

Mork, P. J., Holtermann, A., & Nilsen, T. I. (2012). Effect of body mass index and physical exercise on risk of knee and hip osteoarthritis: Longitudinal data from the Norwegian HUNT Study. *Journal of Epidemiology and Community Health, 66,* 678–683.

Morko, J. P., Soderstrom, M., Saamanen, A. M. K., Salminen, H. J., & Vuorio, E. I. (2004). Up regulation of cathepsin K expression in articular chondrocytes in a transgenic mouse model for osteoarthritis. *Annals of the Rheumatic Diseases, 63,* 649–655.

Moroco, J. R., Hinton, R., Buschang, P., Milam, S. B., & Iacopino, A. M. (1997). Type II collagen and TGF-betas in developing and aging porcine mandibular condylar cartilage: Immunohistochemical studies. *Cell and Tissue Research, 289,* 119–124.

Morrell, K. C., Hodge, W. A., Krebs, D. E., & Mann, R. W. (2005). Corroboration of in vivo cartilage pressures with implications for synovial joint tribology and osteoarthritis causation. *Proceedings of the National Academy of Sciences of the United States of America, 102,* 14819–14824.

Morrison, D. K. (2012). MAP kinase pathways. *Cold Spring Harbor Perspectives in Biology, 4.*

Moseley, H., Foulds, W. S., Allan, D., & Kyle, P. M. (1984). Routes of clearance of radioactive water from the rabbit vitreous. *British Journal of Ophthalmology, 68,* 145–151.

Mott, J. D., & Werb, Z. (2004). Regulation of matrix biology by matrix metalloproteinases. *Current Opinion in Cell Biology, 16,* 558–564.

Mout, R., Ray, M., Lee, Y. W., Scaletti, F., & Rotello, V. M. (2017). In vivo delivery of CRISPR/Cas9 for therapeutic gene editing: Progress and challenges. *Bioconjugate Chemistry, 28,* 880–884.

Moutos, F. T., Glass, K. A., Compton, S. A., Ross, A. K., Gersbach, C. A., Guilak, F., et al. (2016). Anatomically shaped tissue-engineered cartilage with tunable and inducible anticytokine delivery for biological joint resurfacing. *Proceedings of the National Academy of Sciences of the United States of America, 113,* E4513–E4522.

Mow, V. C., Ratcliffe, A., & Poole, A. R. (1992). Cartilage and diarthrodial joints as paradigms for hierarchical materials and structures. *Biomaterials, 13,* 67–97.

Muehleman, C., Bareither, D., Huch, K., Cole, A. A., & Kuettner, K. E. (1997). Prevalence of degenerative morphological changes in the joints of the lower extremity. *Osteoarthritis and Cartilage, 5,* 23–37.

Muinos-Lopez, E., Rendal-Vazquez, M. E., Hermida-Gomez, T., Fuentes-Boquete, I., Diaz-Prado, S., & Blanco, F. J. (2012). Cryopreservation effect on proliferative and chondrogenic potential of human chondrocytes isolated from superficial and deep cartilage. *Open Orthopaedics Journal, 6,* 150–159.

Muir, H. (1995). The chondrocyte, architect of cartilage—Biomechanics, structure, function and molecular-biology of cartilage matrix macromolecules. *BioEssays, 17,* 1039–1048.

Muldrew, K., Chung, M., Novak, K., Schachar, N. S., Zernicke, R. F., McGann, L. E., et al. (2001). Evidence of chondrocyte repopulation in adult ovine articular cartilage following cryoinjury and long-term transplantation. *Osteoarthritis Cartilage, 9,* 432–439.

Munakata, H., Takagaki, K., Majima, M., & Endo, M. (1999). Interaction between collagens and glycosaminoglycans investigated using a surface plasmon resonance biosensor. *Glycobiology, 9,* 1023–1027.

Munugoda, I. P., Wills, K., Cicuttini, F., Graves, S. E., Lorimer, M., Jones, G., et al. (2018). The association between ambulatory activity, body composition and hip or knee joint replacement due to osteoarthritis: A prospective cohort study. *Osteoarthritis Cartilage, 26,* 671–679.

Murakami, T., Nakashima, K., Sawae, Y., Sakai, N., Hosoda, N., & Klein, J. (2009). Roles of adsorbed film and gel layer in hydration lubrication for articular cartilage. *Proceedings of the Institution of Mechanical Engineers Part J—Journal of Engineering Tribology, 223,* 603–604.

Murphy, G. (2011). Tissue inhibitors of metalloproteinases. *Genome Biology, 12,* R233.

Murphy, G., & Lee, M. H. (2005). What are the roles of metalloproteinases in cartilage and bone damage? *Annals of the Rheumatic Diseases, 64,* 44–47.

Murphy, G., & Nagase, H. (2008). Reappraising metalloproteinases in rheumatoid arthritis and osteoarthritis: Destruction or repair? *Nature Clinical Practice Rheumatology, 4,* 128–135.

Murphy-Ryan, M., Psychogios, A., & Lindor, N. M. (2010). Hereditary disorders of connective tissue: A guide to the emerging differential diagnosis. *Genetics in Medicine, 12,* 344–354.

Murray, R. C., Smith, R. K., Henson, F. M., & Goodship, A. (2001). The distribution of cartilage oligomeric matrix protein (COMP) in equine carpal articular cartilage and its variation with exercise and cartilage deterioration. *The Veterinary Journal, 162,* 121–128.

Myers, D. B., Marshall, R. N., & Palmer, D. G. (1983). Morphological and biochemical comparison of convex and concave articular surfaces from adult subtalar and midtarsal joints. *Scandinavian Journal of Rheumatology, 12,* 119–123.

Myllarniemi, M., Lindholm, P., Ryynanen, M. J., Kliment, C. R., Salmenkivi, K., Keski-Oja, J., et al. (2008). Gremlin-mediated decrease in bone morphogenetic protein signaling promotes pulmonary fibrosis. *American Journal of Respiratory and Critical Care Medicine, 177,* 321–329.

Nabeshima, Y., Grood, E. S., Sakurai, A., & Herman, J. H. (1996). Uniaxial tension inhibits tendon collagen degradation by collagenase in vitro. *Journal of Orthopaedic Research, 14,* 123–130.

Nadzir, M. M., Kino-Oka, M., Sugawara, K., & Taya, M. (2013). Modulation of chondrocyte migration and aggregation by insulin-like growth factor-1 in cultured cartilage. *Biotechnology Letters, 35,* 295–300.

Nagai, T., Sato, M., Furukawa, K. S., Kutsuna, T., Ohta, N., Ushida, T., et al. (2008). Optimization of allograft implantation using scaffold-free chondrocyte plates. *Tissue Engineering Part A, 14,* 1225–1235.

Nagase, H. (2001). Substrate specificity of MMPs. In N. J. Clendeninn & K. Appelt (Eds.), *Matrix metalloproteinase inhibitors in cancer therapy.* Humana Press.

Nagase, H., & Kashiwagi, M. (2003). Aggrecanases and cartilage matrix degradation. *Arthritis Research & Therapy, 5,* 94–103.

Nagase, H., Visse, R., & Murphy, G. (2006). Structure and function of matrix metalloproteinases and TIMPs. *Cardiovascular Research, 69,* 562–573.

Naito, S., Shiomi, T., Okada, A., Kimura, T., Chijiiwa, M., Fujita, Y., et al. (2007). Expression of ADAMTS4 (aggrecanase-1) in human osteoarthritic cartilage. *Pathology International, 57,* 703–711.

Naka, M. H., Morita, Y., & Ikeuchi, K. (2005). Influence of proteoglycan contents and of tissue hydration on the frictional characteristics of articular cartilage. *Proceedings of the Institution of Mechanical Engineers Part H—Journal of Engineering in Medicine, 219,* 175–182.

Nakagawa, K., Sakiyama, H., Tsuchida, T., Yamaguchi, K., Toyoguchi, T., Masuda, R., et al. (1999). Complement C1s activation in degenerating articular cartilage of rheumatoid arthritis patients: Immunohistochemical studies with an active form specific antibody. *Annals of the Rheumatic Diseases, 58,* 175–181.

Nakashima, T., Hayashi, M., Fukunaga, T., Kurata, K., Oh-Hora, M., Feng, J. Q., et al. (2011). Evidence for osteocyte regulation of bone homeostasis through RANKL expression. *Nature Medicine, 17,* 1231–1234.

Nap, R. J., & Szleifer, I. (2008). Structure and interactions of aggrecans: Statistical thermodynamic approach. *Biophysical Journal, 95,* 4570–4583.

Nardozza Junior, A., Szelbracikowski Sdos, S., Nardi, A. C., & Almeida, J. C. (2011). Age-related testosterone decline in a Brazilian cohort of healthy military men. *International Brazilian Journal of Urology, 37,* 591–597.

Necas, J., Bartosikova, L., Brauner, P., & Kolar, J. (2008). Hyaluronic acid (hyaluronan): A review. *Veterinarni Medicina, 53,* 397–411.

Neff, P., Eidel, B., & Martin, R. J. (2016). Geometry of logarithmic strain measures in solid mechanics. *Archive for Rational Mechanics and Analysis, 222,* 507–572.

Negoro, K., Kobayashi, S., Takeno, K., Uchida, K., & Baba, H. (2008). Effect of osmolarity on glycosaminoglycan production and cell metabolism of articular chondrocyte under three-dimensional culture system. *Clinical and Experimental Rheumatology, 26,* 534–541.

Nelson, F., Dahlberg, L., Laverty, S., Reiner, A., Pidoux, I., Ionescu, M., et al. (1998). Evidence for altered synthesis of type II collagen in patients with osteoarthritis. *Journal of Clinical Investigation, 102,* 2115–2125.

Nelson, P. H. (2017). Osmosis and thermodynamics explained by solute blocking. *European Biophysics Journal with Biophysics Letters, 46,* 59–64.

Nematnasser, S. (1979). Decomposition of strain measures and their rates in finite deformation elasto-plasticity. *International Journal of Solids and Structures, 15,* 155–166.

Neogi, T. (2013). The epidemiology and impact of pain in osteoarthritis. *Osteoarthritis and Cartilage, 21,* 1145–1153.

Neph, S., Stergachis, A. B., Reynolds, A., Sandstrom, R., Borenstein, E., & Stamatoyannopoulos, J. A. (2012). Circuitry and dynamics of human transcription factor regulatory networks. *Cell, 150,* 1274–1286.

Nerlich, A. G., Boos, N., Wiest, I., & Aebi, M. (1998). Immunolocalization of major interstitial collagen types in human lumbar intervertebral discs of various ages. *Virchows Archiv—An International Journal of Pathology, 432,* 67–76.

Neu, C. P., Khalafi, A., Komvopoulos, K., Schmid, T. M., & Reddi, A. H. (2007). Mechanotransduction of bovine articular cartilage superficial zone protein by transforming growth factor beta signaling. *Arthritis and Rheumatism, 56,* 3706–3714.

Neu, C. P., Komvopoulos, K., & Reddi, A. H. (2008). The interface of functional biotribology and regenerative medicine in synovial joints. *Tissue Engineering Part B: Reviews, 14,* 235–247.

Neu, C. P., Reddi, A. H., Komvopoulos, K., Schmid, T. M., & di Cesare, P. E. (2010). Increased friction coefficient and superficial zone protein expression in patients with advanced osteoarthritis. *Arthritis and Rheumatism, 62,* 2680–2687.

Neuhold, L. A., Killar, L., Zhao, W., Sung, M. L., Warner, L., Kulik, J., et al. (2001). Postnatal expression in hyaline cartilage of constitutively active human collagenase-3 (MMP-13) induces osteoarthritis in mice. *The Journal of Clinical Investigation, 107,* 35–44.

Neuman, P., Dahlberg, L. E., Englund, M., & Struglics, A. (2017). Concentrations of synovial fluid biomarkers and the prediction of knee osteoarthritis 16 years after anterior cruciate ligament injury. *Osteoarthritis and Cartilage, 25,* 492–498.

Newman, A. P. (1998). Articular cartilage repair. *American Journal of Sports Medicine, 26,* 309–324.

Newton, P. T., Xie, M., Medvedeva, E., Savendahl, L., & Chagin, A. S. (2018). Activation of mTORC1 in chondrocytes does not affect proliferation or differentiation, but causes the resting zone of the growth plate to become disordered. *Bone Reports,* 8.

Ng, A. F., Yang, Y. O., Wong, R. W., Hagg, E. U., & Rabie, A. B. (2006). Factors regulating condylar cartilage growth under repeated load application. *Frontiers in Bioscience, 11,* 949–954.

Ng, C. P., & Swartz, M. A. (2003). Fibroblast alignment under interstitial fluid flow using a novel 3-D tissue culture model. *American Journal of Physiology-Heart and Circulatory Physiology, 284,* H1771–H1777.

Ng, K. W., Wanivenhaus, F., Chen, T., Hsu, H. C., Allon, A. A., Abrams, V. D., et al. (2012). A novel macroporous polyvinyl alcohol scaffold promotes chondrocyte migration and interface formation in an in vitro cartilage defect model. *Tissue Engineering Part A, 18,* 1273–1281.

Ng, L., Grodzinsky, A. J., Patwari, P., Sandy, J., Plaas, A., & Ortiz, C. (2003). Individual cartilage aggrecan macromolecules and their constituent glycosaminoglycans visualized via atomic force microscopy. *Journal of Structural Biology, 143,* 242–257.

Nguyen, L. T., Sharma, A. R., Chakraborty, C., Saibaba, B., Ahn, M. E., & Lee, S. S. (2017). Review of prospects of biological fluid biomarkers in osteoarthritis. *International Journal of Molecular Sciences,* 18.

Nguyen, Q. T., Wong, B. L., Chun, J., Yoon, Y. C., Talke, F. E., & Sah, R. L. (2010). Macroscopic assessment of cartilage shear: Effects of counter-surface roughness, synovial fluid lubricant, and compression offset. *Journal of Biomechanics, 43,* 1787–1793.

Nia, H. T., Bozchalooi, I. S., Li, Y., Han, L., Hung, H. H., Frank, E., et al. (2013). High-bandwidth AFM-based rheology reveals that cartilage is most sensitive to high loading rates at early stages of impairment. *Biophysical Journal, 104,* 1529–1537.

Nia, H. T., Han, L., Li, Y., Ortiz, C., & Grodzinsky, A. (2011). Poroelasticity of cartilage at the nanoscale. *Biophysical Journal, 101,* 2304–2313.

Nickien, M., Thambyah, A., & Broom, N. (2013). How changes in fibril-level organization correlate with the macrolevel behavior of articular cartilage. *Wiley Interdisciplinary Reviews: Systems Biology and Medicine, 5,* 495–509.

Niemela, O., Risteli, L., Parkkinen, J., & Risteli, J. (1985). Purification and characterization of the N-terminal propeptide of human type-III procollagen. *Biochemical Journal, 232,* 145–150.

Nieminen, H. J., Ylitalo, T., Karhula, S., Suuronen, J. P., Kauppinen, S., Serimaa, R., et al. (2015). Determining collagen distribution in articular cartilage using contrast-enhanced micro-computed tomography. *Osteoarthritis Cartilage, 23,* 1613–1621.

Nilsson, J., Noborn, F., Toledo, A. G., Nasir, W., Sihlbom, C., & Larson, G. (2017). Characterization of glycan structures of chondroitin sulfate-glycopeptides facilitated by sodium ion-pairing and positive mode LC-MS/MS. *Journal of the American Society for Mass Spectrometry, 28,* 229–241.

Niu, J. B., Clancy, M., Aliabadi, P., Vasan, R., & Felson, D. T. (2017). Metabolic syndrome, its components, and knee osteoarthritis: The framingham osteoarthritis study. *Arthritis & Rheumatology, 69,* 1194–1203.

Northrup, S. H., & Erickson, H. P. (1992). Kinetics of protein protein association explained by Brownian dynamics computer-simulation. *Proceedings of the National Academy of Sciences of the United States of America, 89,* 3338–3342.

Nose, M. (2011). A polygene network model for the complex pathological phenotypes of collagen disease. *Pathology International, 61,* 619–629.

Noskina, Y., & Cole, N. (2002). *Presence of proliferation markers in human articular chondrocytes.* Available: http://www.ors.org/Transactions/48/0375.pdf.

Nudelman, F., Lausch, A. J., Sommerdijk, N. A. J. M., & Sone, E. D. (2013). In vitro models of collagen biomineralization. *Journal of Structural Biology, 183,* 258–269.

Nugent, A. E., Speicher, D. M., Gradisar, I., McBurney, D. L., Baraga, A., Doane, K. J., et al. (2009). Advanced osteoarthritis in humans is associated with altered collagen VI expression and upregulation of ER-stress markers Grp78 and Bag-1. *Journal of Histochemistry and Cytochemistry, 57,* 923–931.

Oates, K. M., Krause, W. E., Jones, R. L., & Colby, R. H. (2006). Rheopexy of synovial fluid and protein aggregation. *Journal of the Royal Society, Interface, 3,* 167–174.

Ofek, G., Natoli, R. M., & Athanasiou, K. A. (2009). In situ mechanical properties of the chondrocyte cytoplasm and nucleus. *Journal of Biomechanics, 42,* 873–877.

Ogawa, H., Kozhemyakina, E., Hung, H. H., Grodzinsky, A. J., & Lassar, A. B. (2014). Mechanical motion promotes expression of Prg4 in articular cartilage via multiple CREB-dependent, fluid flow shear stress-induced signaling pathways. *Genes & Development, 28,* 127–139.

Okada, A., Mochizuki, S., Yatabe, T., Kimura, T., Shiomi, T., Fujita, Y., et al. (2008). ADAM-12 (meltrin alpha) is involved in chondrocyte proliferation via cleavage of insulin-like growth factor binding protein 5 in osteoarthritic cartilage. *Arthritis and Rheumatism, 58,* 778–789.

Okubo, N., Fujiwara, H., Minami, Y., Kunimoto, T., Hosokawa, T., Emura, Y. U., et al. (2015). Parathyroid hormone resets the cartilage circadian clock of the organ-cultured murine femur. *Acta Orthopaedica, 86,* 627–631.

Olin, A. I., Morgelin, M., Sasaki, T., Timpl, R., Heinegard, D., & Aspberg, A. (2001). The proteoglycans aggrecan and versican form networks with fibulin-2 through their lectin domain binding. *Journal of Biological Chemistry, 276,* 1253–1261.

Oloyede, A., & Broom, N. D. (1991). Is classical consolidation theory applicable to articular cartilage deformation? *Clinical Biomechanics, 6,* 206–212.

Oloyede, A., & Broom, N. D. (1994). The generalized consolidation of articular cartilage: An investigation of its near-physiological response to static load. *Connective Tissue Research, 31,* 75–86.

Onnerfjord, P., Khabut, A., Reinholt, F. P., Svensson, O., & Heinegard, D. (2012). Quantitative proteomic analysis of eight cartilaginous tissues reveals characteristic differences as well as similarities between subgroups. *Journal of Biological Chemistry, 287,* 18913–18924.

Orfanidou, T., Iliopoulos, D., Malizos, K. N., & Tsezou, A. (2009). Involvement of SOX-9 and FGF-23 in RUNX-2 regulation in osteoarthritic chondrocytes. *Journal of Cellular and Molecular Medicine, 13,* 3186–3194.

Ornitz, D. M., & Itoh, N. (2015). The fibroblast growth factor signaling pathway. *Wiley Interdisciplinary Reviews-Developmental Biology, 4,* 215–266.

Osborn, J. W., Averina, V. A., & Fink, G. D. (2009). Current computational models do not reveal the importance of the nervous system in long-term control of arterial pressure. *Experimental Physiology, 94,* 389–396.

Ota, N., Takaishi, H., Kosaki, N., Takito, J., Yoda, M., Tohmonda, T., et al. (2009). Accelerated cartilage resorption by chondroclasts during bone fracture healing in osteoprotegerin-deficient mice. *Endocrinology, 150,* 4823–4834.

Otsuka, S., Aoyama, T., Furu, M., Ito, K., Jin, Y., Nasu, A., et al. (2009). PGE2 signal via EP2 receptors evoked by a selective agonist enhances regeneration of injured articular cartilage. *Osteoarthritis and Cartilage, 17,* 529–538.

Otsuki, S., Hanson, S. R., Miyaki, S., Grogan, S. P., Kinoshita, M., Asahara, H., et al. (2010). Extracellular sulfatases support cartilage homeostasis by regulating BMP and FGF signaling pathways. *Proceedings of the National Academy of Sciences of the United States of America, 107,* 10202–10207.

Oungoulian, S. R., Chang, S., Bortz, O., Hehir, K. E., Zhu, K., Willis, C. E., et al. (2013). Articular cartilage wear characterization with a particle sizing and counting analyzer. *Journal of Biomechanical Engineering, 135,* 024501.

Oungoulian, S. R., Durney, K. M., Jones, B. K., Ahmad, C. S., Hung, C. T., & Ateshian, G. A. (2015). Wear and damage of articular cartilage with friction against orthopedic implant materials. *Journal of Biomechanics, 48,* 1957–1964.

Pabbruwe, M. B., Esfandiari, E., Kafienah, W., Tarlton, J. F., & Hollander, A. P. (2009). Induction of cartilage integration by a chondrocyte/collagen-scaffold implant. *Biomaterials, 30,* 4277–4286.

Pabinger, C., Berghold, A., Boehler, N., & Labek, G. (2013). Revision rates after knee replacement. Cumulative results from worldwide clinical studies versus joint registers. *Osteoarthritis Cartilage, 21,* 263–268.

Paetzold, H., Goepfert, C., Huber, G., Hoenig, E., Portner, R., Schilling, A. F., et al. (2012). The development of the collagen fibre network in tissue-engineered cartilage constructs in vivo. Engineered cartilage reorganises fibre network. *European Cells & Materials, 23,* 209–221.

Page-Mccaw, A., Ewald, A. J., & Werb, Z. (2007). Matrix metalloproteinases and the regulation of tissue remodeling. *Nature Reviews Molecular Cell Biology, 8,* 221–233.

Page-Thomas, D. P., Bard, D., King, B., & Dingle, J. T. (1987). Clearance of proteoglycan from joint cavities. *Annals of the Rheumatic Diseases, 46,* 934–937.

Palfi, V. K., & Perczel, A. (2010). Stability of the hydration layer of tropocollagen: A QM study. *Journal of Computational Chemistry, 31,* 764–777.

Palmer, G. D., Chao P. H., Raia, F., Mauck, R. L., Valhmu, W. B., & Hung, C. T. (2001). Time-dependent aggrecan gene expression of articular chondrocytes in response to hyperosmotic loading. *Osteoarthritis Cartilage*, 9, 761–770.

Pan, J., Zhou, X. Z., Li, W., Novotny, J. E., Doty, S. B., & Wang, L. Y. (2009). In situ measurement of transport between subchondral bone and articular cartilage. *Journal of Orthopaedic Research, 27,* 1347–1352.

Pang, X. C., Lin, L. J., & Tang, B. (2017). Unraveling the role of calcium ions in the mechanical properties of individual collagen fibrils. *Scientific Reports, 7.*

Panula, H. E., Hyttinen, M. M., Arokoski, J. P. A., Langsjo, T. K., Pelttari, A., Kiviranta, I., et al. (1998). Articular cartilage superficial zone collagen birefringence reduced and cartilage thickness increased before surface fibrillation in experimental osteoarthritis. *Annals of the Rheumatic Diseases, 57,* 237–245.

Panwar, P., Du, X., Sharma, V., Lamour, G., Castro, M., Li, H. B., et al. (2013). Effects of cysteine proteases on the structural and mechanical properties of collagen fibers. *Journal of Biological Chemistry, 288,* 5940–5950.

Pap, T., & Bertrand, J. (2013). Syndecans in cartilage breakdown and synovial inflammation. *Nature Reviews Rheumatology, 9,* 43–55.

Papalia, R., del Buono, A., Osti, L., Denaro, V., & Maffulli, N. (2011). Meniscectomy as a risk factor for knee osteoarthritis: A systematic review. *British Medical Bulletin, 99,* 89–106.

Park, S., Hung, C. T., & Ateshian, G. A. (2004). Mechanical response of bovine articular cartilage under dynamic unconfined compression loading at physiological stress levels. *Osteoarthritis Cartilage, 12,* 65–73.

Parkar, A. A., Kahmann, J. D., Howat, S. L. T., Bayliss, M. T., & Day, A. J. (1998). TSG-6 interacts with hyaluronan and aggrecan in a pH-dependent manner via a common functional element: Implications for its regulation in inflamed cartilage. *FEBS Letters, 428,* 171–176.

Parkes, M., Cann, P., & Jeffers, J. (2017). Real-time observation of fluid flows in tissue during stress relaxation using Raman spectroscopy. *Journal of Biomechanics, 60,* 261–265.

Parkkinen, J. J., Ikonen, J., Lammi, M. J., Laakkonen, J., Tammi, M., & Helminen, H. J. (1993). Effects of cyclic hydrostatic pressure on proteoglycan synthesis in cultured chondrocytes and articular cartilage explants. *Archives of Biochemistry and Biophysics, 300,* 458–465.

Patel, L., Sun, W., Glasson, S. S., Morris, E. A., Flannery, C. R., & Chockalingam, P. S. (2011). Tenascin-C induces inflammatory mediators and matrix degradation in osteoarthritic cartilage. *BMC Musculoskelet Disord, 12,* 164.

Patwari, P., Kurz, B., Sandy, J. D., & Grodzinsky, A. J. (2000). Mannosamine inhibits aggrecanase-mediated changes in the physical properties and biochemical composition of articular cartilage. *Archives of Biochemistry and Biophysics, 374,* 79–85.

Pearle, A. D., Warren, R. F., & Rodeo, S. A. (2005). Basic science of articular cartilage and osteoarthritis. *Clinics in Sports Medicine, 24,* 1–12.

Pereira, A. M., Tudor, C., Kanger, J. S., Subramaniam, V., & Martin-Blanco, E. (2011). Integrin-dependent activation of the JNK signaling pathway by mechanical stress. *PLoS ONE, 6,* e26182.

Perrin, D. D. (1965). Binding of tetracyclines to bone. *Nature, 208,* 787.

Peters, G., & Smith, D. W. (2002). Solute transport through a deforming porous medium. *International Journal for Numerical and Analytical Methods in Geomechanics, 26,* 683–717.

Pfander, D., Cramer, T., Deuerling, D., Weseloh, G., & Swoboda, B. (2000). Expression of thrombospondin-1 and its receptor CD36 in human osteoarthritic cartilage. *Annals of the Rheumatic Diseases, 59,* 448–454.

Pfander, D., Cramer, T., Schipani, E., & Johnson, R. S. (2003). HIF-1alpha controls extracellular matrix synthesis by epiphyseal chondrocytes. *Journal of Cell Science, 116,* 1819–1826.

Pfander, D., Heinz, N., Rothe, P., Carl, H. D., & Swoboda, B. (2004). Tenascin and aggrecan expression by articular chondrocytes is influenced by interleukin 1beta: A possible explanation for the changes in matrix synthesis during osteoarthritis. *Annals of the Rheumatic Diseases, 63,* 240–244.

Pfeiffer, H., Mornstad, H., & Teivens, A. (1995). Estimation of chronologic age using the aspartic acid racemization method. II. On human cortical bone. *International Journal of Legal Medicine, 108,* 24–26.

Phan, M. N., Leddy, H. A., Votta, B. J., Kumar, S., Levy, D. S., Lipshutz, D. B., et al. (2009). Functional characterization of TRPV4 as an osmotically sensitive ion channel in porcine articular chondrocytes. *Arthritis and Rheumatism, 60,* 3028–3037.

Philipse, A., & Vrij, A. (2011). The Donnan equilibrium: I. On the thermodynamic foundation of the Donnan equation of state. *Journal of Physics: Condensed Matter, 23,* 194106.

Pierce, D. M., Ricken, T., & Holzapfel, G. A. (2013). A hyperelastic biphasic fibre-reinforced model of articular cartilage considering distributed collagen fibre orientations: Continuum basis, computational aspects and applications. *Computer Methods in Biomechanics and Biomedical Engineering, 16,* 1344–1361.

Pilz, G. A., Bottes, S., Betizeau, M., Jorg, D. J., Carta, S., Simons, B. D., et al. (2018). Live imaging of neurogenesis in the adult mouse hippocampus. *Science, 359,* 658.

Plaas, A., Osborn, B., Yoshihara, Y., Bai, Y., Bloom, T., Nelson, F., et al. (2007). Aggrecanolysis in human osteoarthritis: Confocal localization and biochemical characterization of ADAMTS5-hyaluronan complexes in articular cartilages. *Osteoarthritis Cartilage, 15,* 719–734.

Plaas, A., Velasco, J., Gorski, D. J., Li, J., Cole, A., Christopherson, K., et al. (2011). The relationship between fibrogenic TGF beta 1 signaling in the joint and cartilage degradation in post-injury osteoarthritis. *Osteoarthritis and Cartilage, 19,* 1081–1090.

Plochocki, J. H., Ward, C. V., & Smith, D. E. (2009). Evaluation of the chondral modeling theory using fe-simulation and numeric shape optimization. *Journal of Anatomy, 214,* 768–777.

Poole, A. R., Kobayashi, M., Yasuda, T., Laverty, S., Mwale, F., Kojima, T., et al. (2002). Type II collagen degradation and its regulation in articular cartilage in osteoarthritis. *Annals of the Rheumatic Diseases, 61,* 78–81.

Poole, A. R., Rosenberg, L. C., Reiner, A., Ionescu, M., Bogoch, E., & Roughley, P. J. (1996). Contents and distributions of the proteoglycans decorin and biglycan in normal and osteoarthritic human articular cartilage. *Journal of Orthopaedic Research, 14,* 681–689.

Poole, C. A. (1997). Articular cartilage chondrons: Form, function and failure. *Journal of Anatomy, 191*(Pt 1), 1–13.

Poole, C. A., Ayad, S., & Gilbert, R. T. (1992). Chondrons from articular-cartilage. 5. Immunohistochemical evaluation of type-VI collagen organization in isolated chondrons by light, confocal and electron-microscopy. *Journal of Cell Science, 103,* 1101–1110.

Poole, C. A., Ayad, S., & Schofield, J. R. (1988a). Chondrons from articular-cartilage I. Immunohistochemical evaluation of type-VI collagen organization in isolated chondrons by light, confocal and electron-microscopy. *Journal of Cell Science, 90,* 635–643.

Poole, C. A., Flint, M. H., & Beaumont, B. W. (1984). Morphological and functional interrelationships of articular cartilage matrices. *Journal of Anatomy, 138*(Pt 1), 113–138.

Poole, C. A., Flint, M. H., & Beaumont, B. W. (1988b). Chondrons extracted from canine tibial cartilage: Preliminary report on their isolation and structure. *Journal of Orthopaedic Research, 6,* 408–419.

Poole, C. A., Glant, T. T., & Schofield, J. R. (1991a). Chondrons from articular cartilage. IV. Immunolocalization of proteoglycan epitopes in isolated canine tibial chondrons. *Journal of Histochemistry and Cytochemistry, 39,* 1175–1187.

Poole, C. A., Matsuoka, A., & Schofield, J. R. (1991b). Chondrons from articular cartilage. III. Morphologic changes in the cellular microenvironment of chondrons isolated from osteoarthritic cartilage. *Arthritis and Rheumatism, 34,* 22–35.

Povysil, C., Kana, R., Dundr, P., Tvrdik, D., Horak, M., Vaculik, J., et al. (2008). Distribution of chondrocytes containing alpha-smooth muscle actin in human normal, osteoarthrotic, and transplanted articular cartilage. *Pathology, Research and Practice, 204,* 883–890.

Prakash, J., & Vij, S. K. (1974). Effect of velocity slip on porous-walled squeeze films. *Wear, 29,* 363–372.

Pratta, M. A., Yao, W., Decicco, C., Tortorella, M. D., Liu, R. Q., Copeland, R. A., et al. (2003). Aggrecan protects cartilage collagen from proteolytic cleavage. *Journal of Biological Chemistry, 278,* 45539–45545.

Prekasan, D., & Saju, K. K. (2016). Review of the tribological characteristics of synovial fluid. In *1st Global Colloquium on Recent Advancements and Effectual Researches in Engineering, Science and Technology—Raerest 2016, 25,* 1170–1174.

Price, F. M., Levick, J. R., & Mason, R. M. (1996). Glycosaminoglycan concentration in synovium and other tissues of rabbit knee in relation to synovial hydraulic resistance. *Journal of Physiology, 495*(Pt 3), 803–820.

Pritchard, M. H., & Jessop, J. D. (1977). Chondrocalcinosis in primary hyperparathyroidism—Influence of age, metabolic bone-disease, and parathyroidectomy. *Annals of the Rheumatic Diseases, 36,* 146–151.

Pritchard, S., Votta, B. J., Kumar, S., & Guilak, F. (2008). Interleukin-1 inhibits osmotically induced calcium signaling and volume regulation in articular chondrocytes. *Osteoarthritis Cartilage, 16,* 1466–1473.

Provenzano, P. P., Alejandro-Osorio, A. L., Valhmu, W. B., Jensen, K. T., & Vanderby, R., Jr. (2005). Intrinsic fibroblast-mediated remodeling of damaged collagenous matrices in vivo. *Matrix Biology, 23,* 543–555.

Provot, S., Kempf, H., Murtaugh, L. C., Chung, U. I., Kim, D. W., Chyung, J., et al. (2006). Nkx3.2/Bapx1 acts as a negative regulator of chondrocyte maturation. *Development, 133,* 651–662.

Prydz, K. (2015). Determinants of glycosaminoglycan (GAG) structure. *Biomolecules, 5,* 2003–2022.

Qin, Y. X., Rubin, C. T., & McLeod, K. J. (1998). Nonlinear dependence of loading intensity and cycle number in the maintenance of bone mass and morphology. *Journal of Orthopaedic Research, 16,* 482–489.

Qiu, W., Murray, M. M., Shortkroff, S., Lee, C. R., Martin, S. D., & Spector, M. (2000). Outgrowth of chondrocytes from human articular cartilage explants and expression of alpha-smooth muscle actin. *Wound Repair Regen, 8,* 383–391.

Qu, F., Holloway, J. L., Esterhai, J. L., Burdick, J. A., & Mauck, R. L. (2017). Programmed biomolecule delivery to enable and direct cell migration for connective tissue repair. *Nature Communications, 8,* 1780.

Quicke, J. G., Foster, N. E., Thomas, M. J., & Holden, M. A. (2014). Is long-term physical activity safe for older adults with knee pain?: A systematic review. *Osteoarthritis and Cartilage, 22,* S11–S12.

Quinn, M. A., Gough, A. K., Green, M. J., Devlin, J., Hensor, E. M., Greenstein, A., et al. (2006). Anti-CCP antibodies measured at disease onset help identify seronegative rheumatoid arthritis and predict radiological and functional outcome. *Rheumatology (Oxford), 45,* 478–480.

Quinn, T. M., Allen, R. G., Schalet, B. J., Perumbuli, P., & Hunziker, E. B. (2001). Matrix and cell injury due to sub-impact loading of adult bovine articular cartilage explants: Effects of strain rate and peak stress. *Journal of Orthopaedic Research, 19,* 242–249.

Quinn, T. M., Grodzinsky, A. J., Buschmann, M. D., Kim, Y. J., & Hunziker, E. B. (1998). Mechanical compression alters proteoglycan deposition and matrix deformation around individual cells in cartilage explants. *Journal of Cell Science, 111*(Pt 5), 573–583.

Quiroga, J. M. P., Wilson, W., Ito, K., & van Donkelaar, C. C. (2017). The effect of loading rate on the development of early damage in articular cartilage. *Biomechanics and Modeling in Mechanobiology, 16,* 263–273.

Raddatz, G., Hagemann, S., Aran, D., Sohle, J., Kulkarni, P. P., Kaderali, L., et al. (2013). Aging is associated with highly defined epigenetic changes in the human epidermis. *Epigenetics & Chromatin, 6,* 36.

Raichlen, D. A., Armstrong, H., & Lieberman, D. E. (2011). Calcaneus length determines running economy: Implications for endurance running performance in modern humans and neandertals. *Journal of Human Evolution, 60,* 299–308.

Ramos, Y. F. M., & Meulenbelt, I. (2015). Implementation of functional genomics for bench-to-bedside transition in osteoarthritis. *Current Rheumatology Reports, 17,* 53.

Randall, R. M., Shao, Y. Y., Wang, L., & Ballock, R. T. (2012). Activation of Wnt planar cell polarity (PCP) signaling promotes growth plate column formation in vitro. *Journal of Orthopaedic Research, 30,* 1906–1914.

Raspanti, M., Viola, M., Forlino, A., Tenni, R., Gruppi, C., & Tira, M. E. (2008). Glycosaminoglycans show a specific periodic interaction with type I collagen fibrils. *Journal of Structural Biology, 164,* 134–139.

Ray, A., Bal, B. S., & Ray, B. K. (2005). Transcriptional induction of matrix metalloproteinase-9 in the chondrocyte and synoviocyte cells is regulated via a novel mechanism: Evidence for functional cooperation between serum amyloid A-activating factor-1 and AP-1. *Journal of Immunology, 175,* 4039–4048.

Ray, A., Singh, P. N., Sohaskey, M. L., Harland, R. M., & Bandyopadhyay, A. (2015). Precise spatial restriction of BMP signaling is essential for articular cartilage differentiation. *Development, 142,* 1169–1179.

Recklies, A. D., White, C., Melching, L., & Roughley, P. J. (2001). Differential regulation and expression of hyaluronan synthases in human articular chondrocytes, synovial cells and osteosarcoma cells. *Biochemical Journal, 354,* 17–24.

Rees, S. G., Flannery, C. R., Little, C. B., Hughes, C. E., Caterson, B., & Dent, C. M. (2000). Catabolism of aggrecan, decorin and biglycan in tendon. *Biochemical Journal, 350*(Pt 1), 181–188.

Reginato, A. M., & Olsen, B. R. (2002). The role of structural genes in the pathogenesis of osteoarthritic disorders. *Arthritis Research, 4,* 337–345.

Reginster, J. Y., Badurski, J., Bellamy, N., Bensen, W., Chapurlat, R., Chevalier, X., et al. (2013). Efficacy and safety of strontium ranelate in the treatment of knee osteoarthritis: Results of a double-blind, randomised placebo-controlled trial. *Annals of the Rheumatic Diseases, 72,* 179–186.

Reijman, M., Hazes, J. M. W., Bierma-Zeinstra, S. M. A., Koes, B. W., Christgau, S., Christiansen, C., et al. (2004). A new marker for osteoarthritis—Cross-sectional and longitudinal approach. *Arthritis and Rheumatism, 50,* 2471–2478.

Rekhi, R., & Qutub, A. A. (2013). Systems approaches for synthetic biology: A pathway toward mammalian design. *Frontiers in Physiology, 4,* 285.

Responte, D. J., Lee, J. K., Hu, J. C., & Athanasiou, K. A. (2012). Biomechanics-driven chondrogenesis: From embryo to adult. *The FASEB Journal, 26,* 3614–3624.

Revell, P. A., Pirie, C., Amir, G., Rashad, S., & Walker, F. (1990). Metabolic-activity in the calcified zone of cartilage—Observations on tetracycline labeled articular-cartilage in human osteoarthritic hips. *Rheumatology International, 10,* 143–147.

Reynard, L. N., Ratnayake, M., Santibanez-Koref, M., & Loughlin, J. (2016). Functional characterization of the osteoarthritis susceptibility mapping to CHST11—A bioinformatics and molecular study. *PLoS ONE, 11,* e0159024.

Rhee, D. K., Marcelino, J., Al-Mayouf, S., Schelling, D. K., Bartels, C. F., Cui, Y., et al. (2005). Consequences of disease-causing mutations on lubricin protein synthesis, secretion, and post-translational processing. *Journal of Biological Chemistry, 280,* 31325–31332.

Rhee, D. K., Marcelino, J., Baker, M. A., Gong, Y. Q., Smits, P., Lefebvre, V., et al. (2005). The secreted glycoprotein lubricin protects cartilage surfaces and inhibits synovial cell overgrowth. *Journal of Clinical Investigation, 115,* 622–631.

Rice, J. R., & Cleary, M. P. (1976). Some basic stress diffusion solutions for fluid-saturated elastic porous-media with compressible constituents. *Reviews of Geophysics, 14,* 227–241.

RICE, J. R. (1994). Solid Mechanics, School of Engineering and Applied Sciences, and Department of Earth and Planetary Sciences, Harvard University, Cambridge, MA.

Rieppo, J., Hyttinen, M. M., Halmesmaki, E., Ruotsalainen, H., Vasara, A., Kiviranta, I., et al. (2009). Changes in spatial collagen content and collagen network architecture in porcine articular cartilage during growth and maturation. *Osteoarthritis and Cartilage, 17,* 448–455.

Roberts, S., Menage, J., Sandell, L. J., Evans, E. H., & Richardson, J. B. (2009). Immunohistochemical study of collagen types I and II and procollagen IIA in human cartilage repair tissue following autologous chondrocyte implantation. *Knee, 16,* 398–404.

Robins, J. C., Akeno, N., Mukherjee, A., Dalal, R. R., Aronow, B. J., Koopman, P., et al. (2005). Hypoxia induces chondrocyte-specific gene expression in mesenchymal cells in association with transcriptional activation of Sox9. *Bone, 37,* 313–322.

Robins, S. P. (2007). Biochemistry and functional significance of collagen cross-linking. *Biochemical Society Transactions, 35,* 849–852.

Robinson, W. H., Lepus, C. M., Wang, Q., Raghu, H., Mao, R., Lindstrom, T. M., et al. (2016). Low-grade inflammation as a key mediator of the pathogenesis of osteoarthritis. *Nature Reviews Rheumatology, 12,* 580–592.

Robson, M. D., Gatehouse, P. D., Bydder, M., & Bydder, G. M. (2003). Magnetic resonance: An introduction to ultrashort TE (UTE) imaging. *Journal of Computer Assisted Tomography, 27,* 825–846.

Rohl, L., Linde, F., Odgaard, A., & Hvid, I. (1997). Simultaneous measurement of stiffness and energy absorptive properties of articular cartilage and subchondral trabecular bone. *Proceedings of the Institution of Mechanical Engineers, Part H, 211,* 257–264.

Rolauffs, B., Rothdiener, M., Bahrs, C., Badke, A., Weise, K., Kuettner, K. E., et al. (2011). Onset of preclinical osteoarthritis: The angular spatial organization permits early diagnosis. *Arthritis and Rheumatism, 63,* 1637–1647.

Rolauffs, B., Williams, J. M., Aurich, M., Grodzinsky, A. J., Kuettner, K. E., & Cole, A. A. (2010). Proliferative remodeling of the spatial organization of human superficial chondrocytes distant from focal early osteoarthritis. *Arthritis and Rheumatism, 62,* 489–498.

Rolauffs, B., Williams, J. M., Grodzinsky, A. J., Kuettner, K. E., & Cole, A. A. (2008). Distinct horizontal patterns in the spatial organization of superficial zone chondrocytes of human joints. *Journal of Structural Biology, 162,* 335–344.

Rosa, R. G., Collavino, K., Lakhani, A., Delve, E., Weber, J. F., Rosenthal, A. K., et al. (2014). Clodronate exerts an anabolic effect on articular chondrocytes mediated through the purinergic receptor pathway. *Osteoarthritis Cartilage, 22,* 1327–1336.

Rosario, P. W. (2010). Normal values of serum IGF-1 in adults: Results from a Brazilian population. *Arq Bras Endocrinol Metabol, 54,* 477–481.

Rosenberg, K., Olsson, H., Morgelin, M., & Heinegard, D. (1998). Cartilage oligomeric matrix protein shows high affinity zinc-dependent interaction with triple helical collagen. *Journal of Biological Chemistry, 273,* 20397–20403.

Rosenblum, G., van den Steen, P. E., Cohen, S. R., Bitler, A., Brand, D. D., Opdenakker, G., et al. (2010). Direct visualization of protease action on collagen triple helical structure. *PLoS ONE, 5,* e11043.

Rosenthal, A. K., Heinkel, D., & Gohr, C. M. (2003). Thyroxine stimulates transglutaminase activity in articular chondrocytes. *Osteoarthritis Cartilage, 11,* 463–470.

Rosenzweig, D. H., Djap, M. J., Ou, S. J., & Quinn, T. M. (2012). Mechanical injury of bovine cartilage explants induces depth-dependent, transient changes in MAP kinase activity associated with apoptosis. *Osteoarthritis Cartilage, 20,* 1591–1602.

Roth, T. E., Goldberg, J. S., & Behrents, R. G. (1984). Synovial-fluid pressure determination in the temporomandibular-joint. *Oral Surgery, Oral Medicine, Oral Pathology, Oral Radiology and Endodontics, 57,* 583–588.

Roughley, P. J., & Mort, J. S. (2014). The role of aggrecan in normal and osteoarthritic cartilage. *Journal of Experimental Orthopaedics.*

Rountree, R. B., Schoor, M., Chen, H., Marks, M. E., Harley, V., Mishina, Y., et al. (2004). BMP receptor signaling is required for postnatal maintenance of articular cartilage. *PLoS Biology, 2,* 1815–1827.

Roux, E. (2014). The concept of function in modern physiology. *Journal of Physiology-London, 592,* 2245–2249.

Rowe, R. G., Keena, D., Sabeh, F., Willis, A. L., & Weiss, S. J. (2011). Pulmonary fibroblasts mobilize the membrane-tethered matrix metalloprotease, MT1-MMP, to destructively remodel and invade interstitial type I collagen barriers. *American Journal of Physiology. Lung Cellular and Molecular Physiology, 301,* L683–L692.

Rowe, R. K., Quigley, R. M., Brachman, R. W., & Booker, J. R. (2004). *Barrier systems for waste disposal facilities.* Boca Raton, FL: CRC Press.

Rucker, R. B., Kosonen, T., Clegg, M. S., Mitchell, A. E., Rucker, B. R., Uriu-Hare, J. Y., et al. (1998). Copper, lysyl oxidase, and extracellular matrix protein cross-linking. *American Journal of Clinical Nutrition, 67,* 996S–1002S.

Ruiz-Heiland, G., Horn, A., Zerr, P., Hofstetter, W., Baum, W., Stock, M., et al. (2011). Smoothening of periarticular bone: Blockade of the hedgehog pathway inhibits osteophyte formation in arthritis. *Osteoarthritis and Cartilage, 19,* S98.

Ruiz-Heiland, G., Horn, A., Zerr, P., Hofstetter, W., Baum, W., Stock, M., et al. (2012). Blockade of the hedgehog pathway inhibits osteophyte formation in arthritis. *Annals of the Rheumatic Diseases, 71,* 400–407.

Russel, W. B. (1980). Review of the role of colloidal forces in the rheology of suspensions. *Journal of Rheology, 24,* 287–317.

Russel, W. B., & Benzing, D. W. (1981). The viscoelastic properties of ordered lattices—A self-consistent field-theory. *Journal of Colloid and Interface Science, 83,* 163–177.

Ryan, J. A., Eisner, E. A., Duraine, G., You, Z. B., & Reddi, A. H. (2009). Mechanical compression of articular cartilage induces chondrocyte proliferation and inhibits proteoglycan synthesis by activation of the ERK pathway: Implications for tissue engineering and regenerative medicine. *Journal of Tissue Engineering and Regenerative Medicine, 3,* 107–116.

Ryan, M. C., & Sandell, L. J. (1990). Differential expression of a cysteine-rich domain in the amino-terminal propeptide of type II (cartilage) procollagen by alternative splicing of mRNA. *Journal of Biological Chemistry, 265,* 10334–10339.

Ryu, J. H., Yang, S., Shin, Y., Rhee, J., Chun, C. H., & Chun, J. S. (2011). Interleukin-6 plays an essential role in hypoxia-inducible factor 2alpha-induced experimental osteoarthritic cartilage destruction in mice. *Arthritis and Rheumatism, 63,* 2732–2743.

Saadat, E., Lan, H., Majumdar, S., Rempel, D. M., & King, K. B. (2006). Long-term cyclical in vivo loading increases cartilage proteoglycan content in a spatially specific manner: An infrared microspectroscopic imaging and polarized light microscopy study. *Arthritis Research & Therapy, 8,* R147.

Saarakkala, S., Julkunen, P., Kiviranta, P., Makitalo, J., Jurvelin, J. S., & Korhonen, R. K. (2010). Depth-wise progression of osteoarthritis in human articular cartilage: Investigation of composition, structure and biomechanics. *Osteoarthritis Cartilage, 18,* 73–81.

Sabaratnam, S., Coleman, P. J., Mason, R. M., & Levick, J. R. (2007). Interstitial matrix proteins determine hyaluronan reflection and fluid retention in rabbit joints: Effect of protease. *Journal of Physiology-London, 578,* 291–299.

Sah, R. L., Grodzinsky, A. J., Plaas, A. H., & Sandy, J. D. (1990). Effects of tissue compression on the hyaluronate-binding properties of newly synthesized proteoglycans in cartilage explants. *Biochemical Journal, 267,* 803–808.

Sah, R. L., Kim, Y. J., Doong, J. Y., Grodzinsky, A. J., Plaas, A. H., & Sandy, J. D. (1989). Biosynthetic response of cartilage explants to dynamic compression. *Journal of Orthopaedic Research, 7,* 619–636.

Sahay, A., Scobie, K. N., Hill, A. S., O'Carroll, C. M., Kheirbek, M. A., Burghardt, N. S., et al. (2011). Increasing adult hippocampal neurogenesis is sufficient to improve pattern separation. *Nature, 472,* 466–U539.

Sahlman, J., Pitkanen, M. T., Prockop, D. J., Arita, M., Li, S. W., Helminen, H. J., et al. (2004). A human COL2A1 gene with an Arg519Cys mutation causes osteochondrodysplasia in transgenic mice. *Arthritis and Rheumatism, 50,* 3153–3160.

Sahni, M., Ambrosetti, D. C., Mansukhani, A., Gertner, R., Levy, D., & Basilico, C. (1999). FGF signaling inhibits chondrocyte proliferation and regulates bone development through the STAT-1 pathway. *Genes & Development, 13,* 1361–1366.

Sakakura, Y., Hosokawa, Y., Tsuruga, E., Irie, K., Nakamura, M., & Yajima, T. (2007). Contributions of matrix metalloproteinases toward Meckel's cartilage resorption in mice: Immunohistochemical studies, including comparisons with developing endochondral bones. *Cell and Tissue Research, 328,* 137–151.

Salminen, H., Perala, M., Lorenzo, P., Saxne, T., Heinegard, D., Saamanen, A. M., et al. (2000). Up-regulation of cartilage oligomeric matrix protein at the onset of articular cartilage degeneration in a transgenic mouse model of osteoarthritis. *Arthritis and Rheumatism, 43,* 1742–1748.

Sanchez, C., Bay-Jensen, A. C., Pap, T., Dvir-Ginzberg, M., Quasnichka, H., Barrett-Jolley, R., et al. (2017). Chondrocyte secretome: A source of novel insights and exploratory biomarkers of osteoarthritis. *Osteoarthritis Cartilage, 25,* 1199–1209.

Sandell, L. J., & Adler, P. (1999). Developmental patterns of cartilage. *Frontiers in Bioscience, 4,* D731–D742.

Sandell, L. J., & Aigner, T. (2001). Articular cartilage and changes in arthritis. An introduction: Cell biology of osteoarthritis. *Arthritis Research & Therapy, 3,* 107–113.

Sandy, J. D., Flannery, C. R., Neame, P. J., & Lohmander, L. S. (1992). The structure of aggrecan fragments in human synovial fluid. Evidence for the involvement in osteoarthritis of a novel proteinase which cleaves the Glu 373-Ala 374 bond of the interglobular domain. *The Journal of Clinical Investigation, 89,* 1512–1516.

Sandy, J. D., & Verscharen, C. (2001). Analysis of aggrecan in human knee cartilage and synovial fluid indicates that aggrecanase (ADAMTS) activity is responsible for the catabolic turnover and loss of whole aggrecan whereas other protease activity is required for C-terminal processing in vivo. *Biochemical Journal, 358,* 615–626.

Sangiorgio, V., Pitto, M., Palestini, P., & Masserini, M. (2004). GPI-anchored proteins and lipid rafts. *Italian Journal of Biochemistry, 53,* 98–111.

Sariahmetoglu, M., Crawford, B. D., Leon, H., Sawicka, J., Li, L. J., Ballermann, B. J., et al. (2007). Regulation of matrix metalloproteinase-2 (MMP-2) activity by phosphorylation. *The FASEB Journal, 21,* 2486–2495.

Sarrazin, S., Lamanna, W. C., & Esko, J. D. (2011). Heparan sulfate proteoglycans. *Cold Spring Harbor Perspectives in Biology, 3.*

Sato, T., Ovejero, M. D., Hou, P., Heegaard, A. M., Kumegawa, M., Foged, N. T., et al. (1997). Identification of the membrane-type matrix metalloproteinase MT1-MMP in osteoclasts. *Journal of Cell Science, 110,* 589–596.

Sauerland, K., Plaas, A. H., Raiss, R. X., & Steinmeyer, J. (2003). The sulfation pattern of chondroitin sulfate from articular cartilage explants in response to mechanical loading. *Biochimica et Biophysica Acta, 1638,* 241–248.

Scallan, J., Huxley, V. H., & Korthuis, R. J. (2010). Capillary fluid exchange: Regulation, functions, and pathology. Morgan and Claypool Life Sciences.

Schechtman, H., & Bader, D. L. (1997). In vitro fatigue of human tendons. *Journal of Biomechanics, 30,* 829–835.

Scher, J. U., Pillinger, M. H., & Abramson, S. B. (2007). Nitric oxide synthases and osteoarthritis. *Current Rheumatology Reports, 9,* 9–15.

Schett, G., Kleyer, A., Perricone, C., Sahinbegovic, E., Iagnocco, A., Zwerina, J., et al. (2013). Diabetes is an independent predictor for severe osteoarthritis results from a longitudinal cohort study. *Diabetes Care, 36,* 403–409.

Schicht, M., Ernst, J., Nielitz, A., Fester, L., Tsokos, M., Guddat, S. S., et al. (2014). Articular cartilage chondrocytes express aromatase and use enzymes involved in estrogen metabolism. *Arthritis Research & Therapy, 16,* R93.

Schimpl, M., Rush, C. L., Betou, M., Eggleston, I. M., Recklies, A. D., & van Aalten, D. M. F. (2012). Human YKL-39 is a pseudo-chitinase with retained chitooligosaccharide-binding properties. *Biochemical Journal, 446,* 149–157.

Schinagl, R. M., Gurskis, D., Chen, A. C., & Sah, R. L. (1997). Depth-dependent confined compression modulus of full-thickness bovine articular cartilage. *Journal of Orthopaedic Research, 15,* 499–506.

Schinagl, R. M., Ting, M. K., Price, J. H., & Sah, R. L. (1996). Video microscopy to quantitate the inhomogeneous equilibrium strain within articular cartilage during confined compression. *Annals of Biomedical Engineering, 24,* 500–512.

Schipani, E., & Provot, S. (2003). PTHrP, PTH, and the PTH/PTHrP receptor in endochondral bone development. *Birth Defects Research Part C-Embryo Today-Reviews, 69,* 352–362.

Schipani, E., Ryan, H. E., Didrickson, S., Kobayashi, T., Knight, M., & Johnson, R. S. (2001). Hypoxia in cartilage: HIF-1 alpha is essential for chondrocyte growth arrest and survival. *Genes & Development, 15,* 2865–2876.

Schneiderman, R., Keret, D., & Maroudas, A. (1986). Effects of mechanical and osmotic pressure on the rate of glycosaminoglycan synthesis in the human adult femoral head cartilage: An in vitro study. *Journal of Orthopaedic Research, 4,* 393–408.

Scholzen, T., & Gerdes, J. (2000). The Ki-67 protein: From the known and the unknown. *Journal of Cellular Physiology, 182,* 311–322.

Schott, E. M., Lillis, J. A., Farnsworth, C. W., Villani, D., Ketz, J. P., Ashton, J. M., et al. (2018). Early synovial B-cell infiltration as a candidate mechanism of pathogenesis in the OA of obesity. *Osteoarthritis and Cartilage, 26,* S14–S14.

Schwartz, A. G., Lipner, J. H., Pasteris, J. D., Genin, G. M., & Thomopoulos, S. (2013). Muscle loading is necessary for the formation of a functional tendon enthesis. *Bone, 55,* 44–51.

Schwartz, M. (2004). Rho signalling at a glance. *Journal of Cell Science, 117,* 5457–5458.

Scotton, C. J., & Chambers, R. C. (2007). Molecular targets in pulmonary fibrosis—The myofibroblast in focus. *Chest, 132,* 1311–1321.

Seeliger, E., Wronski, T., Ladwig, M., Rebeschke, T., Persson, P. B., & Reinhardt, H. W. (2005). The 'body fluid pressure control system' relies on the Renin-Angiotensin-aldosterone system: Balance studies in freely moving dogs. *Clinical and Experimental Pharmacology and Physiology, 32,* 394–399.

Seifzadeh, A., Wang, J., Oguamanam, D. C. D., & Papini, M. (2011). A nonlinear biphasic fiber-reinforced porohyperviscoelastic model of articular cartilage incorporating fiber reorientation and dispersion. *Journal of Biomechanical Engineering—Transactions of the ASME, 133.*

Seki, K., & Hata, A. (2004). Indian hedgehog gene is a target of the bone morphogenetic protein signaling pathway. *Journal of Biological Chemistry, 279,* 18544–18549.

Sellam, J., & Berenbaum, F. (2013). Is osteoarthritis a metabolic disease? *Joint Bone Spine, 80,* 568–573.

Sengers, B. G., Heywood, H. K., Lee, D. A., Oomens, C. W. J., & Bader, D. L. (2005). Nutrient utilization by bovine articular chondrocytes: A combined experimental and theoretical approach. *Journal of Biomechanical Engineering—Transactions of the ASME, 127,* 758–766.

Seol, D., McCabe, D. J., Choe, H., Zheng, H., Yu, Y., Jang, K., et al. (2012). Chondrogenic progenitor cells respond to cartilage injury. *Arthritis and Rheumatism, 64,* 3626–3637.

Seror, J., Merkher, Y., Kampf, N., Collinson, L., Day, A. J., Maroudas, A., et al. (2011). Articular cartilage proteoglycans as boundary lubricants: Structure and frictional interaction of surface-attached hyaluronan and hyaluronan-aggrecan complexes. *Biomacromolecules, 12,* 3432–3443.

Sessa, L., Gatti, E., Zeni, F., Antonelli, A., Catucci, A., Koch, M., et al. (2014). The receptor for advanced glycation end-products (RAGE) is only present in mammals, and belongs to a family of cell adhesion molecules (CAMs). *PLoS ONE, 9,* e86903.

Sforza, C., Grandi, G., Binelli, M., Tommasi, D. G., Rosati, R., & Ferrario, V. F. (2009). Age- and sex-related changes in the normal human ear. *Forensic Science International, 187.*

Sforza, C., Grandi, G., De Menezes, M., Tartaglia, G. M., & Ferrario, V. F. (2011). Age- and sex-related changes in the normal human external nose. *Forensic Science International, 204* (1-3), 205.e1-9.

Sharif, M., Kirwan, J., Charni, N., Sandell, L. J., Whittles, C., & Garnero, P. (2007). A 5-yr longitudinal study of type IIA collagen synthesis and total type II collagen degradation in patients with knee osteoarthritis—Association with disease progression. *Rheumatology, 46,* 938–943.

Sharif, M., Whitehouse, A., Sharman, P., Perry, M., & Adams, M. (2004). Increased apoptosis in human osteoarthritic cartilage corresponds to reduced cell density and expression of caspase-3. *Arthritis and Rheumatism, 50,* 507–515.

Shark, L. K., Chen, H., & Goodacre, J. (2010). Discovering differences in acoustic emission between healthy and osteoarthritic knees using a four-phase model of sit-stand-sit movements. *Open Medical Informatics Journal, 4,* 116–125.

Sharma, A., Kudesia, P., Shi, Q., & Gandhi, R. (2016). Anxiety and depression in patients with osteoarthritis: Impact and management challenges. *Open Access Rheumatology: Research and Reviews, 8,* 103–113.

Sharma, A. R., Jagga, S., Lee, S. S., & Nam, J. S. (2013). Interplay between cartilage and subchondral bone contributing to pathogenesis of osteoarthritis. *International Journal of Molecular Sciences, 14,* 19805–19830.

Sharma, V., Jaishankar, A., Wang, Y. C., & McKinley, G. H. (2011). Rheology of globular proteins: Apparent yield stress, high shear rate viscosity and interfacial viscoelasticity of bovine serum albumin solutions. *Soft Matter, 7,* 5150–5160.

Shekhawat, V. K., Laurent, M. P., Muehleman, C., & Wimmer, M. A. (2009). Surface topography of viable articular cartilage measured with scanning white light interferometry. *Osteoarthritis Cartilage, 17,* 1197–1203.

Shen, J., Li, J., Wang, B., Jin, H., Wang, M., Zhang, Y., et al. (2013). Deletion of the transforming growth factor beta receptor type II gene in articular chondrocytes leads to a progressive osteoarthritis-like phenotype in mice. *Arthritis and Rheumatism, 65,* 3107–3119.

Shen, J., Li, S., & Chen, D. (2014). TGF-β signaling and the development of osteoarthritis. *Bone Research, 2.*

Sheng, D. C., Sloan, S. W., Gens, A., & Smith, D. W. (2003). Finite element formulation and algorithms for unsaturated soils. Part I: Theory. *International Journal for Numerical and Analytical Methods in Geomechanics, 27,* 745–765.

Shepherd, D. E., & Seedhom, B. B. (1999). Thickness of human articular cartilage in joints of the lower limb. *Annals of the Rheumatic Diseases, 58,* 27–34.

Shepherd, J. H., & Screen, H. R. C. (2013). Fatigue loading of tendon. *International Journal of Experimental Pathology, 94,* 260–270.

Shi, S. L., Mercer, S., Eckert, G. J., & Trippel, S. B. (2013). Growth factor transgenes interactively regulate articular chondrocytes. *Journal of Cellular Biochemistry, 114,* 908–919.

Shimizu, S., Asou, Y., Itoh, S., Chung, U. I., Kawaguchi, H., Shinomiya, K., et al. (2007). Prevention of cartilage destruction with intraarticular osteoclastogenesis inhibitory factor/osteoprotegerin in a murine model of osteoarthritis. *Arthritis and Rheumatism, 56,* 3358–3365.

Siebelt, M., Groen, H. C., Koelewijn, S. J., de Blois, E., Sandker, M., Waarsing, J. H., et al. (2014). Increased physical activity severely induces osteoarthritic changes in knee joints with papain induced sulfate-glycosaminoglycan depleted cartilage. *Arthritis Research & Therapy, 16,* R32.

Silver, F. H., Freeman, J. W., & Seehra, G. P. (2003). Collagen self-assembly and the development of tendon mechanical properties. *Journal of Biomechanics, 36,* 1529–1553.

Silverwood, V., Blagojevic-Bucknall, M., Jinks, C., Jordan, J. L., Protheroe, J., & Jordan, K. P. (2015). Current evidence on risk factors for knee osteoarthritis in older adults: A systematic review and meta-analysis. *Osteoarthritis Cartilage, 23,* 507–515.

Simkin, P. A. (2008). A biography of the chondrocyte. *Annals of the Rheumatic Diseases, 67,* 1064–1068.

Simkin, P. A. (2012). Consider the tidemark. *Journal of Rheumatology, 39,* 890–892.

Simkin, P. A. (2015). The human knee: A window on the microvasculature. *Tissue Barriers, 3,* e970465.

Simon, D., Mascarenhas, R., Saltzman, B. M., Rollins, M., Bach, B. R., Jr., & Macdonald, P. (2015). The relationship between anterior cruciate ligament injury and osteoarthritis of the knee. *Advances in Orthopedics*, 928301.

Simsa-Maziel, S., & Monsonego-Ornan, E. (2012). Interleukin-1beta promotes proliferation and inhibits differentiation of chondrocytes through down-regulation of FGFR-3-P21 pathway and diminishes bone quality. *Bone, 50,* S100–S101.

Singh-Zocchi, M., Andreasen, A., & Zocchi, G. (1999). Osmotic pressure contribution of albumin to colloidal interactions. *Proceedings of the National Academy of Sciences of the United States of America, 96,* 6711–6715.

Sivan, S. S., Tsitron, E., Wachtel, E., Roughley, P. J., Sakkee, N., van der Ham, F., et al. (2006). Aggrecan turnover in human intervertebral disc as determined by the racemization of aspartic acid. *Journal of Biological Chemistry, 281,* 13009–13014.

Sjoberg, A. P., Manderson, G. A., Morgelin, M., Day, A. J., Heinegard, D., & Blom, A. M. (2009). Short leucine-rich glycoproteins of the extracellular matrix display diverse patterns of complement interaction and activation. *Molecular Immunology, 46,* 830–839.

Skempton, A. W. (1960). Terzaghi's discovery of effective stress. In L. Bejerrum, A. Casagrande, R. B. Peck, & A. W. Skemption (Eds.) *From theory to practice in soil mechanics.* New York, London: Wiley.

Skurk, C., Maatz, H., Rocnik, E., Bialik, A., Force, T., & Walsh, K. (2005). Glycogen-Synthase Kinase3beta/beta-catenin axis promotes angiogenesis through activation of vascular endothelial growth factor signaling in endothelial cells. *Circulation Research, 96,* 308–318.

Slater, J. C. (1939). *Introduction to chemical physics.* New York: McGraw-Hill.

Smid, J. K., Faulkes, S., & Rudnicki, M. A. (2015). Periostin induces pancreatic regeneration. *Endocrinology, 156,* 824–836.

Smilgies, D. M., & Folta-Stogniew, E. (2015). Molecular weight-gyration radius relation of globular proteins: A comparison of light scattering, small-angle X-ray scattering and structure-based data. *Journal of Applied Crystallography, 48,* 1604–1606.

Smith, D. W. (2000). One-dimensional contaminant transport through a deforming porous medium: Theory and a solution for a quasi-steady-state problem. *International Journal for Numerical and Analytical Methods in Geomechanics, 24,* 693–722.

Smith, D. W., & Booker, J. R. (1993). Green-functions for a fully coupled thermoporoelastic material. *International Journal for Numerical and Analytical Methods in Geomechanics, 17,* 139–163.
Smith, D. W., Gardiner, B. S., Davidson, J., & Grodzinsky, A. J. (2016). Computational model for the analysis of cartilage and cartilage tissue constructs. *Journal of Tissue Engineering and Regenerative Medicine, 10,* 334–347.
Smith, D. W., Gardiner, B. S., & Dunstan, C. (2012). Bone balance within a cortical BMU: Local controls of bone resorption and formation. *PLoS ONE, 7,* e40268.
Smith, D. W., Pivonka, P., Jungnickel, C., & Fityus, S. (2004). Theoretical analysis of anion exclusion and diffusive transport through platy-clay soils. *Transport in Porous Media, 57,* 251–277.
Smith, M. D. (2011). The normal synovium. *The Open Rheumatology Journal, 5,* 100–106.
Smits, P., Li, P., Mandel, J., Zhang, Z. P., Deng, J. M., Behringer, R. R., et al. (2001). The transcription factors L-sox5 and sox6 are essential for cartilage formation. *Developmental Cell, 1,* 277–290.
Smyth, P. A., Rifkin, R. E., Jackson, R. L., & Hanson, R. R. (2012). A surface roughness comparison of cartilage in different types of synovial joints. *Journal of Biomechanical Engineering, 134,* 021006.
Sophia Fox, A. J., Bedi, A., & Rodeo, S. A. (2009). The basic science of articular cartilage: Structure, composition, and function. *Sports Health, 1,* 461–468.
Sormaala, M. J., Niva, M. H., Kiuru, M. J., Mattila, V. M., & Pihlajamaki, H. K. (2006). Bone stress injuries of the talus in military recruits. *Bone, 39,* 199–204.
Sowers, M. F., Karvonen-Gutierrez, C. A., Yosef, M., Jannausch, M., Jiang, Y., Garnero, P., et al. (2009). Longitudinal changes of serum COMP and urinary CTX-II predict X-ray defined knee osteoarthritis severity and stiffness in women. *Osteoarthritis and Cartilage, 17,* 1609–1614.
Spalding, K. L., Bergmann, O., Alkass, K., Bernard, S., Salehpour, M., Huttner, H. B., et al. (2013). Dynamics of hippocampal neurogenesis in adult humans. *Cell, 153,* 1219–1227.
Spector, T. D., & Macgregor, A. J. (2004). Risk factors for osteoarthritis: Genetics. *Osteoarthritis Cartilage, 12*(Suppl A), S39–S44.
Spitters, T. W., Leijten, J. C., Deus, F. D., Costa, I. B., van Apeldoorn, A. A., van Blitterswijk, C. A., et al. (2013). A dual flow bioreactor with controlled mechanical stimulation for cartilage tissue engineering. *Tissue Engineering Part C: Methods, 19,* 774–783.
St-Pierre, J. P., Gan, L., Wang, J., Pilliar, R. M., Grynpas, M. D., & Kandel, R. A. (2012). The incorporation of a zone of calcified cartilage improves the interfacial shear strength between in vitro-formed cartilage and the underlying substrate. *Acta Biomaterialia, 8,* 1603–1615.
Stachowiak, G. P., Stachowiak, G. W., & Podsiadlo, P. (2006). Automated classification of articular cartilage surfaces based on surface texture. *Proceedings of the Institution of Mechanical Engineers, Part H, 220,* 831–843.
Stangenberg, L., Burzyn, D., Binstadt, B. A., Weissleder, R., Mahmood, U., Benoist, C., et al. (2014). Denervation protects limbs from inflammatory arthritis via an impact on the microvasculature. *Proceedings of the National Academy of Sciences of the United States of America, 111,* 11419–11424.
Stankovich, J., & Carnie, S. L. (1996). Electrical double layer interaction between dissimilar spherical colloidal particles and between a sphere and a plate: Nonlinear Poisson-Boltzmann theory. *Langmuir, 12,* 1453–1461.
Stannus, O., Jones, G., Cicuttini, F., Parameswaran, V., Quinn, S., Burgess, J., et al. (2010). Circulating levels of IL-6 and TNF-alpha are associated with knee radiographic osteoarthritis and knee cartilage loss in older adults. *Osteoarthritis and Cartilage, 18,* 1441–1447.
Stattin, E. L., Wiklund, F., Lindblom, K., Onnerfjord, P., Jonsson, B. A., Tegner, Y., et al. (2010). A missense mutation in the aggrecan C-type lectin domain disrupts extracellular matrix interactions and causes dominant familial osteochondritis dissecans. *American Journal of Human Genetics, 86,* 126–137.

Stewart, C. E. H., & Rotwein, P. (1996). Insulin-like growth factor-II is an autocrine survival factor for differentiating myoblasts. *Journal of Biological Chemistry, 271,* 11330–11338.

Stewart, T., Jin, Z. M., & Fisher, J. (1997). Friction of composite cushion bearings for total knee joint replacements under adverse lubrication conditions. *Proceedings of the Institution of Mechanical Engineers Part H, 211,* 451–465.

Stoppiello, L. A., Mapp, P. I., Wilson, D., Hill, R., Scammell, B. E., & Walsh, D. A. (2014). Structural associations of symptomatic knee osteoarthritis. *Arthritis Rheumatol, 66,* 3018–3027.

Strazzulla, L. C., & Cronstein, B. N. (2016). Regulation of bone and cartilage by adenosine signaling. *Purinergic Signal, 12,* 583–593.

Struglics, A., & Hansson, M. (2010). Calpain is involved in C-terminal truncation of human aggrecan. *Osteoarthritis and Cartilage, 18,* S10.

Struglics, A., Larsson, S., Hansson, M., & Lohmander, L. S. (2009). Western blot quantification of aggrecan fragments in human synovial fluid indicates differences in fragment patterns between joint diseases. *Osteoarthritis Cartilage, 17,* 497–506.

Struglics, A., Larsson, S., Kumahashi, N., Frobell, R., & Lohmander, L. S. (2015). Changes in cytokines and aggrecan ARGS neoepitope in synovial fluid and serum and in C-terminal crosslinking telopeptide of type II collagen and N-terminal crosslinking telopeptide of type I collagen in urine over five years after anterior cruciate ligament rupture: An exploratory analysis in the knee anterior cruciate ligament, nonsurgical versus surgical treatment trial. *Arthritis & Rheumatology, 67,* 1816–1825.

Struglics, A., Larsson, S., Pratta, M. A., Kumar, S., Lark, M. W., & Lohmander, L. S. (2006). Human osteoarthritis synovial fluid and joint cartilage contain both aggrecanase- and matrix metalloproteinase-generated aggrecan fragments. *Osteoarthritis Cartilage, 14,* 101–113.

Stuart, K., Paderi, J., Snyder, P. W., Freeman, L., & Panitch, A. (2011). Collagen-binding peptidoglycans inhibit MMP mediated collagen degradation and reduce dermal scarring. *PLoS ONE, 6,* e22139.

Sun, J., Lei, L., Tsai, C. M., Wang, Y., Shi, Y., Ouyang, M., et al. (2017). Engineered proteins with sensing and activating modules for automated reprogramming of cellular functions. *Nature Communications, 8,* 477.

SUZUKI, T., YAMAZAKI, T. & YAZAKI, Y. 2001. The role of the natriuretic peptides in the cardiovascular system. *Cardiovascular Research*, 51, 489-494.

Svensson, L., Aszodi, A., Heinegard, D., Hunziker, E. B., Reinholt, F. P., Fassler, R., et al. (2002). Cartilage oligomeric matrix protein-deficient mice have normal skeletal development. *Molecular and Cellular Biology, 22,* 4366–4371.

Svensson, L., Aszodi, A., Reinholt, F. P., Fassler, R., Heinegard, D., & Oldberg, A. (1999). Fibromodulin-null mice have abnormal collagen fibrils, tissue organization, and altered lumican deposition in tendon. *Journal of Biological Chemistry, 274,* 9636–9647.

Svensson, R. B., Hassenkam, T., Grant, C. A., & Magnusson, S. P. (2010). Tensile properties of human collagen fibrils and fascicles are insensitive to environmental salts. *Biophysical Journal, 99,* 4020–4027.

Swann, D. A., Sotman, S., Dixon, M., & Brooks, C. (1977). The isolation and partial characterization of the major glycoprotein (LGP-I) from the articular lubricating fraction from bovine synovial fluid. *Biochemical Journal, 161,* 473–485.

Swartz, M. A., & Fleury, M. E. (2007). Interstitial flow and its effects in soft tissues. *Annual Review of Biomedical Engineering, 9,* 229–256.

Szarko, M., & Xia, Y. (2012). Direct visualisation of the depth-dependent mechanical properties of full-thickness articular cartilage. *Open Journal of Orthopedics, 2.*

Taga, Y., Kusubata, M., Ogawa-Goto, K., & Hattori, S. (2013). Site-specific quantitative analysis of overglycosylation of collagen in osteogenesis imperfecta using hydrazide chemistry and SILAC. *Journal of Proteome Research, 12,* 2225–2232.

Taherian, M., & Mousavi, S. M. (2017). Modeling and simulation of forward osmosis process using agent-based model system. *Computers & Chemical Engineering, 100,* 104–118.

Tahimic, C. G., Wang, Y., & Bikle, D. D. (2013). Anabolic effects of IGF-1 signaling on the skeleton. *Frontiers in Endocrinology (Lausanne), 4,* 6.
Takahashi, I., Nuckolls, G. H., Takahashi, K., Tanaka, O., Semba, I., Dashner, R., et al. (1998). Compressive force promotes sox9, type II collagen and aggrecan and inhibits IL-1beta expression resulting in chondrogenesis in mouse embryonic limb bud mesenchymal cells. *Journal of Cell Science, 111*(Pt 14), 2067–2076.
Tam, E. M., Wu, Y. I., Butler, G. S., Stack, M. S., & Overall, C. M. (2002). Collagen binding properties of the membrane type-1 matrix metalloproteinase (MT1-MMP) hemopexin C domain—The ectodomain of the 44-kDa autocatalytic product of MT1-MMP inhibits cell invasion by disrupting native type I collagen cleavage. *Journal of Biological Chemistry, 277,* 39005–39014.
Tanamas, S. K., Wluka, A. E., Pelletier, J. P., Pelletier, J. M., Abram, F., Berry, P. A., et al. (2010). Bone marrow lesions in people with knee osteoarthritis predict progression of disease and joint replacement: A longitudinal study. *Rheumatology (Oxford), 49,* 2413–2419.
Tang, X., Zarebska, J., Curtinha, M., Stott, B., & Vincent, T. (2017). Connective tissue growth factor determines cartilage thickness and osteoarthritic risk by controlling activation of latent transforming growth factor beta in cartilage. *Osteoarthritis and Cartilage, 25,* S17.
Tang, Y., Ballarini, R., Buehler, M. J., & Eppell, S. J. (2010). Deformation micromechanisms of collagen fibrils under uniaxial tension. *Journal of the Royal Society, Interface, 7,* 839–850.
Tardif, G., Hum, D., Pelletier, J. P., Boileau, C., Ranger, P., & Martel-Pelletier, J. (2004). Differential gene expression and regulation of the bone morphogenetic protein antagonists follistatin and gremlin in normal and osteoarthritic human chondrocytes and synovial fibroblasts. *Arthritis and Rheumatism, 50,* 2521–2530.
Tat, S. K., Pelletier, J. P., Velasco, C. R., Padrines, M., & Martel-Pelletier, J. (2009). New perspective in osteoarthritis: The OPG and RANKL system as a potential therapeutic target? *Keio Journal of Medicine, 58,* 29–40.
Ten Dijke, P., & Arthur, H. M. (2007). Extracellular control of TGFbeta signalling in vascular development and disease. *Nature Reviews Molecular Cell Biology, 8,* 857–869.
Terzaghi, K. (1925). *Erdbaumechanik auf bodenphysikalischer grundlage*. Leipzig u. Wien: F. Deuticke.
Tew, S. R., Peffers, M. J., McKay, T. R., Lowe, E. T., Khan, W. S., Hardingham, T. E., et al. (2009). Hyperosmolarity regulates SOX9 mRNA posttranscriptionally in human articular chondrocytes. *American Journal of Physiology. Cell Physiology, 297,* C898–C906.
Thibault, M., Poole, A. R., & Buschmann, M. D. (2002). Cyclic compression of cartilage/bone explants in vitro leads to physical weakening, mechanical breakdown of collagen and release of matrix fragments. *Journal of Orthopaedic Research, 20,* 1265–1273.
Thieszen, S. L., & Rosenquist, T. H. (1995). Expression of collagens and decorin during aortic-arch artery development—Implications for matrix pattern-formation. *Matrix Biology, 14,* 573–582.
Thomas, C. M., Fuller, C. J., Whittles, C. E., & Sharif, M. (2007). Chondrocyte death by apoptosis is associated with cartilage matrix degradation. *Osteoarthritis Cartilage, 15,* 27–34.
Thomas, L. (1956). Reversible collapse of rabbit ears after intravenous papain, and prevention of recovery by cortisone. *Journal of Experimental Medicine, 104,* 245–252.
Thorfve, A., Dehne, T., Lindahl, A., Brittberg, M., Pruss, A., Ringe, J., et al. (2012). Characteristic markers of the Wnt signaling pathways are differentially expressed in osteoarthritic cartilage. *Cartilage, 3,* 43–57.
Thorpe, C. T., Streeter, I., Pinchbeck, G. L., Goodship, A. E., Clegg, P. D., & Birch, H. L. (2010). Aspartic acid racemization and collagen degradation markers reveal an accumulation of damage in tendon collagen that is enhanced with aging. *Journal of Biological Chemistry, 285,* 15674–15681.

Tock, J., van Putten, V., Stenmark, K. R., & Nemenoff, R. A. (2003). Induction of SM-alpha-actin expression by mechanical strain in adult vascular smooth muscle cells is mediated through activation of JNK and p38 MAP kinase. *Biochemical and Biophysical Research Communications, 301,* 1116–1121.

Torzilli, P. A., Bhargava, M., Park, S., & Chen, C. T. (2010). Mechanical load inhibits IL-1 induced matrix degradation in articular cartilage. *Osteoarthritis Cartilage, 18,* 97–105.

Toyoda, T., Seedhom, B. B., Yao, J. Q., Kirkham, J., Brookes, S., & Bonass, W. A. (2003). Hydrostatic pressure modulates proteoglycan metabolism in chondrocytes seeded in agarose. *Arthritis and Rheumatism, 48,* 2865–2872.

Trapnell, C. (2015). Defining cell types and states with single-cell genomics. *Genome Research, 25,* 1491–1498.

Treppo, S., Koepp, H., Quan, E. C., Cole, A. A., Kuettner, K. E., & Grodzinsky, A. J. (2000). Comparison of biomechanical and biochemical properties of cartilage from human knee and ankle pairs. *Journal of Orthopaedic Research, 18,* 739–748.

Trickey, W. R., Lee, G. M., & Guilak, F. (2000). Viscoelastic properties of chondrocytes from normal and osteoarthritic human cartilage. *Journal of Orthopaedic Research, 18,* 891–898.

Troeberg, L., Mulloy, B., Ghosh, P., Lee, M. H., Murphy, G., & Nagase, H. (2012). Pentosan polysulfate increases affinity between ADAMTS-5 and TIMP-3 through formation of an electrostatically driven trimolecular complex. *Biochemical Journal, 443,* 307–315.

Tsai, W. C., Tang, F. T., Wong, M. K., & Pang, J. H. S. (2003). Inhibition of tendon cell migration by dexamethasone is correlated with reduced alpha-smooth muscle actin gene expression: A potential mechanism of delayed tendon healing. *Journal of Orthopaedic Research, 21,* 265–271.

Tseng, T. L., Konada, U., & Kwon, Y. (2016). A novel approach to predict surface roughness in machining operations using fuzzy set theory. *Journal of Computational Design and Engineering, 3,* 1–13.

Tsipouras, P., & Ramirez, F. (1987). Genetic disorders of collagen. *Journal of Medical Genetics, 24,* 2–8.

Tsuchida, A. I., Beekhuizen, M., Rutgers, M., van Osch, G. J. V. M., Bekkers, J. E. J., Bot, A. G. J., et al. (2012). Interleukin-6 is elevated in synovial fluid of patients with focal cartilage defects and stimulates cartilage matrix production in an in vitro regeneration model. *Arthritis Research & Therapy, 14.*

Tsuchida, A. I., Beekhuizen, M., 't Hart, M. C., Radstake, T. R., Dhert, W. J., Saris, D. B., et al. (2014). Cytokine profiles in the joint depend on pathology, but are different between synovial fluid, cartilage tissue and cultured chondrocytes. *Arthritis Research and Theraphy, 16,* 441.

Tsumaki, N., Nakase, T., Miyaji, T., Kakiuchi, M., Kimura, T., Ochi, T., et al. (2002). Bone morphogenetic protein signals are required for cartilage formation and differently regulate joint development during skeletogenesis. *Journal of Bone and Mineral Research, 17,* 898–906.

Udaka, J., Ohmori, S., Terui, T., Ohtsuki, I., Ishiwata, S., Kurihara, S., et al. (2008). Disuse-induced preferential loss of the giant protein titin depresses muscle performance via abnormal sarcomeric organization. *Journal of General Physiology, 131,* 33–41.

Uquillas, J. A., & Akkus, O. (2012). Modeling the electromobility of type-I collagen molecules in the electrochemical fabrication of dense and aligned tissue constructs. *Annals of Biomedical Engineering, 40,* 1641–1653.

Urban, J. P., Hall, A. C., & Gehl, K. A. (1993). Regulation of matrix synthesis rates by the ionic and osmotic environment of articular chondrocytes. *Journal of Cellular Physiology, 154,* 262–270.

Urban, J. P. G., Maroudas, A., Bayliss, M. T., & Dillon, J. (1979). Swelling pressures of proteoglycans at the concentrations found in cartilaginous tissues. *Biorheology, 16,* 447–464.

Usami, Y., Gunawardena, A. T., Iwamoto, M., & Enomoto-Iwamoto, M. (2016). Wnt signaling in cartilage development and diseases: Lessons from animal studies. *Laboratory Investigation, 96,* 186–196.

Utzny, C., Faroudi, M., & Valitutti, S. (2005). Frequency encoding of T-cell receptor engagement dynamics in calcium time series. *Biophysical Journal, 88,* 1–14.

Vadon-Le Goff, S., Kronenberg, D., Bourhis, J. M., Bijakowski, C., Raynal, N., Ruggiero, F., et al. (2011). Procollagen C-proteinase enhancer stimulates procollagen processing by binding to the C-propeptide region only. *Journal of Biological Chemistry, 286,* 38932–38938.

Valdes, A. M., & Spector, T. D. (2008). The contribution of genes to osteoarthritis. *Rheumatic Diseases Clinics of North America, 34,* 581–603.

van Amerongen, R., & Berns, A. (2006). Knockout mouse models to study Wnt signal transduction. *Trends in Genetics, 22,* 678–689.

van den Berg, W. B. (2011). Osteoarthritis year 2010 in review: Pathomechanisms. *Osteoarthritis Cartilage, 19,* 338–341.

van der Kraan, P. M. (2017). The changing role of TGF beta in healthy, ageing and osteoarthritic joints. *Nature Reviews Rheumatology, 13,* 155–163.

van der Kraan, P. M., Goumans, M. J., Blaney Davidson, E., & Ten Dijke, P. (2012). Age-dependent alteration of TGF-beta signalling in osteoarthritis. *Cell and Tissue Research, 347*, 257–265.

van der Kraan, P. M., & van den Berg, W. B. (2007). Osteophytes: Relevance and biology. *Osteoarthritis and Cartilage, 15,* 237–244.

van der Wath, R. C., Gardiner, B. S., Burgess, A. W., & Smith, D. W. (2013). Cell organisation in the colonic crypt: A theoretical comparison of the pedigree and niche concepts. *PLoS ONE, 8,* e73204.

van der Windt, A. E., Haak, E., Das, R. H., Kops, N., Welting, T. J., Caron, M. M., et al. (2010). Physiological tonicity improves human chondrogenic marker expression through nuclear factor of activated T-cells 5 in vitro. *Arthritis Research & Therapy, 12,* R100.

van Dijk, C. N., Reilingh, M. L., Zengerink, M., & van Bergen, C. J. (2010). Osteochondral defects in the ankle: Why painful? *Knee Surgery, Sports Traumatology, Arthroscopy, 18,* 570–580.

van Doren, S. R. (2015). Matrix metalloproteinase interactions with collagen and elastin. *Matrix Biology, 44–46,* 224–231.

van Turnhout, M. C., Haazelager, M. B., Gijsen, M. A. L., Schipper, H., Kranenbarg, S., & van Leeuwen, J. L. (2008). Quantitative description of collagen structure in the articular cartilage of the young and adult equine distal metacarpus. *Animal Biology, 58,* 353–370.

van Turnhout, M. C., Kranenbarg, S., & van Leeuwen, J. L. (2011). Contribution of postnatal collagen reorientation to depth-dependent mechanical properties of articular cartilage. *Biomechanics and Modeling in Mechanobiology, 10,* 269–279.

van Turnhout, M. C., Schipper, H., Engel, B., Buist, W., Kranenbarg, S., & van Leeuwen, J. L. (2010). Postnatal development of collagen structure in ovine articular cartilage. *BMC Developmental Biology, 10,* 62.

van Valburg, A. A., van Roermund, P. M., Marijnissen, A. C., Wenting, M. J., Verbout, A. J., Lafeber, F. P., et al. (2000). Joint distraction in treatment of osteoarthritis (II): Effects on cartilage in a canine model. *Osteoarthritis Cartilage, 8,* 1–8.

Varma, A. K. (2013). Charcot neuroarthropathy of the foot and ankle: A review. *Journal of Foot and Ankle Surgery, 52,* 740–749.

Vaughan, L., Mendler, M., Huber, S., Bruckner, P., Winterhalter, K. H., Irwin, M. I., et al. (1988). D-periodic distribution of collagen type IX along cartilage fibrils. *Journal of Cell Biology, 106,* 991–997.

Veidal, S. S., Karsdal, M. A., Vassiliadis, E., Nawrocki, A., Larsen, M. R., Quoc, H. T. N., Hagglund, P., Luo, Y. Y., Zheng, Q. L., Vainer, B. & Leeming, D. J. (2011). MMP mediated degradation of type VI collagen is highly associated with liver fibrosis—Identification and validation of a novel biochemical marker assay. *PLoS ONE, 6*, e24753.

Velloso, C. P. (2008). Regulation of muscle mass by growth hormone and IGF-I. *British Journal of Pharmacology, 154,* 557–568.

Venalainen, M. S., Mononen, M. E., Salo, J., Rasanen, L. P., Jurvelin, J. S., Toyras, J., et al. (2016). Quantitative evaluation of the mechanical risks caused by focal cartilage defects in the knee. *Scientific Reports, 6*, 37538.

Verbruggen, G., Cornelissen, M., Almqvist, K. F., Wang, L., Elewaut, D., Broddelez, C., et al. (2000). Influence of aging on the synthesis and morphology of the aggrecans synthesized by differentiated human articular chondrocytes. *Osteoarthritis Cartilage, 8,* 170–179.

Veres, S. P., Harrison, J. M., & Lee, J. M. (2014). Mechanically overloading collagen fibrils uncoils collagen molecules, placing them in a stable, denatured state. *Matrix Biology, 33,* 54–59.

Verruijt, A. (1995). *Computational Geomechanics*. Dordrecht: Kluwer Academic Publishers.

Verruijt, A. (2013). *Theory and problems of poroelasticity*, Self-published.

Verzijl, N., Degroot, J., Oldehinkel, E., Bank, R. A., Thorpe, S. R., Baynes, J. W., et al. (2000). Age-related accumulation of Maillard reaction products in human articular cartilage collagen. *Biochemical Journal, 350*(Pt 2), 381–387.

Verzijl, N., Degroot, J., Thorpe, S. R., Bank, R. A., Shaw, J. N., Lyons, T. J., et al. (2000). Effect of collagen turnover on the accumulation of advanced glycation end products. *Journal of Biological Chemistry, 275,* 39027–39031.

Videman, T., Eronen, I., & Candolin, T. (1981). [3H]proline incorporation and hydroxyproline concentration in articular cartilage during the development of osteoarthritis caused by immobilization. A study in vivo with rabbits. *Biochemical Journal, 200,* 435–440.

Vignon, E., & Arlot, M. (1981). Macroscopically normal cartilage from the human osteoarthritic femoral-head. 2. Measurement of cartilage thickness and cell-density. *Journal of Rheumatology, 8,* 447–450.

Vignon, E., Arlot, M., Hartmann, D., Moyen, B., & Ville, G. (1983). Hypertrophic repair of articular cartilage in experimental osteoarthrosis. *Annals of the Rheumatic Diseases, 42,* 82–88.

Vilker, V. L., Colton, C. K., & Smith, K. A. (1981). The osmotic-pressure of concentrated protein solutions—Effect of concentration and pH in saline solutions of bovine serum-albumin. *Journal of Colloid and Interface Science, 79,* 548–566.

Villiger, P. M., & Lotz, M. (1992). Differential expression of TGF beta isoforms by human articular chondrocytes in response to growth factors. *Journal of Cellular Physiology, 151,* 318–325.

Vincent, T., Hermansson, M., Bolton, M., Wait, R., & Saklatvala, J. (2002). Basic FGF mediates an immediate response of articular cartilage to mechanical injury. *Proceedings of the National Academy of Sciences of the United States of America, 99,* 8259–8264.

Vincent, T. L., & Watt, F. E. (2018). Osteoarthritis. *Medicine, 46,* 7.

Vincourt, J. B., Etienne, S., Cottet, J., Delaunay, C., Malanda, B., Lionneton, F., et al. (2010). C-propeptides of procollagens I alpha 1 and II that differentially accumulate in enchondromas versus chondrosarcomas regulate tumor cell survival and migration. *Cancer Research, 70,* 4739–4748.

Visse, R., & Nagase, H. (2003). Matrix metalloproteinases and tissue inhibitors of metalloproteinases: Structure, function, and biochemistry. *Circulation Research, 92,* 827–839.

Vortkamp, A., Lee, K., Lanske, B., Segre, G. V., Kronenberg, H. M., & Tabin, C. J. (1996). Regulation of rate of cartilage differentiation by Indian hedgehog and PTH-related protein. *Science, 273,* 613–622.

Vortkamp, A., Pathi, S., Peretti, G. M., Caruso, E. M., Zaleske, D. J., & Tabin, C. J. (1998). Recapitulation of signals regulating embryonic bone formation during postnatal growth and in fracture repair. *Mechanisms of Development, 71,* 65–76.

Vynios, D. H., Papageorgakopoulou, N., Sazakli, H., & Tsiganos, C. P. (2001). The interactions of cartilage proteoglycans with collagens are determined by their structures. *Biochimie, 83,* 899–906.

Wadhwa, S., & Kapila, S. (2008). TMJ disorders: Future innovations in diagnostics and therapeutics. *Journal of Dental Education, 72,* 930–947.

Wagenaar-Miller, R. A., Engelholm, L. H., Gavard, J., Yamada, S. S., Gutkind, J. S., Behrendt, N., et al. (2007). Complementary roles of intracellular and pericellular collagen degradation pathways in vivo. *Molecular and Cellular Biology, 27,* 6309–6322.

Waggett, A. D., Benjamin, M., & Ralphs, J. R. (2006). Connexin 32 and 43 gap junctions differentially modulate tenocyte response to cyclic mechanical load. *European Journal of Cell Biology, 85,* 1145–1154.

Waggett, A. D., Rees, S. G., & Caterson, B. (2004). Increased turnover of proteoglycan aggregates in tendon vs. cartilage. *International Journal of Experimental Pathology, 85,* A76–A77.

Waller, K. A., Zhang, L. X., Elsaid, K. A., Fleming, B. C., Warman, M. L., & Jay, G. D. (2013). Role of lubricin and boundary lubrication in the prevention of chondrocyte apoptosis. *Proceedings of the National Academy of Sciences of the United States of America, 110,* 5852–5857.

Walker, P. S., Dowson, D., Longfield, M. D., & Wright, V. (1968). "Boosted lubrication" in synovial joints by fluid entrapment and enrichment. *Annals of the Rheumatic Diseases, 27,* 512–520.

Wang, B. L., Jin, H. T., Shu, B., Mira, R. R., & Chen, D. (2015a). Chondrocytes-specific expression of osteoprotegerin modulates osteoclast formation in metaphyseal bone. *Scientific Reports, 5,* 13667.

Wang, H., Casaburi, R., Taylor, W. E., Aboellail, H., Storer, T. W., & Kopple, J. D. (2005). Skeletal muscle mRNA for IGF-IEa, IGF-II, and IGF-I receptor is decreased in sedentary chronic hemodialysis patients. *Kidney International, 68,* 352–361.

Wang, L., Shao, Y. Y., & Ballock, R. T. (2007). Thyroid hormone interacts with the Wnt/β-catenin signaling pathway in the terminal differentiation of growth plate chondrocytes. *Journal of Bone and Mineral Research, 22*(12), 1988–1995.

Wang, L., Shao, Y. Y., & Ballock, R. T. (2010). Thyroid hormone-mediated growth and differentiation of growth plate chondrocytes involves IGF-1 modulation of beta-catenin signaling. *Journal of Bone and Mineral Research, 25,* 1138–1146.

Wang, L. M., & Detamore, M. S. (2009). Effects of growth factors and glucosamine on porcine mandibular condylar cartilage cells and hyaline cartilage cells for tissue engineering applications. *Archives of Oral Biology, 54,* 1–5.

Wang, Q., Breinan, H. A., Hsu, H. P., & Spector, M. (2000). Healing of defects in canine articular cartilage: Distribution of nonvascular alpha-smooth muscle actin-containing cells. *Wound Repair and Regeneration, 8,* 145–158.

Wang, W. G., Rigueur, D., & Lyons, K. M. (2014). TGF beta signaling in cartilage development and maintenance. *Birth Defects Research Part C-Embryo Today-Reviews, 102,* 37–51.

Wang, X., Brouillette, M. J., Ayati, B. P., & Martin, J. A. (2015b). A validated model of the pro- and anti-inflammatory cytokine balancing act in articular cartilage lesion formation. *Frontiers in Bioengineering and Biotechnology, 3,* 25.

Wang, X. W., Liu, J., Wang, Z., Huang, Y. M., Liu, W. P., Zhu, X., et al. (2013). Periostin contributes to the acquisition of multipotent stem cell-like properties in human mammary epithelial cells and breast cancer cells. *PLoS ONE, 8,* e72962.

Wang, Y., Ding, C., Wluka, A. E., Davis, S., Ebeling, P. R., Jones, G., et al. (2006). Factors affecting progression of knee cartilage defects in normal subjects over 2 years. *Rheumatology (Oxford), 45,* 79–84.

Wang, Y., Wluka, A. E., & Cicuttini, F. M. (2005). The determinants of change in tibial plateau bone area in osteoarthritic knees: A cohort study. *Arthritis Research & Therapy, 7,* R687–R693.

Warde, N. (2011). Osteoarthritis: Sclerostin inhibits Wnt signaling in OA chondrocytes and protects against inflammation-induced cartilage damage. *Nature Reviews Rheumatology, 7,* 438.

Warth, L. C., Callaghan, J. J., Liu, S. S., Klaassen, A. L., Goetz, D. D., & Johnston, R. C. (2014). Thirty-five-year results after charnley total hip arthroplasty in patients less than fifty years old: A concise follow-up of previous reports. *Journal of Bone and Joint Surgery. American Volume, 96,* 1814–1819.

Watanabe, H., Gao, L., Sugiyama, S., Doege, K., Kimata, K., & Yamada, Y. (1995). Mouse aggrecan, a large cartilage proteoglycan: Protein sequence, gene structure and promoter sequence. *Biochemical Journal, 308*(Pt 2), 433–440.

Watkins, P. J., & Edmonds, M. E. (1983). Sympathetic-nerve failure in diabetes. *Diabetologia, 25,* 73–77.

Waugh, C. M., Blazevich, A. J., Fath, F., & Korff, T. (2012). Age-related changes in mechanical properties of the Achilles tendon. *Journal of Anatomy, 220,* 144–155.

Wayne, G. J., Deng, S. J., Amour, A., Borman, S., Matico, R., Carter, H. L., et al. (2007). TIMP-3 inhibition of ADAMTS-4 (Aggrecanase-1) is modulated by interactions between aggrecan and the C-terminal domain of ADAMTS-4. *Journal of Biological Chemistry, 282,* 20991–20998.

Wedig, M. L., Bae, W. C., Temple, M. M., Sah, R. L., & Gray, M. L. (2005). GAG profiles in normal human articular cartilage. In *51st Annual Meeting of the Orthopaedic Research Society.*

Wegrowski, J., Bellon, G., & Randoux, A. (1990). Human type-Vi collagen—Isolation and characterization of alpha-1 and alpha-2 chains by 2-dimensional preparative gel-electrophoresis. *Biochimica et Biophysica Acta, 1039,* 189–196.

Wei, F., Zhou, J., Wei, X., Zhang, J., Fleming, B. C., Terek, R., et al. (2012). Activation of Indian hedgehog promotes chondrocyte hypertrophy and upregulation of MMP-13 in human osteoarthritic cartilage. *Osteoarthritis Cartilage, 20,* 755–763.

Weightman, B., Chappell, D. J., & Jenkins, E. A. (1978). A second study of tensile fatigue properties of human articular cartilage. *Annals of the Rheumatic Diseases, 37,* 58–63.

Weisberg, H. F. (1978). Osmotic pressure of the serum proteins. *Annals of Clinical and Laboratory Science, 8,* 155–164.

Wellard, R. M., Ravasio, J. P., Guesne, S., Bell, C., Oloyede, A., Tevelen, G., et al. (2014). Simultaneous magnetic resonance imaging and consolidation measurement of articular cartilage. *Sensors (Basel), 14,* 7940–7958.

Wells, J. A. (1996). Binding in the growth hormone receptor complex. *Proceedings of the National Academy of Sciences of the United States of America, 93,* 1–6.

Wennerstrom, H., Jonsson, B., & Linse, P. (1982). The cell model for poly-electrolyte systems—Exact statistical mechanical relations, Monte-Carlo simulations, and the Poisson-Boltzmann approximation. *Journal of Chemical Physics, 76,* 4665–4670.

Wesseling-Perry, K. (2014). The BRC canopy an important player in bone remodeling. *American Journal of Pathology, 184,* 924–926.

Whittaker, J. L., Woodhouse, L. J., Nettel-Aguirre, A., & Emery, C. A. (2015). Outcomes associated with early post-traumatic osteoarthritis and other negative health consequences 3-10 years following knee joint injury in youth sport. *Osteoarthritis and Cartilage, 23,* 1122–1129.

Wiberg, C., Hedbom, E., Khairullina, A., Lamande, S. R., Oldberg, A., Timpl, R., et al. (2001). Biglycan and decorin bind close to the N-terminal region of the collagen VI triple helix. *Journal of Biological Chemistry, 276,* 18947–18952.

Wiberg, C., Klatt, A. R., Wagener, R., Paulsson, M., Bateman, J. F., Heinegard, D., et al. (2003). Complexes of matrilin-1 and biglycan or decorin connect collagen VI microfibrils to both collagen II and aggrecan. *Journal of Biological Chemistry, 278,* 37698–37704.

Wiederhielm, C. A., Woodbury, J. W., Kirk, S., & Rushmer, R. F. (1964). Pulsatile pressures in the microcirculation of frog's mesentery. *American Journal of Physiology, 207,* 173–176.

Wiegant, K., van Roermund, P. M., Intema, F., Cotofana, S., Eckstein, F., Mastbergen, S. C., et al. (2013). Sustained clinical and structural benefit after joint distraction in the treatment of severe knee osteoarthritis. *Osteoarthritis Cartilage, 21,* 1660–1667.

Wiig, H., Reed, R. K., & Aukland, K. (1981). Micropuncture measurement of interstitial fluid pressure in rat subcutis and skeletal-muscle—Comparison to wick-in-needle technique. *Microvascular Research, 21,* 308–319.

Williams, C. J., & Jimenez, S. A. (1995). Heritable diseases of cartilage caused by mutations in collagen genes. *Journal of Rheumatology. Supplement, 43,* 28–33.

Williams, G. R. (2013). Thyroid hormone actions in cartilage and bone. *European Thyroid Journal, 2,* 3–13.

Williams, R. P. W., & Comper, W. D. (1990). Osmotic flow caused by polyelectrolytes. *Biophysical Chemistry, 36,* 12.

Wilson, S. L., Yang, Y., & el Haj, A. J. (2014). Corneal stromal cell plasticity: In vitro regulation of cell phenotype through cell-cell interactions in a three-dimensional model. *Tissue Engineering Part A, 20,* 225–238.

Wilson, S. R., Peters, C., Saftig, P., & Bromme, D. (2009). Cathepsin K activity-dependent regulation of osteoclast actin ring formation and bone resorption. *Journal of Biological Chemistry, 284,* 2584–2592.

Wilson, W., Driessen, N. J. B., van Donkelaar, C. C., & Ito, K. (2006). Prediction of collagen orientation in articular cartilage by a collagen remodeling algorithm. *Osteoarthritis and Cartilage, 14,* 1196–1202.

Wilson, W., van Burken, C., van Donkelaar, C., Buma, P., van Rietbergen, B., & Huiskes, R. (2006). Causes of mechanically induced collagen damage in articular cartilage. *Journal of Orthopaedic Research, 24,* 220–228.

Winby, C. R., Lloyd, D. G., Besier, T. F., & Kirk, T. B. (2009). Muscle and external load contribution to knee joint contact loads during normal gait. *Journal of Biomechanics, 42,* 2294–2300.

Wirrig, E. E., Snarr, B. S., Chintalapudi, M. R., O'Neal, J, L., Phelps, A. L., Barth, J. L., et al. (2007). Cartilage link protein 1 (Crtl1), an extracellular matrix component playing an important role in heart development. *Developmental Biology, 310,* 291–303.

Wirth, W., Hunter, D. J., Nevitt, M. C., Kwoh, C. K., Sharma, L., Guermazi, A., et al. (2016). Predictive versus concurrent validity of cartilage thickness change as a marker of progression of knee osteoarthritis—Data from the osteoarthritis initiative. *Osteoarthritis and Cartilage, 24,* S241–S241.

Witzeneder, K., Lindenmair, A., Gabriel, C., Holler, K., Theiss, D., Redl, H., et al. (2013). Human-derived alternatives to fetal bovine serum in cell culture. *Transfusion Medicine and Hemotheraphy, 40,* 417–423.

Wlodarski, K. H., Brodzikowska, A., & Kuzaka, B. (2014). Are chondroclasts and osteoclasts identical? *Folia Biologica-Krakow, 62,* 143–147.

Wolf, A., Ackermann, B., & Steinmeyer, J. (2007). Collagen synthesis of articular cartilage explants in response to frequency of cyclic mechanical loading. *Cell and Tissue Research, 327,* 155–166.

Wong, M., & Carter, D. R. (2003). Articular cartilage functional histomorphology and mechanobiology: A research perspective. *Bone, 33,* 1–13.

Wong, M., Ponticiello, M., Kovanen, V., & Jurvelin, J. S. (2000). Volumetric changes of articular cartilage during stress relaxation in unconfined compression. *Journal of Biomechanics, 33,* 1049–1054.

Wood, L., Ferrell, W. R., & Baxendale, R. H. (1988). Pressures in normal and acutely distended human knee joints and effects on quadriceps maximal voluntary contractions. *Quarterly Journal of Experimental Physiology and Cognate Medical Sciences, 73,* 305–314.

Woods, A., Wang, G., & Beier, F. (2005). RhoA/ROCK signaling regulates Sox9 expression and actin organization during chondrogenesis. *Journal of Biological Chemistry, 280,* 11626–11634.

Workman, J., Thambyah, A., & Broom, N. (2017). The influence of early degenerative changes on the vulnerability of articular cartilage to impact-induced injury. *Clinical Biomechanics, 43,* 40–49.

Wright, V., & Dowson, D. (1976). Lubrication and cartilage. *Journal of Anatomy, 121,* 107–118.

Wu, D. Q., & Pan, W. J. (2010). GSK3: A multifaceted kinase in Wnt signaling. *Trends in Biochemical Sciences, 35,* 161–168.

Wu, J. J., Weis, M. A., Kim, L. S., & Eyre, D. R. (2010). Type III collagen, a fibril network modifier in articular cartilage. *Journal of Biological Chemistry, 285,* 18537–18544.

Wu, J. Z., & Prausnitz, J. M. (1999). Osmotic pressures of aqueous bovine serum albumin solutions at high ionic strength. *Fluid Phase Equilibria, 155,* 139–154.

Wu, Q., Zhang, Y., & Chen, Q. (2001). Indian hedgehog is an essential component of mechanotransduction complex to stimulate chondrocyte proliferation. *The Journal of Biological Chemistry, 276,* 35290–35296.

Wyatt, K. E., Bourne, J. W., & Torzilli, P. A. (2009). Deformation-dependent enzyme mechanokinetic cleavage of type I collagen. *Journal of Biomechanical Engineering, 131,* 051004.

Xia, Y. (2013). MRI of articular cartilage at microscopic resolution. *Bone and Joint Research, 2,* 9–17.

Xu, H., Edwards, J., Banerji, S., Prevo, R., Jackson, D. G., & Athanasou, N. A. (2003). Distribution of lymphatic vessels in normal and arthritic human synovial tissues. *Annals of the Rheumatic Diseases, 62,* 1227–1229.

Xu, P., & Yu, B. M. (2008). Developing a new form of permeability and Kozeny-Carman constant for homogeneous porous media by means of fractal geometry. *Advances in Water Resources, 31,* 74–81.

Xu, X., Urban, J. P., Tirlapur, U. K., & Cui, Z. (2010). Osmolarity effects on bovine articular chondrocytes during three-dimensional culture in alginate beads. *Osteoarthritis Cartilage, 18,* 433–439.

Yamashita, A., Morioka, M., Yahara, Y., Okada, M., Kobayashi, T., Kuriyama, S., et al. (2015). Generation of scaffoldless hyaline cartilaginous tissue from human iPSCs. *Stem Cell Reports, 4,* 404–418.

Yamashita, S., Miyaki, S., Kato, Y., Yokoyama, S., Sato, T., Barrionuevo, F., et al. (2012). L-Sox5 and Sox6 proteins enhance chondrogenic miR-140 microRNA expression by strengthening dimeric Sox9 activity. *Journal of Biological Chemistry, 287,* 22206–22215.

Yan, B., Zhang, Z. M., Jin, D. D., Cai, C., Jia, C. H., Liu, W., et al. (2016). mTORC1 regulates PTHrP to coordinate chondrocyte growth, proliferation and differentiation. *Nature Communications, 7,* 11151.

Yanai, T., Ishii, T., Chang, F., & Ochiai, N. (2005). Repair of large full-thickness articular cartilage defects in the rabbit: The effects of joint distraction and autologous bone-marrow-derived mesenchymal cell transplantation. *Journal of Bone and Joint Surgery. British Volume, 87,* 721–729.

Yang, K., Wang, X., Zhang, H. M., Wang, Z. L., Nan, G. X., Li, Y. S., et al. (2016). The evolving roles of canonical WNT signaling in stem cells and tumorigenesis: Implications in targeted cancer therapies. *Laboratory Investigation, 96,* 116–136.

Yao, M. X., Qiu, W., Liu, R. C., Efremov, A. K., Cong, P. W., Seddiki, R., et al. (2014). Force-dependent conformational switch of alpha-catenin controls vinculin binding. *Nature Communications, 5,* 4525.

Yasuda, K., & Hayashi, K. (1999). Changes in biomechanical properties of tendons and ligaments from joint disuse. *Osteoarthritis Cartilage, 7,* 122–129.

Yasuhara, R., Ohta, Y., Yuasa, T., Kondo, N., Hoang, T., Addya, S., et al. (2011). Roles of beta-catenin signaling in phenotypic expression and proliferation of articular cartilage superficial zone cells. *Laboratory Investigation, 91,* 1739–1752.

Yates, M. P., Settle S. L., & Al, E. (2010). IGFBP-5 Metabolism is disrupted in the rat medial meniscal tear model of osteoarthritis. *Cartilage, 1,* 43–54.

Ye, Z., Chen, Y., Zhang, R., Dai, H., Zeng, C., Zeng, H., et al. (2014). c-Jun N-terminal kinase—c-Jun pathway transactivates Bim to promote osteoarthritis. *Canadian Journal of Physiology and Pharmacology, 92,* 132–139.

Yeh, A. T., Hammer-Wilson, M. J., van Sickle, D. C., Benton, H. P., Zoumi, A., Tromberg, B. J., et al. (2005). Nonlinear optical microscopy of articular cartilage. *Osteoarthritis Cartilage, 13,* 345–352.

Yeung, C. Y. C., Gossan, N., Lu, Y. H., Hughes, A., Hensman, J. J., Bayer, M. L., et al. (2014). Gremlin-2 is a BMP antagonist that is regulated by the circadian clock. *Scientific Reports,* 4.

Yin, J., Xia, Y., & Lu, M. (2012). Concentration profiles of collagen and proteoglycan in articular cartilage by Fourier transform infrared imaging and principal component regression. *Spectrochimica Acta Part A: Molecular and Biomolecular Spectroscopy, 88,* 90–96.

Young, S. R., Gardiner, B., Mehdizadeh, A., Rubenson, J., Umberger, B. & Smith, D. W. (2016). adaptive remodeling of Achilles tendon: A multi-scale computational model. *PLoS Computational Biology, 12,* e1005106.

Yu, D. G., Ding, H. F., Mao, Y. Q., Liu, M., Yu, B., Zhao, X., et al. (2013). Strontium ranelate reduces cartilage degeneration and subchondral bone remodeling in rat osteoarthritis model. *Acta Pharmacologica Sinica, 34,* 393–402.

Yu, W. H., Yu, S. S. C., Meng, Q., Brew, K., & Woessner, J. F. (2000). TIMP-3 binds to sulfated glycosaminoglycans of the extracellular matrix. *Journal of Biological Chemistry, 275,* 31226–31232.

YU, H., & ROHAN, T. 2000. Role of the insulin-like growth factor family in cancer development and progression. *Jnci-Journal of the National Cancer Institute, 92,* 1472–1489.

Yuan, X. L., Meng, H. Y., Wang, Y. C., Peng, J., Guo, Q. Y., Wang, A. Y., et al. (2014). Bone-cartilage interface crosstalk in osteoarthritis: Potential pathways and future therapeutic strategies. *Osteoarthritis and Cartilage, 22,* 1077–1089.

Zamparo, O., & Comper, W. D. (1989). hydraulic conductivity of chondroitin sulfate proteoglycan solutions. *Archives of Biochemistry and Biophysics, 274,* 259–269.

Zanotti, S., & Canalis, E. (2013). Interleukin 6 mediates selected effects of Notch in chondrocytes. *Osteoarthritis Cartilage, 21,* 1766–1773.

Zappone, B., Greene, G. W., Oroudjev, E., Jay, G. D., & Israelachvili, J. N. (2008). Molecular aspects of boundary lubrication by human lubricin: Effect of disulfide bonds and enzymatic digestion. *Langmuir, 24,* 1495–1508.

Zareian, R., Church, K. P., Saeidi, N., Flynn, B. P., Beale, J. W., & Ruberti, J. W. (2010). Probing collagen/enzyme mechanochemistry in native tissue with dynamic, enzyme-induced creep. *Langmuir, 26,* 9917–9926.

Zbyn, S., Mlynarik, V., Juras, V., Szomolanyi, P., & Trattnig, S. (2014). Sodium MR imaging of articular cartilage pathologies. *Current Radiology Reports, 2,* 41.

Zehnder, A. F., Kristiansen, A. G., Adams, J. C., Kujawa, S. G., Merchant, S. N., & McKenna, M. J. (2006). Osteoprotegrin knockout mice demonstrate abnormal remodeling of the otic capsule and progressive hearing loss. *Laryngoscope, 116,* 201–206.

Zhang, J., Chen, Z. X., Wang, L., Li, D. C., & Jin, Z. M. (2017). Load application for the contact mechanics analysis and wear prediction of total knee replacement. *Proceedings of the Institution of Mechanical Engineers Part H-Journal of Engineering in Medicine, 231,* 444–454.

Zhang, L., Gardiner, B. S., Smith, D. W., Pivona, P., & Grodzinsky, A. J. (2010a). The transport of insulin-like growth factor through cartilage. In K. Vafai (Ed.), *Applications in biological systems and biotechnology*. CRC Press.

Zhang, L., Gardiner, B. S., Smith, D. W., Pivonka, P., & Grodzinsky, A. (2007). The effect of cyclic deformation and solute binding on solute transport in cartilage. *Archives of Biochemistry and Biophysics, 457,* 47–56.

Zhang, L., Gardiner, B. S., Smith, D. W., Pivonka, P., & Grodzinsky, A. (2008a). A fully coupled poroelastic reactive-transport model of cartilage. *Molecular and Cellular Biomechanics, 5,* 133–153.

Zhang, L., Gardiner, B. S., Smith, D. W., Pivonka, P., & Grodzinsky, A. J. (2010b). On the role of diffusible binding partners in modulating the transport and concentration of proteins in tissues. *Journal of Theoretical Biology, 263,* 20–29.

Zhang, L., Smith, D. W., Gardiner, B. S., & Grodzinsky, A. J. (2013a). Modeling the insulin-like growth factor system in articular cartilage. *PLoS ONE, 8,* e66870.

Zhang, L. H., Gardiner, B. S., Smith, D. W., Pivonka, P., & Grodzinsky, A. J. (2008b). IGF uptake with competitive binding in articular cartilage. *Journal of Biological Systems, 16,* 175–195.

Zhang, L. H., Gardiner, B. S., Smith, D. W., Pivonka, P., & Grodzinsky, A. J. (2009). Integrated model of IGF-I mediated biosynthesis in a deformed articular cartilage. *Journal of Engineering Mechanics-ASCE, 135,* 439–449.

Zhang, L. H., Miramini, S., Smith, D. W., Gardiner, B. S., & Grodzinsky, A. J. (2015a). Time evolution of deformation in a human cartilage under cyclic loading. *Annals of Biomedical Engineering, 43,* 1166–1177.

Zhang, L. Z., Zheng, H. A., Jiang, Y., Tu, Y. H., Jiang, P. H., & Yang, A. L. (2012a). Mechanical and biologic link between cartilage and subchondral bone in osteoarthritis. *Arthritis Care & Research, 64,* 960–967.

Zhang, R., Fang, H., Chen, Y., Shen, J., Lu, H., Zeng, C., et al. (2012b). Gene expression analyses of subchondral bone in early experimental osteoarthritis by microarray. *PLoS ONE, 7,* e32356.

Zhang, W., Chen, J., Zhang, S., & Ouyang, H. W. (2012c). Inhibitory function of parathyroid hormone-related protein on chondrocyte hypertrophy: The implication for articular cartilage repair. *Arthritis Research & Therapy, 14,* 221.

Zhang, X. X., Yin, J. H., Mao, Z. H., & Xia, Y. (2015b). Discrimination of healthy and osteoarthritic articular cartilages by Fourier transform infrared imaging and partial least squares-discriminant analysis. *Journal of Biomedical Optics, 20,* 060501.

Zhang, Y., Wang, F., Tan, H., Chen, G., Guo, L., & Yang, L. (2012). Analysis of the mineral composition of the human calcified cartilage zone. *International Journal of Medical Sciences, 9,* 353–360.

Zhen, E. Y., Brittain, I. J., Laska, D. A., Mitchell, P. G., Sumer, E. U., Karsdal, M. A., et al. (2008). Characterization of metalloprotease cleavage products of human articular cartilage. *Arthritis and Rheumatism, 58,* 2420–2431.

Zhen, G. H., & Cao, X. (2014). Targeting TGF beta signaling in subchondral bone and articular cartilage homeostasis. *Trends in Pharmacological Sciences, 35,* 227–236.

Zhong, L., Huang, X., Karperien, M., & Post, J. N. (2015). The regulatory role of signaling crosstalk in hypertrophy of MSCs and human articular chondrocytes. *International Journal of Molecular Sciences, 16,* 19225–19247.

Zhou, J. M., Wei, X. C., & Wei, L. (2014). Indian hedgehog, a critical modulator in osteoarthritis, could be a potential therapeutic target for attenuating cartilage degeneration disease. *Connective Tissue Research, 55,* 257–261.

Zhou, Q., Guo, R., Wood, R., Boyce, B. F., Liang, Q., Wang, Y. J., et al. (2011). Vascular endothelial growth factor C attenuates joint damage in chronic inflammatory arthritis by accelerating local lymphatic drainage in mice. *Arthritis and Rheumatism, 63,* 2318–2328.

Zhou, S., Cui, Z., & Urban, J. P. (2004). Factors influencing the oxygen concentration gradient from the synovial surface of articular cartilage to the cartilage-bone interface: A modeling study. *Arthritis and Rheumatism, 50,* 3915–3924.

Zhou, X., Wu, W., Li, H., Cheng, Y., Wei, N., Zong, J., et al. (2014). Transcriptome analysis of alternative splicing events regulated by SRSF10 reveals position-dependent splicing modulation. *Nucleic Acids Research, 42,* 4019–4030.

Zhou, Z. Y., Liu, Y. K., Chen, H. L., & Liu, F. (2014). Body mass index and knee osteoarthritis risk: A dose-response meta-analysis. *Obesity (Silver Spring), 22,* 2180–2185.

Zhu, J., & Mogilner, A. (2016). Comparison of cell migration mechanical strategies in three-dimensional matrices: A computational study. *Interface Focus, 6,* 20160040.

Zhu, M., Chen, M., Zuscik, M., Wu, Q., Wang, Y. J., Rosier, R. N., et al. (2008). Inhibition of beta-catenin signaling in articular chondrocytes results in articular cartilage destruction. *Arthritis and Rheumatism, 58,* 2053–2064.

Zimmerman, N. B., Smith, D. G., Pottenger, L. A., & Cooperman, D. R. (1988). Mechanical disruption of human patellar cartilage by repetitive loading in vitro. *Clinical Orthopaedics and Related* Research, 302–307.

Zuo, X. Y., Pan, W., Feng, T. T., Shi, X. H., & Dai, J. F. (2014). Matrix metalloproteinase 3 promotes cellular anti-dengue virus response via interaction with transcription factor NF kappa B in cell nucleus. *PLoS ONE, 9,* e84748.